RSMeans

Mechanical Cost Data
24th Annual Edition

2001

Senior Editor
Melville J. Mossman, PE

Contributing Editors
Barbara Balboni
Robert A. Bastoni
Howard M. Chandler
John H. Chiang, PE
Paul C. Crosscup
Jennifer L. Curran
Stephen E. Donnelly
J. Robert Lang
Robert C. McNichols
Robert W. Mewis
Melville J. Mossman, PE
John J. Moylan
Jeannene D. Murphy
Peter T. Nightingale
Stephen C. Plotner
Michael J. Regan
Stephen J. Sechovicz
Phillip R. Waier, PE

Manager, Engineering Operations
John H. Ferguson, PE

Group Vice President
Durwood S. Snead

Vice President and General Manager
Roger J. Grant

Vice President, Sales and Marketing
John M. Shea

Production Manager
Michael Kokernak

Production Coordinator
Marion E. Schofield

Technical Support
Thomas J. Dion
Michael H. Donelan
Jonathan Forgit
Mary Lou Geary
Gary L. Hoitt
Paula Reale-Camelio
Robin Richardson
Kathryn S. Rodriguez
Sheryl A. Rose
Elizabeth A. Ryan
James N. Wills

Book & Cover Design
Norman R. Forgit

Editorial Advisory Board

James E. Armstrong
Energy Consultant
EnergyVision

William R. Barry, CCC
Cost Consultant

Robert F. Cox, PhD
Assistant Professor
ME Rinker Sr. School of Bldg. Constr.
University of Florida

Roy F. Gilley, AIA
Principal
Gilley-Hinkel Architects

Kenneth K. Humphreys, PhD, PE, CCE
Secretary-Treasurer
International Cost Engineering Council

Patricia L. Jackson, PE
President
Jackson A&E Assoc., Inc.

Martin F. Joyce
Vice President, Utility Division
Bond Brothers, Inc.

R.S. Means Company, Inc. ("R.S. Means"), its authors, editors and engineers, apply diligence and judgment in locating and using reliable sources for the information published. **However, R.S. Means makes no express or implied warranty or guarantee in connection with the content of the information contained herein, including the accuracy, correctness, value, sufficiency, or completeness of the data, methods and other information contained herein. R.S. Means makes no express or implied warranty of merchantability or fitness for a particular purpose.** R.S. Means shall have no liability to any customer or third party for any loss, expense, or damage including consequential, incidental, special or punitive damages, including lost profits or lost revenue, caused directly or indirectly by any error or omission, or arising out of, or in connection with, the information contained herein.

No part of this publication may be reproduced, stored in a retrieval system, or transmitted in any form or by any means without prior written permission of R.S. Means Company, Inc.

Second Printing

Foreword

R.S. Means Co., Inc. is a subsidiary of CMD Group (Construction Market Data), a leading provider of construction information, products, and services in North America and globally. CMD's flagship product line includes more than 100 regional editions, national construction project information, sales leads, and over 70 local plan room/libraries in major business centers. CMD also publishes ProFile, a directory of more than 20,000 architectural firms in the U.S. Architect's First Source (AFS) provides Publishers First Source for Products, CSI's SPEC-DATA® and CSI's MANU-SPEC® building product directories for the selection of nationally available building products. R.S. Means provides Construction cost data and training, and consulting services in print, CD-ROM and online. Associated Construction Publications (ACP) provides regional magazines covering the U.S. highways heavy construction, and heavy equipment trader. Manufacturers' Survey Associates (MSA) provides surveys, plans, and specifications. CMD Group is headquartered in Atlanta and has 1,400 employees worldwide. CMD Group is owned by Cahners Business Information (www.cahners.com), a member of the Reed Elsevier plc group (NYSE: RUK and ENL). Cahners is the leading U.S. provider of business information to 16 vertical markets, including building & construction, entertainment, manufacturing, and retail. Cahners' content portfolio encompasses 140 Web sites as well as *Variety, Publishers Weekly, Design News* and 127 other market-leading business-to-business magazines. Cahners developed the leading business-to-business Web portals e-inSITE in the electronics industry, Manufacturing.net in manufacturing as well as Buildingteam.com and HousingZone.com in construction, and maintains an active program of Internet and print launches.

Our Mission

Since 1942, R.S. Means Company, Inc. has been actively engaged in construction cost publishing and consulting throughout North America.

Today, over 50 years after the company began, our primary objective remains the same: to provide you, the construction and facilities professional, with the most current and comprehensive construction cost data possible.

Whether you are a contractor, an owner, an architect, an engineer, a facilities manager, or anyone else who needs a fast and reliable construction cost estimate, you'll find this publication to be a highly useful and necessary tool.

Today, with the constant flow of new construction methods and materials, it's difficult to find the time to look at and evaluate all the different construction cost possibilities. In addition, because labor and material costs keep changing, last year's cost information is not a reliable basis for today's estimate or budget.

That's why so many construction professionals turn to R.S. Means. We keep track of the costs for you, along with a wide range of other key information, from city cost indexes . . . to productivity rates . . . to crew composition . . . to contractor's overhead and profit rates.

R.S. Means performs these functions by collecting data from all facets of the industry, and organizing it in a format that is instantly accessible to you. From the preliminary budget to the detailed unit price estimate, you'll find the data in this book useful for all phases of construction cost determination.

The Staff, the Organization, and Our Services

When you purchase one of R.S. Means' publications, you are in effect hiring the services of a full-time staff of construction and engineering professionals.

Our thoroughly experienced and highly qualified staff works daily at collecting, analyzing, and disseminating comprehensive cost information for your needs. These staff members have years of practical construction experience and engineering training prior to joining the firm. As a result, you can count on them not only for the cost figures, but also for additional background reference information that will help you create a realistic estimate.

The Means organization is always prepared to help you solve construction problems through its five major divisions: Construction and Cost Data Publishing, Electronic Products and Services, Consulting Services, Insurance Division, and Educational Services.

Besides a full array of construction cost estimating books, Means also publishes a number of other reference works for the construction industry. Subjects include construction estimating and project and business management; special topics such as HVAC, roofing, plumbing, and hazardous waste remediation; and a library of facility management references.

In addition, you can access all of our construction cost data through your computer with Means CostWorks 2001 CD-ROM, an electronic tool that offers over 50,000 lines of Means detailed construction cost data, along with assembly and whole building cost data. You can also access Means cost information from our Web site at www.rsmeans.com

What's more, you can increase your knowledge and improve your construction estimating and management performance with a Means Construction Seminar or In-House Training Program. These two-day seminar programs offer unparalleled opportunities for everyone in your organization to get updated on a wide variety of construction-related issues.

Means also is a worldwide provider of construction cost management and analysis services for commercial and government owners and of claims and valuation services for insurers.

In short, R.S. Means can provide you with the tools and expertise for constructing accurate and dependable construction estimates and budgets in a variety of ways.

Robert Snow Means Established a Tradition of Quality That Continues Today

Robert Snow Means spent years building his company, making certain he always delivered a quality product.

Today, at R.S. Means, we do more than talk about the quality of our data and the usefulness of our books. We stand behind all of our data, from historical cost indexes... to construction materials and techniques... to current costs.

If you have any questions about our products or services, please call us toll-free at 1-800-334-3509. Our customer service representatives will be happy to assist you or visit our Web site at www.rsmeans.com

Table of Contents

Foreword	ii
How the Book Is Built: An Overview	iv
Quick Start	v
How To Use the Book: The Details	vi
Unit Price Section	1
Assemblies Section	407
Reference Section	513
Reference Numbers	514
Change Orders	548
Crew Listings	552
Historical Cost Indexes	577
City Cost Indexes	578
Location Factors	598
Abbreviations	604
Index	607
Other Means Publications and Reference Works	Yellow Pages
Installing Contractor's Overhead & Profit	Inside Back Cover

UNIT PRICES

GENERAL REQUIREMENTS	1
SITE CONSTRUCTION	2
CONCRETE	3
MASONRY	4
METALS	5
WOOD & PLASTICS	6
THERMAL & MOISTURE PROTECTION	7
DOORS & WINDOWS	8
FINISHES	9
SPECIALTIES	10
EQUIPMENT	11
SPECIAL CONSTRUCTION	13
CONVEYING SYSTEMS	14
MECHANICAL	15
ELECTRICAL	16
SQUARE FOOT	17

ASSEMBLIES

MECHANICAL	8
SPECIAL CONSTRUCTION	11
SITE WORK	12

REFERENCE INFORMATION

- REFERENCE NUMBERS
- CREWS
- COST INDEXES
- INDEX

How the Book Is Built: An Overview

A Powerful Construction Tool

You have in your hands one of the most powerful construction tools available today. A successful project is built on the foundation of an accurate and dependable estimate. This book will enable you to construct just such an estimate.

For the casual user the book is designed to be:

- quickly and easily understood so you can get right to your estimate
- filled with valuable information so you can understand the necessary factors that go into the cost estimate

For the regular user, the book is designed to be:

- a handy desk reference that can be quickly referred to for key costs
- a comprehensive, fully reliable source of current construction costs and productivity rates, so you'll be prepared to estimate any project
- a source book for preliminary project cost, product selections, and alternate materials and methods

To meet all of these requirements we have organized the book into the following clearly defined sections.

Quick Start

This one-page section (see following page) can quickly get you started on your estimate.

How To Use the Book: The Details

This section contains an in-depth explanation of how the book is arranged... and how you can use it to determine a reliable construction cost estimate. It includes information about how we develop our cost figures and how to completely prepare your estimate.

Unit Price Section

All cost data has been divided into the 16 divisions according to the MasterFormat system of classification and numbering as developed by the Construction Specifications Institute (CSI) and Construction Specifications Canada (CSC). For a listing of these divisions and an outline of their subdivisions, see the Unit Price Section Table of Contents.

Estimating tips are included at the beginning of each division.

Division 17: Quick Project Estimates
In addition to the 16 Unit Price Divisions there is a S.F. (Square Foot) and C.F. (Cubic Foot) Cost Division, Division 17. It contains costs for 58 different building types that allow you to quickly make a rough estimate for the overall cost of a project or its major components.

Assemblies Section

The cost data in this section has been organized in an "Assemblies" format. These assemblies are the functional elements of a building and are arranged according to the 12 divisions of the UniFormat classification system. For a complete explanation of a typical "Assemblies" page, see "How To Use the Assemblies Cost Tables."

Reference Section

This section includes information on Reference Numbers, Change Orders, Crew Listings, Historical Cost Indexes, City Cost Indexes, Location Factors and a listing of Abbreviations. It is visually identified by a vertical gray bar on the edge of pages.

Reference Numbers: Following selected major classifications throughout the book are "reference numbers" shown in bold squares. These numbers refer you to related information in other sections.

In these other sections, you'll find related tables, explanations, and estimating information. Also included are alternate pricing methods and technical data, along with information on design and economy in construction. You'll also find helpful tips on estimating and construction.

Change Orders: This section includes information on the factors that influence the pricing of change orders.

Crew Listings: This section lists all the crews referenced. For the purposes of this book, a crew is composed of more than one trade classification and/or the addition of equipment to any trade classification. Power equipment is included in the cost of the crew. Costs are shown both with bare labor rates and with the installing contractor's overhead and profit added. For each, the total crew cost per eight-hour day and the composite cost per labor-hour are listed.

Historical Cost Indexes: These indexes provide you with data to adjust construction costs over time. If you know costs for a past project, you can use these indexes to estimate the cost to construct the same project today.

City Cost Indexes: Obviously, costs vary depending on the regional economy. You can adjust the "national average" costs in this book to 305 major cities throughout the U.S. and Canada by using the data in this section.

Location Factors, to quickly adjust the data to over 930 zip code areas, are included.

Abbreviations: A listing of the abbreviations and the terms they represent, is included.

Index

A comprehensive listing of all terms and subjects in this book to help you find what you need quickly.

The Scope of This Book

This book is designed to be comprehensive and easy to use. To that end we have made certain assumptions:

1. We have established material prices based on a "national average."
2. We have computed labor costs based on a 30-city "national average" of union wage rates.
3. We have targeted the data for projects of a certain size range.

For a more detailed explanation of how the cost data is developed, see "How To Use the Book: The Details."

Project Size

This book is aimed primarily at commercial and industrial projects costing $1,000,000 and up, or large multi-family housing projects. Costs are primarily for new construction or major renovation of buildings rather than repairs or minor alterations.

With reasonable exercise of judgment the figures can be used for any building work. *However, for civil engineering structures such as bridges, dams, highways, or the like, please refer to* **Means Heavy Construction Cost Data.**

Quick Start

If you feel you are ready to use this book and don't think you need the detailed instructions that begin on the following page, this Quick Start section is for you.

These steps will allow you to get started estimating in a matter of minutes.

1 First, decide whether you require a Unit Price or Assemblies type estimate. Unit price estimating requires a breakdown of the work to individual items. Assemblies estimates combine individual items or components into building systems.

If you need to estimate each line item separately, follow the instructions for **Unit Prices.**

If you can use an estimate for the entire assembly or system, follow the instructions for **Assemblies.**

2 Find each cost data section you need in the Table of Contents (either for Unit Prices or Assemblies).

Unit Prices: The cost data for Unit Prices has been divided into 16 divisions according to the CSI MasterFormat.

Assemblies: The cost data for Assemblies has been divided into 12 divisions according to the UniFormat.

3 Turn to the indicated section and locate the line item or assemblies table you need for your estimate. Portions of a sample page layout from both the Unit Price Listings and the Assemblies Cost Tables appear below.

Unit Prices: If there is a reference number listed at the beginning of the section, it refers to additional information you may find useful. See the referenced section for additional information.

- Note the crew code designation. You'll find full descriptions of crews in the Crew Listings including labor-hour and equipment costs.

Assemblies: The Assemblies (*not* shown in full here) are generally separated into three parts: 1) an illustration of the system to be estimated; 2) the components and related costs of a typical system; and 3) the costs for similar systems with dimensional and/or size variations. The Assemblies Section also contains reference numbers for additional useful information.

4 Determine the total number of units your job will require.

Unit Prices: Note the unit of measure for the material you're using is listed under "Unit."

- **Bare Costs:** These figures show unit costs for materials and installation. Labor and equipment costs are calculated according to crew costs and average daily output. Bare costs do not contain allowances for overhead, profit or taxes.
- "Labor-hours" allows you to calculate the total labor-hours to complete that task. Just multiply the quantity of work by this figure for an estimate of activity duration.

Assemblies: Note the unit of measure for the assembly or system you're estimating is listed in the Assemblies Table.

5 Then multiply the total units by . . .

Unit Prices: "Total Incl. O&P" which stands for the total cost including the installing contractor's overhead and profit. (See the "How To Use the Unit Price Pages" for a complete explanation.)

Assemblies: The "Total" in the right-hand column, which is the total cost **including the installing contractor's overhead and profit.** (See the "How To Use the Assemblies Cost Tables" section for a complete explanation.)

Material and equipment cost figures include a 10% markup. For labor markups, see the inside back cover of this book. If the work is to be subcontracted, add the general contractor's markup, approximately 10%.

6 The price you calculate will be an estimate for either an individual item of work or a completed assembly or *system*.

7 Compile a list of all items or assemblies included in the total project.

Summarize cost information, and add project overhead.

Localize costs by using the City Cost Indexes or Location Factors found in the Reference Section.

For a more complete explanation of the way costs are derived, please see the following sections.

Editors' Note: We urge you to spend time reading and understanding all of the supporting material and to take into consideration the reference material such as Crews Listing and the "reference numbers."

Unit Price Pages

Assemblies Pages

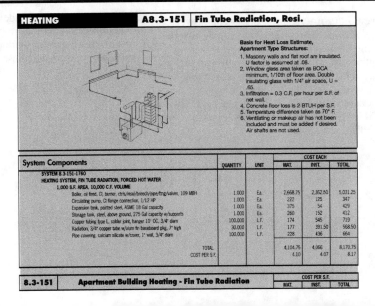

How to Use the Book: The Details

What's Behind the Numbers? The Development of Cost Data

The staff at R.S. Means continuously monitors developments in the construction industry in order to ensure reliable, thorough and up-to-date cost information.

While *overall* construction costs may vary relative to general economic conditions, price fluctuations within the industry are dependent upon many factors. Individual price variations may, in fact, be opposite to overall economic trends. Therefore, costs are continually monitored and complete updates are published yearly. Also, new items are frequently added in response to changes in materials and methods.

Costs—$ (U.S.)

All costs represent U.S. national averages and are given in U.S. dollars. The Means City Cost Indexes can be used to adjust costs to a particular location. The City Cost Indexes for Canada can be used to adjust U.S. national averages to local costs in Canadian dollars.

Material Costs

The R.S. Means staff contacts manufacturers, dealers, distributors, and contractors all across the U.S. and Canada to determine national average material costs. If you have access to current material costs for your specific location, you may wish to make adjustments to reflect differences from the national average. Included within material costs are fasteners for a normal installation. R.S. Means engineers use manufacturers' recommendations, written specifications and/or standard construction practice for size and spacing of fasteners. Adjustments to material costs may be required for your specific application or location. Material costs do not include sales tax.

Labor Costs

Labor costs are based on the average of wage rates from 30 major U.S. cities. Rates are determined from labor union agreements or prevailing wages for construction trades for the current year. Rates along with overhead and profit markups are listed on the inside back cover of this book.

- If wage rates in your area vary from those used in this book, or if rate increases are expected within a given year, labor costs should be adjusted accordingly.

Labor costs reflect productivity based on actual working conditions. These figures include time spent during a normal workday on tasks other than actual installation, such as material receiving and handling, mobilization at site, site movement, breaks, and cleanup.

Productivity data is developed over an extended period so as not to be influenced by abnormal variations and reflects a typical average.

Equipment Costs

Equipment costs include not only rental, but also operating costs for equipment under normal use. The operating costs include parts and labor for routine servicing such as repair and replacement of pumps, filters and worn lines. Normal operating expendables such as fuel, lubricants, tires and electricity (where applicable) are also included. Extraordinary operating expendables with highly variable wear patterns such as diamond bits and blades are excluded. These costs are included under materials. Equipment rental rates are obtained from industry sources throughout North America—contractors, suppliers, dealers, manufacturers, and distributors.

Crew Equipment Cost/Day—The power equipment required for each crew is included in the crew cost. The daily cost for crew equipment is based on dividing the weekly bare rental rate by 5 (number of working days per week), and then adding the hourly operating cost times 8 (hours per day). This "Crew Equipment Cost/Day" is listed in Subdivision 01590.

General Conditions

Cost data in this book is presented in two ways: Bare Costs and Total Cost including O&P (Overhead and Profit). General Conditions, when applicable, should also be added to the Total Cost including O&P. The costs for General Conditions are listed in Division 1 of the Unit Price Section and the Reference Section of this book. General Conditions for the *Installing Contractor* may range from 0% to 10% of the Total Cost including O&P. For the *General* or *Prime Contractor*, costs for General Conditions may range from 5% to 15% of the Total Cost including O&P, with a figure of 10% as the most typical allowance.

Overhead and Profit

Total Cost including O&P for the *Installing Contractor* is shown in the last column on both the Unit Price and the Assemblies pages of this book. This figure is the sum of the bare material cost plus 10% for profit, the base labor cost plus total overhead and profit, and the bare equipment cost plus 10% for profit. Details for the calculation of Overhead and Profit on labor are shown on the inside back cover and in the Reference Section of this book. (See the "How to Use the Unit Price Pages" for an example of this calculation.)

Factors Affecting Costs

Costs can vary depending upon a number of variables. Here's how we have handled the main factors affecting costs.

Quality—The prices for materials and the workmanship upon which productivity is based represent sound construction work. They are also in line with U.S. government specifications.

Overtime—We have made no allowance for overtime. If you anticipate premium time or work beyond normal working hours, be sure to make an appropriate adjustment to your labor costs.

Productivity—The productivity, daily output, and labor-hour figures for each line item are based on working an eight-hour day in daylight hours in moderate temperatures. For work that extends beyond normal work hours or is performed under adverse conditions, productivity may decrease. (See the section in "How To Use the Unit Price Pages" for more on productivity.)

Size of Project—The size, scope of work, and type of construction project will have a significant impact on cost. Economies of scale can reduce costs for large projects. Unit costs can often run higher for small projects. Costs in this book are intended for the size and type of project as previously described in "How the Book Is Built: An Overview." Costs for projects of a significantly different size or type should be adjusted accordingly.

Location—Material prices in this book are for metropolitan areas. However, in dense urban areas, traffic and site storage limitations may increase costs. Beyond a 20-mile radius of large cities, extra trucking or transportation charges may also increase the material costs slightly. On the other hand, lower wage rates may be in effect. Be sure to consider both these factors when preparing an estimate, particularly if the job site is located in a central city or remote rural location.

In addition, highly specialized subcontract items may require travel and per diem expenses for mechanics.

Other factors —
- season of year
- contractor management
- weather conditions
- local union restrictions
- building code requirements
- availability of:
  - adequate energy
  - skilled labor
  - building materials
- owner's special requirements/restrictions
- safety requirements
- environmental considerations

Unpredictable Factors—General business conditions influence "in-place" costs of all items. Substitute materials and construction methods may have to be employed. These may affect the installed cost and/or life cycle costs. Such factors may be difficult to evaluate and cannot necessarily be predicted on the basis of the job's location in a particular section of the country. Thus, where these factors apply, you may find significant, but unavoidable cost variations for which you will have to apply a measure of judgment to your estimate.

Rounding of Costs

In general, all unit prices in excess of $5.00 have been rounded to make them easier to use and still maintain adequate precision of the results. The rounding rules we have chosen are in the following table.

Prices from . . .	Rounded to the nearest . . .
$.01 to $5.00	$.01
$5.01 to $20.00	$.05
$20.01 to $100.00	$.50
$100.01 to $300.00	$1.00
$300.01 to $1,000.00	$5.00
$1,000.01 to $10,000.00	$25.00
$10,000.01 to $50,000.00	$100.00
$50,000.01 and above	$500.00

Contingencies

Contingencies: The allowance for contingencies generally is to provide for unforeseen construction difficulties. On alterations or repair jobs 20% is none too much. If drawings are final and only field contingencies are being considered, 2% or 3% is probably sufficient and often nothing need be added. As far as the contract is concerned, future changes in plans will be covered by extras. The contractor should consider inflationary price trends and possible material shortages during the course of the job. If drawings are not complete or approved or a budget cost wanted, it is wise to add 5% to 10%. Contingencies, then, are a matter of judgment. Additional allowances are shown in Division 01250-200 for contingencies and job conditions and Division 01250-400 for factors to convert prices for repair and remodeling jobs.

Note: For an explanation of "Selective Demolition" versus "Removal for Replacement" see Reference number R15050-720.

Important Estimating Considerations

One reason for listing a job size "minimum" is to ensure that the construction craftsmen are productive for 8 hours a day on a continuous basis. Otherwise, tasks that only require 6 to 7 hours could be billed as an 8-hour day, thus providing an erroneous productivity estimate. The "productivity," or daily output of each craftsman, includes mobilization and cleanup time, break time, plan layout time, as well as an allowance to carry stock from the storage trailer or location on the job site up to 200' into the building and on the first or second floor. If material has to be transported over greater distances or to higher floors, an additional allowance should be considered by the estimator. An allowance has also been included in the piping and fittings installation time for a leak check and minor tightening.

Equipment installation time includes all applicable items following: positioning, leveling and securing the unit in place, connecting all associated piping, ducts, vents, etc., which shall have been estimated separately, connecting to an adjacent power source, filling/bleeding, startup, adjusting the controls up and down to assure proper response, setting the integral controls/valves/regulators/thermostats for proper operation (does not include external building type control systems, DDC systems, etc.), explaining/training owner's operator, and warranty. A reasonable breakdown of the labor costs would be as follows:

1. Movement into building, installation/ setting of equipment — 35%
2. Connecting to piping/duct/power, etc. — 25%
3. Filling/flushing/cleaning/touchup, etc. — 15%
4. Startup/running adjustments — 5%
5. Training owner's rep. — 5%
6. Warranty/call back/service — 15%

Major Sections of this Book	CSI Format		Assemblies Format	
	Division	Description	Division	Description
Mechanical Related Items	1	General Requirements	*1	Foundations
	2	Site Work	*2	Substructures
	3	Concrete	*3	Superstructures
	4	Masonry	*4	Exterior Closure
	5	Metals	*5	Roofing
	6	Wood and Plastics	*6	Interior Construction
	7	Thermal and Moisture Protection	*7	Conveying Systems
	8	Doors and Windows		
	*9	Finishes		
Items have been rearranged to be consistent with major sections of this book.	10	Specialties	*10	General Conditions & Profit
	11	Equipment	*11	Special Construction
	*12	Furnishings	12	Site Work
	13	Special Construction		
	14	Conveying Systems		
Mechanical	15	Mechanical	8	Mechanical
Electrical	16	Electrical	*9	Electrical
Cost Modifications	17	Square Foot	14	Square Foot Costs
	Appendix	City Cost Index		
*Division does not appear in this book.				

Note that cranes or other lifting equipment are not included on any lines in Division 15. For example, if a crane is required to lift a heavy piece of pipe into place high above a gym floor, or to put a rooftop unit on the roof of a four-story building, etc., it must be added. Due to the potential for extreme variation—from nothing additional required, to a major crane or helicopter—we feel that including a nominal amount for "lifting contingency" would be useless and detract from the accuracy of the estimate. When using equipment rental from Means do not forget to include the cost of the operator(s).

Organization of the Book

Format Division Numbers—Note that the approximate relation between the CSI format numbers and the Assemblies format are given above. Not all the divisions appear in the book.

Estimating Labor-Hours

The labor-hours expressed in this publication are based on Average Installation time, using an efficiency level of approximately 60-65% (see item 7) which has been found reasonable and acceptable by many contractors.

The book uses this national efficiency average to establish a consistent benchmark.

For bid situations, adjustments to this efficiency level should be the responsibility of the contractor bidding the project.

The unit labor-hour is divided in the following manner. A typical day for a journeyman might be:

1.	Study Plans	3%	14.4 min.
2.	Material Procurement	3%	14.4 min.
3.	Receiving and Storing	3%	14.4 min
4.	Mobilization	5%	24.0 min.
5.	Site Movement	5%	24.0 min.
6.	Layout and Marking	8%	38.4 min.
7.	Actual Installation	64%	307.2 min.
8.	Cleanup	3%	14.4 min.
9.	Breaks, Non-Productive	6%	28.8 min.
		100%	480.0 min.

If any of the percentages expressed in this breakdown do not apply to the particular work or project situation, then that percentage or a portion of it may be deducted from or added to labor hours.

Final Checklist

Estimating can be a straightforward process provided you remember the basics. Here's a checklist of some of the items you should remember to do before completing your estimate.

Did you remember to . . .

- factor in the City Cost Index for your locale
- take into consideration which items have been marked up and by how much
- mark up the entire estimate sufficiently for your purposes
- read the background information on techniques and technical matters that could impact your project time span and cost
- include all components of your project in the final estimate
- double check your figures to be sure of your accuracy
- for more information, please see "Tips for Accurate Estimating," R01100-005 in the Reference Section.
- call R.S. Means if you have any questions about your estimate or the data you've found in our publications

Remember, R.S. Means stands behind its publications. If you have any questions about your estimate . . . about the costs you've used from our books . . . or even about the technical aspects of the job that may affect your estimate, feel free to call the R.S. Means editors at 1-800-334-3509.

Unit Price Section

Table of Contents

Div No.

	General Requirements	**5**
01100	Summary	6
01200	Price & Payment Procedures	6
01300	Administrative Requirements	7
01500	Temporary Facilities & Controls	8
01700	Execution Requirements	18
01800	Facility Operation	18
	Site Construction	**19**
02050	Basic Site Materials & Methods	20
02100	Site Remediation	21
02200	Site Preparation	22
02300	Earthwork	25
02400	Tunneling, Boring & Jacking	28
02500	Utility Services	28
02600	Drainage & Containment	40
02700	Bases, Ballasts, Pavements, & Appurtenances	43
02800	Site Improvements & Amenities	44
02950	Site Restoration & Rehab.	45
	Concrete	**49**
03100	Concrete Forms & Accessories	50
03200	Concrete Reinforcement	50
03300	Cast-In-Place Concrete	50
	Masonry	**53**
04050	Basic Masonry Mat. & Methods	54
04800	Masonry Assemblies	54
	Metals	**55**
05050	Basic Materials & Methods	56
05100	Structural Metal Framing	58
	Wood & Plastics	**61**
06050	Basic Wd. & Plastic Mat. & Meth.	62
06100	Rough Carpentry	63
06400	Architectural Woodwork	63
	Thermal & Moisture Protection	**65**
07600	Flashing & Sheet Metal	66
07700	Roof Specialties & Accessories	66
07800	Fire & Smoke Protection	67
07900	Joint Sealers	68

Div No.

	Doors & Windows	**71**
08300	Specialty Doors	72
	Finishes	**75**
09900	Paints & Coatings	76
	Specialties	**77**
10200	Louvers & Vents	78
10270	Access Flooring	78
10300	Fireplaces & Stoves	78
	Equipment	**81**
11010	Maintenance Equipment	82
11100	Mercantile Equipment	82
11110	Coml. Laun. & Dry Cln. Eq.	82
11140	Vehicle Service Equip.	83
11170	Solid Waste Handling Equip.	83
11300	Fluid Waste Treat. & Disp. Equip.	84
11400	Food Service Equipment	85
11450	Residential Equipment	86
11470	Darkroom Equipment	87
11500	Industrial & Process Equipment	88
11600	Laboratory Equipment	90
11700	Medical Equipment	90
	Special Construction	**93**
13030	Special Purpose Rooms	94
13120	Pre-Engineered Structures	96
13150	Swimming Pools	96
13175	Ice Rinks	97
13200	Storage Tanks	97
13280	Hazardous Mat. Remediation	101
13600	Solar & Wind Energy Equipment	104
13800	Building Automation & Control	107
13850	Detection & Alarm	116
	Conveying Systems	**117**
14500	Material Handling	118

Div No.

	Mechanical	**119**
15050	Basic Materials & Methods	120
15100	Building Services Piping	141
15200	Process Piping	276
15400	Plumbing Fixtures & Equipment	282
15500	Heat Generation Equipment	286
15600	Refrigeration Equipment	311
15700	Heating, Ventilation, & A/C Equip.	319
15800	Air Distribution	347
15900	HVAC Instrumentation Controls	383
15950	Testing/Adjusting/Balancing	384
	Electrical	**387**
16100	Wiring Methods	388
16200	Electrical Power	396
16400	Low-Voltage Distribution	396
	Square Foot	**399**
17100	S.F., C.F. & % of Total Costs	400

How to Use the Unit Price Pages

The following is a detailed explanation of a sample entry in the Unit Price Section. Next to each bold number below is the item being described with appropriate component of the sample entry following in parenthesis. Some prices are listed as bare costs, others as costs that include overhead and profit of the installing contractor. In most cases, if the work is to be subcontracted, the general contractor will need to add an additional markup (R.S. Means suggests using 10%) to the figures in the column "Total Incl. O&P."

1 Division Number/Title (15100/Building Services Piping)

Use the Unit Price Section Table of Contents to locate specific items. The sections are classified according to the CSI MasterFormat.

2 Line Numbers (15110 200 1090)

Each unit price line item has been assigned a unique 12-digit code based on the CSI MasterFormat classification.

- Level One - CSI-MasterFormat Division
- Level Two - CSI
- **15100**
- **15110-200-1090**
- Means 12 digit Line Number
- Level Four - Means
- Level Three - CSI

3 Description (Valves, Iron Body)

Each line item is described in detail. Sub-items and additional sizes are indented beneath the appropriate line items. The first line or two after the main item (in boldface) may contain descriptive information that pertains to all line items beneath this boldface listing.

4 Reference Number Information

R15100-090 You'll see reference numbers shown in bold rectangles at the beginning of some sections. These refer to related items in the Reference Section, visually identified by a vertical gray bar on the edge of pages.

The relation may be: (1) an estimating procedure that should be read before estimating, (2) an alternate pricing method, or (3) technical information.

The "R" designates the Reference Section. The numbers refer to the MasterFormat classification system. An "A" denotes the Assemblies Section and the numbers refer to the UniFormat classification system.

Example: The rectangle number above is directing you to refer to the reference number R15100-090. This particular reference number provides technical data on valve selection.

Crew (Q-2)

The "Crew" column designates the typical trade or crew used to install the item. If an installation can be accomplished by one trade and requires no power equipment, that trade and the number of workers are listed (for example, "2 Plum"). If an installation requires a composite crew, a crew code designation is listed (for example, "Q-2"). You'll find full details on all composite crews in the Crew Listings.
- For a complete list of all trades utilized in this book and their abbreviations, see the inside back cover.

Crews

Crew No.	Bare Costs		Incl. Subs O & P		Cost Per Labor-Hour	
Crew Q-2	Hr.	Daily	Hr.	Daily	Bare Costs	Incl. O&P
2 Plumbers	$34.45	$551.20	$52.00	$832.00	$32.15	$48.53
1 Plumber Apprentice	27.55	220.40	41.60	332.80		
24 L.H., Daily Totals		$771.60		$1164.80	$32.15	$48.53

 ## Productivity: Daily Output (4.50)/ Labor-Hours (5.333)

The "Daily Output" represents the typical number of units the designated crew will install in a normal 8-hour day. To find out the number of days the given crew would require to complete the installation, divide your quantity by the daily output. For example:

Quantity	÷	Daily Output	=	Duration
10 Ea.	÷	4.50 Ea./ Crew Day	=	2.22 Crew Days

The "Labor-Hours" figure represents the number of labor-hours required to install one unit of work. To find out the number of labor-hours required for your particular task, multiply the quantity of the item times the number of labor-hours shown. For example:

Quantity	x	Productivity Rate	=	Duration
10 Ea.	x	5.333 Labor-Hours/Ea.	=	53.33 Labor-Hours

 ## Unit (Ea.)

The abbreviated designation indicates the unit of measure upon which the price, production, and crew are based (Ea. = Each). For a complete listing of abbreviations refer to the Abbreviations Listing in the Reference Section of this book.

 ## Bare Costs:
Mat. (Bare Material Cost) (264)

The unit material cost is the "bare" material cost with no overhead and profit included. *Costs shown reflect national average material prices for January of the current year and include delivery to the job site. No sales taxes are included.*

Labor (171)

The unit labor cost is derived by multiplying bare labor-hour costs for Crew Q-2 by labor-hour units. The bare labor-hour cost is found in the Crew Section under Q-2. (If a trade is listed, the hourly labor cost—the wage rate—is found on the inside back cover.)

Labor-Hour Cost Crew Q-2	x	Labor-Hour Units	=	Labor
$32.15	x	5.333	=	$171.46

Equip. (Equipment) (0)

Equipment costs for each crew are listed in the description of each crew. Tools or equipment whose value justifies purchase or ownership by a contractor are considered overhead as shown on the inside back cover. The unit equipment cost is derived by multiplying the bare equipment hourly cost by the labor-hour units.

Equipment Cost Crew Q-2	x	Labor-Hour Units	=	Equip.
$0	x	5.333	=	$0

Total (435)

The total of the bare costs is the arithmetic total of the three previous columns: mat., labor, and equip.

Material	+	Labor	+	Equip.	=	Total
$264	+	$171	+	$0	=	$435

 ## Total Costs Including O&P

This figure is the sum of the bare material cost plus 10% for profit; the bare labor cost plus total overhead and profit (per the inside back cover or, if a crew is listed, from the crew listings); and the bare equipment cost plus 10% for profit.

Material is Bare Material Cost + 10% = 264 + 26.40	=	$290
Labor for Crew Q-2 = Labor-Hour Cost (48.53) x Labor-Hour Units (5.333)	=	$258.81
Equip. is Bare Equip. Cost + 10% = 0 + 0	=	$0
Total (Rounded)	=	$550

Division 1
General Requirements

Estimating Tips

The General Requirements of any contract are very important to both the bidder and the owner. These lay the ground rules under which the contract will be executed and have a significant influence on the cost of operations. Therefore, it is extremely important to thoroughly read and understand the General Requirements both before preparing an estimate and when the estimate is complete, to ascertain that nothing in the contract is overlooked. Caution should be exercised when applying items listed in Division 1 to an estimate. Many of the items are included in the unit prices listed in the other divisions such as mark-ups on labor and company overhead.

01300 Administrative Requirements
- Before determining a final cost estimate, it is a good practice to review all the items listed in subdivision 01300 to make final adjustments for items that may need customizing to specific job conditions.

01330 Submittal Procedures
- Requirements for initial and periodic submittals can represent a significant cost to the General Requirements of a job. Thoroughly check the submittal specifications when estimating a project to determine any costs that should be included.

01400 Quality Requirements
- All projects will require some degree of Quality Control. This cost is not included in the unit cost of construction listed in each division. Depending upon the terms of the contract, the various costs of inspection and testing can be the responsibility of either the owner or the contractor. Be sure to include the required costs in your estimate.

01500 Temporary Facilities & Controls
- Barricades, access roads, safety nets, scaffolding, security and many more requirements for the execution of a safe project are elements of direct cost. These costs can easily be overlooked when preparing an estimate. When looking through the major classifications of this subdivision, determine which items apply to each division in your estimate.

01590 Equipment Rental
- This subdivision contains transportation, handling, storage, protection and product options and substitutions. Listed in this cost manual are average equipment rental rates for all types of equipment. This is useful information when estimating the time and materials requirement of any particular operation in order to establish a unit or total cost.
- A good rule of thumb is that weekly rental is 3 times daily rental and that monthly rental is 3 times weekly rental.

- The figures in the column for Crew Equipment Cost represent the rental rate used in determining the daily cost of equipment in a crew. It is calculated by dividing the weekly rate by 5 days and adding the hourly operating cost times 8 hours.

01770 Closeout Procedures
- When preparing an estimate, read the specifications to determine the requirements for Contract Closeout thoroughly. Final cleaning, record documentation, operation and maintenance data, warranties and bonds, and spare parts and maintenance materials can all be elements of cost for the completion of a contract. Do not overlook these in your estimate.

01830 Operations & Maintenance
- If maintenance and repair are included in your contract, they require special attention. To estimate the cost to remove and replace any unit usually requires a site visit to determine the accessibility and the specific difficulty at that location. Obstructions, dust control, safety, and often overtime hours must be considered when preparing your estimate.

Reference Numbers
Reference numbers are shown in bold squares at the beginning of some major classifications. These numbers refer to related items in the Reference Section. The reference information may be an estimating procedure, an alternate pricing method or technical information.

Note: Not all subdivisions listed here necessarily appear in this publication.

01100 | Summary

01107 | Professional Consultant

			CREW	DAILY OUTPUT	LABOR-HOURS	UNIT	MAT.	LABOR	EQUIP.	TOTAL	TOTAL INCL O&P	
300	0010	**ENGINEERING FEES**										300
	0020	Educational planning consultant, minimum				Project					.50%	
	0100	Maximum				"					2.50%	
	0200	Electrical, minimum				Contrct					4.10%	
	0300	Maximum									10.10%	
	0400	Elevator & conveying systems, minimum									2.50%	
	0500	Maximum									5%	
	0600	Food service & kitchen equipment, minimum									8%	
	0700	Maximum									12%	
	1000	Mechanical (plumbing & HVAC), minimum									4.10%	
	1100	Maximum									10.10%	

01200 | Price & Payment Procedures

01250 | Contract Modification Procedures

			CREW	DAILY OUTPUT	LABOR-HOURS	UNIT	MAT.	LABOR	EQUIP.	TOTAL	TOTAL INCL O&P	
200	0010	**CONTINGENCIES** for estimate at conceptual stage				Project					20%	200
	0050	Schematic stage									15%	
	0100	Preliminary working drawing stage (Design Dev.)									10%	
	0150	Final working drawing stage									3%	
300	0010	**CREWS** For building construction, see How To Use This Book										300
400	0010	**FACTORS** Cost adjustments										400
	0100	Add to construction costs for particular job requirements										
	0500	Cut & patch to match existing construction, add, minimum				Costs	2%	3%				
	0550	Maximum					5%	9%				
	0800	Dust protection, add, minimum					1%	2%				
	0850	Maximum					4%	11%				
	1100	Equipment usage curtailment, add, minimum					1%	1%				
	1150	Maximum					3%	10%				
	1400	Material handling & storage limitation, add, minimum					1%	1%				
	1450	Maximum					6%	7%				
	1700	Protection of existing work, add, minimum					2%	2%				
	1750	Maximum					5%	7%				
	2000	Shift work requirements, add, minimum						5%				
	2050	Maximum						30%				
	2300	Temporary shoring and bracing, add, minimum					2%	5%				
	2350	Maximum					5%	12%				
	2400	Work inside prisons and high security areas, add, minimum						30%				
	2450	Maximum						50%				
500	0010	**JOB CONDITIONS** Modifications to total										500
	0020	project cost summaries										
	0100	Economic conditions, favorable, deduct				Project					2%	
	0200	Unfavorable, add									5%	
	0300	Hoisting conditions, favorable, deduct									2%	
	0400	Unfavorable, add									5%	
	0700	Labor availability, surplus, deduct									1%	
	0800	Shortage, add									10%	
	0900	Material storage area, available, deduct									1%	
	1000	Not available, add									2%	

01200 | Price & Payment Procedures

01250 | Contract Modification Procedures

			CREW	DAILY OUTPUT	LABOR-HOURS	UNIT	2001 BARE COSTS MAT.	LABOR	EQUIP.	TOTAL	TOTAL INCL O&P	
500	1100	Subcontractor availability, surplus, deduct				Project					5%	500
	1200	Shortage, add									12%	
	1300	Work space, available, deduct									2%	
	1400	Not available, add									5%	
600	0010	**OVERTIME** For early completion of projects or where	R01100-110			Costs		100%				600
	0020	labor shortages exist, add to usual labor, up to										

01255 | Cost Indexes

			CREW	DAILY OUTPUT	LABOR-HOURS	UNIT	MAT.	LABOR	EQUIP.	TOTAL	INCL O&P	
200	0010	**CONSTRUCTION COST INDEX** (Reference) over 930 zip code locations in				%						200
	0020	The U.S. and Canada, total bldg cost, min. (Clarksdale, MS)									66.80%	
	0050	Average									100%	
	0100	Maximum (New York, NY)									134.50%	
400	0010	**HISTORICAL COST INDEXES** (Reference) Back to 1951										400
500	0010	**LABOR INDEX** (Reference) For over 930 zip code locations in				%		35%				500
	0020	the U.S. and Canada, minimum (Clarksdale, MS)										
	0050	Average						100%				
	0100	Maximum (New York, NY)						161.90%				
600	0011	**MATERIAL INDEX** For over 930 zip code locations in				%	92.20%					600
	0020	the U.S. and Canada, minimum (Elizabethtown, KY)										
	0040	Average					100%					
	0060	Maximum (Ketchikan, AK)					144.20%					

01290 | Payment Procedures

			CREW	DAILY OUTPUT	LABOR-HOURS	UNIT	MAT.	LABOR	EQUIP.	TOTAL	INCL O&P	
800	0010	**TAXES** Sales tax, State, average	R01100-090			%	4.71%					800
	0050	Maximum					7.25%					
	0200	Social Security, on first $76,200 of wages	R01100-100					7.65%				
	0300	Unemployment, MA, combined Federal and State, minimum						2.10%				
	0350	Average						7%				
	0400	Maximum						8%				

01300 | Administrative Requirements

01310 | Project Management/Coordination

			CREW	DAILY OUTPUT	LABOR-HOURS	UNIT	MAT.	LABOR	EQUIP.	TOTAL	INCL O&P	
150	0010	**PERMITS** Rule of thumb, most cities, minimum				Job					.50%	150
	0100	Maximum				"					2%	
200	0010	**PERFORMANCE BOND** For buildings, minimum	R01100-080			Job					.60%	200
	0100	Maximum				"					2.50%	
350	0010	**INSURANCE** Builders risk, standard, minimum	R01100-040			Job					.22%	350
	0050	Maximum									.59%	
	0200	All-risk type, minimum	R01100-060								.25%	
	0250	Maximum									.62%	
	0400	Contractor's equipment floater, minimum				Value					.50%	
	0450	Maximum				"					1.50%	
	0600	Public liability, average				Job					1.55%	
	0800	Workers' compensation & employer's liability, average										
	0850	by trade, carpentry, general				Payroll		19.01%				
	1000	Electrical						6.79%				
	1150	Insulation						17.65%				
	1450	Plumbing						8.47%				

01300 | Administrative Requirements

01310 | Project Management/Coordination

			CREW	DAILY OUTPUT	LABOR-HOURS	UNIT	2001 BARE COSTS MAT.	LABOR	EQUIP.	TOTAL	TOTAL INCL O&P	
350	1550	Sheet metal work (HVAC)	R01100-040			Payroll		11.93%				350
500	0010	**MARK-UP** For General Contractors for change	R01100-070									500
	0100	of scope of job as bid										
	0200	Extra work, by subcontractors, add				%					10%	
	0250	By General Contractor, add									15%	
	0400	Omitted work, by subcontractors, deduct all but									5%	
	0450	By General Contractor, deduct all but									7.50%	
	0600	Overtime work, by subcontractors, add									15%	
	0650	By General Contractor, add									10%	
	1000	Installing contractors, on his own labor, minimum						47.80%				
	1100	Maximum						87.50%				
600	0010	**OVERHEAD** As percent of direct costs, minimum	R01100-070			%				5%		600
	0050	Average								13%		
	0100	Maximum								30%		
620	0010	**OVERHEAD & PROFIT** Allowance to add to items in this	R01100-070									620
	0020	book that do not include Subs O&P, average				%					25%	
	0100	Allowance to add to items in this book that										
	0110	do include Subs O&P, minimum				%					5%	
	0150	Average									10%	
	0200	Maximum									15%	
	0300	Typical, by size of project, under $100,000									30%	
	0350	$500,000 project									25%	
	0400	$2,000,000 project									20%	
	0450	Over $10,000,000 project									15%	

01321 | Construction Photos

			CREW	DAILY OUTPUT	LABOR-HOURS	UNIT	MAT.	LABOR	EQUIP.	TOTAL	TOTAL INCL O&P	
500	0010	**PHOTOGRAPHS** 8" x 10", 4 shots, 2 prints ea., std. mounting				Set	263			263	289	500
	0100	Hinged linen mounts					290			290	320	
	0200	8" x 10", 4 shots, 2 prints each, in color					310			310	340	
	0300	For I.D. slugs, add to all above					3.75			3.75	4.13	
	1500	Time lapse equipment, camera and projector, buy					3,575			3,575	3,925	
	1550	Rent per month					530			530	585	
	1700	Cameraman and film, including processing, B.&W.				Day	1,225			1,225	1,350	
	1720	Color				"	1,225			1,225	1,350	

01500 | Temporary Facilities & Controls

01510 | Temporary Utilities

			CREW	DAILY OUTPUT	LABOR-HOURS	UNIT	2001 BARE COSTS MAT.	LABOR	EQUIP.	TOTAL	TOTAL INCL O&P	
800	0010	**TEMPORARY UTILITIES**										800
	0100	Heat, incl. fuel and operation, per week, 12 hrs. per day	1 Skwk	100	.080	CSF Flr	5.10	2.39		7.49	9.40	
	0200	24 hrs. per day	"	60	.133		7.70	3.98		11.68	14.70	
	0350	Lighting, incl. service lamps, wiring & outlets, minimum	1 Elec	34	.235		2.07	8.05		10.12	14.35	
	0360	Maximum	"	17	.471		4.51	16.15		20.66	29	
	0400	Power for temp lighting only, per month, min/month 6.6 KWH								.75	1.18	
	0450	Maximum/month 23.6 KWH								2.85	3.14	
	0600	Power for job duration incl. elevator, etc., minimum								47	51.70	
	0650	Maximum								110	121	
	1000	Toilet, portable, see division 01590-400										

01500 | Temporary Facilities & Controls

01520 | Construction Facilities

			CREW	DAILY OUTPUT	LABOR-HOURS	UNIT	2001 BARE COSTS MAT.	LABOR	EQUIP.	TOTAL	TOTAL INCL O&P	
500	0010	**OFFICE** Trailer, furnished, no hookups, 20' x 8', buy	2 Skwk	1	16	Ea.	5,325	480		5,805	6,625	500
	0250	Rent per month					150			150	165	
	0300	32' x 8', buy	2 Skwk	.70	22.857		7,825	680		8,505	9,700	
	0350	Rent per month					165			165	182	
	0400	50' x 10', buy	2 Skwk	.60	26.667		13,400	795		14,195	16,000	
	0450	Rent per month					295			295	325	
	0500	50' x 12', buy	2 Skwk	.50	32		15,800	955		16,755	18,900	
	0550	Rent per month					345			345	380	
	0700	For air conditioning, rent per month, add					35.50			35.50	39	
	0800	For delivery, add per mile				Mile	1.50			1.50	1.65	
	1000	Portable buildings, prefab, on skids, economy, 8' x 8'	2 Carp	265	.060	S.F.	80	1.76		81.76	91	
	1100	Deluxe, 8' x 12'	"	150	.107	"	87	3.11		90.11	100	
	1200	Storage boxes, 20' x 8', buy	2 Skwk	1.80	8.889	Ea.	3,300	265		3,565	4,050	
	1250	Rent per month					72.50			72.50	80	
	1300	40' x 8', buy	2 Skwk	1.40	11.429		3,950	340		4,290	4,875	
	1350	Rent per month					103			103	113	
550	0010	**FIELD OFFICE EXPENSE**										550
	0100	Field office expense, office equipment rental average				Month	136			136	150	
	0120	Office supplies, average				"	82			82	90	
	0125	Office trailer rental, see division 01520-500										
	0140	Telephone bill; avg. bill/month incl. long dist.				Month	200			200	220	
	0160	Field office lights & HVAC				"	92			92	101	

01530 | Temporary Construction

700	0010	**PROTECTION** Stair tread, 2" x 12" planks, 1 use	1 Carp	75	.107	Tread	4.43	3.11		7.54	9.75	700
	0100	Exterior plywood, 1/2" thick, 1 use		65	.123		1.48	3.59		5.07	7.25	
	0200	3/4" thick, 1 use	↓	60	.133	↓	2.25	3.89		6.14	8.55	
900	0010	**WINTER PROTECTION** Reinforced plastic on wood										900
	0100	framing to close openings	2 Clab	750	.021	S.F.	.35	.49		.84	1.15	
	0200	Tarpaulins hung over scaffolding, 8 uses, not incl. scaffolding		1,500	.011		.16	.24		.40	.56	
	0250	Tarpaulin polyester reinf. w/ integral fastening system 11 mils thick		1,600	.010		.73	.23		.96	1.16	
	0300	Prefab fiberglass panels, steel frame, 8 uses	↓	1,200	.013	↓	.70	.30		1	1.25	

01540 | Construction Aids

750	0010	**SCAFFOLDING**										750
	0015	Steel tubular, rent, 1 use/mo, no plank, erect & dismantle	R01540-100									
	0090	Building exterior, wall face, 1 to 5 stories	3 Carp	24	1	C.S.F.	24.50	29		53.50	72	
	0200	6 to 12 stories	4 Carp	21.20	1.509		24.50	44		68.50	95.50	
	0310	13 to 20 stories	5 Carp	20	2		24.50	58.50		83	118	
	0460	Building interior, wall face area, up to 16' high	3 Carp	25	.960		24.50	28		52.50	70.50	
	0560	16' to 40' high		23	1.043	↓	24.50	30.50		55	74	
	0800	Building interior floor area, up to 30' high		312	.077	C.C.F.	1.87	2.24		4.11	5.55	
	0900	Over 30' high	4 Carp	275	.116	"	1.87	3.39		5.26	7.35	
	0910	Steel tubular, heavy duty shoring, buy										
	0920	Frames 5' high 2' wide				Ea.	75			75	82.50	
	0925	5' high 4' wide					85			85	93.50	
	0930	6' high 2' wide					86			86	94.50	
	0935	6' high 4' wide				↓	101			101	111	
	0940	Accessories										
	0945	Cross braces				Ea.	16			16	17.60	
	0950	U-head, 8" x 8"					16			16	17.60	
	0955	J-head, 4" x 8"					12			12	13.20	
	0960	Base plate, 8" x 8"					13			13	14.30	
	0965	Leveling jack	↓				30.50			30.50	33.50	

GENERAL REQUIREMENTS

01500 | Temporary Facilities & Controls

01540 | Construction Aids

			CREW	DAILY OUTPUT	LABOR-HOURS	UNIT	2001 BARE COSTS MAT.	LABOR	EQUIP.	TOTAL	TOTAL INCL O&P	
750	1000	Steel tubular, regular, buy										750
	1100	Frames 3' high 5' wide				Ea.	58			58	64	
	1150	5' high 5' wide					67			67	73.50	
	1200	6'-4" high 5' wide					84			84	92.50	
	1350	7'-6" high 6' wide					145			145	160	
	1500	Accessories cross braces					15			15	16.50	
	1550	Guardrail post					15			15	16.50	
	1600	Guardrail 7' section					7.25			7.25	8	
	1650	Screw jacks & plates					24			24	26.50	
	1700	Sidearm brackets					28			28	31	
	1750	8" casters					33			33	36.50	
	1800	Plank 2" x 10" x 16'-0"					42.50			42.50	47	
	1900	Stairway section					235			235	259	
	1910	Stairway starter bar					21			21	23.50	
	1920	Stairway inside handrail					53			53	58.50	
	1930	Stairway outside handrail					73			73	80.50	
	1940	Walk-thru frame guardrail					28			28	31	
	2000	Steel tubular, regular, rent/mo.										
	2100	Frames 3' high 5' wide				Ea.	3.75			3.75	4.13	
	2150	5' high 5' wide					3.75			3.75	4.13	
	2200	6'-4" high 5' wide					3.75			3.75	4.13	
	2250	7'-6" high 6' wide					7			7	7.70	
	2500	Accessories, cross braces					.60			.60	.66	
	2550	Guardrail post					1			1	1.10	
	2600	Guardrail 7' section					.75			.75	.83	
	2650	Screw jacks & plates					1.50			1.50	1.65	
	2700	Sidearm brackets					1.50			1.50	1.65	
	2750	8" casters					6			6	6.60	
	2800	Outrigger for rolling tower					3			3	3.30	
	2850	Plank 2" x 10" x 16'-0"					5			5	5.50	
	2900	Stairway section					10			10	11	
	2910	Stairway starter bar					.10			.10	.11	
	2920	Stairway inside handrail					5			5	5.50	
	2930	Stairway outside handrail					5			5	5.50	
	2940	Walk-thru frame guardrail					2			2	2.20	
	3000	Steel tubular, heavy duty shoring, rent/mo.										
	3250	5' high 2' & 4' wide				Ea.	5			5	5.50	
	3300	6' high 2' & 4' wide					5			5	5.50	
	3500	Accessories, cross braces					1			1	1.10	
	3600	U - head, 8" x 8"					1			1	1.10	
	3650	J - head, 4" x 8"					1			1	1.10	
	3700	Base plate, 8" x 8"					1			1	1.10	
	3750	Leveling jack					2			2	2.20	
	5700	Planks, 2"x10"x16'-0" in place, up to 50' high	3 Carp	144	.167			4.86		4.86	7.60	
	5800	In place over 50' high	4 Carp	160	.200			5.85		5.85	9.10	
	6000	Heavy duty shoring for elevated slab forms to 8'-2" high, floor area										
	6010	Set up and take down										
	6100	1 use/month	4 Carp	36	.889	C.S.F.	29.50	26		55.50	73	
	6150	2 uses/month	"	36	.889	"	14.80	26		40.80	57	
	6500	To 14'-8" high										
	6600	1 use/month	4 Carp	18	1.778	C.S.F.	43	52		95	129	
	6650	2 uses/month	"	18	1.778	"	21.50	52		73.50	105	
755	0011	**SCAFFOLDING SPECIALTIES**										755
	1200	Sidewalk bridge, heavy duty steel posts & beams, including										
	1210	parapet protection & waterproofing										
	1220	8' to 10' wide, 2 posts	3 Carp	15	1.600	L.F.	42	46.50		88.50	119	

01500 | Temporary Facilities & Controls

01540 | Construction Aids

		CREW	DAILY OUTPUT	LABOR-HOURS	UNIT	2001 BARE COSTS MAT.	LABOR	EQUIP.	TOTAL	TOTAL INCL O&P	
755	1230	3 posts	3 Carp	10	2.400	L.F.	63	70		133	179	755
	1500	Sidewalk bridge using tubular steel										
	1510	scaffold frames, including planking	3 Carp	45	.533	L.F.	4.72	15.55		20.27	29.50	
	1600	For 2 uses per month, deduct from all above					50%					
	1700	For 1 use every 2 months, add to all above					100%					
	1900	Catwalks, 32" wide, no guardrails, 6' span, buy				Ea.	120			120	132	
	2000	10' span, buy				"	190			190	209	
	2800	Hand winch-operated masons										
	2810	scaffolding, no plank moving required										
	2900	98' long, 10'-6" high, buy				Ea.	19,700			19,700	21,700	
	3000	Rent per month					790			790	865	
	3100	28'-6" high, buy					25,700			25,700	28,200	
	3200	Rent per month					1,025			1,025	1,125	
	3400	196' long, 28'-6" high, buy					49,700			49,700	54,500	
	3500	Rent per month					2,000			2,000	2,175	
	3600	64'-6" high, buy					73,000			73,000	80,000	
	3700	Rent per month					2,900			2,900	3,200	
	3720	Putlog, standard, 8' span, with hangers, buy					61			61	67	
	3730	Rent per month					10			10	11	
	3750	12' span, buy					92			92	101	
	3755	Rent per month					15			15	16.50	
	3760	Trussed type, 16' span, buy					210			210	231	
	3770	Rent per month					20			20	22	
	3790	22' span, buy					252			252	277	
	3795	Rent per month					30			30	33	
	3800	Rolling ladders with handrails, 30" wide, buy, 2 step					144			144	158	
	4000	7 step					430			430	475	
	4050	10 step					645			645	710	
	4100	Rolling towers, buy, 5' wide, 7' long, 10' high					1,150			1,150	1,250	
	4200	For 5' high added sections, to buy, add					188			188	207	
	4300	Complete incl. wheels, railings, outriggers,										
	4350	21' high, to buy				Ea.	1,925			1,925	2,125	
	4400	Rent/month				"	138			138	152	
800	0010	**TARPAULINS** Cotton duck, 10 oz. to 13.13 oz. per S.Y., minimum				S.F.	.46			.46	.51	800
	0050	Maximum					.55			.55	.61	
	0100	Polyvinyl coated nylon, 14 oz. to 18 oz., minimum					.45			.45	.50	
	0150	Maximum					.65			.65	.72	
	0200	Reinforced polyethylene 3 mils thick, white					.10			.10	.11	
	0300	4 mils thick, white, clear or black					.12			.12	.13	
	0400	5.5 mils thick, clear					.09			.09	.10	
	0500	White, fire retardant					.16			.16	.18	
	0600	7.5 mils, oil resistant, fire retardant					.17			.17	.19	
	0700	8.5 mils, black					.22			.22	.24	
	0710	Woven polyethylene, 6 mils thick					.45			.45	.50	
	0730	Polyester reinforced w/ integral fastening system 11 mils thick					1			1	1.10	
	0740	Mylar polyester, non-reinforced, 7 mils thick					1.10			1.10	1.21	
820	0010	**SMALL TOOLS** As % of contractor's work, minimum				Total					.50%	820
	0100	Maximum				"					2%	

01550 | Vehicular Access & Parking

		CREW	DAILY OUTPUT	LABOR-HOURS	UNIT	MAT.	LABOR	EQUIP.	TOTAL	TOTAL INCL O&P	
700	0010	**ROADS AND SIDEWALKS** Temporary										700
	0050	Roads, gravel fill, no surfacing, 4" gravel depth	B-14	715	.067	S.Y.	2.72	1.62	.25	4.59	5.80	
	0100	8" gravel depth	"	615	.078	"	5.45	1.89	.29	7.63	9.25	
	1000	Ramp, 3/4" plywood on 2" x 6" joists, 16" O.C.	2 Carp	300	.053	S.F.	1.34	1.55		2.89	3.90	

01500 | Temporary Facilities & Controls

01550 | Vehicular Access & Parking

			CREW	DAILY OUTPUT	LABOR-HOURS	UNIT	MAT.	LABOR	EQUIP.	TOTAL	TOTAL INCL O&P	
700	1100	On 2" x 10" joists, 16" O.C.	2 Carp	275	.058	S.F.	1.96	1.70		3.66	4.81	700
	2200	Sidewalks, 2" x 12" planks, 2 uses	1 Carp	350	.023		.74	.67		1.41	1.85	
	2300	Exterior plywood, 2 uses, 1/2" thick		750	.011		.25	.31		.56	.76	
	2400	5/8" thick		650	.012		.31	.36		.67	.90	
	2500	3/4" thick	↓	600	.013	↓	.37	.39		.76	1.02	

01560 | Barriers & Enclosures

			CREW	DAILY OUTPUT	LABOR-HOURS	UNIT	MAT.	LABOR	EQUIP.	TOTAL	TOTAL INCL O&P	
100	0010	**BARRICADES** 5' high, 3 rail @ 2" x 8", fixed	2 Carp	30	.533	L.F.	10.80	15.55		26.35	36.50	100
	0150	Movable		20	.800		10.80	23.50		34.30	48.50	
	1000	Guardrail, wooden, 3' high, 1" x 6", on 2" x 4" posts		200	.080		1.05	2.33		3.38	4.80	
	1100	2" x 6", on 4" x 4" posts	↓	165	.097		1.65	2.83		4.48	6.25	
	1200	Portable metal with base pads, buy					15			15	16.50	
	1250	Typical installation, assume 10 reuses	2 Carp	600	.027	↓	1.58	.78		2.36	2.96	
	1300	Barricade tape, polyethelyne, 7 mil, 3" wide x 500' long roll				Ea.	25			25	27.50	
250	0010	**FENCING** Chain link, 11 ga, 5' high	2 Clab	100	.160	L.F.	3.58	3.66		7.24	9.65	250
	0100	6' high		75	.213		3.28	4.87		8.15	11.25	
	0200	Rented chain link, 6' high, to 500' (up to 12 mo.)		100	.160		2.26	3.66		5.92	8.20	
	0250	Over 1000' (up to 12 mo.)	↓	110	.145		1.64	3.32		4.96	7	
	0350	Plywood, painted, 2" x 4" frame, 4' high	A-4	135	.178		4.44	5		9.44	12.65	
	0400	4" x 4" frame, 8' high	"	110	.218		7.90	6.15		14.05	18.20	
	0500	Wire mesh on 4" x 4" posts, 4' high	2 Carp	100	.160		6	4.66		10.66	13.85	
	0550	8' high	"	80	.200		9	5.85		14.85	19	
400	0010	**TEMPORARY CONSTRUCTION** See also division 01530										400

01580 | Project Signs

			CREW	DAILY OUTPUT	LABOR-HOURS	UNIT	MAT.	LABOR	EQUIP.	TOTAL	TOTAL INCL O&P	
700	0010	**SIGNS** Hi-intensity reflectorized, no posts, buy				S.F.	14			14	15.40	700

01500 | Temporary Facilities & Controls

01590 | Equipment Rental

			UNIT	HOURLY OPER. COST	RENT PER DAY	RENT PER WEEK	RENT PER MONTH	CREW EQUIPMENT COST/DAY
0010	**EARTHWORK EQUIPMENT RENTAL** Without operators	R01590-100						
0050	Augers for truck or trailer mounting, vertical drilling							
0060	4" to 36" diam., 54 H.P., gas, 10' spindle travel		Ea.	16.30	650	1,945	5,825	519.40
0070	14' spindle travel			18.75	810	2,430	7,300	636
0100	Excavator, diesel hydraulic, crawler mounted, 1/2 C.Y. cap.			8.80	415	1,240	3,725	318.40
0120	5/8 C.Y. capacity			10.50	460	1,380	4,150	360
0140	3/4 C.Y. capacity			12.95	515	1,550	4,650	413.60
0150	1 C.Y. capacity			16.35	600	1,800	5,400	490.80
0200	1-1/2 C.Y. capacity			19.40	730	2,190	6,575	593.20
0300	2 C.Y. capacity			26.85	1,075	3,200	9,600	854.80
0320	2-1/2 C.Y. capacity			49.60	1,925	5,800	17,400	1,557
0340	3-1/2 C.Y. capacity			61.75	2,475	7,400	22,200	1,974
0341	Attachments							
0342	Bucket thumbs			1.10	172	515	1,550	111.80
0345	Grapples			1	202	605	1,825	129
0350	Gradall type, truck mounted, 3 ton @ 15' radius, 5/8 C.Y.			21.05	800	2,400	7,200	648.40
0370	1 C.Y. capacity			24.60	965	2,890	8,675	774.80
0400	Backhoe-loader, 40 to 45 H.P., 5/8 C.Y. capacity			4.60	220	660	1,975	168.80
0450	45 H.P. to 60 H.P., 3/4 C.Y. capacity			5.50	228	685	2,050	181
0460	80 H.P., 1-1/4 C.Y. capacity			6.30	297	890	2,675	228.40
0470	112 H.P., 1-1/2 C.Y. capacity			10.30	370	1,105	3,325	303.40
0480	Attachments							
0482	Compactor, 20,000 lb			1.75	196	588	1,775	131.60
0485	Hydraulic hammer, 750 ft-lbs			1.30	243	729	2,175	156.20
0486	Hydraulic hammer, 1200 ft-lbs			1.75	294	882	2,650	190.40
0750	Bucket, clamshell, general purpose, 3/8 C.Y.			.75	61.50	185	555	43
0800	1/2 C.Y.			.75	70	210	630	48
0850	3/4 C.Y.			.85	80	240	720	54.80
0900	1 C.Y.			.90	106	319	955	71
0950	1-1/2 C.Y.			1.45	133	400	1,200	91.60
1000	2 C.Y.			1.55	167	500	1,500	112.40
1200	Compactor, roller, 2 drum 2000 lb., operator walking			3.75	135	405	1,225	111
1250	Rammer compactor, gas, 1000 lb. blow			.75	46.50	140	420	34
1300	Vibratory plate, gas, 13" plate, 1000 lb. blow			.75	43.50	130	390	32
1350	24" plate, 5000 lb. blow			1.95	75	225	675	60.60
3710	Screening plant 110 hp. w / 5' x 10' screen			13	415	1,250	3,750	354
3720	5' x 16' screen			14.30	495	1,480	4,450	410.40
4891	Attachments for all skid steer loaders							
4892	Auger		Ea.	.49	81.50	245	735	52.90
4893	Backhoe			.63	105	314	940	67.85
4894	Broom			.62	103	309	925	66.75
4895	Forks			.20	33.50	101	305	21.80
4896	Grapple			.47	78	234	700	50.55
4897	Concrete hammer			1.02	171	512	1,525	110.55
4898	Tree spade			.92	153	460	1,375	99.35
4899	Trencher			.73	121	364	1,100	78.65
4900	Trencher, chain, boom type, gas, operator walking, 12 H.P.			1.60	140	420	1,250	96.80
4910	Operator riding, 40 H.P.			4.55	267	800	2,400	196.40
5000	Wheel type, diesel, 4' deep, 12" wide			10.60	565	1,700	5,100	424.80
5100	Diesel, 6' deep, 20" wide			19.10	805	2,420	7,250	636.80
5150	Ladder type, diesel, 5' deep, 8" wide			9.15	425	1,275	3,825	328.20
5200	Diesel, 8' deep, 16" wide			27.05	825	2,475	7,425	711.40
5250	Truck, dump, tandem, 12 ton payload			18.05	375	1,130	3,400	370.40
5300	Three axle dump, 16 ton payload			24.85	475	1,420	4,250	482.80
5350	Dump trailer only, rear dump, 16-1/2 C.Y.			4	180	540	1,625	140
5400	20 C.Y.			4.45	183	550	1,650	145.60
5450	Flatbed, single axle, 1-1/2 ton rating			9.75	133	400	1,200	158
5500	3 ton rating			12.40	142	425	1,275	184.20

01500 | Temporary Facilities & Controls

01590 | Equipment Rental

			UNIT	HOURLY OPER. COST	RENT PER DAY	RENT PER WEEK	RENT PER MONTH	CREW EQUIPMENT COST/DAY	
200	5550	Off highway rear dump, 25 ton capacity	Ea.	37.80	1,000	2,995	8,975	901.40	200
	5600	35 ton capacity		38.80	1,325	3,950	11,900	1,100	
400	0010	**GENERAL EQUIPMENT RENTAL** Without operators							400
	0150	Aerial lift, scissor type, to 15' high, 1000 lb. cap., electric	Ea.	1.15	83.50	250	750	59.20	
	0160	To 25' high, 2000 lb. capacity		2	128	385	1,150	93	
	0170	Telescoping boom to 40' high, 750 lb. capacity, gas		6.50	265	795	2,375	211	
	0180	1000 lb. capacity		8.20	390	1,175	3,525	300.60	
	0190	To 60' high, 750 lb. capacity		8.30	405	1,220	3,650	310.40	
	0200	Air compressor, portable, gas engine, 60 C.F.M.		5.41	58.50	175	525	78.30	
	0300	160 C.F.M.		6.80	73.50	220	660	98.40	
	0400	Diesel engine, rotary screw, 250 C.F.M.		6.90	117	350	1,050	125.20	
	0500	365 C.F.M.		8.10	150	450	1,350	154.80	
	0600	600 C.F.M.		13.25	242	725	2,175	251	
	0700	750 C.F.M.		17.40	258	775	2,325	294.20	
	0800	For silenced models, small sizes, add		3%	5%	5%	5%		
	0900	Large sizes, add		5%	7%	7%	7%		
	0920	Air tools and accessories							
	0930	Breaker, pavement, 60 lb.	Ea.	.20	30	90	270	19.60	
	0940	80 lb.		.20	40	120	360	25.60	
	0980	Dust control per drill		1.10	16	48	144	18.40	
	1000	Hose, air with couplings, 50' long, 3/4" diameter		.02	5	15	45	3.15	
	1100	1" diameter		.03	5.65	17	51	3.65	
	1200	1-1/2" diameter		.04	9.35	28	84	5.90	
	1300	2" diameter		.11	17.65	53	159	11.50	
	1400	2-1/2" diameter		.12	20.50	62	186	13.35	
	1410	3" diameter		.16	26.50	80	240	17.30	
	1560	Tamper, single, 35 lb.		.45	29.50	89	267	21.40	
	1570	Triple, 140 lb.		.90	46.50	140	420	35.20	
	1580	Wrenches, impact, air powered, up to 3/4" bolt		.10	21.50	65	195	13.80	
	1590	Up to 1-1/4" bolt		.15	40	120	360	25.20	
	2100	Generator, electric, gas engine, 1.5 KW to 3 KW		1.13	40	120	360	33.05	
	2200	5 KW		1.49	58.50	175	525	46.90	
	2300	10 KW		2.40	133	400	1,200	99.20	
	2400	25 KW		6.80	153	460	1,375	146.40	
	2500	Diesel engine, 20 KW		4.35	117	350	1,050	104.80	
	2600	50 KW		6.90	133	400	1,200	135.20	
	2700	100 KW		11.88	193	580	1,750	211.05	
	2800	250 KW		29.05	335	1,000	3,000	432.40	
	2900	Heaters, space, oil or electric, 50 MBH		.59	27.50	82	246	21.10	
	3000	100 MBH		1.06	35	105	315	29.50	
	3100	300 MBH		3.41	53.50	160	480	59.30	
	3150	500 MBH		6.77	70	210	630	96.15	
	3200	Hose, water, suction with coupling, 20' long, 2" diameter		.06	10	30	90	6.50	
	3210	3" diameter		.06	15	45	135	9.50	
	3220	4" diameter		.06	20	60	180	12.50	
	3230	6" diameter		.06	36.50	110	330	22.50	
	3240	8" diameter		.36	59.50	178	535	38.50	
	3250	Discharge hose with coupling, 50' long, 2" diameter		.05	8.35	25	75	5.40	
	3260	3" diameter		.05	10	30	90	6.40	
	3270	4" diameter		.06	13.35	40	120	8.50	
	3280	6" diameter		.07	33.50	100	300	20.55	
	3290	8" diameter		.33	54.50	163	490	35.25	
	3300	Ladders, extension type, 16' to 36' long		.14	18.35	55	165	12.10	
	3400	40' to 60' long		.18	29.50	89	267	19.25	
	3500	Light towers, towable, with diesel generator, 2000 watt		1.50	108	325	975	77	
	3600	4000 watt		1.70	130	390	1,175	91.60	
	4100	Pump, centrifugal gas pump, 1-1/2", 4 MGPH		.85	31.50	95	285	25.80	
	4200	2", 8 MGPH		1.20	33.50	100	300	29.60	

01500 | Temporary Facilities & Controls
01590 | Equipment Rental

			UNIT	HOURLY OPER. COST	RENT PER DAY	RENT PER WEEK	RENT PER MONTH	CREW EQUIPMENT COST/DAY	
400	4300	3", 15 MGPH	Ea.	1.40	46.50	140	420	39.20	400
	4400	6", 90 MGPH		9.30	150	450	1,350	164.40	
	4500	Submersible electric pump, 1-1/4", 55 GPM		.35	36.50	110	330	24.80	
	4600	1-1/2", 83 GPM		.41	40	120	360	27.30	
	4700	2", 120 GPM		.42	46.50	140	420	31.35	
	4800	3", 300 GPM		.75	58.50	175	525	41	
	4900	4", 560 GPM		5.30	82	246	740	91.60	
	5000	6", 1590 GPM		8	233	700	2,100	204	
	5100	Diaphragm pump, gas, single, 1-1/2" diameter		.52	29.50	88	264	21.75	
	5200	2" diameter		.80	40.50	122	365	30.80	
	5300	3" diameter		.75	47.50	143	430	34.60	
	5400	Double, 4" diameter		1.75	80	240	720	62	
	5500	Trash pump, self-priming, gas, 2" diameter		1.65	40.50	122	365	37.60	
	5600	Diesel, 4" diameter		2.65	73.50	220	660	65.20	
	5650	Diesel, 6" diameter		6.40	122	365	1,100	124.20	
	5700	Salamanders, L.P. gas fired, 100,000 B.T.U.		1.06	20	60	180	20.50	
	5800	Saw, chain, gas engine, 18" long		.15	41.50	125	375	26.20	
	5900	36" long		.15	70	210	630	43.20	
	5950	60" long		.15	83.50	250	750	51.20	
	6000	Masonry, table mounted, 14" diameter, 5 H.P.		1.30	57	171	515	44.60	
	6100	Circular, hand held, electric, 7-1/4" diameter		.16	23.50	70	210	15.30	
	6200	12" diameter		.25	33.50	100	300	22	
	6350	Torch, cutting, acetylene-oxygen, 150' hose		1.50	33.50	100	300	32	
	6360	Hourly operating cost includes tips and gas		7.80				62.40	
	6410	Toilet, portable chemical		.10	16.35	49	147	10.60	
	6420	Recycle flush type		.12	20.50	61	183	13.15	
	6430	Toilet, fresh water flush, garden hose,		.14	23	69	207	14.90	
	6440	Hoisted, non-flush, for high rise		.12	20	60	180	12.95	
	6450	Toilet, trailers, minimum		.20	34	102	305	22	
	6460	Maximum		.61	102	306	920	66.10	
	6500	Trailers, platform, flush deck, 2 axle, 25 ton capacity		3.35	123	370	1,100	100.80	
	6600	40 ton capacity		4.55	223	670	2,000	170.40	
	6700	3 axle, 50 ton capacity		5	247	740	2,225	188	
	6800	75 ton capacity		6.45	325	975	2,925	246.60	
	7100	Truck, pickup, 3/4 ton, 2 wheel drive		4.75	75	225	675	83	
	7200	4 wheel drive		4.90	83.50	250	750	89.20	
	7300	Tractor, 4 x 2, 30 ton capacity, 195 H.P.		12.55	365	1,100	3,300	320.40	
	7410	250 H.P.		17.20	435	1,300	3,900	397.60	
	7620	Vacuum truck, hazardous material, 2500 gallon		7.10	305	920	2,750	240.80	
	7625	5,000 gallon		7.15	410	1,225	3,675	302.20	
	7700	Welder, electric, 200 amp		4.97	43.50	130	390	65.75	
	7800	300 amp		7.83	63.50	190	570	100.65	
	7900	Gas engine, 200 amp		4.40	55	165	495	68.20	
	8000	300 amp		5.25	70	210	630	84	
	8100	Wheelbarrow, any size		.06	9.65	29	87	6.30	
600	0010	**LIFTING AND HOISTING EQUIPMENT RENTAL**	R01590 -100						600
	0100	without operators							
	0200	Crane, climbing, 106' jib, 6000 lb. capacity, 410 FPM	Ea.	38.75	1,250	3,750	11,300	1,060	
	0300	101' jib, 10,250 lb. capacity, 270 FPM	"	43.75	1,575	4,750	14,300	1,300	
	0400	Tower, static, 130' high, 106' jib,	Ea.	41.73	1,450	4,345	13,000	1,203	
	0500	6200 lb. capacity at 400 FPM							
	0600	Crawler, cable, 1/2 C.Y., 15 tons at 12' radius		4.97	530	1,590	4,775	357.75	
	0700	3/4 C.Y., 20 tons at 12' radius		18.55	540	1,625	4,875	473.40	
	0800	1 C.Y., 25 tons at 12' radius		25.90	565	1,700	5,100	547.20	
	0900	Crawler mounted, lattice boom, 1-1/2 C.Y., 40 tons at 12' radius		27	835	2,500	7,500	716	
	1000	2 C.Y., 50 tons at 12' radius		36.35	990	2,975	8,925	885.80	
	1100	3 C.Y., 75 tons at 12' radius		39.15	1,025	3,100	9,300	933.20	

GENERAL REQUIREMENTS 1

01500 | Temporary Facilities & Controls

01590 | Equipment Rental

			UNIT	HOURLY OPER. COST	RENT PER DAY	RENT PER WEEK	RENT PER MONTH	CREW EQUIPMENT COST/DAY
600	1200	100 ton capacity, standard boom	Ea.	47.35	1,375	4,150	12,500	1,209
	1300	165 ton capacity, standard boom	R01590 -100	71.50	2,225	6,700	20,100	1,912
	1400	200 ton capacity, 150' boom		73.25	2,575	7,700	23,100	2,126
	1500	450' boom		105.30	3,550	10,680	32,000	2,978
	1600	Truck mounted, lattice boom, 6 x 4, 20 tons at 10' radius		18.10	685	2,050	6,150	554.80
	1700	25 tons at 10' radius		23.15	1,050	3,150	9,450	815.20
	1800	8 x 4, 30 tons at 10' radius		25.95	650	1,950	5,850	597.60
	1900	40 tons at 12' radius		31.45	835	2,500	7,500	751.60
	2000	8 x 4, 60 tons at 15' radius		38.65	1,000	3,000	9,000	909.20
	2050	82 tons at 15' radius		41.57	1,525	4,600	13,800	1,253
	2100	90 tons at 15' radius		43.08	1,575	4,725	14,200	1,290
	2200	115 tons at 15' radius		45.40	1,800	5,400	16,200	1,443
	2300	150 tons at 18' radius		46	1,850	5,550	16,700	1,478
	2350	165 tons at 18' radius		51.75	2,175	6,500	19,500	1,714
	2400	Truck mounted, hydraulic, 12 ton capacity		27.35	495	1,480	4,450	514.80
	2500	25 ton capacity		28.70	665	2,000	6,000	629.60
	2550	33 ton capacity		31.85	915	2,750	8,250	804.80
	2600	55 ton capacity		38.95	985	2,950	8,850	901.60
	2700	80 ton capacity		56.80	1,450	4,355	13,100	1,325
	2800	Self-propelled, 4 x 4, with telescoping boom, 5 ton		11.50	320	960	2,875	284
	2900	12-1/2 ton capacity		17.15	480	1,440	4,325	425.20
	3000	15 ton capacity		19.45	600	1,800	5,400	515.60
	3100	25 ton capacity		22.25	705	2,120	6,350	602
	3200	Derricks, guy, 20 ton capacity, 60' boom, 75' mast		11.44	305	920	2,750	275.50
	3300	100' boom, 115' mast		18.63	530	1,590	4,775	467.05
	3400	Stiffleg, 20 ton capacity, 70' boom, 37' mast		13.30	395	1,185	3,550	343.40
	3500	100' boom, 47' mast		20.98	640	1,925	5,775	552.85
	3550	Helicopter, small, lift to 1250 lbs. maximum, w/pilot		62.12	2,475	7,445	22,300	1,986
	3600	Hoists, chain type, overhead, manual, 3/4 ton		.06	6.65	20	60	4.50
	3900	10 ton		.25	28.50	85	255	19
	4000	Hoist and tower, 5000 lb. cap., portable electric, 40' high		3.75	177	530	1,600	136
	4100	For each added 10' section, add		.08	13.65	41	123	8.85
	4200	Hoist and single tubular tower, 5000 lb. electric, 100' high		5.08	247	740	2,225	188.65
	4300	For each added 6'-6" section, add		.14	23.50	71	213	15.30
	4400	Hoist and double tubular tower, 5000 lb., 100' high		5.45	272	815	2,450	206.60
	4500	For each added 6'-6" section, add		.15	25.50	77	231	16.60
	4550	Hoist and tower, mast type, 6000 lb., 100' high		5.88	282	845	2,525	216.05
	4570	For each added 10' section, add		.10	17	51	153	11
	4600	Hoist and tower, personnel, electric, 2000 lb., 100' @ 125 FPM		12.33	750	2,245	6,725	547.65
	4700	3000 lb., 100' @ 200 FPM		14.13	850	2,550	7,650	623.05
	4800	3000 lb., 150' @ 300 FPM		15.65	950	2,855	8,575	696.20
	4900	4000 lb., 100' @ 300 FPM		16.20	970	2,910	8,725	711.60
	5000	6000 lb., 100' @ 275 FPM		17.50	1,025	3,060	9,175	752
	5100	For added heights up to 500', add	L.F.	.01	1.70	5.10	15.30	1.10
	5200	Jacks, hydraulic, 20 ton	Ea.	.15	14.65	44	132	10
	5500	100 ton	"	.20	40.50	122	365	26
	6000	Jacks, hydraulic, climbing with 50' jackrods						
	6010	and control consoles, minimum 3 mo. rental						
	6100	30 ton capacity	Ea.	1.53	102	305	915	73.25
	6150	For each added 10' jackrod section, add		.05	3.33	10	30	2.40
	6300	50 ton capacity		2.45	163	490	1,475	117.60
	6350	For each added 10' jackrod section, add		.06	4	12	36	2.90
	6500	125 ton capacity		6.38	425	1,275	3,825	306.05
	6550	For each added 10' jackrod section, add		.44	29	87	261	20.90
	6600	Cable jack, 10 ton capacity with 200' cable		1.28	85	255	765	61.25
	6650	For each added 50' of cable, add		.13	8.65	26	78	6.25
700	0010	**WELLPOINT EQUIPMENT RENTAL** See also division 02240						
	0020	Based on 2 months rental						

01500 | Temporary Facilities & Controls

01590 | Equipment Rental

			UNIT	HOURLY OPER. COST	RENT PER DAY	RENT PER WEEK	RENT PER MONTH	CREW EQUIPMENT COST/DAY	
700	0100	Combination jetting & wellpoint pump, 60 H.P. diesel	Ea.	8.29	252	755	2,275	217.30	700
	0200	High pressure gas jet pump, 200 H.P., 300 psi	"	14.52	215	645	1,925	245.15	
	0300	Discharge pipe, 8" diameter	L.F.	.01	.41	1.22	3.66	.30	
	0350	12" diameter		.01	.60	1.80	5.40	.45	
	0400	Header pipe, flows up to 150 G.P.M., 4" diameter		.01	.37	1.12	3.36	.30	
	0500	400 G.P.M., 6" diameter		.01	.43	1.30	3.90	.35	
	0600	800 G.P.M., 8" diameter		.01	.60	1.80	5.40	.45	
	0700	1500 G.P.M., 10" diameter		.01	.63	1.90	5.70	.45	
	0800	2500 G.P.M., 12" diameter		.02	1.20	3.60	10.80	.90	
	0900	4500 G.P.M., 16" diameter		.02	1.53	4.60	13.80	1.10	
	0950	For quick coupling aluminum and plastic pipe, add	▼	.02	1.58	4.75	14.25	1.10	
	1100	Wellpoint, 25' long, with fittings & riser pipe, 1-1/2" or 2" diameter	Ea.	.05	3.17	9.50	28.50	2.30	
	1200	Wellpoint pump, diesel powered, 4" diameter, 20 H.P.		4.05	145	435	1,300	119.40	
	1300	6" diameter, 30 H.P.		5.28	180	540	1,625	150.25	
	1400	8" suction, 40 H.P.		7.18	247	740	2,225	205.45	
	1500	10" suction, 75 H.P.		9.81	288	865	2,600	251.50	
	1600	12" suction, 100 H.P.		14.70	460	1,385	4,150	394.60	
	1700	12" suction, 175 H.P.	▼	19.46	510	1,530	4,600	461.70	

GENERAL REQUIREMENTS 1

01700 | Execution Requirements

01740 | Cleaning

			CREW	DAILY OUTPUT	LABOR-HOURS	UNIT	2001 BARE COSTS MAT.	LABOR	EQUIP.	TOTAL	TOTAL INCL O&P	
500	0010	**CLEANING UP** After job completion, allow, minimum				Job					.30%	500
	0040	Maximum				"					1%	
	0050	Cleanup of floor area, continuous, per day	A-5	24	.750	M.S.F.	1.65	17.20	1.64	20.49	30.50	
	0100	Final	"	11.50	1.565	"	2.63	36	3.43	42.06	62.50	
	1000	Mechanical demolition, see division 02225										

01800 | Facility Operation

01810 | Commissioning

			CREW	DAILY OUTPUT	LABOR-HOURS	UNIT	2001 BARE COSTS MAT.	LABOR	EQUIP.	TOTAL	TOTAL INCL O&P	
100	0010	**COMMISSIONING**										100
	0100	As-built drawings, punchlist, training, manuals, minimum				Project					.50%	
	0150	Maximum				"					2%	

For information about Means Estimating Seminars, see yellow pages 11 and 12 in back of book

Division 2
Site Construction

Estimating Tips

02200 Site Preparation
- If possible visit the site and take an inventory of the type, quantity and size of the trees. Certain trees may have a landscape resale value or firewood value. Stump disposal can be very expensive, particularly if they cannot be buried at the site. Consider using a bulldozer in lieu of hand cutting trees.
- Estimators should visit the site to determine the need for haul road, access, storage of materials, and security considerations. When estimating for access roads on unstable soil, consider using a geotextile stabilization fabric. It can greatly reduce the quantity of crushed stone or gravel. Sites of limited size and access can cause cost overruns due to lost productivity. Theft and damage is another consideration if the location is isolated. A temporary fence or security guards may be required. Investigate the site thoroughly.

02210 Subsurface Investigation
In preparing estimates on structures involving earthwork or foundations, all information concerning soil characteristics should be obtained. Look particularly for hazardous waste, evidence of prior dumping of debris, and previous stream beds.
- The costs shown for selective demolition do not include rubbish handling or disposal. These items should be estimated separately using Means data or other sources.

02300 Earthwork
- Estimating the actual cost of performing earthwork requires careful consideration of the variables involved. This includes items such as type of soil, whether or not water will be encountered, dewatering, whether or not banks need bracing, disposal of excavated earth, length of haul to fill or spoil sites, etc. If the project has large quantities of cut or fill, consider raising or lowering the site to reduce costs while paying close attention to the effect on site drainage and utilities if doing this.
- If the project has large quantities of fill, creating a borrow pit on the site can significantly lower the costs. It is very important to consider what time of year the project is scheduled for completion. Bad weather can create large cost overruns from dewatering, site repair and lost productivity from cold weather.

02700 Bases, Ballasts, Pavements/Appurtenances
- When estimating paving, keep in mind the project schedule. If an asphaltic paving project is in a colder climate and runs through to the spring, consider placing the base course in the autumn, then topping it in the spring just prior to completion. This could save considerable costs in spring repair. Keep in mind that prices for asphalt and concrete are generally higher in the cold seasons.

02500 Utility Services
02600 Drainage & Containment
- Never assume that the water, sewer and drainage lines will go in at the early stages of the project. Consider the site access needs before dividing the site in half with open trenches, loose pipe, and machinery obstructions. Always inspect the site to establish that the site drawings are complete. Check off all existing utilities on your drawing as you locate them. If you find any discrepancies, mark up the site plan for further research. Differing site conditions can be very costly if discovered later in the project.

02900 Planting
- The timing of planting and guarantee specifications often dictate the costs for establishing tree and shrub growth and a stand of grass or ground cover. Establish the work performance schedule to coincide with the local planting season. Maintenance and growth guarantees can add from 20% to 100% to the total landscaping cost. The cost to replace trees and shrubs can be as high as 5% of the total cost depending on the planting zone, soil conditions and time of year.

Reference Numbers
Reference numbers are shown in bold squares at the beginning of some major classifications. These numbers refer to related items in the Reference Section. The reference information may be an estimating procedure, an alternate pricing method or technical information.

Note: Not all subdivisions listed here necessarily appear in this publication.

02050 | Basic Site Materials & Methods

02065 | Cement & Concrete

			CREW	DAILY OUTPUT	LABOR-HOURS	UNIT	MAT.	LABOR	EQUIP.	TOTAL	TOTAL INCL O&P	
300	0010	ASPHALTIC CONCRETE plant mix (145 lb. per C.F.)				Ton	33			33	36.50	300
	0200	All weather patching mix, hot					31.50			31.50	34.50	
	0250	Cold patch					36			36	39.50	
	0300	Berm mix					31.50			31.50	34.50	

02080 | Utility Materials

			CREW	DAILY OUTPUT	LABOR-HOURS	UNIT	MAT.	LABOR	EQUIP.	TOTAL	TOTAL INCL O&P	
790	0010	UNDERGROUND MARKING TAPE A12.3-110										790
	0400	Underground tape, detectable aluminum, 2"	1 Clab	150	.053	C.L.F.	3.70	1.22		4.92	6	
	0500	6"	"	140	.057	"	9.25	1.31		10.56	12.25	
900	0010	PIPING VALVES Water distribution, see also div. 15110										900
	3000	Butterfly valves with boxes, cast iron, mech. jt.										
	3100	4" diameter	B-20	6	4	Ea.	228	103		331	410	
	3140	6" diameter	"	5	4.800		262	124		386	480	
	3180	8" diameter	B-21	4	7		315	186	35.50	536.50	680	
	3300	10" diameter		3.50	8		465	213	40.50	718.50	890	
	3340	12" diameter		3	9.333		610	248	47.50	905.50	1,100	
	3400	14" diameter		2	14		1,050	375	71	1,496	1,800	
	3440	16" diameter		2	14		1,400	375	71	1,846	2,200	
	3460	18" diameter		1.50	18.667		1,750	495	94.50	2,339.50	2,800	
	3480	20" diameter		1	28		2,200	745	142	3,087	3,725	
	3500	24" diameter		.50	56		3,075	1,500	284	4,859	6,000	
	3600	With lever operator										
	3610	4" diameter	B-20	6	4	Ea.	121	103		224	295	
	3614	6" diameter	"	5	4.800		169	124		293	380	
	3616	8" diameter	B-21	4	7		245	186	35.50	466.50	600	
	3618	10" diameter		3.50	8		395	213	40.50	648.50	805	
	3620	12" diameter		3	9.333		730	248	47.50	1,025.50	1,225	
	3622	14" diameter		2	14		1,175	375	71	1,621	1,925	
	3624	16" diameter		2	14		1,575	375	71	2,021	2,400	
	3626	18" diameter		1.50	18.667		1,825	495	94.50	2,414.50	2,875	
	3628	20" diameter		1	28		2,425	745	142	3,312	3,950	
	3630	24" diameter		.50	56		2,600	1,500	284	4,384	5,500	
	3700	Check valves, flanged										
	3710	4" diameter	B-20	6	4	Ea.	395	103		498	595	
	3714	6" diameter	"	5	4.800		665	124		789	930	
	3716	8" diameter	B-21	4	7		1,350	186	35.50	1,571.50	1,825	
	3718	10" diameter		3.50	8		2,175	213	40.50	2,428.50	2,775	
	3720	12" diameter		3	9.333		3,325	248	47.50	3,620.50	4,100	
	3722	14" diameter		2	14		5,325	375	71	5,771	6,500	
	3724	16" diameter		2	14		7,325	375	71	7,771	8,725	
	3726	18" diameter		1.50	18.667		10,900	495	94.50	11,489.50	12,800	
	3728	20" diameter		1	28		12,600	745	142	13,487	15,200	
	3730	24" diameter		.50	56		18,200	1,500	284	19,984	22,600	
	3800	Gate valves, C.I., 250 PSI, mechanical joint, w/boxes										
	3810	4" diameter	B-21A	8	5	Ea.	160	141	53	354	450	
	3814	6" diameter		6.80	5.882		193	166	62.50	421.50	535	
	3816	8" diameter		5.60	7.143		285	201	76	562	710	
	3818	10" diameter		4.80	8.333		460	235	88.50	783.50	970	
	3820	12" diameter		4.08	9.804		635	276	104	1,015	1,225	
	3822	14" diameter	B-21	2	14		735	375	71	1,181	1,475	
	3824	16" diameter		1	28		865	745	142	1,752	2,250	
	3826	18" diameter		.80	35		1,075	930	177	2,182	2,825	
	3828	20" diameter		.80	35		1,350	930	177	2,457	3,150	
	3830	24" diameter		.50	56		1,675	1,500	284	3,459	4,450	
	3831	30" diameter		.35	80		2,000	2,125	405	4,530	5,975	

02050 | Basic Site Materials & Methods

02080 | Utility Materials

		CREW	DAILY OUTPUT	LABOR-HOURS	UNIT	2001 BARE COSTS MAT.	LABOR	EQUIP.	TOTAL	TOTAL INCL O&P	
3832	36" diameter	B-21	.30	93.333	Ea.	2,400	2,475	475	5,350	7,050	900
3900	Globe valves, flanged, iron body, class 125										
3910	4" diameter	B-20	10	2.400	Ea.	865	62		927	1,050	
3914	6" diameter	"	9	2.667		1,625	69		1,694	1,875	
3916	8" diameter	B-21	6	4.667		3,000	124	23.50	3,147.50	3,525	
3918	10" diameter		5	5.600		4,725	149	28.50	4,902.50	5,475	
3920	12" diameter		4	7		6,625	186	35.50	6,846.50	7,600	
3922	14" diameter		3	9.333		7,100	248	47.50	7,395.50	8,225	
3924	16" diameter		2	14		7,000	375	71	7,446	8,350	
3926	18" diameter		.80	35		7,700	930	177	8,807	10,100	
3928	20" diameter		.60	46.667		7,600	1,250	237	9,087	10,500	
3930	24" diameter		.50	56		9,000	1,500	284	10,784	12,500	

02100 | Site Remediation

02115 | Underground Tank Removal

		CREW	DAILY OUTPUT	LABOR-HOURS	UNIT	2001 BARE COSTS MAT.	LABOR	EQUIP.	TOTAL	TOTAL INCL O&P	
0010	**REMOVAL OF UNDERGROUND STORAGE TANKS** R02115-200										200
0011	Petroleum storage tanks, non-leaking										
0100	Excavate & load onto trailer										
0110	3000 gal. to 5000 gal. tank	B-14	4	12	Ea.		290	45	335	500	
0120	6000 gal to 8000 gal tank	B-3A	3	13.333			325	198	523	720	
0130	9000 gal to 12000 gal tank	"	2	20			485	297	782	1,075	
0190	Known leaking tank add				%				100%	100%	
0200	Remove sludge, water and remaining product from tank bottom										
0201	of tank with vacuum truck										
0300	3000 gal to 5000 gal tank	A-13	5	1.600	Ea.		46	104	150	185	
0310	6000 gal to 8000 gal tank		4	2			57.50	130	187.50	230	
0320	9000 gal to 12000 gal tank		3	2.667			76.50	174	250.50	305	
0390	Dispose of sludge off-site, average				Gal.					4	
0400	Insert inert solid CO2 "dry ice" into tank										
0401	For cleaning/transporting tanks (1.5 lbs./100 gal. cap)	1 Clab	500	.016	Lb.	1.20	.37		1.57	1.89	
1020	Haul tank to certified salvage dump, 100 miles round trip										
1023	3000 gal. to 5000 gal. tank				Ea.				550	630	
1026	6000 gal. to 8000 gal. tank								650	750	
1029	9,000 gal. to 12,000 gal. tank								875	1,000	
1100	Disposal of contaminated soil to landfill										
1110	Minimum				C.Y.					100	
1111	Maximum				"					300	
1120	Disposal of contaminated soil to										
1121	bituminous concrete batch plant										
1130	Minimum				C.Y.					50	
1131	Maximum				"					100	
2010	Decontamination of soil on site incl poly tarp on top/bottom										
2011	Soil containment berm, and chemical treatment										
2020	Minimum	B-11C	100	.160	C.Y.	5.60	4.26	1.81	11.67	14.70	
2021	Maximum	"	100	.160		7.25	4.26	1.81	13.32	16.55	
2050	Disposal of decontaminated soil, minimum									60	
2055	Maximum									125	

For expanded coverage of these items see *Means Heavy Construction Cost Data 2001*

02200 | Site Preparation

02210 | Subsurface Investigation

		CREW	DAILY OUTPUT	LABOR-HOURS	UNIT	MAT.	2001 BARE COSTS LABOR	EQUIP.	TOTAL	TOTAL INCL O&P	
320	0010 **DRILLING, CORE** Reinforced concrete slab, up to 6" thick slab										320
	0020 Including bit, layout and set up										
	0100 1" diameter core	B-89A	28	.571	Ea.	2.30	15.05	3.62	20.97	30	
	0150 Each added inch thick, add		300	.053		.41	1.41	.34	2.16	3.02	
	0300 3" diameter core		23	.696		5.10	18.35	4.41	27.86	39	
	0350 Each added inch thick, add		186	.086		.92	2.27	.55	3.74	5.15	
	0500 4" diameter core		19	.842		5.10	22	5.35	32.45	46.50	
	0550 Each added inch thick, add		170	.094		1.17	2.48	.60	4.25	5.85	
	0700 6" diameter core		14	1.143		8.45	30	7.25	45.70	64.50	
	0750 Each added inch thick, add		140	.114		1.43	3.01	.72	5.16	7.10	
	0900 8" diameter core		11	1.455		11.55	38.50	9.20	59.25	83	
	0950 Each added inch thick, add		95	.168		1.94	4.44	1.07	7.45	10.25	
	1100 10" diameter core		10	1.600		15.40	42	10.15	67.55	94	
	1150 Each added inch thick, add		80	.200		2.55	5.25	1.27	9.07	12.45	
	1300 12" diameter core		9	1.778		18.45	47	11.25	76.70	106	
	1350 Each added inch thick, add		68	.235		3.06	6.20	1.49	10.75	14.70	
	1500 14" diameter core		7	2.286		22.50	60	14.50	97	135	
	1550 Each added inch thick, add		55	.291		3.88	7.65	1.84	13.37	18.30	
	1700 18" diameter core		4	4		29	105	25.50	159.50	225	
	1750 Each added inch thick, add		28	.571		5.10	15.05	3.62	23.77	33	
	1760 For horizontal holes, add to above								30%	30%	
	1770 Prestressed hollow core plank, 6" thick										
	1780 1" diameter core	B-89A	52	.308	Ea.	1.53	8.10	1.95	11.58	16.50	
	1790 Each added inch thick, add		350	.046		.26	1.20	.29	1.75	2.50	
	1800 3" diameter core		50	.320		3.37	8.45	2.03	13.85	19.15	
	1810 Each added inch thick, add		240	.067		.56	1.76	.42	2.74	3.83	
	1820 4" diameter core		48	.333		4.49	8.80	2.11	15.40	21	
	1830 Each added inch thick, add		216	.074		.77	1.95	.47	3.19	4.43	
	1840 6" diameter core		44	.364		5.55	9.60	2.31	17.46	23.50	
	1850 Each added inch thick, add		175	.091		.92	2.41	.58	3.91	5.45	
	1860 8" diameter core		32	.500		7.45	13.20	3.17	23.82	32	
	1870 Each added inch thick, add		118	.136		1.28	3.57	.86	5.71	7.95	
	1880 10" diameter core		28	.571		10.05	15.05	3.62	28.72	38.50	
	1890 Each added inch thick, add		99	.162		1.38	4.26	1.02	6.66	9.30	
	1900 12" diameter core		22	.727		12.25	19.15	4.61	36.01	48.50	
	1910 Each added inch thick, add		85	.188		2.04	4.96	1.19	8.19	11.30	
	1950 Minimum charge for above, 3" diameter core		7	2.286	Total		60	14.50	74.50	110	
	2000 4" diameter core		6.80	2.353			62	14.90	76.90	113	
	2050 6" diameter core		6	2.667			70.50	16.90	87.40	129	
	2100 8" diameter core		5.50	2.909			76.50	18.45	94.95	141	
	2150 10" diameter core		4.75	3.368			89	21.50	110.50	163	
	2200 12" diameter core		3.90	4.103			108	26	134	198	
	2250 14" diameter core		3.38	4.734			125	30	155	229	
	2300 18" diameter core		3.15	5.079			134	32	166	246	

02220 | Site Demolition

		CREW	DAILY OUTPUT	LABOR-HOURS	UNIT	MAT.	LABOR	EQUIP.	TOTAL	TOTAL INCL O&P	
875	0010 **SITE DEMOLITION** No hauling, abandon catch basin or manhole	B-6	7	3.429	Ea.		85	26	111	160	875
	0020 Remove existing catch basin or manhole, masonry		4	6			149	45	194	280	
	0030 Catch basin or manhole frames and covers, stored		13	1.846			46	13.90	59.90	86.50	
	0040 Remove and reset		7	3.429			85	26	111	160	
	0900 Hydrants, fire, remove only	B-21A	5	8			225	85	310	440	
	0950 Remove and reset	"	2	20			565	213	778	1,100	
	1710 Pavement removal, bituminous roads, 3" thick	B-38	690	.058	S.Y.		1.50	1.36	2.86	3.81	
	1750 4" to 6" thick		420	.095			2.47	2.23	4.70	6.25	
	1800 Bituminous driveways		640	.063			1.62	1.46	3.08	4.11	
	1900 Concrete to 6" thick, hydraulic hammer, mesh reinforced		255	.157			4.07	3.67	7.74	10.30	

Important: See the Reference Section for critical supporting data - Reference Nos., Crews, & City Cost Indexes

02200 | Site Preparation

02220 | Site Demolition

		CREW	DAILY OUTPUT	LABOR-HOURS	UNIT	MAT.	LABOR	EQUIP.	TOTAL	TOTAL INCL O&P	
875	2000 Rod reinforced	B-38	200	.200	S.Y.		5.20	4.68	9.88	13.15	875
	2100 Concrete, 7" to 24" thick, plain		33	1.212	C.Y.		31.50	28.50	60	79.50	
	2200 Reinforced		24	1.667	"		43	39	82	110	
	2300 With hand held air equipment, bituminous, to 6" thick	B-39	1,900	.025	S.F.		.61	.09	.70	1.05	
	2320 Concrete to 6" thick, no reinforcing		1,200	.040			.97	.15	1.12	1.66	
	2340 Mesh reinforced		1,400	.034			.83	.13	.96	1.43	
	2360 Rod reinforced		765	.063			1.52	.23	1.75	2.61	
	4000 Sidewalk removal, bituminous, 2-1/2" thick	B-6	325	.074	S.Y.		1.83	.56	2.39	3.44	
	4050 Brick, set in mortar		185	.130			3.22	.98	4.20	6.05	
	4100 Concrete, plain, 4"		160	.150			3.72	1.13	4.85	7	
	4200 Mesh reinforced		150	.160			3.97	1.21	5.18	7.50	

02225 | Selective Demolition

		CREW	DAILY OUTPUT	LABOR-HOURS	UNIT	MAT.	LABOR	EQUIP.	TOTAL	TOTAL INCL O&P	
730	0010 **RUBBISH HANDLING** The following are to be added to the										730
	0020 demolition prices										
	0400 Chute, circular, prefabricated steel, 18" diameter	B-1	40	.600	L.F.	19.60	14.10		33.70	43.50	
	0440 30" diameter	"	30	.800	"	26.50	18.80		45.30	59	
	0725 Dumpster, weekly rental, 1 dump/week, 20 C.Y. capacity (8Tons)				Week					425	
	0800 30 C.Y. capacity (10 Tons)									640	
	0840 40 C.Y. capacity (13 Tons)									775	
	1000 Dust partition, 6 mil polyethylene, 4' x 8' panels, 1" x 3" frame	2 Carp	2,000	.008	S.F.	.17	.23		.40	.54	
	1080 2" x 4" frame	"	2,000	.008	"	.33	.23		.56	.72	
	2000 Load, haul to chute & dumping into chute, 50' haul	2 Clab	24	.667	C.Y.		15.25		15.25	24	
	2040 100' haul		16.50	.970			22		22	34.50	
	2080 Over 100' haul, add per 100 L.F.		35.50	.451			10.30		10.30	16.10	
	2120 In elevators, per 10 floors, add		140	.114			2.61		2.61	4.09	
	3000 Loading & trucking, including 2 mile haul, chute loaded	B-16	45	.711			16.80	10.75	27.55	38	
	3040 Hand loading truck, 50' haul	"	48	.667			15.75	10.05	25.80	35.50	
	3080 Machine loading truck	B-17	120	.267			6.55	4.59	11.14	15.15	
	5000 Haul, per mile, up to 8 C.Y. truck	B-34B	1,165	.007			.16	.41	.57	.71	
	5100 Over 8 C.Y. truck	"	1,550	.005			.12	.31	.43	.53	
740	0010 **DUMP CHARGES** Typical urban city, tipping fees only										740
	0100 Building construction materials				Ton					55	
	0200 Trees, brush, lumber									45	
	0300 Rubbish only									50	
	0500 Reclamation station, usual charge									80	
790	0010 **TORCH CUTTING** Steel, 1" thick plate	1 Clab	32	.250	L.F.		5.70		5.70	8.95	790
	0040 1" diameter bar	"	210	.038	Ea.		.87		.87	1.36	
	1000 Oxygen lance cutting, reinforced concrete walls										
	1040 12" to 16" thick walls	1 Clab	10	.800	L.F.		18.30		18.30	28.50	
	1080 24" thick walls	"	6	1.333	"		30.50		30.50	47.50	

02240 | Dewatering

		CREW	DAILY OUTPUT	LABOR-HOURS	UNIT	MAT.	LABOR	EQUIP.	TOTAL	TOTAL INCL O&P	
500	0010 **DEWATERING** Excavate drainage trench, 2' wide, 2' deep	B-11C	90	.178	C.Y.		4.73	2.01	6.74	9.45	500
	0100 2' wide, 3' deep, with backhoe loader	"	135	.119			3.15	1.34	4.49	6.30	
	0200 Excavate sump pits by hand, light soil	1 Clab	7.10	1.127			26		26	40.50	
	0300 Heavy soil	"	3.50	2.286			52		52	81.50	
	0500 Pumping 8 hr., attended 2 hrs. per day, including 20 L.F.										
	0550 of suction hose & 100 L.F. discharge hose										
	0600 2" diaphragm pump used for 8 hours	B-10H	4	3	Day		83.50	12.05	95.55	141	
	0650 4" diaphragm pump used for 8 hours	B-10I	4	3			83.50	23	106.50	153	
	0800 8 hrs. attended, 2" diaphragm pump	B-10H	1	12			335	48	383	565	
	0900 3" centrifugal pump	B-10J	1	12			335	61.50	396.50	580	

For expanded coverage of these items see *Means Heavy Construction Cost Data 2001*

02200 | Site Preparation

02240 | Dewatering

			CREW	DAILY OUTPUT	LABOR-HOURS	UNIT	MAT.	LABOR	EQUIP.	TOTAL	TOTAL INCL O&P
500	1000	4" diaphragm pump	B-10I	1	12	Day		335	91.50	426.50	610
	1100	6" centrifugal pump	B-10K	1	12	↓		335	228	563	760
	1300	CMP, incl. excavation 3' deep, 12" diameter	B-6	115	.209	L.F.	6.35	5.20	1.57	13.12	16.70
	1400	18" diameter		100	.240	"	7.90	5.95	1.81	15.66	19.90
	1600	Sump hole construction, incl. excavation and gravel, pit		1,250	.019	C.F.	.67	.48	.14	1.29	1.63
	1700	With 12" gravel collar, 12" pipe, corrugated, 16 ga.		70	.343	L.F.	12.10	8.50	2.59	23.19	29.50
	1800	15" pipe, corrugated, 16 ga.		55	.436		15.45	10.85	3.29	29.59	37.50
	1900	18" pipe, corrugated, 16 ga.		50	.480		18.15	11.90	3.62	33.67	42.50
	2000	24" pipe, corrugated, 14 ga.		40	.600	↓	22	14.90	4.52	41.42	52
	2200	Wood lining, up to 4' x 4', add		300	.080	SFCA	12.30	1.99	.60	14.89	17.25
	9950	See div. 02240-900 for wellpoints									
	9960	See div. 02240-700 for deep well systems									
700	0010	**WELLS** For dewatering 10' to 20' deep, 2' diameter									
	0020	with steel casing, minimum	B-6	165	.145	V.L.F.	2	3.61	1.10	6.71	9
	0050	Average		98	.245		4	6.10	1.85	11.95	15.85
	0100	Maximum		49	.490	↓	10	12.15	3.69	25.84	34
	0300	For pumps for dewatering, see division 01590-400-4100 to 4400									
	0500	For domestic water wells, see division 02520-900									
900	0010	**WELLPOINTS** For wellpoint equipment rental, see div. 01590-700 R02240-900									
	0100	Installation and removal of single stage system									
	0110	Labor only, .75 labor-hours per L.F., minimum	1 Clab	10.70	.748	LF Hdr		17.10		17.10	26.50
	0200	2.0 labor-hours per L.F., maximum	"	4	2	"		45.50		45.50	71.50
	0400	Pump operation, 4 @ 6 hr. shifts									
	0410	Per 24 hour day	4 Eqlt	1.27	25.197	Day		725		725	1,100
	0500	Per 168 hour week, 160 hr. straight, 8 hr. double time		.18	177	Week		5,100		5,100	7,750
	0550	Per 4.3 week month	↓	.04	800	Month		23,000		23,000	34,800
	0600	Complete installation, operation, equipment rental, fuel &									
	0610	removal of system with 2" wellpoints 5' O.C.									
	0700	100' long header, 6" diameter, first month	4 Eqlt	3.23	9.907	LF Hdr	112	285		397	555
	0800	Thereafter, per month		4.13	7.748		89.50	223		312.50	435
	1000	200' long header, 8" diameter, first month		6	5.333		100	153		253	340
	1100	Thereafter, per month		8.39	3.814		50.50	110		160.50	222
	1300	500' long header, 8" diameter, first month		10.63	3.010		39	86.50		125.50	174
	1400	Thereafter, per month		20.91	1.530		28	44		72	97
	1600	1,000' long header, 10" diameter, first month		11.62	2.754		33.50	79		112.50	157
	1700	Thereafter, per month	↓	41.81	.765	↓	16.75	22		38.75	52
	1900	Note: above figures include pumping 168 hrs. per week									
	1910	and include the pump operator and one stand-by pump.									

02250 | Shoring & Underpinning

			CREW	DAILY OUTPUT	LABOR-HOURS	UNIT	MAT.	LABOR	EQUIP.	TOTAL	TOTAL INCL O&P
400	0010	**SHEET PILING** Steel, not incl. wales, 22 psf, 15' excav., left in place	B-40	10.81	5.920	Ton	810	173	185	1,168	1,375
	0100	Drive, extract & salvage		6	10.667		340	310	335	985	1,250
	0300	20' deep excavation, 27 psf, left in place R02250-450		12.95	4.942		810	145	154	1,109	1,300
	0400	Drive, extract & salvage		6.55	9.771	↓	340	286	305	931	1,175
	1200	15' deep excavation, 22 psf, left in place		983	.065	S.F.	9.45	1.91	2.03	13.39	15.75
	1300	Drive, extract & salvage		656	.098		3.84	2.85	3.05	9.74	12.20
	1500	20' deep excavation, 27 psf, left in place		960	.067		11.85	1.95	2.08	15.88	18.50
	1600	Drive, extract & salvage	↓	640	.100	↓	4.99	2.93	3.12	11.04	13.70
	2100	Rent steel sheet piling and wales, first month				Ton	180			180	198
	2200	Per added month					18			18	19.80
	2300	Rental piling left in place, add to rental					580			580	635
	2500	Wales, connections & struts, 2/3 salvage				↓	187			187	206
	3900	Wood, solid sheeting, incl. wales, braces and spacers,									
	3910	drive, extract & salvage, 8' deep excavation	B-31	330	.121	S.F.	1.73	2.97	.45	5.15	7.05
	4000	10' deep, 50 S.F./hr. in & 150 S.F./hr. out		300	.133		1.78	3.27	.50	5.55	7.60
	4100	12' deep, 45 S.F./hr. in & 135 S.F./hr. out	↓	270	.148	↓	1.83	3.63	.55	6.01	8.30

02200 | Site Preparation

02250 | Shoring & Underpinning

		CREW	DAILY OUTPUT	LABOR-HOURS	UNIT	MAT.	LABOR	EQUIP.	TOTAL	TOTAL INCL O&P
4200	14' deep, 42 S.F./hr. in & 126 S.F./hr. out	B-31	250	.160	S.F.	1.88	3.92	.60	6.40	8.90
4300	16' deep, 40 S.F./hr. in & 120 S.F./hr. out		240	.167		1.94	4.09	.62	6.65	9.25
4400	18' deep, 38 S.F./hr. in & 114 S.F./hr. out		230	.174		2.01	4.26	.65	6.92	9.55
4520	Left in place, 8' deep, 55 S.F./hr.		440	.091		3.11	2.23	.34	5.68	7.30
4540	10' deep, 50 S.F./hr.		400	.100		3.27	2.45	.37	6.09	7.85
4560	12' deep, 45 S.F./hr.		360	.111		3.46	2.72	.42	6.60	8.50
4565	14' deep, 42 S.F./hr.		335	.119		3.66	2.93	.45	7.04	9.10
4570	16' deep, 40 S.F./hr.		320	.125		3.89	3.06	.47	7.42	9.60
4700	Alternate pricing, left in place, 8' deep		1.76	22.727	M.B.F.	700	555	85	1,340	1,725
4800	Drive, extract and salvage, 8' deep		1.32	30.303	"	620	745	113	1,478	1,950

02260 | Excavation Support/Protection

		CREW	DAILY OUTPUT	LABOR-HOURS	UNIT	MAT.	LABOR	EQUIP.	TOTAL	TOTAL INCL O&P
0010	**SOLDIER BEAMS & LAGGING** H piles with 3" wood sheeting									
0020	horizontal between piles, including removal of wales & braces									
0100	No hydrostatic head, 15' deep, 1 line of braces, minimum	B-50	545	.205	S.F.	7.40	5.75	3.06	16.21	21
0200	Maximum		495	.226		8.20	6.30	3.37	17.87	23
1300	No hydrostatic head, left in place, 15' dp., 1 line of braces, min.		635	.176		9.85	4.93	2.63	17.41	21.50
1400	Maximum		575	.195		10.55	5.45	2.90	18.90	23.50
2350	Lagging only, 3" thick wood between piles 8' O.C., minimum	B-46	400	.120		1.64	3.14	.11	4.89	7.05
2370	Maximum		250	.192		2.46	5	.17	7.63	11.15
2400	Open sheeting no bracing, for trenches to 10' deep, min.		1,736	.028		.74	.72	.02	1.48	2.03
2450	Maximum		1,510	.032		.82	.83	.03	1.68	2.30
2500	Tie-back method, add to open sheeting, add, minimum								20%	20%
2550	Maximum								60%	60%
2700	Tie-backs only, based on tie-backs total length, minimum	B-46	86.80	.553	L.F.	8.60	14.45	.50	23.55	34
2750	Maximum		38.50	1.247	"	15.10	32.50	1.12	48.72	71.50
3500	Tie-backs only, typical average, 25' long		2	24	Ea.	380	630	21.50	1,031.50	1,475
3600	35' long		1.58	30.380	"	505	795	27.50	1,327.50	1,875

02300 | Earthwork

02315 | Excavation and Fill

		CREW	DAILY OUTPUT	LABOR-HOURS	UNIT	MAT.	LABOR	EQUIP.	TOTAL	TOTAL INCL O&P
0010	**BACKFILL** By hand, no compaction, light soil	1 Clab	14	.571	C.Y.		13.05		13.05	20.50
0100	Heavy soil		11	.727			16.60		16.60	26
0300	Compaction in 6" layers, hand tamp, add to above		20.60	.388			8.85		8.85	13.90
0400	Roller compaction operator walking, add	B-10A	100	.120			3.34	1.11	4.45	6.30
0500	Air tamp, add	B-9C	190	.211			4.89	.93	5.82	8.65
0600	Vibrating plate, add	A-1	60	.133			3.05	1.08	4.13	5.95
0800	Compaction in 12" layers, hand tamp, add to above	1 Clab	34	.235			5.40		5.40	8.40
0900	Roller compaction operator walking, add	B-10A	150	.080			2.23	.74	2.97	4.21
1000	Air tamp, add	B-9	285	.140			3.26	.62	3.88	5.80
1100	Vibrating plate, add	A-1	90	.089			2.03	.72	2.75	3.97
1300	Dozer backfilling, bulk, up to 300' haul, no compaction	B-10B	1,200	.010			.28	.61	.89	1.10
1400	Air tamped	B-11B	240	.067			1.77	3.85	5.62	6.95
1600	Compacting backfill, 6" to 12" lifts, vibrating roller	B-10C	800	.015			.42	1.13	1.55	1.88
1700	Sheepsfoot roller	B-10D	750	.016			.45	1.07	1.52	1.86
1900	Dozer backfilling, trench, up to 300' haul, no compaction	B-10B	900	.013			.37	.82	1.19	1.47
2000	Air tamped	B-11B	235	.068			1.81	3.93	5.74	7.10
2200	Compacting backfill, 6" to 12" lifts, vibrating roller	B-10C	700	.017			.48	1.29	1.77	2.15
2300	Sheepsfoot roller	B-10D	650	.018			.51	1.24	1.75	2.15

For expanded coverage of these items see *Means Heavy Construction Cost Data 2001*

02300 | Earthwork

02315 | Excavation and Fill

			CREW	DAILY OUTPUT	LABOR-HOURS	UNIT	MAT.	LABOR	EQUIP.	TOTAL	TOTAL INCL O&P	
130	0010	**BEDDING** For pipe and conduit, not incl. compaction										130
	0050	Crushed or screened bank run gravel	B-6	150	.160	C.Y.	17	3.97	1.21	22.18	26	
	0100	Crushed stone 3/4" to 1/2"		150	.160		18.80	3.97	1.21	23.98	28	
	0200	Sand, dead or bank		150	.160		3.55	3.97	1.21	8.73	11.40	
	0500	Compacting bedding in trench	A-1	90	.089			2.03	.72	2.75	3.97	
340	0010	**DRILLING AND BLASTING** Only, rock, open face, under 1500 C.Y.	B-47	225	.107	C.Y.	1.80	2.72	3.29	7.81	9.80	340
	0100	Over 1500 C.Y.		300	.080		1.80	2.04	2.47	6.31	7.85	
	0200	Areas where blasting mats are required, under 1500 C.Y.		175	.137		1.80	3.49	4.23	9.52	12.05	
	0250	Over 1500 C.Y.		250	.096		1.80	2.45	2.96	7.21	9	
	2200	Trenches, up to 1500 C.Y.		22	1.091		5.20	28	33.50	66.70	86	
	2300	Over 1500 C.Y.		26	.923		5.20	23.50	28.50	57.20	74	
440	0010	**EXCAVATING, STRUCTURAL** Hand, pits to 6' deep, sandy soil	1 Clab	8	1	C.Y.		23		23	36	440
	0100	Heavy soil or clay		4	2			45.50		45.50	71.50	
	0300	Pits 6' to 12' deep, sandy soil		5	1.600			36.50		36.50	57	
	0500	Heavy soil or clay		3	2.667			61		61	95.50	
	0700	Pits 12' to 18' deep, sandy soil		4	2			45.50		45.50	71.50	
	0900	Heavy soil or clay		2	4			91.50		91.50	143	
	1500	For wet or muck hand excavation, add to above				%				50%	50%	
505	0010	**FILL** Spread dumped material, by dozer, no compaction	B-10B	1,000	.012	C.Y.		.33	.74	1.07	1.32	505
	0100	By hand	1 Clab	12	.667	"		15.25		15.25	24	
	0500	Gravel fill, compacted, under floor slabs, 4" deep	B-37	10,000	.005	S.F.	.15	.12	.01	.28	.36	
	0600	6" deep		8,600	.006		.23	.13	.02	.38	.48	
	0700	9" deep		7,200	.007		.38	.16	.02	.56	.69	
	0800	12" deep		6,000	.008		.53	.19	.02	.74	.90	
	1000	Alternate pricing method, 4" deep		120	.400	C.Y.	11.35	9.65	1.11	22.11	29	
	1100	6" deep		160	.300		11.35	7.25	.83	19.43	24.50	
	1200	9" deep		200	.240		11.35	5.80	.67	17.82	22	
	1300	12" deep		220	.218		11.35	5.25	.61	17.21	21.50	
	1400	Granular fill					7.60			7.60	8.35	
900	0010	**EXCAVATING, TRENCH** or continuous footing, common earth	A12.3 -110									900
	0020	No sheeting or dewatering included										
	0050	1' to 4' deep, 3/8 C.Y. tractor loader/backhoe	B-11C	150	.107	C.Y.		2.84	1.21	4.05	5.70	
	0060	1/2 C.Y. tractor loader/backhoe	B-11M	200	.080			2.13	1.14	3.27	4.53	
	0090	4' to 6' deep, 1/2 C.Y. tractor loader/backhoe	"	200	.080			2.13	1.14	3.27	4.53	
	0100	5/8 C.Y. hydraulic backhoe	B-12Q	250	.064			1.82	1.44	3.26	4.34	
	0110	3/4 C.Y. hydraulic backhoe	B-12F	300	.053			1.52	1.38	2.90	3.82	
	0300	1/2 C.Y. hydraulic excavator, truck mounted	B-12J	200	.080			2.28	3.24	5.52	7	
	0500	6' to 10' deep, 3/4 C.Y. hydraulic backhoe, 6' to 10' deep	B-12F	225	.071			2.02	1.84	3.86	5.10	
	0600	1 C.Y. hydraulic excavator, truck mounted	B-12K	400	.040			1.14	1.94	3.08	3.85	
	0900	10' to 14' deep, 3/4 C.Y. hydraulic backhoe	B-12F	200	.080			2.28	2.07	4.35	5.75	
	1000	1-1/2 C.Y. hydraulic backhoe	B-12B	540	.030			.84	1.10	1.94	2.49	
	1300	14' to 20' deep, 1 C.Y. hydraulic backhoe	B-12A	320	.050			1.42	1.53	2.95	3.84	
	1400	By hand with pick and shovel 2' to 6' deep, light soil	1 Clab	8	1			23		23	36	
	1500	Heavy soil	"	4	2			45.50		45.50	71.50	
	1700	For tamping backfilled trenches, air tamp, add	A-1	100	.080			1.83	.65	2.48	3.57	
	1900	Vibrating plate, add	B-18	230	.104			2.45	.26	2.71	4.13	
	2100	Trim sides and bottom for concrete pours, common earth		1,500	.016	S.F.		.38	.04	.42	.63	
	2300	Hardpan		600	.040	"		.94	.10	1.04	1.58	
940	0010	**EXCAVATING, UTILITY TRENCH** Common earth										940
	0050	Trenching with chain trencher, 12 H.P., operator walking										
	0100	4" wide trench, 12" deep	B-53	800	.010	L.F.		.29	.12	.41	.57	
	0150	18" deep		750	.011			.31	.13	.44	.60	

02300 | Earthwork

02315 | Excavation and Fill

		CREW	DAILY OUTPUT	LABOR-HOURS	UNIT	MAT.	LABOR	EQUIP.	TOTAL	TOTAL INCL O&P
0200	24" deep	B-53	700	.011	L.F.		.33	.14	.47	.65
0300	6" wide trench, 12" deep		650	.012			.35	.15	.50	.70
0350	18" deep		600	.013			.38	.16	.54	.76
0400	24" deep		550	.015			.42	.18	.60	.82
0450	36" deep		450	.018			.51	.22	.73	1.01
0600	8" wide trench, 12" deep		475	.017			.48	.20	.68	.95
0650	18" deep		400	.020			.58	.24	.82	1.14
0700	24" deep		350	.023			.66	.28	.94	1.30
0750	36" deep		300	.027			.77	.32	1.09	1.52
1000	Backfill by hand including compaction, add									
1050	4" wide trench, 12" deep	A-1	800	.010	L.F.		.23	.08	.31	.45
1100	18" deep		530	.015			.34	.12	.46	.67
1150	24" deep		400	.020			.46	.16	.62	.90
1300	6" wide trench, 12" deep		540	.015			.34	.12	.46	.66
1350	18" deep		405	.020			.45	.16	.61	.89
1400	24" deep		270	.030			.68	.24	.92	1.32
1450	36" deep		180	.044			1.02	.36	1.38	1.99
1600	8" wide trench, 12" deep		400	.020			.46	.16	.62	.90
1650	18" deep		265	.030			.69	.24	.93	1.35
1700	24" deep		200	.040			.91	.32	1.23	1.79
1750	36" deep		135	.059			1.35	.48	1.83	2.65
2000	Chain trencher, 40 H.P. operator riding									
2050	6" wide trench and backfill, 12" deep	B-54	1,200	.007	L.F.		.19	.16	.35	.47
2100	18" deep		1,000	.008			.23	.20	.43	.57
2150	24" deep		975	.008			.24	.20	.44	.58
2200	36" deep		900	.009			.26	.22	.48	.63
2250	48" deep		750	.011			.31	.26	.57	.75
2300	60" deep		650	.012			.35	.30	.65	.87
2400	8" wide trench and backfill, 12" deep		1,000	.008			.23	.20	.43	.57
2450	18" deep		950	.008			.24	.21	.45	.60
2500	24" deep		900	.009			.26	.22	.48	.63
2550	36" deep		800	.010			.29	.25	.54	.71
2600	48" deep		650	.012			.35	.30	.65	.87
2700	12" wide trench and backfill, 12" deep		975	.008			.24	.20	.44	.58
2750	18" deep		860	.009			.27	.23	.50	.66
2800	24" deep		800	.010			.29	.25	.54	.71
2850	36" deep		725	.011			.32	.27	.59	.78
3000	16" wide trench and backfill, 12" deep		835	.010			.28	.24	.52	.68
3050	18" deep		750	.011			.31	.26	.57	.75
3100	24" deep		700	.011			.33	.28	.61	.81
3200	Compaction with vibratory plate, add								50%	50%
5100	Hand excavate and trim for pipe bells after trench excavation									
5200	8" pipe	1 Clab	155	.052	L.F.		1.18		1.18	1.85
5300	18" pipe	"	130	.062	"		1.41		1.41	2.20

02320 | Hauling

		CREW	DAILY OUTPUT	LABOR-HOURS	UNIT	MAT.	LABOR	EQUIP.	TOTAL	TOTAL INCL O&P
0011	HAULING Excavated or borrow material, loose cubic yards									
0015	no loading included, highway haulers									
0020	6 C.Y. dump truck, 1/4 mile round trip, 5.0 loads/hr.	B-34A	195	.041	C.Y.		.98	1.90	2.88	3.59
0030	1/2 mile round trip, 4.1 loads/hr.		160	.050			1.20	2.32	3.52	4.38
0040	1 mile round trip, 3.3 loads/hr.		130	.062			1.47	2.85	4.32	5.40
0100	2 mile round trip, 2.6 loads/hr.		100	.080			1.92	3.70	5.62	7
0150	3 mile round trip, 2.1 loads/hr.		80	.100			2.40	4.63	7.03	8.75
0200	4 mile round trip, 1.8 loads/hr.		70	.114			2.74	5.30	8.04	10
0310	12 C.Y. dump truck, 1/4 mile round trip 3.7 loads/hr.	B-34B	288	.028			.67	1.68	2.35	2.86
0400	2 mile round trip, 2.2 loads/hr.		180	.044			1.06	2.68	3.74	4.58

For expanded coverage of these items see *Means Heavy Construction Cost Data 2001*

02300 | Earthwork

02320 | Hauling

			CREW	DAILY OUTPUT	LABOR-HOURS	UNIT	2001 BARE COSTS				TOTAL INCL O&P
							MAT.	LABOR	EQUIP.	TOTAL	
200	0450	3 mile round trip, 1.9 loads/hr.	B-34B	170	.047	C.Y.		1.13	2.84	3.97	4.84
	0500	4 mile round trip, 1.6 loads/hr.		125	.064			1.53	3.86	5.39	6.60
	1300	Hauling in medium traffic, add								20%	20%
	1400	Heavy traffic, add								30%	30%
	1600	Grading at dump, or embankment if required, by dozer	B-10B	1,000	.012			.33	.74	1.07	1.32
	1800	Spotter at fill or cut, if required	1 Clab	8	1	Hr.		23		23	36

R01590-100

02400 | Tunneling, Boring & Jacking

02441 | Microtunneling

			CREW	DAILY OUTPUT	LABOR-HOURS	UNIT	2001 BARE COSTS				TOTAL INCL O&P
							MAT.	LABOR	EQUIP.	TOTAL	
400	0010	MICROTUNNELING Not including excavation, backfill, shoring,									
	0020	or dewatering, average 50'/day, slurry method									
	0100	24" to 48" outside diameter, minimum				L.F.					600
	0110	Adverse conditions, add				%					50%
	1000	Rent microtunneling machine, average monthly lease				Month					80,000
	1010	Operating technician				Day					600
	1100	Mobilization and demobilization, minimum				Job					40,000
	1110	Maximum				"					400,000

02445 | Boring or Jacking Conduits

			CREW	DAILY OUTPUT	LABOR-HOURS	UNIT	MAT.	LABOR	EQUIP.	TOTAL	TOTAL INCL O&P
300	0010	HORIZONTAL BORING Casing only, 100' minimum,										300
	0020	not incl. jacking pits or dewatering									
	0100	Roadwork, 1/2" thick wall, 24" diameter casing	B-42	20	3.200	L.F.	47.50	82.50	62	192	251
	0200	36" diameter		16	4		75	103	77.50	255.50	330
	0300	48" diameter		15	4.267		110	110	82.50	302.50	385
	0500	Railroad work, 24" diameter		15	4.267		47.50	110	82.50	240	320
	0600	36" diameter		14	4.571		75	118	88.50	281.50	365
	0700	48" diameter		12	5.333		110	137	103	350	450
	0900	For ledge, add								145	175
	1000	Small diameter boring, 3", sandy soil	B-82	900	.018		13	.46	.05	13.51	15.05
	1040	Rocky soil	"	500	.032		13	.83	.09	13.92	15.65
	1100	Prepare jacking pits, incl. mobilization & demobilization, minimum				Ea.				2,650	3,150
	1101	Maximum				"				15,000	18,000

02500 | Utility Services

02510 | Water Distribution

			CREW	DAILY OUTPUT	LABOR-HOURS	UNIT	MAT.	LABOR	EQUIP.	TOTAL	TOTAL INCL O&P
350	0010	DISTRIBUTION CONNECTION										350
	7020	Crosses, 4" x 4"	B-21	37	.757	Ea.	335	20	3.84	358.84	400
	7040	6" x 6"		25	1.120		395	30	5.70	430.70	490
	7060	8" x 6"		21	1.333		485	35.50	6.75	527.25	600
	7080	8" x 8"		21	1.333		525	35.50	6.75	567.25	645
	7100	10" x 6"		21	1.333		965	35.50	6.75	1,007.25	1,125

02500 | Utility Services

02510 | Water Distribution

		Crew	Daily Output	Labor-Hours	Unit	Mat.	Labor	Equip.	Total	Total Incl O&P	
350	7120	10" x 10"	B-21	21	1.333	Ea.	1,050	35.50	6.75	1,092.25	1,225	350
	7140	12" x 6"		18	1.556		965	41.50	7.90	1,014.40	1,125	
	7160	12" x 12"		18	1.556		1,225	41.50	7.90	1,274.40	1,425	
	7180	14" x 6"		16	1.750		2,425	46.50	8.85	2,480.35	2,725	
	7200	14" x 14"		16	1.750		2,550	46.50	8.85	2,605.35	2,875	
	7220	16" x 6"		14	2		2,575	53	10.15	2,638.15	2,925	
	7240	16" x 10"		14	2		2,625	53	10.15	2,688.15	2,975	
	7260	16" x 16"		14	2		2,750	53	10.15	2,813.15	3,125	
	7280	18" x 6"		10	2.800		3,850	74.50	14.20	3,938.70	4,375	
	7300	18" x 12"		10	2.800		3,875	74.50	14.20	3,963.70	4,400	
	7320	18" x 18"		10	2.800		4,000	74.50	14.20	4,088.70	4,525	
	7340	20" x 6"		8	3.500		3,125	93	17.75	3,235.75	3,600	
	7360	20" x 12"		8	3.500		3,350	93	17.75	3,460.75	3,850	
	7380	20" x 20"		8	3.500		4,875	93	17.75	4,985.75	5,550	
	7400	24" x 6"		6	4.667		4,025	124	23.50	4,172.50	4,650	
	7420	24" x 12"		6	4.667		4,100	124	23.50	4,247.50	4,725	
	7440	24" x 18"		6	4.667		6,100	124	23.50	6,247.50	6,925	
	7460	24" x 24"		6	4.667		6,350	124	23.50	6,497.50	7,200	
	7600	Cut-in sleeves with rubber gaskets, 4"		18	1.556		111	41.50	7.90	160.40	195	
	7620	6"		12	2.333		141	62	11.85	214.85	265	
	7640	8"		10	2.800		191	74.50	14.20	279.70	340	
	7660	10"		10	2.800		265	74.50	14.20	353.70	425	
	7680	12"		9	3.111		315	83	15.75	413.75	490	
	7800	Cut-in valves with rubber gaskets, 4"		18	1.556		289	41.50	7.90	338.40	395	
	7820	6"		12	2.333		390	62	11.85	463.85	540	
	7840	8"		10	2.800		605	74.50	14.20	693.70	795	
	7860	10"		10	2.800		925	74.50	14.20	1,013.70	1,150	
	7880	12"		9	3.111		1,175	83	15.75	1,273.75	1,425	
	8000	Sleeves with rubber gaskets, 4" x 4"		37	.757		267	20	3.84	290.84	330	
	8020	6" x 6"		25	1.120		315	30	5.70	350.70	405	
	8040	8" x 6"		21	1.333		395	35.50	6.75	437.25	500	
	8060	8" x 8"		21	1.333		420	35.50	6.75	462.25	530	
	8080	10" x 6"		21	1.333		770	35.50	6.75	812.25	915	
	8100	10" x 10"		21	1.333		835	35.50	6.75	877.25	985	
	8120	12" x 6"		18	1.556		770	41.50	7.90	819.40	925	
	8140	12" x 12"		18	1.556		975	41.50	7.90	1,024.40	1,150	
	8160	14" x 6"		16	1.750		1,925	46.50	8.85	1,980.35	2,200	
	8180	14" x 14"		16	1.750		2,025	46.50	8.85	2,080.35	2,325	
	8200	16" x 6"		14	2		2,050	53	10.15	2,113.15	2,375	
	8220	16" x 10"		14	2		2,100	53	10.15	2,163.15	2,400	
	8240	16" x 16"		14	2		2,200	53	10.15	2,263.15	2,525	
	8260	18" x 6"		10	2.800		3,075	74.50	14.20	3,163.70	3,525	
	8280	18" x 12"		10	2.800		3,100	74.50	14.20	3,188.70	3,550	
	8300	18" x 18"		10	2.800		3,200	74.50	14.20	3,288.70	3,650	
	8320	20" x 6"		8	3.500		2,500	93	17.75	2,610.75	2,925	
	8340	20" x 12"		8	3.500		2,675	93	17.75	2,785.75	3,125	
	8360	20" x 20"		8	3.500		3,900	93	17.75	4,010.75	4,475	
	8380	24" x 6"		6	4.667		3,225	124	23.50	3,372.50	3,775	
	8400	24" x 12"		6	4.667		3,275	124	23.50	3,422.50	3,825	
	8420	24" x 18"		6	4.667		4,875	124	23.50	5,022.50	5,600	
	8440	24" x 24"		6	4.667		5,075	124	23.50	5,222.50	5,800	
	8800	Curb box, 6' long	B-20	20	1.200		132	31		163	194	
	8820	8' long	"	18	1.333		133	34.50		167.50	200	
	9000	Valves, gate valve, N.R.S. post type, 4" diameter	B-21	32	.875		233	23.50	4.44	260.94	297	
	9020	6" diameter		20	1.400		297	37.50	7.10	341.60	390	
	9040	8" diameter		16	1.750		475	46.50	8.85	530.35	600	

For expanded coverage of these items see *Means Heavy Construction Cost Data 2001*

02500 | Utility Services

02510 | Water Distribution

			Crew	Daily Output	Labor-Hours	Unit	Mat.	Labor	Equip.	Total	Total Incl O&P
350	9060	10" diameter	B-21	16	1.750	Ea.	735	46.50	8.85	790.35	890
	9080	12" diameter		13	2.154		935	57.50	10.90	1,003.40	1,125
	9100	14" diameter		11	2.545		2,200	67.50	12.90	2,280.40	2,550
	9120	O.S.&Y., 4" diameter		32	.875		330	23.50	4.44	357.94	400
	9140	6" diameter		20	1.400		425	37.50	7.10	469.60	535
	9160	8" diameter		16	1.750		650	46.50	8.85	705.35	795
	9180	10" diameter		16	1.750		935	46.50	8.85	990.35	1,100
	9200	12" diameter		13	2.154		1,250	57.50	10.90	1,318.40	1,450
	9220	14" diameter		11	2.545		2,200	67.50	12.90	2,280.40	2,550
	9400	Check valves, rubber disc, 2-1/2" diameter	B-20	44	.545		258	14.10		272.10	305
	9420	3" diameter	"	38	.632		258	16.35		274.35	310
	9440	4" diameter	B-21	32	.875		320	23.50	4.44	347.94	390
	9480	6" diameter		20	1.400		480	37.50	7.10	524.60	590
	9500	8" diameter		16	1.750		700	46.50	8.85	755.35	850
	9520	10" diameter		16	1.750		1,250	46.50	8.85	1,305.35	1,450
	9540	12" diameter		13	2.154		1,975	57.50	10.90	2,043.40	2,275
	9542	14" diameter		11	2.545		2,775	67.50	12.90	2,855.40	3,175
	9700	Detector check valves, reducing, 4" diameter		32	.875		640	23.50	4.44	667.94	745
	9720	6" diameter		20	1.400		985	37.50	7.10	1,029.60	1,150
	9740	8" diameter		16	1.750		1,600	46.50	8.85	1,655.35	1,850
	9760	10" diameter		16	1.750		3,175	46.50	8.85	3,230.35	3,575
	9800	Galvanized, 4" diameter		32	.875		835	23.50	4.44	862.94	960
	9820	6" diameter		20	1.400		1,225	37.50	7.10	1,269.60	1,400
	9840	8" diameter		16	1.750		1,950	46.50	8.85	2,005.35	2,225
	9860	10" diameter		16	1.750		3,850	46.50	8.85	3,905.35	4,325
800	0010	**PIPING, WATER DISTRIBUTION SYSTEMS** Pipe, laid									
	0020	in trench, excavation and backfill not included (R02510-800)									
	1400	Ductile Iron, cement lined, class 50 water pipe, 18' lengths									
	1410	Mechanical joint, 4" diameter	B-20	144	.167	L.F.	8.75	4.31		13.06	16.40
	1420	6" diameter		126	.190		9.10	4.92		14.02	17.70
	1430	8" diameter		108	.222		11.65	5.75		17.40	22
	1440	10" diameter		90	.267		15.15	6.90		22.05	27.50
	1450	12" diameter	B-21	72	.389		19.35	10.35	1.97	31.67	40
	1460	14" diameter		54	.519		24	13.80	2.63	40.43	50.50
	1470	16" diameter		46	.609		26	16.20	3.09	45.29	57
	1480	18" diameter		42	.667		34.50	17.75	3.38	55.63	68.50
	1490	24" diameter		35	.800		51	21.50	4.06	76.56	94
	1550	Push on joint, 4" diameter	B-20	155	.155		6.45	4		10.45	13.35
	1560	6" diameter		135	.178		7.45	4.60		12.05	15.35
	1570	8" diameter		115	.209		10.50	5.40		15.90	20
	1580	10" diameter		98	.245		15.95	6.35		22.30	27.50
	1590	12" diameter		78	.308		16.80	7.95		24.75	31
	1600	14" diameter	B-21	58	.483		18.50	12.85	2.45	33.80	43
	1610	16" diameter		52	.538		26	14.35	2.73	43.08	54
	1620	18" diameter		43	.651		29	17.35	3.30	49.65	62.50
	1630	20" diameter		41	.683		32	18.15	3.46	53.61	67.50
	1640	24" diameter		40	.700		41	18.65	3.55	63.20	78
	1950	Butterfly valves with boxes, cast iron									
	1970	4" diameter	B-20	6	4	Ea.	360	103		463	555
	1990	6" diameter	"	5	4.800		485	124		609	730
	2010	8" diameter	B-21	4	7		725	186	35.50	946.50	1,125
	2030	10" diameter		3.50	8		955	213	40.50	1,208.50	1,425
	2050	12" diameter		3	9.333		1,325	248	47.50	1,620.50	1,875
	2070	14" diameter		2	14		1,800	375	71	2,246	2,625
	2090	16" diameter		2	14		2,225	375	71	2,671	3,100

02500 | Utility Services

02510 | Water Distribution

			CREW	DAILY OUTPUT	LABOR-HOURS	UNIT	2001 BARE COSTS MAT.	LABOR	EQUIP.	TOTAL	TOTAL INCL O&P	
800	2650	Polyvinyl chloride pipe, class 160, S.D.R.-26, 1-1/2" diameter R02510-800	B-20	300	.080	L.F.	.59	2.07		2.66	3.89	800
	2700	2" diameter		250	.096		.89	2.48		3.37	4.87	
	2750	2-1/2" diameter		250	.096		1.32	2.48		3.80	5.35	
	2800	3" diameter		200	.120		1.89	3.10		4.99	6.95	
	2850	4" diameter		200	.120		3.09	3.10		6.19	8.25	
	2900	6" diameter		180	.133		6.65	3.45		10.10	12.75	
	2950	8" diameter	B-21	160	.175		11.30	4.66	.89	16.85	20.50	
	8000	Fittings, ductile iron, mechanical joint										
	8010	90° bend 4" diameter	B-20	37	.649	Ea.	145	16.75		161.75	187	
	8020	6" diameter		25	.960		195	25		220	253	
	8040	8" diameter		21	1.143		275	29.50		304.50	350	
	8060	10" diameter	B-21	21	1.333		435	35.50	6.75	477.25	545	
	8080	12" diameter		18	1.556		595	41.50	7.90	644.40	730	
	8100	14" diameter		16	1.750		610	46.50	8.85	665.35	750	
	8120	16" diameter		14	2		660	53	10.15	723.15	825	
	8140	18" diameter		10	2.800		1,525	74.50	14.20	1,613.70	1,800	
	8160	20" diameter		8	3.500		2,300	93	17.75	2,410.75	2,700	
	8180	24" diameter		6	4.667		2,600	124	23.50	2,747.50	3,075	
	8200	Wye or tee, 4" diameter	B-20	25	.960		213	25		238	273	
	8220	6" diameter		17	1.412		160	36.50		196.50	233	
	8240	8" diameter		14	1.714		234	44.50		278.50	330	
	8260	10" diameter	B-21	14	2		800	53	10.15	863.15	975	
	8280	12" diameter		12	2.333		975	62	11.85	1,048.85	1,175	
	8300	14" diameter		10	2.800		910	74.50	14.20	998.70	1,125	
	8320	16" diameter		8	3.500		1,050	93	17.75	1,160.75	1,325	
	8340	18" diameter		6	4.667		1,975	124	23.50	2,122.50	2,400	
	8360	20" diameter		4	7		2,600	186	35.50	2,821.50	3,175	
	8380	24" diameter		3	9.333		3,025	248	47.50	3,320.50	3,750	
820	0010	**PIPING, WATER DISTRIBUTION, DUCTILE IRON** cement lined										820
	4000	Drill and tap pressurized main (labor only)										
	4100	6" main, 1" to 2" service	Q-1	3	5.333	Ea.		165		165	250	
	4150	8" main, 1" to 2" service	"	2.75	5.818	"		180		180	272	
	4500	Tap and insert gate valve										
	4600	8" main, 4" branch	B-21	3.20	8.750	Ea.	233	44.50		277.50	415	
	4650	6" branch		2.70	10.370		276	52.50		328.50	490	
	4700	10" main, 4" branch		2.70	10.370		276	52.50		328.50	490	
	4750	6" branch		2.35	11.915		315	60.50		375.50	560	
	4800	12" main, 6" branch		2.35	11.915		315	60.50		375.50	560	
	4850	8" branch		2.35	11.915		315	60.50		375.50	560	
850	0010	**PIPING, HDPE BUTT FUSION JOINTS**, SDR 21, 40' lengths										850
	0100	4" diameter	B-22A	400	.095	L.F.	1.75	2.48	1.41	5.64	7.35	
	0200	6" diameter		380	.100		2.87	2.61	1.48	6.96	8.85	
	0300	8" diameter		320	.119		4.87	3.09	1.76	9.72	12.10	
	0400	10" diameter		300	.127		7.60	3.30	1.87	12.77	15.55	
	0500	12" diameter		260	.146		10.65	3.81	2.16	16.62	20	
	0600	14" diameter		220	.173		12.80	4.50	2.55	19.85	24	
	0700	16" diameter		180	.211		16.75	5.50	3.12	25.37	30.50	
	0800	18" diameter		140	.271		21	7.05	4.01	32.06	39	
	0900	24" diameter		100	.380		37.50	9.90	5.60	53	63	
	1000	Fittings										
	1100	Elbows, 90 degrees										
	1200	4" diameter	B-22B	32	.500	Ea.	27	13.20	3.71	43.91	54.50	
	1300	6" diameter		28	.571		72.50	15.05	4.23	91.78	108	
	1400	8" diameter		24	.667		150	17.55	4.94	172.49	198	
	1500	10" diameter		18	.889		264	23.50	6.60	294.10	335	

For expanded coverage of these items see *Means Heavy Construction Cost Data 2001*

02500 | Utility Services

02510 | Water Distribution

		Crew	Daily Output	Labor-Hours	Unit	Mat.	Labor	Equip.	Total	Total Incl O&P
1600	12" diameter	B-22B	12	1.333	Ea.	450	35	9.90	494.90	560
1700	14" diameter		9	1.778		435	47	13.15	495.15	570
1800	16" diameter		6	2.667		525	70.50	19.75	615.25	705
1900	18" diameter		4	4		705	105	29.50	839.50	980
2000	24" diameter		3	5.333		1,425	141	39.50	1,605.50	1,850
2100	Tees									
2200	4" diameter	B-22B	30	.533	Ea.	31	14.05	3.95	49	60.50
2300	6" diameter		26	.615		85	16.20	4.56	105.76	124
2400	8" diameter		22	.727		215	19.15	5.40	239.55	273
2500	10" diameter		15	1.067		284	28	7.90	319.90	365
2600	12" diameter		10	1.600		390	42	11.85	443.85	505
2700	14" diameter		8	2		460	52.50	14.80	527.30	605
2800	16" diameter		6	2.667		540	70.50	19.75	630.25	725
2900	18" diameter		4	4		710	105	29.50	844.50	980
3000	24" diameter		2	8		1,475	211	59.50	1,745.50	2,025

02520 | Wells

		Crew	Daily Output	Labor-Hours	Unit	Mat.	Labor	Equip.	Total	Total Incl O&P
0010	**WELLS** Domestic water									
0100	Drilled, 4" to 6" diameter	B-23	120	.333	L.F.		7.75	17.10	24.85	31
0200	8" diameter	"	95.20	.420	"		9.75	21.50	31.25	39
0400	Gravel pack well, 40' deep, incl. gravel & casing, complete									
0500	24" diameter casing x 18" diameter screen	B-23	.13	307	Total	21,000	7,150	15,800	43,950	51,500
0600	36" diameter casing x 18" diameter screen		.12	333	"	22,600	7,750	17,100	47,450	55,500
0800	Observation wells, 1-1/4" riser pipe		163	.245	V.L.F.	11.55	5.70	12.60	29.85	35.50
0900	For flush Buffalo roadway box, add	1 Skwk	16.60	.482	Ea.	31.50	14.40		45.90	57
1200	Test well, 2-1/2" diameter, up to 50' deep (15 to 50 GPM)	B-23	1.51	26.490	"	475	615	1,350	2,440	2,975
1300	Over 50' deep, add	"	121.80	.328	L.F.	12.60	7.65	16.85	37.10	44.50
1500	Pumps, installed in wells to 100' deep, 4" submersible									
1510	1/2 H.P.	Q-1	3.22	4.969	Ea.	355	154		509	625
1520	3/4 H.P.		2.66	6.015		425	186		611	750
1600	1 H.P.		2.29	6.987		480	217		697	850
1700	1-1/2 H.P.	Q-22	1.60	10		620	310	320	1,250	1,500
1800	2 H.P.		1.33	12.030		840	375	385	1,600	1,925
1900	3 H.P.		1.14	14.035		1,050	435	450	1,935	2,300
2000	5 H.P.		1.14	14.035		1,550	435	450	2,435	2,850
2050	Remove and install motor only, 4 H.P.		1.14	14.035		615	435	450	1,500	1,825
3000	Pump, 6" submersible, 25' to 150' deep, 25 H.P., 249 to 297 GPM		.89	17.978		3,900	555	580	5,035	5,775
3100	25' to 500' deep, 30 H.P., 100 to 300 GPM		.73	21.918		3,950	680	705	5,335	6,125
8110	Well screen assembly, stainless steel, 2" diameter	B-23A	273	.088	L.F.	34	2.29	7.15	43.44	49
8120	3" diameter		253	.095		47.50	2.47	7.70	57.67	65
8130	4" diameter		200	.120		54	3.12	9.75	66.87	75
8140	5" diameter		168	.143		63.50	3.72	11.65	78.87	88.50
8150	6" diameter		126	.190		76	4.96	15.50	96.46	108
8160	8" diameter		98.50	.244		99.50	6.35	19.85	125.70	141
8170	10" diameter		73	.329		125	8.55	27	160.55	180
8180	12" diameter		62.50	.384		147	10	31.50	188.50	211
8190	14" diameter		54.30	.442		166	11.50	36	213.50	240
8200	16" diameter		48.30	.497		184	12.95	40.50	237.45	267
8210	18" diameter		39.20	.612		228	15.95	50	293.95	330
8220	20" diameter		31.20	.769		261	20	62.50	343.50	385
8230	24" diameter		23.80	1.008		325	26	82	433	485
8240	26" diameter		21	1.143		355	29.50	93	477.50	540
8300	Slotted PVC, 1-1/4" diameter		521	.046		1.26	1.20	3.75	6.21	7.35
8310	1-1/2" diameter		488	.049		1.88	1.28	4	7.16	8.45
8320	2" diameter		273	.088		2.75	2.29	7.15	12.19	14.40

02500 | Utility Services

02520 | Wells

		Crew	Daily Output	Labor-Hours	Unit	Mat.	Labor	Equip.	Total	Total Incl O&P	
900	8330	3" diameter	B-23A	253	.095	L.F.	2.73	2.47	7.70	12.90	15.30	900
	8340	4" diameter		200	.120		3.39	3.12	9.75	16.26	19.30	
	8350	5" diameter		168	.143		3.56	3.72	11.65	18.93	22.50	
	8360	6" diameter		126	.190		4.65	4.96	15.50	25.11	30	
	8370	8" diameter		98.50	.244		6.70	6.35	19.85	32.90	39	
	8400	Artificial gravel pack, 2" screen, 6" casing	B-23B	174	.138		1.89	3.59	12.15	17.63	21	
	8405	8" casing		111	.216		2.58	5.65	19.10	27.33	32.50	
	8410	10" casing		74.50	.322		3.25	8.40	28.50	40.15	48	
	8415	12" casing		60	.400		4.38	10.40	35.50	50.28	60	
	8420	14" casing		50.20	.478		5.65	12.45	42	60.10	72	
	8425	16" casing		40.70	.590		7.80	15.35	52	75.15	89	
	8430	18" casing		36	.667		9.10	17.35	59	85.45	102	
	8435	20" casing		29.50	.814		10.45	21	72	103.45	123	
	8440	24" casing		25.70	.934		11.60	24.50	82.50	118.60	141	
	8445	26" casing		24.60	.976		13	25.50	86	124.50	148	
	8450	30" casing		20	1.200		14.90	31	106	151.90	180	
	8455	36" casing		16.40	1.463		16	38	129	183	219	
	8500	Develop well		8	3	Hr.	165	78	265	508	595	
	8550	Pump test well		8	3		44.50	78	265	387.50	460	
	8560	Standby well	B-23A	8	3		43.50	78	244	365.50	440	
	8570	Standby, drill rig		8	3			78	244	322	390	
	8580	Surface seal well, concrete filled		1	24	Ea.	410	625	1,950	2,985	3,575	
	8590	Well test pump, install & remove	B-23	1	40			930	2,050	2,980	3,700	
	8600	Well sterilization, chlorine	2 Clab	1	16		345	365		710	950	
	9950	See div. 02240-900 for wellpoints										
	9960	See div. 02240-700 for drainage wells										
910	0010	**PUMPS, WELL** Water system, with pressure control										910
	1000	Deep well, jet, 42 gal. galvanized tank										
	1040	3/4 HP	1 Plum	.80	10	Ea.	545	345		890	1,125	
	3000	Shallow well, jet, 30 gal. galvanized tank										
	3040	1/2 HP	1 Plum	2	4	Ea.	550	138		688	815	

02530 | Sanitary Sewerage

		Crew	Daily Output	Labor-Hours	Unit	Mat.	Labor	Equip.	Total	Total Incl O&P	
730	0010	**PIPING, DRAINAGE & SEWAGE, CONCRETE**										730
	0020	Not including excavation or backfill										
	1000	Non-reinforced pipe, extra strength, B&S or T&G joints										
	1010	6" diameter	B-14	265.04	.181	L.F.	3.48	4.38	.68	8.54	11.40	
	1020	8" diameter		224	.214		3.83	5.20	.81	9.84	13.15	
	1030	10" diameter		216	.222		4.25	5.35	.84	10.44	13.95	
	1040	12" diameter		200	.240		5.20	5.80	.90	11.90	15.75	
	1050	15" diameter		180	.267		6.10	6.45	1.01	13.56	17.80	
	1060	18" diameter		144	.333		7.50	8.05	1.26	16.81	22	
	1070	21" diameter		112	.429		9.20	10.35	1.62	21.17	28	
	1080	24" diameter		100	.480		11.30	11.60	1.81	24.71	32.50	
	2000	Reinforced culvert, class 3, no gaskets										
	2010	12" diameter	B-14	210	.229	L.F.	9.80	5.50	.86	16.16	20.50	
	2020	15" diameter		175	.274		12.10	6.65	1.03	19.78	24.50	
	2030	18" diameter		130	.369		12.20	8.90	1.39	22.49	29	
	2035	21" diameter		120	.400		14.85	9.65	1.51	26.01	33	
	2040	24" diameter		100	.480		17.80	11.60	1.81	31.21	39.50	
	2045	27" diameter	B-13	92	.609		21.50	15.05	6.85	43.40	54.50	
	2050	30" diameter		88	.636		24.50	15.75	7.15	47.40	59.50	
	2060	36" diameter		72	.778		35	19.25	8.75	63	78	
	2070	42" diameter	B-13B	72	.778		47.50	19.25	12.50	79.25	96.50	
	2080	48" diameter		64	.875		59.50	21.50	14.10	95.10	115	

For expanded coverage of these items see Means Heavy Construction Cost Data 2001

02500 | Utility Services

02530 | Sanitary Sewerage

		CREW	DAILY OUTPUT	LABOR-HOURS	UNIT	MAT.	LABOR	EQUIP.	TOTAL	TOTAL INCL O&P	
730	2090	60" diameter	B-13B	48	1.167	L.F.	95	29	18.80	142.80	170	730
	2100	72" diameter		40	1.400		135	34.50	22.50	192	227	
	2120	84" diameter		32	1.750		245	43.50	28	316.50	365	
	2140	96" diameter		24	2.333		294	57.50	37.50	389	455	
	2200	With gaskets, class 3, 12" diameter	B-21	168	.167		10.80	4.44	.85	16.09	19.75	
	2220	15" diameter		160	.175		12.95	4.66	.89	18.50	22.50	
	2230	18" diameter		152	.184		16.20	4.90	.93	22.03	26.50	
	2240	24" diameter		136	.206		24.50	5.50	1.04	31.04	36.50	
	2260	30" diameter	B-13	88	.636		32.50	15.75	7.15	55.40	68	
	2270	36" diameter	"	72	.778		48.50	19.25	8.75	76.50	93	
	2290	48" diameter	B-13B	64	.875		79.50	21.50	14.10	115.10	137	
	2310	72" diameter	"	40	1.400		205	34.50	22.50	262	305	
	2330	Flared ends, 6'-1" long, 12" diameter	B-21	190	.147		37	3.92	.75	41.67	47.50	
	2340	15" diameter		155	.181		43	4.81	.92	48.73	55.50	
	2400	6'-2" long, 18" diameter		122	.230		45.50	6.10	1.16	52.76	61	
	2420	24" diameter		88	.318		55	8.45	1.61	65.06	75.50	
	2440	36" diameter	B-13	60	.933		84	23	10.50	117.50	140	
	3040	Vitrified plate lined, add to above, 30" to 36" diameter				SFCA	3.18			3.18	3.50	
	3050	42" to 54" diameter, add					3.40			3.40	3.74	
	3060	60" to 72" diameter, add					3.98			3.98	4.38	
	3070	Over 72" diameter, add					4.24			4.24	4.66	
	3080	Radius pipe, add to pipe prices, 12" to 60" diameter				L.F.	50%					
	3090	Over 60" diameter, add				"	20%					
	3500	Reinforced elliptical, 8' lengths, C507 class 3										
	3520	14" x 23" inside, round equivalent 18" diameter	B-21	82	.341	L.F.	22.50	9.10	1.73	33.33	40.50	
	3530	24" x 38" inside, round equivalent 30" diameter	B-13	58	.966		41	24	10.85	75.85	94	
	3540	29" x 45" inside, round equivalent 36" diameter		52	1.077		50.50	26.50	12.10	89.10	110	
	3550	38" x 60" inside, round equivalent 48" diameter		38	1.474		77.50	36.50	16.55	130.55	160	
	3560	48" x 76" inside, round equivalent 60" diameter		26	2.154		139	53.50	24	216.50	262	
	3570	58" x 91" inside, round equivalent 72" diameter		22	2.545		187	63	28.50	278.50	335	
	3780	Concrete slotted pipe, class 4 mortar joint										
	3800	12" diameter	B-21	168	.167	L.F.	11.15	4.44	.85	16.44	20	
	3840	18" diameter	"	152	.184	"	17.25	4.90	.93	23.08	27.50	
	3900	Class 4 O-ring										
	3940	12" diameter	B-21	168	.167	L.F.	11.75	4.44	.85	17.04	20.50	
	3960	18" diameter	"	152	.184	"	16.45	4.90	.93	22.28	26.50	
780	0010	**PIPING, DRAINAGE & SEWAGE, POLYVINYL CHLORIDE**										780
	0020	Not including excavation or backfill										
	2000	10' lengths, S.D.R. 35, B&S, 4" diameter	B-20	375	.064	L.F.	1.41	1.65		3.06	4.14	
	2040	6" diameter		350	.069		2.41	1.77		4.18	5.45	
	2080	8" diameter		335	.072		4.40	1.85		6.25	7.75	
	2120	10" diameter	B-21	330	.085		6.55	2.26	.43	9.24	11.20	
	2160	12" diameter		320	.087		7.35	2.33	.44	10.12	12.20	
	2200	15" diameter		190	.147		11.15	3.92	.75	15.82	19.20	
790	0010	**PIPING, DRAINAGE & SEWAGE, VITRIFIED CLAY** C700										790
	0020	Not including excavation or backfill,										
	4030	Extra strength, compression joints, C425										
	5000	4" diameter x 4' long	B-20	265	.091	L.F.	1.54	2.34		3.88	5.35	
	5020	6" diameter x 5' long	"	200	.120		2.61	3.10		5.71	7.75	
	5040	8" diameter x 5' long	B-21	200	.140		3.70	3.73	.71	8.14	10.65	
	5060	10" diameter x 5' long		190	.147		6.20	3.92	.75	10.87	13.70	
	5080	12" diameter x 6' long		150	.187		8.25	4.97	.95	14.17	17.90	
	5100	15" diameter x 7' long		110	.255		14.45	6.75	1.29	22.49	28	
	5120	18" diameter x 7' long		88	.318		21.50	8.45	1.61	31.56	38.50	
	5140	24" diameter x 7' long		45	.622		39	16.55	3.15	58.70	72.50	
	5160	30" diameter x 7' long	B-22	31	.968		72.50	26	6.85	105.35	128	

Important: See the Reference Section for critical supporting data - Reference Nos., Crews, & City Cost Indexes

02500 | Utility Services

02530 | Sanitary Sewerage

			CREW	DAILY OUTPUT	LABOR-HOURS	UNIT	2001 BARE COSTS MAT.	LABOR	EQUIP.	TOTAL	TOTAL INCL O&P	
790	5180	36" diameter x 7' long	B-22	20	1.500	L.F.	113	40.50	10.65	164.15	200	790
	6000	For 3' lengths, add					30%	30%				
	6020	For 2' lengths, add					40%	60%				
	6060	For plain joints, deduct					25%					
	7060	2' lengths, add to above					40%					

02550 | Piped Energy Distribution

			CREW	DAILY OUTPUT	LABOR-HOURS	UNIT	MAT.	LABOR	EQUIP.	TOTAL	TOTAL INCL O&P	
450	0010	**GAS STATION PRODUCT LINE**										450
	0020	Primary containment pipe, fiberglass-reinforced										
	0030	Plastic pipe 15' & 30' lengths										
	0040	2" diameter	Q-6	425	.056	L.F.	2.91	1.84		4.75	5.95	
	0050	3" diameter		400	.060		3.92	1.95		5.87	7.25	
	0060	4" diameter		375	.064		5.05	2.08		7.13	8.70	
	0100	Fittings										
	0110	Elbows, 90° & 45°, bell-ends, 2"	Q-6	24	1	Ea.	29.50	32.50		62	81.50	
	0120	3" diameter		22	1.091		31.50	35.50		67	88.50	
	0130	4" diameter		20	1.200		42	39		81	105	
	0200	Tees, bell ends, 2"		21	1.143		36.50	37		73.50	96.50	
	0210	3" diameter		18	1.333		37	43.50		80.50	106	
	0220	4" diameter		15	1.600		48.50	52		100.50	132	
	0230	Flanges bell ends, 2"		24	1		8.65	32.50		41.15	58.50	
	0240	3" diameter		22	1.091		10.90	35.50		46.40	65.50	
	0250	4" diameter		20	1.200		14.85	39		53.85	75.50	
	0260	Sleeve couplings, 2"		21	1.143		7.45	37		44.45	64	
	0270	3" diameter		18	1.333		10.30	43.50		53.80	77	
	0280	4" diameter		15	1.600		14.25	52		66.25	94	
	0290	Threaded adapters 2"		21	1.143		10.15	37		47.15	67	
	0300	3" diameter		18	1.333		14.80	43.50		58.30	82	
	0310	4" diameter		15	1.600		21.50	52		73.50	102	
	0320	Reducers, 2"		27	.889		12.50	29		41.50	57.50	
	0330	3" diameter		22	1.091		14.80	35.50		50.30	70	
	0340	4" diameter		20	1.200		21.50	39		60.50	82.50	
	1010	Gas station product line for secondary containment (double wall)										
	1100	Fiberglass reinforced plastic pipe 25' lengths										
	1120	Pipe, plain end, 3"	Q-6	375	.064	L.F.	4.04	2.08		6.12	7.60	
	1130	4" diameter		350	.069		5.05	2.23		7.28	8.90	
	1140	5" diameter		325	.074		7	2.40		9.40	11.35	
	1150	6" diameter		300	.080		8.90	2.60		11.50	13.75	
	1200	Fittings										
	1230	Elbows, 90° & 45°, 3"	Q-6	18	1.333	Ea.	36	43.50		79.50	105	
	1240	4" diameter		16	1.500		64.50	49		113.50	145	
	1250	5" diameter		14	1.714		149	56		205	248	
	1260	6" diameter		12	2		151	65		216	264	
	1270	Tees, 3"		15	1.600		54.50	52		106.50	139	
	1280	4" diameter		12	2		80.50	65		145.50	187	
	1290	5" diameter		9	2.667		163	87		250	310	
	1300	6" diameter		6	4		170	130		300	385	
	1310	Couplings, 3"		18	1.333		25	43.50		68.50	93	
	1320	4" diameter		16	1.500		67	49		116	147	
	1330	5" diameter		14	1.714		139	56		195	237	
	1340	6" diameter		12	2		144	65		209	257	
	1350	Cross-over nipples, 3"		18	1.333		5.90	43.50		49.40	72	
	1360	4" diameter		16	1.500		6.95	49		55.95	81	
	1370	5" diameter		14	1.714		10.30	56		66.30	95.50	
	1380	6" diameter		12	2		10.80	65		75.80	110	

For expanded coverage of these items see Means Heavy Construction Cost Data 2001

02500 | Utility Services

02550 | Piped Energy Distribution

			CREW	DAILY OUTPUT	LABOR-HOURS	UNIT	MAT.	LABOR	EQUIP.	TOTAL	TOTAL INCL O&P	
450	1400	Telescoping, reducers, concentric 4" x 3"	Q-6	18	1.333	Ea.	19.75	43.50		63.25	87	450
	1410	5" x 4"		17	1.412		51.50	46		97.50	126	
	1420	6" x 5"	↓	16	1.500	↓	124	49		173	210	
464	0010	**PIPING, GAS SERVICE & DISTRIBUTION, POLYETHYLENE**										464
	0020	not including excavation or backfill										
	1000	60 psi coils, comp cplg @ 100', 1/2" diameter, SDR 9.3	B-20A	608	.053	L.F.	.38	1.44		1.82	2.63	
	1040	1-1/4" diameter, SDR 11		544	.059		.65	1.61		2.26	3.18	
	1100	2" diameter, SDR 11		488	.066		.80	1.80		2.60	3.63	
	1160	3" diameter, SDR 11	↓	408	.078		1.68	2.15		3.83	5.15	
	1500	60 PSI 40' joints with coupling, 3" diameter, SDR 11	B-21A	408	.098		1.74	2.76	1.04	5.54	7.30	
	1540	4" diameter, SDR 11		352	.114		3.84	3.20	1.21	8.25	10.45	
	1600	6" diameter, SDR 11		328	.122		11.95	3.44	1.30	16.69	19.85	
	1640	8" diameter, SDR 11	↓	272	.147	↓	16.30	4.14	1.56	22	26	
550	0010	**PIPE CONDUIT, PREFABRICATED**										550
	0020	Does not include trenching, fittings or crane.										
	0300	For cathodic protection, add 12 to 14%										
	0310	of total built-up price (casing plus service pipe)										
	0580	Polyurethane insulated system, 250°F. max. temp.										
	0620	Black steel service pipe, standard wt., 1/2" insulation										
	0660	3/4" diam. pipe size	Q-17	54	.296	L.F.	16.55	9.30	1.86	27.71	34.50	
	0670	1" diam. pipe size		50	.320		18.20	10.05	2.01	30.26	37.50	
	0680	1-1/4" diam. pipe size		47	.340		20.50	10.70	2.14	33.34	41	
	0690	1-1/2" diam. pipe size		45	.356		22	11.15	2.24	35.39	44	
	0700	2" diam. pipe size		42	.381		23	11.95	2.40	37.35	45.50	
	0710	2-1/2" diam. pipe size		34	.471		23	14.75	2.96	40.71	51.50	
	0720	3" diam. pipe size		28	.571		27	17.95	3.59	48.54	60.50	
	0730	4" diam. pipe size		22	.727		34	23	4.57	61.57	77	
	0740	5" diam. pipe size	↓	18	.889		43.50	28	5.60	77.10	95.50	
	0750	6" diam. pipe size	Q-18	23	1.043		50.50	34	4.37	88.87	112	
	0760	8" diam. pipe size		19	1.263		73.50	41	5.30	119.80	149	
	0770	10" diam. pipe size		16	1.500		97	49	6.30	152.30	187	
	0780	12" diam. pipe size		13	1.846		121	60	7.75	188.75	232	
	0790	14" diam. pipe size		11	2.182		135	71	9.15	215.15	265	
	0800	16" diam. pipe size		10	2.400		154	78	10.05	242.05	299	
	0810	18" diam. pipe size		8	3		178	97.50	12.55	288.05	355	
	0820	20" diam. pipe size		7	3.429		198	112	14.35	324.35	400	
	0830	24" diam. pipe size	↓	6	4		242	130	16.75	388.75	480	
	0900	For 1" thick insulation, add					10%					
	0940	For 1-1/2" thick insulation, add					13%					
	0980	For 2" thick insulation, add				↓	20%					
	1500	Gland seal for system, 3/4" diam. pipe size	Q-17	32	.500	Ea.	269	15.70	3.15	287.85	325	
	1510	1" diam. pipe size		32	.500		269	15.70	3.15	287.85	325	
	1540	1-1/4" diam. pipe size		30	.533		288	16.75	3.35	308.10	345	
	1550	1-1/2" diam. pipe size		30	.533		288	16.75	3.35	308.10	345	
	1560	2" diam. pipe size		28	.571		340	17.95	3.59	361.54	405	
	1570	2-1/2" diam. pipe size		26	.615		370	19.30	3.87	393.17	440	
	1580	3" diam. pipe size		24	.667		395	21	4.19	420.19	470	
	1590	4" diam. pipe size		22	.727		470	23	4.57	497.57	555	
	1600	5" diam. pipe size	↓	19	.842		575	26.50	5.30	606.80	680	
	1610	6" diam. pipe size	Q-18	26	.923		610	30	3.87	643.87	725	
	1620	8" diam. pipe size		25	.960		710	31	4.02	745.02	830	
	1630	10" diam. pipe size		23	1.043		865	34	4.37	903.37	1,000	
	1640	12" diam. pipe size		21	1.143		955	37	4.79	996.79	1,100	
	1650	14" diam. pipe size		19	1.263		1,075	41	5.30	1,121.30	1,250	
	1660	16" diam. pipe size	↓	18	1.333		1,250	43.50	5.60	1,299.10	1,450	

02500 | Utility Services

02550 | Piped Energy Distribution

		Crew	Daily Output	Labor-Hours	Unit	Mat.	Labor	Equip.	Total	Total Incl O&P
1670	18" diam. pipe size	Q-18	16	1.500	Ea.	1,350	49	6.30	1,405.30	1,550
1680	20" diam. pipe size		14	1.714		1,525	56	7.20	1,588.20	1,775
1690	24" diam. pipe size		12	2		1,700	65	8.40	1,773.40	1,950
2000	Elbow, 45° for system									
2020	3/4" diam. pipe size	Q-17	14	1.143	Ea.	170	36	7.20	213.20	249
2040	1" diam. pipe size		13	1.231		174	38.50	7.75	220.25	259
2050	1-1/4" diam. pipe size		11	1.455		198	45.50	9.15	252.65	297
2060	1-1/2" diam. pipe size		9	1.778		225	56	11.20	292.20	345
2070	2" diam. pipe size		6	2.667		216	83.50	16.75	316.25	380
2080	2-1/2" diam. pipe size		4	4		234	126	25	385	475
2090	3" diam. pipe size		3.50	4.571		270	143	29	442	545
2100	4" diam. pipe size		3	5.333		315	167	33.50	515.50	635
2110	5" diam. pipe size		2.80	5.714		405	179	36	620	755
2120	6" diam. pipe size	Q-18	4	6		460	195	25	680	830
2130	8" diam. pipe size		3	8		665	260	33.50	958.50	1,150
2140	10" diam. pipe size		2.40	10		850	325	42	1,217	1,475
2150	12" diam. pipe size		2	12		1,125	390	50.50	1,565.50	1,875
2160	14" diam. pipe size		1.80	13.333		1,375	435	56	1,866	2,250
2170	16" diam. pipe size		1.60	15		1,650	490	63	2,203	2,625
2180	18" diam. pipe size		1.30	18.462		2,075	600	77.50	2,752.50	3,275
2190	20" diam. pipe size		1	24		2,600	780	101	3,481	4,150
2200	24" diam. pipe size		.70	34.286		3,275	1,125	144	4,544	5,425
2260	For elbow, 90°, add					25%				
2300	For tee, straight, add					85%	30%			
2340	For tee, reducing, add					170%	30%			
2380	For weldolet, straight, add					50%				
2400	Polyurethane insulation, 1"									
2410	FRP carrier and casing									
2420	4"	Q-5	18	.889	L.F.	42.50	28		70.50	88.50
2422	6"	Q-6	23	1.043		73.50	34		107.50	133
2424	8"		19	1.263		120	41		161	194
2426	10"		16	1.500		161	49		210	251
2428	12"		13	1.846		210	60		270	320
2430	FRP carrier and PVC casing									
2440	4"	Q-5	18	.889	L.F.	13.70	28		41.70	57
2444	8"	Q-6	19	1.263		28.50	41		69.50	93.50
2446	10"		16	1.500		41	49		90	119
2448	12"		13	1.846		52.50	60		112.50	149
2450	PVC carrier and casing									
2460	4"	Q-1	36	.444	L.F.	6.75	13.80		20.55	28.50
2462	6"	"	29	.552		9.45	17.10		26.55	36.50
2464	8"	Q-2	36	.667		13.15	21.50		34.65	47
2466	10"		32	.750		18	24		42	56.50
2468	12"		31	.774		21	25		46	60.50
2800	Calcium silicate insulated system, high temp. (1200°F)									
2840	Steel casing with protective exterior coating									
2850	6-5/8" diameter	Q-18	52	.462	L.F.	31	15	1.93	47.93	58.50
2860	8-5/8" diameter		50	.480		34	15.60	2.01	51.61	62.50
2870	10-3/4" diameter		47	.511		39.50	16.60	2.14	58.24	71
2880	12-3/4" diameter		44	.545		43	17.75	2.29	63.04	77
2890	14" diameter		41	.585		48.50	19.05	2.45	70	85
2900	16" diameter		39	.615		52	20	2.58	74.58	90.50
2910	18" diameter		36	.667		57.50	21.50	2.79	81.79	99.50
2920	20" diameter		34	.706		65	23	2.96	90.96	109
2930	22" diameter		32	.750		90	24.50	3.14	117.64	139
2940	24" diameter		29	.828		103	27	3.47	133.47	157

For expanded coverage of these items see *Means Heavy Construction Cost Data 2001*

02500 | Utility Services

02550 | Piped Energy Distribution

		CREW	DAILY OUTPUT	LABOR-HOURS	UNIT	2001 BARE COSTS MAT.	LABOR	EQUIP.	TOTAL	TOTAL INCL O&P
2950	26" diameter	Q-18	26	.923	L.F.	117	30	3.87	150.87	179
2960	28" diameter		23	1.043		146	34	4.37	184.37	216
2970	30" diameter		21	1.143		155	37	4.79	196.79	231
2980	32" diameter		19	1.263		174	41	5.30	220.30	260
2990	34" diameter		18	1.333		176	43.50	5.60	225.10	266
3000	36" diameter		16	1.500		189	49	6.30	244.30	288
3040	For multi-pipe casings, add					10%				
3060	For oversize casings, add					2%				
3400	Steel casing gland seal, single pipe									
3420	6-5/8" diameter	Q-18	25	.960	Ea.	430	31	4.02	465.02	525
3440	8-5/8" diameter		23	1.043		505	34	4.37	543.37	610
3450	10-3/4" diameter		21	1.143		565	37	4.79	606.79	685
3460	12-3/4" diameter		19	1.263		675	41	5.30	721.30	810
3470	14" diameter		17	1.412		735	46	5.90	786.90	885
3480	16" diameter		16	1.500		865	49	6.30	920.30	1,025
3490	18" diameter		15	1.600		960	52	6.70	1,018.70	1,125
3500	20" diameter		13	1.846		1,050	60	7.75	1,117.75	1,275
3510	22" diameter		12	2		1,200	65	8.40	1,273.40	1,425
3520	24" diameter		11	2.182		1,350	71	9.15	1,430.15	1,600
3530	26" diameter		10	2.400		1,525	78	10.05	1,613.05	1,800
3540	28" diameter		9.50	2.526		1,725	82	10.60	1,817.60	2,025
3550	30" diameter		9	2.667		1,775	87	11.15	1,873.15	2,100
3560	32" diameter		8.50	2.824		1,975	92	11.85	2,078.85	2,325
3570	34" diameter		8	3		2,150	97.50	12.55	2,260.05	2,525
3580	36" diameter		7	3.429		2,300	112	14.35	2,426.35	2,700
3620	For multi-pipe casings, add					5%				
4000	Steel casing anchors, single pipe									
4020	6-5/8" diameter	Q-18	8	3	Ea.	385	97.50	12.55	495.05	585
4040	8-5/8" diameter		7.50	3.200		405	104	13.40	522.40	615
4050	10-3/4" diameter		7	3.429		530	112	14.35	656.35	770
4060	12-3/4" diameter		6.50	3.692		565	120	15.45	700.45	825
4070	14" diameter		6	4		665	130	16.75	811.75	945
4080	16" diameter		5.50	4.364		775	142	18.30	935.30	1,075
4090	18" diameter		5	4.800		870	156	20	1,046	1,225
4100	20" diameter		4.50	5.333		960	173	22.50	1,155.50	1,325
4110	22" diameter		4	6		1,075	195	25	1,295	1,500
4120	24" diameter		3.50	6.857		1,175	223	28.50	1,426.50	1,650
4130	26" diameter		3	8		1,325	260	33.50	1,618.50	1,875
4140	28" diameter		2.50	9.600		1,450	310	40	1,800	2,125
4150	30" diameter		2	12		1,550	390	50.50	1,990.50	2,350
4160	32" diameter		1.50	16		1,850	520	67	2,437	2,875
4170	34" diameter		1	24		2,075	780	101	2,956	3,550
4180	36" diameter		1	24		2,250	780	101	3,131	3,750
4220	For multi-pipe, add					5%	20%			
4800	Steel casing elbow									
4820	6-5/8" diameter	Q-18	15	1.600	Ea.	530	52	6.70	588.70	670
4830	8-5/8" diameter		15	1.600		565	52	6.70	623.70	710
4850	10-3/4" diameter		14	1.714		685	56	7.20	748.20	840
4860	12-3/4" diameter		13	1.846		810	60	7.75	877.75	990
4870	14" diameter		12	2		865	65	8.40	938.40	1,050
4880	16" diameter		11	2.182		945	71	9.15	1,025.15	1,175
4890	18" diameter		10	2.400		1,075	78	10.05	1,163.05	1,300
4900	20" diameter		9	2.667		1,150	87	11.15	1,248.15	1,425
4910	22" diameter		8	3		1,250	97.50	12.55	1,360.05	1,525
4920	24" diameter		7	3.429		1,375	112	14.35	1,501.35	1,675
4930	26" diameter		6	4		1,500	130	16.75	1,646.75	1,875

02500 | Utility Services

02550 | Piped Energy Distribution

		CREW	DAILY OUTPUT	LABOR-HOURS	UNIT	MAT.	LABOR	EQUIP.	TOTAL	TOTAL INCL O&P
4940	28" diameter	Q-18	5	4.800	Ea.	1,600	156	20	1,776	2,025
4950	30" diameter		4	6		1,625	195	25	1,845	2,100
4960	32" diameter		3	8		1,850	260	33.50	2,143.50	2,450
4970	34" diameter		2	12		2,025	390	50.50	2,465.50	2,875
4980	36" diameter		2	12		2,150	390	50.50	2,590.50	3,025
5500	Black steel service pipe, std. wt., 1" thick insulation									
5510	3/4" diameter pipe size	Q-17	54	.296	L.F.	13.50	9.30	1.86	24.66	31
5540	1" diameter pipe size		50	.320		14.05	10.05	2.01	26.11	33
5550	1-1/4" diameter pipe size		47	.340		15.75	10.70	2.14	28.59	36
5560	1-1/2" diameter pipe size		45	.356		17.15	11.15	2.24	30.54	38
5570	2" diameter pipe size		42	.381		19.05	11.95	2.40	33.40	41.50
5580	2-1/2" diameter pipe size		34	.471		20	14.75	2.96	37.71	48
5590	3" diameter pipe size		28	.571		23	17.95	3.59	44.54	56
5600	4" diameter pipe size		22	.727		29	23	4.57	56.57	71.50
5610	5" diameter pipe size		18	.889		39.50	28	5.60	73.10	91.50
5620	6" diameter pipe size	Q-18	23	1.043		43	34	4.37	81.37	104
6000	Black steel service pipe, std. wt., 1-1/2" thick insul.									
6010	3/4" diameter pipe size	Q-17	54	.296	L.F.	13.70	9.30	1.86	24.86	31
6040	1" diameter pipe size		50	.320		15.30	10.05	2.01	27.36	34
6050	1-1/4" diameter pipe size		47	.340		17.15	10.70	2.14	29.99	37.50
6060	1-1/2" diameter pipe size		45	.356		18.70	11.15	2.24	32.09	40
6070	2" diameter pipe size		42	.381		20.50	11.95	2.40	34.85	43
6080	2-1/2" diameter pipe size		34	.471		21.50	14.75	2.96	39.21	50
6090	3" diameter pipe size		28	.571		24.50	17.95	3.59	46.04	57.50
6100	4" diameter pipe size		22	.727		31	23	4.57	58.57	73.50
6110	5" diameter pipe size		18	.889		39.50	28	5.60	73.10	91.50
6120	6" diameter pipe size	Q-18	23	1.043		45	34	4.37	83.37	106
6130	8" diameter pipe size		19	1.263		65	41	5.30	111.30	139
6140	10" diameter pipe size		16	1.500		84.50	49	6.30	139.80	173
6150	12" diameter pipe size		13	1.846		101	60	7.75	168.75	210
6190	For 2" thick insulation, add					15%				
6220	For 2-1/2" thick insulation, add					25%				
6260	For 3" thick insulation, add					30%				
6800	Black steel service pipe, ex. hvy. wt., 1" thick insul.									
6820	3/4" diameter pipe size	Q-17	50	.320	L.F.	14.20	10.05	2.01	26.26	33
6840	1" diameter pipe size		47	.340		15.10	10.70	2.14	27.94	35
6850	1-1/4" diameter pipe size		44	.364		17.45	11.40	2.29	31.14	39
6860	1-1/2" diameter pipe size		42	.381		18.55	11.95	2.40	32.90	41
6870	2" diameter pipe size		40	.400		19.60	12.55	2.52	34.67	43
6880	2-1/2" diameter pipe size		31	.516		24.50	16.20	3.25	43.95	54.50
6890	3" diameter pipe size		27	.593		27.50	18.60	3.73	49.83	62
6900	4" diameter pipe size		21	.762		36	24	4.79	64.79	81
6910	5" diameter pipe size		17	.941		50.50	29.50	5.90	85.90	107
6920	6" diameter pipe size	Q-18	22	1.091		56	35.50	4.57	96.07	120
7400	Black steel service pipe, ex. hvy. wt., 1-1/2" thick insul.									
7420	3/4" diameter pipe size	Q-17	50	.320	L.F.	14.20	10.05	2.01	26.26	33
7440	1" diameter pipe size		47	.340		16.20	10.70	2.14	29.04	36.50
7450	1-1/4" diameter pipe size		44	.364		18.80	11.40	2.29	32.49	40.50
7460	1-1/2" diameter pipe size		42	.381		20.50	11.95	2.40	34.85	43
7470	2" diameter pipe size		40	.400		21	12.55	2.52	36.07	44.50
7480	2-1/2" diameter pipe size		31	.516		24.50	16.20	3.25	43.95	55
7490	3" diameter pipe size		27	.593		29	18.60	3.73	51.33	64
7500	4" diameter pipe size		21	.762		38	24	4.79	66.79	83
7510	5" diameter pipe size		17	.941		52	29.50	5.90	87.40	109
7520	6" diameter pipe size	Q-18	22	1.091		57.50	35.50	4.57	97.57	122
7530	8" diameter pipe size		18	1.333		86.50	43.50	5.60	135.60	167

For expanded coverage of these items see *Means Heavy Construction Cost Data 2001*

02500 | Utility Services

02550 | Piped Energy Distribution

		CREW	DAILY OUTPUT	LABOR-HOURS	UNIT	2001 BARE COSTS				TOTAL INCL O&P
						MAT.	LABOR	EQUIP.	TOTAL	
7540	10" diameter pipe size	Q-18	15	1.600	L.F.	103	52	6.70	161.70	199
7550	12" diameter pipe size	↓	13	1.846		126	60	7.75	193.75	237
7590	For 2" thick insulation, add					13%				
7640	For 2-1/2" thick insulation, add					18%				
7680	For 3" thick insulation, add				↓	24%				
9000	Combined steam pipe with condensate return									
9010	8" and 4" in 24" case	Q-18	100	.240	L.F.	185	7.80	1.01	193.81	217
9020	6" and 3" in 20" case		104	.231		127	7.50	.97	135.47	151
9030	3" and 1-1/2" in 16" case		110	.218		95	7.10	.91	103.01	116
9040	2" and 1-1/4" in 12-3/4" case		114	.211		82.50	6.85	.88	90.23	102
9050	1-1/2" and 1-1/4" in 10-3/4" case	↓	116	.207	↓	72	6.75	.87	79.62	90.50
9100	Steam pipe only (no return)									
9110	6" in 18" case	Q-18	108	.222	L.F.	104	7.25	.93	112.18	126
9120	4" in 14" case		112	.214		82.50	6.95	.90	90.35	102
9130	3" in 14" case		114	.211		76	6.85	.88	83.73	95.50
9140	2-1/2" in 12-3/4" case		118	.203		71	6.60	.85	78.45	89
9150	2" in 12-3/4" case	↓	122	.197	↓	68	6.40	.82	75.22	85.50

02600 | Drainage & Containment

02620 | Subdrainage

		CREW	DAILY OUTPUT	LABOR-HOURS	UNIT	2001 BARE COSTS				TOTAL INCL O&P
						MAT.	LABOR	EQUIP.	TOTAL	
0010	**PIPING, SUBDRAINAGE, CONCRETE**									
0021	Not including excavation and backfill									
3000	Porous wall concrete underdrain, std. strength, 4" diameter	B-20	335	.072	L.F.	1.78	1.85		3.63	4.86
3020	6" diameter	"	315	.076		2.31	1.97		4.28	5.65
3040	8" diameter	B-21	310	.090		2.86	2.40	.46	5.72	7.40
3100	18" diameter	"	165	.170		9.15	4.52	.86	14.53	18.10
4000	Extra strength, 6" diameter	B-20	315	.076		2.36	1.97		4.33	5.70
4020	8" diameter	B-21	310	.090		3.54	2.40	.46	6.40	8.15
4040	10" diameter		285	.098		7.10	2.61	.50	10.21	12.40
4060	12" diameter		230	.122		7.65	3.24	.62	11.51	14.20
4080	15" diameter		200	.140		8.50	3.73	.71	12.94	15.95
4100	18" diameter	↓	165	.170	↓	12.40	4.52	.86	17.78	21.50
0010	**PIPING, SUBDRAINAGE, CORRUGATED METAL**									
0021	Not including excavation and backfill									
2010	Aluminum, perforated									
2020	6" diameter, 18 ga.	B-14	380	.126	L.F.	2.62	3.05	.48	6.15	8.15
2200	8" diameter, 16 ga.		370	.130		3.81	3.14	.49	7.44	9.60
2220	10" diameter, 16 ga.		360	.133		4.76	3.22	.50	8.48	10.80
2240	12" diameter, 16 ga.		285	.168		5.35	4.07	.63	10.05	12.90
2260	18" diameter, 16 ga.	↓	205	.234	↓	8	5.65	.88	14.53	18.55
3000	Uncoated galvanized, perforated									
3020	6" diameter, 18 ga.	B-20	380	.063	L.F.	4.08	1.63		5.71	7.05
3200	8" diameter, 16 ga.	"	370	.065		5.60	1.68		7.28	8.80
3220	10" diameter, 16 ga.	B-21	360	.078		8.40	2.07	.39	10.86	12.90
3240	12" diameter, 16 ga.		285	.098		8.80	2.61	.50	11.91	14.30
3260	18" diameter, 16 ga.	↓	205	.137		13.45	3.63	.69	17.77	21
4000	Steel, perforated, asphalt coated									
4020	6" diameter 18 ga.	B-20	380	.063	L.F.	3.26	1.63		4.89	6.15

02600 | Drainage & Containment

02620 | Subdrainage

			CREW	DAILY OUTPUT	LABOR-HOURS	UNIT	2001 BARE COSTS MAT.	LABOR	EQUIP.	TOTAL	TOTAL INCL O&P	
240	4030	8" diameter 18 ga	B-20	370	.065	L.F.	5.10	1.68		6.78	8.25	240
	4040	10" diameter 16 ga	B-21	360	.078		5.85	2.07	.39	8.31	10.10	
	4050	12" diameter 16 ga	↓	285	.098		6.75	2.61	.50	9.86	12	
	4060	18" diameter 16 ga		205	.137	↓	9.20	3.63	.69	13.52	16.50	
250	0010	**PIPING, SUBDRAINAGE, PLASTIC**										250
	0020	Not including excavation and backfill										
	1110	10"				Ea.	3.61			3.61	3.97	
	2100	Perforated PVC, 4" diameter	B-14	314	.153	L.F.	1.44	3.69	.58	5.71	7.95	
	2110	6" diameter		300	.160		2.66	3.87	.60	7.13	9.60	
	2120	8" diameter		290	.166		3.02	4	.62	7.64	10.20	
	2130	10" diameter		280	.171		4.68	4.14	.65	9.47	12.30	
	2140	12" diameter	↓	270	.178	↓	6.50	4.30	.67	11.47	14.60	
280	0010	**PIPING, SUBDRAINAGE, VITRIFIED CLAY**										280
	0020	Not including excavation and backfill										
	3000	Perforated, 5' lengths, C700, 4" diameter	B-14	400	.120	L.F.	1.91	2.90	.45	5.26	7.10	
	3020	6" diameter		315	.152		3.25	3.68	.57	7.50	9.95	
	3040	8" diameter		290	.166		4.62	4	.62	9.24	12	
	3060	12" diameter	↓	275	.175		9.60	4.22	.66	14.48	17.80	
	4000	Channel pipe, 4" diameter	B-20	430	.056		1.79	1.44		3.23	4.23	
	4020	6" diameter		335	.072		3.15	1.85		5	6.35	
	4060	8" diameter	↓	295	.081		4.81	2.10		6.91	8.60	
	4080	12" diameter	B-21	280	.100	↓	10.50	2.66	.51	13.67	16.25	

02630 | Storm Drainage

			CREW	DAILY OUTPUT	LABOR-HOURS	UNIT	MAT.	LABOR	EQUIP.	TOTAL	TOTAL INCL O&P	
100	0010	**PIPING, STORM DRAINAGE, CORRUGATED METAL**										100
	0020	Not including excavation or backfill										
	2000	Corrugated metal pipe, galvanized and coated										
	2020	Bituminous coated with paved invert, 20' lengths										
	2040	8" diameter, 16 ga.	B-14	330	.145	L.F.	5.85	3.52	.55	9.92	12.50	
	2060	10" diameter, 16 ga.		260	.185		7	4.46	.70	12.16	15.40	
	2080	12" diameter, 16 ga.		210	.229		8.45	5.50	.86	14.81	18.85	
	2100	15" diameter, 16 ga.		200	.240		10.25	5.80	.90	16.95	21.50	
	2120	18" diameter, 16 ga.		190	.253		13.15	6.10	.95	20.20	25	
	2140	24" diameter, 14 ga.	↓	160	.300		16.15	7.25	1.13	24.53	30.50	
	2160	30" diameter, 14 ga.	B-13	120	.467		21	11.55	5.25	37.80	46.50	
	2180	36" diameter, 12 ga.		120	.467		31	11.55	5.25	47.80	57.50	
	2200	48" diameter, 12 ga.	↓	100	.560		47	13.85	6.30	67.15	80.50	
	2220	60" diameter, 10 ga.	B-13B	75	.747		61	18.45	12	91.45	109	
	2240	72" diameter, 8 ga.	"	45	1.244	↓	93	31	20	144	172	
	2500	Galvanized, uncoated, 20' lengths										
	2520	8" diameter, 16 ga.	B-14	355	.135	L.F.	4.61	3.27	.51	8.39	10.70	
	2540	10" diameter, 16 ga.		280	.171		5.10	4.14	.65	9.89	12.75	
	2560	12" diameter, 16 ga.		220	.218		5.75	5.25	.82	11.82	15.45	
	2580	15" diameter, 16 ga.		220	.218		7.65	5.25	.82	13.72	17.50	
	2600	18" diameter, 16 ga.		205	.234		10.30	5.65	.88	16.83	21	
	2620	24" diameter, 14 ga.	↓	175	.274		14.10	6.65	1.03	21.78	27	
	2640	30" diameter, 14 ga.	B-13	130	.431		17.90	10.65	4.84	33.39	41.50	
	2660	36" diameter, 12 ga.		130	.431		29.50	10.65	4.84	44.99	54.50	
	2680	48" diameter, 12 ga.	↓	110	.509		39.50	12.60	5.70	57.80	69.50	
	2690	60" diameter, 10 ga.	B-13B	78	.718		62	17.75	11.55	91.30	108	
	2780	End sections, 8" diameter	B-14	24	2	Ea.	55	48.50	7.55	111.05	144	
	2785	10" diameter		22	2.182		56.50	52.50	8.25	117.25	153	
	2790	12" diameter		35	1.371		42.50	33	5.15	80.65	104	
	2800	18" diameter	↓	30	1.600		65.50	38.50	6.05	110.05	139	
	2810	24" diameter	B-13	25	2.240		95.50	55.50	25	176	219	
	2820	30" diameter	↓	25	2.240		181	55.50	25	261.50	315	

For expanded coverage of these items see *Means Heavy Construction Cost Data 2001*

02600 | Drainage & Containment

02630 | Storm Drainage

		CREW	DAILY OUTPUT	LABOR-HOURS	UNIT	2001 BARE COSTS MAT.	LABOR	EQUIP.	TOTAL	TOTAL INCL O&P	
2825	36" diameter	B-13	20	2.800	Ea.	240	69.50	31.50	341	405	100
2830	48" diameter	↓	10	5.600		495	139	63	697	830	
2835	60" diameter	B-13B	5	11.200		650	277	180	1,107	1,350	
2840	72" diameter	"	4	14	↓	1,175	345	225	1,745	2,050	
3000	Corrugated galvanized or alum. oval arch culverts, coated & paved										
3020	17" x 13", 16 ga., 15" equivalent	B-14	200	.240	L.F.	17.35	5.80	.90	24.05	29	
3040	21" x 15", 16 ga., 18" equivalent		150	.320		22.50	7.75	1.21	31.46	38	
3060	28" x 20", 14 ga., 24" equivalent		125	.384		32	9.30	1.45	42.75	51.50	
3080	35" x 24", 14 ga., 30" equivalent	↓	100	.480		39.50	11.60	1.81	52.91	63.50	
3100	42" x 29", 12 ga., 36" equivalent	B-13	100	.560		58.50	13.85	6.30	78.65	92.50	
3120	49" x 33", 12 ga., 42" equivalent		90	.622		70.50	15.40	7	92.90	109	
3140	57" x 38", 12 ga., 48" equivalent	↓	75	.747	↓	79.50	18.45	8.40	106.35	125	
3160	Steel, plain oval arch culverts, plain										
3180	17" x 13", 16 ga., 15" equivalent	B-14	225	.213	L.F.	9.55	5.15	.80	15.50	19.40	
3200	21" x 15", 16 ga., 18" equivalent		175	.274		11.30	6.65	1.03	18.98	24	
3220	28" x 20", 14 ga., 24" equivalent		150	.320		18.15	7.75	1.21	27.11	33.50	
3240	35" x 24", 14 ga., 30" equivalent	B-13	108	.519		23	12.85	5.85	41.70	51.50	
3260	42" x 29", 12 ga., 36" equivalent		108	.519		37.50	12.85	5.85	56.20	68	
3280	49" x 33", 12 ga., 42" equivalent		92	.609		44.50	15.05	6.85	66.40	80	
3300	57" x 38", 12 ga., 48" equivalent		75	.747	↓	39	18.45	8.40	65.85	81	
3320	End sections, 17" x 13"		22	2.545	Ea.	52	63	28.50	143.50	187	
3340	42" x 29"	↓	17	3.294	"	214	81.50	37	332.50	405	
3360	Multi-plate arch, steel	B-20	1,690	.014	Lb.	.60	.37		.97	1.24	
0010	**CATCH BASINS OR MANHOLES** not including footing, excavation,										200
0020	backfill, frame and cover										
0050	Brick, 4' inside diameter, 4' deep	D-1	1	16	Ea.	271	420		691	950	
0100	6' deep		.70	22.857		380	600		980	1,350	
0150	8' deep		.50	32	↓	485	840		1,325	1,825	
0200	For depths over 8', add		4	4	V.L.F.	105	105		210	278	
0400	Concrete blocks (radial), 4' I.D., 4' deep		1.50	10.667	Ea.	225	280		505	680	
0500	6' deep		1	16		298	420		718	980	
0600	8' deep		.70	22.857		370	600		970	1,350	
0700	For depths over 8', add	↓	5.50	2.909	V.L.F.	38.50	76.50		115	160	
0800	Concrete, cast in place, 4' x 4', 8" thick, 4' deep	C-14H	2	24	Ea.	345	690	18.95	1,053.95	1,500	
0900	6' deep		1.50	32		500	925	25.50	1,450.50	2,025	
1000	8' deep		1	48		715	1,375	38	2,128	3,000	
1100	For depths over 8', add		8	6	V.L.F.	85.50	173	4.74	263.24	375	
1110	Precast, 4' I.D., 4' deep	B-22	4.10	7.317	Ea.	305	197	52	554	695	
1120	6' deep		3	10		425	269	71	765	965	
1130	8' deep		2	15	↓	455	405	107	967	1,250	
1140	For depths over 8', add		16	1.875	V.L.F.	78.50	50.50	13.30	142.30	180	
1150	5' I.D., 4' deep	B-6	3	8	Ea.	445	199	60.50	704.50	860	
1160	6' deep		2	12		605	298	90.50	993.50	1,225	
1170	8' deep		1.50	16		760	395	121	1,276	1,575	
1180	For depths over 8', add		12	2	V.L.F.	99	49.50	15.10	163.60	202	
1190	6' I.D., 4' deep		2	12	Ea.	730	298	90.50	1,118.50	1,375	
1200	6' deep		1.50	16		950	395	121	1,466	1,800	
1210	8' deep		1	24		1,175	595	181	1,951	2,400	
1220	For depths over 8', add		8	3	V.L.F.	153	74.50	22.50	250	310	
1250	Slab tops, precast, 8" thick										
1300	4' diameter manhole	B-6	8	3	Ea.	156	74.50	22.50	253	310	
1400	5' diameter manhole		7.50	3.200		275	79.50	24	378.50	450	
1500	6' diameter manhole		7	3.429		335	85	26	446	530	
1600	Frames & covers, C.I., 24" square, 500 lb.		7.80	3.077		214	76.50	23	313.50	380	
1700	26" D shape, 600 lb.	↓	7	3.429		340	85	26	451	535	

02600 | Drainage & Containment

02630 | Storm Drainage

		CREW	DAILY OUTPUT	LABOR-HOURS	UNIT	2001 BARE COSTS MAT.	LABOR	EQUIP.	TOTAL	TOTAL INCL O&P
200	1800	Light traffic, 18" diameter, 100 lb.	B-6	10	2.400	Ea.	114	59.50	18.10	191.60	238
	1900	24" diameter, 300 lb.		8.70	2.759		138	68.50	21	227.50	280
	2000	36" diameter, 900 lb.		5.80	4.138		385	103	31	519	615
	2100	Heavy traffic, 24" diameter, 400 lb.		7.80	3.077		158	76.50	23	257.50	320
	2200	36" diameter, 1150 lb.		3	8		460	199	60.50	719.50	875
	2300	Mass. State standard, 26" diameter, 475 lb.		7	3.429		275	85	26	386	465
	2400	30" diameter, 620 lb.		7	3.429		300	85	26	411	490
	2500	Watertight, 24" diameter, 350 lb.		7.80	3.077		260	76.50	23	359.50	430
	2600	26" diameter, 500 lb.		7	3.429		261	85	26	372	445
	2700	32" diameter, 575 lb.		6	4		500	99.50	30	629.50	735
	2800	3 piece cover & frame, 10" deep,									
	2900	1200 lbs., for heavy equipment	B-6	3	8	Ea.	795	199	60.50	1,054.50	1,250
	3000	Raised for paving 1-1/4" to 2" high,									
	3100	4 piece expansion ring									
	3200	20" to 26" diameter	1 Clab	3	2.667	Ea.	105	61		166	212
	3300	30" to 36" diameter	"	3	2.667	"	147	61		208	258
	3320	Frames and covers, existing, raised for paving, 2", including									
	3340	row of brick, concrete collar, up to 12" wide frame	B-6	18	1.333	Ea.	32	33	10.05	75.05	97.50
	3360	20" to 26" wide frame		11	2.182		42	54	16.45	112.45	148
	3380	30" to 36" wide frame		9	2.667		52.50	66	20	138.50	182
	3400	Inverts, single channel brick	D-1	3	5.333		59	140		199	282
	3500	Concrete		5	3.200		46	84		130	181
	3600	Triple channel, brick		2	8		89.50	210		299.50	425
	3700	Concrete		3	5.333		61.50	140		201.50	285
	3800	Steps, heavyweight cast iron, 7" x 9"	1 Bric	40	.200		9.45	5.90		15.35	19.55
	3900	8" x 9"		40	.200		14.20	5.90		20.10	25
	3928	12" x 10-1/2"		40	.200		13	5.90		18.90	23.50
	4000	Standard sizes, galvanized steel		40	.200		11.75	5.90		17.65	22
	4100	Aluminum		40	.200		15	5.90		20.90	25.50

02700 | Bases, Ballasts, Pavements & Appurtenances

02740 | Flexible Pavement

		CREW	DAILY OUTPUT	LABOR-HOURS	UNIT	2001 BARE COSTS MAT.	LABOR	EQUIP.	TOTAL	TOTAL INCL O&P
300	0010	**ASPHALTIC CONCRETE PAVEMENT** for highways									
	0020	and large paved areas									
	0080	Binder course, 1-1/2" thick	B-25	7,725	.011	S.Y.	1.97	.29	.22	2.48	2.85
	0120	2" thick		6,345	.014		2.63	.35	.27	3.25	3.73
	0160	3" thick		4,905	.018		3.90	.45	.35	4.70	5.35
	0200	4" thick		4,140	.021		5.20	.53	.41	6.14	7.05
	0300	Wearing course, 1" thick	B-25B	10,575	.009		1.54	.23	.18	1.95	2.25
	0340	1-1/2" thick		7,725	.012		2.34	.32	.25	2.91	3.33
	0380	2" thick		6,345	.015		3.14	.39	.30	3.83	4.39
	0420	2-1/2" thick		5,480	.018		3.88	.45	.35	4.68	5.35
	0460	3" thick		4,900	.020		4.62	.50	.39	5.51	6.30
	0800	Alternate method of figuring paving costs									
	0810	Binder course, 1-1/2" thick	B-25	630	.140	Ton	26	3.50	2.71	32.21	37
	0811	2" thick		690	.128		26	3.20	2.47	31.67	36
	0812	3" thick		800	.110		26	2.76	2.13	30.89	35
	0813	4" thick		850	.104		26	2.60	2.01	30.61	34.50

For expanded coverage of these items see *Means Heavy Construction Cost Data 2001*

02700 | Bases, Ballasts, Pavements & Appurtenances

02740 | Flexible Pavement

		Description	CREW	DAILY OUTPUT	LABOR-HOURS	UNIT	MAT.	LABOR	EQUIP.	TOTAL	TOTAL INCL O&P	
300	0850	Wearing course, 1" thick	B-25B	575	.167	Ton	28.50	4.26	3.33	36.09	42	300
	0851	1-1/2" thick		630	.152		28.50	3.89	3.04	35.43	41	
	0852	2" thick		690	.139		28.50	3.55	2.78	34.83	40	
	0853	2-1/2" thick		745	.129		28.50	3.29	2.57	34.36	39.50	
	0854	3" thick		800	.120		28.50	3.06	2.40	33.96	39	
	1000	Pavement replacement over trench, 2" thick	B-37	90	.533	S.Y.	3.20	12.90	1.48	17.58	25	
	1050	4" thick		70	.686		6.35	16.55	1.91	24.81	35	
	1080	6" thick		55	.873		10.05	21	2.43	33.48	47	
315	0011	PAVING Asphaltic concrete, parking lots & driveways										315
	0020	6" stone base, 2" binder course, 1" topping	B-25C	9,000	.005	S.F.	1.13	.14	.17	1.44	1.63	
	0300	Binder course, 1-1/2" thick		35,000	.001		.30	.04	.04	.38	.43	
	0400	2" thick		25,000	.002		.39	.05	.06	.50	.58	
	0500	3" thick		15,000	.003		.60	.08	.10	.78	.90	
	0600	4" thick		10,800	.004		.79	.11	.14	1.04	1.19	
	0800	Sand finish course, 3/4" thick		41,000	.001		.17	.03	.04	.24	.27	
	0900	1" thick		34,000	.001		.21	.04	.04	.29	.34	
	1000	Fill pot holes, hot mix, 2" thick	B-16	4,200	.008		.39	.18	.12	.69	.83	
	1100	4" thick		3,500	.009		.56	.22	.14	.92	1.11	
	1120	6" thick		3,100	.010		.76	.24	.16	1.16	1.38	
	1140	Cold patch, 2" thick	B-51	3,000	.016		.45	.37	.05	.87	1.14	
	1160	4" thick		2,700	.018		.86	.41	.06	1.33	1.65	
	1180	6" thick		1,900	.025		1.33	.59	.08	2	2.48	

02800 | Site Improvements and Amenities

02810 | Irrigation System

		Description	CREW	DAILY OUTPUT	LABOR-HOURS	UNIT	MAT.	LABOR	EQUIP.	TOTAL	TOTAL INCL O&P	
800	0010	SPRINKLER IRRIGATION SYSTEM For lawns										800
	0100	Golf course with fully automatic system	C-17	.05	1,600	9 holes	79,000	48,400		127,400	162,500	
	0200	24' diam. head at 15' O.C incl. piping, auto oper., minimum	B-20	70	.343	Head	17.35	8.85		26.20	33	
	0300	Maximum		40	.600		40	15.50		55.50	68.50	
	0500	60' diam. head at 40' O.C. incl. piping, auto oper., minimum		28	.857		52.50	22		74.50	92.50	
	0600	Maximum		23	1.043		147	27		174	205	
	0800	Residential system, custom, 1" supply		2,000	.012	S.F.	.26	.31		.57	.78	
	0900	1-1/2" supply		1,800	.013	"	.29	.34		.63	.86	

02815 | Fountains

		Description	CREW	DAILY OUTPUT	LABOR-HOURS	UNIT	MAT.	LABOR	EQUIP.	TOTAL	TOTAL INCL O&P	
225	0010	FOUNTAINS/AERATORS										225
	0100	Pump w/controls										
	0200	Single phase, 100' chord, 1/2 H.P. pump	2 Skwk	4.40	3.636	Ea.	2,550	109		2,659	3,000	
	0300	3/4 H.P. pump		4.30	3.721		2,925	111		3,036	3,375	
	0400	1 H.P. pump		4.20	3.810		2,950	114		3,064	3,425	
	0500	1-1/2 H.P. pump		4.10	3.902		3,050	116		3,166	3,525	
	0600	2 H.P. pump		4	4		3,100	119		3,219	3,575	
	0700	Three phase, 200' chord, 5 H.P. pump		3.90	4.103		7,475	122		7,597	8,425	
	0800	7-1/2 H.P. pump		3.80	4.211		8,450	126		8,576	9,500	
	0900	10 H.P. pump		3.70	4.324		9,425	129		9,554	10,600	
	1000	15 H.P. pump		3.60	4.444		11,200	133		11,333	12,500	
	1100	Nozzles, minimum		8	2		187	59.50		246.50	300	
	1200	Maximum		8	2		246	59.50		305.50	365	
	1300	Lights w/mounting kits, 200 watt		18	.889		310	26.50		336.50	380	

02800 | Site Improvements and Amenities

02815 | Fountains

		CREW	DAILY OUTPUT	LABOR-HOURS	UNIT	2001 BARE COSTS				TOTAL INCL O&P
						MAT.	LABOR	EQUIP.	TOTAL	
1400	300 watt	2 Skwk	18	.889	Ea.	345	26.50		371.50	415
1500	500 watt		18	.889		375	26.50		401.50	455
1600	Color blender	▼	12	1.333	▼	292	40		332	385

02950 | Site Restoration & Rehabilitation

02955 | Restoration of Underground Piping

		CREW	DAILY OUTPUT	LABOR-HOURS	UNIT	2001 BARE COSTS				TOTAL INCL O&P
						MAT.	LABOR	EQUIP.	TOTAL	
0010	**PIPE INTERNAL CLEANING & INSPECTION**									
0100	Cleaning, pressure pipe systems									
0120	Pig method, lengths 1000' to 10,000'									
0140	4" diameter thru 24" diameter, minimum				L.F.				2	2.30
0160	Maximum				"				5	6
6000	Sewage/sanitary systems									
6100	Power rodder with header & cutters									
6110	Mobilization charge, minimum				Total				300	350
6120	Mobilization charge, maximum				"				700	800
6140	4" diameter				L.F.				1	1.15
6150	6" diameter								1.25	1.45
6160	8" diameter								1.50	1.75
6170	10" diameter								1.75	2
6180	12" diameter								1.90	2.20
6190	14" diameter								2	2.30
6200	16" diameter								2.25	2.60
6210	18" diameter								2.50	2.85
6220	20" diameter								2.60	3
6230	24" diameter								2.90	3.35
6240	30" diameter								3.50	4
6250	36" diameter								4	4.60
6260	48" diameter								4.30	4.95
6270	60" diameter								5.60	6.45
6280	72" diameter				▼				6.20	7.15
9000	Inspection, television camera with film									
9060	500 linear feet				Total				1,000	1,150
0010	**PIPE REPAIR**									
0020	Not including excavation or backfill									
0100	Clamp, stainless steel, lightweight, for steel pipe									
0110	3" long, 1/2" diameter pipe	1 Plum	34	.235	Ea.	7.60	8.10		15.70	20.50
0120	3/4" diameter pipe		32	.250		7.80	8.60		16.40	21.50
0130	1" diameter pipe		30	.267		8.40	9.20		17.60	23
0140	1-1/4" diameter pipe		28	.286		8.90	9.85		18.75	24.50
0150	1-1/2" diameter pipe		26	.308		9.20	10.60		19.80	26
0160	2" diameter pipe		24	.333		10.05	11.50		21.55	28.50
0170	2-1/2" diameter pipe		23	.348		10.95	12		22.95	30
0180	3" diameter pipe		22	.364		12.10	12.55		24.65	32
0190	3-1/2" diameter pipe	▼	21	.381		12.95	13.10		26.05	34
0200	4" diameter pipe	B-20	56	.429		13.50	11.10		24.60	32
0210	5" diameter pipe		53	.453		15.55	11.70		27.25	35.50
0220	6" diameter pipe		48	.500		17.25	12.95		30.20	39.50
0230	8" diameter pipe	▼	30	.800	▼	20	20.50		40.50	54.50

For expanded coverage of these items see *Means Site Work and Landscape Cost Data 2001*

02950 | Site Restoration & Rehabilitation

02955 | Restoration of Underground Piping

		CREW	DAILY OUTPUT	LABOR-HOURS	UNIT	MAT.	LABOR	EQUIP.	TOTAL	TOTAL INCL O&P
0240	10" diameter pipe	B-20	28	.857	Ea.	52	22		74	91.50
0250	12" diameter pipe		24	1		55	26		81	101
0260	14" diameter pipe		22	1.091		57.50	28		85.50	108
0270	16" diameter pipe		20	1.200		62	31		93	117
0280	18" diameter pipe		18	1.333		65.50	34.50		100	126
0290	20" diameter pipe		16	1.500		72.50	39		111.50	141
0300	24" diameter pipe		14	1.714		80.50	44.50		125	158
0360	For 6" long, add					100%	40%			
0370	For 9" long, add					200%	100%			
0380	For 12" long, add					300%	150%			
0390	For 18" long, add					500%	200%			
1000	Clamp, stainless steel, with threaded service tap									
1040	Full seal for iron, steel, PVC pipe									
1100	6" long, 2" diameter pipe	1 Plum	17	.471	Ea.	98	16.20		114.20	133
1110	2-1/2" diameter pipe		16	.500		104	17.25		121.25	140
1120	3" diameter pipe		15.60	.513		110	17.65		127.65	148
1130	3-1/2" diameter pipe		15	.533		115	18.35		133.35	155
1140	4" diameter pipe	B-20	40	.600		127	15.50		142.50	164
1150	6" diameter pipe		34	.706		146	18.25		164.25	190
1160	8" diameter pipe		21	1.143		182	29.50		211.50	247
1170	10" diameter pipe		20	1.200		201	31		232	270
1180	12" diameter pipe		17	1.412		230	36.50		266.50	310
1205	For 9" long, add					20%	45%			
1210	For 12" long, add					40%	80%			
1220	For 18" long, add					70%	110%			
1600	Clamp, stainless steel, single section									
1640	Full seal for iron, steel, PVC pipe									
1700	6" long, 2" diameter pipe	1 Plum	17	.471	Ea.	50.50	16.20		66.70	80
1710	2-1/2" diameter pipe		16	.500		55	17.25		72.25	86.50
1720	3" diameter pipe		15.60	.513		57.50	17.65		75.15	90
1730	3-1/2" diameter pipe		15	.533		57	18.35		75.35	90
1740	4" diameter pipe	B-20	40	.600		69	15.50		84.50	101
1750	6" diameter pipe		34	.706		86.50	18.25		104.75	124
1760	8" diameter pipe		21	1.143		115	29.50		144.50	174
1770	10" diameter pipe		20	1.200		132	31		163	194
1780	12" diameter pipe		17	1.412		144	36.50		180.50	215
1800	For 9" long, add					40%	45%			
1810	For 12" long, add					60%	80%			
1820	For 18" long, add					120%	110%			
2000	Clamp, stainless steel, two section									
2040	Full seal, for iron, steel, PVC pipe									
2100	6" long, 4" diameter pipe	B-20	24	1	Ea.	98	26		124	149
2110	6" diameter pipe		20	1.200		115	31		146	176
2120	8" diameter pipe		13	1.846		127	47.50		174.50	214
2130	10" diameter pipe		12	2		184	51.50		235.50	283
2140	12" diameter pipe		10	2.400		201	62		263	320
2200	9" long, 4" diameter pipe		16	1.500		127	39		166	200
2210	6" diameter pipe		13	1.846		144	47.50		191.50	233
2220	8" diameter pipe		9	2.667		150	69		219	272
2230	10" diameter pipe		8	3		207	77.50		284.50	350
2240	12" diameter pipe		7	3.429		230	88.50		318.50	390
2250	14" diameter pipe		6.40	3.750		345	97		442	530
2260	16" diameter pipe		6	4		360	103		463	560
2270	18" diameter pipe		5	4.800		375	124		499	605
2280	20" diameter pipe		4.60	5.217		405	135		540	655
2290	24" diameter pipe		4	6		690	155		845	1,000

02950 | Site Restoration & Rehabilitation

02955 | Restoration of Underground Piping

		CREW	DAILY OUTPUT	LABOR-HOURS	UNIT	MAT.	LABOR	EQUIP.	TOTAL	TOTAL INCL O&P	
680	2320 For 12" long, add to 9"				Ea.	15%	25%				680
	2330 For 18" long, add to 9"				↓	70%	55%				
	8000 For internal cleaning and inspection, see Div. 02955-100										
	8100 For pipe testing, see Div. 15955-700										
800	0010 **CORROSION RESISTANCE** Wrap & coat, add to pipe, 4" dia.				L.F.	1.35			1.35	1.49	800
	0020 5" diameter					1.42			1.42	1.56	
	0040 6" diameter					1.42			1.42	1.56	
	0060 8" diameter					2.19			2.19	2.41	
	0080 10" diameter					3.27			3.27	3.60	
	0100 12" diameter					3.27			3.27	3.60	
	0120 14" diameter					4.79			4.79	5.25	
	0140 16" diameter					4.79			4.79	5.25	
	0160 18" diameter					4.84			4.84	5.30	
	0180 20" diameter					5.05			5.05	5.60	
	0200 24" diameter					6.20			6.20	6.80	
	0220 Small diameter pipe, 1" diameter, add					1.02			1.02	1.12	
	0240 2" diameter					1.12			1.12	1.23	
	0260 2-1/2" diameter					1.23			1.23	1.35	
	0280 3" diameter				↓	1.23			1.23	1.35	
	0300 Fittings, field covered, add				S.F.	6.50			6.50	7.15	
	0500 Coating, bituminous, per diameter inch, 1 coat, add				L.F.	.24			.24	.26	
	0540 3 coat					.38			.38	.42	
	0560 Coal tar epoxy, per diameter inch, 1 coat, add					.15			.15	.17	
	0600 3 coat					.30			.30	.33	
	1000 Polyethylene H.D. extruded, .025" thk., 1/2" diameter add					.22			.22	.24	
	1020 3/4" diameter					.23			.23	.25	
	1040 1" diameter					.26			.26	.29	
	1060 1-1/4" diameter					.31			.31	.34	
	1080 1-1/2" diameter					.35			.35	.39	
	1100 .030" thk., 2" diameter					.38			.38	.42	
	1120 2-1/2" diameter					.45			.45	.50	
	1140 .035" thk., 3" diameter					.55			.55	.61	
	1160 3-1/2" diameter					.63			.63	.69	
	1180 4" diameter					.70			.70	.77	
	1200 .040" thk, 5" diameter					.97			.97	1.07	
	1220 6" diameter					1.01			1.01	1.11	
	1240 8" diameter					1.32			1.32	1.45	
	1260 10" diameter					1.61			1.61	1.77	
	1280 12" diameter					1.92			1.92	2.11	
	1300 .060" thk., 14" diameter					2.65			2.65	2.92	
	1320 16" diameter					3.11			3.11	3.42	
	1340 18" diameter					3.67			3.67	4.04	
	1360 20" diameter				↓	4.13			4.13	4.54	
	1380 Fittings, field wrapped, add				S.F.	4.25			4.25	4.68	

For information about Means Estimating Seminars, see yellow pages 11 and 12 in back of book

For expanded coverage of these items see Means Site Work and Landscape Cost Data 2001

Division Notes

	CREW	DAILY OUTPUT	LABOR-HOURS	UNIT	MAT.	LABOR	EQUIP.	TOTAL	TOTAL INCL O&P

(2001 BARE COSTS)

Division 3 Concrete

Estimating Tips
General
- Carefully check all the plans and specifications. Concrete often appears on drawings other than structural drawings, including mechanical and electrical drawings for equipment pads. The cost of cutting and patching is often difficult to estimate. See Subdivision 02225 for demolition costs.
- Always obtain concrete prices from suppliers near the job site. A volume discount can often be negotiated depending upon competition in the area. Remember to add for waste, particularly for slabs and footings on grade.

03100 Concrete Forms & Accessories
- A primary cost for concrete construction is forming. Most jobs today are constructed with prefabricated forms. The selection of the forms best suited for the job and the total square feet of forms required for efficient concrete forming and placing are key elements in estimating concrete construction. Enough forms must be available for erection to make efficient use of the concrete placing equipment and crew.
- Concrete accessories for forming and placing depend upon the systems used. Study the plans and specifications to assure that all special accessory requirements have been included in the cost estimate such as anchor bolts, inserts and hangers.

03200 Concrete Reinforcement
- Ascertain that the reinforcing steel supplier has included all accessories, cutting, bending and an allowance for lapping, splicing and waste. A good rule of thumb is 10% for lapping, splicing and waste. Also, 10% waste should be allowed for welded wire fabric.

03300 Cast-in-Place Concrete
- When estimating structural concrete, pay particular attention to requirements for concrete additives, curing methods and surface treatments. Special consideration for climate, hot or cold, must be included in your estimate. Be sure to include requirements for concrete placing equipment and concrete finishing.

03400 Precast Concrete
03500 Cementitious Decks & Toppings
- The cost of hauling precast concrete structural members is often an important factor. For this reason, it is important to get a quote from the nearest supplier. It may become economically feasible to set up precasting beds on the site if the hauling costs are prohibitive.

Reference Numbers
Reference numbers are shown in bold squares at the beginning of some major classifications. These numbers refer to related items in the Reference Section. The reference information may be an estimating procedure, an alternate pricing method or technical information.

Note: Not all subdivisions listed here necessarily appear in this publication.

03100 | Concrete Forms & Accessories

03110 | Structural C.I.P. Forms

			CREW	DAILY OUTPUT	LABOR-HOURS	UNIT	2001 BARE COSTS MAT.	LABOR	EQUIP.	TOTAL	TOTAL INCL O&P	
425	0010	FORMS IN PLACE, EQUIPMENT FOUNDATIONS 1 use	C-2	160	.300	SFCA	2.57	8.55		11.12	16.20	425
	0050	2 use		190	.253		1.42	7.20		8.62	12.80	
	0100	3 use		200	.240		1.03	6.80		7.83	11.85	
	0150	4 use	↓	205	.234	↓	.84	6.65		7.49	11.30	
620	0010	ACCESSORIES, SLEEVES AND CHASES										620
	0100	Plastic, 1 use, 9" long, 2" diameter	1 Carp	100	.080	Ea.	.53	2.33		2.86	4.23	
	0150	4" diameter		90	.089		1.56	2.59		4.15	5.75	
	0200	6" diameter		75	.107		2.75	3.11		5.86	7.90	
	0250	12" diameter		60	.133		18.05	3.89		21.94	26	
	5000	Sheet metal, 2" diameter		100	.080		.68	2.33		3.01	4.40	
	5100	4" diameter		90	.089		.85	2.59		3.44	4.99	
	5150	6" diameter		75	.107		1.23	3.11		4.34	6.20	
	5200	12" diameter		60	.133		2.46	3.89		6.35	8.80	
	6000	Steel pipe, 2" diameter		100	.080		1.52	2.33		3.85	5.30	
	6100	4" diameter		90	.089		4.48	2.59		7.07	9	
	6150	6" diameter		75	.107		7.60	3.11		10.71	13.20	
	6200	12" diameter	↓	60	.133	↓	27	3.89		30.89	36	

03200 | Concrete Reinforcement

03210 | Reinforcing Steel

			CREW	DAILY OUTPUT	LABOR-HOURS	UNIT	2001 BARE COSTS MAT.	LABOR	EQUIP.	TOTAL	TOTAL INCL O&P	
600	0010	REINFORCING IN PLACE A615 Grade 60										600
	0602	Slab on grade, #3 to #7	4 Rodm	4,200	.008	Lb.	.28	.25		.53	.72	

03300 | Cast-In-Place Concrete

03310 | Structural Concrete

			CREW	DAILY OUTPUT	LABOR-HOURS	UNIT	2001 BARE COSTS MAT.	LABOR	EQUIP.	TOTAL	TOTAL INCL O&P	
220	0010	CONCRETE, READY MIX Regular weight										220
	0020	2000 psi				C.Y.	62.50			62.50	69	
	0100	2500 psi					61.50			61.50	67.50	
	0150	3000 psi					65			65	71	
	0200	3500 psi					66			66	72.50	
	0300	4000 psi					68.50			68.50	75.50	
	0350	4500 psi					70.50			70.50	78	
	0400	5000 psi					73			73	80.50	
	0411	6000 psi					83			83	91.50	
	0412	8000 psi					136			136	149	
	0413	10,000 psi					193			193	212	
	0414	12,000 psi					233			233	256	
	1000	For high early strength cement, add					10%					
	2000	For all lightweight aggregate, add				↓	45%					
240	0010	CONCRETE IN PLACE Including forms (4 uses), reinforcing										240
	0050	steel, including finishing unless otherwise indicated										

03300 | Cast-In-Place Concrete

03310 | Structural Concrete

			CREW	DAILY OUTPUT	LABOR-HOURS	UNIT	MAT.	LABOR	EQUIP.	TOTAL	TOTAL INCL O&P	
240	0500	Chimney foundations, industrial, minimum	C-14C	32.22	3.476	C.Y.	133	97	1.18	231.18	300	240
	0510	Maximum		23.71	4.724		155	132	1.61	288.61	380	
	3800	Footings, spread under 1 C.Y.		38.07	2.942		96	82	1	179	237	
	3850	Over 5 C.Y.		81.04	1.382		87	38.50	.47	125.97	158	
	3900	Footings, strip, 18" x 9", plain		41.04	2.729		86	76	.93	162.93	216	
	3950	36" x 12", reinforced		61.55	1.820		88	51	.62	139.62	178	
	4000	Foundation mat, under 10 C.Y.		38.67	2.896		118	81	.98	199.98	259	
	4050	Over 20 C.Y.		56.40	1.986		105	55.50	.68	161.18	204	
	4650	Slab on grade, not including finish, 4" thick	C-14E	60.75	1.449		79.50	41.50	.62	121.62	155	
	4700	6" thick	"	92	.957		76	27.50	.41	103.91	128	
	4751	Slab on grade, incl. troweled finish, not incl. forms										
	4760	or reinforcing, over 10,000 S.F., 4" thick slab	C-14F	3,425	.021	S.F.	.86	.56	.01	1.43	1.79	
	4820	6" thick slab		3,350	.021		1.25	.57	.01	1.83	2.25	
	4840	8" thick slab		3,184	.023		1.72	.60	.01	2.33	2.80	
	4900	12" thick slab		2,734	.026		2.57	.70	.01	3.28	3.90	
	4950	15" thick slab		2,505	.029		3.23	.76	.02	4.01	4.73	
700	0010	**PLACING CONCRETE** and vibrating, including labor & equipment										700
	1900	Footings, continuous, shallow, direct chute	C-6	120	.400	C.Y.		9.60	.63	10.23	15.60	
	1950	Pumped	C-20	150	.427			10.55	4.37	14.92	21	
	2000	With crane and bucket	C-7	90	.800			19.85	11.15	31	43	
	2100	Footings, continuous, deep, direct chute	C-6	140	.343			8.25	.54	8.79	13.35	
	2150	Pumped	C-20	160	.400			9.85	4.10	13.95	19.75	
	2200	With crane and bucket	C-7	110	.655			16.25	9.15	25.40	35	
	2900	Foundation mats, over 20 C.Y., direct chute	C-6	350	.137			3.30	.22	3.52	5.35	
	2950	Pumped	C-20	400	.160			3.95	1.64	5.59	7.90	
	3000	With crane and bucket	C-7	300	.240			5.95	3.35	9.30	12.90	

03350 | Concrete Finishing

			CREW	DAILY OUTPUT	LABOR-HOURS	UNIT	MAT.	LABOR	EQUIP.	TOTAL	TOTAL INCL O&P	
300	0010	**FINISHING FLOORS** Monolithic, screed finish	1 Cefi	900	.009	S.F.		.25		.25	.37	300
	0100	Screed and bull float (darby) finish		725	.011			.31		.31	.46	
	0150	Screed, float, and broom finish		630	.013			.36		.36	.53	
600	0010	**PATCHING CONCRETE**										600
	0100	Floors, 1/4" thick, small areas, regular grout	1 Cefi	170	.047	S.F.	.06	1.32		1.38	2.03	
	0150	Epoxy grout	"	100	.080	"	3.19	2.24		5.43	6.85	
	0300	Slab on Grade, cut outs, up to 50 C.F.	2 Cefi	50	.320	C.F.	4.80	8.95		13.75	18.60	
	2000	Walls, including chipping, cleaning and epoxy grout										
	2100	Minimum	1 Cefi	65	.123	S.F.	3.19	3.44		6.63	8.60	
	2150	Average		50	.160		4.79	4.47		9.26	11.90	
	2200	Maximum		40	.200		6.40	5.60		12	15.30	
	2510	Underlayment, P.C based self-leveling, 4100 psi, pumped, 1/4"	C-8	20,000	.003		1.20	.07	.03	1.30	1.46	
	2520	1/2"		19,000	.003		2.16	.08	.03	2.27	2.53	
	2530	3/4"		18,000	.003		3.36	.08	.03	3.47	3.86	
	2540	1"		17,000	.003		4.56	.08	.03	4.67	5.15	
	2550	1-1/2"		15,000	.004		6.95	.10	.04	7.09	7.85	
	2560	Hand mix, 1/2"	C-18	4,000	.002		2.16	.05	.01	2.22	2.47	
	2610	Topping, P.C. based self-level/dry 6100 psi, pumped, 1/4"	C-8	20,000	.003		1.80	.07	.03	1.90	2.12	
	2620	1/2"		19,000	.003		3.24	.08	.03	3.35	3.71	
	2630	3/4"		18,000	.003		5.05	.08	.03	5.16	5.70	
	2660	1"		17,000	.003		6.85	.08	.03	6.96	7.65	
	2670	1-1/2"		15,000	.004		10.45	.10	.04	10.59	11.70	
	2680	Hand mix, 1/2"	C-18	4,000	.002		3.24	.05	.01	3.30	3.65	

For information about Means Estimating Seminars, see yellow pages 11 and 12 in back of book

For expanded coverage of these items see *Means Concrete & Masonry Cost Data 2001*

Division Notes

	CREW	DAILY OUTPUT	LABOR-HOURS	UNIT	2001 BARE COSTS				TOTAL INCL O&P
					MAT.	LABOR	EQUIP.	TOTAL	

Division 4 Masonry

Estimating Tips

04050 Basic Masonry Materials & Methods

- The terms *mortar* and *grout* are often used interchangeably, and incorrectly. Mortar is used to bed masonry units, seal the entry of air and moisture, provide architectural appearance, and allow for size variations in the units. Grout is used primarily in reinforced masonry construction and is used to bond the masonry to the reinforcing steel. Common mortar types are M(2500 psi), S(1800 psi), N(750 psi), and O(350 psi), and conform to ASTM C270. Grout is either fine or coarse, conforms to ASTM C476, and in-place strengths generally exceed 2500 psi. Mortar and grout are different components of masonry construction and are placed by entirely different methods. An estimator should be aware of their unique uses and costs.

04200 Masonry Units

- The most common types of unit masonry are brick and concrete masonry. The major classifications of brick are building brick (ASTM C62), facing brick (ASTM C216) and glazed brick, fire brick and pavers. Many varieties of texture and appearance can exist within these classifications, and the estimator would be wise to check local custom and availability within the project area. On repair and remodeling jobs, matching the existing brick may be the most important criteria.
- Brick and concrete block are priced by the piece and then converted into a price per square foot of wall. Openings less than two square feet are generally ignored by the estimator because any savings in units used is offset by the cutting and trimming required.
- All masonry walls, whether interior or exterior, require bracing. The cost of bracing walls during construction should be included by the estimator and this bracing must remain in place until permanent bracing is complete. Permanent bracing of masonry walls is accomplished by masonry itself, in the form of pilasters or abutting wall corners, or by anchoring the walls to the structural frame. Accessories in the form of anchors, anchor slots and ties are used, but their supply and installation can be by different trades. For instance, anchor slots on spandrel beams and columns are supplied and welded in place by the steel fabricator, but the ties from the slots into the masonry are installed by the bricklayer. Regardless of the installation method the estimator must be certain that these accessories are accounted for in pricing.

Reference Numbers

Reference numbers are shown in bold squares at the beginning of some major classifications. These numbers refer to related items in the Reference Section. The reference information may be an estimating procedure, an alternate pricing method or technical information.

Note: Not all subdivisions listed here necessarily appear in this publication.

04050 | Basic Masonry Materials & Methods

04090 | Masonry Accessories

			CREW	DAILY OUTPUT	LABOR-HOURS	UNIT	MAT.	LABOR	EQUIP.	TOTAL	TOTAL INCL O&P	
860	0010	VENT BOX Extruded aluminum, 4" deep, 2-3/8" x 8-1/8"	1 Bric	30	.267	Ea.	30.50	7.90		38.40	45.50	860
	0050	5" x 8-1/8"		25	.320		42.50	9.45		51.95	61	
	0100	2-1/4" x 25"		25	.320		60	9.45		69.45	80.50	
	0150	5" x 16-1/2"		22	.364		44	10.75		54.75	65	
	0200	6" x 16-1/2"		22	.364		85	10.75		95.75	110	
	0250	7-3/4" x 16-1/2"		20	.400		80	11.85		91.85	106	
	0400	For baked enamel finish, add					35%					
	0500	For cast aluminum, painted, add					60%					
	1000	Stainless steel ventilators, 6" x 6"	1 Bric	25	.320		90	9.45		99.45	114	
	1050	8" x 8"		24	.333		95	9.85		104.85	120	
	1100	12" x 12"		23	.348		110	10.30		120.30	137	
	1150	12" x 6"		24	.333		95	9.85		104.85	120	
	1200	Foundation block vent, galv., 1-1/4" thk, 8" high, 16" long, no damper		30	.267		16	7.90		23.90	30	
	1250	For damper, add					5			5	5.50	

04800 | Masonry Assemblies

04810 | Unit Masonry Assemblies

			CREW	DAILY OUTPUT	LABOR-HOURS	UNIT	MAT.	LABOR	EQUIP.	TOTAL	TOTAL INCL O&P	
160	0010	CHIMNEY See Div. 03310 for foundation, add to prices below R04210-050										160
	1800	Metal, high temp. steel jacket, factory lining, 24" diam.	E-2	65	.862	V.L.F.	174	27.50	19.85	221.35	261	
	1900	60" diameter	"	30	1.867		630	59.50	43	732.50	845	
	2100	Poured concrete, brick lining, 200' high x 10' diam.					5,450			5,450	6,000	
	2200	200' high x 18' diameter									4,995	
	2400	250' high x 9' diameter									3,920	
	2500	300' x 14' diameter									4,690	
	2700	400' high x 18' diameter									5,255	
	2800	500' x 20' diameter					9,700			9,700	10,700	

For information about Means Estimating Seminars, see yellow pages 11 and 12 in back of book

Division 5
Metals

Estimating Tips

05050 Basic Metal Materials & Methods

- Nuts, bolts, washers, connection angles and plates can add a significant amount to both the tonnage of a structural steel job as well as the estimated cost. As a rule of thumb add 10% to the total weight to account for these accessories.
- Type 2 steel construction, commonly referred to as "simple construction," consists generally of field bolted connections with lateral bracing supplied by other elements of the building, such as masonry walls or x-bracing. The estimator should be aware, however, that shop connections may be accomplished by welding or bolting. The method may be particular to the fabrication shop and may have an impact on the estimated cost.

05200 Metal Joists

- In any given project the total weight of open web steel joists is determined by the loads to be supported and the design. However, economies can be realized in minimizing the amount of labor used to place the joists. This is done by maximizing the joist spacing and therefore minimizing the number of joists required to be installed on the job. Certain spacings and locations may be required by the design, but in other cases maximizing the spacing and keeping it as uniform as possible will keep the costs down.

05300 Metal Deck

- The takeoff and estimating of metal deck involves more than simply the area of the floor or roof and the type of deck specified or shown on the drawings. Many different sizes and types of openings may exist. Small openings for individual pipes or conduits may be drilled after the floor/roof is installed, but larger openings may require special deck lengths as well as reinforcing or structural support. The estimator should determine who will be supplying this reinforcing. Additionally, some deck terminations are part of the deck package, such as screed angles and pour stops, and others will be part of the steel contract, such as angles attached to structural members and cast-in-place angles and plates. The estimator must ensure that all pieces are accounted for in the complete estimate.

05500 Metal Fabrications

- The most economical steel stairs are those that use common materials, standard details and most importantly, a uniform and relatively simple method of field assembly. Commonly available A36 channels and plates are very good choices for the main stringers of the stairs, as are angles and tees for the carrier members. Risers and treads are usually made by specialty shops, and it is most economical to use a typical detail in as many places as possible. The stairs should be pre-assembled and shipped directly to the site. The field connections should be simple and straightforward to be accomplished efficiently and with a minimum of equipment and labor.

Reference Numbers

Reference numbers are shown in bold squares at the beginning of some major classifications. These numbers refer to related items in the Reference Section. The reference information may be an estimating procedure, an alternate pricing method or technical information.

Note: Not all subdivisions listed here necessarily appear in this publication.

05050 | Basic Materials & Methods

05090 | Metal Fastenings

			CREW	DAILY OUTPUT	LABOR-HOURS	UNIT	MAT.	LABOR	EQUIP.	TOTAL	TOTAL INCL O&P
340	0010	**DRILLING** For anchors, up to 4" deep, incl. bit and layout									
	0050	in concrete or brick walls and floors, no anchor									
	0100	Holes, 1/4" diameter	1 Carp	75	.107	Ea.	.09	3.11		3.20	4.96
	0150	For each additional inch of depth, add		430	.019		.02	.54		.56	.87
	0200	3/8" diameter		63	.127		.08	3.70		3.78	5.90
	0250	For each additional inch of depth, add		340	.024		.02	.69		.71	1.09
	0300	1/2" diameter		50	.160		.07	4.66		4.73	7.40
	0350	For each additional inch of depth, add		250	.032		.02	.93		.95	1.48
	0400	5/8" diameter		48	.167		.14	4.86		5	7.75
	0450	For each additional inch of depth, add		240	.033		.03	.97		1	1.56
	0500	3/4" diameter		45	.178		.14	5.20		5.34	8.25
	0550	For each additional inch of depth, add		220	.036		.04	1.06		1.10	1.70
	0600	7/8" diameter		43	.186		.18	5.40		5.58	8.70
	0650	For each additional inch of depth, add		210	.038		.04	1.11		1.15	1.79
	0700	1" diameter		40	.200		.21	5.85		6.06	9.35
	0750	For each additional inch of depth, add		190	.042		.05	1.23		1.28	1.98
	0800	1-1/4" diameter		38	.211		.30	6.15		6.45	9.95
	0850	For each additional inch of depth, add		180	.044		.08	1.30		1.38	2.11
	0900	1-1/2" diameter		35	.229		.48	6.65		7.13	10.95
	0950	For each additional inch of depth, add		165	.048		.12	1.41		1.53	2.34
	1000	For ceiling installations, add						40%			
	1100	Drilling & layout for drywall or plaster walls, no anchor									
	1200	Holes, 1/4" diameter	1 Carp	150	.053	Ea.	.01	1.55		1.56	2.44
	1300	3/8" diameter		140	.057		.01	1.67		1.68	2.62
	1400	1/2" diameter		130	.062		.01	1.79		1.80	2.82
	1500	3/4" diameter		120	.067		.02	1.94		1.96	3.06
	1600	1" diameter		110	.073		.03	2.12		2.15	3.35
	1700	1-1/4" diameter		100	.080		.04	2.33		2.37	3.69
	1800	1-1/2" diameter		90	.089		.06	2.59		2.65	4.12
	1900	For ceiling installations, add						40%			
380	0010	**EXPANSION ANCHORS** & shields									
	0100	Bolt anchors for concrete, brick or stone, no layout and drilling									
	0200	Expansion shields, zinc, 1/4" diameter, 1" long, single	1 Carp	90	.089	Ea.	.92	2.59		3.51	5.05
	0300	1-3/8" long, double		85	.094		1.01	2.74		3.75	5.40
	0400	3/8" diameter, 2" long, single		85	.094		1.52	2.74		4.26	5.95
	0500	2" long, double		80	.100		1.88	2.92		4.80	6.65
	0600	1/2" diameter, 2-1/2" long, single		80	.100		2.51	2.92		5.43	7.30
	0700	2-1/2" long, double		75	.107		2.43	3.11		5.54	7.55
	0800	5/8" diameter, 2-5/8" long, single		75	.107		3.59	3.11		6.70	8.80
	0900	3" long, double		70	.114		3.59	3.33		6.92	9.15
	1000	3/4" diameter, 2-3/4" long, single		70	.114		5.35	3.33		8.68	11.05
	1100	4" long, double		65	.123		7.15	3.59		10.74	13.45
	1410	Concrete anchor, w/rod & epoxy cartridge, 1-3/4" diameter x 15" long	E-22	20	1.200		77	36.50		113.50	143
	1415	18" long		17	1.412		92.50	43		135.50	170
	1420	2" diameter x 18" long		16	1.500		118	46		164	202
	1425	24" long		15	1.600		154	49		203	247
	1430	Chemical anchor, w/rod & epoxy cartridge, 3/4" diam. x 9-1/2" long		27	.889		11.35	27		38.35	55
	1435	1" diameter x 11-3/4" long		24	1		21.50	30.50		52	71.50
	1440	1-1/4" diameter x 14" long		21	1.143		41	35		76	99.50
	1500	Self drilling anchor, snap-off, for 1/4" diameter bolt	1 Carp	26	.308		.93	8.95		9.88	15.05
	1600	3/8" diameter bolt		23	.348		1.39	10.15		11.54	17.40
	1700	1/2" diameter bolt		20	.400		2.09	11.65		13.74	20.50
	1800	5/8" diameter bolt		18	.444		3.57	12.95		16.52	24.50
	1900	3/4" diameter bolt		16	.500		6.35	14.60		20.95	30
	2100	Hollow wall anchors for gypsum wall board, plaster or tile									
	2300	1/8" diameter, short	1 Carp	160	.050	Ea.	.26	1.46		1.72	2.57

05050 | Basic Materials & Methods

05090 | Metal Fastenings

		CREW	DAILY OUTPUT	LABOR-HOURS	UNIT	2001 BARE COSTS MAT.	LABOR	EQUIP.	TOTAL	TOTAL INCL O&P
2400	Long	1 Carp	160	.050	Ea.	.29	1.46		1.75	2.60
2500	3/16" diameter, short		150	.053		.60	1.55		2.15	3.09
2600	Long		150	.053		.66	1.55		2.21	3.16
2700	1/4" diameter, short		140	.057		.80	1.67		2.47	3.49
2800	Long		140	.057		.89	1.67		2.56	3.59
3000	Toggle bolts, bright steel, 1/8" diameter, 2" long		85	.094		.21	2.74		2.95	4.52
3100	4" long		80	.100		.33	2.92		3.25	4.92
3200	3/16" diameter, 3" long		80	.100		.34	2.92		3.26	4.93
3300	6" long		75	.107		.49	3.11		3.60	5.40
3400	1/4" diameter, 3" long		75	.107		.37	3.11		3.48	5.25
3500	6" long		70	.114		.53	3.33		3.86	5.80
3600	3/8" diameter, 3" long		70	.114		.66	3.33		3.99	5.95
3700	6" long		60	.133		1.20	3.89		5.09	7.40
3800	1/2" diameter, 4" long		60	.133		1.91	3.89		5.80	8.20
3900	6" long		50	.160		3.89	4.66		8.55	11.60
5000	Screw anchors for concrete, masonry,									
5100	stone & tile, no layout or drilling included									
5200	Jute fiber, #6, #8, & #10, 1" long	1 Carp	240	.033	Ea.	.20	.97		1.17	1.74
5300	#12, 1-1/2" long		200	.040		.29	1.17		1.46	2.14
5400	#14, 2" long		160	.050		.46	1.46		1.92	2.79
5500	#16, 2" long		150	.053		.48	1.55		2.03	2.96
5600	#20, 2" long		140	.057		.77	1.67		2.44	3.46
5700	Lag screw shields, 1/4" diameter, short		90	.089		.40	2.59		2.99	4.49
5800	Long		85	.094		.45	2.74		3.19	4.79
5900	3/8" diameter, short		85	.094		.65	2.74		3.39	5
6000	Long		80	.100		.80	2.92		3.72	5.45
6100	1/2" diameter, short		80	.100		1.11	2.92		4.03	5.80
6200	Long		75	.107		1.25	3.11		4.36	6.25
6300	3/4" diameter, short		70	.114		2.47	3.33		5.80	7.90
6400	Long		65	.123		3.16	3.59		6.75	9.10
6600	Lead, #6 & #8, 3/4" long		260	.031		.18	.90		1.08	1.60
6700	#10 - #14, 1-1/2" long		200	.040		.28	1.17		1.45	2.13
6800	#16 & #18, 1-1/2" long		160	.050		.38	1.46		1.84	2.70
6900	Plastic, #6 & #8, 3/4" long		260	.031		.04	.90		.94	1.44
7000	#8 & #10, 7/8" long		240	.033		.04	.97		1.01	1.56
7100	#10 & #12, 1" long		220	.036		.05	1.06		1.11	1.72
7200	#14 & #16, 1-1/2" long		160	.050		.07	1.46		1.53	2.36
8000	Wedge anchors, not including layout or drilling									
8050	Carbon steel, 1/4" diameter, 1-3/4" long	1 Carp	150	.053	Ea.	.32	1.55		1.87	2.78
8100	3 1/4" long		145	.055		.42	1.61		2.03	2.98
8150	3/8" diameter, 2-1/4" long		150	.053		.60	1.55		2.15	3.09
8200	5" long		145	.055		1.06	1.61		2.67	3.69
8250	1/2" diameter, 2-3/4" long		140	.057		.84	1.67		2.51	3.53
8300	7" long		130	.062		1.44	1.79		3.23	4.39
8350	5/8" diameter, 3-1/2" long		130	.062		1.55	1.79		3.34	4.52
8400	8-1/2" long		115	.070		3.29	2.03		5.32	6.80
8450	3/4" diameter, 4-1/4" long		115	.070		1.99	2.03		4.02	5.35
8500	10" long		100	.080		4.50	2.33		6.83	8.60
8550	1" diameter, 6" long		100	.080		6.80	2.33		9.13	11.15
8575	9" long		80	.100		8.80	2.92		11.72	14.20
8600	12" long		80	.100		9.50	2.92		12.42	15
8650	1-1/4" diameter, 9" long		70	.114		13.25	3.33		16.58	19.80
8700	12" long		60	.133		13.30	3.89		17.19	21
8750	For type 303 stainless steel, add					350%				
8800	For type 316 stainless steel, add					450%				

05050 | Basic Materials & Methods

05090 | Metal Fastenings

			CREW	DAILY OUTPUT	LABOR-HOURS	UNIT	2001 BARE COSTS MAT.	LABOR	EQUIP.	TOTAL	TOTAL INCL O&P	
460	0005	**LAG SCREWS**										460
	0010	Steel, 1/4" diameter, 2" long	1 Carp	200	.040	Ea.	.05	1.17		1.22	1.88	
	0100	3/8" diameter, 3" long		150	.053		.15	1.55		1.70	2.60	
	0200	1/2" diameter, 3" long		130	.062		.25	1.79		2.04	3.09	
	0300	5/8" diameter, 3" long	▼	120	.067	▼	.62	1.94		2.56	3.72	
500	0005	**MACHINE SCREWS**										500
	0010	Steel, round head, #8 x 1" long				C	1.85			1.85	2.04	
	0110	#8 x 2" long					4.39			4.39	4.83	
	0200	#10 x 1" long					2.79			2.79	3.07	
	0300	#10 x 2" long				▼	5.30			5.30	5.85	
540	0010	**MACHINERY ANCHORS**, heavy duty, incl. sleeve, floating base nut,										540
	0020	lower stud & coupling nut, fiber plug, connecting stud, washer & nut.										
	0030	For flush mounted embedment in poured concrete heavy equip. pads.										
	0200	Material only, 1/2" diameter stud & bolt				Ea.	45			45	49.50	
	0300	5/8" diameter					50			50	55	
	0500	3/4" diameter					57.50			57.50	63	
	0600	7/8" diameter					62.50			62.50	69	
	0800	1" diameter					66			66	72.50	
	0900	1-1/4" diameter				▼	87.50			87.50	96.50	
580	0005	**POWDER ACTUATED** Tools & fasteners										580
	0010	Stud driver, .22 caliber, buy, minimum				Ea.	310			310	340	
	0100	Maximum				"	540			540	595	
	0300	Powder charges for above, low velocity				C	16.55			16.55	18.20	
	0400	Standard velocity					25			25	27.50	
	0600	Drive pins & studs, 1/4" & 3/8" diam., to 3" long, minimum					21.50			21.50	24	
	0700	Maximum				▼	53			53	58.50	
	0800	Pneumatic stud driver for 1/8" diameter studs				Ea.	1,875			1,875	2,050	
	0900	Drive pins for above, 1/2" to 3/4" long				M	410			410	450	
600	0010	**RIVETS**										600
	0100	Aluminum rivet & mandrel, 1/2" grip length x 1/8" diameter				C	4.53			4.53	4.98	
	0200	3/16" diameter					7			7	7.70	
	0300	Aluminum rivet, steel mandrel, 1/8" diameter					4.23			4.23	4.65	
	0400	3/16" diameter					8.90			8.90	9.80	
	0500	Copper rivet, steel mandrel, 1/8" diameter					5.50			5.50	6.05	
	1200	Steel rivet and mandrel, 1/8" diameter					4.74			4.74	5.20	
	1300	3/16" diameter				▼	7.15			7.15	7.85	
	1400	Hand riveting tool, minimum				Ea.	88.50			88.50	97.50	
	1500	Maximum					465			465	510	
	1600	Power riveting tool, minimum					880			880	965	
	1700	Maximum				▼	2,000			2,000	2,200	

05100 | Structural Metal Framing

05120 | Structural Steel

			CREW	DAILY OUTPUT	LABOR-HOURS	UNIT	2001 BARE COSTS MAT.	LABOR	EQUIP.	TOTAL	TOTAL INCL O&P	
440	0010	**LIGHTWEIGHT FRAMING**										440
	0400	Angle framing, field fabricated, 4" and larger	E-3	440	.055	Lb.	.36	1.83	.19	2.38	3.92	

Important: See the Reference Section for critical supporting data - Reference Nos., Crews, & City Cost Indexes

05100 | Structural Metal Framing

05120 | Structural Steel

			CREW	DAILY OUTPUT	LABOR-HOURS	UNIT	2001 BARE COSTS MAT.	LABOR	EQUIP.	TOTAL	TOTAL INCL O&P	
440	0450	Less than 4" angles	E-3	265	.091	Lb.	.37	3.04	.32	3.73	6.25	440
	0600	Channel framing, field fabricated, 8" and larger		500	.048		.37	1.61	.17	2.15	3.50	
	0650	Less than 8" channels		335	.072		.37	2.40	.25	3.02	5.05	
	1000	Continuous slotted channel framing system, minimum	2 Sswk	220	.073		1.90	2.39		4.29	6.40	
	1200	Maximum	"	135	.119		2.14	3.89		6.03	9.40	
	1250	Plate & bar stock for reinforcing beams and trusses					.67			.67	.74	
	1300	Cross bracing, rods, 3/4" diameter	E-3	700	.034		.74	1.15	.12	2.01	3.02	
	1310	7/8" diameter		850	.028		.74	.95	.10	1.79	2.64	
	1320	1" diameter		1,000	.024		.74	.80	.08	1.62	2.36	
	1330	Angle, 5" x 5" x 3/8"		440	.055		.74	1.83	.19	2.76	4.34	
	1350	Hanging lintels, average		350	.069		.74	2.30	.24	3.28	5.25	
	1380	Roof frames, 3'-0" square, 5' span	E-2	4,200	.013		.74	.43	.31	1.48	1.89	
	1400	Tie rod, not upset, 1-1/2" to 4" diameter, with turnbuckle	2 Sswk	800	.020		.80	.66		1.46	2.07	
	1420	No turnbuckle		700	.023		.77	.75		1.52	2.20	
	1500	Upset, 1-3/4" to 4" diameter, with turnbuckle		800	.020		.80	.66		1.46	2.07	
	1520	No turnbuckle		700	.023		.77	.75		1.52	2.20	
520	0005	**PIPE SUPPORT FRAMING**										520
	0010	Under 10#/L.F.	E-4	3,900	.008	Lb.	.82	.27	.02	1.11	1.42	
	0200	10.1 to 15#/L.F.		4,300	.007		.81	.25	.02	1.08	1.36	
	0400	15.1 to 20#/L.F.		4,800	.007		.80	.22	.02	1.04	1.30	
	0600	Over 20#/L.F.		5,400	.006		.78	.20	.02	1	1.24	

For information about Means Estimating Seminars, see yellow pages 11 and 12 in back of book

Division Notes

	CREW	DAILY OUTPUT	LABOR-HOURS	UNIT	2001 BARE COSTS				TOTAL INCL O&P
					MAT.	LABOR	EQUIP.	TOTAL	

Division 6
Wood & Plastics

Estimating Tips
06050 Basic Wood & Plastic Materials & Methods
- Common to any wood framed structure are the accessory connector items such as screws, nails, adhesives, hangers, connector plates, straps, angles and holdowns. For typical wood framed buildings, such as residential projects, the aggregate total for these items can be significant, especially in areas where seismic loading is a concern. For floor and wall framing, nail quantities can be figured on a "pounds per thousand board feet basis", with 10 to 25 lbs. per MBF the range. Holdowns, hangers and other connectors should be taken off by the piece.

06100 Rough Carpentry
- Lumber is a traded commodity and therefore sensitive to supply and demand in the marketplace. Even in "budgetary" estimating of wood framed projects, it is advisable to call local suppliers for the latest market pricing.
- Common quantity units for wood framed projects are "thousand board feet" (MBF). A board foot is a volume of wood, 1" x 1' x 1', or 144 cubic inches. Board foot quantities are generally calculated using nominal material dimensions—dressed sizes are ignored. Board foot per lineal foot of any stick of lumber can be calculated by dividing the nominal cross sectional area by 12. As an example, 2,000 lineal feet of 2 x 12 equates to 4 MBF by dividing the nominal area, 2 x 12, by 12, which equals 2, and multiplying by 2,000 to give 4,000 board feet. This simple rule applies to all nominal dimensioned lumber.
- Waste is an issue of concern at the quantity takeoff for any area of construction. Framing lumber is sold in even foot lengths, i.e., 10', 12', 14', 16', and depending on spans, wall heights and the grade of lumber, waste is inevitable. A rule of thumb for lumber waste is 5% to 10% depending on material quality and the complexity of the framing.
- Wood in various forms and shapes is used in many projects, even where the main structural framing is steel, concrete or masonry. Plywood as a back-up partition material and 2x boards used as blocking and cant strips around roof edges are two common examples. The estimator should ensure that the costs of all wood materials are included in the final estimate.

06200 Finish Carpentry
- It is necessary to consider the grade of workmanship when estimating labor costs for erecting millwork and interior finish. In practice, there are three grades: premium, custom and economy. The Means daily output for base and case moldings is in the range of 200 to 250 L.F. per carpenter per day. This is appropriate for most average custom grade projects. For premium projects an adjustment to productivity of 25% to 50% should be made depending on the complexity of the job.

Reference Numbers
Reference numbers are shown in bold squares at the beginning of some major classifications. These numbers refer to related items in the Reference Section. The reference information may be an estimating procedure, an alternate pricing method or technical information.

Note: Not all subdivisions listed here necessarily appear in this publication.

06050 | Basic Wood / Plastic Materials / Methods

06055 | Wood & Plastic Laminate

			CREW	DAILY OUTPUT	LABOR-HOURS	UNIT	2001 BARE COSTS MAT.	LABOR	EQUIP.	TOTAL	TOTAL INCL O&P	
720	0010	**CONVECTOR COVERS** Laminated plastic on 3/4"										720
	0020	thick particle board, 12" wide, minimum	1 Carp	16	.500	L.F.	20	14.60		34.60	45	
	0050	Average		13	.615		25.50	17.95		43.45	56	
	0100	Maximum		13	.615		35	17.95		52.95	66.50	
	0300	Add to above for grille, minimum		150	.053	S.F.	1.60	1.55		3.15	4.19	
	0400	Maximum		75	.107	"	4.50	3.11		7.61	9.80	

06090 | Wood & Plastic Fastenings

			CREW	DAILY OUTPUT	LABOR-HOURS	UNIT	MAT.	LABOR	EQUIP.	TOTAL	TOTAL INCL O&P	
600	0010	**NAILS** Prices of material only, based on 50# box purchase, copper, plain				Lb.	8.35			8.35	9.15	600
	0400	Stainless steel, plain					4.87			4.87	5.35	
	0500	Box, 3d to 20d, bright					1.13			1.13	1.24	
	0520	Galvanized					1.34			1.34	1.47	
	0600	Common, 3d to 60d, plain					.77			.77	.85	
	0700	Galvanized					1.18			1.18	1.30	
	0800	Aluminum					3.31			3.31	3.64	
	1000	Annular or spiral thread, 4d to 60d, plain					1.67			1.67	1.84	
	1200	Galvanized					1.67			1.67	1.84	
	1400	Drywall nails, plain					1.33			1.33	1.46	
	1600	Galvanized					1.33			1.33	1.46	
	1800	Finish nails, 4d to 10d, plain					.94			.94	1.03	
	2000	Galvanized					1.04			1.04	1.14	
	2100	Aluminum					4.85			4.85	5.35	
	2300	Flooring nails, hardened steel, 2d to 10d, plain					1.22			1.22	1.34	
	2400	Galvanized					2.17			2.17	2.39	
	2500	Gypsum lath nails, 1-1/8", 13 ga. flathead, blued					1.33			1.33	1.46	
	2600	Masonry nails, hardened steel, 3/4" to 3" long, plain					1.71			1.71	1.88	
	2700	Galvanized					1.71			1.71	1.88	
	5000	Add to prices above for cement coating					.07			.07	.08	
	5200	Zinc or tin plating					.12			.12	.13	
	5500	Vinyl coated sinkers, 8d to 16d					.55			.55	.61	
650	0010	**NAILS** mat. only, for pneumatic tools, framing, per carton of 5000, 2"				Ea.	36.50			36.50	40	650
	0100	2-3/8"					41.50			41.50	45.50	
	0200	Per carton of 4000, 3"					37			37	40.50	
	0300	3-1/4"					39			39	43	
	0400	Per carton of 5000, 2-3/8", galv.					56.50			56.50	62	
	0500	Per carton of 4000, 3", galv.					63.50			63.50	70	
	0600	3-1/4", galv.					79			79	87	
	0700	Roofing, per carton of 7200, 1"					34.50			34.50	38	
	0800	1-1/4"					32			32	35.50	
	0900	1-1/2"					37			37	40.50	
	1000	1-3/4"					44.50			44.50	49	
700	0010	**SHEET METAL SCREWS** Steel, standard, #8 x 3/4", plain				C	2.69			2.69	2.96	700
	0100	Galvanized					3.37			3.37	3.71	
	0300	#10 x 1", plain					3.69			3.69	4.06	
	0400	Galvanized					4.26			4.26	4.69	
	0600	With washers, #14 x 1", plain					10			10	11	
	0700	Galvanized					10			10	11	
	0900	#14 x 2", plain					10			10	11	
	1000	Galvanized					17.80			17.80	19.55	
	1500	Self-drilling, with washers, (pinch point) #8 x 3/4", plain					6			6	6.60	
	1600	Galvanized					6			6	6.60	
	1800	#10 x 3/4", plain					6			6	6.60	
	1900	Galvanized					6			6	6.60	
	3000	Stainless steel w/aluminum or neoprene washers, #14 x 1", plain					18			18	19.80	
	3100	#14 x 2", plain					24.50			24.50	27	

06050 | Basic Wood / Plastic Materials / Methods

06090 | Wood & Plastic Fastenings

			CREW	DAILY OUTPUT	LABOR-HOURS	UNIT	2001 BARE COSTS MAT.	LABOR	EQUIP.	TOTAL	TOTAL INCL O&P	
750	0010	WOOD SCREWS #8, 1" long, steel				C	3.29			3.29	3.62	750
	0100	Brass					11			11	12.10	
	0200	#8, 2" long, steel					3.69			3.69	4.06	
	0300	Brass					11.50			11.50	12.65	
	0400	#10, 1" long, steel					4.22			4.22	4.64	
	0500	Brass					22.50			22.50	24.50	
	0600	#10, 2" long, steel					7.50			7.50	8.25	
	0700	Brass					39.50			39.50	43.50	
	0800	#10, 3" long, steel					11.75			11.75	12.90	
	1000	#12, 2" long, steel					4.79			4.79	5.25	
	1100	Brass					16			16	17.60	
	1500	#12, 3" long, steel					15.90			15.90	17.50	
	2000	#12, 4" long, steel					28.50			28.50	31	

06100 | Rough Carpentry

06160 | Sheathing

			CREW	DAILY OUTPUT	LABOR-HOURS	UNIT	2001 BARE COSTS MAT.	LABOR	EQUIP.	TOTAL	TOTAL INCL O&P	
800	0010	SHEATHING Plywood on roof, CDX										800
	0030	5/16" thick	2 Carp	1,600	.010	S.F.	.33	.29		.62	.83	
	0035	Pneumatic nailed		1,952	.008		.33	.24		.57	.74	
	0050	3/8" thick		1,525	.010		.38	.31		.69	.89	
	0055	Pneumatic nailed		1,860	.009		.38	.25		.63	.80	
	0100	1/2" thick		1,400	.011		.49	.33		.82	1.06	
	0105	Pneumatic nailed		1,708	.009		.49	.27		.76	.97	
	0200	5/8" thick		1,300	.012		.62	.36		.98	1.24	
	0205	Pneumatic nailed		1,586	.010		.62	.29		.91	1.14	
	0300	3/4" thick		1,200	.013		.75	.39		1.14	1.43	
	0305	Pneumatic nailed		1,464	.011		.75	.32		1.07	1.32	
	0500	Plywood on walls with exterior CDX, 3/8" thick		1,200	.013		.38	.39		.77	1.02	
	0505	Pneumatic nailed		1,488	.011		.38	.31		.69	.90	
	0600	1/2" thick		1,125	.014		.49	.41		.90	1.19	
	0605	Pneumatic nailed		1,395	.011		.49	.33		.82	1.06	
	0700	5/8" thick		1,050	.015		.62	.44		1.06	1.37	
	0705	Pneumatic nailed		1,302	.012		.62	.36		.98	1.24	
	0800	3/4" thick		975	.016		.75	.48		1.23	1.57	
	0805	Pneumatic nailed		1,209	.013		.75	.39		1.14	1.42	

06400 | Architectural Woodwork

06440 | Wood Ornaments

			CREW	DAILY OUTPUT	LABOR-HOURS	UNIT	2001 BARE COSTS MAT.	LABOR	EQUIP.	TOTAL	TOTAL INCL O&P	
400	0010	LOUVERS Redwood, 2'-0" diameter, full circle	1 Carp	16	.500	Ea.	105	14.60		119.60	139	400
	0100	Half circle		16	.500		104	14.60		118.60	137	
	0200	Octagonal		16	.500		86	14.60		100.60	118	
	0300	Triangular, 5/12 pitch, 5'-0" at base		16	.500		180	14.60		194.60	221	

For expanded coverage of these items see *Means Interior Cost Data 2001*

Division Notes

		CREW	DAILY OUTPUT	LABOR-HOURS	UNIT	2001 BARE COSTS				TOTAL INCL O&P
						MAT.	LABOR	EQUIP.	TOTAL	

Division 7
Thermal & Moisture Protection

Estimating Tips
07100 Dampproofing & Waterproofing
- Be sure of the job specifications before pricing this subdivision. The difference in cost between waterproofing and dampproofing can be great. Waterproofing will hold back standing water. Dampproofing prevents the transmission of water vapor. Also included in this section are vapor retarding membranes.

07200 Thermal Protection
- Insulation and fireproofing products are measured by area, thickness, volume or R value. Specifications may only give what the specific R value should be in a certain situation. The estimator may need to choose the type of insulation to meet that R value.

07300 Shingles, Roof Tiles & Roof Coverings
07400 Roofing & Siding Panels
- Many roofing and siding products are bought and sold by the square. One square is equal to an area that measures 100 square feet. This simple change in unit of measure could create a large error if the estimator is not observant. Accessories and fasteners necessary for a complete installation must be figured into any calculations for both material and labor.

07500 Membrane Roofing
07600 Flashing & Sheet Metal
07700 Roof Specialties & Accessories
- The items in these subdivisions compose a roofing system. No one component completes the installation and all must be estimated. Built-up or single ply membrane roofing systems are made up of many products and installation trades. Wood blocking at roof perimeters or penetrations, parapet coverings, reglets, roof drains, gutters, downspouts, sheet metal flashing, skylights, smoke vents or roof hatches all need to be considered along with the roofing material. Several different installation trades will need to work together on the roofing system. Inherent difficulties in the scheduling and coordination of various trades must be accounted for when estimating labor costs.

07900 Joint Sealers
- To complete the weather-tight shell the sealants and caulkings must be estimated. Where different materials meet—at expansion joints, at flashing penetrations, and at hundreds of other locations throughout a construction project—they provide another line of defense against water penetration. Often, an entire system is based on the proper location and placement of caulking or sealants. The detail drawings that are included as part of a set of architectural plans, show typical locations for these materials. When caulking or sealants are shown at typical locations, this means the estimator must include them for all the locations where this detail is applicable. Be careful to keep different types of sealants separate, and remember to consider backer rods and primers if necessary.

Reference Numbers
Reference numbers are shown in bold squares at the beginning of some major classifications. These numbers refer to related items in the Reference Section. The reference information may be an estimating procedure, an alternate pricing method or technical information.

Note: Not all subdivisions listed here necessarily appear in this publication.

07600 | Flashing & Sheet Metal

07650 | Flexible Flashing

			Crew	Daily Output	Labor-Hours	Unit	Mat.	Labor	Equip.	Total	Total Incl O&P	
600	0010	**FLASHING** Aluminum, mill finish, .013" thick	1 Shee	145	.055	S.F.	.35	1.85		2.20	3.25	600
	0030	.016" thick		145	.055		.51	1.85		2.36	3.42	
	0060	.019" thick		145	.055		.77	1.85		2.62	3.71	
	0100	.032" thick		145	.055		1.04	1.85		2.89	4	
	0200	.040" thick		145	.055		1.45	1.85		3.30	4.46	
	0300	.050" thick		145	.055		1.84	1.85		3.69	4.88	
	0400	Painted finish, add					.24			.24	.26	
	1300	Asphalt flashing cement, 5 gallon				Gal.	3.34			3.34	3.67	
	1600	Copper, 16 oz, sheets, under 1000 lbs.	1 Shee	115	.070	S.F.	2.18	2.33		4.51	6	
	1700	Over 4000 lbs.		155	.052		2.91	1.73		4.64	5.85	
	1900	20 oz sheets, under 1000 lbs.		110	.073		3.90	2.44		6.34	8.05	
	2000	Over 4000 lbs.		145	.055		3.62	1.85		5.47	6.85	
	2200	24 oz sheets, under 1000 lbs.		105	.076		4.69	2.56		7.25	9.10	
	2300	Over 4000 lbs.		135	.059		4.34	1.99		6.33	7.85	
	2500	32 oz sheets, under 1000 lbs.		100	.080		6.20	2.68		8.88	11	
	2600	Over 4000 lbs.		130	.062		5.80	2.06		7.86	9.60	
	4300	Copper-clad stainless steel, .015" thick, under 500 lbs.		115	.070		3.16	2.33		5.49	7.10	
	4400	Over 2000 lbs.		155	.052		3.05	1.73		4.78	6.05	
	4600	.018" thick, under 500 lbs.		100	.080		4.18	2.68		6.86	8.75	
	4700	Over 2000 lbs.		145	.055		3.06	1.85		4.91	6.25	
	5800	Lead, 2.5 lb. per SF, up to 12" wide	1 Rofc	135	.059		2.80	1.52		4.32	5.70	
	5900	Over 12" wide	"	135	.059		3.25	1.52		4.77	6.20	
	8500	Shower pan, bituminous membrane, 7 oz	1 Shee	155	.052		1.08	1.73		2.81	3.86	
	8550	3 ply copper and fabric, 3 oz		155	.052		1.60	1.73		3.33	4.43	
	8600	7 oz		155	.052		3.30	1.73		5.03	6.30	
	8650	Copper, 16 oz		100	.080		3.05	2.68		5.73	7.50	
	8700	Lead on copper and fabric, 5 oz		155	.052		1.62	1.73		3.35	4.45	
	8800	7 oz		155	.052		2.87	1.73		4.60	5.85	
	8850	Polyvinyl chloride, .030" thick		160	.050		.30	1.68		1.98	2.92	
	8900	Stainless steel sheets, 32 ga, .010" thick		155	.052		2.16	1.73		3.89	5.05	
	9000	28 ga, .015" thick		155	.052		2.55	1.73		4.28	5.50	
	9100	26 ga, .018" thick		155	.052		3.25	1.73		4.98	6.25	
	9200	24 ga, .025" thick		155	.052		4.22	1.73		5.95	7.30	
	9400	Terne coated stainless steel, .015" thick, 28 ga		155	.052		4.05	1.73		5.78	7.15	
	9500	.018" thick, 26 ga		155	.052		4.57	1.73		6.30	7.70	
	9600	Zinc and copper alloy (brass), .020" thick		155	.052		3.26	1.73		4.99	6.25	
	9700	.027" thick		155	.052		4.37	1.73		6.10	7.50	
	9800	.032" thick		155	.052		5.10	1.73		6.83	8.25	
	9900	.040" thick		155	.052		6.20	1.73		7.93	9.50	

07700 | Roof Specialties & Accessories

07710 | Manufactured Roof Specialties

			Crew	Daily Output	Labor-Hours	Unit	Mat.	Labor	Equip.	Total	Total Incl O&P	
400	0010	**DOWNSPOUTS** Aluminum 2" x 3", .020" thick, embossed	1 Shee	190	.042	L.F.	.74	1.41		2.15	2.99	400
	0100	Enameled		190	.042		.90	1.41		2.31	3.17	
	4800	Steel, galvanized, round, corrugated, 2" or 3" diam, 28 ga		190	.042		.73	1.41		2.14	2.98	
	4900	4" diameter, 28 gauge		145	.055		1.07	1.85		2.92	4.04	
	5100	5" diameter, 28 gauge		130	.062		1.40	2.06		3.46	4.73	
	5200	26 gauge		130	.062		1.36	2.06		3.42	4.69	

07700 | Roof Specialties & Accessories

07710 | Manufactured Roof Specialties

		Crew	Daily Output	Labor-Hours	Unit	Mat.	Labor	Equip.	Total	Total Incl O&P	
400	5400 6" diameter, 28 gauge	1 Shee	105	.076	L.F.	1.94	2.56		4.50	6.10	400
	5500 26 gauge		105	.076		1.42	2.56		3.98	5.50	
	7500 Steel pipe, black, extra heavy, 4" diameter		20	.400		4.48	13.40		17.88	25.50	
	7600 6" diameter		18	.444		7.60	14.90		22.50	31.50	

07800 | Fire & Smoke Protection

07840 | Firestopping

		Crew	Daily Output	Labor-Hours	Unit	Mat.	Labor	Equip.	Total	Total Incl O&P	
100	0010 FIRESTOPPING										100
	0100 Metallic piping, non insulated R07800-030										
	0110 Through walls, 2" diameter	1 Carp	16	.500	Ea.	9.60	14.60		24.20	33.50	
	0120 4" diameter		14	.571		14.65	16.65		31.30	42	
	0130 6" diameter		12	.667		19.70	19.45		39.15	52	
	0140 12" diameter		10	.800		35	23.50		58.50	75	
	0150 Through floors, 2" diameter		32	.250		5.80	7.30		13.10	17.80	
	0160 4" diameter		28	.286		8.35	8.35		16.70	22.50	
	0170 6" diameter		24	.333		11	9.70		20.70	27.50	
	0180 12" diameter		20	.400		18.50	11.65		30.15	39	
	0190 Metallic piping, insulated										
	0200 Through walls, 2" diameter	1 Carp	16	.500	Ea.	13.60	14.60		28.20	38	
	0210 4" diameter		14	.571		18.65	16.65		35.30	46.50	
	0220 6" diameter		12	.667		23.50	19.45		42.95	56.50	
	0230 12" diameter		10	.800		39	23.50		62.50	79	
	0240 Through floors, 2" diameter		32	.250		9.80	7.30		17.10	22	
	0250 4" diameter		28	.286		12.35	8.35		20.70	26.50	
	0260 6" diameter		24	.333		15	9.70		24.70	31.50	
	0270 12" diameter		20	.400		18.50	11.65		30.15	39	
	0280 Non metallic piping, non insulated										
	0290 Through walls, 2" diameter	1 Carp	12	.667	Ea.	39.50	19.45		58.95	74	
	0300 4" diameter		10	.800		49.50	23.50		73	91	
	0310 6" diameter		8	1		69	29		98	122	
	0330 Through floors, 2" diameter		16	.500		31	14.60		45.60	57	
	0340 4" diameter		6	1.333		38.50	39		77.50	103	
	0350 6" diameter		6	1.333		46	39		85	112	
	0370 Ductwork, insulated & non insulated, round										
	0380 Through walls, 6" diameter	1 Carp	12	.667	Ea.	20	19.45		39.45	52.50	
	0390 12" diameter		10	.800		40	23.50		63.50	80.50	
	0400 18" diameter		8	1		65	29		94	117	
	0410 Through floors, 6" diameter		16	.500		11	14.60		25.60	35	
	0420 12" diameter		14	.571		20	16.65		36.65	48	
	0430 18" diameter		12	.667		35	19.45		54.45	69	
	0440 Ductwork, insulated & non insulated, rectangular										
	0450 With stiffener/closure angle, through walls, 6" x 12"	1 Carp	8	1	Ea.	16.65	29		45.65	64	
	0460 12" x 24"		6	1.333		22	39		61	85.50	
	0470 24" x 48"		4	2		63	58.50		121.50	161	
	0480 With stiffener/closure angle, through floors, 6" x 12"		10	.800		9	23.50		32.50	46.50	
	0490 12" x 24"		8	1		16.20	29		45.20	63.50	
	0500 24" x 48"		6	1.333		32	39		71	96	
	0510 Multi trade openings										
	0520 Through walls, 6" x 12"	1 Carp	2	4	Ea.	35	117		152	221	

07800 | Fire & Smoke Protection

07840 | Firestopping

			Crew	Daily Output	Labor-Hours	Unit	Mat.	Labor	Equip.	Total	Total Incl O&P
0530		12" x 24"	1 Carp	1	8	Ea.	141	233		374	520
0540		24" x 48"	2 Carp	1	16		565	465		1,030	1,350
0550		48" x 96"	"	.75	21.333		2,275	620		2,895	3,475
0560	Through floors, 6" x 12"		1 Carp	2	4		35	117		152	221
0570		12" x 24"	"	1	8		141	233		374	520
0580		24" x 48"	2 Carp	.75	21.333		565	620		1,185	1,600
0590		48" x 96"	"	.50	32		2,275	935		3,210	3,950
0600	Structural penetrations, through walls										
0610		Steel beams, W8 x 10	1 Carp	8	1	Ea.	22	29		51	69.50
0620		W12 x 14		6	1.333		35	39		74	99.50
0630		W21 x 44		5	1.600		70	46.50		116.50	150
0640		W36 x 135		3	2.667		170	77.50		247.50	310
0650		Bar joists, 18" deep		6	1.333		32	39		71	96
0660		24" deep		6	1.333		40	39		79	105
0670		36" deep		5	1.600		60	46.50		106.50	139
0680		48" deep		4	2		70	58.50		128.50	168
0690	Construction joints, floor slab at exterior wall										
0700	Precast, brick, block or drywall exterior										
0710		2" wide joint	1 Carp	125	.064	L.F.	5	1.87		6.87	8.40
0720		4" wide joint	"	75	.107	"	10	3.11		13.11	15.85
0730	Metal panel, glass or curtain wall exterior										
0740		2" wide joint	1 Carp	40	.200	L.F.	11.85	5.85		17.70	22
0750		4" wide joint	"	25	.320	"	16.15	9.35		25.50	32.50
0760	Floor slab to drywall partition										
0770		Flat joint	1 Carp	100	.080	L.F.	4.90	2.33		7.23	9.05
0780		Fluted joint		50	.160		10	4.66		14.66	18.30
0790		Etched fluted joint		75	.107		6.50	3.11		9.61	12
0800	Floor slab to concrete/masonry partition										
0810		Flat joint	1 Carp	75	.107	L.F.	11	3.11		14.11	16.95
0820		Fluted joint	"	50	.160	"	13	4.66		17.66	21.50
0830	Concrete/CMU wall joints										
0840		1" wide	1 Carp	100	.080	L.F.	6	2.33		8.33	10.25
0850		2" wide		75	.107		11	3.11		14.11	16.95
0860		4" wide		50	.160		21	4.66		25.66	30.50
0870	Concrete/CMU floor joints										
0880		1" wide	1 Carp	200	.040	L.F.	3	1.17		4.17	5.10
0890		2" wide		150	.053		5.50	1.55		7.05	8.50
0900		4" wide		100	.080		10.50	2.33		12.83	15.20

07900 | Joint Sealers

07920 | Joint Sealants

		Crew	Daily Output	Labor-Hours	Unit	Mat.	Labor	Equip.	Total	Total Incl O&P
0010	**CAULKING AND SEALANTS**									
0020	Acoustical sealant, elastomeric, cartridges				Ea.	2.05			2.05	2.26
0030	Backer rod, polyethylene, 1/4" diameter	1 Bric	4.60	1.739	C.L.F.	1.28	51.50		52.78	81
0050	1/2" diameter		4.60	1.739		2.62	51.50		54.12	82.50
0070	3/4" diameter		4.60	1.739		5.05	51.50		56.55	85
0090	1" diameter		4.60	1.739		8	51.50		59.50	88.50
0100	Acrylic latex caulk, white									
0200	11 fl. oz cartridge				Ea.	1.80			1.80	1.98

07900 | Joint Sealers

07920 | Joint Sealants

		Crew	Daily Output	Labor-Hours	Unit	2001 Bare Costs Mat.	Labor	Equip.	Total	Total Incl O&P
0500	1/4" x 1/2"	1 Bric	248	.032	L.F.	.15	.95		1.10	1.64
0600	1/2" x 1/2"		250	.032		.29	.95		1.24	1.78
0800	3/4" x 3/4"		230	.035		.66	1.03		1.69	2.32
0900	3/4" x 1"		200	.040		.88	1.18		2.06	2.80
1000	1" x 1"	↓	180	.044	↓	1.10	1.32		2.42	3.24
1400	Butyl based, bulk				Gal.	22			22	24
1500	Cartridges				"	26.50			26.50	29.50
1700	Bulk, in place 1/4" x 1/2", 154 L.F./gal.	1 Bric	230	.035	L.F.	.14	1.03		1.17	1.75
1800	1/2" x 1/2", 77 L.F./gal.	"	180	.044	"	.29	1.32		1.61	2.34
2000	Latex acrylic based, bulk				Gal.	23			23	25.50
2100	Cartridges					26			26	28.50
2300	Polysulfide compounds, 1 component, bulk					43.50			43.50	47.50
2600	1 or 2 component, in place, 1/4" x 1/4", 308 L.F./gal.	1 Bric	145	.055	L.F.	.14	1.63		1.77	2.67
2700	1/2" x 1/4", 154 L.F./gal.		135	.059		.28	1.75		2.03	3.02
2900	3/4" x 3/8", 68 L.F./gal.		130	.062		.64	1.82		2.46	3.52
3000	1" x 1/2", 38 L.F./gal.	↓	130	.062	↓	1.14	1.82		2.96	4.07
3200	Polyurethane, 1 or 2 component				Gal.	47			47	51.50
3300	Cartridges				"	46.50			46.50	51
3500	Bulk, in place, 1/4" x 1/4"	1 Bric	150	.053	L.F.	.15	1.58		1.73	2.61
3600	1/2" x 1/4"		145	.055		.31	1.63		1.94	2.86
3800	3/4" x 3/8", 68 L.F./gal.		130	.062		.69	1.82		2.51	3.58
3900	1" x 1/2"		110	.073	↓	1.22	2.15		3.37	4.67
4100	Silicone rubber, bulk				Gal.	34			34	37.50
4200	Cartridges				"	40			40	44
4400	Neoprene gaskets, closed cell, adhesive, 1/8" x 3/8"	1 Bric	240	.033	L.F.	.20	.99		1.19	1.74
4500	1/4" x 3/4"		215	.037		.48	1.10		1.58	2.23
4700	1/2" x 1"		200	.040		1.40	1.18		2.58	3.37
4800	3/4" x 1-1/2"	↓	165	.048	↓	2.91	1.44		4.35	5.40
5500	Resin epoxy coating, 2 component, heavy duty				Gal.	26			26	28.50
5800	Tapes, sealant, P.V.C. foam adhesive, 1/16" x 1/4"				C.L.F.	4.60			4.60	5.05
5900	1/16" x 1/2"					6.80			6.80	7.50
5950	1/16" x 1"					11.30			11.30	12.45
6000	1/8" x 1/2"				↓	7.65			7.65	8.40
6200	Urethane foam, 2 component, handy pack, 1 C.F.				Ea.	27.50			27.50	30.50
6300	50.0 C.F. pack				C.F.	14.05			14.05	15.45

For information about Means Estimating Seminars, see yellow pages 11 and 12 in back of book

Division Notes

		CREW	DAILY OUTPUT	LABOR-HOURS	UNIT	2001 BARE COSTS				TOTAL INCL O&P
						MAT.	LABOR	EQUIP.	TOTAL	

Division 8
Doors & Windows

Estimating Tips

08100 Metal Doors & Frames
- Most metal doors and frames look alike, but there may be significant differences among them. When estimating these items be sure to choose the line item that most closely compares to the specification or door schedule requirements regarding:
  - type of metal
  - metal gauge
  - door core material
  - fire rating
  - finish

08200 Wood & Plastic Doors
- Wood and plastic doors vary considerably in price. The primary determinant is the veneer material. Lauan, birch and oak are the most common veneers. Other variables include the following:
  - hollow or solid core
  - fire rating
  - flush or raised panel
  - finish
- If the specifications require compliance with AWI (Architectural Woodwork Institute) standards or acoustical standards, the cost of the door may increase substantially. All wood doors are priced pre-mortised for hinges and predrilled for cylindrical locksets.

08300 Specialty Doors
- There are many varieties of special doors, and they are usually priced per each. Add frames, hardware or operators required for a complete installation.

08510 Steel Windows
- Most metal windows are delivered preglazed. However, some metal windows are priced without glass. Refer to 08800 Glazing for glass pricing. The grade C indicates commercial grade windows, usually ASTM C-35.

08550 Wood Windows
- All wood windows are priced preglazed. The two glazing options priced are single pane float glass and insulating glass 1/2" thick. Add the cost of screens and grills if required.

08700 Hardware
- Hardware costs add considerably to the cost of a door. The most efficient method to determine the hardware requirements for a project is to review the door schedule. This schedule, in conjunction with the specifications, is all you should need to take off the door hardware.
- Door hinges are priced by the pair, with most doors requiring 1-1/2 pairs per door. The hinge prices do not include installation labor because it is included in door installation. Hinges are classified according to the frequency of use.

08800 Glazing
- Different openings require different types of glass. The three most common types are:
  - float
  - tempered
  - insulating
- Most exterior windows are glazed with insulating glass. Entrance doors and window walls, where the glass is less than 18" from the floor, are generally glazed with tempered glass. Interior windows and some residential windows are glazed with float glass.

08900 Glazed Curtain Wall
- Glazed curtain walls consist of the metal tube framing and the glazing material. The cost data in this subdivision is presented for the metal tube framing alone or the composite wall. If your estimate requires a detailed takeoff of the framing, be sure to add the glazing cost.

Reference Numbers
Reference numbers are shown in bold squares at the beginning of some major classifications. These numbers refer to related items in the Reference Section. The reference information may be an estimating procedure, an alternate pricing method or technical information.

Note: Not all subdivisions listed here necessarily appear in this publication.

08300 | Specialty Doors

08341 | Cold Storage Doors

			CREW	DAILY OUTPUT	LABOR-HOURS	UNIT	2001 BARE COSTS MAT.	LABOR	EQUIP.	TOTAL	TOTAL INCL O&P
200	0010	**COLD STORAGE**									200
	0020	Single, 20 ga. galvanized steel									
	0300	Horizontal sliding, 5' x 7', manual operation, 2" thick	2 Carp	2	8	Ea.	2,075	233		2,308	2,650
	0400	4" thick		2	8		2,725	233		2,958	3,375
	0500	6" thick		2	8		2,650	233		2,883	3,300
	0800	5' x 7', power operation, 2" thick		1.90	8.421		4,425	245		4,670	5,225
	0900	4" thick		1.90	8.421		4,475	245		4,720	5,300
	1000	6" thick		1.90	8.421		5,100	245		5,345	6,000
	1300	9' x 10', manual operation, 2" insulation		1.70	9.412		3,575	274		3,849	4,350
	1400	4" insulation		1.70	9.412		3,675	274		3,949	4,475
	1500	6" insulation		1.70	9.412		4,425	274		4,699	5,300
	1800	Power operation, 2" insulation		1.60	10		6,125	292		6,417	7,200
	1900	4" insulation		1.60	10		6,275	292		6,567	7,350
	2000	6" insulation		1.70	9.412		7,125	274		7,399	8,250
	2300	For stainless steel face, add					20%				
	3000	Hinged, lightweight, 3' x 7'-0", galvanized 1 face, 2" thick	2 Carp	2	8	Ea.	1,025	233		1,258	1,500
	3050	4" thick		1.90	8.421		1,175	245		1,420	1,650
	3300	Aluminum doors, 3' x 7'-0", 4" thick		1.90	8.421		1,050	245		1,295	1,525
	3350	6" thick		1.40	11.429		1,875	335		2,210	2,575
	3600	Stainless steel, 3' x 7'-0", 4" thick		1.90	8.421		1,350	245		1,595	1,850
	3650	6" thick		1.40	11.429		2,275	335		2,610	3,025
	3900	Painted, 3' x 7'-0", 4" thick		1.90	8.421		965	245		1,210	1,425
	3950	6" thick		1.40	11.429		1,850	335		2,185	2,550
	5000	Bi-parting, electric operated									
	5010	6' x 8' opening, galv. faces, 4" thick for cooler	2 Carp	.80	20	Opng.	5,775	585		6,360	7,250
	5050	For freezer, 4" thick		.80	20		6,350	585		6,935	7,900
	5300	For door buck framing and door protection, add		2.50	6.400		470	187		657	810
	6000	Galvanized batten door, galvanized hinges, 4' x 7'		2	8		1,425	233		1,658	1,925
	6050	6' x 8'		1.80	8.889		1,950	259		2,209	2,525
	6500	Fire door, 3 hr., 6' x 8', single slide		.80	20		6,675	585		7,260	8,225
	6550	Double, bi-parting		.70	22.857		8,650	665		9,315	10,600

08344 | Industrial Doors

			CREW	DAILY OUTPUT	LABOR-HOURS	UNIT	MAT.	LABOR	EQUIP.	TOTAL	TOTAL INCL O&P
120	0010	**AIR CURTAINS** Incl. motor starters, transformers,									120
	0050	door switches & temperature controls									
	0100	Shipping and receiving doors, unheated, minimal wind stoppage									
	0150	8' high, multiples of 3' wide	2 Shee	3.60	4.444	L.F.	175	149		324	420
	0160	5' wide		5	3.200		186	107		293	370
	0210	10' high, multiples of 4' wide		3.60	4.444		174	149		323	420
	0250	12' high, 3'-6" wide		3.30	4.848		172	163		335	440
	0260	12' wide		3.30	4.848		203	163		366	475
	0350	16' high, 3'-6" wide		3.30	4.848		335	163		498	620
	0360	12' wide		3.30	4.848		168	163		331	435
	1500	Customer entrance doors, unheated, minimal wind stoppage									
	1550	10' high, multiples of 3' wide	2 Shee	3.30	4.848	L.F.	175	163		338	445
	1560	5' wide		5	3.200		189	107		296	375
	1650	Maximum wind stoppage, 12' high, multiples of 4' wide		3.30	4.848		187	163		350	455
	1700	Heated, minimal wind stoppage, electric heat									
	1750	8' high, multiples of 3' wide	2 Shee	3.10	5.161	L.F.	480	173		653	790
	1850	10' high, multiples of 3' wide		3.10	5.161		460	173		633	770
	1860	Multiples of 5' wide		3.10	5.161		365	173		538	665
	1950	Maximum wind stoppage, steam heat									
	1960	12' high, multiples of 4' wide	2 Shee	2.80	5.714	L.F.	291	192		483	615
	2000	Walk-in coolers and freezers, ambient air, minimal wind stoppage									
	2050	8' high, multiples of 3' wide	2 Shee	3.60	4.444	L.F.	156	149		305	400

08300 | Specialty Doors

08344	Industrial Doors	CREW	DAILY OUTPUT	LABOR-HOURS	UNIT	2001 BARE COSTS MAT.	LABOR	EQUIP.	TOTAL	TOTAL INCL O&P	
120	2060	Multiples of 5' wide	2 Shee	3.60	4.444	L.F.	156	149		305	400	120
	2250	Maximum wind stoppage, 12' high, multiples of 3' wide		3.30	4.848		167	163		330	435	
	2450	Conveyor openings or service windows, unheated, 5' high		4	4		202	134		336	430	
	2460	Heated, electric, 5' high, 2'-6" wide		3.30	4.848		207	163		370	480	

For information about Means Estimating Seminars, see yellow pages 11 and 12 in back of book

For expanded coverage of these items see *Means Interior Cost Data 2001*

Division Notes

		CREW	DAILY OUTPUT	LABOR-HOURS	UNIT	2001 BARE COSTS				TOTAL INCL O&P
						MAT.	LABOR	EQUIP.	TOTAL	

Division 9 Finishes

Estimating Tips

General
- Room Finish Schedule: A complete set of plans should contain a room finish schedule. If one is not available, it would be well worth the time and effort to put one together. A room finish schedule should contain the room number, room name (for clarity), floor materials, base materials, wainscot materials, wainscot height, wall materials (for each wall), ceiling materials, ceiling height and special instructions.
- Surplus Finishes: Review the specifications to determine if there is any requirement to provide certain amounts of extra materials for the owner's maintenance department. In some cases the owner may require a substantial amount of materials, especially when it is a special order item or long lead time item.

09200 Plaster & Gypsum Board
- Lath is estimated by the square yard for both gypsum and metal lath, plus usually 5% allowance for waste. Furring, channels and accessories are measured by the linear foot. An extra foot should be allowed for each accessory miter or stop.
- Plaster is also estimated by the square yard. Deductions for openings vary by preference, from zero deduction to 50% of all openings over 2 feet in width. Some estimators deduct a percentage of the total yardage for openings. The estimator should allow one extra square foot for each linear foot of horizontal interior or exterior angle located below the ceiling level. Also, double the areas of small radius work.
- Each room should be measured, perimeter times maximum wall height. Ceiling areas are equal to length times width.
- Drywall accessories, studs, track, and acoustical caulking are all measured by the linear foot. Drywall taping is figured by the square foot. Gypsum wallboard is estimated by the square foot. No material deductions should be made for door or window openings under 32 S.F. Coreboard can be obtained in a 1" thickness for solid wall and shaft work. Additions should be made to price out the inside or outside corners.
- Different types of partition construction should be listed separately on the quantity sheets. There may be walls with studs of various widths, double studded, and similar or dissimilar surface materials. Shaft work is usually different construction from surrounding partitions requiring separate quantities and pricing of the work.

09300 Tile
09400 Terrazzo
- Tile and terrazzo areas are taken off on a square foot basis. Trim and base materials are measured by the linear foot. Accent tiles are listed per each. Two basic methods of installation are used. Mud set is approximately 30% more expensive than the thin set. In terrazzo work, be sure to include the linear footage of embedded decorative strips, grounds, machine rubbing and power cleanup.

09600 Flooring
- Wood flooring is available in strip, parquet, or block configuration. The latter two types are set in adhesives with quantities estimated by the square foot. The laying pattern will influence labor costs and material waste. In addition to the material and labor for laying wood floors, the estimator must make allowances for sanding and finishing these areas unless the flooring is prefinished.
- Most of the various types of flooring are all measured on a square foot basis. Base is measured by the linear foot. If adhesive materials are to be quantified, they are estimated at a specified coverage rate by the gallon depending upon the specified type and the manufacturer's recommendations.
- Sheet flooring is measured by the square yard. Roll widths vary, so consideration should be given to use the most economical width, as waste must be figured into the total quantity. Consider also the installation methods available, direct glue down or stretched.

09700 Wall Finishes
- Wall coverings are estimated by the square foot. The area to be covered is measured, length by height of wall above baseboards, to calculate the square footage of each wall. This figure is divided by the number of square feet in the single roll which is being used. Deduct, in full, the areas of openings such as doors and windows. Where a pattern match is required allow 25%-30% waste. One gallon of paste should be sufficient to hang 12 single rolls of light to medium weight paper.

09800 Acoustical Treatment
- Acoustical systems fall into several categories. The takeoff of these materials is by the square foot of area with a 5% allowance for waste. Do not forget about scaffolding, if applicable, when estimating these systems.

09900 Paints & Coatings
- Painting is one area where bids vary to a greater extent than almost any other section of a project. This arises from the many methods of measuring surfaces to be painted. The estimator should check the plans and specifications carefully to be sure of the required number of coats.
- Protection of adjacent surfaces is not included in painting costs. When considering the method of paint application, an important factor is the amount of protection and masking required. These must be estimated separately and may be the determining factor in choosing the method of application.

Reference Numbers
Reference numbers are shown in bold squares at the beginning of some major classifications. These numbers refer to related items in the Reference Section. The reference information may be an estimating procedure, an alternate pricing method or technical information.

Note: Not all subdivisions listed here necessarily appear in this publication.

09900 | Paints & Coatings

09910 | Paints & Coatings

			CREW	DAILY OUTPUT	LABOR-HOURS	UNIT	MAT.	LABOR	EQUIP.	TOTAL	TOTAL INCL O&P
630	0010	**MISCELLANEOUS, INTERIOR**									
	3800	Grilles, per side, oil base, primer coat, brushwork	1 Pord	520	.015	S.F.	.10	.40		.50	.72
	3850	Spray		1,140	.007		.10	.18		.28	.39
	3880	Paint 1 coat, brushwork		520	.015		.10	.40		.50	.72
	3900	Spray		1,140	.007		.12	.18		.30	.41
	3920	Paint 2 coats, brushwork		325	.025		.20	.64		.84	1.20
	3940	Spray		650	.012		.23	.32		.55	.74
	3950	Prime & paint 1 coat		325	.025		.20	.64		.84	1.20
	3960	Prime & paint 2 coats	▼	270	.030	▼	.20	.77		.97	1.40
	4250	Paint 1 coat, brushwork	2 Pord	1,300	.012	L.F.	.11	.32		.43	.61
	4500	Louvers, one side, primer, brushwork	1 Pord	524	.015	S.F.	.06	.40		.46	.67
	4520	Paint one coat, brushwork		520	.015		.05	.40		.45	.67
	4530	Spray		1,140	.007		.06	.18		.24	.35
	4540	Paint two coats, brushwork		325	.025		.10	.64		.74	1.10
	4550	Spray		650	.012		.12	.32		.44	.62
	4560	Paint three coats, brushwork		270	.030		.16	.77		.93	1.35
	4570	Spray	▼	500	.016	▼	.17	.42		.59	.83
	5000	Pipe, to 4" diameter, primer or sealer coat, oil base, brushwork	2 Pord	1,250	.013	L.F.	.06	.33		.39	.57
	5100	Spray		2,165	.007		.06	.19		.25	.35
	5200	Paint 1 coat, brushwork		1,250	.013		.06	.33		.39	.58
	5300	Spray		2,165	.007		.05	.19		.24	.35
	5350	Paint 2 coats, brushwork		775	.021		.11	.54		.65	.94
	5400	Spray		1,240	.013		.12	.34		.46	.64
	5420	Paint 3 coats, brushwork		775	.021		.16	.54		.70	1
	5450	To 8" diameter, primer or sealer coat, brushwork		620	.026		.12	.67		.79	1.16
	5500	Spray		1,085	.015		.19	.39		.58	.80
	5550	Paint 1 coat, brushwork		620	.026		.17	.67		.84	1.21
	5600	Spray		1,085	.015		.19	.39		.58	.79
	5650	Paint 2 coats, brushwork		385	.042		.22	1.08		1.30	1.89
	5700	Spray		620	.026		.24	.67		.91	1.30
	5720	Paint 3 coats, brushwork		385	.042		.33	1.08		1.41	2.01
	5750	To 12" diameter, primer or sealer coat, brushwork		415	.039		.17	1.01		1.18	1.72
	5800	Spray		725	.022		.20	.58		.78	1.10
	5850	Paint 1 coat, brushwork		415	.039		.17	1.01		1.18	1.72
	6000	Spray		725	.022		.19	.58		.77	1.09
	6200	Paint 2 coats, brushwork		260	.062		.33	1.61		1.94	2.81
	6250	Spray		415	.039		.36	1.01		1.37	1.93
	6270	Paint 3 coats, brushwork		260	.062		.49	1.61		2.10	2.99
	6300	To 16" diameter, primer or sealer coat, brushwork		310	.052		.23	1.35		1.58	2.31
	6350	Spray		540	.030		.26	.77		1.03	1.47
	6400	Paint 1 coat, brushwork		310	.052		.23	1.35		1.58	2.30
	6450	Spray		540	.030		.25	.77		1.02	1.46
	6500	Paint 2 coats, brushwork		195	.082		.44	2.14		2.58	3.74
	6550	Spray	▼	310	.052	▼	.49	1.35		1.84	2.59
	6600	Radiators, per side, primer, brushwork	1 Pord	520	.015	S.F.	.06	.40		.46	.67
	6620	Paint one coat, brushwork		520	.015		.05	.40		.45	.67
	6640	Paint two coats, brushwork		340	.024		.10	.61		.71	1.06
	6660	Paint three coats, brushwork	▼	283	.028	▼	.16	.74		.90	1.29

For information about Means Estimating Seminars, see yellow pages 11 and 12 in back of book

Division 10 Specialties

Estimating Tips
General
- The items in this division are usually priced per square foot or each.
- Many items in Division 10 require some type of support system that is not usually furnished with the item. Examples of these systems include blocking for the attachment of grab bars and support angles for ceiling hung toilet partitions. The required blocking or supports must be added to the estimate in the appropriate division.
- Some items in Division 10, such as lockers, may require assembly before installation. Verify the amount of assembly required. Assembly can often exceed installation time.

10100 Visual Display Boards
- Toilet partitions are priced by the stall. A stall consists of a side wall, pilaster and door with hardware. Toilet tissue holders and grab bars are extra.

10600 Partitions
- The required acoustical rating of a folding partition can have a significant impact on costs. Verify the sound transmission coefficient rating of the panel priced to the specification requirements.

Reference Numbers
Reference numbers are shown in bold squares at the beginning of some major classifications. These numbers refer to related items in the Reference Section. The reference information may be an estimating procedure, an alternate pricing method or technical information.

Note: Not all subdivisions listed here necessarily appear in this publication.

10200 | Louvers & Vents

10210 | Wall Louvers

			CREW	DAILY OUTPUT	LABOR-HOURS	UNIT	2001 BARE COSTS MAT.	LABOR	EQUIP.	TOTAL	TOTAL INCL O&P	
800	0010	**LOUVERS** Aluminum with screen, residential, 8" x 8"	1 Carp	38	.211	Ea.	6.85	6.15		13	17.15	800
	0100	12" x 12"		38	.211		8.20	6.15		14.35	18.60	
	0200	12" x 18"		35	.229		12	6.65		18.65	23.50	
	0250	14" x 24"		30	.267		15.50	7.75		23.25	29	
	0300	18" x 24"		27	.296		18	8.65		26.65	33.50	
	0500	24" x 30"		24	.333		25	9.70		34.70	42.50	
	0700	Triangle, adjustable, small		20	.400		22	11.65		33.65	42.50	
	0800	Large		15	.533		39.50	15.55		55.05	68	
	1200	Extruded aluminum, see division 15850-600										
	2100	Midget, aluminum, 3/4" deep, 1" diameter	1 Carp	85	.094	Ea.	.63	2.74		3.37	4.98	
	2150	3" diameter		60	.133		1.32	3.89		5.21	7.55	
	2200	4" diameter		50	.160		1.89	4.66		6.55	9.40	
	2250	6" diameter		30	.267		2.25	7.75		10	14.65	
	2300	Ridge vent strip, mill finish	1 Shee	155	.052	L.F.	2.14	1.73		3.87	5	
	2400	Under eaves vent, aluminum, mill finish, 16" x 4"	1 Carp	48	.167	Ea.	1.24	4.86		6.10	8.95	
	2500	16" x 8"	"	48	.167	"	1.69	4.86		6.55	9.45	

10270 | Access Flooring

10275 | Access Flooring

			CREW	DAILY OUTPUT	LABOR-HOURS	UNIT	2001 BARE COSTS MAT.	LABOR	EQUIP.	TOTAL	TOTAL INCL O&P	
150	0010	**PEDESTAL ACCESS FLOORS** Computer room application, metal										150
	0020	Particle board or steel panels, no covering, under 6,000 S.F.	2 Carp	400	.040	S.F.	6.25	1.17		7.42	8.70	
	0300	Metal covered, over 6,000 S.F.		450	.036		6.20	1.04		7.24	8.40	
	0400	Aluminum, 24" panels		500	.032		21	.93		21.93	24.50	
	0600	For carpet covering, add					4.44			4.44	4.88	
	0700	For vinyl floor covering, add					4.30			4.30	4.73	
	0900	For high pressure laminate covering, add					3.01			3.01	3.31	
	0910	For snap on stringer system, add	2 Carp	1,000	.016		1.02	.47		1.49	1.85	
	0950	Office applications, to 8" high, steel panels,										
	0960	no covering, over 6,000 S.F.	2 Carp	500	.032	S.F.	6.55	.93		7.48	8.70	
	1000	Machine cutouts after initial installation	1 Carp	10	.800	Ea.	33.50	23.50		57	73.50	
	1050	Pedestals, 6" to 12"	2 Carp	85	.188		6.35	5.50		11.85	15.55	
	1100	Air conditioning grilles, 4" x 12"	1 Carp	17	.471		42	13.70		55.70	67.50	
	1150	4" x 18"	"	14	.571		57.50	16.65		74.15	89	
	1200	Approach ramps, minimum	2 Carp	85	.188	S.F.	17.45	5.50		22.95	28	
	1300	Maximum	"	60	.267	"	26.50	7.75		34.25	41	
	1500	Handrail, 2 rail, aluminum	1 Carp	15	.533	L.F.	75	15.55		90.55	107	

10300 | Fireplaces & Stoves

10305 | Manufactured Fireplaces

			CREW	DAILY OUTPUT	LABOR-HOURS	UNIT	2001 BARE COSTS MAT.	LABOR	EQUIP.	TOTAL	TOTAL INCL O&P	
100	0010	**FIREPLACE, PREFABRICATED** Free standing or wall hung										100
	0100	with hood & screen, minimum	1 Carp	1.30	6.154	Ea.	960	179		1,139	1,325	

10300 | Fireplaces & Stoves

10305	Manufactured Fireplaces	CREW	DAILY OUTPUT	LABOR-HOURS	UNIT	2001 BARE COSTS MAT.	LABOR	EQUIP.	TOTAL	TOTAL INCL O&P	
100	0150	Average	1 Carp	1	8	Ea.	1,150	233		1,383	1,650	100
	0200	Maximum		.90	8.889	↓	2,775	259		3,034	3,450	
	0500	Chimney dbl. wall, all stainless, over 8'-6", 7" diam., add		33	.242	V.L.F.	46	7.05		53.05	62	
	0600	10" diameter, add		32	.250		46	7.30		53.30	62.50	
	0700	12" diameter, add		31	.258		60.50	7.50		68	78.50	
	0800	14" diameter, add		30	.267	↓	76.50	7.75		84.25	96	
	1000	Simulated brick chimney top, 4' high, 16" x 16"		10	.800	Ea.	164	23.50		187.50	218	
	1100	24" x 24"		7	1.143	"	305	33.50		338.50	390	
	1500	Simulated logs, gas fired, 40,000 BTU, 2' long, minimum		7	1.143	Set	420	33.50		453.50	510	
	1600	Maximum		6	1.333		590	39		629	710	
	1700	Electric, 1,500 BTU, 1'-6" long, minimum		7	1.143		114	33.50		147.50	177	
	1800	11,500 BTU, maximum	↓	6	1.333	↓	246	39		285	330	
10320	**Stoves**											
100	0010	**WOODBURNING STOVES** Cast iron, minimum	2 Carp	1.30	12.308	Ea.	725	360		1,085	1,350	100
	0020	Average		1	16		1,150	465		1,615	2,000	
	0030	Maximum	↓	.80	20	↓	2,175	585		2,760	3,300	
	0050	For gas log lighter, add					36			36	39.50	

For information about Means Estimating Seminars, see yellow pages 11 and 12 in back of book

Division Notes

		CREW	DAILY OUTPUT	LABOR-HOURS	UNIT	2001 BARE COSTS				TOTAL INCL O&P
						MAT.	LABOR	EQUIP.	TOTAL	

Division 11 Equipment

Estimating Tips

General
- The items in this division are usually priced per square foot or each. Many of these items are purchased by the owner for installation by the contractor. Check the specifications for responsibilities, and include time for receiving, storage, installation and mechanical and electrical hook-ups in the appropriate divisions.
- Many items in Division 11 require some type of support system that is not usually furnished with the item. Examples of these systems include blocking for the attachment of casework and support angles for ceiling hung projection screens. The required blocking or supports must be added to the estimate in the appropriate division.
- Some items in Division 11 may require assembly or electrical hook-ups. Verify the amount of assembly required or the need for a hard electrical connection and add the appropriate costs.

Reference Numbers
Reference numbers are shown in bold squares at the beginning of some major classifications. These numbers refer to related items in the Reference Section. The reference information may be an estimating procedure, an alternate pricing method or technical information.

Note: Not all subdivisions listed here necessarily appear in this publication.

11010 | Maintenance Equipment

11013 | Floor/Wall Cleaning Equipment

		CREW	DAILY OUTPUT	LABOR-HOURS	UNIT	MAT.	LABOR	EQUIP.	TOTAL	TOTAL INCL O&P
800	0010 **VACUUM CLEANING**									800
	0020 Central, 3 inlet, residential	1 Skwk	.90	8.889	Total	555	265		820	1,025
	0200 Commercial		.70	11.429		960	340		1,300	1,575
	0400 5 inlet system, residential		.50	16		840	480		1,320	1,675
	0600 7 inlet system, commercial		.40	20		1,050	595		1,645	2,075
	0800 9 inlet system, residential	▼	.30	26.667		1,200	795		1,995	2,575
	4010 Rule of thumb: First 1200 S.F., installed								1,044	
	4020 For each additional S.F., add				S.F.				.17	

11100 | Mercantile Equipment

11104 | Display Cases & Systems

		CREW	DAILY OUTPUT	LABOR-HOURS	UNIT	MAT.	LABOR	EQUIP.	TOTAL	TOTAL INCL O&P
700	0010 **REFRIGERATED FOOD CASES**									700
	0030 Dairy, multi-deck, 12' long	Q-5	3	5.333	Ea.	7,075	167		7,242	8,025
	0100 For rear sliding doors, add					1,000			1,000	1,100
	0200 Delicatessen case, service deli, 12' long, single deck	Q-5	3.90	4.103		4,750	129		4,879	5,425
	0300 Multi-deck, 18 S.F. shelf display		3	5.333		5,825	167		5,992	6,650
	0400 Freezer, self-contained, chest-type, 30 C.F.		3.90	4.103		3,475	129		3,604	4,025
	0500 Glass door, upright, 78 C.F.		3.30	4.848		6,600	152		6,752	7,500
	0600 Frozen food, chest type, 12' long		3.30	4.848		4,775	152		4,927	5,475
	0700 Glass door, reach-in, 5 door		3	5.333		9,150	167		9,317	10,400
	0800 Island case, 12' long, single deck		3.30	4.848		5,400	152		5,552	6,175
	0900 Multi-deck		3	5.333		11,400	167		11,567	12,900
	1000 Meat case, 12' long, single deck		3.30	4.848		3,925	152		4,077	4,550
	1050 Multi-deck		3.10	5.161		6,775	162		6,937	7,700
	1100 Produce, 12' long, single deck		3.30	4.848		5,200	152		5,352	5,950
	1200 Multi-deck	▼	3.10	5.161		5,800	162		5,962	6,625

11110 | Commercial Laundry & Dry Cleaning Equipment

11119 | Laundry Cleaning

		CREW	DAILY OUTPUT	LABOR-HOURS	UNIT	MAT.	LABOR	EQUIP.	TOTAL	TOTAL INCL O&P
450	0010 **LAUNDRY EQUIPMENT** Not incl. rough-in									450
	0500 Dryers, gas fired residential, 16 lb. capacity, average	1 Plum	3	2.667	Ea.	520	92		612	710
	1000 Commercial, 30 lb. capacity, coin operated, single		3	2.667		2,375	92		2,467	2,775
	1100 Double stacked		2	4		5,100	138		5,238	5,800
	1500 Industrial, 30 lb. capacity		2	4		2,025	138		2,163	2,425
	1600 50 lb. capacity		1.70	4.706		2,650	162		2,812	3,175
	5000 Washers, residential, 4 cycle, average		3	2.667		595	92		687	795
	5300 Commercial, coin operated, average	▼	3	2.667		955	92		1,047	1,200
	6000 Combination washer/extractor, 20 lb. capacity	L-6	1.50	8		3,425	275		3,700	4,175
	6100 30 lb. capacity		.80	15		6,600	515		7,115	8,025
	6200 50 lb. capacity		.68	17.647		8,225	605		8,830	9,975
	6300 75 lb. capacity	▼	.30	40	▼	17,000	1,375		18,375	20,800

11110 | Commercial Laundry & Dry Cleaning Equipment

	11119	Laundry Cleaning	CREW	DAILY OUTPUT	LABOR-HOURS	UNIT	2001 BARE COSTS MAT.	LABOR	EQUIP.	TOTAL	TOTAL INCL O&P	
450	6350	125 lb. capacity	L-6	.16	75	Ea.	20,600	2,575		23,175	26,500	450

11140 | Vehicle Service Equipment

	11141	Service Station Equipment	CREW	DAILY OUTPUT	LABOR-HOURS	UNIT	2001 BARE COSTS MAT.	LABOR	EQUIP.	TOTAL	TOTAL INCL O&P	
150	0010	**AUTOMOTIVE**										150
	0030	Compressors, electric, 1-1/2 H.P., standard controls	L-4	1.50	16	Ea.	315	435		750	1,025	
	0550	Dual controls		1.50	16		460	435		895	1,200	
	0600	5 H.P., 115/230 volt, standard controls		1	24		1,525	655		2,180	2,725	
	0650	Dual controls		1	24		1,625	655		2,280	2,825	
	1100	Product dispenser with vapor recovery for 6 nozzles, installed, not										
	1110	including piping to storage tanks				Ea.	15,000			15,000	16,500	
	2200	Hoists, single post, 8,000# capacity, swivel arms	L-4	.40	60		3,500	1,625		5,125	6,425	
	2400	Two posts, adjustable frames, 11,000# capacity		.25	96		4,500	2,625		7,125	9,050	
	2500	24,000# capacity		.15	160		6,000	4,375		10,375	13,500	
	2700	7,500# capacity, frame supports		.50	48		5,000	1,300		6,300	7,550	
	2800	Four post, roll on ramp		.50	48		4,500	1,300		5,800	7,000	
	2810	Hydraulic lifts, above ground, 2 post, clear floor, 6000 lb cap		2.67	8.989		4,700	245		4,945	5,550	
	2815	9000 lb capacity		2.29	10.480		11,200	286		11,486	12,800	
	2820	15,000 lb capacity		2	12		12,600	325		12,925	14,400	
	2825	30,000 lb capacity		1.60	15		27,900	410		28,310	31,200	
	2830	4 post, ramp style, 25,000 lb capacity		2	12		10,900	325		11,225	12,500	
	2835	35,000 lb capacity		1	24		51,000	655		51,655	57,000	
	2840	50,000 lb capacity		1	24		57,000	655		57,655	63,500	
	2845	75,000 lb capacity		1	24		66,000	655		66,655	73,500	
	2850	For drive thru tracks, add, minimum					700			700	770	
	2855	Maximum					1,200			1,200	1,325	
	2860	Ramp extensions, 3'(set of 2)					575			575	635	
	2865	Rolling jack platform					2,000			2,000	2,200	
	2870	Elec/hyd jacking beam					5,350			5,350	5,875	
	2880	Scissor lift, portable, 6000 lb capacity					5,250			5,250	5,775	
	3000	Lube equipment, 3 reel type, with pumps, not including piping	L-4	.50	48	Set	6,000	1,300		7,300	8,650	
	3700	Pump lubrication, pneumatic, not incl. air compressor										
	3710	Oil/gear lube	Q-1	9.60	1.667	Ea.	665	51.50		716.50	815	
	3720	Grease	"	9.60	1.667		615	51.50		666.50	760	
	4000	Spray painting booth, 26' long, complete	L-4	.40	60		12,000	1,625		13,625	15,800	

11170 | Solid Waste Handling Equipment

	11179	Waste Handling Equipment	CREW	DAILY OUTPUT	LABOR-HOURS	UNIT	2001 BARE COSTS MAT.	LABOR	EQUIP.	TOTAL	TOTAL INCL O&P	
150	0010	**WASTE HANDLING**										150
	0020	Compactors, 115 volt, 250#/hr., chute fed	L-4	1	24	Ea.	8,225	655		8,880	10,100	

11170 | Solid Waste Handling Equipment

11179 | Waste Handling Equipment

		CREW	DAILY OUTPUT	LABOR-HOURS	UNIT	MAT.	LABOR	EQUIP.	TOTAL	TOTAL INCL O&P	
150	0100	Hand fed	L-4	2.40	10	Ea.	5,675	273		5,948	6,675	150
	0300	Multi-bag, 230 volt, 600#/hr, chute fed		1	24		7,025	655		7,680	8,750	
	0400	Hand fed		1	24		6,375	655		7,030	8,050	
	0500	Containerized, hand fed, 2 to 6 C.Y. containers, 250#/hr.		1	24		7,475	655		8,130	9,250	
	0550	For chute fed, add per floor		1	24		860	655		1,515	1,975	
	1000	Heavy duty industrial compactor, 0.5 C.Y. capacity		1	24		5,050	655		5,705	6,575	
	1050	1.0 C.Y. capacity		1	24		7,525	655		8,180	9,300	
	1100	3 C.Y. capacity		.50	48		10,500	1,300		11,800	13,700	
	1150	5.0 C.Y. capacity		.50	48		14,000	1,300		15,300	17,500	
	1200	Combination shredder/compactor (5,000 lbs./hr.)		.50	48		25,200	1,300		26,500	29,800	
	1400	For handling hazardous waste materials, 55 gallon drum packer, std.					11,700			11,700	12,900	
	1410	55 gallon drum packer w/HEPA filter					14,900			14,900	16,400	
	1420	55 gallon drum packer w/charcoal & HEPA filter					19,000			19,000	20,900	
	1430	All of the above made explosion proof, add					9,075			9,075	9,975	
	1500	Crematory, not including building, 1 place	Q-3	.20	160		42,400	5,250		47,650	54,500	
	1750	2 place		.10	320		60,500	10,500		71,000	82,500	
	4400	Incinerator, gas, not incl. chimney, elec. or pipe, 50#/hr., minimum		.80	40		16,200	1,325		17,525	19,800	
	4420	Maximum		.70	45.714		21,100	1,500		22,600	25,500	
	4440	200 lb. per hr., minimum (batch type)		.60	53.333		21,100	1,750		22,850	25,900	
	4460	Maximum (with feeder)		.50	64		41,100	2,100		43,200	48,400	
	4480	400 lb. per hr., minimum (batch type)		.30	106		25,000	3,500		28,500	32,800	
	4500	Maximum (with feeder)		.25	128		47,200	4,200		51,400	58,500	
	4520	800 lb. per hr., with feeder, minimum		.20	160		61,500	5,250		66,750	75,500	
	4540	Maximum		.17	188		84,000	6,175		90,175	102,000	
	4560	1,200 lb. per hr., with feeder, minimum		.15	213		89,000	7,000		96,000	108,500	
	4580	Maximum		.11	290		107,000	9,550		116,550	132,000	
	4600	2,000 lb. per hr., with feeder, minimum		.10	320		155,500	10,500		166,000	187,000	
	4620	Maximum		.05	640		261,500	21,000		282,500	319,000	
	4700	For heat recovery system, add, minimum		.25	128		50,500	4,200		54,700	62,000	
	4710	Add, maximum		.11	290		161,500	9,550		171,050	192,000	
	4720	For automatic ash conveyer, add		.50	64		20,200	2,100		22,300	25,400	
	4750	Large municipal incinerators, incl. stack, minimum		.25	128	Ton/day	12,900	4,200		17,100	20,600	
	4850	Maximum		.10	320	"	34,300	10,500		44,800	53,500	
	5800	Shredder, industrial, minimum				Ea.	15,100			15,100	16,600	
	5850	Maximum					81,000			81,000	89,000	
	5900	Baler, industrial, minimum					6,050			6,050	6,650	
	5950	Maximum					353,500			353,500	389,000	

11300 | Fluid Waste Treatment & Disposal Equipment

11310 | Sewage & Sludge Pumps

			CREW	DAILY OUTPUT	LABOR-HOURS	UNIT	MAT.	LABOR	EQUIP.	TOTAL	TOTAL INCL O&P	
350	0010	**PUMPS, PNEUMATIC EJECTOR**											350
	0020	With steel receiver, level controls, inlet/outlet gate/check valves											
	0030	Cross connect. not incl. compressor, fittings or piping											
	0040	Duplex											
	0050	30 GPM		Q-2	1.70	14.118	Ea.	13,800	455		14,255	15,900	
	0060	50 GPM			1.56	15.385		19,500	495		19,995	22,100	
	0070	100 GPM			1.30	18.462		23,600	595		24,195	26,800	
	0080	150 GPM			1.10	21.818		45,400	700		46,100	51,000	

11300 | Fluid Waste Treatment & Disposal Equipment

11310 | Sewage & Sludge Pumps

		CREW	DAILY OUTPUT	LABOR-HOURS	UNIT	2001 BARE COSTS MAT.	LABOR	EQUIP.	TOTAL	TOTAL INCL O&P	
350	0090	200 GPM	Q-2	.85	28.235	Ea.	58,000	910		58,910	65,000	350
	0100	250 GPM	Q-3	.91	35.165		74,500	1,150		75,650	84,000	
	0110	300 GPM	"	.57	56.140	↓	91,000	1,850		92,850	103,000	

11400 | Food Service Equipment

11410 | Food Prep Equipment

			CREW	DAILY OUTPUT	LABOR-HOURS	UNIT	2001 BARE COSTS MAT.	LABOR	EQUIP.	TOTAL	TOTAL INCL O&P	
150	0010	**COMMERCIAL KITCHEN EQUIPMENT**										150
	0020	Bake oven, gas, one section	Q-1	8	2	Ea.	3,750	62		3,812	4,225	
	0300	Two sections		7	2.286		7,475	71		7,546	8,325	
	0600	Three sections	↓	6	2.667		10,700	82.50		10,782.50	11,800	
	0900	Electric convection, single deck	L-7	4	7		5,075	197		5,272	5,875	
	2350	Cooler, reach-in, beverage, 6' long	Q-1	6	2.667	↓	2,800	82.50		2,882.50	3,200	
	2700	Dishwasher, commercial, rack type										
	2720	10 to 12 racks per hour	Q-1	3.20	5	Ea.	2,550	155		2,705	3,050	
	2750	Semi-automatic 38 to 50 racks per hour	"	1.30	12.308		5,375	380		5,755	6,475	
	2800	Automatic, 190 to 230 racks per hour	Q-2	.70	34.286		11,000	1,100		12,100	13,800	
	2820	235 to 275 racks per hour		.50	48		22,200	1,550		23,750	26,700	
	2840	8,750 to 12,500 dishes per hour	↓	.20	120		41,700	3,850		45,550	51,500	
	4300	Freezers, reach-in, 44 C.F.	Q-1	4	4		7,400	124		7,524	8,325	
	4500	68 C.F.		3	5.333		8,725	165		8,890	9,825	
	5800	Ice cube maker, 50 pounds per day		6	2.667		1,525	82.50		1,607.50	1,800	
	6050	500 pounds per day	↓	4	4		2,825	124		2,949	3,300	
	6350	20 Gal kettle w/steam jacket, tilting w/positive lock, S/S	L-7	7	4		5,675	112		5,787	6,400	
	6600	60 gallons	"	6	4.667		6,350	131		6,481	7,175	
	7950	Hood fire protection system, minimum	Q-1	3	5.333		2,900	165		3,065	3,425	
	8050	Maximum		1	16		21,100	495		21,595	24,000	
	8300	Refrigerators, reach-in type, 44 C.F.		5	3.200		5,075	99		5,174	5,725	
	8310	With glass doors, 68 C.F.	↓	4	4	↓	6,450	124		6,574	7,275	

11420 | Food Cooking Equipment

			CREW	DAILY OUTPUT	LABOR-HOURS	UNIT	2001 BARE COSTS MAT.	LABOR	EQUIP.	TOTAL	TOTAL INCL O&P	
550	0010	**KITCHEN VENTILATION,** Commercial										550
	1010	Heat reclaim unit, air to air										
	1100	Heat pipe exchanger										
	1110	Combined supply/exhaust air volume										
	1120	2.5 to 6.0 MCFM	Q-10	2.80	8.571	MCFM	4,725	268		4,993	5,625	
	1130	6 to 16 MCFM		5	4.800		2,900	150		3,050	3,400	
	1140	16 to 22 MCFM	↓	6	4	↓	2,250	125		2,375	2,700	
	2010	Packaged ventilation										
	2100	Combined supply/exhaust air volume										
	2110	With ambient supply										
	2120	3 to 7 MCFM	Q-10	5	4.800	MCFM	855	150		1,005	1,175	
	2130	7 to 12 MCFM		7.10	3.380		570	106		676	790	
	2140	12 to 22 MCFM	↓	10.70	2.243	↓	400	70.50		470.50	550	
	2400	With tempered supply										
	2410	3 to 7 MCFM	Q-10	5	4.800	MCFM	1,475	150		1,625	1,850	
	2420	7 to 12 MCFM		7.10	3.380		955	106		1,061	1,225	
	2430	12 to 22 MCFM	↓	10.70	2.243	↓	700	70.50		770.50	880	
	4000	Electrostatic precipitators										

11400 | Food Service Equipment

11420 | Food Cooking Equipment

		CREW	DAILY OUTPUT	LABOR-HOURS	UNIT	2001 BARE COSTS MAT.	LABOR	EQUIP.	TOTAL	TOTAL INCL O&P
4100	Smoke pollution control									
4110	Exhaust air volume									
4120	1 to 2.4 MCFM	Q-10	2.15	11.163	MCFM	5,525	350		5,875	6,625
4130	2.4 to 4.8 MCFM		2.70	8.889		6,300	278		6,578	7,350
4140	4.8 to 7.2 MCFM		3.60	6.667		6,400	209		6,609	7,375
4150	7.2 to 12 MCFM	↓	5.40	4.444	↓	6,700	139		6,839	7,600
4400	Electrostatic precipitators with built-in exhaust fans									
4410	Smoke pollution control									
4420	Exhaust air volume									
4440	1 to 2.4 MCFM	Q-10	2.15	11.163	MCFM	6,650	350		7,000	7,875
4450	2.4 to 4.8 MCFM		2.70	8.889		5,275	278		5,553	6,225
4460	4.8 to 7.2 MCFM		3.60	6.667		4,525	209		4,734	5,300
4470	7.2 to 12 MCFM	↓	5.40	4.444	↓	3,900	139		4,039	4,500
6000	Exhaust hoods									
6100	Centrifugal grease extraction									
6110	Water wash type									
6120	Per foot of length, minimum	Q-10	7.10	3.380	L.F.	1,125	106		1,231	1,425
6130	Maximum	"	7.10	3.380	"	1,925	106		2,031	2,300
6200	Non water wash type									
6210	Per foot of length	Q-10	13.40	1.791	L.F.	825	56		881	995
6400	Island style ventilator									
6410	Water wash type									
6420	Per foot of length, minimum	Q-10	7.10	3.380	L.F.	1,500	106		1,606	1,825
6430	Maximum	"	7.10	3.380	"	2,500	106		2,606	2,925
6460	Non water wash type									
6470	Per foot of length, minimum	Q-10	13.40	1.791	L.F.	650	56		706	800
6480	Maximum	"	13.40	1.791	"	1,325	56		1,381	1,525

11450 | Residential Equipment

11454 | Residential Appliances

		CREW	DAILY OUTPUT	LABOR-HOURS	UNIT	2001 BARE COSTS MAT.	LABOR	EQUIP.	TOTAL	TOTAL INCL O&P
0010	**RESIDENTIAL APPLIANCES**									
0020	Cooking range, 30" free standing, 1 oven, minimum	2 Clab	10	1.600	Ea.	249	36.50		285.50	330
0050	Maximum		4	4		1,200	91.50		1,291.50	1,475
0150	2 oven, minimum		10	1.600		1,300	36.50		1,336.50	1,475
0200	Maximum	↓	10	1.600		1,425	36.50		1,461.50	1,600
1500	Combination range, refrigerator and sink, 30" wide, minimum	L-1	2	8		700	275		975	1,175
1550	Maximum		1	16		1,400	550		1,950	2,375
1570	60" wide, average		1.40	11.429		2,325	395		2,720	3,150
1590	72" wide, average		1.20	13.333		2,625	460		3,085	3,575
1600	Office model, 48" wide		2	8		1,975	275		2,250	2,600
1620	Refrigerator and sink only	↓	2.40	6.667	↓	1,975	229		2,204	2,525
1640	Combination range, refrigerator, sink, microwave									
1660	oven and ice maker	L-1	.80	20	Ea.	3,875	690		4,565	5,275
2450	Dehumidifier, portable, automatic, 15 pint					195			195	214
2550	40 pint					226			226	249
2750	Dishwasher, built-in, 2 cycles, minimum	L-1	4	4		246	138		384	475
2800	Maximum		2	8		271	275		546	715
2950	4 or more cycles, minimum	↓	4	4	↓	242	138		380	475

11450 | Residential Equipment

11454 | Residential Appliances

		CREW	DAILY OUTPUT	LABOR-HOURS	UNIT	MAT.	LABOR	EQUIP.	TOTAL	TOTAL INCL O&P	
500	2960	Average	L-1	4	4	Ea.	320	138		458	560	500
	3000	Maximum	↓	2	8		510	275		785	975	
	3200	Dryer, automatic, minimum	L-2	3	5.333		251	137		388	490	
	3250	Maximum	"	2	8		555	205		760	930	
	3300	Garbage disposer, sink type, minimum	L-1	10	1.600		37.50	55		92.50	124	
	3350	Maximum	"	10	1.600		178	55		233	279	
	3550	Heater, electric, built-in, 1250 watt, ceiling type, minimum	1 Elec	4	2		68	68.50		136.50	177	
	3600	Maximum		3	2.667		111	91.50		202.50	259	
	3700	Wall type, minimum		4	2		96.50	68.50		165	208	
	3750	Maximum		3	2.667		128	91.50		219.50	278	
	3900	1500 watt wall type, with blower		4	2		119	68.50		187.50	233	
	3950	3000 watt	↓	3	2.667		244	91.50		335.50	405	
	4150	Hood for range, 2 speed, vented, 30" wide, minimum	L-3	5	3.200		35.50	101		136.50	194	
	4200	Maximum		3	5.333		445	168		613	750	
	4300	42" wide, minimum		5	3.200		168	101		269	340	
	4330	Custom		5	3.200		600	101		701	815	
	4350	Maximum	↓	3	5.333		730	168		898	1,050	
	4500	For ventless hood, 2 speed, add					14.05			14.05	15.50	
	4650	For vented 1 speed, deduct from maximum					27			27	30	
	4850	Humidifier, portable, 8 gallons per day					97.50			97.50	107	
	5000	15 gallons per day					157			157	173	
	5200	Icemaker, automatic, 20 lb. per day	1 Plum	7	1.143		530	39.50		569.50	645	
	5350	51 lb. per day		2	4		790	138		928	1,075	
	6400	Sump pump cellar drainer, pedestal, 1/3 H.P., molded PVC base		3	2.667		86.50	92		178.50	234	
	6450	Solid brass	↓	2	4		179	138		317	405	
	6460	Sump pump, see also division 15440-940										
	6650	Washing machine, automatic, minimum	1 Plum	3	2.667	Ea.	325	92		417	500	
	6700	Maximum	"	1	8		640	276		916	1,125	
	6900	Water heater, electric, glass lined, 30 gallon, minimum	L-1	5	3.200		221	110		331	410	
	6950	Maximum		3	5.333		305	183		488	615	
	7100	80 gallon, minimum		2	8		420	275		695	875	
	7150	Maximum	↓	1	16		580	550		1,130	1,475	
	7180	Water heater, gas, glass lined, 30 gallon, minimum	2 Plum	5	3.200		298	110		408	495	
	7220	Maximum		3	5.333		415	184		599	730	
	7260	50 gallon, minimum		2.50	6.400		430	220		650	810	
	7300	Maximum	↓	1.50	10.667	↓	600	365		965	1,225	
	7310	Water heater, see also division 15480-200										
	7350	Water softener, automatic, to 30 grains per gallon	2 Plum	5	3.200	Ea.	420	110		530	625	
	7400	To 100 grains per gallon	"	4	4		550	138		688	815	
	7450	Vent kits for dryers	1 Carp	10	.800	↓	10.30	23.50		33.80	48	

11470 | Darkroom Equipment

11471 | Darkroom Processing

		CREW	DAILY OUTPUT	LABOR-HOURS	UNIT	MAT.	LABOR	EQUIP.	TOTAL	TOTAL INCL O&P	
700	0010	**DARKROOM EQUIPMENT**										700
	0020	Developing sink, 5" deep, 24" x 48"	Q-1	2	8	Ea.	3,950	248		4,198	4,725	
	0050	48" x 52"		1.70	9.412		4,000	292		4,292	4,850	
	0200	10" deep, 24" x 48"		1.70	9.412		4,975	292		5,267	5,925	
	0250	24" x 108"		1.50	10.667		1,800	330		2,130	2,475	
	3500	Washers, round, minimum sheet 11" x 14"	↓	2	8		2,425	248		2,673	3,050	

11470 | Darkroom Equipment

		11471	Darkroom Processing	CREW	DAILY OUTPUT	LABOR-HOURS	UNIT	2001 BARE COSTS MAT.	LABOR	EQUIP.	TOTAL	TOTAL INCL O&P	
700	3550		Maximum sheet 20" x 24"	Q-1	1	16	Ea.	2,825	495		3,320	3,850	700
	3800		Square, minimum sheet 20" x 24"		1	16		2,525	495		3,020	3,525	
	3900		Maximum sheet 50" x 56"	↓	.80	20	↓	3,950	620		4,570	5,275	
	4500		Combination tank sink, tray sink, washers, with										
	4510		dry side tables, average	Q-1	.45	35.556	Ea.	7,575	1,100		8,675	10,000	

11500 | Industrial & Process Equipment

		11520	Industrial Equipment	CREW	DAILY OUTPUT	LABOR-HOURS	UNIT	2001 BARE COSTS MAT.	LABOR	EQUIP.	TOTAL	TOTAL INCL O&P	
250	0010	**DUST COLLECTION SYSTEMS** Commercial / industrial											250
	0120		Central vacuum units										
	0130		Includes stand, filters and motorized shaker										
	0200		500 CFM, 10" inlet, 2 HP	Q-20	2.40	8.333	Ea.	3,075	259		3,334	3,800	
	0220		1000 CFM, 10" inlet, 3 HP		2.20	9.091		3,225	282		3,507	3,975	
	0240		1500 CFM, 10" inlet, 5 HP		2	10		3,375	310		3,685	4,200	
	0260		3000 CFM, 13" inlet, 10 HP		1.50	13.333		8,625	415		9,040	10,100	
	0280		5000 CFM, 16" inlet, 2 @ 10 HP	↓	1	20	↓	9,125	620		9,745	11,000	
	1000		Vacuum tubing, galvanized										
	1100		2-1/8" OD, 16 ga.	Q-9	440	.036	L.F.	1.76	1.10		2.86	3.64	
	1110		2-1/2" OD, 16 ga.		420	.038		2.04	1.15		3.19	4.02	
	1120		3" OD, 16 ga.		400	.040		2.59	1.21		3.80	4.72	
	1130		3-1/2" OD, 16 ga.		380	.042		3.68	1.27		4.95	6	
	1140		4" OD, 16 ga.		360	.044		3.76	1.34		5.10	6.20	
	1150		5" OD, 14 ga.		320	.050		6.95	1.51		8.46	10	
	1160		6" OD, 14 ga.		280	.057		7.80	1.73		9.53	11.25	
	1170		8" OD, 14 ga.		200	.080		11.80	2.42		14.22	16.75	
	1180		10" OD, 12 ga.		160	.100		11.80	3.02		14.82	17.65	
	1190		12" OD, 12 ga.		120	.133		34	4.03		38.03	43	
	1200		14" OD, 12 ga.	↓	80	.200	↓	37.50	6.05		43.55	50.50	
	1940		Hose, flexible wire reinforced rubber										
	1956		3" dia.	Q-9	400	.040	L.F.	5.25	1.21		6.46	7.60	
	1960		4" dia.		360	.044		7.10	1.34		8.44	9.85	
	1970		5" dia.		320	.050		8.15	1.51		9.66	11.30	
	1980		6" dia.	↓	280	.057	↓	9.50	1.73		11.23	13.10	
	2000		90° Elbow, slip fit										
	2110		2-1/8" dia.	Q-9	70	.229	Ea.	7.35	6.90		14.25	18.75	
	2120		2-1/2" dia.		65	.246		10.30	7.45		17.75	23	
	2130		3" dia.		60	.267		13.95	8.05		22	28	
	2140		3-1/2" dia.		55	.291		17.25	8.80		26.05	32.50	
	2150		4" dia.		50	.320		21.50	9.65		31.15	38.50	
	2160		5" dia.		45	.356		40.50	10.75		51.25	61	
	2170		6" dia.		40	.400		56	12.10		68.10	80	
	2180		8" dia.	↓	30	.533	↓	106	16.10		122.10	142	
	2400		45° Elbow, slip fit										
	2410		2-1/8" dia.	Q-9	70	.229	Ea.	6.45	6.90		13.35	17.75	
	2420		2-1/2" dia.		65	.246		9.50	7.45		16.95	22	
	2430		3" dia.		60	.267		11.45	8.05		19.50	25	
	2440		3-1/2" dia.		55	.291		14.40	8.80		23.20	29.50	
	2450		4" dia.	↓	50	.320	↓	18.70	9.65		28.35	35.50	

11500 | Industrial & Process Equipment

11520 | Industrial Equipment

		CREW	DAILY OUTPUT	LABOR-HOURS	UNIT	2001 BARE COSTS				TOTAL INCL O&P
						MAT.	LABOR	EQUIP.	TOTAL	
2460	5" dia.	Q-9	45	.356	Ea.	31.50	10.75		42.25	51
2470	6" dia.		40	.400		42.50	12.10		54.60	65.50
2480	8" dia.	▼	35	.457	▼	83.50	13.80		97.30	114
2800	90° TY, slip fit thru 6" dia									
2810	2-1/8" dia.	Q-9	42	.381	Ea.	14.40	11.50		25.90	33.50
2820	2-1/2" dia.		39	.410		18.70	12.40		31.10	39.50
2830	3" dia.		36	.444		24.50	13.40		37.90	47.50
2840	3-1/2" dia.		33	.485		31.50	14.65		46.15	57
2850	4" dia.		30	.533		45.50	16.10		61.60	75
2860	5" dia.		27	.593		86	17.90		103.90	122
2870	6" dia.	▼	24	.667		117	20		137	160
2880	8" dia., butt end					249			249	274
2890	10" dia., butt end					450			450	495
2900	12" dia., butt end					450			450	495
2910	14" dia., butt end					815			815	895
2920	6 x 4" dia., butt end					71.50			71.50	78.50
2930	8" x 4" dia., butt end					135			135	149
2940	10" x 4" dia., butt end					177			177	195
2950	12" x 4" dia., butt end				▼	204			204	224
3100	90° Elbow, butt end segmented									
3110	8" dia., butt end, segmented				Ea.	161			161	177
3120	10" dia., butt end, segmented					264			264	291
3130	12" dia., butt end, segmented					335			335	370
3140	14" dia., butt end, segmented				▼	420			420	460
3200	45° Elbow, butt end segmented									
3210	8" dia., butt end, segmented				Ea.	137			137	151
3220	10" dia., butt end, segmented					191			191	210
3230	12" dia., butt end, segmented					195			195	215
3240	14" dia., butt end, segmented				▼	240			240	264
3400	All butt end fittings require one coupling per joint.									
3410	Labor for fitting included with couplings.									
3460	Compression coupling, galvanized, neoprene gasket									
3470	2-1/8" dia.	Q-9	44	.364	Ea.	11.20	11		22.20	29.50
3480	2-1/2" dia.		44	.364		11.20	11		22.20	29.50
3490	3" dia.		38	.421		14.70	12.70		27.40	36
3500	3-1/2" dia.		35	.457		16.45	13.80		30.25	39.50
3510	4" dia.		33	.485		17.90	14.65		32.55	42
3520	5" dia.		29	.552		20.50	16.65		37.15	48
3530	6" dia.		26	.615		24	18.60		42.60	54.50
3540	8" dia.		22	.727		43.50	22		65.50	81.50
3550	10" dia.		20	.800		63.50	24		87.50	108
3560	12" dia.		18	.889		79	27		106	129
3570	14" dia.	▼	16	1	▼	112	30		142	170
3800	Air gate valves, galvanized									
3810	2-1/8" dia.	Q-9	30	.533	Ea.	85	16.10		101.10	119
3820	2-1/2" dia.		28	.571		89.50	17.25		106.75	125
3830	3" dia.		26	.615		104	18.60		122.60	143
3840	4" dia.		23	.696		136	21		157	183
3850	6" dia.	▼	18	.889	▼	188	27		215	248

11600 | Laboratory Equipment

11620 | Laboratory Equipment

		CREW	DAILY OUTPUT	LABOR-HOURS	UNIT	MAT.	LABOR	EQUIP.	TOTAL	TOTAL INCL O&P
350	0010 **LABORATORY EQUIPMENT**									350
	0020 Cabinets, base, door units, metal	2 Carp	18	.889	L.F.	127	26		153	181
	0300 Drawer units	"	18	.889	"	275	26		301	345
	2000 Fume hood, with countertop & base, not including HVAC									
	2020 Simple, minimum	2 Carp	5.40	2.963	L.F.	505	86.50		591.50	690
	2050 Complex, including fixtures		2.40	6.667		1,200	194		1,394	1,600
	2100 Special, maximum	↓	1.70	9.412	↓	1,250	274		1,524	1,800
	2200 Ductwork, minimum	2 Shee	1	16	Hood	835	535		1,370	1,750
	2250 Maximum	"	.50	32	"	4,225	1,075		5,300	6,300
	2300 Service fixtures, average				Ea.	43			43	47
	2500 For sink assembly with hot and cold water, add	1 Plum	1.40	5.714		525	197		722	870
	2550 Glassware washer, undercounter, minimum	L-1	1.80	8.889		6,200	305		6,505	7,275
	2600 Maximum	"	1	16		7,525	550		8,075	9,100
	2800 Sink, one piece plastic, flask wash, hose, free standing	1 Plum	1.60	5		1,325	172		1,497	1,700
	2850 Epoxy resin sink, 25" x 16" x 10"	"	2	4	↓	147	138		285	370
	4200 Alternate pricing method: as percent of lab furniture									
	4400 Installation, not incl. plumbing & duct work				% Furn.					22%
	4800 Plumbing, final connections, simple system									10%
	5000 Moderately complex system									15%
	5200 Complex system									20%
	5400 Electrical, simple system									10%
	5600 Moderately complex system									20%
	5800 Complex system				↓					35%
	6000 Safety equipment, eye wash, hand held				Ea.	210			210	231
	6200 Deluge shower				"	500			500	550
	6300 Rule of thumb: lab furniture including installation & connection									
	6320 High school				S.F.					29
	6340 College									42.75
	6360 Clinical, health care									36.75
	6380 Industrial				↓					59.25

11700 | Medical Equipment

11710 | Medical Equipment

		CREW	DAILY OUTPUT	LABOR-HOURS	UNIT	MAT.	LABOR	EQUIP.	TOTAL	TOTAL INCL O&P
500	0010 **MEDICAL EQUIPMENT**									500
	0015 Autopsy table, standard	1 Plum	1	8	Ea.	6,500	276		6,776	7,575
	0020 Deluxe	↓	.60	13.333		13,000	460		13,460	15,100
	0700 Distiller, water, steam heated, 50 gal. capacity		1.40	5.714		13,500	197		13,697	15,200
	1800 Heat therapy unit, humidified, 26" x 78" x 28"					2,475			2,475	2,725
	3200 Mortuary refrigerator, end operated, 2 capacity					10,300			10,300	11,400
	3300 6 capacity					19,200			19,200	21,100
	5600 Sterilizers, floor loading, 26" x 62" x 42", single door, steam					118,000			118,000	129,500
	5650 Double door, steam					151,500			151,500	166,500
	5800 General purpose, 20" x 20" x 38", single door					11,400			11,400	12,500
	6000 Portable, counter top, steam, minimum					2,825			2,825	3,100
	6020 Maximum					4,425			4,425	4,850
	6050 Portable, counter top, gas, 17"x15"x32-1/2"					29,200			29,200	32,100
	6150 Manual washer/sterilizer, 16"x16"x26"	1 Plum	2	4	↓	40,000	138		40,138	44,200
	6200 Steam generators, electric 10 KW to 180 KW, freestanding									
	6250 Minimum	1 Elec	3	2.667	Ea.	5,950	91.50		6,041.50	6,675

11700 | Medical Equipment

11710 | Medical Equipment

			CREW	DAILY OUTPUT	LABOR-HOURS	UNIT	2001 BARE COSTS MAT.	LABOR	EQUIP.	TOTAL	TOTAL INCL O&P	
500	6300	Maximum	1 Elec	.70	11.429	Ea.	21,400	390		21,790	24,200	500
	8100	Utensil washer-sanitizer	1 Plum	2	4		8,100	138		8,238	9,125	
	8200	Bed pan washer-sanitizer	"	2	4		5,300	138		5,438	6,025	
	8400	Whirlpool bath, mobile, sst, 18" x 24" x 60"					3,500			3,500	3,850	
	8450	Fixed, incl. mixing valves	1 Plum	2	4	↓	6,850	138		6,988	7,725	

11740 | Dental Equipment

			CREW	DAILY OUTPUT	LABOR-HOURS	UNIT	MAT.	LABOR	EQUIP.	TOTAL	INCL O&P	
200	0010	**DENTAL EQUIPMENT**										200
	0020	Central suction system, minimum	1 Plum	1.20	6.667	Ea.	2,700	230		2,930	3,325	
	0100	Maximum	"	.90	8.889		20,900	305		21,205	23,500	
	0600	Chair, electric or hydraulic, minimum	1 Skwk	.50	16		3,400	480		3,880	4,500	
	0700	Maximum		.25	32		7,200	955		8,155	9,425	
	2000	Light, ceiling mounted, minimum		8	1		1,775	30		1,805	2,000	
	2100	Maximum	↓	8	1		3,575	30		3,605	3,975	
	2200	Unit light, minimum	2 Skwk	5.33	3.002		1,325	89.50		1,414.50	1,600	
	2210	Maximum		5.33	3.002		4,325	89.50		4,414.50	4,900	
	2220	Track light, minimum		3.20	5		2,350	149		2,499	2,800	
	2230	Maximum	↓	3.20	5		5,150	149		5,299	5,900	
	2300	Sterilizers, steam portable, minimum					2,975			2,975	3,275	
	2350	Maximum					10,300			10,300	11,300	
	2600	Steam, institutional					12,800			12,800	14,100	
	2650	Dry heat, electric, portable, 3 trays				↓	1,000			1,000	1,100	

For information about Means Estimating Seminars, see yellow pages 11 and 12 in back of book

Division Notes

		CREW	DAILY OUTPUT	LABOR-HOURS	UNIT	2001 BARE COSTS				TOTAL INCL O&P
						MAT.	LABOR	EQUIP.	TOTAL	

Division 13
Special Construction

Estimating Tips
General
- The items and systems in this division are usually estimated, purchased, supplied and installed as a unit by one or more subcontractors. The estimator must ensure that all parties are operating from the same set of specifications and assumptions and that all necessary items are estimated and will be provided. Many times the complex items and systems are covered but the more common ones such as excavation or a crane are overlooked for the very reason that everyone assumes nobody could miss them. The estimator should be the central focus and be able to ensure that all systems are complete.
- Another area where problems can develop in this division is at the interface between systems. The estimator must ensure, for instance, that anchor bolts, nuts and washers are estimated and included for the air-supported structures and pre-engineered buildings to be bolted to their foundations. Utility supply is a common area where essential items or pieces of equipment can be missed or overlooked due to the fact that each subcontractor may feel it is the others' responsibility. The estimator should also be aware of certain items which may be supplied as part of a package but installed by others, and ensure that the installing contractor's estimate includes the cost of installation. Conversely, the estimator must also ensure that items are not costed by two different subcontractors, resulting in an inflated overall estimate.

13120 Pre-Engineered Structures
- The foundations and floor slab, as well as rough mechanical and electrical, should be estimated, as this work is required for the assembly and erection of the structure. Generally, as noted in the book, the pre-engineered building comes as a shell and additional features must be included by the estimator. Here again, the estimator must have a clear understanding of the scope of each portion of the work and all the necessary interfaces.

13200 Storage Tanks
- The prices in this subdivision for above and below ground storage tanks do not include foundations or hold-down slabs. The estimator should refer to Divisions 2 and 3 for foundation system pricing. In addition to the foundations, required tank accessories such as tank gauges, leak detection devices, and additional manholes and piping must be added to the tank prices.

Reference Numbers
Reference numbers are shown in bold squares at the beginning of some major classifications. These numbers refer to related items in the Reference Section. The reference information may be an estimating procedure, an alternate pricing method or technical information.

Note: Not all subdivisions listed here necessarily appear in this publication.

13030 | Special Purpose Rooms

13035 | Special Purpose Rooms

		CREW	DAILY OUTPUT	LABOR-HOURS	UNIT	2001 BARE COSTS MAT.	LABOR	EQUIP.	TOTAL	TOTAL INCL O&P
100	0010 **INTEGRATED CEILINGS** Lighting, ventilating & acoustical										100
	0100 Luminaire, incl. connections, 5' x 5' modules, 50% lighted	L-3	90	.178	S.F.	3.53	5.60		9.13	12.55
	0200 100% lighted	"	50	.320		5.25	10.10		15.35	21.50
	0400 For ventilating capacity with perforations, add				↓					.27
	0600 For supply air diffuser, add				Ea.					64
	0700 Dimensionaire, 2' x 4' board system, see also div. 09510-760	L-3	50	.320	L.F.	8.20	10.10		18.30	24.50
	0900 Tile system	1 Carp	250	.032	S.F.	2.14	.93		3.07	3.81
	1000 For air bar suspension, add, minimum									.61
	1100 Maximum				↓					1.05
	1200 For vaulted coffer, including fixture, stock, add				Ea.	80			80	88
	1300 Custom, add, minimum									133
	1400 Average									345
	1500 Maximum				↓					575
	1800 Radiant hot water system with finished acoustic ceiling,									
	1810 not including supply piping. Heating only (gross S.F.)									
	2100 Elementary schools, minimum				S.F.					5.30
	2200 Maximum									6.65
	2400 High schools and colleges, minimum									4.75
	2500 Maximum									6.65
	2700 Libraries, minimum									4.80
	2800 Maximum									5.75
	3000 Hospitals, minimum									6.50
	3100 Maximum									8.15
	3300 Office buildings, minimum									4.60
	3400 Maximum									6.15
	3600 For combined heating and cooling, add, minimum									30%
	3700 Maximum									40%
	4000 Radiant electric ceiling board, strapped between joists	1 Elec	250	.032		1.74	1.10		2.84	3.55
	4100 2' x 4' heating panel for grid system, manila finish		25	.320	Ea.	80	11		91	104
	4200 Textured epoxy finish		22	.364		82	12.45		94.45	109
	4300 Vinyl finish		19	.421		84	14.45		98.45	114
	4400 Hair cell, ABS plastic finish	↓	13	.615	↓	94	21		115	135
	4500 2' x 4' alternate blank panel, for use with above									
	4600 Manila finish	1 Elec	50	.160	Ea.	25	5.50		30.50	35.50
	4700 Textured epoxy finish		45	.178		30	6.10		36.10	42
	4800 Vinyl finish		40	.200		32	6.85		38.85	45.50
	4900 Hair cell, ABS plastic finish	↓	25	.320	↓	64	11		75	87
200	0010 **CLEAN ROOMS**										200
	1100 Clean room, soft wall, 12' x 12', Class 100	1 Carp	.18	44.444	Ea.	12,200	1,300		13,500	15,400
	1110 Class 1,000		.18	44.444		9,450	1,300		10,750	12,400
	1120 Class 10,000		.21	38.095		8,050	1,100		9,150	10,600
	1130 Class 100,000	↓	.21	38.095	↓	8,050	1,100		9,150	10,600
	2800 Ceiling grid support, slotted channel struts 4'-0" O.C., ea. way				S.F.					6.50
	3000 Ceiling panel, vinyl coated foil on mineral substrate									
	3020 Sealed, non-perforated				S.F.					1.40
	4000 Ceiling panel seal, silicone sealant, 150 L.F./gal.	1 Carp	150	.053	L.F.	.22	1.55		1.77	2.67
	4100 Two sided adhesive tape	"	240	.033	"	.10	.97		1.07	1.63
	4200 Clips, one per panel				Ea.	.85			.85	.94
	6000 HEPA filter, 2'x4', 99.97% eff., 3" dp beveled frame(silicone seal)					260			260	286
	6040 6" deep skirted frame (channel seal)					300			300	330
	6100 99.99% efficient, 3" deep beveled frame (silicone seal)					280			280	310
	6140 6" deep skirted frame (channel seal)					320			320	350
	6200 99.999% efficient, 3" deep beveled frame (silicone seal)					320			320	350
	6240 6" deep skirted frame (channel seal)				↓	360			360	395
	7000 Wall panel systems, including channel strut framing									

13030 | Special Purpose Rooms

13035 | Special Purpose Rooms

			Crew	Daily Output	Labor-Hours	Unit	Mat.	Labor	Equip.	Total	Total Incl O&P	
200	7020	Polyester coated aluminum, particle board				S.F.					20	200
	7100	Porcelain coated aluminum, particle board									35	
	7400	Wall panel support, slotted channel struts, to 12' high				↓					18	
700	0010	**REFRIGERATION** Curbs, 12" high, 4" thick, concrete	2 Carp	58	.276	L.F.	3.02	8.05		11.07	15.90	700
	1000	Doors, see division 08341-200										
	2400	Finishes, 2 coat portland cement plaster, 1/2" thick	1 Plas	48	.167	S.F.	.83	4.55		5.38	7.85	
	2500	For galvanized reinforcing mesh, add	1 Lath	335	.024		.56	.67		1.23	1.63	
	2700	3/16" thick latex cement	1 Plas	88	.091		1.42	2.48		3.90	5.35	
	2900	For glass cloth reinforced ceilings, add	"	450	.018		.34	.49		.83	1.11	
	3100	Fiberglass panels, 1/8" thick	1 Carp	149.45	.054		1.88	1.56		3.44	4.51	
	3200	Polystyrene, plastic finish ceiling, 1" thick		274	.029		1.74	.85		2.59	3.24	
	3400	2" thick		274	.029		1.97	.85		2.82	3.50	
	3500	4" thick	↓	219	.037		2.20	1.06		3.26	4.09	
	3800	Floors, concrete, 4" thick	1 Cefi	93	.086		.82	2.40		3.22	4.47	
	3900	6" thick	"	85	.094	↓	1.24	2.63		3.87	5.25	
	4000	Insulation, 1" to 6" thick, cork				B.F.	.80			.80	.88	
	4100	Urethane					.79			.79	.87	
	4300	Polystyrene, regular					.54			.54	.59	
	4400	Bead board				↓	.41			.41	.45	
	4600	Installation of above, add per layer	2 Carp	657.60	.024	S.F.	.25	.71		.96	1.39	
	4700	Wall and ceiling juncture		298.90	.054	L.F.	1.28	1.56		2.84	3.85	
	4900	Partitions, galvanized sandwich panels, 4" thick, stock		219.20	.073	S.F.	5.40	2.13		7.53	9.30	
	5000	Aluminum or fiberglass	↓	219.20	.073	"	5.95	2.13		8.08	9.85	
	5200	Prefab walk-in, 7'-6" high, aluminum, incl. door & floors,										
	5210	not incl. partitions or refrigeration, 6' x 6' O.D. nominal	2 Carp	54.80	.292	SF Flr.	97	8.50		105.50	120	
	5500	10' x 10' O.D. nominal		82.20	.195		78	5.65		83.65	95	
	5700	12' x 14' O.D. nominal		109.60	.146		70	4.26		74.26	83.50	
	5800	12' x 20' O.D. nominal	↓	109.60	.146		61	4.26		65.26	73.50	
	6100	For 8'-6" high, add					5%					
	6300	Rule of thumb for complete units, w/o doors & refrigeration, cooler	2 Carp	146	.110		88	3.19		91.19	102	
	6400	Freezer		109.60	.146		104	4.26		108.26	121	
	6600	Shelving, plated or galvanized, steel wire type		360	.044	SF Hor.	7.70	1.30		9	10.50	
	6700	Slat shelf type	↓	375	.043		9.50	1.24		10.74	12.40	
	6900	For stainless steel shelving, add				↓	300%					
	7000	Vapor barrier, on wood walls	2 Carp	1,644	.010	S.F.	.11	.28		.39	.56	
	7200	On masonry walls	"	1,315	.012	"	.29	.35		.64	.88	
	7500	For air curtain doors, see division 08344-120										
800	0010	**SAUNA** Prefabricated, incl. heater & controls, 7' high, 6' x 4', C/C	L-7	2.20	12.727	Ea.	3,200	360		3,560	4,075	800
	0050	6' x 4', C/P		2	14		2,525	395		2,920	3,375	
	0400	6' x 5', C/C		2	14		3,225	395		3,620	4,125	
	0450	6' x 5', C/P		2	14		2,875	395		3,270	3,750	
	0600	6' x 6', C/C		1.80	15.556		3,800	435		4,235	4,850	
	0650	6' x 6', C/P		1.80	15.556		3,000	435		3,435	3,975	
	0800	6' x 9', C/C		1.60	17.500		4,175	490		4,665	5,350	
	0850	6' x 9', C/P		1.60	17.500		4,175	490		4,665	5,350	
	1000	8' x 12', C/C		1.10	25.455		7,350	715		8,065	9,175	
	1050	8' x 12', C/P		1.10	25.455		5,775	715		6,490	7,450	
	1200	8' x 8', C/C		1.40	20		4,925	560		5,485	6,300	
	1250	8' x 8', C/P		1.40	20		4,350	560		4,910	5,650	
	1400	8' x 10', C/C		1.20	23.333		5,400	655		6,055	6,950	
	1450	8' x 10', C/P		1.20	23.333		4,725	655		5,380	6,225	
	1600	10' x 12', C/C		1	28		7,200	785		7,985	9,125	
	1650	10' x 12', C/P	↓	1	28		6,375	785		7,160	8,250	
	2500	Heaters only (incl. above), wall mounted, to 200 C.F.					505			505	555	
	2750	To 300 C.F.				↓	530			530	580	

13030 | Special Purpose Rooms

13035 | Special Purpose Rooms

			CREW	DAILY OUTPUT	LABOR-HOURS	UNIT	2001 BARE COSTS MAT.	LABOR	EQUIP.	TOTAL	TOTAL INCL O&P	
800	3000	Floor standing, to 720 C.F., 10,000 watts, w/controls	1 Elec	3	2.667	Ea.	925	91.50		1,016.50	1,150	800
	3250	To 1,000 C.F., 16,000 watts	"	3	2.667	↓	1,575	91.50		1,666.50	1,850	
940	0010	**STEAM BATH** Heater, timer & head, single, to 140 C.F.	1 Plum	1.20	6.667	Ea.	840	230		1,070	1,275	940
	0500	To 300 C.F.		1.10	7.273		970	251		1,221	1,450	
	1000	Commercial size, with blow-down assembly, to 800 C.F.		.90	8.889		3,800	305		4,105	4,625	
	1500	To 2500 C.F.	↓	.80	10		4,000	345		4,345	4,925	
	2000	Multiple, motels, apts., 2 baths, w/ blow-down assm., 500 C.F.	Q-1	1.30	12.308		3,025	380		3,405	3,900	
	2500	4 baths	"	.70	22.857		3,675	710		4,385	5,100	
	2700	Conversion unit for residential tub, including door				↓	1,775			1,775	1,950	

13120 | Pre-Engineered Structures

13128 | Pre-Eng. Structures

			CREW	DAILY OUTPUT	LABOR-HOURS	UNIT	2001 BARE COSTS MAT.	LABOR	EQUIP.	TOTAL	TOTAL INCL O&P	
200	0010	**COMFORT STATIONS** Prefab., stock, w/doors, windows & fixt.										200
	0100	Not incl. interior finish or electrical										
	0300	Mobile, on steel frame, minimum				S.F.	37.50			37.50	41	
	0350	Maximum					62.50			62.50	69	
	0400	Permanent, including concrete slab, minimum	B-12J	50	.320		138	9.10	12.95	160.05	180	
	0500	Maximum	"	43	.372	↓	200	10.60	15.10	225.70	253	
	0600	Alternate pricing method, mobile, minimum				Fixture	1,725			1,725	1,900	
	0650	Maximum					2,575			2,575	2,825	
	0700	Permanent, minimum	B-12J	.70	22.857		9,900	650	925	11,475	12,900	
	0750	Maximum	"	.50	32	↓	16,500	910	1,300	18,710	21,000	

13150 | Swimming Pools

13151 | Swimming Pools

			CREW	DAILY OUTPUT	LABOR-HOURS	UNIT	2001 BARE COSTS MAT.	LABOR	EQUIP.	TOTAL	TOTAL INCL O&P	
200	0010	**SWIMMING POOLS** Residential in-ground, vinyl lined, concrete sides										200
	0020	Sides including equipment, sand bottom	B-52	300	.187	SF Surf	10.40	5	1.28	16.68	20.50	
	0100	Metal or polystyrene sides	B-14	410	.117		8.70	2.83	.44	11.97	14.45	
	0200	Add for vermiculite bottom				↓	.67			.67	.74	
	0500	Gunite bottom and sides, white plaster finish										
	0600	12' x 30' pool	B-52	145	.386	SF Surf	17.25	10.40	2.66	30.31	38	
	0720	16' x 32' pool		155	.361		15.50	9.70	2.49	27.69	35	
	0750	20' x 40' pool	↓	250	.224	↓	13.85	6.05	1.54	21.44	26.50	
	0810	Concrete bottom and sides, tile finish										
	0820	12' x 30' pool	B-52	80	.700	SF Surf	17.40	18.85	4.82	41.07	54	
	0830	16' x 32' pool		95	.589		14.40	15.85	4.06	34.31	45.50	
	0840	20' x 40' pool	↓	130	.431	↓	11.45	11.60	2.96	26.01	34	
	1100	Motel, gunite with plaster finish, incl. medium										
	1150	capacity filtration & chlorination	B-52	115	.487	SF Surf	21	13.10	3.35	37.45	47.50	
	1200	Municipal, gunite with plaster finish, incl. high										
	1250	capacity filtration & chlorination	B-52	100	.560	SF Surf	27.50	15.05	3.85	46.40	57.50	

13150 | Swimming Pools

13151 | Swimming Pools

		CREW	DAILY OUTPUT	LABOR-HOURS	UNIT	2001 BARE COSTS MAT.	LABOR	EQUIP.	TOTAL	TOTAL INCL O&P	
200	1350 Add for formed gutters				L.F.	47.50			47.50	52.50	200
	1360 Add for stainless steel gutters				"	141			141	155	
	1700 Filtration and deck equipment only, as % of total				Total				20%	20%	
	1800 Deck equipment, rule of thumb, 20' x 40' pool				SF Pool					1.30	
	1900 5000 S.F. pool				"					1.90	
	3000 Painting pools, preparation + 3 coats, 20' x 40' pool, epoxy	2 Pord	.33	48.485	Total	595	1,275		1,870	2,575	
	3100 Rubber base paint, 18 gallons	"	.33	48.485		450	1,275		1,725	2,425	
	3500 42' x 82' pool, 75 gallons, epoxy paint	3 Pord	.14	171		2,500	4,475		6,975	9,575	
	3600 Rubber base paint	"	.14	171	↓	1,925	4,475		6,400	8,950	
700	0010 SWIMMING POOL EQUIPMENT Diving stand, stainless steel, 3 meter	2 Carp	.40	40	Ea.	4,625	1,175		5,800	6,925	700
	0300 1 meter	"	2.70	5.926	"	3,750	173		3,923	4,400	
	0900 Filter system, sand or diatomite type, incl. pump, 6,000 gal./hr.	2 Plum	1.80	8.889	Total	1,050	305		1,355	1,600	
	1020 Add for chlorination system, 800 S.F. pool		3	5.333	Ea.	196	184		380	495	
	1040 5,000 S.F. pool	↓	3	5.333	"	1,150	184		1,334	1,525	
	1100 Gutter system, stainless steel, with grating, stock,										
	1110 contains supply and drainage system	E-1	20	1.200	L.F.	160	38.50	4.20	202.70	247	
	1120 Integral gutter and 5' high wall system, stainless steel	"	10	2.400	"	240	77	8.40	325.40	405	
	2100 Lights, underwater, 12 volt, with transformer, 300 watt	1 Elec	.40	20	Ea.	116	685		801	1,150	
	2200 110 volt, 500 watt, standard		.40	20		124	685		809	1,150	
	2400 Low water cutoff type	↓	.40	20	↓	128	685		813	1,175	
	2800 Heaters, see division 15510-880										

13175 | Ice Rinks

13176 | Ice Rinks

		CREW	DAILY OUTPUT	LABOR-HOURS	UNIT	2001 BARE COSTS MAT.	LABOR	EQUIP.	TOTAL	TOTAL INCL O&P	
500	0010 ICE SKATING Equipment incl. refrigeration, plumbing & cooling										500
	0020 coils & concrete slab, 85' x 200' rink										
	0300 55° system, 5 mos., 100 ton				Total					451,000	
	0700 90° system, 12 mos., 135 ton				"					500,500	
	1000 Dasher boards, 1/2" H.D. polyethylene faced steel frame, 3' acrylic										
	1020 screen at sides, 5' acrylic ends, 85' x 200'	F-5	.06	533	Ea.	115,000	15,800		130,800	151,000	
	1100 Fiberglass & aluminum construction, same sides and ends	"	.06	533		160,000	15,800		175,800	200,500	
	1200 Subsoil heating system (recycled from compressor), 85' x 200'	Q-7	.27	118	↓	25,000	3,950		28,950	33,500	
	1300 Subsoil insulation, 2 lb. polystyrene with vapor barrier, 85' x 200'	2 Carp	.14	114	↓	30,000	3,325		33,325	38,200	

13200 | Storage Tanks

13201 | Storage Tanks

		CREW	DAILY OUTPUT	LABOR-HOURS	UNIT	2001 BARE COSTS MAT.	LABOR	EQUIP.	TOTAL	TOTAL INCL O&P	
200	0010 ELEVATED STORAGE TANKS Not incl. pipe or pumps										200
	3000 Elevated water tanks, 100' to bottom capacity line, incl painting										
	3010 50,000 gallons				Ea.					166,000	
	3300 100,000 gallons				↓					229,000	

13200 | Storage Tanks

13201	Storage Tanks	CREW	DAILY OUTPUT	LABOR-HOURS	UNIT	2001 BARE COSTS MAT.	LABOR	EQUIP.	TOTAL	TOTAL INCL O&P
3400	250,000 gallons				Ea.					316,000
3600	500,000 gallons									510,000
3700	750,000 gallons									703,000
3900	1,000,000 gallons									812,000
0010	**GROUND TANKS** Not incl. pipe or pumps, prestress conc., 250,000 gal.				Ea.					280,000
0100	500,000 gallons									380,000
0300	1,000,000 gallons									540,000
0400	2,000,000 gallons									816,000
0600	4,000,000 gallons									1,290,000
0700	6,000,000 gallons									1,760,000
0750	8,000,000 gallons									2,240,000
0800	10,000,000 gallons									2,700,000
0900	Steel, ground level, ht/dia less than 1, not incl. fdn, 100,000 gallons									98,000
1000	250,000 gallons									137,000
1200	500,000 gallons									209,000
1250	750,000 gallons									249,000
1300	1,000,000 gallons									336,000
1500	2,000,000 gallons									615,000
1600	4,000,000 gallons									955,000
1800	6,000,000 gallons									1,325,000
1850	8,000,000 gallons									1,735,000
1900	10,000,000 gallons									1,937,500
2100	Steel standpipes, hgt/dia more than 1, 100' to overflow, no fdn									
2200	500,000 gallons				Ea.					275,000
2400	750,000 gallons									335,000
2500	1,000,000 gallons									408,000
2700	1,500,000 gallons									570,000
2800	2,000,000 gallons									700,000
3000	Steel, storage, above ground, including cradles, coating,									
3020	fittings, not including fdn, pumps or piping									
3040	Single wall, interior, 275 gallon	Q-5	5	3.200	Ea.	236	100		336	410
3060	550 gallon	"	2.70	5.926		825	186		1,011	1,175
3080	1,000 gallon	Q-7	5	6.400		1,050	213		1,263	1,475
3100	1,500 gallon		4.75	6.737		1,975	224		2,199	2,525
3120	2,000 gallon		4.60	6.957		2,000	231		2,231	2,550
3140	5,000 gallon		3.20	10		4,200	330		4,530	5,125
3150	10,000 gallon		2	16		6,900	530		7,430	8,400
3160	15,000 gallon		1.70	18.824		9,650	625		10,275	11,500
3170	20,000 gallon		1.45	22.069		11,300	735		12,035	13,500
3180	25,000 gallon capacity		1.30	24.615		14,100	820		14,920	16,700
3190	30,000 gallon		1.10	29.091		17,000	965		17,965	20,200
3320	Double wall, 500 gallon capacity	Q-5	2.40	6.667		2,625	209		2,834	3,200
3330	2000 gallon capacity	Q-7	4.15	7.711		4,800	256		5,056	5,650
3340	4000 gallon capacity		3.60	8.889		8,525	295		8,820	9,825
3350	6000 gallon capacity		2.40	13.333		10,200	445		10,645	11,900
3360	8000 gallon capacity		2	16		11,600	530		12,130	13,600
3370	10000 gallon capacity		1.80	17.778		12,800	590		13,390	15,000
3380	15000 gallon capacity		1.50	21.333		19,100	710		19,810	22,100
3390	20000 gallon capacity		1.30	24.615		21,700	820		22,520	25,100
3400	25000 gallon capacity		1.15	27.826		26,400	925		27,325	30,400
3410	30000 gallon capacity		1	32		28,900	1,075		29,975	33,400
4000	Fixed roof oil storage tanks, steel, (1 BBL=42 GAL w/ fdn 3'D x 1'W)									
4200	5,000 barrels				Ea.					144,000
4300	25,000 barrels									202,000
4500	55,000 barrels									342,000
4600	100,000 barrels									550,000

13200 | Storage Tanks

13201 | Storage Tanks

		Crew	Daily Output	Labor-Hours	Unit	2001 Bare Costs Mat.	Labor	Equip.	Total	Total Incl O&P
4800	150,000 barrels				Ea.					775,000
4900	225,000 barrels									1,168,000
5100	Floating roof gasoline tanks, steel, 5,000 barrels									122,000
5200	25,000 barrels									273,000
5400	55,000 barrels									413,000
5500	100,000 barrels									625,000
5700	150,000 barrels									846,000
5800	225,000 barrels									1,173,000
6000	Wood tanks, ground level, 2" cypress, 3,000 gallons	C-1	.19	168		5,775	4,650		10,425	13,600
6100	2-1/2" cypress, 10,000 gallons		.12	266		15,600	7,350		22,950	28,700
6300	3" redwood or 3" fir, 20,000 gallons		.10	320		23,800	8,825		32,625	40,000
6400	30,000 gallons		.08	400		29,500	11,000		40,500	49,700
6600	45,000 gallons		.07	457		44,600	12,600		57,200	69,000
6700	Larger sizes, minimum				Gal.					.66
6900	Maximum				"					.81
7000	Vinyl coated fabric pillow tanks, freestanding, 5,000 gallons	4 Clab	4	8	Ea.	1,900	183		2,083	2,375
7100	Supporting embankment not included, 25,000 gallons	6 Clab	2	24		6,700	550		7,250	8,225
7200	50,000 gallons	8 Clab	1.50	42.667		9,750	975		10,725	12,200
7300	100,000 gallons	9 Clab	.90	80		15,100	1,825		16,925	19,500
7400	150,000 gallons		.50	144		19,100	3,300		22,400	26,200
7500	200,000 gallons		.40	180		22,300	4,125		26,425	30,900
7600	250,000 gallons		.30	240		26,800	5,475		32,275	38,100
0010	**UNDERGROUND STORAGE TANKS**									
0210	Fiberglass, underground, single wall, U.L. listed, not including									
0220	manway or hold-down strap									
0225	550 gallon capacity	Q-5	2.67	5.993	Ea.	1,450	188		1,638	1,875
0230	1,000 gallon capacity	"	2.46	6.504		1,825	204		2,029	2,300
0240	2,000 gallon capacity	Q-7	4.57	7.002		2,375	233		2,608	2,950
0250	4,000 gallon capacity		3.55	9.014		3,425	300		3,725	4,225
0260	6,000 gallon capacity		2.67	11.985		3,825	400		4,225	4,800
0270	8,000 gallon capacity		2.29	13.974		4,800	465		5,265	5,975
0280	10,000 gallon capacity		2	16		5,200	530		5,730	6,525
0282	12,000 gallon capacity		1.88	17.021		7,100	565		7,665	8,650
0284	15,000 gallon capacity		1.68	19.048		7,575	635		8,210	9,300
0290	20,000 gallon capacity		1.45	22.069		10,500	735		11,235	12,600
0300	25,000 gallon capacity		1.28	25		17,400	830		18,230	20,400
0320	30,000 gallon capacity		1.14	28.070		21,700	935		22,635	25,300
0340	40,000 gallon capacity		.89	35.955		30,500	1,200		31,700	35,300
0360	48,000 gallon capacity		.81	39.506		50,000	1,325		51,325	57,000
0500	For manway, fittings and hold-downs, add					20%	15%			
0600	For manways, add					840			840	925
1000	For helical heating coil, add	Q-5	2.50	6.400		2,125	201		2,326	2,650
1020	Fiberglass, underground, double wall, U.L. listed									
1030	includes manways, not incl. hold-down straps									
1040	600 gallon capacity	Q-5	2.42	6.612	Ea.	3,200	207		3,407	3,825
1050	1,000 gallon capacity	"	2.25	7.111		4,250	223		4,473	5,000
1060	2,500 gallon capacity	Q-7	4.16	7.692		6,050	256		6,306	7,025
1070	3,000 gallon capacity		3.90	8.205		6,500	273		6,773	7,550
1080	4,000 gallon capacity		3.64	8.791		7,100	292		7,392	8,250
1090	6,000 gallon capacity		2.42	13.223		8,225	440		8,665	9,725
1100	8,000 gallon capacity		2.08	15.385		10,400	510		10,910	12,200
1110	10,000 gallon capacity		1.82	17.582		10,800	585		11,385	12,800
1120	12,000 gallon capacity		1.70	18.824		13,200	625		13,825	15,400
1122	15,000 gallon capacity		1.52	21.053		18,600	700		19,300	21,600
1124	20,000 gallon capacity		1.33	24.060		22,600	800		23,400	26,100
1126	25,000 gallon capacity		1.16	27.586		29,600	915		30,515	34,000

13200 | Storage Tanks

13201 | Storage Tanks

			CREW	DAILY OUTPUT	LABOR-HOURS	UNIT	2001 BARE COSTS				TOTAL INCL O&P
							MAT.	LABOR	EQUIP.	TOTAL	
800	1128	30,000 gallon capacity	Q-7	1.03	31.068	Ea.	35,400	1,025		36,425	40,600	800
	1140	For hold-down straps, add					2%	10%			
	1150	For hold-downs 500-4000 gal, add	Q-7	16	2	Set	162	66.50		228.50	278
	1160	For hold-downs 5000-15000 gal, add		8	4		325	133		458	555
	1170	For hold-downs 20,000 gal, add		5.33	6.004		485	200		685	835
	1180	For hold-downs 25,000 gal, add		4	8		650	266		916	1,125
	1190	For hold-downs 30,000 gal, add		2.60	12.308		975	410		1,385	1,700
	2210	Fiberglass, underground, single wall, U.L. listed, including									
	2220	hold-down straps, no manways									
	2225	550 gallon capacity	Q-5	2	8	Ea.	1,625	251		1,876	2,150
	2230	1,000 gallon capacity	"	1.88	8.511		1,975	267		2,242	2,575
	2240	2,000 gallon capacity	Q-7	3.55	9.014		2,525	300		2,825	3,225
	2250	4,000 gallon capacity		2.90	11.034		3,600	365		3,965	4,500
	2260	6,000 gallon capacity		2	16		4,150	530		4,680	5,350
	2270	8,000 gallon capacity		1.78	17.978		5,125	600		5,725	6,525
	2280	10,000 gallon capacity		1.60	20		5,525	665		6,190	7,075
	2282	12,000 gallon capacity		1.52	21.053		7,425	700		8,125	9,200
	2284	15,000 gallon capacity		1.39	23.022		7,900	765		8,665	9,850
	2290	20,000 gallon capacity		1.14	28.070		11,000	935		11,935	13,500
	2300	25,000 gallon capacity		.96	33.333		18,000	1,100		19,100	21,500
	2320	30,000 gallon capacity		.80	40		22,700	1,325		24,025	27,000
	3020	Fiberglass, underground, double wall, U.L. listed									
	3030	includes manways and hold-down straps									
	3040	600 gallon capacity	Q-5	1.86	8.602	Ea.	3,350	270		3,620	4,075
	3050	1,000 gallon capacity	"	1.70	9.412		4,425	295		4,720	5,300
	3060	2,500 gallon capacity	Q-7	3.29	9.726		6,225	325		6,550	7,325
	3070	3,000 gallon capacity		3.13	10.224		6,675	340		7,015	7,875
	3080	4,000 gallon capacity		2.93	10.921		7,275	365		7,640	8,550
	3090	6,000 gallon capacity		1.86	17.204		8,550	570		9,120	10,300
	3100	8,000 gallon capacity		1.65	19.394		10,700	645		11,345	12,800
	3110	10,000 gallon capacity		1.48	21.622		11,100	720		11,820	13,300
	3120	12,000 gallon capacity		1.40	22.857		13,500	760		14,260	16,100
	3122	15,000 gallon capacity		1.28	25		18,900	830		19,730	22,100
	3124	20,000 gallon capacity		1.06	30.189		23,100	1,000		24,100	26,900
	3126	25,000 gallon capacity		.90	35.556		30,200	1,175		31,375	35,100
	3128	30,000 gallon capacity		.74	43.243		36,400	1,425		37,825	42,200
	5000	Steel underground, sti-P3, set in place, not incl. hold-down bars.									
	5500	Excavation, pad, pumps and piping not included									
	5510	Single wall, 500 gallon capacity, 7 gauge shell	Q-5	2.70	5.926	Ea.	820	186		1,006	1,175
	5520	1,000 gallon capacity, 7 gauge shell	"	2.50	6.400		1,425	201		1,626	1,875
	5530	2,000 gallon capacity, 1/4" thick shell	Q-7	4.60	6.957		2,300	231		2,531	2,875
	5535	2,500 gallon capacity, 7 gauge shell	Q-5	3	5.333		2,500	167		2,667	3,000
	5540	5,000 gallon capacity, 1/4" thick shell	Q-7	3.20	10		4,675	330		5,005	5,650
	5560	10,000 gallon capacity, 1/4" thick shell		2	16		7,550	530		8,080	9,100
	5580	15,000 gallon capacity, 5/16" thick shell		1.70	18.824		10,200	625		10,825	12,200
	5600	20,000 gallon capacity, 5/16" thick shell		1.50	21.333		14,700	710		15,410	17,200
	5610	25,000 gallon capacity, 3/8" thick shell		1.30	24.615		17,600	820		18,420	20,600
	5620	30,000 gallon capacity, 3/8" thick shell		1.10	29.091		21,500	965		22,465	25,100
	5630	40,000 gallon capacity, 3/8" thick shell		.90	35.556		30,700	1,175		31,875	35,500
	5640	50,000 gallon capacity, 3/8" thick shell		.80	40		38,300	1,325		39,625	44,200
	6200	Steel, underground, 360°, double wall, U.L. listed,									
	6210	with sti-P3 corrosion protection,									
	6220	(dielectric coating, cathodic protection, electrical									
	6230	isolation) 30 year warranty,									
	6240	not incl. manholes or hold-downs.									
	6250	500 gallon capacity	Q-5	2.40	6.667	Ea.	2,300	209		2,509	2,850

13200 | Storage Tanks

13201	Storage Tanks	CREW	DAILY OUTPUT	LABOR-HOURS	UNIT	2001 BARE COSTS MAT.	LABOR	EQUIP.	TOTAL	TOTAL INCL O&P
6260	1,000 gallon capactiy	Q-5	2.25	7.111	Ea.	3,350	223		3,573	4,000
6270	2,000 gallon capacity	Q-7	4.16	7.692		4,400	256		4,656	5,200
6280	3,000 gallon capacity		3.90	8.205		5,250	273		5,523	6,175
6290	4,000 gallon capacity		3.64	8.791		6,425	292		6,717	7,500
6300	5,000 gallon capacity		2.91	10.997		8,100	365		8,465	9,450
6310	6,000 gallon capacity		2.42	13.223		9,600	440		10,040	11,300
6320	8,000 gallon capacity		2.08	15.385		10,000	510		10,510	11,800
6330	10,000 gallon capacity		1.82	17.582		11,300	585		11,885	13,300
6340	12,000 gallon capacity		1.70	18.824		13,600	625		14,225	15,900
6350	15,000 gallon capacity		1.33	24.060		16,300	800		17,100	19,100
6360	20,000 gallon capacity		1.33	24.060		20,000	800		20,800	23,100
6370	25,000 gallon capacity		1.16	27.586		30,400	915		31,315	34,800
6380	30,000 gallon capacity		1.03	31.068		36,600	1,025		37,625	41,900
6390	40,000 gallon capacity		.80	40		47,000	1,325		48,325	53,500
6395	50,000 gallon capacity		.73	43.836		57,000	1,450		58,450	64,500
6400	For hold-downs 500-2000 gal, add		16	2	Set	268	66.50		334.50	395
6410	For hold-downs 3000-6000 gal, add		12	2.667		350	88.50		438.50	520
6420	For hold-downs 8000-12,000 gal, add		11	2.909		850	96.50		946.50	1,075
6430	For hold-downs 15,000 gal, add		9	3.556		1,075	118		1,193	1,350
6440	For hold-downs 20,000 gal, add		8	4		1,175	133		1,308	1,500
6450	For hold-downs 20,000 gal plus, add		6	5.333		2,525	177		2,702	3,050
6500	For manways, add				Ea.	555			555	610
6600	In place with hold-downs									
6652	550 gallon capacity	Q-5	1.84	8.696	Ea.	2,550	273		2,823	3,225

13280 | Hazardous Material Remediation

13281	Hazardous Material Remediation	CREW	DAILY OUTPUT	LABOR-HOURS	UNIT	2001 BARE COSTS MAT.	LABOR	EQUIP.	TOTAL	TOTAL INCL O&P
0010	**BULK ASBESTOS REMOVAL**									
0020	Includes disposable tools and 1 suit and respirator/day/worker									
0200	Boiler insulation	A-9	480	.133	S.F.	.06	4.29		4.35	6.95
0210	With metal lath add				%				50%	
0300	Boiler breeching or flue insulation	A-9	520	.123	S.F.		3.96		3.96	6.35
0310	For active boiler, add				%				100%	
0400	Duct or AHU insulation	A-10B	440	.073	S.F.		2.35		2.35	3.76
0500	Duct vibration isolation joints, up to 24 Sq. In. duct	A-9	56	1.143	Ea.		37		37	59
0520	25 Sq. In. to 48 Sq. In. duct		48	1.333			43		43	69
0530	49 Sq. In. to 76 Sq. In. duct		40	1.600			51.50		51.50	82.50
0600	Pipe insulation, air cell type, up to 4" diameter pipe		900	.071	L.F.		2.29		2.29	3.67
0610	4" to 8" diameter pipe		800	.080			2.58		2.58	4.13
0620	10" to 12" diameter pipe		700	.091			2.95		2.95	4.72
0630	14" to 16" diameter pipe		550	.116			3.75		3.75	6
0650	Over 16" diameter pipe		650	.098	S.F.		3.17		3.17	5.10
0700	With glove bag up to 3" diameter pipe		100	.640	L.F.	3.15	20.50		23.65	36.50
1000	Pipe fitting insulation up to 4" diameter pipe		320	.200	Ea.		6.45		6.45	10.30
1100	6" to 8" diameter pipe		304	.211			6.80		6.80	10.85
1110	10" to 12" diameter pipe		192	.333			10.75		10.75	17.20
1120	14" to 16" diameter pipe		128	.500			16.10		16.10	26
1130	Over 16" diameter pipe		176	.364	S.F.		11.70		11.70	18.75
1200	With glove bag, up to 8" diameter pipe		40	1.600	Ea.	6.55	51.50		58.05	89.50

13280 | Hazardous Material Remediation

13281 | Hazardous Material Remediation

			CREW	DAILY OUTPUT	LABOR-HOURS	UNIT	MAT.	LABOR	EQUIP.	TOTAL	TOTAL INCL O&P	
120	2000	Scrape foam fireproofing from flat surface	A-9	2,400	.027	S.F.		.86		.86	1.38	120
	2100	Irregular surfaces		1,200	.053			1.72		1.72	2.75	
	3000	Remove cementitious material from flat surface		1,800	.036			1.15		1.15	1.83	
	3100	Irregular surface		1,400	.046			1.47		1.47	2.36	
	6000	Remove contaminated soil from crawl space by hand		400	.160	C.F.		5.15		5.15	8.25	
	6100	With large production vacuum loader	A-12	700	.091	"		2.95	.74	3.69	5.55	
	7000	Radiator backing, not including radiator removal	A-9	1,200	.053	S.F.		1.72		1.72	2.75	
	9000	For type C (supplied air) respirator equipment, add				%					10%	
125	0010	**ASBESTOS ABATEMENT WORK AREA** Containment and preparation.										125
	0100	Pre-cleaning, HEPA vacuum and wet wipe, flat surfaces	A-10	12,000	.005	S.F.		.17		.17	.28	
	0200	Protect carpeted area, 2 layers 6 mil poly on 3/4" plywood	"	1,000	.064		1.50	2.06		3.56	4.95	
	0300	Separation barrier, 2" x 4" @ 16", 1/2" plywood ea. side, 8' high	2 Carp	400	.040		1.25	1.17		2.42	3.20	
	0310	12' high		320	.050		1.40	1.46		2.86	3.82	
	0320	16' high		200	.080		1.50	2.33		3.83	5.30	
	0400	Personnel decontam. chamber, 2" x 4" @ 16", 3/4" ply ea. side		280	.057		2.50	1.67		4.17	5.35	
	0450	Waste decontam. chamber, 2" x 4" studs @ 16", 3/4" ply ea. side		360	.044		3	1.30		4.30	5.35	
	0500	Cover surfaces with polyethelene sheeting										
	0501	Including glue and tape										
	0550	Floors, each layer, 6 mil	A-10	8,000	.008	S.F.	.10	.26		.36	.52	
	0551	4 mil		9,000	.007		.06	.23		.29	.44	
	0560	Walls, each layer, 6 mil		6,000	.011		.09	.34		.43	.65	
	0561	4 mil		7,000	.009		.07	.29		.36	.55	
	0570	For heights above 12', add						20%				
	0575	For heights above 20', add						30%				
	0580	For fire retardant poly, add					100%					
	0590	For large open areas, deduct					10%	20%				
	0600	Seal floor penetrations with foam firestop to 36 Sq. In.	2 Carp	200	.080	Ea.	6.25	2.33		8.58	10.55	
	0610	36 Sq. In. to 72 Sq. In.		125	.128		12.50	3.73		16.23	19.60	
	0615	72 Sq. In. to 144 Sq. In.		80	.200		25	5.85		30.85	36.50	
	0620	Wall penetrations, to 36 square inches		180	.089		6.25	2.59		8.84	10.95	
	0630	36 Sq. In. to 72 Sq. In.		100	.160		12.50	4.66		17.16	21	
	0640	72 Sq. In. to 144 Sq. In.		60	.267		25	7.75		32.75	39.50	
	0800	Caulk seams with latex	1 Carp	230	.035	L.F.	.15	1.01		1.16	1.76	
	0900	Set up neg. air machine, 1-2k C.F.M. /25 M.C.F. volume	1 Asbe	4.30	1.860	Ea.		60		60	96	
130	0010	**DEMOLITION IN ASBESTOS CONTAMINATED AREA**										130
	0200	Ceiling, including suspension system, plaster and lath	A-9	2,100	.030	S.F.		.98		.98	1.57	
	0210	Finished plaster, leaving wire lath		585	.109			3.52		3.52	5.65	
	0220	Suspended acoustical tile		3,500	.018			.59		.59	.94	
	0230	Concealed tile grid system		3,000	.021			.69		.69	1.10	
	0240	Metal pan grid system		1,500	.043			1.37		1.37	2.20	
	0250	Gypsum board		2,500	.026			.82		.82	1.32	
	0260	Lighting fixtures up to 2' x 4'		72	.889	Ea.		28.50		28.50	46	
	0400	Partitions, non load bearing										
	0410	Plaster, lath, and studs	A-9	690	.093	S.F.	.55	2.99		3.54	5.40	
	0450	Gypsum board and studs	"	1,390	.046	"		1.48		1.48	2.38	
	9000	For type C (supplied air) respirator equipment, add				%					10%	
135	0010	**ASBESTOS ABATEMENT EQUIPMENT** and supplies, buy										135
	0200	Air filtration device, 2000 C.F.M.				Ea.	2,500			2,500	2,750	
	0250	Large volume air sampling pump, minimum					450			450	495	
	0260	Maximum					1,000			1,000	1,100	
	0300	Airless sprayer unit, 2 gun					2,000			2,000	2,200	
	0350	Light stand, 500 watt					250			250	275	
	0400	Personal respirators										
	0410	Negative pressure, 1/2 face, dual operation, min.				Ea.	22.50			22.50	25	
	0420	Maximum					24			24	26.50	
	0450	P.A.P.R., full face, minimum					400			400	440	

13280 | Hazardous Material Remediation

13281 | Hazardous Material Remediation

		CREW	DAILY OUTPUT	LABOR-HOURS	UNIT	2001 BARE COSTS MAT.	LABOR	EQUIP.	TOTAL	TOTAL INCL O&P
0460	Maximum				Ea.	700			700	770
0470	Supplied air, full face, incl. air line, minimum	R13281-120				450			450	495
0480	Maximum					600			600	660
0500	Personnel sampling pump, minimum					450			450	495
0510	Maximum					750			750	825
1500	Power panel, 20 unit, incl. G.F.I.					1,800			1,800	1,975
1600	Shower unit, including pump and filters					1,125			1,125	1,250
1700	Supplied air system (type C)					10,000			10,000	11,000
1750	Vacuum cleaner, HEPA, 16 gal., stainless steel, wet/dry					1,000			1,000	1,100
1760	55 gallon					2,200			2,200	2,425
1800	Vacuum loader, 9-18 ton/hr					90,000			90,000	99,000
1900	Water atomizer unit, including 55 gal. drum					230			230	253
2000	Worker protection, whole body, foot, head cover & gloves, plastic					30			30	33
2500	Respirator, single use					10.50			10.50	11.55
2550	Cartridge for respirator					37.50			37.50	41.50
2570	Glove bag, 7 mil, 50" x 64"					8.50			8.50	9.35
2580	10 mil, 44" x 60"					8.40			8.40	9.25
3000	HEPA vacuum for work area, minimum					1,000			1,000	1,100
3050	Maximum					2,775			2,775	3,050
6000	Disposable polyethelene bags, 6 mil, 3 C.F.					1.15			1.15	1.27
6300	Disposable fiber drums, 3 C.F.					6.50			6.50	7.15
6400	Pressure sensitive caution lables, 3" x 5"					8.90			8.90	9.80
6450	11" x 17"					6.50			6.50	7.15
6500	Negative air machine, 1800 C.F.M.					775			775	855
140 0010	**DECONTAMINATION CONTAINMENT AREA DEMOLITION** and clean-up									
0100	Spray exposed substrate with surfactant (bridging)									
0200	Flat surfaces	A-9	6,000	.011	S.F.	.35	.34		.69	.94
0250	Irregular surfaces		4,000	.016	"	.30	.52		.82	1.16
0300	Pipes, beams, and columns		2,000	.032	L.F.	.55	1.03		1.58	2.26
1000	Spray encapsulate polyethelene sheeting		8,000	.008	S.F.	.30	.26		.56	.74
1100	Roll down polyethelene sheeting		8,000	.008	"		.26		.26	.41
1500	Bag polyethelene sheeting		400	.160	Ea.	.75	5.15		5.90	9.10
2000	Fine clean exposed substrate, with nylon brush		2,400	.027	S.F.		.86		.86	1.38
2500	Wet wipe substrate		4,800	.013			.43		.43	.69
2600	Vacuum surfaces, fine brush		6,400	.010			.32		.32	.52
3000	Structural demolition									
3100	Wood stud walls	A-9	2,800	.023	S.F.		.74		.74	1.18
3500	Window manifolds, not incl. window replacement		4,200	.015			.49		.49	.79
3600	Plywood carpet protection		2,000	.032			1.03		1.03	1.65
4000	Remove custom decontamination facility	A-10A	8	3	Ea.	15	97		112	172
4100	Remove portable decontamination facility	3 Asbe	12	2	"	12.50	64.50		77	117
5000	HEPA vacuum, shampoo carpeting	A-9	4,800	.013	S.F.	.05	.43		.48	.75
9000	Final cleaning of protected surfaces	A-10A	8,000	.003	"		.10		.10	.16
145 0010	**OSHA TESTING**									
0100	Certified technician, minimum				Day					300
0110	Maximum				"					500
0200	Personal sampling, PCM analysis, minimum	1 Asbe	8	1	Ea.	2.75	32		34.75	54.50
0210	Maximum	"	4	2	"	3	64.50		67.50	106
0300	Industrial hygienist, minimum				Day					400
0310	Maximum				"					550
1000	Cleaned area samples	1 Asbe	8	1	Ea.	2.63	32		34.63	54.50
1100	PCM air sample analysis, minimum		8	1		30	32		62	84.50
1110	Maximum		4	2		3.09	64.50		67.59	106
1200	TEM air sample analysis, minimum								100	125
1210	Maximum								400	500

13280 | Hazardous Material Remediation

13281 | Hazardous Material Remediation

			CREW	DAILY OUTPUT	LABOR-HOURS	UNIT	MAT.	LABOR	EQUIP.	TOTAL	TOTAL INCL O&P	
150	0010	**ENCAPSULATION WITH SEALANTS**										150
	0100	Ceilings and walls, minimum	A-9	21,000	.003	S.F.	.25	.10		.35	.44	
	0110	Maximum		10,600	.006	"	.40	.19		.59	.75	
	0300	Pipes to 12" diameter including minor repairs, minimum		800	.080	L.F.	.35	2.58		2.93	4.52	
	0310	Maximum		400	.160	"	1	5.15		6.15	9.35	
155	0010	**WASTE PACKAGING, HANDLING, & DISPOSAL**										155
	0100	Collect and bag bulk material, 3 C.F. bags, by hand	A-9	400	.160	Ea.	1.15	5.15		6.30	9.50	
	0200	Large production vacuum loader	A-12	880	.073		.80	2.34	.59	3.73	5.30	
	1000	Double bag and decontaminate	A-9	960	.067		2.30	2.15		4.45	5.95	
	2000	Containerize bagged material in drums, per 3 C.F. drum	"	800	.080		6.50	2.58		9.08	11.30	
	3000	Cart bags 50' to dumpster	2 Asbe	400	.040			1.29		1.29	2.06	
	5000	Disposal charges, not including haul, minimum				C.Y.					50	
	5020	Maximum				"					175	
	5100	Remove refrigerant from system	1 Plum	40	.200	Lb.		6.90		6.90	10.40	
	9000	For type C (supplied air) respirator equipment, add				%					10%	

13600 | Solar and Wind Energy Equipment

13630 | Solar Collector Components

			CREW	DAILY OUTPUT	LABOR-HOURS	UNIT	MAT.	LABOR	EQUIP.	TOTAL	TOTAL INCL O&P	
200	0010	**SOLAR ENERGY**	A8.3-610									200
	0020	System/Package prices, not including connecting										
	0030	pipe, insulation, or special heating/plumbing fixtures	A8.3-620									
	0150	For solar ultraviolet pipe insulation see Div. 15086										
	0500	Hot water, standard package, low temperature	A8.3-630									
	0540	1 collector, circulator, fittings, 65 gal. tank	Q-1	.50	32	Ea.	1,350	990		2,340	3,000	
	0580	2 collectors, circulator, fittings, 120 gal. tank	A8.3-640	.40	40		1,925	1,250		3,175	4,000	
	0620	3 collectors, circulator, fittings, 120 gal. tank		.34	47.059		2,350	1,450		3,800	4,775	
	0700	Medium temperature package	A8.3-650									
	0720	1 collector, circulator, fittings, 80 gal. tank	Q-1	.50	32	Ea.	1,450	990		2,440	3,100	
	0740	2 collectors, circulator, fittings, 120 gal. tank	A8.3-660	.40	40		2,075	1,250		3,325	4,175	
	0780	3 collectors, circulator, fittings, 120 gal. tank		.30	53.333		3,775	1,650		5,425	6,650	
	0980	For each additional 120 gal. tank, add					660			660	730	
	2250	Controller, liquid temperature	A8.3-710	1 Plum	5	1.600		92	55		147	184
	2300	Circulators, air										
	2310	Blowers	A8.3-720									
	2330	100-300 S.F. system, 1/10 HP		Q-9	16	1	Ea.	128	30		158	188
	2340	300-500 S.F. system, 1/5 HP	A8.3-740	15	1.067		177	32		209	244	
	2350	Two speed, 100-300 S.F., 1/10 HP	R13600-610	14	1.143		128	34.50		162.50	195	
	2400	Reversible fan, 20" diameter, 2 speed		18	.889		102	27		129	154	
	2520	Space & DHW system, less duct work		.50	32		1,400	965		2,365	3,050	
	2550	Booster fan 6" diameter, 120 CFM		16	1		33	30		63	83	
	2570	6" diameter, 225 CFM		16	1		40.50	30		70.50	91	
	2580	8" diameter, 150 CFM		16	1		37	30		67	87.50	
	2590	8" diameter, 310 CFM		14	1.143		57	34.50		91.50	117	
	2600	8" diameter, 425 CFM		14	1.143		64	34.50		98.50	124	
	2650	Rheostat		32	.500		13.95	15.10		29.05	39	
	2660	Shutter/damper		12	1.333		45	40.50		85.50	112	
	2670	Shutter motor		16	1		97	30		127	154	
	2800	Circulators, liquid, 1/25 HP, 5.3 GPM		Q-1	14	1.143		139	35.50		174.50	207

13600 | Solar and Wind Energy Equipment

13630 | Solar Collector Components

			CREW	DAILY OUTPUT	LABOR-HOURS	UNIT	MAT.	LABOR	EQUIP.	TOTAL	TOTAL INCL O&P	
200	2820	1/20 HP, 17 GPM	A8.3-610	Q-1	12	1.333	Ea.	146	41.50		187.50	224	200
	2850	1/20 HP, 17 GPM, stainless steel			12	1.333		230	41.50		271.50	315	
	2870	1/12 HP, 30 GPM	A8.3-620		10	1.600		315	49.50		364.50	420	
	3000	Collector panels, air with aluminum absorber plate											
	3010	Wall or roof mount	A8.3-630										
	3040	Flat black, plastic glazing											
	3080	4' x 8'	A8.3-640	Q-9	6	2.667	Ea.	600	80.50		680.50	785	
	3100	4' x 10'			5	3.200	"	740	96.50		836.50	965	
	3200	Flush roof mount, 10' to 16' x 22" wide	A8.3-650		96	.167	L.F.	146	5.05		151.05	169	
	3210	Manifold, by L.F. width of collectors			160	.100	"	27	3.02		30.02	34	
	3300	Collector panels, liquid with copper absorber plate	A8.3-660										
	3320	Black chrome, tempered glass glazing											
	3330	Alum. frame, 4' x 8', 5/32" single glazing	A8.3-710	Q-1	9.50	1.684	Ea.	545	52		597	680	
	3390	Alum. frame, 4' x 10', 5/32" single glazing		"	6	2.667	"	650	82.50		732.50	835	
	3440	Flat black	A8.3-720										
	3450	Alum. frame, 3' x 8'		Q-1	9	1.778	Ea.	430	55		485	560	
	3500	Alum. frame, 4' x 8.5'	A8.3-740		5.50	2.909		490	90		580	675	
	3520	Alum. frame, 4' x 10.5'			10	1.600		595	49.50		644.50	730	
	3540	Alum. frame, 4' x 12.5'	R13600-610		5	3.200		735	99		834	960	
	3550	Liquid with fin tube absorber plate											
	3560	Alum. frame 4' x 8' tempered glass		Q-1	10	1.600	Ea.	495	49.50		544.50	620	
	3580	Liquid with vacuum tubes, 4' x 6'-10"			9	1.778		845	55		900	1,025	
	3600	Liquid, full wetted, plastic, alum. frame, 3' x 10'			5	3.200		195	99		294	365	
	3650	Collector panel mounting, flat roof or ground rack			7	2.286		64.50	71		135.50	178	
	3670	Roof clamps			70	.229	Set	1.61	7.10		8.71	12.45	
	3700	Roof strap, teflon		1 Plum	205	.039	L.F.	9.50	1.34		10.84	12.50	
	3900	Differential controller with two sensors											
	3930	Thermostat, hard wired		1 Plum	8	1	Ea.	92	34.50		126.50	153	
	3950	Line cord and receptacle			12	.667		77.50	23		100.50	120	
	4050	Pool valve system			2.50	3.200		199	110		309	385	
	4070	With 12 VAC actuator			2	4		256	138		394	490	
	4080	Pool pump system, 2" pipe size			6	1.333		160	46		206	246	
	4100	Five station with digital read-out			3	2.667		202	92		294	360	
	4150	Sensors											
	4200	Brass plug, 1/2" MPT		1 Plum	32	.250	Ea.	16.60	8.60		25.20	31.50	
	4210	Brass plug, reversed			32	.250		22.50	8.60		31.10	38	
	4220	Freeze prevention			32	.250		20	8.60		28.60	35	
	4240	Screw attached			32	.250		10.15	8.60		18.75	24	
	4250	Brass, immersion			32	.250		24.50	8.60		33.10	40	
	4300	Heat exchanger											
	4330	Fluid to air coil, up flow, 45 MBH		Q-1	4	4	Ea.	279	124		403	490	
	4380	70 MBH			3.50	4.571		315	142		457	560	
	4400	80 MBH			3	5.333		420	165		585	710	
	4580	Fluid to fluid package includes two circulating pumps											
	4590	expansion tank, check valve, relief valve											
	4600	controller, high temperature cutoff and sensors		Q-1	2.50	6.400	Ea.	695	198		893	1,075	
	4650	Heat transfer fluid											
	4700	Propylene glycol, inhibited anti-freeze		1 Plum	28	.286	Gal.	8.80	9.85		18.65	24.50	
	4800	Solar storage tanks, knocked down											
	4810	Air, galvanized steel clad, double wall, 4" fiberglass											
	4820	insulation, 20 Mil PVC lining											
	4870	4' high, 4' x 4', = 64 C.F./450 gallons		Q-9	2	8	Ea.	1,725	242		1,967	2,275	
	4880	4' x 8' = 128 C.F./900 gallons			1.50	10.667		2,250	320		2,570	2,975	
	4890	4' x 12' = 190 C.F./1300 gallons			1.30	12.308		3,000	370		3,370	3,875	
	4900	8' x 8' = 250 C.F./1700 gallons			1	16		3,350	485		3,835	4,425	
	5010	6'-3" high, 7' x 7' = 306 C.F./2000 gallons		Q-10	1.20	20		8,150	625		8,775	9,925	

13600 | Solar and Wind Energy Equipment

13630 | Solar Collector Components

			CREW	DAILY OUTPUT	LABOR-HOURS	UNIT	2001 BARE COSTS MAT.	LABOR	EQUIP.	TOTAL	TOTAL INCL O&P	
200	5020	7' x 10'-6" = 459 C.F./3000 gallons	A8.3 -610	Q-10	.80	30	Ea.	10,300	940		11,240	12,800	200
	5030	7' x 14' = 613 C.F./4000 gallons			.60	40		12,000	1,250		13,250	15,100	
	5040	10'-6" x 10'-6" = 689 C.F./4500 gallons	A8.3 -620		.50	48		12,400	1,500		13,900	15,900	
	5050	10'-6" x 14' = 919 C.F./6000 gallons			.40	60		14,600	1,875		16,475	18,900	
	5060	14' x 14' = 1225 C.F./8000 gallons		Q-11	.40	80		17,100	2,550		19,650	22,800	
	5070	14' x 17'-6" = 1531 C.F./10,000 gallons	A8.3 -630		.30	106		19,200	3,425		22,625	26,400	
	5080	17'-6" x 17'-6" = 1914 C.F./12,500 gallons	A8.3 -640		.25	128		22,600	4,100		26,700	31,200	
	5090	17'-6" x 21' = 2297 C.F./15,000 gallons			.20	160		25,200	5,125		30,325	35,700	
	5100	21' x 21' = 2756 C.F./18,000 gallons	A8.3 -650		.18	177		28,900	5,700		34,600	40,600	
	5120	30 Mil reinforced Chemflex lining,											
	5140	4' high, 4' x 4' = 64 C.F./450 gallons	A8.3 -660	Q-9	2	8	Ea.	1,750	242		1,992	2,300	
	5150	4' x 8' = 128 C.F./900 gallons			1.50	10.667		2,650	320		2,970	3,425	
	5160	4' x 12' = 190 C.F./1300 gallons	A8.3 -710		1.30	12.308		3,525	370		3,895	4,450	
	5170	8' x 8' = 250 C.F./1700 gallons			1	16		3,925	485		4,410	5,075	
	5190	6'-3" high, 7' x 7' = 306 C.F./2000 gallons	A8.3 -720	Q-10	1.20	20		8,700	625		9,325	10,500	
	5200	7' x 10'-6" = 459 C.F./3000 gallons			.80	30		10,600	940		11,540	13,200	
	5210	7' x 14' = 613 C.F./4000 gallons	A8.3 -740		.60	40		12,300	1,250		13,550	15,500	
	5220	10'-6" x 10'-6" = 689 C.F./4500 gallons			.50	48		12,700	1,500		14,200	16,300	
	5230	10'-6" x 14' = 919 C.F./6000 gallons	R13600 -610		.40	60		15,000	1,875		16,875	19,400	
	5240	14' x 14' = 1225 C.F./8000 gallons		Q-11	.40	80		17,500	2,550		20,050	23,300	
	5250	14' x 17'-6" = 1531 C.F./10,000 gallons			.30	106		19,600	3,425		23,025	26,900	
	5260	17'-6" x 17'-6" = 1914 C.F./12,500 gallons			.25	128		23,100	4,100		27,200	31,700	
	5270	17'-6" x 21' = 2297 C.F./15,000 gallons			.20	160		26,000	5,125		31,125	36,500	
	5280	21' x 21' = 2756 C.F./18,000 gallons			.18	177		29,800	5,700		35,500	41,600	
	5290	30 Mil reinforced Hypalon lining, add						.02%					
	7000	Solar control valves and vents											
	7050	Air purger, 1" pipe size		1 Plum	12	.667	Ea.	41	23		64	79.50	
	7070	Air eliminator, automatic 3/4" size			32	.250		27	8.60		35.60	42.50	
	7090	Air vent, automatic, 1/8" fitting			32	.250		8.80	8.60		17.40	22.50	
	7100	Manual, 1/8" NPT			32	.250		1.94	8.60		10.54	15.15	
	7120	Backflow preventer, 1/2" pipe size			16	.500		58.50	17.25		75.75	90	
	7130	3/4" pipe size			16	.500		59	17.25		76.25	91	
	7150	Balancing valve, 3/4" pipe size			20	.400		22.50	13.80		36.30	45.50	
	7180	Draindown valve, 1/2" copper tube			9	.889		187	30.50		217.50	252	
	7200	Flow control valve, 1/2" pipe size			22	.364		49.50	12.55		62.05	73.50	
	7220	Expansion tank, up to 5 gal.			32	.250		58	8.60		66.60	77	
	7250	Hydronic controller (aquastat)			8	1		41.50	34.50		76	97.50	
	7400	Pressure gauge, 2" dial			32	.250		21	8.60		29.60	36	
	7450	Relief valve, temp. and pressure 3/4" pipe size			30	.267		7.90	9.20		17.10	22.50	
	7500	Solenoid valve, normally closed											
	7520	Brass, 3/4" NPT, 24V		1 Plum	9	.889	Ea.	212	30.50		242.50	279	
	7530	1" NPT, 24V			9	.889		212	30.50		242.50	279	
	7750	Vacuum relief valve, 3/4" pipe size			32	.250		25	8.60		33.60	40.50	
	7800	Thermometers											
	7820	Digital temperature monitoring, 4 locations		1 Plum	2.50	3.200	Ea.	135	110		245	315	
	7900	Upright, 1/2" NPT			8	1		25.50	34.50		60	80	
	7970	Remote probe, 2" dial			8	1		29	34.50		63.50	84	
	7990	Stem, 2" dial, 9" stem			16	.500		19	17.25		36.25	47	
	8250	Water storage tank with heat exchanger and electric element											
	8270	66 gal. with 2" x 2 lb. density insulation		1 Plum	1.60	5	Ea.	650	172		822	975	
	8300	80 gal. with 2" x 2 lb. density insulation			1.60	5		745	172		917	1,075	
	8380	120 gal. with 2" x 2 lb. density insulation			1.40	5.714		840	197		1,037	1,225	
	8400	120 gal. with 2" x 2 lb. density insul., 40 S.F. heat coil			1.40	5.714		965	197		1,162	1,375	
	8500	Water storage module, plastic											
	8600	Tubular, 12" diameter, 4' high		1 Carp	48	.167	Ea.	76	4.86		80.86	91	
	8610	12" diameter, 8' high			40	.200		116	5.85		121.85	137	

13600 | Solar and Wind Energy Equipment

13630 | Solar Collector Components

			CREW	DAILY OUTPUT	LABOR-HOURS	UNIT	MAT.	LABOR	EQUIP.	TOTAL	TOTAL INCL O&P	
200	8620	18" diameter, 5' high	A8.3-610	1 Carp	38	.211	Ea.	127	6.15		133.15	150	200
	8630	18" diameter, 10' high		↓	32	.250		167	7.30		174.30	195	
	8640	58" diameter, 5' high	A8.3-620	2 Carp	32	.500		460	14.60		474.60	535	
	8650	Cap, 12" diameter						12			12	13.20	
	8660	18" diameter	A8.3-630				↓	15			15	16.50	

Column headers under "2001 BARE COSTS": MAT., LABOR, EQUIP., TOTAL

13800 | Building Automation & Control

13832 | Direct Digital Controls

			CREW	DAILY OUTPUT	LABOR-HOURS	UNIT	MAT.	LABOR	EQUIP.	TOTAL	TOTAL INCL O&P	
200	0010	**CONTROL COMPONENTS / DDC SYSTEMS** (Sub's quote incl. M & L)											200
	0100	Analog inputs											
	0110	Sensors (avg. 50' run in 1/2" EMT)											
	0120	Duct temperature					Ea.					346.80	
	0130	Space temperature										622.20	
	0140	Duct humidity, +/- 3%										652.80	
	0150	Space humidity, +/- 2%										999.60	
	0160	Duct static pressure										530.40	
	0170	C.F.M./Transducer										714	
	0172	Water temp. (see 15120 for well tap add)										612	
	0174	Water flow (see 15120 for circuit sensor add)										2,244	
	0176	Water pressure differential (see 15120 for tap add)										918	
	0177	Steam flow (see 15120 for circuit sensor add)										2,244	
	0178	Steam pressure (see 15120 for tap add)										958.80	
	0180	K.W./Transducer										1,275	
	0182	K.W.H. totalization (not incl. elec. meter pulse xmtr.)										586	
	0190	Space static pressure					↓					1,000	
	1000	Analog outputs (avg. 50' run in 1/2" EMT)											
	1010	P/I Transducer					Ea.					591.60	
	1020	Analog output, matl. in MUX										285.60	
	1030	Pneumatic (not incl. control device)										601.80	
	1040	Electric (not incl control device)					↓					357	
	2000	Status (Alarms)											
	2100	Digital inputs (avg. 50' run in 1/2" EMT)											
	2110	Freeze					Ea.					408	
	2120	Fire										367.20	
	2130	Differential pressure, (air)										561	
	2140	Differential pressure, (water)										795.60	
	2150	Current sensor										408	
	2160	Duct high temperature thermostat										535.50	
	2170	Duct smoke detector					↓					663	
	2200	Digital output (avg. 50' run in 1/2" EMT)											
	2210	Start/stop					Ea.					326.40	
	2220	On/off (maintained contact)					"					561	
	3000	Controller M.U.X. panel, incl. function boards											
	3100	48 point					Ea.					4,947	
	3110	128 point					"					6,783	
	3200	D.D.C. controller (avg. 50' run in conduit)											
	3210	Mechanical room											
	3214	16 point controller (incl. 120v/1ph power supply)					Ea.					3,060	

13800 | Building Automation & Control

13832 | Direct Digital Controls

		CREW	DAILY OUTPUT	LABOR-HOURS	UNIT	MAT.	LABOR	EQUIP.	TOTAL	TOTAL INCL O&P
3229	32 point controller (incl. 120v/1ph power supply)				Ea.					5,100
3230	Includes software programming and checkout									
3260	Space									
3266	V.A.V. terminal box (incl. space temp. sensor)				Ea.					790.50
3280	Host computer (avg. 50' run in conduit)									
3281	Package complete with PC, keyboard,									
3282	printer, color CRT, modem, basic software				Ea.					9,180
4000	Front end costs									
4100	Computer (P.C.)/software program				Ea.					6,120
4200	Color graphics software									3,672
4300	Color graphics slides									459
4350	Additional dot matrix printer									918
4400	Communications trunk cable				L.F.					3.57
4500	Engineering labor, (not incl. dftg.)				Point					77.52
4600	Calibration labor									77.52
4700	Start-up, checkout labor									117.30
4800	Drafting labor, as req'd									
5000	Communications bus (data transmission cable)									
5010	#18 twisted shielded pair in 1/2" EMT conduit				C.L.F.					357
8000	Applications software									
8050	Basic maintenance manager software (not incl. data base entry)				Ea.					1,836
8100	Time program				Point					6.43
8120	Duty cycle									12.80
8140	Optimum start/stop									38.76
8160	Demand limiting									19.18
8180	Enthalpy program									38.76
8200	Boiler optimization				Ea.					1,147.50
8220	Chiller optimization				"					1,530
8240	Custom applications									
8260	Cost varies with complexity									

13834 | Electric/Electronic Control

		CREW	DAILY OUTPUT	LABOR-HOURS	UNIT	MAT.	LABOR	EQUIP.	TOTAL	TOTAL INCL O&P
0010	**CONTROL SYSTEMS, ELECTRONIC** R15050-710									
0020	For electronic costs, add to division 13836-200				Ea.					15%

13836 | Pneumatic Controls

		CREW	DAILY OUTPUT	LABOR-HOURS	UNIT	MAT.	LABOR	EQUIP.	TOTAL	TOTAL INCL O&P
0010	**CONTROL SYSTEMS, PNEUMATIC** (Sub's quote incl. mat. & labor)									
0011	and nominal 50 Ft. of tubing. Add control panelboard if req'd.									
0100	Heating and Ventilating, split system									
0200	Mixed air control, economizer cycle, panel readout, tubing R15050-710									
0220	Up to 10 tons	Q-19	.68	35.294	Ea.	2,750	1,150		3,900	4,775
0240	For 10 to 20 tons		.63	37.915		2,950	1,225		4,175	5,100
0260	For over 20 tons		.58	41.096		3,200	1,325		4,525	5,525
0270	Enthalpy cycle, up to 10 tons		.50	48.387		3,050	1,575		4,625	5,700
0280	For 10 to 20 tons		.46	52.174		3,300	1,700		5,000	6,175
0290	For over 20 tons		.42	56.604		3,575	1,825		5,400	6,675
0300	Heating coil, hot water, 3 way valve,									
0320	Freezestat, limit control on discharge, readout	Q-5	.69	23.088	Ea.	2,050	725		2,775	3,350
0500	Cooling coil, chilled water, room									
0520	Thermostat, 3 way valve	Q-5	2	8	Ea.	915	251		1,166	1,375
0600	Cooling tower, fan cycle, damper control,									
0620	Control system including water readout in/out at panel	Q-19	.67	35.821	Ea.	3,625	1,150		4,775	5,750
1000	Unit ventilator, day/night operation,									
1100	freezestat, ASHRAE, cycle 2	Q-19	.91	26.374	Ea.	2,000	855		2,855	3,500
2000	Compensated hot water from boiler, valve control,									
2100	readout and reset at panel, up to 60 GPM	Q-19	.55	43.956	Ea.	3,750	1,425		5,175	6,275

13800 | Building Automation & Control

13836 | Pneumatic Controls

			CREW	DAILY OUTPUT	LABOR-HOURS	UNIT	MAT.	LABOR	EQUIP.	TOTAL	TOTAL INCL O&P
2120	For 120 GPM	R15050-710	Q-19	.51	47.059	Ea.	4,025	1,525		5,550	6,725
2140	For 240 GPM			.49	49.180		4,225	1,600		5,825	7,025
3000	Boiler room combustion air, damper to 5 SF, controls			1.36	17.582		1,825	570		2,395	2,850
3500	Fan coil, heating and cooling valves, 4 pipe control system			3	8		820	259		1,079	1,300
3600	Heat exchanger system controls			.86	27.907		1,775	905		2,680	3,300
3900	Multizone control (one per zone), includes thermostat, damper										
3910	motor and reset of discharge temperature		Q-5	.51	31.373	Ea.	1,825	985		2,810	3,475
4000	Pneumatic thermostat, including controlling room radiator valve		"	2.43	6.593		545	207		752	910
4040	Program energy saving optimizer		Q-19	1.21	19.786		4,550	640		5,190	5,975
4060	Pump control system		"	3	8		840	259		1,099	1,325
4080	Reheat coil control system, not incl coil		Q-5	2.43	6.593		710	207		917	1,100
4500	Air supply for pneumatic control system										
4600	Tank mounted duplex compressor, starter, alternator,										
4620	piping, dryer, PRV station and filter										
4630	1/2 HP		Q-19	.68	35.139	Ea.	6,750	1,125		7,875	9,125
4640	3/4 HP			.64	37.383		7,075	1,200		8,275	9,600
4650	1 HP			.61	39.539		7,725	1,275		9,000	10,400
4660	1-1/2 HP			.57	41.739		8,225	1,350		9,575	11,100
4680	3 HP			.55	43.956		11,200	1,425		12,625	14,500
4690	5 HP			.42	57.143		19,600	1,850		21,450	24,300
4800	Main air supply, includes 3/8" copper main and labor		Q-5	1.82	8.791	C.L.F.	227	276		503	665
4810	If poly tubing used, deduct										30%
7000	Static pressure control for air handling unit, includes pressure										
7010	sensor, receiver controller, readout and damper motors		Q-19	.64	37.383	Ea.	5,375	1,200		6,575	7,725
7020	If return air fan requires control, add										70%
8600	VAV boxes, incl. thermostat, damper motor, reheat coil & tubing		Q-5	1.46	10.989		845	345		1,190	1,450
8610	If no reheat coil, deduct										204
9400	Sub assemblies for assembly systems										

13838 | Pneumatic/Electric Controls

		CREW	DAILY OUTPUT	LABOR-HOURS	UNIT	MAT.	LABOR	EQUIP.	TOTAL	TOTAL INCL O&P
0010	CONTROL COMPONENTS									
0500	Aquastats									
0508	Immersion type									
0510	High/low limit, breaks contact w/temp rise	1 Stpi	4	2	Ea.	105	69.50		174.50	220
0514	Sequencing, break 2 switches w/temp rise		4	2		85.50	69.50		155	199
0518	Circulating, makes contact w/temp rise		4	2		77.50	69.50		147	190
0600	Carbon monoxide detector system									
0606	Panel	1 Stpi	4	2	Ea.	2,300	69.50		2,369.50	2,625
0610	Sensor	"	7.30	1.096	"	310	38		348	400
0700	Controller, receiver									
0730	Pneumatic, panel mount, single input	1 Plum	8	1	Ea.	261	34.50		295.50	340
0740	With conversion mounting bracket		8	1		270	34.50		304.50	350
0750	Dual input, with control point adjustment		7	1.143		360	39.50		399.50	455
0850	Electric, single snap switch	1 Elec	4	2		330	68.50		398.50	465
0860	Dual snap switches		3	2.667		420	91.50		511.50	595
0870	Humidity controller		8	1		196	34.50		230.50	267
0880	Load limiting controller		8	1		430	34.50		464.50	525
0890	Temperature controller		8	1		245	34.50		279.50	320
0900	Control panel readout									
0910	Panel up to 12 indicators	1 Stpi	1.20	6.667	Ea.	130	232		362	495
0914	Panel up to 24 indicators		.86	9.302		163	325		488	670
0918	Panel up to 48 indicators		.50	16		385	560		945	1,275
1000	Enthalpy control, boiler water temperature control									
1010	governed by outdoor temperature, with timer	1 Elec	3	2.667	Ea.	195	91.50		286.50	350
1300	Energy control/monitor									
1320	BTU computer meter and controller	1 Stpi	1	8	Ea.	1,375	279		1,654	1,925

13800 | Building Automation & Control

13838 | Pneumatic/Electric Controls

		CREW	DAILY OUTPUT	LABOR-HOURS	UNIT	2001 BARE COSTS MAT.	LABOR	EQUIP.	TOTAL	TOTAL INCL O&P
1600	Flow meters									
1610	Gas	1 Stpi	4	2	Ea.	73	69.50		142.50	186
1620	Liquid	"	4	2	"	73	69.50		142.50	186
1640	Freezestat									
1644	20' sensing element, adjustable	1 Stpi	3.50	2.286	Ea.	80.50	79.50		160	209
2000	Gauges, pressure or vacuum									
2100	2" diameter dial	1 Stpi	32	.250	Ea.	14.20	8.70		22.90	29
2200	2-1/2" diameter dial		32	.250		16.55	8.70		25.25	31.50
2300	3-1/2" diameter dial		32	.250		20.50	8.70		29.20	35.50
2400	4-1/2" diameter dial		32	.250		31	8.70		39.70	47
2700	Flanged iron case, black ring									
2800	3-1/2" diameter dial	1 Stpi	32	.250	Ea.	65	8.70		73.70	84.50
2900	4-1/2" diameter dial		32	.250		88.50	8.70		97.20	111
3000	6" diameter dial		32	.250		70.50	8.70		79.20	90.50
3010	Steel case, 0 - 300 psi									
3012	2" diam. dial	1 Stpi	16	.500	Ea.	20.50	17.45		37.95	49
3014	4" diam. dial	"	16	.500	"	34	17.45		51.45	64
3020	Aluminum case, 0 - 300 psi									
3022	3-1/2" diam. dial	1 Stpi	16	.500	Ea.	47	17.45		64.45	78
3024	4-1/2" diam. dial		16	.500		69.50	17.45		86.95	103
3026	6" diam. dial		16	.500		116	17.45		133.45	154
3028	8-1/2" diam. dial		16	.500		274	17.45		291.45	325
3030	Brass case, 0 - 300 psi									
3032	2" diam. dial	1 Stpi	16	.500	Ea.	23	17.45		40.45	51.50
3034	4-1/2" diam. dial	"	16	.500	"	52.50	17.45		69.95	84
3040	Steel case, high pressure, 0 -10,000 psi									
3042	4-1/2" diam. dial	1 Stpi	16	.500	Ea.	131	17.45		148.45	171
3044	6-1/2" diam. dial		16	.500		210	17.45		227.45	257
3046	8-1/2" diam. dial		16	.500		380	17.45		397.45	445
3080	Pressure gauge, differential, magnehelic									
3084	0 - 2" W.C., with air filter kit	1 Stpi	6	1.333	Ea.	72	46.50		118.50	149
3300	For compound pressure-vacuum, add					18%				
3350	Humidistat									
3360	Pneumatic operation									
3361	Room humidistat, direct acting	1 Stpi	12	.667	Ea.	154	23		177	204
3362	Room humidistat, reverse acting		12	.667		154	23		177	204
3363	Room humidity transmitter		17	.471		166	16.40		182.40	208
3364	Duct mounted controller		12	.667		203	23		226	258
3365	Duct mounted transmitter		12	.667		183	23		206	236
3366	Humidity indicator, 3-1/2"		28	.286		73.50	9.95		83.45	96
3390	Electric operated	1 Shee	8	1		29.50	33.50		63	84.50
3400	Relays									
3430	Pneumatic/electric	1 Plum	16	.500	Ea.	179	17.25		196.25	223
3440	Pneumatic proportioning		8	1		97.50	34.50		132	159
3450	Pneumatic switching		12	.667		86.50	23		109.50	130
3460	Selector, 3 point		6	1.333		62	46		108	138
3470	Pneumatic time delay		8	1		229	34.50		263.50	305
3500	Sensor, air operated									
3520	Humidity	1 Plum	16	.500	Ea.	203	17.25		220.25	249
3540	Pressure		16	.500		46	17.25		63.25	76.50
3560	Temperature		12	.667		97	23		120	142
3600	Electric operated									
3620	Humidity	1 Elec	8	1	Ea.	25.50	34.50		60	79
3650	Pressure		8	1		810	34.50		844.50	940
3680	Temperature		10	.800		93.50	27.50		121	144
3700	Switches									

13800 | Building Automation & Control

13838 | Pneumatic/Electric Controls

		CREW	DAILY OUTPUT	LABOR-HOURS	UNIT	MAT.	LABOR	EQUIP.	TOTAL	TOTAL INCL O&P
3710	Minimum position									
3720	Electrical	1 Stpi	4	2	Ea.	25	69.50		94.50	133
3730	Pneumatic	"	4	2	"	35.50	69.50		105	144
4000	Thermometers									
4100	Dial type, 3-1/2" diameter, vapor type, union connection	1 Stpi	32	.250	Ea.	95	8.70		103.70	118
4120	Liquid type, union connection		32	.250		133	8.70		141.70	159
4130	Remote reading, 15' capillary		32	.250		104	8.70		112.70	127
4500	Stem type, 6-1/2" case, 2" stem, 1/2" NPT		32	.250		54	8.70		62.70	72.50
4520	4" stem, 1/2" NPT		32	.250		74.50	8.70		83.20	94.50
4600	9" case, 3-1/2" stem, 3/4" NPT		28	.286		90	9.95		99.95	114
4620	6" stem, 3/4" NPT		28	.286		96	9.95		105.95	120
4640	8" stem, 3/4" NPT		28	.286		118	9.95		127.95	145
4660	12" stem, 1" NPT		26	.308		130	10.70		140.70	159
4670	Bi-metal, dial type, steel case brass stem									
4672	2" dial, 4" - 9" stem	1 Stpi	16	.500	Ea.	29.50	17.45		46.95	59
4673	2-1/2" dial, 4" - 9" stem		16	.500		32	17.45		49.45	61.50
4674	3-1/2" dial, 4" - 9" stem		16	.500		32	17.45		49.45	61.50
4680	Mercury filled, industrial, union connection type									
4682	Angle stem, 7" scale	1 Stpi	16	.500	Ea.	74.50	17.45		91.95	109
4683	9" scale		16	.500		109	17.45		126.45	147
4684	12" scale		16	.500		121	17.45		138.45	160
4686	Straight stem, 7" scale		16	.500		96	17.45		113.45	132
4687	9" scale		16	.500		118	17.45		135.45	157
4688	12" scale		16	.500		129	17.45		146.45	169
4690	Mercury filled, industrial, separable socket type									
4692	Angle stem, with socket, 7" scale	1 Stpi	16	.500	Ea.	62.50	17.45		79.95	95.50
4693	9" scale		16	.500		75	17.45		92.45	109
4694	12" scale		16	.500		152	17.45		169.45	194
4696	Straight stem, with socket, 7" scale		16	.500		62.50	17.45		79.95	95.50
4697	9" scale		16	.500		75	17.45		92.45	109
4698	12" scale		16	.500		88.50	17.45		105.95	124
5000	Thermostats									
5030	Manual	1 Shee	8	1	Ea.	26.50	33.50		60	81
5040	1 set back, electric, timed		8	1		79	33.50		112.50	139
5050	2 set back, electric, timed		8	1		108	33.50		141.50	171
5100	Locking cover					13.70			13.70	15.10
5200	24 hour, automatic, clock	1 Shee	8	1		101	33.50		134.50	163
5220	Electric, low voltage, 2 wire	1 Elec	13	.615		17.05	21		38.05	50.50
5230	3 wire		10	.800		21	27.50		48.50	64
5236	Heating/cooling, low voltage, with clock		8	1		119	34.50		153.50	182
5240	Pneumatic									
5250	Single temp., single pressure	1 Stpi	8	1	Ea.	117	35		152	182
5251	Dual pressure		8	1		165	35		200	235
5252	Dual temp., dual pressure		8	1		157	35		192	226
5253	Reverse acting w/averaging element		8	1		144	35		179	211
5254	Heating-cooling w/deadband		8	1		165	35		200	235
5255	Integral w/piston top valve actuator		8	1		154	35		189	222
5256	Dual temp., dual pressure		8	1		157	35		192	226
5257	Low limit, 8' averaging element		8	1		109	35		144	173
5258	Room single temp. proportional		8	1		48.50	35		83.50	106
5260	Dual temp, direct acting for VAV fan		8	1		143	35		178	210
5262	Capillary tube type, 20', -30 to +100°F		8	1		93	35		128	156
5300	Transmitter, pneumatic									
5320	Temperature averaging element	Q-1	8	2	Ea.	93	62		155	196
5350	Pressure differential	1 Plum	7	1.143		545	39.50		584.50	660
5370	Humidity, duct		8	1		183	34.50		217.50	253

13800 | Building Automation & Control

13838 | Pneumatic/Electric Controls

		CREW	DAILY OUTPUT	LABOR-HOURS	UNIT	2001 BARE COSTS MAT.	LABOR	EQUIP.	TOTAL	TOTAL INCL O&P
5380	Room	1 Plum	12	.667	Ea.	166	23		189	218
5390	Temperature, with averaging element	↓	6	1.333		104	46		150	185
5420	Electric operated, humidity	1 Elec	8	1		49.50	34.50		84	106
5430	DPST	"	8	1	↓	69.50	34.50		104	128
5500	Timer/time clocks									
5510	7 day, 12 hr. battery	1 Stpi	2.60	3.077	Ea.	208	107		315	390
5600	Recorders									
5610	Hydrograph humidity, wall mtd., aluminum case, grad. chart									
5612	8"	1 Stpi	1.60	5	Ea.	650	174		824	980
5614	10"	"	1.60	5	"	960	174		1,134	1,325
5620	Thermo-hydrograph, wall mtd., grad. chart, 5' SS probe									
5622	8"	1 Stpi	1.60	5	Ea.	440	174		614	745
5624	10"	"	1.60	5	"	770	174		944	1,125
5630	Time of operation, wall mtd., 1 pen cap, 24 hr. clock									
5632	6"	1 Stpi	1.60	5	Ea.	380	174		554	680
5634	8"	"	1.60	5	"	380	174		554	680
5640	Pressure & vacuum, bourdon tube type									
5642	Flush mount, 1 pen									
5644	6"	1 Stpi	1.60	5	Ea.	350	174		524	650
5646	8"	"	1.60	5	"	700	174		874	1,025
6000	Valves, motorized zone									
6100	Sweat connections, 1/2" C x C	1 Stpi	20	.400	Ea.	66	13.95		79.95	93.50
6110	3/4" C x C		20	.400		70.50	13.95		84.45	98.50
6120	1" C x C		19	.421		82.50	14.65		97.15	113
6140	1/2" C x C, with end switch, 2 wire		20	.400		69.50	13.95		83.45	97.50
6150	3/4" C x C, with end switch, 2 wire		20	.400		74	13.95		87.95	103
6160	1" C x C, with end switch, 2 wire		19	.421		84	14.65		98.65	115
6180	1-1/4" C x C, w/end switch, 2 wire	↓	15	.533	↓	101	18.60		119.60	139
7090	Valves, motor controlled, including actuator									
7100	Electric motor actuated									
7200	Brass, two way, screwed									
7210	1/2" pipe size	L-6	36	.333	Ea.	192	11.45		203.45	228
7220	3/4" pipe size		30	.400		226	13.75		239.75	270
7230	1" pipe size		28	.429		244	14.75		258.75	291
7240	1-1/2" pipe size		19	.632		340	21.50		361.50	410
7250	2" pipe size	↓	16	.750	↓	435	26		461	515
7350	Brass, three way, screwed									
7360	1/2" pipe size	L-6	33	.364	Ea.	207	12.50		219.50	247
7370	3/4" pipe size		27	.444		242	15.30		257.30	289
7380	1" pipe size		25.50	.471		254	16.20		270.20	305
7384	1-1/4" pipe size		21	.571		271	19.65		290.65	330
7390	1-1/2" pipe size		17	.706		365	24.50		389.50	440
7400	2" pipe size	↓	14	.857	↓	480	29.50		509.50	575
7550	Iron body, two way, flanged									
7560	2-1/2" pipe size	L-6	4	3	Ea.	340	103		443	530
7570	3" pipe size		3	4		450	138		588	700
7580	4" pipe size	↓	2	6		640	206		846	1,025
7850	Iron body, three way, flanged									
7860	2-1/2" pipe size	L-6	3	4	Ea.	835	138		973	1,125
7870	3" pipe size		2.50	4.800		980	165		1,145	1,325
7880	4" pipe size	↓	2	6	↓	1,250	206		1,456	1,675
8000	Pneumatic, air operated									
8050	Brass, two way, screwed									
8060	1/2" pipe size, class 250	1 Plum	24	.333	Ea.	96.50	11.50		108	123
8070	3/4" pipe size, class 250		20	.400		129	13.80		142.80	163
8080	1" pipe size, class 250	↓	19	.421	↓	147	14.50		161.50	184

13800 | Building Automation & Control

13838 | Pneumatic/Electric Controls

		CREW	DAILY OUTPUT	LABOR-HOURS	UNIT	2001 BARE COSTS MAT.	LABOR	EQUIP.	TOTAL	TOTAL INCL O&P	
200	8090	1-1/4" pipe size, class 125	1 Plum	15	.533	Ea.	214	18.35		232.35	263	200
	8100	1-1/2" pipe size, class 125		13	.615		245	21		266	300	
	8110	2" pipe size, class 125	↓	11	.727	↓	278	25		303	345	
	8180	Brass, three way, screwed										
	8190	1/2" pipe size, class 250	1 Plum	22	.364	Ea.	148	12.55		160.55	181	
	8200	3/4" pipe size, class 250		18	.444		152	15.30		167.30	190	
	8210	1" pipe size, class 250		17	.471		167	16.20		183.20	209	
	8214	1-1/4" pipe size, class 250		14	.571		239	19.70		258.70	293	
	8220	1-1/2" pipe size, class 125		11	.727		269	25		294	335	
	8230	2" pipe size, class 125	↓	9	.889	↓	480	30.50		510.50	570	
	8250	Iron body, two way, screwed										
	8260	1/2" pipe size, class 300	1 Plum	24	.333	Ea.	720	11.50		731.50	805	
	8270	3/4" pipe size, class 300		20	.400		740	13.80		753.80	835	
	8280	1" pipe size, class 300		19	.421		1,000	14.50		1,014.50	1,125	
	8290	1-1/4" pipe size, class 300		15	.533		1,450	18.35		1,468.35	1,600	
	8300	1-1/2" pipe size, class 300		13	.615		1,475	21		1,496	1,650	
	8310	2" pipe size, class 300		11	.727		1,475	25		1,500	1,675	
	8320	1-1/4" pipe size, class 125		15	.533		735	18.35		753.35	835	
	8330	1-1/2" pipe size, class 125		13	.615		370	21		391	440	
	8340	2" pipe size, class 125	↓	11	.727	↓	445	25		470	530	
	8450	Iron body, two way, flanged										
	8460	2-1/2" pipe size, 250 lb. flanges	Q-1	5	3.200	Ea.	2,700	99		2,799	3,125	
	8470	3" pipe size, 250 lb. flanges		4.50	3.556		2,875	110		2,985	3,350	
	8480	4" pipe size, 250 lb. flanges		3	5.333		3,500	165		3,665	4,125	
	8510	2-1/2" pipe size, class 125		5	3.200		565	99		664	770	
	8520	3" pipe size, class 125		4.50	3.556		675	110		785	905	
	8530	4" pipe size, class 125	↓	3	5.333		865	165		1,030	1,200	
	8540	5" pipe size, class 125	Q-2	3.40	7.059		1,800	227		2,027	2,325	
	8550	6" pipe size, class 125	"	3	8	↓	2,125	257		2,382	2,750	
	8560	Iron body, three way, flanged										
	8570	2-1/2" pipe size, class 125	Q-1	4.50	3.556	Ea.	645	110		755	875	
	8580	3" pipe size, class 125		4	4		800	124		924	1,075	
	8590	4" pipe size, class 125	↓	2.50	6.400		1,600	198		1,798	2,075	
	8600	6" pipe size, class 125	Q-2	3	8	↓	2,200	257		2,457	2,800	
	9005	Pneumatic system misc. components										
	9010	Adding/subtracting repeater	1 Stpi	20	.400	Ea.	96	13.95		109.95	127	
	9020	Adjustable ratio network		16	.500		90	17.45		107.45	126	
	9030	Comparator	↓	20	.400	↓	184	13.95		197.95	223	
	9040	Cumulator										
	9041	Air switching	1 Stpi	20	.400	Ea.	86.50	13.95		100.45	116	
	9042	Averaging		20	.400		90	13.95		103.95	120	
	9043	2:1 ratio		20	.400		81	13.95		94.95	110	
	9044	Sequencing		20	.400		55	13.95		68.95	81.50	
	9045	Two-position		16	.500		99	17.45		116.45	136	
	9046	Two-position pilot	↓	16	.500	↓	93	17.45		110.45	130	
	9050	Damper actuator										
	9051	For smoke control	1 Stpi	8	1	Ea.	125	35		160	191	
	9052	For ventilation	"	8	1	"	136	35		171	203	
	9053	Series duplex pedestal mounted										
	9054	For inlet vanes on fans, compressors	1 Stpi	6	1.333	Ea.	625	46.50		671.50	755	
	9060	Enthalpy logic center		16	.500		195	17.45		212.45	241	
	9070	High/low pressure selector		22	.364		35	12.65		47.65	57.50	
	9080	Signal limiter		20	.400		55	13.95		68.95	81.50	
	9081	Transmitter	↓	28	.286	↓	21	9.95		30.95	38	
	9090	Optimal start										
	9091	Warmup or cooldown programmer	1 Stpi	10	.800	Ea.	208	28		236	271	

13800 | Building Automation & Control

13838 | Pneumatic/Electric Controls

			CREW	DAILY OUTPUT	LABOR-HOURS	UNIT	MAT.	LABOR	EQUIP.	TOTAL	TOTAL INCL O&P	
200	9092	Mass temperature transmitter	1 Stpi	18	.444	Ea.	90.50	15.50		106	123	200
	9100	Step controller with positioner, time delay restrictor										
	9110	recycler air valve, switches and cam settings										
	9112	With cabinet										
	9113	6 points	1 Stpi	4	2	Ea.	425	69.50		494.50	575	
	9114	6 points, 6 manual sequences		2	4		450	139		589	705	
	9115	8 points	▼	3	2.667	▼	540	93		633	735	
	9200	Pressure controller and switches										
	9210	High static pressure limit	1 Stpi	8	1	Ea.	125	35		160	191	
	9220	Pressure transmitter		8	1		136	35		171	202	
	9230	Differential pressure transmitter		8	1		545	35		580	655	
	9240	Static pressure transmitter		8	1		315	35		350	405	
	9250	Proportional-only control, single input		6	1.333		261	46.50		307.50	355	
	9260	Proportional plus integral, single input		6	1.333		390	46.50		436.50	500	
	9270	Proportional-only, dual input		6	1.333		360	46.50		406.50	465	
	9280	Proportional plus integral, dual input		6	1.333		475	46.50		521.50	595	
	9281	Time delay for above proportional units		16	.500		103	17.45		120.45	140	
	9290	Proportional/2 position controller units		6	1.333		276	46.50		322.50	375	
	9300	Differential press. control dir./rev.	▼	6	1.333	▼	186	46.50		232.50	274	
	9310	Air pressure reducing valve										
	9311	1/8" size	1 Stpi	18	.444	Ea.	20	15.50		35.50	45.50	
	9315	Precision valve, 1/4" size		17	.471		106	16.40		122.40	142	
	9320	Air flow controller		12	.667		107	23		130	153	
	9330	Booster relay volume amplifier		18	.444		59	15.50		74.50	88.50	
	9340	Series restrictor, straight		60	.133		3.16	4.65		7.81	10.50	
	9341	T-fitting		40	.200		4.63	6.95		11.58	15.60	
	9342	In-line adjustable		48	.167		20	5.80		25.80	31	
	9350	Diode tee		40	.200		10.65	6.95		17.60	22	
	9351	Restrictor tee		40	.200		10.65	6.95		17.60	22	
	9360	Pneumatic gradual switch		8	1		141	35		176	208	
	9361	Selector switch		8	1		62	35		97	121	
	9370	Electro-pneumatic motor driven servo		10	.800		665	28		693	775	
	9380	Regulated power supply		10	.800		460	28		488	550	
	9390	Fan control switch and mounting base		16	.500		62.50	17.45		79.95	95.50	
	9400	Circulating pump sequencer	▼	14	.571	▼	645	19.90		664.90	740	
	9410	Pneumatic tubing, fittings and accessories										
	9414	Tubing, urethane										
	9415	1/8" OD x 1/16" ID	1 Stpi	120	.067	L.F.	.42	2.32		2.74	3.97	
	9416	1/4" OD x 1/8" ID		115	.070		.83	2.42		3.25	4.57	
	9417	5/32" OD x 3/32" ID	▼	110	.073	▼	.47	2.53		3	4.35	
	9420	Coupling, straight										
	9422	Barb x barb										
	9423	1/4" x 5/32"	1 Stpi	160	.050	Ea.	.47	1.74		2.21	3.15	
	9424	1/4" x 1/4"		158	.051		.25	1.76		2.01	2.94	
	9425	3/8" x 3/8"		154	.052		.32	1.81		2.13	3.08	
	9426	3/8" x 1/4"		150	.053		.32	1.86		2.18	3.16	
	9427	1/2" x 1/4"		148	.054		.70	1.88		2.58	3.61	
	9428	1/2" x 3/8"		144	.056		.63	1.94		2.57	3.61	
	9429	1/2" x 1/2"	▼	140	.057	▼	.47	1.99		2.46	3.53	
	9440	Tube x tube										
	9441	1/4" x 1/4"	1 Stpi	100	.080	Ea.	1.30	2.79		4.09	5.65	
	9442	3/8" x 1/4"		96	.083		1.61	2.90		4.51	6.15	
	9443	3/8" x 3/8"		92	.087		1.61	3.03		4.64	6.35	
	9444	1/2" x 3/8"		88	.091		2.27	3.17		5.44	7.30	
	9445	1/2" x 1/2"	▼	84	.095	▼	2.95	3.32		6.27	8.25	
	9450	Elbow coupling										

13800 | Building Automation & Control

13838 | Pneumatic/Electric Controls

		CREW	DAILY OUTPUT	LABOR-HOURS	UNIT	2001 BARE COSTS MAT.	LABOR	EQUIP.	TOTAL	TOTAL INCL O&P
9452	Barb x barb									
9454	1/4" x 1/4"	1 Stpi	158	.051	Ea.	.52	1.76		2.28	3.23
9455	3/8" x 3/8"		154	.052		.70	1.81		2.51	3.50
9456	1/2" x 1/2"		140	.057		.93	1.99		2.92	4.03
9460	Tube x tube									
9462	1/2" x 1/2"	1 Stpi	84	.095	Ea.	3.62	3.32		6.94	9
9470	Tee coupling									
9472	Barb x barb x barb									
9474	1/4" x 1/4" x 5/32"	1 Stpi	108	.074	Ea.	.84	2.58		3.42	4.82
9475	1/4" x 1/4" x 1/4"		104	.077		.56	2.68		3.24	4.67
9476	3/8" x 3/8" x 5/32"		102	.078		1.81	2.73		4.54	6.10
9477	3/8" x 3/8" x 1/4"		99	.081		.84	2.82		3.66	5.15
9478	3/8" x 3/8" x 3/8"		98	.082		.90	2.84		3.74	5.30
9479	1/2" x 1/2" x 1/4"		96	.083		1.73	2.90		4.63	6.30
9480	1/2" x 1/2" x 3/8"		95	.084		.97	2.93		3.90	5.50
9481	1/2" x 1/2" x 1/2"		92	.087		1.01	3.03		4.04	5.70
9484	Tube x tube x tube									
9485	1/4" x 1/4" x 1/4"	1 Stpi	66	.121	Ea.	2.51	4.22		6.73	9.15
9486	3/8" x 1/4" x 1/4"		64	.125		4.75	4.36		9.11	11.85
9487	3/8" x 3/8" x 1/4"		61	.131		3.72	4.57		8.29	11
9488	3/8" x 3/8" x 3/8"		60	.133		3.91	4.65		8.56	11.30
9489	1/2" x 1/2" x 3/8"		58	.138		6.05	4.81		10.86	13.90
9490	1/2" x 1/2" x 1/2"		56	.143		5.40	4.98		10.38	13.40
9492	Needle valve									
9494	Tube x tube									
9495	1/4" x 1/4"	1 Stpi	86	.093	Ea.	4.57	3.24		7.81	9.95
9496	3/8" x 3/8"	"	82	.098	"	5.10	3.40		8.50	10.75
9600	Electronic system misc. components									
9610	Electric motor damper actuator	1 Elec	8	1	Ea.	237	34.50		271.50	310
9620	Damper position indicator		10	.800		75.50	27.50		103	124
9630	Pneumatic-electronic transducer		12	.667		276	23		299	340
9640	Electro-pneumatic transducer		12	.667		276	23		299	340
9650	Remote reset control with motor drive		6	1.333		315	45.50		360.50	415
9660	Manual remote set point control		16	.500		60.50	17.15		77.65	92.50
9670	Alarm unit with two adjustable points		12	.667		291	23		314	355
9700	Modulating step controller									
9701	2-5 steps	1 Elec	5.30	1.509	Ea.	1,125	52		1,177	1,325
9702	6-10 steps		3.20	2.500		1,450	86		1,536	1,725
9703	11-15 steps		2.70	2.963		1,775	102		1,877	2,100
9704	16-20 steps		1.60	5		2,075	172		2,247	2,550
9710	Staging network									
9711	Master one stage	1 Elec	10	.800	Ea.	380	27.50		407.50	460
9712	Two stage		9	.889		500	30.50		530.50	595
9720	Sequencing network		8	1		575	34.50		609.50	680
9730	Regulated DC power supply		16	.500		350	17.15		367.15	410
9740	Accessory unit regulator									
9741	Three function	1 Elec	9	.889	Ea.	189	30.50		219.50	254
9742	Five function		8	1		365	34.50		399.50	450
9750	Accessory power supply		18	.444		73	15.25		88.25	103
9760	Step down transformer		16	.500		123	17.15		140.15	161

13850 | Detection & Alarm

13851 | Detection & Alarm

			DAILY CREW	LABOR-HOURS	UNIT	MAT.	2001 BARE COSTS LABOR	EQUIP.	TOTAL	TOTAL INCL O&P	
065	0010	**DETECTION SYSTEMS**, not including wires & conduits									065
	0100	Burglar alarm, battery operated, mechanical trigger	1 Elec	4	2	Ea.	249	68.50		317.50	375	
	0200	Electrical trigger		4	2		297	68.50		365.50	425	
	0400	For outside key control, add		8	1		70.50	34.50		105	129	
	0600	For remote signaling circuitry, add		8	1		112	34.50		146.50	174	
	0800	Card reader, flush type, standard		2.70	2.963		835	102		937	1,075	
	1000	Multi-code		2.70	2.963		1,075	102		1,177	1,325	
	3594	Fire, alarm control panel										
	3600	4 zone	2 Elec	2	8	Ea.	920	274		1,194	1,400	
	3800	8 zone		1	16		1,400	550		1,950	2,375	
	4000	12 zone		.67	23.988		1,825	825		2,650	3,225	
	4020	Alarm device	1 Elec	8	1		122	34.50		156.50	185	
	4050	Actuating device		8	1		292	34.50		326.50	370	
	4200	Battery and rack		4	2		690	68.50		758.50	860	
	4400	Automatic charger		8	1		445	34.50		479.50	540	
	4600	Signal bell		8	1		49.50	34.50		84	106	
	4800	Trouble buzzer or manual station		8	1		37	34.50		71.50	91.50	
	5000	Detector, rate of rise		8	1		33.50	34.50		68	88	
	5100	Fixed temperature		8	1		28	34.50		62.50	82	
	5200	Smoke detector, ceiling type		6.20	1.290		75	44.50		119.50	149	
	5400	Duct type		3.20	2.500		250	86		336	405	
	5600	Strobe and horn		5.30	1.509		95	52		147	183	
	5800	Fire alarm horn		6.70	1.194		36.50	41		77.50	101	
	6000	Door holder, electro-magnetic		4	2		77.50	68.50		146	188	
	6200	Combination holder and closer		3.20	2.500		430	86		516	605	
	6400	Code transmitter		4	2		690	68.50		758.50	860	
	6600	Drill switch		8	1		86.50	34.50		121	146	
	6800	Master box		2.70	2.963		3,100	102		3,202	3,550	
	7000	Break glass station		8	1		50	34.50		84.50	106	
	7800	Remote annunciator, 8 zone lamp		1.80	4.444		175	152		327	420	
	8000	12 zone lamp	2 Elec	2.60	6.154		300	211		511	645	
	8200	16 zone lamp	"	2.20	7.273		300	249		549	700	
350	0010	**TANK LEAK DETECTION SYSTEMS** Liquid and vapor										350
	0100	For hydrocarbons and hazardous liquids/vapors										
	0120	Controller, data acquisition, incl. printer, modem, RS232 port										
	0140	24 channel, for use with all probes				Ea.	2,050			2,050	2,250	
	0160	9 channel, for external monitoring				"	1,575			1,575	1,750	
	0200	Probes										
	0210	Well monitoring										
	0220	Liquid phase detection				Ea.	710			710	785	
	0230	Hydrocarbon vapor, fixed position					655			655	720	
	0240	Hydrocarbon vapor, float mounted					580			580	640	
	0250	Both liquid and vapor hydrocarbon					880			880	970	
	0300	Secondary containment, liquid phase										
	0310	Pipe trench/manway sump				Ea.	320			320	350	
	0320	Double wall pipe and manual sump					310			310	340	
	0330	Double wall fiberglass annular space					286			286	315	
	0340	Double wall steel tank annular space					286			286	315	
	0500	Accessories										
	0510	Modem, non-dedicated phone line				Ea.	256			256	282	
	0600	Monitoring, internal										
	0610	Automatic tank gauge, incl. overfill				Ea.	1,025			1,025	1,125	
	0620	Product line				"	1,100			1,100	1,200	
	0700	Monitoring, special										
	0710	Cathodic protection				Ea.	610			610	670	
	0720	Annular space chemical monitor				"	830			830	915	

For information about Means Estimating Seminars, see yellow pages 11 and 12 in back of book

Important: See the Reference Section for critical supporting data - Reference Nos., Crews, & City Cost Indexes

Division 14
Conveying Systems

Estimating Tips
General
- Many products in Division 14 will require some type of support or blocking for installation not included with the item itself. Examples are supports for conveyors or tube systems, attachment points for lifts, and footings for hoists or cranes. Add these supports in the appropriate division.

14100 Dumbwaiters
14200 Elevators
- Dumbwaiters and elevators are estimated and purchased in a method similar to buying a car. The manufacturer has a base unit with standard features. Added to this base unit price will be whatever options the owner or specifications require. Increased load capacity, additional vertical travel, additional stops, higher speed, and cab finish options are items to be considered. When developing an estimate for dumbwaiters and elevators, remember that some items needed by the installers may have to be included as part of the general contract.

Examples are:
- shaftway
- rail support brackets
- machine room
- electrical supply
- sill angles
- electrical connections
- pits
- roof penthouses
- pit ladders

Check the job specifications and drawings before pricing.

14300 Escalators & Moving Walks
- Escalators and moving walks are specialty items installed by specialty contractors. There are numerous options associated with these items. For specific options contact a manufacturer or contractor. In a method similar to estimating dumbwaiters and elevators, you should verify the extent of general contract work and add items as necessary.

14400 Lifts
14500 Material Handling
14600 Hoists & Cranes
- Products such as correspondence lifts, conveyors, chutes, pneumatic tube systems, material handling cranes and hoists as well as other items specified in this subdivision may require trained installers. The general contractor might not have any choice as to who will perform the installation or when it will be performed. Long lead times are often required for these products, making early decisions in scheduling necessary.

Reference Numbers
Reference numbers are shown in bold squares at the beginning of some major classifications. These numbers refer to related items in the Reference Section. The reference information may be an estimating procedure, an alternate pricing method or technical information.

Note: Not all subdivisions listed here necessarily appear in this publication.

14500 | Material Handling

14560 | Chutes

			CREW	DAILY OUTPUT	LABOR-HOURS	UNIT	MAT.	LABOR	EQUIP.	TOTAL	TOTAL INCL O&P	
250	0010	**CHUTES** Linen or refuse, incl. sprinklers, 12' floor height										250
	0050	Aluminized steel, 16 ga., 18" diameter	2 Shee	3.50	4.571	Floor	710	153		863	1,025	
	0100	24" diameter		3.20	5		755	168		923	1,100	
	0200	30" diameter		3	5.333		870	179		1,049	1,225	
	0300	36" diameter		2.80	5.714		1,000	192		1,192	1,400	
	0400	Galvanized steel, 16 ga., 18" diameter		3.50	4.571		700	153		853	1,000	
	0500	24" diameter		3.20	5		790	168		958	1,125	
	0600	30" diameter		3	5.333		885	179		1,064	1,250	
	0700	36" diameter		2.80	5.714		1,050	192		1,242	1,450	
	0800	Stainless steel, 18" diameter		3.50	4.571		1,300	153		1,453	1,675	
	0900	24" diameter		3.20	5		1,350	168		1,518	1,750	
	1000	30" diameter		3	5.333		1,700	179		1,879	2,150	
	1005	36" diameter		2.80	5.714		1,725	192		1,917	2,200	
	1200	Linen chute bottom collector, aluminized steel		4	4	Ea.	980	134		1,114	1,275	
	1300	Stainless steel		4	4		1,250	134		1,384	1,575	
	1500	Refuse, bottom hopper, aluminized steel, 18" diameter		3	5.333		600	179		779	935	
	1600	24" diameter		3	5.333		665	179		844	1,000	
	1800	36" diameter		3	5.333		1,125	179		1,304	1,500	
	2900	Package chutes, spiral type, minimum		4.50	3.556	Floor	1,700	119		1,819	2,050	
	3000	Maximum		1.50	10.667	"	4,425	360		4,785	5,425	

14580 | Pneumatic Tube Systems

			CREW	DAILY OUTPUT	LABOR-HOURS	UNIT	MAT.	LABOR	EQUIP.	TOTAL	TOTAL INCL O&P	
800	0010	**PNEUMATIC TUBE SYSTEM** Single tube, 2 stations, blower										800
	0020	100' long, stock										
	0100	3" diameter	2 Stpi	.12	133	Total	4,800	4,650		9,450	12,300	
	0300	4" diameter	"	.09	177	"	5,400	6,200		11,600	15,300	
	0400	Twin tube, two stations or more, conventional system										
	0600	2-1/2" round	2 Stpi	62.50	.256	L.F.	9.65	8.90		18.55	24	
	0700	3" round		46	.348		11.10	12.10		23.20	30.50	
	0900	4" round		49.60	.323		12.30	11.25		23.55	30.50	
	1000	4" x 7" oval		37.60	.426		17.90	14.85		32.75	42	
	1050	Add for blower		2	8	System	3,775	279		4,054	4,600	
	1110	Plus for each round station, add		7.50	2.133	Ea.	425	74.50		499.50	580	
	1150	Plus for each oval station, add		7.50	2.133	"	425	74.50		499.50	580	
	1200	Alternate pricing method: base cost, minimum		.75	21.333	Total	4,650	745		5,395	6,225	
	1300	Maximum		.25	64	"	9,200	2,225		11,425	13,500	
	1500	Plus total system length, add, minimum		93.40	.171	L.F.	5.95	5.95		11.90	15.55	
	1600	Maximum		37.60	.426	"	17.65	14.85		32.50	42	
	1800	Completely automatic system, 4" round, 15 to 50 stations		.29	55.172	Station	14,600	1,925		16,525	19,000	
	2200	51 to 144 stations		.32	50		11,300	1,750		13,050	15,100	
	2400	6" round or 4" x 7" oval, 15 to 50 stations		.24	66.667		18,300	2,325		20,625	23,600	
	2800	51 to 144 stations		.23	69.565		15,400	2,425		17,825	20,600	

For information about Means Estimating Seminars, see yellow pages 11 and 12 in back of book

Division 15 Mechanical

Estimating Tips

15100 Building Services Piping
This subdivision is primarily basic pipe and related materials. The pipe may be used by any of the mechanical disciplines, i.e., plumbing, fire protection, heating, and air conditioning.

- The piping section lists the add to labor for elevated pipe installation. These adds apply to all elevated pipe, fittings, valves, insulation, etc., that are placed above 10' high. CAUTION: the correct percentage may vary for the same pipe. For example, the percentage add for the basic pipe installation should be based on the maximum height that the craftsman must install for that particular section. If the pipe is to be located 14' above the floor but it is suspended on threaded rod from beams, the bottom flange of which is 18' high (4' rods), then the height is actually 18' and the add is 20%. The pipe coverer, however, does not have to go above the 14' and so his add should be 10%.
- Most pipe is priced first as straight pipe with a joint (coupling, weld, etc.) every 10' and a hanger usually every 10'. There are exceptions with hanger spacing such as: for cast iron pipe (5') and plastic pipe (3 per 10'). Following each type of pipe there are several lines listing sizes and the amount to be subtracted to delete couplings and hangers. This is for pipe that is to be buried or supported together on trapeze hangers. The reason that the couplings are deleted is that these runs are usually long and frequently longer lengths of pipe are used. By deleting the couplings the estimator is expected to look up and add back the correct reduced number of couplings.
- When preparing an estimate it may be necessary to approximate the fittings. Fittings usually run between 25% and 50% of the cost of the pipe. The lower percentage is for simpler runs, and the higher number is for complex areas like mechanical rooms.

15400 Plumbing Fixtures & Equipment
- Plumbing fixture costs usually require two lines, the fixture itself and its "rough-in, supply and waste".
- Remember that gas and oil fired units need venting.

15500 Heat Generation Equipment
- When estimating the cost of an HVAC system, check to see who is responsible for providing and installing the temperature control system. It is possible to overlook controls, assuming that they would be included in the electrical estimate.
- When looking up a boiler be careful on specified capacity. Some manufacturers rate their products on output while others use input.
- Include HVAC insulation for pipe, boiler and duct (wrap and liner).
- Be careful when looking up mechanical items to get the correct pressure rating and connection type (thread, weld, flange).

15700 Heating/Ventilation/Air Conditioning Equipment
- Combination heating and cooling units are sized by the air conditioning requirements. (See Reference No. R15710-020 for preliminary sizing guide.)
- A ton of air conditioning is nominally 400 CFM.
- Rectangular duct is taken off by the linear foot for each size, but its cost is usually estimated by the pound. Remember that SMACNA standards now base duct on internal pressure.
- Prefabricated duct is estimated and purchased like pipe: straight sections and fittings.
- Note that cranes or other lifting equipment are not included on any lines in Division 15. For example, if a crane is required to lift a heavy piece of pipe into place high above a gym floor, or to put a rooftop unit on the roof of a four-story building, etc., it must be added. Due to the potential for extreme variation—from nothing additional required, to a major crane or helicopter—we feel that including a nominal amount for "lifting contingency" would be useless and detract from the accuracy of the estimate. When using equipment rental from Means do not forget to include the cost of the operator(s).

Reference Numbers
Reference numbers are shown in bold squares at the beginning of some major classifications. These numbers refer to related items in the Reference Section. The reference information may be an estimating procedure, an alternate pricing method or technical information.

Note: Not all subdivisions listed here necessarily appear in this publication.

15050 | Basic Materials & Methods

15051 | Mechanical General

			CREW	DAILY OUTPUT	LABOR-HOURS	UNIT	2001 BARE COSTS MAT.	LABOR	EQUIP.	TOTAL	TOTAL INCL O&P	
100	0010	**BOILERS, GENERAL** Prices do not include flue piping, elec. wiring,										100
	0020	gas or oil piping, boiler base, pad, or tankless unless noted										
	0100	Boiler H.P.: 10 KW = 34 lbs/steam/hr = 33,475 BTU/hr. R15500-020										
	0120											
	0150	To convert SFR to BTU rating: Hot water, 150 x SFR;										
	0160	Forced hot water, 180 x SFR; steam, 240 x SFR										
700	0010	**PIPING** See also divisions 02500 & 02600 for site work R15100-070										700
	1000	Add to labor for elevated installation										
	1080	10' to 15' high							10%			
	1100	15' to 20' high							20%			
	1120	20' to 25' high							25%			
	1140	25' to 30' high							35%			
	1160	30' to 35' high							40%			
	1180	35' to 40' high							50%			
	1200	Over 40' high							55%			

15055 | Selective Mech Demolition

			CREW	DAILY OUTPUT	LABOR-HOURS	UNIT	MAT.	LABOR	EQUIP.	TOTAL	TOTAL INCL O&P	
300	0010	**HVAC DEMOLITION** R15050-720										300
	0100	Air conditioner, split unit, 3 ton	Q-5	2	8	Ea.		251		251	380	
	0150	Package unit, 3 ton	Q-6	3	8			260		260	395	
	0300	Boiler, electric	Q-19	2	12			390		390	585	
	0340	Gas or oil, steel, under 150 MBH	Q-6	3	8			260		260	395	
	0380	Over 150 MBH	"	2	12			390		390	590	
	1000	Ductwork, 4" high, 8" wide	1 Clab	200	.040	L.F.		.91		.91	1.43	
	1020	10" wide		190	.042			.96		.96	1.51	
	1040	14" wide		180	.044			1.02		1.02	1.59	
	1100	6" high, 8" wide		165	.048			1.11		1.11	1.73	
	1120	12" wide		150	.053			1.22		1.22	1.91	
	1140	18" wide		135	.059			1.35		1.35	2.12	
	1200	10" high, 12" wide		125	.064			1.46		1.46	2.29	
	1220	18" wide		115	.070			1.59		1.59	2.49	
	1240	24" wide		110	.073			1.66		1.66	2.60	
	1300	12"-14" high, 16"-18" wide		85	.094			2.15		2.15	3.36	
	1320	24" wide		75	.107			2.44		2.44	3.81	
	1340	48" wide		71	.113			2.57		2.57	4.03	
	1400	18" high, 24" wide		67	.119			2.73		2.73	4.27	
	1420	36" wide		63	.127			2.90		2.90	4.54	
	1440	48" wide		59	.136			3.10		3.10	4.85	
	1500	30" high, 36" wide		56	.143			3.26		3.26	5.10	
	1520	48" wide		53	.151			3.45		3.45	5.40	
	1540	72" wide		50	.160			3.66		3.66	5.70	
	1550	Duct heater, electric strip	1 Elec	8	1	Ea.		34.50		34.50	51	
	2200	Furnace, electric	Q-20	2	10			310		310	475	
	2300	Gas or oil, under 120 MBH	Q-9	4	4			121		121	187	
	2340	Over 120 MBH	"	3	5.333			161		161	249	
	2800	Heat pump, package unit, 3 ton	Q-5	2.40	6.667			209		209	315	
	2840	Split unit, 3 ton		2	8			251		251	380	
	3000	Mechanical equipment, light items. Unit is weight, not cooling.		.90	17.778	Ton		560		560	840	
	3600	Heavy items		1.10	14.545	"		455		455	690	
	3700	Deduct for salvage (when applicable), minimum				Job					52	
	3750	Maximum				"					420	
600	0010	**PLUMBING DEMOLITION** R15050-720										600
	1020	Fixtures, including 10' piping										
	1100	Bath tubs, cast iron	1 Plum	4	2	Ea.		69		69	104	
	1120	Fiberglass		6	1.333			46		46	69.50	

15050 | Basic Materials & Methods

15055 | Selective Mech Demolition

		CREW	DAILY OUTPUT	LABOR-HOURS	UNIT	MAT.	LABOR	EQUIP.	TOTAL	TOTAL INCL O&P
600 1140	Steel	1 Plum	5	1.600	Ea.		55		55	83
1200	Lavatory, wall hung		10	.800			27.50		27.50	41.50
1220	Counter top		8	1			34.50		34.50	52
1300	Sink, steel or cast iron, single		8	1			34.50		34.50	52
1320	Double		7	1.143			39.50		39.50	59.50
1400	Water closet, floor mounted		8	1			34.50		34.50	52
1420	Wall mounted		7	1.143			39.50		39.50	59.50
1500	Urinal, floor mounted		4	2			69		69	104
1520	Wall mounted		7	1.143			39.50		39.50	59.50
1600	Water fountains, free standing		8	1			34.50		34.50	52
1620	Recessed		6	1.333			46		46	69.50
2000	Piping, metal, to 1-1/2" diameter		200	.040	L.F.		1.38		1.38	2.08
2050	2" to 3-1/2" diameter		150	.053			1.84		1.84	2.77
2100	4" to 6" diameter	2 Plum	100	.160			5.50		5.50	8.30
2150	8" to 14" diameter		60	.267			9.20		9.20	13.85
2153	16" diameter		55	.291			10		10	15.15
2160	Deduct for salvage, aluminum scrap				Ton					525
2170	Brass scrap									420
2180	Copper scrap									1,025
2200	Lead scrap									235
2220	Steel scrap									65
2250	Water heater, 40 gal.	1 Plum	6	1.333	Ea.		46		46	69.50
6000	Remove and reset fixtures, minimum		6	1.333			46		46	69.50
6100	Maximum		4	2			69		69	104

15060 | Hangers & Supports

		CREW	DAILY OUTPUT	LABOR-HOURS	UNIT	MAT.	LABOR	EQUIP.	TOTAL	TOTAL INCL O&P
300 0010	**PIPE HANGERS AND SUPPORTS**, TYPE numbers per MSS-SP58									
0050	Brackets									
0060	Beam side or wall, malleable iron, TYPE 34									
0070	3/8" threaded rod size	1 Plum	48	.167	Ea.	1.73	5.75		7.48	10.55
0080	1/2" threaded rod size		48	.167		3.33	5.75		9.08	12.30
0090	5/8" threaded rod size		48	.167		4.43	5.75		10.18	13.50
0100	3/4" threaded rod size		48	.167		4.73	5.75		10.48	13.85
0110	7/8" threaded rod size		48	.167		6.10	5.75		11.85	15.35
0120	For concrete installation, add						30%			
0150	Wall, welded steel, medium, TYPE 32									
0160	0 size, 12" wide, 18" deep	1 Plum	34	.235	Ea.	126	8.10		134.10	151
0170	1 size, 18" wide 24" deep		34	.235		150	8.10		158.10	177
0180	2 size, 24" wide, 30" deep		34	.235		198	8.10		206.10	230
0200	Beam attachment, welded, TYPE 22									
0202	3/8"	Q-15	80	.200	Ea.	11.60	6.20	1.26	19.06	23.50
0203	1/2"		76	.211		12.35	6.55	1.32	20.22	25
0204	5/8"		72	.222		13.10	6.90	1.40	21.40	26.50
0205	3/4"		68	.235		18.20	7.30	1.48	26.98	32.50
0206	7/8"		64	.250		23	7.75	1.57	32.32	38.50
0207	1"		56	.286		24.50	8.85	1.80	35.15	42
0300	Clamps									
0310	C-clamp, for mounting on steel beam flange, w/locknut, TYPE 23									
0320	3/8" threaded rod size	1 Plum	160	.050	Ea.	2.02	1.72		3.74	4.82
0330	1/2" threaded rod size		160	.050		2.42	1.72		4.14	5.25
0340	5/8" threaded rod size		160	.050		3.46	1.72		5.18	6.40
0350	3/4" threaded rod size		160	.050		5	1.72		6.72	8.10
0352	7/8" threaded rod size		140	.057		13.10	1.97		15.07	17.35
0400	High temperature to 1050°F, alloy steel									
0410	4" pipe size	Q-1	106	.151	Ea.	31	4.68		35.68	41
0420	6" pipe size		106	.151		52	4.68		56.68	64.50

15050 | Basic Materials & Methods

15060 | Hangers & Supports

		CREW	DAILY OUTPUT	LABOR-HOURS	UNIT	MAT.	LABOR	EQUIP.	TOTAL	TOTAL INCL O&P
0430	8" pipe size	Q-1	97	.165	Ea.	58.50	5.10		63.60	72
0440	10" pipe size		84	.190		79.50	5.90		85.40	96.50
0450	12" pipe size		72	.222		92.50	6.90		99.40	112
0460	14" pipe size		64	.250		219	7.75		226.75	253
0470	16" pipe size		56	.286		253	8.85		261.85	291
0480	Beam clamp, flange type, TYPE 25									
0482	For 3/8" bolt	1 Plum	48	.167	Ea.	5.60	5.75		11.35	14.85
0483	For 1/2" bolt		44	.182		6.30	6.25		12.55	16.40
0484	For 5/8" bolt		40	.200		8.15	6.90		15.05	19.35
0485	For 3/4" bolt		36	.222		12.60	7.65		20.25	25.50
0486	For 1" bolt		32	.250		36.50	8.60		45.10	53
0500	I-beam, for mounting on bottom flange, strap iron, TYPE 21									
0510	2" flange size	1 Plum	96	.083	Ea.	3.72	2.87		6.59	8.40
0520	3" flange size		95	.084		3.84	2.90		6.74	8.60
0530	4" flange size		93	.086		3.93	2.96		6.89	8.80
0540	5" flange size		92	.087		4.08	3		7.08	9
0550	6" flange size		90	.089		4.17	3.06		7.23	9.20
0560	7" flange size		88	.091		4.31	3.13		7.44	9.45
0570	8" flange size		86	.093		4.46	3.20		7.66	9.75
0600	One hole, vertical mounting, malleable iron									
0610	1/2" pipe size	1 Plum	160	.050	Ea.	.71	1.72		2.43	3.38
0620	3/4" pipe size		145	.055		.88	1.90		2.78	3.84
0630	1" pipe size		136	.059		1.33	2.03		3.36	4.52
0640	1-1/4" pipe size		128	.063		1.66	2.15		3.81	5.10
0650	1-1/2" pipe size		120	.067		1.87	2.30		4.17	5.55
0660	2" pipe size		112	.071		2.82	2.46		5.28	6.80
0670	2-1/2" pipe size		104	.077		7.70	2.65		10.35	12.45
0680	3" pipe size		96	.083		10.65	2.87		13.52	16.10
0690	3-1/2" pipe size		90	.089		10.65	3.06		13.71	16.35
0700	4" pipe size		84	.095		41.50	3.28		44.78	50.50
0750	Riser or extension pipe, carbon steel, TYPE 8									
0756	1/2" pipe size	1 Plum	52	.154	Ea.	2.61	5.30		7.91	10.85
0760	3/4" pipe size		48	.167		2.61	5.75		8.36	11.50
0770	1" pipe size		47	.170		2.63	5.85		8.48	11.75
0780	1-1/4" pipe size		46	.174		3.31	6		9.31	12.70
0790	1-1/2" pipe size		45	.178		3.56	6.10		9.66	13.15
0800	2" pipe size		43	.186		3.62	6.40		10.02	13.65
0810	2-1/2" pipe size		41	.195		3.80	6.70		10.50	14.35
0820	3" pipe size		40	.200		4.05	6.90		10.95	14.85
0830	3-1/2" pipe size		39	.205		5	7.05		12.05	16.15
0840	4" pipe size		38	.211		5.10	7.25		12.35	16.55
0850	5" pipe size		37	.216		7.45	7.45		14.90	19.40
0860	6" pipe size		36	.222		8.55	7.65		16.20	21
0870	8" pipe size		34	.235		13.85	8.10		21.95	27.50
0880	10" pipe size		32	.250		22.50	8.60		31.10	38
0890	12" pipe size		28	.286		27	9.85		36.85	44.50
0900	For plastic coating 3/4" to 4", add					190%				
0910	For copper plating 3/4" to 4", add					58%				
0950	Two piece, complete, carbon steel, medium weight, TYPE 4									
0960	1/2" pipe size	Q-1	137	.117	Ea.	2.10	3.62		5.72	7.75
0970	3/4" pipe size		134	.119		2.10	3.70		5.80	7.90
0980	1" pipe size		132	.121		2.22	3.76		5.98	8.10
0990	1-1/4" pipe size		130	.123		2.76	3.82		6.58	8.80
1000	1-1/2" pipe size		126	.127		2.76	3.94		6.70	9
1010	2" pipe size		124	.129		3.41	4		7.41	9.80
1020	2-1/2" pipe size		120	.133		3.67	4.13		7.80	10.30

15050 | Basic Materials & Methods

15060 | Hangers & Supports

		CREW	DAILY OUTPUT	LABOR-HOURS	UNIT	2001 BARE COSTS MAT.	LABOR	EQUIP.	TOTAL	TOTAL INCL O&P
1030	3" pipe size	Q-1	117	.137	Ea.	4.09	4.24		8.33	10.90
1040	3-1/2" pipe size		114	.140		5.80	4.35		10.15	12.95
1050	4" pipe size		110	.145		5.80	4.51		10.31	13.20
1060	5" pipe size		106	.151		8.40	4.68		13.08	16.30
1070	6" pipe size		104	.154		14.15	4.77		18.92	23
1080	8" pipe size		100	.160		16.90	4.96		21.86	26
1090	10" pipe size		96	.167		30.50	5.15		35.65	41.50
1100	12" pipe size		89	.180		35.50	5.55		41.05	47.50
1110	14" pipe size		82	.195		61.50	6.05		67.55	76.50
1120	16" pipe size		68	.235		66.50	7.30		73.80	84
1130	For galvanized, add					45%				
1150	Insert, concrete									
1160	Wedge type, carbon steel body, malleable iron nut									
1170	1/4" threaded rod size	1 Plum	96	.083	Ea.	4.75	2.87		7.62	9.60
1180	3/8" threaded rod size		96	.083		4.99	2.87		7.86	9.85
1190	1/2" threaded rod size		96	.083		5.15	2.87		8.02	10.05
1200	5/8" threaded rod size		96	.083		5.25	2.87		8.12	10.10
1210	3/4" threaded rod size		96	.083		5.75	2.87		8.62	10.70
1220	7/8" threaded rod size		96	.083		6.70	2.87		9.57	11.70
1230	For galvanized, add					1.07			1.07	1.18
1250	Pipe guide sized for insulation									
1260	No. 1, 1" pipe size, 1" thick insulation	1 Stpi	26	.308	Ea.	125	10.70		135.70	154
1270	No. 2, 1-1/4"-2" pipe size, 1" thick insulation		23	.348		148	12.10		160.10	181
1280	No. 3, 1-1/4"-2" pipe size, 1-1/2" thick insulation		21	.381		148	13.30		161.30	183
1290	No. 4, 2-1/2"-3-1/2" pipe size, 1-1/2" thick insulation		18	.444		148	15.50		163.50	187
1300	No. 5, 4"-5" pipe size, 1-1/2" thick insulation		16	.500		164	17.45		181.45	208
1310	No. 6, 5"-6" pipe size, 2" thick insulation	Q-5	21	.762		182	24		206	237
1320	No. 7, 8" pipe size, 2" thick insulation		16	1		320	31.50		351.50	400
1330	No. 8, 10" pipe size, 2" thick insulation		12	1.333		605	42		647	730
1340	No. 9, 12" pipe size, 2" thick insulation	Q-6	17	1.412		605	46		651	735
1350	No. 10, 12"-14" pipe size, 2-1/2" thick insulation		16	1.500		785	49		834	940
1360	No. 11, 16" pipe size, 2-1/2" thick insulation		10.50	2.286		785	74.50		859.50	975
1370	No. 12, 16"-18" pipe size, 3" thick insulation		9	2.667		830	87		917	1,050
1380	No. 13, 20" pipe size, 3" thick insulation		7.50	3.200		830	104		934	1,075
1390	No. 14, 24" pipe size, 3" thick insulation		7	3.429		1,050	112		1,162	1,350
1400	Bands									
1410	Adjustable band, carbon steel, for non-insulated pipe, TYPE 7									
1420	1/2" pipe size	Q-1	142	.113	Ea.	.82	3.49		4.31	6.15
1430	3/4" pipe size		140	.114		.82	3.54		4.36	6.25
1440	1" pipe size		137	.117		.93	3.62		4.55	6.45
1450	1-1/4" pipe size		134	.119		.97	3.70		4.67	6.65
1460	1-1/2" pipe size		131	.122		1.09	3.79		4.88	6.90
1470	2" pipe size		129	.124		1.50	3.84		5.34	7.45
1480	2-1/2" pipe size		125	.128		1.58	3.97		5.55	7.75
1490	3" pipe size		122	.131		1.76	4.07		5.83	8.10
1500	3-1/2" pipe size		119	.134		2.73	4.17		6.90	9.30
1510	4" pipe size		114	.140		2.73	4.35		7.08	9.55
1520	5" pipe size		110	.145		2.98	4.51		7.49	10.10
1530	6" pipe size		108	.148		4	4.59		8.59	11.35
1540	8" pipe size		104	.154		6.25	4.77		11.02	14.05
1550	For copper plated, add					50%				
1560	For galvanized, add					30%				
1570	For plastic coating, add					30%				
1600	Adjusting nut malleable iron, steel band, TYPE 9									
1610	1/2" pipe size, galvanized band	Q-1	137	.117	Ea.	.49	3.62		4.11	6
1620	3/4" pipe size, galvanized band		135	.119		.49	3.67		4.16	6.10

15050 | Basic Materials & Methods

15060 | Hangers & Supports

			CREW	DAILY OUTPUT	LABOR-HOURS	UNIT	2001 BARE COSTS MAT.	LABOR	EQUIP.	TOTAL	TOTAL INCL O&P	
300	1630	1" pipe size, galvanized band	Q-1	132	.121	Ea.	.49	3.76		4.25	6.20	300
	1640	1-1/4" pipe size, galvanized band		129	.124		.54	3.84		4.38	6.40	
	1650	1-1/2" pipe size, galvanized band		126	.127		.54	3.94		4.48	6.55	
	1660	2" pipe size, galvanized band		124	.129		.57	4		4.57	6.70	
	1670	2-1/2" pipe size, galvanized band		120	.133		1.08	4.13		5.21	7.45	
	1680	3" pipe size, galvanized band		117	.137		1.17	4.24		5.41	7.70	
	1690	3-1/2" pipe size, galvanized band		114	.140		1.45	4.35		5.80	8.15	
	1700	4" pipe size, cadmium plated band		110	.145		2.17	4.51		6.68	9.20	
	1710	5" pipe size, cadmium plated band		106	.151		2.94	4.68		7.62	10.30	
	1720	6" pipe size, cadmium plated band		104	.154		3.29	4.77		8.06	10.80	
	1730	8" pipe size, cadmium plated band		100	.160		3.70	4.96		8.66	11.55	
	1740	For plastic coated band, add					35%					
	1750	For completely copper coated, add					45%					
	1800	Clevis, adjustable, carbon steel, for non-insulated pipe, TYPE 1										
	1810	1/2" pipe size	Q-1	137	.117	Ea.	.81	3.62		4.43	6.35	
	1820	3/4" pipe size		135	.119		.81	3.67		4.48	6.45	
	1830	1" pipe size		132	.121		.86	3.76		4.62	6.60	
	1840	1-1/4" pipe size		129	.124		.97	3.84		4.81	6.85	
	1850	1-1/2" pipe size		126	.127		1.07	3.94		5.01	7.15	
	1860	2" pipe size		124	.129		1.18	4		5.18	7.35	
	1870	2-1/2" pipe size		120	.133		1.92	4.13		6.05	8.35	
	1880	3" pipe size		117	.137		2.42	4.24		6.66	9.05	
	1890	3-1/2" pipe size		114	.140		2.53	4.35		6.88	9.35	
	1900	4" pipe size		110	.145		2.92	4.51		7.43	10	
	1910	5" pipe size		106	.151		4.03	4.68		8.71	11.50	
	1920	6" pipe size		104	.154		4.64	4.77		9.41	12.30	
	1930	8" pipe size		100	.160		7.65	4.96		12.61	15.90	
	1940	10" pipe size		96	.167		20.50	5.15		25.65	30.50	
	1950	12" pipe size		89	.180		26	5.55		31.55	37	
	1960	14" pipe size		82	.195		40	6.05		46.05	53	
	1970	16" pipe size		68	.235		56.50	7.30		63.80	73.50	
	1971	18" pipe size		54	.296		67.50	9.20		76.70	88.50	
	1972	20" pipe size		38	.421		133	13.05		146.05	166	
	1980	For galvanized, add					66%					
	1990	For copper plated 1/2" to 4", add					77%					
	2000	For light weight 1/2" to 4", deduct					13%					
	2010	Insulated pipe type, 3/4" to 12" pipe, add					180%					
	2020	Insulated pipe type, chrome-moly U-strap, add					530%					
	2250	Split ring, malleable iron, for non-insulated pipe, TYPE 11										
	2260	1/2" pipe size	Q-1	137	.117	Ea.	2.94	3.62		6.56	8.70	
	2270	3/4" pipe size		135	.119		2.96	3.67		6.63	8.80	
	2280	1" pipe size		132	.121		3.13	3.76		6.89	9.10	
	2290	1-1/4" pipe size		129	.124		3.88	3.84		7.72	10.05	
	2300	1-1/2" pipe size		126	.127		4.72	3.94		8.66	11.15	
	2310	2" pipe size		124	.129		5.35	4		9.35	11.95	
	2320	2-1/2" pipe size		120	.133		7.70	4.13		11.83	14.75	
	2330	3" pipe size		117	.137		9.65	4.24		13.89	17	
	2340	3-1/2" pipe size		114	.140		13.35	4.35		17.70	21.50	
	2350	4" pipe size		110	.145		13.35	4.51		17.86	21.50	
	2360	5" pipe size		106	.151		30	4.68		34.68	40	
	2370	6" pipe size		104	.154		30	4.77		34.77	40	
	2380	8" pipe size		100	.160		53	4.96		57.96	65.50	
	2390	For copper plated, add					8%					
	2500	Washer, flat steel										
	2502	3/8"	1 Plum	240	.033	Ea.	.04	1.15		1.19	1.77	
	2503	1/2"		220	.036		.07	1.25		1.32	1.97	

15050 | Basic Materials & Methods

15060 | Hangers & Supports

		Crew	Daily Output	Labor-Hours	Unit	2001 Bare Costs Mat.	Labor	Equip.	Total	Total Incl O&P
2504	5/8"	1 Plum	200	.040	Ea.	.15	1.38		1.53	2.25
2505	3/4"		180	.044		.11	1.53		1.64	2.43
2506	7/8"		160	.050		.15	1.72		1.87	2.77
2507	1"		140	.057		.16	1.97		2.13	3.15
2508	1-1/8"		120	.067		.20	2.30		2.50	3.69
2520	Nut, steel, hex									
2522	3/8"	1 Plum	200	.040	Ea.	.22	1.38		1.60	2.32
2523	1/2"		180	.044		.39	1.53		1.92	2.74
2524	5/8"		160	.050		.82	1.72		2.54	3.50
2525	3/4"		140	.057		1.10	1.97		3.07	4.18
2526	7/8"		120	.067		1.85	2.30		4.15	5.50
2527	1"		100	.080		2	2.76		4.76	6.35
2528	1-1/8"		80	.100		2.36	3.45		5.81	7.80
2532	Turnbuckle, TYPE 13									
2534	3/8"	1 Plum	80	.100	Ea.	13.50	3.45		16.95	20
2535	1/2"		72	.111		16.90	3.83		20.73	24.50
2536	5/8"		64	.125		17.55	4.31		21.86	26
2538	7/8"		48	.167		32.50	5.75		38.25	44
2539	1"		40	.200		39	6.90		45.90	53.50
2540	1-1/8"		32	.250		70	8.60		78.60	90
2544	Eye rod, welded									
2546	3/8"	1 Plum	56	.143	Ea.	17.95	4.92		22.87	27
2547	1/2"		48	.167		26.50	5.75		32.25	37.50
2548	5/8"		40	.200		40	6.90		46.90	55
2549	3/4"		32	.250		51	8.60		59.60	69
2550	7/8"		24	.333		62.50	11.50		74	86.50
2551	1"		16	.500		79	17.25		96.25	113
2556	Eye rod, welded, linked									
2558	3/8"	1 Plum	56	.143	Ea.	55.50	4.92		60.42	68.50
2559	1/2"		48	.167		75.50	5.75		81.25	91.50
2560	5/8"		40	.200		108	6.90		114.90	129
2561	3/4"		32	.250		140	8.60		148.60	167
2562	7/8"		24	.333		171	11.50		182.50	205
2563	1"		16	.500		220	17.25		237.25	267
2650	Rods, carbon steel									
2660	Continuous thread									
2670	1/4" thread size	1 Plum	144	.056	L.F.	.45	1.91		2.36	3.39
2680	3/8" thread size		144	.056		.50	1.91		2.41	3.44
2690	1/2" thread size		144	.056		.78	1.91		2.69	3.75
2700	5/8" thread size		144	.056		1.11	1.91		3.02	4.11
2710	3/4" thread size		144	.056		1.97	1.91		3.88	5.05
2720	7/8" thread size		144	.056		2.47	1.91		4.38	5.60
2721	1" thread size	Q-1	160	.100		4.19	3.10		7.29	9.30
2722	1-1/8" thread size	"	120	.133		5.35	4.13		9.48	12.15
2725	1/4" thread size, bright finish	1 Plum	144	.056		.84	1.91		2.75	3.81
2726	1/2" thread size, bright finish	"	144	.056		1.47	1.91		3.38	4.51
2730	For galvanized, add					40%				
2750	Both ends machine threaded 18" length									
2760	3/8" thread size	1 Plum	240	.033	Ea.	3.58	1.15		4.73	5.65
2770	1/2" thread size		240	.033		5.75	1.15		6.90	8.05
2780	5/8" thread size		240	.033		8.25	1.15		9.40	10.80
2790	3/4" thread size		240	.033		12.65	1.15		13.80	15.65
2800	7/8" thread size		240	.033		22	1.15		23.15	25.50
2810	1" thread size		240	.033		31.50	1.15		32.65	36
2820	Rod couplings									
2821	3/8"	1 Plum	60	.133	Ea.	1.26	4.59		5.85	8.35

15050 | Basic Materials & Methods

15060 | Hangers & Supports

			CREW	DAILY OUTPUT	LABOR-HOURS	UNIT	MAT.	LABOR	EQUIP.	TOTAL	TOTAL INCL O&P	
300	2822	1/2"	1 Plum	54	.148	Ea.	1.60	5.10		6.70	9.45	300
	2823	5/8"		48	.167		2.46	5.75		8.21	11.35	
	2824	3/4"		44	.182		3.59	6.25		9.84	13.40	
	2825	7/8"		40	.200		5.15	6.90		12.05	16.05	
	2826	1"		34	.235		6.90	8.10		15	19.85	
	2827	1-1/8"		30	.267		12.90	9.20		22.10	28	
	2860	Pipe hanger assy, adj. clevis, saddle, rod, clamp, insul. allowance										
	2864	1/2" pipe size	Q-5	35	.457	Ea.	21.50	14.35		35.85	45	
	2868	1" pipe size		34.60	.462		27	14.50		41.50	51.50	
	2872	2" pipe size		33.30	.480		30.50	15.10		45.60	56.50	
	2876	3" pipe size		31.20	.513		46	16.10		62.10	75	
	2880	4" pipe size		30.70	.521		48.50	16.35		64.85	78	
	2884	6" pipe size		29.80	.537		60.50	16.85		77.35	92.50	
	2888	8" pipe size		28	.571		96.50	17.95		114.45	133	
	2892	10" pipe size		25.20	.635		132	19.90		151.90	175	
	2896	12" pipe size		23.20	.690		146	21.50		167.50	194	
	2900	Rolls										
	2910	Adjustable yoke, carbon steel with CI roll, TYPE 43										
	2918	2" pipe size	Q-1	140	.114	Ea.	11.75	3.54		15.29	18.30	
	2920	2-1/2" pipe size		137	.117		11.75	3.62		15.37	18.40	
	2930	3" pipe size		131	.122		13.10	3.79		16.89	20	
	2940	3-1/2" pipe size		124	.129		17.05	4		21.05	25	
	2950	4" pipe size		117	.137		17.65	4.24		21.89	26	
	2960	5" pipe size		110	.145		20.50	4.51		25.01	29.50	
	2970	6" pipe size		104	.154		25.50	4.77		30.27	35.50	
	2980	8" pipe size		96	.167		38.50	5.15		43.65	50	
	2990	10" pipe size		80	.200		45.50	6.20		51.70	59.50	
	3000	12" pipe size		68	.235		67	7.30		74.30	85	
	3010	14" pipe size		56	.286		180	8.85		188.85	211	
	3020	16" pipe size		48	.333		231	10.35		241.35	270	
	3050	Chair, carbon steel with CI roll										
	3060	2" pipe size	1 Plum	68	.118	Ea.	15.90	4.05		19.95	23.50	
	3070	2-1/2" pipe size		65	.123		17.20	4.24		21.44	25.50	
	3080	3" pipe size		62	.129		18.60	4.45		23.05	27	
	3090	3-1/2" pipe size		60	.133		22.50	4.59		27.09	31.50	
	3100	4" pipe size		58	.138		25	4.75		29.75	34.50	
	3110	5" pipe size		56	.143		27.50	4.92		32.42	37.50	
	3120	6" pipe size		53	.151		38	5.20		43.20	50	
	3130	8" pipe size		50	.160		51.50	5.50		57	65	
	3140	10" pipe size		48	.167		64.50	5.75		70.25	79.50	
	3150	12" pipe size		46	.174		94	6		100	112	
	3170	Single pipe roll, (see line 2650 for rods), TYPE 41, 1" pipe size	Q-1	137	.117		8.50	3.62		12.12	14.80	
	3180	1-1/4" pipe size		131	.122		8.50	3.79		12.29	15.05	
	3190	1-1/2" pipe size		129	.124		8.60	3.84		12.44	15.30	
	3200	2" pipe size		124	.129		9.05	4		13.05	16	
	3210	2-1/2" pipe size		118	.136		9.70	4.20		13.90	17.05	
	3220	3" pipe size		115	.139		10.25	4.31		14.56	17.80	
	3230	3-1/2" pipe size		113	.142		11.95	4.39		16.34	19.75	
	3240	4" pipe size		112	.143		13.05	4.43		17.48	21	
	3250	5" pipe size		110	.145		14.45	4.51		18.96	22.50	
	3260	6" pipe size		101	.158		18.75	4.91		23.66	28	
	3270	8" pipe size		90	.178		29	5.50		34.50	40.50	
	3280	10" pipe size		80	.200		34.50	6.20		40.70	47.50	
	3290	12" pipe size		68	.235		50	7.30		57.30	66	
	3291	14" pipe size		56	.286		82.50	8.85		91.35	104	
	3292	16" pipe size		48	.333		112	10.35		122.35	139	

15050 | Basic Materials & Methods
15060 | Hangers & Supports

			CREW	DAILY OUTPUT	LABOR-HOURS	UNIT	2001 BARE COSTS MAT.	LABOR	EQUIP.	TOTAL	TOTAL INCL O&P	
300	3293	18" pipe size	Q-1	40	.400	Ea.	120	12.40		132.40	151	300
	3294	20" pipe size		35	.457		165	14.15		179.15	203	
	3296	24" pipe size		30	.533		261	16.55		277.55	310	
	3297	30" pipe size		25	.640		515	19.85		534.85	595	
	3298	36" pipe size		20	.800		530	25		555	620	
	3300	Saddles (add vertical pipe riser, usually 3" diameter)										
	3310	Pipe support, complete, adjust., CI saddle, TYPE 36										
	3320	2-1/2" pipe size	1 Plum	96	.083	Ea.	127	2.87		129.87	144	
	3330	3" pipe size		88	.091		129	3.13		132.13	147	
	3340	3-1/2" pipe size		79	.101		130	3.49		133.49	148	
	3350	4" pipe size		68	.118		168	4.05		172.05	191	
	3360	5" pipe size		64	.125		174	4.31		178.31	198	
	3370	6" pipe size		59	.136		174	4.67		178.67	198	
	3380	8" pipe size		53	.151		184	5.20		189.20	210	
	3390	10" pipe size		50	.160		203	5.50		208.50	232	
	3400	12" pipe size		48	.167		222	5.75		227.75	253	
	3450	For standard pipe support, one piece, CI deduct					34%					
	3460	For stanchion support, CI with steel yoke, deduct					60%					
	3500	Multiple pipe support 2' long strut	1 Plum	30	.267	Ea.	4.40	9.20		13.60	18.70	
	3510	3' long strut		20	.400		6.60	13.80		20.40	28.50	
	3520	4' long strut		17	.471		8.80	16.20		25	34	
	3550	Insulation shield 1" thick, 1/2" pipe size, TYPE 40	1 Asbe	100	.080		5.55	2.57		8.12	10.25	
	3560	3/4" pipe size		100	.080		5.55	2.57		8.12	10.25	
	3570	1" pipe size		98	.082		6.30	2.62		8.92	11.15	
	3580	1-1/4" pipe size		98	.082		6.90	2.62		9.52	11.80	
	3590	1-1/2" pipe size		96	.083		6.95	2.68		9.63	11.95	
	3600	2" pipe size		96	.083		7.30	2.68		9.98	12.35	
	3610	2-1/2" pipe size		94	.085		7.65	2.74		10.39	12.85	
	3620	3" pipe size		94	.085		8	2.74		10.74	13.20	
	3630	2" thick, 3-1/2" pipe size		92	.087		11	2.80		13.80	16.60	
	3640	4" pipe size		92	.087		11	2.80		13.80	16.60	
	3650	5" pipe size		90	.089		16.55	2.86		19.41	23	
	3660	6" pipe size		90	.089		18.65	2.86		21.51	25	
	3670	8" pipe size		88	.091		28	2.92		30.92	35	
	3680	10" pipe size		88	.091		41.50	2.92		44.42	50	
	3690	12" pipe size		86	.093		47.50	2.99		50.49	57	
	3700	14" pipe size		86	.093		47.50	2.99		50.49	57.50	
	3710	16" pipe size		84	.095		64	3.06		67.06	75.50	
	3720	18" pipe size		84	.095		64	3.06		67.06	75.50	
	3730	20" pipe size		82	.098		71	3.14		74.14	83	
	3732	24" pipe size		80	.100		84.50	3.22		87.72	98	
	3733	30" pipe size		78	.103		92	3.30		95.30	106	
	3735	36" pipe size		75	.107		101	3.43		104.43	117	
	3750	Covering protection saddle, TYPE 39										
	3760	1" covering size										
	3770	3/4" pipe size	1 Plum	68	.118	Ea.	10	4.05		14.05	17.10	
	3780	1" pipe size		68	.118		10	4.05		14.05	17.10	
	3790	1-1/4" pipe size		68	.118		10	4.05		14.05	17.10	
	3800	1-1/2" pipe size		66	.121		10.85	4.18		15.03	18.25	
	3810	2" pipe size		66	.121		10.85	4.18		15.03	18.25	
	3820	2-1/2" pipe size		64	.125		10.85	4.31		15.16	18.45	
	3830	3" pipe size		64	.125		15.65	4.31		19.96	23.50	
	3840	3-1/2" pipe size		62	.129		16.70	4.45		21.15	25	
	3850	4" pipe size		62	.129		16.70	4.45		21.15	25	
	3860	5" pipe size		60	.133		16.70	4.59		21.29	25.50	
	3870	6" pipe size		60	.133		19.90	4.59		24.49	29	

15050 | Basic Materials & Methods

15060 | Hangers & Supports

		CREW	DAILY OUTPUT	LABOR-HOURS	UNIT	2001 BARE COSTS MAT.	LABOR	EQUIP.	TOTAL	TOTAL INCL O&P
3900	1-1/2" covering size									
3910	3/4" pipe size	1 Plum	68	.118	Ea.	12.50	4.05		16.55	19.85
3920	1" pipe size		68	.118		12.50	4.05		16.55	19.85
3930	1-1/4" pipe size		68	.118		12.50	4.05		16.55	19.85
3940	1-1/2" pipe size		66	.121		12.50	4.18		16.68	20
3950	2" pipe size		66	.121		13.55	4.18		17.73	21.50
3960	2-1/2" pipe size		64	.125		16	4.31		20.31	24
3970	3" pipe size		64	.125		16	4.31		20.31	24
3980	3-1/2" pipe size		62	.129		16.25	4.45		20.70	24.50
3990	4" pipe size		62	.129		16.25	4.45		20.70	24.50
4000	5" pipe size		60	.133		16.25	4.59		20.84	25
4010	6" pipe size		60	.133		23.50	4.59		28.09	32.50
4020	8" pipe size		58	.138		29.50	4.75		34.25	39
4022	10" pipe size		56	.143		29.50	4.92		34.42	39.50
4024	12" pipe size		54	.148		43	5.10		48.10	54.50
4028	2" covering size									
4029	2-1/2" pipe size	1 Plum	62	.129	Ea.	15.35	4.45		19.80	23.50
4032	3" pipe size		60	.133		18.05	4.59		22.64	27
4033	4" pipe size		58	.138		18.05	4.75		22.80	27
4034	6" pipe size		56	.143		26	4.92		30.92	36
4035	8" pipe size		54	.148		30.50	5.10		35.60	41
4080	10" pipe size		58	.138		32.50	4.75		37.25	42.50
4090	12" pipe size		56	.143		48	4.92		52.92	60.50
4100	14" pipe size		56	.143		48	4.92		52.92	60.50
4110	16" pipe size		54	.148		67.50	5.10		72.60	81.50
4120	18" pipe size		54	.148		67.50	5.10		72.60	81.50
4130	20" pipe size		52	.154		75	5.30		80.30	90.50
4150	24" pipe size		50	.160		87.50	5.50		93	104
4160	30" pipe size		48	.167		96	5.75		101.75	115
4180	36" pipe size		45	.178		106	6.10		112.10	126
4186	2-1/2" covering size									
4187	3" pipe size	1 Plum	58	.138	Ea.	17.45	4.75		22.20	26.50
4188	4" pipe size		56	.143		17.45	4.92		22.37	26.50
4189	6" pipe size		52	.154		29	5.30		34.30	40
4190	8" pipe size		48	.167		35.50	5.75		41.25	48
4191	10" pipe size		44	.182		39	6.25		45.25	52.50
4192	12" pipe size		40	.200		52	6.90		58.90	68
4193	14" pipe size		36	.222		52	7.65		59.65	69
4194	16" pipe size		32	.250		77	8.60		85.60	97.50
4195	18" pipe size		28	.286		82	9.85		91.85	105
4200	Sockets									
4210	Rod end, malleable iron, TYPE 16									
4220	1/4" thread size	1 Plum	240	.033	Ea.	.92	1.15		2.07	2.74
4230	3/8" thread size		240	.033		1.49	1.15		2.64	3.37
4240	1/2" thread size		230	.035		1.78	1.20		2.98	3.77
4250	5/8" thread size		225	.036		3.52	1.23		4.75	5.70
4260	3/4" thread size		220	.036		5.75	1.25		7	8.20
4270	7/8" thread size		210	.038		11.25	1.31		12.56	14.35
4290	Strap, 1/2" pipe size, TYPE 26	Q-1	142	.113		2.23	3.49		5.72	7.70
4300	3/4" pipe size		140	.114		2.27	3.54		5.81	7.85
4310	1" pipe size		137	.117		2.32	3.62		5.94	8
4320	1-1/4" pipe size		134	.119		2.61	3.70		6.31	8.45
4330	1-1/2" pipe size		131	.122		3.27	3.79		7.06	9.30
4340	2" pipe size		129	.124		3.37	3.84		7.21	9.50
4350	2-1/2" pipe size		125	.128		5.90	3.97		9.87	12.45
4360	3" pipe size		122	.131		6.45	4.07		10.52	13.25

15050 | Basic Materials & Methods

15060 | Hangers & Supports

		CREW	DAILY OUTPUT	LABOR-HOURS	UNIT	2001 BARE COSTS MAT.	LABOR	EQUIP.	TOTAL	TOTAL INCL O&P	
300	4370	3-1/2" pipe size	Q-1	119	.134	Ea.	10.80	4.17		14.97	18.20	300
	4380	4" pipe size	↓	114	.140	↓	12.10	4.35		16.45	19.85	
	4400	U-bolt, carbon steel										
	4410	Standard, with nuts, TYPE 42										
	4420	1/2" pipe size	1 Plum	160	.050	Ea.	1.04	1.72		2.76	3.74	
	4430	3/4" pipe size		158	.051		1.04	1.74		2.78	3.77	
	4450	1" pipe size		152	.053		1.14	1.81		2.95	3.99	
	4460	1-1/4" pipe size		148	.054		1.26	1.86		3.12	4.20	
	4470	1-1/2" pipe size		143	.056		1.31	1.93		3.24	4.35	
	4480	2" pipe size		139	.058		1.35	1.98		3.33	4.48	
	4490	2-1/2" pipe size		134	.060		1.87	2.06		3.93	5.15	
	4500	3" pipe size		128	.063		3.42	2.15		5.57	7	
	4510	3-1/2" pipe size		122	.066		3.54	2.26		5.80	7.30	
	4520	4" pipe size		117	.068		3.66	2.36		6.02	7.60	
	4530	5" pipe size		114	.070		3.92	2.42		6.34	7.95	
	4540	6" pipe size		111	.072		3.96	2.48		6.44	8.10	
	4550	8" pipe size		109	.073		4.74	2.53		7.27	9	
	4560	10" pipe size		107	.075		8.30	2.58		10.88	13	
	4570	12" pipe size	↓	104	.077	↓	10.60	2.65		13.25	15.70	
	4580	For plastic coating on 1/2" thru 6" size, add					150%					
	4700	U-hook, carbon steel, requires mounting screws or bolts										
	4710	3/4" thru 2" pipe size										
	4720	6" long	1 Plum	96	.083	Ea.	3.20	2.87		6.07	7.85	
	4730	8" long		96	.083		3.42	2.87		6.29	8.10	
	4740	10" long		96	.083		4.32	2.87		7.19	9.10	
	4750	12" long	↓	96	.083		4.79	2.87		7.66	9.60	
	4760	For copper plated, add					50%					
	8000	Pipe clamp, plastic, 1/2" CTS	1 Plum	80	.100		.20	3.45		3.65	5.40	
	8010	3/4" CTS		73	.110		.21	3.78		3.99	5.95	
	8020	1" CTS		68	.118		.47	4.05		4.52	6.60	
	8080	Economy clamp, 1/4" CTS		175	.046		.05	1.57		1.62	2.44	
	8090	3/8" CTS		168	.048		.06	1.64		1.70	2.55	
	8100	1/2" CTS		160	.050		.06	1.72		1.78	2.67	
	8110	3/4" CTS		145	.055		.12	1.90		2.02	3	
	8200	Half clamp, 1/2" CTS		80	.100		.18	3.45		3.63	5.40	
	8210	3/4" CTS		73	.110		.19	3.78		3.97	5.90	
	8300	Suspension clamp, 1/2" CTS		80	.100		.20	3.45		3.65	5.40	
	8310	3/4" CTS		73	.110		.21	3.78		3.99	5.95	
	8320	1" CTS		68	.118		.58	4.05		4.63	6.75	
	8400	Insulator, 1/2" CTS		80	.100		.34	3.45		3.79	5.55	
	8410	3/4" CTS		73	.110		.35	3.78		4.13	6.10	
	8420	1" CTS		68	.118		.37	4.05		4.42	6.50	
	8500	J hook clamp with nail 1/2" CTS		240	.033		.10	1.15		1.25	1.84	
	8501	3/4" CTS	↓	240	.033	↓	.10	1.15		1.25	1.84	

15070 | Mech Sound/Vibration/Seismic Ctrl

		CREW	DAILY OUTPUT	LABOR-HOURS	UNIT	MAT.	LABOR	EQUIP.	TOTAL	TOTAL INCL O&P	
800	0010	**VIBRATION ABSORBERS**										800
	0100	Hangers, neoprene flex										
	0200	10 - 120 lb. capacity				Ea.	9.05			9.05	9.95	
	0220	75 - 550 lb. capacity				↓	13.50			13.50	14.85	
	0240	250 - 1100 lb. capacity					25			25	27.50	
	0260	1000 - 4000 lb. capacity					56			56	61.50	
	0500	Spring flex, 60 lb. capacity					15.80			15.80	17.35	
	0520	450 lb. capacity					22.50			22.50	25	
	0540	900 lb. capacity					22.50			22.50	25	
	0560	1100-1300 lb. capacity					26			26	28.50	

15050 | Basic Materials & Methods

15070 | Mech Sound/Vibration/Seismic Ctrl

			CREW	DAILY OUTPUT	LABOR-HOURS	UNIT	MAT.	LABOR	EQUIP.	TOTAL	TOTAL INCL O&P	
800	0600	Rubber in shear										800
	0610	45 - 340 lb., up to 1/2" rod size	1 Stpi	22	.364	Ea.	11	12.65		23.65	31.50	
	0620	130 - 700 lb., up to 3/4" rod size		20	.400		26	13.95		39.95	49.50	
	0630	50 - 1000 lb., up to 3/4" rod size		18	.444		31	15.50		46.50	57.50	
	1000	Mounts, neoprene, 45 - 380 lb. capacity					7.55			7.55	8.30	
	1020	250 - 1100 lb. capacity					22.50			22.50	25	
	1040	1000 - 4000 lb. capacity					50.50			50.50	55.50	
	1100	Spring flex, 60 lb. capacity					30.50			30.50	33.50	
	1120	165 lb. capacity					30.50			30.50	33.50	
	1140	260 lb. capacity					30.50			30.50	33.50	
	1160	450 lb. capacity					40			40	44.50	
	1180	600 lb. capacity					40			40	44.50	
	1200	750 lb. capacity					40			40	44.50	
	1220	900 lb. capacity					40			40	44.50	
	1240	1100 lb. capacity					44.50			44.50	49	
	1260	1300 lb. capacity					44.50			44.50	49	
	1280	1500 lb. capacity					65			65	71.50	
	1300	1800 lb. capacity					65			65	71.50	
	1320	2200 lb. capacity					72			72	79.50	
	1340	2600 lb. capacity					72			72	79.50	
	1399	Spring type										
	1400	2 Piece										
	1410	50 - 1000 lb.	1 Stpi	12	.667	Ea.	97	23		120	142	
	1420	1100 - 1600 lb.	"	12	.667	"	106	23		129	152	
	1500	Double spring open										
	1510	150 - 450 lb.	1 Stpi	24	.333	Ea.	66	11.60		77.60	90	
	1520	500 - 1000 lb.		24	.333		66	11.60		77.60	90	
	1530	1100 - 1600 lb.		24	.333		66	11.60		77.60	90	
	1540	1700 - 2400 lb.		24	.333		73	11.60		84.60	98	
	1550	2500 - 3400 lb.		24	.333		121	11.60		132.60	151	
	2000	Pads, cork rib, 18" x 18" x 1", 10-50 psi					83.50			83.50	91.50	
	2020	18" x 36" x 1", 10-50 psi					167			167	183	
	2100	Shear flexible pads, 18" x 18" x 3/8", 20-70 psi					39.50			39.50	43.50	
	2120	18" x 36" x 3/8", 20-70 psi					78			78	85.50	
	2150	Laminated neoprene and cork										
	2160	1" thick	1 Stpi	16	.500	S.F.	43	17.45		60.45	74	
	3000	Note overlap in capacities due to deflections										

15075 | Mech Identification

			CREW	DAILY OUTPUT	LABOR-HOURS	UNIT	MAT.	LABOR	EQUIP.	TOTAL	TOTAL INCL O&P	
400	0010	**PIPING SYSTEM IDENTIFICATION LABELS**										400
	0100	Indicate contents and flow direction										
	0106	Pipe markers										
	0110	Plastic snap around										
	0114	1/2" pipe	1 Plum	80	.100	Ea.	3	3.45		6.45	8.50	
	0116	3/4" pipe		80	.100		3	3.45		6.45	8.50	
	0118	1" pipe		80	.100		3	3.45		6.45	8.50	
	0120	2" pipe		75	.107		3.95	3.67		7.62	9.90	
	0122	3" pipe		70	.114		7.40	3.94		11.34	14.10	
	0124	4" pipe		60	.133		7.40	4.59		11.99	15.10	
	0126	6" pipe		60	.133		8	4.59		12.59	15.75	
	0128	8" pipe		56	.143		11	4.92		15.92	19.55	
	0130	10" pipe		56	.143		11	4.92		15.92	19.55	
	0200	Over 10" pipe size		50	.160		13.50	5.50		19	23	
	1110	Self adhesive										
	1114	1" pipe	1 Plum	80	.100	Ea.	2.10	3.45		5.55	7.50	

15050 | Basic Materials & Methods

15075 | Mech Identification

		CREW	DAILY OUTPUT	LABOR-HOURS	UNIT	2001 BARE COSTS MAT.	LABOR	EQUIP.	TOTAL	TOTAL INCL O&P
1116	2" pipe	1 Plum	75	.107	Ea.	2.10	3.67		5.77	7.85
1118	3" pipe		70	.114		3.40	3.94		7.34	9.70
1120	4" pipe		60	.133		3.40	4.59		7.99	10.70
1122	6" pipe		60	.133		3.40	4.59		7.99	10.70
1124	8" pipe		56	.143		4.70	4.92		9.62	12.60
1126	10" pipe		56	.143		4.70	4.92		9.62	12.60
1200	Over 10" pipe size		50	.160		6.55	5.50		12.05	15.50
2000	Valve tags									
2010	Numbered plus identifying legend									
2100	Brass, 2" diameter	1 Plum	40	.200	Ea.	1.75	6.90		8.65	12.35
2200	Plastic, 1-1/2" Dia.	"	40	.200	"	1.60	6.90		8.50	12.15

15082 | Duct Insulation

		CREW	DAILY OUTPUT	LABOR-HOURS	UNIT	MAT.	LABOR	EQUIP.	TOTAL	TOTAL INCL O&P
0010	**INSULATION**									
0100	Rule of thumb, as a percentage of total mechanical costs				Job				10%	
0110	Insulation req'd is based on the surface size/area to be covered									
1000	Boiler, 1-1/2" calcium silicate, 1/2" cement finish	Q-14	50	.320	S.F.	2.53	9.25		11.78	17.65
1020	2" fiberglass	"	80	.200	"	1.49	5.80		7.29	10.90
2000	Breeching, 2" calcium silicate with 1/2" cement finish, no lath									
2020	Rectangular	Q-14	42	.381	S.F.	2.86	11		13.86	21
2040	Round	"	38.70	.413	"	3.21	11.95		15.16	22.50
2300	Calcium silicate block, + 200°F to + 1200°F									
2310	On irregular surfaces, valves and fittings									
2340	1" thick	Q-14	30	.533	S.F.	2	15.45		17.45	26.50
2360	1-1/2" thick		25	.640		2.34	18.50		20.84	32
2380	2" thick		22	.727		3.22	21		24.22	37
2400	3" thick		18	.889		4.71	25.50		30.21	46
2410	On plane surfaces									
2420	1" thick	Q-14	126	.127	S.F.	2	3.67		5.67	8.10
2430	1-1/2" thick		120	.133		2.34	3.86		6.20	8.75
2440	2" thick		100	.160		3.22	4.63		7.85	10.95
2450	3" thick		70	.229		4.71	6.60		11.31	15.80
2900	Domestic water heater wrap kit									
2920	1-1/2" with vinyl jacket, 20-60 gal.	1 Plum	8	1	Ea.	16.25	34.50		50.75	70
2930	Insulated protectors, (ADA)									
2935	For exposed piping under sinks or lavatories.									
2940	Vinyl coated foam, velcro tabs									
2945	P Trap, 1-1/4" or 1-1/2"	1 Plum	32	.250	Ea.	12.85	8.60		21.45	27
2960	Valve and supply cover									
2965	1/2", 3/8", and 7/16" pipe size	1 Plum	32	.250	Ea.	12.85	8.60		21.45	27
2970	Extension drain cover									
2975	1-1/4", or 1-1/2" pipe size	1 Plum	32	.250	Ea.	13.15	8.60		21.75	27.50
2980	Tailpiece offset (wheelchair)									
2985	1-1/4" pipe size	1 Plum	32	.250	Ea.	13.50	8.60		22.10	28
3000	Ductwork									
3020	Blanket type, fiberglass, flexible									
3030	Fire resistant liner, black coating one side									
3050	1/2" thick, 2 lb. density	Q-14	380	.042	S.F.	.29	1.22		1.51	2.27
3060	1" thick, 1-1/2 lb. density		350	.046		.39	1.32		1.71	2.55
3070	1-1/2" thick, 1-1/2 lb. density		320	.050		.68	1.45		2.13	3.07
3080	2" thick, 1-1/2 lb. density		300	.053		.80	1.54		2.34	3.35
3140	FRK vapor barrier wrap, .75 lb. density									
3160	1" thick	Q-14	350	.046	S.F.	.27	1.32		1.59	2.42
3170	1-1/2" thick		320	.050		.24	1.45		1.69	2.58
3180	2" thick		300	.053		.29	1.54		1.83	2.79

15050 | Basic Materials & Methods

15082 | Duct Insulation

		CREW	DAILY OUTPUT	LABOR-HOURS	UNIT	2001 BARE COSTS MAT.	LABOR	EQUIP.	TOTAL	TOTAL INCL O&P
3190	3" thick	Q-14	260	.062	S.F.	.43	1.78		2.21	3.32
3200	4" thick	↓	242	.066	↓	.58	1.91		2.49	3.70
3210	Vinyl jacket, same as FRK									
3280	Unfaced, 1 lb. density									
3310	1" thick	Q-14	360	.044	S.F.	.24	1.29		1.53	2.32
3320	1-1/2" thick		330	.048		.33	1.40		1.73	2.61
3330	2" thick	↓	310	.052	↓	.39	1.49		1.88	2.82
3340	Board type fiberglass liner, 1-1/2 lb. density									
3344	1" thick	Q-14	150	.107	S.F.	.35	3.09		3.44	5.35
3345	1-1/2" thick		130	.123		.57	3.56		4.13	6.35
3346	2" thick	↓	120	.133	↓	.68	3.86		4.54	6.95
3490	Board type, fiberglass liner, 3 lb. density									
3500	Fire resistant, black pigmented, 1 side									
3520	1" thick	Q-14	150	.107	S.F.	1.16	3.09		4.25	6.20
3540	1-1/2" thick		130	.123		1.43	3.56		4.99	7.25
3560	2" thick	↓	120	.133	↓	1.72	3.86		5.58	8.10
3600	FRK vapor barrier									
3620	1" thick	Q-14	150	.107	S.F.	1.27	3.09		4.36	6.35
3630	1-1/2" thick		130	.123		1.64	3.56		5.20	7.50
3640	2" thick	↓	120	.133	↓	1.99	3.86		5.85	8.40
3680	No finish									
3700	1" thick	Q-14	170	.094	S.F.	.68	2.72		3.40	5.10
3710	1-1/2" thick		140	.114		1.02	3.31		4.33	6.40
3720	2" thick	↓	130	.123	↓	1.30	3.56		4.86	7.15
3730	Sheet insulation									
3760	Polyethylene foam, closed cell, UV resistant									
3770	Standard temperature (-90° to +212° F)									
3771	1/4" thick	Q-14	450	.036	S.F.	1.10	1.03		2.13	2.86
3772	3/8" thick		440	.036		1.62	1.05		2.67	3.46
3773	1/2" thick		420	.038		2.02	1.10		3.12	3.99
3774	3/4" thick		400	.040		2.95	1.16		4.11	5.10
3775	1" thick	↓	380	.042	↓	3.75	1.22		4.97	6.10
3779	Adhesive (see line 7878)									
3780	Foam, rubber									
3782	1" thick	1 Stpi	50	.160	S.F.	2.50	5.60		8.10	11.15
3795	Finishes									
3800	1/2" cement over 1" wire mesh, incl. corner bead	Q-14	100	.160	S.F.	1.23	4.63		5.86	8.75
3820	Glass cloth, pasted on		170	.094		2	2.72		4.72	6.55
3900	Weatherproof, non-metallic, 2 lb. per S.F.	↓	100	.160	↓	2.86	4.63		7.49	10.55
4000	Pipe covering (price copper tube one size less than IPS)									
4040	Air cell, asbestos free corrugated felt, with cover									
4100	3 ply, 1/2" iron pipe size	Q-14	145	.110	L.F.	.75	3.19		3.94	5.95
4130	3/4" iron pipe size		145	.110		.85	3.19		4.04	6.05
4140	1" iron pipe size		145	.110		.90	3.19		4.09	6.10
4150	1-1/4" iron pipe size		140	.114		1	3.31		4.31	6.40
4160	1-1/2" iron pipe size		140	.114		1.10	3.31		4.41	6.50
4170	2" iron pipe size		135	.119		1.20	3.43		4.63	6.80
4180	2-1/2" iron pipe size		135	.119		1.35	3.43		4.78	7
4190	3" iron pipe size		130	.123		1.50	3.56		5.06	7.35
4200	3-1/2" iron pipe size		130	.123		1.76	3.56		5.32	7.65
4210	4" iron pipe size		125	.128		1.95	3.70		5.65	8.10
4220	5" iron pipe size		120	.133		2.25	3.86		6.11	8.70
4230	6" iron pipe size		115	.139		2.40	4.03		6.43	9.10
4240	8" iron pipe size		105	.152		3.30	4.41		7.71	10.70
4250	10" iron pipe size		95	.168		3.90	4.87		8.77	12.10
4260	12" iron pipe size	↓	90	.178	↓	4.50	5.15		9.65	13.20

Important: See the Reference Section for critical supporting data - Reference Nos., Crews, & City Cost Indexes

15050 | Basic Materials & Methods

15082 | Duct Insulation

		Crew	Daily Output	Labor-Hours	Unit	Mat.	Labor	Equip.	Total	Total Incl O&P
4280	Cellular glass, closed cell foam, all service jacket, sealant,									
4281	working temp. (-450°F to +900°F), 0 water vapor xmission									
4284	1" wall,									
4286	1/2" iron pipe size	Q-14	120	.133	L.F.	1.73	3.86		5.59	8.10
4300	1-1/2" wall,									
4301	1" iron pipe size	Q-14	105	.152	L.F.	2.78	4.41		7.19	10.10
4304	2-1/2" iron pipe size		90	.178		4.68	5.15		9.83	13.40
4306	3" iron pipe size		85	.188		4.73	5.45		10.18	13.90
4308	4" iron pipe size		70	.229		6.30	6.60		12.90	17.50
4310	5" iron pipe size		65	.246		7.25	7.10		14.35	19.35
4320	2" wall,									
4322	1" iron pipe size	Q-14	100	.160	L.F.	4.13	4.63		8.76	11.95
4324	2-1/2" iron pipe size		85	.188		6.15	5.45		11.60	15.45
4326	3" iron pipe size		80	.200		6.30	5.80		12.10	16.15
4328	4" iron pipe size		65	.246		7.05	7.10		14.15	19.15
4330	5" iron pipe size		60	.267		8.45	7.70		16.15	21.50
4332	6" iron pipe size		50	.320		10	9.25		19.25	26
4336	8" iron pipe size		40	.400		12.95	11.55		24.50	33
4338	10" iron pipe size		35	.457		13.35	13.25		26.60	35.50
4350	2-1/2" wall,									
4360	12" iron pipe size	Q-14	32	.500	L.F.	21	14.45		35.45	46.50
4362	14" iron pipe size	"	28	.571	"	22.50	16.55		39.05	51.50
4370	3" wall,									
4378	6" iron pipe size	Q-14	48	.333	L.F.	13.35	9.65		23	30
4380	8" iron pipe size		38	.421		15.60	12.20		27.80	36.50
4382	10" iron pipe size		33	.485		20.50	14.05		34.55	45
4384	16" iron pipe size		25	.640		26.50	18.50		45	59
4386	18" iron pipe size		22	.727		29	21		50	65.50
4388	20" iron pipe size		20	.800		34.50	23		57.50	74.50
4400	3-1/2" wall,									
4412	12" iron pipe size	Q-14	27	.593	L.F.	30	17.15		47.15	60.50
4414	14" iron pipe size	"	25	.640	"	32	18.50		50.50	64.50
4430	4" wall,									
4446	16" iron pipe size	Q-14	22	.727	L.F.	32.50	21		53.50	69
4448	18" iron pipe size		20	.800		42	23		65	83
4450	20" iron pipe size		18	.889		43.50	25.50		69	89
4480	Fittings, average with fabric and mastic									
4484	1" wall,									
4486	1/2" iron pipe size	1 Asbe	40	.200	Ea.	3.07	6.45		9.52	13.70
4500	1-1/2" wall,									
4502	1" iron pipe size	1 Asbe	38	.211	Ea.	5.30	6.75		12.05	16.65
4504	2-1/2" iron pipe size		32	.250		6.60	8.05		14.65	20
4506	3" iron pipe size		30	.267		7.25	8.55		15.80	21.50
4508	4" iron pipe size		28	.286		11.15	9.20		20.35	27
4510	5" iron pipe size		24	.333		12.25	10.70		22.95	30.50
4520	2" wall,									
4522	1" iron pipe size	1 Asbe	36	.222	Ea.	6.15	7.15		13.30	18.20
4524	2-1/2" iron pipe size		30	.267		7.25	8.55		15.80	21.50
4526	3" iron pipe size		28	.286		8.70	9.20		17.90	24.50
4528	4" iron pipe size		24	.333		11.70	10.70		22.40	30
4530	5" iron pipe size		22	.364		14.75	11.70		26.45	35
4532	6" iron pipe size		20	.400		17.25	12.85		30.10	39.50
4536	8" iron pipe size		12	.667		26	21.50		47.50	63.50
4538	10" iron pipe size		8	1		32.50	32		64.50	87
4550	2-1/2" wall,									
4560	12" iron pipe size	1 Asbe	6	1.333	Ea.	55.50	43		98.50	130

15050 | Basic Materials & Methods

15082 | Duct Insulation

		Crew	Daily Output	Labor-Hours	Unit	Mat.	Labor	Equip.	Total	Total Incl O&P
4562	14" iron pipe size	1 Asbe	4	2	Ea.	64.50	64.50		129	174
4570	3" wall,									
4578	6" iron pipe size	1 Asbe	16	.500	Ea.	20	16.10		36.10	48
4580	8" iron pipe size		10	.800		29.50	25.50		55	73.50
4582	10" iron pipe size		6	1.333		42	43		85	115
4584	16" iron pipe size		2	4		100	129		229	315
4586	18" iron pipe size		2	4		129	129		258	350
4588	20" iron pipe size		2	4		153	129		282	375
4600	3-1/2" wall,									
4612	12" iron pipe size	1 Asbe	4	2	Ea.	72.50	64.50		137	183
4614	14" iron pipe size	"	2	4	"	80	129		209	294
4630	4" wall,									
4646	16" iron pipe size	1 Asbe	2	4	Ea.	120	129		249	340
4648	18" iron pipe size		2	4		147	129		276	370
4650	20" iron pipe size		2	4		172	129		301	395
4900	Calcium silicate, with cover									
5100	1" wall, 1/2" iron pipe size	Q-14	170	.094	L.F.	2.07	2.72		4.79	6.65
5130	3/4" iron pipe size		170	.094		2.08	2.72		4.80	6.65
5140	1" iron pipe size		170	.094		1.92	2.72		4.64	6.45
5150	1-1/4" iron pipe size		165	.097		1.98	2.81		4.79	6.65
5160	1-1/2" iron pipe size		165	.097		2.03	2.81		4.84	6.70
5170	2" iron pipe size		160	.100		2.41	2.89		5.30	7.30
5180	2-1/2" iron pipe size		160	.100		2.58	2.89		5.47	7.45
5190	3" iron pipe size		150	.107		2.74	3.09		5.83	7.95
5200	4" iron pipe size		140	.114		3.18	3.31		6.49	8.80
5210	5" iron pipe size		135	.119		3.51	3.43		6.94	9.35
5220	6" iron pipe size		130	.123		3.81	3.56		7.37	9.90
5280	1-1/2" wall, 1/2" iron pipe size		150	.107		2.27	3.09		5.36	7.45
5310	3/4" iron pipe size		150	.107		2.28	3.09		5.37	7.45
5320	1" iron pipe size		150	.107		2.49	3.09		5.58	7.70
5330	1-1/4" iron pipe size		145	.110		2.68	3.19		5.87	8.05
5340	1-1/2" iron pipe size		145	.110		2.81	3.19		6	8.20
5350	2" iron pipe size		140	.114		3.07	3.31		6.38	8.70
5360	2-1/2" iron pipe size		140	.114		3.35	3.31		6.66	9
5370	3" iron pipe size		135	.119		3.51	3.43		6.94	9.35
5380	4" iron pipe size		125	.128		4.10	3.70		7.80	10.45
5390	5" iron pipe size		120	.133		4.60	3.86		8.46	11.25
5400	6" iron pipe size		110	.145		4.74	4.21		8.95	11.95
5402	8" iron pipe size		95	.168		6.50	4.87		11.37	14.95
5404	10" iron pipe size		85	.188		8.40	5.45		13.85	17.90
5406	12" iron pipe size		80	.200		9.90	5.80		15.70	20
5408	14" iron pipe size		75	.213		11.15	6.15		17.30	22
5410	16" iron pipe size		70	.229		12.50	6.60		19.10	24.50
5412	18" iron pipe size		65	.246		14.25	7.10		21.35	27
5460	2" wall, 1/2" iron pipe size		135	.119		3.62	3.43		7.05	9.50
5490	3/4" iron pipe size		135	.119		3.63	3.43		7.06	9.50
5500	1" iron pipe size		135	.119		3.83	3.43		7.26	9.70
5510	1-1/4" iron pipe size		130	.123		4.26	3.56		7.82	10.40
5520	1-1/2" iron pipe size		130	.123		4.30	3.56		7.86	10.45
5530	2" iron pipe size		125	.128		4.52	3.70		8.22	10.90
5540	2-1/2" iron pipe size		125	.128		5.45	3.70		9.15	11.90
5550	3" iron pipe size		120	.133		6	3.86		9.86	12.80
5560	4" iron pipe size		115	.139		6.10	4.03		10.13	13.20
5570	5" iron pipe size		110	.145		6.95	4.21		11.16	14.40
5580	6" iron pipe size		105	.152		7.80	4.41		12.21	15.65
5581	8" iron pipe size		95	.168		9.35	4.87		14.22	18.05

Important: See the Reference Section for critical supporting data - Reference Nos., Crews, & City Cost Indexes

15050 | Basic Materials & Methods

15082 | Duct Insulation

			CREW	DAILY OUTPUT	LABOR-HOURS	UNIT	MAT.	LABOR	EQUIP.	TOTAL	TOTAL INCL O&P
200	5582	10" iron pipe size	Q-14	85	.188	L.F.	11.70	5.45		17.15	21.50	200
	5583	12" iron pipe size		80	.200		13	5.80		18.80	23.50
	5584	14" iron pipe size		75	.213		14.30	6.15		20.45	25.50
	5585	16" iron pipe size		70	.229		16.30	6.60		22.90	28.50
	5586	18" iron pipe size		65	.246		17.75	7.10		24.85	31
	5587	3" wall, 1-1/4" iron pipe size		100	.160		6.15	4.63		10.78	14.15
	5588	1-1/2" iron pipe size		100	.160		6.50	4.63		11.13	14.55
	5589	2" iron pipe size		95	.168		6.90	4.87		11.77	15.40
	5590	2-1/2" iron pipe size		95	.168		8.20	4.87		13.07	16.85
	5591	3" iron pipe size		90	.178		7.80	5.15		12.95	16.85
	5592	4" iron pipe size		85	.188		10	5.45		15.45	19.70
	5593	6" iron pipe size		80	.200		12.65	5.80		18.45	23
	5594	8" iron pipe size		75	.213		14.95	6.15		21.10	26.50
	5595	10" iron pipe size		65	.246		18.05	7.10		25.15	31.50
	5596	12" iron pipe size		60	.267		20	7.70		27.70	34.50
	5597	14" iron pipe size		55	.291		22.50	8.40		30.90	38
	5598	16" iron pipe size		50	.320		24.50	9.25		33.75	42
	5599	18" iron pipe size		45	.356		27	10.30		37.30	46.50
	5600	Calcium silicate, no cover									
	5720	1" wall, 1/2" iron pipe size	Q-14	180	.089	L.F.	1.93	2.57		4.50	6.25
	5740	3/4" iron pipe size		180	.089		1.93	2.57		4.50	6.25
	5750	1" iron pipe size		180	.089		1.80	2.57		4.37	6.10
	5760	1-1/4" iron pipe size		175	.091		1.81	2.65		4.46	6.25
	5770	1-1/2" iron pipe size		175	.091		1.83	2.65		4.48	6.25
	5780	2" iron pipe size		170	.094		2.18	2.72		4.90	6.75
	5790	2-1/2" iron pipe size		170	.094		2.34	2.72		5.06	6.95
	5800	3" iron pipe size		160	.100		2.47	2.89		5.36	7.35
	5810	4" iron pipe size		150	.107		2.83	3.09		5.92	8.05
	5820	5" iron pipe size		145	.110		3.13	3.19		6.32	8.55
	5830	6" iron pipe size		140	.114		3.38	3.31		6.69	9
	5900	1-1/2" wall, 1/2" iron pipe size		160	.100		2.07	2.89		4.96	6.90
	5920	3/4" iron pipe size		160	.100		2.07	2.89		4.96	6.90
	5930	1" iron pipe size		160	.100		2.26	2.89		5.15	7.10
	5940	1-1/4" iron pipe size		155	.103		2.54	2.99		5.53	7.55
	5950	1-1/2" iron pipe size		155	.103		2.55	2.99		5.54	7.60
	5960	2" iron pipe size		150	.107		2.79	3.09		5.88	8
	5970	2-1/2" iron pipe size		150	.107		3.07	3.09		6.16	8.30
	5980	3" iron pipe size		145	.110		3.20	3.19		6.39	8.60
	5990	4" iron pipe size		135	.119		3.72	3.43		7.15	9.60
	6000	5" iron pipe size		130	.123		4.17	3.56		7.73	10.30
	6010	6" iron pipe size		120	.133		4.27	3.86		8.13	10.90
	6020	7" iron pipe size		115	.139		5.15	4.03		9.18	12.10
	6030	8" iron pipe size		105	.152		5.80	4.41		10.21	13.40
	6040	9" iron pipe size		100	.160		6.95	4.63		11.58	15.05
	6050	10" iron pipe size		95	.168		7.70	4.87		12.57	16.30
	6060	12" iron pipe size		90	.178		9.10	5.15		14.25	18.25
	6070	14" iron pipe size		85	.188		10.35	5.45		15.80	20
	6080	16" iron pipe size		80	.200		11.60	5.80		17.40	22
	6090	18" iron pipe size		75	.213		13.25	6.15		19.40	24.50
	6120	2" wall, 1/2" iron pipe size		145	.110		3.38	3.19		6.57	8.80
	6140	3/4" iron pipe size		145	.110		3.38	3.19		6.57	8.80
	6150	1" iron pipe size		145	.110		3.56	3.19		6.75	9
	6160	1-1/4" iron pipe size		140	.114		3.98	3.31		7.29	9.70
	6170	1-1/2" iron pipe size		140	.114		4.01	3.31		7.32	9.70
	6180	2" iron pipe size		135	.119		4.22	3.43		7.65	10.15
	6190	2-1/2" iron pipe size		135	.119		4.59	3.43		8.02	10.55

15050 | Basic Materials & Methods

15082 | Duct Insulation

		CREW	DAILY OUTPUT	LABOR-HOURS	UNIT	2001 BARE COSTS MAT.	LABOR	EQUIP.	TOTAL	TOTAL INCL O&P	
200	6200	3" iron pipe size	Q-14	130	.123	L.F.	4.84	3.56		8.40	11	200
	6210	4" iron pipe size		125	.128		5.70	3.70		9.40	12.20	
	6220	5" iron pipe size		120	.133		6.50	3.86		10.36	13.35	
	6230	6" iron pipe size		115	.139		7.30	4.03		11.33	14.50	
	6240	7" iron pipe size		110	.145		7.95	4.21		12.16	15.50	
	6250	8" iron pipe size		105	.152		8.70	4.41		13.11	16.65	
	6260	9" iron pipe size		100	.160		9.85	4.63		14.48	18.25	
	6270	10" iron pipe size		95	.168		10.90	4.87		15.77	19.80	
	6280	12" iron pipe size		90	.178		12.20	5.15		17.35	21.50	
	6290	14" iron pipe size		85	.188		13.40	5.45		18.85	23.50	
	6300	16" iron pipe size		80	.200		15.30	5.80		21.10	26	
	6310	18" iron pipe size		75	.213		16.70	6.15		22.85	28.50	
	6320	20" iron pipe size		65	.246		19.45	7.10		26.55	33	
	6330	22" iron pipe size		60	.267		21	7.70		28.70	35.50	
	6340	24" iron pipe size		55	.291		21.50	8.40		29.90	37.50	
	6360	3" wall, 1/2" iron pipe size		115	.139		6.15	4.03		10.18	13.20	
	6380	3/4" iron pipe size		115	.139		6.15	4.03		10.18	13.20	
	6390	1" iron pipe size		115	.139		5.55	4.03		9.58	12.55	
	6400	1-1/4" iron pipe size		110	.145		5.75	4.21		9.96	13.05	
	6410	1-1/2" iron pipe size		110	.145		6.10	4.21		10.31	13.45	
	6420	2" iron pipe size		105	.152		6.50	4.41		10.91	14.20	
	6430	2-1/2" iron pipe size		105	.152		7.75	4.41		12.16	15.60	
	6440	3" iron pipe size		100	.160		8.50	4.63		13.13	16.75	
	6450	4" iron pipe size		95	.168		9.80	4.87		14.67	18.60	
	6460	5" iron pipe size		90	.178		10.65	5.15		15.80	20	
	6470	6" iron pipe size		90	.178		12.05	5.15		17.20	21.50	
	6480	7" iron pipe size		85	.188		13.75	5.45		19.20	24	
	6490	8" iron pipe size		85	.188		14.25	5.45		19.70	24.50	
	6500	9" iron pipe size		80	.200		16.10	5.80		21.90	27	
	6510	10" iron pipe size		75	.213		17.25	6.15		23.40	29	
	6520	12" iron pipe size		70	.229		19.10	6.60		25.70	31.50	
	6530	14" iron pipe size		65	.246		21.50	7.10		28.60	35	
	6540	16" iron pipe size		60	.267		23.50	7.70		31.20	38.50	
	6550	18" iron pipe size		55	.291		26	8.40		34.40	42	
	6560	20" iron pipe size		50	.320		29	9.25		38.25	47	
	6570	22" iron pipe size		45	.356		31.50	10.30		41.80	51.50	
	6580	24" iron pipe size		40	.400		34	11.55		45.55	56	
	6600	Fiberglass, with all service jacket										
	6640	1/2" wall, 1/2" iron pipe size	Q-14	250	.064	L.F.	.88	1.85		2.73	3.94	
	6660	3/4" iron pipe size		240	.067		.98	1.93		2.91	4.17	
	6670	1" iron pipe size		230	.070		1.06	2.01		3.07	4.39	
	6680	1-1/4" iron pipe size		220	.073		1.16	2.10		3.26	4.65	
	6690	1-1/2" iron pipe size		220	.073		1.24	2.10		3.34	4.73	
	6700	2" iron pipe size		210	.076		1.33	2.20		3.53	4.99	
	6710	2-1/2" iron pipe size		200	.080		1.53	2.31		3.84	5.40	
	6720	3" iron pipe size		190	.084		1.70	2.44		4.14	5.75	
	6730	3-1/2" iron pipe size		180	.089		1.83	2.57		4.40	6.15	
	6740	4" iron pipe size		160	.100		2.17	2.89		5.06	7	
	6750	5" iron pipe size		150	.107		2.51	3.09		5.60	7.70	
	6760	6" iron pipe size		120	.133		2.64	3.86		6.50	9.10	
	6840	1" wall, 1/2" iron pipe size		240	.067		1.06	1.93		2.99	4.26	
	6860	3/4" iron pipe size		230	.070		1.21	2.01		3.22	4.55	
	6870	1" iron pipe size		220	.073		1.24	2.10		3.34	4.73	
	6880	1-1/4" iron pipe size		210	.076		1.40	2.20		3.60	5.05	
	6890	1-1/2" iron pipe size		210	.076		1.51	2.20		3.71	5.20	
	6900	2" iron pipe size		200	.080		1.65	2.31		3.96	5.55	

15050 | Basic Materials & Methods

15082 | Duct Insulation

		CREW	DAILY OUTPUT	LABOR-HOURS	UNIT	MAT.	LABOR	EQUIP.	TOTAL	TOTAL INCL O&P	
200	6910	2-1/2" iron pipe size	Q-14	190	.084	L.F.	1.81	2.44		4.25	5.90	200
	6920	3" iron pipe size		180	.089		2.05	2.57		4.62	6.40	
	6930	3-1/2" iron pipe size		170	.094		2.26	2.72		4.98	6.85	
	6940	4" iron pipe size		150	.107		2.64	3.09		5.73	7.85	
	6950	5" iron pipe size		140	.114		3.04	3.31		6.35	8.65	
	6960	6" iron pipe size		120	.133		3.23	3.86		7.09	9.75	
	6970	7" iron pipe size		110	.145		3.98	4.21		8.19	11.15	
	6980	8" iron pipe size		100	.160		4.61	4.63		9.24	12.45	
	6990	9" iron pipe size		90	.178		5	5.15		10.15	13.75	
	7000	10" iron pipe size		90	.178		5.40	5.15		10.55	14.20	
	7010	12" iron pipe size		80	.200		6.15	5.80		11.95	16	
	7020	14" iron pipe size		80	.200		7.10	5.80		12.90	17.05	
	7030	16" iron pipe size		70	.229		8.45	6.60		15.05	19.90	
	7040	18" iron pipe size		70	.229		10.30	6.60		16.90	22	
	7050	20" iron pipe size		60	.267		11.50	7.70		19.20	25	
	7060	24" iron pipe size		60	.267		14	7.70		21.70	28	
	7080	1-1/2" wall, 1/2" iron pipe size		230	.070		2.16	2.01		4.17	5.60	
	7100	3/4" iron pipe size		220	.073		2.26	2.10		4.36	5.85	
	7110	1" iron pipe size		210	.076		2.37	2.20		4.57	6.15	
	7120	1-1/4" iron pipe size		200	.080		2.54	2.31		4.85	6.50	
	7130	1-1/2" iron pipe size		200	.080		2.68	2.31		4.99	6.65	
	7140	2" iron pipe size		190	.084		2.93	2.44		5.37	7.10	
	7150	2-1/2" iron pipe size		180	.089		3.21	2.57		5.78	7.65	
	7160	3" iron pipe size		170	.094		3.31	2.72		6.03	8	
	7170	3-1/2" iron pipe size		160	.100		3.63	2.89		6.52	8.60	
	7180	4" iron pipe size		140	.114		3.78	3.31		7.09	9.45	
	7190	5" iron pipe size		130	.123		4.22	3.56		7.78	10.35	
	7200	6" iron pipe size		110	.145		4.34	4.21		8.55	11.50	
	7210	7" iron pipe size		100	.160		5.05	4.63		9.68	12.95	
	7220	8" iron pipe size		90	.178		5.45	5.15		10.60	14.25	
	7230	9" iron pipe size		85	.188		6.30	5.45		11.75	15.65	
	7240	10" iron pipe size		80	.200		6.85	5.80		12.65	16.80	
	7250	12" iron pipe size		75	.213		7.45	6.15		13.60	18.10	
	7260	14" iron pipe size		70	.229		8.65	6.60		15.25	20	
	7270	16" iron pipe size		65	.246		9.70	7.10		16.80	22	
	7280	18" iron pipe size		60	.267		12.55	7.70		20.25	26	
	7290	20" iron pipe size		55	.291		13.30	8.40		21.70	28	
	7300	24" iron pipe size		50	.320		15.55	9.25		24.80	32	
	7320	2" wall, 1/2" iron pipe size		220	.073		3.49	2.10		5.59	7.20	
	7340	3/4" iron pipe size		210	.076		3.49	2.20		5.69	7.35	
	7350	1" iron pipe size		200	.080		3.73	2.31		6.04	7.80	
	7360	1-1/4" iron pipe size		190	.084		3.95	2.44		6.39	8.25	
	7370	1-1/2" iron pipe size		190	.084		4.17	2.44		6.61	8.50	
	7380	2" iron pipe size		180	.089		4.33	2.57		6.90	8.90	
	7390	2-1/2" iron pipe size		170	.094		4.65	2.72		7.37	9.45	
	7400	3" iron pipe size		160	.100		5	2.89		7.89	10.15	
	7410	3-1/2" iron pipe size		150	.107		5.30	3.09		8.39	10.75	
	7420	4" iron pipe size		130	.123		5.75	3.56		9.31	12.05	
	7430	5" iron pipe size		120	.133		5.75	3.86		9.61	12.50	
	7440	6" iron pipe size		100	.160		6.65	4.63		11.28	14.75	
	7450	7" iron pipe size		90	.178		7.65	5.15		12.80	16.65	
	7460	8" iron pipe size		80	.200		8.25	5.80		14.05	18.35	
	7470	9" iron pipe size		75	.213		9.10	6.15		15.25	19.95	
	7480	10" iron pipe size		70	.229		9.85	6.60		16.45	21.50	
	7490	12" iron pipe size		65	.246		10.85	7.10		17.95	23.50	
	7500	14" iron pipe size		60	.267		12.25	7.70		19.95	26	

15050 | Basic Materials & Methods

15082 | Duct Insulation

			CREW	DAILY OUTPUT	LABOR-HOURS	UNIT	MAT.	LABOR	EQUIP.	TOTAL	TOTAL INCL O&P
200	7510	16" iron pipe size	Q-14	55	.291	L.F.	14	8.40		22.40	29
	7520	18" iron pipe size		50	.320		16.70	9.25		25.95	33
	7530	20" iron pipe size		45	.356		18.50	10.30		28.80	37
	7540	24" iron pipe size	↓	40	.400	↓	21	11.55		32.55	41.50
	7600	For fiberglass with standard canvas jacket, deduct					5%				
	7660	For fittings, add 3 L.F. for each fitting									
	7680	plus 4 L.F. for each flange of the fitting									
	7700	For equipment or congested areas, add						20%			
	7720	Finishes, for .010" aluminum jacket, add	Q-14	200	.080	S.F.	.31	2.31		2.62	4.05
	7740	For .016" aluminum jacket, add		200	.080		.40	2.31		2.71	4.15
	7760	For .010" stainless steel, add	↓	160	.100	↓	1.20	2.89		4.09	5.95
	7780	For single layer of felt, add					10%	10%			
	7782	For roofing paper, 45 lb to 55 lb, add					25%	10%			
	7784	Polyethylene tubing flexible closed cell foam, UV resistant									
	7828	Standard temperature (-90° to +212° F)									
	7830	3/8" wall, 1/8" iron pipe size	1 Asbe	130	.062	L.F.	.25	1.98		2.23	3.45
	7831	1/4" iron pipe size		130	.062		.28	1.98		2.26	3.48
	7832	3/8" iron pipe size		130	.062		.29	1.98		2.27	3.49
	7833	1/2" iron pipe size		126	.063		.33	2.04		2.37	3.63
	7834	3/4" iron pipe size		122	.066		.36	2.11		2.47	3.78
	7835	1" iron pipe size		120	.067		.46	2.14		2.60	3.94
	7836	1-1/4" iron pipe size		118	.068		.59	2.18		2.77	4.14
	7837	1-1/2" iron pipe size		118	.068		.66	2.18		2.84	4.22
	7838	2" iron pipe size		116	.069		.88	2.22		3.10	4.52
	7839	2-1/2" iron pipe size		114	.070		1.20	2.26		3.46	4.93
	7840	3" iron pipe size		112	.071		1.34	2.30		3.64	5.15
	7842	1/2" wall, 1/8" iron pipe size		120	.067		.39	2.14		2.53	3.86
	7843	1/4" iron pipe size		120	.067		.41	2.14		2.55	3.88
	7844	3/8" iron pipe size		120	.067		.42	2.14		2.56	3.89
	7845	1/2" iron pipe size		118	.068		.51	2.18		2.69	4.05
	7846	3/4" iron pipe size		116	.069		.53	2.22		2.75	4.13
	7847	1" iron pipe size		114	.070		.68	2.26		2.94	4.36
	7848	1-1/4" iron pipe size		112	.071		.83	2.30		3.13	4.59
	7849	1-1/2" iron pipe size		110	.073		1.01	2.34		3.35	4.86
	7850	2" iron pipe size		108	.074		1.25	2.38		3.63	5.20
	7851	2-1/2" iron pipe size		106	.075		1.62	2.43		4.05	5.65
	7852	3" iron pipe size		104	.077		1.76	2.47		4.23	5.90
	7853	3-1/2" iron pipe size		102	.078		2.55	2.52		5.07	6.85
	7854	4" iron pipe size		100	.080		2.79	2.57		5.36	7.20
	7855	3/4" wall, 1/8" iron pipe size		110	.073		.91	2.34		3.25	4.75
	7856	1/4" iron pipe size		110	.073		.91	2.34		3.25	4.75
	7857	3/8" iron pipe size		108	.074		1.05	2.38		3.43	4.97
	7858	1/2" iron pipe size		106	.075		1.26	2.43		3.69	5.30
	7859	3/4" iron pipe size		104	.077		1.50	2.47		3.97	5.60
	7860	1" iron pipe size		102	.078		1.75	2.52		4.27	5.95
	7861	1-1/4" iron pipe size		100	.080		2.37	2.57		4.94	6.75
	7862	1-1/2" iron pipe size		100	.080		2.70	2.57		5.27	7.10
	7863	2" iron pipe size		98	.082		3.22	2.62		5.84	7.75
	7864	2-1/2" iron pipe size		96	.083		3.99	2.68		6.67	8.70
	7865	3" iron pipe size		94	.085		5.10	2.74		7.84	10.05
	7866	3-1/2" iron pipe size		92	.087		6.30	2.80		9.10	11.45
	7867	4" iron pipe size		90	.089		6.70	2.86		9.56	12
	7868	1" wall, 1/4" iron pipe size		100	.080		1.96	2.57		4.53	6.30
	7869	3/8" iron pipe size		98	.082		1.96	2.62		4.58	6.35
	7870	1/2" iron pipe size		96	.083		2.45	2.68		5.13	7
	7871	3/4" iron pipe size	↓	94	.085	↓	2.48	2.74		5.22	7.10

Important: See the Reference Section for critical supporting data - Reference Nos., Crews, & City Cost Indexes

15050 | Basic Materials & Methods

15082 | Duct Insulation

		CREW	DAILY OUTPUT	LABOR-HOURS	UNIT	MAT.	LABOR	EQUIP.	TOTAL	TOTAL INCL O&P
7872	1" iron pipe size	1 Asbe	92	.087	L.F.	2.87	2.80		5.67	7.65
7873	1-1/4" iron pipe size		90	.089		4.02	2.86		6.88	9
7874	1-1/2" iron pipe size		90	.089		4.31	2.86		7.17	9.30
7875	2" iron pipe size		88	.091		5.65	2.92		8.57	10.95
7876	2-1/2" iron pipe size		86	.093		7.85	2.99		10.84	13.45
7877	3" iron pipe size	▼	84	.095	▼	8.20	3.06		11.26	13.95
7878	Contact cement, quart can				Ea.	18.55			18.55	20.50
7879	Rubber tubing, flexible closed cell foam									
7880	3/8" wall, 1/4" iron pipe size	1 Asbe	120	.067	L.F.	.28	2.14		2.42	3.74
7900	3/8" iron pipe size		120	.067		.31	2.14		2.45	3.77
7910	1/2" iron pipe size		115	.070		.36	2.24		2.60	3.98
7920	3/4" iron pipe size		115	.070		.44	2.24		2.68	4.06
7930	1" iron pipe size		110	.073		.49	2.34		2.83	4.29
7940	1-1/4" iron pipe size		110	.073		.60	2.34		2.94	4.41
7950	1-1/2" iron pipe size		110	.073		.72	2.34		3.06	4.54
8100	1/2" wall, 1/4" iron pipe size		90	.089		.58	2.86		3.44	5.20
8120	3/8" iron pipe size		90	.089		.57	2.86		3.43	5.20
8130	1/2" iron pipe size		89	.090		.45	2.89		3.34	5.15
8140	3/4" iron pipe size		89	.090		.51	2.89		3.40	5.20
8150	1" iron pipe size		88	.091		.55	2.92		3.47	5.30
8160	1-1/4" iron pipe size		87	.092		.69	2.96		3.65	5.50
8170	1-1/2" iron pipe size		87	.092		.83	2.96		3.79	5.65
8180	2" iron pipe size		86	.093		1.59	2.99		4.58	6.55
8190	2-1/2" iron pipe size		86	.093		1.58	2.99		4.57	6.55
8200	3" iron pipe size		85	.094		1.66	3.03		4.69	6.70
8210	3-1/2" iron pipe size		85	.094		2.22	3.03		5.25	7.30
8220	4" iron pipe size		80	.100		2.62	3.22		5.84	8.05
8230	5" iron pipe size		80	.100		3.61	3.22		6.83	9.10
8300	3/4" wall, 1/4" iron pipe size		90	.089		.65	2.86		3.51	5.30
8320	3/8" iron pipe size		90	.089		.72	2.86		3.58	5.35
8330	1/2" iron pipe size		89	.090		.78	2.89		3.67	5.50
8340	3/4" iron pipe size		89	.090		.86	2.89		3.75	5.60
8350	1" iron pipe size		88	.091		1.14	2.92		4.06	5.95
8360	1-1/4" iron pipe size		87	.092		1.30	2.96		4.26	6.15
8370	1-1/2" iron pipe size		87	.092		1.76	2.96		4.72	6.70
8380	2" iron pipe size		86	.093		2.34	2.99		5.33	7.35
8390	2-1/2" iron pipe size		86	.093		3.12	2.99		6.11	8.20
8400	3" iron pipe size		85	.094		3.19	3.03		6.22	8.35
8410	3-1/2" iron pipe size		85	.094		3.90	3.03		6.93	9.15
8420	4" iron pipe size		80	.100		4.51	3.22		7.73	10.10
8430	5" iron pipe size		80	.100		5.45	3.22		8.67	11.15
8440	6" iron pipe size		80	.100		6.55	3.22		9.77	12.40
8444	1" wall, 1/2" iron pipe size		86	.093		1.63	2.99		4.62	6.60
8445	3/4" iron pipe size		84	.095		1.82	3.06		4.88	6.90
8446	1" iron pipe size		84	.095		2.21	3.06		5.27	7.35
8447	1-1/4" iron pipe size		82	.098		2.57	3.14		5.71	7.85
8448	1-1/2" iron pipe size		82	.098		3.38	3.14		6.52	8.70
8449	2" iron pipe size		80	.100		4.51	3.22		7.73	10.10
8450	2-1/2" iron pipe size	▼	80	.100	▼	5.90	3.22		9.12	11.60
8456	Rubber insulation tape, 1/8" x 2" x 30'				Ea.	10.90			10.90	12
8460	Polyolefin tubing, flexible closed cell foam, UV stabilized, work									
8462	temp.-165° F to +210° F, 0 water vapor transmission.									
8464	3/8" wall, 1/8" iron pipe size	1 Asbe	140	.057	L.F.	.24	1.84		2.08	3.20
8466	1/4" iron pipe size		140	.057		.25	1.84		2.09	3.22
8468	3/8" iron pipe size		140	.057		.27	1.84		2.11	3.24
8470	1/2" iron pipe size	▼	136	.059	▼	.30	1.89		2.19	3.36

15050 | Basic Materials & Methods

15082 | Duct Insulation

		CREW	DAILY OUTPUT	LABOR-HOURS	UNIT	MAT.	LABOR	EQUIP.	TOTAL	TOTAL INCL O&P
8472	3/4" iron pipe size	1 Asbe	132	.061	L.F.	.35	1.95		2.30	3.51
8474	1" iron pipe size		130	.062		.38	1.98		2.36	3.59
8476	1-1/4" iron pipe size		128	.063		.50	2.01		2.51	3.77
8478	1-1/2" iron pipe size		128	.063		.62	2.01		2.63	3.90
8480	2" iron pipe size		126	.063		.69	2.04		2.73	4.03
8482	2-1/2" iron pipe size		123	.065		1	2.09		3.09	4.45
8484	3" iron pipe size		121	.066		1.56	2.13		3.69	5.15
8486	4" iron pipe size		118	.068		2.17	2.18		4.35	5.90
8500	1/2" wall, 1/8" iron pipe size		130	.062		.35	1.98		2.33	3.56
8502	1/4" iron pipe size		130	.062		.37	1.98		2.35	3.58
8504	3/8" iron pipe size		130	.062		.40	1.98		2.38	3.61
8506	1/2" iron pipe size		128	.063		.44	2.01		2.45	3.70
8508	3/4" iron pipe size		126	.063		.51	2.04		2.55	3.83
8510	1" iron pipe size		123	.065		.55	2.09		2.64	3.96
8512	1-1/4" iron pipe size		121	.066		.69	2.13		2.82	4.17
8514	1-1/2" iron pipe size		119	.067		.83	2.16		2.99	4.37
8516	2" iron pipe size		117	.068		.97	2.20		3.17	4.59
8518	2-1/2" iron pipe size		114	.070		1.31	2.26		3.57	5.05
8520	3" iron pipe size		112	.071		1.84	2.30		4.14	5.70
8522	4" iron pipe size		110	.073		2.28	2.34		4.62	6.25
8534	3/4" wall, 1/8" iron pipe size		120	.067		.53	2.14		2.67	4.01
8536	1/4" iron pipe size		120	.067		.56	2.14		2.70	4.05
8538	3/8" iron pipe size		117	.068		.65	2.20		2.85	4.24
8540	1/2" iron pipe size		114	.070		.72	2.26		2.98	4.40
8542	3/4" iron pipe size		112	.071		.94	2.30		3.24	4.71
8544	1" iron pipe size		110	.073		1.09	2.34		3.43	4.95
8546	1-1/4" iron pipe size		108	.074		1.45	2.38		3.83	5.40
8548	1-1/2" iron pipe size		108	.074		1.71	2.38		4.09	5.70
8550	2" iron pipe size		106	.075		1.99	2.43		4.42	6.10
8552	2-1/2" iron pipe size		104	.077		2.51	2.47		4.98	6.70
8554	3" iron pipe size		102	.078		3.24	2.52		5.76	7.60
8556	4" iron pipe size		100	.080		3.98	2.57		6.55	8.50
8570	1" wall, 1/8" iron pipe size		110	.073		.96	2.34		3.30	4.81
8572	1/4" iron pipe size		108	.074		.99	2.38		3.37	4.90
8574	3/8" iron pipe size		106	.075		1.02	2.43		3.45	5
8576	1/2" iron pipe size		104	.077		1.07	2.47		3.54	5.15
8578	3/4" iron pipe size		102	.078		1.40	2.52		3.92	5.60
8580	1" iron pipe size		100	.080		1.62	2.57		4.19	5.90
8582	1-1/4" iron pipe size		97	.082		1.84	2.65		4.49	6.25
8584	1-1/2" iron pipe size		97	.082		2.11	2.65		4.76	6.55
8586	2" iron pipe size		95	.084		2.76	2.71		5.47	7.40
8588	2-1/2" iron pipe size		93	.086		3.60	2.77		6.37	8.40
8590	3" iron pipe size		91	.088		3.95	2.83		6.78	8.90
8600	Seam Roller				Ea.	6.25			6.25	6.85
8602	Fuse Seal applicator (Butane)				"	152			152	167
8604	Fuse Seal 1/2" sticks				Lb.	12.20			12.20	13.45
8606	Leaktite adhesive				Qt.	18.30			18.30	20
8608	Leaktite adhesive				Gal.	51.50			51.50	56.50
8610	NOTE: Preslit/preglued vs unslit, same price									

15100 | Building Services Piping

15107 | Metal Pipe & Fittings

		CREW	DAILY OUTPUT	LABOR-HOURS	UNIT	2001 BARE COSTS MAT.	LABOR	EQUIP.	TOTAL	TOTAL INCL O&P	
220	0010 **PIPE, BRASS** Plain end,										220
	0900 Field threaded, coupling & clevis hanger 10′ O.C.										
	0920 Regular weight										
	0980 1/8″ diameter	1 Plum	62	.129	L.F.	2.55	4.45		7	9.50	
	1000 1/4″ diameter		57	.140		3.18	4.84		8.02	10.80	
	1100 3/8″ diameter		52	.154		3.79	5.30		9.09	12.15	
	1120 1/2″ diameter		48	.167		4.51	5.75		10.26	13.60	
	1140 3/4″ diameter		46	.174		6	6		12	15.65	
	1160 1″ diameter		43	.186		8.50	6.40		14.90	19	
	1180 1-1/4″ diameter	Q-1	72	.222		12.70	6.90		19.60	24.50	
	1200 1-1/2″ diameter		65	.246		14.95	7.65		22.60	28	
	1220 2″ diameter		53	.302		20	9.35		29.35	36	
	1240 2-1/2″ diameter		41	.390		30.50	12.10		42.60	52	
	1260 3″ diameter		31	.516		43	16		59	71.50	
	1280 3-1/2″ diameter	Q-2	39	.615		71.50	19.80		91.30	109	
	1300 4″ diameter	"	37	.649		71.50	21		92.50	110	
	1930 To delete coupling & hanger, subtract										
	1940 1/8″ diam. to 1/2″ diam.					14%	46%				
	1950 3/4″ diam. to 1-1/2″ diam.					8%	47%				
	1960 2″ diam. to 4″ diam.					10%	37%				
260	0010 **PIPE, BRASS, FITTINGS** Rough bronze, threaded										260
	1000 Standard wt., 90° Elbow										
	1040 1/8″	1 Plum	13	.615	Ea.	7	21		28	39.50	
	1060 1/4″		13	.615		7	21		28	39.50	
	1080 3/8″		13	.615		7	21		28	39.50	
	1100 1/2″		12	.667		7	23		30	42	
	1120 3/4″		11	.727		9.35	25		34.35	48.50	
	1140 1″		10	.800		15.20	27.50		42.70	58	
	1160 1-1/4″	Q-1	17	.941		24.50	29		53.50	71	
	1180 1-1/2″		16	1		30.50	31		61.50	80.50	
	1200 2″		14	1.143		49	35.50		84.50	108	
	1220 2-1/2″		11	1.455		119	45		164	199	
	1240 3″		8	2		181	62		243	293	
	1260 4″	Q-2	11	2.182		365	70		435	510	
	1280 5″		8	3		1,025	96.50		1,121.50	1,275	
	1300 6″		7	3.429		1,500	110		1,610	1,825	
	1500 45° Elbow, 1/8″	1 Plum	13	.615		8.60	21		29.60	41.50	
	1540 1/4″		13	.615		8.60	21		29.60	41.50	
	1560 3/8″		13	.615		8.60	21		29.60	41.50	
	1580 1/2″		12	.667		8.60	23		31.60	44	
	1600 3/4″		11	.727		12.20	25		37.20	51.50	
	1620 1″		10	.800		21	27.50		48.50	64.50	
	1640 1-1/4″	Q-1	17	.941		33	29		62	80.50	
	1660 1-1/2″		16	1		41.50	31		72.50	93	
	1680 2″		14	1.143		67.50	35.50		103	128	
	1700 2-1/2″		11	1.455		128	45		173	209	
	1720 3″		8	2		198	62		260	310	
	1740 4″	Q-2	11	2.182		500	70		570	655	
	1760 5″		8	3		745	96.50		841.50	960	
	1780 6″		7	3.429		745	110		855	980	
	2000 Tee, 1/8″	1 Plum	9	.889		8.20	30.50		38.70	55	
	2040 1/4″		9	.889		8.20	30.50		38.70	55	
	2060 3/8″		9	.889		8.20	30.50		38.70	55	
	2080 1/2″		8	1		8.20	34.50		42.70	61	
	2100 3/4″		7	1.143		11.65	39.50		51.15	72.50	
	2120 1″		6	1.333		21	46		67	92.50	

15100 | Building Services Piping

15107 | Metal Pipe & Fittings

		CREW	DAILY OUTPUT	LABOR-HOURS	UNIT	2001 BARE COSTS MAT.	LABOR	EQUIP.	TOTAL	TOTAL INCL O&P
2140	1-1/4"	Q-1	10	1.600	Ea.	36	49.50		85.50	115
2160	1-1/2"		9	1.778		41	55		96	128
2180	2"		8	2		68	62		130	168
2200	2-1/2"		7	2.286		161	71		232	285
2220	3"		5	3.200		247	99		346	420
2240	4"	Q-2	7	3.429		610	110		720	840
2260	5"		5	4.800		1,275	154		1,429	1,625
2280	6"		4	6		2,000	193		2,193	2,500
2500	Coupling, 1/8"	1 Plum	26	.308		5.85	10.60		16.45	22.50
2540	1/4"		22	.364		5.85	12.55		18.40	25.50
2560	3/8"		18	.444		5.85	15.30		21.15	29.50
2580	1/2"		15	.533		5.85	18.35		24.20	34
2600	3/4"		14	.571		8.20	19.70		27.90	38.50
2620	1"		13	.615		14	21		35	47.50
2640	1-1/4"	Q-1	22	.727		23.50	22.50		46	59.50
2660	1-1/2"		20	.800		30.50	25		55.50	71
2680	2"		18	.889		50.50	27.50		78	97
2700	2-1/2"		14	1.143		85.50	35.50		121	148
2720	3"		10	1.600		119	49.50		168.50	206
2740	4"	Q-2	12	2		244	64.50		308.50	365
2760	5"		10	2.400		460	77		537	620
2780	6"		9	2.667		650	85.50		735.50	845
3000	Union, 125 lb									
3020	1/8"	1 Plum	12	.667	Ea.	19.90	23		42.90	56.50
3040	1/4"		12	.667		19.90	23		42.90	56.50
3060	3/8"		12	.667		19.90	23		42.90	56.50
3080	1/2"		11	.727		19.90	25		44.90	60
3100	3/4"		10	.800		27.50	27.50		55	71.50
3120	1"		9	.889		36	30.50		66.50	85.50
3140	1-1/4"	Q-1	16	1		52	31		83	105
3160	1-1/2"		15	1.067		62	33		95	119
3180	2"		13	1.231		95.50	38		133.50	163
3200	2-1/2"		10	1.600		267	49.50		316.50	370
3220	3"		7	2.286		470	71		541	625
3240	4"	Q-2	10	2.400		1,025	77		1,102	1,250
3320	For 250 lb. (navy pattern), add					100%				
0010	**PIPE, COPPER** Solder joints									
0100	Solder									
0120	Solder, lead free, roll				Lb.	8.70			8.70	9.55
1000	Type K tubing, couplings & clevis hangers 10' O.C.									
1100	1/4" diameter	1 Plum	84	.095	L.F.	1.15	3.28		4.43	6.20
1120	3/8" diameter		82	.098		1.27	3.36		4.63	6.45
1140	1/2" diameter		78	.103		1.36	3.53		4.89	6.85
1160	5/8" diameter		77	.104		1.52	3.58		5.10	7.05
1180	3/4" diameter		74	.108		1.98	3.72		5.70	7.75
1200	1" diameter		66	.121		2.40	4.18		6.58	8.95
1220	1-1/4" diameter		56	.143		2.88	4.92		7.80	10.60
1240	1-1/2" diameter		50	.160		3.53	5.50		9.03	12.20
1260	2" diameter		40	.200		5.25	6.90		12.15	16.15
1280	2-1/2" diameter	Q-1	60	.267		7.60	8.25		15.85	21
1300	3" diameter		54	.296		10.30	9.20		19.50	25
1320	3-1/2" diameter		42	.381		12.85	11.80		24.65	32
1330	4" diameter		38	.421		15.95	13.05		29	37.50
1340	5" diameter		32	.500		41.50	15.50		57	69.50
1360	6" diameter	Q-2	38	.632		60	20.50		80.50	96.50
1380	8" diameter		34	.706		107	22.50		129.50	152

15100 | Building Services Piping

15107 | Metal Pipe & Fittings

			DAILY	LABOR-		2001 BARE COSTS				TOTAL	
		CREW	OUTPUT	HOURS	UNIT	MAT.	LABOR	EQUIP.	TOTAL	INCL O&P	
420	1390	For other than full hard temper, add				L.F.	13%					420
	1440	For silver solder, add	R15100 -050					15%				
	1800	For medical clean, (oxygen class), add					12%					
	1950	To delete cplgs. & hngrs., 1/4"-1" pipe, subtract					27%	60%				
	1960	1-1/4"-3" pipe, subtract					14%	52%				
	1970	3-1/2"-5" pipe, subtract					10%	60%				
	1980	6"-8" pipe, subtract					19%	53%				
	2000	Type L tubing, couplings & hangers 10' O.C.										
	2100	1/4" diameter	1 Plum	88	.091	L.F.	.97	3.13		4.10	5.80	
	2120	3/8" diameter		84	.095		1.13	3.28		4.41	6.20	
	2140	1/2" diameter		81	.099		1.25	3.40		4.65	6.55	
	2160	5/8" diameter		79	.101		1.60	3.49		5.09	7	
	2180	3/4" diameter		76	.105		1.59	3.63		5.22	7.20	
	2200	1" diameter		68	.118		2.02	4.05		6.07	8.35	
	2220	1-1/4" diameter		58	.138		2.54	4.75		7.29	9.95	
	2240	1-1/2" diameter		52	.154		3.07	5.30		8.37	11.35	
	2260	2" diameter		42	.190		4.47	6.55		11.02	14.80	
	2280	2-1/2" diameter	Q-1	62	.258		6.50	8		14.50	19.25	
	2300	3" diameter		56	.286		8.65	8.85		17.50	23	
	2320	3-1/2" diameter		43	.372		11.05	11.55		22.60	29.50	
	2340	4" diameter		39	.410		13.55	12.70		26.25	34	
	2360	5" diameter		34	.471		38	14.60		52.60	63.50	
	2380	6" diameter	Q-2	40	.600		47.50	19.30		66.80	81	
	2400	8" diameter	"	36	.667		82	21.50		103.50	123	
	2410	For other than full hard temper, add					21%					
	2590	For silver solder, add						15%				
	2900	For medical clean, (oxygen class), add					12%					
	2940	To delete cplgs. & hngrs., 1/4"-1" pipe, subtract					37%	63%				
	2960	1-1/4"-3" pipe, subtract					12%	53%				
	2970	3-1/2"-5" pipe, subtract					12%	63%				
	2980	6"-8" pipe, subtract					24%	55%				
	3000	Type M tubing, couplings & hangers 10' O.C.										
	3100	1/4" diameter	1 Plum	90	.089	L.F.	1.01	3.06		4.07	5.75	
	3120	3/8" diameter		87	.092		1.02	3.17		4.19	5.90	
	3140	1/2" diameter		84	.095		1.05	3.28		4.33	6.10	
	3160	5/8" diameter		81	.099		1.26	3.40		4.66	6.55	
	3180	3/4" diameter		78	.103		1.34	3.53		4.87	6.80	
	3200	1" diameter		70	.114		1.70	3.94		5.64	7.80	
	3220	1-1/4" diameter		60	.133		2.17	4.59		6.76	9.35	
	3240	1-1/2" diameter		54	.148		2.78	5.10		7.88	10.75	
	3260	2" diameter		44	.182		4.17	6.25		10.42	14.05	
	3280	2-1/2" diameter	Q-1	64	.250		5.75	7.75		13.50	18.05	
	3300	3" diameter		58	.276		7.30	8.55		15.85	21	
	3320	3-1/2" diameter		45	.356		9.30	11		20.30	27	
	3340	4" diameter		40	.400		12.15	12.40		24.55	32	
	3360	5" diameter		36	.444		36.50	13.80		50.30	61	
	3370	6" diameter	Q-2	42	.571		46	18.35		64.35	78.50	
	3380	8" diameter	"	38	.632		77	20.50		97.50	115	
	3440	For silver solder, add						15%				
	3960	To delete cplgs. & hngrs., 1/4"-1" pipe, subtract					35%	65%				
	3970	1-1/4"-3" pipe, subtract					19%	56%				
	3980	3-1/2"-5" pipe, subtract					13%	65%				
	3990	6"-8" pipe, subtract					28%	58%				
	4000	Type DWV tubing, couplings & hangers 10' O.C.										
	4100	1-1/4" diameter	1 Plum	60	.133	L.F.	2.14	4.59		6.73	9.30	
	4120	1-1/2" diameter		54	.148		2.54	5.10		7.64	10.50	

15100 | Building Services Piping

15107 | Metal Pipe & Fittings

		CREW	DAILY OUTPUT	LABOR-HOURS	UNIT	2001 BARE COSTS MAT.	LABOR	EQUIP.	TOTAL	TOTAL INCL O&P
4140	2" diameter	1 Plum	44	.182	L.F.	3.19	6.25		9.44	12.95
4160	3" diameter	Q-1	58	.276		5.25	8.55		13.80	18.65
4180	4" diameter		40	.400		8.95	12.40		21.35	28.50
4200	5" diameter		36	.444		26	13.80		39.80	49.50
4220	6" diameter	Q-2	42	.571		36.50	18.35		54.85	67.50
4240	8" diameter	"	38	.632		80	20.50		100.50	119
4730	To delete cplgs. & hngrs., 1-1/4"-2" pipe, subtract					16%	53%			
4740	2-1/2"-4" pipe, subtract					13%	60%			
4750	5"-8" pipe, subtract					23%	58%			
5200	ACR tubing, type L, hard temper, cleaned and									
5220	capped, no couplings or hangers									
5240	3/8" OD				L.F.	.40			.40	.44
5250	1/2" OD					.54			.54	.59
5260	5/8" OD					.75			.75	.83
5270	3/4" OD					.99			.99	1.09
5280	7/8" OD					1.17			1.17	1.29
5290	1-1/8" OD					1.66			1.66	1.83
5300	1-3/8" OD					2.29			2.29	2.52
5310	1-5/8" OD					2.89			2.89	3.18
5320	2-1/8" OD					4.65			4.65	5.10
5330	2-5/8" OD					6.60			6.60	7.25
5340	3-1/8" OD					9.15			9.15	10.05
5350	3-5/8" OD					11.55			11.55	12.70
5360	4-1/8" OD					15			15	16.50
5380	ACR tubing, type L, annealed, cleaned and capped									
5381	No couplings or hangers									
5384	3/8"	1 Stpi	160	.050	L.F.	.40	1.74		2.14	3.07
5385	1/2"		160	.050		.54	1.74		2.28	3.22
5386	5/8"		160	.050		.75	1.74		2.49	3.46
5387	3/4"		130	.062		.99	2.14		3.13	4.33
5388	7/8"		130	.062		1.17	2.14		3.31	4.53
5389	1-1/8"		115	.070		1.66	2.42		4.08	5.50
5390	1-3/8"		100	.080		2.29	2.79		5.08	6.75
5391	1-5/8"		90	.089		2.89	3.10		5.99	7.85
5392	2-1/8"		80	.100		4.65	3.49		8.14	10.35
5393	2-5/8"	Q-5	125	.128		6.60	4.02		10.62	13.30
5394	3-1/8"		105	.152		9.15	4.78		13.93	17.25
5395	4-1/8"		95	.168		15	5.30		20.30	24.50
5800	Refrigeration tubing, dryseal, 50' coils									
5840	1/8" OD				Coil	10.35			10.35	11.40
5850	3/16" OD					11.35			11.35	12.50
5860	1/4" OD					14			14	15.40
5870	5/16" OD					17.95			17.95	19.75
5880	3/8" OD					21			21	23
5890	1/2" OD					26.50			26.50	29.50
5900	5/8" OD					36			36	39.50
5910	3/4" OD					42			42	46
5920	7/8" OD					62			62	68.50
5930	1-1/8" OD					89			89	97.50
5940	1-3/8" OD					124			124	136
5950	1-5/8" OD					159			159	175
9400	Sub assemblies used in assembly systems									
9410	Chilled water unit, coil connections per unit under 10 ton	Q-5	.80	20	System	470	630		1,100	1,475
9420	Chilled water unit, coil connections per unit 10 ton and up		1	16		905	500		1,405	1,750
9430	Chilled water dist. piping per ton, less than 61 ton systems		26	.615		8.85	19.30		28.15	39
9440	Chilled water dist. piping per ton, 61 through 120 ton systems	Q-6	31	.774		21	25		46	61

15100 | Building Services Piping

15107 | Metal Pipe & Fittings

			CREW	DAILY OUTPUT	LABOR-HOURS	UNIT	2001 BARE COSTS MAT.	LABOR	EQUIP.	TOTAL	TOTAL INCL O&P	
420	9450	Chilled water dist. piping/ton, 135 ton systems and up	Q-8	25.40	1.260	System	29	42	3.97	74.97	99.50	420
	9510	Refrigerant piping/ton of cooling for remote condensers R15100-050	Q-5	2	8		89.50	251		340.50	480	
	9520	Refrigerant piping per ton up to 10 ton w/remote condensing unit		2.40	6.667		52.50	209		261.50	375	
	9530	Refrigerant piping per ton, 20 ton w/remote condensing unit		2	8		67	251		318	455	
	9540	Refrigerant piping per ton, 40 ton w/remote condensing unit		1.90	8.421		84	264		348	495	
	9550	Refrigerant piping per ton, 75-80 ton w/remote condensing unit	Q-6	2.40	10		115	325		440	615	
	9560	Refrigerant piping per ton, 100 ton w/remote condensing unit	"	2.20	10.909		149	355		504	700	
460	0010	**PIPE, COPPER, FITTINGS** Wrought unless otherwise noted										460
	0040	Solder joints, copper x copper										
	0070	90° elbow, 1/4"	1 Plum	22	.364	Ea.	.60	12.55		13.15	19.55	
	0090	3/8"		22	.364		.60	12.55		13.15	19.55	
	0100	1/2"		20	.400		.22	13.80		14.02	21	
	0110	5/8"		19	.421		.48	14.50		14.98	22.50	
	0120	3/4"		19	.421		.48	14.50		14.98	22.50	
	0130	1"		16	.500		1.12	17.25		18.37	27	
	0140	1-1/4"		15	.533		1.69	18.35		20.04	29.50	
	0150	1-1/2"		13	.615		2.64	21		23.64	35	
	0160	2"		11	.727		4.80	25		29.80	43.50	
	0170	2-1/2"	Q-1	13	1.231		9.15	38		47.15	67.50	
	0180	3"		11	1.455		12.80	45		57.80	82	
	0190	3-1/2"		10	1.600		42.50	49.50		92	122	
	0200	4"		9	1.778		33	55		88	119	
	0210	5"		6	2.667		121	82.50		203.50	258	
	0220	6"	Q-2	9	2.667		169	85.50		254.50	315	
	0230	8"	"	8	3		680	96.50		776.50	895	
	0250	45° elbow, 1/4"	1 Plum	22	.364		1.20	12.55		13.75	20	
	0270	3/8"		22	.364		.98	12.55		13.53	20	
	0280	1/2"		20	.400		.98	13.80		14.78	22	
	0290	5/8"		19	.421		1.86	14.50		16.36	24	
	0300	3/4"		19	.421		.38	14.50		14.88	22.50	
	0310	1"		16	.500		.65	17.25		17.90	26.50	
	0320	1-1/4"		15	.533		1.16	18.35		19.51	29	
	0330	1-1/2"		13	.615		1.63	21		22.63	34	
	0340	2"		11	.727		2.66	25		27.66	41	
	0350	2-1/2"	Q-1	13	1.231		4.45	38		42.45	62.50	
	0360	3"		13	1.231		14.05	38		52.05	73	
	0370	3-1/2"		10	1.600		24.50	49.50		74	102	
	0380	4"		9	1.778		30	55		85	116	
	0390	5"		6	2.667		116	82.50		198.50	253	
	0400	6"	Q-2	9	2.667		174	85.50		259.50	320	
	0410	8"	"	8	3		620	96.50		716.50	830	
	0450	Tee, 1/4"	1 Plum	14	.571		1.35	19.70		21.05	31	
	0470	3/8"		14	.571		1.03	19.70		20.73	30.50	
	0480	1/2"		13	.615		.36	21		21.36	32.50	
	0490	5/8"		12	.667		2.23	23		25.23	37	
	0500	3/4"		12	.667		.89	23		23.89	35.50	
	0510	1"		10	.800		2.59	27.50		30.09	44.50	
	0520	1-1/4"		9	.889		3.92	30.50		34.42	50.50	
	0530	1-1/2"		8	1		5.45	34.50		39.95	58	
	0540	2"		7	1.143		8.45	39.50		47.95	69	
	0550	2-1/2"	Q-1	8	2		17.10	62		79.10	112	
	0560	3"		7	2.286		26	71		97	136	
	0570	3-1/2"		6	2.667		75.50	82.50		158	208	
	0580	4"		5	3.200		63	99		162	220	
	0590	5"		4	4		197	124		321	405	

15100 | Building Services Piping

15107 | Metal Pipe & Fittings

		CREW	DAILY OUTPUT	LABOR-HOURS	UNIT	MAT.	LABOR	EQUIP.	TOTAL	TOTAL INCL O&P
0600	6"	Q-2	6	4	Ea.	269	129		398	490
0610	8"	"	5	4.800		1,100	154		1,254	1,425
0612	Tee, reducing on the outlet, 1/4"	1 Plum	15	.533		2.17	18.35		20.52	30
0613	3/8"		15	.533		2.04	18.35		20.39	29.50
0614	1/2"		14	.571		1.78	19.70		21.48	31.50
0615	5/8"		13	.615		3.14	21		24.14	35.50
0616	3/4"		12	.667		.81	23		23.81	35.50
0617	1"		11	.727		2.59	25		27.59	41
0618	1-1/4"		10	.800		3.81	27.50		31.31	45.50
0619	1-1/2"		9	.889		4.04	30.50		34.54	50.50
0620	2"		8	1		6.40	34.50		40.90	59
0621	2-1/2"	Q-1	9	1.778		19.95	55		74.95	105
0622	3"		8	2		24	62		86	120
0623	4"		6	2.667		42.50	82.50		125	172
0624	5"		5	3.200		197	99		296	365
0625	6"	Q-2	7	3.429		269	110		379	460
0626	8"	"	6	4		1,100	129		1,229	1,400
0630	Tee, reducing on the run, 1/4"	1 Plum	15	.533		2.70	18.35		21.05	30.50
0631	3/8"		15	.533		3.62	18.35		21.97	31.50
0632	1/2"		14	.571		2.45	19.70		22.15	32
0633	5/8"		13	.615		3.62	21		24.62	36
0634	3/4"		12	.667		.99	23		23.99	35.50
0635	1"		11	.727		3.08	25		28.08	41.50
0636	1-1/4"		10	.800		5	27.50		32.50	47
0637	1-1/2"		9	.889		8.30	30.50		38.80	55
0638	2"		8	1		11	34.50		45.50	64
0639	2-1/2"	Q-1	9	1.778		27	55		82	113
0640	3"		8	2		36.50	62		98.50	134
0641	4"		6	2.667		84.50	82.50		167	218
0642	5"		5	3.200		197	99		296	365
0643	6"	Q-2	7	3.429		269	110		379	460
0644	8"	"	6	4		1,100	129		1,229	1,400
0650	Coupling, 1/4"	1 Plum	24	.333		.15	11.50		11.65	17.50
0670	3/8"		24	.333		.18	11.50		11.68	17.55
0680	1/2"		22	.364		.15	12.55		12.70	19.05
0690	5/8"		21	.381		.46	13.10		13.56	20.50
0700	3/4"		21	.381		.32	13.10		13.42	20
0710	1"		18	.444		.63	15.30		15.93	23.50
0715	1-1/4"		17	.471		1.10	16.20		17.30	25.50
0716	1-1/2"		15	.533		1.45	18.35		19.80	29
0718	2"		13	.615		2.41	21		23.41	34.50
0721	2-1/2"	Q-1	15	1.067		5.10	33		38.10	55.50
0722	3"		13	1.231		7.85	38		45.85	66
0724	3-1/2"		8	2		14.75	62		76.75	110
0726	4"		7	2.286		15.40	71		86.40	124
0728	5"		6	2.667		36	82.50		118.50	165
0731	6"	Q-2	8	3		56	96.50		152.50	208
0732	8"	"	7	3.429		193	110		303	380
0741	Coupling, reducing, concentric									
0743	1/2"	1 Plum	23	.348	Ea.	.35	12		12.35	18.50
0745	3/4"		21.50	.372		.62	12.80		13.42	20
0747	1"		19.50	.410		1.13	14.15		15.28	22.50
0749	1-1/2"		16	.500		1.42	17.25		18.67	27.50
0751	2"		14	.571		3.64	19.70		23.34	33.50
0753	3"	Q-1	14	1.143		10.60	35.50		46.10	65
0755	4"	"	8	2		24.50	62		86.50	121

15100 | Building Services Piping

15107 | Metal Pipe & Fittings

		CREW	DAILY OUTPUT	LABOR-HOURS	UNIT	MAT.	LABOR	EQUIP.	TOTAL	TOTAL INCL O&P
0757	5"	Q-2	7.50	3.200	Ea.	78	103		181	241
0759	6"		7	3.429		133	110		243	310
0761	8"		6.50	3.692		219	119		338	420
0771	Cap, sweat									
0773	1/2"	1 Plum	40	.200	Ea.	.15	6.90		7.05	10.55
0775	3/4"		38	.211		.26	7.25		7.51	11.25
0777	1"		32	.250		.62	8.60		9.22	13.70
0779	1-1/2"		26	.308		1.28	10.60		11.88	17.40
0781	2"		22	.364		2.34	12.55		14.89	21.50
0791	Flange, sweat									
0793	3"	Q-1	22	.727	Ea.	26	22.50		48.50	62.50
0795	4"		18	.889		36.50	27.50		64	82
0797	5"		12	1.333		70.50	41.50		112	140
0799	6"	Q-2	18	1.333		74	43		117	146
0801	8"	"	16	1.500		193	48		241	285
0850	Unions, 1/4"	1 Plum	21	.381		3.83	13.10		16.93	24
0870	3/8"		21	.381		3.65	13.10		16.75	24
0880	1/2"		19	.421		2.10	14.50		16.60	24.50
0890	5/8"		18	.444		8.75	15.30		24.05	32.50
0900	3/4"		18	.444		2.63	15.30		17.93	26
0910	1"		15	.533		4.72	18.35		23.07	32.50
0920	1-1/4"		14	.571		8.10	19.70		27.80	38.50
0930	1-1/2"		12	.667		10.65	23		33.65	46
0940	2"		10	.800		18.15	27.50		45.65	61.50
0950	2-1/2"	Q-1	12	1.333		37.50	41.50		79	104
0960	3"	"	10	1.600		97.50	49.50		147	182
0980	Adapter, copper x male IPS, 1/4"	1 Plum	20	.400		2.11	13.80		15.91	23.50
0990	3/8"		20	.400		1.05	13.80		14.85	22
1000	1/2"		18	.444		.46	15.30		15.76	23.50
1010	3/4"		17	.471		.76	16.20		16.96	25.50
1020	1"		15	.533		1.84	18.35		20.19	29.50
1030	1-1/4"		13	.615		2.72	21		23.72	35
1040	1-1/2"		12	.667		3.12	23		26.12	38
1050	2"		11	.727		5.30	25		30.30	44
1060	2-1/2"	Q-1	10.50	1.524		16.85	47		63.85	90
1070	3"		10	1.600		23	49.50		72.50	101
1080	3-1/2"		9	1.778		38	55		93	125
1090	4"		8	2		38	62		100	136
1200	5"		6	2.667		182	82.50		264.50	325
1210	6"	Q-2	8.50	2.824		203	91		294	360
1214	Adapter, copper x female IPS									
1216	1/2"	1 Plum	18	.444	Ea.	.67	15.30		15.97	23.50
1218	3/4"		17	.471		.93	16.20		17.13	25.50
1220	1"		15	.533		2.13	18.35		20.48	30
1222	1-1/2"		12	.667		4.89	23		27.89	40
1224	2"		11	.727		6.65	25		31.65	45.50
1250	Cross, 1/2"		10	.800		2.78	27.50		30.28	44.50
1260	3/4"		9.50	.842		5.40	29		34.40	50
1270	1"		8	1		9.15	34.50		43.65	62
1280	1-1/4"		7.50	1.067		12.50	37		49.50	69.50
1290	1-1/2"		6.50	1.231		18.70	42.50		61.20	84.50
1300	2"		5.50	1.455		35.50	50		85.50	115
1310	2-1/2"	Q-1	6.50	2.462		81.50	76.50		158	205
1320	3"		5.50	2.909		107	90		197	254
1340	4"		4.70	3.404		150	106		256	325
1350	5"		3.20	5		213	155		368	470

15100 | Building Services Piping

15107	Metal Pipe & Fittings	CREW	DAILY OUTPUT	LABOR-HOURS	UNIT	2001 BARE COSTS MAT.	LABOR	EQUIP.	TOTAL	TOTAL INCL O&P
1360	6"	Q-2	4.70	5.106	Ea.	298	164		462	580
1500	Tee fitting, mechanically formed, (Type 1).									
1520	1/2" run size, 3/8" to 1/2" branch size	1 Plum	80	.100	Ea.		3.45		3.45	5.20
1530	3/4" run size, 3/8" to 3/4" branch size		60	.133			4.59		4.59	6.95
1540	1" run size, 3/8" to 1" branch size		54	.148			5.10		5.10	7.70
1550	1-1/4" run size, 3/8" to 1-1/4" branch size		48	.167			5.75		5.75	8.65
1560	1-1/2" run size, 3/8" to 1-1/2" branch size		40	.200			6.90		6.90	10.40
1570	2" run size, 3/8" to 2" branch size		35	.229			7.85		7.85	11.90
1580	2-1/2" run size, 1/2" to 2" branch size		32	.250			8.60		8.60	13
1590	3" run size, 1" to 2" branch size		26	.308			10.60		10.60	16
1600	4" run size, 1" to 2" branch size		24	.333			11.50		11.50	17.35
1640	Tee fitting, mechanically formed, (Type 2)									
1650	2-1/2" run size, 2-1/2" branch size	1 Plum	12.50	.640	Ea.		22		22	33.50
1660	3" run size, 2-1/2" to 3" branch size		12	.667			23		23	34.50
1670	3-1/2" run size, 2-1/2" to 3-1/2" branch size		11	.727			25		25	38
1680	4" run size, 2-1/2" to 4" branch size		10.50	.762			26.50		26.50	39.50
1698	5" run size, 2" to 4" branch size		9.50	.842			29		29	44
1700	6" run size, 2" to 4" branch size		8.50	.941			32.50		32.50	49
1710	8" run size, 2" to 4" branch size		7	1.143			39.50		39.50	59.50
1800	ACR fittings									
1802	Tee, straight									
1808	5/8"	1 Stpi	12	.667	Ea.	2.23	23		25.23	37.50
1810	3/4"		12	.667		.89	23		23.89	36
1812	7/8"		10	.800		.89	28		28.89	43
1814	1-1/8"		10	.800		5.30	28		33.30	48
1816	1-3/8"		9	.889		8.05	31		39.05	56
1818	1-5/8"		8	1		11.15	35		46.15	65
1820	2-1/8"		7	1.143		17.35	40		57.35	79
1822	2-5/8"	Q-5	8	2		35	63		98	134
1824	3-1/8"		7	2.286		53.50	71.50		125	167
1826	4-1/8"		5	3.200		129	100		229	294
1830	90° elbow									
1836	5/8"	1 Stpi	19	.421	Ea.	1.40	14.65		16.05	23.50
1838	3/4"		19	.421		2.25	14.65		16.90	24.50
1840	7/8"		16	.500		2.36	17.45		19.81	29
1842	1-1/8"		16	.500		3.36	17.45		20.81	30
1844	1-3/8"		15	.533		3.47	18.60		22.07	32
1846	1-5/8"		13	.615		2.64	21.50		24.14	35.50
1848	2-1/8"		11	.727		4.80	25.50		30.30	44
1850	2-5/8"	Q-5	13	1.231		18.75	38.50		57.25	79
1852	3-1/8"		11	1.455		12.80	45.50		58.30	83
1854	4-1/8"		9	1.778		33	56		89	120
1860	Coupling									
1866	5/8"	1 Stpi	21	.381	Ea.	.32	13.30		13.62	20.50
1868	3/4"		21	.381		.46	13.30		13.76	20.50
1870	7/8"		18	.444		.32	15.50		15.82	24
1872	1-1/8"		18	.444		.63	15.50		16.13	24
1874	1-3/8"		17	.471		1.10	16.40		17.50	26
1876	1-5/8"		15	.533		1.45	18.60		20.05	29.50
1878	2-1/8"		13	.615		2.41	21.50		23.91	35
1880	2-5/8"	Q-5	15	1.067		5.10	33.50		38.60	56
1882	3-1/8"		13	1.231		7.85	38.50		46.35	67
1884	4-1/8"		7	2.286		15.40	71.50		86.90	125
2000	DWV, solder joints, copper x copper									
2030	90° Elbow, 1-1/4"	1 Plum	13	.615	Ea.	1.97	21		22.97	34
2050	1-1/2"		12	.667		2.64	23		25.64	37.50

15100 | Building Services Piping

15107 | Metal Pipe & Fittings

			CREW	DAILY OUTPUT	LABOR-HOURS	UNIT	MAT.	LABOR	EQUIP.	TOTAL	TOTAL INCL O&P
2070		2"	1 Plum	10	.800	Ea.	3.83	27.50		31.33	45.50
2090		3"	Q-1	10	1.600		9.45	49.50		58.95	85.50
2100		4"	"	9	1.778		44.50	55		99.50	132
2150	45° Elbow, 1-1/4"		1 Plum	13	.615		1.80	21		22.80	34
2170	1-1/2"			12	.667		1.50	23		24.50	36
2180	2"			10	.800		3.43	27.50		30.93	45.50
2190	3"		Q-1	10	1.600		7.30	49.50		56.80	83
2200	4"		"	9	1.778		31	55		86	117
2250	Tee, Sanitary, 1-1/4"		1 Plum	9	.889		3.87	30.50		34.37	50.50
2270	1-1/2"			8	1		4.82	34.50		39.32	57.50
2290	2"			7	1.143		5.60	39.50		45.10	65.50
2310	3"		Q-1	7	2.286		20	71		91	129
2330	4"		"	6	2.667		51	82.50		133.50	181
2400	Coupling, 1-1/4"		1 Plum	14	.571		.92	19.70		20.62	30.50
2420	1-1/2"			13	.615		1.14	21		22.14	33.50
2440	2"			11	.727		1.58	25		26.58	39.50
2460	3"		Q-1	11	1.455		3.06	45		48.06	71.50
2480	4"		"	10	1.600		7.35	49.50		56.85	83
2600	Traps, see division 15155										
3500	Compression joint fittings										
3510	As used for plumbing and oil burner work										
3520	Fitting price includes nuts and sleeves										
3540	Sleeve, 1/8"					Ea.	.02			.02	.02
3550	3/16"						.02			.02	.02
3560	1/4"						.02			.02	.02
3570	5/16"						.03			.03	.03
3580	3/8"						.04			.04	.04
3600	1/2"						.12			.12	.13
3620	Nut, 1/8"						.08			.08	.09
3630	3/16"						.10			.10	.11
3640	1/4"						.11			.11	.12
3650	5/16"						.12			.12	.13
3660	3/8"						.14			.14	.15
3670	1/2"						.33			.33	.36
3710	Union, 1/8"		1 Plum	26	.308		.56	10.60		11.16	16.60
3720	3/16"			24	.333		.54	11.50		12.04	17.95
3730	1/4"			24	.333		.50	11.50		12	17.90
3740	5/16"			23	.348		.64	12		12.64	18.80
3750	3/8"			22	.364		.88	12.55		13.43	19.85
3760	1/2"			22	.364		1.39	12.55		13.94	20.50
3780	5/8"			21	.381		3.83	13.10		16.93	24
3820	Union tee, 1/8"			17	.471		1.50	16.20		17.70	26
3830	3/16"			16	.500		1.06	17.25		18.31	27
3840	1/4"			15	.533		1.04	18.35		19.39	28.50
3850	5/16"			15	.533		1.29	18.35		19.64	29
3860	3/8"			15	.533		2.07	18.35		20.42	30
3870	1/2"			15	.533		3.28	18.35		21.63	31
3910	Union elbow, 1/4"			24	.333		.72	11.50		12.22	18.15
3920	5/16"			23	.348		1.06	12		13.06	19.25
3930	3/8"			22	.364		1.53	12.55		14.08	20.50
3940	1/2"			22	.364		2.49	12.55		15.04	21.50
3980	Female connector, 1/8"			26	.308		.46	10.60		11.06	16.50
4000	3/16" x 1/8"			24	.333		.40	11.50		11.90	17.80
4010	1/4" x 1/8"			24	.333		.41	11.50		11.91	17.80
4020	1/4"			24	.333		.57	11.50		12.07	18
4030	3/8" x 1/4"			22	.364		.70	12.55		13.25	19.65

15100 | Building Services Piping

15107 | Metal Pipe & Fittings

		CREW	DAILY OUTPUT	LABOR-HOURS	UNIT	2001 BARE COSTS MAT.	LABOR	EQUIP.	TOTAL	TOTAL INCL O&P	
460	4040	1/2" x 3/8"	1 Plum	22	.364	Ea.	1.22	12.55		13.77	20	460
	4050	5/8" x 1/2"		21	.381		1.76	13.10		14.86	21.50	
	4090	Male connector, 1/8"		26	.308		.36	10.60		10.96	16.40	
	4100	3/16" x 1/8"		24	.333		.39	11.50		11.89	17.80	
	4110	1/4" x 1/8"		24	.333		.39	11.50		11.89	17.80	
	4120	1/4"		24	.333		.39	11.50		11.89	17.80	
	4130	5/16" x 1/8"		23	.348		.41	12		12.41	18.55	
	4140	5/16" x 1/4"		23	.348		.45	12		12.45	18.60	
	4150	3/8" x 1/8"		22	.364		.29	12.55		12.84	19.20	
	4160	3/8" x 1/4"		22	.364		.47	12.55		13.02	19.40	
	4170	3/8"		22	.364		.77	12.55		13.32	19.75	
	4180	3/8" x 1/2"		22	.364		.82	12.55		13.37	19.80	
	4190	1/2" x 3/8"		22	.364		.75	12.55		13.30	19.75	
	4200	1/2"		22	.364		1.13	12.55		13.68	20	
	4210	5/8" x 1/2"		21	.381		1.46	13.10		14.56	21.50	
	4240	Male elbow, 1/8"		26	.308		.54	10.60		11.14	16.60	
	4250	3/16" x 1/8"		24	.333		.55	11.50		12.05	17.95	
	4260	1/4" x 1/8"		24	.333		.47	11.50		11.97	17.85	
	4270	1/4"		24	.333		.58	11.50		12.08	18	
	4280	3/8" x 1/4"		22	.364		.70	12.55		13.25	19.65	
	4290	3/8"		22	.364		.90	12.55		13.45	19.90	
	4300	1/2" x 1/4"		22	.364		1.53	12.55		14.08	20.50	
	4310	1/2" x 3/8"		22	.364		1.25	12.55		13.80	20.50	
	4340	Female elbow, 1/8"		26	.308		1.03	10.60		11.63	17.15	
	4350	1/4" x 1/8"		24	.333		.68	11.50		12.18	18.10	
	4360	1/4"		24	.333		.86	11.50		12.36	18.30	
	4370	3/8" x 1/4"		22	.364		1.04	12.55		13.59	20	
	4380	1/2" x 3/8"		22	.364		1.57	12.55		14.12	20.50	
	4390	1/2"		22	.364		1.58	12.55		14.13	20.50	
	4420	Male run tee, 1/4" x 1/8"		15	.533		.93	18.35		19.28	28.50	
	4430	5/16" x 1/8"		15	.533		1.15	18.35		19.50	29	
	4440	3/8" x 1/4"		15	.533		1.47	18.35		19.82	29	
	4480	1/4" x 1/8"		15	.533		.90	18.35		19.25	28.50	
	4490	1/4"		15	.533		1.30	18.35		19.65	29	
	4500	3/8" x 1/4"		15	.533		1.42	18.35		19.77	29	
	4510	1/2" x 3/8"		15	.533		2.39	18.35		20.74	30	
	4520	1/2"	▼	15	.533	▼	2.92	18.35		21.27	30.50	
	4800	Flare joint fittings										
	4810	Refrigeration fittings										
	4820	Flare joint nuts and labor not incl. in price. Add 1 nut per jnt.										
	4830	90° Elbow, 1/4"				Ea.	1.96			1.96	2.16	
	4840	3/8"					3.03			3.03	3.33	
	4850	1/2"					4.44			4.44	4.88	
	4860	5/8"					6.10			6.10	6.70	
	4870	3/4"					11.30			11.30	12.45	
	5030	Tee, 1/4"					3.15			3.15	3.47	
	5040	5/16"					3.32			3.32	3.65	
	5050	3/8"					4.27			4.27	4.70	
	5060	1/2"					7.80			7.80	8.60	
	5070	5/8"					8.80			8.80	9.70	
	5080	3/4"					11.20			11.20	12.35	
	5140	Union, 3/16"					1.97			1.97	2.17	
	5150	1/4"					.60			.60	.66	
	5160	5/16"					.67			.67	.74	
	5170	3/8"					1.61			1.61	1.77	
	5180	1/2"				▼	2.37			2.37	2.61	

15100 | Building Services Piping

15107 | Metal Pipe & Fittings

		CREW	DAILY OUTPUT	LABOR-HOURS	UNIT	MAT.	LABOR	EQUIP.	TOTAL	TOTAL INCL O&P
5190	5/8"				Ea.	3.71			3.71	4.08
5200	3/4"					5.20			5.20	5.70
5260	Long flare nut, 3/16"	1 Stpi	42	.190		.42	6.65		7.07	10.45
5270	1/4"		41	.195		.48	6.80		7.28	10.80
5280	5/16"		40	.200		.73	6.95		7.68	11.30
5290	3/8"		39	.205		1.13	7.15		8.28	12.05
5300	1/2"		38	.211		1.70	7.35		9.05	12.90
5310	5/8"		37	.216		3.42	7.55		10.97	15.10
5320	3/4"		34	.235		5.80	8.20		14	18.75
5380	Short flare nut, 3/16"		42	.190		.29	6.65		6.94	10.30
5390	1/4"		41	.195		.39	6.80		7.19	10.70
5400	5/16"		40	.200		.57	6.95		7.52	11.15
5410	3/8"		39	.205		.47	7.15		7.62	11.30
5420	1/2"		38	.211		.68	7.35		8.03	11.80
5430	5/8"		36	.222		.96	7.75		8.71	12.75
5440	3/4"		34	.235		3.36	8.20		11.56	16.10
5500	90° Elbow flare by MIPS, 1/4"					1.38			1.38	1.52
5510	3/8"					1.92			1.92	2.11
5520	1/2"					3.08			3.08	3.39
5530	5/8"					3.91			3.91	4.30
5540	3/4"					5.60			5.60	6.15
5600	Flare by FIPS, 1/4"					4.09			4.09	4.50
5610	3/8"					3.57			3.57	3.93
5620	1/2"					6.75			6.75	7.45
5670	Flare by sweat, 1/4"					1.42			1.42	1.56
5680	3/8"					1.38			1.38	1.52
5690	1/2"					2.51			2.51	2.76
5700	5/8"					3.47			3.47	3.82
5760	Tee flare by IPS, 1/4"					2.19			2.19	2.41
5770	3/8"					2.16			2.16	2.38
5780	1/2"					4.45			4.45	4.90
5790	5/8"					4.53			4.53	4.98
5850	Connector, 1/4"					.43			.43	.47
5860	3/8"					.80			.80	.88
5870	1/2"					.82			.82	.90
5880	5/8"					.94			.94	1.03
5890	3/4"					5.30			5.30	5.85
5950	Seal cap, 1/4"					.21			.21	.23
5960	3/8"					.26			.26	.29
5970	1/2"					.42			.42	.46
5980	5/8"					1.26			1.26	1.39
5990	3/4"					4.39			4.39	4.83
6000	Water service fittings									
6010	Flare joints nut and labor are included in the fitting price.									
6020	90° Elbow, cxc, 3/8"	1 Plum	19	.421	Ea.	5.80	14.50		20.30	28.50
6030	1/2"		18	.444		5.80	15.30		21.10	29.50
6040	3/4"		16	.500		8.20	17.25		25.45	35
6050	1"		15	.533		15.55	18.35		33.90	44.50
6060	1-1/4"		12	.667		47	23		70	86.50
6070	1-1/2"		12	.667		50.50	23		73.50	90
6080	2"		10	.800		53.50	27.50		81	101
6090	C x MPT, 3/8"		19	.421		9.05	14.50		23.55	32
6100	1/2"		18	.444		9.05	15.30		24.35	33
6110	3/4"		16	.500		10.80	17.25		28.05	38
6120	1"		15	.533		13.95	18.35		32.30	43
6130	1-1/4"		13	.615		30	21		51	65

15100 | Building Services Piping

15107 | Metal Pipe & Fittings

			CREW	DAILY OUTPUT	LABOR-HOURS	UNIT	MAT.	LABOR	EQUIP.	TOTAL	TOTAL INCL O&P	
460	6140	1-1/2"	1 Plum	12	.667	Ea.	30	23		53	67.50	460
	6150	2"		10	.800		77.50	27.50		105	127	
	6160	C x FPT, 3/8"		19	.421		4.49	14.50		18.99	27	
	6170	1/2"		18	.444		4.49	15.30		19.79	28	
	6180	3/4"		16	.500		6.20	17.25		23.45	33	
	6190	1"		15	.533		13.50	18.35		31.85	42.50	
	6200	1-1/4"		13	.615		42	21		63	78	
	6210	1-1/2"		12	.667		97	23		120	142	
	6220	2"		10	.800		148	27.50		175.50	205	
	6230	Tee, cxcxc, 3/8"		13	.615		10.70	21		31.70	44	
	6240	1/2"		12	.667		10.70	23		33.70	46.50	
	6250	3/4"		11	.727		11.65	25		36.65	51	
	6260	1"		10	.800		23	27.50		50.50	66.50	
	6270	1-1/4"		8	1		76	34.50		110.50	136	
	6280	1-1/2"		8	1		91	34.50		125.50	152	
	6290	2"		7	1.143		163	39.50		202.50	240	
	6300	Tee, cxcxFPT, 3/8"		13	.615		8.60	21		29.60	41.50	
	6310	1/2"		12	.667		12.70	23		35.70	48.50	
	6320	3/4"		11	.727		18	25		43	58	
	6330	Tube nut, cxnut seat, 3/8"		40	.200		1.13	6.90		8.03	11.65	
	6340	1/2"		38	.211		1.13	7.25		8.38	12.20	
	6350	3/4"		34	.235		1.91	8.10		10.01	14.35	
	6360	1"		32	.250		3.47	8.60		12.07	16.80	
	6370	1-1/4"		25	.320		20.50	11		31.50	39	
	6380	Coupling, cxc, 3/8"		19	.421		5.80	14.50		20.30	28.50	
	6390	1/2"		18	.444		5.80	15.30		21.10	29.50	
	6400	3/4"		16	.500		6.80	17.25		24.05	33.50	
	6410	1"		15	.533		13.10	18.35		31.45	42	
	6420	1-1/4"		12	.667		21	23		44	58	
	6430	1-1/2"		12	.667		33	23		56	71	
	6440	2"		10	.800		50	27.50		77.50	96.50	
	6450	Adapter, cxFPT, 3/8"		19	.421		4.23	14.50		18.73	26.50	
	6460	1/2"		18	.444		4.23	15.30		19.53	27.50	
	6470	3/4"		16	.500		5	17.25		22.25	31.50	
	6480	1"		15	.533		11.25	18.35		29.60	40	
	6490	1-1/4"		13	.615		21.50	21		42.50	55.50	
	6500	1-1/2"		12	.667		21.50	23		44.50	58	
	6510	2"		10	.800		30	27.50		57.50	74.50	
	6520	Adapter, cxMPT, 3/8"		19	.421		3.64	14.50		18.14	26	
	6530	1/2"		18	.444		3.64	15.30		18.94	27	
	6540	3/4"		16	.500		5.10	17.25		22.35	31.50	
	6550	1"		15	.533		9.05	18.35		27.40	37.50	
	6560	1-1/4"		13	.615		24	21		45	58	
	6570	1-1/2"		12	.667		24	23		47	60.50	
	6580	2"		10	.800		32	27.50		59.50	77	
	6990	Polybutylene/polyethylene pipe, See 15108 for plastic ftng.										
	7000	Insert type Brass/copper, 100 psi @ 180°F, CTS										
	7010	Adapter MPT 3/8" x 3/8" CTS	1 Plum	29	.276	Ea.	.97	9.50		10.47	15.40	
	7020	1/2" x 1/2"		26	.308		.97	10.60		11.57	17.05	
	7030	3/4" x 1/2"		26	.308		1.44	10.60		12.04	17.60	
	7040	3/4" x 3/4"		25	.320		1.42	11		12.42	18.20	
	7050	Adapter CTS 1/2" x 1/2" sweat		24	.333		.36	11.50		11.86	17.75	
	7060	3/4" x 3/4" sweat		22	.364		.47	12.55		13.02	19.40	
	7070	Coupler center set 3/8" CTS		25	.320		.49	11		11.49	17.20	
	7080	1/2" CTS		23	.348		.38	12		12.38	18.50	
	7090	3/4" CTS		22	.364		.47	12.55		13.02	19.40	

15100 | Building Services Piping

15107 | Metal Pipe & Fittings

		Description	CREW	DAILY OUTPUT	LABOR-HOURS	UNIT	2001 BARE COSTS MAT.	LABOR	EQUIP.	TOTAL	TOTAL INCL O&P	
460	7100	Elbow 90°, copper 3/8"	1 Plum	25	.320	Ea.	.65	11		11.65	17.35	460
	7110	1/2" CTS		23	.348		.69	12		12.69	18.85	
	7120	3/4" CTS		22	.364		.84	12.55		13.39	19.80	
	7130	Tee copper 3/8" CTS		17	.471		.85	16.20		17.05	25.50	
	7140	1/2" CTS		15	.533		.66	18.35		19.01	28	
	7150	3/4" CTS		14	.571		.95	19.70		20.65	30.50	
	7160	3/8" x 3/8" x 1/2"		16	.500		1.30	17.25		18.55	27.50	
	7170	1/2" x 3/8" x 1/2"		15	.533		1.30	18.35		19.65	29	
	7180	3/4" x 1/2" x 3/4"		14	.571		1.30	19.70		21	31	
480	0010	**PIPE/TUBE, GROOVED-JOINT FOR COPPER**										480
	4000	Fittings: coupling and labor required at joints not incl.										
	4001	in fitting price. add 1 per joint for installed price.										
	4010	Coupling, rigid style										
	4018	2" diameter	1 Plum	50	.160	Ea.	8.05	5.50		13.55	17.15	
	4020	2-1/2" diameter	Q-1	80	.200		9.15	6.20		15.35	19.40	
	4022	3" diameter		67	.239		10.25	7.40		17.65	22.50	
	4024	4" diameter		50	.320		15.75	9.90		25.65	32.50	
	4026	5" diameter		40	.400		27	12.40		39.40	48	
	4028	6" diameter	Q-2	50	.480		36	15.45		51.45	63	
	4100	Elbow, 90° or 45°										
	4108	2" diameter	1 Plum	25	.320	Ea.	12.50	11		23.50	30.50	
	4110	2-1/2" diameter	Q-1	40	.400		13.60	12.40		26	33.50	
	4112	3" diameter		33	.485		19.25	15.05		34.30	43.50	
	4114	4" diameter		25	.640		45	19.85		64.85	79.50	
	4116	5" diameter		20	.800		129	25		154	180	
	4118	6" diameter	Q-2	25	.960		206	31		237	274	
	4200	Tee										
	4208	2" diameter	1 Plum	17	.471	Ea.	21	16.20		37.20	47.50	
	4210	2-1/2" diameter	Q-1	27	.593		22	18.35		40.35	52	
	4212	3" diameter		22	.727		33.50	22.50		56	70.50	
	4214	4" diameter		17	.941		74	29		103	125	
	4216	5" diameter		13	1.231		208	38		246	287	
	4218	6" diameter	Q-2	17	1.412		257	45.50		302.50	350	
	4300	Reducer, concentric										
	4310	3" x 2-1/2" diameter	Q-1	35	.457	Ea.	18.20	14.15		32.35	41.50	
	4312	4" x 2-1/2" diameter		32	.500		38	15.50		53.50	65	
	4314	4" x 3" diameter		29	.552		38	17.10		55.10	67.50	
	4316	5" x 3" diameter		25	.640		108	19.85		127.85	149	
	4318	5" x 4" diameter		22	.727		108	22.50		130.50	153	
	4320	6" x 3" diameter	Q-2	28	.857		117	27.50		144.50	170	
	4322	6" x 4" diameter		26	.923		117	29.50		146.50	173	
	4324	6" x 5" diameter		24	1		117	32		149	177	
	4350	Flange, w/groove gasket										
	4351	ANSI class 125 and 150										
	4355	2" diameter	1 Plum	23	.348	Ea.	65.50	12		77.50	90	
	4356	2-1/2" diameter	Q-1	37	.432		68.50	13.40		81.90	95.50	
	4358	3" diameter		31	.516		71.50	16		87.50	103	
	4360	4" diameter		23	.696		78.50	21.50		100	119	
	4362	5" diameter		19	.842		106	26		132	156	
	4364	6" diameter	Q-2	23	1.043		115	33.50		148.50	178	
620	0010	**PIPE, STEEL**	R15100-050									620
	0020	All pipe sizes are to Spec. A-53 unless noted otherwise										
	0030	Schedule 10, see 15107-690-0500										
	0050	Schedule 40, threaded, with couplings, and clevis type										

MECHANICAL 15

15100 | Building Services Piping

15107 | Metal Pipe & Fittings

		CREW	DAILY OUTPUT	LABOR-HOURS	UNIT	2001 BARE COSTS MAT.	LABOR	EQUIP.	TOTAL	TOTAL INCL O&P
620	0060 hangers sized for covering, 10' O.C.									620
	0540 Black, 1/4" diameter	1 Plum	66	.121	L.F.	1.29	4.18		5.47	7.70
	0550 3/8" diameter		65	.123		1.40	4.24		5.64	7.95
	0560 1/2" diameter		63	.127		1.22	4.37		5.59	7.95
	0570 3/4" diameter		61	.131		1.36	4.52		5.88	8.30
	0580 1" diameter		53	.151		1.73	5.20		6.93	9.75
	0590 1-1/4" diameter	Q-1	89	.180		2.04	5.55		7.59	10.65
	0600 1-1/2" diameter		80	.200		2.29	6.20		8.49	11.85
	0610 2" diameter		64	.250		2.89	7.75		10.64	14.90
	0620 2-1/2" diameter		50	.320		4.59	9.90		14.49	20
	0630 3" diameter		43	.372		5.80	11.55		17.35	24
	0640 3-1/2" diameter		40	.400		7.70	12.40		20.10	27
	0650 4" diameter		36	.444		8.55	13.80		22.35	30.50
	0660 5" diameter		26	.615		12.55	19.10		31.65	43
	0670 6" diameter	Q-2	31	.774		16.50	25		41.50	55.50
	0680 8" diameter		27	.889		22.50	28.50		51	67.50
	0690 10" diameter		23	1.043		39	33.50		72.50	93.50
	0700 12" diameter		18	1.333		49.50	43		92.50	119
	1220 To delete coupling & hanger, subtract									
	1230 1/4" diam. to 3/4" diam.					31%	56%			
	1240 1" diam. to 1-1/2" diam.					23%	51%			
	1250 2" diam. to 4" diam.					23%	41%			
	1260 5" diam. to 12" diam.					21%	45%			
	1290 Galvanized, 1/4" diameter	1 Plum	66	.121	L.F.	1.49	4.18		5.67	7.95
	1300 3/8" diameter		65	.123		1.60	4.24		5.84	8.15
	1310 1/2" diameter		63	.127		1.37	4.37		5.74	8.10
	1320 3/4" diameter		61	.131		1.53	4.52		6.05	8.50
	1330 1" diameter		53	.151		1.81	5.20		7.01	9.85
	1340 1-1/4" diameter	Q-1	89	.180		2.16	5.55		7.71	10.75
	1350 1-1/2" diameter		80	.200		2.39	6.20		8.59	12
	1360 2" diameter		64	.250		3.03	7.75		10.78	15.05
	1370 2-1/2" diameter		50	.320		4.52	9.90		14.42	19.95
	1380 3" diameter		43	.372		5.60	11.55		17.15	23.50
	1390 3-1/2" diameter		40	.400		7.10	12.40		19.50	26.50
	1400 4" diameter		36	.444		8.05	13.80		21.85	30
	1410 5" diameter		26	.615		14.90	19.10		34	45.50
	1420 6" diameter	Q-2	31	.774		20.50	25		45.50	60
	1430 8" diameter		27	.889		33	28.50		61.50	79.50
	1440 10" diameter		23	1.043		48.50	33.50		82	104
	1450 12" diameter		18	1.333		66.50	43		109.50	138
	1750 To delete coupling & hanger, subtract									
	1760 1/4" diam. to 3/4" diam.					31%	56%			
	1770 1" diam. to 1-1/2" diam.					23%	51%			
	1780 2" diam. to 4" diam.					23%	41%			
	1790 5" diam. to 12" diam.					21%	45%			
	2000 Welded, sch. 40, on yoke & roll hangers, sized for covering,									
	2010 10' O.C. (no hangers incl. for 14" diam. and up)									
	2040 Black, 1" diameter	Q-15	93	.172	L.F.	2.60	5.35	1.08	9.03	12.10
	2050 1-1/4" diameter		84	.190		3.23	5.90	1.20	10.33	13.75
	2060 1-1/2" diameter		76	.211		3.41	6.55	1.32	11.28	15.05
	2070 2" diameter		61	.262		3.88	8.15	1.65	13.68	18.40
	2080 2-1/2" diameter		47	.340		5.05	10.55	2.14	17.74	24
	2090 3" diameter		43	.372		5.45	11.55	2.34	19.34	26
	2100 3-1/2" diameter		39	.410		6.85	12.70	2.58	22.13	29.50
	2110 4" diameter		37	.432		7.90	13.40	2.72	24.02	31.50
	2120 5" diameter		32	.500		13.10	15.50	3.15	31.75	41.50

15100 | Building Services Piping

15107 | Metal Pipe & Fittings

			CREW	DAILY OUTPUT	LABOR-HOURS	UNIT	MAT.	LABOR	EQUIP.	TOTAL	TOTAL INCL O&P
2130	6" diameter	R15100-050	Q-16	36	.667	L.F.	16.80	21.50	2.79	41.09	54
2140	8" diameter			29	.828		24.50	26.50	3.47	54.47	71
2150	10" diameter			24	1		43	32	4.19	79.19	101
2160	12" diameter			19	1.263		53	40.50	5.30	98.80	125
2170	14" diameter			15	1.600		34	51.50	6.70	92.20	122
2180	16" diameter			13	1.846		32	59.50	7.75	99.25	133
2190	18" diameter			11	2.182		56.50	70	9.15	135.65	179
2200	20" diameter			9	2.667		64	85.50	11.15	160.65	212
2220	24" diameter			8	3		71.50	96.50	12.55	180.55	238
2230	26" diameter			7.50	3.200		77.50	103	13.40	193.90	255
2240	30" diameter			6	4		84.50	129	16.75	230.25	305
2260	36" diameter			4.50	5.333		96.50	171	22.50	290	390
2323	.375" wall, A-53, gr. A/B, seamless, welded w/hngrs.										
2324	12" diameter		Q-16	20	1.200	L.F.	36	38.50	5.05	79.55	103
2325	14" diameter			17	1.412		43	45.50	5.90	94.40	122
2326	16" diameter			14	1.714		54	55	7.20	116.20	150
2334	.50" wall, 8" diameter			25	.960		23	31	4.02	58.02	76.50
2335	10" diameter			20	1.200		33.50	38.50	5.05	77.05	100
2336	12" diameter			16	1.500		39	48	6.30	93.30	123
2337	14" diameter			13	1.846		54.50	59.50	7.75	121.75	158
2345	Sch. 40, A-53, gr. A/B, ERW, welded w/hngrs.										
2346	2" diameter		Q-15	61	.262	L.F.	2.45	8.15	1.65	12.25	16.80
2347	2-1/2" diameter			47	.340		3.54	10.55	2.14	16.23	22
2348	3" diameter			43	.372		4.42	11.55	2.34	18.31	25
2349	4" diameter			38	.421		6.05	13.05	2.65	21.75	29.50
2350	6" diameter		Q-16	37	.649		8.95	21	2.72	32.67	44.50
2351	8" diameter			29	.828		13.15	26.50	3.47	43.12	58.50
2352	10" diameter			24	1		25	32	4.19	61.19	80.50
2363	.375" wall, A-53, gr. A/B, ERW, welded w/hngrs.										
2364	12" diameter		Q-16	20	1.200	L.F.	27.50	38.50	5.05	71.05	93.50
2365	14" diameter			17	1.412		31	45.50	5.90	82.40	109
2366	16" diameter			14	1.714		43.50	55	7.20	105.70	138
2560	To delete hanger, subtract										
2570	1" diam. to 1-1/2" diam.						15%	34%			
2580	2" diam. to 3-1/2" diam.						9%	21%			
2590	4" diam. to 12" diam.						5%	12%			
2640	Galvanized, 1" diameter		Q-15	93	.172	L.F.	2.74	5.35	1.08	9.17	12.25
2650	1-1/4" diameter			84	.190		3.41	5.90	1.20	10.51	13.95
2660	1-1/2" diameter			76	.211		3.63	6.55	1.32	11.50	15.30
2670	2" diameter			61	.262		4.18	8.15	1.65	13.98	18.70
2680	2-1/2" diameter			47	.340		5.20	10.55	2.14	17.89	24
2690	3" diameter			43	.372		6.05	11.55	2.34	19.94	26.50
2700	3-1/2" diameter			39	.410		7.60	12.70	2.58	22.88	30.50
2710	4" diameter			37	.432		8.55	13.40	2.72	24.67	32.50
2720	5" diameter			32	.500		16.05	15.50	3.15	34.70	44.50
2730	6" diameter		Q-16	36	.667		21	21.50	2.79	45.29	58.50
2740	8" diameter			29	.828		30.50	26.50	3.47	60.47	77.50
2750	10" diameter			24	1		51.50	32	4.19	87.69	110
2760	12" diameter			19	1.263		64.50	40.50	5.30	110.30	138
3160	To delete hanger, subtract										
3170	1" diam. to 1-1/2" diam.						30%	34%			
3180	2" diam. to 3-1/2" diam.						19%	21%			
3190	4" diam. to 12" diam.						10%	12%			
3250	Flanged, 150 lb. weld neck, on yoke & roll hangers										
3260	sized for covering, 10' O.C.										
3290	Black, 1" diameter		Q-15	70	.229	L.F.	6.05	7.10	1.44	14.59	18.95

15100 | Building Services Piping

15107 | Metal Pipe & Fittings

			CREW	DAILY OUTPUT	LABOR-HOURS	UNIT	MAT.	LABOR	EQUIP.	TOTAL	TOTAL INCL O&P
3300		1-1/4" diameter	Q-15	64	.250	L.F.	6.70	7.75	1.57	16.02	21
3310		1-1/2" diameter	R15100-050	58	.276		6.85	8.55	1.74	17.14	22.50
3320		2" diameter		45	.356		7.70	11	2.24	20.94	27.50
3330		2-1/2" diameter		36	.444		9.45	13.80	2.80	26.05	34.50
3340		3" diameter		32	.500		10.30	15.50	3.15	28.95	38.50
3350		3-1/2" diameter		29	.552		12.60	17.10	3.47	33.17	43.50
3360		4" diameter		26	.615		13.55	19.10	3.87	36.52	48
3370		5" diameter		21	.762		22	23.50	4.79	50.29	65.50
3380		6" diameter	Q-16	25	.960		26	31	4.02	61.02	79.50
3390		8" diameter		19	1.263		39	40.50	5.30	84.80	110
3400		10" diameter		16	1.500		67.50	48	6.30	121.80	154
3410		12" diameter		14	1.714		92	55	7.20	154.20	192
3470		For 300 lb. flanges, add					63%				
3480		For 600 lb. flanges, add					310%				
3960		To delete flanges & hanger, subtract									
3970		1" diam. to 2" diam.					76%	65%			
3980		2-1/2" diam. to 4" diam.					62%	59%			
3990		5" diam. to 12" diam.					60%	46%			
4040		Galvanized, 1" diameter	Q-15	70	.229	L.F.	6.20	7.10	1.44	14.74	19.10
4050		1-1/4" diameter		64	.250		6.85	7.75	1.57	16.17	21
4060		1-1/2" diameter		58	.276		7.10	8.55	1.74	17.39	22.50
4070		2" diameter		45	.356		7.75	11	2.24	20.99	27.50
4080		2-1/2" diameter		36	.444		9.35	13.80	2.80	25.95	34.50
4090		3" diameter		32	.500		10.55	15.50	3.15	29.20	38.50
4100		3-1/2" diameter		29	.552		12.60	17.10	3.47	33.17	43.50
4110		4" diameter		26	.615		14.20	19.10	3.87	37.17	49
4120		5" diameter		21	.762		25	23.50	4.79	53.29	68.50
4130		6" diameter	Q-16	25	.960		30	31	4.02	65.02	84
4140		8" diameter		19	1.263		39.50	40.50	5.30	85.30	111
4150		10" diameter		16	1.500		68	48	6.30	122.30	154
4160		12" diameter		14	1.714		86.50	55	7.20	148.70	186
4220		For 300 lb. flanges, add					65%				
4240		For 600 lb. flanges, add					350%				
4660		To delete flanges & hanger, subtract									
4670		1" diam. to 2" diam.					73%	65%			
4680		2-1/2" diam. to 4" diam.					58%	59%			
4690		5" diam. to 12" diam.					43%	46%			
4750		Schedule 80, threaded, with couplings, and clevis type hangers									
4760		sized for covering, 10' O.C.									
4790		Black, 1/4" diameter	1 Plum	54	.148	L.F.	1.52	5.10		6.62	9.35
4800		3/8" diameter		53	.151		1.56	5.20		6.76	9.55
4810		1/2" diameter		52	.154		1.69	5.30		6.99	9.85
4820		3/4" diameter		50	.160		1.90	5.50		7.40	10.40
4830		1" diameter		45	.178		2.30	6.10		8.40	11.80
4840		1-1/4" diameter	Q-1	75	.213		3	6.60		9.60	13.30
4850		1-1/2" diameter		69	.232		3.28	7.20		10.48	14.45
4860		2" diameter		56	.286		4.38	8.85		13.23	18.15
4870		2-1/2" diameter		44	.364		6.35	11.25		17.60	24
4880		3" diameter		38	.421		7.85	13.05		20.90	28.50
4890		3-1/2" diameter		35	.457		9.70	14.15		23.85	32
4900		4" diameter		32	.500		11.40	15.50		26.90	36
4910		5" diameter		23	.696		20	21.50		41.50	54.50
4920		6" diameter	Q-2	28	.857		32	27.50		59.50	76.50
4930		8" diameter		23	1.043		52.50	33.50		86	108
4940		10" diameter		20	1.200		80.50	38.50		119	147
4950		12" diameter		15	1.600		93	51.50		144.50	180

15100 | Building Services Piping

15107 | Metal Pipe & Fittings

		Crew	Daily Output	Labor-Hours	Unit	Mat.	Labor	Equip.	Total	Total Incl O&P
5061	A-106, gr. A/B seamless w/ cplgs. & hgrs, 1/4" dia. (R15100-050)	1 Plum	63	.127	L.F.	2.12	4.37		6.49	8.95
5062	3/8" diameter		62	.129		2.77	4.45		7.22	9.75
5063	1/2" diameter		61	.131		3.45	4.52		7.97	10.60
5064	3/4" diameter		57	.140		3.99	4.84		8.83	11.70
5065	1" diameter		51	.157		4.30	5.40		9.70	12.90
5066	1-1/4" diameter	Q-1	85	.188		5.75	5.85		11.60	15.10
5067	1-1/2" diameter		77	.208		6.55	6.45		13	16.90
5071	A-53 2" diameter		61	.262		3.69	8.15		11.84	16.35
5072	2-1/2" diameter		48	.333		5.85	10.35		16.20	22
5073	3" diameter		41	.390		7.25	12.10		19.35	26
5074	4" diameter		35	.457		10.50	14.15		24.65	33
5430	To delete coupling & hanger, subtract									
5440	1/4" diam. to 1/2" diam.					31%	54%			
5450	3/4" diam. to 1-1/2" diam.					28%	49%			
5460	2" diam. to 4" diam.					21%	40%			
5470	5" diam. to 12" diam.					17%	43%			
5510	Galvanized, 1/4" diameter	1 Plum	54	.148	L.F.	1.72	5.10		6.82	9.60
5520	3/8" diameter		53	.151		1.76	5.20		6.96	9.80
5530	1/2" diameter		52	.154		1.89	5.30		7.19	10.10
5540	3/4" diameter		50	.160		2.15	5.50		7.65	10.65
5550	1" diameter		45	.178		2.49	6.10		8.59	12
5560	1-1/4" diameter	Q-1	75	.213		3.27	6.60		9.87	13.60
5570	1-1/2" diameter		69	.232		3.61	7.20		10.81	14.80
5580	2" diameter		56	.286		4.84	8.85		13.69	18.65
5590	2-1/2" diameter		44	.364		7.05	11.25		18.30	25
5600	3" diameter		38	.421		8.75	13.05		21.80	29.50
5610	3-1/2" diameter		35	.457		10.85	14.15		25	33.50
5620	4" diameter		32	.500		12.75	15.50		28.25	37.50
5630	5" diameter		23	.696		24.50	21.50		46	59
5640	6" diameter	Q-2	28	.857		91	27.50		118.50	142
5650	8" diameter		23	1.043		51.50	33.50		85	107
5660	10" diameter		20	1.200		79.50	38.50		118	146
5670	12" diameter		15	1.600		91.50	51.50		143	178
5930	To delete coupling & hanger, subtract									
5940	1/4" diam. to 1/2" diam.					31%	54%			
5950	3/4" diam. to 1-1/2" diam.					28%	49%			
5960	2" diam. to 4" diam.					21%	40%			
5970	5" diam. to 12" diam.					17%	43%			
6000	Welded, on yoke & roller hangers									
6010	sized for covering, 10' O.C.									
6040	Black, 1" diameter	Q-15	85	.188	L.F.	2.82	5.85	1.18	9.85	13.20
6050	1-1/4" diameter		79	.203		3.58	6.30	1.27	11.15	14.85
6060	1-1/2" diameter		72	.222		3.86	6.90	1.40	12.16	16.20
6070	2" diameter		57	.281		4.52	8.70	1.77	14.99	20
6080	2-1/2" diameter		44	.364		6.05	11.25	2.29	19.59	26
6090	3" diameter		40	.400		7.15	12.40	2.52	22.07	29.50
6100	3-1/2" diameter		34	.471		8.65	14.60	2.96	26.21	35
6110	4" diameter		33	.485		9.80	15.05	3.05	27.90	36.50
6120	5" diameter		26	.615		17.85	19.10	3.87	40.82	53
6130	6" diameter	Q-16	30	.800		29.50	25.50	3.35	58.35	75
6140	8" diameter		25	.960		50.50	31	4.02	85.52	106
6150	10" diameter		20	1.200		83	38.50	5.05	126.55	155
6160	12" diameter		15	1.600		94.50	51.50	6.70	152.70	189
6540	To delete hanger, subtract									
6550	1" diam. to 1-1/2" diam.					30%	14%			
6560	2" diam. to 3" diam.					23%	9%			

15100 | Building Services Piping

15107 | Metal Pipe & Fittings

			CREW	DAILY OUTPUT	LABOR-HOURS	UNIT	MAT.	LABOR	EQUIP.	TOTAL	TOTAL INCL O&P	
620	6570	3-1/2" diam. to 5" diam.	R15100-050				12%	6%				620
	6580	6" diam. to 12" diam.					10%	4%				
	6610	Galvanized, 1" diameter	Q-15	85	.188	L.F.	3.01	5.85	1.18	10.04	13.40	
	6650	1-1/4" diameter		79	.203		3.85	6.30	1.27	11.42	15.15	
	6660	1-1/2" diameter		72	.222		4.19	6.90	1.40	12.49	16.55	
	6670	2" diameter		57	.281		4.98	8.70	1.77	15.45	20.50	
	6680	2-1/2" diameter		44	.364		6.70	11.25	2.29	20.24	27	
	6690	3" diameter		40	.400		8.05	12.40	2.52	22.97	30.50	
	6700	3-1/2" diameter		34	.471		9.75	14.60	2.96	27.31	36	
	6710	4" diameter		33	.485		11.15	15.05	3.05	29.25	38	
	6720	5" diameter		26	.615		22	19.10	3.87	44.97	57.50	
	6730	6" diameter	Q-16	30	.800		30	25.50	3.35	58.85	75.50	
	6740	8" diameter		25	.960		49.50	31	4.02	84.52	105	
	6750	10" diameter		20	1.200		81.50	38.50	5.05	125.05	154	
	6760	12" diameter		15	1.600		93	51.50	6.70	151.20	187	
	7150	To delete hanger, subtract										
	7160	1" diam. to 1-1/2" diam.					26%	14%				
	7170	2" diam. to 3" diam.					16%	9%				
	7180	3-1/2" diam. to 5" diam.					10%	6%				
	7190	6" diam. to 12" diam.					7%	4%				
	7250	Flanged, 300 lb. weld neck, on yoke & roll hangers										
	7260	sized for covering, 10' O.C.										
	7290	Black, 1" diameter	Q-15	66	.242	L.F.	7.15	7.50	1.52	16.17	21	
	7300	1-1/4" diameter		61	.262		7.90	8.15	1.65	17.70	23	
	7310	1-1/2" diameter		54	.296		8.20	9.20	1.86	19.26	25	
	7320	2" diameter		42	.381		9.60	11.80	2.40	23.80	31	
	7330	2-1/2" diameter		33	.485		11.70	15.05	3.05	29.80	38.50	
	7340	3" diameter		29	.552		12.85	17.10	3.47	33.42	44	
	7350	3-1/2" diameter		24	.667		18.15	20.50	4.19	42.84	55.50	
	7360	4" diameter		23	.696		19.30	21.50	4.38	45.18	58.50	
	7370	5" diameter		19	.842		32.50	26	5.30	63.80	81	
	7380	6" diameter	Q-16	23	1.043		42	33.50	4.37	79.87	101	
	7390	8" diameter		17	1.412		72	45.50	5.90	123.40	154	
	7400	10" diameter		14	1.714		128	55	7.20	190.20	232	
	7410	12" diameter		12	2		151	64.50	8.40	223.90	272	
	7470	For 600 lb. flanges, add					100%					
	7940	To delete flanges & hanger, subtract										
	7950	1" diam. to 1-1/2" diam.					75%	66%				
	7960	2" diam. to 3" diam.					62%	60%				
	7970	3-1/2" diam. to 5" diam.					54%	66%				
	7980	6" diam. to 12" diam.					55%	62%				
	8040	Galvanized, 1" diameter	Q-15	66	.242	L.F.	7.35	7.50	1.52	16.37	21	
	8050	1-1/4" diameter		61	.262		8.20	8.15	1.65	18	23	
	8060	1-1/2" diameter		54	.296		8.55	9.20	1.86	19.61	25.50	
	8070	2" diameter		42	.381		10.05	11.80	2.40	24.25	31.50	
	8080	2-1/2" diameter		33	.485		12.35	15.05	3.05	30.45	39.50	
	8090	3" diameter		29	.552		13.80	17.10	3.47	34.37	45	
	8100	3-1/2" diameter		24	.667		19.25	20.50	4.19	43.94	56.50	
	8110	4" diameter		23	.696		20.50	21.50	4.38	46.38	60	
	8120	5" diameter		19	.842		36.50	26	5.30	67.80	86	
	8130	6" diameter	Q-16	23	1.043		42	33.50	4.37	79.87	102	
	8140	8" diameter		17	1.412		71	45.50	5.90	122.40	153	
	8150	10" diameter		14	1.714		126	55	7.20	188.20	230	
	8160	12" diameter		12	2		150	64.50	8.40	222.90	271	
	8240	For 600 lb. flanges, add					100%					
	8900	To delete flanges & hangers, subtract										

15 MECHANICAL

15100 | Building Services Piping

15107 | Metal Pipe & Fittings

		Crew	Daily Output	Labor-Hours	Unit	2001 Bare Costs Mat.	Labor	Equip.	Total	Total Incl O&P
8910	1" diam. to 1-1/2" diam.	R15100 -050				72%	66%			
8920	2" diam. to 3" diam.					59%	60%			
8930	3-1/2" diam. to 5" diam.					51%	66%			
8940	6" diam. to 12" diam.					49%	62%			
9000	Threading pipe labor, one end, all schedules through 80									
9010	1/4" through 3/4" pipe size	1 Plum	80	.100	Ea.		3.45		3.45	5.20
9020	1" through 2" pipe size		73	.110			3.78		3.78	5.70
9030	2-1/2" pipe size		53	.151			5.20		5.20	7.85
9040	3" pipe size		50	.160			5.50		5.50	8.30
9050	3-1/2" pipe size	Q-1	89	.180			5.55		5.55	8.40
9060	4" pipe size		73	.219			6.80		6.80	10.25
9070	5" pipe size		53	.302			9.35		9.35	14.15
9080	6" pipe size		46	.348			10.80		10.80	16.30
9090	8" pipe size		29	.552			17.10		17.10	26
9100	10" pipe size		21	.762			23.50		23.50	35.50
9110	12" pipe size		13	1.231			38		38	57.50
9200	Welding labor per joint									
9210	Schedule 40,									
9230	1/2" pipe size	Q-15	32	.500	Ea.		15.50	3.15	18.65	27
9240	3/4" pipe size		27	.593			18.35	3.73	22.08	31.50
9250	1" pipe size		23	.696			21.50	4.38	25.88	37.50
9260	1-1/4" pipe size		20	.800			25	5.05	30.05	43
9270	1-1/2" pipe size		19	.842			26	5.30	31.30	45.50
9280	2" pipe size		16	1			31	6.30	37.30	54
9290	2-1/2" pipe size		13	1.231			38	7.75	45.75	66
9300	3" pipe size		12	1.333			41.50	8.40	49.90	72
9310	4" pipe size		10	1.600			49.50	10.05	59.55	86
9320	5" pipe size		9	1.778			55	11.20	66.20	95.50
9330	6" pipe size		8	2			62	12.60	74.60	107
9340	8" pipe size		5	3.200			99	20	119	172
9350	10" pipe size		4	4			124	25	149	215
9360	12" pipe size		3	5.333			165	33.50	198.50	287
9370	14" pipe size		2.60	6.154			191	38.50	229.50	330
9380	16" pipe size		2.20	7.273			225	46	271	390
9390	18" pipe size		2	8			248	50.50	298.50	430
9400	20" pipe size		1.80	8.889			276	56	332	475
9410	22" pipe size		1.70	9.412			292	59	351	505
9420	24" pipe size		1.50	10.667			330	67	397	575
9450	Schedule 80,									
9460	1/2" pipe size	Q-15	27	.593	Ea.		18.35	3.73	22.08	31.50
9470	3/4" pipe size		23	.696			21.50	4.38	25.88	37.50
9480	1" pipe size		20	.800			25	5.05	30.05	43
9490	1-1/4" pipe size		19	.842			26	5.30	31.30	45.50
9500	1-1/2" pipe size		18	.889			27.50	5.60	33.10	47.50
9510	2" pipe size		15	1.067			33	6.70	39.70	57.50
9520	2-1/2" pipe size		12	1.333			41.50	8.40	49.90	72
9530	3" pipe size		11	1.455			45	9.15	54.15	78
9540	4" pipe size		8	2			62	12.60	74.60	107
9550	5" pipe size		6	2.667			82.50	16.75	99.25	143
9560	6" pipe size		5	3.200			99	20	119	172
9570	8" pipe size		4	4			124	25	149	215
9580	10" pipe size		3	5.333			165	33.50	198.50	287
9590	12" pipe size		2	8			248	50.50	298.50	430
9600	14" pipe size	Q-16	2.60	9.231			297	38.50	335.50	495
9610	16" pipe size		2.30	10.435			335	43.50	378.50	555
9620	18" pipe size		2	12			385	50.50	435.50	635

15100 | Building Services Piping

15107 | Metal Pipe & Fittings

			CREW	DAILY OUTPUT	LABOR-HOURS	UNIT	2001 BARE COSTS MAT.	LABOR	EQUIP.	TOTAL	TOTAL INCL O&P	
620	9630	20" pipe size	Q-16	1.80	13.333	Ea.	430		56	486	705	620
	9640	22" pipe size	R15100 -050	1.60	15		480		63	543	800	
	9650	24" pipe size		1.50	16		515		67	582	850	
640	0010	**PIPE, STEEL, FITTINGS** Threaded										640
	0020	Cast Iron										
	0040	Standard weight, black										
	0060	90° Elbow, straight										
	0070	1/4"	1 Plum	16	.500	Ea.	3.34	17.25		20.59	29.50	
	0080	3/8"		16	.500		4.71	17.25		21.96	31	
	0090	1/2"		15	.533		2.12	18.35		20.47	30	
	0100	3/4"		14	.571		2.21	19.70		21.91	32	
	0110	1"		13	.615		2.61	21		23.61	35	
	0120	1-1/4"	Q-1	22	.727		3.71	22.50		26.21	38	
	0130	1-1/2"		20	.800		5.15	25		30.15	43	
	0140	2"		18	.889		8	27.50		35.50	50.50	
	0150	2-1/2"		14	1.143		19.20	35.50		54.70	74.50	
	0160	3"		10	1.600		31.50	49.50		81	110	
	0170	3-1/2"		8	2		65	62		127	165	
	0180	4"		6	2.667		58.50	82.50		141	189	
	0190	5"		5	3.200		113	99		212	275	
	0200	6"	Q-2	7	3.429		144	110		254	325	
	0210	8"	"	6	4		299	129		428	525	
	0250	45° Elbow, straight										
	0260	1/4"	1 Plum	16	.500	Ea.	3.57	17.25		20.82	30	
	0270	3/8"		16	.500		3.83	17.25		21.08	30	
	0280	1/2"		15	.533		3.24	18.35		21.59	31	
	0300	3/4"		14	.571		3.26	19.70		22.96	33	
	0320	1"		13	.615		3.82	21		24.82	36	
	0330	1-1/4"	Q-1	22	.727		5.15	22.50		27.65	39.50	
	0340	1-1/2"		20	.800		8.50	25		33.50	47	
	0350	2"		18	.889		9.85	27.50		37.35	52.50	
	0360	2-1/2"		14	1.143		25	35.50		60.50	81	
	0370	3"		10	1.600		40.50	49.50		90	120	
	0380	3-1/2"		8	2		82	62		144	184	
	0400	4"		6	2.667		82	82.50		164.50	216	
	0420	5"		5	3.200		140	99		239	305	
	0440	6"	Q-2	7	3.429		180	110		290	365	
	0460	8"	"	6	4		360	129		489	590	
	0500	Tee, straight										
	0510	1/4"	1 Plum	10	.800	Ea.	4.66	27.50		32.16	46.50	
	0520	3/8"		10	.800		4.96	27.50		32.46	47	
	0530	1/2"		9	.889		3.29	30.50		33.79	49.50	
	0540	3/4"		9	.889		3.85	30.50		34.35	50	
	0550	1"		8	1		3.42	34.50		37.92	56	
	0560	1-1/4"	Q-1	14	1.143		6.25	35.50		41.75	60.50	
	0570	1-1/2"		13	1.231		8.15	38		46.15	66.50	
	0580	2"		11	1.455		11.30	45		56.30	80.50	
	0590	2-1/2"		9	1.778		29.50	55		84.50	116	
	0600	3"		6	2.667		45	82.50		127.50	175	
	0610	3-1/2"		5	3.200		88	99		187	247	
	0620	4"		4	4		88	124		212	284	
	0630	5"		3	5.333		178	165		343	445	
	0640	6"	Q-2	4	6		202	193		395	515	
	0650	8"	"	3	8		455	257		712	890	
	0660	Tee, reducing run and outlet										

15100 | Building Services Piping

15107 | Metal Pipe & Fittings

		CREW	DAILY OUTPUT	LABOR-HOURS	UNIT	2001 BARE COSTS MAT.	LABOR	EQUIP.	TOTAL	TOTAL INCL O&P	
640	0661	1/2"	1 Plum	9	.889	Ea.	7.05	30.50		37.55	54	640
	0662	3/4"		9	.889		6.10	30.50		36.60	52.50	
	0663	1"		8	1		6.30	34.50		40.80	59	
	0664	1-1/4"	Q-1	14	1.143		6.20	35.50		41.70	60.50	
	0665	1-1/2"		13	1.231		13.25	38		51.25	72	
	0666	2"		11	1.455		17.55	45		62.55	87.50	
	0667	2-1/2"		9	1.778		32	55		87	118	
	0668	3"		6	2.667		56.50	82.50		139	188	
	0669	3-1/2"		5	3.200		100	99		199	260	
	0670	4"		4	4		100	124		224	297	
	0671	5"		3	5.333		207	165		372	480	
	0672	6"	Q-2	4	6		255	193		448	570	
	0673	8"	"	3	8		289	257		546	705	
	0674	Reducer, concentric										
	0675	3/4"	1 Plum	18	.444	Ea.	5.40	15.30		20.70	29	
	0676	1"	"	15	.533		4	18.35		22.35	32	
	0677	1-1/4"	Q-1	26	.615		10.40	19.10		29.50	40.50	
	0678	1-1/2"		24	.667		14.70	20.50		35.20	47	
	0679	2"		21	.762		19.25	23.50		42.75	56.50	
	0680	2-1/2"		18	.889		27.50	27.50		55	71.50	
	0681	3"		14	1.143		37.50	35.50		73	95	
	0682	3-1/2"		12	1.333		65.50	41.50		107	135	
	0683	4"		10	1.600		65.50	49.50		115	148	
	0684	5"		6	2.667		104	82.50		186.50	240	
	0685	6"	Q-2	8	3		119	96.50		215.50	277	
	0686	8"	"	7	3.429		148	110		258	330	
	0687	Reducer, eccentric										
	0688	3/4"	1 Plum	16	.500	Ea.	12.30	17.25		29.55	39.50	
	0689	1"	"	14	.571		12.30	19.70		32	43	
	0690	1-1/4"	Q-1	25	.640		18.30	19.85		38.15	50	
	0691	1-1/2"		22	.727		26	22.50		48.50	62.50	
	0692	2"		20	.800		33	25		58	73.50	
	0693	2-1/2"		16	1		49	31		80	101	
	0694	3"		12	1.333		60.50	41.50		102	129	
	0695	3-1/2"		10	1.600		119	49.50		168.50	205	
	0696	4"		9	1.778		119	55		174	213	
	0697	5"		5	3.200		145	99		244	310	
	0698	6"	Q-2	8	3		198	96.50		294.50	365	
	0699	8"	"	7	3.429		740	110		850	980	
	0700	Standard weight, galvanized cast iron										
	0720	90° Elbow, straight										
	0730	1/4"	1 Plum	16	.500	Ea.	5.55	17.25		22.80	32	
	0740	3/8"		16	.500		5.55	17.25		22.80	32	
	0750	1/2"		15	.533		5.55	18.35		23.90	33.50	
	0760	3/4"		14	.571		6.20	19.70		25.90	36.50	
	0770	1"		13	.615		7.15	21		28.15	40	
	0780	1-1/4"	Q-1	22	.727		11.10	22.50		33.60	46	
	0790	1-1/2"		20	.800		15.15	25		40.15	54	
	0800	2"		18	.889		23	27.50		50.50	66.50	
	0810	2-1/2"		14	1.143		46.50	35.50		82	105	
	0820	3"		10	1.600		70.50	49.50		120	153	
	0830	3-1/2"		8	2		111	62		173	216	
	0840	4"		6	2.667		130	82.50		212.50	267	
	0850	5"		5	3.200		130	99		229	292	
	0860	6"	Q-2	7	3.429		289	110		399	480	
	0870	8"	"	6	4		590	129		719	845	

15100 | Building Services Piping

15107 | Metal Pipe & Fittings

		CREW	DAILY OUTPUT	LABOR-HOURS	UNIT	2001 BARE COSTS MAT.	LABOR	EQUIP.	TOTAL	TOTAL INCL O&P
0900	45° Elbow, straight									
0910	1/4"	1 Plum	16	.500	Ea.	6.30	17.25		23.55	33
0920	3/8"		16	.500		6.30	17.25		23.55	33
0930	1/2"		15	.533		6.30	18.35		24.65	34.50
0940	3/4"		14	.571		7.20	19.70		26.90	37.50
0950	1"		13	.615		8.45	21		29.45	41.50
0960	1-1/4"	Q-1	22	.727		12.60	22.50		35.10	48
0970	1-1/2"		20	.800		18.85	25		43.85	58.50
0980	2"		18	.889		26.50	27.50		54	70.50
0990	2-1/2"		14	1.143		52	35.50		87.50	111
1000	3"		10	1.600		83	49.50		132.50	166
1010	3-1/2"		8	2		152	62		214	261
1020	4"		6	2.667		152	82.50		234.50	292
1030	5"		5	3.200		189	99		288	360
1040	6"	Q-2	7	3.429		244	110		354	435
1050	8"	"	6	4		485	129		614	730
1100	Tee, straight									
1110	1/4"	1 Plum	10	.800	Ea.	7.05	27.50		34.55	49.50
1120	3/8"		10	.800		7.05	27.50		34.55	49.50
1130	1/2"		9	.889		6.15	30.50		36.65	53
1140	3/4"		9	.889		9.10	30.50		39.60	56
1150	1"		8	1		9.65	34.50		44.15	62.50
1160	1-1/4"	Q-1	14	1.143		16.85	35.50		52.35	72
1170	1-1/2"		13	1.231		22	38		60	82
1180	2"		11	1.455		28	45		73	98.50
1190	2-1/2"		9	1.778		60	55		115	149
1200	3"		6	2.667		155	82.50		237.50	296
1210	3-1/2"		5	3.200		159	99		258	325
1220	4"		4	4		178	124		302	385
1230	5"		3	5.333		241	165		406	515
1240	6"	Q-2	4	6		272	193		465	590
1250	8"	"	3	8		620	257		877	1,075
1300	Extra heavy weight, black									
1310	Couplings, steel straight									
1320	1/4"	1 Plum	19	.421	Ea.	2.72	14.50		17.22	25
1330	3/8"		19	.421		2.97	14.50		17.47	25.50
1340	1/2"		19	.421		4.02	14.50		18.52	26.50
1350	3/4"		18	.444		4.28	15.30		19.58	27.50
1360	1"		15	.533		5.45	18.35		23.80	33.50
1370	1-1/4"	Q-1	26	.615		8.75	19.10		27.85	38.50
1380	1-1/2"		24	.667		8.75	20.50		29.25	40.50
1390	2"		21	.762		13.35	23.50		36.85	50
1400	2-1/2"		18	.889		19.80	27.50		47.30	63.50
1410	3"		14	1.143		23.50	35.50		59	79.50
1420	3-1/2"		12	1.333		31.50	41.50		73	97.50
1430	4"		10	1.600		37	49.50		86.50	116
1440	5"		6	2.667		53	82.50		135.50	184
1450	6"	Q-2	8	3		65	96.50		161.50	218
1460	8"		7	3.429		74	110		184	248
1470	10"		6	4		116	129		245	320
1480	12"		4	6		155	193		348	460
1510	90° Elbow, straight									
1520	1/2"	1 Plum	15	.533	Ea.	9.60	18.35		27.95	38
1530	3/4"		14	.571		10.15	19.70		29.85	40.50
1540	1"		13	.615		12.25	21		33.25	45.50
1550	1-1/4"	Q-1	22	.727		18.15	22.50		40.65	54

Important: See the Reference Section for critical supporting data - Reference Nos., Crews, & City Cost Indexes

15100 | Building Services Piping

15107 | Metal Pipe & Fittings

		CREW	DAILY OUTPUT	LABOR-HOURS	UNIT	2001 BARE COSTS MAT.	LABOR	EQUIP.	TOTAL	TOTAL INCL O&P
1560	1-1/2"	Q-1	20	.800	Ea.	22.50	25		47.50	62.50
1580	2"		18	.889		28	27.50		55.50	72.50
1590	2-1/2"		14	1.143		54.50	35.50		90	114
1600	3"		10	1.600		100	49.50		149.50	185
1610	4"		6	2.667		167	82.50		249.50	310
1620	6"	Q-2	7	3.429		435	110		545	640
1650	45° Elbow, straight									
1660	1/2"	1 Plum	15	.533	Ea.	11.35	18.35		29.70	40
1670	3/4"		14	.571		12.90	19.70		32.60	43.50
1680	1"		13	.615		15.55	21		36.55	49
1690	1-1/4"	Q-1	22	.727		22	22.50		44.50	58
1700	1-1/2"		20	.800		28	25		53	68.50
1710	2"		18	.889		40	27.50		67.50	85.50
1720	2-1/2"		14	1.143		60	35.50		95.50	120
1800	Tee, straight									
1810	1/2"	1 Plum	9	.889	Ea.	14.80	30.50		45.30	62.50
1820	3/4"		9	.889		15.10	30.50		45.60	62.50
1830	1"		8	1		18.50	34.50		53	72.50
1840	1-1/4"	Q-1	14	1.143		27.50	35.50		63	83.50
1850	1-1/2"		13	1.231		35	38		73	96
1860	2"		11	1.455		43.50	45		88.50	116
1870	2-1/2"		9	1.778		77	55		132	168
1880	3"		6	2.667		133	82.50		215.50	272
1890	4"		4	4		211	124		335	420
1900	6"	Q-2	4	6		435	193		628	770
4000	Standard weight, black									
4010	Couplings, steel straight, merchants									
4030	1/4"	1 Plum	19	.421	Ea.	.73	14.50		15.23	23
4040	3/8"		19	.421		.88	14.50		15.38	23
4050	1/2"		19	.421		.93	14.50		15.43	23
4060	3/4"		18	.444		1.19	15.30		16.49	24.50
4070	1"		15	.533		1.67	18.35		20.02	29.50
4080	1-1/4"	Q-1	26	.615		2.13	19.10		21.23	31.50
4090	1-1/2"		24	.667		2.70	20.50		23.20	34
4100	2"		21	.762		3.86	23.50		27.36	40
4110	2-1/2"		18	.889		11	27.50		38.50	53.50
4120	3"		14	1.143		15.45	35.50		50.95	70.50
4130	3-1/2"		12	1.333		27.50	41.50		69	92.50
4140	4"		10	1.600		27.50	49.50		77	105
4150	5"		6	2.667		50	82.50		132.50	180
4160	6"	Q-2	8	3		60	96.50		156.50	212
4166	Plug, 1/4"	1 Plum	38	.211		1.06	7.25		8.31	12.10
4167	3/8"		38	.211		1.06	7.25		8.31	12.10
4168	1/2"		38	.211		.84	7.25		8.09	11.85
4169	3/4"		32	.250		1.06	8.60		9.66	14.15
4170	1"		30	.267		1.06	9.20		10.26	15
4171	1-1/4"	Q-1	52	.308		1.70	9.55		11.25	16.25
4172	1-1/2"		48	.333		2.40	10.35		12.75	18.25
4173	2"		42	.381		2.80	11.80		14.60	21
4176	2-1/2"		36	.444		6.80	13.80		20.60	28.50
4200	Standard weight, galvanized									
4210	Couplings, steel straight, merchants									
4230	1/4"	1 Plum	19	.421	Ea.	.57	14.50		15.07	22.50
4240	3/8"		19	.421		.73	14.50		15.23	23
4250	1/2"		19	.421		.78	14.50		15.28	23
4260	3/4"		18	.444		.97	15.30		16.27	24

15100 | Building Services Piping

15107 | Metal Pipe & Fittings

		CREW	DAILY OUTPUT	LABOR-HOURS	UNIT	MAT.	LABOR	EQUIP.	TOTAL	TOTAL INCL O&P
4270	1"	1 Plum	15	.533	Ea.	1.35	18.35		19.70	29
4280	1-1/4"	Q-1	26	.615		1.73	19.10		20.83	31
4290	1-1/2"		24	.667		2.15	20.50		22.65	33.50
4300	2"		21	.762		3.22	23.50		26.72	39
4310	2-1/2"		18	.889		6.85	27.50		34.35	49
4320	3"		14	1.143		9.10	35.50		44.60	63.50
4330	3-1/2"		12	1.333		16.05	41.50		57.55	80
4340	4"		10	1.600		16.05	49.50		65.55	92.50
4350	5"		6	2.667		30	82.50		112.50	158
4360	6"	Q-2	8	3		36.50	96.50		133	187
4370	Plug, galvanized, square head									
4374	1/2"	1 Plum	38	.211	Ea.	2.15	7.25		9.40	13.30
4375	3/4"		32	.250		2.15	8.60		10.75	15.35
4376	1"		30	.267		2.15	9.20		11.35	16.20
4377	1-1/4"	Q-1	52	.308		3.59	9.55		13.14	18.35
4378	1-1/2"		48	.333		4.85	10.35		15.20	21
4379	2"		42	.381		6.05	11.80		17.85	24.50
4380	2-1/2"		36	.444		12.70	13.80		26.50	35
4381	3"		28	.571		16.30	17.70		34	44.50
4382	4"		20	.800		39.50	25		64.50	81
4384	6"		8	2		69.50	62		131.50	170
5000	Malleable iron, 150 lb.									
5020	Black									
5040	90° elbow, straight									
5060	1/4"	1 Plum	16	.500	Ea.	1.75	17.25		19	28
5070	3/8"		16	.500		1.75	17.25		19	28
5080	1/2"		15	.533		1.21	18.35		19.56	29
5090	3/4"		14	.571		1.47	19.70		21.17	31
5100	1"		13	.615		2.56	21		23.56	35
5110	1-1/4"	Q-1	22	.727		4.20	22.50		26.70	38.50
5120	1-1/2"		20	.800		5.55	25		30.55	43.50
5130	2"		18	.889		9.55	27.50		37.05	52
5140	2-1/2"		14	1.143		21	35.50		56.50	77
5150	3"		10	1.600		31	49.50		80.50	109
5160	3-1/2"		8	2		67	62		129	167
5170	4"		6	2.667		67	82.50		149.50	199
5180	5"		5	3.200		206	99		305	375
5190	6"	Q-2	7	3.429		206	110		316	390
5250	45° elbow, straight									
5270	1/4"	1 Plum	16	.500	Ea.	2.64	17.25		19.89	29
5280	3/8"		16	.500		2.64	17.25		19.89	29
5290	1/2"		15	.533		2.01	18.35		20.36	29.50
5300	3/4"		14	.571		2.47	19.70		22.17	32
5310	1"		13	.615		3.13	21		24.13	35.50
5320	1-1/4"	Q-1	22	.727		5.55	22.50		28.05	40
5330	1-1/2"		20	.800		6.85	25		31.85	45
5340	2"		18	.889		10.35	27.50		37.85	53
5350	2-1/2"		14	1.143		30	35.50		65.50	86.50
5360	3"		10	1.600		39	49.50		88.50	118
5370	3-1/2"		8	2		76.50	62		138.50	178
5380	4"		6	2.667		76.50	82.50		159	209
5390	5"		5	3.200		274	99		373	450
5400	6"	Q-2	7	3.429		274	110		384	465
5450	Tee, straight									
5470	1/4"	1 Plum	10	.800	Ea.	2.56	27.50		30.06	44.50
5480	3/8"		10	.800		2.56	27.50		30.06	44.50

15100 | Building Services Piping

15107 | Metal Pipe & Fittings

		Crew	Daily Output	Labor-Hours	Unit	Mat.	Labor	Equip.	Total	Total Incl O&P
5490	1/2"	1 Plum	9	.889	Ea.	1.64	30.50		32.14	48
5500	3/4"		9	.889		2.34	30.50		32.84	48.50
5510	1"		8	1		3.99	34.50		38.49	56.50
5520	1-1/4"	Q-1	14	1.143		6.45	35.50		41.95	60.50
5530	1-1/2"		13	1.231		8.05	38		46.05	66.50
5540	2"		11	1.455		13.70	45		58.70	83
5550	2-1/2"		9	1.778		29.50	55		84.50	116
5560	3"		6	2.667		43.50	82.50		126	173
5570	3-1/2"		5	3.200		105	99		204	266
5580	4"		4	4		105	124		229	305
5590	5"		3	5.333		235	165		400	510
5600	6"	Q-2	4	6		305	193		498	625
5601	Tee, reducing, on outlet									
5602	1/2"	1 Plum	9	.889	Ea.	4.33	30.50		34.83	51
5603	3/4"		9	.889		3.61	30.50		34.11	50
5604	1"		8	1		4.55	34.50		39.05	57
5605	1-1/4"	Q-1	14	1.143		7.90	35.50		43.40	62
5606	1-1/2"		13	1.231		11.35	38		49.35	70
5607	2"		11	1.455		15.50	45		60.50	85
5608	2-1/2"		9	1.778		41	55		96	128
5609	3"		6	2.667		51.50	82.50		134	182
5610	3-1/2"		5	3.200		124	99		223	286
5611	4"		4	4		124	124		248	325
5612	5"		3	5.333		137	165		302	400
5613	6"	Q-2	5	4.800		144	154		298	390
5650	Coupling									
5670	1/4"	1 Plum	19	.421	Ea.	2.17	14.50		16.67	24.50
5680	3/8"		19	.421		2.17	14.50		16.67	24.50
5690	1/2"		19	.421		1.68	14.50		16.18	24
5700	3/4"		18	.444		1.97	15.30		17.27	25
5710	1"		15	.533		2.95	18.35		21.30	31
5720	1-1/4"	Q-1	26	.615		3.82	19.10		22.92	33
5730	1-1/2"		24	.667		5.15	20.50		25.65	36.50
5740	2"		21	.762		7.65	23.50		31.15	44
5750	2-1/2"		18	.889		21	27.50		48.50	64.50
5760	3"		14	1.143		28.50	35.50		64	85
5770	3-1/2"		12	1.333		57.50	41.50		99	126
5780	4"		10	1.600		57.50	49.50		107	138
5790	5"		6	2.667		125	82.50		207.50	262
5800	6"	Q-2	8	3		125	96.50		221.50	283
5840	Reducer, concentric, 1/4"	1 Plum	19	.421		2.29	14.50		16.79	24.50
5850	3/8"		19	.421		2.29	14.50		16.79	24.50
5860	1/2"		19	.421		2.32	14.50		16.82	24.50
5870	3/4"		16	.500		2.75	17.25		20	29
5880	1"		15	.533		4.34	18.35		22.69	32.50
5890	1-1/4"	Q-1	26	.615		5.20	19.10		24.30	34.50
5900	1-1/2"		24	.667		7.45	20.50		27.95	39
5910	2"		21	.762		10.75	23.50		34.25	47.50
5911	2-1/2"		18	.889		22	27.50		49.50	66
5912	3"		14	1.143		30	35.50		65.50	86.50
5913	3-1/2"		12	1.333		62	41.50		103.50	131
5914	4"		10	1.600		62	49.50		111.50	144
5915	5"		6	2.667		116	82.50		198.50	253
5916	6"	Q-2	8	3		150	96.50		246.50	310
5981	Bushing, 1/4"	1 Plum	19	.421		2.18	14.50		16.68	24.50
5982	3/8"		19	.421		2.18	14.50		16.68	24.50

15100 | Building Services Piping

15107 | Metal Pipe & Fittings

		CREW	DAILY OUTPUT	LABOR-HOURS	UNIT	2001 BARE COSTS MAT.	LABOR	EQUIP.	TOTAL	TOTAL INCL O&P	
640	5983	1/2"	1 Plum	19	.421	Ea.	2.18	14.50		16.68	24.50	640
	5984	3/4"		16	.500		2.18	17.25		19.43	28.50	
	5985	1"	↓	15	.533		3.20	18.35		21.55	31	
	5986	1-1/4"	Q-1	26	.615		3.97	19.10		23.07	33.50	
	5987	1-1/2"		24	.667		4.81	20.50		25.31	36.50	
	5988	2"		21	.762		5.25	23.50		28.75	41.50	
	5989	Cap, 1/4"	1 Plum	38	.211		1.61	7.25		8.86	12.70	
	5991	3/8"		38	.211		1.36	7.25		8.61	12.45	
	5992	1/2"		38	.211		1.23	7.25		8.48	12.30	
	5993	3/4"		32	.250		1.67	8.60		10.27	14.85	
	5994	1"	↓	30	.267		2.01	9.20		11.21	16.05	
	5995	1-1/4"	Q-1	52	.308		2.66	9.55		12.21	17.35	
	5996	1-1/2"		48	.333		3.64	10.35		13.99	19.60	
	5997	2"		42	.381		5.30	11.80		17.10	23.50	
	6000	For galvanized elbows, tees, and couplings add				↓	20%					
	6058	For galvanized reducers, caps and plugs add					20%					
	6100	90° elbow, galvanized, 150 lb., reducing										
	6110	3/4" x 1/2"	1 Plum	15.40	.519	Ea.	3.24	17.90		21.14	30.50	
	6112	1" x 3/4"		14	.571		4.24	19.70		23.94	34	
	6114	1" x 1/2"	↓	14.50	.552		4.50	19		23.50	33.50	
	6116	1-1/4" x 1"	Q-1	24.20	.661		7.10	20.50		27.60	39	
	6118	1-1/4" x 3/4"		25.40	.630		8.55	19.55		28.10	39	
	6120	1-1/4" x 1/2"		26.20	.611		9.05	18.95		28	38.50	
	6122	1-1/2" x 1-1/4"		21.60	.741		11.35	23		34.35	47	
	6124	1-1/2" x 1"		23.50	.681		11.35	21		32.35	44.50	
	6126	1-1/2" x 3/4"		24.60	.650		11.35	20		31.35	43	
	6128	2" x 1-1/2"		20.50	.780		13.30	24		37.30	51	
	6130	2" x 1-1/4"		21	.762		15.20	23.50		38.70	52.50	
	6132	2" x 1"		22.80	.702		15.80	22		37.80	50.50	
	6134	2" x 3/4"		23.90	.669		16.10	21		37.10	49	
	6136	2-1/2" x 2"		12.30	1.301		45	40.50		85.50	111	
	6138	2-1/2" x 1-1/2"		12.50	1.280		50	39.50		89.50	115	
	6140	3" x 2-1/2"		8.60	1.860		72.50	57.50		130	167	
	6142	3" x 2"		11.80	1.356		71	42		113	142	
	6144	4" x 3"	↓	8.20	1.951	↓	140	60.50		200.50	246	
	6160	90° elbow, black, 150 lb., reducing										
	6170	1" x 3/4"	1 Plum	14	.571	Ea.	3.06	19.70		22.76	33	
	6174	1-1/2" x 1"	Q-1	23.50	.681		7.25	21		28.25	40	
	6178	1-1/2" x 3/4"		24.60	.650		8.30	20		28.30	39.50	
	6182	2" x 1-1/2"		20.50	.780		11.95	24		35.95	49.50	
	6186	2" x 1"		22.80	.702		11.45	22		33.45	45.50	
	6190	2" x 3/4"		23.90	.669		12.70	21		33.70	45.50	
	6194	2-1/2" x 2"	↓	12.30	1.301		29	40.50		69.50	93	
	7000	Union, with brass seat										
	7010	1/4"	1 Plum	15	.533	Ea.	8.60	18.35		26.95	37	
	7020	3/8"		15	.533		5.95	18.35		24.30	34	
	7030	1/2"		14	.571		5.40	19.70		25.10	35.50	
	7040	3/4"		13	.615		6.15	21		27.15	39	
	7050	1"		12	.667		8.05	23		31.05	43.50	
	7060	1-1/4"	Q-1	21	.762		11.55	23.50		35.05	48	
	7070	1-1/2"		19	.842		14.40	26		40.40	55.50	
	7080	2"		17	.941		16.75	29		45.75	62.50	
	7090	2-1/2"		13	1.231		50	38		88	113	
	7100	3"	↓	9	1.778	↓	60	55		115	149	
	7120	Union, galvanized										
	7124	1/2"	1 Plum	14	.571	Ea.	6.75	19.70		26.45	37	

15100 | Building Services Piping

15107 | Metal Pipe & Fittings

		CREW	DAILY OUTPUT	LABOR-HOURS	UNIT	MAT.	LABOR	EQUIP.	TOTAL	TOTAL INCL O&P
7125	3/4"	1 Plum	13	.615	Ea.	7.75	21		28.75	40.50
7126	1"	↓	12	.667		10.15	23		33.15	45.50
7127	1-1/4"	Q-1	21	.762		14.65	23.50		38.15	51.50
7128	1-1/2"		19	.842		17.80	26		43.80	59
7129	2"		17	.941		20.50	29		49.50	66.50
7130	2-1/2"		13	1.231		71	38		109	136
7131	3"	↓	9	1.778	↓	99.50	55		154.50	192
7500	Malleable iron, 300 lb									
7520	Black									
7540	90° Elbow, straight, 1/4"	1 Plum	16	.500	Ea.	6.15	17.25		23.40	33
7560	3/8"		16	.500		5.45	17.25		22.70	32
7570	1/2"		15	.533		7.10	18.35		25.45	35.50
7580	3/4"		14	.571		8	19.70		27.70	38.50
7590	1"	↓	13	.615		10.25	21		31.25	43.50
7600	1-1/4"	Q-1	22	.727		14.95	22.50		37.45	50.50
7610	1-1/2"		20	.800		17.60	25		42.60	57
7620	2"		18	.889		25.50	27.50		53	69.50
7630	2-1/2"		14	1.143		60	35.50		95.50	120
7640	3"		10	1.600		75.50	49.50		125	158
7650	4"	↓	6	2.667		157	82.50		239.50	298
7700	45° Elbow, straight, 1/4"	1 Plum	16	.500		9.15	17.25		26.40	36
7720	3/8"		16	.500		9.15	17.25		26.40	36
7730	1/2"		15	.533		10.20	18.35		28.55	38.50
7740	3/4"		14	.571		11.35	19.70		31.05	42
7750	1"	↓	13	.615		12.55	21		33.55	46
7760	1-1/4"	Q-1	22	.727		20	22.50		42.50	56
7770	1-1/2"		20	.800		26	25		51	66.50
7780	2"		18	.889		39.50	27.50		67	85
7790	2-1/2"		14	1.143		78.50	35.50		114	140
7800	3"		10	1.600		103	49.50		152.50	189
7810	4"	↓	6	2.667		200	82.50		282.50	345
7850	Tee, straight, 1/4"	1 Plum	10	.800		7.75	27.50		35.25	50
7870	3/8"		10	.800		8.20	27.50		35.70	50.50
7880	1/2"		9	.889		10.45	30.50		40.95	57.50
7890	3/4"		9	.889		11.30	30.50		41.80	58.50
7900	1"	↓	8	1		13.60	34.50		48.10	67
7910	1-1/4"	Q-1	14	1.143		19.45	35.50		54.95	75
7920	1-1/2"		13	1.231		22.50	38		60.50	82.50
7930	2"		11	1.455		33.50	45		78.50	105
7940	2-1/2"		9	1.778		72.50	55		127.50	163
7950	3"		6	2.667		116	82.50		198.50	253
7960	4"	↓	4	4		250	124		374	460
7980	6"	Q-2	4	6		296	193		489	615
7990	8"	"	3	8		465	257		722	900
8050	Couplings, straight, 1/4"	1 Plum	19	.421		5.55	14.50		20.05	28
8070	3/8"		19	.421		5.55	14.50		20.05	28
8080	1/2"		19	.421		6.20	14.50		20.70	29
8090	3/4"		18	.444		7.15	15.30		22.45	31
8100	1"	↓	15	.533		8.15	18.35		26.50	36.50
8110	1-1/4"	Q-1	26	.615		9.70	19.10		28.80	39.50
8120	1-1/2"		24	.667		14.50	20.50		35	47
8130	2"		21	.762		21	23.50		44.50	58.50
8140	2-1/2"		18	.889		34	27.50		61.50	79
8150	3"		14	1.143		49.50	35.50		85	108
8160	4"	↓	10	1.600	↓	69	49.50		118.50	151
8200	Galvanized									

15100 | Building Services Piping

15107 | Metal Pipe & Fittings

			CREW	DAILY OUTPUT	LABOR-HOURS	UNIT	MAT.	LABOR	EQUIP.	TOTAL	TOTAL INCL O&P
640	8220	90° Elbow, straight, 1/4"	1 Plum	16	.500	Ea.	10.45	17.25		27.70	37.50
	8222	3/8"		16	.500		10.65	17.25		27.90	37.50
	8224	1/2"		15	.533		12.65	18.35		31	41.50
	8226	3/4"		14	.571		14.20	19.70		33.90	45
	8228	1"	↓	13	.615		18.25	21		39.25	52
	8230	1-1/4"	Q-1	22	.727		28.50	22.50		51	65.50
	8232	1-1/2"		20	.800		31	25		56	71.50
	8234	2"		18	.889		52	27.50		79.50	99
	8236	2-1/2"		14	1.143		102	35.50		137.50	166
	8238	3"		10	1.600		137	49.50		186.50	226
	8240	4"	↓	6	2.667	↓	310	82.50		392.50	465
	8280	45° Elbow, straight									
	8282	1/2"	1 Plum	15	.533	Ea.	18.85	18.35		37.20	48
	8284	3/4"		14	.571		21.50	19.70		41.20	53.50
	8286	1"	↓	13	.615		24	21		45	58
	8288	1-1/4"	Q-1	22	.727		37	22.50		59.50	74.50
	8290	1-1/2"		20	.800		47.50	25		72.50	90
	8292	2"	↓	18	.889		66	27.50		93.50	114
	8310	Tee, straight, 1/4"	1 Plum	10	.800		14.85	27.50		42.35	58
	8312	3/8"		10	.800		15.10	27.50		42.60	58
	8314	1/2"		9	.889		18.90	30.50		49.40	67
	8316	3/4"		9	.889		21	30.50		51.50	69
	8318	1"	↓	8	1		25.50	34.50		60	80.50
	8320	1-1/4"	Q-1	14	1.143		36.50	35.50		72	94
	8322	1-1/2"		13	1.231		38	38		76	99.50
	8324	2"		11	1.455		61	45		106	135
	8326	2-1/2"		9	1.778		146	55		201	243
	8328	3"		6	2.667		218	82.50		300.50	365
	8330	4"	↓	4	4		425	124		549	650
	8380	Couplings, straight, 1/4"	1 Plum	19	.421		9.85	14.50		24.35	33
	8382	3/8"		19	.421		9.85	14.50		24.35	33
	8384	1/2"		19	.421		10.10	14.50		24.60	33
	8386	3/4"		18	.444		11.40	15.30		26.70	35.50
	8388	1"	↓	15	.533		15.10	18.35		33.45	44
	8390	1-1/4"	Q-1	26	.615		19.65	19.10		38.75	50.50
	8392	1-1/2"		24	.667		28	20.50		48.50	61.50
	8394	2"		21	.762		33.50	23.50		57	72.50
	8396	2-1/2"		18	.889		67	27.50		94.50	115
	8398	3"		14	1.143		93.50	35.50		129	157
	8399	4"	↓	10	1.600	↓	100	49.50		149.50	185
	8529	Black									
	8530	Reducer, concentric, 1/4"	1 Plum	19	.421	Ea.	7.35	14.50		21.85	30
	8531	3/8"		19	.421		7.35	14.50		21.85	30
	8532	1/2"		17	.471		8.85	16.20		25.05	34
	8533	3/4"		16	.500		9.05	17.25		26.30	36
	8534	1"	↓	15	.533		11	18.35		29.35	39.50
	8535	1-1/4"	Q-1	26	.615		14.60	19.10		33.70	45
	8536	1-1/2"		24	.667		21	20.50		41.50	54
	8537	2"	↓	21	.762		21	23.50		44.50	59
	8550	Cap, 1/4"	1 Plum	38	.211		5.70	7.25		12.95	17.20
	8551	3/8"		38	.211		5.70	7.25		12.95	17.20
	8552	1/2"		34	.235		5.65	8.10		13.75	18.50
	8553	3/4"		32	.250		7.15	8.60		15.75	21
	8554	1"	↓	30	.267		9.35	9.20		18.55	24
	8555	1-1/4"	Q-1	52	.308		10.35	9.55		19.90	26
	8556	1-1/2"	↓	48	.333		15.15	10.35		25.50	32.50

15100 | Building Services Piping

15107 | Metal Pipe & Fittings

		CREW	DAILY OUTPUT	LABOR-HOURS	UNIT	2001 BARE COSTS MAT.	LABOR	EQUIP.	TOTAL	TOTAL INCL O&P
8557	2"	Q-1	42	.381	Ea.	21	11.80		32.80	41
8570	Plug, 1/4"	1 Plum	38	.211		.10	7.25		7.35	11.05
8571	3/8"		38	.211		.12	7.25		7.37	11.10
8572	1/2"		34	.235		.16	8.10		8.26	12.45
8573	3/4"		32	.250		.28	8.60		8.88	13.30
8574	1"		30	.267		.61	9.20		9.81	14.50
8575	1-1/4"	Q-1	52	.308		1.19	9.55		10.74	15.70
8576	1-1/2"		48	.333		2.17	10.35		12.52	18
8577	2"		42	.381		3.03	11.80		14.83	21
9500	Union with brass seat, 1/4"	1 Plum	15	.533		12.50	18.35		30.85	41.50
9530	3/8"		15	.533		12.75	18.35		31.10	41.50
9540	1/2"		14	.571		10.30	19.70		30	41
9550	3/4"		13	.615		11.40	21		32.40	44.50
9560	1"		12	.667		14.90	23		37.90	51
9570	1-1/4"	Q-1	21	.762		24.50	23.50		48	62
9580	1-1/2"		19	.842		25.50	26		51.50	67.50
9590	2"		17	.941		31.50	29		60.50	78.50
9600	2-1/2"		13	1.231		102	38		140	170
9610	3"		9	1.778		131	55		186	228
9620	4"		5	3.200		325	99		424	505
9630	Union, all iron, 1/4"	1 Plum	15	.533		19.50	18.35		37.85	49
9650	3/8"		15	.533		20.50	18.35		38.85	50
9660	1/2"		14	.571		20.50	19.70		40.20	52
9670	3/4"		13	.615		22	21		43	56
9680	1"		12	.667		27	23		50	64
9690	1-1/4"	Q-1	21	.762		41.50	23.50		65	81
9700	1-1/2"		19	.842		50.50	26		76.50	95
9710	2"		17	.941		64	29		93	115
9720	2-1/2"		13	1.231		216	38		254	295
9730	3"		9	1.778		216	55		271	320
9750	For galvanized unions, add					15%				
9757	Forged steel, 3000 lb.									
9758	Black									
9760	90° Elbow, 1/4"	1 Plum	16	.500	Ea.	8.10	17.25		25.35	35
9761	3/8"		16	.500		11.15	17.25		28.40	38.50
9762	1/2"		15	.533		8.25	18.35		26.60	36.50
9763	3/4"		14	.571		9.90	19.70		29.60	40.50
9764	1"		13	.615		14.35	21		35.35	48
9765	1-1/4"	Q-1	22	.727		28	22.50		50.50	65
9766	1-1/2"		20	.800		36.50	25		61.50	77.50
9767	2"		18	.889		44	27.50		71.50	90
9780	45° Elbow 1/4"	1 Plum	16	.500		12.20	17.25		29.45	39.50
9781	3/8"		16	.500		12.20	17.25		29.45	39.50
9782	1/2"		15	.533		12.20	18.35		30.55	41
9783	3/4"		14	.571		14.10	19.70		33.80	45
9784	1"		13	.615		19.25	21		40.25	53
9785	1-1/4"	Q-1	22	.727		26.50	22.50		49	63.50
9786	1-1/2"		20	.800		40	25		65	81.50
9787	2"		18	.889		55	27.50		82.50	102
9800	Tee, 1/4"	1 Plum	10	.800		10.90	27.50		38.40	53.50
9801	3/8"		10	.800		12.60	27.50		40.10	55.50
9802	1/2"		9	.889		10.90	30.50		41.40	58
9803	3/4"		9	.889		15.55	30.50		46.05	63
9804	1"		8	1		20	34.50		54.50	74
9805	1-1/4"	Q-1	14	1.143		38	35.50		73.50	95.50
9806	1-1/2"		13	1.231		45	38		83	107

15100 | Building Services Piping

15107 | Metal Pipe & Fittings

			CREW	DAILY OUTPUT	LABOR-HOURS	UNIT	MAT.	LABOR	EQUIP.	TOTAL	TOTAL INCL O&P	
640	9807	2"	Q-1	11	1.455	Ea.	60	45		105	134	640
	9820	Reducer, concentric, 1/4"	1 Plum	19	.421		6.60	14.50		21.10	29.50	
	9821	3/8"		19	.421		6.60	14.50		21.10	29.50	
	9822	1/2"		17	.471		6.60	16.20		22.80	32	
	9823	3/4"		16	.500		7.40	17.25		24.65	34	
	9824	1"		15	.533		10.15	18.35		28.50	38.50	
	9825	1-1/4"	Q-1	26	.615		15.90	19.10		35	46.50	
	9826	1-1/2"		24	.667		18.50	20.50		39	51.50	
	9827	2"		21	.762		26.50	23.50		50	65	
	9840	Cap, 1/4"	1 Plum	38	.211		4.07	7.25		11.32	15.45	
	9841	3/8"		38	.211		4.07	7.25		11.32	15.45	
	9842	1/2"		34	.235		4.07	8.10		12.17	16.75	
	9843	3/4"		32	.250		5.65	8.60		14.25	19.25	
	9844	1"		30	.267		8.75	9.20		17.95	23.50	
	9845	1-1/4"	Q-1	52	.308		13.10	9.55		22.65	29	
	9846	1-1/2"		48	.333		15.15	10.35		25.50	32.50	
	9847	2"		42	.381		23.50	11.80		35.30	43.50	
	9860	Plug, 1/4"	1 Plum	38	.211		1.27	7.25		8.52	12.35	
	9861	3/8"		38	.211		1.58	7.25		8.83	12.70	
	9862	1/2"		34	.235		1.58	8.10		9.68	14	
	9863	3/4"		32	.250		2.75	8.60		11.35	16.05	
	9864	1"		30	.267		3.61	9.20		12.81	17.80	
	9865	1-1/4"	Q-1	52	.308		6.25	9.55		15.80	21.50	
	9866	1-1/2"		48	.333		8.10	10.35		18.45	24.50	
	9867	2"		42	.381		15.15	11.80		26.95	34.50	
	9880	Union, bronze seat, 1/4"	1 Plum	15	.533		14.60	18.35		32.95	43.50	
	9881	3/8"		15	.533		14.60	18.35		32.95	43.50	
	9882	1/2"		14	.571		14.60	19.70		34.30	45.50	
	9883	3/4"		13	.615		17.55	21		38.55	51.50	
	9884	1"		12	.667		23	23		46	59.50	
	9885	1-1/4"	Q-1	21	.762		38.50	23.50		62	78	
	9886	1-1/2"		19	.842		40	26		66	83.50	
	9887	2"		17	.941		51.50	29		80.50	101	
	9900	Coupling, 1/4"	1 Plum	19	.421		3.03	14.50		17.53	25.50	
	9901	3/8"		19	.421		3.37	14.50		17.87	25.50	
	9902	1/2"		17	.471		2.87	16.20		19.07	27.50	
	9903	3/4"		16	.500		4.07	17.25		21.32	30.50	
	9904	1"		15	.533		5.95	18.35		24.30	34	
	9905	1-1/4"	Q-1	26	.615		11.05	19.10		30.15	41	
	9906	1-1/2"		24	.667		15.15	20.50		35.65	47.50	
	9907	2"		21	.762		18.85	23.50		42.35	56.50	
660	0010	**PIPE, STEEL, FITTINGS** Flanged, welded and special type										660
	0020	flanged joints, C.I., standard weight, black. One gasket & bolt										
	0040	set required at each joint, not included (see line 0620)										
	0060	90° Elbow, straight, 1-1/2" pipe size	Q-1	14	1.143	Ea.	127	35.50		162.50	194	
	0080	2" pipe size		13	1.231		100	38		138	168	
	0090	2-1/2" pipe size		12	1.333		108	41.50		149.50	182	
	0100	3" pipe size		11	1.455		129	45		174	209	
	0110	4" pipe size		8	2		111	62		173	217	
	0120	5" pipe size		7	2.286		206	71		277	335	
	0130	6" pipe size	Q-2	9	2.667		176	85.50		261.50	320	
	0140	8" pipe size		8	3		273	96.50		369.50	445	
	0150	10" pipe size		7	3.429		605	110		715	830	
	0160	12" pipe size		6	4		1,050	129		1,179	1,350	
	0171	90° Elbow, reducing										

15100 | Building Services Piping

15107 | Metal Pipe & Fittings

		CREW	DAILY OUTPUT	LABOR-HOURS	UNIT	2001 BARE COSTS				TOTAL INCL O&P
						MAT.	LABOR	EQUIP.	TOTAL	
0172	2-1/2" by 2" pipe size	Q-1	12	1.333	Ea.	246	41.50		287.50	335
0173	3" by 2-1/2" pipe size		11	1.455		260	45		305	355
0174	4" by 3" pipe size		8	2		294	62		356	420
0175	5" by 3" pipe size		7	2.286		380	71		451	520
0176	6" by 4" pipe size	Q-2	9	2.667		420	85.50		505.50	595
0177	8" by 6" pipe size		8	3		620	96.50		716.50	830
0178	10" by 8" pipe size		7	3.429		980	110		1,090	1,250
0179	12" by 10" pipe size		6	4		1,700	129		1,829	2,050
0200	45° Elbow, straight, 1-1/2" pipe size	Q-1	14	1.143		155	35.50		190.50	225
0220	2" pipe size		13	1.231		152	38		190	225
0230	2-1/2" pipe size		12	1.333		155	41.50		196.50	234
0240	3" pipe size		11	1.455		163	45		208	248
0250	4" pipe size		8	2		169	62		231	280
0260	5" pipe size		7	2.286		315	71		386	450
0270	6" pipe size	Q-2	9	2.667		277	85.50		362.50	435
0280	8" pipe size		8	3		405	96.50		501.50	590
0290	10" pipe size		7	3.429		860	110		970	1,100
0300	12" pipe size		6	4		1,325	129		1,454	1,650
0310	Cross, straight									
0311	2-1/2" pipe size	Q-1	6	2.667	Ea.	335	82.50		417.50	495
0312	3" pipe size		5	3.200		335	99		434	520
0313	4" pipe size		4	4		435	124		559	665
0314	5" pipe size		3	5.333		710	165		875	1,025
0315	6" pipe size	Q-2	5	4.800		710	154		864	1,025
0316	8" pipe size		4	6		1,150	193		1,343	1,550
0317	10" pipe size		3	8		1,750	257		2,007	2,325
0318	12" pipe size		2	12		2,650	385		3,035	3,500
0350	Tee, straight, 1-1/2" pipe size	Q-1	10	1.600		109	49.50		158.50	195
0370	2" pipe size		9	1.778		109	55		164	203
0380	2-1/2" pipe size		8	2		159	62		221	268
0390	3" pipe size		7	2.286		165	71		236	289
0400	4" pipe size		5	3.200		169	99		268	335
0410	5" pipe size		4	4		355	124		479	575
0420	6" pipe size	Q-2	6	4		246	129		375	465
0430	8" pipe size		5	4.800		385	154		539	660
0440	10" pipe size		4	6		880	193		1,073	1,250
0450	12" pipe size		3	8		1,375	257		1,632	1,925
0459	Tee, reducing on outlet									
0460	2-1/2" by 2" pipe size	Q-1	8	2	Ea.	236	62		298	355
0461	3" by 2-1/2" pipe size		7	2.286		270	71		341	405
0462	4" by 3" pipe size		5	3.200		660	99		759	880
0463	5" by 4" pipe size		4	4		845	124		969	1,125
0464	6" by 4" pipe size	Q-2	6	4		845	129		974	1,125
0465	8" by 6" pipe size		5	4.800		910	154		1,064	1,225
0466	10" by 8" pipe size		4	6		1,425	193		1,618	1,850
0467	12" by 10" pipe size		3	8		2,100	257		2,357	2,700
0476	Reducer, concentric									
0477	3" by 2-1/2"	Q-1	12	1.333	Ea.	250	41.50		291.50	340
0478	4" by 3"		9	1.778		267	55		322	375
0479	5" by 4"		8	2		420	62		482	555
0480	6" by 4"	Q-2	10	2.400		470	77		547	630
0481	8" by 6"		9	2.667		535	85.50		620.50	715
0482	10" by 8"		8	3		860	96.50		956.50	1,100
0483	12" by 10"		7	3.429		1,550	110		1,660	1,875
0492	Reducer, eccentric									
0493	4" by 3"	Q-1	8	2	Ea.	325	62		387	455

15100 | Building Services Piping

15107 | Metal Pipe & Fittings

		CREW	DAILY OUTPUT	LABOR-HOURS	UNIT	2001 BARE COSTS MAT.	LABOR	EQUIP.	TOTAL	TOTAL INCL O&P
0494	5" by 4"	Q-1	7	2.286	Ea.	420	71		491	565
0495	6" by 4"	Q-2	9	2.667		450	85.50		535.50	625
0496	8" by 6"		8	3		590	96.50		686.50	795
0497	10" by 8"		7	3.429		1,025	110		1,135	1,300
0498	12" by 10"	↓	6	4		1,575	129		1,704	1,925
0500	For galvanized elbows and tees, add					100%				
0520	For extra heavy weight elbows and tees, add					140%				
0620	Gasket and bolt set, 150#, 1/2" thru 1-1/2" pipe size	1 Plum	20	.400		3.05	13.80		16.85	24.50
0630	2" pipe size		13	.615		4.87	21		25.87	37.50
0640	2-1/2" pipe size		12	.667		4.64	23		27.64	39.50
0650	3" pipe size		11	.727		5.40	25		30.40	44
0660	3-1/2" pipe size		9	.889		9.80	30.50		40.30	57
0670	4" pipe size		8	1		9.80	34.50		44.30	63
0680	5" pipe size		7	1.143		14.55	39.50		54.05	75.50
0690	6" pipe size		6	1.333		15.25	46		61.25	86.50
0700	8" pipe size		5	1.600		16.25	55		71.25	101
0710	10" pipe size		4.50	1.778		33.50	61		94.50	130
0720	12" pipe size		4.20	1.905		36	65.50		101.50	139
0730	14" pipe size		4	2		51	69		120	161
0740	16" pipe size		3	2.667		67	92		159	213
0750	18" pipe size		2.70	2.963		93.50	102		195.50	257
0760	20" pipe size		2.30	3.478		125	120		245	320
0780	24" pipe size		1.90	4.211		167	145		312	405
0790	26" pipe size		1.60	5		189	172		361	470
0810	30" pipe size		1.40	5.714		234	197		431	555
0830	36" pipe size	↓	1.10	7.273	↓	288	251		539	695
0850	For 300 lb gasket set, add					40%				
2000	Flanged unions, 125 lb., black, 1/2" pipe size	1 Plum	17	.471	Ea.	25.50	16.20		41.70	52.50
2040	3/4" pipe size		17	.471		33.50	16.20		49.70	61.50
2050	1" pipe size		16	.500		32.50	17.25		49.75	61.50
2060	1-1/4" pipe size	Q-1	28	.571		38.50	17.70		56.20	69
2070	1-1/2" pipe size		27	.593		35.50	18.35		53.85	66.50
2080	2" pipe size		26	.615		42.50	19.10		61.60	76
2090	2-1/2" pipe size		24	.667		57.50	20.50		78	94.50
2100	3" pipe size		22	.727		65	22.50		87.50	106
2110	3-1/2" pipe size		18	.889		133	27.50		160.50	188
2120	4" pipe size		16	1		89	31		120	145
2130	5" pipe size	↓	14	1.143		157	35.50		192.50	227
2140	6" pipe size	Q-2	19	1.263		190	40.50		230.50	271
2150	8" pipe size	"	16	1.500		325	48		373	435
2200	For galvanized unions, add				↓	150%				
2290	Threaded flange									
2300	Cast iron									
2310	Black, 125 lb., per flange									
2320	1" pipe size	1 Plum	27	.296	Ea.	14.55	10.20		24.75	31.50
2330	1-1/4" pipe size	Q-1	44	.364		17.55	11.25		28.80	36.50
2340	1-1/2" pipe size		40	.400		16.20	12.40		28.60	36.50
2350	2" pipe size		36	.444		16.20	13.80		30	39
2360	2-1/2" pipe size		28	.571		18.90	17.70		36.60	47.50
2370	3" pipe size		20	.800		24.50	25		49.50	64
2380	3-1/2" pipe size		16	1		32.50	31		63.50	82.50
2390	4" pipe size		12	1.333		32.50	41.50		74	98.50
2400	5" pipe size	↓	10	1.600		46	49.50		95.50	126
2410	6" pipe size	Q-2	14	1.714		52	55		107	140
2420	8" pipe size		12	2		82	64.50		146.50	188
2430	10" pipe size	↓	10	2.400		142	77		219	273

15100 | Building Services Piping

15107 | Metal Pipe & Fittings

			CREW	DAILY OUTPUT	LABOR-HOURS	UNIT	MAT.	LABOR	EQUIP.	TOTAL	TOTAL INCL O&P
2440		12" pipe size	Q-2	8	3	Ea.	237	96.50		333.50	405
2460		For galvanized flanges, add					95%				
2580	Forged steel,										
2590		Black 150 lb., per flange									
2600		1/2" pipe size	1 Plum	30	.267	Ea.	21.50	9.20		30.70	38
2610		3/4" pipe size		28	.286		21.50	9.85		31.35	39
2620		1" pipe size		27	.296		21.50	10.20		31.70	39.50
2630		1-1/4" pipe size	Q-1	44	.364		21.50	11.25		32.75	41
2640		1-1/2" pipe size		40	.400		21.50	12.40		33.90	42.50
2650		2" pipe size		36	.444		21.50	13.80		35.30	45
2660		2-1/2" pipe size		28	.571		26.50	17.70		44.20	55.50
2670		3" pipe size		20	.800		26.50	25		51.50	66.50
2690		4" pipe size		12	1.333		31	41.50		72.50	96.50
2700		5" pipe size		10	1.600		45	49.50		94.50	125
2710		6" pipe size	Q-2	14	1.714		51	55		106	139
2720		8" pipe size		12	2		86	64.50		150.50	192
2730		10" pipe size		10	2.400		156	77		233	288
2860		Black 300 lb., per flange									
2870		1/2" pipe size	1 Plum	30	.267	Ea.	24	9.20		33.20	40.50
2880		3/4" pipe size		28	.286		24	9.85		33.85	41.50
2890		1" pipe size		27	.296		24	10.20		34.20	42
2900		1-1/4" pipe size	Q-1	44	.364		24	11.25		35.25	43.50
2910		1-1/2" pipe size		40	.400		24	12.40		36.40	45
2920		2" pipe size		36	.444		25	13.80		38.80	48.50
2930		2-1/2" pipe size		28	.571		35	17.70		52.70	65
2940		3" pipe size		20	.800		37.50	25		62.50	79
2960		4" pipe size		12	1.333		53	41.50		94.50	121
2970		6" pipe size	Q-2	14	1.714		93.50	55		148.50	186
3000	Weld joint, butt, carbon steel, standard weight										
3040		90° elbow, long radius									
3050		1/2" pipe size	Q-15	16	1	Ea.	12.50	31	6.30	49.80	67.50
3060		3/4" pipe size		16	1		12.50	31	6.30	49.80	67.50
3070		1" pipe size		16	1		6.40	31	6.30	43.70	61
3080		1-1/4" pipe size		14	1.143		6.40	35.50	7.20	49.10	68.50
3090		1-1/2" pipe size		13	1.231		6.40	38	7.75	52.15	73
3100		2" pipe size		10	1.600		6.40	49.50	10.05	65.95	93
3110		2-1/2" pipe size		8	2		8.65	62	12.60	83.25	117
3120		3" pipe size		7	2.286		9.90	71	14.40	95.30	134
3130		4" pipe size		5	3.200		16.95	99	20	135.95	191
3136		5" pipe size		4	4		35.50	124	25	184.50	254
3140		6" pipe size	Q-16	5	4.800		40.50	154	20	214.50	300
3150		8" pipe size		3.75	6.400		76	206	27	309	425
3160		10" pipe size		3	8		139	257	33.50	429.50	580
3170		12" pipe size		2.50	9.600		199	310	40	549	730
3180		14" pipe size		2	12		275	385	50.50	710.50	940
3190		16" pipe size		1.50	16		375	515	67	957	1,275
3191		18" pipe size		1.25	19.200		485	615	80.50	1,180.50	1,550
3192		20" pipe size		1.15	20.870		715	670	87.50	1,472.50	1,900
3194		24" pipe size		1.02	23.529		945	755	98.50	1,798.50	2,300
3200		45° Elbow, long									
3210		1/2" pipe size	Q-15	16	1	Ea.	11.85	31	6.30	49.15	67
3220		3/4" pipe size		16	1		11.85	31	6.30	49.15	67
3230		1" pipe size		16	1		6.10	31	6.30	43.40	60.50
3240		1-1/4" pipe size		14	1.143		6.10	35.50	7.20	48.80	68
3250		1-1/2" pipe size		13	1.231		6.10	38	7.75	51.85	72.50
3260		2" pipe size		10	1.600		6.10	49.50	10.05	65.65	93

15100 | Building Services Piping

15107 | Metal Pipe & Fittings

		CREW	DAILY OUTPUT	LABOR-HOURS	UNIT	MAT.	LABOR	EQUIP.	TOTAL	TOTAL INCL O&P
3270	2-1/2" pipe size	Q-15	8	2	Ea.	7.35	62	12.60	81.95	115
3280	3" pipe size		7	2.286		8.65	71	14.40	94.05	132
3290	4" pipe size		5	3.200		15.05	99	20	134.05	189
3296	5" pipe size		4	4		22	124	25	171	239
3300	6" pipe size	Q-16	5	4.800		32	154	20	206	290
3310	8" pipe size		3.75	6.400		54	206	27	287	400
3320	10" pipe size		3	8		98	257	33.50	388.50	535
3330	12" pipe size		2.50	9.600		139	310	40	489	660
3340	14" pipe size		2	12		189	385	50.50	624.50	845
3341	16" pipe size		1.50	16		258	515	67	840	1,125
3342	18" pipe size		1.25	19.200		345	615	80.50	1,040.50	1,400
3343	20" pipe size		1.15	20.870		455	670	87.50	1,212.50	1,625
3345	24" pipe size		1.05	22.857		660	735	96	1,491	1,925
3346	26" pipe size		.85	28.235		1,325	910	118	2,353	2,975
3347	30" pipe size		.45	53.333		1,475	1,725	223	3,423	4,475
3349	36" pipe size		.38	63.158		2,450	2,025	265	4,740	6,075
3350	Tee, straight									
3352	For reducing tees and concentrics see starting line 4600									
3360	1/2" pipe size	Q-15	10	1.600	Ea.	30.50	49.50	10.05	90.05	120
3370	3/4" pipe size		10	1.600		16.95	49.50	10.05	76.50	105
3380	1" pipe size		10	1.600		16.95	49.50	10.05	76.50	105
3390	1-1/4" pipe size		9	1.778		19.50	55	11.20	85.70	117
3400	1-1/2" pipe size		8	2		19.50	62	12.60	94.10	129
3410	2" pipe size		6	2.667		16.95	82.50	16.75	116.20	162
3420	2-1/2" pipe size		5	3.200		21.50	99	20	140.50	196
3430	3" pipe size		4	4		23.50	124	25	172.50	241
3440	4" pipe size		3	5.333		33.50	165	33.50	232	325
3446	5" pipe size		2.50	6.400		61	198	40.50	299.50	410
3450	6" pipe size	Q-16	3	8		61	257	33.50	351.50	495
3460	8" pipe size		2.50	9.600		111	310	40	461	635
3470	10" pipe size		2	12		186	385	50.50	621.50	840
3480	12" pipe size		1.60	15		280	480	63	823	1,100
3481	14" pipe size		1.30	18.462		470	595	77.50	1,142.50	1,500
3482	16" pipe size		1	24		565	770	101	1,436	1,900
3483	18" pipe size		.80	30		895	965	126	1,986	2,575
3484	20" pipe size		.75	32		1,375	1,025	134	2,534	3,225
3486	24" pipe size		.70	34.286		1,725	1,100	144	2,969	3,725
3487	26" pipe size		.55	43.636		2,400	1,400	183	3,983	4,975
3488	30" pipe size		.30	80		2,600	2,575	335	5,510	7,100
3490	36" pipe size		.20	120		4,100	3,850	505	8,455	10,900
3491	Eccentric reducer, 1-1/2" pipe size	Q-15	14	1.143		17.30	35.50	7.20	60	80.50
3492	2" pipe size		11	1.455		17.30	45	9.15	71.45	97
3493	2-1/2" pipe size		9	1.778		14.70	55	11.20	80.90	112
3494	3" pipe size		8	2		16.65	62	12.60	91.25	126
3495	4" pipe size		6	2.667		17.90	82.50	16.75	117.15	163
3496	6" pipe size	Q-16	5	4.800		36	154	20	210	295
3497	8" pipe size		4	6		70	193	25	288	395
3498	10" pipe size		3	8		91	257	33.50	381.50	525
3499	12" pipe size		2.50	9.600		148	310	40	498	675
3501	Cap, 1-1/2" pipe size	Q-15	28	.571		7.05	17.70	3.59	28.34	38
3502	2" pipe size		22	.727		6.40	22.50	4.57	33.47	46
3503	2-1/2" pipe size		18	.889		6.40	27.50	5.60	39.50	54.50
3504	3" pipe size		16	1		6.40	31	6.30	43.70	61
3505	4" pipe size		12	1.333		8.30	41.50	8.40	58.20	81
3506	6" pipe size	Q-16	10	2.400		15.70	77	10.05	102.75	144
3507	8" pipe size		8	3		24	96.50	12.55	133.05	186

15100 | Building Services Piping

15107 | Metal Pipe & Fittings

		CREW	DAILY OUTPUT	LABOR-HOURS	UNIT	MAT.	LABOR	EQUIP.	TOTAL	TOTAL INCL O&P
3508	10" pipe size	Q-16	6	4	Ea.	42	129	16.75	187.75	259
3509	12" pipe size		5	4.800		52.50	154	20	226.50	315
3511	14" pipe size		4	6		96.50	193	25	314.50	425
3512	16" pipe size		4	6		113	193	25	331	445
3513	18" pipe size	↓	3	8	↓	130	257	33.50	420.50	570
3517	Weld joint, butt, carbon steel, extra strong									
3519	90° elbow, long									
3520	1/2" pipe size	Q-15	13	1.231	Ea.	15.70	38	7.75	61.45	83.50
3530	3/4" pipe size		12	1.333		15.70	41.50	8.40	65.60	89
3540	1" pipe size		11	1.455		8.30	45	9.15	62.45	87
3550	1-1/4" pipe size		10	1.600		8.30	49.50	10.05	67.85	95
3560	1-1/2" pipe size		9	1.778		8.30	55	11.20	74.50	104
3570	2" pipe size		8	2		8.30	62	12.60	82.90	117
3580	2-1/2" pipe size		7	2.286		12.15	71	14.40	97.55	136
3590	3" pipe size		6	2.667		15.05	82.50	16.75	114.30	160
3600	4" pipe size		4	4		25.50	124	25	174.50	243
3606	5" pipe size	↓	3.50	4.571		50.50	142	29	221.50	300
3610	6" pipe size	Q-16	4.50	5.333		61	171	22.50	254.50	350
3620	8" pipe size		3.50	6.857		115	220	28.50	363.50	495
3630	10" pipe size		2.50	9.600		206	310	40	556	735
3640	12" pipe size	↓	2.25	10.667	↓	291	345	44.50	680.50	890
3650	45° Elbow, long									
3660	1/2" pipe size	Q-15	13	1.231	Ea.	15.70	38	7.75	61.45	83.50
3670	3/4" pipe size		12	1.333		15.70	41.50	8.40	65.60	89
3680	1" pipe size		11	1.455		8.30	45	9.15	62.45	87
3690	1-1/4" pipe size		10	1.600		8.30	49.50	10.05	67.85	95
3700	1-1/2" pipe size		9	1.778		8.30	55	11.20	74.50	104
3710	2" pipe size		8	2		7.70	62	12.60	82.30	116
3720	2-1/2" pipe size		7	2.286		16.30	71	14.40	101.70	141
3730	3" pipe size		6	2.667		18.25	82.50	16.75	117.50	163
3740	4" pipe size		4	4		17.90	124	25	166.90	234
3746	5" pipe size	↓	3.50	4.571		41.50	142	29	212.50	292
3750	6" pipe size	Q-16	4.50	5.333		47.50	171	22.50	241	335
3760	8" pipe size		3.50	6.857		77.50	220	28.50	326	450
3770	10" pipe size		2.50	9.600		125	310	40	475	645
3780	12" pipe size	↓	2.25	10.667		186	345	44.50	575.50	775
3800	Tee, straight									
3810	1/2" pipe size	Q-15	9	1.778	Ea.	38	55	11.20	104.20	137
3820	3/4" pipe size		8.50	1.882		19.50	58.50	11.85	89.85	123
3830	1" pipe size		8	2		20	62	12.60	94.60	129
3840	1-1/4" pipe size		7	2.286		20	71	14.40	105.40	145
3850	1-1/2" pipe size		6	2.667		23.50	82.50	16.75	122.75	169
3860	2" pipe size		5	3.200		20	99	20	139	194
3870	2-1/2" pipe size		4	4		35.50	124	25	184.50	254
3880	3" pipe size		3.50	4.571		42.50	142	29	213.50	293
3890	4" pipe size		2.50	6.400		54	198	40.50	292.50	405
3896	5" pipe size	↓	2.25	7.111		110	220	44.50	374.50	505
3900	6" pipe size	Q-16	2.25	10.667		84.50	345	44.50	474	660
3910	8" pipe size		2	12		172	385	50.50	607.50	825
3920	10" pipe size		1.75	13.714		270	440	57.50	767.50	1,025
3930	12" pipe size	↓	1.50	16		405	515	67	987	1,300
4000	Eccentric reducer, 1-1/2" pipe size	Q-15	10	1.600		23.50	49.50	10.05	83.05	112
4010	2" pipe size		9	1.778		25.50	55	11.20	91.70	123
4020	2-1/2" pipe size		8	2		29	62	12.60	103.60	139
4030	3" pipe size		7	2.286		25.50	71	14.40	110.90	151
4040	4" pipe size	↓	5	3.200		26	99	20	145	201

15100 | Building Services Piping

15107 | Metal Pipe & Fittings

		CREW	DAILY OUTPUT	LABOR-HOURS	UNIT	MAT.	LABOR	EQUIP.	TOTAL	TOTAL INCL O&P
4046	5" pipe size	Q-15	4.70	3.404	Ea.	71	106	21.50	198.50	261
4050	6" pipe size	Q-16	4.50	5.333		70.50	171	22.50	264	360
4060	8" pipe size		3.50	6.857		184	220	28.50	432.50	570
4070	10" pipe size		2.50	9.600		182	310	40	532	710
4080	12" pipe size		2.25	10.667		209	345	44.50	598.50	800
4090	14" pipe size		2.10	11.429		365	365	48	778	1,000
4100	16" pipe size		1.90	12.632		405	405	53	863	1,125
4151	Cap, 1-1/2" pipe size	Q-15	24	.667		7.35	20.50	4.19	32.04	43.50
4152	2" pipe size		18	.889		6.40	27.50	5.60	39.50	54.50
4153	2-1/2" pipe size		16	1		6.70	31	6.30	44	61.50
4154	3" pipe size		14	1.143		8	35.50	7.20	50.70	70
4155	4" pipe size		10	1.600		11.20	49.50	10.05	70.75	98.50
4156	6" pipe size	Q-16	9	2.667		29	85.50	11.15	125.65	173
4157	8" pipe size		7	3.429		38.50	110	14.35	162.85	224
4158	10" pipe size		5	4.800		61	154	20	235	320
4159	12" pipe size		4	6		86	193	25	304	415
4190	Weld fittings, reducing, standard weight									
4200	Welding ring w/spacer pins, 2" pipe size				Ea.	2			2	2.20
4210	2-1/2" pipe size					2.10			2.10	2.31
4220	3" pipe size					2.50			2.50	2.75
4230	4" pipe size					2.70			2.70	2.97
4236	5" pipe size					2.95			2.95	3.25
4240	6" pipe size					3.40			3.40	3.74
4250	8" pipe size					5.65			5.65	6.20
4260	10" pipe size					6.80			6.80	7.50
4270	12" pipe size					7.75			7.75	8.55
4280	14" pipe size					8.60			8.60	9.45
4290	16" pipe size					9.85			9.85	10.85
4300	18" pipe size					10.35			10.35	11.40
4310	20" pipe size					12			12	13.20
4330	24" pipe size					13.95			13.95	15.35
4340	26" pipe size					14			14	15.40
4350	30" pipe size					17.65			17.65	19.40
4370	36" pipe size					24.50			24.50	27
4600	Tee, reducing on outlet									
4601	2-1/2" x 2" pipe size	Q-15	5	3.200	Ea.	30.50	99	20	149.50	206
4602	3" x 2-1/2" pipe size		4	4		30.50	124	25	179.50	248
4603	3-1/2" x 3" pipe size		3.50	4.571		19.85	142	29	190.85	268
4604	4" x 3" pipe size		3	5.333		38.50	165	33.50	237	330
4605	5" x 4" pipe size		2.50	6.400		84	198	40.50	322.50	435
4606	6" x 5" pipe size	Q-16	3	8		76	257	33.50	366.50	510
4607	8" x 6" pipe size		2.50	9.600		145	310	40	495	670
4608	10" x 8" pipe size		2	12		245	385	50.50	680.50	905
4609	12" x 10" pipe size		1.60	15		253	480	63	796	1,075
4610	16" x 12" pipe size		1.50	16		590	515	67	1,172	1,500
4611	14" x 12" pipe size		1.52	15.789		505	510	66	1,081	1,400
4618	Reducer, concentric									
4619	2-1/2" by 2" pipe size	Q-15	10	1.600	Ea.	10.90	49.50	10.05	70.45	98
4620	3" by 2-1/2" pipe size		9	1.778		8.65	55	11.20	74.85	105
4621	3-1/2" by 3" pipe size		8	2		20.50	62	12.60	95.10	130
4622	4" by 2-1/2" pipe size		7	2.286		13.45	71	14.40	98.85	138
4623	5" by 3" pipe size		7	2.286		22	71	14.40	107.40	147
4624	6" by 4" pipe size	Q-16	6	4		26	129	16.75	171.75	241
4625	8" by 6" pipe size		5	4.800		35	154	20	209	294
4626	10" by 8" pipe size		4	6		52.50	193	25	270.50	375
4627	12" by 10" pipe size		3	8		82	257	33.50	372.50	515

Important: See the Reference Section for critical supporting data - Reference Nos., Crews, & City Cost Indexes

15100 | Building Services Piping

15107 | Metal Pipe & Fittings

		CREW	DAILY OUTPUT	LABOR-HOURS	UNIT	MAT.	LABOR	EQUIP.	TOTAL	TOTAL INCL O&P
4660	Reducer, eccentric									
4662	3" x 2"	Q-15	8	2	Ea.	14.40	62	12.60	89	123
4664	4" x 3"		6	2.667		16.65	82.50	16.75	115.90	162
4666	4" x 2"		6	2.667		22	82.50	16.75	121.25	167
4670	6" x 4"	Q-16	5	4.800		33.50	154	20	207.50	292
4672	6" x 3"		5	4.800		62.50	154	20	236.50	325
4676	8" x 6"		4	6		50.50	193	25	268.50	375
4678	8" x 4"		4	6		88	193	25	306	415
4682	10" x 8"		3	8		60	257	33.50	350.50	495
4684	10" x 6"		3	8		107	257	33.50	397.50	545
4688	12" x 10"		2.50	9.600		94.50	310	40	444.50	615
4690	12" x 8"		2.50	9.600		148	310	40	498	675
4691	14" x 12"		2.20	10.909		178	350	45.50	573.50	775
4693	16" x 14"		1.80	13.333		242	430	56	728	975
4694	16" x 12"		2	12		299	385	50.50	734.50	965
4696	18" x 16"		1.60	15		395	480	63	938	1,225
5000	Weld joint, socket, forged steel, 3000 lb., schedule 40 pipe									
5010	90° elbow, straight									
5020	1/4" pipe size	Q-15	22	.727	Ea.	7.45	22.50	4.57	34.52	47.50
5030	3/8" pipe size		22	.727		9.20	22.50	4.57	36.27	49
5040	1/2" pipe size		20	.800		7.45	25	5.05	37.50	51.50
5050	3/4" pipe size		20	.800		7.65	25	5.05	37.70	51.50
5060	1" pipe size		20	.800		9.90	25	5.05	39.95	54
5070	1-1/4" pipe size		18	.889		18.90	27.50	5.60	52	68.50
5080	1-1/2" pipe size		16	1		21	31	6.30	58.30	77
5090	2" pipe size		12	1.333		31.50	41.50	8.40	81.40	107
5100	2-1/2" pipe size		10	1.600		93	49.50	10.05	152.55	188
5110	3" pipe size		8	2		157	62	12.60	231.60	280
5120	4" pipe size		6	2.667		425	82.50	16.75	524.25	615
5130	45° Elbow, straight									
5134	1/4" pipe size	Q-15	22	.727	Ea.	12.10	22.50	4.57	39.17	52.50
5135	3/8" pipe size		22	.727		12.10	22.50	4.57	39.17	52.50
5136	1/2" pipe size		20	.800		9.40	25	5.05	39.45	53.50
5137	3/4" pipe size		20	.800		10.95	25	5.05	41	55
5140	1" pipe size		20	.800		14.50	25	5.05	44.55	59
5150	1-1/4" pipe size		18	.889		19.30	27.50	5.60	52.40	68.50
5160	1-1/2" pipe size		16	1		24	31	6.30	61.30	80.50
5170	2" pipe size		12	1.333		41	41.50	8.40	90.90	117
5180	2-1/2" pipe size		10	1.600		106	49.50	10.05	165.55	202
5190	3" pipe size		8	2		178	62	12.60	252.60	300
5200	4" pipe size		6	2.667		350	82.50	16.75	449.25	530
5250	Tee, straight									
5254	1/4" pipe size	Q-15	15	1.067	Ea.	8.55	33	6.70	48.25	67
5255	3/8" pipe size		15	1.067		10.55	33	6.70	50.25	69
5256	1/2" pipe size		13	1.231		8.55	38	7.75	54.30	75.50
5257	3/4" pipe size		13	1.231		10.80	38	7.75	56.55	78
5260	1" pipe size		13	1.231		14.85	38	7.75	60.60	82.50
5270	1-1/4" pipe size		12	1.333		23.50	41.50	8.40	73.40	97.50
5280	1-1/2" pipe size		11	1.455		31.50	45	9.15	85.65	113
5290	2" pipe size		8	2		44	62	12.60	118.60	156
5300	2-1/2" pipe size		6	2.667		130	82.50	16.75	229.25	286
5310	3" pipe size		5	3.200		315	99	20	434	515
5320	4" pipe size		4	4		500	124	25	649	765
5350	For reducing sizes, add					60%				
5450	Couplings									
5451	1/4" pipe size	Q-15	23	.696	Ea.	6.55	21.50	4.38	32.43	44.50

15100 | Building Services Piping

15107 | Metal Pipe & Fittings

		CREW	DAILY OUTPUT	LABOR-HOURS	UNIT	MAT.	LABOR	EQUIP.	TOTAL	TOTAL INCL O&P
5452	3/8" pipe size	Q-15	23	.696	Ea.	6.55	21.50	4.38	32.43	44.50
5453	1/2" pipe size		21	.762		3.31	23.50	4.79	31.60	44.50
5454	3/4" pipe size		21	.762		4.41	23.50	4.79	32.70	45.50
5460	1" pipe size		20	.800		5.15	25	5.05	35.20	49
5470	1-1/4" pipe size		20	.800		10.10	25	5.05	40.15	54
5480	1-1/2" pipe size		18	.889		11.15	27.50	5.60	44.25	60
5490	2" pipe size		14	1.143		17.55	35.50	7.20	60.25	80.50
5500	2-1/2" pipe size		12	1.333		37	41.50	8.40	86.90	113
5510	3" pipe size		9	1.778		86.50	55	11.20	152.70	190
5520	4" pipe size		7	2.286		130	71	14.40	215.40	266
5570	Union, 1/4" pipe size		21	.762		14.35	23.50	4.79	42.64	56.50
5571	3/8" pipe size		21	.762		14.35	23.50	4.79	42.64	56.50
5572	1/2" pipe size		19	.842		14.35	26	5.30	45.65	61
5573	3/4" pipe size		19	.842		16.30	26	5.30	47.60	63.50
5574	1" pipe size		19	.842		21	26	5.30	52.30	68.50
5575	1-1/4" pipe size		17	.941		34	29	5.90	68.90	88
5576	1-1/2" pipe size		15	1.067		36	33	6.70	75.70	97
5577	2" pipe size		11	1.455		50	45	9.15	104.15	133
5600	Reducer, 1/4" pipe size		23	.696		17.65	21.50	4.38	43.53	57
5601	3/8" pipe size		23	.696		17.65	21.50	4.38	43.53	57
5602	1/2" pipe size		21	.762		12	23.50	4.79	40.29	54
5603	3/4" pipe size		21	.762		13.95	23.50	4.79	42.24	56
5604	1" pipe size		21	.762		15.30	23.50	4.79	43.59	57.50
5605	1-1/4" pipe size		19	.842		27.50	26	5.30	58.80	76
5607	1-1/2" pipe size		17	.941		29.50	29	5.90	64.40	83
5608	2" pipe size		13	1.231		32.50	38	7.75	78.25	102
5612	Cap, 1/4" pipe size		46	.348		6.95	10.80	2.19	19.94	26.50
5613	3/8" pipe size		46	.348		6.95	10.80	2.19	19.94	26.50
5614	1/2" pipe size		42	.381		4.91	11.80	2.40	19.11	26
5615	3/4" pipe size		42	.381		5.70	11.80	2.40	19.90	27
5616	1" pipe size		42	.381		8.80	11.80	2.40	23	30
5617	1-1/4" pipe size		38	.421		10.25	13.05	2.65	25.95	34
5618	1-1/2" pipe size		34	.471		13.95	14.60	2.96	31.51	40.50
5619	2" pipe size		26	.615		22	19.10	3.87	44.97	57.50
5630	T-O-L, 1/4" pipe size, nozzle		23	.696		4.31	21.50	4.38	30.19	42
5631	3/8" pipe size, nozzle		23	.696		4.31	21.50	4.38	30.19	42
5632	1/2" pipe size, nozzle		22	.727		4.31	22.50	4.57	31.38	44
5633	3/4" pipe size, nozzle		21	.762		5.05	23.50	4.79	33.34	46.50
5634	1" pipe size, nozzle		20	.800		5.70	25	5.05	35.75	49.50
5635	1-1/4" pipe size, nozzle		18	.889		8.05	27.50	5.60	41.15	56.50
5636	1-1/2" pipe size, nozzle		16	1		8.75	31	6.30	46.05	63.50
5637	2" pipe size, nozzle		12	1.333		9.05	41.50	8.40	58.95	81.50
5638	2-1/2" pipe size, nozzle		10	1.600		28.50	49.50	10.05	88.05	118
5639	4" pipe size, nozzle		6	2.667		64.50	82.50	16.75	163.75	214
5640	W-O-L, 1/4" pipe size, nozzle		23	.696		7.40	21.50	4.38	33.28	45.50
5641	3/8" pipe size, nozzle		23	.696		7.40	21.50	4.38	33.28	45.50
5642	1/2" pipe size, nozzle		22	.727		7.40	22.50	4.57	34.47	47
5643	3/4" pipe size, nozzle		21	.762		7.60	23.50	4.79	35.89	49
5644	1" pipe size, nozzle		20	.800		7.95	25	5.05	38	52
5645	1-1/4" pipe size, nozzle		18	.889		9.65	27.50	5.60	42.75	58.50
5646	1-1/2" pipe size, nozzle		16	1		9.65	31	6.30	46.95	64.50
5647	2" pipe size, nozzle		12	1.333		10.10	41.50	8.40	60	83
5648	2-1/2" pipe size, nozzle		10	1.600		23	49.50	10.05	82.55	111
5649	3" pipe size, nozzle		8	2		26.50	62	12.60	101.10	137
5650	4" pipe size, nozzle		6	2.667		34	82.50	16.75	133.25	181
5651	5" pipe size, nozzle		5	3.200		82.50	99	20	201.50	263

15100 | Building Services Piping

15107 | Metal Pipe & Fittings

		CREW	DAILY OUTPUT	LABOR-HOURS	UNIT	MAT.	LABOR	EQUIP.	TOTAL	TOTAL INCL O&P
5652	6" pipe size, nozzle	Q-15	4	4	Ea.	105	124	25	254	330
5653	8" pipe size, nozzle		3	5.333		189	165	33.50	387.50	495
5654	10" pipe size, nozzle		2.60	6.154		275	191	38.50	504.50	630
5655	12" pipe size, nozzle		2.20	7.273		430	225	46	701	865
5656	14" pipe size, nozzle		2	8		800	248	50.50	1,098.50	1,300
5660	L-O-L, 1/4" pipe size, outlet		23	.696		47	21.50	4.38	72.88	89
5661	3/8" pipe size, outlet		23	.696		47	21.50	4.38	72.88	89
5662	1/2" pipe size, outlet		22	.727		47	22.50	4.57	74.07	90.50
5663	3/4" pipe size, outlet		21	.762		51.50	23.50	4.79	79.79	98
5664	1" pipe size, outlet		20	.800		69.50	25	5.05	99.55	120
5665	1-1/4" pipe size, outlet		18	.889		108	27.50	5.60	141.10	167
5666	1-1/2" pipe size, outlet		16	1		108	31	6.30	145.30	173
5667	2" pipe size, outlet		12	1.333		171	41.50	8.40	220.90	261
5674	S-O-L, 1/4" pipe size, outlet		23	.696		4.81	21.50	4.38	30.69	42.50
5675	3/8" pipe size, outlet		23	.696		4.81	21.50	4.38	30.69	42.50
5676	1/2" pipe size, outlet		22	.727		4.81	22.50	4.57	31.88	44.50
5677	3/4" pipe size, outlet		21	.762		4.88	23.50	4.79	33.17	46
5678	1" pipe size, outlet		20	.800		5.30	25	5.05	35.35	49
5679	1-1/4" pipe size, outlet		18	.889		8.20	27.50	5.60	41.30	56.50
5680	1-1/2" pipe size, outlet		16	1		8.60	31	6.30	45.90	63.50
5681	2" pipe size, outlet		12	1.333		9.65	41.50	8.40	59.55	82.50
6000	Weld-on flange, forged steel									
6020	Slip-on, 150 lb. flange (welded front and back)									
6050	1/2" pipe size	Q-15	18	.889	Ea.	12.15	27.50	5.60	45.25	61
6060	3/4" pipe size		18	.889		12.15	27.50	5.60	45.25	61
6070	1" pipe size		17	.941		11.20	29	5.90	46.10	63
6080	1-1/4" pipe size		16	1		11.20	31	6.30	48.50	66
6090	1-1/2" pipe size		15	1.067		11.20	33	6.70	50.90	69.50
6100	2" pipe size		12	1.333		11.20	41.50	8.40	61.10	84
6110	2-1/2" pipe size		10	1.600		13.45	49.50	10.05	73	101
6120	3" pipe size		9	1.778		15.70	55	11.20	81.90	113
6130	3-1/2" pipe size		7	2.286		19.50	71	14.40	104.90	144
6140	4" pipe size		6	2.667		19.50	82.50	16.75	118.75	165
6150	5" pipe size		5	3.200		35	99	20	154	211
6160	6" pipe size	Q-16	6	4		31.50	129	16.75	177.25	247
6170	8" pipe size		5	4.800		48	154	20	222	310
6180	10" pipe size		4	6		83.50	193	25	301.50	410
6190	12" pipe size		3	8		124	257	33.50	414.50	565
6191	14" pipe size		2.50	9.600		164	310	40	514	690
6192	16" pipe size		1.80	13.333		271	430	56	757	1,000
6200	300 lb. flange									
6210	1/2" pipe size	Q-15	17	.941	Ea.	20	29	5.90	54.90	72.50
6220	3/4" pipe size		17	.941		20	29	5.90	54.90	72.50
6230	1" pipe size		16	1		20	31	6.30	57.30	76
6240	1-1/4" pipe size		13	1.231		20	38	7.75	65.75	88
6250	1-1/2" pipe size		12	1.333		20	41.50	8.40	69.90	94
6260	2" pipe size		11	1.455		23	45	9.15	77.15	104
6270	2-1/2" pipe size		9	1.778		26	55	11.20	92.20	124
6280	3" pipe size		7	2.286		32	71	14.40	117.40	158
6290	4" pipe size		6	2.667		42.50	82.50	16.75	141.75	190
6300	5" pipe size		4	4		54	124	25	203	274
6310	6" pipe size	Q-16	5	4.800		53	154	20	227	315
6320	8" pipe size		4	6		88.50	193	25	306.50	415
6330	10" pipe size		3.40	7.059		163	227	29.50	419.50	555
6340	12" pipe size		2.80	8.571		200	276	36	512	675
6400	Welding neck, 150 lb. flange									

15100 | Building Services Piping

15107 | Metal Pipe & Fittings

			CREW	DAILY OUTPUT	LABOR-HOURS	UNIT	MAT.	LABOR	EQUIP.	TOTAL	TOTAL INCL O&P
660	6410	1/2" pipe size	Q-15	40	.400	Ea.	16	12.40	2.52	30.92	39
	6420	3/4" pipe size		36	.444		16	13.80	2.80	32.60	41.50
	6430	1" pipe size		32	.500		16	15.50	3.15	34.65	44.50
	6440	1-1/4" pipe size		29	.552		16	17.10	3.47	36.57	47.50
	6450	1-1/2" pipe size		26	.615		16	19.10	3.87	38.97	51
	6460	2" pipe size		20	.800		17.30	25	5.05	47.35	62
	6470	2-1/2" pipe size		16	1		18.55	31	6.30	55.85	74.50
	6480	3" pipe size		14	1.143		20.50	35.50	7.20	63.20	84
	6500	4" pipe size		10	1.600		24.50	49.50	10.05	84.05	113
	6510	5" pipe size		8	2		38.50	62	12.60	113.10	150
	6520	6" pipe size	Q-16	10	2.400		37.50	77	10.05	124.55	168
	6530	8" pipe size		7	3.429		65.50	110	14.35	189.85	254
	6540	10" pipe size		6	4		105	129	16.75	250.75	330
	6550	12" pipe size		5	4.800		153	154	20	327	425
	6551	14" pipe size		4.50	5.333		221	171	22.50	414.50	525
	6552	16" pipe size		3	8		375	257	33.50	665.50	840
	6553	18" pipe size		2.50	9.600		530	310	40	880	1,100
	6554	20" pipe size		2.30	10.435		645	335	43.50	1,023.50	1,275
	6556	24" pipe size		2	12		865	385	50.50	1,300.50	1,575
	6557	26" pipe size		1.70	14.118		1,475	455	59	1,989	2,375
	6558	30" pipe size		.90	26.667		1,925	855	112	2,892	3,550
	6559	36" pipe size		.75	32		2,200	1,025	134	3,359	4,125
	6560	300 lb. flange									
	6570	1/2" pipe size	Q-15	36	.444	Ea.	20	13.80	2.80	36.60	46
	6580	3/4" pipe size		34	.471		20	14.60	2.96	37.56	47.50
	6590	1" pipe size		30	.533		20	16.55	3.35	39.90	50.50
	6600	1-1/4" pipe size		28	.571		20	17.70	3.59	41.29	52.50
	6610	1-1/2" pipe size		24	.667		20	20.50	4.19	44.69	57.50
	6620	2" pipe size		18	.889		23	27.50	5.60	56.10	73
	6630	2-1/2" pipe size		14	1.143		26	35.50	7.20	68.70	90
	6640	3" pipe size		12	1.333		26	41.50	8.40	75.90	100
	6650	4" pipe size		8	2		42.50	62	12.60	117.10	154
	6660	5" pipe size		7	2.286		66	71	14.40	151.40	195
	6670	6" pipe size	Q-16	9	2.667		62	85.50	11.15	158.65	210
	6680	8" pipe size		6	4		107	129	16.75	252.75	330
	6690	10" pipe size		5	4.800		207	154	20	381	485
	6700	12" pipe size		4	6		265	193	25	483	610
	6710	14" pipe size		3.50	6.857		590	220	28.50	838.50	1,025
	6720	16" pipe size		2	12		700	385	50.50	1,135.50	1,400
	7740	Plain ends for plain end pipe, mechanically coupled									
	7750	Cplg & labor required at joints not included add 1 per									
	7760	joint for installed price, see line 9180									
	7770	Malleable iron, painted, unless noted otherwise									
	7800	90° Elbow 1"				Ea.	22.50			22.50	25
	7810	1-1/2"					22.50			22.50	25
	7820	2"					22.50			22.50	25
	7830	2-1/2"					22.50			22.50	25
	7840	3"					27			27	30
	7860	4"					39.50			39.50	43.50
	7870	5" welded steel					137			137	151
	7880	6"					109			109	120
	7890	8" welded steel					230			230	253
	7900	10" welded steel					289			289	320
	7910	12" welded steel					370			370	405
	7970	45° Elbow 1"					22.50			22.50	25
	7980	1-1/2"					22.50			22.50	25

15100 | Building Services Piping

15107 | Metal Pipe & Fittings

		CREW	DAILY OUTPUT	LABOR-HOURS	UNIT	MAT.	LABOR	EQUIP.	TOTAL	TOTAL INCL O&P
7990	2"				Ea.	22.50			22.50	25
8000	2-1/2"					22.50			22.50	25
8010	3"					27			27	30
8030	4"					39.50			39.50	43.50
8040	5" welded steel					137			137	151
8050	6"					109			109	120
8060	8"					126			126	139
8070	10" welded steel					176			176	194
8080	12" welded steel					228			228	251
8140	Tee, straight 1"					30.50			30.50	33.50
8150	1-1/2"					30.50			30.50	33.50
8160	2"					30.50			30.50	33.50
8170	2-1/2"					30.50			30.50	33.50
8180	3"					44			44	48
8200	4"					61			61	67
8210	5" welded steel					134			134	147
8220	6"					165			165	182
8230	8" welded steel					229			229	252
8240	10" welded steel					350			350	385
8250	12" welded steel					475			475	520
8340	Segmentally welded steel, painted									
8390	Wye 2"				Ea.	89			89	98
8400	2-1/2"					89			89	98
8410	3"					100			100	110
8430	4"					149			149	164
8440	5"					181			181	199
8450	6"					249			249	274
8460	8"					325			325	355
8470	10"					475			475	525
8480	12"					735			735	805
8540	Wye, lateral 2"					95.50			95.50	105
8550	2-1/2"					95.50			95.50	105
8560	3"					106			106	117
8580	4"					161			161	177
8590	5"					209			209	230
8600	6"					267			267	294
8610	8"					390			390	430
8620	10"					550			550	605
8630	12"					745			745	820
8690	Cross, 2"					101			101	111
8700	2-1/2"					101			101	111
8710	3"					120			120	132
8730	4"					165			165	182
8740	5"					235			235	258
8750	6"					310			310	340
8760	8"					395			395	435
8770	10"					580			580	640
8780	12"					835			835	920
8800	Tees, reducing 2" x 1"					60			60	66
8810	2" x 1-1/2"					60			60	66
8820	3" x 1"					60			60	66
8830	3" x 1-1/2"					60			60	66
8840	3" x 2"					60			60	66
8850	4" x 1"					87.50			87.50	96.50
8860	4" x 1-1/2"					87.50			87.50	96.50
8870	4" x 2"					87.50			87.50	96.50

15100 | Building Services Piping

15107 | Metal Pipe & Fittings

		CREW	DAILY OUTPUT	LABOR-HOURS	UNIT	2001 BARE COSTS MAT.	LABOR	EQUIP.	TOTAL	TOTAL INCL O&P
8880	4" x 2-1/2"				Ea.	90			90	99
8890	4" x 3"					90			90	99
8900	6" x 2"					146			146	160
8910	6" x 3"					154			154	169
8920	6" x 4"					154			154	169
8930	8" x 2"					187			187	206
8940	8" x 3"					196			196	216
8950	8" x 4"					200			200	220
8960	8" x 5"					205			205	226
8970	8" x 6"					205			205	226
8980	10" x 4"					294			294	325
8990	10" x 6"					310			310	345
9000	10" x 8"					310			310	345
9010	12" x 6"					455			455	500
9020	12" x 8"					455			455	500
9030	12" x 10"					470			470	520
9080	Adapter nipples 3" long									
9090	1"				Ea.	7.80			7.80	8.60
9100	1-1/2"					7.80			7.80	8.60
9110	2"					7.80			7.80	8.60
9120	2-1/2"					9.15			9.15	10.10
9130	3"					11.30			11.30	12.45
9140	4"					18.95			18.95	21
9150	6"					51			51	56.50
9180	Coupling, mechanical, plain end pipe to plain end pipe or fitting									
9190	1"	Q-1	29	.552	Ea.	24	17.10		41.10	52.50
9200	1-1/2"		28	.571		28.50	17.70		46.20	58
9210	2"		27	.593		28.50	18.35		46.85	59
9220	2-1/2"		26	.615		28.50	19.10		47.60	60.50
9230	3"		25	.640		42.50	19.85		62.35	76.50
9240	3-1/2"		24	.667		49	20.50		69.50	85
9250	4"		22	.727		49	22.50		71.50	88
9260	5"	Q-2	28	.857		70	27.50		97.50	119
9270	6"		24	1		85.50	32		117.50	143
9280	8"		19	1.263		152	40.50		192.50	229
9290	10"		16	1.500		197	48		245	290
9300	12"		12	2		248	64.50		312.50	370
9310	Outlets for precut holes through pipe wall									
9331	Strapless type, with gasket									
9332	4" to 8" pipe x 1/2"	1 Plum	13	.615	Ea.	28.50	21		49.50	63.50
9333	4" to 8" pipe x 3/4"		13	.615		30.50	21		51.50	65.50
9334	10" pipe and larger x 1/2"		11	.727		28.50	25		53.50	69.50
9335	10" pipe and larger x 3/4"		11	.727		30.50	25		55.50	71.50
9341	Thermometer wells with gasket									
9342	4" to 8" pipe, 6" stem	1 Plum	14	.571	Ea.	43.50	19.70		63.20	77.50
9343	8" pipe and larger, 6" stem	"	13	.615	"	43.50	21		64.50	80
9400	Mechanical joint ends for plain end pipe									
9410	Malleable iron, black									
9420	90° Elbows, 1-1/4"	Q-1	29	.552	Ea.	10.40	17.10		27.50	37.50
9430	1-1/2"		27	.593		11.40	18.35		29.75	40
9440	2"		24	.667		13.45	20.50		33.95	46
9490	Tee, reducing outlet									
9510	1-1/4" x 1/2"	Q-1	18	.889	Ea.	8.30	27.50		35.80	50.50
9520	1-1/4" x 3/4"		18	.889		8.30	27.50		35.80	50.50
9530	1-1/4" x 1"		18	.889		8.30	27.50		35.80	50.50
9540	1-1/2" x 1/2"		17	.941		8.80	29		37.80	53.50

15100 | Building Services Piping

15107 | Metal Pipe & Fittings

			CREW	DAILY OUTPUT	LABOR-HOURS	UNIT	MAT.	LABOR	EQUIP.	TOTAL	TOTAL INCL O&P	
660	9550	1-1/2" x 3/4"	Q-1	17	.941	Ea.	8.80	29		37.80	53.50	660
	9560	1-1/2" x 1"		17	.941		8.80	29		37.80	53.50	
	9570	2" x 1/2"		15	1.067		10.80	33		43.80	62	
	9580	2" x 3/4"		15	1.067		10.80	33		43.80	62	
	9590	2" x 1"		15	1.067		10.80	33		43.80	62	
	9640	Tee, reducing run and outlet										
	9660	1-1/4" x 1" x 1/2"	Q-1	18	.889	Ea.	8.90	27.50		36.40	51.50	
	9670	1-1/4" x 1" x 3/4"		18	.889		8.90	27.50		36.40	51.50	
	9680	1-1/4" x 1" x 1"		18	.889		8.90	27.50		36.40	51.50	
	9690	1-1/2" x 1-1/4" x 1/2"		17	.941		9.45	29		38.45	54.50	
	9700	1-1/2" x 1-1/4" x 3/4"		17	.941		9.45	29		38.45	54.50	
	9710	1-1/2" x 1-1/4" x 1"		17	.941		9.45	29		38.45	54.50	
	9720	2" x 1-1/2" x 1/2"		15	1.067		11.40	33		44.40	62.50	
	9730	2" x 1-1/2" x 3/4"		15	1.067		11.40	33		44.40	62.50	
	9740	2" x 1-1/2" x 1"		15	1.067		11.40	33		44.40	62.50	
	9790	Tee, outlet										
	9810	3" x 1-1/4"	Q-1	29	.552	Ea.	20	17.10		37.10	48	
	9820	3" x 1-1/2"		28	.571		20	17.70		37.70	48.50	
	9830	3" x 2"		26	.615		22.50	19.10		41.60	54	
	9840	4" x 1-1/4"		28	.571		25	17.70		42.70	54	
	9850	4" x 1-1/2"		26	.615		26.50	19.10		45.60	58	
	9860	4" x 2"		24	.667		26.50	20.50		47	60	
	9940	For galvanized fittings for plain end pipe, add					20%					
690	0010	**PIPE, GROOVED-JOINT STEEL FITTINGS & VALVES**										690
	0020	Pipe includes coupling & clevis type hanger 10' O.C.										
	0500	Schedule 10, black										
	0550	2" diameter	1 Plum	43	.186	L.F.	3.04	6.40		9.44	13	
	0560	2-1/2" diameter	Q-1	61	.262		3.80	8.15		11.95	16.50	
	0570	3" diameter		55	.291		4.35	9		13.35	18.40	
	0580	3-1/2" diameter		53	.302		5.60	9.35		14.95	20.50	
	0590	4" diameter		49	.327		5.70	10.10		15.80	21.50	
	0600	5" diameter		40	.400		9.15	12.40		21.55	29	
	0610	6" diameter	Q-2	46	.522		10.05	16.75		26.80	36.50	
	0620	8" diameter	"	41	.585		19.15	18.80		37.95	49.50	
	0700	To delete couplings & hangers, subtract										
	0710	2" diam. to 5" diam.					25%	20%				
	0720	6" diam. to 8" diam.					27%	15%				
	1000	Schedule 40, black										
	1040	3/4" diameter	1 Plum	71	.113	L.F.	1.92	3.88		5.80	7.95	
	1050	1" diameter		63	.127		2.03	4.37		6.40	8.85	
	1060	1-1/4" diameter		58	.138		2.51	4.75		7.26	9.90	
	1070	1-1/2" diameter		51	.157		2.79	5.40		8.19	11.20	
	1080	2" diameter		40	.200		3.21	6.90		10.11	13.95	
	1090	2-1/2" diameter	Q-1	57	.281		4.71	8.70		13.41	18.35	
	1100	3" diameter		50	.320		5.30	9.90		15.20	21	
	1110	4" diameter		45	.356		7.55	11		18.55	25	
	1120	5" diameter		37	.432		12.75	13.40		26.15	34	
	1130	6" diameter	Q-2	42	.571		16.55	18.35		34.90	45.50	
	1140	8" diameter		37	.649		25	21		46	59	
	1150	10" diameter		31	.774		37.50	25		62.50	78.50	
	1160	12" diameter		27	.889		49	28.50		77.50	97	
	1170	14" diameter		20	1.200		53.50	38.50		92	117	
	1180	16" diameter		17	1.412		57.50	45.50		103	132	
	1190	18" diameter		14	1.714		86.50	55		141.50	178	
	1200	20" diameter		12	2		113	64.50		177.50	221	

15100 | Building Services Piping

15107 | Metal Pipe & Fittings

		CREW	DAILY OUTPUT	LABOR-HOURS	UNIT	2001 BARE COSTS				TOTAL INCL O&P
						MAT.	LABOR	EQUIP.	TOTAL	
1210	24" diameter	Q-2	10	2.400	L.F.	124	77		201	252
1740	To delete coupling & hanger, subtract									
1750	3/4" diam. to 2" diam.					65%	27%			
1760	2-1/2" diam. to 5" diam.					41%	18%			
1770	6" diam. to 12" diam.					31%	13%			
1780	14" diam. to 24" diam.					35%	10%			
1800	Galvanized									
1840	3/4" diameter	1 Plum	71	.113	L.F.	2.11	3.88		5.99	8.15
1850	1" diameter		63	.127		2.17	4.37		6.54	9
1860	1-1/4" diameter		58	.138		2.69	4.75		7.44	10.10
1870	1-1/2" diameter		51	.157		3.01	5.40		8.41	11.45
1880	2" diameter		40	.200		3.51	6.90		10.41	14.25
1890	2-1/2" diameter	Q-1	57	.281		4.87	8.70		13.57	18.50
1900	3" diameter		50	.320		5.90	9.90		15.80	21.50
1910	4" diameter		45	.356		8.20	11		19.20	25.50
1920	5" diameter		37	.432		15.65	13.40		29.05	37.50
1930	6" diameter	Q-2	42	.571		20.50	18.35		38.85	50
1940	8" diameter		37	.649		30.50	21		51.50	65.50
1950	10" diameter		31	.774		46	25		71	88
1960	12" diameter		27	.889		61	28.50		89.50	110
2540	To delete coupling & hanger, subtract									
2550	3/4" diam. to 2" diam.					36%	27%			
2560	2-1/2" diam. to 5" diam.					19%	18%			
2570	6" diam. to 12" diam.					14%	13%			
2600	Schedule 80, black									
2610	3/4" diameter	1 Plum	65	.123	L.F.	2.15	4.24		6.39	8.75
2650	1" diameter		61	.131		2.31	4.52		6.83	9.35
2660	1-1/4" diameter		55	.145		2.92	5		7.92	10.75
2670	1-1/2" diameter		49	.163		3.30	5.60		8.90	12.10
2680	2" diameter		38	.211		3.96	7.25		11.21	15.30
2690	2-1/2" diameter	Q-1	54	.296		5.45	9.20		14.65	19.85
2700	3" diameter		48	.333		6.75	10.35		17.10	23
2710	4" diameter		44	.364		9.60	11.25		20.85	27.50
2720	5" diameter		35	.457		17.65	14.15		31.80	41
2730	6" diameter	Q-2	40	.600		27.50	19.30		46.80	59.50
2740	8" diameter		35	.686		49.50	22		71.50	87.50
2750	10" diameter		29	.828		77.50	26.50		104	125
2760	12" diameter		24	1		86	32		118	143
3240	To delete coupling & hanger, subtract									
3250	3/4" diam. to 2" diam.					30%	25%			
3260	2-1/2" diam. to 5" diam.					14%	17%			
3270	6" diam. to 12" diam.					12%	12%			
3300	Galvanized									
3310	3/4" diameter	1 Plum	65	.123	L.F.	2.40	4.24		6.64	9.05
3350	1" diameter		61	.131		2.50	4.52		7.02	9.55
3360	1-1/4" diameter		55	.145		3.19	5		8.19	11.05
3370	1-1/2" diameter		46	.174		3.63	6		9.63	13.05
3380	2" diameter		38	.211		4.35	7.25		11.60	15.75
3390	2-1/2" diameter	Q-1	54	.296		6	9.20		15.20	20.50
3400	3" diameter		48	.333		7.60	10.35		17.95	24
3410	4" diameter		44	.364		10.80	11.25		22.05	29
3420	5" diameter		35	.457		21	14.15		35.15	44.50
3430	6" diameter	Q-2	40	.600		27.50	19.30		46.80	59.50
3440	8" diameter		35	.686		47.50	22		69.50	85.50
3450	10" diameter		29	.828		75	26.50		101.50	123
3460	12" diameter		24	1		83.50	32		115.50	140

15100 | Building Services Piping

15107 | Metal Pipe & Fittings

		CREW	DAILY OUTPUT	LABOR-HOURS	UNIT	2001 BARE COSTS MAT.	LABOR	EQUIP.	TOTAL	TOTAL INCL O&P
3920	To delete coupling & hanger, subtract									
3930	3/4" diam. to 2" diam.					30%	25%			
3940	2-1/2" diam. to 5" diam.					15%	17%			
3950	6" diam. to 12" diam.					11%	12%			
3990	Fittings: cplg. & labor required at joints not incl. in fitting									
3994	price. Add 1 per joint for installed price.									
4000	Elbow, 90° or 45°, painted									
4030	3/4" diameter	1 Plum	50	.160	Ea.	13.65	5.50		19.15	23.50
4040	1" diameter		50	.160		13.65	5.50		19.15	23.50
4050	1-1/4" diameter		40	.200		13.65	6.90		20.55	25.50
4060	1-1/2" diameter		33	.242		13.65	8.35		22	27.50
4070	2" diameter		25	.320		13.65	11		24.65	31.50
4080	2-1/2" diameter	Q-1	40	.400		13.65	12.40		26.05	33.50
4090	3" diameter		33	.485		24.50	15.05		39.55	49
4100	4" diameter		25	.640		26.50	19.85		46.35	59
4110	5" diameter		20	.800		64	25		89	108
4120	6" diameter	Q-2	25	.960		75	31		106	129
4130	8" diameter		21	1.143		157	36.50		193.50	229
4140	10" diameter		18	1.333		286	43		329	380
4150	12" diameter		15	1.600		455	51.50		506.50	585
4170	14" diameter		12	2		820	64.50		884.50	995
4180	16" diameter		11	2.182		1,075	70		1,145	1,275
4190	18" diameter	Q-3	15	2.133		1,350	70		1,420	1,575
4200	20" diameter		13	2.462		1,775	81		1,856	2,075
4210	24" diameter		11	2.909		2,575	95.50		2,670.50	2,975
4250	For galvanized elbows, add					26%				
4690	Tee, painted									
4700	3/4" diameter	1 Plum	38	.211	Ea.	21	7.25		28.25	34
4740	1" diameter		33	.242		21	8.35		29.35	35.50
4750	1-1/4" diameter		27	.296		21	10.20		31.20	38.50
4760	1-1/2" diameter		22	.364		21	12.55		33.55	42
4770	2" diameter		17	.471		21	16.20		37.20	47.50
4780	2-1/2" diameter	Q-1	27	.593		21	18.35		39.35	50.50
4790	3" diameter		22	.727		29.50	22.50		52	66.50
4800	4" diameter		17	.941		45	29		74	93.50
4810	5" diameter		13	1.231		105	38		143	174
4820	6" diameter	Q-2	17	1.412		122	45.50		167.50	203
4830	8" diameter		14	1.714		268	55		323	375
4840	10" diameter		12	2		555	64.50		619.50	705
4850	12" diameter		10	2.400		775	77		852	970
4851	14" diameter		9	2.667		820	85.50		905.50	1,025
4852	16" diameter		8	3		930	96.50		1,026.50	1,175
4853	18" diameter	Q-3	11	2.909		1,150	95.50		1,245.50	1,425
4854	20" diameter		10	3.200		1,675	105		1,780	1,975
4855	24" diameter		8	4		2,525	131		2,656	3,000
4900	For galvanized tees, add					24%				
4906	Couplings, rigid style, painted									
4908	1" diameter	1 Plum	100	.080	Ea.	9.90	2.76		12.66	15.05
4909	1-1/4" diameter		100	.080		9.90	2.76		12.66	15.05
4910	1-1/2" diameter		67	.119		9.90	4.11		14.01	17.10
4912	2" diameter		50	.160		10.15	5.50		15.65	19.50
4914	2-1/2" diameter	Q-1	80	.200		11.70	6.20		17.90	22
4916	3" diameter		67	.239		13.65	7.40		21.05	26.50
4918	4" diameter		50	.320		19.35	9.90		29.25	36.50
4920	5" diameter		40	.400		25	12.40		37.40	46
4922	6" diameter	Q-2	50	.480		33.50	15.45		48.95	60.50

15100 | Building Services Piping

15107 | Metal Pipe & Fittings

		CREW	DAILY OUTPUT	LABOR-HOURS	UNIT	MAT.	LABOR	EQUIP.	TOTAL	TOTAL INCL O&P
4924	8" diameter	Q-2	42	.571	Ea.	53	18.35		71.35	85.50
4926	10" diameter		35	.686		95	22		117	138
4928	12" diameter		32	.750		106	24		130	154
4930	14" diameter		24	1		139	32		171	202
4931	16" diameter		20	1.200		181	38.50		219.50	257
4932	18" diameter		18	1.333		209	43		252	295
4933	20" diameter	▼	16	1.500	▼	249	48		297	345
4934	24" diameter	Q-9	13	1.231		365	37		402	465
4940	Flexible, standard, painted									
4950	3/4" diameter	1 Plum	100	.080	Ea.	7.10	2.76		9.86	12
4960	1" diameter		100	.080		7.10	2.76		9.86	12
4970	1-1/4" diameter		80	.100		9.45	3.45		12.90	15.60
4980	1-1/2" diameter		67	.119		10.35	4.11		14.46	17.55
4990	2" diameter	▼	50	.160		10.90	5.50		16.40	20.50
5000	2-1/2" diameter	Q-1	80	.200		13.05	6.20		19.25	24
5010	3" diameter		67	.239		14.45	7.40		21.85	27
5020	3-1/2" diameter		57	.281		21	8.70		29.70	36
5030	4" diameter		50	.320		21	9.90		30.90	38
5040	5" diameter	▼	40	.400		32.50	12.40		44.90	54
5050	6" diameter	Q-2	50	.480		38	15.45		53.45	65.50
5070	8" diameter		42	.571		62	18.35		80.35	96
5090	10" diameter		35	.686		103	22		125	148
5110	12" diameter		32	.750		118	24		142	166
5120	14" diameter		24	1		143	32		175	206
5130	16" diameter		20	1.200		188	38.50		226.50	264
5140	18" diameter		18	1.333		220	43		263	305
5150	20" diameter		16	1.500		345	48		393	455
5160	24" diameter	▼	13	1.846	▼	380	59.50		439.50	505
5176	Lightweight style, painted									
5178	1-1/2" diameter	1 Plum	67	.119	Ea.	9	4.11		13.11	16.10
5180	2" diameter	"	50	.160		9.20	5.50		14.70	18.45
5182	2-1/2" diameter	Q-1	80	.200		10.70	6.20		16.90	21
5184	3" diameter		67	.239		12.45	7.40		19.85	25
5186	3-1/2" diameter		57	.281		17.60	8.70		26.30	32.50
5188	4" diameter		50	.320		17.60	9.90		27.50	34.50
5190	5" diameter	▼	40	.400		25.50	12.40		37.90	46.50
5192	6" diameter	Q-2	50	.480		30.50	15.45		45.95	57
5194	8" diameter		42	.571		48	18.35		66.35	80.50
5196	10" diameter		35	.686		129	22		151	176
5198	12" diameter	▼	32	.750		144	24		168	196
5200	For galvanized couplings, add				▼	33%				
5220	Tee, reducing, painted									
5225	2" x 1-1/2" diameter	Q-1	38	.421	Ea.	45.50	13.05		58.55	69.50
5226	2-1/2" x 2" diameter		28	.571		45.50	17.70		63.20	76.50
5227	3" x 2-1/2" diameter		23	.696		40	21.50		61.50	76.50
5228	4" x 3" diameter		18	.889		54	27.50		81.50	101
5229	5" x 4" diameter	▼	15	1.067		116	33		149	178
5230	6" x 4" diameter	Q-2	18	1.333		128	43		171	206
5231	8" x 6" diameter		15	1.600		268	51.50		319.50	370
5232	10" x 8" diameter		13	1.846		350	59.50		409.50	475
5233	12" x 10" diameter		11	2.182		535	70		605	695
5234	14" x 12" diameter		10	2.400		600	77		677	775
5235	16" x 12" diameter	▼	9	2.667		740	85.50		825.50	945
5236	18" x 12" diameter	Q-3	12	2.667		890	87.50		977.50	1,100
5237	18" x 16" diameter		11	2.909		1,125	95.50		1,220.50	1,375
5238	20" x 16" diameter	▼	10	3.200	▼	1,475	105		1,580	1,775

Important: See the Reference Section for critical supporting data - Reference Nos., Crews, & City Cost Indexes

15100 | Building Services Piping

15107 | Metal Pipe & Fittings

		CREW	DAILY OUTPUT	LABOR-HOURS	UNIT	2001 BARE COSTS MAT.	LABOR	EQUIP.	TOTAL	TOTAL INCL O&P
5239	24" x 20" diameter	Q-3	9	3.556	Ea.	2,325	117		2,442	2,725
5240	Reducer, concentric, painted									
5241	2-1/2" x 2" diameter	Q-1	43	.372	Ea.	15.85	11.55		27.40	35
5242	3" x 2-1/2" diameter		35	.457		19.20	14.15		33.35	42.50
5243	4" x 3" diameter		29	.552		23	17.10		40.10	51.50
5244	5" x 4" diameter	↓	22	.727		32.50	22.50		55	69.50
5245	6" x 4" diameter	Q-2	26	.923		37.50	29.50		67	86.50
5246	8" x 6" diameter		23	1.043		98.50	33.50		132	159
5247	10" x 8" diameter		20	1.200		201	38.50		239.50	279
5248	12" x 10" diameter	↓	16	1.500	↓	360	48		408	470
5255	Eccentric, painted									
5256	2-1/2" x 2" diameter	Q-1	42	.381	Ea.	33.50	11.80		45.30	55
5257	3" x 2-1/2" diameter		34	.471		38.50	14.60		53.10	64.50
5258	4" x 3" diameter		28	.571		47	17.70		64.70	78.50
5259	5" x 4" diameter	↓	21	.762		64	23.50		87.50	106
5260	6" x 4" diameter	Q-2	25	.960		75	31		106	129
5261	8" x 6" diameter		22	1.091		152	35		187	220
5262	10" x 8" diameter		19	1.263		415	40.50		455.50	520
5263	12" x 10" diameter	↓	15	1.600	↓	575	51.50		626.50	710
5270	Coupling, reducing, painted									
5272	2" x 1-1/2" diameter	1 Plum	52	.154	Ea.	16	5.30		21.30	25.50
5274	2-1/2" x 2" diameter	Q-1	82	.195		21	6.05		27.05	32
5276	3" x 2" diameter		69	.232		24	7.20		31.20	37.50
5278	4" x 2" diameter		52	.308		38	9.55		47.55	56.50
5280	5" x 4" diameter	↓	42	.381		43	11.80		54.80	65.50
5282	6" x 4" diameter	Q-2	52	.462		65.50	14.85		80.35	94.50
5284	8" x 6" diameter	"	44	.545	↓	98	17.55		115.55	135
5290	Outlet coupling, painted									
5294	1-1/2" x 1" pipe size	1 Plum	65	.123	Ea.	19.95	4.24		24.19	28.50
5296	2" x 1" pipe size	"	48	.167		20.50	5.75		26.25	31
5298	2-1/2" x 1" pipe size	Q-1	78	.205		32	6.35		38.35	44.50
5300	2-1/2" x 1" pipe size	1 Plum	70	.114		36	3.94		39.94	45.50
5302	3" x 1" pipe size	Q-1	65	.246		30.50	7.65		38.15	45
5304	4" x 3/4" pipe size		48	.333		45	10.35		55.35	65.50
5306	4" x 1-1/2" pipe size	↓	46	.348		64.50	10.80		75.30	87.50
5308	6" x 1-1/2" pipe size	Q-2	44	.545	↓	91	17.55		108.55	127
5750	Flange, w/groove gasket, black steel									
5752	See 15107-660-0620 for gasket & bolt set									
5760	ANSI class 125 and 150, painted									
5780	2" pipe size	1 Plum	23	.348	Ea.	45.50	12		57.50	68
5790	2-1/2" pipe size	Q-1	37	.432		56.50	13.40		69.90	82
5800	3" pipe size		31	.516		61	16		77	91
5820	4" pipe size		23	.696		81.50	21.50		103	123
5830	5" pipe size	↓	19	.842		95	26		121	144
5840	6" pipe size	Q-2	23	1.043		103	33.50		136.50	165
5850	8" pipe size		17	1.412		117	45.50		162.50	197
5860	10" pipe size		14	1.714		185	55		240	287
5870	12" pipe size		12	2		242	64.50		306.50	365
5880	14" pipe size		10	2.400		555	77		632	725
5890	16" pipe size		9	2.667		645	85.50		730.50	840
5900	18" pipe size		6	4		795	129		924	1,075
5910	20" pipe size		5	4.800		955	154		1,109	1,275
5920	24" pipe size	↓	4.50	5.333	↓	1,225	171		1,396	1,600
5940	ANSI class 350, painted									
5946	2" pipe size	1 Plum	23	.348	Ea.	57	12		69	80.50
5948	2-1/2" pipe size	Q-1	37	.432	↓	66	13.40		79.40	92.50

15100 | Building Services Piping

15107 | Metal Pipe & Fittings

		CREW	DAILY OUTPUT	LABOR-HOURS	UNIT	2001 BARE COSTS MAT.	LABOR	EQUIP.	TOTAL	TOTAL INCL O&P
5950	3" pipe size	Q-1	31	.516	Ea.	90	16		106	123
5952	4" pipe size		23	.696		120	21.50		141.50	165
5954	5" pipe size	↓	19	.842		137	26		163	190
5956	6" pipe size	Q-2	23	1.043		159	33.50		192.50	226
5958	8" pipe size		17	1.412		183	45.50		228.50	270
5960	10" pipe size	↓	14	1.714		291	55		346	405
5962	12" pipe size	1 Plum	12	.667	↓	310	23		333	375
6100	Coupling, for PVC plastic pipe									
6110	2" diameter	1 Plum	50	.160	Ea.	12.20	5.50		17.70	21.50
6112	2-1/2" diameter	Q-1	80	.200		17.10	6.20		23.30	28
6114	3" diameter		67	.239		21	7.40		28.40	34
6116	4" diameter	↓	50	.320		27	9.90		36.90	45
6118	6" diameter	Q-2	50	.480		45	15.45		60.45	73
6120	8" diameter		42	.571		74	18.35		92.35	109
6122	10" diameter		35	.686		121	22		143	167
6124	12" diameter	↓	32	.750	↓	153	24		177	206
7400	Suction diffuser									
7402	Grooved end inlet x flanged outlet									
7410	3" x 3"	Q-1	50	.320	Ea.	480	9.90		489.90	540
7412	4" x 4"		38	.421		650	13.05		663.05	735
7414	5" x 5"	↓	30	.533		760	16.55		776.55	860
7416	6" x 6"	Q-2	38	.632		955	20.50		975.50	1,075
7418	8" x 8"		27	.889		1,775	28.50		1,803.50	2,000
7420	10" x 10"		20	1.200		2,425	38.50		2,463.50	2,725
7422	12" x 12"		16	1.500		4,000	48		4,048	4,475
7424	14" x 14"		15	1.600		5,025	51.50		5,076.50	5,625
7426	16" x 14"	↓	14	1.714	↓	5,175	55		5,230	5,775
7500	Strainer, tee type, painted									
7506	2" pipe size	1 Plum	38	.211	Ea.	298	7.25		305.25	340
7508	2-1/2" pipe size	Q-1	62	.258		315	8		323	355
7510	3" pipe size		50	.320		350	9.90		359.90	400
7512	4" pipe size		38	.421		400	13.05		413.05	460
7514	5" pipe size	↓	30	.533		575	16.55		591.55	660
7516	6" pipe size	Q-2	38	.632		625	20.50		645.50	720
7518	8" pipe size		27	.889		960	28.50		988.50	1,100
7520	10" pipe size		20	1.200		1,400	38.50		1,438.50	1,600
7522	12" pipe size		16	1.500		1,825	48		1,873	2,075
7524	14" pipe size		15	1.600		6,475	51.50		6,526.50	7,200
7526	16" pipe size	↓	14	1.714	↓	8,050	55		8,105	8,925
7570	Expansion joint, max. 3" travel									
7572	2" diameter	1 Plum	38	.211	Ea.	245	7.25		252.25	280
7574	3" diameter	Q-1	50	.320		288	9.90		297.90	330
7576	4" diameter	"	38	.421		360	13.05		373.05	415
7578	6" diameter	Q-2	38	.632	↓	625	20.50		645.50	720
7800	Ball valve w/handle, carbon steel trim									
7810	1-1/2" pipe size	1 Plum	50	.160	Ea.	73	5.50		78.50	88.50
7812	2" pipe size	"	38	.211		100	7.25		107.25	121
7814	2-1/2" pipe size	Q-1	62	.258		207	8		215	240
7816	3" pipe size		50	.320		325	9.90		334.90	375
7818	4" pipe size	↓	38	.421		505	13.05		518.05	575
7820	6" pipe size	Q-2	30	.800	↓	1,475	25.50		1,500.50	1,675
7830	With gear operator									
7834	2-1/2" pipe size	Q-1	62	.258	Ea.	365	8		373	410
7836	3" pipe size		50	.320		480	9.90		489.90	545
7838	4" pipe size	↓	38	.421		650	13.05		663.05	735
7840	6" pipe size	Q-2	30	.800	↓	1,650	25.50		1,675.50	1,875

Important: See the Reference Section for critical supporting data - Reference Nos., Crews, & City Cost Indexes

15100 | Building Services Piping

15107 | Metal Pipe & Fittings

		CREW	DAILY OUTPUT	LABOR-HOURS	UNIT	MAT.	LABOR	EQUIP.	TOTAL	TOTAL INCL O&P
7870	Check valve									
7874	2-1/2" pipe size	Q-1	62	.258	Ea.	131	8		139	156
7876	3" pipe size		50	.320		154	9.90		163.90	185
7878	4" pipe size		38	.421		163	13.05		176.05	199
7880	5" pipe size		30	.533		272	16.55		288.55	325
7882	6" pipe size	Q-2	38	.632		320	20.50		340.50	385
7884	8" pipe size		27	.889		440	28.50		468.50	530
7886	10" pipe size		20	1.200		1,275	38.50		1,313.50	1,450
7888	12" pipe size		16	1.500		1,500	48		1,548	1,725
7900	Plug valve, balancing, w/lever operator									
7906	3" pipe size	Q-1	50	.320	Ea.	268	9.90		277.90	310
7908	4" pipe size	"	38	.421		295	13.05		308.05	345
7909	6" pipe size	Q-2	30	.800		530	25.50		555.50	620
7916	With gear operator									
7920	3" pipe size	Q-1	50	.320	Ea.	485	9.90		494.90	550
7922	4" pipe size	"	38	.421		510	13.05		523.05	580
7924	6" pipe size	Q-2	38	.632		745	20.50		765.50	850
7926	8" pipe size		27	.889		995	28.50		1,023.50	1,150
7928	10" pipe size		20	1.200		1,575	38.50		1,613.50	1,775
7930	12" pipe size		16	1.500		2,275	48		2,323	2,600
8000	Butterfly valve, 2 position handle, with standard trim									
8010	1-1/2" pipe size	1 Plum	50	.160	Ea.	115	5.50		120.50	135
8020	2" pipe size	"	38	.211		115	7.25		122.25	138
8030	3" pipe size	Q-1	50	.320		166	9.90		175.90	197
8050	4" pipe size	"	38	.421		182	13.05		195.05	221
8070	6" pipe size	Q-2	38	.632		370	20.50		390.50	435
8080	8" pipe size		27	.889		710	28.50		738.50	830
8090	10" pipe size		20	1.200		1,125	38.50		1,163.50	1,275
8200	With stainless steel trim									
8240	1-1/2" pipe size	1 Plum	50	.160	Ea.	146	5.50		151.50	169
8250	2" pipe size	"	38	.211		146	7.25		153.25	172
8270	3" pipe size	Q-1	50	.320		197	9.90		206.90	232
8280	4" pipe size	"	38	.421		214	13.05		227.05	255
8300	6" pipe size	Q-2	38	.632		400	20.50		420.50	470
8310	8" pipe size		27	.889		825	28.50		853.50	950
8320	10" pipe size		20	1.200		1,225	38.50		1,263.50	1,400
8322	12" pipe size		16	1.500		1,375	48		1,423	1,575
8324	14" pipe size		15	1.600		1,800	51.50		1,851.50	2,075
8326	16" pipe size		14	1.714		2,475	55		2,530	2,800
8328	18" pipe size	Q-3	12	2.667		3,050	87.50		3,137.50	3,475
8330	20" pipe size		11	2.909		4,125	95.50		4,220.50	4,675
8332	24" pipe size		10	3.200		5,250	105		5,355	5,925
8336	Note: sizes 8" up w/manual gear operator									
9000	Cut one groove, labor									
9010	3/4" pipe size	Q-1	152	.105	Ea.		3.26		3.26	4.93
9020	1" pipe size		140	.114			3.54		3.54	5.35
9030	1-1/4" pipe size		124	.129			4		4	6.05
9040	1-1/2" pipe size		114	.140			4.35		4.35	6.55
9050	2" pipe size		104	.154			4.77		4.77	7.20
9060	2-1/2" pipe size		96	.167			5.15		5.15	7.80
9070	3" pipe size		88	.182			5.65		5.65	8.50
9080	3-1/2" pipe size		83	.193			6		6	9
9090	4" pipe size		78	.205			6.35		6.35	9.60
9100	5" pipe size		72	.222			6.90		6.90	10.40
9110	6" pipe size		70	.229			7.10		7.10	10.70
9120	8" pipe size		54	.296			9.20		9.20	13.85

15100 | Building Services Piping

15107 | Metal Pipe & Fittings

		CREW	DAILY OUTPUT	LABOR-HOURS	UNIT	2001 BARE COSTS				TOTAL INCL O&P
						MAT.	LABOR	EQUIP.	TOTAL	
9130	10" pipe size	Q-1	38	.421	Ea.		13.05		13.05	19.70
9140	12" pipe size		30	.533			16.55		16.55	25
9150	14" pipe size		20	.800			25		25	37.50
9160	16" pipe size		19	.842			26		26	39.50
9170	18" pipe size		18	.889			27.50		27.50	41.50
9180	20" pipe size		17	.941			29		29	44
9190	24" pipe size		15	1.067			33		33	50
9210	Roll one groove									
9220	3/4" pipe size	Q-1	266	.060	Ea.		1.86		1.86	2.82
9230	1" pipe size		228	.070			2.18		2.18	3.28
9240	1-1/4" pipe size		200	.080			2.48		2.48	3.74
9250	1-1/2" pipe size		178	.090			2.79		2.79	4.21
9260	2" pipe size		116	.138			4.28		4.28	6.45
9270	2-1/2" pipe size		110	.145			4.51		4.51	6.80
9280	3" pipe size		100	.160			4.96		4.96	7.50
9290	3-1/2" pipe size		94	.170			5.30		5.30	7.95
9300	4" pipe size		86	.186			5.75		5.75	8.70
9310	5" pipe size		84	.190			5.90		5.90	8.90
9320	6" pipe size		80	.200			6.20		6.20	9.35
9330	8" pipe size		66	.242			7.50		7.50	11.35
9340	10" pipe size		58	.276			8.55		8.55	12.90
9350	12" pipe size		46	.348			10.80		10.80	16.30
9360	14" pipe size		30	.533			16.55		16.55	25
9370	16" pipe size		28	.571			17.70		17.70	26.50
9380	18" pipe size		27	.593			18.35		18.35	27.50
9390	20" pipe size		25	.640			19.85		19.85	30
9400	24" pipe size		23	.696			21.50		21.50	32.50
0010	**PIPE, STAINLESS STEEL**									
0020	Welded, with clevis type hangers 10' O.C.									
0500	Schedule 5, type 304									
0540	1/2" diameter	Q-15	128	.125	L.F.	4.67	3.88	.79	9.34	11.85
0550	3/4" diameter		116	.138		5.25	4.28	.87	10.40	13.20
0560	1" diameter		103	.155		6.40	4.82	.98	12.20	15.40
0570	1-1/4" diameter		93	.172		7.30	5.35	1.08	13.73	17.30
0580	1-1/2" diameter		85	.188		8.40	5.85	1.18	15.43	19.30
0590	2" diameter		69	.232		10.05	7.20	1.46	18.71	23.50
0600	2-1/2" diameter		53	.302		14.05	9.35	1.90	25.30	31.50
0610	3" diameter		48	.333		16.90	10.35	2.10	29.35	36.50
0620	4" diameter		44	.364		21.50	11.25	2.29	35.04	43
0630	5" diameter		36	.444		43	13.80	2.80	59.60	71.50
0640	6" diameter	Q-16	42	.571		40	18.35	2.39	60.74	74
0650	8" diameter		34	.706		61	22.50	2.96	86.46	105
0660	10" diameter		26	.923		85.50	29.50	3.87	118.87	143
0670	12" diameter		21	1.143		112	36.50	4.79	153.29	184
0700	To delete hangers, subtract									
0710	1/2" diam. to 1-1/2" diam.					8%	19%			
0720	2" diam. to 5" diam.					4%	9%			
0730	6" diam. to 12" diam.					3%	4%			
0750	For small quantities, add				L.F.	10%				
1250	Schedule 5, type 316									
1290	1/2" diameter	Q-15	128	.125	L.F.	5.85	3.88	.79	10.52	13.15
1300	3/4" diameter		116	.138		6.75	4.28	.87	11.90	14.80
1310	1" diameter		103	.155		7.65	4.82	.98	13.45	16.80
1320	1-1/4" diameter		93	.172		9.15	5.35	1.08	15.58	19.35
1330	1-1/2" diameter		85	.188		10.40	5.85	1.18	17.43	21.50

Important: See the Reference Section for critical supporting data - Reference Nos., Crews, & City Cost Indexes

15100 | Building Services Piping

15107 | Metal Pipe & Fittings

		CREW	DAILY OUTPUT	LABOR-HOURS	UNIT	MAT.	LABOR	EQUIP.	TOTAL	TOTAL INCL O&P
1340	2" diameter	Q-15	69	.232	L.F.	12.60	7.20	1.46	21.26	26.50
1350	2-1/2" diameter		53	.302		18.10	9.35	1.90	29.35	36
1360	3" diameter		48	.333		23	10.35	2.10	35.45	43
1370	4" diameter		44	.364		28.50	11.25	2.29	42.04	50.50
1380	5" diameter		36	.444		53.50	13.80	2.80	70.10	83
1390	6" diameter	Q-16	42	.571		52.50	18.35	2.39	73.24	87.50
1400	8" diameter		34	.706		78.50	22.50	2.96	103.96	124
1410	10" diameter		26	.923		113	29.50	3.87	146.37	173
1420	12" diameter		21	1.143		146	36.50	4.79	187.29	221
1490	For small quantities, add					10%				
1940	To delete hanger, subtract									
1950	1/2" diam. to 1-1/2" diam.					5%	19%			
1960	2" diam. to 5" diam.					3%	9%			
1970	6" diam. to 12" diam.					2%	4%			
2000	Schedule 10, type 304									
2040	1/4" diameter	Q-15	131	.122	L.F.	3.69	3.79	.77	8.25	10.60
2050	3/8" diameter		128	.125		3.89	3.88	.79	8.56	11
2060	1/2" diameter		125	.128		4.77	3.97	.81	9.55	12.15
2070	3/4" diameter		113	.142		5.50	4.39	.89	10.78	13.70
2080	1" diameter		100	.160		7.85	4.96	1.01	13.82	17.25
2090	1-1/4" diameter		91	.176		9.45	5.45	1.11	16.01	19.85
2100	1-1/2" diameter		83	.193		10.40	6	1.21	17.61	21.50
2110	2" diameter		67	.239		12.85	7.40	1.50	21.75	27
2120	2-1/2" diameter		51	.314		16	9.75	1.97	27.72	34.50
2130	3" diameter		46	.348		19.60	10.80	2.19	32.59	40
2140	4" diameter		42	.381		25.50	11.80	2.40	39.70	48.50
2150	5" diameter		35	.457		34	14.15	2.88	51.03	62
2160	6" diameter	Q-16	40	.600		40	19.30	2.51	61.81	76
2170	8" diameter		33	.727		61	23.50	3.05	87.55	106
2180	10" diameter		25	.960		85.50	31	4.02	120.52	145
2190	12" diameter		21	1.143		107	36.50	4.79	148.29	179
2250	For small quantities, add					10%				
2650	To delete hanger, subtract									
2660	1/4" diam. to 3/4" diam.					9%	22%			
2670	1" diam. to 2" diam.					4%	15%			
2680	2-1/2" diam. to 5" diam.					3%	8%			
2690	6" diam. to 12" diam.					3%	4%			
2750	Schedule 10, type 316									
2790	1/4" diameter	Q-15	131	.122	L.F.	4.14	3.79	.77	8.70	11.10
2800	3/8" diameter		128	.125		4.59	3.88	.79	9.26	11.75
2810	1/2" diameter		125	.128		5.85	3.97	.81	10.63	13.30
2820	3/4" diameter		113	.142		6.75	4.39	.89	12.03	15.10
2830	1" diameter		100	.160		8.85	4.96	1.01	14.82	18.35
2840	1-1/4" diameter		91	.176		11.85	5.45	1.11	18.41	22.50
2850	1-1/2" diameter		83	.193		13.30	6	1.21	20.51	25
2860	2" diameter		67	.239		16.35	7.40	1.50	25.25	31
2870	2-1/2" diameter		51	.314		21	9.75	1.97	32.72	40.50
2880	3" diameter		46	.348		27	10.80	2.19	39.99	48
2890	4" diameter		42	.381		32.50	11.80	2.40	46.70	56.50
2900	5" diameter		35	.457		48.50	14.15	2.88	65.53	77.50
2910	6" diameter	Q-16	40	.600		52	19.30	2.51	73.81	89.50
2920	8" diameter		33	.727		85.50	23.50	3.05	112.05	133
2930	10" diameter		25	.960		113	31	4.02	148.02	176
2940	12" diameter		21	1.143		142	36.50	4.79	183.29	217
2990	For small quantities, add					10%				
3430	To delete hanger, subtract									

15100 | Building Services Piping

15107 | Metal Pipe & Fittings

		CREW	DAILY OUTPUT	LABOR-HOURS	UNIT	2001 BARE COSTS MAT.	LABOR	EQUIP.	TOTAL	TOTAL INCL O&P
3440	1/4" diam. to 3/4" diam.					6%	22%			
3450	1" diam. to 2" diam.					3%	15%			
3460	2-1/2" diam. to 5" diam.					2%	8%			
3470	6" diam. to 12" diam.					2%	4%			
3500	Threaded, couplings and hangers 10' O.C.									
3520	Schedule 40, type 304									
3540	1/4" diameter	1 Plum	54	.148	L.F.	4.87	5.10		9.97	13.05
3550	3/8" diameter		53	.151		5.55	5.20		10.75	13.95
3560	1/2" diameter		52	.154		6.55	5.30		11.85	15.25
3570	3/4" diameter		51	.157		8.50	5.40		13.90	17.50
3580	1" diameter		45	.178		10.70	6.10		16.80	21
3590	1-1/4" diameter	Q-1	76	.211		12.60	6.55		19.15	23.50
3600	1-1/2" diameter		69	.232		14.70	7.20		21.90	27
3610	2" diameter		57	.281		19.35	8.70		28.05	34.50
3620	2-1/2" diameter		44	.364		31	11.25		42.25	51
3630	3" diameter		38	.421		40	13.05		53.05	63.50
3640	4" diameter	Q-2	51	.471		55	15.15		70.15	84
3740	For small quantities, add					10%				
4200	To delete couplings & hangers, subtract									
4210	1/4" diam. to 3/4" diam.					15%	56%			
4220	1" diam. to 2" diam.					18%	49%			
4230	2-1/2" diam. to 4" diam.					34%	40%			
4250	Schedule 40, type 316									
4290	1/4" diameter	1 Plum	54	.148	L.F.	5.80	5.10		10.90	14.05
4300	3/8" diameter		53	.151		7.50	5.20		12.70	16.10
4310	1/2" diameter		52	.154		8.25	5.30		13.55	17.05
4320	3/4" diameter		51	.157		10.35	5.40		15.75	19.50
4330	1" diameter		45	.178		13.60	6.10		19.70	24
4340	1-1/4" diameter	Q-1	76	.211		16	6.55		22.55	27.50
4350	1-1/2" diameter		69	.232		18.75	7.20		25.95	31.50
4360	2" diameter		57	.281		25	8.70		33.70	40.50
4370	2-1/2" diameter		44	.364		40.50	11.25		51.75	61.50
4380	3" diameter		38	.421		54.50	13.05		67.55	79.50
4390	4" diameter	Q-2	51	.471		73.50	15.15		88.65	104
4490	For small quantities, add					10%				
4900	To delete couplings & hangers, subtract									
4910	1/4" diam. to 3/4" diam.					12%	56%			
4920	1" diam. to 2" diam.					14%	49%			
4930	2-1/2" diam. to 4" diam.					27%	40%			
5000	Schedule 80, type 304									
5040	1/4" diameter	1 Plum	53	.151	L.F.	6.40	5.20		11.60	14.90
5050	3/8" diameter		52	.154		7.30	5.30		12.60	16
5060	1/2" diameter		51	.157		9.10	5.40		14.50	18.15
5070	3/4" diameter		48	.167		11.65	5.75		17.40	21.50
5080	1" diameter		43	.186		15.90	6.40		22.30	27
5090	1-1/4" diameter	Q-1	73	.219		22.50	6.80		29.30	35.50
5100	1-1/2" diameter		67	.239		26.50	7.40		33.90	40
5110	2" diameter		54	.296		33.50	9.20		42.70	50.50
5190	For small quantities, add					10%				
5700	To delete couplings & hangers, subtract									
5710	1/4" diam. to 3/4" diam.					10%	53%			
5720	1" diam. to 2" diam.					14%	47%			
5750	Schedule 80, type 316									
5790	1/4" diameter	1 Plum	53	.151	L.F.	7.90	5.20		13.10	16.50
5800	3/8" diameter		52	.154		8.95	5.30		14.25	17.80
5810	1/2" diameter		51	.157		12.75	5.40		18.15	22

15100 | Building Services Piping

15107 | Metal Pipe & Fittings

		CREW	DAILY OUTPUT	LABOR-HOURS	UNIT	2001 BARE COSTS MAT.	LABOR	EQUIP.	TOTAL	TOTAL INCL O&P
5820	3/4" diameter	1 Plum	48	.167	L.F.	14.45	5.75		20.20	24.50
5830	1" diameter	▼	43	.186		19.90	6.40		26.30	31.50
5840	1-1/4" diameter	Q-1	73	.219		29	6.80		35.80	42.50
5850	1-1/2" diameter		67	.239		33.50	7.40		40.90	47.50
5860	2" diameter	▼	54	.296	▼	42.50	9.20		51.70	61
5950	For small quantities, add					10%				
7000	To delete couplings & hangers, subtract									
7010	1/4" diam. to 3/4" diam.					9%	53%			
7020	1" diam. to 2" diam.					14%	47%			
8000	Weld joints with clevis type hangers 10' O.C.									
8010	Schedule 40, type 304									
8050	1/8" pipe size	Q-15	126	.127	L.F.	3.44	3.94	.80	8.18	10.60
8060	1/4" pipe size		125	.128		4.28	3.97	.81	9.06	11.60
8070	3/8" pipe size		122	.131		4.84	4.07	.82	9.73	12.35
8080	1/2" pipe size		118	.136		6.10	4.20	.85	11.15	14
8090	3/4" pipe size		109	.147		7.25	4.55	.92	12.72	15.85
8100	1" pipe size		95	.168		9.05	5.20	1.06	15.31	19
8110	1-1/4" pipe size		86	.186		11.15	5.75	1.17	18.07	22.50
8120	1-1/2" pipe size		78	.205		13	6.35	1.29	20.64	25.50
8130	2" pipe size		62	.258		16.65	8	1.62	26.27	32
8140	2-1/2" pipe size		49	.327		25	10.10	2.05	37.15	45
8150	3" pipe size		44	.364		32	11.25	2.29	45.54	54.50
8160	3-1/2" pipe size		44	.364		37.50	11.25	2.29	51.04	61
8170	4" pipe size		39	.410		43	12.70	2.58	58.28	69.50
8180	5" pipe size	▼	32	.500		62.50	15.50	3.15	81.15	95.50
8190	6" pipe size	Q-16	37	.649		75	21	2.72	98.72	117
8200	8" pipe size		29	.828		121	26.50	3.47	150.97	177
8210	10" pipe size		24	1		204	32	4.19	240.19	278
8220	12" pipe size	▼	20	1.200	▼	231	38.50	5.05	274.55	320
8300	Schedule 40, type 316									
8310	1/8" pipe size	Q-15	126	.127	L.F.	4.31	3.94	.80	9.05	11.55
8320	1/4" pipe size		125	.128		5.10	3.97	.81	9.88	12.50
8330	3/8" pipe size		122	.131		6.70	4.07	.82	11.59	14.40
8340	1/2" pipe size		118	.136		7.65	4.20	.85	12.70	15.75
8350	3/4" pipe size		109	.147		8.85	4.55	.92	14.32	17.60
8360	1" pipe size		95	.168		11.60	5.20	1.06	17.86	22
8370	1-1/4" pipe size		86	.186		14.30	5.75	1.17	21.22	25.50
8380	1-1/2" pipe size		78	.205		16.70	6.35	1.29	24.34	29.50
8390	2" pipe size		62	.258		21.50	8	1.62	31.12	38
8400	2-1/2" pipe size		49	.327		33	10.10	2.05	45.15	54
8410	3" pipe size		44	.364		44.50	11.25	2.29	58.04	68.50
8420	3-1/2" pipe size		44	.364		51.50	11.25	2.29	65.04	76.50
8430	4" pipe size		39	.410		58.50	12.70	2.58	73.78	86.50
8440	5" pipe size	▼	32	.500		88.50	15.50	3.15	107.15	124
8450	6" pipe size	Q-16	37	.649		99.50	21	2.72	123.22	143
8460	8" pipe size		29	.828		185	26.50	3.47	214.97	248
8470	10" pipe size		24	1		248	32	4.19	284.19	325
8480	12" pipe size	▼	20	1.200	▼	335	38.50	5.05	378.55	435
8500	Schedule 80, type 304									
8510	1/4" pipe size	Q-15	110	.145	L.F.	5.75	4.51	.91	11.17	14.10
8520	3/8" pipe size		109	.147		6.50	4.55	.92	11.97	15
8530	1/2" pipe size		106	.151		8.10	4.68	.95	13.73	17.05
8540	3/4" pipe size		96	.167		10.35	5.15	1.05	16.55	20.50
8550	1" pipe size		87	.184		13.75	5.70	1.16	20.61	25
8560	1-1/4" pipe size		81	.198		17.70	6.10	1.24	25.04	30
8570	1-1/2" pipe size		74	.216		20	6.70	1.36	28.06	33.50

15100 | Building Services Piping

15107 | Metal Pipe & Fittings

		CREW	DAILY OUTPUT	LABOR-HOURS	UNIT	2001 BARE COSTS MAT.	LABOR	EQUIP.	TOTAL	TOTAL INCL O&P
8580	2" pipe size	Q-15	58	.276	L.F.	25	8.55	1.74	35.29	42.50
8590	2-1/2" pipe size		46	.348		51.50	10.80	2.19	64.49	75
8600	3" pipe size		41	.390		69	12.10	2.45	83.55	97
8610	4" pipe size		33	.485		108	15.05	3.05	126.10	145
8630	6" pipe size	Q-16	30	.800		237	25.50	3.35	265.85	305
8640	Schedule 80, type 316									
8650	1/4" pipe size	Q-15	110	.145	L.F.	7.15	4.51	.91	12.57	15.65
8660	3/8" pipe size		109	.147		8.15	4.55	.92	13.62	16.80
8670	1/2" pipe size		106	.151		11.60	4.68	.95	17.23	21
8680	3/4" pipe size		96	.167		12.80	5.15	1.05	19	23
8690	1" pipe size		87	.184		17.10	5.70	1.16	23.96	28.50
8700	1-1/4" pipe size		81	.198		23	6.10	1.24	30.34	36
8710	1-1/2" pipe size		74	.216		26	6.70	1.36	34.06	40.50
8720	2" pipe size		58	.276		32.50	8.55	1.74	42.79	51
8730	2-1/2" pipe size		46	.348		85	10.80	2.19	97.99	112
8740	3" pipe size		41	.390		99	12.10	2.45	113.55	130
8760	4" pipe size		33	.485		137	15.05	3.05	155.10	177
8770	6" pipe size	Q-16	30	.800		370	25.50	3.35	398.85	450
8790	Schedule 160, type 304									
8800	1/2" pipe size	Q-15	96	.167	L.F.	11.40	5.15	1.05	17.60	21.50
8810	3/4" pipe size		88	.182		13.35	5.65	1.14	20.14	24.50
8820	1" pipe size		80	.200		22.50	6.20	1.26	29.96	35.50
8840	1-1/2" pipe size		67	.239		31	7.40	1.50	39.90	47
8850	2" pipe size		53	.302		44.50	9.35	1.90	55.75	65
8870	3" pipe size		37	.432		72.50	13.40	2.72	88.62	103
8900	Schedule 160, type 316									
8910	1/2" pipe size	Q-15	96	.167	L.F.	16.85	5.15	1.05	23.05	27.50
8920	3/4" pipe size		88	.182		17.70	5.65	1.14	24.49	29.50
8930	1" pipe size		80	.200		25	6.20	1.26	32.46	38
8940	1-1/4" pipe size		73	.219		35.50	6.80	1.38	43.68	51
8950	1-1/2" pipe size		67	.239		41	7.40	1.50	49.90	58
8960	2" pipe size		53	.302		56.50	9.35	1.90	67.75	78.50
8990	4" pipe size		30	.533		143	16.55	3.35	162.90	186
9100	Threading pipe labor, sst, one end, schedules 40 & 80									
9110	1/4" through 3/4" pipe size	1 Plum	61.50	.130	Ea.		4.48		4.48	6.75
9120	1" through 2" pipe size		55.90	.143			4.93		4.93	7.45
9130	2-1/2" pipe size		41.50	.193			6.65		6.65	10
9140	3" pipe size		38.50	.208			7.15		7.15	10.80
9150	3-1/2" pipe size	Q-1	68.40	.234			7.25		7.25	10.95
9160	4" pipe size		73	.219			6.80		6.80	10.25
9170	5" pipe size		40.70	.393			12.20		12.20	18.40
9180	6" pipe size		35.40	.452			14		14	21
9190	8" pipe size		22.30	.717			22		22	33.50
9200	10" pipe size		16.10	.994			31		31	46.50
9210	12" pipe size		12.30	1.301			40.50		40.50	61
9250	Welding labor per joint for stainless steel									
9260	Schedule 5 and 10									
9270	1/4" pipe size	Q-15	36	.444	Ea.		13.80	2.80	16.60	24
9280	3/8" pipe size		35	.457			14.15	2.88	17.03	24.50
9290	1/2" pipe size		35	.457			14.15	2.88	17.03	24.50
9300	3/4" pipe size		28	.571			17.70	3.59	21.29	30.50
9310	1" pipe size		25	.640			19.85	4.03	23.88	34.50
9320	1-1/4" pipe size		22	.727			22.50	4.57	27.07	39
9330	1-1/2" pipe size		21	.762			23.50	4.79	28.29	41
9340	2" pipe size		18	.889			27.50	5.60	33.10	47.50
9350	2-1/2" pipe size		14	1.143			35.50	7.20	42.70	61.50

15100 | Building Services Piping

15107 | Metal Pipe & Fittings

		CREW	DAILY OUTPUT	LABOR-HOURS	UNIT	MAT.	LABOR	EQUIP.	TOTAL	TOTAL INCL O&P	
920	9360	3" pipe size	Q-15	13	1.231	Ea.		38	7.75	45.75	66	920
	9370	4" pipe size		12	1.333			41.50	8.40	49.90	72	
	9380	5" pipe size		10	1.600			49.50	10.05	59.55	86	
	9390	6" pipe size		9	1.778			55	11.20	66.20	95.50	
	9400	8" pipe size		6	2.667			82.50	16.75	99.25	143	
	9410	10" pipe size		4	4			124	25	149	215	
	9420	12" pipe size		3	5.333			165	33.50	198.50	287	
	9500	Schedule 40										
	9510	1/4" pipe size	Q-15	35	.457	Ea.		14.15	2.88	17.03	24.50	
	9520	3/8" pipe size		35	.457			14.15	2.88	17.03	24.50	
	9530	1/2" pipe size		34	.471			14.60	2.96	17.56	25.50	
	9540	3/4" pipe size		28	.571			17.70	3.59	21.29	30.50	
	9550	1" pipe size		24	.667			20.50	4.19	24.69	35.50	
	9560	1-1/4" pipe size		21	.762			23.50	4.79	28.29	41	
	9570	1-1/2" pipe size		20	.800			25	5.05	30.05	43	
	9580	2" pipe size		17	.941			29	5.90	34.90	50.50	
	9590	2-1/2" pipe size		14	1.143			35.50	7.20	42.70	61.50	
	9600	3" pipe size		13	1.231			38	7.75	45.75	66	
	9610	4" pipe size		11	1.455			45	9.15	54.15	78	
	9620	5" pipe size		9	1.778			55	11.20	66.20	95.50	
	9630	6" pipe size		8	2			62	12.60	74.60	107	
	9640	8" pipe size		5	3.200			99	20	119	172	
	9650	10" pipe size		4	4			124	25	149	215	
	9660	12" pipe size		3	5.333			165	33.50	198.50	287	
	9750	Schedule 80										
	9760	1/4" pipe size	Q-15	28	.571	Ea.		17.70	3.59	21.29	30.50	
	9770	3/8" pipe size		28	.571			17.70	3.59	21.29	30.50	
	9780	1/2" pipe size		28	.571			17.70	3.59	21.29	30.50	
	9790	3/4" pipe size		24	.667			20.50	4.19	24.69	35.50	
	9800	1" pipe size		21	.762			23.50	4.79	28.29	41	
	9810	1-1/4" pipe size		20	.800			25	5.05	30.05	43	
	9820	1-1/2" pipe size		19	.842			26	5.30	31.30	45.50	
	9830	2" pipe size		16	1			31	6.30	37.30	54	
	9840	2-1/2" pipe size		13	1.231			38	7.75	45.75	66	
	9850	3" pipe size		12	1.333			41.50	8.40	49.90	72	
	9860	4" pipe size		8	2			62	12.60	74.60	107	
	9870	5" pipe size		6	2.667			82.50	16.75	99.25	143	
	9880	6" pipe size		5	3.200			99	20	119	172	
	9890	8" pipe size		3	5.333			165	33.50	198.50	287	
	9900	10" pipe size		2	8			248	50.50	298.50	430	
	9910	12" pipe size		1	16			495	101	596	860	
	9920	Schedule 160, 1/2" pipe size		25	.640			19.85	4.03	23.88	34.50	
	9930	3/4" pipe size		22	.727			22.50	4.57	27.07	39	
	9940	1" pipe size		19	.842			26	5.30	31.30	45.50	
	9950	1-1/4" pipe size		18	.889			27.50	5.60	33.10	47.50	
	9960	1-1/2" pipe size		17	.941			29	5.90	34.90	50.50	
	9970	2" pipe size		14	1.143			35.50	7.20	42.70	61.50	
	9980	3" pipe size		11	1.455			45	9.15	54.15	78	
	9990	4" pipe size		7	2.286			71	14.40	85.40	123	
960	0010	**PIPE, STAINLESS STEEL, FITTINGS**										960
	0100	Butt weld joint, schedule 5, type 304										
	0120	90° Elbow, long										
	0140	1/2"	Q-15	16	1	Ea.	9.55	31	6.30	46.85	64.50	
	0150	3/4"		16	1		15	31	6.30	52.30	70.50	
	0160	1"		15	1.067		10.95	33	6.70	50.65	69.50	

15100 | Building Services Piping

15107 | Metal Pipe & Fittings

		CREW	DAILY OUTPUT	LABOR-HOURS	UNIT	2001 BARE COSTS MAT.	LABOR	EQUIP.	TOTAL	TOTAL INCL O&P	
0170	1-1/4"	Q-15	14	1.143	Ea.	17.50	35.50	7.20	60.20	80.50	960
0180	1-1/2"		13	1.231		13.75	38	7.75	59.50	81	
0190	2"		11	1.455		18.15	45	9.15	72.30	98	
0200	2-1/2"		8	2		26.50	62	12.60	101.10	136	
0210	3"		6	2.667		33	82.50	16.75	132.25	180	
0220	3-1/2"		5	3.200		175	99	20	294	365	
0230	4"		4	4		61.50	124	25	210.50	282	
0240	5"		3	5.333		440	165	33.50	638.50	765	
0250	6"	Q-16	4	6		175	193	25	393	510	
0260	8"		3	8		298	257	33.50	588.50	750	
0270	10"		2	12		615	385	50.50	1,050.50	1,300	
0280	12"		1.57	15.287		970	490	64	1,524	1,875	
0320	For schedule 5, type 316, add					30%					
0600	45° Elbow, long										
0620	1/2"	Q-15	16	1	Ea.	13.75	31	6.30	51.05	69	
0630	3/4"		16	1		13.75	31	6.30	51.05	69	
0640	1"		15	1.067		15.50	33	6.70	55.20	74.50	
0650	1-1/4"		14	1.143		26.50	35.50	7.20	69.20	90.50	
0660	1-1/2"		13	1.231		14.10	38	7.75	59.85	81.50	
0670	2"		11	1.455		16.20	45	9.15	70.35	96	
0680	2-1/2"		8	2		27	62	12.60	101.60	137	
0690	3"		6	2.667		27.50	82.50	16.75	126.75	174	
0700	3-1/2"		5	3.200		175	99	20	294	365	
0710	4"		4	4		49	124	25	198	269	
0720	5"		3	5.333		305	165	33.50	503.50	620	
0730	6"	Q-16	4	6		105	193	25	323	435	
0740	8"		3	8		183	257	33.50	473.50	630	
0750	10"		2	12		410	385	50.50	845.50	1,075	
0760	12"		1.57	15.287		655	490	64	1,209	1,525	
0800	For schedule 5, type 316, add					25%					
1100	Tee, straight										
1130	1/2"	Q-15	10	1.600	Ea.	38.50	49.50	10.05	98.05	129	
1140	3/4"		10	1.600		38.50	49.50	10.05	98.05	129	
1150	1"		10	1.600		37	49.50	10.05	96.55	127	
1160	1-1/4"		9	1.778		63	55	11.20	129.20	165	
1170	1-1/2"		8	2		37	62	12.60	111.60	148	
1180	2"		7	2.286		38.50	71	14.40	123.90	165	
1190	2-1/2"		5	3.200		110	99	20	229	293	
1200	3"		4	4		72	124	25	221	294	
1210	3-1/2"		3	5.333		192	165	33.50	390.50	500	
1220	4"		2	8		102	248	50.50	400.50	545	
1230	5"		2	8		410	248	50.50	708.50	880	
1240	6"	Q-16	2.60	9.231		305	297	38.50	640.50	830	
1250	8"		2	12		700	385	50.50	1,135.50	1,400	
1260	10"		1.80	13.333		1,100	430	56	1,586	1,900	
1270	12"		1.40	17.143		1,400	550	72	2,022	2,425	
1320	For schedule 5, type 316, add					25%					
2000	Butt weld joint, schedule 10, type 304										
2020	90° elbow, long										
2040	1/2"	Q-15	16	1	Ea.	9	31	6.30	46.30	64	
2050	3/4"		16	1		9	31	6.30	46.30	64	
2060	1"		15	1.067		9	33	6.70	48.70	67.50	
2070	1-1/4"		14	1.143		16.80	35.50	7.20	59.50	80	
2080	1-1/2"		13	1.231		13.10	38	7.75	58.85	80.50	
2090	2"		11	1.455		17.50	45	9.15	71.65	97.50	
2100	2-1/2"		8	2		40	62	12.60	114.60	151	

15100 | Building Services Piping

15107 | Metal Pipe & Fittings

		CREW	DAILY OUTPUT	LABOR-HOURS	UNIT	MAT.	LABOR	EQUIP.	TOTAL	TOTAL INCL O&P
2110	3"	Q-15	6	2.667	Ea.	47.50	82.50	16.75	146.75	195
2120	3-1/2"		5	3.200		202	99	20	321	395
2130	4"		4	4		77.50	124	25	226.50	300
2140	5"		3	5.333		415	165	33.50	613.50	740
2150	6"	Q-16	4	6		168	193	25	386	505
2160	8"		3	8		350	257	33.50	640.50	810
2170	10"		2	12		755	385	50.50	1,190.50	1,475
2180	12"		1.57	15.287		1,175	490	64	1,729	2,075
2500	45° elbow, long									
2520	1/2"	Q-15	16	1	Ea.	16.50	31	6.30	53.80	72
2530	3/4"		16	1		16.50	31	6.30	53.80	72
2540	1"		15	1.067		14.70	33	6.70	54.40	73.50
2550	1-1/4"		14	1.143		31.50	35.50	7.20	74.20	96
2560	1-1/2"		13	1.231		13.10	38	7.75	58.85	80.50
2570	2"		11	1.455		15.05	45	9.15	69.20	94.50
2580	2-1/2"		8	2		36	62	12.60	110.60	147
2590	3"		6	2.667		40.50	82.50	16.75	139.75	188
2600	3-1/2"		5	3.200		202	99	20	321	395
2610	4"		4	4		45.50	124	25	194.50	265
2620	5"		3	5.333		315	165	33.50	513.50	630
2630	6"	Q-16	4	6		98	193	25	316	425
2640	8"		3	8		233	257	33.50	523.50	685
2650	10"		2	12		540	385	50.50	975.50	1,225
2660	12"		1.57	15.287		825	490	64	1,379	1,725
3000	Tee, straight									
3030	1/2"	Q-15	10	1.600	Ea.	41.50	49.50	10.05	101.05	132
3040	3/4"		10	1.600		41.50	49.50	10.05	101.05	132
3050	1"		10	1.600		31.50	49.50	10.05	91.05	121
3060	1-1/4"		9	1.778		63	55	11.20	129.20	165
3070	1-1/2"		8	2		34.50	62	12.60	109.10	145
3080	2"		7	2.286		35	71	14.40	120.40	161
3090	2-1/2"		5	3.200		116	99	20	235	299
3100	3"		4	4		81	124	25	230	305
3110	3-1/2"		3	5.333		122	165	33.50	320.50	420
3120	4"		2	8		96	248	50.50	394.50	535
3130	5"		2	8		390	248	50.50	688.50	855
3140	6"	Q-16	2.60	9.231		290	297	38.50	625.50	815
3150	8"		2	12		705	385	50.50	1,140.50	1,400
3151	10"		1.80	13.333		1,175	430	56	1,661	1,975
3152	12"		1.40	17.143		1,625	550	72	2,247	2,675
3154	For schedule 10, type 316, add					25%				
3281	Butt weld joint, schedule 40, type 304									
3284	90° Elbow, long, 1/2"	Q-15	14	1.143	Ea.	15.75	35.50	7.20	58.45	79
3288	3/4"		14	1.143		15.75	35.50	7.20	58.45	79
3289	1"		12	1.333		17.25	41.50	8.40	67.15	91
3290	1-1/4"		11	1.455		25.50	45	9.15	79.65	106
3300	1-1/2"		10	1.600		20.50	49.50	10.05	80.05	109
3310	2"		9	1.778		28.50	55	11.20	94.70	127
3320	2-1/2"		7	2.286		59.50	71	14.40	144.90	188
3330	3"		6	2.667		76	82.50	16.75	175.25	227
3340	3-1/2"		5.50	2.909		315	90	18.30	423.30	505
3350	4"		5	3.200		140	99	20	259	325
3360	5"		4	4		495	124	25	644	760
3370	6"	Q-16	6	4		440	129	16.75	585.75	695
3380	8"		4	6		740	193	25	958	1,125
3390	10"		3	8		1,550	257	33.50	1,840.50	2,125

15100 | Building Services Piping

15107 | Metal Pipe & Fittings

		CREW	DAILY OUTPUT	LABOR-HOURS	UNIT	MAT.	LABOR	EQUIP.	TOTAL	TOTAL INCL O&P
3400	12"	Q-16	2	12	Ea.	2,175	385	50.50	2,610.50	3,025
3410	For schedule 40, type 316, add					25%				
3460	45° Elbow, long, 1/2"	Q-15	14	1.143	Ea.	21	35.50	7.20	63.70	84.50
3470	3/4"		14	1.143		21	35.50	7.20	63.70	84.50
3480	1"		12	1.333		21	41.50	8.40	70.90	95
3490	1-1/4"		11	1.455		31	45	9.15	85.15	112
3500	1-1/2"		10	1.600		24	49.50	10.05	83.55	113
3510	2"		9	1.778		29.50	55	11.20	95.70	127
3520	2-1/2"		7	2.286		59.50	71	14.40	144.90	188
3530	3"		6	2.667		55	82.50	16.75	154.25	203
3540	3-1/2"		5.50	2.909		320	90	18.30	428.30	510
3550	4"		5	3.200		99	99	20	218	281
3560	5"		4	4		345	124	25	494	595
3570	6"	Q-16	6	4		258	129	16.75	403.75	495
3580	8"		4	6		490	193	25	708	860
3590	10"		3	8		920	257	33.50	1,210.50	1,425
3600	12"		2	12		1,275	385	50.50	1,710.50	2,025
3610	For schedule 40, type 316, add					25%				
3660	Tee, straight 1/2"	Q-15	11	1.455	Ea.	44.50	45	9.15	98.65	127
3670	3/4"		9	1.778		44.50	55	11.20	110.70	144
3680	1"		8	2		44.50	62	12.60	119.10	156
3690	1-1/4"		7	2.286		73.50	71	14.40	158.90	204
3700	1-1/2"		6.50	2.462		44.50	76.50	15.50	136.50	181
3710	2"		6	2.667		48	82.50	16.75	147.25	196
3720	2-1/2"		5	3.200		120	99	20	239	305
3730	3"		4	4		97	124	25	246	320
3740	3-1/2"		3.70	4.324		139	134	27	300	385
3750	4"		3.50	4.571		139	142	29	310	400
3760	5"		3	5.333		515	165	33.50	713.50	855
3770	6"	Q-16	4	6		445	193	25	663	805
3780	8"		3	8		1,025	257	33.50	1,315.50	1,550
3790	10"		2	12		2,175	385	50.50	2,610.50	3,000
3800	12"		1.50	16		2,200	515	67	2,782	3,250
3810	For schedule 40, type 316, add					25%				
3820	Tee, reducing on outlet, 3/4" x 1/2"	Q-15	11	1.455	Ea.	48	45	9.15	102.15	131
3822	1" x 1/2"		9	1.778		66.50	55	11.20	132.70	169
3824	1" x 3/4"		9	1.778		55.50	55	11.20	121.70	156
3826	1-1/4" x 1"		8	2		176	62	12.60	250.60	300
3828	1-1/2" x 1/2"		7	2.286		92.50	71	14.40	177.90	225
3830	1-1/2" x 3/4"		7	2.286		92.50	71	14.40	177.90	225
3832	1-1/2" x 1"		7	2.286		66.50	71	14.40	151.90	196
3834	2" x 1"		6.50	2.462		65	76.50	15.50	157	203
3836	2" x 1-1/2"		6.50	2.462		61	76.50	15.50	153	199
3838	2-1/2" x 2"		6	2.667		157	82.50	16.75	256.25	315
3840	3" x 1-1/2"		5	3.200		134	99	20	253	320
3842	3" x 2"		5	3.200		113	99	20	232	296
3844	4" x 2"		4	4		204	124	25	353	440
3846	4" x 3"		4	4		167	124	25	316	400
3848	5" x 4"		3.50	4.571		525	142	29	696	825
3850	6" x 3"	Q-16	4.60	5.217		420	168	22	610	735
3852	6" x 4"		4.60	5.217		380	168	22	570	695
3854	8" x 4"		4	6		700	193	25	918	1,100
3856	10" x 8"		3	8		1,950	257	33.50	2,240.50	2,550
3858	12" x 10"		2	12		2,400	385	50.50	2,835.50	3,275
3950	Reducer, concentric, 3/4" x 1/2"	Q-15	14	1.143		22	35.50	7.20	64.70	86
3952	1" x 3/4"		13	1.231		26	38	7.75	71.75	94.50

15100 | Building Services Piping

15107 | Metal Pipe & Fittings

		CREW	DAILY OUTPUT	LABOR-HOURS	UNIT	MAT.	LABOR	EQUIP.	TOTAL	TOTAL INCL O&P
3954	1-1/4" x 3/4"	Q-15	12	1.333	Ea.	74	41.50	8.40	123.90	153
3956	1-1/4" x 1"		12	1.333		40	41.50	8.40	89.90	116
3958	1-1/2" x 3/4"		11	1.455		46	45	9.15	100.15	129
3960	1-1/2" x 1"		11	1.455		15.45	45	9.15	69.60	95
3962	2" x 1"		10	1.600		18.50	49.50	10.05	78.05	107
3964	2" x 1-1/2"		10	1.600		15.45	49.50	10.05	75	103
3966	2-1/2" x 1"		8	2		111	62	12.60	185.60	229
3968	2-1/2" x 2"		8	2		55.50	62	12.60	130.10	168
3970	3" x 1"		7	2.286		74	71	14.40	159.40	204
3972	3" x 1-1/2"		7	2.286		34.50	71	14.40	119.90	160
3974	3" x 2"		7	2.286		28	71	14.40	113.40	153
3976	4" x 2"		6	2.667		38	82.50	16.75	137.25	185
3978	4" x 3"		6	2.667		33.50	82.50	16.75	132.75	180
3980	5" x 3"		4	4		278	124	25	427	520
3982	5" x 4"		4	4		222	124	25	371	460
3984	6" x 3"	Q-16	7	3.429		106	110	14.35	230.35	298
3986	6" x 4"		7	3.429		88	110	14.35	212.35	279
3988	8" x 4"		5	4.800		335	154	20	509	620
3990	8" x 6"		5	4.800		213	154	20	387	490
3992	10" x 6"		4	6		445	193	25	663	810
3994	10" x 8"		4	6		263	193	25	481	610
3995	12" x 6"		3	8		665	257	33.50	955.50	1,150
3996	12" x 8"		3	8		610	257	33.50	900.50	1,100
3997	12" x 10"		3	8		365	257	33.50	655.50	830
4000	Socket weld joint, 3000 lb., type 304									
4100	90° Elbow									
4140	1/4"	Q-15	20	.800	Ea.	21.50	25	5.05	51.55	66.50
4150	3/8"		18	.889		27.50	27.50	5.60	60.60	78
4160	1/2"		17	.941		29	29	5.90	63.90	82.50
4170	3/4"		16	1		35	31	6.30	72.30	92.50
4180	1"		15	1.067		53	33	6.70	92.70	115
4190	1-1/4"		14	1.143		92	35.50	7.20	134.70	162
4200	1-1/2"		13	1.231		114	38	7.75	159.75	191
4210	2"		11	1.455		184	45	9.15	238.15	280
4300	45° Elbow									
4340	1/4"	Q-15	20	.800	Ea.	41	25	5.05	71.05	88
4350	3/8"		18	.889		41	27.50	5.60	74.10	92.50
4360	1/2"		17	.941		41	29	5.90	75.90	95.50
4370	3/4"		16	1		47	31	6.30	84.30	106
4380	1"		15	1.067		68	33	6.70	107.70	132
4390	1-1/4"		14	1.143		110	35.50	7.20	152.70	182
4400	1-1/2"		13	1.231		111	38	7.75	156.75	188
4410	2"		11	1.455		201	45	9.15	255.15	300
4500	Tee									
4540	1/4"	Q-15	13	1.231	Ea.	26	38	7.75	71.75	95
4550	3/8"		12	1.333		31	41.50	8.40	80.90	106
4560	1/2"		12	1.333		41.50	41.50	8.40	91.40	117
4570	3/4"		11	1.455		48	45	9.15	102.15	131
4580	1"		10	1.600		64.50	49.50	10.05	124.05	157
4590	1-1/4"		9	1.778		114	55	11.20	180.20	221
4600	1-1/2"		8	2		167	62	12.60	241.60	290
4610	2"		7	2.286		253	71	14.40	338.40	400
5000	Socket weld joint, 3000 lb., type 316									
5100	90° Elbow									
5140	1/4"	Q-15	20	.800	Ea.	25	25	5.05	55.05	70.50
5150	3/8"		18	.889		29	27.50	5.60	62.10	79.50

15100 | Building Services Piping

15107 | Metal Pipe & Fittings

		CREW	DAILY OUTPUT	LABOR-HOURS	UNIT	MAT.	LABOR	EQUIP.	TOTAL	TOTAL INCL O&P
5160	1/2"	Q-15	17	.941	Ea.	35	29	5.90	69.90	89
5170	3/4"		16	1		46	31	6.30	83.30	105
5180	1"		15	1.067		68.50	33	6.70	108.20	133
5190	1-1/4"		14	1.143		122	35.50	7.20	164.70	196
5200	1-1/2"		13	1.231		141	38	7.75	186.75	221
5210	2"		11	1.455		240	45	9.15	294.15	340
5300	45° Elbow									
5340	1/4"	Q-15	20	.800	Ea.	52.50	25	5.05	82.55	101
5350	3/8"		18	.889		52.50	27.50	5.60	85.60	106
5360	1/2"		17	.941		52.50	29	5.90	87.40	109
5370	3/4"		16	1		59.50	31	6.30	96.80	119
5380	1"		15	1.067		90	33	6.70	129.70	156
5390	1-1/4"		14	1.143		126	35.50	7.20	168.70	200
5400	1-1/2"		13	1.231		141	38	7.75	186.75	222
5410	2"		11	1.455		211	45	9.15	265.15	310
5500	Tee									
5540	1/4"	Q-15	13	1.231	Ea.	31.50	38	7.75	77.25	101
5550	3/8"		12	1.333		38.50	41.50	8.40	88.40	114
5560	1/2"		12	1.333		48	41.50	8.40	97.90	125
5570	3/4"		11	1.455		59.50	45	9.15	113.65	144
5580	1"		10	1.600		91	49.50	10.05	150.55	186
5590	1-1/4"		9	1.778		145	55	11.20	211.20	254
5600	1-1/2"		8	2		208	62	12.60	282.60	335
5610	2"		7	2.286		330	71	14.40	415.40	490
5700	For socket weld joint, 6000 lb., type 304 and 316, add					100%				
6000	Threaded companion flange									
6010	Stainless steel, 150 lb., type 304									
6020	1/2" diam.	1 Plum	30	.267	Ea.	45.50	9.20		54.70	64
6030	3/4" diam.		28	.286		50.50	9.85		60.35	71
6040	1" diam.		27	.296		56.50	10.20		66.70	78
6050	1-1/4" diam.	Q-1	44	.364		77	11.25		88.25	102
6060	1-1/2" diam.		40	.400		79	12.40		91.40	106
6070	2" diam.		36	.444		88	13.80		101.80	118
6080	2-1/2" diam.		28	.571		135	17.70		152.70	175
6090	3" diam.		20	.800		147	25		172	199
6110	4" diam.		12	1.333		231	41.50		272.50	315
6130	6" diam.	Q-2	14	1.714		440	55		495	570
6140	8" diam.	"	12	2		825	64.50		889.50	1,000
6150	For type 316 add					40%				
6260	Weld flanges, stainless steel, type 304									
6270	Slip on, 150 lb. (welded, front and back)									
6280	1/2" diam.	Q-15	16	1	Ea.	45.50	31	6.30	82.80	104
6290	3/4" diam.		16	1		50.50	31	6.30	87.80	110
6300	1" diam.		13	1.231		34.50	38	7.75	80.25	104
6310	1-1/4" diam.		12	1.333		77	41.50	8.40	126.90	157
6320	1-1/2" diam.		11	1.455		51	45	9.15	105.15	134
6330	2" diam.		10	1.600		66.50	49.50	10.05	126.05	159
6340	2-1/2" diam.		9	1.778		135	55	11.20	201.20	243
6350	3" diam.		8	2		147	62	12.60	221.60	268
6370	4" diam.		6	2.667		140	82.50	16.75	239.25	297
6390	6" diam.	Q-16	6	4		220	129	16.75	365.75	455
6400	8" diam.	"	4	6		825	193	25	1,043	1,225
6410	For type 316, add					40%				
6530	Weld neck 150 lb.									
6540	1/2" diam.	Q-15	35	.457	Ea.	32.50	14.15	2.88	49.53	60.50
6550	3/4" diam.		32	.500		37	15.50	3.15	55.65	67.50

15100 | Building Services Piping

15107 | Metal Pipe & Fittings

		CREW	DAILY OUTPUT	LABOR-HOURS	UNIT	MAT.	LABOR	EQUIP.	TOTAL	TOTAL INCL O&P
6560	1" diam.	Q-15	24	.667	Ea.	44	20.50	4.19	68.69	83.50
6570	1-1/4" diam.		23	.696		52.50	21.50	4.38	78.38	95.50
6580	1-1/2" diam.		20	.800		61.50	25	5.05	91.55	111
6590	2" diam.		18	.889		79	27.50	5.60	112.10	134
6600	2-1/2" diam.		14	1.143		131	35.50	7.20	173.70	205
6610	3" diam.		12	1.333		149	41.50	8.40	198.90	236
6630	4" diam.		9	1.778		193	55	11.20	259.20	305
6640	5" diam.		9	1.778		256	55	11.20	322.20	375
6650	6" diam.	Q-16	9	2.667		298	85.50	11.15	394.65	470
6652	8" diam.		8	3		420	96.50	12.55	529.05	620
6654	10" diam.		7	3.429		575	110	14.35	699.35	810
6656	12" diam.		6	4		910	129	16.75	1,055.75	1,200
6670	For type 316 add					23%				
7000	Threaded joint, 150 lb., type 304									
7030	90° elbow									
7040	1/8"	1 Plum	13	.615	Ea.	12.80	21		33.80	46
7050	1/4"		13	.615		12.80	21		33.80	46
7070	3/8"		13	.615		15.45	21		36.45	49
7080	1/2"		12	.667		7.90	23		30.90	43
7090	3/4"		11	.727		8.95	25		33.95	48
7100	1"		10	.800		11.90	27.50		39.40	54.50
7110	1-1/4"	Q-1	17	.941		33	29		62	80
7120	1-1/2"		16	1		20	31		51	69
7130	2"		14	1.143		29	35.50		64.50	85.50
7140	2-1/2"		11	1.455		144	45		189	227
7150	3"		8	2		182	62		244	295
7160	4"	Q-2	11	2.182		201	70		271	325
7180	45° elbow									
7190	1/8"	1 Plum	13	.615	Ea.	17.60	21		38.60	51.50
7200	1/4"		13	.615		17.60	21		38.60	51.50
7210	3/8"		13	.615		19.60	21		40.60	53.50
7220	1/2"		12	.667		9.50	23		32.50	45
7230	3/4"		11	.727		11.65	25		36.65	51
7240	1"		10	.800		12.85	27.50		40.35	55.50
7250	1-1/4"	Q-1	17	.941		31	29		60	78
7260	1-1/2"		16	1		20.50	31		51.50	69.50
7270	2"		14	1.143		30.50	35.50		66	87
7280	2-1/2"		11	1.455		204	45		249	292
7290	3"		8	2		298	62		360	425
7300	4"	Q-2	11	2.182		254	70		324	385
7320	Tee, straight									
7330	1/8"	1 Plum	9	.889	Ea.	19.70	30.50		50.20	67.50
7340	1/4"		9	.889		19.70	30.50		50.20	67.50
7350	3/8"		9	.889		22.50	30.50		53	71
7360	1/2"		8	1		12.15	34.50		46.65	65.50
7370	3/4"		7	1.143		12.75	39.50		52.25	73.50
7380	1"		6.50	1.231		15.25	42.50		57.75	81
7390	1-1/4"	Q-1	11	1.455		45	45		90	117
7400	1-1/2"		10	1.600		31	49.50		80.50	109
7410	2"		9	1.778		37	55		92	124
7420	2-1/2"		7	2.286		208	71		279	335
7430	3"		5	3.200		315	99		414	500
7440	4"	Q-2	7	3.429		298	110		408	495
7460	Coupling, straight									
7470	1/8"	1 Plum	19	.421	Ea.	4.86	14.50		19.36	27.50
7480	1/4"		19	.421		5.85	14.50		20.35	28.50

15100 | Building Services Piping

15107 | Metal Pipe & Fittings

		CREW	DAILY OUTPUT	LABOR-HOURS	UNIT	2001 BARE COSTS MAT.	LABOR	EQUIP.	TOTAL	TOTAL INCL O&P
7490	3/8"	1 Plum	19	.421	Ea.	6.90	14.50		21.40	29.50
7500	1/2"		19	.421		4.81	14.50		19.31	27.50
7510	3/4"		18	.444		12.45	15.30		27.75	36.50
7520	1"		15	.533		16.50	18.35		34.85	45.50
7530	1-1/4"	Q-1	26	.615		14.15	19.10		33.25	44.50
7540	1-1/2"		24	.667		17.05	20.50		37.55	50
7550	2"		21	.762		27	23.50		50.50	65.50
7560	2-1/2"		18	.889		61.50	27.50		89	109
7570	3"		14	1.143		82.50	35.50		118	144
7580	4"	Q-2	16	1.500		123	48		171	208
7600	Reducer, concentric, 1/2"	1 Plum	12	.667		4.23	23		27.23	39
7610	3/4"		11	.727		5.50	25		30.50	44
7612	1"		10	.800		9.60	27.50		37.10	52
7614	1-1/4"	Q-1	17	.941		19.60	29		48.60	65.50
7616	1-1/2"		16	1		21.50	31		52.50	71
7618	2"		14	1.143		34	35.50		69.50	90.50
7620	2-1/2"		11	1.455		71	45		116	146
7622	3"		8	2		103	62		165	208
7624	4"	Q-2	11	2.182		148	70		218	268
7710	Union									
7720	1/8"	1 Plum	12	.667	Ea.	23.50	23		46.50	60.50
7730	1/4"		12	.667		23.50	23		46.50	60.50
7740	3/8"		12	.667		27.50	23		50.50	64.50
7750	1/2"		11	.727		16.35	25		41.35	56
7760	3/4"		10	.800		23	27.50		50.50	66.50
7770	1"		9	.889		33	30.50		63.50	82.50
7780	1-1/4"	Q-1	16	1		148	31		179	209
7790	1-1/2"		15	1.067		160	33		193	226
7800	2"		13	1.231		202	38		240	280
7810	2-1/2"		10	1.600		435	49.50		484.50	550
7820	3"		7	2.286		575	71		646	740
7830	4"	Q-2	10	2.400		795	77		872	990
7838	Caps									
7840	1/2"	1 Plum	24	.333	Ea.	4.61	11.50		16.11	22.50
7841	3/4"		22	.364		6.70	12.55		19.25	26.50
7842	1"		20	.400		10.70	13.80		24.50	33
7843	1-1/2"	Q-1	32	.500		28	15.50		43.50	54.50
7844	2"	"	28	.571		36	17.70		53.70	66
7845	4"	Q-2	22	1.091		141	35		176	208
7850	For 150 lb., type 316, add					25%				
8750	Threaded joint, 2000 lb., type 304									
8770	90° Elbow									
8780	1/8"	1 Plum	13	.615	Ea.	15.40	21		36.40	49
8790	1/4"		13	.615		15.40	21		36.40	49
8800	3/8"		13	.615		19.05	21		40.05	53
8810	1/2"		12	.667		26	23		49	63
8820	3/4"		11	.727		30	25		55	71
8830	1"		10	.800		39.50	27.50		67	85
8840	1-1/4"	Q-1	17	.941		63.50	29		92.50	114
8850	1-1/2"		16	1		102	31		133	159
8860	2"		14	1.143		143	35.50		178.50	211
8880	45° Elbow									
8890	1/8"	1 Plum	13	.615	Ea.	32	21		53	67.50
8900	1/4"		13	.615		32	21		53	67.50
8910	3/8"		13	.615		40	21		61	76
8920	1/2"		12	.667		40	23		63	78.50

15100 | Building Services Piping

15107 | Metal Pipe & Fittings

		Crew	Daily Output	Labor-Hours	Unit	Mat.	Labor	Equip.	Total	Total Incl O&P
8930	3/4"	1 Plum	11	.727	Ea.	44.50	25		69.50	86.50
8940	1"	↓	10	.800		52.50	27.50		80	99.50
8950	1-1/4"	Q-1	17	.941		90.50	29		119.50	144
8960	1-1/2"		16	1		121	31		152	180
8970	2"	↓	14	1.143	↓	140	35.50		175.50	208
8990	Tee, straight									
9000	1/8"	1 Plum	9	.889	Ea.	19.90	30.50		50.40	68
9010	1/4"		9	.889		19.90	30.50		50.40	68
9020	3/8"		9	.889		25.50	30.50		56	74
9030	1/2"		8	1		33	34.50		67.50	88
9040	3/4"		7	1.143		49.50	39.50		89	114
9050	1"	↓	6.50	1.231		51.50	42.50		94	121
9060	1-1/4"	Q-1	11	1.455		85	45		130	162
9070	1-1/2"		10	1.600		141	49.50		190.50	231
9080	2"	↓	9	1.778	↓	262	55		317	370
9100	For couplings and unions use 3000 lb., type 304									
9120	2000 lb., type 316									
9130	90° Elbow									
9140	1/8"	1 Plum	13	.615	Ea.	18.70	21		39.70	52.50
9150	1/4"		13	.615		18.70	21		39.70	52.50
9160	3/8"		13	.615		22	21		43	56
9170	1/2"		12	.667		26.50	23		49.50	63.50
9180	3/4"		11	.727		34.50	25		59.50	76
9190	1"	↓	10	.800		51.50	27.50		79	98
9200	1-1/4"	Q-1	17	.941		92	29		121	145
9210	1-1/2"		16	1		106	31		137	163
9220	2"	↓	14	1.143	↓	180	35.50		215.50	252
9240	45° Elbow									
9250	1/8"	1 Plum	13	.615	Ea.	39.50	21		60.50	75.50
9260	1/4"		13	.615		39.50	21		60.50	75.50
9270	3/8"		13	.615		39.50	21		60.50	75.50
9280	1/2"		12	.667		39.50	23		62.50	78
9300	3/4"		11	.727		44.50	25		69.50	87
9310	1"	↓	10	.800		67.50	27.50		95	116
9320	1-1/4"	Q-1	17	.941		94.50	29		123.50	148
9330	1-1/2"		16	1		129	31		160	189
9340	2"	↓	14	1.143	↓	159	35.50		194.50	228
9360	Tee, straight									
9370	1/8"	1 Plum	9	.889	Ea.	24	30.50		54.50	72
9380	1/4"		9	.889		24	30.50		54.50	72
9390	3/8"		9	.889		29	30.50		59.50	78
9400	1/2"		8	1		36	34.50		70.50	91.50
9410	3/4"		7	1.143		44.50	39.50		84	109
9420	1"	↓	6.50	1.231		68.50	42.50		111	139
9430	1-1/4"	Q-1	11	1.455		109	45		154	187
9440	1-1/2"		10	1.600		156	49.50		205.50	246
9450	2"	↓	9	1.778	↓	249	55		304	355
9470	For couplings and unions use 3000 lb., type 316									
9490	3000 lb., type 304									
9510	Coupling									
9520	1/8"	1 Plum	19	.421	Ea.	6.20	14.50		20.70	29
9530	1/4"		19	.421		6.80	14.50		21.30	29.50
9540	3/8"		19	.421		7.90	14.50		22.40	30.50
9550	1/2"		19	.421		9.55	14.50		24.05	32.50
9560	3/4"		18	.444		12.70	15.30		28	37
9570	1"	↓	15	.533	↓	21.50	18.35		39.85	51

15100 | Building Services Piping

15107 | Metal Pipe & Fittings

			CREW	DAILY OUTPUT	LABOR-HOURS	UNIT	MAT.	LABOR	EQUIP.	TOTAL	TOTAL INCL O&P	
960	9580	1-1/4"	Q-1	26	.615	Ea.	48.50	19.10		67.60	82.50	960
	9590	1-1/2"		24	.667		61	20.50		81.50	98.50	
	9600	2"	↓	21	.762	↓	80.50	23.50		104	124	
	9620	Union										
	9630	1/8"	1 Plum	12	.667	Ea.	48	23		71	87	
	9640	1/4"		12	.667		64	23		87	105	
	9650	3/8"		12	.667		51	23		74	91	
	9660	1/2"		11	.727		51	25		76	94.50	
	9670	3/4"		10	.800		62.50	27.50		90	111	
	9680	1"	↓	9	.889		96.50	30.50		127	152	
	9690	1-1/4"	Q-1	16	1		162	31		193	225	
	9700	1-1/2"		15	1.067		182	33		215	251	
	9710	2"	↓	13	1.231	↓	244	38		282	325	
	9730	3000 lb., type 316										
	9750	Coupling										
	9770	1/8"	1 Plum	19	.421	Ea.	6.55	14.50		21.05	29	
	9780	1/4"		19	.421		7.40	14.50		21.90	30	
	9790	3/8"		19	.421		7.90	14.50		22.40	30.50	
	9800	1/2"		19	.421		11.20	14.50		25.70	34.50	
	9810	3/4"		18	.444		16.50	15.30		31.80	41	
	9820	1"	↓	15	.533		27.50	18.35		45.85	58	
	9830	1-1/4"	Q-1	26	.615		57.50	19.10		76.60	92.50	
	9840	1-1/2"		24	.667		72	20.50		92.50	111	
	9850	2"	↓	21	.762	↓	100	23.50		123.50	146	
	9870	Union										
	9880	1/8"	1 Plum	12	.667	Ea.	49	23		72	88.50	
	9890	1/4"		12	.667		49	23		72	88.50	
	9900	3/8"		12	.667		56.50	23		79.50	97	
	9910	1/2"		11	.727		56.50	25		81.50	101	
	9920	3/4"		10	.800		74	27.50		101.50	123	
	9930	1"	↓	9	.889		117	30.50		147.50	174	
	9940	1-1/4"	Q-1	16	1		185	31		216	250	
	9950	1-1/2"		15	1.067		228	33		261	300	
	9960	2"	↓	13	1.231	↓	290	38		328	380	

15108 | Plastic Pipe & Fittings

			CREW	DAILY OUTPUT	LABOR-HOURS	UNIT	MAT.	LABOR	EQUIP.	TOTAL	TOTAL INCL O&P	
520	0010	**PIPE, PLASTIC**										520
	0020	Fiberglass reinforced, couplings 10' O.C., hangers 3 per 10'										
	0080	General service										
	0120	2" diameter	Q-1	59	.271	L.F.	10.15	8.40		18.55	24	
	0140	3" diameter		52	.308		12.80	9.55		22.35	28.50	
	0150	4" diameter		48	.333		14.55	10.35		24.90	31.50	
	0160	6" diameter	↓	39	.410		22	12.70		34.70	43	
	0170	8" diameter	Q-2	49	.490		33.50	15.75		49.25	61	
	0180	10" diameter		41	.585		51	18.80		69.80	84.50	
	0190	12" diameter	↓	36	.667	↓	65	21.50		86.50	104	
	0200	High strength										
	0240	2" diameter	Q-1	58	.276	L.F.	10.45	8.55		19	24.50	
	0260	3" diameter		51	.314		14.70	9.75		24.45	31	
	0280	4" diameter		47	.340		17.40	10.55		27.95	35	
	0300	6" diameter	↓	38	.421		26.50	13.05		39.55	48.50	
	0320	8" diameter	Q-2	48	.500		41	16.10		57.10	69.50	
	0340	10" diameter		40	.600		60	19.30		79.30	95	
	0360	12" diameter	↓	36	.667	↓	74	21.50		95.50	114	
	0550	To delete coupling & hangers, subtract										
	0560	2" diam. to 6" diam.					33%	56%				

15100 | Building Services Piping

15108 | Plastic Pipe & Fittings

		CREW	DAILY OUTPUT	LABOR-HOURS	UNIT	2001 BARE COSTS MAT.	LABOR	EQUIP.	TOTAL	TOTAL INCL O&P
0570	8" diam. to 12" diam.					31%	52%			
0600	PVC, high impact/pressure, cplgs. 10' O.C., hangers 3 per 10'									
1020	Schedule 80									
1040	1/4" diameter	1 Plum	58	.138	L.F.	2.65	4.75		7.40	10.05
1060	3/8" diameter		55	.145		2.84	5		7.84	10.65
1070	1/2" diameter		50	.160		2.53	5.50		8.03	11.10
1080	3/4" diameter		47	.170		2.68	5.85		8.53	11.80
1090	1" diameter		43	.186		3.07	6.40		9.47	13
1100	1-1/4" diameter		39	.205		3.55	7.05		10.60	14.55
1110	1-1/2" diameter		34	.235		3.84	8.10		11.94	16.45
1120	2" diameter	Q-1	55	.291		4.44	9		13.44	18.50
1130	2-1/2" diameter		52	.308		5.70	9.55		15.25	20.50
1140	3" diameter		50	.320		7.80	9.90		17.70	23.50
1150	4" diameter		46	.348		10.25	10.80		21.05	27.50
1160	5" diameter		42	.381		13.80	11.80		25.60	33
1170	6" diameter		38	.421		18	13.05		31.05	39.50
1180	8" diameter	Q-2	47	.511		25.50	16.40		41.90	53
1190	10" diameter		42	.571		38.50	18.35		56.85	70
1200	12" diameter		38	.632		50	20.50		70.50	85.50
1730	To delete coupling & hangers, subtract									
1740	1/4" diam. to 1/2" diam.					62%	80%			
1750	3/4" diam. to 1-1/4" diam.					58%	73%			
1760	1-1/2" diam. to 6" diam.					40%	57%			
1770	8" diam. to 12" diam.					34%	50%			
1800	PVC, couplings 10' O.C., hangers 3 per 10'									
1820	Schedule 40									
1860	1/2" diameter	1 Plum	54	.148	L.F.	2.23	5.10		7.33	10.15
1870	3/4" diameter		51	.157		2.28	5.40		7.68	10.65
1880	1" diameter		46	.174		2.48	6		8.48	11.75
1890	1-1/4" diameter		42	.190		2.72	6.55		9.27	12.90
1900	1-1/2" diameter		36	.222		2.85	7.65		10.50	14.70
1910	2" diameter	Q-1	59	.271		3.17	8.40		11.57	16.20
1920	2-1/2" diameter		56	.286		3.97	8.85		12.82	17.70
1930	3" diameter		53	.302		4.96	9.35		14.31	19.60
1940	4" diameter		48	.333		6.05	10.35		16.40	22.50
1950	5" diameter		43	.372		7.75	11.55		19.30	26
1960	6" diameter		39	.410		10.10	12.70		22.80	30.50
1970	8" diameter	Q-2	48	.500		13.10	16.10		29.20	39
1980	10" diameter		43	.558		25.50	17.95		43.45	55.50
1990	12" diameter		42	.571		33.50	18.35		51.85	64
2000	14" diameter		31	.774		52.50	25		77.50	95.50
2010	16" diameter		23	1.043		83	33.50		116.50	142
2340	To delete coupling & hangers, subtract									
2360	1/2" diam. to 1-1/4" diam.					65%	74%			
2370	1-1/2" diam. to 6" diam.					44%	57%			
2380	8" diam. to 12" diam.					41%	53%			
2390	14" diam. to 16" diam.					48%	45%			
2420	Schedule 80									
2440	1/4" diameter	1 Plum	58	.138	L.F.	2.10	4.75		6.85	9.45
2450	3/8" diameter		55	.145		2.10	5		7.10	9.85
2460	1/2" diameter		50	.160		2.27	5.50		7.77	10.80
2470	3/4" diameter		47	.170		2.34	5.85		8.19	11.45
2480	1" diameter		43	.186		2.57	6.40		8.97	12.45
2490	1-1/4" diameter		39	.205		2.85	7.05		9.90	13.80
2500	1-1/2" diameter		34	.235		3.01	8.10		11.11	15.55
2510	2" diameter	Q-1	55	.291		3.30	9		12.30	17.25

15100 | Building Services Piping

15108 | Plastic Pipe & Fittings

		CREW	DAILY OUTPUT	LABOR-HOURS	UNIT	2001 BARE COSTS MAT.	LABOR	EQUIP.	TOTAL	TOTAL INCL O&P
2520	2-1/2" diameter	Q-1	52	.308	L.F.	3.86	9.55		13.41	18.65
2530	3" diameter		50	.320		5.45	9.90		15.35	21
2540	4" diameter		46	.348		6.80	10.80		17.60	24
2550	5" diameter		42	.381		10	11.80		21.80	29
2560	6" diameter	↓	38	.421		11.75	13.05		24.80	32.50
2570	8" diameter	Q-2	47	.511		15.65	16.40		32.05	42
2580	10" diameter		42	.571		31.50	18.35		49.85	62
2590	12" diameter	↓	38	.632	↓	40	20.50		60.50	74.50
2830	To delete coupling & hangers, subtract									
2840	1/4" diam. to 1/2" diam.					66%	80%			
2850	3/4" diam. to 1-1/4" diam.					61%	73%			
2860	1-1/2" diam. to 6" diam.					41%	57%			
2870	8" diam. to 12" diam.					31%	50%			
2900	Schedule 120									
2910	1/2" diameter	1 Plum	50	.160	L.F.	2.66	5.50		8.16	11.20
2950	3/4" diameter		47	.170		2.84	5.85		8.69	12
2960	1" diameter		43	.186		3.31	6.40		9.71	13.30
2970	1-1/4" diameter		39	.205		3.89	7.05		10.94	14.95
2980	1-1/2" diameter	↓	33	.242		4.26	8.35		12.61	17.30
2990	2" diameter	Q-1	54	.296		5.05	9.20		14.25	19.45
3000	2-1/2" diameter		52	.308		7.35	9.55		16.90	22.50
3010	3" diameter		49	.327		9.05	10.10		19.15	25.50
3020	4" diameter		45	.356		12.85	11		23.85	31
3030	6" diameter	↓	37	.432	↓	23	13.40		36.40	45.50
3240	To delete coupling & hangers, subtract									
3250	1/2" diam. to 1-1/4" diam.					52%	74%			
3260	1-1/2" diam. to 4" diam.					30%	57%			
3270	6" diam.					17%	50%			
3300	PVC, pressure, couplings 10' O.C., hangers 3 per 10'									
3310	SDR 26, 160 psi									
3350	1-1/4" diameter	1 Plum	42	.190	L.F.	2.55	6.55		9.10	12.70
3360	1-1/2" diameter	"	36	.222		2.66	7.65		10.31	14.50
3370	2" diameter	Q-1	59	.271		2.83	8.40		11.23	15.80
3380	2-1/2" diameter		56	.286		3.60	8.85		12.45	17.30
3390	3" diameter		53	.302		4.56	9.35		13.91	19.15
3400	4" diameter		48	.333		5.65	10.35		16	22
3420	6" diameter	↓	39	.410		9.85	12.70		22.55	30
3430	8" diameter	Q-2	48	.500	↓	13.10	16.10		29.20	39
3660	To delete coupling & hangers, subtract									
3670	1-1/4" diam.					63%	68%			
3680	1-1/2" diam. to 4" diam.					48%	57%			
3690	6" diam. to 8" diam.					60%	54%			
3720	SDR 21, 200 psi, 1/2" diameter	1 Plum	54	.148	L.F.	2.19	5.10		7.29	10.10
3740	3/4" diameter		51	.157		2.20	5.40		7.60	10.60
3750	1" diameter		46	.174		2.36	6		8.36	11.65
3760	1-1/4" diameter		42	.190		2.58	6.55		9.13	12.75
3770	1-1/2" diameter	↓	36	.222		2.70	7.65		10.35	14.50
3780	2" diameter	Q-1	59	.271		2.91	8.40		11.31	15.90
3790	2-1/2" diameter		56	.286		4.29	8.85		13.14	18.05
3800	3" diameter		53	.302		4.72	9.35		14.07	19.35
3810	4" diameter		48	.333		5.90	10.35		16.25	22
3830	6" diameter	↓	39	.410		10.40	12.70		23.10	30.50
3840	8" diameter	Q-2	48	.500	↓	14.25	16.10		30.35	40
4000	To delete coupling & hangers, subtract									
4010	1/2" diam. to 3/4" diam.					71%	77%			
4020	1" diam. to 1-1/4" diam.					63%	70%			

15100 | Building Services Piping

15108 | Plastic Pipe & Fittings

		CREW	DAILY OUTPUT	LABOR-HOURS	UNIT	MAT.	LABOR	EQUIP.	TOTAL	TOTAL INCL O&P
4030	1-1/2" diam. to 6" diam.					44%	57%			
4040	8" diam.					46%	54%			
4100	DWV type, schedule 40, couplings 10' O.C., hangers 3 per 10'									
4120	ABS									
4140	1-1/4" diameter	1 Plum	42	.190	L.F.	2.33	6.55		8.88	12.45
4150	1-1/2" diameter	"	36	.222		2.36	7.65		10.01	14.15
4160	2" diameter	Q-1	59	.271		2.49	8.40		10.89	15.45
4170	3" diameter		53	.302		3.62	9.35		12.97	18.15
4180	4" diameter		48	.333		4.07	10.35		14.42	20
4190	6" diameter	▼	39	.410	▼	12.90	12.70		25.60	33.50
4360	To delete coupling & hangers, subtract									
4370	1-1/4" diam.					64%	68%			
4380	1-1/2" diam. to 6" diam.					54%	57%			
4400	PVC									
4410	1-1/4" diameter	1 Plum	42	.190	L.F.	2.33	6.55		8.88	12.45
4420	1-1/2" diameter	"	36	.222		2.44	7.65		10.09	14.25
4460	2" diameter	Q-1	59	.271		2.51	8.40		10.91	15.45
4470	3" diameter		53	.302		4.34	9.35		13.69	18.95
4480	4" diameter		48	.333		4.74	10.35		15.09	21
4490	6" diameter	▼	39	.410		8.15	12.70		20.85	28
4500	8" diameter	Q-2	48	.500	▼	17.35	16.10		33.45	43.50
4750	To delete coupling & hangers, subtract									
4760	1-1/4" diam. to 1-1/2" diam.					71%	64%			
4770	2" diam. to 8" diam.					60%	57%			
4800	PVC, clear pipe, cplgs. 10' O.C., hangers 3 per 10', Sched. 40									
4840	1/4" diameter	1 Plum	59	.136	L.F.	2.49	4.67		7.16	9.80
4850	3/8" diameter		56	.143		2.72	4.92		7.64	10.45
4860	1/2" diameter		54	.148		3.10	5.10		8.20	11.10
4870	3/4" diameter		51	.157		3.44	5.40		8.84	11.95
4880	1" diameter		46	.174		4.31	6		10.31	13.80
4890	1-1/4" diameter		42	.190		5.20	6.55		11.75	15.60
4900	1-1/2" diameter	▼	36	.222		5.85	7.65		13.50	17.95
4910	2" diameter	Q-1	59	.271		7.20	8.40		15.60	20.50
4920	2-1/2" diameter		56	.286		10.60	8.85		19.45	25
4930	3" diameter		53	.302		13.30	9.35		22.65	29
4940	3-1/2" diameter		50	.320		20	9.90		29.90	37
4950	4" diameter	▼	48	.333	▼	19.05	10.35		29.40	36.50
5250	To delete coupling & hangers, subtract									
5260	1/4" diam. to 3/8" diam.					60%	81%			
5270	1/2" diam. to 3/4" diam.					41%	77%			
5280	1" diam. to 1-1/2" diam.					26%	67%			
5290	2" diam. to 4" diam.					16%	58%			
5360	CPVC, couplings 10' O.C., hangers 3 per 10'									
5380	Schedule 40									
5460	1/2" diameter	1 Plum	54	.148	L.F.	2.88	5.10		7.98	10.85
5470	3/4" diameter		51	.157		3.18	5.40		8.58	11.65
5480	1" diameter		46	.174		3.75	6		9.75	13.15
5490	1-1/4" diameter		42	.190		4.35	6.55		10.90	14.70
5500	1-1/2" diameter	▼	36	.222		5.35	7.65		13	17.45
5510	2" diameter	Q-1	59	.271		6.20	8.40		14.60	19.55
5520	2-1/2" diameter		56	.286		9.95	8.85		18.80	24.50
5530	3" diameter		53	.302		12	9.35		21.35	27.50
5540	4" diameter		48	.333		17.35	10.35		27.70	34.50
5550	6" diameter	▼	43	.372	▼	24.50	11.55		36.05	44.50
5730	To delete coupling & hangers, subtract									
5740	1/2" diam. to 3/4" diam.					37%	77%			

15100 | Building Services Piping

15108 | Plastic Pipe & Fittings

		CREW	DAILY OUTPUT	LABOR-HOURS	UNIT	2001 BARE COSTS MAT.	LABOR	EQUIP.	TOTAL	TOTAL INCL O&P
5750	1" diam. to 1-1/4" diam.					27%	70%			
5760	1-1/2" diam. to 3" diam.					21%	57%			
5770	4" diam. to 6" diam.					16%	57%			
5800	Schedule 80									
5860	1/2" diameter	1 Plum	50	.160	L.F.	3.27	5.50		8.77	11.90
5870	3/4" diameter		47	.170		3.36	5.85		9.21	12.55
5880	1" diameter		43	.186		4.02	6.40		10.42	14.05
5890	1-1/4" diameter		39	.205		4.76	7.05		11.81	15.90
5900	1-1/2" diameter		34	.235		5.90	8.10		14	18.75
5910	2" diameter	Q-1	55	.291		7.05	9		16.05	21.50
5920	2-1/2" diameter		52	.308		11.05	9.55		20.60	26.50
5930	3" diameter		50	.320		13.65	9.90		23.55	30
5940	4" diameter		46	.348		19.95	10.80		30.75	38.50
5950	6" diameter		38	.421		30.50	13.05		43.55	53
5960	8" diameter	Q-2	47	.511		64.50	16.40		80.90	96
6060	To delete couplings & hangers, subtract									
6070	1/2" diam. to 3/4" diam.					44%	77%			
6080	1" diam. to 1-1/4" diam.					32%	71%			
6090	1-1/2" diam. to 4" diam.					25%	58%			
6100	6" diam. to 8" diam.					20%	53%			
6240	CTS, 1/2" diameter	1 Plum	54	.148	L.F.	2.12	5.10		7.22	10.05
6250	3/4" diameter		51	.157		2.60	5.40		8	11
6260	1" diameter		46	.174		2.90	6		8.90	12.25
6270	1 1/4"		42	.190		3.40	6.55		9.95	13.65
6280	1 1/2" diameter		36	.222		3.92	7.65		11.57	15.85
6290	2" diameter	Q-1	59	.271		5.30	8.40		13.70	18.55
6370	To delete coupling & hangers, subtract									
6380	1/2" diam.					51%	79%			
6390	3/4" diam.					40%	76%			
6392	1" thru 2" diam.					72%	68%			
7280	Polyethylene, flexible, no couplings or hangers									
7282	Note: For labor costs add 25% to the couplings and fittings labor total.									
7300	SDR 15, 100 psi									
7310	3/4" diameter				L.F.	.17			.17	.19
7350	1" diameter					.22			.22	.24
7360	1-1/4" diameter					.41			.41	.45
7370	1-1/2" diameter					.49			.49	.54
7380	2" diameter					.80			.80	.88
7700	SIDR 9, 160 psi									
7710	1/2" diameter				L.F.	.22			.22	.24
7750	3/4" diameter					.22			.22	.24
7760	1" diameter					.34			.34	.37
7770	1-1/4" diameter					.60			.60	.66
7780	1-1/2" diameter					.84			.84	.92
7790	2" diameter					1.38			1.38	1.52
8120	SDR 9, 200 psi									
8150	3/4" diameter				L.F.	.20			.20	.22
8160	1" diameter					.32			.32	.35
8170	1-1/4" diameter					.49			.49	.54
8420	SDR 7, 250 psi									
8440	3/4" diameter				L.F.	.28			.28	.31
8450	1" diameter					.46			.46	.51
8460	1-1/4" diameter					.80			.80	.88
8800	PVC, type PSP, drain & sewer, belled end gasket jnt., no hngr.									
8840	3" diameter				L.F.	.32			.32	.35
8850	4" diameter					.43			.43	.47

15100 | Building Services Piping

15108 | Plastic Pipe & Fittings

			CREW	DAILY OUTPUT	LABOR-HOURS	UNIT	2001 BARE COSTS MAT.	LABOR	EQUIP.	TOTAL	TOTAL INCL O&P	
520	8860	6" diameter				L.F.	.89			.89	.98	520
	9000	Perforated										
	9040	4" diameter				L.F.	.57			.57	.63	
560	0010	**PIPE, PLASTIC, FITTINGS**										560
	0030	Epoxy resin, fiberglass reinforced, general service										
	0090	Elbow, 90°, 2"	Q-1	33.10	.483	Ea.	45.50	15		60.50	72.50	
	0100	3"		20.80	.769		64.50	24		88.50	107	
	0110	4"		16.50	.970		88	30		118	143	
	0120	6"	↓	10.10	1.584		134	49		183	222	
	0130	8"	Q-2	9.30	2.581		248	83		331	395	
	0140	10"		8.50	2.824		310	91		401	475	
	0150	12"	↓	7.60	3.158	↓	445	102		547	645	
	0160	45° Elbow, same as 90°										
	0170	Elbow, 90°, flanged										
	0172	2"	Q-1	23	.696	Ea.	91	21.50		112.50	133	
	0173	3"		16	1		105	31		136	163	
	0174	4"		13	1.231		138	38		176	209	
	0176	6"	↓	8	2		249	62		311	370	
	0177	8"	Q-2	9	2.667		450	85.50		535.50	620	
	0178	10"		7	3.429		615	110		725	840	
	0179	12"	↓	5	4.800	↓	835	154		989	1,150	
	0186	Elbow, 45°, flanged										
	0188	2"	Q-1	23	.696	Ea.	91.50	21.50		113	134	
	0189	3"		16	1		105	31		136	163	
	0190	4"		13	1.231		137	38		175	209	
	0192	6"	↓	8	2		251	62		313	370	
	0193	8"	Q-2	9	2.667		450	85.50		535.50	625	
	0194	10"		7	3.429		615	110		725	840	
	0195	12"	↓	5	4.800		835	154		989	1,150	
	0290	Tee, 2"	Q-1	20	.800		55	25		80	98	
	0300	3"		13.90	1.151		64.50	35.50		100	125	
	0310	4"		11	1.455		76.50	45		121.50	152	
	0320	6"	↓	6.70	2.388		205	74		279	340	
	0330	8"	Q-2	6.20	3.871		237	124		361	450	
	0340	10"		5.70	4.211		380	135		515	620	
	0350	12"	↓	5.10	4.706	↓	460	151		611	735	
	0352	Tee, flanged										
	0354	2"	Q-1	17	.941	Ea.	124	29		153	180	
	0355	3"		10	1.600		166	49.50		215.50	258	
	0356	4"		8	2		185	62		247	297	
	0358	6"	↓	5	3.200		365	99		464	550	
	0359	8"	Q-2	6	4		625	129		754	880	
	0360	10"		5	4.800		905	154		1,059	1,225	
	0361	12"	↓	4	6	↓	1,150	193		1,343	1,575	
	0365	Wye, flanged										
	0367	2"	Q-1	17	.941	Ea.	320	29		349	395	
	0368	3"		10	1.600		355	49.50		404.50	465	
	0369	4"		8	2		440	62		502	580	
	0371	6"	↓	5	3.200		530	99		629	730	
	0372	8"	Q-2	6	4		690	129		819	950	
	0373	10"		5	4.800		1,175	154		1,329	1,500	
	0374	12"	↓	4	6	↓	1,425	193		1,618	1,850	
	0380	Couplings										
	0410	2"	Q-1	33.10	.483	Ea.	9.85	15		24.85	33.50	
	0420	3"	↓	20.80	.769	↓	16.95	24		40.95	54.50	

MECHANICAL 15

15100 | Building Services Piping

15108 | Plastic Pipe & Fittings

		CREW	DAILY OUTPUT	LABOR-HOURS	UNIT	MAT.	LABOR	EQUIP.	TOTAL	TOTAL INCL O&P
0430	4"	Q-1	16.50	.970	Ea.	20	30		50	67.50
0440	6"	↓	10.10	1.584		38	49		87	116
0450	8"	Q-2	9.30	2.581		58	83		141	189
0460	10"		8.50	2.824		87.50	91		178.50	233
0470	12"	↓	7.60	3.158	↓	117	102		219	282
0473	High corrosion resistant couplings, add					30%				
0474	Reducer, concentric, flanged									
0475	2" x 1-1/2"	Q-1	30	.533	Ea.	132	16.55		148.55	170
0476	3" x 2"		24	.667		150	20.50		170.50	196
0477	4" x 3"		19	.842		158	26		184	214
0479	6" x 4"	↓	15	1.067		206	33		239	277
0480	8" x 6"	Q-2	16	1.500		289	48		337	395
0481	10" x 8"		13	1.846		385	59.50		444.50	515
0482	12" x 10"	↓	11	2.182	↓	555	70		625	715
0486	Adapter, bell x male or female									
0488	2"	Q-1	28	.571	Ea.	17.70	17.70		35.40	46
0489	3"		20	.800		22.50	25		47.50	62.50
0491	4"		17	.941		31	29		60	78
0492	6"	↓	12	1.333		61.50	41.50		103	130
0493	8"	Q-2	15	1.600		86	51.50		137.50	172
0494	10"	"	11	2.182	↓	151	70		221	272
0528	Flange									
0532	2"	Q-1	46	.348	Ea.	17.40	10.80		28.20	35.50
0533	3"		32	.500		21	15.50		36.50	47
0534	4"		26	.615		29.50	19.10		48.60	61.50
0536	6"	↓	16	1		50	31		81	102
0537	8"	Q-2	18	1.333		82	43		125	155
0538	10"		14	1.714		114	55		169	209
0539	12"	↓	10	2.400	↓	161	77		238	293
2100	PVC schedule 80									
2110	90° elbow, 1/2"	1 Plum	30.30	.264	Ea.	.98	9.10		10.08	14.85
2130	3/4"		26	.308		1.27	10.60		11.87	17.40
2140	1"		22.70	.352		2.03	12.15		14.18	20.50
2150	1-1/4"		20.20	.396		2.71	13.65		16.36	23.50
2160	1-1/2"	↓	18.20	.440		2.91	15.15		18.06	26
2170	2"	Q-1	33.10	.483		3.51	15		18.51	26.50
2180	3"		20.80	.769		9.25	24		33.25	46
2190	4"		16.50	.970		19.15	30		49.15	66.50
2200	6"	↓	10.10	1.584		54.50	49		103.50	134
2210	8"	Q-2	9.30	2.581		150	83		233	290
2250	45° elbow, 1/2"	1 Plum	30.30	.264		1.86	9.10		10.96	15.80
2270	3/4"		26	.308		2.83	10.60		13.43	19.10
2280	1"		22.70	.352		4.24	12.15		16.39	23
2290	1-1/4"		20.20	.396		5.40	13.65		19.05	26.50
2300	1-1/2"	↓	18.20	.440		6.40	15.15		21.55	30
2310	2"	Q-1	33.10	.483		8.30	15		23.30	31.50
2320	3"		20.80	.769		21	24		45	59.50
2330	4"		16.50	.970		38	30		68	87.50
2340	6"	↓	10.10	1.584		48	49		97	127
2350	8"	Q-2	9.30	2.581		92.50	83		175.50	227
2400	Tee, 1/2"	1 Plum	20.20	.396		2.79	13.65		16.44	23.50
2420	3/4"		17.30	.462		2.91	15.95		18.86	27
2430	1"		15.20	.526		3.63	18.15		21.78	31.50
2440	1-1/4"		13.50	.593		10	20.50		30.50	42
2450	1-1/2"	↓	12.10	.661		10	23		33	45.50
2460	2"	Q-1	20	.800		12.50	25		37.50	51.50

15100 | Building Services Piping

15108 | Plastic Pipe & Fittings

		CREW	DAILY OUTPUT	LABOR-HOURS	UNIT	2001 BARE COSTS MAT.	LABOR	EQUIP.	TOTAL	TOTAL INCL O&P	
2470	3"	Q-1	13.90	1.151	Ea.	17	35.50		52.50	72.50	
2480	4"		11	1.455		19.65	45		64.65	89.50	
2490	6"		6.70	2.388		61	74		135	180	
2500	8"	Q-2	6.20	3.871		198	124		322	405	
2510	Flange, socket, 150 lb., 1/2"	1 Plum	55.60	.144		5.40	4.96		10.36	13.40	
2514	3/4"		47.60	.168		5.75	5.80		11.55	15.10	
2518	1"		41.70	.192		6.40	6.60		13	17.05	
2522	1-1/2"		33.30	.240		6.75	8.30		15.05	19.95	
2526	2"	Q-1	60.60	.264		9	8.20		17.20	22.50	
2530	4"		30.30	.528		19.40	16.35		35.75	46	
2534	6"		18.50	.865		30.50	27		57.50	74	
2538	8"	Q-2	17.10	1.404		50	45		95	123	
2550	Coupling, 1/2"	1 Plum	30.30	.264		1.78	9.10		10.88	15.70	
2570	3/4"		26	.308		2.44	10.60		13.04	18.70	
2580	1"		22.70	.352		2.47	12.15		14.62	21	
2590	1-1/4"		20.20	.396		3.77	13.65		17.42	24.50	
2600	1-1/2"		18.20	.440		4.07	15.15		19.22	27.50	
2610	2"	Q-1	33.10	.483		4.36	15		19.36	27.50	
2620	3"		20.80	.769		12.35	24		36.35	49.50	
2630	4"		16.50	.970		15.45	30		45.45	62.50	
2640	6"		10.10	1.584		33.50	49		82.50	111	
2650	8"	Q-2	9.30	2.581		35	83		118	164	
2660	10"		8.50	2.824		41	91		132	182	
2670	12"		7.60	3.158		44	102		146	202	
2700	PVC (white), schedule 40, socket joints										
2760	90° elbow, 1/2"	1 Plum	33.30	.240	Ea.	.26	8.30		8.56	12.80	
2770	3/4"		28.60	.280		.29	9.65		9.94	14.85	
2780	1"		25	.320		.53	11		11.53	17.25	
2790	1-1/4"		22.20	.360		.94	12.40		13.34	19.80	
2800	1-1/2"		20	.400		1	13.80		14.80	22	
2810	2"	Q-1	36.40	.440		1.56	13.65		15.21	22	
2820	2-1/2"		26.70	.599		4.75	18.60		23.35	33.50	
2830	3"		22.90	.699		5.70	21.50		27.20	39	
2840	4"		18.20	.879		10.20	27.50		37.70	52	
2850	5"		12.10	1.322		26.50	41		67.50	91	
2860	6"		11.10	1.441		32.50	44.50		77	103	
2870	8"	Q-2	10.30	2.330		84.50	75		159.50	206	
2980	45° elbow, 1/2"	1 Plum	33.30	.240		.43	8.30		8.73	12.95	
2990	3/4"		28.60	.280		.66	9.65		10.31	15.30	
3000	1"		25	.320		.80	11		11.80	17.55	
3010	1-1/4"		22.20	.360		1.12	12.40		13.52	20	
3020	1-1/2"		20	.400		.90	13.80		14.70	22	
3030	2"	Q-1	36.40	.440		1.83	13.65		15.48	22.50	
3040	2-1/2"		26.70	.599		4.76	18.60		23.36	33.50	
3050	3"		22.90	.699		7.40	21.50		28.90	40.50	
3060	4"		18.20	.879		13.25	27.50		40.75	55.50	
3070	5"		12.10	1.322		26.50	41		67.50	91	
3080	6"		11.10	1.441		32.50	44.50		77	104	
3090	8"	Q-2	10.30	2.330		93	75		168	215	
3180	Tee, 1/2"	1 Plum	22.20	.360		.32	12.40		12.72	19.10	
3190	3/4"		19	.421		.36	14.50		14.86	22.50	
3200	1"		16.70	.479		.70	16.50		17.20	26	
3210	1-1/4"		14.80	.541		1.10	18.60		19.70	29	
3220	1-1/2"		13.30	.601		1.33	20.50		21.83	33	
3230	2"	Q-1	24.20	.661		1.93	20.50		22.43	33	
3240	2-1/2"		17.80	.899		6.35	28		34.35	49	

15100 | Building Services Piping

15108 | Plastic Pipe & Fittings

		CREW	DAILY OUTPUT	LABOR-HOURS	UNIT	MAT.	LABOR	EQUIP.	TOTAL	TOTAL INCL O&P
3250	3"	Q-1	15.20	1.053	Ea.	8.35	32.50		40.85	58.50
3260	4"		12.10	1.322		15.10	41		56.10	78.50
3270	5"		8.10	1.975		36.50	61		97.50	133
3280	6"		7.40	2.162		51	67		118	157
3290	8"	Q-2	6.80	3.529		60.50	113		173.50	238
3380	Coupling, 1/2"	1 Plum	33.30	.240		.17	8.30		8.47	12.70
3390	3/4"		28.60	.280		.24	9.65		9.89	14.80
3400	1"		25	.320		.41	11		11.41	17.10
3410	1-1/4"		22.20	.360		.56	12.40		12.96	19.35
3420	1-1/2"		20	.400		.60	13.80		14.40	21.50
3430	2"	Q-1	36.40	.440		.94	13.65		14.59	21.50
3440	2-1/2"		26.70	.599		2.06	18.60		20.66	30.50
3450	3"		22.90	.699		3.24	21.50		24.74	36
3460	4"		18.20	.879		5.05	27.50		32.55	46.50
3470	5"		12.10	1.322		11.40	41		52.40	74.50
3480	6"		11.10	1.441		14.75	44.50		59.25	84
3490	8"	Q-2	10.30	2.330		28	75		103	144
3710	Reducing insert, schedule 40, socket weld									
3712	3/4"	1 Plum	31.50	.254	Ea.	.28	8.75		9.03	13.50
3713	1"		27.50	.291		.51	10		10.51	15.70
3715	1-1/2"		22	.364		.71	12.55		13.26	19.70
3716	2"	Q-1	40	.400		1.17	12.40		13.57	20
3717	4"		20	.800		6.25	25		31.25	44.50
3718	6"		12.20	1.311		15.50	40.50		56	78.50
3719	8"	Q-2	11.30	2.124		54	68.50		122.50	163
3730	Reducing insert, socket weld x female/male thread									
3732	1/2"	1 Plum	38.30	.209	Ea.	1.25	7.20		8.45	12.25
3733	3/4"		32.90	.243		.77	8.40		9.17	13.50
3734	1"		28.80	.278		1.09	9.55		10.64	15.65
3736	1-1/2"		23	.348		1.95	12		13.95	20.50
3737	2"	Q-1	41.90	.382		1.04	11.85		12.89	19
3738	4"	"	20.90	.766		7.80	23.50		31.30	44.50
3742	Male adapter, socket weld x male thread									
3744	1/2"	1 Plum	38.30	.209	Ea.	.24	7.20		7.44	11.10
3745	3/4"		32.90	.243		.36	8.40		8.76	13.05
3746	1"		28.80	.278		.47	9.55		10.02	14.95
3748	1-1/2"		23	.348		.77	12		12.77	18.95
3749	2"	Q-1	41.90	.382		1.01	11.85		12.86	18.95
3750	4"	"	20.90	.766		5.60	23.50		29.10	42
3754	Female adapter, socket weld x female thread									
3756	1/2"	1 Plum	38.30	.209	Ea.	.39	7.20		7.59	11.30
3757	3/4"		32.90	.243		.50	8.40		8.90	13.20
3758	1"		28.80	.278		.58	9.55		10.13	15.10
3760	1-1/2"		23	.348		.77	12		12.77	18.95
3761	2"	Q-1	41.90	.382		1.04	11.85		12.89	19
3762	4"	"	20.90	.766		5.85	23.50		29.35	42.50
3800	PVC, schedule 80, socket joints									
3810	Reducing insert									
3812	3/4"	1 Plum	28.60	.280	Ea.	.78	9.65		10.43	15.40
3813	1"		25	.320		3.03	11		14.03	20
3815	1-1/2"		20	.400		4.79	13.80		18.59	26.50
3816	2"	Q-1	36.40	.440		6.80	13.65		20.45	28
3817	4"		18.20	.879		26	27.50		53.50	69.50
3818	6"		11.10	1.441		36	44.50		80.50	108
3819	8"	Q-2	10.20	2.353		56	75.50		131.50	176
3830	Reducing insert, socket weld x female/male thread									

Important: See the Reference Section for critical supporting data - Reference Nos., Crews, & City Cost Indexes

15100 | Building Services Piping

15108 | Plastic Pipe & Fittings

		CREW	DAILY OUTPUT	LABOR-HOURS	UNIT	MAT.	LABOR	EQUIP.	TOTAL	TOTAL INCL O&P
3832	1/2"	1 Plum	34.80	.230	Ea.	2.28	7.90		10.18	14.45
3833	3/4"		29.90	.268		1.98	9.20		11.18	16.10
3834	1"		26.10	.307		3.43	10.55		13.98	19.70
3836	1-1/2"		20.90	.383		5.75	13.20		18.95	26.50
3837	2"	Q-1	38	.421		8.35	13.05		21.40	29
3838	4"	"	19	.842		23.50	26		49.50	65.50
3844	Adapter, male socket x male thread									
3846	1/2"	1 Plum	34.80	.230	Ea.	1.80	7.90		9.70	13.95
3847	3/4"		29.90	.268		1.98	9.20		11.18	16.10
3848	1"		26.10	.307		3.43	10.55		13.98	19.70
3850	1-1/2"		20.90	.383		5.75	13.20		18.95	26.50
3851	2"	Q-1	38	.421		8.35	13.05		21.40	29
3852	4"	"	19	.842		11.60	26		37.60	52.50
3860	Adapter, female socket x female thread									
3862	1/2"	1 Plum	34.80	.230	Ea.	1.44	7.90		9.34	13.55
3863	3/4"		29.90	.268		3.37	9.20		12.57	17.60
3864	1"		26.10	.307		4.97	10.55		15.52	21.50
3866	1-1/2"		20.90	.383		9.90	13.20		23.10	31
3867	2"	Q-1	38	.421		17.25	13.05		30.30	38.50
3868	4"	"	19	.842		52.50	26		78.50	97.50
3872	Union, socket joints									
3874	1/2"	1 Plum	25.80	.310	Ea.	4.97	10.70		15.67	21.50
3875	3/4"		22.10	.362		7.20	12.45		19.65	27
3876	1"		19.30	.415		6.15	14.30		20.45	28.50
3878	1-1/2"		15.50	.516		16.20	17.80		34	45
3879	2"	Q-1	28.10	.569		22	17.65		39.65	50.50
3888	Cap									
3890	1/2"	1 Plum	54.50	.147	Ea.	2.38	5.05		7.43	10.25
3891	3/4"		46.70	.171		3.33	5.90		9.23	12.55
3892	1"		41	.195		4.44	6.70		11.14	15.05
3894	1-1/2"		32.80	.244		5.35	8.40		13.75	18.60
3895	2"	Q-1	59.50	.269		14.10	8.35		22.45	28
3896	4"		30	.533		42.50	16.55		59.05	72
3897	6"		18.20	.879		108	27.50		135.50	160
3898	8"	Q-2	16.70	1.437		136	46		182	220
4500	DWV, ABS, non pressure, socket joints									
4540	1/4 Bend, 1-1/4"	1 Plum	20.20	.396	Ea.	1.90	13.65		15.55	22.50
4560	1-1/2"	"	18.20	.440		.85	15.15		16	24
4570	2"	Q-1	33.10	.483		1.39	15		16.39	24
4580	3"		20.80	.769		3.31	24		27.31	39.50
4590	4"		16.50	.970		5.75	30		35.75	52
4600	6"		10.10	1.584		36.50	49		85.50	114
4650	1/8 Bend, same as 1/4 Bend									
4800	Tee, sanitary									
4820	1-1/4"	1 Plum	13.50	.593	Ea.	2.40	20.50		22.90	33.50
4830	1-1/2"	"	12.10	.661		1.25	23		24.25	36
4840	2"	Q-1	20	.800		1.77	25		26.77	39.50
4850	3"		13.90	1.151		4.50	35.50		40	59
4860	4"		11	1.455		11.15	45		56.15	80.50
4862	Tee, sanitary, reducing, 2" x 1-1/2"		22	.727		2.63	22.50		25.13	37
4864	3" x 2"		15.30	1.046		3.80	32.50		36.30	53
4868	4" x 3"		12.10	1.322		11.65	41		52.65	75
4870	Combination Y and 1/8 bend									
4872	1-1/2"	1 Plum	12.10	.661	Ea.	4.51	23		27.51	39.50
4874	2"	Q-1	20	.800		5.45	25		30.45	43.50
4876	3"		13.90	1.151		9.65	35.50		45.15	64.50

15100 | Building Services Piping

15108 | Plastic Pipe & Fittings

		CREW	DAILY OUTPUT	LABOR-HOURS	UNIT	MAT.	LABOR	EQUIP.	TOTAL	TOTAL INCL O&P
4878	4"	Q-1	11	1.455	Ea.	18.95	45		63.95	89
4880	3" x 1-1/2"		15.50	1.032		11.25	32		43.25	61
4882	4" x 3"	↓	12.10	1.322		16.95	41		57.95	80.50
4900	Wye, 1-1/4"	1 Plum	13.50	.593		2.77	20.50		23.27	34
4902	1-1/2"	"	12.10	.661		2.83	23		25.83	37.50
4904	2"	Q-1	20	.800		2.90	25		27.90	40.50
4906	3"		13.90	1.151		6.70	35.50		42.20	61.50
4908	4"		11	1.455		15.80	45		60.80	85.50
4910	6"		6.70	2.388		60	74		134	178
4918	3" x 1-1/2"		15.50	1.032		6.60	32		38.60	56
4920	4" x 3"		12.10	1.322		10.95	41		51.95	74
4922	6" x 4"	↓	6.90	2.319		35.50	72		107.50	148
4930	Double Wye, 1-1/2"	1 Plum	9.10	.879		4.51	30.50		35.01	50.50
4932	2"	Q-1	16.60	.964		5.80	30		35.80	51.50
4934	3"		10.40	1.538		15	47.50		62.50	88.50
4936	4"		8.25	1.939		30.50	60		90.50	125
4940	2" x 1-1/2"		16.80	.952		5.80	29.50		35.30	51
4942	3" x 2"		10.60	1.509		11.10	47		58.10	83
4944	4" x 3"		8.45	1.893		32	58.50		90.50	124
4946	6" x 4"		7.25	2.207		67	68.50		135.50	177
4950	Reducer bushing, 2" x 1-1/2"		36.40	.440		.99	13.65		14.64	21.50
4952	3" x 1-1/2"		27.30	.586		3.57	18.15		21.72	31.50
4954	4" x 2"		18.20	.879		8.20	27.50		35.70	50
4956	6" x 4"	↓	11.10	1.441		22	44.50		66.50	91.50
4960	Couplings, 1-1/2"	1 Plum	18.20	.440		2.07	15.15		17.22	25.50
4962	2"	Q-1	33.10	.483		.66	15		15.66	23
4963	3"		20.80	.769		2.22	24		26.22	38.50
4964	4"		16.50	.970		3.40	30		33.40	49
4966	6"		10.10	1.584		18	49		67	94
4970	2" x 1-1/2"		33.30	.480		1.76	14.90		16.66	24.50
4972	3" x 1-1/2"		21	.762		5.05	23.50		28.55	41
4974	4" x 3"	↓	16.70	.958		8.70	29.50		38.20	54.50
4978	Closet flange, 4"	1 Plum	32	.250		3.23	8.60		11.83	16.55
4980	4" x 3"	"	34	.235	↓	3.03	8.10		11.13	15.60
5000	DWV, PVC, schedule 40, socket joints									
5040	1/4 bend, 1-1/4"	1 Plum	20.20	.396	Ea.	1.64	13.65		15.29	22.50
5060	1-1/2"	"	18.20	.440		.82	15.15		15.97	24
5070	2"	Q-1	33.10	.483		1.28	15		16.28	24
5080	3"		20.80	.769		3.19	24		27.19	39.50
5090	4"		16.50	.970		5.70	30		35.70	52
5100	6"	↓	10.10	1.584		35	49		84	113
5105	8"	Q-2	9.30	2.581		70.50	83		153.50	203
5106	10"	"	8.50	2.824		70.50	91		161.50	215
5110	1/4 bend, long sweep, 1-1/2"	1 Plum	18.20	.440		2.61	15.15		17.76	26
5112	2"	Q-1	33.10	.483		2.11	15		17.11	25
5114	3"		20.80	.769		5.55	24		29.55	42
5116	4"	↓	16.50	.970		10.95	30		40.95	57.50
5150	1/8 bend, 1-1/4"	1 Plum	20.20	.396		1.67	13.65		15.32	22.50
5170	1-1/2"	"	18.20	.440		.92	15.15		16.07	24
5180	2"	Q-1	33.10	.483		1.35	15		16.35	24
5190	3"		20.80	.769		2.75	24		26.75	39
5200	4"		16.50	.970		4.56	30		34.56	50.50
5210	6"	↓	10.10	1.584		32	49		81	109
5215	8"	Q-2	9.30	2.581		47.50	83		130.50	177
5216	10"		8.50	2.824		71.50	91		162.50	216
5217	12"	↓	7.60	3.158	↓	114	102		216	279

15100 | Building Services Piping

15108 | Plastic Pipe & Fittings

		CREW	DAILY OUTPUT	LABOR-HOURS	UNIT	2001 BARE COSTS MAT.	LABOR	EQUIP.	TOTAL	TOTAL INCL O&P
5250	Tee, sanitary 1-1/4"	1 Plum	13.50	.593	Ea.	2.94	20.50		23.44	34
5254	1-1/2"	"	12.10	.661		1.24	23		24.24	36
5255	2"	Q-1	20	.800		1.69	25		26.69	39.50
5256	3"		13.90	1.151		4.65	35.50		40.15	59
5257	4"		11	1.455		8.85	45		53.85	78
5259	6"		6.70	2.388		51.50	74		125.50	169
5261	8"	Q-2	6.20	3.871		154	124		278	360
5264	2" x 1-1/2"	Q-1	22	.727		4.46	22.50		26.96	39
5266	3" x 1-1/2"		15.50	1.032		3.56	32		35.56	52.50
5268	4" x 3"		12.10	1.322		16.35	41		57.35	80
5271	6" x 4"		6.90	2.319		55.50	72		127.50	170
5276	Tee, sanitary, reducing									
5281	2" x 1-1/2" x 1-1/2"	Q-1	23	.696	Ea.	2.46	21.50		23.96	35
5282	2" x 1-1/2" x 2"		22	.727		3.38	22.50		25.88	37.50
5283	2" x 2" x 1-1/2"		22	.727		1.95	22.50		24.45	36
5284	3" x 3" x 1-1/2"		15.50	1.032		3.96	32		35.96	53
5285	3" x 3" x 2"		15.30	1.046		4.22	32.50		36.72	53.50
5286	4" x 4" x 1-1/2"		12.30	1.301		13.60	40.50		54.10	76
5287	4" x 4" x 2"		12.20	1.311		13.35	40.50		53.85	76
5288	4" x 4" x 3"		12.10	1.322		16.35	41		57.35	80
5291	6" x 6" x 4"		6.90	2.319		55.50	72		127.50	170
5294	Tee, double sanitary									
5295	1-1/2"	1 Plum	9.10	.879	Ea.	4.10	30.50		34.60	50
5296	2"	Q-1	16.60	.964		4.52	30		34.52	50
5297	3"		10.40	1.538		12.65	47.50		60.15	86
5298	4"		8.25	1.939		20.50	60		80.50	114
5303	Wye, reducing									
5304	2" x 1-1/2" x 1-1/2"	Q-1	23	.696	Ea.	3.81	21.50		25.31	36.50
5305	2" x 2" x 1-1/2"		22	.727		4.80	22.50		27.30	39.50
5306	3" x 3" x 2"		15.30	1.046		9	32.50		41.50	59
5307	4" x 4" x 2"		12.20	1.311		9.25	40.50		49.75	71.50
5309	4" x 4" x 3"		12.10	1.322		10.80	41		51.80	74
5314	Combination Y & 1/8 bend, 1-1/2"	1 Plum	12.10	.661		3.92	23		26.92	39
5315	2"	Q-1	20	.800		5.05	25		30.05	43
5317	3"		13.90	1.151		8.35	35.50		43.85	63
5318	4"		11	1.455		16.40	45		61.40	86
5319	6"		6.70	2.388		72	74		146	191
5320	8"	Q-2	6.20	3.871		115	124		239	315
5324	Combination Y & 1/8 bend, reducing									
5325	2" x 2" x 1-1/2"	Q-1	22	.727	Ea.	5.60	22.50		28.10	40
5327	3" x 3" x 1-1/2"		15.50	1.032		9.75	32		41.75	59
5328	3" x 3" x 2"		15.30	1.046		6.10	32.50		38.60	55.50
5329	4" x 4" x 2"		12.20	1.311		13.55	40.50		54.05	76.50
5331	Wye, 1-1/4"	1 Plum	13.50	.593		3.13	20.50		23.63	34.50
5332	1-1/2"	"	12.10	.661		2.44	23		25.44	37
5333	2"	Q-1	20	.800		2.45	25		27.45	40
5334	3"		13.90	1.151		5.55	35.50		41.05	60
5335	4"		11	1.455		10.35	45		55.35	79.50
5336	6"		6.70	2.388		38	74		112	154
5337	8"	Q-2	6.20	3.871		39	124		163	231
5338	10"		5.70	4.211		61	135		196	271
5339	12"		5.10	4.706		90	151		241	325
5341	2" x 1-1/2"	Q-1	22	.727		4.31	22.50		26.81	38.50
5342	3" x 1-1/2"		15.50	1.032		5.70	32		37.70	55
5343	4" x 3"		12.10	1.322		6.70	41		47.70	69.50
5344	6" x 4"		6.90	2.319		38	72		110	151

15100 | Building Services Piping

15108 | Plastic Pipe & Fittings

		CREW	DAILY OUTPUT	LABOR-HOURS	UNIT	2001 BARE COSTS MAT.	LABOR	EQUIP.	TOTAL	TOTAL INCL O&P
5345	8" x 6"	Q-2	6.40	3.750	Ea.	52	121		173	239
5347	Double wye, 1-1/2"	1 Plum	9.10	.879		5.20	30.50		35.70	51
5348	2"	Q-1	16.60	.964		6.70	30		36.70	52.50
5349	3"		10.40	1.538		17.25	47.50		64.75	91
5350	4"		8.25	1.939		35	60		95	130
5354	2" x 1-1/2"		16.80	.952		6.10	29.50		35.60	51.50
5355	3" x 2"		10.60	1.509		12.85	47		59.85	84.50
5356	4" x 3"		8.45	1.893		27.50	58.50		86	119
5357	6" x 4"		7.25	2.207		58	68.50		126.50	167
5374	Coupling, 1-1/4"	1 Plum	20.20	.396		1.58	13.65		15.23	22
5376	1-1/2"	"	18.20	.440		.53	15.15		15.68	23.50
5378	2"	Q-1	33.10	.483		.55	15		15.55	23
5380	3"		20.80	.769		1.58	24		25.58	37.50
5390	4"		16.50	.970		2.61	30		32.61	48.50
5400	6"		10.10	1.584		13.95	49		62.95	89.50
5402	8"	Q-2	9.30	2.581		31.50	83		114.50	160
5404	2" x 1-1/2"	Q-1	33.30	.480		1.37	14.90		16.27	24
5406	3" x 1-1/2"		21	.762		4.37	23.50		27.87	40.50
5408	4" x 3"		16.70	.958		7.55	29.50		37.05	53.50
5410	Reducer bushing, 2" x 1-1/4"		36.50	.438		.92	13.60		14.52	21.50
5411	2" x 1-1/2"		36.40	.440		1.37	13.65		15.02	22
5412	3" x 1-1/2"		27.30	.586		4.37	18.15		22.52	32.50
5413	3" x 2"		27.10	.590		3.65	18.30		21.95	31.50
5414	4" x 2"		18.20	.879		8.05	27.50		35.55	50
5415	4" x 3"		16.70	.958		8.40	29.50		37.90	54.50
5416	6" x 4"		11.10	1.441		22.50	44.50		67	92.50
5418	8" x 6"	Q-2	10.20	2.353		45.50	75.50		121	164
5425	Closet flange 4"	Q-1	32	.500		10	15.50		25.50	34.50
5426	4" x 3"	"	34	.471		7.20	14.60		21.80	30
5450	Solvent cement for PVC, industrial grade, per quart				Qt.	10.80			10.80	11.90
5500	CPVC, Schedule 80, threaded joints									
5540	90° Elbow, 1/4"	1 Plum	32	.250	Ea.	6.25	8.60		14.85	19.90
5560	1/2"		30.30	.264		2.45	9.10		11.55	16.45
5570	3/4"		26	.308		3.13	10.60		13.73	19.45
5580	1"		22.70	.352		4.96	12.15		17.11	24
5590	1-1/4"		20.20	.396		10.75	13.65		24.40	32.50
5600	1-1/2"		18.20	.440		12	15.15		27.15	36
5610	2"	Q-1	33.10	.483		14.45	15		29.45	38.50
5620	2-1/2"		24.20	.661		33.50	20.50		54	67.50
5630	3"		20.80	.769		37.50	24		61.50	77.50
5640	4"		16.50	.970		68	30		98	120
5650	6"		10.10	1.584		136	49		185	224
5700	45° Elbow same as 90° Elbow									
5850	Tee, 1/4"	1 Plum	22	.364	Ea.	5.75	12.55		18.30	25
5870	1/2"		20.20	.396		5.75	13.65		19.40	27
5880	3/4"		17.30	.462		5.85	15.95		21.80	30.50
5890	1"		15.20	.526		7.15	18.15		25.30	35.50
5900	1-1/4"		13.50	.593		15.10	20.50		35.60	47.50
5910	1-1/2"		12.10	.661		17.30	23		40.30	53.50
5920	2"	Q-1	20	.800		19.25	25		44.25	58.50
5930	2-1/2"		16.20	.988		49	30.50		79.50	100
5940	3"		13.90	1.151		49	35.50		84.50	108
5950	4"		11	1.455		65	45		110	140
5960	6"		6.70	2.388		170	74		244	299
6000	Coupling, 1/4"	1 Plum	32	.250		3.88	8.60		12.48	17.25
6020	1/2"		30.30	.264		2.72	9.10		11.82	16.75

15100 | Building Services Piping

15108 | Plastic Pipe & Fittings

		Crew	Daily Output	Labor-Hours	Unit	2001 Bare Costs Mat.	Labor	Equip.	Total	Total Incl O&P
6030	3/4"	1 Plum	26	.308	Ea.	3.87	10.60		14.47	20.50
6040	1"		22.70	.352		4.16	12.15		16.31	23
6050	1-1/4"		20.20	.396		5.15	13.65		18.80	26
6060	1-1/2"		18.20	.440		11.45	15.15		26.60	35.50
6070	2"	Q-1	33.10	.483		12	15		27	35.50
6080	2-1/2"		24.20	.661		28.50	20.50		49	62.50
6090	3"		20.80	.769		33.50	24		57.50	73
6100	4"		16.50	.970		58	30		88	110
6110	6"		10.10	1.584		45.50	49		94.50	124
6120	8"	Q-2	9.30	2.581		215	83		298	360
6200	CTS, 100 psi at 180°F, hot and cold water									
6230	90° Elbow, 1/2"	1 Plum	20	.400	Ea.	.07	13.80		13.87	21
6250	3/4"		19	.421		.16	14.50		14.66	22
6251	1"		16	.500		.52	17.25		17.77	26.50
6252	1-1/4"		15	.533		.89	18.35		19.24	28.50
6253	1-1/2"		14	.571		1.43	19.70		21.13	31
6254	2"	Q-1	23	.696		3.08	21.50		24.58	36
6260	45° Elbow, 1/2"	1 Plum	20	.400		.12	13.80		13.92	21
6280	3/4"		19	.421		.18	14.50		14.68	22
6281	1"		16	.500		.48	17.25		17.73	26.50
6282	1-1/4"		15	.533		.94	18.35		19.29	28.50
6283	1-1/2"		14	.571		1.49	19.70		21.19	31
6284	2"	Q-1	23	.696		3.14	21.50		24.64	36
6290	Tee, 1/2"	1 Plum	13	.615		.10	21		21.10	32
6310	3/4"		12	.667		.23	23		23.23	35
6311	1"		11	.727		1.18	25		26.18	39.50
6312	1-1/4"		10	.800		1.80	27.50		29.30	43.50
6313	1-1/2"		10	.800		2.34	27.50		29.84	44
6314	2"	Q-1	17	.941		3.80	29		32.80	48
6320	Coupling, 1/2"	1 Plum	22	.364		.09	12.55		12.64	19
6340	3/4"		21	.381		.10	13.10		13.20	19.90
6341	1"		18	.444		.45	15.30		15.75	23.50
6342	1-1/4"		17	.471		.56	16.20		16.76	25
6343	1-1/2"		16	.500		.79	17.25		18.04	27
6344	2"	Q-1	28	.571		1.58	17.70		19.28	28
6360	Solvent cement for CPVC, commercial grade, per quart				Qt.	10.45			10.45	11.50
7990	Polybutyl/polyethyl pipe, for copper fittings see 15107-460-7000									
8000	Compression type, PVC, 160 psi cold water									
8010	Coupling, 3/4" CTS	1 Plum	21	.381	Ea.	2.11	13.10		15.21	22
8020	1" CTS		18	.444		2.61	15.30		17.91	26
8030	1-1/4" CTS		17	.471		3.63	16.20		19.83	28.50
8040	1-1/2" CTS		16	.500		5	17.25		22.25	31.50
8050	2" CTS		15	.533		6.95	18.35		25.30	35
8060	Female adapter, 3/4" FPT x 3/4" CTS		23	.348		3.35	12		15.35	22
8070	3/4" FPT x 1" CTS		21	.381		5	13.10		18.10	25.50
8080	1" FPT x 1" CTS		20	.400		5.05	13.80		18.85	26.50
8090	1-1/4" FPT x 1-1/4" CTS		18	.444		6.60	15.30		21.90	30.50
8100	1-1/2" FPT x 1-1/2" CTS		16	.500		7.70	17.25		24.95	34.50
8110	2" FPT x 2" CTS		13	.615		11.45	21		32.45	44.50
8130	Male adapter, 3/4" MPT x 3/4" CTS		23	.348		3.02	12		15.02	21.50
8140	3/4" MPT x 1" CTS		21	.381		3.71	13.10		16.81	24
8150	1" MPT x 1" CTS		20	.400		3.79	13.80		17.59	25
8160	1-1/4" MPT x 1-1/4" CTS		18	.444		5.80	15.30		21.10	29.50
8170	1-1/2" MPT x 1-1/2" CTS		16	.500		6.75	17.25		24	33.50
8180	2" MPT x 2" CTS		13	.615		10.75	21		31.75	44
8200	Spigot adapter, 3/4" IPS x 3/4" CTS		23	.348		3.02	12		15.02	21.50

15100 | Building Services Piping

15108 | Plastic Pipe & Fittings

		CREW	DAILY OUTPUT	LABOR-HOURS	UNIT	MAT.	LABOR	EQUIP.	TOTAL	TOTAL INCL O&P
8210	3/4" IPS x 1" CTS	1 Plum	21	.381	Ea.	3.02	13.10		16.12	23
8220	1" IPS x 1" CTS		20	.400		3.26	13.80		17.06	24.50
8230	1-1/4" IPS x 1-1/4" CTS		18	.444		5.80	15.30		21.10	29.50
8240	1-1/2" IPS x 1-1/2" CTS		16	.500		6.75	17.25		24	33.50
8250	2" IPS x 2" CTS		13	.615		10.75	21		31.75	44
8270	Price includes insert stiffeners									
8280	250 psi is same price as 160 psi									
8300	Insert type, nylon, 160 & 250 psi, cold water									
8310	Clamp ring stainless steel, 3/4" IPS	1 Plum	115	.070	Ea.	.62	2.40		3.02	4.30
8320	1" IPS		107	.075		.63	2.58		3.21	4.58
8330	1-1/4" IPS		101	.079		.63	2.73		3.36	4.81
8340	1-1/2" IPS		95	.084		.64	2.90		3.54	5.10
8350	2" IPS		85	.094		.64	3.24		3.88	5.60
8370	Coupling, 3/4" IPS		22	.364		.62	12.55		13.17	19.60
8390	1-1/4" IPS		18	.444		1.49	15.30		16.79	24.50
8400	1-1/2" IPS		17	.471		1.94	16.20		18.14	26.50
8410	2" IPS		16	.500		2.33	17.25		19.58	28.50
8430	Elbow, 90°, 3/4" IPS		22	.364		.91	12.55		13.46	19.90
8440	1" IPS		19	.421		1	14.50		15.50	23
8450	1-1/4" IPS		18	.444		1.12	15.30		16.42	24
8460	1-1/2" IPS		17	.471		1.33	16.20		17.53	26
8470	2" IPS		16	.500		1.85	17.25		19.10	28
8490	Male adapter, 3/4" IPS x 3/4" MPT		25	.320		.63	11		11.63	17.35
8500	1" IPS x 1" MPT		21	.381		.83	13.10		13.93	20.50
8510	1-1/4" IPS x 1-1/4" MPT		20	.400		1.37	13.80		15.17	22.50
8520	1-1/2" IPS x 1-1/2" MPT		18	.444		1.81	15.30		17.11	25
8530	2" IPS x 2" MPT		15	.533		2.33	18.35		20.68	30
8550	Tee, 3/4" IPS		14	.571		1.19	19.70		20.89	31
8560	1" IPS		13	.615		1.94	21		22.94	34
8570	1-1/4" IPS		12	.667		3.50	23		26.50	38.50
8580	1-1/2" IPS		11	.727		4.40	25		29.40	43
8590	2" IPS		10	.800		5.95	27.50		33.45	48
8610	Insert type, PVC, 100 psi @ 180°F, hot & cold water									
8620	Coupler, male, 3/8" CTS x 3/8" MPT	1 Plum	29	.276	Ea.	.38	9.50		9.88	14.75
8630	3/8" CTS x 1/2" MPT		28	.286		.38	9.85		10.23	15.25
8640	1/2" CTS x 1/2" MPT		27	.296		.38	10.20		10.58	15.80
8650	1/2" CTS x 3/4" MPT		26	.308		1.07	10.60		11.67	17.20
8660	3/4" CTS x 1/2" MPT		25	.320		1.07	11		12.07	17.85
8670	3/4" CTS x 3/4" MPT		25	.320		.45	11		11.45	17.15
8700	Coupling, 3/8" CTS x 1/2" CTS		25	.320		2.15	11		13.15	19
8710	1/2" CTS		23	.348		2.50	12		14.50	21
8720	1/2" CTS x stub		23	.348		2.52	12		14.52	21
8730	3/4" CTS		22	.364		3.96	12.55		16.51	23.50
8750	Elbow 90°, 3/8" CTS		25	.320		2.67	11		13.67	19.60
8760	1/2" CTS		23	.348		2.83	12		14.83	21
8770	3/4" CTS		22	.364		4.61	12.55		17.16	24
8800	Rings, crimp, copper, 3/8" CTS		120	.067		.11	2.30		2.41	3.59
8810	1/2" CTS		117	.068		.12	2.36		2.48	3.69
8820	3/4" CTS		115	.070		.15	2.40		2.55	3.79
8850	Reducer tee, 3/8" x 3/8" x 1/2" CTS		17	.471		1.46	16.20		17.66	26
8860	1/2" x 3/8" x 1/2" CTS		15	.533		1.25	18.35		19.60	29
8870	3/4" x 1/2" x 1/2" CTS		14	.571		1.31	19.70		21.01	31
8890	3/4" x 3/4" x 1/2" CTS		14	.571		1.24	19.70		20.94	31
8900	3/4" x 1/2" x 3/8" CTS		14	.571		1.09	19.70		20.79	30.50
8930	Tee, 3/8" CTS		17	.471		1.05	16.20		17.25	25.50
8940	1/2" CTS		15	.533		1.19	18.35		19.54	29

15100 | Building Services Piping

15108 | Plastic Pipe & Fittings

			CREW	DAILY OUTPUT	LABOR-HOURS	UNIT	2001 BARE COSTS MAT.	LABOR	EQUIP.	TOTAL	TOTAL INCL O&P	
560	8950	3/4" CTS	1 Plum	14	.571	Ea.	1.29	19.70		20.99	31	560
	8960	Copper rings included in fitting price										
	9000	Flare type, assembled, acetal, hot & cold water										
	9010	Coupling, 1/4" & 3/8" CTS	1 Plum	24	.333	Ea.	2.14	11.50		13.64	19.70	
	9020	1/2" CTS		22	.364		2.45	12.55		15	21.50	
	9030	3/4" CTS		21	.381		3.61	13.10		16.71	24	
	9040	1" CTS		18	.444		4.61	15.30		19.91	28	
	9050	Elbow 90°, 1/4" CTS		26	.308		2.38	10.60		12.98	18.60	
	9060	3/8" CTS		24	.333		2.55	11.50		14.05	20	
	9070	1/2" CTS		22	.364		3.01	12.55		15.56	22	
	9080	3/4" CTS		21	.381		4.61	13.10		17.71	25	
	9090	1" CTS		18	.444		5.80	15.30		21.10	29.50	
	9110	Tee, 1/4" & 3/8" CTS		15	.533		2.62	18.35		20.97	30.50	
	9120	1/2" CTS		14	.571		3.44	19.70		23.14	33.50	
	9130	3/4" CTS		13	.615		5.25	21		26.25	38	
	9140	1" CTS		12	.667		7	23		30	42	
	9550	For plastic hangers see 15060-300-8000										
	9560	For copper/brass fittings see 15107-460-7000										
590	0010	**PIPE, HIGH DENSITY POLYETHYLENE PLASTIC (HDPE)**										590
	0020	Not incl. hangers, trenching, backfill, hoisting or digging equipment.										
	0030	Standard length is 40', add a weld for each joint.										
	0050	Straight										
	0054	1" diameter DR 11				L.F.	.28			.28	.31	
	0058	1-1/2" diameter DR 11					.62			.62	.68	
	0062	2" diameter DR 11					.79			.79	.87	
	0066	3" diameter DR 11					1.72			1.72	1.89	
	0070	3" diameter DR 17					1.15			1.15	1.27	
	0074	4" diameter DR 11					2.84			2.84	3.12	
	0078	4" diameter DR 17					1.91			1.91	2.10	
	0082	6" diameter DR 11					6.15			6.15	6.75	
	0086	6" diameter DR 17					4.13			4.13	4.54	
	0090	8" diameter DR 11					10.40			10.40	11.45	
	0094	8" diameter DR 26					4.69			4.69	5.15	
	0098	10" diameter DR 11					16.20			16.20	17.85	
	0102	10" diameter DR 26					7.25			7.25	8	
	0106	12" diameter DR 11					23			23	25	
	0110	12" diameter DR 26					10.25			10.25	11.25	
	0114	16" diameter DR 11					34			34	37	
	0118	16" diameter DR 26					15.20			15.20	16.70	
	0122	18" diameter DR 11					43			43	47	
	0126	18" diameter DR 26					19.20			19.20	21	
	0130	20" diameter DR 11					53			53	58	
	0134	20" diameter DR 26					23.50			23.50	26	
	0138	22" diameter DR 11					64			64	70.50	
	0142	22" diameter DR 26					28.50			28.50	31.50	
	0146	24" diameter DR 11					76			76	84	
	0150	24" diameter DR 26					34			34	37.50	
	0154	28" diameter DR 17					69.50			69.50	76.50	
	0158	28" diameter DR 26					46.50			46.50	51	
	0162	30" diameter DR 21					65.50			65.50	72	
	0166	30" diameter DR 26					43			43	47.50	
	0170	36" diameter DR 26					43			43	47.50	
	0174	42" diameter DR 26					105			105	115	
	0178	48" diameter DR 26					132			132	146	
	0182	54" diameter DR 26					173			173	190	
	0300	90° Elbow										

15100 | Building Services Piping

15108 | Plastic Pipe & Fittings

		CREW	DAILY OUTPUT	LABOR-HOURS	UNIT	2001 BARE COSTS				TOTAL INCL O&P
						MAT.	LABOR	EQUIP.	TOTAL	
0304	1" diameter DR 11				Ea.	6.45			6.45	7.10
0308	1-1/2" diameter DR 11					10.75			10.75	11.85
0312	2" diameter DR 11					12			12	13.20
0316	3" diameter DR 11					23			23	25.50
0320	3" diameter DR 17					23			23	25.50
0324	4" diameter DR 11					27			27	29.50
0328	4" diameter DR 17					27			27	29.50
0332	6" diameter DR 11					79			79	87
0336	6" diameter DR 17					79			79	87
0340	8" diameter DR 11					260			260	286
0344	8" diameter DR 26					177			177	195
0348	10" diameter DR 11					645			645	710
0352	10" diameter DR 26					325			325	360
0356	12" diameter DR 11					645			645	710
0360	12" diameter DR 26					355			355	390
0364	16" diameter DR 11					720			720	795
0368	16" diameter DR 26					580			580	640
0372	18" diameter DR 11					755			755	830
0376	18" diameter DR 26					755			755	830
0380	20" diameter DR 11					1,125			1,125	1,225
0384	20" diameter DR 26					1,125			1,125	1,225
0388	22" diameter DR 11					1,400			1,400	1,550
0392	22" diameter DR 26					1,400			1,400	1,550
0396	24" diameter DR 11					1,625			1,625	1,775
0400	24" diameter DR 26					1,625			1,625	1,775
0404	28" diameter DR 17					2,450			2,450	2,700
0408	28" diameter DR 26					2,450			2,450	2,700
0412	30" diameter DR 17					2,250			2,250	2,475
0416	30" diameter DR 26					2,250			2,250	2,475
0420	36" diameter DR 26					3,450			3,450	3,775
0424	42" diameter DR 26					3,875			3,875	4,250
0428	48" diameter DR 26					4,075			4,075	4,500
0432	54" diameter DR 26				▼	9,475			9,475	10,400
0500	45° Elbow									
0512	2" diameter DR 11				Ea.	24			24	26.50
0516	3" diameter DR 11					21			21	23
0520	3" diameter DR 17					34			34	37.50
0524	4" diameter DR 11					35			35	38.50
0528	4" diameter DR 17					42			42	46
0532	6" diameter DR 11					74			74	81.50
0536	6" diameter DR 17					70			70	77
0540	8" diameter DR 11					175			175	193
0544	8" diameter DR 26					105			105	116
0548	10" diameter DR 11					490			490	540
0552	10" diameter DR 26					142			142	156
0556	12" diameter DR 11					590			590	650
0560	12" diameter DR 26					224			224	246
0564	16" diameter DR 11					282			282	310
0568	16" diameter DR 26					282			282	310
0572	18" diameter DR 11					415			415	455
0576	18" diameter DR 26					415			415	455
0580	20" diameter DR 11					645			645	710
0584	20" diameter DR 26					645			645	710
0588	22" diameter DR 11					755			755	830
0592	22" diameter DR 26					755			755	830
0596	24" diameter DR 11				▼	925			925	1,025

15100 | Building Services Piping

15108 | Plastic Pipe & Fittings

		CREW	DAILY OUTPUT	LABOR-HOURS	UNIT	2001 BARE COSTS MAT.	LABOR	EQUIP.	TOTAL	TOTAL INCL O&P
0600	24" diameter DR 26				Ea.	925			925	1,025
0604	28" diameter DR 17					1,350			1,350	1,500
0608	28" diameter DR 26					1,350			1,350	1,500
0612	30" diameter DR 17					1,300			1,300	1,425
0616	30" diameter DR 26					1,300			1,300	1,425
0620	36" diameter DR 26					2,125			2,125	2,350
0624	42" diameter DR 26					2,600			2,600	2,850
0628	48" diameter DR 26					2,675			2,675	2,925
0632	54" diameter DR 26				↓	4,350			4,350	4,775
0700	Tee									
0704	1" diameter DR 11				Ea.	6.45			6.45	7.10
0708	1-1/2" diameter DR 11					10.75			10.75	11.85
0712	2" diameter DR 11					15			15	16.50
0716	3" diameter DR 11					24			24	26.50
0720	3" diameter DR 17					24			24	26.50
0724	4" diameter DR 11					38			38	42
0728	4" diameter DR 17					38			38	42
0732	6" diameter DR 11					94			94	103
0736	6" diameter DR 17					94			94	103
0740	8" diameter DR 11					217			217	239
0744	8" diameter DR 17					197			197	217
0748	10" diameter DR 11					675			675	745
0752	10" diameter DR 17					242			242	266
0756	12" diameter DR 11					975			975	1,075
0760	12" diameter DR 17					276			276	305
0764	16" diameter DR 11					645			645	710
0768	16" diameter DR 17					645			645	710
0772	18" diameter DR 11					685			685	755
0776	18" diameter DR 17					685			685	755
0780	20" diameter DR 11					850			850	935
0784	20" diameter DR 17					850			850	935
0788	22" diameter DR 11					1,175			1,175	1,300
0792	22" diameter DR 17					1,175			1,175	1,300
0796	24" diameter DR 11					1,175			1,175	1,300
0800	24" diameter DR 17					1,175			1,175	1,300
0804	28" diameter DR 17					2,100			2,100	2,300
0812	30" diameter DR 17					2,350			2,350	2,575
0820	36" diameter DR 17					4,675			4,675	5,125
0824	42" diameter DR 26					5,125			5,125	5,650
0828	48" diameter DR 26				↓	6,000			6,000	6,600
1000	Flange adptr, w/back-up ring and 1/2 cost of plated bolt set									
1004	1" diameter DR 11				Ea.	59			59	65
1008	1-1/2" diameter DR 11					60.50			60.50	66.50
1012	2" diameter DR 11					33.50			33.50	36.50
1016	3" diameter DR 11					41			41	45
1020	3" diameter DR 17					41			41	45
1024	4" diameter DR 11					54			54	59
1028	4" diameter DR 17					54			54	59
1032	6" diameter DR 11					78.50			78.50	86
1036	6" diameter DR 17					78.50			78.50	86
1040	8" diameter DR 11					112			112	123
1044	8" diameter DR 26					112			112	123
1048	10" diameter DR 11					168			168	185
1052	10" diameter DR 26					168			168	185
1056	12" diameter DR 11					243			243	267
1060	12" diameter DR 26				↓	243			243	267

15100 | Building Services Piping

15108 | Plastic Pipe & Fittings

		CREW	DAILY OUTPUT	LABOR-HOURS	UNIT	MAT.	LABOR	EQUIP.	TOTAL	TOTAL INCL O&P
1064	16" diameter DR 11				Ea.	520			520	570
1068	16" diameter DR 26					520			520	570
1072	18" diameter DR 11					665			665	730
1076	18" diameter DR 26					665			665	730
1080	20" diameter DR 11					950			950	1,050
1084	20" diameter DR 26					950			950	1,050
1088	22" diameter DR 11					1,125			1,125	1,225
1092	22" diameter DR 26					1,125			1,125	1,225
1096	24" diameter DR 17					1,175			1,175	1,300
1100	24" diameter DR 32.5					1,175			1,175	1,300
1104	28" diameter DR 15.5					2,025			2,025	2,225
1108	28" diameter DR 32.5					1,975			1,975	2,175
1112	30" diameter DR 11					2,175			2,175	2,375
1116	30" diameter DR 21					1,875			1,875	2,075
1120	36" diameter DR 26					2,800			2,800	3,075
1124	42" diameter DR 26					3,150			3,150	3,475
1128	48" diameter DR 26					4,450			4,450	4,900
1132	54" diameter DR 26					4,975			4,975	5,475
1200	Reducer									
1204	1-1/2" x 1" diameter DR 11				Ea.	6.90			6.90	7.55
1208	2" x 1-1/2" diameter DR 11					10.75			10.75	11.85
1212	3" x 2" diameter DR 11					14			14	15.40
1216	4" x 2" diameter DR 11					21			21	23
1220	4" x 3" diameter DR 11					21			21	23
1224	6" x 4" diameter DR 11					49			49	54
1228	8" x 6" diameter DR 11					86			86	94.50
1232	10" x 8" diameter DR 11					115			115	127
1236	12" x 8" diameter DR 11					215			215	237
1240	12" x 10" diameter DR 11					154			154	169
1244	14" x 12" diameter DR 11					173			173	190
1248	16" x 14" diameter DR 11					208			208	229
1252	18" x 16" diameter DR 11					415			415	460
1256	20" x 18" diameter DR 11					565			565	620
1260	22" x 20" diameter DR 11					1,150			1,150	1,275
1264	24" x 22" diameter DR 11					740			740	810
1268	26" x 24" diameter DR 11					865			865	950
1272	28" x 24" diameter DR 11					995			995	1,100
1276	32" x 28" diameter DR 17					1,200			1,200	1,325
1280	36" x 32" diameter DR 17					1,425			1,425	1,550
4000	Welding labor per joint, not including welding machine									
4010	Pipe joint size (cost based on thickest wall for each dia.)									
4030	1" pipe size	4 Skwk	273	.117	Ea.		3.50		3.50	5.50
4040	1-1/2" pipe size		175	.183			5.45		5.45	8.55
4050	2" pipe size		128	.250			7.45		7.45	11.70
4060	3" pipe size		100	.320			9.55		9.55	15
4070	4" pipe size		77	.416			12.40		12.40	19.45
4080	6" pipe size	5 Skwk	63	.635			18.95		18.95	30
4090	8" pipe size		48	.833			25		25	39
4100	10" pipe size		40	1			30		30	47
4110	12" pipe size	6 Skwk	41	1.171			35		35	55
4120	16" pipe size		34	1.412			42		42	66
4130	18" pipe size		32	1.500			45		45	70.50
4140	20" pipe size	8 Skwk	37	1.730			51.50		51.50	81
4150	22" pipe size		35	1.829			54.50		54.50	85.50
4160	24" pipe size		34	1.882			56		56	88
4170	28" pipe size		33	1.939			58		58	91

15100 | Building Services Piping

15108 | Plastic Pipe & Fittings

		CREW	DAILY OUTPUT	LABOR-HOURS	UNIT	2001 BARE COSTS MAT.	LABOR	EQUIP.	TOTAL	TOTAL INCL O&P
4180	30" pipe size	8 Skwk	32	2	Ea.		59.50		59.50	93.50
4190	36" pipe size		31	2.065			61.50		61.50	96.50
4200	42" pipe size	↓	30	2.133			63.50		63.50	100
4210	48" pipe size	9 Skwk	33	2.182			65		65	102
4220	54" pipe size	"	31	2.323	↓		69.50		69.50	109
4300	Note: Cost for set up each time welder is moved.									
4301	Add 50% of a weld cost									
4310	Welder usually remains stationary with pipe moved through it.									
4340	Weld machine, rental per day based on dia. capacity									
4350	1" thru 2" diameter				Ea.			35		38.50
4360	3" thru 4" diameter							40		44
4370	6" thru 8" diameter							90		99
4380	10" thru 12" diameter							155		170.50
4390	16" thru 18" diameter							225		247.50
4400	20" thru 24" diameter							435		478.50
4410	28" thru 32" diameter							475		525
4420	36" diameter							495		545
4430	42" thru 54" diameter				↓			775		855
5000	Dual wall contained pipe									
5040	Straight									
5054	1" DR 11 x 3" DR 11				L.F.	5.05			5.05	5.55
5058	1" DR 11 x 4" DR 11					5.55			5.55	6.10
5062	1-1/2" DR 11 x 4" DR 17					5.85			5.85	6.45
5066	2" DR 11 x 4" DR 17					5.90			5.90	6.50
5070	2" DR 11 x 6" DR 17					9.25			9.25	10.15
5074	3" DR 11 x 6" DR 17					10.15			10.15	11.20
5078	3" DR 11 x 6" DR 26					8.15			8.15	8.95
5082	3" DR 17 x 8" DR 11					16.80			16.80	18.45
5086	3" DR 17 x 8" DR 17					13.25			13.25	14.60
5090	4" DR 11 x 8" DR 17					15			15	16.50
5094	4" DR 17 x 8" DR 26					11.85			11.85	13.05
5098	6" DR 11 x 10" DR 17					24			24	26.50
5102	6" DR 17 x 10" DR 26					18.25			18.25	20
5106	6" DR 26 x 10" DR 26					16.85			16.85	18.50
5110	8" DR 17 x 12" DR 26					26.50			26.50	29
5114	8" DR 26 x 12" DR 32.5					22			22	24
5118	10" DR 17 x 14" DR 26					34.50			34.50	38
5122	10" DR 17 x 16" DR 26					39			39	43
5126	10" DR 26 x 16" DR 26					35.50			35.50	39
5130	12" DR 26 x 16" DR 26					40			40	44
5134	12" DR 17 x 18" DR 26					51.50			51.50	56.50
5138	12" DR 26 x 18" DR 26					46			46	50.50
5142	14" DR 26 x 20" DR 32.5					50.50			50.50	55.50
5146	16" DR 26 x 22" DR 32.5					76			76	83.50
5150	18" DR 26 x 24" DR 32.5					70			70	77
5154	20" DR 32.5 x 28" DR 32.5					90.50			90.50	99.50
5158	22" DR 32.5 x 30" DR 32.5					90.50			90.50	99.50
5162	24" DR 32.5 x 32" DR 32.5					115			115	127
5166	36" DR 32.5 x 42" DR 32.5				↓	195			195	215
5300	Force transfer coupling									
5354	1" DR 11 x 3" DR 11				Ea.	158			158	174
5358	1" DR 11 x 4" DR 17					166			166	183
5362	1-1/2" DR 11 x 4" DR 17					174			174	192
5366	2" DR 11 x 4" DR 17					182			182	200
5370	2" DR 11 x 6" DR 17					264			264	291
5374	3" DR 11 x 6" DR 17				↓	264			264	291

15100 | Building Services Piping

15108 | Plastic Pipe & Fittings

		CREW	DAILY OUTPUT	LABOR-HOURS	UNIT	2001 BARE COSTS MAT.	LABOR	EQUIP.	TOTAL	TOTAL INCL O&P	
5378	3" DR 11 x 6" DR 26				Ea.	264			264	291	
5382	3" DR 11 x 8" DR 11					279			279	305	
5386	3" DR 11 x 8" DR 17					279			279	305	
5390	4" DR 11 x 8" DR 17					279			279	305	
5394	4" DR 17 x 8" DR 26					245			245	270	
5398	6" DR 11 x 10" DR 17					355			355	390	
5402	6" DR 17 x 10" DR 26					292			292	320	
5406	6" DR 26 x 10" DR 26					292			292	320	
5410	8" DR 17 x 12" DR 26					390			390	425	
5414	8" DR 26 x 12" DR 32.5					315			315	345	
5418	10" DR 17 x 14" DR 26					510			510	560	
5422	10" DR 17 x 16" DR 26					510			510	560	
5426	10" DR 26 x 16" DR 26					510			510	560	
5430	12" DR 26 x 16" DR 26					635			635	700	
5434	12" DR 17 x 18" DR 26					655			655	720	
5438	12" DR 26 x 18" DR 26					655			655	720	
5442	14" DR 26 x 20" DR 32.5					690			690	760	
5446	16" DR 26 x 22" DR 32.5					710			710	780	
5450	18" DR 26 x 24" DR 32.5					770			770	845	
5454	20" DR 32.5 x 28" DR 32.5					940			940	1,025	
5458	22" DR 32.5 x 30" DR 32.5					1,025			1,025	1,150	
5462	24" DR 32.5 x 32" DR 32.5					1,125			1,125	1,225	
5466	36" DR 32.5 x 42" DR 32.5				▼	1,850			1,850	2,050	
5600	90° Elbow										
5654	1" DR 11 x 3" DR 11				Ea.	131			131	144	
5658	1" DR 11 x 4" DR 17					125			125	138	
5662	1-1/2" DR 11 x 4" DR 17					137			137	151	
5666	2" DR 11 x 4" DR 17					145			145	160	
5670	2" DR 11 x 6" DR 17					184			184	203	
5674	3" DR 11 x 6" DR 17					211			211	232	
5678	3" DR 17 x 6" DR 26					168			168	185	
5682	3" DR 17 x 8" DR 11					340			340	375	
5686	3" DR 17 x 8" DR 17					265			265	292	
5690	4" DR 11 x 8" DR 17					310			310	345	
5694	4" DR 17 x 8" DR 26					248			248	273	
5698	6" DR 11 x 10" DR 17					415			415	455	
5702	6" DR 17 x 10" DR 26					320			320	350	
5706	6" DR 26 x 10" DR 26					298			298	330	
5710	8" DR 17 x 12" DR 26					475			475	520	
5714	8" DR 26 x 12" DR 32.5					435			435	480	
5718	10" DR 17 x 14" DR 26					745			745	820	
5722	10" DR 17 x 16" DR 26					710			710	785	
5726	10" DR 26 x 16" DR 26					660			660	725	
5730	12" DR 26 x 16" DR 26					730			730	805	
5734	12" DR 17 x 18" DR 26					870			870	960	
5738	12" DR 26 x 18" DR 26					800			800	880	
5742	14" DR 26 x 20" DR 32.5					1,050			1,050	1,150	
5746	16" DR 26 x 22" DR 32.5					1,075			1,075	1,175	
5750	18" DR 26 x 24" DR 32.5					1,300			1,300	1,425	
5754	20" DR 32.5 x 28" DR 32.5					1,500			1,500	1,650	
5758	22" DR 32.5 x 30" DR 32.5					2,050			2,050	2,250	
5762	24" DR 32.5 x 32" DR 32.5					2,000			2,000	2,200	
5766	36" DR 32.5 x 42" DR 32.5				▼	3,250			3,250	3,550	
5800	45° Elbow										
5804	1" DR 11 x 3" DR 11				Ea.	78.50			78.50	86	
5808	1" DR 11 x 4" DR 17				▼	78			78	85.50	

15100 | Building Services Piping

15108 | Plastic Pipe & Fittings

		CREW	DAILY OUTPUT	LABOR-HOURS	UNIT	MAT.	LABOR	EQUIP.	TOTAL	TOTAL INCL O&P
5812	1-1/2" DR 11 x 4" DR 17				Ea.	83.50			83.50	92
5816	2" DR 11 x 4" DR 17					91			91	100
5820	2" DR 11 x 6" DR 17					114			114	126
5824	3" DR 11 x 6" DR 17					127			127	140
5828	3" DR 17 x 6" DR 26					105			105	116
5832	3" DR 17 x 8" DR 11					194			194	214
5836	3" DR 17 x 8" DR 17					155			155	171
5840	4" DR 11 x 8" DR 17					182			182	200
5844	4" DR 17 x 8" DR 26					148			148	163
5848	6" DR 11 x 10" DR 17					240			240	264
5852	6" DR 17 x 10" DR 26					188			188	207
5856	6" DR 26 x 10" DR 26					177			177	195
5860	8" DR 17 x 12" DR 26					286			286	315
5864	8" DR 26 x 12" DR 32.5					267			267	294
5868	10" DR 17 x 14" DR 26					450			450	495
5872	10" DR 17 x 16" DR 26					440			440	480
5876	10" DR 26 x 16" DR 26					410			410	450
5880	12" DR 26 x 16" DR 26					465			465	510
5884	12" DR 17 x 18" DR 26					545			545	600
5888	12" DR 26 x 18" DR 26					510			510	560
5892	14" DR 26 x 20" DR 32.5					655			655	720
5896	16" DR 26 x 22" DR 32.5					685			685	755
5900	18" DR 26 x 24" DR 32.5					810			810	895
5904	20" DR 32.5 x 28" DR 32.5					995			995	1,100
5908	22" DR 32.5 x 30" DR 32.5					1,275			1,275	1,400
5912	24" DR 32.5 x 32" DR 32.5					1,250			1,250	1,375
5916	36" DR 32.5 x 42" DR 32.5				▼	1,950			1,950	2,150
6000	Access port with 4" riser									
6050	1" DR 11 x 4" DR 17				Ea.	171			171	188
6054	1-1/2" DR 11 x 4" DR 17					175			175	192
6058	2" DR 11 x 6" DR 17					213			213	234
6062	3" DR 11 x 6" DR 17					214			214	235
6066	3" DR 17 x 6" DR 26					210			210	231
6070	3" DR 17 x 8" DR 11					240			240	264
6074	3" DR 17 x 8" DR 17					229			229	252
6078	4" DR 11 x 8" DR 17					237			237	261
6082	4" DR 17 x 8" DR 26					227			227	250
6086	6" DR 11 x 10" DR 17					297			297	325
6090	6" DR 17 x 10" DR 26					273			273	300
6094	6" DR 26 x 10" DR 26					268			268	295
6098	8" DR 17 x 12" DR 26					287			287	315
6102	8" DR 26 x 12" DR 32.5				▼	274			274	300
6200	End termination with vent plug									
6204	1" DR 11 x 3" DR 11				Ea.	142			142	156
6208	1" DR 11 x 4" DR 17					151			151	167
6212	1-1/2" DR 11 x 4" DR 17					151			151	167
6216	2" DR 11 x 4" DR 17					165			165	181
6220	2" DR 11 x 6" DR 17					188			188	206
6224	3" DR 11 x 6" DR 17					188			188	206
6228	3" DR 17 x 6" DR 26					188			188	206
6232	3" DR 17 x 8" DR 11					260			260	285
6236	3" DR 17 x 8" DR 17					260			260	285
6240	4" DR 11 x 8" DR 17					260			260	285
6244	4" DR 17 x 8" DR 26					225			225	248
6248	6" DR 11 x 10" DR 17					310			310	340
6252	6" DR 17 x 10" DR 26				▼	253			253	278

15100 | Building Services Piping

15108 | Plastic Pipe & Fittings

		CREW	DAILY OUTPUT	LABOR-HOURS	UNIT	2001 BARE COSTS MAT.	LABOR	EQUIP.	TOTAL	TOTAL INCL O&P	
6256	6" DR 26 x 10" DR 26				Ea.	253			253	278	590
6260	8" DR 17 x 12" DR 26					375			375	410	
6264	8" DR 26 x 12" DR 32.5					300			300	330	
6268	10" DR 17 x 14" DR 26					390			390	430	
6272	10" DR 17 x 16" DR 26					390			390	430	
6276	10" DR 26 x 16" DR 26					390			390	430	
6280	12" DR 26 x 16" DR 26					390			390	430	
6284	12" DR 17 x 18" DR 26					515			515	565	
6288	12" DR 26 x 18" DR 26					515			515	565	
6292	14" DR 26 x 20" DR 32.5					525			525	575	
6296	16" DR 26 x 22" DR 32.5					605			605	665	
6300	18" DR 26 x 24" DR 32.5					625			625	690	
6304	20" DR 32.5 x 28" DR 32.5					800			800	880	
6308	22" DR 32.5 x 30" DR 32.5					880			880	965	
6312	24" DR 32.5 x 32" DR 32.5					970			970	1,075	
6316	36" DR 32.5 x 42" DR 32.5					1,425			1,425	1,550	
6600	Tee										
6604	1" DR 11 x 3" DR 11				Ea.	144			144	159	
6608	1" DR 11 x 4" DR 17					179			179	197	
6612	1-1/2" DR 11 x 4" DR 17					199			199	219	
6616	2" DR 11 x 4" DR 17					217			217	239	
6620	2" DR 11 x 6" DR 17					264			264	290	
6624	3" DR 11 x 6" DR 17					305			305	335	
6628	3" DR 17 x 6" DR 26					296			296	325	
6632	3" DR 17 x 8" DR 11					360			360	395	
6636	3" DR 17 x 8" DR 17					340			340	375	
6640	4" DR 11 x 8" DR 17					370			370	410	
6644	4" DR 17 x 8" DR 26					355			355	390	
6648	6" DR 11 x 10" DR 17					465			465	510	
6652	6" DR 17 x 10" DR 26					440			440	485	
6656	6" DR 26 x 10" DR 26					435			435	480	
6660	8" DR 17 x 12" DR 26					435			435	480	
6664	8" DR 26 x 12" DR 32.5					570			570	625	
6668	10" DR 17 x 14" DR 26					655			655	725	
6672	10" DR 17 x 16" DR 26					690			690	760	
6676	10" DR 26 x 16" DR 26					690			690	760	
6680	12" DR 26 x 16" DR 26					745			745	815	
6684	12" DR 17 x 18" DR 26					745			745	815	
6688	12" DR 26 x 18" DR 26					825			825	905	
6692	14" DR 26 x 20" DR 32.5					1,100			1,100	1,200	
6696	16" DR 26 x 22" DR 32.5					1,325			1,325	1,475	
6700	18" DR 26 x 24" DR 32.5					1,450			1,450	1,600	
6704	20" DR 32.5 x 28" DR 32.5					1,825			1,825	2,025	
6708	22" DR 32.5 x 30" DR 32.5					2,175			2,175	2,400	
6712	24" DR 32.5 x 32" DR 32.5					2,600			2,600	2,875	
6716	36" DR 32.5 x 42" DR 32.5					5,300			5,300	5,850	
6800	Wye										
6816	2" DR 11 x 4" DR 17				Ea.	274			274	300	
6820	2" DR 11 x 6" DR 17					310			310	340	
6824	3" DR 11 x 6" DR 17					315			315	345	
6828	3" DR 17 x 6" DR 26					305			305	335	
6832	3" DR 17 x 8" DR 11					380			380	420	
6836	3" DR 17 x 8" DR 17					355			355	395	
6840	4" DR 11 x 8" DR 17					390			390	430	
6844	4" DR 17 x 8" DR 26					370			370	405	
6848	6" DR 11 x 10" DR 17					480			480	530	

Important: See the Reference Section for critical supporting data - Reference Nos., Crews, & City Cost Indexes

15100 | Building Services Piping

15108 | Plastic Pipe & Fittings

		CREW	DAILY OUTPUT	LABOR-HOURS	UNIT	MAT.	LABOR	EQUIP.	TOTAL	TOTAL INCL O&P	
590											590
6852	6" DR 17 x 10" DR 26				Ea.	450			450	495	
6856	6" DR 26 x 10" DR 26					445			445	490	
6860	8" DR 17 x 12" DR 26					640			640	700	
6864	8" DR 26 x 12" DR 32.5					615			615	680	
6868	10" DR 17 x 14" DR 26					770			770	845	
6872	10" DR 17 x 16" DR 26					815			815	900	
6876	10" DR 26 x 16" DR 26					805			805	885	
6880	12" DR 26 x 16" DR 26					905			905	995	
6884	12" DR 17 x 18" DR 26					1,025			1,025	1,125	
6888	12" DR 26 x 18" DR 26					1,025			1,025	1,125	
6892	14" DR 26 x 20" DR 32.5					1,250			1,250	1,375	
6896	16" DR 26 x 22" DR 32.5					1,800			1,800	2,000	
6900	18" DR 26 x 24" DR 32.5					2,125			2,125	2,350	
6904	20" DR 32.5 x 28" DR 32.5					2,675			2,675	2,925	
6908	22" DR 32.5 x 30" DR 32.5					3,025			3,025	3,350	
6912	24" DR 32.5 x 32" DR 32.5					3,450			3,450	3,800	
9000	Welding labor per joint, not including welding machine										
9010	Pipe joint size, outer pipe (cost based on the thickest walls)										
9020	Straight pipe										
9050	3" pipe size	4 Skwk	96	.333	Ea.		9.95		9.95	15.60	
9060	4" pipe size	"	77	.416			12.40		12.40	19.45	
9070	6" pipe size	5 Skwk	60	.667			19.90		19.90	31	
9080	8" pipe size	"	40	1			30		30	47	
9090	10" pipe size	6 Skwk	41	1.171			35		35	55	
9100	12" pipe size		39	1.231			36.50		36.50	57.50	
9110	14" pipe size		38	1.263			37.50		37.50	59	
9120	16" pipe size		35	1.371			41		41	64.50	
9130	18" pipe size	8 Skwk	45	1.422			42.50		42.50	66.50	
9140	20" pipe size		42	1.524			45.50		45.50	71.50	
9150	22" pipe size		40	1.600			48		48	75	
9160	24" pipe size		38	1.684			50.50		50.50	79	
9170	28" pipe size		37	1.730			51.50		51.50	81	
9180	30" pipe size		36	1.778			53		53	83.50	
9190	32" pipe size		35	1.829			54.50		54.50	85.50	
9200	42" pipe size		32	2			59.50		59.50	93.50	
9300	Note: Cost for set up each time welder is moved.										
9301	Add 100% of weld labor cost										
9310	For handling between fitting welds add 50% of weld labor cost										
9320	Welder usually remains stationary with pipe moved through it										
9360	Weld machine, rental per day based on dia. capacity										
9380	3" thru 4" diameter				Ea.			55		60.50	
9390	6" thru 8" diameter							180		198	
9400	10" thru 12" diameter							270		297	
9410	14" thru 18" diameter							365		402	
9420	20" thru 24" diameter							650		715	
9430	28" thru 32" diameter							765		840	
9440	42" thru 58" diameter							790		870	

15110 | Valves

		CREW	DAILY OUTPUT	LABOR-HOURS	UNIT	MAT.	LABOR	EQUIP.	TOTAL	TOTAL INCL O&P	
100	0010	**VALVES, BRASS**									100
0030	For motorized valves, see Division 13838-200										
0500	Gas cocks, threaded										
0510	1/4" size	1 Plum	26	.308	Ea.	7.80	10.60		18.40	24.50	
0520	3/8" size		24	.333		8.25	11.50		19.75	26.50	
0530	1/2" size		24	.333		7.65	11.50		19.15	26	

15100 | Building Services Piping

15110 | Valves

			Crew	Daily Output	Labor-Hours	Unit	2001 Bare Costs Mat.	Labor	Equip.	Total	Total Incl O&P	
100	0540	3/4" size	1 Plum	22	.364	Ea.	10.30	12.55		22.85	30.50	100
	0550	1" size		19	.421		12.60	14.50		27.10	36	
	0560	1-1/4" size		15	.533		36	18.35		54.35	67	
	0570	1-1/2" size		13	.615		45	21		66	81.50	
	0580	2" size	↓	11	.727	↓	65.50	25		90.50	110	
	0670	For larger sizes use lubricated plug valve, Section 15110-600										
160	0010	**VALVES, BRONZE**										160
	1020	Angle, 150 lb., rising stem, threaded	R15100 -090									
	1030	1/8" size	1 Plum	24	.333	Ea.	42	11.50		53.50	63.50	
	1040	1/4" size		24	.333		42	11.50		53.50	63.50	
	1050	3/8" size		24	.333		42	11.50		53.50	63.50	
	1060	1/2" size		22	.364		42	12.55		54.55	65	
	1070	3/4" size		20	.400		56.50	13.80		70.30	83	
	1080	1" size		19	.421		81	14.50		95.50	111	
	1090	1-1/4" size		15	.533		104	18.35		122.35	143	
	1100	1-1/2" size		13	.615		135	21		156	181	
	1110	2" size	↓	11	.727	↓	219	25		244	279	
	1380	Ball, 150 psi, threaded										
	1400	1/4" size	1 Plum	24	.333	Ea.	6.40	11.50		17.90	24.50	
	1430	3/8" size		24	.333		6.40	11.50		17.90	24.50	
	1450	1/2" size		22	.364		6.40	12.55		18.95	26	
	1460	3/4" size		20	.400		10.60	13.80		24.40	32.50	
	1470	1" size		19	.421		13.30	14.50		27.80	36.50	
	1480	1-1/4" size		15	.533		22.50	18.35		40.85	52.50	
	1490	1-1/2" size		13	.615		28.50	21		49.50	63.50	
	1500	2" size		11	.727		35.50	25		60.50	77.50	
	1510	2-1/2" diameter		9	.889		169	30.50		199.50	232	
	1520	3" diameter	↓	8	1	↓	196	34.50		230.50	267	
	1600	Butterfly, 175 psi, full port, solder or threaded ends										
	1610	Stainless steel disc and stem										
	1620	1/4" size	1 Plum	24	.333	Ea.	5.75	11.50		17.25	23.50	
	1630	3/8" size		24	.333		5.75	11.50		17.25	23.50	
	1640	1/2" size		22	.364		6.10	12.55		18.65	25.50	
	1650	3/4" size		20	.400		9.85	13.80		23.65	32	
	1660	1" size		19	.421		12.25	14.50		26.75	35.50	
	1670	1-1/4" size		15	.533		19.45	18.35		37.80	49	
	1680	1-1/2" size		13	.615		25	21		46	59.50	
	1690	2" size	↓	11	.727	↓	31.50	25		56.50	72.50	
	1750	Check, swing, class 150, regrinding disc, threaded										
	1800	1/8" size	1 Plum	24	.333	Ea.	22.50	11.50		34	42.50	
	1830	1/4" size		24	.333		22.50	11.50		34	42.50	
	1840	3/8" size		24	.333		22.50	11.50		34	42.50	
	1850	1/2" size		24	.333		22.50	11.50		34	42.50	
	1860	3/4" size		20	.400		30	13.80		43.80	54	
	1870	1" size		19	.421		45	14.50		59.50	71.50	
	1880	1-1/4" size		15	.533		62.50	18.35		80.85	96	
	1890	1-1/2" size		13	.615		73.50	21		94.50	113	
	1900	2" size	↓	11	.727		108	25		133	157	
	1910	2-1/2" size	Q-1	15	1.067		179	33		212	247	
	1920	3" size	"	13	1.231	↓	265	38		303	350	
	2000	For 200 lb, add					5%	10%				
	2040	For 300 lb, add					15%	15%				
	2060	Check swing, 300#, sweat, 3/8" size	1 Plum	24	.333	Ea.	12.90	11.50		24.40	31.50	
	2070	1/2" size		24	.333		12.90	11.50		24.40	31.50	
	2080	3/4" size		20	.400		16.45	13.80		30.25	39	
	2090	1" size	↓	19	.421		23	14.50		37.50	47	

15100 | Building Services Piping

15110 | Valves

			Crew	Daily Output	Labor-Hours	Unit	Mat.	Labor	Equip.	Total	Total Incl O&P
160	2100	1-1/4" size	1 Plum	15	.533	Ea.	29.50	18.35		47.85	60
	2110	1-1/2" size		13	.615		39.50	21		60.50	75.50
	2120	2" size		11	.727		59	25		84	103
	2130	2-1/2" size	Q-1	15	1.067		141	33		174	205
	2140	3" size	"	13	1.231		188	38		226	264
	2350	Check, lift, class 150, horizontal composition disc, threaded									
	2430	1/4" size	1 Plum	24	.333	Ea.	50.50	11.50		62	73
	2440	3/8" size		24	.333		77	11.50		88.50	102
	2450	1/2" size		24	.333		56	11.50		67.50	79
	2460	3/4" size		20	.400		68.50	13.80		82.30	96
	2470	1" size		19	.421		97	14.50		111.50	128
	2480	1-1/4" size		15	.533		130	18.35		148.35	171
	2490	1-1/2" size		13	.615		155	21		176	203
	2500	2" size		11	.727		259	25		284	325
	2850	Gate, N.R.S., soldered, 125 psi									
	2900	3/8" size	1 Plum	24	.333	Ea.	16.85	11.50		28.35	36
	2920	1/2" size		24	.333		14.75	11.50		26.25	33.50
	2940	3/4" size		20	.400		16.85	13.80		30.65	39.50
	2950	1" size		19	.421		24	14.50		38.50	48.50
	2960	1-1/4" size		15	.533		40	18.35		58.35	71
	2970	1-1/2" size		13	.615		45	21		66	81.50
	2980	2" size		11	.727		57.50	25		82.50	102
	2990	2-1/2" size	Q-1	15	1.067		145	33		178	209
	3000	3" size	"	13	1.231		189	38		227	266
	3350	Threaded, class 150									
	3410	1/4" size	1 Plum	24	.333	Ea.	25	11.50		36.50	45
	3420	3/8" size		24	.333		25	11.50		36.50	45
	3430	1/2" size		24	.333		23	11.50		34.50	42.50
	3440	3/4" size		20	.400		25.50	13.80		39.30	49
	3450	1" size		19	.421		33	14.50		47.50	58.50
	3460	1-1/4" size		15	.533		44.50	18.35		62.85	76.50
	3470	1-1/2" size		13	.615		56	21		77	94
	3480	2" size		11	.727		77	25		102	123
	3490	2-1/2" size	Q-1	15	1.067		178	33		211	246
	3500	3" size	"	13	1.231		251	38		289	335
	3600	Gate, flanged, 150 lb.									
	3610	1" size	1 Plum	7	1.143	Ea.	415	39.50		454.50	520
	3620	1-1/2" size		6	1.333		515	46		561	640
	3630	2" size		5	1.600		765	55		820	925
	3640	3" size	Q-1	4.50	3.556		1,375	110		1,485	1,675
	3850	Rising stem, soldered, 300 psi									
	3900	3/8" size	1 Plum	24	.333	Ea.	37.50	11.50		49	59
	3920	1/2" size		24	.333		35	11.50		46.50	56
	3940	3/4" size		20	.400		44.50	13.80		58.30	70
	3950	1" size		19	.421		55	14.50		69.50	82.50
	3960	1-1/4" size		15	.533		76	18.35		94.35	111
	3970	1-1/2" size		13	.615		93	21		114	134
	3980	2" size		11	.727		137	25		162	189
	3990	2-1/2" size	Q-1	15	1.067		330	33		363	415
	4000	3" size	"	13	1.231		385	38		423	485
	4250	Threaded, class 150									
	4310	1/4" size	1 Plum	24	.333	Ea.	21.50	11.50		33	41.50
	4320	3/8" size		24	.333		21.50	11.50		33	41.50
	4330	1/2" size		24	.333		20.50	11.50		32	40
	4340	3/4" size		20	.400		24	13.80		37.80	47.50
	4350	1" size		19	.421		32	14.50		46.50	57

15100 | Building Services Piping

15110 | Valves

			CREW	DAILY OUTPUT	LABOR-HOURS	UNIT	MAT.	LABOR	EQUIP.	TOTAL	TOTAL INCL O&P	
160	4360	1-1/4" size	1 Plum	15	.533	Ea.	43	18.35		61.35	75	160
	4370	1-1/2" size		13	.615		54	21		75	91.50	
	4380	2" size		11	.727		73.50	25		98.50	119	
	4390	2-1/2" size	Q-1	15	1.067		171	33		204	238	
	4400	3" size	"	13	1.231		240	38		278	320	
	4500	For 300 psi, threaded, add					100%	15%				
	4540	For chain operated type, add					15%					
	4850	Globe, class 150, rising stem, threaded										
	4920	1/4" size	1 Plum	24	.333	Ea.	31.50	11.50		43	52	
	4940	3/8" size		24	.333		31.50	11.50		43	52	
	4950	1/2" size		24	.333		31.50	11.50		43	52	
	4960	3/4" size		20	.400		43	13.80		56.80	68	
	4970	1" size		19	.421		67	14.50		81.50	95.50	
	4980	1-1/4" size		15	.533		105	18.35		123.35	144	
	4990	1-1/2" size		13	.615		128	21		149	172	
	5000	2" size		11	.727		191	25		216	248	
	5010	2-1/2" size	Q-1	15	1.067		385	33		418	475	
	5020	3" size	"	13	1.231		550	38		588	660	
	5120	For 300 lb threaded, add					50%	15%				
	5130	Globe, 300 lb., sweat, 3/8" size	1 Plum	24	.333	Ea.	33.50	11.50		45	54.50	
	5140	1/2" size		24	.333		33	11.50		44.50	53.50	
	5150	3/4" size		20	.400		47.50	13.80		61.30	73.50	
	5160	1" size		19	.421		71.50	14.50		86	101	
	5170	1-1/4" size		15	.533		110	18.35		128.35	149	
	5180	1-1/2" size		13	.615		131	21		152	176	
	5190	2" size		11	.727		197	25		222	254	
	5200	2-1/2" size	Q-1	15	1.067		435	33		468	530	
	5210	3" size	"	13	1.231		630	38		668	755	
	5600	Relief, pressure & temperature, self-closing, ASME, threaded										
	5640	3/4" size	1 Plum	28	.286	Ea.	65	9.85		74.85	86.50	
	5650	1" size		24	.333		94.50	11.50		106	121	
	5660	1-1/4" size		20	.400		189	13.80		202.80	229	
	5670	1-1/2" size		18	.444		360	15.30		375.30	425	
	5680	2" size		16	.500		395	17.25		412.25	460	
	5950	Pressure, poppet type, threaded										
	6000	1/2" size	1 Plum	30	.267	Ea.	13.15	9.20		22.35	28.50	
	6040	3/4" size	"	28	.286	"	44	9.85		53.85	63.50	
	6400	Pressure, water, ASME, threaded										
	6440	3/4" size	1 Plum	28	.286	Ea.	43	9.85		52.85	62.50	
	6450	1" size		24	.333		91	11.50		102.50	117	
	6460	1-1/4" size		20	.400		149	13.80		162.80	185	
	6470	1-1/2" size		18	.444		206	15.30		221.30	249	
	6480	2" size		16	.500		298	17.25		315.25	350	
	6490	2-1/2" size		15	.533		875	18.35		893.35	990	
	6900	Reducing, water pressure										
	6920	300 psi to 25-75 psi, threaded or sweat										
	6940	1/2" size	1 Plum	24	.333	Ea.	108	11.50		119.50	136	
	6950	3/4" size		20	.400		108	13.80		121.80	140	
	6960	1" size		19	.421		167	14.50		181.50	206	
	6970	1-1/4" size		15	.533		300	18.35		318.35	360	
	6980	1-1/2" size		13	.615		450	21		471	530	
	6990	2" size		11	.727		655	25		680	760	
	7100	For built-in by-pass or 10-35 psi, add					7			7	7.70	
	7700	High capacity, 250 psi to 25-75 psi, threaded										
	7740	1/2" size	1 Plum	24	.333	Ea.	108	11.50		119.50	136	
	7780	3/4" size		20	.400		108	13.80		121.80	140	

15100 | Building Services Piping

15110 | Valves

		CREW	DAILY OUTPUT	LABOR-HOURS	UNIT	2001 BARE COSTS MAT.	LABOR	EQUIP.	TOTAL	TOTAL INCL O&P
7790	1" size	1 Plum R15100-090	19	.421	Ea.	167	14.50		181.50	206
7800	1-1/4" size		15	.533		300	18.35		318.35	360
7810	1-1/2" size		13	.615		450	21		471	530
7820	2" size		11	.727		655	25		680	760
7830	2-1/2" size		9	.889		870	30.50		900.50	1,000
7840	3" size		8	1		1,025	34.50		1,059.50	1,175
7850	3" flanged (iron body)	Q-1	10	1.600		1,050	49.50		1,099.50	1,225
7860	4" flanged (iron body)	"	8	2		1,750	62		1,812	2,000
7920	For higher pressure, add					25%				
8000	Silent check, bronze trim									
8010	Compact wafer type, for 125 or 150 lb. flanges									
8020	1-1/2" size	1 Plum	11	.727	Ea.	255	25		280	320
8021	2" size	"	9	.889		274	30.50		304.50	345
8022	2-1/2" size	Q-1	9	1.778		288	55		343	400
8023	3" size		8	2		320	62		382	450
8024	4" size		5	3.200		550	99		649	755
8025	5" size	Q-2	6	4		760	129		889	1,025
8026	6" size		5	4.800		865	154		1,019	1,175
8027	8" size		4.50	5.333		1,850	171		2,021	2,275
8028	10" size		4	6		3,250	193		3,443	3,875
8029	12" size		3	8		7,600	257		7,857	8,750
8050	For 250 or 300 lb flanges, thru 6" no change									
8051	For 8" and 10", add				Ea.	40%	10%			
8060	Full flange wafer type, 150 lb.									
8061	1" size	1 Plum	14	.571	Ea.	272	19.70		291.70	330
8062	1-1/4" size		12	.667		284	23		307	345
8063	1-1/2" size		11	.727		340	25		365	415
8064	2" size		9	.889		460	30.50		490.50	550
8065	2-1/2" size	Q-1	9	1.778		520	55		575	655
8066	3" size		8	2		575	62		637	725
8067	4" size		5	3.200		850	99		949	1,075
8068	5" size	Q-2	6	4		1,100	129		1,229	1,400
8069	6" size	"	5	4.800		1,450	154		1,604	1,800
8080	For 300 lb., add					40%	10%			
8100	Globe type, 150 lb.									
8110	2" size	1 Plum	9	.889	Ea.	535	30.50		565.50	635
8111	2-1/2" size	Q-1	9	1.778		660	55		715	815
8112	3" size		8	2		790	62		852	965
8113	4" size		5	3.200		1,100	99		1,199	1,350
8114	5" size	Q-2	6	4		1,350	129		1,479	1,675
8115	6" size	"	5	4.800		1,875	154		2,029	2,275
8130	For 300 lb., add					20%	10%			
8140	Screwed end type, 250 lb.									
8141	1/2" size	1 Plum	24	.333	Ea.	48	11.50		59.50	70.50
8142	3/4" size		20	.400		48	13.80		61.80	74
8143	1" size		19	.421		55	14.50		69.50	82.50
8144	1-1/4" size		15	.533		76	18.35		94.35	111
8145	1-1/2" size		13	.615		85	21		106	126
8146	2" size		11	.727		114	25		139	164
8350	Tempering, water, sweat connections									
8400	1/2" size	1 Plum	24	.333	Ea.	40	11.50		51.50	61.50
8440	3/4" size	"	20	.400	"	49	13.80		62.80	75
8650	Threaded connections									
8700	1/2" size	1 Plum	24	.333	Ea.	49	11.50		60.50	71.50
8740	3/4" size		20	.400		187	13.80		200.80	227
8750	1" size		19	.421		206	14.50		220.50	249

15100 | Building Services Piping

15110 | Valves

			CREW	DAILY OUTPUT	LABOR-HOURS	UNIT	MAT.	LABOR	EQUIP.	TOTAL	TOTAL INCL O&P	
160	8760	1-1/4" size	1 Plum	15	.533	Ea.	330	18.35		348.35	390	160
	8770	1-1/2" size		13	.615		355	21		376	425	
	8780	2" size		11	.727		540	25		565	630	
200	0010	**VALVES, IRON BODY**										200
	0020	For grooved joint, see Division 15107-690										
	0100	Angle, 125 lb.										
	0110	Flanged										
	0116	2" size	1 Plum	5	1.600	Ea.	630	55		685	780	
	0118	4" size	Q-1	3	5.333		1,050	165		1,215	1,400	
	0120	6" size	Q-2	3	8		2,050	257		2,307	2,650	
	0122	8" size	"	2.50	9.600		3,650	310		3,960	4,500	
	0560	Butterfly, lug type, pneumatic operator, 2" size	1 Stpi	14	.571		221	19.90		240.90	273	
	0570	3" size	Q-1	8	2		229	62		291	345	
	0580	4" size	"	5	3.200		250	99		349	425	
	0590	6" size	Q-2	5	4.800		335	154		489	605	
	0600	8" size		4.50	5.333		405	171		576	705	
	0610	10" size		4	6		510	193		703	850	
	0620	12" size		3	8		685	257		942	1,150	
	0630	14" size		2.30	10.435		985	335		1,320	1,575	
	0640	18" size		1.50	16		1,700	515		2,215	2,650	
	0650	20" size		1	24		2,050	770		2,820	3,425	
	0800	Butterfly, lug type, gear operated, 2-1/2" size, 200 lb. except noted	Q-1	9	1.778		174	55		229	274	
	0810	3" size		8	2		183	62		245	295	
	0820	4" size		5	3.200		206	99		305	375	
	0830	5" size	Q-2	5	4.800		257	154		411	515	
	0840	6" size		5	4.800		284	154		438	550	
	0850	8" size		4.50	5.333		370	171		541	665	
	0860	10" size		4	6		475	193		668	815	
	0870	12" size		3	8		690	257		947	1,150	
	0880	14" size, 150 lb.		2.30	10.435		815	335		1,150	1,400	
	0890	16" size, 150 lb.		1.75	13.714		1,150	440		1,590	1,950	
	0900	18" size, 150 lb.		1.50	16		1,500	515		2,015	2,425	
	0910	20" size, 150 lb.		1	24		1,775	770		2,545	3,125	
	0930	24" size, 150 lb.		.75	32		2,475	1,025		3,500	4,275	
	1020	Butterfly, wafer type, gear actuator, 200 lb.										
	1030	2" size	1 Plum	14	.571	Ea.	109	19.70		128.70	149	
	1040	2-1/2" size	Q-1	9	1.778		112	55		167	206	
	1050	3" size		8	2		116	62		178	222	
	1060	4" size		5	3.200		145	99		244	310	
	1070	5" size	Q-2	5	4.800		175	154		329	425	
	1080	6" size		5	4.800		198	154		352	450	
	1090	8" size		4.50	5.333		264	171		435	550	
	1100	10" size		4	6		330	193		523	655	
	1110	12" size		3	8		460	257		717	895	
	1200	Wafer type, lever actuator, 200 lb.										
	1220	2" size	1 Plum	14	.571	Ea.	76	19.70		95.70	113	
	1230	2-1/2" size	Q-1	9	1.778		79	55		134	170	
	1240	3" size		8	2		83.50	62		145.50	185	
	1250	4" size		5	3.200		102	99		201	262	
	1260	5" size	Q-2	5	4.800		141	154		295	390	
	1270	6" size		5	4.800		170	154		324	420	
	1280	8" size		4.50	5.333		251	171		422	535	
	1290	10" size		4	6		325	193		518	650	
	1300	12" size		3	8		505	257		762	945	
	1400	Diverter, 150 lb. flanged, bronze or iron plugs										

Important: See the Reference Section for critical supporting data - Reference Nos., Crews, & City Cost Indexes

15100 | Building Services Piping

15110 | Valves

			Crew	Daily Output	Labor-Hours	Unit	Mat.	2001 Bare Costs Labor	Equip.	Total	Total Incl O&P
1440		2" pipe size	Q-1	2	8	Ea.	2,725	248		2,973	3,375
1450		3" pipe size	"	1.50	10.667	"	3,900	330		4,230	4,800
1600	Gate, threaded, 125 lb.										
1604		1-1/2"	1 Plum	13	.615	Ea.	150	21		171	196
1605		2"		11	.727		530	25		555	625
1606		3"		8	1		420	34.50		454.50	515
1607		4"		5	1.600		955	55		1,010	1,125
1650	Gate, 125 lb., N.R.S.										
2150	Flanged										
2200		2" size	1 Plum	5	1.600	Ea.	276	55		331	390
2240		2-1/2" size	Q-1	5	3.200		284	99		383	460
2260		3" size		4.50	3.556		320	110		430	515
2280		4" size		3	5.333		455	165		620	750
2290		5" size	Q-2	3.40	7.059		775	227		1,002	1,200
2300		6" size		3	8		775	257		1,032	1,250
2320		8" size		2.50	9.600		1,325	310		1,635	1,950
2340		10" size		2.20	10.909		2,350	350		2,700	3,100
2360		12" size		1.70	14.118		3,225	455		3,680	4,225
2370		14" size		1.30	18.462		5,500	595		6,095	6,950
2380		16" size		1	24		7,675	770		8,445	9,625
2420	For 250 lb flanged, add						200%	10%			
3500	OS&Y, 125 lb., threaded										
3504		3/4"	1 Plum	18	.444	Ea.	385	15.30		400.30	450
3505		1"		16	.500		440	17.25		457.25	505
3506		1-1/2" size		13	.615		465	21		486	540
3507		2" size		11	.727		475	25		500	565
3508		2-1/2" size		9	.889		490	30.50		520.50	585
3510		4" size		7	1.143		790	39.50		829.50	925
3550	OS&Y, 125 lb., flanged										
3600		2" size	1 Plum	5	1.600	Ea.	209	55		264	310
3640		2-1/2" size	Q-1	5	3.200		217	99		316	390
3660		3" size		4.50	3.556		243	110		353	435
3670		3-1/2" size		3	5.333		271	165		436	550
3680		4" size		3	5.333		345	165		510	630
3690		5" size	Q-2	3.40	7.059		570	227		797	975
3700		6" size		3	8		570	257		827	1,025
3720		8" size		2.50	9.600		1,025	310		1,335	1,600
3740		10" size		2.20	10.909		1,850	350		2,200	2,575
3760		12" size		1.70	14.118		2,475	455		2,930	3,400
3770		14" size		1.30	18.462		4,725	595		5,320	6,100
3780		16" size		1	24		7,075	770		7,845	8,975
3790		18" size		.80	30		9,500	965		10,465	12,000
3800		20" size		.60	40		13,300	1,275		14,575	16,600
3830		24" size		.50	48		19,700	1,550		21,250	23,900
3900	For 250 lb, flanged, add						200%	10%			
4350	Globe, OS&Y										
4352	Class 125, threaded										
4358		4" size	Q-1	12	1.333	Ea.	1,025	41.50		1,066.50	1,200
4540	Class 125, flanged										
4550		2" size	1 Plum	5	1.600	Ea.	430	55		485	555
4560		2-1/2" size	Q-1	5	3.200		605	99		704	815
4570		3" size		4.50	3.556		525	110		635	745
4580		4" size		3	5.333		750	165		915	1,075
4590		5" size	Q-2	3.40	7.059		1,375	227		1,602	1,850
4600		6" size		3	8		1,375	257		1,632	1,900
4610		8" size		2.50	9.600		2,675	310		2,985	3,425

15100 | Building Services Piping

15110 | Valves

			CREW	DAILY OUTPUT	LABOR-HOURS	UNIT	MAT.	LABOR	EQUIP.	TOTAL	TOTAL INCL O&P	
200	4612	10" size	Q-2	2.20	10.909	Ea.	4,200	350		4,550	5,125	200
	4614	12" size	R15100 -090	1.70	14.118		4,850	455		5,305	6,000	
	5040	Class 250, flanged										
	5050	2" size	1 Plum	4.50	1.778	Ea.	645	61		706	805	
	5060	2-1/2" size	Q-1	4.50	3.556		840	110		950	1,100	
	5070	3" size		4	4		870	124		994	1,150	
	5080	4" size		2.70	5.926		1,275	184		1,459	1,675	
	5090	5" size	Q-2	3	8		2,300	257		2,557	2,925	
	5100	6" size		2.70	8.889		2,300	286		2,586	2,950	
	5110	8" size		2.20	10.909		3,875	350		4,225	4,800	
	5120	10" size		2	12		5,250	385		5,635	6,350	
	5130	12" size		1.60	15		8,175	480		8,655	9,725	
	5240	Valve sprocket rim w/chain, for 2" valve	1 Stpi	30	.267		71.50	9.30		80.80	92.50	
	5250	2-1/2" valve		27	.296		71.50	10.35		81.85	94	
	5260	3-1/2" valve		25	.320		71.50	11.15		82.65	95.50	
	5270	6" valve		20	.400		71.50	13.95		85.45	99.50	
	5280	8" valve		18	.444		108	15.50		123.50	143	
	5290	12" valve		16	.500		108	17.45		125.45	146	
	5300	16" valve		12	.667		147	23		170	196	
	5310	20" valve		10	.800		147	28		175	203	
	5320	36" valve		8	1		272	35		307	350	
	5450	Swing check, 125 lb., threaded										
	5500	2" size	1 Plum	11	.727	Ea.	340	25		365	415	
	5540	2-1/2" size	Q-1	15	1.067		400	33		433	490	
	5550	3" size		13	1.231		445	38		483	550	
	5560	4" size		10	1.600		710	49.50		759.50	860	
	5950	Flanged										
	5994	1" size	1 Plum	7	1.143	Ea.	78	39.50		117.50	146	
	5998	1-1/2" size		6	1.333		114	46		160	195	
	6000	2" size		5	1.600		144	55		199	241	
	6040	2-1/2" size	Q-1	5	3.200		174	99		273	340	
	6050	3" size		4.50	3.556		187	110		297	370	
	6060	4" size		3	5.333		294	165		459	575	
	6070	6" size	Q-2	3	8		505	257		762	945	
	6080	8" size		2.50	9.600		945	310		1,255	1,525	
	6090	10" size		2.20	10.909		1,625	350		1,975	2,300	
	6100	12" size		1.70	14.118		2,525	455		2,980	3,450	
	6110	18" size		1.30	18.462		16,600	595		17,195	19,100	
	6114	24" size		.75	32		30,800	1,025		31,825	35,400	
	6160	For 250 lb flanged, add					200%	20%				
	6600	Silent check, bronze trim										
	6610	Compact wafer type, for 125 or 150 lb. flanges										
	6630	1-1/2" size	1 Plum	11	.727	Ea.	118	25		143	167	
	6640	2" size	"	9	.889		140	30.50		170.50	200	
	6650	2-1/2" size	Q-1	9	1.778		154	55		209	252	
	6660	3" size		8	2		165	62		227	275	
	6670	4" size		5	3.200		210	99		309	380	
	6680	5" size	Q-2	6	4		275	129		404	500	
	6690	6" size		6	4		370	129		499	600	
	6700	8" size		4.50	5.333		660	171		831	985	
	6710	10" size		4	6		1,150	193		1,343	1,550	
	6720	12" size		3	8		2,175	257		2,432	2,800	
	6740	For 250 or 300 lb. flanges, thru 6" no change										
	6741	For 8" and 10", add				Ea.	11%	10%				
	6750	Twin disc										
	6752	2" size	1 Plum	9	.889	Ea.	168	30.50		198.50	231	

15100 | Building Services Piping

15110 | Valves

			Crew	Daily Output	Labor-Hours	Unit	2001 Bare Costs Mat.	Labor	Equip.	Total	Total Incl O&P	
200	6754	4" size	Q-1	5	3.200	Ea.	282	99		381	460	200
	6756	6" size	Q-2	5	4.800		415	154		569	690	
	6758	8" size		4.50	5.333		645	171		816	965	
	6760	10" size		4	6		1,025	193		1,218	1,425	
	6762	12" size		3	8		1,250	257		1,507	1,750	
	6764	18" size		1.50	16		5,050	515		5,565	6,325	
	6766	24" size		.75	32		7,475	1,025		8,500	9,775	
	6800	Full flange type, 150 lb.										
	6810	1" size	1 Plum	14	.571	Ea.	138	19.70		157.70	181	
	6811	1-1/4" size		12	.667		145	23		168	194	
	6812	1-1/2" size		11	.727		168	25		193	223	
	6813	2" size		9	.889		193	30.50		223.50	258	
	6814	2-1/2" size	Q-1	9	1.778		234	55		289	340	
	6815	3" size		8	2		242	62		304	360	
	6816	4" size		5	3.200		299	99		398	480	
	6817	5" size	Q-2	6	4		470	129		599	715	
	6818	6" size		5	4.800		620	154		774	915	
	6819	8" size		4.50	5.333		1,050	171		1,221	1,400	
	6820	10" size		4	6		1,325	193		1,518	1,750	
	6840	For 250 lb., add					30%	10%				
	6900	Globe type, 125 lb.										
	6910	2" size	1 Plum	9	.889	Ea.	267	30.50		297.50	340	
	6911	2-1/2" size	Q-1	9	1.778		294	55		349	410	
	6912	3" size		8	2		315	62		377	445	
	6913	4" size		5	3.200		425	99		524	620	
	6914	5" size	Q-2	6	4		550	129		679	800	
	6915	6" size		5	4.800		685	154		839	990	
	6916	8" size		4.50	5.333		1,250	171		1,421	1,625	
	6917	10" size		4	6		1,550	193		1,743	2,000	
	6918	12" size		3	8		2,625	257		2,882	3,300	
	6919	14" size		2.30	10.435		3,600	335		3,935	4,450	
	6920	16" size		1.75	13.714		5,175	440		5,615	6,375	
	6921	18" size		1.50	16		7,475	515		7,990	9,000	
	6922	20" size		1	24		9,075	770		9,845	11,200	
	6923	24" size		.75	32		13,200	1,025		14,225	16,200	
	6940	For 250 lb., add					40%	10%				
	6980	Screwed end type, 125 lb.										
	6981	1" size	1 Plum	19	.421	Ea.	50.50	14.50		65	77.50	
	6982	1-1/4" size		15	.533		65.50	18.35		83.85	99.50	
	6983	1-1/2" size		13	.615		81	21		102	121	
	6984	2" size		11	.727		110	25		135	159	
400	0010	**MULTIPURPOSE VALVES**										400
	0100	Functions as a shut off, balancing, check & metering valve										
	1000	Cast iron body										
	1010	Threaded										
	1020	1-1/2" size	1 Stpi	11	.727	Ea.	280	25.50		305.50	350	
	1030	2" size	"	8	1		340	35		375	425	
	1040	2-1/2" size	Q-5	5	3.200		365	100		465	555	
	1050	3" size	"	4.50	3.556		390	112		502	595	
	1200	Flanged										
	1210	3" size	Q-5	4.20	3.810	Ea.	425	120		545	645	
	1220	4" size	"	3	5.333		855	167		1,022	1,200	
	1230	5" size	Q-6	3.80	6.316		1,000	205		1,205	1,400	
	1240	6" size		3	8		1,250	260		1,510	1,775	
	1250	8" size		2.50	9.600		1,825	310		2,135	2,500	

15100 | Building Services Piping

15110 | Valves

		CREW	DAILY OUTPUT	LABOR-HOURS	UNIT	MAT.	LABOR	EQUIP.	TOTAL	TOTAL INCL O&P	
400	1260	10" size	Q-6	2.20	10.909	Ea.	2,650	355		3,005	3,450	400
	1270	12" size		2.10	11.429		7,225	370		7,595	8,500	
	1280	14" size	↓	2	12	↓	10,200	390		10,590	11,900	
500	0010	**VALVES, PLASTIC** R15100-090										500
	1100	Angle, PVC, threaded										
	1110	1/4" size	1 Plum	26	.308	Ea.	47.50	10.60		58.10	68	
	1120	1/2" size		26	.308		47.50	10.60		58.10	68	
	1130	3/4" size		25	.320		55.50	11		66.50	77.50	
	1140	1" size	↓	23	.348	↓	66.50	12		78.50	91	
	1150	Ball, PVC, socket or threaded, single union										
	1200	1/4" size	1 Plum	26	.308	Ea.	24	10.60		34.60	42.50	
	1220	3/8" size		26	.308		24	10.60		34.60	42.50	
	1230	1/2" size		26	.308		24	10.60		34.60	42.50	
	1240	3/4" size		25	.320		28.50	11		39.50	48	
	1250	1" size		23	.348		34	12		46	55.50	
	1260	1-1/4" size		21	.381		45	13.10		58.10	69.50	
	1270	1-1/2" size		20	.400		57	13.80		70.80	83.50	
	1280	2" size	↓	17	.471		81.50	16.20		97.70	114	
	1290	2-1/2" size	Q-1	26	.615		204	19.10		223.10	253	
	1300	3" size		24	.667		204	20.50		224.50	255	
	1310	4" size	↓	20	.800		262	25		287	325	
	1360	For PVC, flanged, add					100%	15%				
	1450	Double union 1/2" size	1 Plum	26	.308		13.80	10.60		24.40	31	
	1460	3/4" size		25	.320		16.20	11		27.20	34.50	
	1470	1" size		23	.348		17.75	12		29.75	37.50	
	1480	1-1/4" size		21	.381		27	13.10		40.10	49.50	
	1490	1-1/2" size		20	.400		30	13.80		43.80	54	
	1500	2" size	↓	17	.471	↓	42.50	16.20		58.70	71	
	1650	CPVC, socket or threaded, single union										
	1700	1/2" size	1 Plum	26	.308	Ea.	33.50	10.60		44.10	52.50	
	1720	3/4" size		25	.320		42	11		53	62.50	
	1730	1" size		23	.348		50	12		62	73	
	1750	1-1/4" size		21	.381		83.50	13.10		96.60	112	
	1760	1-1/2" size		20	.400		83.50	13.80		97.30	113	
	1770	2" size		17	.471		116	16.20		132.20	153	
	1780	3" size	Q-1	24	.667		345	20.50		365.50	410	
	1840	For CPVC, flanged, add					65%	15%				
	1880	For true union, socket or threaded, add				↓	50%	5%				
	2050	Polypropylene, threaded										
	2100	1/4" size	1 Plum	26	.308	Ea.	30	10.60		40.60	49	
	2120	3/8" size		26	.308		30	10.60		40.60	49	
	2130	1/2" size		26	.308		30	10.60		40.60	49	
	2140	3/4" size		25	.320		37.50	11		48.50	57.50	
	2150	1" size		23	.348		44.50	12		56.50	67	
	2160	1-1/4" size		21	.381		64.50	13.10		77.60	91	
	2170	1-1/2" size		20	.400		74	13.80		87.80	103	
	2180	2" size	↓	17	.471		101	16.20		117.20	136	
	2190	3" size	Q-1	24	.667		266	20.50		286.50	325	
	2200	4" size	"	20	.800	↓	445	25		470	530	
	2550	PVC, three way, socket or threaded										
	2600	1/2" size	1 Plum	26	.308	Ea.	56.50	10.60		67.10	78	
	2640	3/4" size		25	.320		64	11		75	87	
	2650	1" size		23	.348		69.50	12		81.50	94.50	
	2660	1-1/2" size		20	.400		140	13.80		153.80	175	
	2670	2" size	↓	17	.471	↓	188	16.20		204.20	232	

15100 | Building Services Piping

15110 | Valves

			Crew	Daily Output	Labor-Hours	Unit	Mat.	Labor	Equip.	Total	Total Incl O&P	
500	2680	3" size	Q-1	24	.667	Ea.	410	20.50		430.50	480	500
	2740	For flanged, add				↓	60%	15%				
	3150	Ball check, PVC, socket or threaded										
	3200	1/4" size	1 Plum	26	.308	Ea.	29	10.60		39.60	48	
	3220	3/8" size		26	.308		29	10.60		39.60	48	
	3240	1/2" size		26	.308		29	10.60		39.60	48	
	3250	3/4" size		25	.320		32.50	11		43.50	52	
	3260	1" size		23	.348		40.50	12		52.50	63	
	3270	1-1/4" size		21	.381		68	13.10		81.10	95	
	3280	1-1/2" size		20	.400		68	13.80		81.80	96	
	3290	2" size	↓	17	.471		93	16.20		109.20	127	
	3310	3" size	Q-1	24	.667		257	20.50		277.50	315	
	3320	4" size	"	20	.800		365	25		390	440	
	3360	For PVC, flanged, add				↓	50%	15%				
	3750	CPVC, socket or threaded										
	3800	1/2" size	1 Plum	26	.308	Ea.	42.50	10.60		53.10	62.50	
	3840	3/4" size		25	.320		50.50	11		61.50	72	
	3850	1" size		23	.348		60	12		72	83.50	
	3860	1-1/2" size		20	.400		103	13.80		116.80	134	
	3870	2" size	↓	17	.471		138	16.20		154.20	177	
	3880	3" size	Q-1	24	.667		375	20.50		395.50	440	
	3920	4" size	"	20	.800		505	25		530	595	
	3930	For CPVC, flanged, add				↓	40%	15%				
	4340	Polypropylene, threaded										
	4360	1/2" size	1 Plum	26	.308	Ea.	25	10.60		35.60	43.50	
	4400	3/4" size		25	.320		29	11		40	48.50	
	4440	1" size		23	.348		37.50	12		49.50	59	
	4450	1-1/2" size		20	.400		72.50	13.80		86.30	101	
	4460	2" size	↓	17	.471		91	16.20		107.20	125	
	4500	For polypropylene flanged, add				↓	200%	15%				
	4850	Foot valve, PVC, socket or threaded										
	4900	1/2" size	1 Plum	34	.235	Ea.	40.50	8.10		48.60	57	
	4930	3/4" size		32	.250		46	8.60		54.60	64	
	4940	1" size		28	.286		60	9.85		69.85	81	
	4950	1-1/4" size		27	.296		115	10.20		125.20	142	
	4960	1-1/2" size		26	.308		115	10.60		125.60	143	
	4970	2" size		24	.333		134	11.50		145.50	164	
	4980	3" size		20	.400		320	13.80		333.80	370	
	4990	4" size	↓	18	.444		560	15.30		575.30	640	
	5000	For flanged, add				↓	25%	10%				
	5050	CPVC, socket or threaded										
	5060	1/2" size	1 Plum	34	.235	Ea.	89.50	8.10		97.60	111	
	5070	3/4" size		32	.250		108	8.60		116.60	132	
	5080	1" size		28	.286		127	9.85		136.85	155	
	5090	1-1/4" size		27	.296		200	10.20		210.20	235	
	5100	1-1/2" size		26	.308		208	10.60		218.60	245	
	5110	2" size		24	.333		267	11.50		278.50	310	
	5120	3" size		20	.400		505	13.80		518.80	575	
	5130	4" size	↓	18	.444		1,075	15.30		1,090.30	1,225	
	5140	For flanged, add				↓	25%	10%				
	5280	Needle valve, PVC, threaded										
	5300	1/4" size	1 Plum	26	.308	Ea.	32.50	10.60		43.10	52	
	5340	3/8" size		26	.308		38	10.60		48.60	58	
	5360	1/2" size		26	.308		38	10.60		48.60	58	
	5380	For polypropylene, add				↓	10%					
	5800	Y check, PVC, socket or threaded										

15100 | Building Services Piping

15110 | Valves

		CREW	DAILY OUTPUT	LABOR-HOURS	UNIT	2001 BARE COSTS MAT.	LABOR	EQUIP.	TOTAL	TOTAL INCL O&P	
500	5820	1/2" size	1 Plum	26	.308	Ea.	52.50	10.60		63.10	73.50	500
	5840	3/4" size		25	.320		56.50	11		67.50	78.50	
	5850	1" size		23	.348		62	12		74	86	
	5860	1-1/4" size		21	.381		97	13.10		110.10	127	
	5870	1-1/2" size		20	.400		105	13.80		118.80	137	
	5880	2" size		17	.471		131	16.20		147.20	170	
	5890	2-1/2" size		15	.533		277	18.35		295.35	335	
	5900	3" size	Q-1	24	.667		260	20.50		280.50	315	
	5910	4" size	"	20	.800		455	25		480	540	
	5960	For PVC flanged, add					45%	15%				
	6350	Y sediment strainer, PVC, socket or threaded										
	6400	1/2" size	1 Plum	26	.308	Ea.	29.50	10.60		40.10	48.50	
	6440	3/4" size		24	.333		32	11.50		43.50	53	
	6450	1" size		23	.348		39	12		51	61	
	6460	1-1/4" size		21	.381		64.50	13.10		77.60	91	
	6470	1-1/2" size		20	.400		67.50	13.80		81.30	95.50	
	6480	2" size		17	.471		78.50	16.20		94.70	111	
	6490	2-1/2" size		15	.533		192	18.35		210.35	239	
	6500	3" size	Q-1	24	.667		192	20.50		212.50	242	
	6510	4" size	"	20	.800		320	25		345	390	
	6560	For PVC, flanged, add					55%	15%				
600	0010	**VALVES, SEMI-STEEL**										600
	1020	Lubricated plug valve, threaded, 200 psi										
	1030	1/2" pipe size	1 Plum	18	.444	Ea.	61.50	15.30		76.80	90.50	
	1040	3/4" pipe size		16	.500		61.50	17.25		78.75	93.50	
	1050	1" pipe size		14	.571		79	19.70		98.70	116	
	1060	1-1/4" pipe size		12	.667		94.50	23		117.50	139	
	1070	1-1/2" pipe size		11	.727		102	25		127	150	
	1080	2" pipe size		8	1		120	34.50		154.50	184	
	1090	2-1/2" pipe size	Q-1	5	3.200		186	99		285	355	
	1100	3" pipe size	"	4.50	3.556		228	110		338	415	
	6990	Flanged, 200 psi										
	7000	2" pipe size	1 Plum	8	1	Ea.	145	34.50		179.50	211	
	7010	2-1/2" pipe size	Q-1	5	3.200		218	99		317	390	
	7020	3" pipe size		4.50	3.556		263	110		373	455	
	7030	4" pipe size		3	5.333		335	165		500	615	
	7036	5" pipe size		2.50	6.400		500	198		698	850	
	7040	6" pipe size	Q-2	3	8		655	257		912	1,100	
	7050	8" pipe size		2.50	9.600		1,175	310		1,485	1,775	
	7060	10" pipe size		2.20	10.909		1,725	350		2,075	2,425	
	7070	12" pipe size		1.70	14.118		2,975	455		3,430	3,925	
700	0010	**VALVES, STEEL**										700
	0800	Cast										
	1350	Check valve, swing type, 150 lb., flanged										
	1400	2" size	1 Plum	8	1	Ea.	465	34.50		499.50	565	
	1440	2-1/2" size	Q-1	5	3.200		595	99		694	805	
	1450	3" size		4.50	3.556		595	110		705	820	
	1460	4" size		3	5.333		875	165		1,040	1,200	
	1470	6" size	Q-2	3	8		1,350	257		1,607	1,900	
	1480	8" size		2.50	9.600		2,225	310		2,535	2,925	
	1490	10" size		2.20	10.909		3,300	350		3,650	4,150	
	1500	12" size		1.70	14.118		4,725	455		5,180	5,875	
	1510	14" size		1.30	18.462		6,750	595		7,345	8,325	
	1520	16" size		1	24		8,225	770		8,995	10,200	
	1548	For 600 lb., flanged, add					110%	20%				

15100 | Building Services Piping

15110 | Valves

			CREW	DAILY OUTPUT	LABOR-HOURS	UNIT	MAT.	LABOR	EQUIP.	TOTAL	TOTAL INCL O&P	
700	1571	300 lb., 2" size	1 Plum	7.40	1.081	Ea.	545	37		582	655	700
	1572	2-1/2" size	Q-1	4.20	3.810		825	118		943	1,075	
	1573	3" size		4	4		825	124		949	1,100	
	1574	4" size		2.80	5.714		1,175	177		1,352	1,550	
	1575	6" size	Q-2	2.90	8.276		2,100	266		2,366	2,725	
	1576	8" size		2.40	10		3,475	320		3,795	4,275	
	1577	10" size		2.10	11.429		4,400	365		4,765	5,400	
	1578	12" size		1.60	15		5,975	480		6,455	7,300	
	1579	14" size		1.20	20		9,725	645		10,370	11,700	
	1581	16" size		.90	26.667		13,500	855		14,355	16,100	
	1700	Diverter, 150 lb., flanged, bronze or iron plugs										
	1720	1-1/2" pipe size	Q-1	3	5.333	Ea.	2,450	165		2,615	2,950	
	1730	2" pipe size		2	8		2,775	248		3,023	3,425	
	1740	3" pipe size		1.50	10.667		3,950	330		4,280	4,850	
	1750	4" pipe size		1	16		6,025	495		6,520	7,375	
	1760	6" pipe size	Q-2	1	24		11,500	770		12,270	13,900	
	1790	For 300 lb., add					20%	5%				
	1810	Diverter, 150 lb., flanged, stainless steel plugs										
	1830	1-1/2" pipe size	Q-1	3	5.333	Ea.	2,400	165		2,565	2,900	
	1840	2" pipe size		2	8		3,200	248		3,448	3,900	
	1850	3" pipe size		1.50	10.667		4,325	330		4,655	5,250	
	1860	4" pipe size		1	16		6,525	495		7,020	7,925	
	1870	6" pipe size	Q-2	1	24		13,200	770		13,970	15,700	
	1890	For 300 lb., add					10%	5%				
	1950	Gate valve, 150 lb., flanged										
	2000	2" size	1 Plum	8	1	Ea.	540	34.50		574.50	645	
	2040	2-1/2" size	Q-1	5	3.200		765	99		864	990	
	2050	3" size		4.50	3.556		765	110		875	1,000	
	2060	4" size		3	5.333		930	165		1,095	1,275	
	2070	6" size	Q-2	3	8		1,475	257		1,732	2,025	
	2080	8" size		2.50	9.600		2,350	310		2,660	3,075	
	2090	10" size		2.20	10.909		3,700	350		4,050	4,600	
	2100	12" size		1.70	14.118		5,050	455		5,505	6,225	
	2110	14" size		1.30	18.462		8,400	595		8,995	10,100	
	2120	16" size		1	24		10,200	770		10,970	12,400	
	2130	18" size		.80	30		11,300	965		12,265	13,900	
	2140	20" size		.60	40		13,100	1,275		14,375	16,400	
	2650	300 lb., flanged										
	2700	2" size	1 Plum	7.40	1.081	Ea.	740	37		777	865	
	2740	2-1/2" size	Q-1	4.20	3.810		995	118		1,113	1,275	
	2750	3" size		4	4		995	124		1,119	1,275	
	2760	4" size		2.80	5.714		1,375	177		1,552	1,800	
	2770	6" size	Q-2	2.90	8.276		2,350	266		2,616	2,975	
	2780	8" size		2.40	10		3,675	320		3,995	4,500	
	2790	10" size		2.10	11.429		5,425	365		5,790	6,525	
	2800	12" size		1.60	15		7,200	480		7,680	8,650	
	2810	14" size		1.20	20		11,300	645		11,945	13,500	
	2820	16" size		.90	26.667		14,200	855		15,055	16,900	
	2830	18" size		.70	34.286		17,800	1,100		18,900	21,200	
	2840	20" size		.50	48		21,500	1,550		23,050	26,000	
	3200	600 lb., flanged										
	3220	2" size	1 Plum	7	1.143	Ea.	1,175	39.50		1,214.50	1,325	
	3240	2-1/2" size	Q-1	4	4		1,650	124		1,774	2,000	
	3250	3" size		3.60	4.444		1,650	138		1,788	2,025	
	3260	4" size		2.50	6.400		2,675	198		2,873	3,250	
	3270	6" size	Q-2	2.60	9.231		5,050	297		5,347	6,025	

Note for 1571: R15100-090

15100 | Building Services Piping

15110 | Valves

			CREW	DAILY OUTPUT	LABOR-HOURS	UNIT	MAT.	LABOR	EQUIP.	TOTAL	TOTAL INCL O&P
3280	8" size	R15100-090	Q-2	2.10	11.429	Ea.	8,150	365		8,515	9,525
3290	10" size			1.70	14.118		11,900	455		12,355	13,800
3300	12" size			1.20	20		15,300	645		15,945	17,800
3310	14" size			.80	30		27,900	965		28,865	32,200
3320	16" size			.60	40		34,500	1,275		35,775	40,000
3330	18" size			.50	48		42,500	1,550		44,050	49,100
3340	20" size			.40	60		53,000	1,925		54,925	61,500
3650	Globe valve, 150 lb., flanged										
3700	2" size		1 Plum	8	1	Ea.	675	34.50		709.50	795
3740	2-1/2" size		Q-1	5	3.200		850	99		949	1,075
3750	3" size			4.50	3.556		850	110		960	1,100
3760	4" size			3	5.333		1,250	165		1,415	1,625
3770	6" size		Q-2	3	8		1,950	257		2,207	2,550
3780	8" size			2.50	9.600		3,625	310		3,935	4,450
3790	10" size			2.20	10.909		8,925	350		9,275	10,400
3800	12" size			1.70	14.118		11,400	455		11,855	13,200
4080	300 lb., flanged										
4100	2" size		1 Plum	7.40	1.081	Ea.	935	37		972	1,075
4140	2-1/2" size		Q-1	4.20	3.810		1,250	118		1,368	1,575
4150	3" size			4	4		1,250	124		1,374	1,575
4160	4" size			2.80	5.714		1,750	177		1,927	2,200
4170	6" size		Q-2	2.90	8.276		3,150	266		3,416	3,850
4180	8" size			2.40	10		5,175	320		5,495	6,175
4190	10" size			2.10	11.429		12,700	365		13,065	14,500
4200	12" size			1.60	15		14,900	480		15,380	17,100
4680	600 lb., flanged										
4700	2" size		1 Plum	7	1.143	Ea.	1,325	39.50		1,364.50	1,500
4740	2-1/2" size		Q-1	4	4		1,975	124		2,099	2,350
4750	3" size			3.60	4.444		1,975	138		2,113	2,375
4760	4" size			2.50	6.400		3,200	198		3,398	3,800
4770	6" size		Q-2	2.60	9.231		6,600	297		6,897	7,725
4780	8" size		"	2.10	11.429		10,600	365		10,965	12,300
4800	Silent check, 316 S.S. trim										
4810	Full flange type, 150 lb.										
4811	1" size		1 Plum	14	.571	Ea.	305	19.70		324.70	365
4812	1-1/4" size			12	.667		305	23		328	370
4813	1-1/2" size			11	.727		335	25		360	405
4814	2" size			9	.889		400	30.50		430.50	490
4815	2-1/2" size		Q-1	9	1.778		485	55		540	620
4816	3" size			8	2		565	62		627	715
4817	4" size			5	3.200		770	99		869	995
4818	5" size		Q-2	6	4		965	129		1,094	1,250
4819	6" size			5	4.800		1,250	154		1,404	1,600
4820	8" size			4.50	5.333		1,800	171		1,971	2,225
4821	10" size			4	6		2,550	193		2,743	3,125
4840	For 300 lb., add						20%	10%			
4860	For 600 lb., add						50%	15%			
4900	Globe type, 150 lb.										
4910	2" size		1 Plum	9	.889	Ea.	465	30.50		495.50	560
4911	2-1/2" size		Q-1	9	1.778		575	55		630	715
4912	3" size			8	2		640	62		702	800
4913	4" size			5	3.200		850	99		949	1,075
4914	5" size		Q-2	6	4		1,075	129		1,204	1,375
4915	6" size			5	4.800		1,325	154		1,479	1,675
4916	8" size			4.50	5.333		1,850	171		2,021	2,275
4917	10" size			4	6		3,200	193		3,393	3,800

Important: See the Reference Section for critical supporting data - Reference Nos., Crews, & City Cost Indexes

15100 | Building Services Piping

15110 | Valves

		CREW	DAILY OUTPUT	LABOR-HOURS	UNIT	MAT.	LABOR	EQUIP.	TOTAL	TOTAL INCL O&P
4918	12" size	Q-2	3	8	Ea.	3,925	257		4,182	4,700
4919	14" size		2.30	10.435		5,925	335		6,260	7,025
4920	16" size		1.75	13.714		8,275	440		8,715	9,800
4921	18" size	↓	1.50	16		8,700	515		9,215	10,400
4940	For 300 lb., add					30%	10%			
4960	For 600 lb., add				↓	60%	15%			
5150	Forged									
5340	Ball valve, 800 psi, threaded, 1/4" size	1 Plum	24	.333	Ea.	34	11.50		45.50	54.50
5350	3/8" size		24	.333		34	11.50		45.50	54.50
5360	1/2" size		24	.333		40	11.50		51.50	61.50
5370	3/4" size		20	.400		52.50	13.80		66.30	79
5380	1" size		19	.421		67	14.50		81.50	95.50
5390	1-1/4" size		15	.533		88.50	18.35		106.85	125
5400	1-1/2" size		13	.615		116	21		137	160
5410	2" size	↓	11	.727		154	25		179	207
5460	Ball valve, 800 lb., socket weld, 1/4" size	Q-15	19	.842		33.50	26	5.30	64.80	82
5470	3/8" size		19	.842		33.50	26	5.30	64.80	82
5480	1/2" size		19	.842		40	26	5.30	71.30	89.50
5490	3/4" size		19	.842		52.50	26	5.30	83.80	103
5500	1" size		15	1.067		67	33	6.70	106.70	131
5510	1-1/4" size		13	1.231		88.50	38	7.75	134.25	164
5520	1-1/2" size		11	1.455		116	45	9.15	170.15	206
5530	2" size	↓	8.50	1.882		155	58.50	11.85	225.35	272
5550	Ball valve, 150 lb., flanged									
5560	4" size	Q-1	3	5.333	Ea.	415	165		580	710
5570	6" size	Q-2	3	8		1,000	257		1,257	1,500
5580	8" size	"	2.50	9.600	↓	1,800	310		2,110	2,450
5650	Check valve, class 800, horizontal, socket									
5652	1/4" size	Q-15	19	.842	Ea.	41	26	5.30	72.30	90.50
5654	3/8" size		19	.842		41	26	5.30	72.30	90.50
5656	1/2" size		19	.842		41	26	5.30	72.30	90.50
5658	3/4" size		19	.842		44.50	26	5.30	75.80	94.50
5660	1" size		15	1.067		52	33	6.70	91.70	114
5662	1-1/4" size		13	1.231		99.50	38	7.75	145.25	175
5664	1-1/2" size		11	1.455		99.50	45	9.15	153.65	187
5666	2" size	↓	8.50	1.882	↓	139	58.50	11.85	209.35	254
5698	Threaded									
5700	1/4" size	1 Plum	24	.333	Ea.	41	11.50		52.50	62.50
5720	3/8" size		24	.333		41	11.50		52.50	62.50
5730	1/2" size		24	.333		41	11.50		52.50	62.50
5740	3/4" size		20	.400		44.50	13.80		58.30	70
5750	1" size		19	.421		52	14.50		66.50	79
5760	1-1/4" size		15	.533		99.50	18.35		117.85	137
5770	1-1/2" size		13	.615		99.50	21		120.50	141
5780	2" size	↓	11	.727		139	25		164	191
5840	For class 150, flanged, add					100%	15%			
5860	For class 300, flanged, add				↓	120%	20%			
6100	Gate, class 800, inside screw stem, socket									
6102	3/8" size	Q-15	19	.842	Ea.	43	26	5.30	74.30	93
6103	1/2" size		19	.842		43	26	5.30	74.30	93
6104	3/4" size		19	.842		50.50	26	5.30	81.80	101
6105	1" size		15	1.067		57.50	33	6.70	97.20	121
6106	1-1/4" size		13	1.231		115	38	7.75	160.75	193
6107	1-1/2" size		11	1.455		115	45	9.15	169.15	205
6108	2" size	↓	8.50	1.882	↓	144	58.50	11.85	214.35	259
6110	For OS&Y, socket, add					50%				

15100 | Building Services Piping

15110 | Valves

		Crew	Daily Output	Labor-Hours	Unit	Mat.	2001 Bare Costs Labor	Equip.	Total	Total Incl O&P
6118	Threaded									
6120	3/8" size	1 Plum	24	.333	Ea.	43	11.50		54.50	65
6130	1/2" size		24	.333		43	11.50		54.50	65
6140	3/4" size		20	.400		50.50	13.80		64.30	76.50
6150	1" size		19	.421		57.50	14.50		72	85.50
6160	1-1/4" size		15	.533		115	18.35		133.35	155
6170	1-1/2" size		13	.615		115	21		136	159
6180	2" size		11	.727		144	25		169	196
6240	For OS&Y, threaded, add					50%				
6260	For OS&Y, flanged, add					100%	20%			
6700	Globe, OS&Y, class 800, socket									
6710	1/4" size	Q-15	19	.842	Ea.	49.50	26	5.30	80.80	100
6720	3/8" size		19	.842		49.50	26	5.30	80.80	100
6730	1/2" size		19	.842		49.50	26	5.30	80.80	100
6740	3/4" size		19	.842		52.50	26	5.30	83.80	103
6750	1" size		15	1.067		66	33	6.70	105.70	130
6760	1-1/4" size		13	1.231		126	38	7.75	171.75	205
6770	1-1/2" size		11	1.455		126	45	9.15	180.15	217
6780	2" size		8.50	1.882		162	58.50	11.85	232.35	279
6860	For OS&Y, flanged, add					300%	20%			
6880	Threaded									
6882	1/4" size	1 Plum	24	.333	Ea.	49.50	11.50		61	72
6884	3/8" size		24	.333		49.50	11.50		61	72
6886	1/2" size		24	.333		49.50	11.50		61	72
6888	3/4" size		20	.400		52.50	13.80		66.30	79
6890	1" size		19	.421		66	14.50		80.50	95
6892	1-1/4" size		15	.533		126	18.35		144.35	167
6894	1-1/2" size		13	.615		126	21		147	171
6896	2" size		11	.727		162	25		187	216
0010	**VALVES, STAINLESS STEEL**									
1610	Ball, threaded 1/4" size	1 Stpi	24	.333	Ea.	28	11.60		39.60	48
1620	3/8" size		24	.333		28	11.60		39.60	48
1630	1/2" size		22	.364		28	12.65		40.65	49.50
1640	3/4" size		20	.400		46.50	13.95		60.45	72
1650	1" size		19	.421		56.50	14.65		71.15	84
1660	1-1/4 size		15	.533		107	18.60		125.60	146
1670	1-1/2" size		13	.615		112	21.50		133.50	156
1680	2" size		11	.727		146	25.50		171.50	199
1700	Check, 200 lb., threaded									
1710	1/4" size	1 Plum	24	.333	Ea.	72	11.50		83.50	96.50
1720	1/2" size		22	.364		72	12.55		84.55	98
1730	3/4" size		20	.400		79	13.80		92.80	108
1750	1" size		19	.421		101	14.50		115.50	133
1760	1-1/2" size		13	.615		191	21		212	242
1770	2" size		11	.727		320	25		345	390
1800	150 lb., flanged									
1810	2-1/2" size	Q-1	5	3.200	Ea.	765	99		864	990
1820	3" size		4.50	3.556		765	110		875	1,000
1830	4" size		3	5.333		1,150	165		1,315	1,525
1840	6" size	Q-2	3	8		2,050	257		2,307	2,650
1850	8" size	"	2.50	9.600		3,425	310		3,735	4,225
1900	Diverter, 150 lb. flanged, stainless steel plugs									
1910	1-1/2" pipe size	Q-1	3	5.333	Ea.	4,750	165		4,915	5,475
1920	2" pipe size		2	8		6,175	248		6,423	7,150
1940	3" pipe size		1.50	10.667		7,850	330		8,180	9,125

15100 | Building Services Piping

15110 | Valves

			CREW	DAILY OUTPUT	LABOR-HOURS	UNIT	MAT.	LABOR	EQUIP.	TOTAL	TOTAL INCL O&P
800	1950	4" pipe size	Q-1	1	16	Ea.	13,200	495		13,695	15,300
	2100	Gate, OS&Y, 150 lb., flanged	R15100-090								
	2120	1/2" size	1 Plum	18	.444	Ea.	310	15.30		325.30	365
	2140	3/4" size		16	.500		335	17.25		352.25	395
	2150	1" size		14	.571		375	19.70		394.70	440
	2155	1-1/4" size		12	.667		470	23		493	550
	2160	1-1/2" size		11	.727		470	25		495	555
	2170	2" size		8	1		520	34.50		554.50	625
	2180	2-1/2" size	Q-1	5	3.200		810	99		909	1,050
	2190	3" size		4.50	3.556		885	110		995	1,150
	2200	4" size		3	5.333		1,225	165		1,390	1,600
	2205	5" size		2.80	5.714		2,200	177		2,377	2,700
	2210	6" size	Q-2	3	8		2,200	257		2,457	2,825
	2220	8" size		2.50	9.600		3,400	310		3,710	4,225
	2230	10" size		2.30	10.435		4,950	335		5,285	5,950
	2240	12" size		1.90	12.632		6,900	405		7,305	8,225
	2260	For 300 lb., flanged, add					120%	15%			
	2600	600 lb., flanged									
	2620	1/2" size	1 Plum	16	.500	Ea.	720	17.25		737.25	815
	2640	3/4" size		14	.571		880	19.70		899.70	995
	2650	1" size		12	.667		1,100	23		1,123	1,250
	2660	1-1/2" size		10	.800		1,550	27.50		1,577.50	1,750
	2670	2" size		7	1.143		2,775	39.50		2,814.50	3,100
	2680	2-1/2" size	Q-1	4	4		7,975	124		8,099	8,950
	2690	3" size	"	3.60	4.444		8,050	138		8,188	9,050
	3100	Globe, OS&Y, 150 lb., flanged									
	3120	1/2" size	1 Plum	18	.444	Ea.	345	15.30		360.30	405
	3140	3/4" size		16	.500		380	17.25		397.25	445
	3150	1" size		14	.571		470	19.70		489.70	550
	3160	1-1/2" size		11	.727		700	25		725	810
	3170	2" size		8	1		855	34.50		889.50	990
	3180	2-1/2" size	Q-1	5	3.200		1,800	99		1,899	2,125
	3190	3" size		4.50	3.556		1,850	110		1,960	2,200
	3200	4" size		3	5.333		2,150	165		2,315	2,625
	3210	6" size	Q-2	3	8		4,075	257		4,332	4,875
	5000	Silent check, 316 S.S. body and trim									
	5010	Compact wafer type, 300 lb.									
	5020	1" size	1 Plum	14	.571	Ea.	555	19.70		574.70	640
	5021	1-1/4" size		12	.667		600	23		623	695
	5022	1-1/2" size		11	.727		620	25		645	725
	5023	2" size		9	.889		685	30.50		715.50	795
	5024	2-1/2" size	Q-1	9	1.778		975	55		1,030	1,150
	5025	3" size	"	8	2		1,075	62		1,137	1,275
	5100	Full flange wafer type, 150 lb.									
	5110	1" size	1 Plum	14	.571	Ea.	390	19.70		409.70	460
	5111	1-1/4" size		12	.667		420	23		443	495
	5112	1-1/2" size		11	.727		480	25		505	570
	5113	2" size		9	.889		625	30.50		655.50	735
	5114	2-1/2" size	Q-1	9	1.778		750	55		805	905
	5115	3" size		8	2		845	62		907	1,025
	5116	4" size		5	3.200		1,250	99		1,349	1,525
	5117	5" size	Q-2	6	4		1,475	129		1,604	1,800
	5118	6" size		5	4.800		1,825	154		1,979	2,250
	5119	8" size		4.50	5.333		3,050	171		3,221	3,600
	5120	10" size		4	6		4,225	193		4,418	4,950
	5200	Globe type, 300 lb.									

15100 | Building Services Piping

15110 | Valves

		CREW	DAILY OUTPUT	LABOR-HOURS	UNIT	MAT.	LABOR	EQUIP.	TOTAL	TOTAL INCL O&P
5210	2" size	1 Plum	9	.889	Ea.	715	30.50		745.50	830
5211	2-1/2" size	Q-1	9	1.778		880	55		935	1,050
5212	3" size		8	2		1,100	62		1,162	1,300
5213	4" size		5	3.200		1,500	99		1,599	1,800
5214	5" size	Q-2	6	4		2,125	129		2,254	2,525
5215	6" size		5	4.800		2,400	154		2,554	2,850
5216	8" size		4.50	5.333		3,375	171		3,546	3,975
5217	10" size		4	6		4,775	193		4,968	5,550
5300	Screwed end type, 300 lb.									
5310	1/2" size	1 Plum	24	.333	Ea.	60.50	11.50		72	84
5311	3/4" size		20	.400		72	13.80		85.80	101
5312	1" size		19	.421		84	14.50		98.50	114
5313	1-1/4" size		15	.533		111	18.35		129.35	150
5314	1-1/2" size		13	.615		120	21		141	164
5315	2" size		11	.727		149	25		174	202

Reference: R15100-090

15120 | Piping Specialties

		CREW	DAILY OUTPUT	LABOR-HOURS	UNIT	MAT.	LABOR	EQUIP.	TOTAL	TOTAL INCL O&P
0010	**AIR CONTROL**									
0030	Air separator, with strainer									
0040	2" diameter	Q-5	6	2.667	Ea.	515	83.50		598.50	695
0080	2-1/2" diameter		5	3.200		585	100		685	795
0100	3" diameter		4	4		890	126		1,016	1,175
0120	4" diameter		3	5.333		1,300	167		1,467	1,675
0130	5" diameter	Q-6	3.60	6.667		1,650	217		1,867	2,125
0140	6" diameter		3.40	7.059		1,975	230		2,205	2,525
0160	8" diameter		3	8		2,925	260		3,185	3,625
0180	10" diameter		2.20	10.909		4,575	355		4,930	5,550
0200	12" diameter		1.70	14.118		7,725	460		8,185	9,200
0210	14" diameter		1.30	18.462		9,200	600		9,800	11,000
0220	16" diameter		1	24		13,900	780		14,680	16,400
0230	18" diameter		.80	30		18,200	975		19,175	21,500
0240	20" diameter		.60	40		20,400	1,300		21,700	24,400
0300	Without strainer									
0310	2" diameter	Q-5	6	2.667	Ea.	400	83.50		483.50	565
0320	2-1/2" diameter		5	3.200		480	100		580	675
0330	3" diameter		4	4		670	126		796	930
0340	4" diameter		3	5.333		1,050	167		1,217	1,400
0350	5" diameter		2.40	6.667		1,425	209		1,634	1,900
0360	6" diameter	Q-6	3.40	7.059		1,650	230		1,880	2,150
0370	8" diameter		3	8		2,250	260		2,510	2,875
0380	10" diameter		2.20	10.909		3,325	355		3,680	4,175
0390	12" diameter		1.70	14.118		5,150	460		5,610	6,375
0400	14" diameter		1.30	18.462		7,725	600		8,325	9,400
0410	16" diameter		1	24		10,800	780		11,580	13,100
0420	18" diameter		.80	30		14,100	975		15,075	17,000
0430	20" diameter		.60	40		16,900	1,300		18,200	20,600
1000	Micro-bubble separator for total air removal, closed loop system									
1010	Requires bladder type tank in system.									
1020	Water (hot or chilled) or glycol system									
1030	Threaded									
1040	3/4" diameter	1 Stpi	20	.400	Ea.	67	13.95		80.95	94.50
1050	1" diameter		19	.421		75	14.65		89.65	105
1060	1-1/4" diameter		16	.500		103	17.45		120.45	141
1070	1-1/2" diameter		13	.615		135	21.50		156.50	181
1080	2" diameter		11	.727		655	25.50		680.50	760

15100 | Building Services Piping

15120 | Piping Specialties

		Crew	Daily Output	Labor-Hours	Unit	Mat.	Labor	Equip.	Total	Total Incl O&P	
120	1090 2-1/2" diameter	Q-5	15	1.067	Ea.	725	33.50		758.50	850	120
	1100 3" diameter		13	1.231		1,050	38.50		1,088.50	1,200	
	1110 4" diameter	▼	10	1.600	▼	1,125	50		1,175	1,300	
	1230 Flanged										
	1250 2" diameter	1 Stpi	8	1	Ea.	835	35		870	970	
	1260 2-1/2" diameter	Q-5	5	3.200		905	100		1,005	1,150	
	1270 3" diameter		4.50	3.556		1,225	112		1,337	1,525	
	1280 4" diameter	▼	3	5.333		1,350	167		1,517	1,725	
	1290 5" diameter	Q-6	3.40	7.059		2,525	230		2,755	3,125	
	1300 6" diameter	"	3	8	▼	2,625	260		2,885	3,275	
	1400 Larger sizes available										
	1590 With extended tank dirt catcher										
	1600 Flanged										
	1620 2" diameter	1 Stpi	7.40	1.081	Ea.	1,100	37.50		1,137.50	1,275	
	1630 2-1/2" diameter	Q-5	4.20	3.810		1,150	120		1,270	1,450	
	1640 3" diameter		4	4		1,525	126		1,651	1,875	
	1650 4" diameter	▼	2.80	5.714		1,650	179		1,829	2,075	
	1660 5" diameter	Q-6	3.10	7.742		2,775	252		3,027	3,425	
	1670 6" diameter	"	2.90	8.276	▼	2,950	269		3,219	3,650	
	1800 With drain/dismantle/cleaning access flange										
	1810 Flanged										
	1820 2" diameter	1 Stpi	7	1.143	Ea.	2,100	40		2,140	2,375	
	1830 2-1/2" diameter	Q-5	4	4		2,175	126		2,301	2,600	
	1840 3" diameter		3.60	4.444		2,850	139		2,989	3,325	
	1850 4" diameter	▼	2.50	6.400		3,000	201		3,201	3,600	
	1860 5" diameter	Q-6	2.80	8.571		5,550	279		5,829	6,550	
	1870 6" diameter	"	2.60	9.231	▼	5,925	300		6,225	6,975	
	2000 Boiler air fitting, in line separator										
	2010 1-1/4", 19 GPM	Q-5	22	.727	Ea.	77.50	23		100.50	120	
	2020 1-1/2", 29 GPM		20	.800		88	25		113	135	
	2021 2", 50 GPM	▼	18	.889	▼	410	28		438	490	
	2040 Boiler fitting, top outlet										
	2050 2", 12 GPM	Q-5	18	.889	Ea.	116	28		144	169	
	2060 2", 19 GPM		17	.941		77.50	29.50		107	130	
	2070 2", 26 GPM		16	1		88	31.50		119.50	145	
	2100 3", 46 GPM		8	2		116	63		179	222	
	2110 3", 68 GPM		6	2.667		123	83.50		206.50	261	
	2130 4", 125 GPM		4.50	3.556		144	112		256	325	
	2150 6", 220 GPM	Q-6	3.80	6.316		263	205		468	600	
	2160 6", 640 GPM	"	3.60	6.667	▼	375	217		592	735	
	2400 Compression tank air fitting										
	2410 For tanks 9" to 24" diameter	1 Stpi	15	.533	Ea.	33	18.60		51.60	64.50	
	2420 For tanks 100 gallon or larger	"	12	.667	"	137	23		160	185	
140	0010 **AIR PURGING SCOOP** with tappings										140
	0020 for air vent and expansion tank connection										
	0100 1" pipe size, threaded	1 Stpi	19	.421	Ea.	13.25	14.65		27.90	36.50	
	0110 1-1/4" pipe size, threaded		15	.533		13.30	18.60		31.90	42.50	
	0120 1-1/2" pipe size, threaded		13	.615		29	21.50		50.50	64.50	
	0130 2" pipe size, threaded	▼	11	.727		33	25.50		58.50	75	
	0140 2-1/2" pipe size, threaded	Q-5	15	1.067		70.50	33.50		104	129	
	0150 3" pipe size, threaded		13	1.231		92.50	38.50		131	161	
	0160 4" 150 lb. flanges	▼	3	5.333	▼	207	167		374	480	
160	0010 **AUTOMATIC AIR VENT**										160
	0020 Cast iron body, stainless steel internals, float type										

15100 | Building Services Piping

15120 | Piping Specialties

			CREW	DAILY OUTPUT	LABOR-HOURS	UNIT	MAT.	LABOR	EQUIP.	TOTAL	TOTAL INCL O&P	
160	0060	1/2" NPT inlet, 300 psi	1 Stpi	12	.667	Ea.	78.50	23		101.50	121	160
	0140	3/4" NPT inlet, 300 psi		12	.667		78.50	23		101.50	121	
	0180	1/2" NPT inlet, 250 psi		10	.800		251	28		279	320	
	0220	3/4" NPT inlet, 250 psi		10	.800		251	28		279	320	
	0260	1" NPT inlet, 250 psi		10	.800		370	28		398	445	
	0340	1-1/2" NPT inlet, 250 psi	Q-5	12	1.333		780	42		822	925	
	0380	2" NPT inlet, 250 psi	"	12	1.333		780	42		822	925	
	0600	Forged steel body, stainless steel internals, float type										
	0640	1/2" NPT inlet, 750 psi	1 Stpi	12	.667	Ea.	770	23		793	880	
	0680	3/4" NPT inlet, 750 psi		12	.667		770	23		793	880	
	0760	3/4" NPT inlet, 1000 psi		10	.800		1,150	28		1,178	1,325	
	0800	1" NPT inlet, 1000 psi	Q-5	12	1.333		1,150	42		1,192	1,350	
	0880	1-1/2" NPT inlet, 1000 psi		10	1.600		3,250	50		3,300	3,650	
	0920	2" NPT inlet, 1000 psi		10	1.600		3,375	50		3,425	3,775	
	1100	Formed steel body, non corrosive										
	1110	1/8" NPT inlet 150 psi	1 Stpi	32	.250	Ea.	6.70	8.70		15.40	20.50	
	1120	1/4" NPT inlet 150 psi		32	.250		22.50	8.70		31.20	38	
	1130	3/4" NPT inlet 150 psi		32	.250		22.50	8.70		31.20	38	
	1300	Chrome plated brass, automatic/manual, for radiators										
	1310	1/8" NPT inlet, nickel plated brass	1 Stpi	32	.250	Ea.	3.92	8.70		12.62	17.45	
180	0010	**CIRCUIT SENSOR** Flow meter										180
	0020	Metering stations										
	0040	Wafer orifice insert type										
	0060	2-1/2" pipe size	Q-5	12	1.333	Ea.	119	42		161	194	
	0100	3" pipe size		11	1.455		133	45.50		178.50	215	
	0140	4" pipe size		8	2		151	63		214	261	
	0180	5" pipe size		7.30	2.192		193	69		262	315	
	0220	6" pipe size		6.40	2.500		228	78.50		306.50	370	
	0260	8" pipe size	Q-6	5.30	4.528		320	147		467	570	
	0280	10" pipe size		4.60	5.217		375	170		545	665	
	0360	12" pipe size		4.20	5.714		615	186		801	955	
	2000	In-line probe type										
	2200	Copper, soldered										
	2210	1/2" size	1 Stpi	24	.333	Ea.	104	11.60		115.60	132	
	2220	3/4" size		20	.400		108	13.95		121.95	140	
	2230	1" size		19	.421		122	14.65		136.65	156	
	2240	1-1/4" size		15	.533		128	18.60		146.60	169	
	2250	1-1/2" size		13	.615		151	21.50		172.50	199	
	2260	2" size		11	.727		169	25.50		194.50	225	
	2270	2-1/2" size	Q-5	15	1.067		179	33.50		212.50	248	
	2280	3" size	"	13	1.231		194	38.50		232.50	273	
	2500	Brass, threaded										
	2510	1" size	1 Stpi	19	.421	Ea.	128	14.65		142.65	163	
	2520	1-1/4" size		15	.533		142	18.60		160.60	184	
	2530	1-1/2" size		13	.615		170	21.50		191.50	220	
	2540	2" size		11	.727		179	25.50		204.50	236	
	2550	2-1/2" size	Q-5	15	1.067		195	33.50		228.50	265	
	2560	3" size	"	13	1.231		199	38.50		237.50	278	
	2700	Steel, threaded										
	2710	1" size	1 Stpi	19	.421	Ea.	123	14.65		137.65	157	
	2720	1-1/4" size		15	.533		134	18.60		152.60	175	
	2730	1-1/2" size		13	.615		151	21.50		172.50	199	
	2740	2" size		11	.727		169	25.50		194.50	225	
	2750	2-1/2" size	Q-5	15	1.067		179	33.50		212.50	248	
	2760	3" size	"	13	1.231		198	38.50		236.50	277	
	3000	Pitot tube probe type										

15100 | Building Services Piping

15120 | Piping Specialties

		Crew	Daily Output	Labor-Hours	Unit	Mat.	Labor	Equip.	Total	Total Incl O&P
3100	Weld-on mounting with probe									
3110	2" size	Q-17	18	.889	Ea.	139	28	5.60	172.60	201
3120	2-1/2" size		18	.889		142	28	5.60	175.60	204
3130	3" size		18	.889		151	28	5.60	184.60	214
3140	4" size		18	.889		199	28	5.60	232.60	267
3150	5" size		17	.941		223	29.50	5.90	258.40	296
3160	6" size		17	.941		288	29.50	5.90	323.40	365
3170	8" size		16	1		410	31.50	6.30	447.80	505
3180	10" size		15	1.067		445	33.50	6.70	485.20	550
3190	12" size		15	1.067		475	33.50	6.70	515.20	585
3200	14" size		14	1.143		545	36	7.20	588.20	660
3210	16" size		14	1.143		645	36	7.20	688.20	770
3220	18" size		13	1.231		690	38.50	7.75	736.25	820
3230	20" size		13	1.231		745	38.50	7.75	791.25	885
3240	24" size		12	1.333		830	42	8.40	880.40	985
3400	Weld-on wet tap with probe									
3410	2" size, 5/16" probe dia.	Q-17	14	1.143	Ea.	335	36	7.20	378.20	430
3420	2-1/2" size, 5/16" probe dia.		14	1.143		345	36	7.20	388.20	440
3430	3" size, 5/16" probe dia.		14	1.143		365	36	7.20	408.20	460
3440	4" size, 5/16" probe dia.		14	1.143		370	36	7.20	413.20	470
3450	5" size, 3/8" probe dia.		13	1.231		390	38.50	7.75	436.25	490
3460	6" size, 3/8" probe dia.		13	1.231		415	38.50	7.75	461.25	520
3470	8" size, 1/2" probe dia.		12	1.333		430	42	8.40	480.40	545
3480	10" size, 3/4" probe dia.		11	1.455		460	45.50	9.15	514.65	585
3490	12" size, 3/4" probe dia.		11	1.455		500	45.50	9.15	554.65	630
3500	14" size, 1" probe dia.		10	1.600		565	50	10.05	625.05	705
3510	16" size, 1" probe dia.		10	1.600		675	50	10.05	735.05	825
3520	18" size, 1" probe dia.		9	1.778		725	56	11.20	792.20	890
3530	20" size, 1" probe dia.		8	2		775	63	12.60	850.60	960
3540	24" size, 1" probe dia.		8	2		870	63	12.60	945.60	1,075
3700	Clamp-on wet tap with probe									
3710	2" size, 5/16" probe dia.	Q-5	16	1	Ea.	400	31.50		431.50	490
3720	2-1/2" size, 5/16" probe dia.		16	1		410	31.50		441.50	500
3730	3" size, 5/16" probe dia.		15	1.067		420	33.50		453.50	510
3740	4" size, 5/16" probe dia.		15	1.067		485	33.50		518.50	585
3750	5" size, 3/8" probe dia.		14	1.143		515	36		551	620
3760	6" size, 3/8" probe dia.		14	1.143		545	36		581	655
3770	8" size, 1/2" probe dia.		13	1.231		600	38.50		638.50	720
3780	10" size, 3/4" probe dia.		12	1.333		625	42		667	755
3790	12" size, 3/4" probe dia.		12	1.333		700	42		742	835
3800	14" size, 1" probe dia.		11	1.455		945	45.50		990.50	1,125
3810	16" size, 1" probe dia.		11	1.455		1,100	45.50		1,145.50	1,275
3820	18" size, 1" probe dia.		10	1.600		1,175	50		1,225	1,375
3830	20" size, 1" probe dia		9	1.778		1,275	56		1,331	1,475
3840	24" size, 1" probe dia.		9	1.778		1,375	56		1,431	1,600
4000	Wet tap drills									
4010	3/8" dia. for 5/16" probe				Ea.	80.50			80.50	88.50
4020	1/2" dia. for 3/8" probe					84			84	92.50
4030	5/8" dia for 1/2" probe					99.50			99.50	109
4040	7/8" dia for 3/4" probe					130			130	143
4050	1-3/16" dia for 1" probe					139			139	152
4100	Wet tap punch									
4110	5/16" dia for 5/16" probe				Ea.	84			84	92.50
4120	3/8" dia for 3/8" probe					90.50			90.50	99.50
4130	1/2" dia for 1/2" probe					108			108	119
4150	Note: labor for wet tap drill or punch									

15100 | Building Services Piping

15120 | Piping Specialties

			CREW	DAILY OUTPUT	LABOR-HOURS	UNIT	2001 BARE COSTS				TOTAL INCL O&P
							MAT.	LABOR	EQUIP.	TOTAL	
180	4160	is included with the mounting and probe assembly										180
	9000	Readout instruments									
	9200	Gauges									
	9220	GPM gauge				Ea.	1,350			1,350	1,500
	9230	Dual gauge kit				"	1,400			1,400	1,550
200	0010	**CIRCUIT SETTER** Balance valve										200
	0012	Bronze body, soldered									
	0013	1/2" size	1 Stpi	24	.333	Ea.	41.50	11.60		53.10	63
	0014	3/4" size	"	20	.400	"	44	13.95		57.95	69
	0018	Threaded									
	0019	1/2" pipe size	1 Stpi	22	.364	Ea.	43	12.65		55.65	66.50
	0020	3/4" pipe size		20	.400		39	13.95		52.95	64
	0040	1" pipe size		18	.444		59.50	15.50		75	89
	0050	1-1/4" pipe size		15	.533		87.50	18.60		106.10	125
	0060	1-1/2" pipe size		12	.667		103	23		126	149
	0080	2" pipe size		10	.800		148	28		176	204
	0100	2-1/2" pipe size	Q-5	15	1.067		335	33.50		368.50	420
	0120	3" pipe size	"	10	1.600		470	50		520	595
	0130	Cast iron body, flanged									
	0136	3" pipe size, flanged	Q-5	4	4	Ea.	495	126		621	730
	0140	4" pipe size	"	3	5.333		705	167		872	1,025
	0200	For differential meter, accurate to 1%, add					575			575	630
	0300	For differential meter 400 psi/200° F continuous, add					1,400			1,400	1,550
220	0010	**COCKS, DRAINS & SPECIALTIES**										220
	1000	Boiler drain									
	1010	Pipe thread to hose									
	1020	Bronze									
	1030	1/2" size	1 Stpi	36	.222	Ea.	4.10	7.75		11.85	16.20
	1040	3/4" size	"	34	.235	"	4.44	8.20		12.64	17.30
	1100	Solder to hose									
	1110	Bronze									
	1120	1/2" size	1 Stpi	46	.174	Ea.	4.10	6.05		10.15	13.65
	1130	3/4" size	"	44	.182	"	4.44	6.35		10.79	14.45
	1600	With built-in vacuum breaker									
	1610	1/2" IP or solder	1 Stpi	36	.222	Ea.	12.60	7.75		20.35	25.50
	1630	With tamper proof vacuum breaker									
	1640	1/2" IP or solder	1 Stpi	36	.222	Ea.	15.65	7.75		23.40	29
	1650	3/4" IP or solder	"	34	.235	"	15.95	8.20		24.15	30
	3000	Cocks									
	3010	Air, lever or tee handle									
	3020	Bronze, single thread									
	3030	1/8" size	1 Stpi	52	.154	Ea.	5.45	5.35		10.80	14.10
	3040	1/4" size		46	.174		5.90	6.05		11.95	15.65
	3050	3/8" size		40	.200		6.65	6.95		13.60	17.80
	3060	1/2" size		36	.222		6.65	7.75		14.40	19.05
	3100	Bronze, double thread									
	3110	1/8" size	1 Stpi	26	.308	Ea.	7.30	10.70		18	24.50
	3120	1/4" size		22	.364		7.05	12.65		19.70	27
	3130	3/8" size		18	.444		7.80	15.50		23.30	32
	3140	1/2" size		15	.533		9.25	18.60		27.85	38
	4000	Steam									
	4010	Bronze									
	4020	1/8" size	1 Stpi	26	.308	Ea.	14.25	10.70		24.95	32
	4030	1/4" size		22	.364		13.10	12.65		25.75	33.50
	4040	3/8" size		18	.444		12.80	15.50		28.30	37.50

15100 | Building Services Piping

15120 | Piping Specialties

		CREW	DAILY OUTPUT	LABOR-HOURS	UNIT	MAT.	LABOR	EQUIP.	TOTAL	TOTAL INCL O&P	
220	4050	1/2" size	1 Stpi	15	.533	Ea.	12.05	18.60		30.65	41.50	220
	4060	3/4" size		14	.571		13.20	19.90		33.10	44.50	
	4070	1" size		13	.615		24.50	21.50		46	59	
	4080	1-1/4" size		11	.727		37	25.50		62.50	79	
	4090	1-1/2" size		10	.800		46.50	28		74.50	93	
	4100	2" size		9	.889		68.50	31		99.50	122	
	4300	Bronze, 3 way										
	4320	1/4" size	1 Stpi	15	.533	Ea.	27.50	18.60		46.10	58	
	4330	3/8" size		12	.667		28.50	23		51.50	66.50	
	4340	1/2" size		10	.800		16.85	28		44.85	60.50	
	4350	3/4" size		9.50	.842		29	29.50		58.50	76.50	
	4360	1" size		8.50	.941		35	33		68	88	
	4370	1-1/4" size		7.50	1.067		52.50	37		89.50	114	
	4380	1-1/2" size		6.50	1.231		61	43		104	132	
	4390	2" size		6	1.333		90	46.50		136.50	169	
	4500	Gauge cock, brass										
	4510	1/4" FPT	1 Stpi	24	.333	Ea.	6.15	11.60		17.75	24.50	
	4512	1/4" MPT	"	24	.333	"	7.15	11.60		18.75	25.50	
	4600	Pigtail, steam syphon										
	4604	1/4"	1 Stpi	24	.333	Ea.	9.30	11.60		20.90	28	
	4650	Snubber valve										
	4654	1/4"	1 Stpi	22	.364	Ea.	9.40	12.65		22.05	29.50	
	4660	Nipple, black steel										
	4664	1/4" x 3"	1 Stpi	37	.216	Ea.	.44	7.55		7.99	11.85	
250	0010	**DIELECTRIC UNIONS** Standard gaskets for water and air										250
	0020	250 psi maximum pressure										
	0280	Female IPT to sweat, straight										
	0300	1/2" pipe size	1 Plum	24	.333	Ea.	3.15	11.50		14.65	21	
	0340	3/4" pipe size		20	.400		3.15	13.80		16.95	24.50	
	0360	1" pipe size		19	.421		6.10	14.50		20.60	28.50	
	0380	1-1/4" pipe size		15	.533		10	18.35		28.35	38.50	
	0400	1-1/2" pipe size		13	.615		15	21		36	48.50	
	0420	2" pipe size		11	.727		21	25		46	61	
	0580	Female IPT to brass pipe thread, straight										
	0600	1/2" pipe size	1 Plum	24	.333	Ea.	5.95	11.50		17.45	24	
	0640	3/4" pipe size		20	.400		7.60	13.80		21.40	29.50	
	0660	1" pipe size		19	.421		11.40	14.50		25.90	34.50	
	0680	1-1/4" pipe size		15	.533		17.05	18.35		35.40	46.50	
	0700	1-1/2" pipe size		13	.615		23.50	21		44.50	58	
	0720	2" pipe size		11	.727		30.50	25		55.50	72	
	0780	Female IPT to female IPT, straight										
	0800	1/2" pipe size	1 Plum	24	.333	Ea.	4.75	11.50		16.25	22.50	
	0840	3/4" pipe size		20	.400		5.30	13.80		19.10	27	
	0860	1" pipe size		19	.421		7	14.50		21.50	29.50	
	0880	1-1/4" pipe size		15	.533		9.75	18.35		28.10	38.50	
	0900	1-1/2" pipe size		13	.615		14.95	21		35.95	48.50	
	0920	2" pipe size		11	.727		21.50	25		46.50	61.50	
	2000	175 psi maximum pressure										
	2180	Female IPT to sweat										
	2200	1-1/2" pipe size	1 Plum	11	.727	Ea.	38.50	25		63.50	80	
	2240	2" pipe size	"	9	.889		44	30.50		74.50	94.50	
	2260	2-1/2" pipe size	Q-1	15	1.067		48	33		81	103	
	2280	3" pipe size		14	1.143		65.50	35.50		101	126	
	2300	4" pipe size		11	1.455		174	45		219	260	
	2320	5" pipe size	Q-2	14	1.714		284	55		339	395	
	2340	6" pipe size		12	2		330	64.50		394.50	455	

15100 | Building Services Piping

15120 | Piping Specialties

			CREW	DAILY OUTPUT	LABOR-HOURS	UNIT	2001 BARE COSTS MAT.	LABOR	EQUIP.	TOTAL	TOTAL INCL O&P	
250	2360	8" pipe size	Q-2	10	2.400	Ea.	910	77		987	1,125	250
	2480	Female IPT to brass pipe										
	2500	1-1/2" pipe size	1 Plum	11	.727	Ea.	58	25		83	102	
	2540	2" pipe size	"	9	.889		69	30.50		99.50	122	
	2560	2-1/2" pipe size	Q-1	15	1.067		100	33		133	160	
	2580	3" pipe size		14	1.143		116	35.50		151.50	182	
	2600	4" pipe size	▼	11	1.455		199	45		244	287	
	2620	5" pipe size	Q-2	14	1.714		224	55		279	330	
	2640	6" pipe size		12	2		267	64.50		331.50	390	
	2680	8" pipe size	▼	10	2.400	▼	435	77		512	595	
	2880	Female IPT to female IPT										
	2900	1-1/2" pipe size	1 Plum	11	.727	Ea.	36	25		61	77.50	
	2940	2" pipe size	"	9	.889		38.50	30.50		69	88.50	
	2960	2-1/2" pipe size	Q-1	15	1.067		48	33		81	103	
	2980	3" pipe size		14	1.143		60	35.50		95.50	120	
	3000	4" pipe size	▼	11	1.455		91	45		136	168	
	3020	5" pipe size	Q-2	14	1.714		105	55		160	199	
	3040	6" pipe size		12	2		141	64.50		205.50	252	
	3060	8" pipe size	▼	10	2.400		192	77		269	325	
	3380	Copper to copper										
	3400	1-1/2" pipe size	1 Plum	11	.727	Ea.	47.50	25		72.50	90.50	
	3440	2" pipe size	"	9	.889		54	30.50		84.50	106	
	3460	2-1/2" pipe size	Q-1	15	1.067		58	33		91	114	
	3480	3" pipe size		14	1.143		72.50	35.50		108	133	
	3500	4" pipe size	▼	11	1.455		230	45		275	320	
	3520	5" pipe size	Q-2	14	1.714		370	55		425	490	
	3540	6" pipe size		12	2		450	64.50		514.50	590	
	3560	8" pipe size	▼	10	2.400		1,200	77		1,277	1,450	
280	0010	**EXPANSION COUPLINGS** Hydronic										280
	0100	Copper to copper, sweat										
	0120	3/4" diameter	1 Stpi	20	.400	Ea.	25.50	13.95		39.45	49	
	0140	1" diameter		19	.421		32.50	14.65		47.15	57.50	
	0160	1-1/4" diameter	▼	15	.533	▼	44.50	18.60		63.10	76.50	
	1000	Baseboard riser fitting, 5" stub by coupling 12" long										
	1020	1/2" diameter	1 Stpi	24	.333	Ea.	10.25	11.60		21.85	29	
	1040	3/4" diameter		20	.400		14.50	13.95		28.45	37	
	1060	1" diameter		19	.421		20	14.65		34.65	44.50	
	1080	1-1/4" diameter	▼	15	.533		25	18.60		43.60	55.50	
	1180	9" Stub by tubing 8" long										
	1200	1/2" diameter	1 Stpi	24	.333	Ea.	9.25	11.60		20.85	28	
	1220	3/4" diameter		20	.400		12.20	13.95		26.15	34.50	
	1240	1" diameter		19	.421		16.95	14.65		31.60	40.50	
	1260	1-1/4" diameter	▼	15	.533		23	18.60		41.60	53	
300	0010	**EXPANSION JOINTS**										300
	0100	Bellows type, neoprene cover, flanged spool										
	0120	6" face to face, 1" diameter	1 Stpi	13	.615	Ea.	158	21.50		179.50	207	
	0140	1-1/4" diameter		11	.727		161	25.50		186.50	216	
	0160	1-1/2" diameter	▼	10.60	.755		161	26.50		187.50	217	
	0180	2" diameter	Q-5	13.30	1.203		163	38		201	236	
	0190	2-1/2" diameter		12.40	1.290		170	40.50		210.50	248	
	0200	3" diameter		11.40	1.404		193	44		237	279	
	0210	4" diameter		8.40	1.905		206	60		266	320	
	0220	5" diameter		7.60	2.105		254	66		320	380	
	0230	6" diameter		6.80	2.353		261	74		335	400	
	0240	8" diameter	▼	5.40	2.963		299	93		392	470	

15100 | Building Services Piping

15120 | Piping Specialties

		CREW	DAILY OUTPUT	LABOR-HOURS	UNIT	2001 BARE COSTS				TOTAL INCL O&P
						MAT.	LABOR	EQUIP.	TOTAL	
300	0250	10" diameter	Q-5	5	3.200	Ea.	400	100		500	590
	0260	12" diameter		4.60	3.478		455	109		564	665
	0480	10" face to face, 2" diameter		13	1.231		240	38.50		278.50	325
	0500	2-1/2" diameter		12	1.333		253	42		295	340
	0520	3" diameter		11	1.455		257	45.50		302.50	350
	0540	4" diameter		8	2		287	63		350	410
	0560	5" diameter		7	2.286		390	71.50		461.50	540
	0580	6" diameter		6	2.667		355	83.50		438.50	520
	0600	8" diameter		5	3.200		425	100		525	615
	0620	10" diameter		4.60	3.478		455	109		564	665
	0640	12" diameter		4	4		520	126		646	760
	0660	14" diameter		3.80	4.211		650	132		782	915
	0680	16" diameter		2.90	5.517		745	173		918	1,075
	0700	18" diameter		2.50	6.400		840	201		1,041	1,225
	0720	20" diameter		2.10	7.619		880	239		1,119	1,325
	0740	24" diameter		1.80	8.889		1,000	279		1,279	1,550
	0760	26" diameter		1.40	11.429		1,125	360		1,485	1,775
	0780	30" diameter		1.20	13.333		1,225	420		1,645	1,975
	0800	36" diameter		1	16		1,525	500		2,025	2,425
	1000	Bellows with internal sleeves and external covers									
	1010	Stainless steel, 150 lb.									
	1020	With male threads									
	1030	1/2" diameter	1 Stpi	22	.364	Ea.	84	12.65		96.65	112
	1040	3/4" diameter		20	.400		84	13.95		97.95	114
	1050	1" diameter		19	.421		91	14.65		105.65	122
	1060	1-1/4" diameter		16	.500		148	17.45		165.45	190
	1070	1-1/2" diameter		13	.615		297	21.50		318.50	360
	1080	2" diameter		11	.727		320	25.50		345.50	390
	1110	With flanged ends									
	1120	1-1/4" diameter	1 Stpi	12	.667	Ea.	297	23		320	360
	1130	1-1/2" diameter		11	.727		297	25.50		322.50	365
	1140	2" diameter		9	.889		475	31		506	570
	1150	3" diameter	Q-5	8	2		480	63		543	625
	1160	4" diameter	"	5	3.200		510	100		610	715
	1170	5" diameter	Q-6	6	4		550	130		680	800
	1180	6" diameter		5	4.800		600	156		756	895
	1190	8" diameter		4.50	5.333		720	173		893	1,050
	1200	10" diameter		4	6		880	195		1,075	1,250
	1210	12" diameter		3.60	6.667		1,125	217		1,342	1,550
320	0010	**EXPANSION TANKS**									
	1400	Plastic, corrosion resistant, see Plumbing Cost Data									
	1410	Div. 15220									
	1505	Fiberglass and steel single / double wall storage, see Div 13201									
	1510	Tank leak detection systems, see Div 13851-350									
	2000	Steel, liquid expansion, ASME, painted, 15 gallon capacity	Q-5	17	.941	Ea.	340	29.50		369.50	415
	2020	24 gallon capacity		14	1.143		340	36		376	430
	2040	30 gallon capacity		12	1.333		375	42		417	480
	2060	40 gallon capacity		10	1.600		430	50		480	550
	2080	60 gallon capacity		8	2		515	63		578	660
	2100	80 gallon capacity		7	2.286		530	71.50		601.50	690
	2120	100 gallon capacity		6	2.667		685	83.50		768.50	875
	2130	120 gallon capacity		5	3.200		760	100		860	985
	2140	135 gallon capacity		4.50	3.556		790	112		902	1,050
	2150	175 gallon capacity		4	4		1,100	126		1,226	1,400
	2160	220 gallon capacity		3.60	4.444		1,250	139		1,389	1,575

15100 | Building Services Piping

15120 | Piping Specialties

			CREW	DAILY OUTPUT	LABOR-HOURS	UNIT	MAT.	LABOR	EQUIP.	TOTAL	TOTAL INCL O&P	
320	2170	240 gallon capacity	Q-5	3.30	4.848	Ea.	1,400	152		1,552	1,775	**320**
	2180	305 gallon capacity		3	5.333		1,900	167		2,067	2,325	
	2190	400 gallon capacity	↓	2.80	5.714	↓	2,125	179		2,304	2,600	
	2360	Galvanized										
	2370	15 gallon capacity	Q-5	17	.941	Ea.	495	29.50		524.50	590	
	2380	24 gallon capacity		14	1.143		625	36		661	740	
	2390	30 gallon capacity		12	1.333		600	42		642	725	
	2400	40 gallon capacity		10	1.600		705	50		755	850	
	2410	60 gallon capacity		8	2		810	63		873	985	
	2420	80 gallon capacity		7	2.286		1,100	71.50		1,171.50	1,300	
	2430	100 gallon capacity		6	2.667		1,325	83.50		1,408.50	1,600	
	2440	120 gallon capacity		5	3.200		1,550	100		1,650	1,850	
	2450	135 gallon capacity		4.50	3.556		1,650	112		1,762	2,000	
	2460	175 gallon capacity		4	4		2,025	126		2,151	2,425	
	2470	220 gallon capacity		3.60	4.444		2,325	139		2,464	2,775	
	2480	240 gallon capacity		3.30	4.848		2,425	152		2,577	2,900	
	2490	305 gallon capacity		3	5.333		3,300	167		3,467	3,875	
	2500	400 gallon capacity		2.80	5.714		4,000	179		4,179	4,675	
	3000	Steel ASME expansion, rubber diaphragm, 19 gal. cap. accept.		12	1.333		1,275	42		1,317	1,475	
	3020	31 gallon capacity		8	2		1,425	63		1,488	1,650	
	3040	61 gallon capacity		6	2.667		2,000	83.50		2,083.50	2,300	
	3060	79 gallon capacity		5	3.200		2,125	100		2,225	2,475	
	3080	119 gallon capacity		4	4		2,250	126		2,376	2,675	
	3100	158 gallon capacity		3.80	4.211		3,150	132		3,282	3,675	
	3120	211 gallon capacity		3.30	4.848		3,625	152		3,777	4,225	
	3140	317 gallon capacity		2.80	5.714		4,750	179		4,929	5,500	
	3160	422 gallon capacity		2.60	6.154		7,025	193		7,218	8,025	
	3180	528 gallon capacity	↓	2.40	6.667	↓	7,700	209		7,909	8,800	
350	0010	**FLEXIBLE CONNECTORS**, Corrugated, 7/8" O.D., 1/2" I.D.										**350**
	0050	Gas, seamless brass, steel fittings										
	0200	12" long	1 Plum	36	.222	Ea.	9.95	7.65		17.60	22.50	
	0220	18" long		36	.222		12.35	7.65		20	25	
	0240	24" long		34	.235		14.60	8.10		22.70	28.50	
	0260	30" long		34	.235		15.75	8.10		23.85	29.50	
	0280	36" long		32	.250		17.40	8.60		26	32	
	0320	48" long		30	.267		22	9.20		31.20	38.50	
	0340	60" long		30	.267		26.50	9.20		35.70	43	
	0360	72" long	↓	30	.267		30.50	9.20		39.70	47.50	
	2000	Water, copper tubing, dielectric separators										
	2100	12" long	1 Plum	36	.222	Ea.	8.50	7.65		16.15	21	
	2220	15" long		36	.222		9.45	7.65		17.10	22	
	2240	18" long		36	.222		10.30	7.65		17.95	23	
	2260	24" long	↓	34	.235	↓	12.65	8.10		20.75	26	
370	0010	**FLEXIBLE METAL HOSE** Connectors, standard lengths										**370**
	0100	Bronze braided, bronze ends										
	0120	3/8" diameter x 12"	1 Stpi	26	.308	Ea.	14.40	10.70		25.10	32	
	0140	1/2" diameter x 12"		24	.333		13.70	11.60		25.30	32.50	
	0160	3/4" diameter x 12"		20	.400		20	13.95		33.95	43	
	0180	1" diameter x 18"		19	.421		26	14.65		40.65	50.50	
	0200	1-1/2" diameter x 18"		13	.615		41	21.50		62.50	77.50	
	0220	2" diameter x 18"	↓	11	.727	↓	49.50	25.50		75	93	
	1000	Carbon steel ends										
	1020	1/4" diameter x 12"	1 Stpi	28	.286	Ea.	11	9.95		20.95	27	
	1040	3/8" diameter x 12"		26	.308		11.30	10.70		22	28.50	
	1060	1/2" diameter x 12"	↓	24	.333		11	11.60		22.60	29.50	

15100 | Building Services Piping

15120 | Piping Specialties

		CREW	DAILY OUTPUT	LABOR-HOURS	UNIT	2001 BARE COSTS MAT.	LABOR	EQUIP.	TOTAL	TOTAL INCL O&P
1080	1/2" diameter x 24"	1 Stpi	24	.333	Ea.	14.10	11.60		25.70	33
1100	1/2" diameter x 36"		24	.333		17.20	11.60		28.80	36.50
1120	3/4" diameter x 12"		20	.400		14.60	13.95		28.55	37
1140	3/4" diameter x 24"		20	.400		17.90	13.95		31.85	40.50
1160	3/4" diameter x 36"		20	.400		21	13.95		34.95	44.50
1180	1" diameter x 18"		19	.421		19.20	14.65		33.85	43
1200	1" diameter x 30"		19	.421		22.50	14.65		37.15	47
1220	1" diameter x 36"		19	.421		24.50	14.65		39.15	49
1240	1-1/4" diameter x 18"		15	.533		19.60	18.60		38.20	49.50
1260	1-1/4" diameter x 36"		15	.533		26	18.60		44.60	56.50
1280	1-1/2" diameter x 18"		13	.615		21	21.50		42.50	55.50
1300	1-1/2" diameter x 36"		13	.615		28	21.50		49.50	63.50
1320	2" diameter x 24"		11	.727		27.50	25.50		53	69
1340	2" diameter x 36"		11	.727		34	25.50		59.50	76
1360	2-1/2" diameter x 24"		9	.889		39	31		70	90
1380	2-1/2" diameter x 36"		9	.889		46	31		77	97.50
1400	3" diameter x 24"		7	1.143		46.50	40		86.50	111
1420	3" diameter x 36"		7	1.143		54	40		94	120
2000	Carbon steel braid, carbon steel solid ends									
2100	1/2" diameter x 12"	1 Stpi	24	.333	Ea.	11	11.60		22.60	29.50
2120	3/4" diameter x 12"		20	.400		14.60	13.95		28.55	37
2140	1" diameter x 12"		19	.421		17.40	14.65		32.05	41
2160	1-1/4" diameter x 12"		15	.533		17.40	18.60		36	47
2180	1-1/2" diameter x 12"		13	.615		18.80	21.50		40.30	53
3000	Stainless steel braid, welded on carbon steel ends									
3100	1/2" diameter x 12"	1 Stpi	24	.333	Ea.	16.10	11.60		27.70	35.50
3120	3/4" diameter x 12"		20	.400		23	13.95		36.95	46
3140	3/4" diameter x 24"		20	.400		27.50	13.95		41.45	51.50
3160	3/4" diameter x 36"		20	.400		32.50	13.95		46.45	57
3180	1" diameter x 12"		19	.421		26	14.65		40.65	50.50
3200	1" diameter x 24"		19	.421		31.50	14.65		46.15	56.50
3220	1" diameter x 36"		19	.421		37	14.65		51.65	63
3240	1-1/4" diameter x 12"		15	.533		28.50	18.60		47.10	59
3260	1-1/4" diameter x 24"		15	.533		36	18.60		54.60	67.50
3280	1-1/4" diameter x 36"		15	.533		43.50	18.60		62.10	76
3300	1-1/2" diameter x 12"		13	.615		31	21.50		52.50	67
3320	1-1/2" diameter x 24"		13	.615		40	21.50		61.50	76.50
3340	1-1/2" diameter x 36"		13	.615		48.50	21.50		70	86
3400	Metal sst braid, over corrugated stainless steel, flanged ends									
3410	150 PSI									
3420	1/2" diameter x 12"	1 Stpi	24	.333	Ea.	20	11.60		31.60	39.50
3430	1" diameter x 12"		20	.400		32	13.95		45.95	56
3440	1-1/2" diameter x 12"		15	.533		37	18.60		55.60	68.50
3450	2-1/2" diameter x 9"		12	.667		54	23		77	94.50
3460	3" diameter x 9"		9	.889		65	31		96	119
3470	4" diameter x 9"		7	1.143		80	40		120	148
3480	4" diameter x 30"		5	1.600		170	56		226	271
3490	4" diameter x 36"		4.80	1.667		190	58		248	297
3500	6" diameter x 11"		5	1.600		132	56		188	229
3510	6" diameter x 36"		3.80	2.105		251	73.50		324.50	390
3520	8" diameter x 12"		4	2		256	69.50		325.50	385
3530	10" diameter x 13"		3	2.667		325	93		418	500
3540	12" diameter x 14"	Q-5	4	4		655	126		781	910
6000	Molded rubber with helical wire reinforcement									
6010	150 PSI									
6020	1-1/2" diameter x 12"	1 Stpi	15	.533	Ea.	39	18.60		57.60	70.50

15100 | Building Services Piping

15120 | Piping Specialties

			CREW	DAILY OUTPUT	LABOR-HOURS	UNIT	2001 BARE COSTS MAT.	LABOR	EQUIP.	TOTAL	TOTAL INCL O&P	
370	6030	2" diameter x 12"	1 Stpi	12	.667	Ea.	145	23		168	195	370
	6040	3" diameter x 12"		8	1		187	35		222	259	
	6050	4" diameter x 12"		6	1.333		218	46.50		264.50	310	
	6060	6" diameter x 18"		4	2		275	69.50		344.50	410	
	6070	8" diameter x 24"		3	2.667		355	93		448	530	
	6080	10" diameter x 24"		2	4		400	139		539	650	
	6090	12" diameter x 24"	Q-5	3	5.333		455	167		622	755	
	7000	Molded teflon with stainless steel flanges										
	7010	150 PSI										
	7020	2-1/2" diameter x 3-3/16"	Q-1	7.80	2.051	Ea.	325	63.50		388.50	455	
	7030	3" diameter x 3-5/8"		6.50	2.462		415	76.50		491.50	575	
	7040	4" diameter x 3-5/8"		5	3.200		450	99		549	645	
	7050	6" diameter x 4"		4.30	3.721		645	115		760	885	
	7060	8" diameter x 6"		3.80	4.211		860	131		991	1,150	
400	0010	**FLOAT VALVES**										400
	0020	With ball and bracket										
	0030	Single seat, threaded										
	0040	Brass body										
	0050	1/2"	1 Stpi	11	.727	Ea.	37	25.50		62.50	79	
	0060	3/4"		9	.889		44.50	31		75.50	96	
	0070	1"		7	1.143		56.50	40		96.50	123	
	0080	1-1/2"		4.50	1.778		83	62		145	185	
	0090	2"		3.60	2.222		84	77.50		161.50	210	
	0300	For condensate receivers, CI, in-line mount										
	0320	1" inlet	1 Stpi	7	1.143	Ea.	116	40		156	187	
	0360	For condensate receiver, CI, external float, flanged tank mount										
	0370	3/4" inlet	1 Stpi	5	1.600	Ea.	98.50	56		154.50	192	
420	0010	**FLOW CHECK CONTROL**										420
	0100	Bronze body, soldered										
	0110	3/4" size	1 Stpi	20	.400	Ea.	34.50	13.95		48.45	59	
	0120	1" size	"	19	.421	"	41.50	14.65		56.15	67.50	
	0200	Cast iron body, threaded										
	0210	3/4" size	1 Stpi	20	.400	Ea.	27.50	13.95		41.45	51.50	
	0220	1" size		19	.421		31	14.65		45.65	56	
	0230	1-1/4" size		15	.533		37.50	18.60		56.10	69.50	
	0240	1-1/2" size		13	.615		57.50	21.50		79	96	
	0250	2" size		11	.727		82.50	25.50		108	130	
	0300	Flanged inlet, threaded outlet										
	0310	2-1/2" size	1 Stpi	8	1	Ea.	183	35		218	255	
	0320	3" size	Q-5	9	1.778		218	56		274	325	
	0330	4" size	"	7	2.286		425	71.50		496.50	575	
520	0010	**HYDRONIC HEATING CONTROL VALVES**										520
	0050	Hot water, nonelectric, thermostatic										
	0100	Radiator supply, 1/2" diameter	1 Stpi	24	.333	Ea.	36	11.60		47.60	57	
	0120	3/4" diameter		20	.400		36.50	13.95		50.45	61.50	
	0140	1" diameter		19	.421		45.50	14.65		60.15	72	
	0160	1-1/4" diameter		15	.533		64	18.60		82.60	98.50	
	0500	For low pressure steam, add					25%					
	1000	Manual, radiator supply										
	1010	1/2" pipe size, angle union	1 Stpi	24	.333	Ea.	28.50	11.60		40.10	48.50	
	1020	3/4" pipe size, angle union		20	.400		26.50	13.95		40.45	50.50	
	1030	1" pipe size, angle union		19	.421		34.50	14.65		49.15	60	
	1100	Radiator, balancing, straight, sweat connections										
	1110	1/2" pipe size	1 Stpi	24	.333	Ea.	7.70	11.60		19.30	26	
	1120	3/4" pipe size		20	.400		11.30	13.95		25.25	33.50	

15100 | Building Services Piping

15120 | Piping Specialties

			CREW	DAILY OUTPUT	LABOR-HOURS	UNIT	MAT.	LABOR	EQUIP.	TOTAL	TOTAL INCL O&P	
520	1130	1" pipe size	1 Stpi	19	.421	Ea.	18.40	14.65		33.05	42.50	520
	1200	Steam, radiator, supply										
	1210	1/2" pipe size, angle union	1 Stpi	24	.333	Ea.	25.50	11.60		37.10	45.50	
	1220	3/4" pipe size, angle union		20	.400		28	13.95		41.95	52	
	1230	1" pipe size, angle union		19	.421		35.50	14.65		50.15	61	
	1240	1-1/4" pipe size, angle union		15	.533		31.50	18.60		50.10	62.50	
	8000	System balancing and shut-off										
	8020	Butterfly, quarter turn, calibrated, threaded or solder										
	8040	Bronze, -30° F to +350° F, pressure to 175 psi										
	8060	1/2" size	1 Stpi	22	.364	Ea.	9.80	12.65		22.45	30	
	8070	3/4" size		20	.400		10.55	13.95		24.50	32.50	
	8080	1" size		19	.421		11.70	14.65		26.35	35	
	8090	1-1/4" size		15	.533		19.25	18.60		37.85	49	
	8100	1-1/2" size		13	.615		24	21.50		45.50	59	
	8110	2" size		11	.727		28.50	25.50		54	70	
550	0010	**LIQUID DRAINERS**										550
	0100	Guided lever type										
	0110	Cast iron body, threaded										
	0120	1/2" pipe size	1 Stpi	22	.364	Ea.	78.50	12.65		91.15	105	
	0130	3/4" pipe size		20	.400		210	13.95		223.95	252	
	0140	1" pipe size		19	.421		335	14.65		349.65	390	
	0150	1-1/2" pipe size		11	.727		725	25.50		750.50	840	
	0160	2" pipe size		8	1		725	35		760	855	
	0200	Forged steel body, threaded										
	0210	1/2" pipe size	1 Stpi	18	.444	Ea.	725	15.50		740.50	825	
	0220	3/4" pipe size		16	.500		725	17.45		742.45	825	
	0230	1" pipe size		14	.571		1,100	19.90		1,119.90	1,225	
	0240	1-1/2" pipe size	Q-5	10	1.600		3,175	50		3,225	3,575	
	0250	2" pipe size	"	8	2		3,175	63		3,238	3,600	
	0300	Socket weld										
	0310	1/2" pipe size	Q-17	16	1	Ea.	835	31.50	6.30	872.80	975	
	0320	3/4" pipe size		14	1.143		835	36	7.20	878.20	980	
	0330	1" pipe size		12	1.333		1,225	42	8.40	1,275.40	1,425	
	0340	1-1/2" pipe size		9	1.778		3,300	56	11.20	3,367.20	3,725	
	0350	2" pipe size		7	2.286		3,300	71.50	14.40	3,385.90	3,750	
	0400	Flanged										
	0410	1/2" pipe size	1 Stpi	12	.667	Ea.	1,550	23		1,573	1,725	
	0420	3/4" pipe size		10	.800		1,550	28		1,578	1,750	
	0430	1" pipe size		7	1.143		2,025	40		2,065	2,275	
	0440	1-1/2" pipe size	Q-5	9	1.778		4,425	56		4,481	4,950	
	0450	2" pipe size	"	7	2.286		4,425	71.50		4,496.50	4,975	
	1000	Fixed lever and snap action type										
	1100	Cast iron body, threaded										
	1110	1/2" pipe size	1 Stpi	22	.364	Ea.	123	12.65		135.65	155	
	1120	3/4" pipe size		20	.400		325	13.95		338.95	375	
	1130	1" pipe size		19	.421		325	14.65		339.65	375	
	1200	Forged steel body, threaded										
	1210	1/2" pipe size	1 Stpi	18	.444	Ea.	785	15.50		800.50	890	
	1220	3/4" pipe size		16	.500		785	17.45		802.45	890	
	1230	1" pipe size		14	.571		1,950	19.90		1,969.90	2,150	
	1240	1-1/4" pipe size		12	.667		1,950	23		1,973	2,150	
	2000	High pressure high leverage type										
	2100	Forged steel body, threaded										
	2110	1/2" pipe size	1 Stpi	18	.444	Ea.	1,550	15.50		1,565.50	1,725	
	2120	3/4" pipe size		16	.500		1,550	17.45		1,567.45	1,725	

15100 | Building Services Piping

15120 | Piping Specialties

		Crew	Daily Output	Labor-Hours	Unit	Mat.	Labor	Equip.	Total	Total Incl O&P	
550	2130	1" pipe size	1 Stpi	14	.571	Ea.	2,575	19.90		2,594.90	2,875	550
	2140	1-1/4" pipe size	↓	12	.667		2,575	23		2,598	2,875	
	2150	1-1/2" pipe size	Q-5	10	1.600		3,925	50		3,975	4,375	
	2160	2" pipe size	"	8	2	↓	3,925	63		3,988	4,400	
	2200	Socket weld										
	2210	1/2" pipe size	Q-17	16	1	Ea.	1,675	31.50	6.30	1,712.80	1,900	
	2220	3/4" pipe size		14	1.143		1,675	36	7.20	1,718.20	1,900	
	2230	1" pipe size		12	1.333		2,700	42	8.40	2,750.40	3,050	
	2240	1-1/4" pipe size		11	1.455		2,700	45.50	9.15	2,754.65	3,050	
	2250	1-1/2" pipe size		9	1.778		4,050	56	11.20	4,117.20	4,550	
	2260	2" pipe size	↓	7	2.286	↓	4,050	71.50	14.40	4,135.90	4,575	
	2300	Flanged										
	2310	1/2" pipe size	1 Stpi	12	.667	Ea.	2,475	23		2,498	2,750	
	2320	3/4" pipe size		10	.800		2,475	28		2,503	2,775	
	2330	1" pipe size		7	1.143		3,825	40		3,865	4,250	
	2340	1-1/4" pipe size	↓	6	1.333		3,825	46.50		3,871.50	4,275	
	2350	1-1/2" pipe size	Q-5	9	1.778		5,175	56		5,231	5,775	
	2360	2" pipe size	"	7	2.286	↓	5,175	71.50		5,246.50	5,800	
	3000	Guided lever dual gravity type										
	3100	Cast iron body, threaded										
	3110	1/2" pipe size	1 Stpi	22	.364	Ea.	300	12.65		312.65	350	
	3120	3/4" pipe size		20	.400		300	13.95		313.95	350	
	3130	1" pipe size		19	.421		440	14.65		454.65	500	
	3140	1-1/2" pipe size		11	.727		850	25.50		875.50	975	
	3150	2" pipe size	↓	8	1	↓	850	35		885	990	
	3200	Forged steel body, threaded										
	3210	1/2" pipe size	1 Stpi	18	.444	Ea.	845	15.50		860.50	955	
	3220	3/4" pipe size		16	.500		845	17.45		862.45	955	
	3230	1" pipe size	↓	14	.571		1,250	19.90		1,269.90	1,400	
	3240	1-1/2" pipe size	Q-5	10	1.600		3,325	50		3,375	3,750	
	3250	2" pipe size	"	8	2	↓	3,325	63		3,388	3,775	
	3300	Socket weld										
	3310	1/2" pipe size	Q-17	16	1	Ea.	955	31.50	6.30	992.80	1,100	
	3320	3/4" pipe size		14	1.143		955	36	7.20	998.20	1,100	
	3330	1" pipe size		12	1.333		1,375	42	8.40	1,425.40	1,600	
	3340	1-1/2" pipe size		9	1.778		3,450	56	11.20	3,517.20	3,900	
	3350	2" pipe size	↓	7	2.286	↓	3,450	71.50	14.40	3,535.90	3,925	
	3400	Flanged										
	3410	1/2" pipe size	1 Stpi	12	.667	Ea.	1,675	23		1,698	1,875	
	3420	3/4" pipe size		10	.800		1,675	28		1,703	1,900	
	3430	1" pipe size	↓	7	1.143		2,175	40		2,215	2,450	
	3440	1-1/2" pipe size	Q-5	9	1.778		4,600	56		4,656	5,125	
	3450	2" pipe size	"	7	2.286	↓	4,600	71.50		4,671.50	5,150	
580	0010	**MIXING VALVE** Automatic water tempering										580
	0050	3/4" size	1 Stpi	18	.444	Ea.	335	15.50		350.50	390	
	0100	1" size		16	.500		475	17.45		492.45	550	
	0120	1-1/4" size		13	.615		690	21.50		711.50	790	
	0140	1-1/2" size		10	.800		820	28		848	945	
	0160	2" size		8	1		1,025	35		1,060	1,200	
	0180	3" size	↓	4	2	↓	2,375	69.50		2,444.50	2,725	
610	0010	**MONOFLOW TEE FITTING**										610
	1100	For one pipe hydronic, supply and return										
	1110	Copper, soldered										
	1120	3/4" x 1/2" size	1 Stpi	13	.615	Ea.	12	21.50		33.50	45.50	

15100 | Building Services Piping

15120 | Piping Specialties

		Crew	Daily Output	Labor-Hours	Unit	Mat.	2001 Bare Costs Labor	Equip.	Total	Total Incl O&P	
610	1130	1" x 1/2" size	1 Stpi	12	.667	Ea.	12	23		35	48	610
	1140	1" x 3/4" size		11	.727		12	25.50		37.50	51.50	
	1150	1-1/4" x 1/2" size		11	.727		12.75	25.50		38.25	52.50	
	1160	1-1/4" x 3/4" size		10	.800		12.75	28		40.75	56	
	1170	1-1/2" x 3/4" size		10	.800		18	28		46	62	
	1180	1-1/2" x 1" size		9	.889		18	31		49	67	
	1190	2" x 3/4" size		9	.889		27.50	31		58.50	77	
	1200	2" x 1" size		8	1		27.50	35		62.50	82.50	
640	0010	**PRESSURE REDUCING VALVE** Steam, pilot operated										640
	0100	Threaded, iron body										
	0120	1/2" size	1 Stpi	18	.444	Ea.	555	15.50		570.50	635	
	0140	3/4" size		14	.571		580	19.90		599.90	665	
	0160	1" size		13	.615		615	21.50		636.50	710	
	0180	1-1/4" size		10	.800		720	28		748	835	
	0200	1-1/2" size		8	1		780	35		815	915	
	0220	2" size		5	1.600		1,025	56		1,081	1,200	
	1000	Flanged, iron body, 125 lb. flanges										
	1020	2" size	1 Stpi	8	1	Ea.	1,075	35		1,110	1,225	
	1040	2-1/2" size	"	4	2		1,350	69.50		1,419.50	1,600	
	1060	3" size	Q-5	4.50	3.556		1,375	112		1,487	1,700	
	1080	4" size	"	3	5.333		1,750	167		1,917	2,175	
	1500	For 250 lb. flanges, add					5%					
670	0010	**PRESSURE REGULATOR**										670
	0100	Gas appliance regulators										
	0106	Main burner and pilot applications										
	0108	Rubber seat poppet type										
	0109	1/8" pipe size	1 Stpi	24	.333	Ea.	10.60	11.60		22.20	29.50	
	0110	1/4" pipe size		24	.333		12.15	11.60		23.75	31	
	0112	3/8" pipe size		24	.333		13.10	11.60		24.70	32	
	0113	1/2" pipe size		24	.333		14.55	11.60		26.15	33.50	
	0114	3/4" pipe size		20	.400		17.45	13.95		31.40	40	
	0122	Lever action type										
	0123	3/8" pipe size	1 Stpi	24	.333	Ea.	21.50	11.60		33.10	41	
	0124	1/2" pipe size		24	.333		21.50	11.60		33.10	41	
	0125	3/4" pipe size		20	.400		43	13.95		56.95	68.50	
	0126	1" pipe size		19	.421		43	14.65		57.65	69.50	
	0132	Double diaphragm type										
	0133	3/8" pipe size	1 Stpi	24	.333	Ea.	24.50	11.60		36.10	44.50	
	0134	1/2" pipe size		24	.333		37.50	11.60		49.10	58.50	
	0135	3/4" pipe size		20	.400		64	13.95		77.95	91.50	
	0136	1" pipe size		19	.421		78.50	14.65		93.15	108	
	0137	1-1/4" pipe size		15	.533		255	18.60		273.60	310	
	0138	1-1/2" pipe size		13	.615		475	21.50		496.50	555	
	0139	2" pipe size		11	.727		475	25.50		500.50	560	
	0140	2-1/2" pipe size	Q-5	15	1.067		925	33.50		958.50	1,075	
	0141	3" pipe size		13	1.231		925	38.50		963.50	1,075	
	0142	4" pipe size (flanged)		8	2		1,700	63		1,763	1,975	
	0160	Main burner only										
	0162	Straight-thru-flow design										
	0163	1/2" pipe size	1 Stpi	24	.333	Ea.	27.50	11.60		39.10	48	
	0164	3/4" pipe size		20	.400		37.50	13.95		51.45	62.50	
	0165	1" pipe size		19	.421		56	14.65		70.65	83.50	
	0166	1-1/4" pipe size		15	.533		56	18.60		74.60	89.50	
	0200	Oil, light, hot water, ordinary steam, threaded										
	0220	Bronze body, 1/4" size	1 Stpi	24	.333	Ea.	122	11.60		133.60	152	
	0230	3/8" size		24	.333		132	11.60		143.60	163	

15100 | Building Services Piping

15120 | Piping Specialties

			Crew	Daily Output	Labor-Hours	Unit	Mat.	Labor	Equip.	Total	Total Incl O&P	
670	0240	1/2" size	1 Stpi	24	.333	Ea.	163	11.60		174.60	198	670
	0250	3/4" size		20	.400		195	13.95		208.95	235	
	0260	1" size		19	.421		297	14.65		311.65	345	
	0270	1-1/4" size		15	.533		400	18.60		418.60	470	
	0280	1-1/2" size		13	.615		430	21.50		451.50	510	
	0290	2" size		11	.727		780	25.50		805.50	900	
	0320	Iron body, 1/4" size		24	.333		95	11.60		106.60	123	
	0330	3/8" size		24	.333		112	11.60		123.60	142	
	0340	1/2" size		24	.333		119	11.60		130.60	149	
	0350	3/4" size		20	.400		144	13.95		157.95	179	
	0360	1" size		19	.421		188	14.65		202.65	229	
	0370	1-1/4" size		15	.533		261	18.60		279.60	315	
	0380	1-1/2" size		13	.615		290	21.50		311.50	355	
	0390	2" size		11	.727		420	25.50		445.50	505	
	0500	Oil, heavy, viscous fluids, threaded										
	0520	Bronze body, 3/8" size	1 Stpi	24	.333	Ea.	209	11.60		220.60	248	
	0530	1/2" size		24	.333		269	11.60		280.60	315	
	0540	3/4" size		20	.400		299	13.95		312.95	350	
	0550	1" size		19	.421		375	14.65		389.65	430	
	0560	1-1/4" size		15	.533		530	18.60		548.60	610	
	0570	1-1/2" size		13	.615		605	21.50		626.50	700	
	0600	Iron body, 3/8" size		24	.333		169	11.60		180.60	204	
	0620	1/2" size		24	.333		202	11.60		213.60	240	
	0630	3/4" size		20	.400		229	13.95		242.95	273	
	0640	1" size		19	.421		271	14.65		285.65	320	
	0650	1-1/4" size		15	.533		390	18.60		408.60	460	
	0660	1-1/2" size		13	.615		420	21.50		441.50	495	
	0800	Process steam, wet or super heated, monel trim, threaded										
	0820	Bronze body, 1/4" size	1 Stpi	24	.333	Ea.	187	11.60		198.60	223	
	0830	3/8" size		24	.333		206	11.60		217.60	244	
	0840	1/2" size		24	.333		256	11.60		267.60	299	
	0850	3/4" size		20	.400		290	13.95		303.95	340	
	0860	1" size		19	.421		400	14.65		414.65	460	
	0870	1-1/4" size		15	.533		495	18.60		513.60	575	
	0880	1-1/2" size		13	.615		525	21.50		546.50	610	
	0890	2" size		11	.727		865	25.50		890.50	990	
	0920	Iron body, max 125 PSIG press out, 1/4" size		24	.333		95	11.60		106.60	123	
	0930	3/8" size		24	.333		112	11.60		123.60	142	
	0940	1/2" size		24	.333		119	11.60		130.60	149	
	0950	3/4" size		20	.400		144	13.95		157.95	179	
	0960	1" size		19	.421		188	14.65		202.65	229	
	0970	1-1/4" size		15	.533		261	18.60		279.60	315	
	0980	1-1/2" size		13	.615		290	21.50		311.50	355	
	0990	2" size		11	.727		420	25.50		445.50	505	
	1000	Flanged, Class 125										
	1006	2-1/2" size	Q-5	10	1.600	Ea.	1,350	50		1,400	1,575	
	1008	3" size		9	1.778		1,375	56		1,431	1,600	
	1010	4" size		8	2		1,750	63		1,813	2,025	
	1020	6" size		7	2.286		4,450	71.50		4,521.50	4,975	
	3000	Steam, high capacity, bronze body, stainless steel trim										
	3020	Threaded, 1/2" diameter	1 Stpi	24	.333	Ea.	705	11.60		716.60	795	
	3030	3/4" diameter		24	.333		705	11.60		716.60	795	
	3040	1" diameter		19	.421		790	14.65		804.65	890	
	3060	1-1/4" diameter		15	.533		870	18.60		888.60	990	
	3080	1-1/2" diameter		13	.615		995	21.50		1,016.50	1,125	
	3100	2" diameter		11	.727		1,225	25.50		1,250.50	1,400	

15100 | Building Services Piping

15120 | Piping Specialties

		CREW	DAILY OUTPUT	LABOR-HOURS	UNIT	2001 BARE COSTS MAT.	LABOR	EQUIP.	TOTAL	TOTAL INCL O&P	
670	3120	2-1/2" diameter	Q-5	12	1.333	Ea.	1,525	42		1,567	1,750	670
	3140	3" diameter	"	11	1.455	↓	1,750	45.50		1,795.50	2,000	
	3500	Flanged connection, iron body, 125 lb. W.S.P.										
	3520	3" diameter	Q-5	11	1.455	Ea.	1,900	45.50		1,945.50	2,175	
	3540	4" diameter	"	5	3.200	"	2,400	100		2,500	2,800	
	9000	For water pressure regulators, see Div. 15110-160										
700	0010	**PRESSURE & TEMPERATURE SAFETY PLUG**										700
	1000	3/4" external thread, 3/8" diam. element										
	1020	Carbon steel										
	1050	7-1/2" insertion	1 Stpi	32	.250	Ea.	32	8.70		40.70	48	
	1120	304 stainless steel										
	1150	7-1/2" insertion	1 Stpi	32	.250	Ea.	32	8.70		40.70	48	
	1220	316 Stainless steel										
	1250	7-1/2" insertion	1 Stpi	32	.250	Ea.	41.50	8.70		50.20	59	
730	0010	**SLEEVES AND ESCUTCHEONS**										730
	0100	Pipe sleeve										
	0110	Steel, w/water stop, 12" long, with link seal										
	0120	2" diam. for 1/2" carrier pipe	1 Plum	8.40	.952	Ea.	24	33		57	75.50	
	0130	2-1/2" diam. for 3/4" carrier pipe		8	1		28.50	34.50		63	83.50	
	0140	2-1/2" diam. for 1" carrier pipe		8	1		26.50	34.50		61	81	
	0150	3" diam. for 1-1/4" carrier pipe		7.20	1.111		37.50	38.50		76	99	
	0160	3-1/2" diam. for 1-1/2" carrier pipe		6.80	1.176		38.50	40.50		79	103	
	0170	4" diam. for 2" carrier pipe		6	1.333		42.50	46		88.50	117	
	0180	4" diam. for 2-1/2" carrier pipe		6	1.333		42	46		88	116	
	0190	5" diam. for 3" carrier pipe		5.40	1.481		53	51		104	136	
	0200	6" diam. for 4" carrier pipe	↓	4.80	1.667		61.50	57.50		119	154	
	0210	10" diam. for 6" carrier pipe	Q-1	8	2		115	62		177	220	
	0220	12" diam. for 8" carrier pipe		7.20	2.222		165	69		234	286	
	0230	14" diam. for 10" carrier pipe		6.40	2.500		174	77.50		251.50	310	
	0240	16" diam. for 12" carrier pipe		5.80	2.759		203	85.50		288.50	355	
	0250	18" diam. for 14" carrier pipe		5.20	3.077		229	95.50		324.50	395	
	0260	24" diam. for 18" carrier pipe		4	4		380	124		504	600	
	0270	24" diam. for 20" carrier pipe		4	4		310	124		434	525	
	0280	30" diam. for 24" carrier pipe	↓	3.20	5	↓	700	155		855	1,000	
	0500	Wall sleeve										
	0510	Ductile iron with rubber gasket seal										
	0520	3"	1 Plum	8.40	.952	Ea.	96	33		129	156	
	0530	4"		7.20	1.111		120	38.50		158.50	190	
	0540	6"		6	1.333		161	46		207	247	
	0550	8"		4	2		206	69		275	330	
	0560	10"		3	2.667		231	92		323	395	
	0570	12"	↓	2.40	3.333	↓	274	115		389	475	
	5000	Escutcheon										
	5100	Split ring, pipe										
	5110	Chrome plated										
	5120	1/2"	1 Plum	160	.050	Ea.	.88	1.72		2.60	3.57	
	5130	3/4"		160	.050		.92	1.72		2.64	3.61	
	5140	1"		135	.059		.98	2.04		3.02	4.16	
	5150	1-1/2"		115	.070		1.48	2.40		3.88	5.25	
	5160	2"		100	.080		1.52	2.76		4.28	5.85	
	5170	4"		80	.100		3.35	3.45		6.80	8.90	
	5180	6"	↓	68	.118	↓	5.60	4.05		9.65	12.25	
760	0010	**STEAM TRAP**										760
	0030	Cast iron body, threaded										
	0040	Inverted bucket										
	0050	1/2" pipe size	1 Stpi	12	.667	Ea.	100	23		123	145	

15100 | Building Services Piping

15120 | Piping Specialties

		CREW	DAILY OUTPUT	LABOR-HOURS	UNIT	2001 BARE COSTS				TOTAL INCL O&P
						MAT.	LABOR	EQUIP.	TOTAL	
0070	3/4" pipe size	1 Stpi	10	.800	Ea.	172	28		200	231
0100	1" pipe size		9	.889		265	31		296	340
0120	1-1/4" pipe size		8	1		400	35		435	495
0130	1-1/2" pipe size		7	1.143		495	40		535	600
0140	2" pipe size		6	1.333		665	46.50		711.50	800
0200	With thermic vent & check valve									
0202	1/2" pipe size	1 Stpi	12	.667	Ea.	163	23		186	214
0204	3/4" pipe size		10	.800		241	28		269	305
0206	1" pipe size		9	.889		355	31		386	435
0208	1-1/4" pipe size		8	1		525	35		560	635
0210	1-1/2" pipe size		7	1.143		715	40		755	850
0212	2" pipe size		6	1.333		1,100	46.50		1,146.50	1,275
1000	Float & thermostatic, 15 psi									
1010	3/4" pipe size	1 Stpi	16	.500	Ea.	96	17.45		113.45	133
1020	1" pipe size		15	.533		115	18.60		133.60	155
1030	1-1/4" pipe size		13	.615		139	21.50		160.50	186
1040	1-1/2" pipe size		9	.889		254	31		285	325
1060	2" pipe size		6	1.333		470	46.50		516.50	590
1100	For vacuum breaker, add					51.50			51.50	56.50
1110	Differential condensate controller									
1120	With union and manual metering valve									
1130	1/2" pipe size	1 Stpi	24	.333	Ea.	415	11.60		426.60	475
1140	3/4" pipe size		20	.400		480	13.95		493.95	550
1150	1" pipe size		19	.421		650	14.65		664.65	735
1160	1-1/4" pipe size		15	.533		875	18.60		893.60	990
1170	1-1/2" pipe size		13	.615		1,050	21.50		1,071.50	1,175
1180	2" pipe size		11	.727		1,375	25.50		1,400.50	1,575
1290	Brass body, threaded									
1300	Thermostatic, angle union, 25 psi									
1310	1/2" pipe size	1 Stpi	24	.333	Ea.	38.50	11.60		50.10	59.50
1320	3/4" pipe size		20	.400		68.50	13.95		82.45	96
1330	1" pipe size		19	.421		99	14.65		113.65	131
1990	Carbon steel body, threaded									
2000	Controlled disc, integral strainer & blow down									
2010	Threaded									
2020	3/8" pipe size	1 Stpi	24	.333	Ea.	180	11.60		191.60	216
2030	1/2" pipe size		24	.333		180	11.60		191.60	216
2040	3/4" pipe size		20	.400		242	13.95		255.95	288
2050	1" pipe size		19	.421		310	14.65		324.65	360
2100	Socket weld									
2110	3/8" pipe size	Q-17	17	.941	Ea.	215	29.50	5.90	250.40	288
2120	1/2" pipe size		16	1		215	31.50	6.30	252.80	291
2130	3/4" pipe size		14	1.143		281	36	7.20	324.20	370
2140	1" pipe size		12	1.333		350	42	8.40	400.40	455
2170	Flanged, 600 lb. ASA									
2180	3/8" pipe size	1 Stpi	12	.667	Ea.	530	23		553	620
2190	1/2" pipe size		12	.667		530	23		553	620
2200	3/4" pipe size		10	.800		635	28		663	740
2210	1" pipe size		7	1.143		750	40		790	885
5000	Forged steel body									
5010	Inverted bucket									
5020	Threaded									
5030	1/2" pipe size	1 Stpi	12	.667	Ea.	320	23		343	390
5040	3/4" pipe size		10	.800		555	28		583	650
5050	1" pipe size		9	.889		835	31		866	960
5060	1-1/4" pipe size		8	1		1,200	35		1,235	1,375

Important: See the Reference Section for critical supporting data - Reference Nos., Crews, & City Cost Indexes

15100 | Building Services Piping

15120 | Piping Specialties

		Crew	Daily Output	Labor-Hours	Unit	Mat.	Labor	Equip.	Total	Total Incl O&P
5070	1-1/2" pipe size	1 Stpi	7	1.143	Ea.	1,725	40		1,765	1,925
5080	2" pipe size	↓	6	1.333	↓	3,025	46.50		3,071.50	3,400
5100	Socket weld									
5110	1/2" pipe size	Q-17	16	1	Ea.	385	31.50	6.30	422.80	480
5120	3/4" pipe size		14	1.143		625	36	7.20	668.20	745
5130	1" pipe size		12	1.333		905	42	8.40	955.40	1,075
5140	1-1/4" pipe size		10	1.600		1,275	50	10.05	1,335.05	1,475
5150	1-1/2" pipe size		9	1.778		1,775	56	11.20	1,842.20	2,050
5160	2" pipe size	↓	7	2.286	↓	3,100	71.50	14.40	3,185.90	3,525
5200	Flanged									
5210	1/2" pipe size	1 Stpi	12	.667	Ea.	650	23		673	750
5220	3/4" pipe size		10	.800		1,050	28		1,078	1,200
5230	1" pipe size		7	1.143		1,500	40		1,540	1,675
5240	1-1/4" pipe size	↓	6	1.333		2,075	46.50		2,121.50	2,350
5250	1-1/2" pipe size	Q-5	9	1.778		2,575	56		2,631	2,900
5260	2" pipe size	"	7	2.286	↓	3,950	71.50		4,021.50	4,425
6000	Stainless steel body									
6010	Inverted bucket									
6020	Threaded									
6030	1/2" pipe size	1 Stpi	12	.667	Ea.	108	23		131	153
6040	3/4" pipe size		10	.800		140	28		168	196
6050	1" pipe size	↓	9	.889	↓	500	31		531	595
6100	Threaded with thermic vent									
6110	1/2" pipe size	1 Stpi	12	.667	Ea.	123	23		146	170
6120	3/4" pipe size		10	.800		157	28		185	215
6130	1" pipe size	↓	9	.889	↓	520	31		551	620
6200	Threaded with check valve									
6210	1/2" pipe size	1 Stpi	12	.667	Ea.	153	23		176	203
6220	3/4" pipe size	"	10	.800	"	209	28		237	272
6300	Socket weld									
6310	1/2" pipe size	Q-17	16	1	Ea.	125	31.50	6.30	162.80	191
6320	3/4" pipe size	"	14	1.143	"	160	36	7.20	203.20	238
0010	**STRAINERS, BASKET TYPE** Perforated stainless steel basket									
0100	Brass or monel available									
2000	Simplex style									
2300	Bronze body									
2320	Screwed, 3/8" pipe size	1 Stpi	22	.364	Ea.	98	12.65		110.65	127
2340	1/2" pipe size		20	.400		100	13.95		113.95	131
2360	3/4" pipe size		17	.471		103	16.40		119.40	138
2380	1" pipe size		15	.533		171	18.60		189.60	216
2400	1-1/4" pipe size		13	.615		201	21.50		222.50	254
2420	1-1/2" pipe size		12	.667		260	23		283	320
2440	2" pipe size	↓	10	.800		395	28		423	475
2460	2-1/2" pipe size	Q-5	15	1.067		565	33.50		598.50	670
2480	3" pipe size	"	14	1.143		815	36		851	955
2600	Flanged, 2" pipe size	1 Stpi	6	1.333		600	46.50		646.50	730
2620	2-1/2" pipe size	Q-5	4.50	3.556		730	112		842	975
2640	3" pipe size		3.50	4.571		1,075	143		1,218	1,400
2660	4" pipe size	↓	3	5.333		1,800	167		1,967	2,225
2680	5" pipe size	Q-6	3.40	7.059		2,150	230		2,380	2,700
2700	6" pipe size		3	8		3,525	260		3,785	4,275
2710	8" pipe size		2.50	9.600	↓	6,300	310		6,610	7,400
3600	Iron body									
3700	Screwed, 3/8" pipe size	1 Stpi	22	.364	Ea.	112	12.65		124.65	142
3720	1/2" pipe size		20	.400		112	13.95		125.95	144
3740	3/4" pipe size	↓	17	.471	↓	147	16.40		163.40	187

15100 | Building Services Piping

15120 | Piping Specialties

		CREW	DAILY OUTPUT	LABOR-HOURS	UNIT	2001 BARE COSTS MAT.	LABOR	EQUIP.	TOTAL	TOTAL INCL O&P
3760	1" pipe size	1 Stpi	15	.533	Ea.	147	18.60		165.60	190
3780	1-1/4" pipe size		13	.615		197	21.50		218.50	249
3800	1-1/2" pipe size		12	.667		216	23		239	272
3820	2" pipe size	↓	10	.800		258	28		286	325
3840	2-1/2" pipe size	Q-5	15	1.067		370	33.50		403.50	455
3860	3" pipe size	"	14	1.143		420	36		456	515
4000	Flanged, 2" pipe size	1 Stpi	6	1.333		350	46.50		396.50	455
4020	2-1/2" pipe size	Q-5	4.50	3.556		475	112		587	695
4040	3" pipe size		3.50	4.571		505	143		648	770
4060	4" pipe size	↓	3	5.333		760	167		927	1,100
4080	5" pipe size	Q-6	3.40	7.059		1,200	230		1,430	1,675
4100	6" pipe size		3	8		1,475	260		1,735	2,025
4120	8" pipe size		2.50	9.600		2,725	310		3,035	3,475
4140	10" pipe size		2.20	10.909		4,100	355		4,455	5,025
4160	12" pipe size		1.70	14.118		4,875	460		5,335	6,075
4180	14" pipe size		1.40	17.143		6,150	560		6,710	7,600
4200	16" pipe size	↓	1	24	↓	6,425	780		7,205	8,250
6000	Cast steel body									
6400	Screwed, 1" pipe size	1 Stpi	15	.533	Ea.	241	18.60		259.60	293
6410	1-1/4" pipe size		13	.615		365	21.50		386.50	435
6420	1-1/2" pipe size		12	.667		365	23		388	435
6440	2" pipe size	↓	10	.800		515	28		543	605
6460	2-1/2" pipe size	Q-5	15	1.067		720	33.50		753.50	845
6480	3" pipe size	"	14	1.143		945	36		981	1,100
6560	Flanged, 2" pipe size	1 Stpi	6	1.333		750	46.50		796.50	895
6580	2-1/2" pipe size	Q-5	4.50	3.556		1,100	112		1,212	1,375
6600	3" pipe size		3.50	4.571		1,200	143		1,343	1,525
6620	4" pipe size	↓	3	5.333		1,675	167		1,842	2,100
6640	6" pipe size	Q-6	3	8		3,050	260		3,310	3,750
6660	8" pipe size	"	2.50	9.600	↓	4,950	310		5,260	5,925
7000	Stainless steel body									
7200	Screwed, 1" pipe size	1 Stpi	15	.533	Ea.	395	18.60		413.60	465
7210	1-1/4" pipe size		13	.615		575	21.50		596.50	670
7220	1-1/2" pipe size		12	.667		575	23		598	670
7240	2" pipe size	↓	10	.800		845	28		873	970
7260	2-1/2" pipe size	Q-5	15	1.067		1,400	33.50		1,433.50	1,600
7280	3" pipe size	"	14	1.143		1,850	36		1,886	2,100
7400	Flanged, 2" pipe size	1 Stpi	6	1.333		1,150	46.50		1,196.50	1,350
7420	2-1/2" pipe size	Q-5	4.50	3.556		2,100	112		2,212	2,500
7440	3" pipe size		3.50	4.571		2,150	143		2,293	2,600
7460	4" pipe size	↓	3	5.333		3,375	167		3,542	3,975
7480	6" pipe size	Q-6	3	8		5,850	260		6,110	6,850
7500	8" pipe size	"	2.50	9.600	↓	9,925	310		10,235	11,400
8100	Duplex style									
8200	Bronze body									
8240	Screwed, 3/4" pipe size	1 Stpi	16	.500	Ea.	1,050	17.45		1,067.45	1,175
8260	1" pipe size		14	.571		1,050	19.90		1,069.90	1,175
8280	1-1/4" pipe size		12	.667		2,050	23		2,073	2,275
8300	1-1/2" pipe size		11	.727		2,050	25.50		2,075.50	2,300
8320	2" pipe size	↓	9	.889		3,250	31		3,281	3,625
8340	2-1/2" pipe size	Q-5	14	1.143		3,250	36		3,286	3,625
8420	Flanged, 2" pipe size	1 Stpi	6	1.333		3,675	46.50		3,721.50	4,125
8440	2-1/2" pipe size	Q-5	4.50	3.556		3,775	112		3,887	4,325
8460	3" pipe size		3.50	4.571		5,200	143		5,343	5,925
8480	4" pipe size	↓	3	5.333		8,200	167		8,367	9,275
8500	5" pipe size	Q-6	3.40	7.059		14,300	230		14,530	16,000

15100 | Building Services Piping

15120 | Piping Specialties

		CREW	DAILY OUTPUT	LABOR-HOURS	UNIT	2001 BARE COSTS MAT.	LABOR	EQUIP.	TOTAL	TOTAL INCL O&P
8520	6" pipe size	Q-6	3	8	Ea.	18,000	260		18,260	20,200
8700	Iron body									
8740	Screwed, 3/4" pipe size	1 Stpi	16	.500	Ea.	545	17.45		562.45	625
8760	1" pipe size		14	.571		545	19.90		564.90	630
8780	1-1/4" pipe size		12	.667		950	23		973	1,075
8800	1-1/2" pipe size		11	.727		950	25.50		975.50	1,100
8820	2" pipe size	▼	9	.889		1,625	31		1,656	1,825
8840	2-1/2" pipe size	Q-5	14	1.143		1,625	36		1,661	1,825
9000	Flanged, 2" pipe size	1 Stpi	6	1.333		1,700	46.50		1,746.50	1,950
9020	2-1/2" pipe size	Q-5	4.50	3.556		1,775	112		1,887	2,125
9040	3" pipe size		3.50	4.571		2,250	143		2,393	2,700
9060	4" pipe size	▼	3	5.333		3,550	167		3,717	4,175
9080	5" pipe size	Q-6	3.40	7.059		7,150	230		7,380	8,225
9100	6" pipe size		3	8		7,150	260		7,410	8,275
9120	8" pipe size		2.50	9.600		13,700	310		14,010	15,600
9140	10" pipe size		2.20	10.909		19,100	355		19,455	21,500
9160	12" pipe size		1.70	14.118		21,200	460		21,660	24,000
9170	14" pipe size		1.40	17.143		25,700	560		26,260	29,100
9180	16" pipe size	▼	1	24	▼	32,200	780		32,980	36,700
9300	Cast steel body									
9340	Screwed, 1" pipe size	1 Stpi	14	.571	Ea.	1,525	19.90		1,544.90	1,700
9360	1-1/2" pipe size		11	.727		2,250	25.50		2,275.50	2,525
9380	2" pipe size		9	.889		3,825	31		3,856	4,250
9460	Flanged, 2" pipe size	▼	6	1.333		3,825	46.50		3,871.50	4,300
9480	2-1/2" pipe size	Q-5	4.50	3.556		4,225	112		4,337	4,825
9500	3" pipe size		3.50	4.571		5,575	143		5,718	6,375
9520	4" pipe size	▼	3	5.333		8,175	167		8,342	9,250
9540	6" pipe size	Q-6	3	8		14,000	260		14,260	15,800
9560	8" pipe size	"	2.50	9.600	▼	27,700	310		28,010	31,000
9700	Stainless steel body									
9740	Screwed, 1" pipe size	1 Stpi	14	.571	Ea.	2,775	19.90		2,794.90	3,075
9760	1-1/2" pipe size		11	.727		4,200	25.50		4,225.50	4,675
9780	2" pipe size		9	.889		6,775	31		6,806	7,500
9860	Flanged, 2" pipe size	▼	6	1.333		7,050	46.50		7,096.50	7,825
9880	2-1/2" pipe size	Q-5	4.50	3.556		7,400	112		7,512	8,325
9900	3" pipe size		3.50	4.571		11,600	143		11,743	13,000
9920	4" pipe size	▼	3	5.333		14,200	167		14,367	15,900
9940	6" pipe size	Q-6	3	8		25,000	260		25,260	27,900
9960	8" pipe size	"	2.50	9.600	▼	46,300	310		46,610	51,500
0010	**STRAINERS, Y TYPE** Bronze body									
0050	Screwed, 150 lb., 1/4" pipe size	1 Stpi	24	.333	Ea.	11.40	11.60		23	30
0070	3/8" pipe size		24	.333		15.90	11.60		27.50	35
0100	1/2" pipe size		20	.400		15.90	13.95		29.85	38.50
0120	3/4" pipe size		19	.421		17.40	14.65		32.05	41
0140	1" pipe size		17	.471		18.70	16.40		35.10	45.50
0150	1-1/4" pipe size		15	.533		42	18.60		60.60	74
0160	1-1/2" pipe size		14	.571		40.50	19.90		60.40	74.50
0180	2" pipe size		13	.615		53.50	21.50		75	91.50
0182	3" pipe size	▼	12	.667		286	23		309	350
0200	300 lb., 2-1/2" pipe size	Q-5	17	.941		286	29.50		315.50	360
0220	3" pipe size		16	1		565	31.50		596.50	675
0240	4" pipe size	▼	15	1.067	▼	1,300	33.50		1,333.50	1,475
0500	For 300 lb rating 1/4" thru 2", add					15%				
1000	Flanged, 150 lb., 1-1/2" pipe size	1 Stpi	11	.727	Ea.	305	25.50		330.50	375
1020	2" pipe size	"	8	1	▼	390	35		425	485

15100 | Building Services Piping

15120 | Piping Specialties

			CREW	DAILY OUTPUT	LABOR-HOURS	UNIT	MAT.	LABOR	EQUIP.	TOTAL	TOTAL INCL O&P	
820	1030	2-1/2" pipe size	Q-5	5	3.200	Ea.	505	100		605	705	820
	1040	3" pipe size		4.50	3.556		620	112		732	855	
	1060	4" pipe size		3	5.333		945	167		1,112	1,275	
	1080	5" pipe size	Q-6	3.40	7.059		1,875	230		2,105	2,425	
	1100	6" pipe size		3	8		2,500	260		2,760	3,150	
	1106	8" pipe size		2.60	9.231		3,900	300		4,200	4,750	
	1500	For 300 lb rating, add					40%					
840	0010	**STRAINERS, Y TYPE** Iron body										840
	0050	Screwed, 250 lb., 1/4" pipe size	1 Stpi	20	.400	Ea.	6.65	13.95		20.60	28.50	
	0070	3/8" pipe size		20	.400		6.30	13.95		20.25	28	
	0100	1/2" pipe size		20	.400		6.65	13.95		20.60	28.50	
	0120	3/4" pipe size		18	.444		7.95	15.50		23.45	32.50	
	0140	1" pipe size		16	.500		10.75	17.45		28.20	38.50	
	0150	1-1/4" pipe size		15	.533		14.75	18.60		33.35	44	
	0160	1-1/2" pipe size		12	.667		17.75	23		40.75	54.50	
	0180	2" pipe size		8	1		27.50	35		62.50	82.50	
	0200	2-1/2" pipe size	Q-5	12	1.333		140	42		182	217	
	0220	3" pipe size		11	1.455		151	45.50		196.50	235	
	0240	4" pipe size		5	3.200		255	100		355	430	
	0500	For galvanized body, add					50%					
	1000	Flanged, 125 lb., 1-1/2" pipe size	1 Stpi	11	.727	Ea.	86	25.50		111.50	133	
	1020	2" pipe size	"	8	1		64	35		99	123	
	1030	2-1/2" pipe size	Q-5	5	3.200		72.50	100		172.50	232	
	1040	3" pipe size		4.50	3.556		84.50	112		196.50	261	
	1060	4" pipe size		3	5.333		155	167		322	425	
	1080	5" pipe size	Q-6	3.40	7.059		242	230		472	610	
	1100	6" pipe size		3	8		295	260		555	720	
	1120	8" pipe size		2.50	9.600		495	310		805	1,025	
	1140	10" pipe size		2	12		950	390		1,340	1,650	
	1160	12" pipe size		1.70	14.118		1,425	460		1,885	2,275	
	1170	14" pipe size		1.30	18.462		2,625	600		3,225	3,800	
	1180	16" pipe size		1	24		3,725	780		4,505	5,275	
	1500	For 250 lb rating, add					20%					
	2000	For galvanized body, add					50%					
	2500	For steel body, add					40%					
870	0010	**SUCTION DIFFUSERS**										870
	0100	Cast iron body with integral straightening vanes, strainer										
	1000	Flanged										
	1010	2" inlet, 1-1/2" pump side	1 Stpi	6	1.333	Ea.	213	46.50		259.50	305	
	1020	2" pump side	"	5	1.600		221	56		277	325	
	1030	3" inlet, 2" pump side	Q-5	6.50	2.462		270	77		347	415	
	1040	2-1/2" pump side		5.30	3.019		395	94.50		489.50	580	
	1050	3" pump side		4.50	3.556		400	112		512	610	
	1060	4" inlet, 3" pump side		3.50	4.571		465	143		608	730	
	1070	4" pump side		3	5.333		540	167		707	850	
	1080	5" inlet, 4" pump side	Q-6	3.80	6.316		615	205		820	990	
	1090	5" pump side		3.40	7.059		735	230		965	1,150	
	1100	6" inlet, 4" pump side		3.30	7.273		640	237		877	1,050	
	1110	5" pump side		3.10	7.742		770	252		1,022	1,225	
	1120	6" pump side		3	8		810	260		1,070	1,300	
	1130	8" inlet, 6" pump side		2.70	8.889		875	289		1,164	1,400	
	1140	8" pump side		2.50	9.600		1,500	310		1,810	2,150	
	1150	10" inlet, 6" pump side		2.40	10		2,425	325		2,750	3,150	
	1160	8" pump side		2.30	10.435		1,750	340		2,090	2,450	
	1170	10" pump side		2.20	10.909		2,525	355		2,880	3,300	

15100 | Building Services Piping

15120 | Piping Specialties

			CREW	DAILY OUTPUT	LABOR-HOURS	UNIT	MAT.	LABOR	EQUIP.	TOTAL	TOTAL INCL O&P	
870	1180	12" inlet, 8" pump side	Q-6	2	12	Ea.	2,675	390		3,065	3,550	870
	1190	10" pump side		1.80	13.333		2,950	435		3,385	3,900	
	1200	12" pump side		1.70	14.118		3,325	460		3,785	4,375	
	1210	14" inlet, 12" pump side		1.50	16		3,575	520		4,095	4,700	
	1220	14" pump side	↓	1.30	18.462	↓	4,500	600		5,100	5,850	
900	0010	**THERMOFLO INDICATOR** For balancing										900
	1000	Sweat connections, 1-1/4" pipe size	1 Stpi	12	.667	Ea.	345	23		368	415	
	1020	1-1/2" pipe size		10	.800		355	28		383	430	
	1040	2" pipe size		8	1		375	35		410	465	
	1060	2-1/2" pipe size	↓	7	1.143		565	40		605	685	
	2000	Flange connections, 3" pipe size	Q-5	5	3.200		675	100		775	895	
	2020	4" pipe size		4	4		810	126		936	1,075	
	2030	5" pipe size		3.50	4.571		1,025	143		1,168	1,350	
	2040	6" pipe size		3	5.333		1,100	167		1,267	1,450	
	2060	8" pipe size	↓	2	8	↓	1,325	251		1,576	1,850	
920	0010	**VENTURI FLOW** Measuring device										920
	0050	1/2" diameter	1 Stpi	24	.333	Ea.	105	11.60		116.60	134	
	0100	3/4" diameter		20	.400		93	13.95		106.95	123	
	0120	1" diameter		19	.421		94	14.65		108.65	125	
	0140	1-1/4" diameter		15	.533		134	18.60		152.60	175	
	0160	1-1/2" diameter		13	.615		140	21.50		161.50	187	
	0180	2" diameter	↓	11	.727		166	25.50		191.50	222	
	0200	2-1/2" diameter	Q-5	16	1		259	31.50		290.50	335	
	0220	3" diameter		14	1.143		335	36		371	425	
	0240	4" diameter	↓	11	1.455		375	45.50		420.50	485	
	0260	5" diameter	Q-6	4	6		585	195		780	940	
	0280	6" diameter		3.50	6.857		660	223		883	1,050	
	0300	8" diameter		3	8		1,050	260		1,310	1,550	
	0320	10" diameter		2	12		1,925	390		2,315	2,700	
	0330	12" diameter		1.80	13.333		3,075	435		3,510	4,025	
	0340	14" diameter		1.60	15		3,250	490		3,740	4,300	
	0350	16" diameter	↓	1.40	17.143	↓	6,500	560		7,060	8,000	
	0500	For meter, add					1,175			1,175	1,300	
940	0010	**WATER SUPPLY METERS**										940
	1000	Detector, serves dual systems such as fire and domestic or										
	1020	process water, wide range cap., UL and FM approved										
	1100	3" mainline x 2" by-pass, 400 GPM	Q-1	3.60	4.444	Ea.	5,100	138		5,238	5,825	
	1140	4" mainline x 2" by-pass, 700 GPM	"	2.50	6.400		5,500	198		5,698	6,350	
	1180	6" mainline x 3" by-pass, 1600 GPM	Q-2	2.60	9.231		7,475	297		7,772	8,650	
	1220	8" mainline x 4" by-pass, 2800 GPM		2.10	11.429		11,300	365		11,665	13,100	
	1260	10" mainline x 6" by-pass, 4400 GPM		2	12		18,400	385		18,785	20,800	
	1300	10"x12" mainlines x 6" by-pass, 5400 GPM	↓	1.70	14.118	↓	19,100	455		19,555	21,700	
	2000	Domestic/commercial, bronze										
	2020	Threaded										
	2060	5/8" diameter, to 20 GPM	1 Plum	16	.500	Ea.	55.50	17.25		72.75	87	
	2080	3/4" diameter, to 30 GPM		14	.571		96	19.70		115.70	136	
	2100	1" diameter, to 50 GPM	↓	12	.667	↓	131	23		154	179	
	2300	Threaded/flanged										
	2340	1-1/2" diameter, to 100 GPM	1 Plum	8	1	Ea.	405	34.50		439.50	495	
	2360	2" diameter, to 160 GPM	"	6	1.333	"	535	46		581	660	
	2600	Flanged, compound										
	2640	3" diameter, 320 GPM	Q-1	3	5.333	Ea.	2,425	165		2,590	2,900	
	2660	4" diameter, to 500 GPM		1.50	10.667		3,775	330		4,105	4,650	
	2680	6" diameter, to 1,000 GPM		1	16		5,425	495		5,920	6,725	
	2700	8" diameter, to 1,800 GPM	↓	.80	20		10,700	620		11,320	12,700	

MECHANICAL 15

15100 | Building Services Piping

15120 | Piping Specialties

			CREW	DAILY OUTPUT	LABOR-HOURS	UNIT	MAT.	LABOR	EQUIP.	TOTAL	TOTAL INCL O&P	
940	7000	Turbine										940
	7260	Flanged										
	7300	2" diameter, to 160 GPM	1 Plum	7	1.143	Ea.	690	39.50		729.50	820	
	7320	3" diameter, to 450 GPM	Q-1	3.60	4.444		995	138		1,133	1,300	
	7340	4" diameter, to 650 GPM	"	2.50	6.400		1,700	198		1,898	2,175	
	7360	6" diameter, to 1800 GPM	Q-2	2.60	9.231		2,850	297		3,147	3,575	
	7380	8" diameter, to 2500 GPM		2.10	11.429		4,850	365		5,215	5,875	
	7400	10" diameter, to 5500 GPM		1.70	14.118		6,575	455		7,030	7,925	
960	0010	**WELD END BALL JOINTS** Steel										960
	0050	2-1/2" diameter	Q-17	13	1.231	Ea.	435	38.50	7.75	481.25	540	
	0100	3" diameter		12	1.333		455	42	8.40	505.40	570	
	0120	4" diameter		11	1.455		730	45.50	9.15	784.65	880	
	0140	5" diameter	Q-18	14	1.714		1,050	56	7.20	1,113.20	1,250	
	0160	6" diameter		12	2		1,175	65	8.40	1,248.40	1,400	
	0180	8" diameter		9	2.667		1,675	87	11.15	1,773.15	2,000	
	0200	10" diameter		8	3		2,500	97.50	12.55	2,610.05	2,900	
	0220	12" diameter		6	4		3,300	130	16.75	3,446.75	3,850	
	0240	14" diameter		4	6		3,975	195	25	4,195	4,700	
980	0010	**ZONE VALVES**										980
	1000	Bronze body										
	1010	2 way, 125 psi										
	1020	Standard, 65' pump head										
	1030	1/2" soldered	1 Stpi	24	.333	Ea.	69.50	11.60		81.10	94	
	1040	3/4" soldered		20	.400		69.50	13.95		83.45	97.50	
	1050	1" soldered		19	.421		74.50	14.65		89.15	104	
	1060	1-1/4" soldered		15	.533		87	18.60		105.60	124	
	1200	High head, 125' pump head										
	1210	1/2" soldered	1 Stpi	24	.333	Ea.	94	11.60		105.60	121	
	1220	3/4" soldered		20	.400		97.50	13.95		111.45	128	
	1230	1" soldered		19	.421		107	14.65		121.65	139	
	1300	Geothermal, 150' pump head										
	1310	3/4" soldered	1 Stpi	20	.400		112	13.95		125.95	144	
	1320	1" soldered		19	.421		119	14.65		133.65	153	
	1330	3/4" threaded		20	.400		112	13.95		125.95	144	
	3000	3 way, 125 psi										
	3200	By-pass, 65' pump head										
	3210	1/2" soldered	1 Stpi	24	.333	Ea.	98.50	11.60		110.10	126	
	3220	3/4" soldered		20	.400		102	13.95		115.95	133	
	3230	1" soldered		19	.421		116	14.65		130.65	149	
	9000	Transformer, for up to 3 zone valves		20	.400		14.30	13.95		28.25	36.50	

15140 | Domestic Water Piping

			CREW	DAILY OUTPUT	LABOR-HOURS	UNIT	MAT.	LABOR	EQUIP.	TOTAL	TOTAL INCL O&P	
100	0010	**BACKFLOW PREVENTER** Includes valves										100
	0020	and four test cocks, corrosion resistant, automatic operation										
	1000	Double check principle										
	1080	Threaded, with gate valves										
	1100	3/4" pipe size	1 Plum	16	.500	Ea.	595	17.25		612.25	680	
	1120	1" pipe size		14	.571		620	19.70		639.70	710	
	1140	1-1/2" pipe size		10	.800		705	27.50		732.50	815	
	1160	2" pipe size		7	1.143		870	39.50		909.50	1,025	
	1300	Flanged, valves are OS&Y										
	1370	1" pipe size	1 Plum	5	1.600	Ea.	650	55		705	800	
	1374	1-1/2" pipe size		5	1.600		740	55		795	900	
	1378	2" pipe size		4.80	1.667		905	57.50		962.50	1,075	
	1380	3" pipe size	Q-1	4.50	3.556		1,450	110		1,560	1,775	
	1400	4" pipe size	"	3	5.333		2,100	165		2,265	2,550	

15100 | Building Services Piping

15140 | Domestic Water Piping

			CREW	DAILY OUTPUT	LABOR-HOURS	UNIT	2001 BARE COSTS MAT.	LABOR	EQUIP.	TOTAL	TOTAL INCL O&P	
100	1420	6" pipe size	Q-2	3	8	Ea.	3,300	257		3,557	4,025	100
	1430	8" pipe size	"	2	12	↓	5,725	385		6,110	6,875	
	4000	Reduced pressure principle										
	4100	Threaded, valves are ball										
	4120	3/4" pipe size	1 Plum	16	.500	Ea.	685	17.25		702.25	780	
	4140	1" pipe size		14	.571		730	19.70		749.70	830	
	4150	1-1/4" pipe size		12	.667		1,025	23		1,048	1,150	
	4160	1-1/2" pipe size		10	.800		1,025	27.50		1,052.50	1,175	
	4180	2" pipe size	↓	7	1.143	↓	1,050	39.50		1,089.50	1,200	
	5000	Flanged, bronze, valves are OS&Y										
	5060	2-1/2" pipe size	Q-1	5	3.200	Ea.	2,725	99		2,824	3,150	
	5080	3" pipe size		4.50	3.556		3,375	110		3,485	3,900	
	5100	4" pipe size	↓	3	5.333		4,800	165		4,965	5,525	
	5120	6" pipe size	Q-2	3	8	↓	10,200	257		10,457	11,600	
	5600	Flanged, iron, valves are OS&Y										
	5660	2-1/2" pipe size	Q-1	5	3.200	Ea.	1,825	99		1,924	2,150	
	5680	3" pipe size		4.50	3.556		1,950	110		2,060	2,300	
	5700	4" pipe size	↓	3	5.333		2,575	165		2,740	3,075	
	5720	6" pipe size	Q-2	3	8		4,050	257		4,307	4,850	
	5740	8" pipe size		2	12		8,175	385		8,560	9,575	
	5760	10" pipe size	↓	1	24	↓	11,200	770		11,970	13,500	
600	0010	**VACUUM BREAKERS** Hot or cold water										600
	1030	Anti-siphon, brass										
	1040	1/4" size	1 Plum	24	.333	Ea.	18.65	11.50		30.15	38	
	1050	3/8" size		24	.333		18.65	11.50		30.15	38	
	1060	1/2" size		24	.333		21	11.50		32.50	41	
	1080	3/4" size		20	.400		25	13.80		38.80	48.50	
	1100	1" size		19	.421		39	14.50		53.50	65	
	1120	1-1/4" size		15	.533		68.50	18.35		86.85	103	
	1140	1-1/2" size		13	.615		80.50	21		101.50	121	
	1160	2" size		11	.727		125	25		150	176	
	1180	2-1/2" size		9	.889		360	30.50		390.50	440	
	1200	3" size	↓	7	1.143		480	39.50		519.50	585	
	1300	For polished chrome, (1/4" thru 1"), add					50%					
	1900	Vacuum relief, water service, bronze										
	2000	1/2" size	1 Plum	30	.267	Ea.	17.70	9.20		26.90	33.50	
	2040	3/4" size	"	28	.286	"	22	9.85		31.85	39.50	
700	0010	**VACUUM BREAKERS**										700
	0011	See also backflow preventers 15140-100										
	1000	Anti-siphon continuous pressure type										
	1010	Max. 150 PSI - 210°F										
	1020	Bronze body										
	1030	1/2" size	1 Stpi	24	.333	Ea.	80.50	11.60		92.10	106	
	1040	3/4" size		20	.400		80.50	13.95		94.45	110	
	1050	1" size		19	.421		83.50	14.65		98.15	114	
	1060	1-1/4" size		15	.533		164	18.60		182.60	209	
	1070	1-1/2" size		13	.615		202	21.50		223.50	256	
	1080	2" size	↓	11	.727	↓	209	25.50		234.50	269	
	1200	Max. 125 PSI with atmospheric vent										
	1210	Brass, in-line construction										
	1220	1/4" size	1 Stpi	24	.333	Ea.	30.50	11.60		42.10	51.50	
	1230	3/8" size	"	24	.333		30.50	11.60		42.10	51.50	
	1260	For polished chrome finish, add				↓	13%					
	2000	Anti-siphon, non-continuous pressure type										
	2010	Hot or cold water 125 PSI - 210° F										

15100 | Building Services Piping

15140 | Domestic Water Piping

			CREW	DAILY OUTPUT	LABOR-HOURS	UNIT	MAT.	LABOR	EQUIP.	TOTAL	TOTAL INCL O&P	
700	2020	Bronze body										700
	2030	1/4" size	1 Stpi	24	.333	Ea.	18.65	11.60		30.25	38	
	2040	3/8" size		24	.333		18.65	11.60		30.25	38	
	2050	1/2" size		24	.333		21	11.60		32.60	41	
	2060	3/4" size		20	.400		25	13.95		38.95	48.50	
	2070	1" size		19	.421		39	14.65		53.65	65	
	2080	1-1/4" size		15	.533		68.50	18.60		87.10	104	
	2090	1-1/2" size		13	.615		80.50	21.50		102	121	
	2100	2" size		11	.727		125	25.50		150.50	177	
	2110	2-1/2" size		8	1		360	35		395	450	
	2120	3" size		6	1.333		480	46.50		526.50	595	
	2150	For polished chrome finish, add					50%					
800	0010	**WATER HAMMER ARRESTORS / SHOCK ABSORBERS**										800
	0490	Copper										
	0500	3/4" male I.P.S. For 1 to 11 fixtures	1 Plum	12	.667	Ea.	16.05	23		39.05	52	
	0600	1" male I.P.S., For 12 to 32 fixtures		8	1		41	34.50		75.50	97	
	0700	1-1/4" male I.P.S. For 33 to 60 fixtures		8	1		47.50	34.50		82	105	
	0800	1-1/2" male I.P.S. For 61 to 113 fixtures		8	1		65	34.50		99.50	124	
	0900	2" male I.P.S. For 114 to 154 fixtures		8	1		104	34.50		138.50	166	
	1000	2-1/2" male I.P.S. For 155 to 330 fixtures		4	2		231	69		300	360	
	4000	Bellows type										
	4010	3/4" FNPT, to 11 fixture units	1 Plum	10	.800	Ea.	68.50	27.50		96	117	
	4020	1" FNPT, to 32 fixture units		8.80	.909		138	31.50		169.50	200	
	4030	1" FNPT, to 60 fixture units		8.80	.909		206	31.50		237.50	275	
	4040	1" FNPT, to 113 fixture units		8.80	.909		515	31.50		546.50	620	
	4050	1" FNPT, to 154 fixture units		6.60	1.212		585	42		627	705	
	4060	1-1/2" FNPT, to 300 fixture units		5	1.600		720	55		775	875	

15155 | Drainage Specialties

			CREW	DAILY OUTPUT	LABOR-HOURS	UNIT	MAT.	LABOR	EQUIP.	TOTAL	TOTAL INCL O&P	
680	0010	**SEPARATORS** Entrainment eliminator, steel body, 150 PSIG										680
	0100	1/4" size	1 Stpi	24	.333	Ea.	134	11.60		145.60	165	
	0120	1/2" size		24	.333		144	11.60		155.60	176	
	0140	3/4" size		20	.400		155	13.95		168.95	192	
	0160	1" size		19	.421		206	14.65		220.65	249	
	0180	1-1/4" size		15	.533		237	18.60		255.60	289	
	0200	1-1/2" size		13	.615		268	21.50		289.50	330	
	0220	2" size		11	.727		299	25.50		324.50	370	
	0240	2-1/2" size	Q-5	15	1.067		1,175	33.50		1,208.50	1,350	
	0260	3" size		13	1.231		1,400	38.50		1,438.50	1,600	
	0280	4" size		10	1.600		1,650	50		1,700	1,875	
	0300	5" size		6	2.667		1,850	83.50		1,933.50	2,175	
	0320	6" size		3	5.333		2,125	167		2,292	2,575	
	0340	8" size	Q-6	4.40	5.455		2,550	177		2,727	3,075	
	0360	10" size	"	4	6		3,825	195		4,020	4,500	
	1000	For 300 PSIG, add					15%					
780	0010	**TRAPS**										780
	4700	Copper, drainage, drum trap										
	4800	3" x 5" solid, 1-1/2" pipe size	1 Plum	16	.500	Ea.	16.70	17.25		33.95	44.50	
	4840	3" x 6" swivel, 1-1/2" pipe size	"	16	.500	"	24.50	17.25		41.75	53	
	5100	P trap, standard pattern										
	5200	1-1/4" pipe size	1 Plum	18	.444	Ea.	11.90	15.30		27.20	36	
	5240	1-1/2" pipe size		17	.471		11.90	16.20		28.10	37.50	
	5260	2" pipe size		15	.533		18.35	18.35		36.70	47.50	
	5280	3" pipe size		11	.727		44.50	25		69.50	87	
	5340	With cleanout and slip joint										

15100 | Building Services Piping

15155 | Drainage Specialties

		CREW	DAILY OUTPUT	LABOR-HOURS	UNIT	2001 BARE COSTS				TOTAL INCL O&P
						MAT.	LABOR	EQUIP.	TOTAL	
780 5360	1-1/4" pipe size	1 Plum	18	.444	Ea.	16.85	15.30		32.15	41.50 **780**
5400	1-1/2" pipe size		17	.471		17.20	16.20		33.40	43.50
5420	2" pipe size	↓	15	.533		28	18.35		46.35	58
5460	For swivel, add				↓	6.75			6.75	7.45
6710	ABS DWV P trap, solvent weld joint									
6720	1-1/2" pipe size	1 Plum	18	.444	Ea.	31	15.30		46.30	57
6722	2" pipe size		17	.471		50	16.20		66.20	79.50
6724	3" pipe size		15	.533		228	18.35		246.35	279
6726	4" pipe size	↓	14	.571	↓	470	19.70		489.70	545
6732	PVC DWV P trap, solvent weld joint									
6733	1-1/2" pipe size	1 Plum	18	.444	Ea.	2.39	15.30		17.69	25.50
6734	2" pipe size		17	.471		3.26	16.20		19.46	28
6735	3" pipe size		15	.533		16.20	18.35		34.55	45.50
6736	4" pipe size		14	.571		39	19.70		58.70	72.50
6760	PP DWV, dilution trap, 1-1/2" pipe size		16	.500		101	17.25		118.25	137
6770	P trap, 1-1/2" pipe size		17	.471		27	16.20		43.20	54
6780	2" pipe size		16	.500		37	17.25		54.25	66.50
6790	3" pipe size		14	.571		85	19.70		104.70	123
6800	4" pipe size		13	.615		109	21		130	151
6810	Running trap, 1-1/2" pipe size		16	.500		28.50	17.25		45.75	57.50
6820	2" pipe size		15	.533		40	18.35		58.35	71.50
6830	S trap, 1-1/2" pipe size		16	.500		29	17.25		46.25	58
6840	2" pipe size		15	.533		36.50	18.35		54.85	67.50
6850	Universal trap, 1-1/2" pipe size		14	.571		82.50	19.70		102.20	121
6860	PVC DWV hub x hub, basin trap, 1-1/4" pipe size		18	.444		4.22	15.30		19.52	27.50
6870	Sink P trap, 1-1/2" pipe size		18	.444		4.22	15.30		19.52	27.50
6880	Tubular S trap, 1-1/2" pipe size	↓	17	.471	↓	10	16.20		26.20	35.50
6890	PVC sch. 40 DWV, drum trap									
6900	1-1/2" pipe size	1 Plum	16	.500	Ea.	12.40	17.25		29.65	39.50
6910	P trap, 1-1/2" pipe size		18	.444		2.39	15.30		17.69	25.50
6920	2" pipe size		17	.471		3.26	16.20		19.46	28
6930	3" pipe size		15	.533		16.20	18.35		34.55	45.50
6940	4" pipe size		14	.571		39	19.70		58.70	72.50
6950	P trap w/clean out, 1-1/2" pipe size		18	.444		5.80	15.30		21.10	29.50
6960	2" pipe size		17	.471		9.85	16.20		26.05	35.50
6970	P trap adjustable, 1-1/2" pipe size		17	.471		3.12	16.20		19.32	28
6980	P trap adj. w/union & cleanout, 1-1/2" pipe size	↓	16	.500	↓	10.30	17.25		27.55	37.50
940 0010	**VENT FLASHING**									**940**
1000	Aluminum with lead ring									
1020	1-1/4" pipe	1 Plum	20	.400	Ea.	6.40	13.80		20.20	28
1030	1-1/2" pipe		20	.400		6.40	13.80		20.20	28
1040	2" pipe		18	.444		6.85	15.30		22.15	30.50
1050	3" pipe		17	.471		7.60	16.20		23.80	33
1060	4" pipe	↓	16	.500	↓	9.15	17.25		26.40	36
1350	Copper with neoprene ring									
1400	1-1/4" pipe	1 Plum	20	.400	Ea.	14.05	13.80		27.85	36.50
1430	1-1/2" pipe		20	.400		14.05	13.80		27.85	36.50
1440	2" pipe		18	.444		14.85	15.30		30.15	39.50
1450	3" pipe		17	.471		17.40	16.20		33.60	43.50
1460	4" pipe	↓	16	.500	↓	19.20	17.25		36.45	47
2000	Galvanized with neoprene ring									
2020	1-1/4" pipe	1 Plum	20	.400	Ea.	5.10	13.80		18.90	26.50
2030	1-1/2" pipe		20	.400		5.10	13.80		18.90	26.50
2040	2" pipe		18	.444		5.40	15.30		20.70	29
2050	3" pipe	↓	17	.471		5.85	16.20		22.05	31

15100 | Building Services Piping

15155 | Drainage Specialties

		CREW	DAILY OUTPUT	LABOR-HOURS	UNIT	2001 BARE COSTS MAT.	LABOR	EQUIP.	TOTAL	TOTAL INCL O&P	
940	2060	4" pipe	1 Plum	16	.500	Ea.	7.60	17.25		24.85	34.50	940
	2980	Neoprene, one piece										
	3000	1-1/4" pipe	1 Plum	24	.333	Ea.	4.92	11.50		16.42	23	
	3030	1-1/2" pipe		24	.333		4.92	11.50		16.42	23	
	3040	2" pipe		23	.348		5.15	12		17.15	24	
	3050	3" pipe		21	.381		6.25	13.10		19.35	26.50	
	3060	4" pipe	▼	20	.400	▼	7.90	13.80		21.70	29.50	

15180 | Heating and Cooling Piping

		CREW	DAILY OUTPUT	LABOR-HOURS	UNIT	MAT.	LABOR	EQUIP.	TOTAL	TOTAL INCL O&P	
100	0010	**ANTI-FREEZE** Inhibited										100
	0900	Ethylene glycol concentrated										
	1000	55 gallon drums, small quantities				Gal.	8.75			8.75	9.60	
	1200	Large quantities					8.20			8.20	9	
	2000	Propylene glycol, for solar heat, small quantities					9.95			9.95	10.90	
	2100	Large quantities				▼	8.80			8.80	9.70	
200	0010	**PUMPS, CIRCULATING** Heated or chilled water application										200
	0600	Bronze, sweat connections, 1/40 HP, in line										
	0640	3/4" size	Q-1	16	1	Ea.	114	31		145	172	
	1000	Flange connection, 3/4" to 1-1/2" size										
	1040	1/12 HP	Q-1	6	2.667	Ea.	310	82.50		392.50	465	
	1060	1/8 HP		6	2.667		535	82.50		617.50	710	
	1100	1/3 HP		6	2.667		595	82.50		677.50	780	
	1140	2" size, 1/6 HP		5	3.200		765	99		864	990	
	1180	2-1/2" size, 1/4 HP		5	3.200		995	99		1,094	1,250	
	1220	3" size, 1/4 HP		4	4		1,050	124		1,174	1,325	
	1260	1/3 HP		4	4		1,275	124		1,399	1,575	
	1300	1/2 HP		4	4		1,300	124		1,424	1,600	
	1340	3/4 HP		4	4		1,425	124		1,549	1,725	
	1380	1 HP	▼	4	4	▼	2,300	124		2,424	2,700	
	2000	Cast iron, flange connection										
	2040	3/4" to 1-1/2" size, in line, 1/12 HP	Q-1	6	2.667	Ea.	202	82.50		284.50	345	
	2060	1/8 HP		6	2.667		340	82.50		422.50	495	
	2100	1/3 HP		6	2.667		380	82.50		462.50	540	
	2140	2" size, 1/6 HP		5	3.200		415	99		514	605	
	2180	2-1/2" size, 1/4 HP		5	3.200		540	99		639	745	
	2220	3" size, 1/4 HP		4	4		550	124		674	790	
	2260	1/3 HP		4	4		745	124		869	1,000	
	2300	1/2 HP		4	4		770	124		894	1,025	
	2340	3/4 HP		4	4		890	124		1,014	1,175	
	2380	1 HP	▼	4	4	▼	1,275	124		1,399	1,600	
	2600	For non-ferrous impeller, add					3%					
	3000	High head, bronze impeller										
	3030	1-1/2" size 1/2 HP	Q-1	5	3.200	Ea.	660	99		759	875	
	3040	1-1/2" size 3/4 HP		5	3.200		705	99		804	925	
	3050	2" size 1 HP		4	4		845	124		969	1,125	
	3090	2" size 1-1/2 HP	▼	4	4	▼	1,050	124		1,174	1,325	
	4000	Close coupled, end suction, bronze impeller										
	4040	1-1/2" size, 1-1/2 HP, to 40 GPM	Q-1	3	5.333	Ea.	1,050	165		1,215	1,425	
	4090	2" size, 2 HP, to 50 GPM		3	5.333		1,075	165		1,240	1,450	
	4100	2" size, 3 HP, to 90 GPM		2.30	6.957		1,200	216		1,416	1,625	
	4190	2-1/2" size, 3 HP, to 150 GPM		2	8		1,275	248		1,523	1,800	
	4300	3" size, 5 HP, to 225 GPM		1.80	8.889		1,400	276		1,676	1,950	
	4410	3" size, 10 HP, to 350 GPM		1.60	10		2,000	310		2,310	2,675	
	4420	4" size, 7-1/2 HP, to 350 GPM	▼	1.60	10		1,950	310		2,260	2,625	
	4520	4" size, 10 HP, to 600 GPM	Q-2	1.70	14.118		2,400	455		2,855	3,300	

15100 | Building Services Piping

15180 | Heating and Cooling Piping

		CREW	DAILY OUTPUT	LABOR-HOURS	UNIT	MAT.	LABOR	EQUIP.	TOTAL	TOTAL INCL O&P
200	4530	5" size, 15 HP, to 1000 GPM	Q-2	1.70	14.118	Ea.	2,875	455		3,330	3,850
	4610	5" size, 20 HP, to 1350 GPM		1.50	16		3,175	515		3,690	4,275
	4620	5" size, 25 HP, to 1550 GPM	↓	1.50	16	↓	3,775	515		4,290	4,950
	5000	Base mounted, bronze impeller, coupling guard									
	5040	1-1/2" size, 1-1/2 HP, to 40 GPM	Q-1	2.30	6.957	Ea.	1,500	216		1,716	1,975
	5090	2" size, 2 HP, to 50 GPM		2.30	6.957		1,700	216		1,916	2,200
	5100	2" size, 3 HP, to 90 GPM		2	8		1,750	248		1,998	2,300
	5190	2-1/2" size, 3 HP, to 150 GPM		1.80	8.889		1,800	276		2,076	2,400
	5300	3" size, 5 HP, to 225 GPM		1.60	10		1,900	310		2,210	2,575
	5410	4" size, 5 HP, to 350 GPM		1.50	10.667		2,075	330		2,405	2,775
	5420	4" size, 7-1/2 HP, to 350 GPM	↓	1.50	10.667		2,325	330		2,655	3,050
	5520	5" size, 10 HP, to 600 GPM	Q-2	1.60	15		2,375	480		2,855	3,325
	5530	5" size, 15 HP, to 1000 GPM		1.60	15		3,275	480		3,755	4,325
	5610	6" size, 20 HP, to 1350 GPM		1.40	17.143		3,600	550		4,150	4,800
	5620	6" size, 25 HP, to 1550 GPM	↓	1.40	17.143	↓	4,600	550		5,150	5,900
	5800	The above pump capacities are based on 1800 RPM,									
	5810	at a 60 foot head. Increasing the RPM									
	5820	or decreasing the head will increase the GPM.									
300	0010	**PUMPS, CONDENSATE RETURN SYSTEM**									
	2000	Simplex									
	2010	With pump, motor, CI receiver, float switch									
	2020	3/4 HP, 15 GPM	Q-1	1.80	8.889	Ea.	3,150	276		3,426	3,900
	2100	Duplex									
	2110	With 2 pumps and motors, CI receiver, float switch, alternator									
	2120	3/4 HP, 15 GPM, 15 Gal CI rcvr	Q-1	1.40	11.429	Ea.	3,450	355		3,805	4,300
	2130	1 HP, 25 GPM		1.20	13.333		3,750	415		4,165	4,775
	2140	1-1/2 HP, 45 GPM		1	16		4,525	495		5,020	5,725
	2150	1-1/2 HP, 60 GPM		1	16	↓	3,600	495		4,095	4,700
700	0010	**REFRIGERATION SPECIALTIES**									
	0600	Accumulator									
	0610	3/4"	1 Stpi	8.80	.909	Ea.	34.50	31.50		66	86
	0614	7/8"		6.40	1.250		41	43.50		84.50	111
	0618	1-1/8"		4.80	1.667		50	58		108	143
	0622	1-3/8"		4	2		68	69.50		137.50	180
	0626	1-5/8"		3.20	2.500		83.50	87		170.50	224
	0630	2-1/8"	↓	2.40	3.333	↓	278	116		394	480
	1000	Filter dryer									
	1010	Replaceable core type, solder									
	1020	1/2"	1 Stpi	20	.400	Ea.	102	13.95		115.95	134
	1030	5/8"		19	.421		107	14.65		121.65	140
	1040	7/8"		18	.444		107	15.50		122.50	142
	1050	1-1/8"		15	.533		112	18.60		130.60	151
	1060	1-3/8"		14	.571		134	19.90		153.90	178
	1070	1-5/8"		12	.667		134	23		157	183
	1080	2-1/8"		10	.800		148	28		176	205
	1090	2-5/8"		9	.889		275	31		306	345
	1100	3-1/8"	↓	8	1	↓	230	35		265	305
	1200	Sealed in-line, solder									
	1210	1/4", 3 cubic inches	1 Stpi	22	.364	Ea.	8.10	12.65		20.75	28
	1220	3/8", 5 cubic inches		21	.381		7.40	13.30		20.70	28
	1230	1/2", 9 cubic inches		20	.400		12	13.95		25.95	34
	1240	1/2", 16 cubic inches		20	.400		15.60	13.95		29.55	38
	1250	5/8", 16 cubic inches		19	.421		16.35	14.65		31	40
	1260	5/8", 30 cubic inches		19	.421		26.50	14.65		41.15	51
	1270	7/8", 30 cubic inches		18	.444		28	15.50		43.50	54.50
	1280	7/8", 41 cubic inches	↓	17	.471	↓	37.50	16.40		53.90	66.50

15100 | Building Services Piping

15180 | Heating and Cooling Piping

			CREW	DAILY OUTPUT	LABOR-HOURS	UNIT	2001 BARE COSTS MAT.	LABOR	EQUIP.	TOTAL	TOTAL INCL O&P
700	1290	1-1/8", 60 cubic inches	1 Stpi	15	.533	Ea.	52	18.60		70.60	85.50
	4000	P-Trap, suction line, solder									
	4010	5/8"	1 Stpi	19	.421	Ea.	9.30	14.65		23.95	32
	4020	3/4"		19	.421		7.35	14.65		22	30
	4030	7/8"		18	.444		7.35	15.50		22.85	31.50
	4040	1-1/8"		15	.533		10.50	18.60		29.10	39.50
	4050	1-3/8"		14	.571		16.40	19.90		36.30	48
	4060	1-5/8"		12	.667		25	23		48	62.50
	4070	2-1/8"	▼	10	.800	▼	51	28		79	98
	4440	Refrigerant, R-502, 30 lb. disposable cylinder				Lb.	36.50			36.50	40.50
	5000	Sightglass									
	5010	Moisture and liquid indicator, solder									
	5020	1/4"	1 Stpi	22	.364	Ea.	8.95	12.65		21.60	29
	5030	3/8"		21	.381		10	13.30		23.30	31
	5040	1/2"		20	.400		12.70	13.95		26.65	35
	5050	5/8"		19	.421		12.95	14.65		27.60	36.50
	5060	7/8"		18	.444		16.95	15.50		32.45	42
	5070	1-1/8"		15	.533		17.70	18.60		36.30	47.50
	5080	1-3/8"		14	.571		17.55	19.90		37.45	49.50
	5090	1-5/8"		12	.667		20.50	23		43.50	57.50
	5100	2-1/8"	▼	10	.800	▼	25.50	28		53.50	70
	7400	Vacuum pump set									
	7410	Two stage, high vacuum continuous duty	1 Stpi	8	1	Ea.	2,000	35		2,035	2,250
	8000	Valves									
	8100	Check valve, soldered									
	8110	5/8"	1 Stpi	36	.222	Ea.	27	7.75		34.75	41
	8114	7/8"		26	.308		46	10.70		56.70	66.50
	8118	1-1/8"		18	.444		127	15.50		142.50	163
	8122	1-3/8"		14	.571		173	19.90		192.90	220
	8126	1-5/8"		13	.615		235	21.50		256.50	292
	8130	2-1/8"	▼	12	.667		355	23		378	425
	8134	2-5/8"	Q-5	22	.727		875	23		898	1,000
	8138	3-1/8"	"	20	.800	▼	1,075	25		1,100	1,250
	8500	Refrigeration valve, packless, soldered									
	8510	1/2"	1 Stpi	38	.211	Ea.	22	7.35		29.35	35
	8514	5/8"		36	.222		47	7.75		54.75	63
	8518	7/8"	▼	26	.308	▼	84	10.70		94.70	109
	8520	Packed, soldered									
	8522	1-1/8"	1 Stpi	18	.444	Ea.	300	15.50		315.50	355
	8526	1-3/8"		14	.571		172	19.90		191.90	219
	8530	1-5/8"		13	.615		213	21.50		234.50	267
	8534	2-1/8"	▼	12	.667		315	23		338	380
	8538	2-5/8"	Q-5	22	.727		470	23		493	550
	8542	3-1/8"		20	.800		625	25		650	725
	8546	4-1/8"	▼	18	.889	▼	895	28		923	1,025
	8600	Solenoid valve, flange/solder									
	8610	1/2"	1 Stpi	38	.211	Ea.	118	7.35		125.35	141
	8614	5/8"		36	.222		147	7.75		154.75	174
	8618	3/4"		30	.267		185	9.30		194.30	217
	8622	7/8"		26	.308		185	10.70		195.70	219
	8626	1-1/8"		18	.444		305	15.50		320.50	360
	8630	1-3/8"		14	.571		370	19.90		389.90	440
	8634	1-5/8"		13	.615		520	21.50		541.50	610
	8638	2-1/8"	▼	12	.667		555	23		578	645
	8800	Thermostatic valve, flange/solder									
	8810	1/2 - 3 ton, 3/8" x 5/8"	1 Stpi	9	.889	Ea.	78.50	31		109.50	134

15100 | Building Services Piping

15180 | Heating and Cooling Piping

			CREW	DAILY OUTPUT	LABOR-HOURS	UNIT	MAT.	LABOR	EQUIP.	TOTAL	TOTAL INCL O&P	
700	8814	4 - 5 ton, 1/2" x 7/8"	1 Stpi	7	1.143	Ea.	78.50	40		118.50	147	700
	8818	6 - 8 ton, 5/8" x 7/8"		5	1.600		78.50	56		134.50	171	
	8822	7 - 12 ton, 7/8" x 1-1/8"		4	2		95.50	69.50		165	210	
	8826	15 - 20 ton, 7/8" x 1-3/8"	↓	3.20	2.500	↓	187	87		274	340	
800	0010	**STEAM CONDENSATE METER**										800
	0100	500 lb. per hour	1 Stpi	14	.571	Ea.	2,300	19.90		2,319.90	2,550	
	0140	1500 lb. per hour		7	1.143		2,475	40		2,515	2,775	
	0160	3000 lb. per hour	↓	5	1.600		3,325	56		3,381	3,725	
	0200	12,000 lb. per hour	Q-5	3.50	4.571	↓	4,450	143		4,593	5,100	

15195 | Fuel Systems

			CREW	DAILY OUTPUT	LABOR-HOURS	UNIT	MAT.	LABOR	EQUIP.	TOTAL	TOTAL INCL O&P	
490	0010	**FUEL OIL SPECIALTIES**										490
	0020	Foot valve, single poppet, metal to metal construction										
	0040	Bevel seat, 1/2" diameter	1 Stpi	20	.400	Ea.	36.50	13.95		50.45	61.50	
	0060	3/4" diameter		18	.444		38.50	15.50		54	65.50	
	0080	1" diameter		16	.500		41	17.45		58.45	71.50	
	0100	1-1/4" diameter		15	.533		54	18.60		72.60	87.50	
	0120	1-1/2" diameter		13	.615		63	21.50		84.50	102	
	0140	2" diameter	↓	11	.727	↓	81.50	25.50		107	129	
	0160	Foot valve, double poppet, metal to metal construction										
	0164	1" diameter	1 Stpi	15	.533	Ea.	83.50	18.60		102.10	120	
	0166	1-1/2" diameter	"	12	.667	"	79	23		102	122	
	0400	Fuel fill box, flush type										
	0408	Nonlocking, watertight										
	0410	1-1/2" diameter	1 Stpi	12	.667	Ea.	12.85	23		35.85	49	
	0440	2" diameter		10	.800		13.80	28		41.80	57	
	0450	2-1/2" diameter		9	.889		46.50	31		77.50	98	
	0460	3" diameter		7	1.143		45	40		85	110	
	0470	4" diameter	↓	5	1.600	↓	61.50	56		117.50	152	
	0500	Locking inner cover										
	0510	2" diameter	1 Stpi	8	1	Ea.	67	35		102	127	
	0520	2-1/2" diameter		7	1.143		90	40		130	159	
	0530	3" diameter		5	1.600		96.50	56		152.50	190	
	0540	4" diameter	↓	4	2	↓	124	69.50		193.50	241	
	0600	Fuel system components										
	0620	Spill container	1 Stpi	4	2	Ea.	485	69.50		554.50	635	
	0640	Fill adapter, 4", straight drop		8	1		32	35		67	88	
	0680	Fill cap, 4"		30	.267		25.50	9.30		34.80	42	
	0700	Extractor fitting, 4" x 1-1/2"		8	1		325	35		360	410	
	0740	Vapor hose adapter, 4"	↓	8	1		69.50	35		104.50	129	
	0760	Wood gage stick, 10'					10.40			10.40	11.45	
	1000	Oil filters, 3/8" IPT., 20 gal. per hour	1 Stpi	20	.400		17.25	13.95		31.20	40	
	1020	32 gal. per hour		18	.444		32.50	15.50		48	59	
	1040	40 gal. per hour		16	.500		35.50	17.45		52.95	65.50	
	1060	50 gal. per hour	↓	14	.571	↓	38	19.90		57.90	72	
	1800	Pump and motor sets										
	1810	Light fuel and diesel oils, 100 PSI										
	1820	25 GPH 1/4 HP	Q-5	6	2.667	Ea.	765	83.50		848.50	965	
	1830	45 GPH, 1/4 HP		6	2.667		770	83.50		853.50	970	
	1840	90 GPH, 1/4 HP		5	3.200		830	100		930	1,050	
	1850	160 GPH, 1/3 HP		4	4		850	126		976	1,125	
	1860	325 GPH, 3/4 HP		4	4		930	126		1,056	1,225	
	1870	700 GPH, 1-1/2 HP		3	5.333		1,700	167		1,867	2,100	
	1880	1000 GPH, 2 HP		3	5.333		1,725	167		1,892	2,150	
	1890	1800 GPH, 5 HP	↓	1.80	8.889	↓	2,650	279		2,929	3,350	

15100 | Building Services Piping

15195 | Fuel Systems

		CREW	DAILY OUTPUT	LABOR-HOURS	UNIT	MAT.	LABOR	EQUIP.	TOTAL	TOTAL INCL O&P
2000	Remote tank gauging system, self contained									
2020	30' transmission line									
2100	30" pointer travel	1 Stpi	2	4	Ea.	1,750	139		1,889	2,125
2120	5" pointer travel	"	2.50	3.200		1,200	112		1,312	1,500
2500	For each additional 10' transmission line, add					43			43	47
3000	Valve, ball check, globe type, 3/8" diameter	1 Stpi	24	.333		6.30	11.60		17.90	24.50
3500	Fusible, 3/8" diameter		24	.333		5.15	11.60		16.75	23
3600	1/2" diameter		24	.333		13.40	11.60		25	32.50
3610	3/4" diameter		20	.400		26.50	13.95		40.45	50.50
3620	1" diameter		19	.421		81	14.65		95.65	111
4000	Nonfusible, 3/8" diameter		24	.333		9.55	11.60		21.15	28
4500	Shutoff, gate type, lever handle, spring-fusible kit									
4520	1/4" diameter	1 Stpi	14	.571	Ea.	15.15	19.90		35.05	46.50
4540	3/8" diameter		12	.667		14.05	23		37.05	50.50
4560	1/2" diameter		10	.800		19.95	28		47.95	64
4570	3/4" diameter		8	1		31.50	35		66.50	87.50
4580	Lever handle, requires weight and fusible kit									
4600	1" diameter	1 Stpi	9	.889	Ea.	290	31		321	365
4620	1-1/4" diameter		8	1		300	35		335	385
4640	1-1/2" diameter		7	1.143		410	40		450	510
4660	2" diameter		6	1.333		430	46.50		476.50	540
4680	For fusible link, weight and braided wire, add					5%				
5000	Vent alarm, whistling signal					26			26	28.50
5500	Vent protector/breather, 1-1/4" diameter	1 Stpi	32	.250		7.35	8.70		16.05	21
5520	1-1/2" diameter		32	.250		8.15	8.70		16.85	22
5540	2" diameter		32	.250		30	8.70		38.70	46
5560	3" diameter		28	.286		36.50	9.95		46.45	55.50
5580	4" diameter		24	.333		47.50	11.60		59.10	69.50
5600	Dust cap, breather, 2"		40	.200		9.65	6.95		16.60	21
8000	Fuel oil and tank heaters									
8020	Electric, capacity rated at 230 volts									
8040	Immersion element in steel manifold									
8060	96 GPH at 50°F rise	Q-5	6.40	2.500	Ea.	630	78.50		708.50	810
8070	128 GPH at 50°F rise		6.20	2.581		665	81		746	850
8080	160 GPH at 50°F rise		5.90	2.712		695	85		780	895
8090	192 GPH at 50°F rise		5.50	2.909		740	91.50		831.50	955
8100	240 GPH at 50°F rise		5.10	3.137		890	98.50		988.50	1,125
8110	288 GPH at 50°F rise		4.60	3.478		1,050	109		1,159	1,350
8120	384 GPH at 50°F rise		3.10	5.161		1,200	162		1,362	1,575
8130	480 GPH at 50°F rise		2.30	6.957		1,500	218		1,718	1,975
8140	576 GPH at 50°F rise		2.10	7.619		1,625	239		1,864	2,150
8300	Suction stub, immersion type									
8320	75" long, 750 watts	1 Stpi	14	.571	Ea.	530	19.90		549.90	615
8330	99" long, 2000 watts		12	.667		745	23		768	855
8340	123" long, 3000 watts		10	.800		805	28		833	925
8400	Electro/steam, electric capacity									
8420	rated at 230 volts/steam at 5 PSIG									
8440	28/50 GPH	Q-5	3.50	4.571	Ea.	1,900	143		2,043	2,300
8450	64/135 GPH		3	5.333		2,250	167		2,417	2,725
8460	80/257 GPH		2.50	6.400		2,975	201		3,176	3,575
8470	96/303 GPH		2.10	7.619		3,225	239		3,464	3,900
8480	120/384 GPH	Q-6	3	8		3,475	260		3,735	4,225
8490	144/720 GPH		1.80	13.333		5,875	435		6,310	7,100
8500	192/800 GPH		1.60	15		6,725	490		7,215	8,125
8510	240/1000 GPH		1.30	18.462		7,575	600		8,175	9,225
8660	Steam, cross flow, rated at 5 PSIG									

15100 | Building Services Piping

15195 | Fuel Systems

		CREW	DAILY OUTPUT	LABOR-HOURS	UNIT	MAT.	LABOR	EQUIP.	TOTAL	TOTAL INCL O&P
8680	42 GPH	Q-5	7	2.286	Ea.	1,225	71.50		1,296.50	1,450
8690	73 GPH		6.70	2.388		1,425	75		1,500	1,700
8700	112 GPH		6.20	2.581		1,475	81		1,556	1,750
8710	158 GPH		5.80	2.759		1,625	86.50		1,711.50	1,900
8720	187 GPH		4	4		2,225	126		2,351	2,650
8730	270 GPH		3.60	4.444		2,400	139		2,539	2,825
8740	365 GPH	Q-6	4.90	4.898		2,800	159		2,959	3,350
8750	635 GPH		3.70	6.486		3,550	211		3,761	4,225
8760	845 GPH		2.50	9.600		5,200	310		5,510	6,200
8770	1420 GPH		1.60	15		8,475	490		8,965	10,100
8780	2100 GPH		1.10	21.818		11,600	710		12,310	13,900
8860	Steam or water, safety type									
8880	Water capacity at 180°F/steam at 5 PSIG									
8900	70/84 GPH	Q-5	3.70	4.324	Ea.	1,500	136		1,636	1,875
8910	90/108 GPH		3.10	5.161		1,775	162		1,937	2,225
8920	120/144 GPH		2.70	5.926		2,075	186		2,261	2,575
8930	150/180 GPH		2.30	6.957		2,450	218		2,668	3,025
8940	180/216 GPH		1.80	8.889		3,050	279		3,329	3,800
8950	250/300 GPH		1.40	11.429		3,850	360		4,210	4,800
8960	320/384 GPH	Q-6	1.90	12.632		4,625	410		5,035	5,700
8970	400/449 GPH		1.50	16		5,650	520		6,170	7,000
8980	480/536 GPH		1.30	18.462		6,625	600		7,225	8,175
8990	600/670 GPH		1.10	21.818		8,400	710		9,110	10,300
9100	Suction bell type									
9120	160 GPH, 17" high	Q-5	8	2	Ea.	2,175	63		2,238	2,475
9140	240 GPH, 24" high	"	7	2.286	"	2,350	71.50		2,421.50	2,675
9200	Tank heating coil									
9240	73" high by 15-3/8" diameter	Q-5	5	3.200	Ea.	3,450	100		3,550	3,950
9280	Tank outlet heater (large capacity tank)									
9300	rated for steam at 5/125 PSIG									
9320	95/200 GPH	Q-5	4	4	Ea.	1,375	126		1,501	1,725
9330	190/400 GPH	"	2.70	5.926		1,950	186		2,136	2,425
9340	360/810 GPH	Q-6	3.20	7.500		2,750	244		2,994	3,400
9350	490/1070 GPH		2.60	9.231		3,425	300		3,725	4,225
9360	830/1850 GPH		2	12		4,375	390		4,765	5,425
9370	1000/2200 GPH		1.70	14.118		4,950	460		5,410	6,150
9380	1310/2920 GPH		1.30	18.462		6,725	600		7,325	8,300
9390	1500/3300 GPH		1.10	21.818		7,400	710		8,110	9,200
9400	2280/5050 GPH		.83	28.916		10,000	940		10,940	12,400
9410	2500/5550 GPH		.80	30		10,800	975		11,775	13,400
9420	3000/6660 GPH		.72	33.333		12,100	1,075		13,175	14,900
9430	4100/9150 GPH		.60	40		14,400	1,300		15,700	17,800
9440	5450/12,100 GPH		.50	48		17,700	1,550		19,250	21,900
9450	7000/15,500 GPH		.40	60		21,900	1,950		23,850	27,000
9460	8700/19,300 GPH		.33	72.727		26,400	2,375		28,775	32,700
9470	10,400/23,200 GPH		.29	82.759		29,300	2,700		32,000	36,300
0010	**METERS**									
4000	Residential									
4010	Gas meter, residential, 3/4" pipe size	1 Plum	14	.571	Ea.	115	19.70		134.70	157
4020	Gas meter, residential, 1" pipe size		12	.667		138	23		161	187
4030	Gas meter, residential, 1-1/4" pipe size		10	.800		138	27.50		165.50	194

15200 | Process Piping

15210 | Process Air/Gas Piping

			CREW	DAILY OUTPUT	LABOR-HOURS	UNIT	MAT.	LABOR	EQUIP.	TOTAL	TOTAL INCL O&P	
100	0010	**COMPRESSORS**										100
	0100	Refrigeration, hermetic, switches and protective devices										
	0200	High back pressure types for R-22										
	0210	1.25 ton	1 Stpi	3	2.667	Ea.	340	93		433	510	
	0220	1.42 ton		3	2.667		340	93		433	515	
	0230	1.68 ton		2.80	2.857		370	99.50		469.50	560	
	0240	2.00 ton		2.60	3.077		460	107		567	665	
	0250	2.37 ton		2.50	3.200		555	112		667	780	
	0260	2.67 ton		2.40	3.333		565	116		681	795	
	0270	3.53 ton		2.30	3.478		585	121		706	825	
	0280	4.43 ton		2.20	3.636		605	127		732	855	
	0290	5.08 ton		2.10	3.810		720	133		853	995	
	0300	5.22 ton		2	4		715	139		854	1,000	
	0310	7.31 ton	Q-5	3	5.333		1,325	167		1,492	1,700	
	0320	9.95 ton	"	2.60	6.154		1,450	193		1,643	1,900	
	0990	Refrigeration, recip. hermetic, switches & protective devices										
	1000	(Ratings are ARI standard 515 group IV using R-22) 2 ton	Q-5	1	16	Ea.	2,650	500		3,150	3,650	
	1100	15 ton	Q-6	.72	33.333		6,275	1,075		7,350	8,525	
	1200	20 ton		.64	37.500		7,450	1,225		8,675	10,100	
	1300	40 ton		.44	54.545		8,225	1,775		10,000	11,700	
	1400	50 ton		.20	120		19,200	3,900		23,100	27,100	
	1500	75 ton	Q-7	.27	118		22,500	3,950		26,450	30,800	
	1600	130 ton	"	.21	152		28,100	5,075		33,175	38,600	
	5250	Air, reciprocating air cooled, splash lubricated, tank mounted										
	5300	Single stage, 1 phase, 140 psi										
	5303	1/2 HP, 30 gal tank	1 Stpi	3	2.667	Ea.	1,050	93		1,143	1,325	
	5305	3/4 HP, 30 gal tank		2.60	3.077		1,075	107		1,182	1,350	
	5307	1 HP, 30 gal tank		2.20	3.636		1,375	127		1,502	1,700	
	5309	2 HP, 30 gal tank	Q-5	4	4		1,750	126		1,876	2,125	
	5310	3 HP, 30 gal vertical tank		3.60	4.444		2,125	139		2,264	2,550	
	5314	3 HP, 60 gal tank		3.50	4.571		2,825	143		2,968	3,325	
	5320	5 HP, 60 gal vertical tank		3.20	5		2,900	157		3,057	3,400	
	5330	5 HP, 80 gal vertical tank		3	5.333		3,050	167		3,217	3,600	
	5340	7.5 HP, 80 gal tank		2.60	6.154		5,225	193		5,418	6,050	
	5600	2 stage pkg., 3 phase										
	5650	6 CFM at 125 psi 1-1/2 HP, 60 gal tank	Q-5	3	5.333	Ea.	2,275	167		2,442	2,750	
	5670	10.9 CFM at 125 psi, 3 HP, 80 gal tank		1.50	10.667		2,725	335		3,060	3,500	
	5680	38.7 CFM at 125 psi, 10 HP, 120 gal tank		.60	26.667		4,900	835		5,735	6,650	
	5690	105 CFM at 125 psi, 25 HP, 250 gal tank	Q-6	.60	40		9,425	1,300		10,725	12,400	
	5800	With single stage pump										
	5850	8.3 CFM at 125 psi, 2 HP, 80 gal tank	Q-6	3.50	6.857	Ea.	2,575	223		2,798	3,150	
	5860	38.7 CFM at 125 psi, 10 HP, 120 gal tank	"	.90	26.667	"	5,425	865		6,290	7,275	
	6000	Reciprocating, 2 stage, tank mtd, 3 Ph., Cap. rated @175 PSIG										
	6050	Pressure lubcatd, hvy duty, 9.7 CFM, 3 HP, 120 Gal. tank	Q-5	1.30	12.308	Ea.	4,150	385		4,535	5,125	
	6054	5 CFM, 1-1/2 HP, 80 gal tank		2.80	5.714		2,425	179		2,604	2,950	
	6056	6.4 CFM, 2 HP, 80 gal tank		2	8		2,575	251		2,826	3,200	
	6058	8.1 CFM, 3 HP, 80 gal tank		1.70	9.412		2,675	295		2,970	3,400	
	6059	14.8 CFM, 5 HP, 80 gal tank		1	16		2,750	500		3,250	3,800	
	6060	16.5 CFM, 5 HP, 120 gal. tank		1	16		4,200	500		4,700	5,375	
	6063	13 CFM, 6 HP, 80 gal tank		.90	17.778		4,425	560		4,985	5,725	
	6066	19.8 CFM, 7.5 HP, 80 gal tank		.80	20		4,675	630		5,305	6,075	
	6070	25.8 CFM, 7-1/2 HP, 120 gal. tank		.80	20		6,250	630		6,880	7,825	
	6078	34.8 CFM, 10 HP, 80 gal. tank		.70	22.857		5,925	715		6,640	7,575	
	6080	34.8 CFM, 10 HP, 120 gal. tank		.60	26.667		6,275	835		7,110	8,175	
	6090	53.7 CFM, 15 HP, 120 gal. tank	Q-6	.80	30		7,800	975		8,775	10,100	
	6100	76.7 CFM, 20 HP, 120 gal. tank		.70	34.286		9,650	1,125		10,775	12,300	

15200 | Process Piping

15210 | Process Air/Gas Piping

		CREW	DAILY OUTPUT	LABOR-HOURS	UNIT	MAT.	LABOR	EQUIP.	TOTAL	TOTAL INCL O&P	
100	6104	76.7 CFM, 20 HP, 240 gal tank	Q-6	.68	35.294	Ea.	9,450	1,150		10,600	12,100	100
	6110	90.1 CFM, 25 HP, 120 gal. tank		.63	38.095		10,100	1,250		11,350	13,000	
	6120	101 CFM, 30 HP, 120 gal. tank		.57	42.105		10,900	1,375		12,275	14,100	
	6130	101 CFM, 30 HP, 250 gal. tank		.52	46.154		11,800	1,500		13,300	15,200	
	6200	Oil-less, 13.6 CFM, 5 HP, 120 gal. tank	Q-5	.88	18.182		12,400	570		12,970	14,500	
	6210	13.6 CFM, 5 HP, 250 gal. tank		.80	20		13,300	630		13,930	15,600	
	6220	18.2 CFM, 7.5 HP, 120 gal. tank		.73	21.918		12,600	690		13,290	14,900	
	6230	18.2 CFM, 7.5 HP, 250 gal. tank		.67	23.881		13,400	750		14,150	15,900	
	6250	30.5 CFM, 10 HP, 120 gal. tank		.57	28.070		14,700	880		15,580	17,400	
	6260	30.5 CFM, 10 HP, 250 gal. tank		.53	30.189		15,500	945		16,445	18,500	
	6270	41.3 CFM, 15 HP, 120 gal. tank	Q-6	.70	34.286		15,900	1,125		17,025	19,200	
	6280	41.3 CFM, 15 HP, 250 gal. tank	"	.67	35.821		16,700	1,175		17,875	20,200	
200	0010	**COMPRESSOR ACCESSORIES**										200
	0100	Aftercooler										
	0130	Air cooled, based on 100 psig and 250°F										
	0140	Tank mounted single stage, 20 SCFM	Q-5	7.50	2.133	Ea.	625	67		692	790	
	0150	46 SCFM		7	2.286		730	71.50		801.50	910	
	0160	77 SCFM		6.50	2.462		1,225	77		1,302	1,475	
	0170	Floor mounted horizontal draft, 36 SCFM		6.50	2.462		645	77		722	825	
	0180	78 SCFM		6	2.667		800	83.50		883.50	1,000	
	0190	190 SCFM		5.70	2.807		1,225	88		1,313	1,475	
	0200	260 SCFM		5.20	3.077		1,375	96.50		1,471.50	1,675	
	0210	440 SCFM		4	4		1,850	126		1,976	2,225	
	0220	650 SCFM		3.50	4.571		2,000	143		2,143	2,425	
	0230	Floor mounted vertical draft, 850 SCFM		2.50	6.400		3,450	201		3,651	4,100	
	0260	2060 SCFM		2	8		5,675	251		5,926	6,625	
	0280	3640 SCFM		1.70	9.412		13,900	295		14,195	15,700	
	0300	Separator with automatic float trap										
	0330	600 lb/hr drain cap. 125 psi 3/4" conn.	Q-5	7.50	2.133	Ea.	315	67		382	445	
	0340	1200 lb/hr drain cap. 125 psi 1" conn.		7	2.286		330	71.50		401.50	475	
	0350	1200 lb/hr drain cap. 125 psi 1-1/2" conn.		6	2.667		375	83.50		458.50	540	
	0360	1200 lb/hr drain cap. 125 psi 2" conn.		5	3.200		405	100		505	595	
	0370	1500 CFM at 200 psi 2-1/2" conn.		4	4		960	126		1,086	1,250	
	0380	4000 CFM at 200 psi 4" conn.		3	5.333		1,050	167		1,217	1,400	
	0400	Water cooled, fixed tube at 250°F										
	0430	52 CFM	Q-5	5	3.200	Ea.	405	100		505	595	
	0440	104 CFM		4	4		470	126		596	710	
	0450	182 CFM		3.50	4.571		730	143		873	1,025	
	0460	312 CFM		3	5.333		815	167		982	1,150	
	0480	494 CFM		2.80	5.714		1,700	179		1,879	2,150	
	0490	702 CFM		2.40	6.667		2,275	209		2,484	2,825	
	0500	806 CFM		2.20	7.273		2,675	228		2,903	3,300	
	0510	1092 CFM		2	8		3,325	251		3,576	4,050	
	0520	1430 CFM		1.50	10.667		3,925	335		4,260	4,800	
	0600	Separator, horizontal										
	0610	100 CFM	Q-5	15	1.067	Ea.	221	33.50		254.50	294	
	0620	350 CFM		10	1.600		320	50		370	425	
	0630	900 CFM		7	2.286		775	71.50		846.50	965	
	0640	1500 CFM		5	3.200		860	100		960	1,100	
	0650	2800 CFM		3.50	4.571		1,275	143		1,418	1,625	
	0700	Dryers, air, deliquescent at 90 psi										
	0730	21 CFM	Q-5	14	1.143	Ea.	655	36		691	780	
	0740	63 CFM		12	1.333		1,200	42		1,242	1,400	
	0750	140 CFM		8	2		2,150	63		2,213	2,450	
	0760	335 CFM		5.50	2.909		2,900	91.50		2,991.50	3,325	
	0770	525 CFM		5	3.200		3,800	100		3,900	4,325	

15200 | Process Piping

15210 | Process Air/Gas Piping

		CREW	DAILY OUTPUT	LABOR-HOURS	UNIT	MAT.	LABOR	EQUIP.	TOTAL	TOTAL INCL O&P
0780	760 CFM	Q-5	3	5.333	Ea.	5,300	167		5,467	6,075
0790	1050 CFM		2.50	6.400		6,575	201		6,776	7,550
0800	1340 CFM		2	8		7,750	251		8,001	8,900
0810	1700 CFM		1.50	10.667		9,850	335		10,185	11,300
0820	2550 CFM	Q-6	2	12		12,800	390		13,190	14,700
0830	3550 CFM		1.50	16		16,100	520		16,620	18,500
0840	4700 CFM		1	24		20,000	780		20,780	23,200
0850	6100 CFM	Q-7	1	32		23,400	1,075		24,475	27,300
0860	8400 CFM		.80	40		29,000	1,325		30,325	33,900
0870	10200 CFM		.60	53.333		31,500	1,775		33,275	37,400
0880	12100 CFM		.50	64		33,800	2,125		35,925	40,300
0890	14200 CFM		.40	80		35,600	2,650		38,250	43,200
0950	Desiccant 33 lb. bag					30.50			30.50	33.50
0960	55 lb. bag					50			50	55
0980	2200 lb. bulk bag 1 or 2					2,000			2,000	2,200
0990	2200 lb. bulk bag 3 or 4					1,950			1,950	2,150
1250	Refrigeration, 50°F pressure dewpoint									
1280	20 CFM	Q-5	7	2.286	Ea.	1,250	71.50		1,321.50	1,475
1290	60 CFM		6	2.667		1,775	83.50		1,858.50	2,075
1300	130 CFM		5.50	2.909		2,300	91.50		2,391.50	2,675
1310	150 CFM		5	3.200		4,825	100		4,925	5,450
1320	200 CFM		4.50	3.556		4,950	112		5,062	5,625
1330	280 CFM		4	4		5,425	126		5,551	6,150
1340	350 CFM		3.50	4.571		5,900	143		6,043	6,700
1350	450 CFM		3	5.333		6,775	167		6,942	7,700
1360	520 CFM		2.80	5.714		7,475	179		7,654	8,500
1370	650 CFM		2.40	6.667		8,225	209		8,434	9,375
1380	800 CFM		2.20	7.273		8,775	228		9,003	10,000
1390	1000 CFM		2	8		11,300	251		11,551	12,800
1400	1400 CFM		1.50	10.667		13,300	335		13,635	15,100
1410	2000 CFM		1.20	13.333		16,200	420		16,620	18,400
1420	2400 CFM		1	16		23,400	500		23,900	26,500
1430	3000 CFM	Q-6	1.30	18.462		31,700	600		32,300	35,800
1440	3600 CFM		1.20	20		35,700	650		36,350	40,300
1450	4800 CFM		1	24		41,800	780		42,580	47,100
1460	6000 CFM		.80	30		51,500	975		52,475	58,000
1470	7200 CFM		.50	48		60,000	1,550		61,550	68,500
1500	Package includes dryer & filtration systems									
1530	150 CFM	Q-5	4.20	3.810	Ea.	7,050	120		7,170	7,925
1540	200 CFM		3.90	4.103		7,200	129		7,329	8,125
1550	280 CFM		3.50	4.571		8,250	143		8,393	9,300
1560	350 CFM		3.20	5		9,275	157		9,432	10,400
1570	450 CFM		3	5.333		10,300	167		10,467	11,600
1580	520 CFM		2.50	6.400		10,900	201		11,101	12,300
1590	650 CFM		2.20	7.273		11,300	228		11,528	12,800
1600	800 CFM		2	8		12,000	251		12,251	13,600
1610	1000 CFM		1.80	8.889		15,400	279		15,679	17,300
1620	1200 CFM		1.40	11.429		17,300	360		17,660	19,500
1630	1400 CFM		1.20	13.333		18,900	420		19,320	21,300
1640	2000 CFM		1	16		21,700	500		22,200	24,700
1650	2400 CFM		.70	22.857		28,000	715		28,715	31,900
2000	Regenerative adsorption with desiccant									
2030	50 CFM	Q-5	5	3.200	Ea.	4,250	100		4,350	4,825
2040	75 CFM		4.50	3.556		4,525	112		4,637	5,175
2050	100 CFM		4.20	3.810		4,975	120		5,095	5,650
2060	125 CFM		4	4		5,400	126		5,526	6,150

15200 | Process Piping

15210 | Process Air/Gas Piping

		CREW	DAILY OUTPUT	LABOR-HOURS	UNIT	MAT.	LABOR	EQUIP.	TOTAL	TOTAL INCL O&P
2070	170 CFM	Q-5	3.70	4.324	Ea.	5,850	136		5,986	6,625
2080	240 CFM		3.40	4.706		7,100	148		7,248	8,025
2090	350 CFM		3	5.333		7,600	167		7,767	8,625
2100	520 CFM		2.50	6.400		9,100	201		9,301	10,300
2110	650 CFM		2	8		10,400	251		10,651	11,900
2120	800 CFM		1.50	10.667		13,100	335		13,435	14,900
2130	1000 CFM		1.20	13.333		15,200	420		15,620	17,400
2140	1200 CFM		1	16		16,000	500		16,500	18,400
2150	1600 CFM		.80	20		20,000	630		20,630	23,000
2160	2000 CFM		.70	22.857		22,800	715		23,515	26,200
2170	2500 CFM		.60	26.667		28,200	835		29,035	32,300
2180	3000 CFM	Q-6	.90	26.667		29,500	865		30,365	33,700
2190	3500 CFM		.70	34.286		42,700	1,125		43,825	48,600
2200	4000 CFM		.50	48		43,900	1,550		45,450	50,500
2300	Desiccant, 50 lb. bag					122			122	134
2330	350 lb. drum					845			845	930
3000	Filters Air									
3100	Adsorptive, odors by glass fiber & carbon, 90 psi									
3140	110 CFM	Q-5	8	2	Ea.	835	63		898	1,025
3150	190 CFM		7.50	2.133		950	67		1,017	1,150
3160	290 CFM		7	2.286		1,175	71.50		1,246.50	1,400
3170	380 CFM		6	2.667		1,600	83.50		1,683.50	1,875
3180	570 CFM		5	3.200		2,250	100		2,350	2,625
3190	870 CFM		4.50	3.556		2,550	112		2,662	3,000
3200	1160 CFM		4	4		3,275	126		3,401	3,800
3210	1450 CFM		3.50	4.571		3,850	143		3,993	4,450
3220	1740 CFM		3	5.333		4,300	167		4,467	4,975
3260	Coalescent system, liquids & solids 90 psi high purity									
3300	110 CFM	Q-5	7	2.286	Ea.	3,125	71.50		3,196.50	3,525
3310	190 CFM		6.70	2.388		3,400	75		3,475	3,850
3320	290 CFM		6.50	2.462		3,575	77		3,652	4,075
3330	380 CFM		6	2.667		4,200	83.50		4,283.50	4,750
3340	570 CFM		4.50	3.556		5,275	112		5,387	5,975
3350	870 CFM		4	4		5,575	126		5,701	6,325
3360	1160 CFM		3.50	4.571		6,350	143		6,493	7,200
3370	1450 CFM		3	5.333		7,150	167		7,317	8,100
3380	1740 CFM		2.50	6.400		7,550	201		7,751	8,625
3390	2030 CFM		2	8		7,825	251		8,076	8,975
3500	Combination air line filter, regulator and lubricator									
3510	1/4" NPT	1 Stpi	16	.500	Ea.	105	17.45		122.45	142
3520	3/8" NPT		16	.500		117	17.45		134.45	155
3530	1/2" NPT		15	.533		117	18.60		135.60	156
3540	3/4" NPT		14	.571		235	19.90		254.90	289
3550	1" NPT		12	.667		235	23		258	294
4000	Couplers, air line, sleeve type									
4010	Female, connection size NPT									
4020	1/4"	1 Stpi	38	.211	Ea.	3.92	7.35		11.27	15.35
4030	3/8"		36	.222		4.89	7.75		12.64	17.10
4040	1/2"		35	.229		10.25	7.95		18.20	23.50
4050	3/4"		34	.235		11.75	8.20		19.95	25.50
4100	Male									
4110	1/4"	1 Stpi	38	.211	Ea.	3.88	7.35		11.23	15.30
4120	3/8"		36	.222		4.89	7.75		12.64	17.10
4130	1/2"		35	.229		9.70	7.95		17.65	22.50
4140	3/4"		34	.235		11.45	8.20		19.65	25
4150	Coupler, combined male and female halves									

15200 | Process Piping

15210 | Process Air/Gas Piping

		CREW	DAILY OUTPUT	LABOR-HOURS	UNIT	2001 BARE COSTS MAT.	LABOR	EQUIP.	TOTAL	TOTAL INCL O&P	
200	4160	1/2"	1 Stpi	17	.471	Ea.	19.95	16.40		36.35	47	200
	4170	3/4"	"	15	.533	"	23	18.60		41.60	53.50	
900	0010	**VACUUM PUMPS**										900
	0100	Medical, with receiver										
	0110	Duplex										
	0120	20 SCFM	Q-1	1.14	14.035	Ea.	14,100	435		14,535	16,200	
	0130	60 SCFM, 10 HP	Q-2	1.20	20	"	24,000	645		24,645	27,300	
	0200	Triplex										
	0220	180 SCFM	Q-2	.86	27.907	Ea.	46,400	895		47,295	52,500	
	0300	Dental oral										
	0310	Duplex										
	0330	165 SCFM with 77 Gal separator	Q-2	1.30	18.462	Ea.	42,000	595		42,595	47,100	

15230 | Industrial Process Piping

			CREW	DAILY OUTPUT	LABOR-HOURS	UNIT	MAT.	LABOR	EQUIP.	TOTAL	TOTAL INCL O&P	
500	0010	**PUMPS, GENERAL UTILITY** With motor										500
	0200	Multi-stage, horizontal split, for boiler feed applications										
	0300	Two stage, 3" discharge x 4" suction, 75 HP	Q-7	.30	106	Ea.	14,900	3,550		18,450	21,800	
	0340	Four stage, 3" discharge x 4" suction, 150 HP	"	.18	177	"	24,500	5,900		30,400	35,900	
	2000	Single stage										
	2060	End suction, 1"D. x 2"S., 3 HP	Q-1	.50	32	Ea.	3,650	990		4,640	5,500	
	2100	1-1/2"D. x 3"S., 10 HP	"	.40	40		3,925	1,250		5,175	6,200	
	2140	2"D. x 3"S., 15 HP	Q-2	.60	40		4,275	1,275		5,550	6,650	
	2180	3"D. x 4"S., 20 HP		.50	48		4,550	1,550		6,100	7,325	
	2220	4"D. x 6"S., 30 HP		.40	60		4,925	1,925		6,850	8,325	
	3000	Double suction, 2"D. x 2-1/2"S., 10 HP	Q-1	.30	53.333		5,050	1,650		6,700	8,050	
	3060	3"D. x 4"S., 15 HP	Q-2	.46	52.174		5,200	1,675		6,875	8,250	
	3100	4"D. x 5"S., 30 HP		.40	60		7,050	1,925		8,975	10,700	
	3140	5"D. x 6"S., 50 HP		.33	72.727		7,575	2,350		9,925	11,900	
	3180	6"D. x 8"S., 60 HP	Q-3	.30	106		8,475	3,500		11,975	14,600	
	3190	75 HP, to 2500 GPM		.28	114		10,500	3,750		14,250	17,200	
	3220	100 HP, to 3000 GPM		.26	123		13,300	4,050		17,350	20,800	
	3240	150 HP, to 4000 GPM		.24	133		17,900	4,375		22,275	26,300	
	4000	Centrifugal, end suction, mounted on base										
	4010	Horizontal mounted, with drip proof motor, rated @ 100' head										
	4020	Vertical split case, single stage										
	4040	100 GPM, 5 HP, 1-1/2" discharge	Q-1	1.70	9.412	Ea.	1,950	292		2,242	2,600	
	4050	200 GPM, 10 HP, 2" discharge		1.30	12.308		2,325	380		2,705	3,125	
	4060	250 GPM, 10 HP, 3" discharge		1.28	12.500		2,425	390		2,815	3,250	
	4070	300 GPM, 15 HP, 2" discharge	Q-2	1.56	15.385		2,625	495		3,120	3,650	
	4080	500 GPM, 20 HP, 4" discharge		1.44	16.667		3,100	535		3,635	4,200	
	4090	750 GPM, 30 HP, 4" discharge		1.20	20		4,150	645		4,795	5,550	
	4100	1050 GPM, 40 HP, 5" discharge		1	24		4,475	770		5,245	6,100	
	4110	1500 GPM, 60 HP, 6" discharge		.60	40		6,525	1,275		7,800	9,125	
	4120	2000 GPM, 75 HP, 6" discharge		.50	48		7,500	1,550		9,050	10,600	
	4130	3000 GPM, 100 HP, 8" discharge		.40	60		11,300	1,925		13,225	15,300	
	4200	Horizontal split case, single stage										
	4210	100 GPM, 7.5 HP, 1-1/2" discharge	Q-1	1.70	9.412	Ea.	3,200	292		3,492	3,950	
	4220	250 GPM, 15 HP, 2-1/2" discharge	"	1.30	12.308		4,175	380		4,555	5,175	
	4230	500 GPM, 20 HP, 4" discharge	Q-2	1.60	15		5,050	480		5,530	6,275	
	4240	750 GPM, 25 HP, 5" discharge		1.54	15.584		5,225	500		5,725	6,500	
	4250	1000 GPM, 40 HP, 5" discharge		1.20	20		6,825	645		7,470	8,500	
	4260	1500 GPM, 50 HP, 6" discharge	Q-3	1.42	22.535		8,600	740		9,340	10,600	
	4270	2000 GPM, 75 HP, 8" discharge		1.14	28.070		11,700	920		12,620	14,200	
	4280	3000 GPM, 100 HP, 10" discharge		.96	33.333		13,000	1,100		14,100	16,000	
	4290	3500 GPM, 150 HP, 10" discharge		.86	37.209		16,300	1,225		17,525	19,800	
	4300	4000 GPM, 200 HP, 10" discharge		.66	48.485		17,900	1,600		19,500	22,100	

15200 | Process Piping

15230 | Industrial Process Piping

		CREW	DAILY OUTPUT	LABOR-HOURS	UNIT	2001 BARE COSTS MAT.	LABOR	EQUIP.	TOTAL	TOTAL INCL O&P
4330	Horizontal split case, two stage, 500' head									
4340	100 GPM, 40 HP, 1-1/2" discharge	Q-2	1.70	14.118	Ea.	4,150	455		4,605	5,225
4350	200 GPM, 50 HP, 1-1/2" discharge	"	1.44	16.667		4,600	535		5,135	5,875
4360	300 GPM, 75 HP, 2" discharge	Q-3	1.57	20.382		4,900	670		5,570	6,400
4370	400 GPM, 100 HP, 3" discharge		1.14	28.070		6,375	920		7,295	8,425
4380	800 GPM, 200 HP, 4" discharge	↓	.86	37.209	↓	15,000	1,225		16,225	18,400
5000	Centrifugal, in-line									
5006	Vertical mount, iron body, 125 lb. flgd, 3550 RPM TEFC mtr									
5010	Single stage									
5020	50 GPM, 3 HP, 1-1/2" discharge	Q-1	2.30	6.957	Ea.	510	216		726	885
5030	75 GPM, 5 HP, 1-1/2" discharge		1.60	10		865	310		1,175	1,425
5040	100 GPM, 7.5 HP, 1-1/2" discharge	↓	1.30	12.308		930	380		1,310	1,600
5050	125 GPM, 10 HP, 1-1/2" discharge	Q-2	1.70	14.118		1,150	455		1,605	1,950
5060	150 GPM, 15 HP, 1-1/2" discharge		1.60	15		1,475	480		1,955	2,350
5070	200 GPM, 30 HP, 1-1/2" discharge		1.50	16		2,425	515		2,940	3,425
5080	250 GPM, 40 HP, 2" discharge		1.40	17.143		3,200	550		3,750	4,350
5090	300 GPM, 50 HP, 3" discharge	↓	1.30	18.462		3,875	595		4,470	5,175
5100	400 GPM, 75 HP, 3" discharge	Q-3	.60	53.333		6,325	1,750		8,075	9,600
5110	600 GPM, 100 HP, 3" discharge	"	.50	64	↓	8,475	2,100		10,575	12,500
8000	Vertical submerged, with non-submerged motor									
8060	1"D., 3 HP	Q-1	.50	32	Ea.	5,275	990		6,265	7,300
8100	1-1/2"D., 10 HP	"	.30	53.333		5,525	1,650		7,175	8,575
8140	2"D., 15 HP	Q-2	.40	60		5,675	1,925		7,600	9,150
8180	3"D., 25 HP		.30	80		5,800	2,575		8,375	10,300
8220	4"D., 30 HP	↓	.20	120	↓	6,500	3,850		10,350	13,000
600 0010	**PUMPS MISCELLANEOUS**									
0020	Water pump, portable, gasoline powered									
0100	6000 GPH, 2" discharge	Q-1	11	1.455	Ea.	785	45		830	935
0110	8000 GPH, 2" discharge		10.50	1.524		785	47		832	935
0120	10,000 GPH, 2" discharge	↓	10	1.600	↓	1,275	49.50		1,324.50	1,475
0500	Pump, propylene body, housing and impeller									
0510	22 GPM, 1/3 HP, 40' HD	1 Plum	5	1.600	Ea.	485	55		540	620
0520	33 GPM, 1/2 HP, 40' HD	Q-1	5	3.200		495	99		594	695
0530	53 GPM, 3/4 HP, 40' HD	"	4	4	↓	530	124		654	765
0600	Rotary pump, CI									
0610	10 GPM, 1/2 HP, 1" discharge	Q-1	5	3.200	Ea.	910	99		1,009	1,150
0614	10 GPM, 3/4 HP, 1" discharge		4.50	3.556		910	110		1,020	1,175
0618	10 GPM 1 HP, 1" discharge		4	4		945	124		1,069	1,225
0620	25 GPM, 1 HP, 1-1/4" discharge		3.80	4.211		945	131		1,076	1,250
0624	25 GPM, 1.5 HP, 1-1/4" discharge		3.60	4.444		1,025	138		1,163	1,325
0628	25 GPM, 2 HP, 1-1/4" discharge		3.20	5	↓	1,025	155		1,180	1,350
0700	Fuel oil pump, 2 stage, 3450 RPM									
0710	1/4 HP, 100 PSI	Q-1	8	2	Ea.	1,550	62		1,612	1,800
0800	Hydraulic oil pump									
0810	75 HP, 3 PH, 480 V	Q-3	1.20	26.667	Ea.	5,075	875		5,950	6,900
1000	Turbine pump, CI									
1010	50 GPM, 2 HP, 3" discharge	Q-1	.80	20	Ea.	1,050	620		1,670	2,075
1020	100 GPM, 3 HP, 4" discharge	Q-2	.96	25		1,575	805		2,380	2,950
1030	250 GPM, 15 HP, 6" discharge	"	.94	25.532		3,150	820		3,970	4,725
1040	500 GPM, 25 HP, 6" discharge	Q-3	1.30	24.615		5,250	810		6,060	7,000
1050	1000 GPM, 50 HP, 8" discharge		1.14	28.070		8,400	920		9,320	10,700
1060	2000 GPM, 100 HP, 10" discharge		1	32		9,450	1,050		10,500	12,000
1070	3000 GPM, 150 HP, 10" discharge		.80	40		10,500	1,325		11,825	13,600
1080	4000 GPM, 200 HP, 12" discharge		.70	45.714		15,800	1,500		17,300	19,600
1090	6000 GPM, 300 HP, 14" discharge	↓	.60	53.333	↓	17,900	1,750		19,650	22,400

15200 | Process Piping

15230	Industrial Process Piping	CREW	DAILY OUTPUT	LABOR-HOURS	UNIT	2001 BARE COSTS MAT.	LABOR	EQUIP.	TOTAL	TOTAL INCL O&P
600 1100	10,000 GPM, 300 HP, 18" discharge	Q-3	.58	55.172	Ea.	19,500	1,800		21,300	24,100

15400 | Plumbing Fixtures & Equipment

15410	Plumbing Fixtures	CREW	DAILY OUTPUT	LABOR-HOURS	UNIT	2001 BARE COSTS MAT.	LABOR	EQUIP.	TOTAL	TOTAL INCL O&P
300 0010	**FAUCETS/FITTINGS**									
5000	Sillcock, compact, brass, IPS or copper to hose	1 Plum	24	.333	Ea.	4.94	11.50		16.44	23
6000	Stop and waste valves, bronze									
6100	Angle, solder end 1/2"	1 Plum	24	.333	Ea.	4.18	11.50		15.68	22
6110	3/4"		20	.400		4.59	13.80		18.39	26
6300	Straightway, solder end 3/8"		24	.333		3.59	11.50		15.09	21.50
6310	1/2"		24	.333		2.32	11.50		13.82	19.90
6320	3/4"		20	.400		2.68	13.80		16.48	24
6330	1"		19	.421		4	14.50		18.50	26.50
6400	Straightway, threaded 3/8"		24	.333		4.07	11.50		15.57	22
6410	1/2"		24	.333		3.13	11.50		14.63	21
6420	3/4"		20	.400		3.73	13.80		17.53	25
6430	1"		19	.421		10.50	14.50		25	33.50

15411	Commercial/Indust Fixtures									
500 0010	**HYDRANTS**									
0050	Wall type, moderate climate, bronze, encased									
0200	3/4" IPS connection	1 Plum	16	.500	Ea.	226	17.25		243.25	275
0300	1" IPS connection	"	14	.571		226	19.70		245.70	279
0500	Anti-siphon type					30.50			30.50	33.50
1000	Non-freeze, bronze, exposed									
1100	3/4" IPS connection, 4" to 9" thick wall	1 Plum	14	.571	Ea.	153	19.70		172.70	199
1120	10" to 14" thick wall		12	.667		166	23		189	218
1140	15" to 19" thick wall		12	.667		185	23		208	239
1160	20" to 24" thick wall		10	.800		201	27.50		228.50	263
1200	For 1" IPS connection, add					15%	10%			
1240	For 3/4" adapter type vacuum breaker, add				Ea.	19.25			19.25	21
1280	For anti-siphon type				"	40.50			40.50	44.50
2000	Non-freeze bronze, encased									
2100	3/4" IPS connection, 5" to 9" thick wall	1 Plum	14	.571	Ea.	264	19.70		283.70	320
2120	10" to 14" thick wall		12	.667		277	23		300	340
2140	15" to 19" thick wall		12	.667		297	23		320	360
2160	20" to 24" thick wall		10	.800		310	27.50		337.50	380
2200	For 1" IPS connection, add y					10%	10%			
2280	Anti-siphon type				Ea.	124			124	136
3000	Ground box type, bronze frame, 3/4" IPS connection									
3080	Non-freeze, all bronze, polished face, set flush									
3100	2 feet depth of bur	1 Plum	8	1	Ea.	282	34.50		316.50	360
3120	3 feet depth of bury		8	1		305	34.50		339.50	385
3140	4 feet depth of bury		8	1		325	34.50		359.50	405
3160	5 feet depth of bury		7	1.143		350	39.50		389.50	440
3180	6 feet depth of bury		7	1.143		370	39.50		409.50	465
3200	7 feet depth of bury		6	1.333		390	46		436	500

15400 | Plumbing Fixtures & Equipment

15411 | Commercial/Indust Fixtures

		CREW	DAILY OUTPUT	LABOR-HOURS	UNIT	2001 BARE COSTS MAT.	LABOR	EQUIP.	TOTAL	TOTAL INCL O&P
3220	8 feet depth of bury	1 Plum	5	1.600	Ea.	415	55		470	540
3240	9 feet depth of bury		4	2		435	69		504	585
3260	10 feet depth of bury	↓	4	2		455	69		524	605
3400	For 1" IPS connection, add					15%	10%			
3450	For 1-1/4" IPS connection, add					325%	14%			
3500	For 1-1/2" connection, add					370%	18%			
3550	For 2" connection, add					445%	24%			
3600	For tapped drain port in box, add				↓	25.50			25.50	28.50
4000	Non-freeze, CI body, bronze frame & scoriated cover									
4010	with hose storage									
4100	2 feet depth of bury	1 Plum	7	1.143	Ea.	535	39.50		574.50	645
4120	3 feet depth of bury		7	1.143		555	39.50		594.50	670
4140	4 feet depth of bury		7	1.143		575	39.50		614.50	690
4160	5 feet depth of bury		6.50	1.231		585	42.50		627.50	710
4180	6 feet depth of bury		6	1.333		595	46		641	725
4200	7 feet depth of bury		5.50	1.455		615	50		665	755
4220	8 feet depth of bury		5	1.600		640	55		695	785
4240	9 feet depth of bury		4.50	1.778		660	61		721	820
4260	10 feet depth of bury	↓	4	2		680	69		749	855
4280	For 1" IPS connection, add					117			117	129
4300	For tapped drain port in box, add				↓	25.50			25.50	28.50
5000	Moderate climate, all bronze, polished face									
5020	and scoriated cover, set flush									
5100	3/4" IPS connection	1 Plum	16	.500	Ea.	201	17.25		218.25	247
5120	1" IPS connection	"	14	.571		200	19.70		219.70	250
5200	For tapped drain port in box, add					25.50			25.50	28.50
6000	Ground post type, all non-freeze, all bronze, aluminum casing									
6010	guard, exposed head, 3/4" IPS connection									
6100	2 feet depth of bury	1 Plum	8	1	Ea.	283	34.50		317.50	360
6120	3 feet depth of bury		8	1		305	34.50		339.50	385
6140	4 feet depth of bury		8	1		330	34.50		364.50	415
6160	5 feet depth of bury		7	1.143		355	39.50		394.50	450
6180	6 feet depth of bury		7	1.143		375	39.50		414.50	475
6200	7 feet depth of bury		6	1.333		405	46		451	515
6220	8 feet depth of bury		5	1.600		430	55		485	555
6240	9 feet depth of bury		4	2		455	69		524	605
6260	10 feet depth of bury	↓	4	2		480	69		549	635
6300	For 1" IPS connection, add					40%	10%			
6350	For 1-1/4" IPS connection, add					140%	14%			
6400	For 1-1/2" IPS connection, add					225%	18%			
6450	For 2" IPS connection, add					315%	24%			

15440 | Plumbing Pumps

		CREW	DAILY OUTPUT	LABOR-HOURS	UNIT	MAT.	LABOR	EQUIP.	TOTAL	TOTAL INCL O&P
0010	**PUMPS, PRESSURE BOOSTER SYSTEM**									
0200	Pump system, with diaphragm tank, control, press. switch									
0300	1 HP pump	Q-1	1.30	12.308	Ea.	3,050	380		3,430	3,925
0400	1-1/2 HP pump		1.25	12.800		3,075	395		3,470	3,975
0420	2 HP pump		1.20	13.333		3,150	415		3,565	4,075
0440	3 HP pump	↓	1.10	14.545		3,200	450		3,650	4,175
0460	5 HP pump	Q-2	1.50	16		3,525	515		4,040	4,650
0480	7-1/2 HP pump		1.42	16.901		3,925	545		4,470	5,150
0500	10 HP pump	↓	1.34	17.910	↓	4,125	575		4,700	5,400
1000	Pump/ energy storage system, diaphragm tank, 3 HP pump									
1100	motor, PRV, switch, gauge, control center, flow switch									
1200	125 lb. working pressure	Q-2	.70	34.286	Ea.	8,850	1,100		9,950	11,400

15400 | Plumbing Fixtures & Equipment

15440 | Plumbing Pumps

			CREW	DAILY OUTPUT	LABOR-HOURS	UNIT	2001 BARE COSTS MAT.	LABOR	EQUIP.	TOTAL	TOTAL INCL O&P	
240	1300	250 lb. working pressure	Q-2	.64	37.500	Ea.	9,675	1,200		10,875	12,400	240
400	0010	**PUMPS, GRINDER SYSTEM** Complete, incl. check valve, tank, std.										400
	0020	controls incl. alarm/disconnect panel w/wire. Excavation not included										
	0260	Simplex, 9 GPM at 60 PSIG, 70 gal. tank				Ea.	2,150			2,150	2,375	
	0300	For manway, 26" I.D., 18" high, add					405			405	445	
	0340	26" I.D., 36" high, add					475			475	525	
	0380	43" I.D., 4' high, add					525			525	575	
	0600	Simplex, 9 GPM at 60 PSIG, 150 gal. tank					2,450			2,450	2,700	
	0660	For manway, add										
	0700	26" I.D., 36" high, add				Ea.	490			490	540	
	0740	26" I.D., 4' high, add					570			570	625	
	2000	Duplex, 18 GPM at 60 PSIG, 150 gal. tank					4,050			4,050	4,450	
	2060	For manway 43" I.D., 4' high, add					1,300			1,300	1,425	
	2400	For core only					1,625			1,625	1,775	
	2410	For salt water core, add					156			156	172	
	2600	For addit. manway 5' to 10' high, 60 gal tank, per ft., add				V.L.F.	145			145	159	
	2620	For manways 5' to 10', 120 gal tank, add				Ea.	158			158	173	
800	0010	**PUMPS, SEWAGE EJECTOR** With operating and level controls										800
	0100	Simplex system incl. tank, cover, pump 15' head										
	0500	37 gal PE tank, 12 GPM, 1/2 HP, 2" discharge	Q-1	3.20	5	Ea.	365	155		520	635	
	0510	3" discharge		3.10	5.161		390	160		550	670	
	0530	87 GPM, .7 HP, 2" discharge		3.20	5		560	155		715	850	
	0540	3" discharge		3.10	5.161		610	160		770	910	
	0600	45 gal. coated stl tank, 12 GPM, 1/2 HP, 2" discharge		3	5.333		645	165		810	960	
	0610	3" discharge		2.90	5.517		675	171		846	1,000	
	0630	87 GPM, .7 HP, 2" discharge		3	5.333		830	165		995	1,150	
	0640	3" discharge		2.90	5.517		875	171		1,046	1,225	
	0660	134 GPM, 1 HP, 2" discharge		2.80	5.714		895	177		1,072	1,250	
	0680	3" discharge		2.70	5.926		945	184		1,129	1,325	
	0700	70 gal. PE tank, 12 GPM, 1/2 HP, 2" discharge		2.60	6.154		730	191		921	1,100	
	0710	3" discharge		2.40	6.667		780	207		987	1,175	
	0730	87 GPM, 0.7 HP, 2" discharge		2.50	6.400		980	198		1,178	1,375	
	0740	3" discharge		2.30	6.957		1,000	216		1,216	1,425	
	0760	134 GPM, 1 HP, 2" discharge		2.20	7.273		1,025	225		1,250	1,475	
	0770	3" discharge		2	8		1,100	248		1,348	1,575	
	0800	75 gal. coated stl. tank, 12 GPM, 1/2 HP, 2" discharge		2.40	6.667		735	207		942	1,125	
	0810	3" discharge		2.20	7.273		765	225		990	1,175	
	0830	87 GPM, .7 HP, 2" discharge		2.30	6.957		925	216		1,141	1,350	
	0840	3" discharge		2.10	7.619		970	236		1,206	1,425	
	0860	134 GPM, 1 HP, 2" discharge		2	8		990	248		1,238	1,475	
	0880	3" discharge		1.80	8.889		1,050	276		1,326	1,575	
	1040	Duplex system incl. tank, covers, pumps										
	1060	110 gal. fiberglass tank, 24 GPM, 1/2 HP, 2" discharge	Q-1	1.60	10	Ea.	1,425	310		1,735	2,050	
	1080	3" discharge		1.40	11.429		1,475	355		1,830	2,150	
	1100	174 GPM, .7 HP, 2" discharge		1.50	10.667		1,875	330		2,205	2,550	
	1120	3" discharge		1.30	12.308		1,950	380		2,330	2,700	
	1140	268 GPM, 1 HP, 2" discharge		1.20	13.333		2,025	415		2,440	2,850	
	1160	3" discharge		1	16		2,100	495		2,595	3,050	
	1260	135 gal. coated stl. tank, 24 GPM, 1/2 HP, 2" discharge	Q-2	1.70	14.118		1,450	455		1,905	2,275	
	2000	3" discharge		1.60	15		1,525	480		2,005	2,400	
	2640	174 GPM, .7 HP, 2" discharge		1.60	15		1,900	480		2,380	2,800	
	2660	3" discharge		1.50	16		2,000	515		2,515	2,975	
	2700	268 GPM, 1 HP, 2" discharge		1.30	18.462		2,050	595		2,645	3,150	
	3040	3" discharge		1.10	21.818		2,150	700		2,850	3,425	
	3060	275 gal. coated stl. tank, 24 GPM, 1/2 HP, 2" discharge		1.50	16		1,800	515		2,315	2,750	

15400 | Plumbing Fixtures & Equipment

15440 | Plumbing Pumps

		CREW	DAILY OUTPUT	LABOR-HOURS	UNIT	MAT.	LABOR	EQUIP.	TOTAL	TOTAL INCL O&P
3080	3" discharge	Q-2	1.40	17.143	Ea.	1,825	550		2,375	2,825
3100	174 GPM, .7 HP, 2" discharge		1.40	17.143		2,325	550		2,875	3,375
3120	3" discharge		1.30	18.462		2,350	595		2,945	3,475
3140	268 GPM, 1 HP, 2" discharge		1.10	21.818		2,550	700		3,250	3,850
3160	3" discharge		.90	26.667		2,650	855		3,505	4,200
3260	Pump system accessories, add									
3300	Alarm horn and lights, 115V mercury switch	Q-1	8	2	Ea.	82	62		144	184
3340	Switch, mag. contactor, alarm bell, light, 3 level control		5	3.200		390	99		489	580
3380	Alternator, mercury switch activated		4	4		785	124		909	1,050
0010	**PUMPS, PEDESTAL SUMP** With float control									
0400	Molded PVC base, 21 GPM at 15' head, 1/3 HP	1 Plum	5	1.600	Ea.	86.50	55		141.50	178
0800	Iron base, 21 GPM at 15' head, 1/3 HP		5	1.600		110	55		165	204
1200	Solid brass, 21 GPM at 15' head, 1/3 HP		5	1.600		179	55		234	280
0010	**PUMPS, SUBMERSIBLE** Sump									
7000	Sump pump, automatic									
7100	Plastic, 1-1/4" discharge, 1/4 HP	1 Plum	6	1.333	Ea.	105	46		151	186
7140	1/3 HP		5	1.600		127	55		182	223
7160	1/2 HP		5	1.600		153	55		208	251
7180	1-1/2" discharge, 1/2 HP		4	2		171	69		240	292
7500	Cast iron, 1-1/4" discharge, 1/4 HP		6	1.333		125	46		171	207
7540	1/3 HP		6	1.333		140	46		186	224
7560	1/2 HP		5	1.600		176	55		231	276

15450 | Potable Water Tanks

		CREW	DAILY OUTPUT	LABOR-HOURS	UNIT	MAT.	LABOR	EQUIP.	TOTAL	TOTAL INCL O&P
0010	**WATER HEATER STORAGE TANKS** 125 psi ASME									
2000	Galvanized steel, 15 gal., 14" diam., 26" LOA	1 Plum	12	.667	Ea.	490	23		513	570
2060	30 gal., 14" diam. x 49" LOA		11	.727		580	25		605	680
2080	80 gal., 20" diam. x 64" LOA		9	.889		900	30.50		930.50	1,025
2100	135 gal., 24" diam. x 75" LOA		6	1.333		1,325	46		1,371	1,550
2120	240 gal., 30" diam. x 86" LOA		4	2		2,275	69		2,344	2,600
2140	300 gal., 36" diam. x 76" LOA		3	2.667		3,700	92		3,792	4,225
2160	400 gal., 36" diam. x 100" LOA	Q-1	4	4		4,550	124		4,674	5,200
2180	500 gal., 36" diam., x 126" LOA	"	3	5.333		5,625	165		5,790	6,450
3000	Glass lined, P.E., 80 gal., 20" diam. x 60" LOA	1 Plum	9	.889		1,600	30.50		1,630.50	1,800
3060	140 gal., 24" diam. x 80" LOA		6	1.333		2,275	46		2,321	2,575
3080	225 gal., 30" diam. x 78" LOA		4	2		2,650	69		2,719	3,025
3100	325 gal., 36" diam. x 81" LOA		3	2.667		3,400	92		3,492	3,900
3120	460 gal., 42" diam. x 84" LOA	Q-1	4	4		3,575	124		3,699	4,125
3140	605 gal., 48" diam. x 87" LOA		3	5.333		6,850	165		7,015	7,775
3160	740 gal., 54" diam. x 91" LOA		3	5.333		7,700	165		7,865	8,725
3180	940 gal., 60" diam. x 93" LOA		2.50	6.400		8,900	198		9,098	10,100
3200	1330 gal., 66" diam. x 107" LOA		2	8		12,200	248		12,448	13,800
3220	1615 gal., 72" diam. x 110" LOA		1.50	10.667		13,900	330		14,230	15,700
3240	2285 gal., 84" diam. x 128" LOA		1	16		17,100	495		17,595	19,600
3260	3440 gal., 96" diam. x 157" LOA	Q-2	1.50	16		25,400	515		25,915	28,700

15480 | Domestic Water Heaters

		CREW	DAILY OUTPUT	LABOR-HOURS	UNIT	MAT.	LABOR	EQUIP.	TOTAL	TOTAL INCL O&P
0010	**HEAT TRANSFER PACKAGES** Complete, controls,									
0020	expansion tank, converter, air separator									
1000	Hot water, 180°F enter, 200°F leaving, 15# steam									
1010	One pump system, 28 GPM	Q-6	.75	32	Ea.	10,700	1,050		11,750	13,400
1020	35 GPM		.70	34.286		11,800	1,125		12,925	14,700
1040	55 GPM		.65	36.923		13,500	1,200		14,700	16,700

15400 | Plumbing Fixtures & Equipment

15480 | Domestic Water Heaters

			CREW	DAILY OUTPUT	LABOR-HOURS	UNIT	2001 BARE COSTS MAT.	LABOR	EQUIP.	TOTAL	TOTAL INCL O&P	
900	1060	130 GPM	Q-6	.55	43.636	Ea.	16,700	1,425		18,125	20,600	900
	1080	255 GPM		.40	60		23,000	1,950		24,950	28,300	
	1100	550 GPM		.30	80		29,200	2,600		31,800	36,100	
	1120	800 GPM		.25	96		35,800	3,125		38,925	44,100	
	1220	Two pump system, 28 GPM		.70	34.286		14,300	1,125		15,425	17,400	
	1240	35 GPM		.65	36.923		17,100	1,200		18,300	20,600	
	1260	55 GPM		.60	40		17,500	1,300		18,800	21,300	
	1280	130 GPM		.50	48		22,400	1,550		23,950	27,100	
	1300	255 GPM		.35	68.571		30,800	2,225		33,025	37,300	
	1320	550 GPM		.25	96		38,900	3,125		42,025	47,500	
	1340	800 GPM		.20	120		47,500	3,900		51,400	58,000	

15500 | Heat Generation Equipment

15510 | Heating Boilers and Accessories

			CREW	DAILY OUTPUT	LABOR-HOURS	UNIT	2001 BARE COSTS MAT.	LABOR	EQUIP.	TOTAL	TOTAL INCL O&P	
110	0010	**BOILER BLOWDOWN SYSTEMS**										110
	1010	Boiler blowdown, auto/manual to 2000 MBH	Q-5	3.75	4.267	Ea.	2,375	134		2,509	2,825	
	1020	7300 MBH	"	3	5.333	"	2,725	167		2,892	3,250	
120	0010	**BURNERS**										120
	0990	Residential, conversion, gas fired, LP or natural										
	1000	Gun type, atmospheric input 72 to 200 MBH	Q-1	2.50	6.400	Ea.	655	198		853	1,025	
	1020	120 to 360 MBH		2	8		725	248		973	1,175	
	1040	280 to 800 MBH		1.70	9.412		1,400	292		1,692	2,000	
	2000	Commercial and industrial, gas/oil, input										
	2050	400 MBH	Q-1	1.50	10.667	Ea.	1,850	330		2,180	2,525	
	2080	For mounting plate, add		35	.457		134	14.15		148.15	169	
	2090	670 MBH		1.40	11.429		2,275	355		2,630	3,025	
	2110	For mounting plate, add		35	.457		134	14.15		148.15	169	
	2140	1155 MBH		1.30	12.308		2,950	380		3,330	3,825	
	2170	For mounting plate, add		35	.457		134	14.15		148.15	169	
	2200	1800 MBH		1.20	13.333		3,200	415		3,615	4,150	
	2230	For mounting plate, add		35	.457		134	14.15		148.15	169	
	2260	3000 MBH		1.10	14.545		4,400	450		4,850	5,525	
	2290	For mounting plate, add		35	.457		134	14.15		148.15	169	
	2320	4100 MBH		1	16		4,550	495		5,045	5,750	
	2350	For mounting plate, add		35	.457		134	14.15		148.15	169	
	2500	Impinged jet, rectangular, burner only, 3-1/2" WC,										
	2520	Gas fired, input, 420 to 640 MBH	Q-1	2	8	Ea.	540	248		788	965	
	2540	560 to 860 MBH		2	8		765	248		1,013	1,225	
	2560	630 to 950 MBH		1.90	8.421		805	261		1,066	1,275	
	2580	950 to 1200 MBH		1.80	8.889		1,150	276		1,426	1,700	
	2600	1120 to 1700 MBH		1.80	8.889		1,525	276		1,801	2,125	
	2620	1700 to 1900 MBH		1.70	9.412		1,725	292		2,017	2,350	
	2640	1400 to 2100 MBH		1.60	10		1,925	310		2,235	2,575	
	2660	1600 to 2400 MBH		1.60	10		2,025	310		2,335	2,700	
	2680	1700 to 2500 MBH		1.50	10.667		2,300	330		2,630	3,025	
	2700	1880 to 2880 MBH		1.40	11.429		2,425	355		2,780	3,175	
	2720	2260 to 3680 MBH		1.40	11.429		2,825	355		3,180	3,625	

15500 | Heat Generation Equipment

15510 | Heating Boilers and Accessories

			CREW	DAILY OUTPUT	LABOR-HOURS	UNIT	MAT.	LABOR	EQUIP.	TOTAL	TOTAL INCL O&P
2740		2590 to 4320 MBH	Q-1	1.30	12.308	Ea.	3,450	380		3,830	4,375
2760		3020 to 5040 MBH		1.20	13.333		4,025	415		4,440	5,050
2780		3450 to 5760 MBH		1.20	13.333		4,600	415		5,015	5,700
2850		Burner pilot		50	.320		14.35	9.90		24.25	31
2860		Thermocouple		50	.320		14.75	9.90		24.65	31.50
2870		Thermocouple & pilot bracket		45	.356		56	11		67	78
2900		For 7" W.C. pressure, increase MBH 40%									
3000	Flame retention oil fired assembly, input										
3020		.50 to 2.25 GPH	Q-1	2.40	6.667	Ea.	291	207		498	630
3040		2.0 to 5.0 GPH		2	8		525	248		773	955
3060		3.0 to 7.0 GPH		1.80	8.889		535	276		811	1,000
3080		6.0 to 12.0 GPH		1.60	10		835	310		1,145	1,375
4600	Gas safety, shut off valve, 3/4" threaded		1 Stpi	20	.400		272	13.95		285.95	320
4610		1" threaded		19	.421		274	14.65		288.65	320
4620		1-1/4" threaded		15	.533		298	18.60		316.60	360
4630		1-1/2" threaded		13	.615		320	21.50		341.50	390
4640		2" threaded		11	.727		360	25.50		385.50	435
4650		2-1/2" threaded	Q-1	15	1.067		415	33		448	505
4660		3" threaded		13	1.231		570	38		608	685
4670		4" flanged		3	5.333		2,150	165		2,315	2,625
4680		6" flanged	Q-2	3	8		5,000	257		5,257	5,900
8020	Coal stoker, automatic, complete system 1000 MBH boiler		Q-7	.33	96.970		14,200	3,225		17,425	20,500
8040		2000 MBH boiler		.33	96.970		15,300	3,225		18,525	21,700
8060		7300 MBH boiler		.30	106		17,000	3,550		20,550	24,100
0010	**BOILERS, ELECTRIC, ASME** Standard controls and trim										
1000	Steam, 6 KW, 20.5 MBH		Q-19	1.20	20	Ea.	7,900	645		8,545	9,675
1040		9 KW, 30.7 MBH		1.20	20		8,250	645		8,895	10,100
1060		18 KW, 61.4 MBH		1.20	20		8,325	645		8,970	10,100
1080		24 KW, 81.8 MBH		1.10	21.818		8,400	705		9,105	10,300
1120		36 KW, 123 MBH		1.10	21.818		8,575	705		9,280	10,500
1160		60 KW, 205 MBH		1	24		9,775	775		10,550	11,900
1220		112 KW, 382 MBH		.75	32		11,200	1,025		12,225	13,900
1240		148 KW, 505 MBH		.65	36.923		12,600	1,200		13,800	15,700
1260		168 KW, 573 MBH		.60	40		14,700	1,300		16,000	18,200
1280		222 KW, 758 MBH		.55	43.636		15,300	1,400		16,700	18,900
1300		296 KW, 1010 MBH		.45	53.333		15,400	1,725		17,125	19,600
1320		300 KW, 1023 MBH		.40	60		17,700	1,950		19,650	22,300
1340		370 KW, 1263 MBH		.35	68.571		19,100	2,225		21,325	24,300
1360		444 KW, 1515 MBH		.30	80		20,100	2,600		22,700	26,000
1380		518 KW, 1768 MBH	Q-21	.36	88.889		20,600	2,925		23,525	27,000
1400		592 KW, 2020 MBH		.34	94.118		23,800	3,100		26,900	30,900
1420		666 KW, 2273 MBH		.32	100		26,000	3,300		29,300	33,600
1460		740 KW, 2526 MBH		.28	114		29,700	3,775		33,475	38,300
1480		814 KW, 2778 MBH		.25	128		31,500	4,225		35,725	41,100
1500		962 KW, 3283 MBH		.22	145		39,600	4,800		44,400	51,000
1520		1036 KW, 3536 MBH		.20	160		45,300	5,275		50,575	58,000
1540		1110 KW, 3788 MBH		.19	168		52,500	5,550		58,050	66,500
1560		2070 KW, 7063 MBH		.18	177		56,500	5,850		62,350	71,000
1580		2250 KW, 7677 MBH		.17	188		58,500	6,200		64,700	73,500
1600		2,340 KW, 7984 MBH		.16	200		60,000	6,600		66,600	75,500
2000	Hot water, 6 KW, 20.5 MBH		Q-19	1.30	18.462		3,400	595		3,995	4,625
2020		9 KW, 30.7 MBH		1.30	18.462		3,425	595		4,020	4,650
2040		24 KW, 82 MBH		1.20	20		3,625	645		4,270	4,975
2060		30 KW, 103 MBH		1.20	20		3,725	645		4,370	5,075
2070		36 KW, 123 MBH		1.20	20		3,950	645		4,595	5,325
2080		48 KW, 164 MBH		1.10	21.818		3,975	705		4,680	5,425

15500 | Heat Generation Equipment

15510 | Heating Boilers and Accessories

			CREW	DAILY OUTPUT	LABOR-HOURS	UNIT	MAT.	LABOR	EQUIP.	TOTAL	TOTAL INCL O&P	
300	2100	74 KW, 253 MBH	Q-19	1.10	21.818	Ea.	4,725	705		5,430	6,250	300
	2120	93 KW, 317 MBH R15500-030		1	24		5,550	775		6,325	7,275	
	2140	112 KW, 382 MBH		.90	26.667		6,025	865		6,890	7,925	
	2160	148 KW, 505 MBH A8.3-110		.75	32		8,550	1,025		9,575	11,000	
	2180	168 KW, 573 MBH		.65	36.923		9,150	1,200		10,350	11,900	
	2200	222 KW, 758 MBH A8.3-120		.60	40		10,000	1,300		11,300	13,000	
	2220	296 KW, 1010 MBH		.55	43.636		10,600	1,400		12,000	13,800	
	2280	370 KW, 1263 MBH		.40	60		14,600	1,950		16,550	19,000	
	2300	444 KW, 1515 MBH		.35	68.571		16,900	2,225		19,125	21,900	
	2340	518 KW, 1768 MBH	Q-21	.44	72.727		18,900	2,400		21,300	24,400	
	2360	592 KW, 2020 MBH		.43	74.419		20,100	2,450		22,550	25,800	
	2400	666 KW, 2273 MBH		.40	80		23,500	2,650		26,150	29,800	
	2420	740 KW, 2526 MBH		.39	82.051		23,800	2,700		26,500	30,300	
	2440	814 KW, 2778 MBH		.38	84.211		27,200	2,775		29,975	34,100	
	2460	888 KW, 3031 MBH		.37	86.486		29,000	2,850		31,850	36,200	
	2480	962 KW, 3283 MBH		.36	88.889		30,800	2,925		33,725	38,300	
	2500	1,036 KW, 3536 MBH		.34	94.118		33,600	3,100		36,700	41,700	
	2520	1110 KW, 3788 MBH		.33	96.970		38,800	3,200		42,000	47,400	
	2540	1440 KW, 4915 MBH		.32	100		40,900	3,300		44,200	50,000	
	2560	1560 KW, 5323 MBH		.31	103		43,100	3,400		46,500	52,500	
	2580	1680 KW, 5733 MBH		.30	106		44,800	3,525		48,325	54,500	
	2600	1800 KW, 6143 MBH		.29	110		46,800	3,650		50,450	57,000	
	2620	1980 KW, 6757 MBH		.28	114		51,000	3,775		54,775	61,500	
	2640	2100 KW, 7167 MBH		.27	118		52,500	3,900		56,400	64,000	
	2660	2220 KW, 7576 MBH		.26	123		55,000	4,050		59,050	66,500	
	2680	2,400 KW, 8191 MBH		.25	128		57,500	4,225		61,725	69,500	
	2700	2610 KW, 8905 MBH		.24	133		60,500	4,400		64,900	73,000	
	2720	2790 KW, 9519 MBH		.23	139		63,000	4,575		67,575	76,500	
	2740	2970 KW, 10133 MBH		.21	152		65,000	5,025		70,025	79,000	
	2760	3150 KW, 10748 MBH		.19	168		66,500	5,550		72,050	81,500	
	2780	3240 KW, 11055 MBH		.18	177		67,500	5,850		73,350	83,500	
	2800	3420 KW, 11669 MBH		.17	188		69,500	6,200		75,700	85,500	
	2820	3,600 KW, 12,283 MBH		.16	200		72,000	6,600		78,600	89,500	
400	0010	**BOILERS, GAS FIRED** Natural or propane, standard controls										400
	1000	Cast iron, with insulated jacket										
	2000	Steam, gross output, 81 MBH	Q-7	1.40	22.857	Ea.	1,125	760		1,885	2,400	
	2020	102 MBH		1.30	24.615		1,300	820		2,120	2,650	
	2040	122 MBH		1	32		1,400	1,075		2,475	3,150	
	2060	163 MBH		.90	35.556		1,725	1,175		2,900	3,675	
	2080	203 MBH		.90	35.556		1,900	1,175		3,075	3,875	
	2100	240 MBH		.85	37.647		1,900	1,250		3,150	4,000	
	2120	280 MBH		.80	40		2,425	1,325		3,750	4,675	
	2140	320 MBH		.70	45.714		2,650	1,525		4,175	5,200	
	2160	360 MBH		.63	51.200		2,975	1,700		4,675	5,850	
	2180	400 MBH		.56	56.838		3,175	1,900		5,075	6,350	
	2200	440 MBH		.51	62.500		3,400	2,075		5,475	6,850	
	2220	544 MBH		.45	71.588		3,975	2,375		6,350	7,975	
	2240	765 MBH		.43	74.419		5,075	2,475		7,550	9,300	
	2260	892 MBH		.38	84.211		6,075	2,800		8,875	10,900	
	2280	1275 MBH		.34	94.118		7,725	3,125		10,850	13,200	
	2300	1530 MBH		.32	100		9,050	3,325		12,375	15,000	
	2320	1,875 MBH		.30	106		10,700	3,550		14,250	17,200	
	2340	2170 MBH		.26	122		12,300	4,050		16,350	19,600	
	2360	2675 MBH		.20	163		14,000	5,425		19,425	23,600	
	2380	3060 MBH		.19	172		15,200	5,725		20,925	25,300	

15500 | Heat Generation Equipment

15510 | Heating Boilers and Accessories

		CREW	DAILY OUTPUT	LABOR-HOURS	UNIT	2001 BARE COSTS MAT.	LABOR	EQUIP.	TOTAL	TOTAL INCL O&P	
400	2400	3570 MBH	Q-7	.18	181	Ea.	16,900	6,050		22,950	27,700	400
	2420	4207 MBH		.16	205		18,700	6,825		25,525	30,900	
	2440	4,720 MBH		.15	207		20,500	6,900		27,400	33,000	
	2460	5660 MBH		.15	220		42,900	7,325		50,225	58,500	
	2480	6,100 MBH		.13	246		49,000	8,175		57,175	66,500	
	2500	6390 MBH		.12	266		52,000	8,875		60,875	70,500	
	2520	6680 MBH		.11	290		52,500	9,675		62,175	72,000	
	2540	6,970 MBH		.10	320		55,000	10,600		65,600	76,500	
	3000	Hot water, gross output, 80 MBH		1.46	21.918		1,000	730		1,730	2,225	
	3020	100 MBH		1.35	23.704		1,175	790		1,965	2,500	
	3040	122 MBH		1.10	29.091		1,275	965		2,240	2,850	
	3060	163 MBH		1	32		1,600	1,075		2,675	3,350	
	3080	203 MBH		1	32		1,775	1,075		2,850	3,575	
	3100	240 MBH		.95	33.684		1,800	1,125		2,925	3,675	
	3120	280 MBH		.90	35.556		2,250	1,175		3,425	4,250	
	3140	320 MBH		.80	40		2,475	1,325		3,800	4,700	
	3160	360 MBH		.71	45.070		2,800	1,500		4,300	5,325	
	3180	400 MBH		.64	50		3,000	1,650		4,650	5,800	
	3200	440 MBH		.58	54.983		3,225	1,825		5,050	6,300	
	3220	544 MBH		.51	62.992		3,800	2,100		5,900	7,325	
	3240	765 MBH		.46	70.022		4,875	2,325		7,200	8,875	
	3260	1,088 MBH		.40	80		6,375	2,650		9,025	11,100	
	3280	1,275 MBH		.36	89.888		7,500	3,000		10,500	12,800	
	3300	1,530 MBH		.31	104		8,875	3,475		12,350	15,100	
	3320	2,000 MBH		.26	125		11,000	4,150		15,150	18,400	
	3340	2,312 MBH		.22	148		12,700	4,925		17,625	21,400	
	3360	2,856 MBH		.20	160		15,200	5,325		20,525	24,700	
	3380	3,264 MBH		.18	179		16,100	5,975		22,075	26,700	
	3400	3,808 MBH		.16	195		17,800	6,475		24,275	29,400	
	3420	4,488 MBH		.15	210		19,700	7,000		26,700	32,300	
	3440	4,720 MBH		.15	220		43,900	7,325		51,225	59,500	
	3460	5,520 MBH		.14	228		49,500	7,600		57,100	66,000	
	3480	6,100 MBH		.13	250		50,000	8,300		58,300	67,500	
	3500	6,390 MBH		.11	285		51,000	9,500		60,500	70,500	
	3520	6,680 MBH		.10	310		54,000	10,300		64,300	75,000	
	3540	6,970 MBH		.09	359		56,000	12,000		68,000	79,500	
	4000	Steel, insulating jacket										
	4500	Steam, not including burner, gross output										
	5500	1,440 MBH	Q-6	.35	68.571	Ea.	15,300	2,225		17,525	20,200	
	5520	1630 MBH		.30	80		15,900	2,600		18,500	21,400	
	5540	1800 MBH		.26	92.308		16,500	3,000		19,500	22,700	
	5560	2520 MBH		.20	120		20,400	3,900		24,300	28,400	
	5580	3,065 MBH		.18	133		21,900	4,325		26,225	30,600	
	5600	3600 MBH		.16	150		23,200	4,875		28,075	32,900	
	5620	5400 MBH		.14	171		32,300	5,575		37,875	44,000	
	5640	7,200 MBH	Q-7	.18	177		40,100	5,900		46,000	53,000	
	5660	10,800 MBH		.17	188		49,900	6,250		56,150	64,500	
	5680	12,600 MBH		.15	213		55,500	7,100		62,600	71,500	
	5700	16,190 MBH		.12	266		70,000	8,875		78,875	90,500	
	5720	17,990 MBH		.10	320		75,000	10,600		85,600	98,500	
	6000	Hot water, including burner & one zone valve, gross output	Q-6	2	12	Ea.	1,575	390		1,965	2,325	
	6010	51.2 MBH		2	12		1,750	390		2,140	2,525	
	6020	72 MBH		1.90	12.632		1,775	410		2,185	2,600	
	6040	89 MBH		1.80	13.333		2,000	435		2,435	2,850	
	6060	105 MBH		1.70	14.118		2,275	460		2,735	3,200	
	6080	132 MBH										

15500 | Heat Generation Equipment

15510 | Heating Boilers and Accessories

			CREW	DAILY OUTPUT	LABOR-HOURS	UNIT	2001 BARE COSTS MAT.	LABOR	EQUIP.	TOTAL	TOTAL INCL O&P	
400	6100	155 MBH	Q-6	1.50	16	Ea.	2,650	520		3,170	3,675	400
	6110	186 MBH		1.40	17.143		3,200	560		3,760	4,350	
	6120	227 MBH		1.30	18.462		4,050	600		4,650	5,350	
	6140	292 MBH		1.20	20		4,575	650		5,225	6,000	
	6160	400 MBH		.80	30		5,775	975		6,750	7,825	
	6180	480 MBH		.70	34.286		6,600	1,125		7,725	8,950	
	6200	640 MBH		.60	40		7,850	1,300		9,150	10,600	
	6220	800 MBH		.50	48		9,150	1,550		10,700	12,500	
	6240	960 MBH		.45	53.333		11,400	1,725		13,125	15,100	
	6260	1200 MBH		.40	60		14,000	1,950		15,950	18,300	
	6280	1440 MBH		.35	68.571		15,800	2,225		18,025	20,800	
	6300	1960 MBH		.30	80		20,900	2,600		23,500	26,900	
	6320	2400 MBH		.20	120		25,500	3,900		29,400	34,000	
	6340	3,000 MBH		.15	160		31,600	5,200		36,800	42,600	
	7000	For tankless water heater on smaller gas units, add					10%					
	7050	For additional zone valves up to 312 MBH add				Ea.	101			101	111	
	7990	Special feature gas fired boilers										
	8000	Pulse combustion, standard controls / trim, 44,000 BTU	Q-5	1.50	10.667	Ea.	2,275	335		2,610	3,000	
	8050	88,000 BTU		1.40	11.429		2,700	360		3,060	3,500	
	8080	134,000 BTU		1.20	13.333		3,500	420		3,920	4,450	
	9000	Wall hung, C.I., sealed combustion, direct vent										
	9010	Packaged, net output										
	9020	45,000 BTUH	Q-6	1.33	18.045	Ea.	1,375	585		1,960	2,375	
	9030	60,800 BTUH		1.33	18.045		1,525	585		2,110	2,550	
	9040	77,000 BTUH		1.33	18.045		1,650	585		2,235	2,700	
460	0010	**BOILERS, GAS/OIL** Combination with burners and controls										460
	1000	Cast iron with insulated jacket										
	2000	Steam, gross output, 720 MBH	Q-7	.43	74.074	Ea.	6,375	2,450		8,825	10,800	
	2020	810 MBH		.38	83.989		6,850	2,800		9,650	11,800	
	2040	1,084 MBH		.34	93.023		7,800	3,100		10,900	13,300	
	2060	1,360 MBH		.33	98.160		9,250	3,275		12,525	15,100	
	2080	1,600 MBH		.30	107		9,825	3,550		13,375	16,200	
	2100	2,040 MBH		.25	130		12,500	4,325		16,825	20,200	
	2120	2,450 MBH		.20	156		13,800	5,200		19,000	23,000	
	2140	2,700 MBH		.19	165		14,600	5,500		20,100	24,400	
	2160	3,000 MBH		.18	175		15,800	5,850		21,650	26,200	
	2180	3,270 MBH		.17	183		17,100	6,125		23,225	28,000	
	2200	3,770 MBH		.17	191		38,800	6,375		45,175	52,000	
	2220	4,070 MBH		.16	200		40,800	6,650		47,450	55,000	
	2240	4,650 MBH		.15	210		45,200	7,000		52,200	60,500	
	2260	5,230 MBH		.14	223		47,600	7,450		55,050	63,500	
	2280	5,520 MBH		.14	235		52,000	7,825		59,825	69,000	
	2300	5,810 MBH		.13	248		52,500	8,250		60,750	70,000	
	2320	6,100 MBH		.12	260		52,500	8,650		61,150	71,000	
	2340	6,390 MBH		.11	296		55,500	9,850		65,350	76,000	
	2360	6,680 MBH		.10	320		58,500	10,600		69,100	80,500	
	2380	6,970 MBH		.09	372		59,000	12,400		71,400	83,500	
	2900	Hot water, gross output										
	2910	200 MBH	Q-6	.61	39.024	Ea.	4,325	1,275		5,600	6,675	
	2920	300 MBH		.49	49.080		4,325	1,600		5,925	7,150	
	2930	400 MBH		.41	57.971		5,050	1,875		6,925	8,400	
	2940	500 MBH		.36	67.039		5,450	2,175		7,625	9,275	
	3000	584 MBH	Q-7	.44	72.072		5,900	2,400		8,300	10,100	
	3020	876 MBH		.40	79.012		7,950	2,625		10,575	12,700	
	3040	1,168 MBH		.31	103		8,900	3,450		12,350	15,000	

15500 | Heat Generation Equipment

15510 | Heating Boilers and Accessories

		CREW	DAILY OUTPUT	LABOR-HOURS	UNIT	MAT.	LABOR	EQUIP.	TOTAL	TOTAL INCL O&P
3060	1,460 MBH	Q-7	.28	113	Ea.	10,500	3,750		14,250	17,200
3080	2,044 MBH		.26	122		13,600	4,050		17,650	21,000
3100	2,628 MBH		.21	150		35,900	5,000		40,900	47,100
3120	3,210 MBH		.18	174		44,000	5,800		49,800	57,000
3140	3,796 MBH		.17	186		52,000	6,175		58,175	67,000
3160	4,088 MBH		.16	195		56,000	6,475		62,475	71,500
3180	4,672 MBH		.16	203		64,000	6,775		70,775	80,500
3200	5,256 MBH		.15	217		68,000	7,225		75,225	86,000
3220	6,000 MBH, 179 BHP		.13	256		92,500	8,500		101,000	114,500
3240	7,130 MBH, 213 BHP		.08	385		101,500	12,800		114,300	131,500
3260	9,800 MBH, 286 BHP		.06	533		111,000	17,700		128,700	149,000
3280	10,900 MBH, 325.6 BHP		.05	592		114,500	19,700		134,200	155,500
3290	12,200 MBH, 364.5 BHP		.05	666		116,000	22,200		138,200	161,000
3300	13,500 MBH, 403.3 BHP		.04	727		117,000	24,200		141,200	165,500
4000	Steel, insulated jacket, skid base, tubeless									
4100	Steam, 15 psi, gross output 517 MBH, 15 BHP	Q-6	.41	57.971	Ea.	11,700	1,875		13,575	15,700
4110	690 MBH, 20 BHP		.38	62.992		12,000	2,050		14,050	16,300
4120	862 MBH, 25 BHP		.36	67.039		12,800	2,175		14,975	17,400
4130	1,004 MBH, 30 BHP		.32	74.074		15,000	2,400		17,400	20,200
4140	1,340 MBH, 40 BHP		.31	78.431		16,500	2,550		19,050	22,000
4150	1,675 MBH, 50 BHP		.28	85.714		17,100	2,800		19,900	23,000
4160	2,175 MBH, 65 BHP		.23	104		17,600	3,400		21,000	24,500
4170	2,510 MBH, 75 BHP		.19	125		18,900	4,075		22,975	27,000
4180	3,015 MBH, 90 BHP		.17	140		21,500	4,575		26,075	30,500
4190	3,350 MBH, 100 BHP	Q-7	.22	146		24,500	4,875		29,375	34,400
4200	3,880 MBH, 116 BHP		.21	153		26,500	5,125		31,625	36,900
4210	4,685 MBH, 140 BHP		.19	168		28,300	5,600		33,900	39,600
4220	5,025 MBH, 150 BHP		.18	177		29,200	5,900		35,100	41,000
4230	5,550 MBH, 166 BHP		.17	189		37,800	6,300		44,100	51,000
4240	6,700 MBH, 200 BHP		.13	256		41,300	8,500		49,800	58,000
4250	7,760 MBH, 232 BHP		.11	296		46,200	9,850		56,050	66,000
4260	8,870 MBH, 265 BHP		.09	336		50,000	11,200		61,200	72,000
4270	11,100 MBH, 322 BHP		.07	426		53,500	14,200		67,700	80,500
4280	13,400 MBH, 400 BHP		.06	516		61,500	17,200		78,700	93,500
4290	17,750 MBH, 530 BHP		.05	680		77,000	22,600		99,600	118,500
4300	20,750 MBH, 620 BHP		.04	800		86,000	26,600		112,600	134,500
4500	Steam, 150 psi gross output, 335 MBH, 10 BHP	Q-6	.54	44.037		9,425	1,425		10,850	12,600
4520	502 MBH, 15 BHP		.40	60		10,700	1,950		12,650	14,800
4560	670 MBH, 20 BHP		.36	65.934		11,700	2,150		13,850	16,200
4600	1,005 MBH, 30 BHP		.31	77.922		15,300	2,525		17,825	20,600
4640	1,339 MBH, 40 BHP		.29	81.911		18,500	2,675		21,175	24,400
4660	1,674 MBH, 50 BHP		.24	100		20,600	3,250		23,850	27,600
4680	2,009 MBH, 60 BHP		.21	114		24,500	3,725		28,225	32,500
4720	2,511 MBH, 75 BHP		.17	141		25,200	4,600		29,800	34,700
5000	Hot water, gross output, 525 MBH		.45	52.980		9,500	1,725		11,225	13,100
5020	630 MBH		.43	55.944		9,600	1,825		11,425	13,400
5040	735 MBH		.41	58.968		11,900	1,925		13,825	15,900
5060	840 MBH		.38	64		13,100	2,075		15,175	17,600
5080	1,050 MBH		.35	67.989		15,300	2,200		17,500	20,300
5100	1,365 MBH		.33	72.727		19,100	2,375		21,475	24,700
5120	1,680 MBH		.29	82.759		21,900	2,700		24,600	28,200
5140	2,310 MBH		.20	120		28,500	3,900		32,400	37,300
5160	2,835 MBH		.18	129		34,400	4,225		38,625	44,200
5180	3,150 MBH		.16	146		37,900	4,750		42,650	48,900
0010	**BOILERS, OIL FIRED** Standard controls, flame retention burner									
1000	Cast iron, with insulated flush jacket									

15500 | Heat Generation Equipment

15510 | Heating Boilers and Accessories

			CREW	DAILY OUTPUT	LABOR-HOURS	UNIT	2001 BARE COSTS				TOTAL INCL O&P
							MAT.	LABOR	EQUIP.	TOTAL	
500	2000	Steam, gross output, 109 MBH	Q-7	1.20	26.667	Ea.	1,375	885		2,260	2,875
	2020	144 MBH		1.10	29.091		1,600	965		2,565	3,225
	2040	173 MBH		1	32		1,750	1,075		2,825	3,500
	2060	207 MBH		.90	35.556		1,875	1,175		3,050	3,850
	2080	236 MBH		.85	37.647		2,700	1,250		3,950	4,875
	2100	300 MBH		.70	45.714		2,725	1,525		4,250	5,300
	2120	480 MBH		.50	64		3,475	2,125		5,600	7,025
	2140	665 MBH		.45	71.111		4,775	2,375		7,150	8,825
	2160	794 MBH		.41	78.049		5,525	2,600		8,125	10,000
	2180	1,084 MBH		.38	85.106		6,350	2,825		9,175	11,300
	2200	1,360 MBH		.33	98.160		7,450	3,275		10,725	13,100
	2220	1,600 MBH		.26	122		8,400	4,050		12,450	15,400
	2240	2,175 MBH		.24	133		10,700	4,450		15,150	18,500
	2260	2,480 MBH		.20	156		12,500	5,200		17,700	21,500
	2280	3,000 MBH		.19	170		14,400	5,650		20,050	24,400
	2300	3,550 MBH		.17	187		15,300	6,225		21,525	26,200
	2320	3,820 MBH		.16	200		17,000	6,650		23,650	28,700
	2340	4,360 MBH		.15	214		18,700	7,150		25,850	31,300
	2360	4,940 MBH		.14	225		36,700	7,500		44,200	51,500
	2380	5,520 MBH		.14	235		43,100	7,825		50,925	59,000
	2400	6,100 MBH		.13	256		49,400	8,500		57,900	67,500
	2420	6,390 MBH		.11	290		52,000	9,675		61,675	72,000
	2440	6,680 MBH		.10	313		53,500	10,400		63,900	74,500
	2460	6,970 MBH		.09	363		55,500	12,100		67,600	79,500
	3000	Hot water, same price as steam									
	4000	For tankless coil in smaller sizes, add					15%				
	5000	Steel, insulated jacket, burner									
	6000	Steam, full water leg construction, gross output									
	6020	144 MBH	Q-6	1.33	18.045	Ea.	3,275	585		3,860	4,475
	6040	198 MBH		1.20	20		3,550	650		4,200	4,875
	6060	252 MBH		1.13	21.295		4,425	695		5,120	5,925
	6080	324 MBH		1.04	23.011		5,050	750		5,800	6,675
	6100	396 MBH		.78	30.691		5,475	1,000		6,475	7,525
	6120	468 MBH		.70	34.483		6,025	1,125		7,150	8,325
	6140	648 MBH		.52	45.977		6,400	1,500		7,900	9,300
	6160	792 MBH		.44	55.046		7,525	1,800		9,325	11,000
	6180	1,008 MBH		.39	61.381		9,325	2,000		11,325	13,300
	6200	1,260 MBH		.35	68.966		10,300	2,250		12,550	14,800
	6220	1,512 MBH		.31	78.689		12,400	2,550		14,950	17,500
	6240	1,800 MBH		.29	83.624		13,600	2,725		16,325	19,000
	6260	2,100 MBH		.23	106		16,000	3,450		19,450	22,800
	6280	2,400 MBH		.19	125		17,800	4,075		21,875	25,800
	6400	Larger sizes are same as steel, gas fired									
	7000	Hot water, gross output, 103 MBH	Q-6	1.60	15	Ea.	965	490		1,455	1,800
	7020	122 MBH		1.45	16.506		2,100	535		2,635	3,100
	7040	137 MBH		1.36	17.595		2,450	570		3,020	3,575
	7060	168 MBH		1.30	18.405		2,625	600		3,225	3,800
	7080	225 MBH		1.22	19.704		2,725	640		3,365	3,975
	7100	315 MBH		.96	25.105		3,575	815		4,390	5,150
	7120	420 MBH		.70	34.483		4,175	1,125		5,300	6,275
	7140	525 MBH		.57	42.403		4,825	1,375		6,200	7,400
	7160	630 MBH		.52	45.977		5,450	1,500		6,950	8,250
	7180	735 MBH		.48	50.104		5,650	1,625		7,275	8,650
	7200	840 MBH		.43	55.172		9,150	1,800		10,950	12,800
	7220	1,050 MBH		.36	65.753		10,300	2,150		12,450	14,500
	7240	1,365 MBH		.32	74.534		10,300	2,425		12,725	15,000

15500 | Heat Generation Equipment

15510 | Heating Boilers and Accessories

			CREW	DAILY OUTPUT	LABOR-HOURS	UNIT	MAT.	LABOR	EQUIP.	TOTAL	TOTAL INCL O&P	
500	7260	1,680 MBH	Q-6	.29	83.624	Ea.	11,600	2,725		14,325	16,900	500
	7280	2,310 MBH		.21	114		13,100	3,725		16,825	20,100	
	7300	2,835 MBH		.17	141		16,900	4,600		21,500	25,500	
	7320	3,150 MBH		.13	184		18,000	6,000		24,000	28,900	
	7340	For tankless coil in steam or hot water, add					7%					
700	0010	**BOILERS, PACKAGED SCOTCH MARINE** Steam or hot water										700
	1000	Packaged fire tube, #2 oil, gross output										
	1006	15 PSI steam										
	1010	1005 MBH, 30 HP	Q-6	.33	72.072	Ea.	22,100	2,350		24,450	27,900	
	1014	1675 MBH, 50 HP	"	.26	93.023		27,600	3,025		30,625	34,900	
	1020	3348 MBH, 100 HP	Q-7	.21	152		29,200	5,075		34,275	39,900	
	1030	5025 MBH, 150 HP		.17	192		36,900	6,400		43,300	50,500	
	1040	6696 MBH, 200 HP		.14	223		45,800	7,450		53,250	61,500	
	1050	8375 MBH, 250 HP		.14	231		47,100	7,700		54,800	63,500	
	1060	10,044 MBH, 300 HP		.13	251		62,000	8,375		70,375	80,500	
	1080	16,740 MBH, 500 HP		.08	380		90,000	12,700		102,700	118,000	
	1100	23,435 MBH, 700 HP		.07	484		110,500	16,100		126,600	146,000	
	1140	To fire #6, and gas, add		.42	76.190		9,825	2,525		12,350	14,600	
	1180	For duplex package feed system										
	1200	To 3348 MBH boiler, add	Q-7	.54	59.259	Ea.	5,275	1,975		7,250	8,800	
	1220	To 6696 MBH boiler, add		.41	78.049		6,275	2,600		8,875	10,800	
	1240	To 10,044 MBH boiler, add		.38	84.211		7,175	2,800		9,975	12,100	
	1260	To 16,740 MBH boiler, add		.28	114		8,200	3,800		12,000	14,700	
	1280	To 23,435 MBH boiler, add		.25	128		9,675	4,250		13,925	17,000	
	1300	150 PSI steam										
	1310	1005 MBH, 30 HP	Q-6	.33	72.072	Ea.	22,000	2,350		24,350	27,800	
	1320	1675 MBH, 50 HP	"	.26	93.023		27,500	3,025		30,525	34,900	
	1330	3350 MBH, 100 HP	Q-7	.21	152		37,500	5,075		42,575	49,000	
	1340	5025 MBH, 150 HP		.17	192		47,900	6,400		54,300	62,000	
	1350	6700 MBH, 200 HP		.14	223		53,000	7,450		60,450	69,500	
	1360	10,050 MBH, 300 HP		.13	251		78,500	8,375		86,875	98,500	
	1400	Packaged scotch marine, #6 oil										
	1410	15 PSI steam										
	1420	1675 MBH, 50 HP	Q-6	.25	97.166	Ea.	33,900	3,150		37,050	42,100	
	1430	3350 MBH, 100 HP	Q-7	.20	159		36,200	5,300		41,500	47,800	
	1440	5025 MBH, 150 HP		.16	202		45,400	6,725		52,125	60,000	
	1450	6700 MBH, 200 HP		.14	233		49,600	7,775		57,375	66,000	
	1460	8375 MBH, 250 HP		.13	244		63,500	8,125		71,625	82,000	
	1470	10,050 MBH, 300 HP		.12	262		74,500	8,725		83,225	95,000	
	1480	13,400 MBH, 400 HP		.09	351		85,500	11,700		97,200	111,500	
	1500	150 PSI steam										
	1510	1675 MBH, 50 HP	Q-6	.25	97.166	Ea.	40,200	3,150		43,350	49,100	
	1520	3350 MBH, 100 HP	Q-7	.20	159		43,500	5,300		48,800	56,000	
	1530	5025 MBH, 150 HP		.16	202		54,000	6,725		60,725	69,500	
	1540	6700 MBH, 200 HP		.14	233		58,000	7,775		65,775	75,000	
	1550	8375 MBH, 250 HP		.13	244		73,500	8,125		81,625	93,500	
	1560	10,050 MBH, 300 HP		.12	262		83,500	8,725		92,225	105,000	
	1570	13,400 MBH, 400 HP		.09	351		97,000	11,700		108,700	124,000	
	2000	Packaged water tube, #2 oil, steam or hot water, gross output										
	2010	200 MBH	Q-6	1.50	16	Ea.	7,075	520		7,595	8,575	
	2014	275 MBH		1.40	17.143		7,675	560		8,235	9,300	
	2018	360 MBH		1.10	21.818		8,600	710		9,310	10,500	
	2022	520 MBH		.65	36.923		8,800	1,200		10,000	11,500	
	2026	600 MBH		.60	40		8,950	1,300		10,250	11,800	
	2030	720 MBH		.55	43.636		9,325	1,425		10,750	12,500	

15500 | Heat Generation Equipment

15510 | Heating Boilers and Accessories

			Daily Output	Labor-Hours	Unit	Mat.	Labor	Equip.	Total	Total Incl O&P	
700	2034	960 MBH	Q-6	.48	50	Ea.	10,300	1,625		11,925	13,900	700
	2040	1200 MBH	Q-7	.50	64		11,700	2,125		13,825	16,100	
	2044	1440 MBH		.45	71.111		14,300	2,375		16,675	19,300	
	2060	1600 MBH		.40	80		14,700	2,650		17,350	20,100	
	2068	1920 MBH		.35	91.429		15,900	3,050		18,950	22,100	
	2072	2160 MBH		.33	96.970		17,600	3,225		20,825	24,300	
	2080	2400 MBH		.30	106		19,700	3,550		23,250	27,100	
	2100	3200 MBH		.25	128		22,600	4,250		26,850	31,200	
	2120	4800 MBH		.20	160		29,500	5,325		34,825	40,400	
	2200	Gas fired										
	2204	200 MBH	Q-6	1.50	16	Ea.	4,800	520		5,320	6,075	
	2208	275 MBH		1.40	17.143		5,000	560		5,560	6,350	
	2212	360 MBH		1.10	21.818		5,000	710		5,710	6,575	
	2216	520 MBH		.65	36.923		5,900	1,200		7,100	8,325	
	2220	600 MBH		.60	40		5,900	1,300		7,200	8,475	
	2224	720 MBH		.55	43.636		6,525	1,425		7,950	9,325	
	2228	960 MBH		.48	50		7,675	1,625		9,300	10,900	
	2232	1220 MBH	Q-7	.50	64		8,275	2,125		10,400	12,300	
	2236	1440 MBH		.45	71.111		8,800	2,375		11,175	13,300	
	2240	1680 MBH		.40	80		9,825	2,650		12,475	14,800	
	2244	1920 MBH		.35	91.429		10,400	3,050		13,450	16,000	
	2248	2160 MBH		.33	96.970		11,200	3,225		14,425	17,200	
	2252	2400 MBH		.30	106		12,600	3,550		16,150	19,300	
760	0010	**BOILERS, SOLID FUEL**										760
	0020	Wood, coal, wood/paper waste, peat. Steel, 3" diam. horiz.										
	0040	fire tubes, low pressure steam or hot water, natural or										
	0060	mechanical draft. Not including dry base.										
	1000	Natural draft, gross output, 1445 MBH	Q-6	.45	53.333	Ea.	11,600	1,725		13,325	15,400	
	1050	1635 MBH		.43	55.814		12,100	1,825		13,925	16,200	
	1100	1805 MBH		.41	58.537		12,700	1,900		14,600	16,800	
	1150	2160 MBH		.38	63.158		13,600	2,050		15,650	18,100	
	1200	2520 MBH		.32	75		16,400	2,450		18,850	21,800	
	1250	3065 MBH	Q-7	.40	80		17,700	2,650		20,350	23,500	
	1300	3600 MBH		.38	84.211		18,900	2,800		21,700	25,000	
	1350	4500 MBH		.29	110		24,500	3,675		28,175	32,500	
	1400	5400 MBH		.26	123		26,400	4,100		30,500	35,200	
	1450	6295 MBH		.23	139		30,600	4,625		35,225	40,600	
	1500	7200 MBH		.21	152		34,600	5,075		39,675	45,800	
	1550	9000 MBH		.19	168		37,600	5,600		43,200	49,800	
	1600	10,800 MBH		.17	188		43,000	6,250		49,250	57,000	
	1650	12,600 MBH		.15	213		47,100	7,100		54,200	62,500	
	1700	14,390 MBH		.13	246		56,500	8,175		64,675	75,000	
	1750	16,190 MBH		.12	266		61,500	8,875		70,375	81,500	
	1800	17,990 MBH		.10	320		68,500	10,600		79,100	91,500	
	3000	Stoker fired (coal) cast iron with flush jacket and										
	3080	insulation, steam or water, gross output, 148 MBH	Q-6	1.50	16	Ea.	3,450	520		3,970	4,575	
	3100	200 MBH		1.30	18.462		3,950	600		4,550	5,250	
	3120	247 MBH		1.20	20		4,425	650		5,075	5,825	
	3140	300 MBH		1	24		4,950	780		5,730	6,625	
	3160	410 MBH		.96	25		5,675	815		6,490	7,450	
	3180	495 MBH		.88	27.273		6,325	885		7,210	8,300	
	3200	580 MBH		.80	30		7,000	975		7,975	9,175	
	3400	1280 MBH		.36	66.667		17,300	2,175		19,475	22,300	
	3420	1460 MBH		.30	80		19,000	2,600		21,600	24,800	
	3440	1640 MBH		.28	85.714		20,200	2,800		23,000	26,500	

15500 | Heat Generation Equipment

15510 | Heating Boilers and Accessories

			CREW	DAILY OUTPUT	LABOR-HOURS	UNIT	MAT.	LABOR	EQUIP.	TOTAL	TOTAL INCL O&P	
760	3460	1820 MBH	Q-6	.26	92.308	Ea.	21,700	3,000		24,700	28,400	760
	3480	2000 MBH		.25	96		23,100	3,125		26,225	30,100	
	3500	2360 MBH		.23	104		26,100	3,400		29,500	33,800	
	3540	2725 MBH		.20	120		29,000	3,900		32,900	37,800	
	3800	2950 MBH	Q-7	.16	200		48,500	6,650		55,150	63,500	
	3820	3210 MBH		.15	213		51,000	7,100		58,100	66,500	
	3840	3480 MBH		.14	228		53,500	7,600		61,100	70,000	
	3860	3745 MBH		.14	228		56,000	7,600		63,600	73,000	
	3880	4000 MBH		.13	246		58,000	8,175		66,175	76,500	
	3900	4200 MBH		.13	246		60,500	8,175		68,675	79,000	
	3920	4400 MBH		.12	266		63,000	8,875		71,875	82,500	
	3940	4600 MBH		.12	266		65,500	8,875		74,375	85,500	
880	0010	**SWIMMING POOL HEATERS** Not including wiring, external										880
	0020	piping, base or pad,										
	0060	Gas fired, input, 115 MBH	Q-6	3	8	Ea.	1,525	260		1,785	2,075	
	0100	135 MBH		2	12		1,725	390		2,115	2,475	
	0160	155 MBH		1.50	16		1,825	520		2,345	2,775	
	0180	175 MBH		1.30	18.462		2,025	600		2,625	3,125	
	0200	190 MBH		1	24		2,350	780		3,130	3,775	
	0220	235 MBH		.70	34.286		2,875	1,125		4,000	4,850	
	0240	285 MBH		.60	40		3,700	1,300		5,000	6,050	
	0260	365 MBH		.50	48		4,175	1,550		5,725	6,925	
	0280	500 MBH		.40	60		5,700	1,950		7,650	9,225	
	0300	600 MBH		.35	68.571		6,550	2,225		8,775	10,600	
	0320	800 MBH		.33	72.727		7,825	2,375		10,200	12,200	
	0360	1000 MBH		.22	109		9,125	3,550		12,675	15,500	
	0370	1,200 MBH		.21	114		11,400	3,725		15,125	18,100	
	0380	1,500 MBH		.19	126		14,100	4,100		18,200	21,700	
	0400	1,800 MBH		.14	171		15,900	5,575		21,475	25,900	
	0410	2,450 MBH		.13	184		21,200	6,000		27,200	32,400	
	0420	3,000 MBH		.11	218		25,800	7,100		32,900	39,100	
	0440	3,750 MBH		.09	266		32,100	8,675		40,775	48,400	
	1000	Oil fired, gross output, 97 MBH		1.70	14.118		1,750	460		2,210	2,625	
	1020	118 MBH		1.70	14.118		1,775	460		2,235	2,650	
	1040	134 MBH		1.60	15		1,900	490		2,390	2,800	
	1060	161 MBH		1.30	18.462		2,300	600		2,900	3,425	
	1080	187 MBH		1.20	20		2,425	650		3,075	3,650	
	1100	214 MBH		1.10	21.818		2,775	710		3,485	4,125	
	1120	262 MBH		.89	26.966		3,125	875		4,000	4,775	
	1140	340 MBH		.68	35.294		4,150	1,150		5,300	6,275	
	1160	420 MBH		.58	41.379		4,750	1,350		6,100	7,250	
	1180	525 MBH		.46	52.174		6,075	1,700		7,775	9,275	
	1200	630 MBH		.41	58.537		6,775	1,900		8,675	10,400	
	1220	735 MBH		.35	68.571		8,350	2,225		10,575	12,600	
	1240	840 MBH		.33	72.727		9,550	2,375		11,925	14,100	
	1260	1,050 MBH		.27	88.889		11,600	2,900		14,500	17,200	
	1280	1,365 MBH		.21	114		14,900	3,725		18,625	22,000	
	1300	1,680 MBH		.18	133		17,600	4,325		21,925	25,900	
	1320	2,310 MBH		.13	184		22,700	6,000		28,700	34,000	
	1340	2,835 MBH		.11	218		28,200	7,100		35,300	41,700	
	1360	3,150 MBH		.09	266		31,700	8,675		40,375	48,000	
	2000	Electric, 12 KW, 4,800 gallon pool	Q-19	3	8		1,525	259		1,784	2,075	
	2020	15 KW, 7,200 gallon pool		2.80	8.571		1,525	277		1,802	2,100	
	2040	24 KW, 9,600 gallon pool		2.40	10		2,050	325		2,375	2,725	
	2060	30 KW, 12,000 gallon pool		2	12		2,125	390		2,515	2,900	
	2080	35 KW, 14,400 gallon pool		1.60	15		2,125	485		2,610	3,050	

15500 | Heat Generation Equipment

15510 | Heating Boilers and Accessories

			CREW	DAILY OUTPUT	LABOR-HOURS	UNIT	MAT.	LABOR	EQUIP.	TOTAL	TOTAL INCL O&P	
880	2100	55 KW, 24,000 gallon pool	Q-19	1.20	20	Ea.	2,900	645		3,545	4,175	880
	9000	To select pool heater: 12 BTUH x S.F. pool area										
	9010	X temperature differential =required output										
	9050	For electric, KW = gallons x 2.5 divided by 1000										
	9100	For family home type pool, double the										
	9110	Rated gallon capacity = 1/2°F rise per hour										

15520 | Feedwater Equipment

			CREW	DAILY OUTPUT	LABOR-HOURS	UNIT	MAT.	LABOR	EQUIP.	TOTAL	TOTAL INCL O&P	
050	0010	**SHOT CHEMICAL FEEDER**										050
	0200	Shot chem feeder, by pass, in line mount, 125 PSIG, 1.7 gallon	2 Stpi	1.80	8.889	Ea.	215	310		525	705	
	0220	Floor mount, 175 PSIG, 5 gallon		1.24	12.903		285	450		735	995	
	0230	12 gallon		.93	17.204		455	600		1,055	1,400	
	0300	150 lb., ASME, 5 gallon		1.24	12.903		1,200	450		1,650	2,000	
	0320	10 gallon		.93	17.204		1,300	600		1,900	2,325	
	0400	300 lb., ASME, 5 gallon		1.24	12.903		3,225	450		3,675	4,200	
	0420	10 gallon		.93	17.204		4,000	600		4,600	5,300	

15530 | Furnaces

			CREW	DAILY OUTPUT	LABOR-HOURS	UNIT	MAT.	LABOR	EQUIP.	TOTAL	TOTAL INCL O&P	
200	0010	**FURNACE COMPONENTS AND COMBINATIONS**										200
	0080	Coils, A/C evaporator, for gas or oil furnaces										
	0090	Add-on, with holding charge										
	0100	Upflow										
	0120	1-1/2 ton cooling	Q-5	4	4	Ea.	127	126		253	330	
	0130	2 ton cooling		3.70	4.324		153	136		289	375	
	0140	3 ton cooling		3.30	4.848		191	152		343	440	
	0150	4 ton cooling		3	5.333		270	167		437	550	
	0160	5 ton cooling		2.70	5.926		345	186		531	660	
	0300	Downflow										
	0330	2-1/2 ton cooling	Q-5	3	5.333	Ea.	178	167		345	450	
	0340	3-1/2 ton cooling		2.60	6.154		239	193		432	555	
	0350	5 ton cooling		2.20	7.273		345	228		573	725	
	0600	Horizontal										
	0630	2 ton cooling	Q-5	3.90	4.103	Ea.	180	129		309	390	
	0640	3 ton cooling		3.50	4.571		206	143		349	445	
	0650	4 ton cooling		3.20	5		259	157		416	520	
	0660	5 ton cooling		2.90	5.517		345	173		518	640	
	2000	Cased evaporator coils for air handlers										
	2100	1-1/2 ton cooling	Q-5	4.40	3.636	Ea.	182	114		296	370	
	2110	2 ton cooling		4.10	3.902		183	122		305	385	
	2120	2-1/2 ton cooling		3.90	4.103		185	129		314	395	
	2130	3 ton cooling		3.70	4.324		222	136		358	450	
	2140	3-1/2 ton cooling		3.50	4.571		272	143		415	515	
	2150	4 ton cooling		3.20	5		273	157		430	535	
	2160	5 ton cooling		2.90	5.517		340	173		513	630	
	3010	Air handler, modular										
	3100	With cased evaporator cooling coil										
	3120	1-1/2 ton cooling	Q-5	3.80	4.211	Ea.	480	132		612	725	
	3130	2 ton cooling		3.50	4.571		480	143		623	740	
	3140	2-1/2 ton cooling		3.30	4.848		540	152		692	825	
	3150	3 ton cooling		3.10	5.161		540	162		702	840	
	3160	3-1/2 ton cooling		2.90	5.517		605	173		778	925	
	3170	4 ton cooling		2.50	6.400		605	201		806	975	
	3180	5 ton cooling		2.10	7.619		725	239		964	1,150	
	3500	With no cooling coil										
	3520	1-1/2 ton coil size	Q-5	12	1.333	Ea.	296	42		338	390	
	3530	2 ton coil size		10	1.600		345	50		395	450	

15500 | Heat Generation Equipment

15530 | Furnaces

			CREW	DAILY OUTPUT	LABOR-HOURS	UNIT	2001 BARE COSTS MAT.	LABOR	EQUIP.	TOTAL	TOTAL INCL O&P	
200	3540	2-1/2 ton coil size	Q-5	10	1.600	Ea.	320	50		370	425	200
	3554	3 ton coil size		9	1.778		360	56		416	480	
	3560	3-1/2 ton coil size		9	1.778		335	56		391	450	
	3570	4 ton coil size		8.50	1.882		310	59		369	430	
	3580	5 ton coil size	▼	8	2	▼	395	63		458	530	
	4000	Heater for above handlers										
	4120	5 kW, 17.1 MBH	Q-5	16	1	Ea.	257	31.50		288.50	330	
	4130	7.5 kW, 25.6 MBH		15.60	1.026		261	32		293	335	
	4140	10 kW, 34.2 MBH		15.20	1.053		310	33		343	390	
	4150	12.5 KW, 42.7 MBH		14.80	1.081		310	34		344	390	
	4160	15 KW, 51.2 MBH		14.40	1.111		385	35		420	480	
	4170	25 KW, 85.4 MBH		14	1.143		385	36		421	480	
	4180	30 KW, 102 MBH	▼	13	1.231	▼	525	38.50		563.50	640	
400	0010	**FURNACES** Hot air heating, blowers, standard controls										400
	0020	not including gas, oil or flue piping										
	1000	Electric, UL listed										
	1020	10.2 MBH	Q-20	5	4	Ea.	315	124		439	535	
	1040	17.1 MBH		4.80	4.167		330	129		459	560	
	1060	27.3 MBH		4.60	4.348		395	135		530	640	
	1100	34.1 MBH		4.40	4.545		405	141		546	660	
	1120	51.6 MBH		4.20	4.762		505	148		653	780	
	1140	68.3 MBH		4	5		610	155		765	910	
	1160	85.3 MBH	▼	3.80	5.263	▼	670	163		833	985	
	3000	Gas, AGA certified, upflow, direct drive models										
	3020	45 MBH input	Q-9	4	4	Ea.	435	121		556	665	
	3040	60 MBH input		3.80	4.211		580	127		707	835	
	3060	75 MBH input		3.60	4.444		615	134		749	885	
	3100	100 MBH input		3.20	5		655	151		806	955	
	3120	125 MBH input		3	5.333		755	161		916	1,075	
	3130	150 MBH input		2.80	5.714		875	173		1,048	1,225	
	3140	200 MBH input		2.60	6.154		1,475	186		1,661	1,900	
	3160	300 MBH input		2.30	6.957		1,600	210		1,810	2,075	
	3180	400 MBH input	▼	2	8	▼	2,050	242		2,292	2,625	
	3200	Gas fired wall furnace										
	3204	Horizontal, recessed										
	3208	7.7 MBH	Q-9	7	2.286	Ea.	400	69		469	545	
	3212	14 MBH		6.50	2.462		430	74.50		504.50	590	
	3216	24 MBH		5	3.200		520	96.50		616.50	720	
	3220	49 MBH		4	4		515	121		636	750	
	3224	65 MBH	▼	3.60	4.444	▼	695	134		829	965	
	3240	Up-flow, recessed										
	3244	7.7 MBH	Q-9	7	2.286	Ea.	254	69		323	385	
	3248	14 MBH		6.50	2.462		294	74.50		368.50	440	
	3252	24 MBH		5	3.200		360	96.50		456.50	550	
	3256	49 MBH		4	4		705	121		826	960	
	3260	65 MBH	▼	3.60	4.444	▼	730	134		864	1,000	
	3500	Gas furnace										
	3510	Up-flow										
	3514	1200 CFM, 45 MBH	Q-9	4	4	Ea.	685	121		806	940	
	3518	75 MBH		3.60	4.444		755	134		889	1,025	
	3522	100 MBH		3.20	5		865	151		1,016	1,200	
	3526	125 MBH		3	5.333		1,000	161		1,161	1,350	
	3530	150 MBH		2.80	5.714		1,225	173		1,398	1,625	
	3540	2000 CFM, 75 MBH		3.60	4.444		830	134		964	1,125	
	3544	100 MBH	▼	3.20	5	▼	940	151		1,091	1,250	

15500 | Heat Generation Equipment

15530 | Furnaces

		Crew	Daily Output	Labor-Hours	Unit	Mat.	Labor	Equip.	Total	Total Incl O&P
400											400
3548	120 MBH	Q-9	3	5.333	Ea.	1,075	161		1,236	1,450
3552	150 MBH	↓	2.80	5.714	↓	1,300	173		1,473	1,700
3560	Horizontal flow									
3564	1200 CFM, 45 MBH	Q-9	4	4	Ea.	690	121		811	940
3568	60 MBH	↓	3.80	4.211		790	127		917	1,050
3572	70 MBH		3.60	4.444		800	134		934	1,075
3576	2000 CFM, 90 MBH		3.30	4.848		865	146		1,011	1,175
3580	110 MBH	↓	3.10	5.161	↓	905	156		1,061	1,250
6000	Oil, UL listed, atomizing gun type burner									
6020	56 MBH output	Q-9	3.60	4.444	Ea.	745	134		879	1,025
6030	84 MBH output		3.50	4.571		775	138		913	1,075
6040	95 MBH output		3.40	4.706		790	142		932	1,100
6060	134 MBH output		3.20	5		1,100	151		1,251	1,425
6080	151 MBH output		3	5.333		1,225	161		1,386	1,575
6100	200 MBH input		2.60	6.154		1,575	186		1,761	2,000
6120	300 MBH input		2.30	6.957		1,825	210		2,035	2,325
6140	400 MBH input	↓	2	8	↓	2,150	242		2,392	2,725
6200	Up flow									
6210	105 MBH	Q-9	3.40	4.706	Ea.	1,125	142		1,267	1,475
6220	140 MBH		3.30	4.848		1,250	146		1,396	1,600
6230	168 MBH	↓	2.90	5.517	↓	1,075	167		1,242	1,425
6260	Lo-boy style									
6270	105 MBH	Q-9	3.40	4.706	Ea.	1,125	142		1,267	1,475
6280	140 MBH		3.30	4.848		1,350	146		1,496	1,700
6290	189 MBH	↓	2.80	5.714	↓	1,350	173		1,523	1,750
6320	Down flow									
6330	105 MBH	Q-9	3.40	4.706	Ea.	1,175	142		1,317	1,525
6340	140 MBH		3.30	4.848		1,350	146		1,496	1,725
6350	168 MBH	↓	2.90	5.517	↓	1,375	167		1,542	1,750
6380	Horizontal									
6390	105 MBH	Q-9	3.40	4.706	Ea.	1,175	142		1,317	1,525
6400	140 MBH		3.30	4.848		1,350	146		1,496	1,700
6410	189 MBH	↓	2.80	5.714	↓	1,400	173		1,573	1,825
8500	Wood and coal combination complete									
8520	95 MBH output	Q-9	3	5.333	Ea.	4,350	161		4,511	5,050
8540	140 MBH output (oil)		3	5.333		4,350	161		4,511	5,050
8560	163 MBH output (oil)	↓	3	5.333	↓	4,350	161		4,511	5,050
440	0010 **FURNACES, COMBINATION SYSTEMS** Heating, cooling, electric air										440
	0020 cleaner, humidification, dehumidification.									
2000	Gas fired, 80 MBH heat output, 24 MBH cooling	Q-9	1.20	13.333	Ea.	2,775	405		3,180	3,675
2020	80 MBH heat output, 36 MBH cooling		1.20	13.333		3,050	405		3,455	3,975
2040	100 MBH heat output, 29 MBH cooling		1	16		3,100	485		3,585	4,175
2060	100 MBH heat output, 36 MBH cooling		1	16		3,275	485		3,760	4,350
2080	100 MBH heat output, 47 MBH cooling	↓	.90	17.778		3,875	535		4,410	5,075
2100	120 MBH heat output, 29 MBH cooling	Q-10	1.30	18.462		3,300	580		3,880	4,525
2120	120 MBH heat output, 42 MBH cooling		1.30	18.462		3,475	580		4,055	4,725
2140	120 MBH heat output, 47 MBH cooling		1.20	20		3,875	625		4,500	5,250
2160	120 MBH heat output, 55 MBH cooling		1.10	21.818		4,175	685		4,860	5,650
2180	144 MBH heat output, 42 MBH cooling		1.20	20		3,625	625		4,250	4,950
2200	144 MBH heat output, 47 MBH cooling		1.20	20		4,025	625		4,650	5,400
2220	144 MBH heat output, 58 MBH cooling	↓	1	24	↓	4,225	750		4,975	5,800
2500	For precharged tubing with connection									
2520	15 feet				Ea.	116			116	127
2540	25 feet					154			154	170
2560	35 feet					199			199	219

15500 | Heat Generation Equipment

15530 | Furnaces

		CREW	DAILY OUTPUT	LABOR-HOURS	UNIT	2001 BARE COSTS MAT.	LABOR	EQUIP.	TOTAL	TOTAL INCL O&P	
440	3000 Oil fired, 84 MBH heat output, 24 MBH cooling	Q-9	1.20	13.333	Ea.	3,100	405		3,505	4,050	440
	3020 84 MBH heat output, 36 MBH cooling		1.20	13.333		3,425	405		3,830	4,375	
	3040 95.2 MBH heat output, 29 MBH cooling		1	16		3,350	485		3,835	4,425	
	3060 95.2 MBH heat output, 36 MBH cooling		1	16		3,475	485		3,960	4,575	
	3080 95.2 MBH heat output, 47 MBH cooling		.90	17.778		4,025	535		4,560	5,250	
	3100 112 MBH heat output, 29 MBH cooling	Q-10	1.30	18.462		3,575	580		4,155	4,825	
	3120 112 MBH heat output, 42 MBH cooling		1.30	18.462		3,875	580		4,455	5,175	
	3140 112 MBH heat output, 47 MBH cooling		1.20	20		4,300	625		4,925	5,700	
	3160 112 MBH heat output, 58 MBH cooling		1.10	21.818		4,700	685		5,385	6,200	
	3180 151.2 MBH heat output, 42 MBH cooling		1.20	20		4,175	625		4,800	5,575	
	3200 151.2 MBH heat output, 47 MBH cooling		1.20	20		4,575	625		5,200	6,000	
	3240 151.2 MBH heat output, 58 MBH cooling		1	24		4,775	750		5,525	6,400	
	3500 For precharged tubing with connection, add										
	3520 15 feet				Ea.	116			116	127	
	3540 25 feet					154			154	170	
	3560 35 feet					185			185	203	

15540 | Fuel-Fired Heaters

		CREW	DAILY OUTPUT	LABOR-HOURS	UNIT	MAT.	LABOR	EQUIP.	TOTAL	TOTAL INCL O&P	
300	0010 **DUCT FURNACES** Includes burner, controls, stainless steel										300
	0020 heat exchanger. Gas fired, electric ignition										
	0030 Indoor installation										
	0080 100 MBH output	Q-5	5	3.200	Ea.	1,025	100		1,125	1,275	
	0100 120 MBH output		4	4		1,175	126		1,301	1,475	
	0130 200 MBH output		2.70	5.926		1,625	186		1,811	2,050	
	0140 240 MBH output		2.30	6.957		1,625	218		1,843	2,125	
	0160 280 MBH output		2	8		1,875	251		2,126	2,450	
	0180 320 MBH output		1.60	10		2,050	315		2,365	2,725	
	0300 For powered venter and adapter, add					277			277	305	
	0500 For required flue pipe, see 15550										
	1000 Outdoor installation, with vent cap										
	1020 75 MBH output	Q-5	4	4	Ea.	1,600	126		1,726	1,975	
	1040 94 MBH output		4	4		1,725	126		1,851	2,100	
	1060 120 MBH output		4	4		1,975	126		2,101	2,375	
	1080 157 MBH output		3.50	4.571		2,300	143		2,443	2,750	
	1100 187 MBH output		3	5.333		2,825	167		2,992	3,375	
	1120 225 MBH output		2.50	6.400		2,950	201		3,151	3,550	
	1140 300 MBH output		1.80	8.889		3,700	279		3,979	4,500	
	1160 375 MBH output		1.60	10		4,050	315		4,365	4,925	
	1180 450 MBH output		1.40	11.429		4,250	360		4,610	5,225	
	1200 600 MBH output		1	16		5,175	500		5,675	6,425	
	1300 Aluminized exchanger, subtract					15%					
	1400 For powered venter, add					10%					
	1500 For two stage gas valve, add					156			156	172	
900	0010 **SPACE HEATERS** Cabinet, grilles, fan, controls, burner,										900
	0020 thermostat, no piping. For flue see 15550										
	1000 Gas fired, floor mounted										
	1100 60 MBH output	Q-5	10	1.600	Ea.	530	50		580	660	
	1120 80 MBH output		9	1.778		535	56		591	675	
	1140 100 MBH output		8	2		580	63		643	730	
	1160 120 MBH output		7	2.286		650	71.50		721.50	825	
	1180 180 MBH output		6	2.667		790	83.50		873.50	995	
	1500 Rooftop mounted, gravity vent, stainless steel exchanger										
	1520 75 MBH output	Q-6	4	6	Ea.	3,275	195		3,470	3,900	
	1540 95 MBH output		3.60	6.667		3,325	217		3,542	4,000	
	1560 120 MBH output		3.30	7.273		3,575	237		3,812	4,275	

15500 | Heat Generation Equipment

15540 | Fuel-Fired Heaters

			CREW	DAILY OUTPUT	LABOR-HOURS	UNIT	MAT.	LABOR	EQUIP.	TOTAL	TOTAL INCL O&P	
900	1580	159 MBH output	Q-6	3	8	Ea.	3,675	260		3,935	4,450	900
	1600	190 MBH output		2.60	9.231		4,150	300		4,450	5,025	
	1620	225 MBH output		2.30	10.435		4,225	340		4,565	5,150	
	1640	300 MBH output		1.90	12.632		5,025	410		5,435	6,150	
	1660	375 MBH output		1.40	17.143		5,700	560		6,260	7,125	
	1680	450 MBH output		1.20	20		5,800	650		6,450	7,375	
	1700	600 MBH output		1	24		7,000	780		7,780	8,875	
	1720	750 MBH output		.80	30		12,000	975		12,975	14,700	
	1740	900 MBH output		.60	40		13,400	1,300		14,700	16,800	
	1760	1200 MBH output		.30	80		16,300	2,600		18,900	21,900	
	1800	For power vent, add					10%					
	1900	For aluminized steel exchanger, subtract					10%					
	2000	Suspension mounted, propeller fan, 20 MBH output	Q-5	8.50	1.882		410	59		469	540	
	2020	40 MBH output		7.50	2.133		450	67		517	595	
	2040	60 MBH output		7	2.286		505	71.50		576.50	665	
	2060	80 MBH output		6	2.667		560	83.50		643.50	740	
	2080	100 MBH output		5.50	2.909		665	91.50		756.50	875	
	2100	130 MBH output		5	3.200		745	100		845	970	
	2120	140 MBH output		4.50	3.556		800	112		912	1,050	
	2140	160 MBH output		4	4		830	126		956	1,100	
	2160	180 MBH output		3.50	4.571		905	143		1,048	1,200	
	2180	200 MBH output		3	5.333		965	167		1,132	1,325	
	2200	240 MBH output		2.70	5.926		1,150	186		1,336	1,525	
	2220	280 MBH output		2.30	6.957		1,350	218		1,568	1,800	
	2240	320 MBH output		2	8		1,550	251		1,801	2,075	
	2500	For powered venter and adapter, add					248			248	272	
	3000	Suspension mounted, blower type, 40 MBH output	Q-5	6.80	2.353		715	74		789	895	
	3020	60 MBH output		6.60	2.424		790	76		866	980	
	3040	84 MBH output		5.80	2.759		845	86.50		931.50	1,050	
	3060	104 MBH output		5.20	3.077		975	96.50		1,071.50	1,225	
	3080	140 MBH output		4.30	3.721		1,175	117		1,292	1,475	
	3100	180 MBH output		3.30	4.848		1,225	152		1,377	1,575	
	3120	240 MBH output		2.50	6.400		1,775	201		1,976	2,275	
	3140	280 MBH output		2	8		2,000	251		2,251	2,575	
	4000	Suspension mounted, sealed combustion system,										
	4020	Aluminized steel exchanger, powered vent										
	4040	100 MBH output	Q-5	5	3.200	Ea.	1,550	100		1,650	1,875	
	4060	120 MBH output		4.70	3.404		1,700	107		1,807	2,000	
	4080	160 MBH output		3.70	4.324		1,875	136		2,011	2,250	
	4100	200 MBH output		2.90	5.517		2,225	173		2,398	2,700	
	4120	240 MBH output		2.50	6.400		2,375	201		2,576	2,900	
	4140	320 MBH output		1.70	9.412		2,725	295		3,020	3,450	
	5000	Wall furnace, 17.5 MBH output		6	2.667		495	83.50		578.50	670	
	5020	24 MBH output		5	3.200		510	100		610	715	
	5040	35 MBH output		4	4		690	126		816	950	
	6000	Oil fired, suspension mounted, 94 MBH output		4	4		1,925	126		2,051	2,325	
	6040	140 MBH output		3	5.333		2,050	167		2,217	2,500	
	6060	184 MBH output		3	5.333		2,200	167		2,367	2,675	

15550 | Breechings, Chimneys & Stacks

			CREW	DAILY OUTPUT	LABOR-HOURS	UNIT	MAT.	LABOR	EQUIP.	TOTAL	TOTAL INCL O&P	
200	0010	**DRAFT CONTROLS**										200
	1000	Barometric, gas fired system only, 6" size for 5" and 6" pipes	1 Shee	20	.400	Ea.	40	13.40		53.40	64.50	
	1020	7" size, for 6" and 7" pipes		19	.421		44	14.15		58.15	70.50	
	1040	8" size, for 7" and 8" pipes		18	.444		51.50	14.90		66.40	80	
	1060	9" size, for 8" and 9" pipes		16	.500		58	16.80		74.80	89.50	
	2000	All fuel, oil, oil/gas, coal										

15500 | Heat Generation Equipment

15550 | Breechings, Chimneys & Stacks

			CREW	DAILY OUTPUT	LABOR-HOURS	UNIT	MAT.	LABOR	EQUIP.	TOTAL	TOTAL INCL O&P	
200	2020	10" for 9" and 10" pipes	1 Shee	15	.533	Ea.	87	17.90		104.90	123	200
	2040	12" for 11" and 12" pipes		15	.533		115	17.90		132.90	155	
	2060	14" for 13" and 14" pipes		14	.571		144	19.15		163.15	189	
	2080	16" for 15" and 16" pipes		13	.615		202	20.50		222.50	254	
	2100	18" for 17" and 18" pipes		12	.667		260	22.50		282.50	320	
	2120	20" for 19" and 21" pipes		10	.800		315	27		342	390	
	2140	24" for 22" and 25" pipes	Q-9	12	1.333		390	40.50		430.50	485	
	2160	28" for 26" and 30" pipes		10	1.600		465	48.50		513.50	585	
	2180	32" for 31" and 34" pipes		8	2		590	60.50		650.50	745	
	3260	For thermal switch for above, add	1 Shee	24	.333		36.50	11.20		47.70	58	
	5000	Vent damper, bi-metal, gas, 3" diameter	Q-9	24	.667		31.50	20		51.50	65.50	
	5010	4" diameter		24	.667		35.50	20		55.50	70	
	5020	5" diameter		23	.696		42.50	21		63.50	79	
	5030	6" diameter		22	.727		48	22		70	86.50	
	5040	7" diameter		21	.762		55	23		78	95.50	
	5050	8" diameter		20	.800		63	24		87	107	
	5101	Electric, automatic, gas, 4" diameter		24	.667		137	20		157	181	
	5110	5" diameter		23	.696		139	21		160	186	
	5121	6" diameter		22	.727		142	22		164	190	
	5130	7" diameter		21	.762		144	23		167	195	
	5140	8" diameter		20	.800		147	24		171	200	
	5150	9" diameter		20	.800		150	24		174	203	
	5160	10" diameter		19	.842		155	25.50		180.50	210	
	5170	12" diameter		19	.842		157	25.50		182.50	213	
	5250	Automatic, oil, 4" diameter		24	.667		153	20		173	199	
	5260	5" diameter		23	.696		156	21		177	204	
	5270	6" diameter		22	.727		159	22		181	209	
	5280	7" diameter		21	.762		162	23		185	214	
	5290	8" diameter		20	.800		169	24		193	224	
	5300	9" diameter		20	.800		171	24		195	227	
	5310	10" diameter		19	.842		172	25.50		197.50	229	
	5320	12" diameter		19	.842		175	25.50		200.50	232	
	5330	14" diameter		18	.889		234	27		261	299	
	5340	16" diameter		17	.941		271	28.50		299.50	340	
	5350	18" diameter		16	1		305	30		335	380	
440	0010	**VENT CHIMNEY** Prefab metal, U.L. listed										440
	0020	Gas, double wall, galvanized steel										
	0080	3" diameter	Q-9	72	.222	V.L.F.	3.48	6.70		10.18	14.20	
	0100	4" diameter		68	.235		4.35	7.10		11.45	15.75	
	0120	5" diameter		64	.250		5.05	7.55		12.60	17.25	
	0140	6" diameter		60	.267		6.25	8.05		14.30	19.30	
	0160	7" diameter		56	.286		8.65	8.65		17.30	23	
	0180	8" diameter		52	.308		9.65	9.30		18.95	25	
	0200	10" diameter		48	.333		20.50	10.05		30.55	38	
	0220	12" diameter		44	.364		27	11		38	47	
	0240	14" diameter		42	.381		46	11.50		57.50	68.50	
	0260	16" diameter		40	.400		62	12.10		74.10	86.50	
	0280	18" diameter		38	.421		80	12.70		92.70	108	
	0300	20" diameter	Q-10	36	.667		94.50	21		115.50	136	
	0320	22" diameter		34	.706		120	22		142	166	
	0340	24" diameter		32	.750		147	23.50		170.50	199	
	0360	26" diameter		31	.774		177	24.50		201.50	233	
	0380	28" diameter		30	.800		187	25		212	244	
	0400	30" diameter		28	.857		197	27		224	259	
	0420	32" diameter		27	.889		229	28		257	295	

15500 | Heat Generation Equipment

15550 | Breechings, Chimneys & Stacks

		CREW	DAILY OUTPUT	LABOR-HOURS	UNIT	MAT.	LABOR	EQUIP.	TOTAL	TOTAL INCL O&P
0440	34" diameter	Q-10	26	.923	V.L.F.	261	29		290	330
0460	36" diameter		25	.960		275	30		305	350
0480	38" diameter		24	1		300	31.50		331.50	380
0500	40" diameter		23	1.043		335	32.50		367.50	420
0520	42" diameter		22	1.091		350	34		384	440
0540	44" diameter		21	1.143		390	36		426	485
0560	46" diameter		20	1.200		430	37.50		467.50	530
0580	48" diameter		19	1.263		475	39.50		514.50	580
0600	For 4", 5" and 6" oval, add					50%				
0650	Gas, double wall, galvanized steel, fittings									
0660	Elbow 45°, 3" diameter	Q-9	36	.444	Ea.	8.40	13.40		21.80	30
0670	4" diameter		34	.471		10.20	14.20		24.40	33.50
0680	5" diameter		32	.500		12	15.10		27.10	36.50
0690	6" diameter		30	.533		14.35	16.10		30.45	41
0700	7" diameter		28	.571		20.50	17.25		37.75	49.50
0710	8" diameter		26	.615		24	18.60		42.60	54.50
0720	10" diameter		24	.667		59	20		79	96
0730	12" diameter		22	.727		75	22		97	117
0740	14" diameter		21	.762		105	23		128	151
0750	16" diameter		20	.800		131	24		155	183
0760	18" diameter		19	.842		177	25.50		202.50	234
0770	20" diameter	Q-10	18	1.333		194	42		236	278
0780	22" diameter		17	1.412		292	44		336	390
0790	24" diameter		16	1.500		380	47		427	495
0800	26" diameter		16	1.500		750	47		797	900
0810	28" diameter		15	1.600		770	50		820	930
0820	30" diameter		14	1.714		815	53.50		868.50	980
0830	32" diameter		14	1.714		915	53.50		968.50	1,075
0840	34" diameter		13	1.846		1,000	58		1,058	1,200
0850	36" diameter		12	2		1,025	62.50		1,087.50	1,225
0860	38" diameter		12	2		1,125	62.50		1,187.50	1,325
0870	40" diameter		11	2.182		1,175	68.50		1,243.50	1,400
0880	42" diameter		11	2.182		1,225	68.50		1,293.50	1,450
0890	44" diameter		10	2.400		1,325	75		1,400	1,575
0900	46" diameter		10	2.400		1,400	75		1,475	1,675
0910	48" diameter		10	2.400		1,475	75		1,550	1,750
0916	Adjustable length									
0918	3" diameter, to 12"	Q-9	36	.444	Ea.	7.05	13.40		20.45	28.50
0920	4" diameter, to 12"		34	.471		8.35	14.20		22.55	31
0924	6" diameter, to 12"		30	.533		10.80	16.10		26.90	37
0928	8" diameter, to 12"		26	.615		15.85	18.60		34.45	46
0930	10" diameter, to 18"		24	.667		59.50	20		79.50	96.50
0932	12" diameter, to 18"		22	.727		72.50	22		94.50	114
0936	16" diameter, to 18"		20	.800		140	24		164	192
0938	18" diameter, to 18"		19	.842		167	25.50		192.50	224
0944	24" diameter, to 18"	Q-10	16	1.500		299	47		346	405
0950	Elbow 90°, adjustable, 3" diameter	Q-9	36	.444		14.40	13.40		27.80	36.50
0960	4" diameter		34	.471		17.50	14.20		31.70	41.50
0970	5" diameter		32	.500		20.50	15.10		35.60	46
0980	6" diameter		30	.533		24.50	16.10		40.60	52
0990	7" diameter		28	.571		38.50	17.25		55.75	69
1010	8" diameter		26	.615		41	18.60		59.60	73.50
1020	Wall thimble, 4 to 7" adjustable, 3" diameter		36	.444		9.10	13.40		22.50	30.50
1022	4" diameter		34	.471		10.15	14.20		24.35	33
1024	5" diameter		32	.500		11.40	15.10		26.50	36
1026	6" diameter		30	.533		14	16.10		30.10	40.50

15500 | Heat Generation Equipment

15550 | Breechings, Chimneys & Stacks

		CREW	DAILY OUTPUT	LABOR-HOURS	UNIT	2001 BARE COSTS MAT.	LABOR	EQUIP.	TOTAL	TOTAL INCL O&P
1028	7" diameter	Q-9	28	.571	Ea.	16.25	17.25		33.50	44.50
1030	8" diameter		26	.615		16.85	18.60		35.45	47
1040	Roof flashing, 3" diameter		36	.444		5.20	13.40		18.60	26
1050	4" diameter		34	.471		5.85	14.20		20.05	28.50
1060	5" diameter		32	.500		9.10	15.10		24.20	33.50
1070	6" diameter		30	.533		12.75	16.10		28.85	39
1080	7" diameter		28	.571		16.10	17.25		33.35	44
1090	8" diameter		26	.615		17.50	18.60		36.10	48
1100	10" diameter		24	.667		47.50	20		67.50	83.50
1110	12" diameter		22	.727		61	22		83	101
1120	14" diameter		20	.800		75	24		99	120
1130	16" diameter		18	.889		92.50	27		119.50	144
1140	18" diameter		16	1		125	30		155	185
1150	20" diameter	Q-10	18	1.333		163	42		205	244
1160	22" diameter		14	1.714		206	53.50		259.50	310
1170	24" diameter		12	2		247	62.50		309.50	370
1200	Tee, 3" diameter	Q-9	27	.593		21.50	17.90		39.40	51
1210	4" diameter		26	.615		22	18.60		40.60	53
1220	5" diameter		25	.640		24	19.35		43.35	56.50
1230	6" diameter		24	.667		26	20		46	59.50
1240	7" diameter		23	.696		28.50	21		49.50	64
1250	8" diameter		22	.727		39	22		61	77
1260	10" diameter		21	.762		106	23		129	153
1270	12" diameter		20	.800		128	24		152	178
1280	14" diameter		18	.889		208	27		235	271
1290	16" diameter		16	1		295	30		325	370
1300	18" diameter		14	1.143		370	34.50		404.50	460
1310	20" diameter	Q-10	17	1.412		500	44		544	620
1320	22" diameter		13	1.846		620	58		678	775
1330	24" diameter		12	2		705	62.50		767.50	875
1340	26" diameter		12	2		1,025	62.50		1,087.50	1,225
1350	28" diameter		11	2.182		1,150	68.50		1,218.50	1,375
1360	30" diameter		10	2.400		1,250	75		1,325	1,500
1370	32" diameter		10	2.400		1,475	75		1,550	1,750
1380	34" diameter		9	2.667		1,475	83.50		1,558.50	1,750
1390	36" diameter		9	2.667		1,775	83.50		1,858.50	2,075
1400	38" diameter		8	3		1,775	94		1,869	2,100
1410	40" diameter		7	3.429		1,925	107		2,032	2,300
1420	42" diameter		6	4		2,100	125		2,225	2,525
1430	44" diameter		5	4.800		2,225	150		2,375	2,675
1440	46" diameter		5	4.800		2,450	150		2,600	2,900
1450	48" diameter		5	4.800		2,600	150		2,750	3,100
1460	Tee cap, 3" diameter	Q-9	45	.356		1.53	10.75		12.28	18.30
1470	4" diameter		42	.381		1.69	11.50		13.19	19.60
1480	5" diameter		40	.400		2.28	12.10		14.38	21
1490	6" diameter		37	.432		2.63	13.05		15.68	23
1500	7" diameter		35	.457		4.29	13.80		18.09	26
1510	8" diameter		34	.471		5	14.20		19.20	27.50
1520	10" diameter		32	.500		9.60	15.10		24.70	34
1530	12" diameter		30	.533		12.80	16.10		28.90	39
1540	14" diameter		28	.571		26.50	17.25		43.75	55.50
1550	16" diameter		25	.640		30.50	19.35		49.85	63.50
1560	18" diameter		24	.667		34.50	20		54.50	69
1570	20" diameter	Q-10	27	.889		37	28		65	83.50
1580	22" diameter		22	1.091		42.50	34		76.50	99.50
1590	24" diameter		21	1.143		48	36		84	108

15500 | Heat Generation Equipment

15550 | Breechings, Chimneys & Stacks

		CREW	DAILY OUTPUT	LABOR-HOURS	UNIT	MAT.	LABOR	EQUIP.	TOTAL	TOTAL INCL O&P
1600	26" diameter	Q-10	20	1.200	Ea.	83	37.50		120.50	149
1610	28" diameter		19	1.263		103	39.50		142.50	174
1620	30" diameter		18	1.333		113	42		155	189
1630	32" diameter		17	1.412		131	44		175	213
1640	34" diameter		16	1.500		146	47		193	234
1650	36" diameter		15	1.600		163	50		213	257
1660	38" diameter		14	1.714		177	53.50		230.50	277
1670	40" diameter		14	1.714		193	53.50		246.50	295
1680	42" diameter		13	1.846		205	58		263	315
1690	44" diameter		12	2		222	62.50		284.50	340
1700	46" diameter		12	2		235	62.50		297.50	355
1710	48" diameter		11	2.182		265	68.50		333.50	395
1750	Top, 3" diameter	Q-9	46	.348		8	10.50		18.50	25
1760	4" diameter		44	.364		8.40	11		19.40	26
1770	5" diameter		42	.381		11.55	11.50		23.05	30.50
1780	6" diameter		40	.400		14.65	12.10		26.75	35
1790	7" diameter		38	.421		22	12.70		34.70	43.50
1800	8" diameter		36	.444		27.50	13.40		40.90	50.50
1810	10" diameter		34	.471		66	14.20		80.20	94.50
1820	12" diameter		32	.500		99.50	15.10		114.60	133
1830	14" diameter		30	.533		113	16.10		129.10	149
1840	16" diameter		28	.571		204	17.25		221.25	251
1850	18" diameter		26	.615		370	18.60		388.60	435
1860	20" diameter	Q-10	28	.857		400	27		427	480
1870	22" diameter		22	1.091		545	34		579	655
1880	24" diameter		20	1.200		680	37.50		717.50	805
1882	26" diameter		19	1.263		1,025	39.50		1,064.50	1,175
1884	28" diameter		18	1.333		1,175	42		1,217	1,350
1886	30" diameter		17	1.412		1,300	44		1,344	1,525
1888	32" diameter		16	1.500		1,375	47		1,422	1,600
1890	34" diameter		15	1.600		1,475	50		1,525	1,700
1892	36" diameter		14	1.714		1,575	53.50		1,628.50	1,825
1900	Gas, double wall, galvanized steel, oval									
1904	4" x 1'	Q-9	68	.235	V.L.F.	11.20	7.10		18.30	23.50
1906	5"		64	.250		23	7.55		30.55	36.50
1908	5"/6" x 1'		60	.267		23	8.05		31.05	38
1910	Oval fittings									
1912	Adjustable length									
1914	4" diameter to 12" lg	Q-9	34	.471	Ea.	11.85	14.20		26.05	35
1916	5" diameter to 12" lg		32	.500		33	15.10		48.10	59.50
1918	5"/6" diameter to 12"		30	.533		30	16.10		46.10	58
1920	Elbow 45°									
1922	4"	Q-9	34	.471	Ea.	21.50	14.20		35.70	46
1924	5"		32	.500		41	15.10		56.10	68.50
1926	5"/6"		30	.533		39	16.10		55.10	67.50
1930	Elbow 45°, flat									
1932	4"	Q-9	34	.471	Ea.	21.50	14.20		35.70	46
1934	5"		32	.500		41	15.10		56.10	68.50
1936	5"/6"		30	.533		41	16.10		57.10	70
1940	Top									
1942	4"	Q-9	44	.364	Ea.	16.10	11		27.10	34.50
1944	5"		42	.381		24.50	11.50		36	45
1946	5"/6"		40	.400		24.50	12.10		36.60	45.50
1950	Adjustable flashing									
1952	4"	Q-9	34	.471	Ea.	9.70	14.20		23.90	32.50
1954	5"		32	.500		28.50	15.10		43.60	55

Important: See the Reference Section for critical supporting data - Reference Nos., Crews, & City Cost Indexes

15500 | Heat Generation Equipment

15550 | Breechings, Chimneys & Stacks

		CREW	DAILY OUTPUT	LABOR-HOURS	UNIT	MAT.	LABOR	EQUIP.	TOTAL	TOTAL INCL O&P
1956	5"/6"	Q-9	30	.533	Ea.	19.20	16.10		35.30	46
1960	Tee									
1962	4"	Q-9	26	.615	Ea.	32	18.60		50.60	64
1964	5"		25	.640		61.50	19.35		80.85	98
1966	5"/6"	▼	24	.667	▼	62	20		82	99.50
1970	Tee with short snout									
1972	4"	Q-9	26	.615	Ea.	32	18.60		50.60	64
3200	All fuel, pressure tight, double wall, U.L. listed, 1400°F.									
3210	304 stainless steel liner, aluminized steel outer jacket									
3220	6" diameter	Q-9	60	.267	L.F.	35.50	8.05		43.55	51.50
3221	8" diameter		52	.308		43.50	9.30		52.80	62.50
3222	10" diameter		48	.333		48.50	10.05		58.55	69
3223	12" diameter		44	.364		55.50	11		66.50	78
3224	14" diameter		42	.381		63.50	11.50		75	88
3225	16" diameter		40	.400		67.50	12.10		79.60	92.50
3226	18" diameter		38	.421		76	12.70		88.70	103
3227	20" diameter	Q-10	36	.667		87	21		108	128
3228	24" diameter		32	.750		112	23.50		135.50	160
3229	28" diameter		30	.800		128	25		153	180
3230	32" diameter		27	.889		144	28		172	201
3231	36" diameter		25	.960		161	30		191	224
3232	42" diameter		22	1.091		184	34		218	256
3233	48" diameter	▼	19	1.263	▼	217	39.50		256.50	299
3260	For 316 stainless steel liner add					6%				
3280	All fuel, pressure tight, double wall fittings									
3284	304 stainless steel inner, aluminized steel jacket									
3288	Adjustable 18"/30" section									
3291	5" diameter	Q-9	32	.500	Ea.	91	15.10		106.10	124
3292	6" diameter		30	.533		115	16.10		131.10	152
3293	8" diameter		26	.615		133	18.60		151.60	175
3294	10" diameter		24	.667		149	20		169	195
3295	12" diameter		22	.727		170	22		192	221
3296	14" diameter		21	.762		196	23		219	252
3297	16" diameter		20	.800		204	24		228	262
3298	18" diameter	▼	19	.842		251	25.50		276.50	315
3299	20" diameter	Q-10	18	1.333		290	42		332	385
3300	24" diameter		16	1.500		375	47		422	485
3301	28" diameter		15	1.600		415	50		465	540
3302	32" diameter		14	1.714		430	53.50		483.50	560
3303	36" diameter		12	2		450	62.50		512.50	590
3304	42" diameter		11	2.182		515	68.50		583.50	675
3305	48" diameter	▼	10	2.400		605	75		680	780
3326	For 316 stainless steel liner, add				▼	15%				
3350	Elbow 90° fixed									
3353	5" diameter	Q-9	32	.500	Ea.	170	15.10		185.10	211
3354	6" diameter		30	.533		200	16.10		216.10	245
3355	8" diameter		26	.615		221	18.60		239.60	273
3356	10" diameter		24	.667		254	20		274	310
3357	12" diameter		22	.727		287	22		309	350
3358	14" diameter		21	.762		335	23		358	405
3359	16" diameter		20	.800		375	24		399	455
3360	18" diameter	▼	19	.842		440	25.50		465.50	520
3361	20" diameter	Q-10	18	1.333		490	42		532	600
3362	24" diameter		16	1.500		605	47		652	740
3363	28" diameter		15	1.600		755	50		805	910
3364	32" diameter	▼	14	1.714	▼	1,000	53.50		1,053.50	1,175

15500 | Heat Generation Equipment

15550 | Breechings, Chimneys & Stacks

		CREW	DAILY OUTPUT	LABOR-HOURS	UNIT	2001 BARE COSTS MAT.	LABOR	EQUIP.	TOTAL	TOTAL INCL O&P
3365	36" diameter	Q-10	12	2	Ea.	1,125	62.50		1,187.50	1,350
3366	42" diameter		11	2.182		1,725	68.50		1,793.50	2,000
3367	48" diameter	▼	10	2.400	▼	2,175	75		2,250	2,500
3380	For 316 stainless steel liner, add					20%				
3400	Elbow 45°									
3404	6" diameter	Q-9	30	.533	Ea.	127	16.10		143.10	164
3405	8" diameter		26	.615		143	18.60		161.60	187
3406	10" diameter		24	.667		161	20		181	208
3407	12" diameter		22	.727		184	22		206	236
3408	14" diameter		21	.762		211	23		234	268
3409	16" diameter		20	.800		238	24		262	299
3410	18" diameter	▼	19	.842		269	25.50		294.50	335
3411	20" diameter	Q-10	18	1.333		310	42		352	405
3412	24" diameter		16	1.500		395	47		442	510
3413	28" diameter		15	1.600		445	50		495	570
3414	32" diameter		14	1.714		600	53.50		653.50	745
3415	36" diameter		12	2		805	62.50		867.50	980
3416	42" diameter		11	2.182		1,100	68.50		1,168.50	1,300
3417	48" diameter	▼	10	2.400		1,475	75		1,550	1,750
3430	For 316 stainless steel liner, add				▼	20%				
3450	Tee 90°									
3454	6" diameter	Q-9	24	.667	Ea.	132	20		152	176
3455	8" diameter		22	.727		152	22		174	201
3456	10" diameter		21	.762		178	23		201	231
3457	12" diameter		20	.800		208	24		232	266
3458	14" diameter		18	.889		245	27		272	310
3459	16" diameter		16	1		275	30		305	350
3460	18" diameter	▼	14	1.143		325	34.50		359.50	410
3461	20" diameter	Q-10	17	1.412		365	44		409	470
3462	24" diameter		12	2		450	62.50		512.50	590
3463	28" diameter		11	2.182		565	68.50		633.50	725
3464	32" diameter		10	2.400		770	75		845	965
3465	36" diameter		9	2.667		860	83.50		943.50	1,075
3466	42" diameter		6	4		1,375	125		1,500	1,700
3467	48" diameter	▼	5	4.800		1,750	150		1,900	2,150
3480	For Tee Cap, add					35%	20%			
3500	For 316 stainless steel liner, add					22%				
3520	Plate support, galvanized									
3524	6" diameter	Q-9	26	.615	Ea.	83	18.60		101.60	120
3525	8" diameter		22	.727		93.50	22		115.50	137
3526	10" diameter		20	.800		102	24		126	150
3527	12" diameter		18	.889		106	27		133	159
3528	14" diameter		17	.941		128	28.50		156.50	185
3529	16" diameter		16	1		136	30		166	196
3530	18" diameter	▼	15	1.067		143	32		175	207
3531	20" diameter	Q-10	16	1.500		151	47		198	239
3532	24" diameter		14	1.714		159	53.50		212.50	258
3533	28" diameter		13	1.846		196	58		254	305
3534	32" diameter		12	2		254	62.50		316.50	375
3535	36" diameter		10	2.400		315	75		390	465
3536	42" diameter		9	2.667		560	83.50		643.50	745
3537	48" diameter	▼	8	3	▼	675	94		769	890
3570	Bellows, lined, 316 stainless steel only									
3574	6" diameter	Q-9	30	.533	Ea.	284	16.10		300.10	340
3575	8" diameter		26	.615		310	18.60		328.60	370
3576	10" diameter		24	.667		340	20		360	405

15500 | Heat Generation Equipment

15550 | Breechings, Chimneys & Stacks

		CREW	DAILY OUTPUT	LABOR-HOURS	UNIT	MAT.	LABOR	EQUIP.	TOTAL	TOTAL INCL O&P
3577	12" diameter	Q-9	22	.727	Ea.	375	22		397	450
3578	14" diameter		21	.762		425	23		448	500
3579	16" diameter		20	.800		470	24		494	555
3580	18" diameter	↓	19	.842		525	25.50		550.50	615
3581	20" diameter	Q-10	18	1.333		590	42		632	715
3582	24" diameter		16	1.500		740	47		787	890
3583	28" diameter		15	1.600		830	50		880	995
3584	32" diameter		14	1.714		955	53.50		1,008.50	1,125
3585	36" diameter		12	2		1,100	62.50		1,162.50	1,300
3586	42" diameter		11	2.182		1,375	68.50		1,443.50	1,625
3587	48" diameter	↓	10	2.400	↓	1,700	75		1,775	2,000
3600	Ventilated roof thimble, 304 stainless steel									
3620	6" diameter	Q-9	26	.615	Ea.	191	18.60		209.60	239
3624	8" diameter		22	.727		205	22		227	259
3625	10" diameter		20	.800		206	24		230	265
3626	12" diameter		18	.889		222	27		249	286
3627	14" diameter		17	.941		239	28.50		267.50	305
3628	16" diameter		16	1		254	30		284	325
3629	18" diameter	↓	15	1.067		274	32		306	350
3630	20" diameter	Q-10	16	1.500		293	47		340	400
3631	24" diameter		14	1.714		330	53.50		383.50	450
3632	28" diameter		13	1.846		380	58		438	510
3633	32" diameter		12	2		425	62.50		487.50	565
3634	36" diameter		10	2.400		465	75		540	625
3635	42" diameter		9	2.667		620	83.50		703.50	810
3636	48" diameter	↓	8	3	↓	780	94		874	1,000
3650	For 316 stainless steel, add					10%				
3670	Exit cone, 316 stainless steel only									
3674	6" diameter	Q-9	46	.348	Ea.	125	10.50		135.50	154
3675	8" diameter		42	.381		131	11.50		142.50	163
3676	10" diameter		40	.400		142	12.10		154.10	175
3677	12" diameter		38	.421		149	12.70		161.70	184
3678	14" diameter		37	.432		166	13.05		179.05	202
3679	16" diameter		36	.444		182	13.40		195.40	222
3680	18" diameter	↓	35	.457		224	13.80		237.80	269
3681	20" diameter	Q-10	28	.857		267	27		294	335
3682	24" diameter		26	.923		345	29		374	425
3683	28" diameter		25	.960		415	30		445	505
3684	32" diameter		24	1		520	31.50		551.50	620
3685	36" diameter		22	1.091		595	34		629	710
3686	42" diameter		21	1.143		760	36		796	890
3687	48" diameter	↓	20	1.200	↓	980	37.50		1,017.50	1,125
3720	Roof support assy., incl. 30" pipe sect, 304 st. st.									
3724	6" diameter	Q-9	25	.640	Ea.	218	19.35		237.35	270
3725	8" diameter		21	.762		238	23		261	298
3726	10" diameter		19	.842		261	25.50		286.50	325
3727	12" diameter		17	.941		278	28.50		306.50	350
3728	14" diameter		16	1		310	30		340	385
3729	16" diameter		15	1.067		330	32		362	415
3730	18" diameter	↓	14	1.143		355	34.50		389.50	445
3731	20" diameter	Q-10	15	1.600		380	50		430	500
3732	24" diameter		13	1.846		425	58		483	560
3733	28" diameter		12	2		500	62.50		562.50	645
3734	32" diameter		11	2.182		600	68.50		668.50	765
3735	36" diameter		9	2.667		690	83.50		773.50	890
3736	42" diameter	↓	8	3	↓	1,075	94		1,169	1,325

15500 | Heat Generation Equipment

15550 | Breechings, Chimneys & Stacks

		CREW	DAILY OUTPUT	LABOR-HOURS	UNIT	2001 BARE COSTS MAT.	LABOR	EQUIP.	TOTAL	TOTAL INCL O&P
3737	48" diameter	Q-10	7	3.429	Ea.	1,325	107		1,432	1,625
3750	For 316 stainless steel, add				↓	20%				
3770	Stack cap, 316 stainless steel only									
3773	5" diameter	Q-9	48	.333	Ea.	119	10.05		129.05	147
3774	6" diameter		46	.348		140	10.50		150.50	170
3775	8" diameter		42	.381		162	11.50		173.50	196
3776	10" diameter		40	.400		190	12.10		202.10	228
3777	12" diameter		38	.421		221	12.70		233.70	263
3778	14" diameter		37	.432		260	13.05		273.05	305
3779	16" diameter		36	.444		297	13.40		310.40	345
3780	18" diameter	↓	35	.457		335	13.80		348.80	390
3781	20" diameter	Q-10	28	.857		385	27		412	460
3782	24" diameter		26	.923		460	29		489	555
3783	28" diameter		25	.960		505	30		535	600
3784	32" diameter		24	1		580	31.50		611.50	685
3785	36" diameter		22	1.091		675	34		709	800
3786	42" diameter		21	1.143		805	36		841	940
3787	48" diameter	↓	20	1.200	↓	975	37.50		1,012.50	1,125
3800	Round storm collar									
3806	5" diameter	Q-9	32	.500	Ea.	25	15.10		40.10	51
7800	All fuel, double wall, stainless steel, 6" diameter		60	.267	V.L.F.	28.50	8.05		36.55	44
7802	7" diameter		56	.286		37	8.65		45.65	54
7804	8" diameter		52	.308		43	9.30		52.30	62
7806	10" diameter		48	.333		63	10.05		73.05	85
7808	12" diameter		44	.364		84.50	11		95.50	110
7810	14" diameter	↓	42	.381	↓	111	11.50		122.50	140
8000	All fuel, double wall, stainless steel fittings									
8010	Roof support 6" diameter	Q-9	30	.533	Ea.	73.50	16.10		89.60	106
8020	7" diameter		28	.571		83.50	17.25		100.75	118
8030	8" diameter		26	.615		90	18.60		108.60	128
8040	10" diameter		24	.667		119	20		139	162
8050	12" diameter		22	.727		143	22		165	191
8060	14" diameter		21	.762		181	23		204	235
8100	Elbow 15°, 6" diameter		30	.533		64	16.10		80.10	95.50
8120	7" diameter		28	.571		83.50	17.25		100.75	118
8140	8" diameter		26	.615		82.50	18.60		101.10	119
8160	10" diameter		24	.667		108	20		128	149
8180	12" diameter		22	.727		131	22		153	178
8200	14" diameter		21	.762		156	23		179	208
8300	Insulated tee with insulated tee cap, 6" diameter		30	.533		121	16.10		137.10	158
8340	7" diameter		28	.571		159	17.25		176.25	202
8360	8" diameter		26	.615		179	18.60		197.60	226
8380	10" diameter		24	.667		252	20		272	310
8400	12" diameter		22	.727		355	22		377	425
8420	14" diameter		21	.762		465	23		488	545
8500	Joist shield, 6" diameter		30	.533		37	16.10		53.10	65.50
8510	7" diameter		28	.571		40	17.25		57.25	70.50
8520	8" diameter		26	.615		49.50	18.60		68.10	83
8530	10" diameter		24	.667		66.50	20		86.50	105
8540	12" diameter		22	.727		83	22		105	126
8550	14" diameter		21	.762		103	23		126	149
8600	Round top, 6" diameter		30	.533		41	16.10		57.10	70
8620	7" diameter		28	.571		56	17.25		73.25	88
8640	8" diameter		26	.615		75	18.60		93.60	111
8660	10" diameter		24	.667		135	20		155	180
8680	12" diameter	↓	22	.727		193	22		215	246

Important: See the Reference Section for critical supporting data - Reference Nos., Crews, & City Cost Indexes

15500 | Heat Generation Equipment

15550 | Breechings, Chimneys & Stacks

		CREW	DAILY OUTPUT	LABOR-HOURS	UNIT	MAT.	LABOR	EQUIP.	TOTAL	TOTAL INCL O&P
8700	14" diameter	Q-9	21	.762	Ea.	258	23		281	320
8800	Adjustable roof flashing, 6" diameter		30	.533		49	16.10		65.10	78.50
8820	7" diameter		28	.571		56	17.25		73.25	88
8840	8" diameter		26	.615		60.50	18.60		79.10	95
8860	10" diameter		24	.667		78	20		98	117
8880	12" diameter		22	.727		100	22		122	145
8900	14" diameter	▼	21	.762	▼	125	23		148	174
9000	High temp. (2000° F), steel jacket, acid resist. refractory lining									
9010	11 ga. galvanized jacket, U.L. listed									
9020	Straight section, 48" long, 10" diameter	Q-10	13.30	1.805	Ea.	335	56.50		391.50	455
9030	12" diameter		11.20	2.143		360	67		427	505
9040	18" diameter		7.40	3.243		510	102		612	715
9050	24" diameter		4.60	5.217		695	163		858	1,025
9060	30" diameter		3.70	6.486		885	203		1,088	1,300
9070	36" diameter	▼	2.70	8.889		1,075	278		1,353	1,625
9080	42" diameter	Q-11	3.10	10.323		1,350	330		1,680	1,975
9090	48" diameter		2.70	11.852		1,575	380		1,955	2,325
9100	54" diameter		2.20	14.545		1,925	465		2,390	2,850
9110	60" diameter	▼	2	16		2,525	510		3,035	3,575
9120	Tee section, 10" diameter	Q-10	4.40	5.455		705	171		876	1,050
9130	12" diameter		3.70	6.486		745	203		948	1,125
9140	18" diameter		2.40	10		1,000	315		1,315	1,575
9150	24" diameter		1.50	16		1,400	500		1,900	2,300
9160	30" diameter		1.20	20		1,950	625		2,575	3,125
9170	36" diameter	▼	.80	30		3,300	940		4,240	5,075
9180	42" diameter	Q-11	1	32		4,325	1,025		5,350	6,350
9190	48" diameter		.90	35.556		5,600	1,150		6,750	7,925
9200	54" diameter		.70	45.714		8,175	1,475		9,650	11,300
9210	60" diameter	▼	.60	53.333		9,875	1,700		11,575	13,500
9220	Cleanout pier section, 10" diameter	Q-10	3.50	6.857		545	215		760	930
9230	12" diameter		2.50	9.600		570	300		870	1,100
9240	18" diameter		1.90	12.632		730	395		1,125	1,425
9250	24" diameter		1.30	18.462		925	580		1,505	1,925
9260	30" diameter		1	24		1,125	750		1,875	2,400
9270	36" diameter	▼	.75	32		1,350	1,000		2,350	3,025
9280	42" diameter	Q-11	.90	35.556		1,700	1,150		2,850	3,625
9290	48" diameter		.75	42.667		2,050	1,375		3,425	4,350
9300	54" diameter		.60	53.333		2,450	1,700		4,150	5,325
9310	60" diameter	▼	.50	64		3,125	2,050		5,175	6,575
9320	For drain, add					53%				
9330	Elbow, 30° and 45°, 10" diameter	Q-10	6.60	3.636		460	114		574	680
9340	12" diameter		5.60	4.286		495	134		629	750
9350	18" diameter		3.70	6.486		745	203		948	1,125
9360	24" diameter		2.30	10.435		1,100	325		1,425	1,725
9370	30" diameter		1.85	12.973		1,525	405		1,930	2,300
9380	36" diameter	▼	1.30	18.462		1,850	580		2,430	2,925
9390	42" diameter	Q-11	1.60	20		2,525	640		3,165	3,775
9400	48" diameter		1.35	23.704		3,225	760		3,985	4,725
9410	54" diameter		1.10	29.091		4,175	930		5,105	6,025
9420	60" diameter	▼	1	32		5,075	1,025		6,100	7,175
9430	For 60° and 90° elbow, add					112%				
9440	End cap, 10" diameter	Q-10	26	.923		790	29		819	910
9450	12" diameter		22	1.091		900	34		934	1,050
9460	18" diameter		15	1.600		1,125	50		1,175	1,300
9470	24" diameter		9	2.667		1,375	83.50		1,458.50	1,625
9480	30" diameter	▼	7.30	3.288	▼	1,700	103		1,803	2,000

15500 | Heat Generation Equipment

15550 | Breechings, Chimneys & Stacks

		CREW	DAILY OUTPUT	LABOR-HOURS	UNIT	2001 BARE COSTS				TOTAL INCL O&P
						MAT.	LABOR	EQUIP.	TOTAL	
9490	36" diameter	Q-10	5	4.800	Ea.	1,950	150		2,100	2,375
9500	42" diameter	Q-11	6.30	5.079		2,425	163		2,588	2,925
9510	48" diameter		5.30	6.038		2,875	193		3,068	3,450
9520	54" diameter		4.40	7.273		3,275	233		3,508	3,950
9530	60" diameter		3.90	8.205		3,275	263		3,538	4,000
9540	Increaser (1 diameter), 10" diameter	Q-10	6.60	3.636		390	114		504	605
9550	12" diameter		5.60	4.286		420	134		554	665
9560	18" diameter		3.70	6.486		610	203		813	985
9570	24" diameter		2.30	10.435		880	325		1,205	1,475
9580	30" diameter		1.85	12.973		1,075	405		1,480	1,825
9590	36" diameter		1.30	18.462		1,200	580		1,780	2,225
9592	42" diameter	Q-11	1.60	20		1,325	640		1,965	2,450
9594	48" diameter		1.35	23.704		1,575	760		2,335	2,925
9596	54" diameter		1.10	29.091		1,925	930		2,855	3,525
9600	For expansion joints, add to straight section					7%				
9610	For 1/4" hot rolled steel jacket, add					157%				
9620	For 2950°F very high temperature, add					89%				
9630	26 ga. aluminized jacket, straight section, 48" long									
9640	10" diameter	Q-10	15.30	1.569	V.L.F.	197	49		246	292
9650	12" diameter		12.90	1.860		208	58.50		266.50	320
9660	18" diameter		8.50	2.824		300	88.50		388.50	465
9670	24" diameter		5.30	4.528		430	142		572	695
9680	30" diameter		4.30	5.581		550	175		725	875
9690	36" diameter		3.10	7.742		695	242		937	1,150
9700	Accessories (all models)									
9710	Guy band, 10" diameter	Q-10	32	.750	Ea.	61	23.50		84.50	104
9720	12" diameter		30	.800		64.50	25		89.50	110
9730	18" diameter		26	.923		83.50	29		112.50	137
9740	24" diameter		24	1		124	31.50		155.50	185
9750	30" diameter		20	1.200		143	37.50		180.50	215
9760	36" diameter		18	1.333		158	42		200	238
9770	42" diameter	Q-11	22	1.455		287	46.50		333.50	385
9780	48" diameter		20	1.600		315	51		366	425
9790	54" diameter		16	2		350	64		414	480
9800	60" diameter		12	2.667		375	85.50		460.50	545
9810	Draw band, galv. stl., 11 gauge, 10" diameter	Q-10	32	.750		53	23.50		76.50	95
9820	12" diameter		30	.800		57	25		82	101
9830	18" diameter		26	.923		71.50	29		100.50	123
9840	24" diameter		24	1		90.50	31.50		122	148
9850	30" diameter		20	1.200		105	37.50		142.50	174
9860	36" diameter		18	1.333		121	42		163	198
9870	42" diameter	Q-11	22	1.455		162	46.50		208.50	250
9880	48" diameter		20	1.600		207	51		258	305
9890	54" diameter		16	2		241	64		305	365
9900	60" diameter		12	2.667		305	85.50		390.50	465
9910	Draw band, aluminized stl., 26 gauge, 10" diameter	Q-10	32	.750		15.20	23.50		38.70	53
9920	12" diameter		30	.800		15.20	25		40.20	55
9930	18" diameter		26	.923		19	29		48	65.50
9940	24" diameter		24	1		23	31.50		54.50	73.50
9950	30" diameter		20	1.200		26.50	37.50		64	87.50
9960	36" diameter		18	1.333		30.50	42		72.50	98
9962	42" diameter	Q-11	22	1.455		34	46.50		80.50	110
9964	48" diameter		20	1.600		42	51		93	125
9966	54" diameter		16	2		42	64		106	145
9968	60" diameter		12	2.667		45.50	85.50		131	182

15500 | Heat Generation Equipment

15550 | Breechings, Chimneys & Stacks

		CREW	DAILY OUTPUT	LABOR-HOURS	UNIT	2001 BARE COSTS MAT.	LABOR	EQUIP.	TOTAL	TOTAL INCL O&P
0010	**INDUCED DRAFT FANS**									
1000	Breeching installation									
1800	Hot gas, 600°F, variable pitch pulley and motor									
1840	6" diam. inlet, 1/4 H.P., 1 phase, 400 CFM	Q-9	6	2.667	Ea.	1,000	80.50		1,080.50	1,225
1850	7" diam. inlet, 1/4 H.P., 1 phase, 800 CFM		5	3.200		1,075	96.50		1,171.50	1,325
1860	8" diam. inlet, 1/4 H.P., 1phase, 1120 CFM		4	4		1,575	121		1,696	1,900
1870	9" diam. inlet, 3/4 H.P., 1 phase, 1440 CFM		3.60	4.444		1,850	134		1,984	2,225
1880	10" diam. inlet, 3/4 H.P., 1 phase, 2000 CFM		3.30	4.848		2,075	146		2,221	2,500
1900	12" diam. inlet, 3/4 H.P., 3 phase, 2960 CFM		3	5.333		2,150	161		2,311	2,600
1910	14" diam. inlet, 1 H.P., 3 phase, 4160 CFM		2.60	6.154		2,525	186		2,711	3,050
1920	16" diam. inlet, 2 H.P., 3 phase, 6720 CFM		2.30	6.957		2,750	210		2,960	3,350
1940	18" diam. inlet, 3 H.P., 3 phase, 9120 CFM		2	8		3,325	242		3,567	4,025
1950	20" diam. inlet, 3 H.P., 3 phase, 9760 CFM		1.50	10.667		3,625	320		3,945	4,475
1960	22" diam. inlet, 5 H.P., 3 phase, 13,360 CFM		1	16		4,225	485		4,710	5,400
1980	24" diam. inlet, 7-1/2 H.P., 3 phase, 17,760 CFM		.80	20		5,200	605		5,805	6,650
2300	For multi-blade damper at fan inlet, add					20%				
3600	Chimneytop installation									
3700	6" size	1 Shee	8	1	Ea.	440	33.50		473.50	535
3740	8" size		7	1.143		440	38.50		478.50	545
3780	13" size		6	1.333		605	44.50		649.50	735
3880	For speed control switch, add					39.50			39.50	43.50
3920	For thermal fan control, add					39.50			39.50	43.50
5500	Flue shutter damper for draft control,									
5510	parallel blades									
5550	8" size	Q-9	8	2	Ea.	350	60.50		410.50	480
5560	9" size		7.50	2.133		380	64.50		444.50	520
5570	10" size		7	2.286		390	69		459	535
5580	12" size		6.50	2.462		430	74.50		504.50	590
5590	14" size		6	2.667		490	80.50		570.50	660
5600	16" size		5.50	2.909		530	88		618	720
5610	18" size		5	3.200		580	96.50		676.50	790
5620	20" size		4.50	3.556		645	107		752	875
5630	22" size		4	4		710	121		831	965
5640	24" size		3.50	4.571		775	138		913	1,075
5650	27" size		3	5.333		775	161		936	1,100
5660	30" size		2.50	6.400		825	193		1,018	1,200
5670	32" size		2	8		865	242		1,107	1,325
5680	36" size		1.50	10.667		930	320		1,250	1,525

15600 | Refrigeration Equipment

15620 | Packaged Water Chillers

		CREW	DAILY OUTPUT	LABOR-HOURS	UNIT	2001 BARE COSTS MAT.	LABOR	EQUIP.	TOTAL	TOTAL INCL O&P
0010	**ABSORPTION WATER CHILLERS**									
0020	Steam or hot water, water cooled									
0050	100 ton	Q-7	.13	240	Ea.	110,500	7,975		118,475	133,500
0100	148 ton		.12	258		114,000	8,575		122,575	138,000
0200	200 ton		.12	275		126,500	9,175		135,675	153,500
0240	250 ton		.11	290		139,500	9,675		149,175	168,000
0300	354 ton		.10	304		174,000	10,100		184,100	207,000
0400	420 ton		.10	323		190,500	10,700		201,200	225,500

15600 | Refrigeration Equipment

15620 | Packaged Water Chillers

			CREW	DAILY OUTPUT	LABOR-HOURS	UNIT	MAT.	LABOR	EQUIP.	TOTAL	TOTAL INCL O&P
100	0500	665 ton	Q-7	.09	340	Ea.	267,000	11,300		278,300	311,000
	0600	750 ton		.09	363		312,000	12,100		324,100	362,000
	0700	850 ton		.08	385		333,000	12,800		345,800	386,000
	0800	955 ton		.08	410		356,000	13,600		369,600	412,500
	0900	1125 ton		.08	421		411,000	14,000		425,000	473,500
	1000	1250 ton		.07	444		442,000	14,800		456,800	509,000
	1100	1465 ton		.07	463		524,500	15,400		539,900	600,500
	1200	1660 ton		.07	477		617,000	15,900		632,900	702,500
	2000	For two stage unit, add					80%	25%			
	3000	Gas fired, air cooled									
	3180	3 ton	Q-5	1.30	12.308	Ea.	3,100	385		3,485	3,975
	3200	4 ton		.90	17.778		3,675	560		4,235	4,875
	3220	5 ton		.60	26.667		4,100	835		4,935	5,800
	3250	8 ton		.50	32		9,025	1,000		10,025	11,500
	3270	10 ton		.40	40		10,300	1,250		11,550	13,200
	4000	Water cooled, duplex									
	4130	100 ton	Q-7	.13	246	Ea.	110,500	8,175		118,675	134,000
	4140	200 ton		.11	283		163,500	9,425		172,925	194,000
	4150	300 ton		.11	299		187,500	9,950		197,450	221,000
	4160	400 ton		.10	316		250,000	10,500		260,500	291,000
	4170	500 ton		.10	329		312,500	11,000		323,500	360,000
	4180	600 ton		.09	340		375,000	11,300		386,300	429,500
	4190	700 ton		.09	359		437,000	12,000		449,000	499,000
	4200	800 ton		.08	380		499,500	12,700		512,200	568,500
	4210	900 ton		.08	400		562,000	13,300		575,300	638,500
	4220	1000 ton		.08	421		624,500	14,000		638,500	708,000
600	0010	**CENTRIFUGAL/RECIP. WATER CHILLERS,** With standard controls									
	0020	Centrifugal liquid chiller, water cooled									
	0030	not including water tower									
	0100	Open drive, 2000 ton	Q-7	.07	477	Ea.	483,500	15,900		499,400	555,500
	0110	Centrifugal liquid chiller, air cooled, insulated evaporator									
	0120	130 ton	Q-7	.14	228	Ea.	54,000	7,600		61,600	71,000
	0124	160 ton		.13	246		65,500	8,175		73,675	84,500
	0128	180 ton		.13	250		73,500	8,300		81,800	93,500
	0132	210 ton		.12	258		86,000	8,575		94,575	107,500
	0136	270 ton		.12	266		110,500	8,875		119,375	135,000
	0140	320 ton		.12	275		130,500	9,175		139,675	158,000
	0200	Packaged unit, water cooled, not incl. tower									
	0210	80 ton	Q-7	.14	223	Ea.	28,000	7,450		35,450	42,000
	0220	100 ton		.14	230		35,000	7,650		42,650	50,000
	0230	150 ton		.13	240		52,500	8,000		60,500	70,000
	0240	200 ton		.13	251		85,500	8,375		93,875	106,500
	0250	250 ton		.12	260		87,500	8,650		96,150	109,500
	0260	300 ton		.12	266		105,000	8,875		113,875	129,000
	0270	350 ton		.12	275		122,500	9,175		131,675	149,000
	0280	400 ton		.11	283		143,000	9,425		152,425	171,000
	0282	450 ton		.11	290		157,500	9,675		167,175	188,000
	0290	500 ton		.11	296		175,000	9,850		184,850	207,500
	0292	550 ton		.10	304		192,500	10,100		202,600	227,500
	0300	600 ton		.10	310		210,000	10,300		220,300	246,500
	0302	650 ton		.10	320		227,500	10,600		238,100	266,500
	0304	700 ton		.10	326		245,000	10,900		255,900	286,000
	0306	750 ton		.10	333		262,500	11,100		273,600	305,500
	0310	800 ton		.09	340		280,000	11,300		291,300	325,000
	0312	850 ton		.09	351		297,500	11,700		309,200	345,000
	0316	900 ton		.09	359		315,000	12,000		327,000	364,500

Just one CD is packed with massive amounts of co data covering every detail from the ground up

- **18 detailed cost data books** — available individually, in specially priced packages, or in total.
- **Architects' First Source®** — featuring thousands of products listed by manufacturer.
- **The Construction Specifications Institute's SPEC-DATA®** for technica product information.
- **CSI's MANU-SPEC®** for manufacture complete product specifications.

Plus you get the productivity-enhancing tools to really put this valuable information to work.

Get the Means Construction Dictionary with over 17,000 construction terms — a $99.95 value — FREE on the Means CostWorks 2001 CD!

Save hundreds on specially priced packages! Choose the one that's right for you…

Builder's Package
Includes:
- Residential Cost Data
- Light Commercial Cost Data
- Site Work & Landscape Cost Data
- Concrete & Masonry Cost Data
- Open Shop Building Construction Cost Data

Just $299.99
Save over $300 off the price of buying each book separately!

Building Professional's Package
Includes:
- Building Construction Cost Data
- Mechanical Cost Data
- Plumbing Cost Data
- Electrical Cost Data
- Square Foot Model Costs

Just $349.99
Save over $250 off the price of buying each book separately!

Facility Manager's Package
Includes:
- Facilities Construction Cost Data
- Facilities Maintenance & Repair
- Repair & Remodeling Cost Data
- Building Construction Cost Data

Just $449.99
Save over $300 off the price of buying each book separately!

Design Professional's Package
Includes:
- Building Construction Cost Data
- Square Foot Model Costs
- Interior Cost Data
- Assemblies Cost Data

Just $349.99
Save over $200 off the price of buying each book separately!

Stay current with online cost data updates — included in the cost of the Means CostWorks 2001 CD!

Your 2001 Annual Means CostWorks license includes four FREE quarterly updates from the Means CostWorks Web site

Where can you get cost data with just one click?

Means CostWorks 2001 CD-ROM is the answer!

- Search and organize detailed cost data in an instant to simplify estimating

- Test various options quickly and easily to determine best-cost solutions

- Make changes on the fly and calculate the results automatically to save time and headaches

- Localize costs to your geographic area for greater accuracy and more competitive bids

- Stay current with rising costs thanks to online updates available every three months

Here's all the valuable Means cost data you need — with built-in tools and accessories that work for you!

Means Cost Books are a great reference when you need a quick price on an item. But for complex estimates involving numerous cost factors, you need a resource that can go to work in an instant — and produce the results that win more jobs. That's what you get with the **Means CostWorks 2001 CD.**

Easily search through line items.

Selected items are quantified and added to Costlist.

This comprehensive digital library provides 18 detailed cost books that cover every job costing requirement you can imagine — all at your fingertips using your desktop or laptop PC. Each "book" addresses a specific area of the building and maintenance industry, available separately or in specially priced packages. Plus, you have access to regular cost updates and unlimited access to additional valuable resources on the R.S. Means Web site included in the price of the CD.

Job costing has never been easier — or more precise

Check out all you get... see back cover

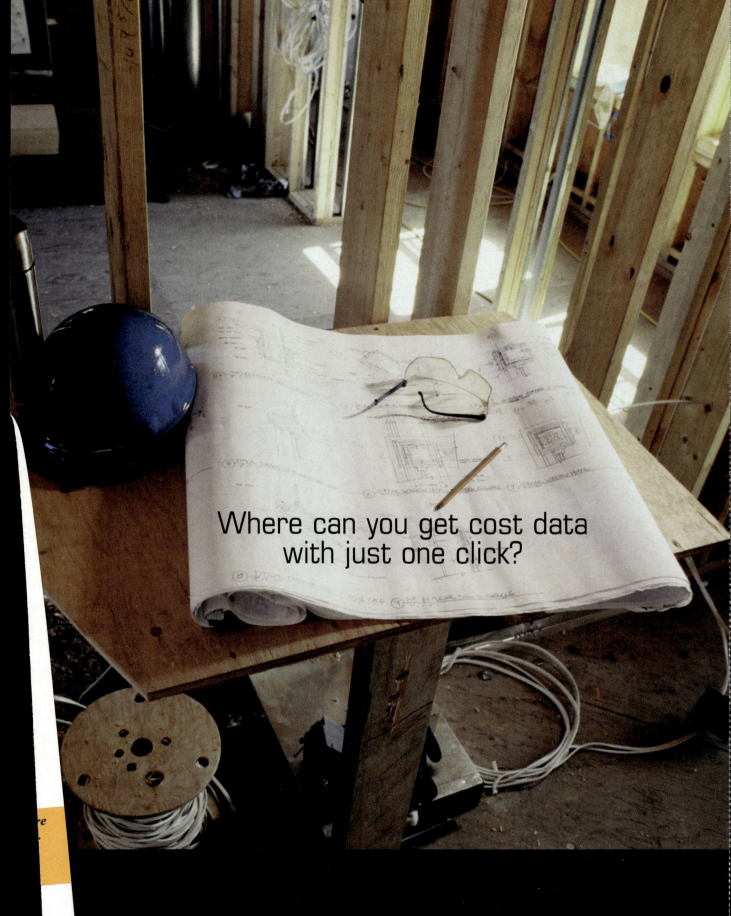

Get your Means CostWorks 2001 CD today!

FAX back the form below Toll Free: 1-800-632-6732 or Order by Phone: 1-800-334-3509

Select the individual digital cost books or packages you need.

Catalog No.	Title	Price	Cost
65061	Assemblies Cost Data	$159.95	
65011	Building Construction Cost Data	$110.95	
65111	Concrete & Masonry Cost Data	$110.95	
65031	Electrical Cost Data	$110.95	
65161	Heavy Construction Cost Data	$110.95	
65091	Interior Cost Data	$110.95	
65181	Light Commercial Cost Data	$139.95	
65021	Mechanical Cost Data	$110.95	
65151	Open Shop BCCD	$110.95	
65211	Plumbing Cost Data	$110.95	
65281	Site Work & Landscape Cost Data	$110.95	
65051	Square Foot Model Costs	$159.95	
65901	Wage Rate Utility (Available for above titles only)	$110.95	
65201	Facilities Construction Cost Data	$259.95	
65041	Repair & Remodeling Cost Data	$110.95	
65311	Building Construction Cost Data/Metric	$110.95	
65361	Heavy Construction Cost Data/Metric	$110.95	
65171	Residential Cost Data	$129.95	
65301	Facilities Maintenance & Repair Cost Data	$259.95	

PACKAGES

Catalog No.	Title	Price	Cost
65511	Builder's Package	$299.99	
65521	Building Professional's Package	$349.99	
65531	Facility Manager's Package	$449.99	
65541	Design Professional's Package	$349.99	

Name: _____

Company: _____

Street: _____

City _____ State: _____ Zip: _____

Phone ( ) _____ FAX: ( ) _____

E-Mail: _____

Subtotal	
MA residents please add 5% sales tax	
Shipping and Handling	$5.00
Grand total	

All items shipped on a single CD with software code to unlock specific selections purchased.

Buy additional keys any time you need to unlock additional cost data books!

☐ Bill me   ☐ Please charge my credit card
☐ Visa   ☐ MasterCard   ☐ American Express   ☐ Discover

Card # _____ Expiration Date: _____

Cardholder Signature: _____

RSMeans CMDGROUP

CWCW-2001

Order on our secure Web site www.rsmeans.com/cwcw2001.html

15600 | Refrigeration Equipment

15620 | Packaged Water Chillers

		CREW	DAILY OUTPUT	LABOR-HOURS	UNIT	MAT.	LABOR	EQUIP.	TOTAL	TOTAL INCL O&P
0318	950 ton	Q-7	.09	363	Ea.	332,500	12,100		344,600	384,500
0320	1000 ton		.09	372		349,500	12,400		361,900	403,000
0324	1100 ton		.08	385		385,000	12,800		397,800	443,000
0330	1200 ton		.08	395		420,000	13,100		433,100	482,500
0340	1300 ton		.08	410		446,000	13,600		459,600	511,000
0350	1400 ton		.08	421		490,500	14,000		504,500	560,500
0360	1500 ton		.07	426		510,000	14,200		524,200	582,500
0370	1600 ton		.07	438		560,500	14,600		575,100	638,500
0380	1700 ton		.07	450		595,500	15,000		610,500	677,500
0390	1800 ton		.07	457		630,500	15,200		645,700	716,500
0400	1900 ton		.07	470		665,500	15,600		681,100	755,500
0490	Reciprocating, packaged w/integral air cooled condenser, 15 ton cool		.37	86.486		12,300	2,875		15,175	17,900
0500	20 ton cooling		.34	94.955		15,500	3,150		18,650	21,900
0510	25 ton cooling		.33	97.859		17,600	3,250		20,850	24,300
0515	30 ton cooling		.31	101		19,000	3,400		22,400	26,000
0517	35 ton cooling		.31	104		21,400	3,475		24,875	28,800
0520	40 ton cooling		.30	108		24,900	3,600		28,500	32,800
0528	45 ton cooling		.29	109		28,000	3,650		31,650	36,300
0536	50 ton cooling		.28	113		30,800	3,750		34,550	39,600
0538	60 ton cooling		.28	115		35,500	3,825		39,325	44,800
0540	65 ton cooling		.27	118		38,800	3,925		42,725	48,600
0546	70 ton cooling		.27	119		40,800	3,975		44,775	51,000
0552	80 ton cooling		.26	123		43,600	4,100		47,700	54,000
0554	90 ton cooling		.25	125		47,400	4,200		51,600	58,500
0600	100 ton cooling		.25	129		53,000	4,300		57,300	64,500
0620	110 ton cooling		.24	132		57,500	4,400		61,900	70,000
0630	130 ton cooling		.24	135		65,500	4,500		70,000	79,000
0640	150 ton cooling		.23	137		75,500	4,575		80,075	90,500
0650	175 ton cooling		.23	140		80,500	4,675		85,175	95,500
0654	190 ton cooling		.22	144		91,500	4,800		96,300	107,500
0660	210 ton cooling		.22	148		100,500	4,925		105,425	118,000
0662	240 ton cooling		.21	151		119,000	5,050		124,050	138,500
0664	270 ton cooling		.20	156		135,000	5,200		140,200	156,500
0666	300 ton cooling		.20	160		146,000	5,325		151,325	168,500
0668	330 ton cooling		.19	164		161,500	5,450		166,950	185,500
0670	360 ton cooling		.19	168		183,000	5,600		188,600	210,000
0672	390 ton cooling		.18	172		192,000	5,750		197,750	219,500
0674	420 ton cooling		.18	177		201,000	5,900		206,900	230,000
0680	Water cooled, single compressor, semi-hermetic, tower not incl.									
0700	2 to 5 ton cooling	Q-5	.57	28.021	Ea.	5,850	880		6,730	7,750
0720	6 ton cooling		.42	38.005		9,150	1,200		10,350	11,900
0740	8 ton cooling		.31	52.117		9,825	1,625		11,450	13,300
0760	10 ton cooling	Q-6	.36	67.039		10,400	2,175		12,575	14,700
0780	15 ton cooling	"	.33	72.727		11,000	2,375		13,375	15,600
0800	20 ton cooling	Q-7	.38	83.989		12,800	2,800		15,600	18,200
0820	30 ton cooling		.33	96.096		15,400	3,200		18,600	21,700
0840	35 ton cooling		.33	97.859		16,100	3,250		19,350	22,600
0980	Water cooled, multiple compress., semi-hermetic, tower not incl.									
1000	15 ton cooling	Q-6	.36	65.934	Ea.	16,000	2,150		18,150	20,800
1020	20 ton cooling	Q-7	.41	78.049		17,400	2,600		20,000	23,100
1040	25 ton cooling		.36	89.888		18,700	3,000		21,700	25,100
1060	30 ton cooling		.31	101		19,900	3,400		23,300	27,000
1080	40 ton cooling		.30	108		21,700	3,600		25,300	29,300
1100	50 ton cooling		.28	113		25,100	3,775		28,875	33,300
1120	60 ton cooling		.25	125		26,700	4,200		30,900	35,700
1130	70 ton cooling		.23	139		34,100	4,625		38,725	44,500

15600 | Refrigeration Equipment

15620 | Packaged Water Chillers

		CREW	DAILY OUTPUT	LABOR-HOURS	UNIT	2001 BARE COSTS				TOTAL INCL O&P
						MAT.	LABOR	EQUIP.	TOTAL	
1140	80 ton cooling	Q-7	.21	151	Ea.	38,400	5,050		43,450	49,800
1150	90 ton cooling		.19	164		47,600	5,475		53,075	61,000
1160	100 ton cooling		.18	179		48,700	5,975		54,675	62,500
1170	110 ton cooling		.17	190		49,600	6,325		55,925	64,000
1180	120 ton cooling		.16	196		52,000	6,525		58,525	67,000
1200	140 ton cooling		.16	202		56,500	6,725		63,225	72,000
1210	160 ton cooling	▼	.15	210	▼	65,500	7,000		72,500	82,500
1300	Water cooled, single compressor, direct drive, tower not incl.									
1320	40 ton cooling,	Q-7	.30	108	Ea.	24,000	3,600		27,600	31,800
1340	50 ton cooling		.28	113		28,000	3,775		31,775	36,500
1360	60 ton cooling		.25	125		30,400	4,200		34,600	39,800
1380	80 ton cooling	▼	.21	151	▼	40,200	5,050		45,250	52,000
1450	Water cooled, dual compressors, direct drive, tower not incl.									
1500	80 ton cooling	Q-7	.14	222	Ea.	46,400	7,375		53,775	62,000
1520	100 ton cooling		.14	228		54,000	7,600		61,600	71,000
1540	120 ton cooling		.14	231		59,500	7,700		67,200	77,000
1560	135 ton cooling		.14	235		68,000	7,825		75,825	86,500
1580	150 ton cooling		.13	240		72,500	8,000		80,500	91,500
1600	175 ton cooling		.13	244		83,000	8,125		91,125	103,500
1620	200 ton cooling		.13	250		88,500	8,300		96,800	109,500
1640	225 ton cooling		.13	256		96,000	8,500		104,500	119,000
1660	250 ton cooling	▼	.12	260	▼	102,000	8,650		110,650	125,000
4000	Packaged chiller, remote air cooled condensers not incl.									
4020	15 ton cooling	Q-7	.30	108	Ea.	13,800	3,600		17,400	20,600
4030	20 ton cooling		.28	115		16,000	3,850		19,850	23,400
4040	25 ton cooling		.25	125		19,000	4,200		23,200	27,300
4050	30 ton cooling		.24	133		21,600	4,450		26,050	30,500
4060	40 ton cooling		.22	144		28,000	4,800		32,800	38,000
4070	50 ton cooling		.21	153		31,400	5,125		36,525	42,200
4080	60 ton cooling		.19	164		34,200	5,450		39,650	45,800
4090	70 ton cooling		.18	173		42,700	5,775		48,475	55,500
4100	80 ton cooling		.17	183		43,800	6,125		49,925	57,500
4110	90 ton cooling		.16	193		57,000	6,450		63,450	72,500
4120	100 ton cooling		.16	203		63,500	6,775		70,275	80,000
4130	110 ton cooling		.15	213		65,000	7,100		72,100	82,000
4140	120 ton cooling		.14	223		67,000	7,450		74,450	84,500
4150	140 ton cooling	▼	.14	233	▼	74,500	7,775		82,275	93,500
8000	Direct expansion, shell and tube type, for built up systems									
8020	1 ton	Q-5	2	8	Ea.	4,500	251		4,751	5,325
8030	5 ton		1.90	8.421		7,475	264		7,739	8,600
8040	10 ton		1.70	9.412		9,250	295		9,545	10,600
8050	20 ton		1.50	10.667		10,400	335		10,735	11,900
8060	30 ton		1	16		12,900	500		13,400	15,000
8070	50 ton	▼	.90	17.778		20,100	560		20,660	22,900
8080	100 ton	Q-6	.90	26.667	▼	37,100	865		37,965	42,200

15640 | Packaged Cooling Towers

			CREW	DAILY OUTPUT	LABOR-HOURS	UNIT	MAT.	LABOR	EQUIP.	TOTAL	TOTAL INCL O&P
0010	**COOLING TOWERS** Packaged units	A8.4 -120									
0070	Galvanized steel										
0080	Draw thru, single flow	A8.4 -230									
0100	Belt drive, 60 tons		Q-6	90	.267	TonAC	78.50	8.65		87.15	99
0150	95 tons			100	.240		69	7.80		76.80	88
0200	110 tons			109	.220		67.50	7.15		74.65	85.50
0250	125 tons		▼	120	.200	▼	65.50	6.50		72	82
1000	For higher capacities, use multiples										

15600 | Refrigeration Equipment

15640 | Packaged Cooling Towers

			CREW	DAILY OUTPUT	LABOR-HOURS	UNIT	MAT.	LABOR	EQUIP.	TOTAL	TOTAL INCL O&P	
400	1500	Induced air, double flow										400
	1900	Gear drive, 150 ton	Q-6	126	.190	TonAC	91	6.20		97.20	109	
	2000	300 ton		129	.186		62	6.05		68.05	77	
	2100	600 ton		132	.182		58.50	5.90		64.40	73	
	2150	840 ton		142	.169		47	5.50		52.50	60	
	2200	Up to 1,000 tons		150	.160		43.50	5.20		48.70	56	
	3000	For higher capacities, use multiples										
	3500	For pumps and piping, add	Q-6	38	.632	TonAC	39	20.50		59.50	74	
	4000	For absorption systems, add				"	75%	75%				
	4100	Cooling water chemical feeder	Q-5	3	5.333	Ea.	263	167		430	540	
	4500	For rigging, see division 01559										
	5000	Fiberglass										
	5010	Draw thru										
	5100	60 tons	Q-6	1.50	16	Ea.	2,775	520		3,295	3,825	
	5120	125 tons		.99	24.242		5,675	790		6,465	7,450	
	5140	300 tons		.43	55.814		13,300	1,825		15,125	17,400	
	5160	600 tons		.22	109		24,200	3,550		27,750	32,000	
	5180	1000 tons		.15	160		41,500	5,200		46,700	53,500	
	6000	Stainless steel										
	6010	Draw thru										
	6100	60 tons	Q-6	1.50	16	Ea.	7,575	520		8,095	9,100	
	6120	110 tons		.99	24.242		12,100	790		12,890	14,500	
	6140	300 tons		.43	55.814		33,500	1,825		35,325	39,700	
	6160	600 tons		.22	109		52,000	3,550		55,550	62,500	
	6180	1000 tons		.15	160		84,500	5,200		89,700	101,000	

15660 | Liquid Coolers/Evap Condensers

			CREW	DAILY OUTPUT	LABOR-HOURS	UNIT	MAT.	LABOR	EQUIP.	TOTAL	TOTAL INCL O&P	
100	0010	**CONDENSERS** Ratings are for 30°F TD, R-22										100
	0080	Air cooled, belt drive, propeller fan										
	0220	45 ton	Q-6	.70	34.286	Ea.	10,100	1,125		11,225	12,800	
	0240	50 ton		.69	34.985		10,400	1,150		11,550	13,200	
	0260	54 ton		.63	37.795		10,800	1,225		12,025	13,700	
	0280	59 ton		.58	41.308		11,500	1,350		12,850	14,600	
	0300	65 ton		.53	45.541		11,900	1,475		13,375	15,300	
	0320	73 ton		.47	51.173		13,400	1,675		15,075	17,200	
	0340	81 ton		.42	56.738		16,100	1,850		17,950	20,500	
	0360	86 ton		.40	60.302		16,700	1,950		18,650	21,400	
	0380	88 ton		.39	61.697		17,800	2,000		19,800	22,600	
	0400	101 ton	Q-7	.45	70.640		20,600	2,350		22,950	26,300	
	0500	159 ton		.31	102		33,100	3,425		36,525	41,600	
	0600	228 ton		.22	148		47,800	4,925		52,725	60,000	
	0700	314 ton		.16	203		58,500	6,775		65,275	74,500	
	0800	471 ton		.11	283		68,500	9,425		77,925	89,000	
	0900	630 ton		.08	376		110,000	12,500		122,500	140,000	
	1000	840 ton		.06	507		147,500	16,900		164,400	187,500	
	1500	May be specified single or multi-circuit										
	1550	Air cooled, direct drive, propeller fan										
	1590	1 ton	Q-5	3.80	4.211	Ea.	545	132		677	800	
	1600	1-1/2 ton		3.60	4.444		650	139		789	925	
	1620	2 ton		3.20	5		715	157		872	1,025	
	1630	3 ton		2.40	6.667		800	209		1,009	1,200	
	1640	5 ton		2	8		1,450	251		1,701	1,975	
	1650	8 ton		1.80	8.889		1,850	279		2,129	2,450	
	1660	10 ton		1.40	11.429		2,250	360		2,610	3,025	
	1670	12 ton		1.30	12.308		2,525	385		2,910	3,350	

15600 | Refrigeration Equipment

15660 | Liquid Coolers/Evap Condensers

			CREW	DAILY OUTPUT	LABOR-HOURS	UNIT	MAT.	LABOR	EQUIP.	TOTAL	TOTAL INCL O&P	
100	1680	14 ton	Q-5	1.20	13.333	Ea.	2,900	420		3,320	3,800	100
	1690	16 ton A8.4-110		1.10	14.545		3,175	455		3,630	4,175	
	1700	21 ton		1	16		3,650	500		4,150	4,775	
	1720	26 ton		.84	19.002		4,125	595		4,720	5,425	
	1740	30 ton		.70	22.792		6,825	715		7,540	8,575	
	1760	41 ton	Q-6	.77	31.008		7,525	1,000		8,525	9,800	
	1780	52 ton		.66	36.419		10,700	1,175		11,875	13,600	
	1800	63 ton		.54	44.037		11,900	1,425		13,325	15,300	
	1820	76 ton		.45	52.980		13,500	1,725		15,225	17,400	
	1840	86 ton		.40	60		16,600	1,950		18,550	21,300	
	1860	97 ton		.35	67.989		18,900	2,200		21,100	24,200	
	1880	105 ton	Q-7	.43	73.563		21,800	2,450		24,250	27,700	
	1890	118 ton		.39	82.687		23,500	2,750		26,250	30,100	
	1900	126 ton		.36	88.154		27,300	2,925		30,225	34,500	
	1910	136 ton		.34	95.238		30,600	3,175		33,775	38,400	
	1920	142 ton		.32	99.379		31,400	3,300		34,700	39,500	
	3400	Evaporative, copper coil, pump, fan motor										
	3440	10 ton	Q-5	.54	29.630	Ea.	4,225	930		5,155	6,050	
	3460	15 ton		.50	32		4,375	1,000		5,375	6,350	
	3480	20 ton		.47	34.043		4,575	1,075		5,650	6,650	
	3500	25 ton		.45	35.556		4,875	1,125		6,000	7,050	
	3520	30 ton		.42	38.095		5,200	1,200		6,400	7,525	
	3540	40 ton	Q-6	.49	48.980		6,875	1,600		8,475	9,975	
	3560	50 ton		.39	61.538		8,625	2,000		10,625	12,500	
	3580	65 ton		.35	68.571		9,575	2,225		11,800	13,900	
	3600	80 ton		.33	72.727		10,300	2,375		12,675	14,900	
	3620	90 ton		.29	82.759		11,400	2,700		14,100	16,600	
	3640	100 ton	Q-7	.36	88.889		12,400	2,950		15,350	18,200	
	3660	110 ton		.33	96.970		13,800	3,225		17,025	20,000	
	3680	125 ton		.30	106		14,900	3,550		18,450	21,800	
	3700	135 ton		.28	114		16,000	3,800		19,800	23,300	
	3720	150 ton		.25	128		17,600	4,250		21,850	25,800	
	3740	165 ton		.23	139		19,400	4,625		24,025	28,300	
	3760	185 ton		.22	145		20,500	4,825		25,325	29,800	
	3860	For fan damper control, add	Q-5	2	8		440	251		691	865	
	8000	Oils										
	8100	Lubricating										
	8120	Oil, lubricating				Oz.	.17			.17	.19	
	8500	Refrigeration										
	8520	Oil, refrigeration				Gal.	14.30			14.30	15.70	

15670 | Refrig Condensing Units

			CREW	DAILY OUTPUT	LABOR-HOURS	UNIT	MAT.	LABOR	EQUIP.	TOTAL	TOTAL INCL O&P	
300	0010	**CONDENSING UNITS**										300
	0030	Air cooled, compressor, standard controls										
	0050	1.5 ton	Q-5	2.50	6.400	Ea.	765	201		966	1,150	
	0100	2 ton		2.10	7.619		810	239		1,049	1,250	
	0200	2.5 ton		1.70	9.412		955	295		1,250	1,500	
	0300	3 ton		1.30	12.308		1,100	385		1,485	1,800	
	0350	3.5 ton		1.10	14.545		1,200	455		1,655	2,025	
	0400	4 ton		.90	17.778		1,450	560		2,010	2,450	
	0500	5 ton		.60	26.667		1,700	835		2,535	3,150	
	0550	7.5 ton		.55	29.091		3,825	915		4,740	5,575	
	0560	8.5 ton		.53	30.189		3,900	945		4,845	5,700	
	0600	10 ton		.50	32		4,400	1,000		5,400	6,375	
	0620	11 ton		.48	33.333		5,300	1,050		6,350	7,425	
	0650	15 ton		.40	40		6,325	1,250		7,575	8,850	

15600 | Refrigeration Equipment

15670 | Refrig Condensing Units

		CREW	DAILY OUTPUT	LABOR-HOURS	UNIT	MAT.	LABOR	EQUIP.	TOTAL	TOTAL INCL O&P
0700	20 ton	Q-6	.40	60	Ea.	11,200	1,950		13,150	15,300
0720	25 ton		.35	68.571		12,500	2,225		14,725	17,100
0750	30 ton		.30	80		14,300	2,600		16,900	19,700
0800	40 ton		.20	120		18,200	3,900		22,100	25,900
0840	50 ton		.18	133		21,800	4,325		26,125	30,600
0860	60 ton		.16	150		25,200	4,875		30,075	35,100
0900	75 ton		.14	171		29,000	5,575		34,575	40,300
1000	80 ton		.12	200		31,600	6,500		38,100	44,500
1100	100 ton		.09	266		43,700	8,675		52,375	61,000
2000	Water cooled, compressor, heat exchanger, controls									
2100	5 ton	Q-5	.70	22.857	Ea.	3,975	715		4,690	5,450
2110	10 ton		.60	26.667		5,825	835		6,660	7,700
2200	15 ton		.50	32		6,825	1,000		7,825	9,050
2300	20 ton	Q-6	.40	60		9,475	1,950		11,425	13,400
2310	30 ton		.30	80		13,100	2,600		15,700	18,300
2400	40 ton		.20	120		17,800	3,900		21,700	25,500
2410	60 ton		.18	133		27,200	4,325		31,525	36,500
2420	80 ton		.14	171		31,400	5,575		36,975	43,000
2500	100 ton		.11	218		38,500	7,100		45,600	53,000
2510	120 ton		.10	240		62,000	7,800		69,800	80,000

15680 | Evaporators, DX Coils

		CREW	DAILY OUTPUT	LABOR-HOURS	UNIT	MAT.	LABOR	EQUIP.	TOTAL	TOTAL INCL O&P
0010	**EVAPORATORS** DX coils, remote compressors not included									
1000	Coolers, reach-in type, above freezing temperatures									
1300	Shallow depth, wall mount, 7 fins per inch, air defrost									
1310	600 BTUH, 8" fan, rust-proof core	Q-5	3.80	4.211	Ea.	258	132		390	485
1320	900 BTUH, 8" fan, rust-proof core		3.30	4.848		320	152		472	580
1330	1200 BTUH, 8" fan, rust-proof core		3	5.333		375	167		542	670
1340	1800 BTUH, 8" fan, rust-proof core		2.40	6.667		440	209		649	795
1350	2500 BTUH, 10" fan, rust-proof core		2.20	7.273		475	228		703	865
1360	3500 BTUH, 10" fan		2	8		555	251		806	990
1370	4500 BTUH, 10" fan		1.90	8.421		590	264		854	1,050
1390	For pan drain, add					34			34	37.50
1600	Undercounter refrigerators, ceiling or wall mount,									
1610	8 fins per inch, air defrost, rust-proof core									
1630	800 BTUH, one 6" fan	Q-5	5	3.200	Ea.	195	100		295	365
1640	1300 BTUH, two 6" fans		4	4		237	126		363	450
1650	1700 BTUH, two 6" fans		3.60	4.444		265	139		404	505
2000	Coolers, reach-in and walk-in types, above freezing temperatures									
2600	Two-way discharge, ceiling mount, 150-4100 CFM,									
2610	above 34°F applications, air defrost									
2630	900 BTUH, 7 fins per inch, 8" fan	Q-5	4	4	Ea.	207	126		333	415
2660	2500 BTUH, 7 fins per inch, 10" fan		2.60	6.154		385	193		578	715
2690	5500 BTUH, 7 fins per inch, 12" fan		1.70	9.412		775	295		1,070	1,300
2720	8500 BTUH, 8 fins per inch, 16" fan		1.20	13.333		1,000	420		1,420	1,725
2750	15,000 BTUH, 7 fins per inch, 18" fan		1.10	14.545		1,150	455		1,605	1,975
2770	24,000 BTUH, 7 fins per inch, two 16" fans		1	16		2,000	500		2,500	2,950
2790	30,000 BTUH, 7 fins per inch, two 18" fans		.90	17.778		2,500	560		3,060	3,575
2850	Two-way discharge, low profile, ceiling mount,									
2860	8 fins per inch, 200-570 CFM, air defrost									
2880	800 BTUH, one 6" fan	Q-5	4	4	Ea.	235	126		361	450
2890	1300 BTUH, two 6" fans		3.50	4.571		280	143		423	525
2900	1800 BTUH, two 6" fans		3.30	4.848		295	152		447	555
2910	2700 BTUH, three 6" fans		2.70	5.926		345	186		531	660
3000	Coolers, walk-in type, above freezing temperatures									
3300	General use, ceiling mount, 108-2080 CFM									

15600 | Refrigeration Equipment

15680 | Evaporators, DX Coils

		CREW	DAILY OUTPUT	LABOR-HOURS	UNIT	MAT.	LABOR	EQUIP.	TOTAL	TOTAL INCL O&P	
300	3320	600 BTUH, 7 fins per inch, 6" fan	Q-5	5.20	3.077	Ea.	186	96.50		282.50	350	300
	3340	1200 BTUH, 7 fins per inch, 8" fan		3.70	4.324		257	136		393	485	
	3360	1800 BTUH, 7 fins per inch, 10" fan		2.70	5.926		360	186		546	675	
	3380	3500 BTUH, 7 fins per inch, 12" fan		2.40	6.667		530	209		739	895	
	3400	5500 BTUH, 8 fins per inch, 12" fan		2	8		590	251		841	1,025	
	3430	8500 BTUH, 8 fins per inch, 16" fan		1.40	11.429		875	360		1,235	1,500	
	3460	15,000 BTUH, 7 fins per inch, two 16" fans		1.10	14.545		1,400	455		1,855	2,250	
	3640	Low velocity, high latent load, ceiling mount,										
	3650	1050-2420 CFM, air defrost, 6 fins per inch										
	3670	6700 BTUH, two 10" fans	Q-5	1.30	12.308	Ea.	1,150	385		1,535	1,850	
	3680	10,000 BTUH, three 10" fans		1	16		1,350	500		1,850	2,225	
	3690	13,500 BTUH, three 10" fans		1	16		1,625	500		2,125	2,525	
	3700	18,000 BTUH, four 10" fans		1	16		2,225	500		2,725	3,175	
	3710	26,500 BTUH, four 10" fans		.80	20		2,650	630		3,280	3,875	
	3730	For electric defrost, add					17%					
	5000	Freezers and coolers, reach-in type, above 34°F										
	5030	to sub-freezing temperature range, low latent load,										
	5050	air defrost, 7 fins per inch										
	5070	1200 BTUH, 8" fan, rustproof core	Q-5	4.40	3.636	Ea.	230	114		344	425	
	5080	1500 BTUH, 8" fan, rustproof core		4.20	3.810		237	120		357	440	
	5090	1800 BTUH, 8" fan, rustproof core		3.70	4.324		244	136		380	475	
	5100	2500 BTUH, 10" fan, rustproof core		2.90	5.517		279	173		452	565	
	5110	3500 BTUH, 10" fan, rustproof core		2.50	6.400		385	201		586	730	
	5120	4500 BTUH, 12" fan, rustproof core		2.10	7.619		500	239		739	910	
	6000	Freezers and coolers, walk-in type										
	6050	1960-18,000 CFM, medium profile										
	6060	Standard motor, 6 fins per inch, to -30°F, all aluminum										
	6080	10,500 BTUH, air defrost, one 18" fan	Q-5	1.40	11.429	Ea.	1,050	360		1,410	1,700	
	6090	12,500 BTUH, air defrost, one 18" fan		1.40	11.429		1,125	360		1,485	1,775	
	6100	16,400 BTUH, air defrost, one 20" fan		1.30	12.308		1,150	385		1,535	1,850	
	6110	20,900 BTUH, air defrost, one 18" fans		1	16		1,350	500		1,850	2,225	
	6120	27,000 BTUH, air defrost, two 18" fans		1	16		2,000	500		2,500	2,950	
	6130	32,900 BTUH, air defrost, two 20" fans	Q-6	1.30	18.462		2,225	600		2,825	3,350	
	6140	39,000 BTUH, air defrost, two 18" fans		1.25	19.200		2,600	625		3,225	3,800	
	6150	44,100 BTUH, air defrost, three 20" fans		1	24		2,950	780		3,730	4,425	
	6155	53,000 BTUH, air defrost, three 20" fans		1	24		2,950	780		3,730	4,425	
	6160	66,200 BTUH, air defrost, four 20" fans		1	24		3,675	780		4,455	5,200	
	6170	78,000 BTUH, air defrost, four 20" fans		.80	30		4,300	975		5,275	6,200	
	6180	88,200 BTUH, air defrost, four 24" fans		.60	40		4,675	1,300		5,975	7,125	
	6190	110,000 BTUH, air defrost, four 24" fans		.50	48		5,500	1,550		7,050	8,400	
	6330	Hot gas defrost, standard motor units, add					35%					
	6370	Electric defrost, 230V, standard motor, add					16%					
	6410	460V, standard or high capacity motor, add					25%					
	6800	Eight fins per inch, increases BTUH 75%										
	6810	Standard capacity, add				Ea.	5%					
	6830	Hot gas & electric defrost not recommended										
	7000	800-4150 CFM, 12" fans										
	7010	Four fins per inch										
	7030	3400 BTUH, 1 fan	Q-5	2.80	5.714	Ea.	365	179		544	670	
	7040	4200 BTUH, 1 fan		2.40	6.667		370	209		579	725	
	7050	5300 BTUH, 1 fan		1.90	8.421		430	264		694	875	
	7060	6800 BTUH, 1 fan		1.70	9.412		485	295		780	975	
	7070	8400 BTUH, 2 fans		1.60	10		645	315		960	1,175	
	7080	10,500 BTUH, 2 fans		1.40	11.429		700	360		1,060	1,300	
	7090	13,000 BTUH, 3 fans		1.30	12.308		905	385		1,290	1,575	
	7100	17,000 BTUH, 4 fans		1.20	13.333		1,150	420		1,570	1,875	

15600 | Refrigeration Equipment

15680 | Evaporators, DX Coils

		CREW	DAILY OUTPUT	LABOR-HOURS	UNIT	2001 BARE COSTS MAT.	LABOR	EQUIP.	TOTAL	TOTAL INCL O&P	
300	7110	21,500 BTUH, 5 fans	Q-5	1	16	Ea.	1,425	500		1,925	2,300	300
	7160	Six fins per inch increases BTUH 35%, add				↓	5%					
	7180	Eight fins per inch increases BTUH 53%,										
	7190	not recommended for below 35°F, add				Ea.	8%					
	7210	For 12°F temperature differential, add					20%					
	7230	For adjustable thermostat control, add				↓	74			74	81	
	7240	For hot gas defrost, except 8 fin models										
	7250	on applications below 35°F, add				Ea.	58%					
	7260	For electric defrost, except 8 fin models										
	7270	on applications below 35°F, add				Ea.	36%					
	8000	Freezers, pass-thru door uprights, walk-in storage										
	8300	Low temperature, thin profile, electric defrost										
	8310	138-1800 CFM, 5 fins per inch										
	8320	900 BTUH, one 6" fan	Q-5	3.30	4.848	Ea.	350	152		502	615	
	8340	1500 BTUH, two 6" fans		2.20	7.273		485	228		713	880	
	8360	2600 BTUH, three 6" fans		1.90	8.421		670	264		934	1,125	
	8370	3300 BTUH, four 6" fans		1.70	9.412		735	295		1,030	1,250	
	8380	4400 BTUH, five 6" fans		1.40	11.429		1,000	360		1,360	1,650	
	8400	7000 BTUH, three 10" fans		1.30	12.308		1,200	385		1,585	1,900	
	8410	8700 BTUH, four 10" fans	↓	1.10	14.545		1,375	455		1,830	2,200	
	8450	For air defrost, deduct				↓	140			140	153	

15700 | Heating/Ventilating/Air Conditioning Equipment

15705 | Curbs/Pads/Stands Prefab

		CREW	DAILY OUTPUT	LABOR-HOURS	UNIT	2001 BARE COSTS MAT.	LABOR	EQUIP.	TOTAL	TOTAL INCL O&P	
600	0010	**CURBS/PADS PREFABRICATED**										600
	6000	Pad, fiberglass reinforced concrete with polystyrene foam core										
	6050	Condenser, 2" thick, 20" x 38"	1 Shee	8	1	Ea.	17.75	33.50		51.25	71.50	
	6070	24" x 24"		16	.500		15.10	16.80		31.90	42.50	
	6090	24" x 36"		12	.667		22.50	22.50		45	59.50	
	6110	24" x 42"	↓	8	1		27	33.50		60.50	82	
	6150	26" x 36"	Q-9	8	2		25	60.50		85.50	121	
	6170	28" x 38"	1 Shee	8	1		29	33.50		62.50	83.50	
	6190	30" x 30"		12	.667		24	22.50		46.50	61	
	6220	30" x 36"	↓	8	1		29.50	33.50		63	84	
	6240	30" x 40"	Q-9	8	2		32.50	60.50		93	129	
	6260	32" x 32"		7	2.286		27	69		96	137	
	6280	36" x 36"		8	2		35	60.50		95.50	132	
	6300	36" x 40"		7	2.286		39	69		108	150	
	6320	36" x 48"		7	2.286		47	69		116	159	
	6340	36" x 54"		6	2.667		53	80.50		133.50	183	
	6360	36" x 60"		7	2.286		59	69		128	172	
	6400	41" x 47"		6	2.667		56.50	80.50		137	186	
	6490	26" round		10	1.600		21.50	48.50		70	98.50	
	6550	30" round		9	1.778		28	53.50		81.50	114	
	6600	36" round	↓	8	2		43.50	60.50		104	142	
	6700	For 3" thick pad, add				↓	20%	5%				
	8800	Stand, corrosion-resistant plastic										
	8850	Heat pump, 6" high	1 Shee	32	.250	Ea.	2.16	8.40		10.56	15.35	

15700 | Heating/Ventilating/Air Conditioning Equipment

15705 | Curbs/Pads/Stands Prefab

			CREW	DAILY OUTPUT	LABOR-HOURS	UNIT	MAT.	LABOR	EQUIP.	TOTAL	TOTAL INCL O&P	
600	8870	12" high	1 Shee	32	.250	Ea.	2.38	8.40		10.78	15.55	600

15710 | Heat Exchangers

			CREW	DAILY OUTPUT	LABOR-HOURS	UNIT	MAT.	LABOR	EQUIP.	TOTAL	TOTAL INCL O&P	
900	0010	**HEAT EXCHANGERS**										900
	0016	Shell & tube type, 4 pass, 3/4" O.D. copper tubes,										
	0020	C.I. heads, C.I. tube sheet, steel shell										
	0100	Hot water 40°F to 180°F, by steam at 10 PSI										
	0120	8 GPM	Q-5	6	2.667	Ea.	895	83.50		978.50	1,100	
	0140	10 GPM		5	3.200		1,350	100		1,450	1,625	
	0160	40 GPM		4	4		2,075	126		2,201	2,500	
	0180	64 GPM		2	8		3,200	251		3,451	3,875	
	0200	96 GPM	▼	1	16		4,225	500		4,725	5,400	
	0220	120 GPM	Q-6	1.50	16		5,550	520		6,070	6,875	
	0240	168 GPM		1	24		6,800	780		7,580	8,675	
	0260	240 GPM		.80	30		10,600	975		11,575	13,200	
	0300	600 GPM	▼	.70	34.286	▼	22,900	1,125		24,025	26,800	
	0500	For bronze head and tube sheet, add					50%					
	1000	Hot water 40°F to 140°F, by water at 200°F										
	1020	7 GPM	Q-5	6	2.667	Ea.	1,100	83.50		1,183.50	1,325	
	1040	16 GPM		5	3.200		1,550	100		1,650	1,850	
	1060	34 GPM		4	4		2,350	126		2,476	2,800	
	1080	55 GPM		3	5.333		3,400	167		3,567	4,000	
	1100	74 GPM		1.50	10.667		4,225	335		4,560	5,150	
	1120	86 GPM	▼	1.40	11.429		5,650	360		6,010	6,775	
	1140	112 GPM	Q-6	2	12		7,025	390		7,415	8,350	
	1160	126 GPM		1.80	13.333		8,775	435		9,210	10,300	
	1180	152 GPM	▼	1	24	▼	11,200	780		11,980	13,500	
	3000	Plate type,										
	3100	400 GPM	Q-6	.80	30	Ea.	24,500	975		25,475	28,500	
	3120	800 GPM	"	.50	48		43,100	1,550		44,650	49,800	
	3140	1200 GPM	Q-7	.34	94.118		88,500	3,125		91,625	102,000	
	3160	1800 GPM	"	.24	133	▼	133,000	4,425		137,425	152,500	
	8000	Heat pipe type, glycol, 50% efficient										
	8010	100 MBH, 1700 CFM	1 Stpi	.80	10	Ea.	15,700	350		16,050	17,800	
	8020	160 MBH, 2700 CFM		.60	13.333		19,800	465		20,265	22,500	
	8030	620 MBH, 4000 CFM	▼	.40	20	▼	26,900	695		27,595	30,700	

15720 | Air Handling Units

			CREW	DAILY OUTPUT	LABOR-HOURS	UNIT	MAT.	LABOR	EQUIP.	TOTAL	TOTAL INCL O&P	
100	0010	**AIR HANDLING UNIT** Built-up										100
	0100	With cooling/heating coil section, filters, mixing box										
	0880	Single zone, horizontal / vertical										
	0890	Constant volume										
	0900	1600 CFM	Q-5	1.20	13.333	Ea.	695	420		1,115	1,400	
	0906	2000 CFM		1.10	14.545		870	455		1,325	1,650	
	0910	3000 CFM		1	16		1,300	500		1,800	2,175	
	0916	4000 CFM	▼	.95	16.842		1,725	530		2,255	2,700	
	0920	5000 CFM	Q-6	1.40	17.143		2,175	560		2,735	3,225	
	0926	6500 CFM		1.30	18.462		2,825	600		3,425	4,000	
	0930	7500 CFM		1.20	20		3,250	650		3,900	4,550	
	0936	9200 CFM		1.10	21.818		4,000	710		4,710	5,475	
	0940	11,500 CFM		1	24		5,000	780		5,780	6,675	
	0946	13,200 CFM		.90	26.667		5,725	865		6,590	7,600	
	0950	16,500 CFM		.80	30		7,150	975		8,125	9,350	
	0960	19,500 CFM	▼	.70	34.286		8,475	1,125		9,600	11,000	

15700 | Heating/Ventilating/Air Conditioning Equipment

15720 | Air Handling Units

			CREW	DAILY OUTPUT	LABOR-HOURS	UNIT	MAT.	LABOR	EQUIP.	TOTAL	TOTAL INCL O&P
100	0970	22,000 CFM	Q-6	.60	40	Ea.	9,550	1,300		10,850	12,500
	0980	27,000 CFM		.50	48		11,700	1,550		13,250	15,300
	0990	34,000 CFM		.40	60		14,800	1,950		16,750	19,200
	1000	40,000 CFM		.30	80		15,400	2,600		18,000	20,800
	1010	47,000 CFM		.20	120		20,400	3,900		24,300	28,300
	1140	60,000 CFM	Q-7	.18	177		44,800	5,900		50,700	58,000
	1160	75,000 CFM	"	.16	200		56,000	6,650		62,650	71,500
	2000	Variable air volume									
	2100	75,000 CFM	Q-7	.12	266	Ea.	58,000	8,875		66,875	77,000
	2140	100,000 CFM		.11	290		85,500	9,675		95,175	108,500
	2160	150,000 CFM		.10	320		128,500	10,600		139,100	157,000
	2300	Multi-zone, horizontal or vertical, blow-thru fan									
	2310	3000 CFM	Q-5	1	16	Ea.	1,325	500		1,825	2,200
	2314	4000 CFM	"	.95	16.842		1,775	530		2,305	2,750
	2318	5000 CFM	Q-6	1.40	17.143		2,225	560		2,785	3,275
	2322	6500 CFM		1.30	18.462		2,875	600		3,475	4,075
	2326	7500 CFM		1.20	20		3,325	650		3,975	4,625
	2330	9200 CFM		1.10	21.818		4,075	710		4,785	5,550
	2334	11,500 CFM		1	24		5,100	780		5,880	6,775
	2336	13,200 CFM		.90	26.667		5,850	865		6,715	7,725
	2340	16,500 CFM		.80	30		7,300	975		8,275	9,500
	2344	19,500 CFM		.70	34.286		8,625	1,125		9,750	11,200
	2348	22,000 CFM		.60	40		9,750	1,300		11,050	12,700
	2352	27,000 CFM		.50	48		12,000	1,550		13,550	15,500
	2356	34,000 CFM		.40	60		15,100	1,950		17,050	19,600
	2360	40,000 CFM		.30	80		17,700	2,600		20,300	23,400
	2364	47,000 CFM		.20	120		20,800	3,900		24,700	28,800
	3000	Rooftop									
	3010	Constant volume									
	3100	2000 CFM	Q-5	1	16	Ea.	3,375	500		3,875	4,450
	3140	5000 CFM	Q-6	.79	30.380		6,050	990		7,040	8,150
	3150	10,000 CFM		.52	46.154		7,650	1,500		9,150	10,700
	3160	15,000 CFM		.39	61.538		11,600	2,000		13,600	15,700
	3170	20,000 CFM		.30	80		14,400	2,600		17,000	19,700
	3200	Variable air volume									
	3240	5000 CFM	Q-6	.70	34.286	Ea.	7,775	1,125		8,900	10,200
	3250	10,000 CFM		.50	48		10,100	1,550		11,650	13,500
	3260	15,000 CFM		.40	60		14,000	1,950		15,950	18,400
	3270	20,000 CFM		.29	82.759		18,300	2,700		21,000	24,200
	3280	30,000 CFM		.20	120		30,000	3,900		33,900	38,900
200	0010	**CENTRAL STATION AIR-HANDLING UNIT** Chilled water									
	1000	Modular, capacity at 700 FPM face velocity									
	1090	Constant volume									
	1100	1300 CFM	Q-5	1.20	13.333	Ea.	2,325	420		2,745	3,200
	1200	1900 CFM		1.10	14.545		2,600	455		3,055	3,550
	1300	3200 CFM		.80	20		3,075	630		3,705	4,350
	1400	5400 CFM	Q-6	.80	30		4,650	975		5,625	6,575
	1500	8000 CFM		.60	40		5,425	1,300		6,725	7,925
	1550	10,000 CFM		.54	44.444		5,875	1,450		7,325	8,650
	1600	12,000 CFM		.46	52.174		6,700	1,700		8,400	9,950
	1650	14,000 CFM		.42	57.143		7,925	1,850		9,775	11,500
	1670	16,000 CFM		.38	63.158		8,875	2,050		10,925	12,900
	1700	20,000 CFM		.31	77.419		11,100	2,525		13,625	16,000
	1800	23,000 CFM		.28	85.714		12,700	2,800		15,500	18,200
	1850	28,000 CFM		.24	100		16,100	3,250		19,350	22,600
	1900	33,500 CFM		.19	126		19,200	4,100		23,300	27,400

15700 | Heating/Ventilating/Air Conditioning Equipment

15720 | Air Handling Units

			CREW	DAILY OUTPUT	LABOR-HOURS	UNIT	2001 BARE COSTS				TOTAL INCL O&P	
							MAT.	LABOR	EQUIP.	TOTAL		
200	1950	42,500 CFM	Q-6	.15	160	Ea.	22,900	5,200		28,100	33,000	200
	2000	52,500 CFM	Q-7	.16	200		31,500	6,650		38,150	44,600	
	2100	63,000 CFM	"	.13	246		37,400	8,175		45,575	53,500	
	2200	For hot water heating coil, add					21%	5%				
	2300	Variable air volume, also incl. heating coil										
	2340	5000 CFM	Q-6	.70	34.286	Ea.	7,300	1,125		8,425	9,700	
	2350	10,000 CFM		.50	48		9,500	1,550		11,050	12,900	
	2360	15,000 CFM		.40	60		13,100	1,950		15,050	17,400	
	2370	20,000 CFM		.29	82.759		17,200	2,700		19,900	23,000	
	2380	30,000 CFM		.20	120		28,000	3,900		31,900	36,700	
	3000	Packaged, 3000 CFM, 7.5 ton		1.70	14.118		1,825	460		2,285	2,700	
	3050	3200 CFM, 8 ton		1.60	15		2,125	490		2,615	3,075	
	3100	4000 CFM, 10 ton		1.40	17.143		2,300	560		2,860	3,400	
	3150	4400 CFM, 11 ton		1.35	17.778		2,725	580		3,305	3,875	
	3200	6000 CFM, 15 ton		1	24		3,250	780		4,030	4,750	
	3250	7000 CFM, 17.5 ton		.90	26.667		4,175	865		5,040	5,900	
	3300	10,000 CFM, 25 ton		.65	36.923		6,000	1,200		7,200	8,425	
	3400	12,000 CFM, 30 ton		.50	48		6,225	1,550		7,775	9,200	
	3500	For hot water heat coil, add					30%	5%				
300	0010	**EVAPORATIVE COOLERS** Ducted, not incl. duct.										300
	0100	Side discharge style, capacities at .25" S.P.										
	0120	1785 CFM, 1/3 HP, 115 V	Q-9	5	3.200	Ea.	310	96.50		406.50	490	
	0140	2740 CFM, 1/3 HP, 115 V		4.50	3.556		355	107		462	555	
	0160	3235 CFM, 1/2 HP, 115 V		4	4		365	121		486	585	
	0180	3615 CFM, 1/2 HP, 230 V		3.60	4.444		370	134		504	615	
	0200	4215 CFM, 3/4 HP, 230 V		3.20	5		470	151		621	750	
	0220	5255 CFM. 1 HP, 115/230 V		3	5.333		750	161		911	1,075	
	0240	6090 CFM, 1 HP, 230/460 V		2.80	5.714		1,050	173		1,223	1,425	
	0260	8300 CFM, 1-1/2 HP, 230/460 V		2.60	6.154		1,075	186		1,261	1,450	
	0280	8360 CFM, 1-1/2 HP, 230/460 V		2.20	7.273		1,475	220		1,695	1,975	
	0300	9725 CFM, 2 HP, 230/460 V		1.80	8.889		1,475	268		1,743	2,050	
	0320	11,715 CFM, 3 HP, 230/460 V		1.40	11.429		1,550	345		1,895	2,225	
	0340	14,410 CFM, 5 HP, 230/460 V		1	16		1,550	485		2,035	2,450	
	0400	For two-speed motor, add					5%					
	0500	For down discharge style, add					10%					
	1000	High capacities at .4" S.P.										
	1040	15,000 CFM, 5 HP, 230/460 V	Q-9	.54	29.630	Ea.	5,550	895		6,445	7,475	
	1080	20,000 CFM, 10HP, 230/460 V	"	.48	33.333		7,150	1,000		8,150	9,400	
	1100	25,000 CFM, 20 HP, 230/460 V	Q-10	.54	44.444		7,875	1,400		9,275	10,800	
	1140	30,000 CFM, 20 HP, 230/460 V		.46	52.174		9,075	1,625		10,700	12,500	
	1160	35,000 CFM, 20 HP, 230/460 V		.44	54.545		9,925	1,700		11,625	13,500	
	1180	40,000 CFM, 30 HP, 230/460 V		.42	57.143		10,400	1,800		12,200	14,200	
500	0010	**MAKE-UP AIR UNIT**										500
	0020	Indoor suspension, natural/LP gas, direct fired,										
	0030	standard control. For flue see division 15550-440										
	0040	70°F temperature rise, MBH is input										
	0100	2000 CFM, 168 MBH	Q-6	3	8	Ea.	4,800	260		5,060	5,700	
	0120	3000 CFM, 252 MBH		2	12		4,800	390		5,190	5,900	
	0140	4000 CFM, 336 MBH		1.80	13.333		5,025	435		5,460	6,175	
	0160	6000 CFM, 502 MBH		1.50	16		6,050	520		6,570	7,450	
	0180	8000 CFM, 670 MBH		1.40	17.143		6,625	560		7,185	8,150	
	0200	10,000 CFM, 838 MBH		1.20	20		7,475	650		8,125	9,200	
	0220	12,000 CFM, 1005 MBH		1	24		7,550	780		8,330	9,475	
	0240	14,000 CFM, 1180 MBH		.94	25.532		7,650	830		8,480	9,650	

T5 MECHANICAL

15700 | Heating/Ventilating/Air Conditioning Equipment

15720 | Air Handling Units

		CREW	DAILY OUTPUT	LABOR-HOURS	UNIT	MAT.	LABOR	EQUIP.	TOTAL	TOTAL INCL O&P	
500	0260	18,000 CFM, 1340 MBH	Q-6	.86	27.907	Ea.	7,700	910		8,610	9,850	500
	0280	20,000 CFM, 1675 MBH	↓	.76	31.579		9,500	1,025		10,525	12,000	
	0300	24,000 CFM, 2007 MBH	Q-7	1	32		9,825	1,075		10,900	12,400	
	0320	30,000 CFM, 2510 MBH		.96	33.333		10,100	1,100		11,200	12,800	
	0340	35,000 CFM, 2930 MBH		.92	34.783		12,700	1,150		13,850	15,800	
	0360	40,000 CFM, 3350 MBH		.88	36.364		13,000	1,200		14,200	16,100	
	0380	45,000 CFM, 3770 MBH		.84	38.095		13,300	1,275		14,575	16,600	
	0400	50,000 CFM, 4180 MBH		.80	40		13,500	1,325		14,825	16,800	
	0420	55,000 CFM, 4600 MBH		.75	42.667		17,200	1,425		18,625	21,100	
	0440	60,000 CFM, 5020 MBH		.70	45.714		17,900	1,525		19,425	22,000	
	0460	65,000 CFM, 5435 MBH		.60	53.333		18,000	1,775		19,775	22,500	
	0480	75,000 CFM, 6275 MBH	↓	.50	64		19,000	2,125		21,125	24,000	
	0600	For discharge louver assembly, add					10%					
	0700	For filters, add					20%					
	0800	For air shut-off damper section, add					10%					
	0900	For vertical unit, add				↓	10%					
	1000	Rooftop unit, natural gas, gravity vent, S.S. exchanger										
	1010	70°F temperature rise, MBH is input										
	1020	995 CFM, 75 MBH	Q-6	4	6	Ea.	3,225	195		3,420	3,850	
	1040	1245 CFM, 95 MBH		3.60	6.667		3,275	217		3,492	3,925	
	1060	1590 CFM, 120 MBH		3.30	7.273		3,525	237		3,762	4,225	
	1080	2090 CFM, 159 MBH		3	8		3,625	260		3,885	4,375	
	1100	2490 CFM, 190 MBH		2.60	9.231		4,150	300		4,450	5,000	
	1120	2985 CFM, 225 MBH		2.30	10.435		4,200	340		4,540	5,150	
	1140	3975 CFM, 300 MBH		1.90	12.632		5,000	410		5,410	6,100	
	1160	4970 CFM, 375 MBH		1.40	17.143		5,775	560		6,335	7,200	
	1180	5950 CFM, 450 MBH		1.20	20		6,100	650		6,750	7,675	
	1200	7940 CFM, 600 MBH		1	24		7,075	780		7,855	8,950	
	1220	9930 CFM, 750 MBH		.80	30		12,100	975		13,075	14,800	
	1240	11,900 CFM, 900 MBH		.60	40		13,500	1,300		14,800	16,800	
	1260	15,880 CFM, 1200 MBH	↓	.30	80	↓	16,200	2,600		18,800	21,700	
	1500	For power vent, add					8%					
	1600	For cleanable filters, add				Ea.	94			94	103	
	1700	For electric modulating gas control, add				"	1,000			1,000	1,100	
	1800	For aluminized steel exchanger, deduct					10%					
	8000	Rooftop unit, electric, with curb, controls										
	8010	33 kW	Q-5	4	4	Ea.	3,175	126		3,301	3,675	
	8020	100 kW		3.60	4.444		5,250	139		5,389	5,975	
	8030	150 kW		3.20	5		6,100	157		6,257	6,925	
	8040	250 kW		2.80	5.714		7,925	179		8,104	8,975	
	8050	400 kW	↓	2.20	7.273	↓	11,600	228		11,828	13,000	

15730 | Unitary Air Conditioning Equip

		CREW	DAILY OUTPUT	LABOR-HOURS	UNIT	MAT.	LABOR	EQUIP.	TOTAL	TOTAL INCL O&P	
200	0010	**COMPUTER ROOM UNITS** A8.4 -810										200
	1000	Air cooled, includes remote condenser but not										
	1020	interconnecting tubing or refrigerant										
	1080	3 ton	Q-5	.50	32	Ea.	7,025	1,000		8,025	9,250	
	1120	5 ton		.45	35.556		9,400	1,125		10,525	12,000	
	1160	6 ton		.30	53.333		17,300	1,675		18,975	21,600	
	1200	8 ton		.27	59.259		19,600	1,850		21,450	24,400	
	1240	10 ton		.25	64		20,400	2,000		22,400	25,400	
	1280	15 ton		.22	72.727		22,800	2,275		25,075	28,500	
	1290	18 ton	↓	.20	80		25,900	2,500		28,400	32,300	
	1320	20 ton	Q-6	.29	82.759		27,200	2,700		29,900	34,000	
	1360	23 ton	"	.28	85.714	↓	28,600	2,800		31,400	35,600	

15700 | Heating/Ventilating/Air Conditioning Equipment

15730 | Unitary Air Conditioning Equip

			CREW	DAILY OUTPUT	LABOR-HOURS	UNIT	MAT.	LABOR	EQUIP.	TOTAL	TOTAL INCL O&P
200	2200	Chilled water, for connection to										200
	2220	existing chiller system of adequate capacity	A8.4 -810								
	2260	5 ton	Q-5	.74	21.622	Ea.	7,925	680		8,605	9,750
	2280	6 ton		.52	30.769		10,500	965		11,465	13,100
	2300	8 ton		.50	32		11,300	1,000		12,300	14,000
	2320	10 ton		.49	32.653		12,600	1,025		13,625	15,400
	2330	12 ton		.48	32.990		13,000	1,025		14,025	15,900
	2360	15 ton		.48	33.333		13,500	1,050		14,550	16,400
	2400	20 ton		.46	34.783		13,800	1,100		14,900	16,900
	2440	23 ton		.42	38.095		14,500	1,200		15,700	17,800
	4000	Glycol system, complete except for interconnecting tubing									
	4060	3 ton	Q-5	.40	40	Ea.	8,600	1,250		9,850	11,400
	4100	5 ton		.38	42.105		11,200	1,325		12,525	14,300
	4120	6 ton		.25	64		22,400	2,000		24,400	27,700
	4140	8 ton		.23	69.565		24,200	2,175		26,375	30,000
	4160	10 ton		.21	76.190		26,300	2,400		28,700	32,500
	4200	15 ton	Q-6	.26	92.308		31,800	3,000		34,800	39,500
	4240	20 ton		.24	100		34,100	3,250		37,350	42,400
	4280	23 ton		.22	109		37,200	3,550		40,750	46,300
	8000	Water cooled system, not including condenser,									
	8020	water supply or cooling tower									
	8060	3 ton	Q-5	.62	25.806	Ea.	6,350	810		7,160	8,200
	8100	5 ton		.54	29.630		8,950	930		9,880	11,300
	8120	6 ton		.35	45.714		16,800	1,425		18,225	20,600
	8140	8 ton		.33	48.485		18,500	1,525		20,025	22,600
	8160	10 ton		.31	51.613		19,600	1,625		21,225	24,100
	8200	15 ton		.27	59.259		22,600	1,850		24,450	27,600
	8240	20 ton	Q-6	.38	63.158		26,600	2,050		28,650	32,400
	8280	23 ton	"	.34	70.588		28,200	2,300		30,500	34,500
500	0010	**PACKAGED TERMINAL AIR CONDITIONER** Cabinet, wall sleeve,										500
	0100	louver, electric heat, thermostat, manual changeover, 208 V									
	0200	6,000 BTUH cooling, 8800 BTU heat	Q-5	6	2.667	Ea.	915	83.50		998.50	1,125
	0220	9,000 BTUH cooling, 13,900 BTU heat		5	3.200		935	100		1,035	1,175
	0240	12,000 BTUH cooling, 13,900 BTU heat		4	4		970	126		1,096	1,275
	0260	15,000 BTUH cooling, 13,900 BTU heat		3	5.333		1,050	167		1,217	1,400
	0280	18,000 BTUH cooling, 10 KW heat		2.40	6.667		1,350	209		1,559	1,825
	0300	24,000 BTUH cooling, 10 KW heat		1.90	8.421		1,375	264		1,639	1,900
	0320	30,000 BTUH cooling, 10 KW heat		1.40	11.429		1,550	360		1,910	2,250
	0340	36,000 BTUH cooling, 10 KW heat		1.25	12.800		1,625	400		2,025	2,400
	0360	42,000 BTUH cooling, 10 KW heat		1	16		2,050	500		2,550	3,025
	0380	48,000 BTUH cooling, 10 KW heat		.90	17.778		2,225	560		2,785	3,300
	0500	For hot water coil, increase heat by 10%, add					5%	10%			
	1000	For steam, increase heat output by 30%, add					8%	10%			
600	0010	**ROOF TOP AIR CONDITIONERS** Standard controls, curb, economizer										600
	1000	Single zone, electric cool, gas heat									
	1090	2 ton cooling, 55 MBH heating	Q-5 A8.4-210	.93	17.204	Ea.	4,050	540		4,590	5,275
	1100	3 ton cooling, 60 MBH heating		.70	22.857		4,225	715		4,940	5,700
	1120	4 ton cooling, 95 MBH heating	A8.4-220	.61	26.403		4,500	830		5,330	6,200
	1140	5 ton cooling, 112 MBH heating		.56	28.520		4,675	895		5,570	6,500
	1145	6 ton cooling, 140 MBH heating	R15700-010	.52	30.769		5,350	965		6,315	7,350
	1150	7.5 ton cooling, 170 MBH heating		.50	32.258		7,325	1,000		8,325	9,575
	1160	10 ton cooling, 200 MBH heating	Q-6 R15700-020	.67	35.982		9,500	1,175		10,675	12,200
	1170	12.5 ton cooling, 230 MBH heating		.63	37.975		10,800	1,225		12,025	13,800
	1180	15 ton cooling, 270 MBH heating		.57	42.032		12,500	1,375		13,875	15,900
	1190	18 ton cooling, 330 MBH heating		.52	45.889		15,100	1,500		16,600	18,900

15700 | Heating/Ventilating/Air Conditioning Equipment

15730 | Unitary Air Conditioning Equip

			CREW	DAILY OUTPUT	LABOR-HOURS	UNIT	MAT.	LABOR	EQUIP.	TOTAL	TOTAL INCL O&P
600	1200	20 ton cooling, 360 MBH heating	Q-7 (A8.4-210)	.67	47.976	Ea.	20,600	1,600		22,200	25,000
	1210	25 ton cooling, 450 MBH heating		.56	57.554		24,600	1,925		26,525	30,000
	1220	30 ton cooling, 540 MBH heating	(A8.4-220)	.47	68.376		29,100	2,275		31,375	35,400
	1240	40 ton cooling, 675 MBH heating		.35	91.168		37,900	3,025		40,925	46,300
	1260	50 ton cooling, 810 MBH heating	(R15700-010)	.28	113		46,000	3,775		49,775	56,000
	1265	60 ton cooling, 900 MBH heating		.23	136		52,000	4,550		56,550	64,500
	1300	Electric cool, electric heat	(R15700-020)								
	1310	5 ton	Q-5	.63	25.600	Ea.	1,475	805		2,280	2,850
	1314	10 ton	Q-6	.74	32.432		2,975	1,050		4,025	4,875
	1318	15 ton	"	.63	37.795		4,450	1,225		5,675	6,750
	1322	20 ton	Q-7	.74	43.185		5,950	1,425		7,375	8,700
	1326	25 ton		.62	51.696		7,425	1,725		9,150	10,800
	1330	30 ton		.52	61.657		8,925	2,050		10,975	12,900
	1334	40 ton		.39	82.051		11,900	2,725		14,625	17,200
	1338	50 ton		.31	102		14,900	3,400		18,300	21,500
	1342	60 ton		.26	123		17,800	4,100		21,900	25,800
	1346	75 ton		.21	153		22,300	5,125		27,425	32,200
	1700	Gas cool, gas heat									
	1720	3 ton cooling, 90 MBH heating	Q-5	.93	17.204	Ea.	4,400	540		4,940	5,675
	1730	4 ton cooling, 112.5 MBH heating		.61	26.403		5,175	830		6,005	6,950
	1750	5 ton cooling, 135 MBH heating		.56	28.571		5,675	895		6,570	7,600
	2000	Multizone, electric cool, gas heat, economizer									
	2100	15 ton cooling, 360 MBH heating	Q-7	.61	52.545	Ea.	39,900	1,750		41,650	46,500
	2120	20 ton cooling, 360 MBH heating		.53	60.038		45,500	2,000		47,500	53,000
	2140	25 ton cooling, 450 MBH heating		.44	71.910		48,800	2,400		51,200	57,000
	2160	28 ton cooling, 450 MBH heating		.40	79.012		52,000	2,625		54,625	61,000
	2180	30 ton cooling, 540 MBH heating		.37	85.562		57,000	2,850		59,850	67,500
	2200	40 ton cooling, 540 MBH heating		.28	113		66,500	3,775		70,275	79,000
	2210	50 ton cooling, 540 MBH heating		.22	142		85,000	4,725		89,725	100,500
	2220	70 ton cooling, 1500 MBH heating		.16	198		105,000	6,600		111,600	125,500
	2240	80 ton cooling, 1500 MBH heating		.14	228		120,000	7,600		127,600	143,500
	2260	90 ton cooling, 1500 MBH heating		.13	256		126,000	8,500		134,500	151,500
	2280	105 ton cooling, 1500 MBH heating		.11	290		147,000	9,675		156,675	176,000
	2400	For hot water heat coil, deduct					5%				
	2500	For steam heat coil, deduct					2%				
	2600	For electric heat, deduct					3%	5%			
	4000	Single zone, electric cool, gas heat, variable air volume									
	4100	12.5 ton cooling, 230 MBH heating	Q-6	.55	43.716	Ea.	13,300	1,425		14,725	16,800
	4120	18 ton cooling, 330 MBH heating	"	.45	52.747		17,700	1,725		19,425	22,100
	4140	25 ton cooling, 450 MBH heating	Q-7	.48	66.116		28,500	2,200		30,700	34,700
	4160	40 ton cooling, 675 MBH heating		.31	104		45,200	3,475		48,675	55,000
	4180	60 ton cooling, 900 MBH heating		.20	157		64,000	5,250		69,250	78,500
	4200	80 ton cooling, 1000 MBH heating		.15	209		100,500	6,950		107,450	121,000
	5000	Single zone, electric cool only									
	5050	3 ton cooling	Q-5	.88	18.203	Ea.	3,675	570		4,245	4,875
	5060	4 ton cooling		.76	21.108		3,925	660		4,585	5,325
	5070	5 ton cooling		.70	22.792		4,125	715		4,840	5,600
	5080	6 ton cooling		.65	24.502		4,700	770		5,470	6,325
	5090	7.5 ton cooling		.62	25.806		6,250	810		7,060	8,100
	5100	8.5 ton cooling		.59	26.981		6,900	845		7,745	8,875
	5110	10 ton cooling	Q-6	.83	28.812		8,325	935		9,260	10,600
	5120	12 ton cooling		.80	30		10,300	975		11,275	12,900
	5130	15 ton cooling		.71	33.613		11,500	1,100		12,600	14,300
	5140	18 ton cooling		.65	36.697		14,100	1,200		15,300	17,300
	5150	20 ton cooling	Q-7	.83	38.415		16,600	1,275		17,875	20,200
	5160	25 ton cooling		.70	45.977		19,600	1,525		21,125	23,900

15700 | Heating/Ventilating/Air Conditioning Equipment

15730 | Unitary Air Conditioning Equip

			CREW	DAILY OUTPUT	LABOR-HOURS	UNIT	2001 BARE COSTS MAT.	LABOR	EQUIP.	TOTAL	TOTAL INCL O&P	
600	5170	30 ton cooling [A8.4-210]	Q-7	.58	54.701	Ea.	23,000	1,825		24,825	28,100	600
	5180	40 ton cooling		.44	73.059		31,500	2,425		33,925	38,400	
	5190	50 ton cooling [A8.4-220]		.35	91.168		38,700	3,025		41,725	47,100	
	5200	60 ton cooling		.29	109		43,800	3,625		47,425	53,500	
	5400	For low heat, add					7%					
	5410	For high heat, add [R15700-010]					10%					
	6000	Single zone electric cooling with variable volume distribution [R15700-020]										
	6020	20 ton cooling	Q-7	.72	44.199	Ea.	22,400	1,475		23,875	26,900	
	6030	25 ton cooling		.60	52.893		26,600	1,750		28,350	32,000	
	6040	30 ton cooling		.51	62.868		29,800	2,100		31,900	36,000	
	6050	40 ton cooling		.38	83.989		37,900	2,800		40,700	45,900	
	6060	50 ton cooling		.31	104		45,700	3,475		49,175	56,000	
	6070	60 ton cooling		.25	125		51,000	4,200		55,200	62,500	
	6200	For low heat, add					7%					
	6210	For high heat, add					10%					
	7000	Multizone, cool/heat, variable volume distribution										
	7100	50 ton cooling	Q-7	.19	164	Ea.	60,000	5,450		65,450	74,000	
	7110	70 ton cooling		.14	228		84,000	7,600		91,600	104,000	
	7120	90 ton cooling		.11	296		92,000	9,850		101,850	116,000	
	7130	105 ton cooling		.10	333		107,000	11,100		118,100	134,500	
	7140	120 ton cooling		.08	380		122,500	12,700		135,200	153,500	
	7150	140 ton cooling		.07	444		141,500	14,800		156,300	178,000	
	7300	Penthouse unit, cool/heat, variable volume distribution										
	7400	150 ton cooling	Q-7	.10	310	Ea.	217,500	10,300		227,800	255,000	
	7410	170 ton cooling		.10	320		238,000	10,600		248,600	278,000	
	7420	200 ton cooling		.10	326		278,000	10,900		288,900	322,500	
	7430	225 ton cooling		.09	340		313,000	11,300		324,300	361,000	
	7440	250 ton cooling		.09	359		312,500	12,000		324,500	362,000	
	7450	270 ton cooling		.09	363		337,500	12,100		349,600	390,000	
	7460	300 ton cooling		.09	372		375,000	12,400		387,400	431,000	
	9400	Sub assemblies for assembly systems										
	9410	Duct work per ton roof top 1 zone units	Q-9	.96	16.667	Ton	262	505		767	1,075	
	9420	Ductwork per ton rooftop multizone units		.50	32		375	965		1,340	1,925	
	9430	Ductwork, VAV cooling/ton rooftop multizone, 1 zone/ton		.32	50		555	1,500		2,055	2,925	
	9440	Duct work per ton packaged water & air cooled units		1.26	12.698		88.50	385		473.50	690	
	9450	Duct work per ton split system remote condensing units		1.32	12.121		83.50	365		448.50	655	
800	0010	**WINDOW UNIT AIR CONDITIONERS**										800
	4000	Portable/window, 15 amp 125V grounded receptacle required										
	4060	5000 BTUH	1 Carp	8	1	Ea.	296	29		325	370	
	4340	6000 BTUH		8	1		335	29		364	415	
	4480	8000 BTUH		6	1.333		385	39		424	485	
	4500	10,000 BTUH		6	1.333		455	39		494	560	
	4520	12,000 BTUH	L-2	8	2		495	51.50		546.50	625	
	4600	Window/thru-the-wall, 15 amp 230V grounded receptacle required										
	4780	17,000 BTUH	L-2	6	2.667	Ea.	715	68.50		783.50	890	
	4940	25,000 BTUH		4	4		1,025	103		1,128	1,275	
	4960	29,000 BTUH		4	4		1,200	103		1,303	1,475	
840	0010	**SELF-CONTAINED SINGLE PACKAGE** [A8.4-230]										840
	0100	Air cooled, for free blow or duct, not incl. remote condenser										
	0110	Constant volume [A8.4-240]										
	0200	3 ton cooling	Q-5	1	16	Ea.	4,400	500		4,900	5,575	
	0210	4 ton cooling	"	.80	20		4,775	630		5,405	6,200	
	0220	5 ton cooling	Q-6	1.20	20		5,525	650		6,175	7,050	
	0230	7.5 ton cooling	"	.90	26.667		7,500	865		8,365	9,550	
	0240	10 ton cooling	Q-7	1	32		9,400	1,075		10,475	11,900	

15700 | Heating/Ventilating/Air Conditioning Equipment

15730 | Unitary Air Conditioning Equip

			CREW	DAILY OUTPUT	LABOR-HOURS	UNIT	MAT.	LABOR	EQUIP.	TOTAL	TOTAL INCL O&P	
840	0250	15 ton cooling	Q-7	.95	33.684	Ea.	13,300	1,125		14,425	16,300	840
	0260	20 ton cooling		.90	35.556		15,300	1,175		16,475	18,600	
	0270	25 ton cooling		.85	37.647		18,400	1,250		19,650	22,200	
	0280	30 ton cooling		.80	40		19,300	1,325		20,625	23,200	
	0300	40 ton cooling		.60	53.333		26,700	1,775		28,475	32,100	
	0320	50 ton cooling		.50	64		32,800	2,125		34,925	39,300	
	0340	60 ton cooling	Q-8	.40	80		38,400	2,650	252	41,302	46,500	
	0380	With hot water heat, variable air volume										
	0390	10 ton cooling	Q-7	.96	33.333	Ea.	12,800	1,100		13,900	15,800	
	0400	20 ton cooling		.86	37.209		20,700	1,225		21,925	24,700	
	0410	30 ton cooling		.76	42.105		26,200	1,400		27,600	30,900	
	0420	40 ton cooling		.58	55.172		36,600	1,825		38,425	43,100	
	0430	50 ton cooling		.48	66.667		44,100	2,225		46,325	52,000	
	0440	60 ton cooling	Q-8	.38	84.211		54,000	2,800	265	57,065	64,000	
	0490	For duct mounting no price change										
	0500	For hot water or steam heating coils, add				Ea.	10%	10%				
	0550	Packaged, with electric heat										
	0552	Constant volume										
	0560	5 ton cooling	Q-6	1.20	20	Ea.	7,825	650		8,475	9,600	
	0570	10 ton cooling	Q-7	1	32		13,000	1,075		14,075	15,900	
	0580	20 ton cooling		.90	35.556		20,400	1,175		21,575	24,200	
	0590	30 ton cooling		.80	40		27,800	1,325		29,125	32,600	
	0600	40 ton cooling		.60	53.333		35,900	1,775		37,675	42,200	
	0610	50 ton cooling		.50	64		45,600	2,125		47,725	53,000	
	0700	Variable air volume										
	0710	10 ton cooling	Q-7	1	32	Ea.	15,400	1,075		16,475	18,500	
	0720	20 ton cooling		.90	35.556		24,300	1,175		25,475	28,500	
	0730	30 ton cooling		.80	40		32,800	1,325		34,125	38,100	
	0740	40 ton cooling		.60	53.333		43,200	1,775		44,975	50,000	
	0750	50 ton cooling		.50	64		53,500	2,125		55,625	62,000	
	0760	60 ton cooling	Q-8	.40	80		62,000	2,650	252	64,902	72,500	
	1000	Water cooled for free blow or duct, not including tower										
	1010	Constant volume										
	1100	3 ton cooling	Q-6	1	24	Ea.	3,075	780		3,855	4,550	
	1120	5 ton cooling		1	24		3,650	780		4,430	5,200	
	1130	7.5 ton cooling		.80	30		5,550	975		6,525	7,575	
	1140	10 ton cooling	Q-7	.90	35.556		7,425	1,175		8,600	9,950	
	1150	15 ton cooling		.85	37.647		9,475	1,250		10,725	12,300	
	1160	20 ton cooling		.80	40		13,300	1,325		14,625	16,600	
	1170	25 ton cooling		.75	42.667		15,700	1,425		17,125	19,400	
	1180	30 ton cooling		.70	45.714		17,600	1,525		19,125	21,700	
	1200	40 ton cooling		.40	80		22,300	2,650		24,950	28,500	
	1220	50 ton cooling		.30	106		27,800	3,550		31,350	35,900	
	1240	60 ton cooling	Q-8	.30	106		32,400	3,550	335	36,285	41,300	
	1300	For hot water or steam heat coils, add					12%	10%				
	2000	Water cooled with electric heat, not including tower										
	2010	Constant volume										
	2100	5 ton cooling	Q-6	1.20	20	Ea.	5,975	650		6,625	7,550	
	2120	10 ton cooling	Q-7	1	32		11,000	1,075		12,075	13,700	
	2130	15 ton cooling	"	.90	35.556		14,500	1,175		15,675	17,800	
	2400	Variable air volume										
	2440	10 ton cooling	Q-7	1	32	Ea.	13,400	1,075		14,475	16,400	
	2450	15 ton cooling	"	.90	35.556	"	18,400	1,175		19,575	22,100	
	2600	Water cooled, hot water coils, not including tower										
	2610	Variable air volume										
	2640	10 ton cooling	Q-7	1	32	Ea.	10,600	1,075		11,675	13,300	

15700 | Heating/Ventilating/Air Conditioning Equipment

15730 | Unitary Air Conditioning Equip

			CREW	DAILY OUTPUT	LABOR-HOURS	UNIT	MAT.	LABOR	EQUIP.	TOTAL	TOTAL INCL O&P	
840	2650	15 ton cooling [A8.4-230]	Q-7	.80	40	Ea.	14,300	1,325		15,625	17,700	840
900	0010	**SPLIT DUCTLESS SYSTEM**										900
	0100	Cooling only, single zone										
	0110	Wall mount										
	0120	3/4 ton cooling	Q-5	2	8	Ea.	1,375	251		1,626	1,900	
	0130	1 ton cooling		1.80	8.889		1,650	279		1,929	2,250	
	0140	1-1/2 ton cooling		1.60	10		2,225	315		2,540	2,925	
	0150	2 ton cooling		1.40	11.429		2,950	360		3,310	3,800	
	1000	Ceiling mount										
	1020	2 ton cooling	Q-5	1.40	11.429	Ea.	3,025	360		3,385	3,875	
	1030	3 ton cooling	"	1.20	13.333	"	4,750	420		5,170	5,850	
	2000	T-Bar mount										
	2010	2 ton cooling	Q-5	1.40	11.429	Ea.	3,450	360		3,810	4,350	
	2020	3 ton cooling		1.20	13.333		4,250	420		4,670	5,300	
	2030	3-1/2 ton cooling		1.10	14.545		5,125	455		5,580	6,325	
	3000	Multizone										
	3010	Wall mount										
	3020	2 @ 3/4 ton cooling	Q-5	1.80	8.889	Ea.	4,250	279		4,529	5,100	
	5000	Cooling / Heating										
	5010	Wall mount										
	5110	1 ton cooling	Q-5	1.70	9.412	Ea.	2,075	295		2,370	2,750	
	5120	1-1/2 ton cooling	"	1.50	10.667	"	2,825	335		3,160	3,600	
	5300	Ceiling mount										
	5310	3 ton cooling	Q-5	1	16	Ea.	5,500	500		6,000	6,825	
	7000	Accessories for all split ductless systems										
	7010	Add for ambient frost control	Q-5	8	2	Ea.	234	63		297	350	
	7020	Add for tube / wiring kit										
	7030	15' kit	Q-5	32	.500	Ea.	87	15.70		102.70	119	
	7040	35' kit	"	24	.667	"	173	21		194	223	

15740 | Heat Pumps

			CREW	DAILY OUTPUT	LABOR-HOURS	UNIT	MAT.	LABOR	EQUIP.	TOTAL	TOTAL INCL O&P	
100	0010	**HEAT PUMPS** (Not including interconnecting tubing)										100
	1000	Air to air, split system, not including curbs, pads, or ductwork										
	1010	For curbs/pads see division 15705										
	1015	1.5 ton cooling, 7 MBH heat @ 0°F	Q-5	1.22	13.115	Ea.	1,875	410		2,285	2,675	
	1020	2 ton cooling, 8.5 MBH heat @ 0°F		1.20	13.333		2,100	420		2,520	2,925	
	1030	2.5 ton cooling, 10 MBH heat @ 0°F		1	16		2,375	500		2,875	3,375	
	1040	3 ton cooling, 13 MBH heat @ 0°F		.80	20		2,700	630		3,330	3,900	
	1050	3.5 ton cooling, 18 MBH heat @ 0°F		.75	21.333		3,275	670		3,945	4,600	
	1054	4 ton cooling, 24 MBH heat @ 0°F		.60	26.667		3,650	835		4,485	5,300	
	1060	5 ton cooling, 27 MBH heat @ 0°F		.50	32		4,275	1,000		5,275	6,225	
	1080	7.5 ton cooling, 33 MBH heat @ 0°F		.30	53.333		6,575	1,675		8,250	9,750	
	1100	10 ton cooling, 50 MBH heat @ 0°F	Q-6	.38	63.158		8,500	2,050		10,550	12,500	
	1120	15 ton cooling, 64 MBH heat @ 0°F		.26	92.308		11,700	3,000		14,700	17,400	
	1130	20 ton cooling, 85 MBH heat @ 0°F		.20	120		17,100	3,900		21,000	24,800	
	1140	25 ton cooling, 119 MBH heat @ 0°F	Q-7	.20	160		20,300	5,325		25,625	30,400	
	1160	30 ton cooling, 163 MBH heat @ 0°F		.17	188		23,700	6,250		29,950	35,600	
	1180	40 ton cooling, 193 MBH heat @ 0°F		.12	266		33,900	8,875		42,775	50,500	
	1200	50 ton cooling, 220 MBH heat @ 0°F		.10	320		47,400	10,600		58,000	68,000	
	1300	Supplementary electric heat coil, included										
	1500	Single package, not including curbs, pads, or plenums										
	1502	1/2 ton cooling, supplementary heat not incl.	Q-5	8	2	Ea.	1,200	63		1,263	1,425	
	1504	3/4 ton cooling, supplementary heat not incl.		6	2.667		1,225	83.50		1,308.50	1,475	
	1506	1 ton cooling, supplementary heat not incl.		4	4		1,325	126		1,451	1,650	
	1510	1.5 ton cooling, 5 MBH heat @ 0°F		1.55	10.323		2,300	325		2,625	3,025	

15700 | Heating/Ventilating/Air Conditioning Equipment

15740 | Heat Pumps

		CREW	DAILY OUTPUT	LABOR-HOURS	UNIT	MAT.	LABOR	EQUIP.	TOTAL	TOTAL INCL O&P
1520	2 ton cooling, 6.5 MBH heat @ 0°F	Q-5	1.50	10.667	Ea.	2,525	335		2,860	3,275
1540	2.5 ton cooling, 8 MBH heat @ 0°F		1.40	11.429		2,875	360		3,235	3,725
1560	3 ton cooling, 10 MBH heat @ 0°F		1.20	13.333		3,025	420		3,445	3,950
1570	3.5 ton cooling, 11 MBH heat @ 0°F		1	16		3,425	500		3,925	4,525
1580	4 ton cooling, 13 MBH heat @ 0°F		.96	16.667		3,800	525		4,325	4,975
1620	5 ton cooling, 27 MBH heat @ 0°F		.65	24.615		4,375	770		5,145	6,000
1640	7.5 ton cooling, 35 MBH heat @ 0°F	↓	.40	40		8,225	1,250		9,475	11,000
1648	10 ton cooling, 45 MBH heat @ 0°F	Q-6	.40	60		10,600	1,950		12,550	14,700
1652	12 ton cooling, 50 MBH heat @ 0°F	"	.36	66.667	↓	13,000	2,175		15,175	17,600
1696	Supplementary electric heat coil incl., except as noted									
1800	Rooftop, with curb, economizer and supplementary electric heat									
1820	10 ton cooling	Q-6	.40	60	Ea.	35,100	1,950		37,050	41,600
1830	20 ton cooling	"	.20	120		50,000	3,900		53,900	61,000
1840	30 ton cooling	Q-7	.19	168		67,500	5,600		73,100	83,000
1850	40 ton cooling		.17	188		75,000	6,250		81,250	92,000
1860	50 ton cooling		.15	213		87,500	7,100		94,600	107,000
1870	60 ton cooling	↓	.13	246	↓	92,500	8,175		100,675	114,500
2000	Water source to air, single package									
2100	1 ton cooling, 13 MBH heat @ 75°F	Q-5	2	8	Ea.	965	251		1,216	1,450
2120	1.5 ton cooling, 17 MBH heat @ 75°F		1.80	8.889		1,150	279		1,429	1,700
2140	2 ton cooling, 19 MBH heat @ 75°F		1.70	9.412		1,225	295		1,520	1,775
2160	2.5 ton cooling, 25 MBH heat @ 75°F		1.60	10		1,300	315		1,615	1,900
2180	3 ton cooling, 27 MBH heat @ 75°F		1.40	11.429		1,350	360		1,710	2,025
2190	3.5 ton cooling, 29 MBH heat @ 75°F		1.30	12.308		1,450	385		1,835	2,150
2200	4 ton cooling, 31 MBH heat @ 75°F		1.20	13.333		1,750	420		2,170	2,550
2220	5 ton cooling, 29 MBH heat @ 75°F		.90	17.778		2,125	560		2,685	3,200
2240	7.5 ton cooling, 35 MBH heat @ 75°F		.60	26.667		2,425	835		3,260	3,925
2250	8.5 ton cooling, 40 MBH heat @ 75°F		.58	27.586		6,900	865		7,765	8,900
2260	10 ton cooling, 50 MBH heat @ 75°F	↓	.53	30.189		7,600	945		8,545	9,775
2280	15 ton cooling, 64 MBH heat @ 75°F	Q-6	.47	51.064		8,000	1,650		9,650	11,300
2300	20 ton cooling, 100 MBH heat @ 75°F		.41	58.537		8,250	1,900		10,150	12,000
2320	30 ton cooling, (twin 15 ton units)		.23	102		16,000	3,325		19,325	22,600
2340	40 ton cooling, (twin 20 ton units)		.20	117		16,500	3,800		20,300	23,900
2360	50 ton cooling, (twin 15 + 20 ton unit)	↓	.15	160	↓	24,300	5,200		29,500	34,600
3960	For supplementary heat coil, add					10%				
4000	For increase in capacity thru use									
4020	of solar collector, size boiler at 60%									

15750 | Humidity Control Equipment

		CREW	DAILY OUTPUT	LABOR-HOURS	UNIT	MAT.	LABOR	EQUIP.	TOTAL	TOTAL INCL O&P
0010	**DEHUMIDIFIERS**									
6000	Selfcontained with filters and standard controls									
6040	1.5 lb/hr, 50 cfm	1 Plum	8	1	Ea.	2,875	34.50		2,909.50	3,225
6060	3 lb/hr, 150 cfm	Q-1	12	1.333		3,275	41.50		3,316.50	3,675
6065	6 lb/hr, 150 cfm		9	1.778		6,050	55		6,105	6,750
6070	16 to 20 lb/hr, 600 cfm		5	3.200		12,100	99		12,199	13,500
6080	30 to 40 lb/hr, 1125 cfm		4	4		20,000	124		20,124	22,100
6090	60 to 75 lb/hr, 2250 cfm		3	5.333		26,200	165		26,365	29,100
6100	120 to 155 lb/hr, 4500 cfm		2	8		47,800	248		48,048	53,000
6110	240 to 310 lb/hr, 9000 cfm	↓	1.50	10.667		67,500	330		67,830	74,500
6120	400 to 515 lb/hr, 15,000 cfm	Q-2	1.60	15		91,500	480		91,980	101,000
6130	530 to 690 lb/hr, 20,000 cfm		1.40	17.143		99,000	550		99,550	110,000
6140	800 to 1030 lb/hr, 30,000 cfm		1.20	20		113,000	645		113,645	125,500
6150	1060 to 1375 lb/hr, 40,000 cfm	↓	1	24	↓	157,500	770		158,270	174,500
0010	**HUMIDIFIERS**									
0520	Steam, room or duct, filter, regulators, auto. controls, 220 V									

15700 | Heating/Ventilating/Air Conditioning Equipment

15750 | Humidity Control Equipment

		CREW	DAILY OUTPUT	LABOR-HOURS	UNIT	2001 BARE COSTS MAT.	LABOR	EQUIP.	TOTAL	TOTAL INCL O&P
0540	11 lb. per hour	Q-5	6	2.667	Ea.	2,000	83.50		2,083.50	2,325
0560	22 lb. per hour		5	3.200		2,200	100		2,300	2,575
0580	33 lb. per hour		4	4		2,250	126		2,376	2,675
0600	50 lb. per hour		4	4		2,750	126		2,876	3,225
0620	100 lb. per hour		3	5.333		3,300	167		3,467	3,900
0640	150 lb. per hour		2.50	6.400		4,375	201		4,576	5,125
0660	200 lb. per hour		2	8		5,425	251		5,676	6,325
0700	With blower									
0720	11 lb. per hour	Q-5	5.50	2.909	Ea.	2,775	91.50		2,866.50	3,200
0740	22 lb. per hour		4.75	3.368		2,975	106		3,081	3,425
0760	33 lb. per hour		3.75	4.267		3,050	134		3,184	3,550
0780	50 lb. per hour		3.50	4.571		3,650	143		3,793	4,250
0800	100 lb. per hour		2.75	5.818		4,125	183		4,308	4,825
0820	150 lb. per hour		2	8		6,100	251		6,351	7,100
0840	200 lb. per hour		1.50	10.667		7,175	335		7,510	8,375
1000	Steam for duct installation, manifold with pneum. controls									
1010	3 - 191 lb/hr, 24" x 24"	Q-5	5	3.200	Ea.	990	100		1,090	1,250
1020	3 - 191 lb/hr, 48" x 48"		4.60	3.478		985	109		1,094	1,250
1030	65 - 334 lb/hr, 36" x 36"		4.80	3.333		1,000	105		1,105	1,250
1040	65 - 334 lb/hr, 72" x 72"		4	4		1,050	126		1,176	1,350
1050	100 - 690 lb/hr, 48" x 48"		4	4		1,650	126		1,776	2,000
1060	100 - 690 lb/hr, 84" x 84"		3.60	4.444		1,750	139		1,889	2,125
1070	220 - 2000 lb/hr, 72" x 72"		3.80	4.211		1,775	132		1,907	2,150
1080	220 - 2000 lb/hr, 144" x 144"		3.40	4.706		2,075	148		2,223	2,500
1090	For electrically controlled unit, add					697				
1092	For additional manifold, add					25%	15%			
1200	Electric steam, self generating, with manifold									
1210	1 - 96 lb/hr, 12" x 72"	Q-5	3.60	4.444	Ea.	4,450	139		4,589	5,100
1220	20 - 168 lb/hr, 12" x 72"		3.60	4.444		4,450	139		4,589	5,100
1290	For room mounting, add		16	1		133	31.50		164.50	195
5000	Furnace type, wheel bypass.									
5020	10 GPD	1 Stpi	4	2	Ea.	124	69.50		193.50	242
5040	14 GPD		3.80	2.105		165	73.50		238.50	292
5060	19 GPD		3.60	2.222		192	77.50		269.50	330

15761 | Air Coils

		CREW	DAILY OUTPUT	LABOR-HOURS	UNIT	MAT.	LABOR	EQUIP.	TOTAL	TOTAL INCL O&P
0010	**DUCT HEATERS** Electric, 480 V, 3 Ph									
0020	Finned tubular insert, 500°F									
0100	8" wide x 6" high, 4.0 kW	Q-20	16	1.250	Ea.	525	39		564	640
0120	12" high, 8.0 kW		15	1.333		840	41.50		881.50	990
0140	18" high, 12.0 kW		14	1.429		1,175	44.50		1,219.50	1,375
0160	24" high, 16.0 kW		13	1.538		1,525	47.50		1,572.50	1,750
0180	30" high, 20.0 kW		12	1.667		1,850	51.50		1,901.50	2,125
0300	12" wide x 6" high, 6.7 kW		15	1.333		540	41.50		581.50	655
0320	12" high, 13.3 kW		14	1.429		870	44.50		914.50	1,025
0340	18" high, 20.0 kW		13	1.538		1,225	47.50		1,272.50	1,425
0360	24" high, 26.7 kW		12	1.667		1,575	51.50		1,626.50	1,800
0380	30" high, 33.3 kW		11	1.818		1,925	56.50		1,981.50	2,175
0500	18" wide x 6" high, 13.3 kW		14	1.429		575	44.50		619.50	705
0520	12" high, 26.7 kW		13	1.538		995	47.50		1,042.50	1,175
0540	18" high, 40.0 kW		12	1.667		1,325	51.50		1,376.50	1,525
0560	24" high, 53.3 kW		11	1.818		1,750	56.50		1,806.50	2,000
0580	30" high, 66.7 kW		10	2		2,200	62		2,262	2,525
0700	24" wide x 6" high, 17.8 kW		13	1.538		650	47.50		697.50	790
0720	12" high, 35.6 kW		12	1.667		1,075	51.50		1,126.50	1,250
0740	18" high, 53.3 kW		11	1.818		1,500	56.50		1,556.50	1,750

15700 | Heating/Ventilating/Air Conditioning Equipment

15761 | Air Coils

		CREW	DAILY OUTPUT	LABOR-HOURS	UNIT	MAT.	LABOR	EQUIP.	TOTAL	TOTAL INCL O&P
0760	24" high, 71.1 kW	Q-20	10	2	Ea.	1,950	62		2,012	2,250
0780	30" high, 88.9 kW		9	2.222		2,400	69		2,469	2,725
0900	30" wide x 6" high, 22.2 kW		12	1.667		675	51.50		726.50	820
0920	12" high, 44.4 kW		11	1.818		1,150	56.50		1,206.50	1,325
0940	18" high, 66.7 kW		10	2		1,600	62		1,662	1,850
0960	24" high, 88.9 kW		9	2.222		2,075	69		2,144	2,375
0980	30" high, 111.0 kW		8	2.500		2,525	77.50		2,602.50	2,925
1400	Note decreased kW available for									
1410	each duct size at same cost									
1420	See line 5000 for modifications and accessories									
2000	Finned tubular flange with insulated									
2020	terminal box, 500°F									
2100	12" wide x 36" high, 54 kW	Q-20	10	2	Ea.	2,500	62		2,562	2,850
2120	40" high, 60 kW		9	2.222		2,875	69		2,944	3,250
2200	24" wide x 36" high, 118.8 kW		9	2.222		2,775	69		2,844	3,150
2220	40" high, 132 kW		8	2.500		3,175	77.50		3,252.50	3,625
2400	36" wide x 8" high, 40 kW		11	1.818		1,275	56.50		1,331.50	1,475
2420	16" high, 80 kW		10	2		1,700	62		1,762	1,975
2440	24" high, 120 kW		9	2.222		2,225	69		2,294	2,550
2460	32" high, 160 kW		8	2.500		2,950	77.50		3,027.50	3,375
2480	36" high, 180 kW		7	2.857		3,500	88.50		3,588.50	3,975
2500	40" high, 200 kW		6	3.333		3,875	103		3,978	4,425
2600	40" wide x 8" high, 45 kW		11	1.818		1,325	56.50		1,381.50	1,525
2620	16" high, 90 kW		10	2		1,825	62		1,887	2,100
2640	24" high, 135 kW		9	2.222		2,325	69		2,394	2,650
2660	32" high, 180 kW		8	2.500		3,100	77.50		3,177.50	3,525
2680	36" high, 202.5 kW		7	2.857		3,550	88.50		3,638.50	4,025
2700	40" high, 225 kW		6	3.333		4,050	103		4,153	4,600
2800	48" wide x 8" high, 54.8 kW		10	2		1,400	62		1,462	1,625
2820	16" high, 109.8 kW		9	2.222		1,925	69		1,994	2,225
2840	24" high, 164.4 kW		8	2.500		2,450	77.50		2,527.50	2,825
2860	32" high, 219.2 kW		7	2.857		3,275	88.50		3,363.50	3,725
2880	36" high, 246.6 kW		6	3.333		3,775	103		3,878	4,300
2900	40" high, 274 kW		5	4		4,250	124		4,374	4,875
3000	56" wide x 8" high, 64 kW		9	2.222		1,575	69		1,644	1,850
3020	16" high, 128 kW		8	2.500		2,200	77.50		2,277.50	2,550
3040	24" high, 192 kW		7	2.857		2,650	88.50		2,738.50	3,050
3060	32" high, 256 kW		6	3.333		3,725	103		3,828	4,250
3080	36" high, 288kW		5	4		4,275	124		4,399	4,900
3100	40" high, 320kW		4	5		4,725	155		4,880	5,450
3200	64" wide x 8" high, 74kW		8	2.500		1,625	77.50		1,702.50	1,925
3220	16" high, 148kW		7	2.857		2,250	88.50		2,338.50	2,600
3240	24" high, 222kW		6	3.333		2,875	103		2,978	3,300
3260	32" high, 296kW		5	4		3,850	124		3,974	4,450
3280	36" high, 333kW		4	5		4,600	155		4,755	5,300
3300	40" high, 370kW		3	6.667		5,075	207		5,282	5,900
3800	Note decreased kW available for									
3820	each duct size at same cost									
5000	Duct heater modifications and accessories									
5120	T.C.O. limit auto or manual reset	Q-20	42	.476	Ea.	83	14.75		97.75	114
5140	Thermostat		28	.714		355	22		377	425
5160	Overheat thermocouple (removable)		7	2.857		520	88.50		608.50	710
5180	Fan interlock relay		18	1.111		122	34.50		156.50	187
5200	Air flow switch		20	1		109	31		140	168
5220	Split terminal box cover		100	.200		36	6.20		42.20	49
8000	To obtain BTU multiply kW by 3413									

15700 | Heating/Ventilating/Air Conditioning Equipment

15761 | Air Coils

		CREW	DAILY OUTPUT	LABOR-HOURS	UNIT	MAT.	LABOR	EQUIP.	TOTAL	TOTAL INCL O&P
0010	**COILS, FLANGED**									
0100	Basic water, DX, or condenser coils									
0110	Copper tubes, alum. fins, galv. end sheets									
0112	H is finned height, L is finned length									
0120	3/8" x .016 tube, .0065 aluminum fins									
0130	2 row, 8 fins per inch									
0140	4" H x 12" L	Q-5	48	.333	Ea.	360	10.45		370.45	410
0150	4" H x 24" L		38.38	.417		375	13.10		388.10	430
0160	4" H x 48" L		19.25	.831		390	26		416	470
0170	4" H x 72" L		12.80	1.250		510	39		549	620
0180	6" H x 12" L		48	.333		370	10.45		380.45	425
0190	6" H x 24" L		25.60	.625		380	19.60		399.60	450
0200	6" H x 48" L		12.80	1.250		400	39		439	500
0210	6" H x 72" L		8.53	1.876		530	59		589	670
0220	10" H x 12" L		30.73	.521		395	16.35		411.35	460
0230	10" H x 24" L		15.33	1.044		410	33		443	500
0240	10" H x 48" L		7.69	2.081		435	65.50		500.50	580
0250	10" H x 72" L		5.12	3.125		580	98		678	790
0260	12" H x 12" L		25.60	.625		400	19.60		419.60	470
0270	12" H x 24" L		12.80	1.250		415	39		454	520
0280	12" H x 48" L		6.40	2.500		450	78.50		528.50	615
0290	12" H x 72" L		4.27	3.747		600	118		718	840
0300	16" H x 16" L		14.38	1.113		420	35		455	515
0310	16" H x 24" L		9.59	1.668		430	52.50		482.50	555
0320	16" H x 48" L		4.80	3.333		525	105		630	740
0330	16" H x 72" L		3.20	5		635	157		792	935
0340	20" H x 24" L		7.69	2.081		460	65.50		525.50	605
0350	20" H x 48" L		3.84	4.167		560	131		691	810
0360	20" H x 72" L		2.56	6.250		690	196		886	1,050
0370	24" H x 24" L		6.40	2.500		475	78.50		553.50	640
0380	24" H x 48" L		3.20	5		595	157		752	890
0390	24" H x 72" L		2.13	7.512		725	236		961	1,150
0400	30" H x 28" L		4.39	3.645		570	114		684	800
0410	30" H x 48" L		2.56	6.250		650	196		846	1,000
0420	30" H x 72" L		1.71	9.357		800	294		1,094	1,325
0430	30" H x 84" L		1.46	10.959		880	345		1,225	1,500
0440	34" H x 32" L		3.39	4.720		585	148		733	865
0450	34" H x 48" L		2.26	7.080		680	222		902	1,075
0460	34" H x 72" L		1.51	10.596		725	335		1,060	1,300
0470	34" H x 84" L		1.29	12.403		810	390		1,200	1,475
0600	For 10 fins per inch, add					1%				
0610	For 12 fins per inch, add					3%				
0620	For 4 row, add				Ea.	30%	50%			
0630	For 6 row, add					60%	100%			
0640	For 8 row, add					80%	150%			
1100	1/2" x .017 tube, .0065 aluminum fins									
1110	2 row, 8 fins per inch									
1120	5" H x 5" L	Q-5	40	.400	Ea.	254	12.55		266.55	299
1130	5" H x 20" L		37.10	.431		266	13.55		279.55	315
1140	5" H x 45" L		16.41	.975		286	30.50		316.50	360
1150	5" H x 90" L		8.21	1.949		485	61		546	625
1160	7.5" H x 5" L		38.50	.416		259	13.05		272.05	305
1170	7.5" H x 20" L		24.62	.650		274	20.50		294.50	330
1180	7.5" H x 45" L		10.94	1.463		305	46		351	405
1190	7.5" H x 90" L		5.46	2.930		510	92		602	705
1200	12.5" H x 10" L		29.49	.543		292	17.05		309.05	345

15700 | Heating/Ventilating/Air Conditioning Equipment

15761 | Air Coils

			CREW	DAILY OUTPUT	LABOR-HOURS	UNIT	2001 BARE COSTS MAT.	LABOR	EQUIP.	TOTAL	TOTAL INCL O&P	
700	1210	12.5" H x 20" L	Q-5	14.75	1.085	Ea.	305	34		339	390	700
	1220	12.5" H x 45" L		6.55	2.443		345	76.50		421.50	495	
	1230	12.5" H x 90" L		3.28	4.878		590	153		743	875	
	1240	15" H x 15" L		16.41	.975		305	30.50		335.50	380	
	1250	15" H x 30" L		8.21	1.949		330	61		391	460	
	1260	15" H x 45" L		5.46	2.930		355	92		447	535	
	1270	15" H x 90" L		2.73	5.861		615	184		799	960	
	1280	20" H x 20" L		9.22	1.735		330	54.50		384.50	445	
	1290	20" H x 45" L		4.10	3.902		440	122		562	670	
	1300	20" H x 90" L		2.05	7.805		675	245		920	1,125	
	1310	25" H x 25" L		5.90	2.712		375	85		460	545	
	1320	25" H x 45" L		3.28	4.878		490	153		643	770	
	1330	25" H x 90" L		1.64	9.756		750	305		1,055	1,275	
	1340	30" H x 30" L		4.10	3.902		460	122		582	690	
	1350	30" H x 45" L		2.73	5.861		525	184		709	855	
	1360	30" H x 90" L		1.37	11.679		810	365		1,175	1,450	
	1370	35" H x 35" L		3.01	5.316		505	167		672	805	
	1380	35" H x 45" L		2.34	6.838		555	215		770	935	
	1390	35" H x 90" L		1.17	13.675		865	430		1,295	1,600	
	1400	42.5" H x 45" L		1.93	8.290		625	260		885	1,075	
	1410	42.5" H x 75" L		1.16	13.793		845	435		1,280	1,575	
	1420	42.5" H x 105" L		.83	19.277		1,050	605		1,655	2,100	
	1600	For 10 fins per inch, add					2%					
	1610	For 12 fins per inch, add					5%					
	1620	For 4 row, add					35%	50%				
	1630	For 6 row, add					65%	100%				
	1640	For 8 row, add					90%	150%				
	2010	5/8" x .020" tube, .0065 aluminum fins										
	2020	2 row, 8 fins per inch										
	2030	6" H x 6" L	Q-5	38	.421	Ea.	256	13.20		269.20	300	
	2040	6" H x 24" L		24.96	.641		274	20		294	330	
	2050	6" H x 48" L		12.48	1.282		297	40		337	385	
	2060	6" H x 96" L		6.24	2.564		515	80.50		595.50	690	
	2070	12" H x 12" L		24.96	.641		282	20		302	340	
	2080	12" H x 24" L		12.48	1.282		305	40		345	395	
	2090	12" H x 48" L		6.24	2.564		335	80.50		415.50	490	
	2100	12" H x 96" L		3.12	5.128		600	161		761	905	
	2110	18" H x 24" L		8.32	1.923		340	60.50		400.50	465	
	2120	18" H x 48" L		4.16	3.846		455	121		576	685	
	2130	18" H x 96" L		2.08	7.692		705	241		946	1,150	
	2140	24" H x 24" L		6.24	2.564		365	80.50		445.50	520	
	2150	24" H x 48" L		3.12	5.128		490	161		651	780	
	2160	24" H x 96" L		1.56	10.256		785	320		1,105	1,350	
	2170	30" H x 30" L		3.99	4.010		480	126		606	715	
	2180	30" H x 48" L		2.50	6.400		545	201		746	905	
	2190	30" H x 96" L		1.25	12.800		890	400		1,290	1,575	
	2200	30" H x 120" L		1	16		1,050	500		1,550	1,900	
	2210	36" H x 36" L		2.77	5.776		540	181		721	870	
	2220	36" H x 48" L		2.08	7.692		610	241		851	1,025	
	2230	36" H x 90" L		1.11	14.414		940	450		1,390	1,700	
	2240	36" H x 120" L		.83	19.277		1,150	605		1,755	2,200	
	2250	42" H x 48" L		1.78	8.989		655	282		937	1,150	
	2260	42" H x 90" L		.95	16.842		1,025	530		1,555	1,925	
	2270	42" H x 120" L		.71	22.535		1,250	705		1,955	2,450	
	2280	51" H x 48" L		1.47	10.884		760	340		1,100	1,350	
	2290	51" H x 96" L		.73	21.918		1,225	690		1,915	2,400	

MECHANICAL 15

333

15700 | Heating/Ventilating/Air Conditioning Equipment

15761 | Air Coils

		CREW	DAILY OUTPUT	LABOR-HOURS	UNIT	2001 BARE COSTS MAT.	LABOR	EQUIP.	TOTAL	TOTAL INCL O&P	
2300	51" H x 120" L	Q-5	.59	27.119	Ea.	1,450	850		2,300	2,850	700
2500	For 10 fins per inch, add					3%					
2510	For 12 fins per inch, add					6%					
2520	For 4 row, add				Ea.	40%	50%				
2530	For 6 row, add					80%	100%				
2540	For 8 row, add					110%	150%				
3000	Hot water booster coils										
3010	Copper tubes, alum. fins, galv. end sheets										
3012	H is finned height, L is finned length										
3020	1/2" x .017" tube, .0065 Al fins										
3030	1 row, 10 fins per inch										
3040	5"H x 10"L	Q-5	40	.400	Ea.	175	12.55		187.55	212	
3050	5"H x 25"L		33.18	.482		180	15.15		195.15	220	
3060	10"H x 10"L		39	.410		181	12.85		193.85	218	
3070	10"H x 20"L		20.73	.772		188	24		212	243	
3080	10"H x 30"L		13.83	1.157		194	36.50		230.50	268	
3090	15"H x 15"L		18.44	.868		193	27		220	253	
3110	15"H x 20"L		13.83	1.157		197	36.50		233.50	272	
3120	15"H x 30"L		9.22	1.735		206	54.50		260.50	310	
3130	20"H x 20"L		10.37	1.543		208	48.50		256.50	300	
3140	20"H x 30"L		6.91	2.315		219	72.50		291.50	350	
3150	20"H x 40"L		5.18	3.089		284	97		381	455	
3160	25"H x 25"L		6.64	2.410		268	75.50		343.50	410	
3170	25"H x 35"L		4.74	3.376		294	106		400	485	
3180	25"H x 45"L		3.69	4.336		315	136		451	550	
3190	30"H x 30"L		4.61	3.471		300	109		409	495	
3200	30"H x 40"L		3.46	4.624		325	145		470	575	
3210	30"H x 50"L		2.76	5.797		350	182		532	660	
3300	For 2 row, add					20%	50%				
3400	1/2" x .020" tube, .008 Al fins										
3410	2 row, 10 fins per inch										
3420	20" H x 27" L	Q-5	7.27	2.201	Ea.	276	69		345	410	
3430	20" H x 36" L		5.44	2.941		350	92.50		442.50	525	
3440	20" H x 45" L		4.35	3.678		380	115		495	595	
3450	25" H x 45" L		3.49	4.585		425	144		569	680	
3460	25" H x 54" L		2.90	5.517		510	173		683	825	
3470	30" H x 54" L		2.42	6.612		565	207		772	935	
3480	35" H x 54" L		2.07	7.729		610	243		853	1,050	
3490	35" H x 63" L		1.78	8.989		880	282		1,162	1,400	
3500	40" H x 72" L		1.36	11.765		995	370		1,365	1,650	
3510	55" H x 63" L		1.13	14.159		1,175	445		1,620	1,975	
3520	55" H x 72" L		.99	16.162		1,225	505		1,730	2,125	
3530	55" H x 81" L		.88	18.182		1,325	570		1,895	2,300	
3540	75" H x 81" L		.64	25		1,650	785		2,435	2,975	
3550	75" H x 90" L		.58	27.586		1,725	865		2,590	3,200	
3560	75" H x 108" L		.48	33.333		2,050	1,050		3,100	3,825	
3800	5/8" x .020 tubing, .0065 aluminum fins										
3810	1 row, 10 fins per inch										
3820	6" H x 12" L	Q-5	38	.421	Ea.	178	13.20		191.20	216	
3830	6" H x 24" L		28.80	.556		184	17.45		201.45	229	
3840	6" H x 30" L		23.04	.694		187	22		209	239	
3850	12" H x 12" L		28.80	.556		189	17.45		206.45	234	
3860	12" H x 24" L		14.40	1.111		198	35		233	271	
3870	12" H x 30" L		11.52	1.389		204	43.50		247.50	290	
3880	18" H x 18" L		12.80	1.250		206	39		245	285	
3890	18" H x 24" L		9.60	1.667		213	52.50		265.50	315	

15700 | Heating/Ventilating/Air Conditioning Equipment

15761 | Air Coils

		CREW	DAILY OUTPUT	LABOR-HOURS	UNIT	MAT.	LABOR	EQUIP.	TOTAL	TOTAL INCL O&P
3900	18" H x 36" L	Q-5	6.40	2.500	Ea.	227	78.50		305.50	370
3910	24" H x 24" L		7.20	2.222		227	69.50		296.50	355
3920	24" H x 30" L		4.80	3.333		287	105		392	475
3930	24" H x 36" L		5.05	3.168		300	99.50		399.50	480
3940	30" H x 30" L		4.61	3.471		310	109		419	505
3950	30" H x 36" L		3.85	4.159		325	131		456	555
3960	30" H x 42" L		3.29	4.863		340	153		493	605
3970	36" H x 36" L		3.20	5		350	157		507	620
3980	36" H x 42" L		2.74	5.839		370	183		553	680
4100	For 2 row, add					30%	50%			
4400	Steam and hot water heating coils									
4410	Copper tubes, alum. fins. galv. end sheets									
4412	H is finned height, L is finned length									
4420	5/8" tubing, to 175 psig working pressure									
4430	1 row, 8 fins per inch									
4440	6"H x 6"L	Q-5	38	.421	Ea.	575	13.20		588.20	655
4450	6"H x 12"L		37	.432		590	13.55		603.55	670
4460	6"H x 24"L		24.48	.654		620	20.50		640.50	710
4470	6"H x 36"L		16.32	.980		615	31		646	720
4480	6"H x 48"L		12.24	1.307		675	41		716	805
4490	6"H x 84"L		6.99	2.289		1,125	72		1,197	1,350
4500	6" H x 108" L		5.44	2.941		1,300	92.50		1,392.50	1,575
4510	12" H x 12" L		24.48	.654		670	20.50		690.50	765
4520	12" H x 24" L		12.24	1.307		720	41		761	850
4530	12" H x 36" L		8.16	1.961		755	61.50		816.50	925
4540	12" H x 48" L		6.12	2.614		790	82		872	995
4550	12" H x 84" L		3.50	4.571		1,175	143		1,318	1,500
4560	12" H x 108" L		2.72	5.882		1,325	185		1,510	1,725
4570	18" H x 18" L		10.88	1.471		690	46		736	830
4580	18" H x 24" L		8.16	1.961		710	61.50		771.50	875
4590	18" H x 48" L		4.08	3.922		845	123		968	1,125
4600	18" H x 84" L		2.33	6.867		1,325	215		1,540	1,775
4610	18" H x 108" L		1.81	8.840		1,525	277		1,802	2,100
4620	24" H x 24" L		6.12	2.614		820	82		902	1,025
4630	24" H x 48" L		3.06	5.229		945	164		1,109	1,300
4640	24" H x 84" L		1.75	9.143		1,475	287		1,762	2,050
4650	24" H x 108" L		1.36	11.765		1,725	370		2,095	2,450
4660	36" H x 36" L		2.72	5.882		1,050	185		1,235	1,425
4670	36" H x 48" L		2.04	7.843		1,125	246		1,371	1,625
4680	36" H x 84" L		1.17	13.675		1,800	430		2,230	2,650
4690	36" H x 108" L		.91	17.582		2,150	550		2,700	3,200
4700	48"H x 48"L		1.53	10.458		1,350	330		1,680	1,975
4710	48"H x 84"L		.87	18.391		2,125	575		2,700	3,225
4720	48"H x 108"L		.68	23.529		2,575	740		3,315	3,950
4730	48"H x 126"L		.58	27.586		2,750	865		3,615	4,325
4910	For 10 fins per inch, add					2%				
4920	For 12 fins per inch, add					5%				
4930	For 2 row, add					30%	50%			
4940	For 3 row, add					50%	100%			
4950	For 4 row, add					80%	123%			
4960	For 230 PSIG heavy duty, add					20%	30%			
6000	Steam, heavy duty, 200 PSIG									
6010	Copper tubes, alum. fins, galv. end sheets									
6012	H is finned height, L is finned length									
6020	1" x .035" tubing, .010 aluminum fins									
6030	1 row, 8 fins per inch									

15700 | Heating/Ventilating/Air Conditioning Equipment

15761 | Air Coils

		CREW	DAILY OUTPUT	LABOR-HOURS	UNIT	2001 BARE COSTS MAT.	LABOR	EQUIP.	TOTAL	TOTAL INCL O&P	
700	6040	12"H x 12"L	Q-5	24.48	.654	Ea.	540	20.50		560.50	625	700
	6050	12"H x 24"L		12.24	1.307		575	41		616	695	
	6060	12"H x 36"L		8.16	1.961		615	61.50		676.50	770	
	6070	12"H x 48"L		6.12	2.614		685	82		767	875	
	6080	12"H x 84"L		3.50	4.571		900	143		1,043	1,200	
	6090	12"H x 108"L		2.72	5.882		1,050	185		1,235	1,425	
	6100	18"H x 24"L		8.16	1.961		655	61.50		716.50	815	
	6110	18"H x 36"L		5.44	2.941		760	92.50		852.50	975	
	6120	18"H x 48"L		4.08	3.922		810	123		933	1,075	
	6130	18"H x 84"L		2.33	6.867		1,075	215		1,290	1,525	
	6140	18"H x 108"L		1.81	8.840		1,275	277		1,552	1,825	
	6150	24"H x 24"L		6.12	2.614		800	82		882	1,000	
	6160	24"H x 36"L		4.08	3.922		875	123		998	1,150	
	6170	24"H x 48"L		3.06	5.229		940	164		1,104	1,275	
	6180	24"H x 84"L		1.75	9.143		1,250	287		1,537	1,800	
	6190	24"H x 108"L		1.36	11.765		1,475	370		1,845	2,175	
	6200	30"H x 36"L		3.26	4.908		990	154		1,144	1,300	
	6210	30"H x 48"L		2.45	6.531		1,075	205		1,280	1,475	
	6220	30"H x 84"L		1.40	11.429		1,425	360		1,785	2,125	
	6230	30"H x 108"L		1.09	14.679		1,700	460		2,160	2,575	
	6240	36"H x 36"L		2.72	5.882		1,100	185		1,285	1,500	
	6250	36"H x 48"L		2.04	7.843		1,200	246		1,446	1,700	
	6260	36"H x 84"L		1.17	13.675		1,600	430		2,030	2,425	
	6270	36"H x 108"L		.91	17.582		1,925	550		2,475	2,950	
	6280	36"H x 120"L		.82	19.512		2,025	610		2,635	3,150	
	6290	48"H x 48"L		1.53	10.458		1,475	330		1,805	2,125	
	6300	48"H x 84"L		.87	18.391		1,975	575		2,550	3,050	
	6310	48"H x 108"L		.68	23.529		2,375	740		3,115	3,725	
	6320	48"H x 120"L		.61	26.230		2,500	825		3,325	4,000	
	6400	For 10 fins per inch, add					4%					
	6410	For 12 fins per inch, add					10%					

15765 | Fan Coil Unit/Unit Ventilators

			CREW	DAILY OUTPUT	LABOR-HOURS	UNIT	MAT.	LABOR	EQUIP.	TOTAL	INCL O&P	
200	0010	**FAN COIL AIR CONDITIONING** Cabinet mounted, filters, controls										200
	0100	Chilled water, 1/2 ton cooling	Q-5	8	2	Ea.	730	63		793	900	
	0110	3/4 ton cooling		7	2.286		775	71.50		846.50	965	
	0120	1 ton cooling		6	2.667		830	83.50		913.50	1,050	
	0140	1.5 ton cooling		5.50	2.909		920	91.50		1,011.50	1,150	
	0150	2 ton cooling		5.25	3.048		1,150	95.50		1,245.50	1,425	
	0160	2.5 ton cooling		5	3.200		1,525	100		1,625	1,825	
	0180	3 ton cooling		4	4		1,675	126		1,801	2,025	
	0190	7.5 ton cooling		2.70	5.926		1,925	186		2,111	2,400	
	0195	8 ton cooling		2.20	7.273		2,225	228		2,453	2,800	
	0200	10 ton cooling	Q-6	2.80	8.571		2,400	279		2,679	3,075	
	0210	12 ton cooling		2.20	10.909		2,825	355		3,180	3,625	
	0220	15 ton cooling		1.50	16		3,350	520		3,870	4,450	
	0240	20 ton cooling		.80	30		4,300	975		5,275	6,200	
	0250	25 ton cooling		.70	34.286		6,150	1,125		7,275	8,450	
	0260	30 ton cooling		.60	40		6,350	1,300		7,650	8,975	
	0262	For hot water coil, add					40%	10%				
	0300	Console, 2 pipe with electric heat										
	0310	1/2 ton cooling	Q-5	8	2	Ea.	905	63		968	1,100	
	0315	3/4 ton cooling		7	2.286		990	71.50		1,061.50	1,175	
	0320	1 ton cooling		6	2.667		1,075	83.50		1,158.50	1,300	
	0330	1.5 ton cooling		5.50	2.909		1,325	91.50		1,416.50	1,600	

15700 | Heating/Ventilating/Air Conditioning Equipment

15765 | Fan Coil Unit/Unit Ventilators

			CREW	DAILY OUTPUT	LABOR-HOURS	UNIT	MAT.	LABOR	EQUIP.	TOTAL	TOTAL INCL O&P	
200	0340	2 ton cooling	Q-5	4.70	3.404	Ea.	1,750	107		1,857	2,075	200
	0345	2-1/2 ton cooling		4.70	3.404		2,525	107		2,632	2,925	
	0350	3 ton cooling		4	4		3,350	126		3,476	3,875	
	0360	4 ton cooling		4	4		3,475	126		3,601	4,025	
	0940	Direct expansion, for use w/air cooled condensing, 1.5 ton cooling		5	3.200		440	100		540	635	
	0950	2 ton cooling		4.80	3.333		475	105		580	680	
	0960	2.5 ton cooling		4.40	3.636		545	114		659	770	
	0970	3 ton cooling		3.80	4.211		620	132		752	885	
	0980	3.5 ton cooling		3.60	4.444		840	139		979	1,125	
	0990	4 ton cooling		3.40	4.706		905	148		1,053	1,225	
	1000	5 ton cooling		3	5.333		1,000	167		1,167	1,350	
	1020	7.5 ton cooling		3	5.333		1,800	167		1,967	2,225	
	1040	10 ton cooling	Q-6	2.60	9.231		2,250	300		2,550	2,925	
	1042	11 ton cooling		2.30	10.435		2,750	340		3,090	3,550	
	1050	15 ton cooling		1.60	15		3,100	490		3,590	4,150	
	1060	20 ton cooling		.70	34.286		4,175	1,125		5,300	6,275	
	1070	25 ton cooling		.65	36.923		5,150	1,200		6,350	7,500	
	1080	30 ton cooling		.60	40		5,775	1,300		7,075	8,325	
	1100	40 ton cooling		.45	53.333		8,775	1,725		10,500	12,300	
	1120	50 ton cooling		.38	63.158		10,300	2,050		12,350	14,400	
	1500	For hot water coil, add					40%	10%				
	1510	For condensing unit add see division 15670.										
	3000	Chilled water, horizontal unit, housing, 2 pipe, controls										
	3100	1/2 ton cooling	Q-5	8	2	Ea.	845	63		908	1,025	
	3110	1 ton cooling		6	2.667		1,025	83.50		1,108.50	1,250	
	3120	1.5 ton cooling		5.50	2.909		1,225	91.50		1,316.50	1,500	
	3130	2 ton cooling		5.25	3.048		1,450	95.50		1,545.50	1,750	
	3140	3 ton cooling		4	4		1,700	126		1,826	2,050	
	3150	3.5 ton cooling		3.80	4.211		1,875	132		2,007	2,250	
	3160	4 ton cooling		3.80	4.211		1,875	132		2,007	2,250	
	3170	5 ton cooling		3.40	4.706		2,175	148		2,323	2,600	
	3180	6 ton cooling		3.40	4.706		2,175	148		2,323	2,600	
	3190	7 ton cooling		2.90	5.517		2,350	173		2,523	2,850	
	3200	8 ton cooling		2.90	5.517		2,350	173		2,523	2,850	
	3210	10 ton cooling	Q-6	2.80	8.571		2,650	279		2,929	3,350	
	3240	For hot water coil, add					40%	10%				
	4000	With electric heat, 2 pipe										
	4100	1/2 ton cooling	Q-5	8	2	Ea.	1,025	63		1,088	1,225	
	4105	3/4 ton cooling		7	2.286		1,125	71.50		1,196.50	1,350	
	4110	1 ton cooling		6	2.667		1,250	83.50		1,333.50	1,500	
	4120	1.5 ton cooling		5.50	2.909		1,375	91.50		1,466.50	1,675	
	4130	2 ton cooling		5.25	3.048		1,750	95.50		1,845.50	2,075	
	4135	2.5 ton cooling		4.80	3.333		2,425	105		2,530	2,800	
	4140	3 ton cooling		4	4		3,200	126		3,326	3,725	
	4150	3.5 ton cooling		3.80	4.211		3,825	132		3,957	4,400	
	4160	4 ton cooling		3.60	4.444		4,400	139		4,539	5,050	
	4170	5 ton cooling		3.40	4.706		4,950	148		5,098	5,650	
	4180	6 ton cooling		3.20	5		4,950	157		5,107	5,650	
	4190	7 ton cooling		2.90	5.517		5,325	173		5,498	6,100	
	4200	8 ton cooling		2.20	7.273		5,850	228		6,078	6,800	
	4210	10 ton cooling	Q-6	2.80	8.571		6,200	279		6,479	7,250	
600	0010	**HEATING & VENTILATING UNITS** Classroom										600
	0020	Includes filter, heating/cooling coils, standard controls										
	0080	750 CFM, 2 tons cooling	Q-6	2	12	Ea.	3,000	390		3,390	3,900	
	0100	1000 CFM, 2-1/2 tons cooling		1.60	15		3,525	490		4,015	4,600	

15700 | Heating/Ventilating/Air Conditioning Equipment

15765 | Fan Coil Unit/Unit Ventilators

		CREW	DAILY OUTPUT	LABOR-HOURS	UNIT	2001 BARE COSTS MAT.	LABOR	EQUIP.	TOTAL	TOTAL INCL O&P	
600	0120	1250 CFM, 3 tons cooling	Q-6	1.40	17.143	Ea.	3,650	560		4,210	4,875	600
	0140	1500 CFM, 4 tons cooling		.80	30		3,900	975		4,875	5,775	
	0160	2000 CFM 5 tons cooling	▼	.50	48		4,575	1,550		6,125	7,400	
	0500	For electric heat, add					35%					
	1000	For no cooling, deduct				▼	25%	10%				

15766 | Fin Tube Radiation

		CREW	DAILY OUTPUT	LABOR-HOURS	UNIT	MAT.	LABOR	EQUIP.	TOTAL	TOTAL INCL O&P	
190	0010	**HYDRONIC HEATING** Terminal units, not incl. main supply pipe										190
	1000	Radiation										
	1100	Panel, baseboard, C.I., including supports, no covers	Q-5	46	.348	L.F.	21	10.90		31.90	39.50	
	1150	Fin tube, wall hung, 14" slope top cover, with damper										
	1200	1-1/4" copper tube, 4-1/4" alum. fin	Q-5	38	.421	L.F.	26	13.20		39.20	48.50	
	1250	1-1/4" steel tube, 4-1/4" steel fin		36	.444		23.50	13.95		37.45	46.50	
	1255	2" steel tube, 4-1/4" steel fin	▼	32	.500	▼	25	15.70		40.70	51	
	1260	21", two tier, slope top, w/damper										
	1262	1-1/4" copper tube, 4-1/4" alum. fin	Q-5	30	.533	L.F.	30.50	16.75		47.25	59	
	1264	1-1/4" steel tube, 4-1/4" steel fin		28	.571		28	17.95		45.95	57.50	
	1266	2" steel tube, 4-1/4" steel fin	▼	25	.640	▼	29.50	20		49.50	63	
	1270	10", flat top, wtih damper										
	1272	1-1/4" copper tube, 4-1/4" alum. fin	Q-5	38	.421	L.F.	26	13.20		39.20	48.50	
	1274	1-1/4" steel tube, 4-1/4" steel fin		36	.444		23.50	13.95		37.45	46.50	
	1276	2" steel tube, 4-1/4" steel fin	▼	32	.500	▼	25	15.70		40.70	51	
	1280	17", two tier, flat top, w/damper										
	1282	1-1/4" copper tube, 4-1/4" alum. fin	Q-5	30	.533	L.F.	30.50	16.75		47.25	59	
	1284	1-1/4" steel tube, 4-1/4" steel fin		28	.571		28	17.95		45.95	57.50	
	1286	2" steel tube, 4-1/4" steel fin	▼	25	.640	▼	29.50	20		49.50	63	
	1301	Rough in wall hung, steel fin w/supply & balance valves	1 Stpi	.91	8.791	Ea.	133	305		438	605	
	1310	Baseboard, pkgd, 1/2" copper tube, alum. fin, 7" high	Q-5	60	.267	L.F.	5.05	8.35		13.40	18.25	
	1320	3/4" copper tube, alum. fin, 7" high		58	.276		5.35	8.65		14	18.95	
	1340	1" copper tube, alum. fin, 8-7/8" high		56	.286		11.10	8.95		20.05	26	
	1360	1-1/4" copper tube, alum. fin, 8-7/8" high		54	.296		16.45	9.30		25.75	32	
	1380	1-1/4" IPS steel tube with steel fins	▼	52	.308	▼	15.80	9.65		25.45	32	
	1381	Rough in baseboard panel & fin tube, supply & balance valves	1 Stpi	1.06	7.547	Ea.	85.50	263		348.50	490	
	1500	Note: fin tube may also require corners, caps, etc.										
	1990	Convector unit, floor recessed, flush, with trim										
	2000	for under large glass wall areas, no damper	Q-5	20	.800	L.F.	24.50	25		49.50	65	
	2100	For unit with damper, add				"	9.95			9.95	10.95	
	3000	Radiators, cast iron										
	3100	Free standing or wall hung, 6 tube, 25" high	Q-5	96	.167	Section	21.50	5.25		26.75	31.50	
	3150	4 tube 25" high		96	.167		15.90	5.25		21.15	25.50	
	3200	4 tube, 19" high	▼	96	.167	▼	14.80	5.25		20.05	24	
	3250	Adj. brackets, 2 per wall radiator up to 30 sections	1 Stpi	32	.250	Ea.	32	8.70		40.70	48	
	3500	Recessed, 20" high x 5" deep, without grille	Q-5	60	.267	Section	17.60	8.35		25.95	32	
	3600	For inlet grille, add				"	1.25			1.25	1.38	
	3950	Unit heaters, propeller, 115 V 2 psi steam, 60°F entering air										
	4000	Horizontal, 12 MBH	Q-5	12	1.333	Ea.	256	42		298	345	
	4020	28.2 MBH		10	1.600		355	50		405	465	
	4040	36.5 MBH		8	2		410	63		473	550	
	4060	43.9 MBH		8	2		430	63		493	565	
	4080	56.5 MBH		7.50	2.133		430	67		497	575	
	4100	65.6 MBH		7	2.286		435	71.50		506.50	585	
	4120	87.6 MBH		6.50	2.462		455	77		532	615	
	4140	96.8 MBH		6	2.667		560	83.50		643.50	740	
	4160	133.3 MBH		5	3.200		575	100		675	780	
	4180	157.6 MBH	▼	4	4	▼	720	126		846	980	

15700 | Heating/Ventilating/Air Conditioning Equipment

15766 | Fin Tube Radiation

			CREW	DAILY OUTPUT	LABOR-HOURS	UNIT	2001 BARE COSTS MAT.	LABOR	EQUIP.	TOTAL	TOTAL INCL O&P	
190	4200	197.7 MBH	Q-5	3	5.333	Ea.	850	167		1,017	1,200	190
	4220	257.2 MBH		2.50	6.400		1,175	201		1,376	1,600	
	4240	286.9 MBH		2	8		1,250	251		1,501	1,750	
	4250	326.0 MBH		1.90	8.421		1,325	264		1,589	1,850	
	4260	364 MBH		1.80	8.889		1,475	279		1,754	2,050	
	4270	404 MBH		1.60	10		1,925	315		2,240	2,600	
	4280	Rough in unit heater, 72.7 MBH, piping, supports & controls	1 Stpi	.47	17.021	Set	310	595		905	1,250	
	4300	For vertical diffuser, add				Ea.	136			136	149	
	4310	Vertical flow, 40 MBH	Q-5	11	1.455		420	45.50		465.50	530	
	4314	58.5 MBH		8	2		430	63		493	565	
	4318	92 MBH		7	2.286		515	71.50		586.50	675	
	4322	109.7 MBH		6	2.667		665	83.50		748.50	860	
	4326	131.0 MBH		4	4		675	126		801	930	
	4330	160.0 MBH		3	5.333		715	167		882	1,050	
	4334	194.0 MBH		2.20	7.273		775	228		1,003	1,200	
	4338	212.0 MBH		2.10	7.619		1,250	239		1,489	1,725	
	4342	247.0 MBH		1.96	8.163		1,300	256		1,556	1,800	
	4346	297.0 MBH		1.80	8.889		1,400	279		1,679	1,975	
	4350	333.0 MBH	Q-6	1.90	12.632		1,825	410		2,235	2,625	
	4354	420 MBH,(460 V)		1.80	13.333		1,925	435		2,360	2,775	
	4358	500 MBH, (460 V)		1.71	14.035		2,375	455		2,830	3,325	
	4362	570 MBH, (460 V)		1.40	17.143		2,850	560		3,410	4,000	
	4366	620 MBH, (460 V)		1.30	18.462		3,275	600		3,875	4,500	
	4370	960 MBH, (460 V)		1.10	21.818		5,750	710		6,460	7,400	
	4400	Unit heaters, propeller, hot water, 115 V										
	4404	Horizontal										
	4408	6 MBH	Q-5	11	1.455	Ea.	240	45.50		285.50	335	
	4412	12 MBH		11	1.455		240	45.50		285.50	335	
	4416	16 MBH		10	1.600		259	50		309	360	
	4420	24 MBH		9	1.778		325	56		381	440	
	4424	29 MBH		8	2		385	63		448	520	
	4428	47 MBH		7	2.286		435	71.50		506.50	585	
	4432	63 MBH		6	2.667		455	83.50		538.50	625	
	4436	81 MBH		5.50	2.909		555	91.50		646.50	750	
	4440	90 MBH		5	3.200		575	100		675	780	
	4460	133 MBH		4	4		720	126		846	980	
	4464	139 MBH		3.50	4.571		850	143		993	1,150	
	4468	198 MBH		2.50	6.400		1,175	201		1,376	1,600	
	4472	224 MBH		2	8		1,225	251		1,476	1,700	
	4476	273 MBH		1.50	10.667		1,300	335		1,635	1,925	
	4500	Vertical										
	4504	8 MBH	Q-5	13	1.231	Ea.	380	38.50		418.50	475	
	4508	12 MBH		13	1.231		380	38.50		418.50	475	
	4512	16 MBH		13	1.231		380	38.50		418.50	475	
	4516	30 MBH		12	1.333		380	42		422	480	
	4520	43 MBH		11	1.455		430	45.50		475.50	540	
	4524	57 MBH		8	2		485	63		548	625	
	4528	68 MBH		7	2.286		515	71.50		586.50	675	
	4540	105 MBH		6	2.667		600	83.50		683.50	785	
	4544	123 MBH		5	3.200		715	100		815	935	
	4550	140 MBH		5	3.200		775	100		875	1,000	
	4554	156 MBH		4	4		875	126		1,001	1,150	
	4560	210 MBH		4	4		955	126		1,081	1,250	
	4564	223 MBH		3	5.333		1,425	167		1,592	1,800	
	4568	257 MBH		2	8		1,425	251		1,676	1,925	
	5400	Cabinet, horizontal, hot water, blower type										

15700 | Heating/Ventilating/Air Conditioning Equipment

15766 | Fin Tube Radiation

		CREW	DAILY OUTPUT	LABOR-HOURS	UNIT	MAT.	LABOR	EQUIP.	TOTAL	TOTAL INCL O&P	
5420	20 MBH	Q-5	10	1.600	Ea.	735	50		785	880	190
5430	60 MBH		7	2.286		1,275	71.50		1,346.50	1,500	
5440	100 MBH		5	3.200		1,300	100		1,400	1,600	
5450	120 MBH		4	4		1,400	126		1,526	1,725	
6000	Valance units, complete with 1/2" cooling coil, enclosure										
6020	2 tube	Q-5	18	.889	L.F.	26	28		54	70.50	
6040	3 tube		16	1		31	31.50		62.50	81.50	
6060	4 tube		16	1		35.50	31.50		67	86.50	
6080	5 tube		15	1.067		42.50	33.50		76	97.50	
6100	6 tube		15	1.067		46	33.50		79.50	102	
6120	8 tube		14	1.143		68.50	36		104.50	130	
6200	For 3/4" cooling coil, add					10%					
9500	To convert SFR to BTU rating: Hot water, 150 x SFR										
9510	Forced hot water, 180 x SFR; steam, 240 x SFR										

15768 | Infrared Heaters

		CREW	DAILY OUTPUT	LABOR-HOURS	UNIT	MAT.	LABOR	EQUIP.	TOTAL	TOTAL INCL O&P	
0010	**INFRA-RED UNIT**										600
0020	Gas fired, unvented, electric ignition, 100% shutoff. Piping and										
0030	wiring not included										
0060	Input, 15 MBH	Q-5	7	2.286	Ea.	310	71.50		381.50	450	
0100	30 MBH		6	2.667		355	83.50		438.50	515	
0120	45 MBH		5	3.200		390	100		490	575	
0140	50 MBH		4.50	3.556		420	112		532	635	
0160	60 MBH		4	4		455	126		581	690	
0180	75 MBH		3	5.333		485	167		652	790	
0200	90 MBH		2.50	6.400		545	201		746	905	
0220	105 MBH		2	8		680	251		931	1,125	
0240	120 MBH		2	8		805	251		1,056	1,275	
2000	Electric, single or three phase										
2050	6 kW, 20,478 BTU	1 Elec	2.30	3.478	Ea.	345	119		464	560	
2100	13.5 KW, 40,956 BTU		2.20	3.636		535	125		660	775	
2150	24 KW, 81,912 BTU		2	4		1,175	137		1,312	1,475	
3000	Oil fired, pump, controls, fusible valve, oil supply tank										
3050	91,000 BTU	Q-5	2.50	6.400	Ea.	5,100	201		5,301	5,925	
3080	105.000 BTU		2.25	7.111		5,350	223		5,573	6,200	
3110	119,000 BTU		2	8		5,625	251		5,876	6,575	

15770 | Floor-Heating & Snow-Melting Eq.

		CREW	DAILY OUTPUT	LABOR-HOURS	UNIT	MAT.	LABOR	EQUIP.	TOTAL	TOTAL INCL O&P	
0010	**ELECTRIC HEATING**, equipments are not including wires & conduits										200
0200	Snow melting for paved surface embedded mat heaters & controls	1 Elec	130	.062	S.F.	6.10	2.11		8.21	9.85	
0400	Cable heating, radiant heat plaster, no controls, in South		130	.062		7.25	2.11		9.36	11.15	
0600	In North		90	.089		7.25	3.05		10.30	12.55	
0800	Cable on 1/2" board not incl. controls, tract housing		90	.089		6.10	3.05		9.15	11.25	
1000	Custom housing		80	.100		7.10	3.43		10.53	12.90	
1100	Rule of thumb: Baseboard units, including control		4.40	1.818	kW	75	62.50		137.50	176	
1200	Duct heaters, including controls		5.30	1.509	"	61.50	52		113.50	146	
1300	Baseboard heaters, 2' long, 375 watt		8	1	Ea.	34	34.50		68.50	88.50	
1400	3' long, 500 watt		8	1		41	34.50		75.50	96	
1600	4' long, 750 watt		6.70	1.194		48	41		89	114	
1800	5' long, 935 watt		5.70	1.404		57	48		105	135	
2000	6' long, 1125 watt		5	1.600		65	55		120	154	
2200	7' long, 1310 watt		4.40	1.818		76	62.50		138.50	177	
2400	8' long, 1500 watt		4	2		81	68.50		149.50	191	
2600	9' long, 1680 watt		3.60	2.222		134	76		210	261	
2800	10' long, 1875 watt		3.30	2.424		134	83		217	271	
2950	Wall heaters with fan, 120 to 277 volt										

15700 | Heating/Ventilating/Air Conditioning Equipment

15770 | Floor-Heating & Snow-Melting Eq.

		CREW	DAILY OUTPUT	LABOR-HOURS	UNIT	MAT.	LABOR	EQUIP.	TOTAL	TOTAL INCL O&P
2970	surface mounted, residential, 750 watt	1 Elec	7	1.143	Ea.	105	39		144	175
2980	1000 watt		7	1.143		107	39		146	177
2990	1250 watt		6	1.333		120	45.50		165.50	201
3000	1500 watt		5	1.600		120	55		175	214
3010	2000 watt		5	1.600		123	55		178	217
3040	2250 watt		4	2		255	68.50		323.50	385
3050	2500 watt		4	2		255	68.50		323.50	385
3070	4000 watt		3.50	2.286		260	78.50		338.50	405
3080	Commercial, 750 watt		7	1.143		173	39		212	249
3090	1000 watt		7	1.143		173	39		212	249
3100	1250 watt		6	1.333		173	45.50		218.50	259
3110	1500 watt		5	1.600		173	55		228	272
3120	2000 watt		5	1.600		173	55		228	272
3130	2500 watt		4.50	1.778		335	61		396	460
3140	3000 watt		4	2		335	68.50		403.50	470
3150	4000 watt		3.50	2.286		360	78.50		438.50	510
3160	Recessed, residential, 750 watt		6	1.333		84	45.50		129.50	161
3170	1000 watt		6	1.333		85	45.50		130.50	162
3180	1250 watt		5	1.600		94	55		149	185
3190	1500 watt		4	2		94	68.50		162.50	205
3210	2000 watt		4	2		105	68.50		173.50	218
3230	2500 watt		3.50	2.286		223	78.50		301.50	360
3240	3000 watt		3	2.667		223	91.50		314.50	380
3250	4000 watt		2.70	2.963		228	102		330	405
3260	Commercial, 750 watt		6	1.333		173	45.50		218.50	259
3270	1000 watt		6	1.333		173	45.50		218.50	259
3280	1250 watt		5	1.600		173	55		228	272
3290	1500 watt		4	2		173	68.50		241.50	292
3300	2000 watt		4	2		173	68.50		241.50	292
3310	2500 watt		3.50	2.286		173	78.50		251.50	305
3320	3000 watt		3	2.667		300	91.50		391.50	465
3330	4000 watt		2.70	2.963		300	102		402	480
3600	Thermostats, integral		16	.500		25	17.15		42.15	53
3800	Line voltage, 1 pole		8	1		25	34.50		59.50	78.50
3810	2 pole		8	1		25	34.50		59.50	78.50
3820	Low voltage, 1 pole		8	1		25	34.50		59.50	78.50
4000	Heat trace system, 400 degree									
4020	115V, 2.5 watts per L.F.	1 Elec	530	.015	L.F.	5.80	.52		6.32	7.15
4030	5 watts per L.F.		530	.015		5.80	.52		6.32	7.15
4050	10 watts per L.F.		530	.015		7.35	.52		7.87	8.85
4060	220V, 4 watts per L.F.		530	.015		5.80	.52		6.32	7.15
4080	480V, 8 watts per L.F.		530	.015		5.80	.52		6.32	7.15
4200	Heater raceway									
4220	5/8" W x 3/8" H	1 Elec	200	.040	L.F.	3.95	1.37		5.32	6.40
4240	5/8" W x 1/2" H	"	190	.042	"	5.50	1.44		6.94	8.20
4260	Heat transfer cement									
4280	1 gallon				Ea.	50			50	55
4300	5 gallon				"	205			205	226
4320	Snap band, clamp									
4340	3/4" pipe size	1 Elec	470	.017	Ea.	.58			.58	.87
4360	1" pipe size		444	.018		.62			.62	.92
4380	1-1/4" pipe size		400	.020		.69			.69	1.02
4400	1-1/2" pipe size		355	.023		.77			.77	1.15
4420	2" pipe size		320	.025		.86			.86	1.28
4440	3" pipe size		160	.050			1.72		1.72	2.56
4460	4" pipe size		100	.080			2.74		2.74	4.10

15700 | Heating/Ventilating/Air Conditioning Equipment

15770 | Floor-Heating & Snow-Melting Eq.

		CREW	DAILY OUTPUT	LABOR-HOURS	UNIT	MAT.	LABOR	EQUIP.	TOTAL	TOTAL INCL O&P
4480	Single pole thermostat NEMA 4, 30 amp	1 Elec	8	1	Ea.	500	34.50		534.50	600
4500	NEMA 7, 30 amp		7	1.143		590	39		629	705
4520	Double pole, NEMA 4, 30 amp		7	1.143		500	39		539	610
4540	NEMA 7, 30 amp		6	1.333		590	45.50		635.50	715
4560	Thermostat/contactor combination, NEMA 4									
4580	30 amp 4 pole	1 Elec	3.60	2.222	Ea.	525	76		601	695
4600	50 amp 4 pole		3	2.667		530	91.50		621.50	720
4620	75 amp 3 pole		2.50	3.200		800	110		910	1,050
4640	75 amp 4 pole		2.30	3.478		1,150	119		1,269	1,425
4680	Control transformer, 50 VA		4	2		81	68.50		149.50	191
4700	75 VA		3.10	2.581		88	88.50		176.50	229
4720	Expediter fitting		11	.727		20	25		45	59
5000	Radiant heating ceiling panels, 2' x 4', 500 watt		16	.500		183	17.15		200.15	227
5050	750 watt		16	.500		202	17.15		219.15	248
5200	For recessed plaster frame, add		32	.250		36	8.60		44.60	52.50
5300	Infra-red quartz heaters, 120 volts, 1000 watts		6.70	1.194		125	41		166	199
5350	1500 watt		5	1.600		125	55		180	220
5400	240 volts, 1500 watt		5	1.600		125	55		180	220
5450	2000 watt		4	2		125	68.50		193.50	240
5500	3000 watt		3	2.667		142	91.50		233.50	293
5550	4000 watt		2.60	3.077		252	106		358	435
5570	Modulating control		.80	10		66	345		411	585
5600	Unit heaters, heavy duty, with fan & mounting bracket									
5650	Single phase, 208-240-277 volt, 3 kW	1 Elec	3.20	2.500	Ea.	350	86		436	515
5700	4 kW		2.80	2.857		360	98		458	540
5750	5 kW		2.40	3.333		360	114		474	565
5800	7 kW		1.90	4.211		560	144		704	830
5850	10 kW		1.30	6.154		610	211		821	985
5900	13 kW		1	8		980	274		1,254	1,475
5950	15 kW		.90	8.889		1,050	305		1,355	1,600
6000	480 volt, 3 kW		3.30	2.424		460	83		543	630
6020	4 kW		3	2.667		525	91.50		616.50	715
6040	5 kW		2.60	3.077		525	106		631	740
6060	7 kW		2	4		650	137		787	920
6080	10 kW		1.40	5.714		650	196		846	1,000
6100	13 kW		1.10	7.273		1,075	249		1,324	1,550
6120	15 kW		1	8		1,075	274		1,349	1,575
6140	20 kW		.90	8.889		1,375	305		1,680	1,975
6300	3 phase, 208-240 volt, 5 kW		2.40	3.333		355	114		469	560
6320	7 kW		1.90	4.211		545	144		689	815
6340	10 kW		1.30	6.154		590	211		801	965
6360	15 kW		.90	8.889		980	305		1,285	1,525
6380	20 kW		.70	11.429		1,400	390		1,790	2,125
6400	25 kW		.50	16		1,675	550		2,225	2,675
6500	480 volt, 5 kW		2.60	3.077		500	106		606	710
6520	7 kW		2	4		625	137		762	895
6540	10 kW		1.40	5.714		660	196		856	1,025
6560	13 kW		1.10	7.273		1,075	249		1,324	1,550
6580	15 kW		1	8		1,075	274		1,349	1,575
6600	20 kW		.90	8.889		1,400	305		1,705	1,975
6620	25 kW		.60	13.333		1,675	455		2,130	2,525
6630	30 kW		.70	11.429		1,925	390		2,315	2,700
6640	40 kW		.60	13.333		2,450	455		2,905	3,375
6650	50 kW		.50	16		2,975	550		3,525	4,100
6800	Vertical discharge heaters, with fan									
6820	Single phase, 208-240-277 volt, 10 kW	1 Elec	1.30	6.154	Ea.	1,150	211		1,361	1,575

15700 | Heating/Ventilating/Air Conditioning Equipment

15770 | Floor-Heating & Snow-Melting Eq.

			CREW	DAILY OUTPUT	LABOR-HOURS	UNIT	2001 BARE COSTS MAT.	LABOR	EQUIP.	TOTAL	TOTAL INCL O&P	
200	6840	15 kW	1 Elec	.90	8.889	Ea.	1,400	305		1,705	1,975	200
	6900	3 phase, 208-240 volt, 10 kW		1.30	6.154		1,075	211		1,286	1,525	
	6920	15 kW		.90	8.889		1,225	305		1,530	1,800	
	6940	20 kW		.70	11.429		1,450	390		1,840	2,175	
	6960	25 kW		.50	16		1,925	550		2,475	2,950	
	6980	30 kW		.40	20		2,100	685		2,785	3,325	
	7000	40 kW		.36	22.222		2,800	760		3,560	4,225	
	7020	50 kW		.32	25		3,200	860		4,060	4,800	
	7100	480 volt, 10 kW		1.40	5.714		1,275	196		1,471	1,700	
	7120	15 kW		1	8		1,400	274		1,674	1,950	
	7140	20 kW		.90	8.889		1,700	305		2,005	2,325	
	7160	25 kW		.60	13.333		1,950	455		2,405	2,825	
	7180	30 kW		.50	16		2,350	550		2,900	3,400	
	7200	40 kW		.40	20		2,800	685		3,485	4,100	
	7220	50 kW		.35	22.857		3,100	785		3,885	4,575	
	7900	Cabinet convector heaters, 240 volt										
	7920	2' long, 1000 watt	1 Elec	5.30	1.509	Ea.	445	52		497	570	
	7940	1500 watt		5.30	1.509		470	52		522	600	
	7960	2000 watt		5.30	1.509		500	52		552	630	
	7980	3' long, 1500 watt		4.60	1.739		515	59.50		574.50	655	
	8000	2250 watt		4.60	1.739		520	59.50		579.50	660	
	8020	3000 watt		4.60	1.739		575	59.50		634.50	725	
	8040	4' long, 2000 watt		4	2		585	68.50		653.50	745	
	8060	3000 watt		4	2		615	68.50		683.50	775	
	8080	4000 watt		4	2		645	68.50		713.50	810	
	8100	Available also in 208 or 277 volt										
	8200	Cabinet unit heaters, 120 to 277 volt, 1 pole,										
	8220	wall mounted, 2 kW	1 Elec	4.60	1.739	Ea.	1,600	59.50		1,659.50	1,850	
	8230	3 kW		4.60	1.739		1,650	59.50		1,709.50	1,925	
	8240	4 kW		4.40	1.818		1,725	62.50		1,787.50	2,000	
	8250	5 kW		4.40	1.818		1,800	62.50		1,862.50	2,075	
	8260	6 kW		4.20	1.905		1,825	65.50		1,890.50	2,100	
	8270	8 kW		4	2		1,925	68.50		1,993.50	2,225	
	8280	10 kW		3.80	2.105		1,950	72		2,022	2,250	
	8290	12 kW		3.50	2.286		2,100	78.50		2,178.50	2,425	
	8300	13.5 kW		2.90	2.759		2,400	94.50		2,494.50	2,800	
	8310	16 kW		2.70	2.963		2,400	102		2,502	2,800	
	8320	20 kW		2.30	3.478		2,650	119		2,769	3,100	
	8330	24 kW		1.90	4.211		2,950	144		3,094	3,450	
	8350	Recessed, 2 kW		4.40	1.818		1,800	62.50		1,862.50	2,075	
	8370	3 kW		4.40	1.818		1,800	62.50		1,862.50	2,075	
	8380	4 kW		4.20	1.905		1,850	65.50		1,915.50	2,125	
	8390	5 kW		4.20	1.905		1,925	65.50		1,990.50	2,225	
	8400	6 kW		4	2		1,950	68.50		2,018.50	2,225	
	8410	8 kW		3.80	2.105		2,075	72		2,147	2,375	
	8420	10 kW		3.50	2.286		2,125	78.50		2,203.50	2,450	
	8430	12 kW		2.90	2.759		2,250	94.50		2,344.50	2,625	
	8440	13.5 kW		2.70	2.963		2,550	102		2,652	2,950	
	8450	16 kW		2.30	3.478		2,575	119		2,694	3,025	
	8460	20 kW		1.90	4.211		2,800	144		2,944	3,300	
	8470	24 kW		1.60	5		3,075	172		3,247	3,625	
	8490	Ceiling mounted, 2 kW		3.20	2.500		1,575	86		1,661	1,875	
	8510	3 kW		3.20	2.500		1,650	86		1,736	1,925	
	8520	4 kW		3	2.667		1,700	91.50		1,791.50	2,000	
	8530	5 kW		3	2.667		1,775	91.50		1,866.50	2,075	
	8540	6 kW		2.80	2.857		1,800	98		1,898	2,150	

15700 | Heating/Ventilating/Air Conditioning Equipment

15770 | Floor-Heating & Snow-Melting Eq.

			CREW	DAILY OUTPUT	LABOR-HOURS	UNIT	MAT.	LABOR	EQUIP.	TOTAL	TOTAL INCL O&P	
200	8550	8 kW	1 Elec	2.40	3.333	Ea.	1,925	114		2,039	2,275	200
	8560	10 kW		2.20	3.636		1,950	125		2,075	2,325	
	8570	12 kW		2	4		2,100	137		2,237	2,500	
	8580	13.5 kW		1.50	5.333		2,225	183		2,408	2,725	
	8590	16 kW		1.30	6.154		2,400	211		2,611	2,975	
	8600	20 kW		.90	8.889		2,650	305		2,955	3,375	
	8610	24 kW	▼	.60	13.333	▼	2,925	455		3,380	3,900	
	8630	208 to 480 V, 3 pole										
	8650	Wall mounted, 2 kW	1 Elec	4.60	1.739	Ea.	1,600	59.50		1,659.50	1,850	
	8670	3 kW		4.60	1.739		1,650	59.50		1,709.50	1,925	
	8680	4 kW		4.40	1.818		1,725	62.50		1,787.50	2,000	
	8690	5 kW		4.40	1.818		1,800	62.50		1,862.50	2,075	
	8700	6 kW		4.20	1.905		1,800	65.50		1,865.50	2,100	
	8710	8 kW		4	2		1,925	68.50		1,993.50	2,225	
	8720	10 kW		3.80	2.105		1,975	72		2,047	2,275	
	8730	12 kW		3.50	2.286		2,100	78.50		2,178.50	2,425	
	8740	13.5 kW		2.90	2.759		2,400	94.50		2,494.50	2,775	
	8750	16 kW		2.70	2.963		2,400	102		2,502	2,800	
	8760	20 kW		2.30	3.478		2,650	119		2,769	3,100	
	8770	24 kW		1.90	4.211		2,950	144		3,094	3,450	
	8790	Recessed, 2 kW		4.40	1.818		2,100	62.50		2,162.50	2,400	
	8810	3 kW		4.40	1.818		1,775	62.50		1,837.50	2,050	
	8820	4 kW		4.20	1.905		1,850	65.50		1,915.50	2,125	
	8830	5 kW		4.20	1.905		1,925	65.50		1,990.50	2,200	
	8840	6 kW		4	2		1,950	68.50		2,018.50	2,225	
	8850	8 kW		3.80	2.105		2,075	72		2,147	2,375	
	8860	10 kW		3.50	2.286		2,125	78.50		2,203.50	2,450	
	8870	12 kW		2.90	2.759		2,225	94.50		2,319.50	2,600	
	8880	13.5 kW		2.70	2.963		2,550	102		2,652	2,950	
	8890	16 kW		2.30	3.478		2,575	119		2,694	3,025	
	8900	20 kW		1.90	4.211		2,800	144		2,944	3,300	
	8920	24 kW		1.60	5		3,075	172		3,247	3,625	
	8940	Ceiling mount, 2 kW		3.20	2.500		1,575	86		1,661	1,875	
	8950	3 kW		3.20	2.500		1,650	86		1,736	1,925	
	8960	4 kW		3	2.667		1,700	91.50		1,791.50	2,000	
	8970	5 kW		3	2.667		1,775	91.50		1,866.50	2,075	
	8980	6 kW		2.80	2.857		1,800	98		1,898	2,150	
	8990	8 kW		2.40	3.333		1,925	114		2,039	2,275	
	9000	10 kW		2.20	3.636		1,950	125		2,075	2,325	
	9020	13.5 kW		1.50	5.333		2,100	183		2,283	2,575	
	9030	16 kW		1.30	6.154		2,400	211		2,611	2,975	
	9040	20 kW		.90	8.889		2,650	305		2,955	3,375	
	9060	24 kW	▼	.60	13.333	▼	2,950	455		3,405	3,900	
500	0010	**RADIANT FLOOR HEATING**										500
	0100	Tubing, PEX (cross-linked polyethylene)										
	0110	Oxygen barrier type for systems with ferrous materials										
	0120	1/2"	Q-5	800	.020	L.F.	.49	.63		1.12	1.49	
	0130	3/4"		535	.030		.62	.94		1.56	2.10	
	0140	1"	▼	400	.040	▼	1.25	1.26		2.51	3.28	
	0200	Non barrier type for ferrous free systems										
	0210	1/2"	Q-5	800	.020	L.F.	.27	.63		.90	1.25	
	0220	3/4"		535	.030		.44	.94		1.38	1.90	
	0230	1"	▼	400	.040	▼	.77	1.26		2.03	2.75	
	1000	Manifolds										
	1110	Brass										

15700 | Heating/Ventilating/Air Conditioning Equipment

15770 | Floor-Heating & Snow-Melting Eq.

		CREW	DAILY OUTPUT	LABOR-HOURS	UNIT	MAT.	LABOR	EQUIP.	TOTAL	TOTAL INCL O&P
1120	Valved									
1130	1", 2 circuit	Q-5	13	1.231	Ea.	36	38.50		74.50	98
1140	1", 3 circuit		11	1.455		49	45.50		94.50	123
1150	1", 4 circuit		9	1.778		66	56		122	157
1160	1-1/4", 3 circuit		10	1.600		55.50	50		105.50	137
1170	1-1/4", 4 circuit		8	2		67.50	63		130.50	170
1300	Valveless									
1310	1", 2 circuit	Q-5	14	1.143	Ea.	18	36		54	74
1320	1", 3 circuit		12	1.333		24	42		66	89.50
1330	1", 4 circuit		10	1.600		31	50		81	110
1340	1-1/4", 3 circuit		11	1.455		30	45.50		75.50	102
1350	1-1/4", 4 circuit		9	1.778		30	56		86	117
1400	Manifold connector kit									
1410	Mounting brackets, couplings, end cap, air vent, valve, plug									
1420	1"	Q-5	8	2	Ea.	61.50	63		124.50	163
1430	1-1/4"	"	7	2.286	"	75	71.50		146.50	191
1610	Copper, (cut to size)									
1620	1" x 1/2" x 72" Lg, 2" sweat drops, 24 circuit	Q-5	3.33	4.805	Ea.	62.50	151		213.50	297
1630	1-1/4" x 1/2" x 72" Lg, 2" sweat drops, 24 circuit		3.20	5		73	157		230	315
1640	1-1/2" x 1/2" x 72" Lg, 2" sweat drops, 24 circuit		3	5.333		88.50	167		255.50	350
1650	1-1/4" x 3/4" x 72" Lg, 2" sweat drops, 24 circuit		3.10	5.161		78	162		240	330
1660	1-1/2" x 3/4" x 72" Lg, 2" sweat drops, 24 circuit		2.90	5.517		78	173		251	345
1710	Copper, (modular)									
1720	1" x 1/2", 2" sweat drops, 2 circuit	Q-5	17.50	.914	Ea.	5.85	28.50		34.35	50
1730	1" x 1/2", 2" sweat drops, 3 circuit		14.50	1.103		8.80	34.50		43.30	62
1740	1" x 1/2", 2" sweat drops, 4 circuit		12.30	1.301		10.15	41		51.15	72.50
1750	1-1/4" x 1/2", 2" sweat drops, 2 circuit		17	.941		6.75	29.50		36.25	52
1760	1-1/4" x 1/2", 2" sweat drops, 3 circuit		14	1.143		10.15	36		46.15	65
1770	1-1/4" x 1/2", 2" sweat drops, 4 circuit		11.80	1.356		13.50	42.50		56	79
3000	Valves									
3110	Motorized zone valve operator, 24V	Q-5	30	.533	Ea.	37	16.75		53.75	66
3120	Motorized zone valve with operator complete, 24V									
3130	3/4"	Q-5	35	.457	Ea.	64.50	14.35		78.85	92.50
3140	1"		32	.500		73.50	15.70		89.20	105
3150	1-1/4"		29.60	.541		90	16.95		106.95	125
3200	Flow control/isolation valve, 1/2"		32	.500		20.50	15.70		36.20	46
3300	Thermostatic mixing valves									
3310	1/2"	Q-5	26.60	.601	Ea.	67.50	18.90		86.40	103
3320	3/4"		25	.640		82.50	20		102.50	122
3330	1"		21.40	.748		88	23.50		111.50	132
3400	3 Way mixing / diverting valve, manual, brass									
3410	1/2"	Q-5	21.30	.751	Ea.	86.50	23.50		110	131
3420	3/4"		20	.800		88.50	25		113.50	136
3430	1"		17.80	.899		104	28		132	157
3500	4 Way mixing / diverting valve, manual, brass									
3510	1/2"	Q-5	16	1	Ea.	92.50	31.50		124	149
3520	3/4"		15	1.067		104	33.50		137.50	165
3530	1"		13.30	1.203		116	38		154	184
3540	1-1/4"		11.40	1.404		127	44		171	206
3550	1-1/2"		10.60	1.509		170	47.50		217.50	259
3560	2"		10	1.600		184	50		234	278
3570	2-1/2"		8	2		505	63		568	650
3800	Motor control, 3 or 4 way, for valves up to 1-1/2"		30	.533		124	16.75		140.75	162
5000	Radiant floor heating, zone control panel									
5110	6 Zone control									
5120	For zone valves	Q-5	16	1	Ea.	146	31.50		177.50	209

15700 | Heating/Ventilating/Air Conditioning Equipment

15770 | Floor-Heating & Snow-Melting Eq.

			CREW	DAILY OUTPUT	LABOR-HOURS	UNIT	MAT.	LABOR	EQUIP.	TOTAL	TOTAL INCL O&P	
500	5130	For circulators	Q-5	16	1	Ea.	156	31.50		187.50	220	500
	7000	PEX tubing fittings, compression type										
	7110	PEX tubing to brass valve conn., 1/2"	Q-5	54	.296	Ea.	2.12	9.30		11.42	16.40	
	7200	PEX x male NPT										
	7210	1/2" x 1/2"	Q-5	54	.296	Ea.	4	9.30		13.30	18.45	
	7220	3/4" x 3/4"		46	.348		6.75	10.90		17.65	24	
	7230	1" x 1"	▼	40	.400	▼	15	12.55		27.55	35.50	
	7300	PEX coupling										
	7310	1/2" x 1/2"	Q-5	50	.320	Ea.	4.88	10.05		14.93	20.50	
	7320	3/4" x 3/4"		42	.381		14.05	11.95		26	33.50	
	7330	1" x 1"	▼	36	.444	▼	18	13.95		31.95	41	
	7400	PEX x copper sweat										
	7410	1/2" x 1/2"	Q-5	38	.421	Ea.	3.50	13.20		16.70	24	
	7420	3/4" x 3/4"		34	.471		7.50	14.75		22.25	31	
	7430	1" x 1"	▼	29	.552	▼	16.50	17.30		33.80	44	

15780 | Energy Recovery Equipment

			CREW	DAILY OUTPUT	LABOR-HOURS	UNIT	MAT.	LABOR	EQUIP.	TOTAL	TOTAL INCL O&P	
100	0010	**HEAT RECOVERY PACKAGES**										100
	0100	Air to air										
	2000	Kitchen exhaust, commercial, heat pipe exchanger										
	2040	Combined supply/exhaust air volume										
	2080	2.5 to 6.0 MCFM	Q-10	2.80	8.571	MCFM	4,725	268		4,993	5,625	
	2120	6 to 16 MCFM		5	4.800		2,900	150		3,050	3,400	
	2160	16 to 22 MCFM	▼	6	4	▼	2,250	125		2,375	2,700	
	2170	For kitchen ventilation see div. 11420										
	4000	Enthalpy recovery wheel										
	4010	1000 max CFM	Q-9	1.20	13.333	Ea.	3,450	405		3,855	4,425	
	4020	2000 max CFM		1	16		3,775	485		4,260	4,900	
	4030	4000 max CFM		.80	20		4,725	605		5,330	6,100	
	4040	6000 max CFM	▼	.70	22.857		5,550	690		6,240	7,175	
	4050	8000 max CFM	Q-10	1	24		6,200	750		6,950	7,975	
	4060	10,000 max CFM		.90	26.667		7,650	835		8,485	9,725	
	4070	20,000 max CFM		.80	30		13,300	940		14,240	16,100	
	4080	25,000 max CFM		.70	34.286		16,200	1,075		17,275	19,500	
	4090	30,000 max CFM		.50	48		17,900	1,500		19,400	22,000	
	4100	40,000 max CFM		.45	53.333		24,400	1,675		26,075	29,400	
	4110	50,000 max CFM	▼	.40	60	▼	29,000	1,875		30,875	34,800	
	8000	Residential ventilation heat exchanger, not incl. duct										
	8010	70 CFM exchanger	1 Shee	8	1	Ea.	595	33.50		628.50	705	
	8020	110 CFM exchanger		7	1.143		895	38.50		933.50	1,050	
	8024	210 CFM exchanger	▼	6.60	1.212		895	40.50		935.50	1,050	
	8028	450 CFM exchanger	Q-9	12	1.333		1,825	40.50		1,865.50	2,075	
	8032	900 CFM exchanger	"	10	1.600	▼	2,275	48.50		2,323.50	2,575	
	8060	For larger applications										
	8062	Use multiple units										

15800 | Air Distribution

15810 | Ducts

			CREW	DAILY OUTPUT	LABOR-HOURS	UNIT	MAT.	LABOR	EQUIP.	TOTAL	TOTAL INCL O&P
600	0010	**DUCTWORK**									600
	0020	Fabricated rectangular, includes fittings, joints, supports,	R15810-050								
	0030	allowance for flexible connections, no insulation	R15810-070								
	0031	NOTE: Fabrication and installation are combined									
	0040	as LABOR cost. Approx. 25% fittings assumed.	R15810-100								
	0042	Fabrication/Inst. is to commercial quality standards									
	0043	(SMACNA or equiv.) for structure, sealing, leak testing, etc.									
	0050	Add to labor for elevated installation									
	0051	of fabricated ductwork									
	0052	10' to 15' high						6%			
	0053	15' to 20' high						12%			
	0054	20' to 25' high						15%			
	0055	25' to 30' high						21%			
	0056	30' to 35' high						24%			
	0057	35' to 40' high						30%			
	0058	Over 40' high						33%			
	0070	For duct insulation and lining see 15082-200-3000									
	0100	Aluminum, alloy 3003-H14, under 100 lb.	Q-10	75	.320	Lb.	3.23	10		13.23	19
	0110	100 to 500 lb.		80	.300		2.23	9.40		11.63	16.95
	0120	500 to 1,000 lb.		95	.253		1.73	7.90		9.63	14.10
	0140	1,000 to 2,000 lb.		120	.200		1.33	6.25		7.58	11.10
	0150	2,000 to 5,000 lb.		130	.185		1.28	5.80		7.08	10.35
	0160	Over 5,000 lb.		145	.166		1.23	5.20		6.43	9.35
	0500	Galvanized steel, under 200 lb.		235	.102		1.32	3.20		4.52	6.40
	0520	200 to 500 lb.		245	.098		.82	3.07		3.89	5.65
	0540	500 to 1,000 lb.		255	.094		.57	2.95		3.52	5.20
	0560	1,000 to 2,000 lb.		265	.091		.42	2.84		3.26	4.84
	0570	2,000 to 5,000 lb.		275	.087		.37	2.73		3.10	4.63
	0580	Over 5,000 lb.		285	.084		.32	2.64		2.96	4.42
	0600	For large quantities special prices available from supplier									
	1000	Stainless steel, type 304, under 100 lb.	Q-10	165	.145	Lb.	3.37	4.56		7.93	10.75
	1020	100 to 500 lb.		175	.137		2.37	4.30		6.67	9.25
	1030	500 to 1,000 lb.		190	.126		1.87	3.96		5.83	8.15
	1040	1,000 to 2,000 lb.		200	.120		1.47	3.76		5.23	7.40
	1050	2,000 to 5,000 lb.		225	.107		1.42	3.34		4.76	6.70
	1060	Over 5,000 lb.		235	.102		1.37	3.20		4.57	6.45
	1080	Note: Minimum order cost exceeds per lb. cost for min. wt.									
	1100	For medium pressure ductwork, add						15%			
	1200	For high pressure ductwork, add						40%			
	1276	Duct tape, 2" x 60 yd roll				Ea.	5.20			5.20	5.70
	1280	Add to labor for elevated installation									
	1282	of prefabricated (purchased) ductwork									
	1283	10' to 15' high						10%			
	1284	15' to 20' high						20%			
	1285	20' to 25' high						25%			
	1286	25' to 30' high						35%			
	1287	30' to 35' high						40%			
	1288	35' to 40' high						50%			
	1289	Over 40' high						55%			
	1300	Flexible, coated fiberglass fabric on corr. resist. metal helix									
	1400	pressure to 12" (WG) UL-181									
	1500	Non-insulated, 3" diameter	Q-9	400	.040	L.F.	.95	1.21		2.16	2.92
	1520	4" diameter		360	.044		.95	1.34		2.29	3.12
	1540	5" diameter		320	.050		1.11	1.51		2.62	3.55
	1560	6" diameter		280	.057		1.25	1.73		2.98	4.04
	1580	7" diameter		240	.067		1.64	2.01		3.65	4.91

15800 | Air Distribution

15810 | Ducts

			CREW	DAILY OUTPUT	LABOR-HOURS	UNIT	2001 BARE COSTS MAT.	LABOR	EQUIP.	TOTAL	TOTAL INCL O&P	
600	1600	8" diameter	Q-9	200	.080	L.F.	1.69	2.42		4.11	5.60	600
	1620	9" diameter		180	.089		1.91	2.68		4.59	6.25	
	1640	10" diameter		160	.100		2.17	3.02		5.19	7.05	
	1660	12" diameter		120	.133		2.57	4.03		6.60	9.05	
	1680	14" diameter		80	.200		3.02	6.05		9.07	12.65	
	1700	16" diameter		60	.267		3.93	8.05		11.98	16.75	
	1800	For adjustable clamps, add				Ea.	.74			.74	.81	
	1900	Insulated, 1" thick with 3/4 lb., PE jacket, 3" diameter	Q-9	380	.042	L.F.	1.64	1.27		2.91	3.76	
	1910	4" diameter		340	.047		1.92	1.42		3.34	4.30	
	1920	5" diameter		300	.053		1.92	1.61		3.53	4.60	
	1940	6" diameter		260	.062		2.15	1.86		4.01	5.25	
	1960	7" diameter		220	.073		2.57	2.20		4.77	6.20	
	1980	8" diameter		180	.089		2.63	2.68		5.31	7.05	
	2000	9" diameter		160	.100		3.16	3.02		6.18	8.15	
	2020	10" diameter		140	.114		3.42	3.45		6.87	9.10	
	2040	12" diameter		100	.160		3.98	4.83		8.81	11.85	
	2060	14" diameter		80	.200		4.82	6.05		10.87	14.65	
	2080	16" diameter		60	.267		5.40	8.05		13.45	18.40	
	2100	18" diameter		45	.356		7.25	10.75		18	24.50	
	2120	20" diameter	Q-10	65	.369		7.60	11.55		19.15	26	
	2500	Insulated, heavy duty, coated fiberglass fabric										
	2520	4" diameter	Q-9	340	.047	L.F.	2.39	1.42		3.81	4.82	
	2540	5" diameter		300	.053		2.85	1.61		4.46	5.65	
	2560	6" diameter		260	.062		3.20	1.86		5.06	6.40	
	2580	7" diameter		220	.073		3.77	2.20		5.97	7.55	
	2600	8" diameter		180	.089		4.07	2.68		6.75	8.60	
	2620	9" diameter		160	.100		4.71	3.02		7.73	9.85	
	2640	10" diameter		140	.114		5.10	3.45		8.55	10.95	
	2660	12" diameter		100	.160		6.05	4.83		10.88	14.10	
	2680	14" diameter		80	.200		7.20	6.05		13.25	17.30	
	2700	16" diameter		60	.267		8	8.05		16.05	21.50	
	2720	18" diameter		45	.356		10	10.75		20.75	27.50	
	2800	Flexible, alum., pressure to 12" (WG) UL-181										
	2820	Non-insulated										
	2830	3" diameter	Q-9	400	.040	L.F.	.94	1.21		2.15	2.90	
	2831	4" diameter		360	.044		1.14	1.34		2.48	3.32	
	2832	5" diameter		320	.050		1.44	1.51		2.95	3.91	
	2833	6" diameter		280	.057		1.77	1.73		3.50	4.61	
	2834	7" diameter		240	.067		2.56	2.01		4.57	5.95	
	2835	8" diameter		200	.080		2.61	2.42		5.03	6.60	
	2836	9" diameter		180	.089		3.03	2.68		5.71	7.45	
	2837	10" diameter		160	.100		3.19	3.02		6.21	8.15	
	2838	12" diameter		120	.133		3.93	4.03		7.96	10.50	
	2839	14" diameter		80	.200		4.69	6.05		10.74	14.50	
	2840	15" diameter		70	.229		5.15	6.90		12.05	16.35	
	2841	16" diameter		60	.267		6.30	8.05		14.35	19.40	
	2842	18" diameter		45	.356		6.85	10.75		17.60	24	
	2843	20" diameter	Q-10	65	.369		7.30	11.55		18.85	26	
	2880	Insulated, 1" thick with 3/4 lb., PE jacket										
	2890	3" diameter	Q-9	380	.042	L.F.	1.73	1.27		3	3.86	
	2891	4" diameter		340	.047		1.95	1.42		3.37	4.34	
	2892	5" diameter		300	.053		2.42	1.61		4.03	5.15	
	2893	6" diameter		260	.062		2.81	1.86		4.67	5.95	
	2894	7" diameter		220	.073		3.19	2.20		5.39	6.90	
	2895	8" diameter		180	.089		3.74	2.68		6.42	8.25	
	2896	9" diameter		160	.100		4.23	3.02		7.25	9.30	

Reference numbers: R15810-050, R15810-070, R15810-100

15800 | Air Distribution

15810 | Ducts

			CREW	DAILY OUTPUT	LABOR-HOURS	UNIT	MAT.	LABOR	EQUIP.	TOTAL	TOTAL INCL O&P
2897	10" diameter	R15810-050	Q-9	140	.114	L.F.	4.78	3.45		8.23	10.60
2898	12" diameter			100	.160		5.35	4.83		10.18	13.30
2899	14" diameter	R15810-070		80	.200		6.25	6.05		12.30	16.20
2900	15" diameter			70	.229		6.90	6.90		13.80	18.25
2901	16" diameter	R15810-100		60	.267		8.15	8.05		16.20	21.50
2902	18" diameter			45	.356		8.85	10.75		19.60	26.50
3000	Transitions, 3" deep, square to round										
3010	6" x 6" to round		Q-9	44	.364	Ea.	6.50	11		17.50	24
3014	6" x 12" to round			42	.381		12.35	11.50		23.85	31.50
3018	8" x 8" to round			40	.400		7.15	12.10		19.25	26.50
3022	9" x 9" to round			38	.421		8.45	12.70		21.15	29
3026	9" x 12" to round			37	.432		13	13.05		26.05	34.50
3030	9" x 15" to round			36	.444		16.25	13.40		29.65	38.50
3034	10" x 10" to round			34	.471		9.75	14.20		23.95	33
3038	10" x 22" to round			32	.500		18.85	15.10		33.95	44
3042	12" x 12" to round			30	.533		10.40	16.10		26.50	36.50
3046	12" x 18" to round			28	.571		19.50	17.25		36.75	48
3050	12" x 24" to round			26	.615		22	18.60		40.60	53
3054	15" x 15" to round			25	.640		13.65	19.35		33	45
3058	15" x 18" to round			24	.667		19.50	20		39.50	52.50
3062	18" x 18" to round			22	.727		16.25	22		38.25	52
3066	21" x 21" to round			21	.762		26.50	23		49.50	65
3070	22" x 22" to round			20	.800		26.50	24		50.50	67
3074	24" x 24" to round			18	.889		26.50	27		53.50	71
3490	Rigid fiberglass duct board, foil reinf. kraft facing										
3500	Rectangular, 1" thick, alum. faced, (FRK), std. weight		Q-10	350	.069	SF Surf	.83	2.15		2.98	4.23
3550	Rigid fiberglass reinforced plastic, FM approved										
3552	for acid fume and smoke exhaust system, nonflammable										
3554	Straight, 4" diameter		Q-9	200	.080	L.F.	10.75	2.42		13.17	15.60
3555	6" diameter			146	.110		14.10	3.31		17.41	20.50
3556	8" diameter			106	.151		17.10	4.56		21.66	26
3557	10" diameter			85	.188		20.50	5.70		26.20	31.50
3558	12" diameter			68	.235		24	7.10		31.10	37
3559	14" diameter			50	.320		35.50	9.65		45.15	54
3560	16" diameter			40	.400		39.50	12.10		51.60	61.50
3561	18" diameter			31	.516		44	15.60		59.60	72.50
3562	20" diameter		Q-10	44	.545		48.50	17.10		65.60	80
3563	22" diameter			40	.600		52.50	18.80		71.30	86.50
3564	24" diameter			37	.649		56.50	20.50		77	94
3565	26" diameter			34	.706		61	22		83	101
3566	28" diameter			28	.857		65.50	27		92.50	114
3567	30" diameter			24	1		69.50	31.50		101	125
3568	32" diameter			22.60	1.062		74	33.50		107.50	133
3569	34" diameter			21.60	1.111		78	35		113	140
3570	36" diameter			20.60	1.165		82.50	36.50		119	147
3571	38" diameter			19.80	1.212		127	38		165	199
3572	42" diameter			18.30	1.311		142	41		183	220
3573	46" diameter			17	1.412		157	44		201	241
3574	50" diameter			15.90	1.509		171	47.50		218.50	261
3575	54" diameter			14.90	1.611		188	50.50		238.50	285
3576	60" diameter			13.60	1.765		215	55.50		270.50	320
3577	64" diameter			12.80	1.875		315	58.50		373.50	435
3578	68" diameter			12.20	1.967		335	61.50		396.50	465
3579	72" diameter			11.90	2.017		360	63		423	495
3584	Note: joints are cemented with										
3586	fiberglass resin, included in material cost.										

15800 | Air Distribution

15810 | Ducts

			CREW	DAILY OUTPUT	LABOR-HOURS	UNIT	MAT.	LABOR	EQUIP.	TOTAL	TOTAL INCL O&P
3590	Elbow, 90°, 4" diameter	R15810-050	Q-9	20.30	.788	Ea.	52	24		76	94
3591	6" diameter			13.80	1.159		72.50	35		107.50	134
3592	8" diameter	R15810-070		10	1.600		80	48.50		128.50	163
3593	10" diameter			7.80	2.051		105	62		167	212
3594	12" diameter	R15810-100		6.50	2.462		133	74.50		207.50	261
3595	14" diameter			5.50	2.909		163	88		251	315
3596	16" diameter			4.80	3.333		194	101		295	370
3597	18" diameter		▼	4.40	3.636		278	110		388	475
3598	20" diameter		Q-10	5.90	4.068		293	127		420	520
3599	22" diameter			5.40	4.444		360	139		499	615
3600	24" diameter			4.90	4.898		415	153		568	690
3601	26" diameter			4.60	5.217		420	163		583	710
3602	28" diameter			4.20	5.714		525	179		704	855
3603	30" diameter			3.90	6.154		585	193		778	945
3604	32" diameter			3.70	6.486		650	203		853	1,025
3605	34" diameter			3.45	6.957		765	218		983	1,175
3606	36" diameter			3.20	7.500		880	235		1,115	1,325
3607	38" diameter			3.07	7.818		1,250	245		1,495	1,750
3608	42" diameter			2.70	8.889		1,500	278		1,778	2,075
3609	46" diameter			2.54	9.449		1,750	296		2,046	2,375
3610	50" diameter			2.33	10.300		2,025	325		2,350	2,725
3611	54" diameter			2.14	11.215		2,325	350		2,675	3,100
3612	60" diameter			1.94	12.371		3,100	385		3,485	4,000
3613	64" diameter			1.82	13.187		4,500	415		4,915	5,600
3614	68" diameter			1.71	14.035		5,000	440		5,440	6,175
3615	72" diameter		▼	1.62	14.815		5,525	465		5,990	6,800
3626	Elbow, 45°, 4" diameter		Q-9	22.30	.717		50	21.50		71.50	88.50
3627	6" diameter			15.20	1.053		69	32		101	125
3628	8" diameter			11	1.455		73.50	44		117.50	149
3629	10" diameter			8.60	1.860		77.50	56		133.50	173
3630	12" diameter			7.20	2.222		85	67		152	198
3631	14" diameter			6.10	2.623		104	79		183	236
3632	16" diameter			5.28	3.030		124	91.50		215.50	277
3633	18" diameter		▼	4.84	3.306		145	100		245	315
3634	20" diameter		Q-10	6.50	3.692		168	116		284	365
3635	22" diameter			5.94	4.040		229	127		356	445
3636	24" diameter			5.40	4.444		261	139		400	505
3637	26" diameter			5.06	4.743		296	149		445	555
3638	28" diameter			4.62	5.195		320	163		483	605
3639	30" diameter			4.30	5.581		370	175		545	675
3640	32" diameter			4.07	5.897		410	185		595	735
3641	34" diameter			3.80	6.316		485	198		683	835
3642	36" diameter			3.52	6.818		560	214		774	945
3643	38" diameter			3.38	7.101		785	222		1,007	1,200
3644	42" diameter			2.97	8.081		940	253		1,193	1,425
3645	46" diameter			2.80	8.571		1,100	268		1,368	1,625
3646	50" diameter			2.56	9.375		1,275	294		1,569	1,850
3647	54" diameter			2.35	10.213		1,450	320		1,770	2,100
3648	60" diameter			2.13	11.268		1,950	355		2,305	2,700
3649	64" diameter			2	12		2,825	375		3,200	3,675
3650	68" diameter			1.88	12.766		3,125	400		3,525	4,075
3651	72" diameter		▼	1.78	13.483		3,450	420		3,870	4,450
3660	Tee, 90°, 4" diameter		Q-9	14.06	1.138		28.50	34.50		63	84.50
3661	6" diameter			9.52	1.681		43.50	51		94.50	127
3662	8" diameter			6.98	2.292		59.50	69		128.50	173
3663	10" diameter		▼	5.45	2.936	▼	77.50	88.50		166	222

Important: See the Reference Section for critical supporting data - Reference Nos., Crews, & City Cost Indexes

15800 | Air Distribution

15810 | Ducts

			CREW	DAILY OUTPUT	LABOR-HOURS	UNIT	MAT.	LABOR	EQUIP.	TOTAL	TOTAL INCL O&P
600	3664	12" diameter	Q-9	4.60	3.478	Ea.	97	105		202	269
	3665	14" diameter R15810-050		3.86	4.145		141	125		266	350
	3666	16" diameter		3.40	4.706		168	142		310	405
	3667	18" diameter R15810-070		3.05	5.246		199	158		357	465
	3668	20" diameter	Q-10	4.12	5.825		233	182		415	540
	3669	22" diameter R15810-100		3.78	6.349		266	199		465	595
	3670	24" diameter		3.46	6.936		300	217		517	670
	3671	26" diameter		3.20	7.500		340	235		575	740
	3672	28" diameter		2.96	8.108		380	254		634	810
	3673	30" diameter		2.75	8.727		425	273		698	885
	3674	32" diameter		2.54	9.449		470	296		766	970
	3675	34" diameter		2.40	10		515	315		830	1,050
	3676	36" diameter		2.27	10.573		565	330		895	1,125
	3677	38" diameter		2.15	11.163		850	350		1,200	1,475
	3678	42" diameter		1.91	12.565		1,025	395		1,420	1,725
	3679	46" diameter		1.78	13.483		1,200	420		1,620	1,975
	3680	50" diameter		1.63	14.724		1,400	460		1,860	2,225
	3681	54" diameter		1.50	16		1,625	500		2,125	2,575
	3682	60" diameter		1.35	17.778		2,025	555		2,580	3,075
	3683	64" diameter		1.27	18.898		3,050	590		3,640	4,275
	3684	68" diameter		1.20	20		3,400	625		4,025	4,725
	3685	72" diameter		1.13	21.239		3,800	665		4,465	5,200
	3690	For Y @ 45°, add					22%				
	3700	Blast gate, 4" diameter	Q-9	22.30	.717		138	21.50		159.50	186
	3701	6" diameter		15.20	1.053		156	32		188	221
	3702	8" diameter		11	1.455		210	44		254	299
	3703	10" diameter		8.60	1.860		240	56		296	350
	3704	12" diameter		7.20	2.222		293	67		360	425
	3705	14" diameter		6.10	2.623		325	79		404	475
	3706	16" diameter		5.28	3.030		360	91.50		451.50	535
	3707	18" diameter		4.84	3.306		395	100		495	590
	3708	20" diameter	Q-10	6.50	3.692		425	116		541	650
	3709	22" diameter		5.94	4.040		460	127		587	705
	3710	24" diameter		5.40	4.444		495	139		634	755
	3711	26" diameter		5.06	4.743		530	149		679	815
	3712	28" diameter		4.62	5.195		570	163		733	875
	3713	30" diameter		4.30	5.581		605	175		780	940
	3714	32" diameter		4.07	5.897		660	185		845	1,025
	3715	34" diameter		3.80	6.316		720	198		918	1,100
	3716	36" diameter		3.52	6.818		775	214		989	1,175
	3717	38" diameter		3.38	7.101		1,050	222		1,272	1,500
	3718	42" diameter		2.97	8.081		1,200	253		1,453	1,725
	3719	46" diameter		2.80	8.571		1,375	268		1,643	1,950
	3720	50" diameter		2.56	9.375		1,575	294		1,869	2,175
	3721	54" diameter		2.35	10.213		1,775	320		2,095	2,450
	3722	60" diameter		2.13	11.268		2,075	355		2,430	2,850
	3723	64" diameter		2	12		2,300	375		2,675	3,125
	3724	68" diameter		1.88	12.766		2,550	400		2,950	3,425
	3725	72" diameter		1.78	13.483		2,800	420		3,220	3,725
	3740	Reducers, 4" diameter	Q-9	24.50	.653		34.50	19.70		54.20	68.50
	3741	6" diameter		16.70	.958		51.50	29		80.50	101
	3742	8" diameter		12	1.333		69.50	40.50		110	139
	3743	10" diameter		9.50	1.684		86	51		137	173
	3744	12" diameter		7.90	2.025		103	61		164	208
	3745	14" diameter		6.70	2.388		120	72		192	243
	3746	16" diameter		5.80	2.759		137	83.50		220.50	280

15800 | Air Distribution

15810 | Ducts

			CREW	DAILY OUTPUT	LABOR-HOURS	UNIT	MAT.	LABOR	EQUIP.	TOTAL	TOTAL INCL O&P	
600	3747	18" diameter	Q-9	5.30	3.019	Ea.	155	91		246	310	600
	3748	20" diameter	Q-10	7.20	3.333		172	104		276	350	
	3749	22" diameter		6.50	3.692		277	116		393	485	
	3750	24" diameter		5.90	4.068		300	127		427	525	
	3751	26" diameter		5.60	4.286		325	134		459	565	
	3752	28" diameter		5.10	4.706		355	147		502	620	
	3753	30" diameter		4.70	5.106		380	160		540	660	
	3754	32" diameter		4.50	5.333		405	167		572	705	
	3755	34" diameter		4.20	5.714		430	179		609	745	
	3756	36" diameter		3.90	6.154		455	193		648	800	
	3757	38" diameter		3.70	6.486		690	203		893	1,075	
	3758	42" diameter		3.30	7.273		760	228		988	1,175	
	3759	46" diameter		3.10	7.742		835	242		1,077	1,300	
	3760	50" diameter		2.80	8.571		905	268		1,173	1,400	
	3761	54" diameter		2.60	9.231		980	289		1,269	1,525	
	3762	60" diameter		2.34	10.256		1,100	320		1,420	1,700	
	3763	64" diameter		2.20	10.909		1,700	340		2,040	2,400	
	3764	68" diameter		2.06	11.650		1,800	365		2,165	2,575	
	3765	72" diameter		1.96	12.245		1,925	385		2,310	2,700	
	3780	Flange, per each, 4" diameter	Q-9	40.60	.394		83.50	11.90		95.40	110	
	3781	6" diameter		27.70	.578		97.50	17.45		114.95	134	
	3782	8" diameter		20.50	.780		122	23.50		145.50	171	
	3783	10" diameter		16.30	.982		146	29.50		175.50	206	
	3784	12" diameter		13.90	1.151		168	35		203	239	
	3785	14" diameter		11.30	1.416		194	43		237	279	
	3786	16" diameter		9.70	1.649		218	50		268	315	
	3787	18" diameter		8.30	1.928		242	58		300	355	
	3788	20" diameter	Q-10	11.30	2.124		266	66.50		332.50	395	
	3789	22" diameter		10.40	2.308		290	72.50		362.50	430	
	3790	24" diameter		9.50	2.526		315	79		394	465	
	3791	26" diameter		8.80	2.727		340	85.50		425.50	500	
	3792	28" diameter		8.20	2.927		365	91.50		456.50	540	
	3793	30" diameter		7.60	3.158		385	99		484	580	
	3794	32" diameter		7.20	3.333		410	104		514	610	
	3795	34" diameter		6.80	3.529		435	111		546	650	
	3796	36" diameter		6.40	3.750		460	117		577	685	
	3797	38" diameter		6.10	3.934		485	123		608	720	
	3798	42" diameter		5.50	4.364		530	137		667	795	
	3799	46" diameter		5	4.800		580	150		730	865	
	3800	50" diameter		4.60	5.217		630	163		793	940	
	3801	54" diameter		4.20	5.714		675	179		854	1,025	
	3802	60" diameter		3.80	6.316		750	198		948	1,125	
	3803	64" diameter		3.60	6.667		795	209		1,004	1,200	
	3804	68" diameter		3.40	7.059		845	221		1,066	1,275	
	3805	72" diameter		3.20	7.500		895	235		1,130	1,350	
	4000	Rigid plastic, corrosive fume resistant PVC										
	4020	Straight, 6" diameter	Q-9	220	.073	L.F.	6.30	2.20		8.50	10.30	
	4030	7" diameter		160	.100		9.35	3.02		12.37	14.90	
	4040	8" diameter		160	.100		9.35	3.02		12.37	14.90	
	4050	9" diameter		140	.114		10.55	3.45		14	16.95	
	4060	10" diameter		120	.133		11.75	4.03		15.78	19.15	
	4070	12" diameter		100	.160		13.75	4.83		18.58	22.50	
	4080	14" diameter		70	.229		17.80	6.90		24.70	30.50	
	4090	16" diameter		62	.258		20.50	7.80		28.30	34.50	
	4100	18" diameter		58	.276		32	8.35		40.35	48	
	4110	20" diameter	Q-10	75	.320		41.50	10		51.50	61	

15800 | Air Distribution

15810 | Ducts

			CREW	DAILY OUTPUT	LABOR-HOURS	UNIT	MAT.	LABOR	EQUIP.	TOTAL	TOTAL INCL O&P
4120	22" diameter	R15810-050	Q-10	60	.400	L.F.	47	12.55		59.55	71
4130	24" diameter			55	.436		53	13.65		66.65	79
4140	26" diameter	R15810-070		42	.571		60	17.90		77.90	93.50
4150	28" diameter			35	.686		64.50	21.50		86	104
4160	30" diameter	R15810-100		28	.857		69.50	27		96.50	118
4170	36" diameter			23	1.043		83.50	32.50		116	142
4180	42" diameter			20	1.200		97	37.50		134.50	165
4200	48" diameter			18	1.333		111	42		153	187
4250	Coupling, 6" diameter		Q-9	88	.182	Ea.	6.55	5.50		12.05	15.75
4260	7" diameter			82	.195		7	5.90		12.90	16.80
4270	8" diameter			78	.205		7.35	6.20		13.55	17.60
4280	9" diameter			74	.216		8	6.55		14.55	18.90
4290	10" diameter			70	.229		8	6.90		14.90	19.45
4300	12" diameter			55	.291		8.55	8.80		17.35	23
4310	14" diameter			44	.364		10.40	11		21.40	28.50
4320	16" diameter			40	.400		12.35	12.10		24.45	32
4330	18" diameter			38	.421		14.30	12.70		27	35.50
4340	20" diameter		Q-10	55	.436		25	13.65		38.65	48.50
4350	22" diameter			50	.480		28.50	15.05		43.55	54.50
4360	24" diameter			45	.533		32.50	16.70		49.20	61.50
4370	26" diameter			39	.615		34	19.25		53.25	67.50
4380	28" diameter			33	.727		35.50	23		58.50	74
4390	30" diameter			27	.889		39	28		67	86
4400	36" diameter			25	.960		61.50	30		91.50	114
4410	42" diameter			23	1.043		69	32.50		101.50	126
4420	48" diameter			21	1.143		80.50	36		116.50	144
4470	Elbow, 90°, 6" diameter		Q-9	44	.364		61.50	11		72.50	84.50
4480	7" diameter			36	.444		66	13.40		79.40	93
4490	8" diameter			28	.571		66	17.25		83.25	99
4500	9" diameter			22	.727		90.50	22		112.50	134
4510	10" diameter			18	.889		107	27		134	159
4520	12" diameter			15	1.067		146	32		178	210
4530	14" diameter			11	1.455		199	44		243	287
4540	16" diameter			10	1.600		240	48.50		288.50	340
4550	18" diameter		Q-10	15	1.600		261	50		311	365
4560	20" diameter			14	1.714		276	53.50		329.50	390
4570	22" diameter			13	1.846		350	58		408	475
4580	24" diameter			12	2		395	62.50		457.50	530
4590	26" diameter			11	2.182		455	68.50		523.50	605
4600	28" diameter			10	2.400		535	75		610	705
4610	30" diameter			9	2.667		630	83.50		713.50	825
4620	36" diameter			7	3.429		870	107		977	1,125
4630	42" diameter			5	4.800		1,200	150		1,350	1,550
4640	48" diameter			4	6		1,575	188		1,763	2,025
4750	Elbow 45°, use 90° and deduct						45%				
5400	Spiral preformed, steel, galv., straight lengths, Max 10" spwg.										
5410	4" diameter, 26 ga.		Q-9	360	.044	L.F.	.85	1.34		2.19	3.01
5416	5" diameter, 26 ga.			320	.050		1.08	1.51		2.59	3.52
5420	6" diameter, 26 ga.			280	.057		1.23	1.73		2.96	4.01
5425	7" diameter, 26 ga.			240	.067		1.46	2.01		3.47	4.72
5430	8" diameter, 26 ga.			200	.080		1.62	2.42		4.04	5.50
5440	10" diameter, 26 ga.			160	.100		2	3.02		5.02	6.85
5450	12" diameter, 26 ga.			120	.133		2.46	4.03		6.49	8.90
5460	14" diameter, 26 ga.			80	.200		2.85	6.05		8.90	12.50
5480	16" diameter, 24 ga.			60	.267		4.24	8.05		12.29	17.10
5490	18" diameter, 24 ga.			50	.320		4.70	9.65		14.35	20

15800 | Air Distribution

15810 | Ducts

			CREW	DAILY OUTPUT	LABOR-HOURS	UNIT	MAT.	LABOR	EQUIP.	TOTAL	TOTAL INCL O&P	
600	5500	20" diameter, 24 ga.	Q-10	65	.369	L.F.	5.25	11.55		16.80	23.50	600
	5510	22" diameter, 24 ga.		60	.400		5.80	12.55		18.35	25.50	
	5520	24" diameter, 24 ga.		55	.436		6.30	13.65		19.95	28	
	5540	30" diameter, 22 ga.		45	.533		9.55	16.70		26.25	36.50	
	5600	36" diameter, 22 ga.		40	.600		11.45	18.80		30.25	41.50	
	5800	Connector, 4" diameter	Q-9	100	.160	Ea.	1.58	4.83		6.41	9.20	
	5810	5" diameter		94	.170		1.69	5.15		6.84	9.80	
	5820	6" diameter		88	.182		1.82	5.50		7.32	10.50	
	5840	8" diameter		78	.205		2.35	6.20		8.55	12.15	
	5860	10" diameter		70	.229		3.71	6.90		10.61	14.75	
	5880	12" diameter		50	.320		4.47	9.65		14.12	19.80	
	5900	14" diameter		44	.364		5.20	11		16.20	22.50	
	5920	16" diameter		40	.400		5.80	12.10		17.90	25	
	5930	18" diameter		37	.432		6.50	13.05		19.55	27	
	5940	20" diameter		34	.471		7.05	14.20		21.25	30	
	5950	22" diameter		31	.516		7.70	15.60		23.30	32.50	
	5960	24" diameter		28	.571		8.25	17.25		25.50	35.50	
	5980	30" diameter		22	.727		10.75	22		32.75	46	
	6000	36" diameter		18	.889		16.10	27		43.10	59.50	
	6300	Elbow, 45°, 4" diameter		60	.267		1.79	8.05		9.84	14.40	
	6310	5" diameter		52	.308		2.30	9.30		11.60	16.90	
	6320	6" diameter		44	.364		2.51	11		13.51	19.70	
	6340	8" diameter		28	.571		3.31	17.25		20.56	30	
	6360	10" diameter		18	.889		5.05	27		32.05	47	
	6380	12" diameter		13	1.231		6.50	37		43.50	64.50	
	6400	14" diameter		11	1.455		12.20	44		56.20	81.50	
	6420	16" diameter		10	1.600		20	48.50		68.50	96.50	
	6430	18" diameter		9.60	1.667		24.50	50.50		75	105	
	6440	20" diameter	Q-10	14	1.714		31	53.50		84.50	117	
	6450	22" diameter		13	1.846		31	58		89	124	
	6460	24" diameter		12	2		41.50	62.50		104	143	
	6480	30" diameter		9	2.667		64	83.50		147.50	200	
	6500	36" diameter		7	3.429		78	107		185	252	
	6600	Elbow, 90°, 4" diameter	Q-9	60	.267		3.05	8.05		11.10	15.80	
	6610	5" diameter		52	.308		3.51	9.30		12.81	18.20	
	6620	6" diameter		44	.364		4.02	11		15.02	21.50	
	6625	7" diameter		36	.444		4.25	13.40		17.65	25	
	6630	8" diameter		28	.571		4.93	17.25		22.18	32	
	6640	10" diameter		18	.889		7.05	27		34.05	49.50	
	6650	12" diameter		13	1.231		9.30	37		46.30	68	
	6660	14" diameter		11	1.455		17.45	44		61.45	87	
	6670	16" diameter		10	1.600		27.50	48.50		76	105	
	6676	18" diameter		9.60	1.667		36	50.50		86.50	117	
	6680	20" diameter	Q-10	14	1.714		48	53.50		101.50	136	
	6684	22" diameter		13	1.846		74.50	58		132.50	172	
	6690	24" diameter		12	2		83.50	62.50		146	188	
	6700	30" diameter		9	2.667		128	83.50		211.50	270	
	6710	36" diameter		7	3.429		156	107		263	335	
	6720	End cap										
	6722	4"	Q-9	110	.145	Ea.	3.49	4.39		7.88	10.65	
	6724	6"		94	.170		4.85	5.15		10	13.30	
	6725	8"		88	.182		6.40	5.50		11.90	15.55	
	6726	10"		78	.205		8.75	6.20		14.95	19.15	
	6727	12"		70	.229		10.45	6.90		17.35	22	
	6728	14"		50	.320		13.95	9.65		23.60	30.50	
	6729	16"		44	.364		19.40	11		30.40	38.50	

Important: See the Reference Section for critical supporting data - Reference Nos., Crews, & City Cost Indexes

15800 | Air Distribution

15810 | Ducts

		CREW	DAILY OUTPUT	LABOR-HOURS	UNIT	MAT.	LABOR	EQUIP.	TOTAL	TOTAL INCL O&P
6730	18"	Q-9	40	.400	Ea.	23.50	12.10		35.60	44
6731	20"		37	.432		27	13.05		40.05	50
6732	22"		34	.471		31	14.20		45.20	56
6733	24"		30	.533		35	16.10		51.10	63.50
6800	Reducing coupling, 6" x 4"		46	.348		4.55	10.50		15.05	21
6820	8" x 6"		40	.400		5.60	12.10		17.70	25
6840	10" x 8"		32	.500		7.05	15.10		22.15	31.50
6860	12" x 10"		24	.667		10.10	20		30.10	42
6880	14" x 12"		20	.800		16.65	24		40.65	56
6900	16" x 14"		18	.889		24.50	27		51.50	68
6920	18" x 16"		16	1		30	30		60	79.50
6940	20" x 18"	Q-10	24	1		36.50	31.50		68	89
6950	22" x 20"		23	1.043		43	32.50		75.50	97.50
6960	24" x 22"		22	1.091		49	34		83	107
6980	30" x 28"		18	1.333		73	42		115	145
7000	36" x 34"		16	1.500		98	47		145	181
7100	Tee, 90°, 4" diameter	Q-9	40	.400		3.99	12.10		16.09	23
7110	5" diameter		34.60	.462		5.30	13.95		19.25	27.50
7120	6" diameter		30	.533		5.75	16.10		21.85	31.50
7130	8" diameter		19	.842		7.50	25.50		33	48
7140	10" diameter		12	1.333		10.70	40.50		51.20	74
7150	12" diameter		9	1.778		13.65	53.50		67.15	98
7160	14" diameter		7	2.286		26	69		95	136
7170	16" diameter		6.70	2.388		36.50	72		108.50	151
7176	18" diameter		6	2.667		50.50	80.50		131	180
7180	20" diameter	Q-10	8.70	2.759		56	86.50		142.50	195
7190	24" diameter		7	3.429		66.50	107		173.50	239
7200	30" diameter		6	4		82	125		207	283
7210	36" diameter		5	4.800		104	150		254	345
7400	Steel, PVC coated both sides, straight lengths									
7410	4" diameter, 26 ga.	Q-9	360	.044	L.F.	1.19	1.34		2.53	3.38
7416	5" diameter, 26 ga.		240	.067		1.48	2.01		3.49	4.74
7420	6" diameter, 26 ga.		280	.057		1.65	1.73		3.38	4.48
7440	8" diameter, 26 ga.		200	.080		2.28	2.42		4.70	6.25
7460	10" diameter, 26 ga.		160	.100		3.07	3.02		6.09	8.05
7480	12" diameter, 26 ga.		120	.133		3.80	4.03		7.83	10.40
7500	14" diameter, 26 ga.		80	.200		5.20	6.05		11.25	15.05
7520	16" diameter, 24 ga.		60	.267		7.35	8.05		15.40	20.50
7540	18" diameter, 24 ga.		45	.356		7.95	10.75		18.70	25.50
7560	20" diameter, 24 ga.	Q-10	65	.369		9.35	11.55		20.90	28
7580	24" diameter, 24 ga.		55	.436		11.90	13.65		25.55	34
7600	30" diameter, 22 ga.		45	.533		17.25	16.70		33.95	45
7620	36" diameter, 22 ga.		40	.600		21	18.80		39.80	52
7890	Connector, 4" diameter	Q-9	100	.160	Ea.	2.49	4.83		7.32	10.20
7896	5" diameter		83	.193		2.66	5.80		8.46	11.95
7900	6" diameter		88	.182		2.80	5.50		8.30	11.60
7920	8" diameter		78	.205		3.54	6.20		9.74	13.45
7940	10" diameter		70	.229		4.87	6.90		11.77	16
7960	12" diameter		50	.320		6.25	9.65		15.90	22
7980	14" diameter		44	.364		8	11		19	26
8000	16" diameter		40	.400		8.90	12.10		21	28.50
8020	18" diameter		37	.432		10	13.05		23.05	31
8040	20" diameter		34	.471		10.75	14.20		24.95	34
8060	24" diameter		28	.571		12.05	17.25		29.30	40
8080	30" diameter		22	.727		15.75	22		37.75	51.50
8100	36" diameter		18	.889		21.50	27		48.50	65

References: R15810-050, R15810-070, R15810-100

15800 | Air Distribution

15810 | Ducts

		CREW	DAILY OUTPUT	LABOR-HOURS	UNIT	MAT.	LABOR	EQUIP.	TOTAL	TOTAL INCL O&P
8390	Elbow, 45°, 4" diameter	Q-9	60	.267	Ea.	3.14	8.05		11.19	15.90
8396	5" diameter		36	.444		3.79	13.40		17.19	24.50
8400	6" diameter		44	.364		4.12	11		15.12	21.50
8420	8" diameter		28	.571		5.45	17.25		22.70	32.50
8440	10" diameter		18	.889		7.90	27		34.90	50
8460	12" diameter		13	1.231		11.65	37		48.65	70.50
8480	14" diameter		11	1.455		21	44		65	91
8500	16" diameter		10	1.600		30	48.50		78.50	108
8520	18" diameter		9	1.778		34.50	53.50		88	121
8540	20" diameter	Q-10	13	1.846		46.50	58		104.50	141
8560	24" diameter		11	2.182		59	68.50		127.50	170
8580	30" diameter		9	2.667		90.50	83.50		174	229
8600	36" diameter		7	3.429		110	107		217	287
8800	Elbow, 90°, 4" diameter	Q-9	60	.267		5.20	8.05		13.25	18.20
8804	5" diameter		52	.308		5.95	9.30		15.25	21
8806	6" diameter		44	.364		6.50	11		17.50	24
8808	8" diameter		28	.571		8.10	17.25		25.35	35.50
8810	10" diameter		18	.889		10.95	27		37.95	53.50
8812	12" diameter		13	1.231		15.10	37		52.10	74
8814	14" diameter		11	1.455		27	44		71	97.50
8816	16" diameter		10	1.600		42	48.50		90.50	121
8818	18" diameter		9.60	1.667		51.50	50.50		102	134
8820	20" diameter	Q-10	14	1.714		65.50	53.50		119	155
8822	24" diameter		12	2		118	62.50		180.50	226
8824	30" diameter		9	2.667		181	83.50		264.50	330
8826	36" diameter		7	3.429		220	107		327	410
8850	Tee, 90°, 4" diameter	Q-9	40	.400		5.85	12.10		17.95	25
8854	5" diameter		34.60	.462		7.70	13.95		21.65	30
8856	6" diameter		30	.533		8.55	16.10		24.65	34.50
8858	8" diameter		19	.842		10.95	25.50		36.45	51.50
8860	10" diameter		12	1.333		14.30	40.50		54.80	78
8862	12" diameter		9	1.778		20.50	53.50		74	106
8864	14" diameter		7	2.286		35.50	69		104.50	146
8866	16" diameter		6.70	2.388		51.50	72		123.50	168
8868	18" diameter		6	2.667		74.50	80.50		155	206
8870	20" diameter	Q-10	8.70	2.759		80.50	86.50		167	222
8872	24" diameter		7	3.429		111	107		218	288
8874	30" diameter		6	4		138	125		263	345
8876	36" diameter		5	4.800		175	150		325	425
9000	Reducing coupling, tapered offset									
9020	Diameter x any smaller diameter									
9040	6" x	Q-9	46	.348	Ea.	6.50	10.50		17	23.50
9060	8" x		40	.400		8	12.10		20.10	27.50
9080	10" x		32	.500		9.65	15.10		24.75	34
9100	12" x		24	.667		13.45	20		33.45	46
9120	14" x		20	.800		21.50	24		45.50	61.50
9140	16" x		18	.889		33	27		60	77.50
9160	18" x		16	1		40	30		70	90.50
9180	20" x	Q-10	24	1		49	31.50		80.50	103
9200	24" x		22	1.091		66.50	34		100.50	126
9220	30" x		18	1.333		94	42		136	168
9240	36" x		16	1.500		122	47		169	207
9400	Lateral 45°, straight barrel									
9410	4" size	Q-9	40	.400	Ea.	6.75	12.10		18.85	26
9416	5" size		34.60	.462		8.85	13.95		22.80	31.50
9420	6" size		30	.533		9.80	16.10		25.90	36

15800 | Air Distribution

15810 | Ducts

			Crew	Daily Output	Labor-Hours	Unit	Mat.	Labor	Equip.	Total	Total Incl O&P	
600	9440	8" size	Q-9	19	.842	Ea.	12.60	25.50		38.10	53.50	600
	9460	10" size	R15810-050	12	1.333		16.45	40.50		56.95	80	
	9480	12" size		9	1.778		20.50	53.50		74	106	
	9500	14" size	R15810-070	7	2.286		27	69		96	137	
	9520	16" size		6.70	2.388		51	72		123	167	
	9540	18" size	R15810-100	6	2.667		72	80.50		152.50	203	
	9560	20" size	Q-10	8.70	2.759		85.50	86.50		172	227	
	9580	24" size		7	3.429		119	107		226	297	
	9600	30" size		6	4		139	125		264	345	
	9620	36" size		5	4.800		181	150		331	430	
	9640	For reducing lateral, add					40%					
	9650	Lateral 45°, reducing										
	9654	6" x 6" x 4"	Q-9	30	.533	Ea.	14.55	16.10		30.65	41	
	9658	8" x 8" x 4"		19	.842		18.40	25.50		43.90	60	
	9662	10" x 10" x 4"		12	1.333		24	40.50		64.50	88.50	
	9666	12" x 12" x 4"		9	1.778		28	53.50		81.50	114	
	9670	14" x 14" x 4"		7	2.286		46.50	69		115.50	158	
	9674	16" x 16" x 4"		6.70	2.388		58	72		130	175	
	9678	18" x 18" x 4"		6	2.667		91	80.50		171.50	224	
	9682	20" x 20" x 4"	Q-10	8.70	2.759		116	86.50		202.50	261	
	9686	22" x 22" x 4"		8	3		132	94		226	290	
	9690	24" x 24" x 4"		7	3.429		142	107		249	320	
	9700	Register boot, elbow or perimeter										
	9710	6" duct diameter	Q-9	60	.267	Ea.	6.50	8.05		14.55	19.60	
	9720	8" duct diameter		56	.286		8.25	8.65		16.90	22.50	
	9730	10" duct diameter		45	.356		10.10	10.75		20.85	27.50	
	9740	12" duct diameter		36	.444		12.30	13.40		25.70	34	
	9800	Steel, stainless, straight lengths										
	9810	3" diameter, 24 Ga.	Q-9	400	.040	L.F.	4.28	1.21		5.49	6.60	
	9820	4" diameter, 24 Ga.		360	.044		5.40	1.34		6.74	8	
	9830	5" diameter, 24 Ga.		320	.050		6.65	1.51		8.16	9.65	
	9840	6" diameter, 24 Ga.		280	.057		7.80	1.73		9.53	11.25	
	9850	7" diameter, 24 Ga.		240	.067		8.95	2.01		10.96	12.95	
	9860	8" diameter, 24 Ga.		200	.080		10.15	2.42		12.57	14.90	
	9870	9" diameter, 24 Ga.		180	.089		11.30	2.68		13.98	16.60	
	9880	10" diameter, 24 Ga.		160	.100		12.45	3.02		15.47	18.35	
	9890	12" diameter, 24 Ga.		120	.133		14.85	4.03		18.88	22.50	
	9900	14" diameter, 24 Ga.		80	.200		17.15	6.05		23.20	28.50	
	9910	16" diameter, 24 Ga.		60	.267		19.50	8.05		27.55	34	
	9920	18" diameter, 24 Ga.		45	.356		22	10.75		32.75	40.50	
	9930	20" diameter, 24 Ga.	Q-10	65	.369		24	11.55		35.55	44.50	
	9940	24" diameter, 24 Ga.		55	.436		29	13.65		42.65	53	
	9950	30" diameter, 22 Ga.		45	.533		44	16.70		60.70	74.50	
	9960	36" diameter, 22 Ga.		40	.600		53.50	18.80		72.30	88	
	9970	40" diameter, 20 Ga.		35	.686		71	21.50		92.50	111	

15820 | Duct Accessories

			Crew	Daily Output	Labor-Hours	Unit	Mat.	Labor	Equip.	Total	Total Incl O&P	
300	0010	**DUCT ACCESSORIES**										300
	0050	Air extractors, 12" x 4"	1 Shee	24	.333	Ea.	14.75	11.20		25.95	33.50	
	0100	8" x 6"		22	.364		14.75	12.20		26.95	35	
	0120	12" x 6"		21	.381		20.50	12.80		33.30	42.50	
	0140	16" x 6"		20	.400		25.50	13.40		38.90	48.50	
	0160	24" x 6"		18	.444		42.50	14.90		57.40	70	
	0180	12" x 8"		20	.400		23	13.40		36.40	45.50	
	0200	20" x 8"		16	.500		33.50	16.80		50.30	62.50	

15800 | Air Distribution

15820 | Duct Accessories

			CREW	DAILY OUTPUT	LABOR-HOURS	UNIT	MAT.	LABOR	EQUIP.	TOTAL	TOTAL INCL O&P	
300	0240	18" x 10"	1 Shee	14	.571	Ea.	32.50	19.15		51.65	65	300
	0260	24" x 10"		12	.667		42	22.50		64.50	80.50	
	0280	24" x 12"		10	.800		45.50	27		72.50	91.50	
	0300	30" x 12"		8	1		56.50	33.50		90	114	
	0350	Duct collar, spin-in type										
	0352	Without damper										
	0354	Round										
	0356	4"	1 Shee	32	.250	Ea.	1.92	8.40		10.32	15.05	
	0357	5"		30	.267		1.92	8.95		10.87	15.90	
	0358	6"		28	.286		2.11	9.60		11.71	17.10	
	0359	7"		26	.308		2.28	10.30		12.58	18.45	
	0360	8"		24	.333		2.41	11.20		13.61	19.90	
	0361	9"		22	.364		2.73	12.20		14.93	22	
	0362	10"		20	.400		2.83	13.40		16.23	23.50	
	0363	12"		17	.471		3.28	15.80		19.08	28	
	1000	Duct access door, insulated, 6" x 6"		14	.571		13.65	19.15		32.80	44.50	
	1020	10" x 10"		11	.727		16.25	24.50		40.75	55.50	
	1040	12" x 12"		10	.800		17.55	27		44.55	61	
	1050	12" x 18"		9	.889		27.50	30		57.50	76	
	1060	16" x 12"		9	.889		24.50	30		54.50	73	
	1070	18" x 18"		8	1		31	33.50		64.50	86.50	
	1074	24" x 18"		8	1		37	33.50		70.50	93	
	1080	24" x 24"		8	1		39.50	33.50		73	95.50	
	1100	Exhaust vent damper, auto., OB with elect. actuator										
	1110	6" x 6"	1 Shee	24	.333	Ea.	67	11.20		78.20	91.50	
	1114	8" x 6"		23	.348		69	11.65		80.65	94	
	1116	8" x 8"		22	.364		72	12.20		84.20	98	
	1118	10" x 6"		23	.348		72	11.65		83.65	97	
	1120	10" x 10"		22	.364		76	12.20		88.20	102	
	1122	12" x 6"		22	.364		73	12.20		85.20	99.50	
	1124	12" x 10"		21	.381		80	12.80		92.80	108	
	1126	12" x 12"		20	.400		84	13.40		97.40	113	
	1128	14" x 6"		21	.381		74	12.80		86.80	101	
	1130	14" x 10"		19	.421		84	14.15		98.15	115	
	1132	16" x 6"		21	.381		74	12.80		86.80	101	
	1134	16" x 8"		20	.400		83	13.40		96.40	112	
	1136	18" x 6"		20	.400		79	13.40		92.40	108	
	1138	18" x 8"		19	.421		84	14.15		98.15	115	
	2000	Fabrics for flexible connections, with metal edge		100	.080	L.F.	1.55	2.68		4.23	5.85	
	2100	Without metal edge		160	.050	"	1.17	1.68		2.85	3.88	
	3000	Fire damper, curtain type, 1-1/2 hr rated, vertical, 6" x 6"		24	.333	Ea.	19.70	11.20		30.90	39	
	3020	8" x 6"		22	.364		19.70	12.20		31.90	40.50	
	3021	8" x 8"		22	.364		19.50	12.20		31.70	40.50	
	3030	10" x 10"		22	.364		19.50	12.20		31.70	40.50	
	3040	12" x 6"		22	.364		19.70	12.20		31.90	40.50	
	3044	12" x 12"		20	.400		20	13.40		33.40	42.50	
	3060	20" x 6"		18	.444		23.50	14.90		38.40	48.50	
	3080	12" x 8"		22	.364		19.50	12.20		31.70	40.50	
	3100	24" x 8"		16	.500		28.50	16.80		45.30	57.50	
	3120	12" x 10"		21	.381		20.50	12.80		33.30	42.50	
	3140	24" x 10"		15	.533		31.50	17.90		49.40	62	
	3160	36" x 10"		12	.667		44.50	22.50		67	83.50	
	3180	16" x 12"		20	.400		26	13.40		39.40	49	
	3200	24" x 12"		13	.615		31	20.50		51.50	66.50	
	3220	48" x 12"		10	.800		48	27		75	94.50	
	3238	14" x 14"		19	.421		25.50	14.15		39.65	50	

15800 | Air Distribution

15820 | Duct Accessories

		CREW	DAILY OUTPUT	LABOR-HOURS	UNIT	MAT.	LABOR	EQUIP.	TOTAL	TOTAL INCL O&P	
300	3240	16" x 14"	1 Shee	18	.444	Ea.	31.50	14.90		46.40	57.50	300
	3260	24" x 14"		12	.667		36.50	22.50		59	75	
	3280	30" x 14"		11	.727		45	24.50		69.50	87	
	3298	16" x 16"		18	.444		31	14.90		45.90	57.50	
	3300	18" x 16"		17	.471		34	15.80		49.80	62	
	3320	24" x 16"		11	.727		36.50	24.50		61	78	
	3340	36" x 16"		10	.800		49	27		76	95.50	
	3356	18" x 18"		16	.500		34	16.80		50.80	63	
	3360	24" x 18"		10	.800		40	27		67	85.50	
	3380	48" x 18"		8	1		64	33.50		97.50	123	
	3398	20" x 20"		10	.800		36.50	27		63.50	81.50	
	3400	24" x 20"		8	1		39.50	33.50		73	95.50	
	3420	36" x 20"		7	1.143		52	38.50		90.50	116	
	3440	24" x 22"		7	1.143		46	38.50		84.50	110	
	3460	30" x 22"		6	1.333		51.50	44.50		96	126	
	3478	24" x 24"		8	1		41.50	33.50		75	98	
	3480	26" x 24"		7	1.143		47	38.50		85.50	111	
	3484	32" x 24"	Q-9	13	1.231		51.50	37		88.50	114	
	3500	48" x 24"		12	1.333		67.50	40.50		108	137	
	3520	28" x 26"		13	1.231		53	37		90	116	
	3540	30" x 28"		12	1.333		58.50	40.50		99	126	
	3560	48" x 30"		11	1.455		81.50	44		125.50	158	
	3580	48" x 36"		10	1.600		88	48.50		136.50	171	
	3590	32" x 40"		10	1.600		69.50	48.50		118	151	
	3600	44" x 44"		10	1.600		91.50	48.50		140	176	
	3620	48" x 48"		8	2		120	60.50		180.50	226	
	3700	U.L. label included in above										
	3800	For horizontal operation, add				Ea.	20%					
	3900	For cap for blades out of air stream, add					20%					
	4000	For 10" 22 ga., U.L. approved sleeve, add					35%					
	4100	For 10" 22 ga., U.L. sleeve, 100% free area, add					65%					
	4200	For oversize openings group dampers										
	4210	Fire damper, round, Type A, 2 Hr										
	4214	4" diam.	1 Shee	20	.400	Ea.	47	13.40		60.40	72	
	4216	5" diam.		16	.500		43	16.80		59.80	73	
	4218	6" diam.		14.30	.559		43	18.75		61.75	76	
	4220	7" diam.		13.30	.601		43	20		63	78	
	4222	8" diam.		11.40	.702		43	23.50		66.50	83.50	
	4224	9" diam.		10.80	.741		47	25		72	90	
	4226	10" diam.		10	.800		49	27		76	95	
	4228	12" diam.		9.50	.842		54	28.50		82.50	103	
	4230	14" diam.		8.20	.976		67	32.50		99.50	124	
	4232	16" diam.		7.60	1.053		67	35.50		102.50	128	
	4234	18" diam.		7.20	1.111		80.50	37.50		118	146	
	4236	20" diam.		6.40	1.250		80.50	42		122.50	154	
	4238	22" diam.		6.10	1.311		96	44		140	174	
	4240	24" diam.		5.70	1.404		96	47		143	179	
	4242	26" diam.		5.50	1.455		114	49		163	201	
	4244	28" diam.		5.40	1.481		114	49.50		163.50	202	
	4246	30" diam.		5.30	1.509		114	50.50		164.50	203	
	4500	Fire/smoke combination damper, louver type, UL										
	4506	6" x 6"	1 Shee	24	.333	Ea.	52.50	11.20		63.70	75.50	
	4510	8" x 8"		22	.364		120	12.20		132.20	151	
	4520	16" x 8"		20	.400		131	13.40		144.40	165	
	4540	18" x 8"		18	.444		131	14.90		145.90	167	
	4560	20" x 8"		16	.500		141	16.80		157.80	181	

15800 | Air Distribution

15820 | Duct Accessories

		CREW	DAILY OUTPUT	LABOR-HOURS	UNIT	MAT.	LABOR	EQUIP.	TOTAL	TOTAL INCL O&P	
300	4580	10" x 10"	1 Shee	21	.381	Ea.	136	12.80		148.80	169	300
	4600	24" x 10"		15	.533		160	17.90		177.90	204	
	4620	30" x 10"		12	.667		165	22.50		187.50	217	
	4640	12" x 12"		20	.400		142	13.40		155.40	177	
	4660	18" x 12"		18	.444		144	14.90		158.90	182	
	4680	24" x 12"		13	.615		152	20.50		172.50	199	
	4700	30" x 12"		11	.727		158	24.50		182.50	212	
	4720	14" x 14"		17	.471		153	15.80		168.80	193	
	4740	16" x 14"		18	.444		157	14.90		171.90	196	
	4760	20" x 14"		14	.571		160	19.15		179.15	206	
	4780	24" x 14"		12	.667		164	22.50		186.50	215	
	4800	30" x 14"		10	.800		173	27		200	233	
	4820	16" x 16"		16	.500		169	16.80		185.80	212	
	4840	20" x 16"		14	.571		174	19.15		193.15	221	
	4860	24" x 16"		11	.727		179	24.50		203.50	235	
	4880	30" x 16"		8	1		187	33.50		220.50	258	
	4900	18" x 18"		15	.533		169	17.90		186.90	214	
	5000	24" x 18"		10	.800		179	27		206	239	
	5020	36" x 18"		7	1.143		196	38.50		234.50	274	
	5040	20" x 20"		13	.615		187	20.50		207.50	238	
	5060	24" x 20"		8	1		191	33.50		224.50	262	
	5080	30" x 20"		8	1		201	33.50		234.50	273	
	5100	36" x 20"		7	1.143		212	38.50		250.50	292	
	5120	24" x 24"		8	1		207	33.50		240.50	279	
	5130	30" x 24"		7.60	1.053		196	35.50		231.50	271	
	5140	36" x 24"	Q-9	12	1.333		226	40.50		266.50	310	
	5141	36" x 30"		10	1.600		238	48.50		286.50	335	
	5142	40" x 36"		8	2		425	60.50		485.50	560	
	5143	48" x 48"		4	4		825	121		946	1,100	
	5150	Damper operator motor, 24 or 120 volt	1 Shee	16	.500		220	16.80		236.80	268	
	5154	Induction air terminal unit, cooling and heating										
	5156	Vertical, cooling w/ electric heating coils										
	5158	60 CFM	Q-20	7.60	2.632	Ea.	885	81.50		966.50	1,100	
	5160	90 CFM		7.20	2.778		1,100	86		1,186	1,325	
	5162	120 CFM		6.80	2.941		1,100	91		1,191	1,350	
	5164	150 CFM		6.40	3.125		1,250	97		1,347	1,525	
	5166	170 CFM		6.20	3.226		1,250	100		1,350	1,525	
	5170	Horizontal, cooling w/ electric heating coils										
	5172	60 CFM	Q-20	7.60	2.632	Ea.	720	81.50		801.50	920	
	5174	90 CFM		7.20	2.778		885	86		971	1,100	
	5176	120 CFM		6.80	2.941		885	91		976	1,125	
	5180	Mixing box, includes electric or pneumatic motor										
	5190	Recommend use with attenuator, see 15820-300-9000										
	5200	Constant volume, 150 to 270 CFM	Q-9	12	1.333	Ea.	485	40.50		525.50	590	
	5210	270 to 600 CFM		11	1.455		495	44		539	615	
	5230	550 to 1000 CFM		9	1.778		495	53.50		548.50	630	
	5240	1000 to 1600 CFM		8	2		510	60.50		570.50	655	
	5250	1300 to 1900 CFM		6	2.667		520	80.50		600.50	695	
	5260	550 to 2640 CFM		5.60	2.857		575	86.50		661.50	765	
	5270	650 to 3120 CFM		5.20	3.077		610	93		703	815	
	5500	VAV Cool only, pneum. press indep. 300 to 600 CFM		11	1.455		320	44		364	420	
	5510	500 to 1000 CFM		9	1.778		325	53.50		378.50	445	
	5520	800 to 1600 CFM		9	1.778		335	53.50		388.50	455	
	5530	1100 to 2000 CFM		8	2		345	60.50		405.50	475	
	5540	1500 to 3000 CFM		7	2.286		370	69		439	515	
	5550	2000 to 4000 CFM		6	2.667		385	80.50		465.50	550	

15800 | Air Distribution

15820 | Duct Accessories

		CREW	DAILY OUTPUT	LABOR-HOURS	UNIT	MAT.	LABOR	EQUIP.	TOTAL	TOTAL INCL O&P
5560	For electric, w/thermostat, press. dependent, add				Ea.	20				
5600	VAV Cool & HW coils, damper, actuator and t'stat									
5610	200 CFM	Q-9	11	1.455	Ea.	555	44		599	680
5620	400 CFM		10	1.600		560	48.50		608.50	690
5630	600 CFM		10	1.600		560	48.50		608.50	690
5640	800 CFM		8	2		575	60.50		635.50	730
5650	1000 CFM		8	2		575	60.50		635.50	730
5660	1250 CFM		6	2.667		630	80.50		710.50	820
5670	1500 CFM		6	2.667		630	80.50		710.50	820
5680	2000 CFM		4	4		685	121		806	940
5684	3000 CFM		3.80	4.211		745	127		872	1,025
5700	VAV Cool only, fan powr'd, damper, actuator, t'stat									
5710	200 CFM	Q-9	10	1.600	Ea.	760	48.50		808.50	910
5720	400 CFM		9	1.778		790	53.50		843.50	950
5730	600 CFM		9	1.778		790	53.50		843.50	950
5740	800 CFM		7	2.286		845	69		914	1,025
5750	1000 CFM		7	2.286		845	69		914	1,025
5760	1250 CFM		5	3.200		910	96.50		1,006.50	1,150
5770	1500 CFM		5	3.200		910	96.50		1,006.50	1,150
5780	2000 CFM		4	4		1,000	121		1,121	1,275
5800	VAV Fan powr'd, cool/HW coils, dampers, actuators, t'stat									
5810	200 CFM	Q-9	10	1.600	Ea.	915	48.50		963.50	1,075
5820	400 CFM		9	1.778		945	53.50		998.50	1,100
5830	600 CFM		9	1.778		945	53.50		998.50	1,100
5840	800 CFM		7	2.286		1,000	69		1,069	1,200
5850	1000 CFM		7	2.286		1,000	69		1,069	1,200
5860	1250 CFM		5	3.200		1,125	96.50		1,221.50	1,400
5870	1500 CFM		5	3.200		1,125	96.50		1,221.50	1,400
5880	2000 CFM		3.50	4.571		1,225	138		1,363	1,575
5990	Multi-blade dampers, opposed blade, 8" x 6"	1 Shee	24	.333		18.40	11.20		29.60	37.50
5994	8" x 8"		22	.364		19.20	12.20		31.40	40
5996	10" x 10"		21	.381		21.50	12.80		34.30	44
6000	12" x 12"		21	.381		25	12.80		37.80	47.50
6020	12" x 18"		18	.444		33.50	14.90		48.40	60
6030	14" x 10"		20	.400		21.50	13.40		34.90	44.50
6031	14" x 14"		17	.471		29.50	15.80		45.30	57
6032	16" x 10"		18	.444		21.50	14.90		36.40	47
6033	16" x 12"		17	.471		29.50	15.80		45.30	57
6035	16" x 16"		16	.500		37	16.80		53.80	66.50
6036	18" x 16"		15	.533		41	17.90		58.90	72.50
6037	18" x 14"		16	.500		37	16.80		53.80	66.50
6038	18" x 18"		15	.533		44	17.90		61.90	76
6040	18" x 24"		12	.667		57	22.50		79.50	97
6060	18" x 28"		10	.800		65.50	27		92.50	114
6068	20" x 6"		18	.444		22.50	14.90		37.40	47.50
6069	20" x 8"		16	.500		25.50	16.80		42.30	54
6070	20" x 16"		14	.571		44	19.15		63.15	78
6071	20" x 18"		13	.615		48	20.50		68.50	85
6072	20" x 20"		13	.615		53	20.50		73.50	90
6073	22" x 6"		20	.400		22.50	13.40		35.90	45
6074	22" x 18"		14	.571		53	19.15		72.15	87.50
6075	22" x 20"		12	.667		57.50	22.50		80	98
6076	24" x 16"		11	.727		52	24.50		76.50	94.50
6077	22" x 22"		10	.800		61.50	27		88.50	110
6078	24" x 20"		8	1		61.50	33.50		95	120
6080	24" x 24"		8	1		72	33.50		105.50	131

15800 | Air Distribution

15820 | Duct Accessories

			CREW	DAILY OUTPUT	LABOR-HOURS	UNIT	MAT.	LABOR	EQUIP.	TOTAL	TOTAL INCL O&P	
300	6100	24" x 28"	1 Shee	6	1.333	Ea.	83	44.50		127.50	161	300
	6110	26" x 26"	↓	6	1.333		84.50	44.50		129	162	
	6120	28" x 28"	Q-9	11	1.455		98	44		142	176	
	6130	30" x 18"		10	1.600		69.50	48.50		118	151	
	6132	30" x 24"		7	2.286		88	69		157	204	
	6133	30" x 30"		6.60	2.424		112	73		185	237	
	6135	32" x 32"		6.40	2.500		128	75.50		203.50	258	
	6150	34" x 34"		6	2.667		144	80.50		224.50	282	
	6151	36" x 12"		10	1.600		59	48.50		107.50	140	
	6152	36" x 16"		8	2		74.50	60.50		135	176	
	6156	36" x 32"		6.30	2.540		144	76.50		220.50	276	
	6157	36" x 34"		6.20	2.581		153	78		231	288	
	6158	36" x 36"		6	2.667		161	80.50		241.50	300	
	6160	44" x 28"		5.80	2.759		154	83.50		237.50	298	
	6180	48" x 36"		5.60	2.857		215	86.50		301.50	370	
	6200	56" x 36"		5.40	2.963		250	89.50		339.50	415	
	6220	60" x 36"		5.20	3.077		268	93		361	440	
	6240	60" x 44"	↓	5	3.200		330	96.50		426.50	510	
	7000	Splitter damper assembly, self-locking, 1' rod	1 Shee	24	.333		15.80	11.20		27	34.50	
	7020	3' rod		22	.364		20.50	12.20		32.70	41.50	
	7040	4' rod		20	.400		22.50	13.40		35.90	45.50	
	7060	6' rod	↓	18	.444	↓	27.50	14.90		42.40	53.50	
	7500	Variable volume modulating motorized damper, incl. elect. mtr.										
	7504	8" x 6"	1 Shee	15	.533	Ea.	101	17.90		118.90	139	
	7506	10" x 6"		14	.571		101	19.15		120.15	141	
	7510	10" x 10"		13	.615		104	20.50		124.50	146	
	7520	12" x 12"		12	.667		106	22.50		128.50	152	
	7522	12" x 16"		11	.727		110	24.50		134.50	159	
	7524	16" x 10"		12	.667		107	22.50		129.50	153	
	7526	16" x 14"		10	.800		112	27		139	165	
	7528	16" x 18"		9	.889		119	30		149	177	
	7540	18" x 12"		10	.800		114	27		141	167	
	7542	18" x 18"		8	1		123	33.50		156.50	188	
	7544	20" x 14"		8	1		126	33.50		159.50	190	
	7546	20" x 18"		7	1.143		134	38.50		172.50	206	
	7560	24" x 12"		8	1		127	33.50		160.50	192	
	7562	24" x 18"		7	1.143		142	38.50		180.50	215	
	7568	28" x 10"		7	1.143		211	38.50		249.50	291	
	7580	28" x 16"		6	1.333		224	44.50		268.50	315	
	7590	30" x 14"		5	1.600		224	53.50		277.50	330	
	7600	30" x 18"		4	2		238	67		305	365	
	7610	30" x 24"	↓	3.80	2.105		258	70.50		328.50	390	
	7690	48" x 48"	Q-9	6	2.667		665	80.50		745.50	860	
	7694	6' x 14' w/2 motors	"	3	5.333		2,600	161		2,761	3,100	
	7700	For thermostat, add	1 Shee	8	1		25.50	33.50		59	80	
	7800	For transformer 40 VA capacity, add	"	16	.500	↓	20	16.80		36.80	48	
	8000	Multi-blade dampers, parallel blade										
	8100	8" x 8"	1 Shee	24	.333	Ea.	52.50	11.20		63.70	75.50	
	8120	12" x 8"		22	.364		52.50	12.20		64.70	77	
	8140	16" x 10"		20	.400		55.50	13.40		68.90	81.50	
	8160	18" x 12"		18	.444		57	14.90		71.90	86	
	8180	22" x 12"		15	.533		59.50	17.90		77.40	93	
	8200	24" x 16"		11	.727		74	24.50		98.50	119	
	8220	28" x 16"		10	.800		77	27		104	127	
	8240	30" x 16"		8	1		79	33.50		112.50	139	
	8260	30" x 18"	↓	7	1.143		86	38.50		124.50	154	

15800 | Air Distribution

15820 | Duct Accessories

		CREW	DAILY OUTPUT	LABOR-HOURS	UNIT	2001 BARE COSTS MAT.	LABOR	EQUIP.	TOTAL	TOTAL INCL O&P
8300	Relief damper, electronic bypass with tight seal									
8310	8" x 6"	1 Shee	22	.364	Ea.	101	12.20		113.20	130
8314	10" x 6"		22	.364		114	12.20		126.20	144
8318	10" x 10"		21	.381		147	12.80		159.80	182
8322	12" x 12"		20	.400		158	13.40		171.40	194
8326	12" x 16"		19	.421		164	14.15		178.15	202
8330	16" x 10"		20	.400		158	13.40		171.40	195
8334	16" x 14"		18	.444		165	14.90		179.90	205
8338	16" x 18"		17	.471		182	15.80		197.80	225
8342	18" x 12"		18	.444		164	14.90		178.90	204
8346	18" x 18"		15	.533		171	17.90		188.90	216
8350	20" x 14"		14	.571		171	19.15		190.15	218
8354	20" x 18"		13	.615		175	20.50		195.50	224
8358	24" x 12"		13	.615		182	20.50		202.50	233
8362	24" x 18"		10	.800		196	27		223	258
8363	24" x 24"		10	.800		204	27		231	266
8364	24" x 36"		9	.889		236	30		266	305
8365	24" x 48"		6	1.333		264	44.50		308.50	360
8366	28" x 10"		13	.615		171	20.50		191.50	220
8370	28" x 16"		10	.800		201	27		228	263
8374	30" x 14"		11	.727		197	24.50		221.50	255
8378	30" x 18"		8	1		199	33.50		232.50	271
8382	30" x 24"		6	1.333		197	44.50		241.50	286
8390	46" x 36"		4	2		277	67		344	410
8394	48" x 48"		3	2.667		320	89.50		409.50	490
8396	54" x 36"	▼	2	4	▼	340	134		474	580
8400	Round damper, butterfly, vol. control w/lever lock reg.									
8410	6" diam.	1 Shee	22	.364	Ea.	21.50	12.20		33.70	43
8412	7" diam.		21	.381		22.50	12.80		35.30	44.50
8414	8" diam.		20	.400		23	13.40		36.40	46
8416	9" diam.		19	.421		24	14.15		38.15	48.50
8418	10" diam.		18	.444		25.50	14.90		40.40	51
8420	12" diam.		16	.500		26.50	16.80		43.30	55
8422	14" diam.		14	.571		33.50	19.15		52.65	66.50
8424	16" diam.		13	.615		36	20.50		56.50	71.50
8426	18" diam.		12	.667		40	22.50		62.50	78.50
8428	20" diam.		11	.727		43	24.50		67.50	85
8430	24" diam.		10	.800		55	27		82	102
8432	30" diam.		9	.889		76	30		106	130
8434	36" diam.	▼	8	1	▼	102	33.50		135.50	164
8500	Round motor operated damper									
8510	6" dia.	1 Shee	22	.364	Ea.	51	12.20		63.20	75.50
8512	7" dia.		21	.381		53	12.80		65.80	78.50
8514	8" dia.		20	.400		56	13.40		69.40	82
8516	10" dia.		18	.444		61.50	14.90		76.40	91
8518	12" dia.		16	.500		66.50	16.80		83.30	99
8520	14" dia.	▼	14	.571	▼	75	19.15		94.15	112
9000	Silencers, noise control for air flow, duct				MCFM	41.50			41.50	45.50
9004	Duct sound trap, packaged									
9010	12" x 18" x 36", 2000 CFM	1 Shee	12	.667	Ea.	248	22.50		270.50	310
9014	24" x 36" x 36", 9000 CFM	Q-9	40	.400		560	12.10		572.10	635
9018	36" x 18" x 36", 6300 CFM		42	.381		560	11.50		571.50	635
9022	24" x 24" x 60", 6000 CFM		3.60	4.444		645	134		779	915
9026	24" x 30" x 60", 7000 CFM		3.50	4.571		725	138		863	1,025
9030	24" x 36" x 60", 8000 CFM		3.40	4.706		800	142		942	1,100
9034	24" x 48" x 60", 11,000 CFM	▼	3.30	4.848		700	146		846	995

15800 | Air Distribution

15820 | Duct Accessories

			CREW	DAILY OUTPUT	LABOR-HOURS	UNIT	MAT.	LABOR	EQUIP.	TOTAL	TOTAL INCL O&P	
300	9038	36" x 18" x 60", 6300 CFM	Q-9	3.50	4.571	Ea.	540	138		678	810	300
	9042	36" x 36" x 60", 12,000 CFM		3.30	4.848		890	146		1,036	1,200	
	9046	36" x 48" x 60", 17,000 CFM		3.10	5.161		1,150	156		1,306	1,500	
	9050	36" x 60" x 60", 20,000 CFM		2.80	5.714		1,375	173		1,548	1,800	
	9054	48" x 48" x 60", 23,000 CFM		2.60	6.154		735	186		921	1,100	
	9100	Louvers, galvanized				MCFM	60			60	66	
	9120	Louvers, mill finish aluminum				"	87.50			87.50	96.50	
	9200	Plenums, measured by panel surface				S.F.	8.25			8.25	9.10	
	9300	Vent, air transfer				Ea.	139			139	153	
	9400	Turning vane components										
	9410	Turning vane rail	1 Shee	160	.050	L.F.	.79	1.68		2.47	3.46	
	9420	Double thick, factory fab. vane		300	.027		.98	.89		1.87	2.46	
	9428	12" high set		170	.047		5.50	1.58		7.08	8.50	
	9432	14" high set		160	.050		6.15	1.68		7.83	9.40	
	9434	16" high set		150	.053		6.80	1.79		8.59	10.25	
	9436	18" high set		144	.056		7.45	1.86		9.31	11.10	
	9438	20" high set		138	.058		8.10	1.94		10.04	11.95	
	9440	22" high set		130	.062		8.75	2.06		10.81	12.85	
	9442	24" high set		124	.065		9.45	2.16		11.61	13.70	
	9444	26" high set		116	.069		10.10	2.31		12.41	14.65	
	9446	30" high set		112	.071		11.40	2.40		13.80	16.25	

15830 | Fans

			CREW	DAILY OUTPUT	LABOR-HOURS	UNIT	MAT.	LABOR	EQUIP.	TOTAL	TOTAL INCL O&P	
100	0010	**FANS** R15700-040										100
	0020	Air conditioning and process air handling										
	0030	Axial flow, compact, low sound, 2.5" S.P. R15800-080										
	0050	3,800 CFM, 5 HP	Q-20	3.40	5.882	Ea.	3,500	182		3,682	4,125	
	0080	6,400 CFM, 5 HP		2.80	7.143		3,900	222		4,122	4,650	
	0100	10,500 CFM, 7-1/2 HP		2.40	8.333		4,875	259		5,134	5,750	
	0120	15,600 CFM, 10 HP		1.60	12.500		6,125	390		6,515	7,350	
	0140	23,000 CFM, 15 HP		.70	28.571		9,400	885		10,285	11,800	
	0160	28,000 CFM, 20 HP		.40	50		10,500	1,550		12,050	13,900	
	0200	In-line centrifugal, supply/exhaust booster										
	0220	aluminum wheel/hub, disconnect switch, 1/4" S.P.										
	0240	500 CFM, 10" diameter connection	Q-20	3	6.667	Ea.	700	207		907	1,075	
	0260	1,380 CFM, 12" diameter connection		2	10		800	310		1,110	1,350	
	0280	1,520 CFM, 16" diameter connection		2	10		915	310		1,225	1,475	
	0300	2,560 CFM, 18" diameter connection		1	20		1,000	620		1,620	2,050	
	0320	3,480 CFM, 20" diameter connection		.80	25		1,200	775		1,975	2,525	
	0326	5,080 CFM, 20" diameter connection		.75	26.667		1,325	825		2,150	2,750	
	0340	7,500 CFM, 22" diameter connection		.70	28.571		1,675	885		2,560	3,200	
	0350	10,000 CFM, 27" diameter connection		.65	30.769		1,975	955		2,930	3,650	
	0500	Axial flow, constant speed										
	0505	Direct drive, 1/8" S.P.										
	0510	12", 1060 CFM, 1/6 HP	Q-20	3	6.667	Ea.	274	207		481	615	
	0514	12", 2095 CFM, 1/2 HP		3	6.667		300	207		507	645	
	0518	16", 2490 CFM, 1/3 HP		2.80	7.143		320	222		542	695	
	0522	20", 4130 CFM, 3/4 HP		2.60	7.692		425	239		664	830	
	0526	22", 4700 CFM, 3/4 HP		2.60	7.692		470	239		709	880	
	0530	24", 5850 CFM, 1 HP		2.50	8		535	248		783	965	
	0534	24", 7925 CFM, 1-1/2 HP		2.40	8.333		540	259		799	985	
	0538	30", 10640 CFM, 1 HP		2.20	9.091		980	282		1,262	1,500	
	0542	30", 14765 CFM, 2-1/2 HP		2.10	9.524		1,025	295		1,320	1,575	
	0546	36", 16780 CFM, 2 HP		2.10	9.524		1,100	295		1,395	1,650	
	0550	36", 22920 CFM, 5 HP		1.80	11.111		1,375	345		1,720	2,025	

15800 | Air Distribution

15830 | Fans

			CREW	DAILY OUTPUT	LABOR-HOURS	UNIT	MAT.	LABOR	EQUIP.	TOTAL	TOTAL INCL O&P
100	0560	Belt drive, 1/8" S.P.										100
	0562	15", 2800 CFM, 1/3 HP R15700-040	Q-20	3.20	6.250	Ea.	475	194		669	815
	0564	15", 3400 CFM, 1/2 HP		3	6.667		485	207		692	850
	0568	18", 3280 CFM, 1/3 HP R15800-080		2.80	7.143		515	222		737	905
	0572	18", 3900 CFM, 1/2 HP		2.80	7.143		520	222		742	910
	0576	18", 5250 CFM, 1 HP		2.70	7.407		600	230		830	1,000
	0584	24", 6430 CFM, 1 HP		2.50	8		745	248		993	1,200
	0588	24", 8860 CFM, 2 HP		2.40	8.333		745	259		1,004	1,225
	0592	30", 9250 CFM, 1 HP		2.20	9.091		905	282		1,187	1,425
	0596	30", 16900 CFM, 5 HP		2	10		1,000	310		1,310	1,575
	0604	36", 14475 CFM, 2 HP		2.30	8.696		1,050	270		1,320	1,575
	0608	36", 20080 CFM, 5 HP		1.80	11.111		1,125	345		1,470	1,775
	0612	36", 14475 CFM, 7-1/2 HP		1.60	12.500		1,250	390		1,640	1,975
	0616	42", 29000 CFM, 7-1/2 HP		1.40	14.286		1,500	445		1,945	2,325
	1000	Return air fan									
	1010	9200 CFM	Q-20	1.40	14.286	Ea.	2,850	445		3,295	3,825
	1020	13,200 CFM		1.30	15.385		4,100	475		4,575	5,225
	1030	16,500 CFM		1.20	16.667		5,125	515		5,640	6,425
	1040	19,500 CFM		1	20		6,050	620		6,670	7,600
	1500	Vaneaxial, low pressure, 2000 CFM, 1/2 HP		3.60	5.556		1,575	172		1,747	2,025
	1520	4,000 CFM, 1 HP		3.20	6.250		1,625	194		1,819	2,075
	1540	8,000 CFM, 2 HP		2.80	7.143		1,900	222		2,122	2,450
	1560	16,000 CFM, 5 HP		2.40	8.333		2,525	259		2,784	3,175
	2000	Blowers, direct drive with motor, complete									
	2020	1045 CFM @ .5" S.P., 1/5 HP	Q-20	18	1.111	Ea.	132	34.50		166.50	198
	2040	1385 CFM @ .5" S.P., 1/4 HP		18	1.111		135	34.50		169.50	202
	2060	1640 CFM @ .5" S.P., 1/3 HP		18	1.111		160	34.50		194.50	229
	2080	1760 CFM @ .5" S.P., 1/2 HP		18	1.111		174	34.50		208.50	244
	2090	4 speed									
	2100	1164 to 1739 CFM @ .5" S.P., 1/3 HP	Q-20	16	1.250	Ea.	236	39		275	320
	2120	1467 to 2218 CFM @ 1.0" S.P., 3/4 HP	"	14	1.429	"	305	44.50		349.50	405
	2500	Ceiling fan, right angle, extra quiet, 0.10" S.P.									
	2520	95 CFM	Q-20	20	1	Ea.	110	31		141	169
	2540	210 CFM		19	1.053		130	32.50		162.50	193
	2560	385 CFM		18	1.111		165	34.50		199.50	235
	2580	885 CFM		16	1.250		325	39		364	420
	2600	1,650 CFM		13	1.538		450	47.50		497.50	570
	2620	2,960 CFM		11	1.818		600	56.50		656.50	745
	2640	For wall or roof cap, add	1 Shee	16	.500		110	16.80		126.80	147
	2660	For straight thru fan, add					10%				
	2680	For speed control switch, add	1 Elec	16	.500		60	17.15		77.15	91.50
	3000	Paddle blade air circulator, 3 speed switch									
	3020	42", 5,000 CFM high, 3000 CFM low	1 Elec	2.40	3.333	Ea.	73	114		187	251
	3040	52", 6,500 CFM high, 4000 CFM low	"	2.20	3.636	"	74	125		199	268
	3100	For antique white motor, same cost									
	3200	For brass plated motor, same cost									
	3300	For light adaptor kit, add				Ea.	26			26	29
	3310	Industrial grade, reversable, 4 blade									
	3312	5500 CFM	Q-20	5	4	Ea.	100	124		224	300
	3314	7600 CFM		5	4		109	124		233	310
	3316	21,015 CFM		4	5		140	155		295	390
	3500	Centrifugal, airfoil, motor and drive, complete									
	3520	1000 CFM, 1/2 HP	Q-20	2.50	8	Ea.	870	248		1,118	1,350
	3540	2,000 CFM, 1 HP		2	10		935	310		1,245	1,500
	3560	4,000 CFM, 3 HP		1.80	11.111		1,200	345		1,545	1,850
	3580	8,000 CFM, 7-1/2 HP		1.40	14.286		1,775	445		2,220	2,625

15800 | Air Distribution

15830 | Fans

			CREW	DAILY OUTPUT	LABOR-HOURS	UNIT	MAT.	LABOR	EQUIP.	TOTAL	TOTAL INCL O&P	
100	3600	12,000 CFM, 10 HP	Q-20	1	20	Ea.	2,600	620		3,220	3,800	100
	4000	Single width, belt drive, not incl. motor, capacities	R15700-040									
	4020	at 2000 FPM, 2.5" S.P. for indicated motor	R15800-080									
	4040	6900 CFM, 5 HP		Q-9	2.40	6.667	Ea.	2,550	201		2,751	3,125
	4060	10,340 CFM, 7-1/2" HP			2.20	7.273		3,550	220		3,770	4,275
	4080	15,320 CFM, 10 HP			2	8		4,200	242		4,442	4,975
	4100	22,780 CFM, 15 HP			1.80	8.889		6,225	268		6,493	7,275
	4120	33,840 CFM, 20 HP			1.60	10		8,275	300		8,575	9,575
	4140	41,400 CFM, 25 HP			1.40	11.429		11,000	345		11,345	12,600
	4160	50,100 CFM, 30 HP			.80	20		13,900	605		14,505	16,200
	4200	Double width wheel, 12,420 CFM, 7.5 HP			2.20	7.273		3,650	220		3,870	4,375
	4220	18,620 CFM, 15 HP			2	8		5,225	242		5,467	6,125
	4240	27,580 CFM, 20 HP			1.80	8.889		6,400	268		6,668	7,450
	4260	40,980 CFM, 25 HP			1.50	10.667		10,000	320		10,320	11,500
	4280	60,920 CFM, 40 HP			1	16		13,500	485		13,985	15,500
	4300	74,520 CFM, 50 HP			.80	20		16,000	605		16,605	18,500
	4320	90,160 CFM, 50 HP			.70	22.857		21,900	690		22,590	25,200
	4340	110,300 CFM, 60 HP			.50	32		29,600	965		30,565	34,100
	4360	134,960 CFM, 75 HP			.40	40		39,300	1,200		40,500	45,200
	4500	Corrosive fume resistant, plastic										
	4600	roof ventilators, centrifugal, V belt drive, motor										
	4620	1/4" S.P., 250 CFM, 1/4 HP		Q-20	6	3.333	Ea.	2,350	103		2,453	2,725
	4640	895 CFM, 1/3 HP			5	4		2,550	124		2,674	3,000
	4660	1630 CFM, 1/2 HP			4	5		3,000	155		3,155	3,550
	4680	2240 CFM, 1 HP			3	6.667		3,150	207		3,357	3,775
	4700	3810 CFM, 2 HP			2	10		3,475	310		3,785	4,300
	4710	5000 CFM, 2 HP			1.80	11.111		7,325	345		7,670	8,575
	4715	8000 CFM, 5 HP			1.40	14.286		8,150	445		8,595	9,650
	4720	11760 CFM, 5 HP			1	20		8,275	620		8,895	10,100
	4740	18810 CFM, 10 HP			.70	28.571		8,575	885		9,460	10,800
	4800	For intermediate capacity, motors may be varied										
	4810	For explosion proof motor, add					Ea.	15%				
	5000	Utility set, centrifugal, V belt drive, motor										
	5020	1/4" S.P., 1200 CFM, 1/4 HP		Q-20	6	3.333	Ea.	2,300	103		2,403	2,675
	5040	1520 CFM, 1/3 HP			5	4		2,300	124		2,424	2,725
	5060	1850 CFM, 1/2 HP			4	5		2,300	155		2,455	2,775
	5080	2180 CFM, 3/4 HP			3	6.667		2,325	207		2,532	2,875
	5100	1/2" S.P., 3600 CFM, 1 HP			2	10		3,375	310		3,685	4,175
	5120	4250 CFM, 1-1/2 HP			1.60	12.500		3,425	390		3,815	4,350
	5140	4800 CFM, 2 HP			1.40	14.286		3,450	445		3,895	4,475
	5160	6920 CFM, 5 HP			1.30	15.385		3,575	475		4,050	4,650
	5180	7700 CFM, 7-1/2 HP			1.20	16.667		3,675	515		4,190	4,850
	5200	For explosion proof motor, add						15%				
	5300	Fume exhauster without hose and intake nozzle										
	5310	630 CFM, 1-1/2 HP		Q-20	2	10	Ea.	635	310		945	1,175
	5320	Hose extension kit 5'		"	20	1	"	47	31		78	99
	5500	Industrial exhauster, for air which may contain granular material										
	5520	1000 CFM, 1-1/2 HP		Q-20	2.50	8	Ea.	2,000	248		2,248	2,575
	5540	2000 CFM, 3 HP			2	10		2,425	310		2,735	3,150
	5560	4000 CFM, 7-1/2 HP			1.80	11.111		3,925	345		4,270	4,825
	5580	8000 CFM, 15 HP			1.40	14.286		6,125	445		6,570	7,400
	5600	12,000 CFM, 30 HP			1	20		8,500	620		9,120	10,300
	6000	Propeller exhaust, wall shutter, 1/4" S.P.										
	6020	Direct drive, two speed										
	6100	375 CFM, 1/10 HP		Q-20	10	2	Ea.	198	62		260	310
	6120	730 CFM, 1/7 HP			9	2.222		220	69		289	350

Important: See the Reference Section for critical supporting data - Reference Nos., Crews, & City Cost Indexes

15800 | Air Distribution

15830 | Fans

			CREW	DAILY OUTPUT	LABOR-HOURS	UNIT	MAT.	LABOR	EQUIP.	TOTAL	TOTAL INCL O&P
6140	1000 CFM, 1/8 HP	R15700-040	Q-20	8	2.500	Ea.	305	77.50		382.50	455
6160	1890 CFM, 1/4 HP			7	2.857		274	88.50		362.50	435
6180	3275 CFM, 1/2 HP	R15800-080		6	3.333		315	103		418	510
6200	4720 CFM, 1 HP			5	4		490	124		614	730
6300	V-belt drive, 3 phase										
6320	6175 CFM, 3/4 HP		Q-20	5	4	Ea.	410	124		534	640
6340	7500 CFM, 3/4 HP			5	4		435	124		559	665
6360	10,100 CFM, 1 HP			4.50	4.444		530	138		668	790
6380	14,300 CFM, 1-1/2 HP			4	5		620	155		775	925
6400	19,800 CFM, 2 HP			3	6.667		710	207		917	1,100
6420	26,250 CFM, 3 HP			2.60	7.692		855	239		1,094	1,300
6440	38,500 CFM, 5 HP			2.20	9.091		1,000	282		1,282	1,525
6460	46,000 CFM, 7-1/2 HP			2	10		1,075	310		1,385	1,675
6480	51,500 CFM, 10 HP			1.80	11.111		1,125	345		1,470	1,750
6490	V-belt drive, 115V., residential, whole house										
6500	Ceiling-wall, 5200 CFM, 1/4 HP, 30" x 30"		1 Shee	6	1.333	Ea.	215	44.50		259.50	305
6510	7500 CFM, 1/3 HP, 36"x36"			5	1.600		225	53.50		278.50	330
6520	10,500 CFM, 1/3 HP, 42"x42"			5	1.600		235	53.50		288.50	340
6530	13,200 CFM, 1/3 HP, 48" x 48"			4	2		260	67		327	390
6540	15,445 CFM, 1/2 HP, 48" x 48"			4	2		270	67		337	400
6550	17,025 CFM, 1/2 HP, 54" x 54"			4	2		520	67		587	680
6560	For two speed motor, add						20%				
6570	Shutter, automatic, ceiling/wall										
6580	30" x 30"		1 Shee	8	1	Ea.	70	33.50		103.50	129
6590	36" x 36"			8	1		80	33.50		113.50	140
6600	42" x 42"			8	1		100	33.50		133.50	162
6610	48" x 48"			7	1.143		100	38.50		138.50	169
6620	54" x 54"			6	1.333		143	44.50		187.50	226
6630	Timer, shut off, to 12 Hr.			20	.400		23	13.40		36.40	46
6650	Residential, bath exhaust, grille, back draft damper										
6660	50 CFM		Q-20	24	.833	Ea.	28.50	26		54.50	71
6670	110 CFM			22	.909		47	28		75	95
6672	180 CFM			22	.909		77	28		105	128
6673	210 CFM			22	.909		103	28		131	157
6674	260 CFM			22	.909		116	28		144	171
6675	300 CFM			22	.909		158	28		186	217
6680	Light combination, squirrel cage, 100 watt, 70 CFM			24	.833		58	26		84	104
6700	Light/heater combination, ceiling mounted										
6710	70 CFM, 1450 watt		Q-20	24	.833	Ea.	70.50	26		96.50	117
6800	Heater combination, recessed, 70 CFM			24	.833		33.50	26		59.50	76
6820	With 2 infrared bulbs			23	.870		50.50	27		77.50	97
6840	Wall mount, 170 CFM			22	.909		104	28		132	157
6846	Ceiling mount, 180 CFM			22	.909		105	28		133	158
6900	Kitchen exhaust, grille, complete, 160 CFM			22	.909		58.50	28		86.50	108
6910	180 CFM			20	1		49.50	31		80.50	102
6920	270 CFM			18	1.111		91	34.50		125.50	153
6930	350 CFM			16	1.250		70	39		109	137
6940	Residential roof jacks and wall caps										
6944	Wall cap with back draft damper										
6946	3" & 4" dia. round duct		1 Shee	11	.727	Ea.	12.05	24.50		36.55	51
6948	6" dia. round duct		"	11	.727	"	29	24.50		53.50	69.50
6958	Roof jack with bird screen and back draft damper										
6960	3" & 4" dia. round duct		1 Shee	11	.727	Ea.	11.60	24.50		36.10	50.50
6962	3-1/4" x 10" rectangular duct		"	10	.800	"	21.50	27		48.50	65
6980	Transition										
6982	3-1/4" x 10" to 6" dia. round		1 Shee	20	.400	Ea.	13.05	13.40		26.45	35

15800 | Air Distribution

15830 | Fans

		CREW	DAILY OUTPUT	LABOR-HOURS	UNIT	MAT.	LABOR	EQUIP.	TOTAL	TOTAL INCL O&P
7000	Roof exhauster, centrifugal, aluminum housing, 12" galvanized									
7020	curb, bird screen, back draft damper, 1/4" S.P.									
7100	Direct drive, 320 CFM, 11" sq. damper	Q-20	7	2.857	Ea.	287	88.50		375.50	450
7120	600 CFM, 11" sq. damper		6	3.333		291	103		394	480
7140	815 CFM, 13" sq. damper		5	4		291	124		415	510
7160	1450 CFM, 13" sq. damper		4.20	4.762		375	148		523	635
7180	2050 CFM, 16" sq. damper		4	5		375	155		530	650
7200	V-belt drive, 1650 CFM, 12" sq. damper		6	3.333		655	103		758	880
7220	2750 CFM, 21" sq. damper		5	4		760	124		884	1,025
7230	3500 CFM, 21" sq. damper		4.50	4.444		840	138		978	1,125
7240	4910 CFM, 23" sq. damper		4	5		1,050	155		1,205	1,400
7260	8525 CFM, 28" sq. damper		3	6.667		1,300	207		1,507	1,750
7280	13,760 CFM, 35" sq. damper		2	10		1,800	310		2,110	2,450
7300	20,558 CFM, 43" sq. damper		1	20		3,800	620		4,420	5,125
7320	For 2 speed winding, add					15%				
7340	For explosionproof motor, add				Ea.	282			282	310
7360	For belt driven, top discharge, add					15%				
7400	Roof mounted kitchen exhaust, aluminum, centrifugal									
7410	Direct drive, 2 speed, temp to 200°F									
7412	1/12 HP, 9-3/4"	Q-20	8	2.500	Ea.	290	77.50		367.50	440
7414	1/3 HP, 12-5/8"		7	2.857		355	88.50		443.50	525
7416	1/2 HP, 13-1/2"		6	3.333		395	103		498	595
7418	3/4 HP, 15"		5	4		560	124		684	810
7424	Belt drive, temp to 250°F									
7426	3/4 HP, 20"	Q-20	5	4	Ea.	1,225	124		1,349	1,550
7428	1-1/2 HP, 24-1/2"		4	5		1,375	155		1,530	1,750
7430	1-1/2 HP, 30"		4	5		1,800	155		1,955	2,225
7450	Upblast, propeller, w/BDD, BS, + C									
7454	30,300 CFM @ 3/8" S.P., 5 HP	Q-20	1.60	12.500	Ea.	1,800	390		2,190	2,575
7458	36,000 CFM @ 1/2" S.P., 15 HP	"	1.60	12.500	"	2,550	390		2,940	3,400
7500	Utility set, steel construction, pedestal, 1/4" S.P.									
7520	Direct drive, 150 CFM, 1/8 HP	Q-20	6.40	3.125	Ea.	585	97		682	795
7540	485 CFM, 1/6 HP		5.80	3.448		740	107		847	980
7560	1950 CFM, 1/2 HP		4.80	4.167		865	129		994	1,150
7580	2410 CFM, 3/4 HP		4.40	4.545		1,600	141		1,741	1,975
7600	3328 CFM, 1-1/2 HP		3	6.667		1,775	207		1,982	2,275
7680	V-belt drive, drive cover, 3 phase									
7700	800 CFM, 1/4 HP	Q-20	6	3.333	Ea.	415	103		518	615
7720	1,300 CFM, 1/3 HP		5	4		435	124		559	670
7740	2,000 CFM, 1 HP		4.60	4.348		515	135		650	770
7760	2,900 CFM, 3/4 HP		4.20	4.762		695	148		843	990
7780	3,600 CFM, 3/4 HP		4	5		855	155		1,010	1,175
7800	4,800 CFM, 1 HP		3.50	5.714		1,000	177		1,177	1,375
7820	6,700 CFM, 1-1/2 HP		3	6.667		1,250	207		1,457	1,700
7830	7,500 CFM, 2 HP		2.50	8		1,250	248		1,498	1,750
7840	11,000 CFM, 3 HP		2	10		2,250	310		2,560	2,950
7860	13,000 CFM, 3 HP		1.60	12.500		2,300	390		2,690	3,125
7880	15,000 CFM, 5 HP		1	20		2,375	620		2,995	3,575
7900	17,000 CFM, 7-1/2 HP		.80	25		2,550	775		3,325	4,000
7920	20,000 CFM, 7-1/2 HP		.80	25		3,025	775		3,800	4,525
8000	Ventilation, residential									
8020	Attic, roof type									
8030	Aluminum dome, damper & curb									
8040	6" diameter, 300 CFM	1 Elec	16	.500	Ea.	213	17.15		230.15	261
8050	7" diameter, 450 CFM		15	.533		233	18.30		251.30	284
8060	9" diameter, 900 CFM		14	.571		375	19.60		394.60	440

15800 | Air Distribution

15830 | Fans

			CREW	DAILY OUTPUT	LABOR-HOURS	UNIT	MAT.	LABOR	EQUIP.	TOTAL	TOTAL INCL O&P	
100	8080	12" diameter, 1000 CFM (gravity) R15700-040	1 Elec	10	.800	Ea.	227	27.50		254.50	291	100
	8090	16" diameter, 1500 CFM (gravity)		9	.889		274	30.50		304.50	345	
	8100	20" diameter, 2500 CFM (gravity) R15800-080		8	1		335	34.50		369.50	420	
	8110	26" diameter, 4000 CFM (gravity)		7	1.143		405	39		444	510	
	8120	32" diameter, 6500 CFM (gravity)		6	1.333		560	45.50		605.50	685	
	8130	38" diameter, 8000 CFM (gravity)		5	1.600		830	55		885	990	
	8140	50" diameter, 13,000 CFM (gravity)		4	2		1,200	68.50		1,268.50	1,425	
	8160	Plastic, ABS dome										
	8180	1050 CFM	1 Elec	14	.571	Ea.	67	19.60		86.60	103	
	8200	1600 CFM	"	12	.667	"	100	23		123	144	
	8240	Attic, wall type, with shutter, one speed										
	8250	12" diameter, 1000 CFM	1 Elec	14	.571	Ea.	168	19.60		187.60	215	
	8260	14" diameter, 1500 CFM		12	.667		182	23		205	234	
	8270	16" diameter, 2000 CFM		9	.889		206	30.50		236.50	273	
	8290	Whole house, wall type, with shutter, one speed										
	8300	30" diameter, 4800 CFM	1 Elec	7	1.143	Ea.	440	39		479	545	
	8310	36" diameter, 7000 CFM		6	1.333		480	45.50		525.50	600	
	8320	42" diameter, 10,000 CFM		5	1.600		540	55		595	670	
	8330	48" diameter, 16,000 CFM		4	2		670	68.50		738.50	835	
	8340	For two speed, add					40			40	44	
	8350	Whole house, lay-down type, with shutter, one speed										
	8360	30" diameter, 4500 CFM	1 Elec	8	1	Ea.	470	34.50		504.50	565	
	8370	36" diameter, 6500 CFM		7	1.143		505	39		544	615	
	8380	42" diameter, 9000 CFM		6	1.333		555	45.50		600.50	680	
	8390	48" diameter, 12,000 CFM		5	1.600		630	55		685	775	
	8440	For two speed, add					30			30	33	
	8450	For 12 hour timer switch, add	1 Elec	32	.250		30	8.60		38.60	46	
	8500	Wall exhausters, centrifugal, auto damper, 1/8" S.P.										
	8520	Direct drive, 610 CFM, 1/20 HP	Q-20	14	1.429	Ea.	160	44.50		204.50	244	
	8540	796 CFM, 1/12 HP		13	1.538		165	47.50		212.50	255	
	8560	822 CFM, 1/6 HP		12	1.667		261	51.50		312.50	365	
	8580	1,320 CFM, 1/4 HP		12	1.667		263	51.50		314.50	370	
	8600	1756 CFM, 1/4 HP		11	1.818		305	56.50		361.50	425	
	8620	1983 CFM, 1/4 HP		10	2		365	62		427	500	
	8640	2900 CFM, 1/2 HP		9	2.222		445	69		514	595	
	8660	3307 CFM, 3/4 HP		8	2.500		490	77.50		567.50	660	
	8670	5940 CFM, 1 HP		7	2.857		800	88.50		888.50	1,025	
	9500	V-belt drive, 3 phase										
	9520	2,800 CFM, 1/4 HP	Q-20	9	2.222	Ea.	725	69		794	905	
	9540	3,740 CFM, 1/2 HP		8	2.500		755	77.50		832.50	950	
	9560	4400 CFM, 3/4 HP		7	2.857		760	88.50		848.50	975	
	9580	5700 CFM, 1-1/2 HP		6	3.333		795	103		898	1,025	

15850 | Air Outlets & Inlets

			CREW	DAILY OUTPUT	LABOR-HOURS	UNIT	MAT.	LABOR	EQUIP.	TOTAL	TOTAL INCL O&P	
300	0010	**DIFFUSERS** Aluminum, opposed blade damper unless noted R15850-060										300
	0100	Ceiling, linear, also for sidewall										
	0120	2" wide R15810-070	1 Shee	32	.250	L.F.	26	8.40		34.40	41.50	
	0140	3" wide		30	.267		29.50	8.95		38.45	46.50	
	0160	4" wide R15800-080		26	.308		34.50	10.30		44.80	54	
	0180	6" wide		24	.333		43	11.20		54.20	64.50	
	0200	8" wide		22	.364		50	12.20		62.20	74	
	0220	10" wide		20	.400		57	13.40		70.40	83.50	
	0240	12" wide		18	.444		64	14.90		78.90	93.50	
	0260	For floor or sill application, add					15%					
	0500	Perforated, 24" x 24" lay-in panel size, 6" x 6"	1 Shee	16	.500	Ea.	79.50	16.80		96.30	113	
	0520	8" x 8"		15	.533		82	17.90		99.90	118	

15800 | Air Distribution

15850 | Air Outlets & Inlets

			CREW	DAILY OUTPUT	LABOR-HOURS	UNIT	MAT.	LABOR	EQUIP.	TOTAL	TOTAL INCL O&P	
300	0530	9" x 9"	1 Shee	14	.571	Ea.	84	19.15		103.15	122	300
	0540	10" x 10"	R15850 -060	14	.571		86	19.15		105.15	124	
	0560	12" x 12"		12	.667		86	22.50		108.50	129	
	0580	15" x 15"	R15810 -070	11	.727		103	24.50		127.50	152	
	0590	16" x 16"		11	.727		107	24.50		131.50	156	
	0600	18" x 18"	R15800 -080	10	.800		114	27		141	168	
	0610	20" x 20"		10	.800		130	27		157	185	
	0620	24" x 24"		9	.889		163	30		193	225	
	1000	Rectangular, 1 to 4 way blow, 6" x 6"		16	.500		46	16.80		62.80	77	
	1010	8" x 8"		15	.533		54	17.90		71.90	87	
	1014	9" x 9"		15	.533		56	17.90		73.90	89	
	1016	10" x 10"		15	.533		67	17.90		84.90	101	
	1020	12" x 6"		15	.533		57	17.90		74.90	90	
	1040	12" x 9"		14	.571		69.50	19.15		88.65	106	
	1060	12" x 12"		12	.667		82	22.50		104.50	125	
	1070	14" x 6"		13	.615		61	20.50		81.50	99.50	
	1074	14" x 14"		12	.667		101	22.50		123.50	146	
	1080	18" x 12"		11	.727		108	24.50		132.50	157	
	1120	15" x 15"		10	.800		114	27		141	167	
	1140	18" x 15"		9	.889		128	30		158	186	
	1150	18" x 18"		9	.889		149	30		179	210	
	1160	21" x 21"		8	1		189	33.50		222.50	259	
	1170	24" x 12"		10	.800		134	27		161	190	
	1180	24" x 24"		7	1.143		236	38.50		274.50	320	
	1500	Round, butterfly damper, 6" diameter		18	.444		15.90	14.90		30.80	40.50	
	1520	8" diameter		16	.500		16.85	16.80		33.65	44.50	
	1540	10" diameter		14	.571		21	19.15		40.15	52.50	
	1560	12" diameter		12	.667		27.50	22.50		50	65	
	1580	14" diameter		10	.800		34.50	27		61.50	79.50	
	1600	18" diameter		9	.889		84	30		114	139	
	1610	20" diameter		9	.889		96.50	30		126.50	152	
	1620	22" diameter		8	1		107	33.50		140.50	170	
	1630	24" diameter		8	1		118	33.50		151.50	182	
	1640	28" diameter		7	1.143		132	38.50		170.50	204	
	1650	32" diameter		6	1.333		144	44.50		188.50	228	
	1700	Round, stl, adjustable core, annular segmented damper										
	1704	6" diameter	1 Shee	18	.444	Ea.	77	14.90		91.90	108	
	1708	8" diameter		16	.500		88.50	16.80		105.30	124	
	1712	10" diameter		14	.571		100	19.15		119.15	140	
	1716	12" diameter		12	.667		127	22.50		149.50	175	
	1720	14" diameter		10	.800		154	27		181	211	
	1724	18" diameter		9	.889		226	30		256	295	
	1728	20" diameter		9	.889		265	30		295	335	
	1732	24" diameter		8	1		410	33.50		443.50	500	
	1736	30" diameter		7	1.143		525	38.50		563.50	635	
	1740	36" diameter		6	1.333		540	44.50		584.50	665	
	2000	T bar mounting, 24" x 24" lay-in frame, 6" x 6"		16	.500		75.50	16.80		92.30	109	
	2020	9" x 9"		14	.571		84	19.15		103.15	122	
	2040	12" x 12"		12	.667		111	22.50		133.50	157	
	2060	15" x 15"		11	.727		143	24.50		167.50	195	
	2080	18" x 18"		10	.800		149	27		176	206	
	2500	Combination supply and return										
	2520	21" x 21" supply, 15" x 15" return	Q-9	10	1.600	Ea.	178	48.50		226.50	270	
	2540	24" x 24" supply, 18" x 18" return		9.50	1.684		224	51		275	325	
	2560	27" x 27" supply, 18" x 18" return		9	1.778		274	53.50		327.50	385	
	2580	30" x 30" supply, 21" x 21" return		8.50	1.882		345	57		402	470	

Important: See the Reference Section for critical supporting data - Reference Nos., Crews, & City Cost Indexes

15800 | Air Distribution

15850 | Air Outlets & Inlets

			CREW	DAILY OUTPUT	LABOR-HOURS	UNIT	MAT.	LABOR	EQUIP.	TOTAL	TOTAL INCL O&P
2600	33" x 33" supply, 24" x 24" return	R15850-060	Q-9	8	2	Ea.	405	60.50		465.50	540
2620	36" x 36" supply, 24" x 24" return		↓	7.50	2.133	↓	475	64.50		539.50	625
3000	Baseboard, white enameled steel	R15810-070									
3100	18" long		1 Shee	20	.400	Ea.	7.80	13.40		21.20	29
3120	24" long	R15800-080		18	.444		15.30	14.90		30.20	40
3140	48" long		↓	16	.500		28	16.80		44.80	57
3400	For matching return, deduct					↓	10%				
4000	Floor, steel, adjustable pattern										
4100	2" x 10"		1 Shee	34	.235	Ea.	10.50	7.90		18.40	24
4120	2" x 12"			32	.250		11.50	8.40		19.90	25.50
4140	2" x 14"			30	.267		13.10	8.95		22.05	28
4200	4" x 10"			28	.286		12.15	9.60		21.75	28
4220	4" x 12"			26	.308		13.40	10.30		23.70	30.50
4240	4" x 14"			25	.320		15.15	10.75		25.90	33.50
4260	6" x 10"			26	.308		18.35	10.30		28.65	36
4280	6" x 12"			24	.333		19.15	11.20		30.35	38.50
4300	6" x 14"		↓	22	.364	↓	19.95	12.20		32.15	41
4500	Linear bar diffuser in wall, sill or ceiling										
4510	2" wide		1 Shee	32	.250	L.F.	25.50	8.40		33.90	41
4512	3" wide			30	.267		30	8.95		38.95	47
4514	4" wide			26	.308		32	10.30		42.30	51.50
4516	6" wide			24	.333		37.50	11.20		48.70	59
4520	10" wide			20	.400		46	13.40		59.40	71
4522	12" wide		↓	18	.444	↓	60	14.90		74.90	89
5000	Sidewall, aluminum, 3 way dispersion										
5100	8" x 4"		1 Shee	24	.333	Ea.	5.75	11.20		16.95	23.50
5110	8" x 6"			24	.333		6.15	11.20		17.35	24
5120	10" x 4"			22	.364		6.05	12.20		18.25	25.50
5130	10" x 6"			22	.364		6.45	12.20		18.65	26
5140	10" x 8"			20	.400		8.85	13.40		22.25	30.50
5160	12" x 4"			18	.444		6.50	14.90		21.40	30
5170	12" x 6"			17	.471		7.20	15.80		23	32.50
5180	12" x 8"			16	.500		9.30	16.80		26.10	36.50
5200	14" x 4"			15	.533		7.25	17.90		25.15	35.50
5220	14" x 6"			14	.571		7.95	19.15		27.10	38
5240	14" x 8"			13	.615		10.40	20.50		30.90	43.50
5260	16" x 6"		↓	12.50	.640	↓	11.45	21.50		32.95	45.50
6000	For steel diffusers instead of aluminum, deduct						10%				
0010	**GRILLES**										
0020	Aluminum										
0100	Air supply, single deflection, adjustable										
0120	8" x 4"		1 Shee	30	.267	Ea.	10.80	8.95		19.75	25.50
0140	8" x 8"			28	.286		13.05	9.60		22.65	29
0160	10" x 4"			24	.333		11.90	11.20		23.10	30.50
0180	10" x 10"			23	.348		15.10	11.65		26.75	34.50
0200	12" x 6"			23	.348		14.05	11.65		25.70	33.50
0220	12" x 12"			22	.364		18.45	12.20		30.65	39.50
0230	14" x 6"			23	.348		15.65	11.65		27.30	35.50
0240	14" x 8"			23	.348		16.75	11.65		28.40	36.50
0260	14" x 14"			22	.364		22.50	12.20		34.70	44
0270	18" x 8"			23	.348		19.30	11.65		30.95	39
0280	18" x 10"			23	.348		22	11.65		33.65	42.50
0300	18" x 18"			21	.381		34.50	12.80		47.30	58
0320	20" x 12"			22	.364		25	12.20		37.20	46.50
0340	20" x 20"			21	.381		42	12.80		54.80	66
0350	24" x 8"		↓	20	.400	↓	24	13.40		37.40	47

15800 | Air Distribution

15850 | Air Outlets & Inlets

			CREW	DAILY OUTPUT	LABOR-HOURS	UNIT	MAT.	LABOR	EQUIP.	TOTAL	TOTAL INCL O&P	
500	0360	24" x 14"	1 Shee	18	.444	Ea.	34	14.90		48.90	60.50	500
	0380	24" x 24"		15	.533		59	17.90		76.90	92.50	
	0400	30" x 8"		20	.400		31	13.40		44.40	54.50	
	0420	30" x 10"		19	.421		34.50	14.15		48.65	60	
	0440	30" x 12"		18	.444		37.50	14.90		52.40	64.50	
	0460	30" x 16"		17	.471		49	15.80		64.80	78.50	
	0480	30" x 18"		17	.471		55	15.80		70.80	85	
	0500	30" x 30"		14	.571		97.50	19.15		116.65	137	
	0520	36" x 12"		17	.471		47	15.80		62.80	76	
	0540	36" x 16"		16	.500		59.50	16.80		76.30	91.50	
	0560	36" x 18"		15	.533		66	17.90		83.90	100	
	0570	36" x 20"		15	.533		72.50	17.90		90.40	107	
	0580	36" x 24"		14	.571		94	19.15		113.15	133	
	0600	36" x 28"		13	.615		115	20.50		135.50	158	
	0620	36" x 30"		12	.667		127	22.50		149.50	174	
	0640	36" x 32"		12	.667		139	22.50		161.50	188	
	0660	36" x 34"		11	.727		151	24.50		175.50	204	
	0680	36" x 36"	▼	11	.727		164	24.50		188.50	219	
	0700	For double deflecting, add					70%					
	1000	Air return, 6" x 6"	1 Shee	26	.308		12.35	10.30		22.65	29.50	
	1020	10" x 6"		24	.333		14.95	11.20		26.15	33.50	
	1040	14" x 6"		23	.348		16.90	11.65		28.55	36.50	
	1060	10" x 8"		23	.348		14.95	11.65		26.60	34.50	
	1080	16" x 8"		22	.364		21.50	12.20		33.70	42.50	
	1100	12" x 12"		22	.364		21.50	12.20		33.70	42.50	
	1120	24" x 12"		18	.444		38.50	14.90		53.40	65	
	1140	30" x 12"		16	.500		45.50	16.80		62.30	76	
	1160	14" x 14"		22	.364		26	12.20		38.20	47.50	
	1180	16" x 16"		22	.364		30.50	12.20		42.70	52.50	
	1200	18" x 18"		21	.381		36.50	12.80		49.30	60	
	1220	24" x 18"		16	.500		45.50	16.80		62.30	76	
	1240	36" x 18"		15	.533		76	17.90		93.90	111	
	1260	24" x 24"		15	.533		60	17.90		77.90	93.50	
	1280	36" x 24"		14	.571		90.50	19.15		109.65	129	
	1300	48" x 24"		12	.667		148	22.50		170.50	197	
	1320	48" x 30"		11	.727		190	24.50		214.50	247	
	1340	36" x 36"		13	.615		153	20.50		173.50	200	
	1360	48" x 36"		11	.727		218	24.50		242.50	278	
	1380	48" x 48"		8	1		285	33.50		318.50	365	
	2000	Door grilles, 12" x 12"		22	.364		30	12.20		42.20	52	
	2020	18" x 12"		22	.364		37	12.20		49.20	60	
	2040	24" x 12"		18	.444		44	14.90		58.90	71.50	
	2060	18" x 18"		18	.444		56	14.90		70.90	84.50	
	2080	24" x 18"		16	.500		69.50	16.80		86.30	103	
	2100	24" x 24"		15	.533		92.50	17.90		110.40	130	
	3000	Filter grille with filter, 12" x 12"		24	.333		35	11.20		46.20	56	
	3020	18" x 12"		20	.400		51.50	13.40		64.90	77	
	3040	24" x 18"		18	.444		74	14.90		88.90	105	
	3060	24" x 24"		16	.500		97	16.80		113.80	133	
	3080	30" x 24"		14	.571		125	19.15		144.15	167	
	3100	30" x 30"		13	.615		157	20.50		177.50	205	
	3950	Eggcrate, framed, 6" x 6" opening		26	.308		16.90	10.30		27.20	34.50	
	3954	8" x 8" opening		24	.333		16.90	11.20		28.10	36	
	3960	10" x 10" opening		23	.348		21.50	11.65		33.15	41.50	
	3970	12" x 12" opening		22	.364		24.50	12.20		36.70	46	
	3980	14" x 14" opening	▼	22	.364	▼	32	12.20		44.20	54	

15800 | Air Distribution

15850 | Air Outlets & Inlets

			CREW	DAILY OUTPUT	LABOR-HOURS	UNIT	MAT.	LABOR	EQUIP.	TOTAL	TOTAL INCL O&P	
500	3984	16" x 16" opening	1 Shee	21	.381	Ea.	38.50	12.80		51.30	62	500
	3990	18" x 18" opening		21	.381		45	12.80		57.80	69.50	
	4020	22" x 22" opening		17	.471		62.50	15.80		78.30	93	
	4040	24" x 24" opening		15	.533		76.50	17.90		94.40	112	
	4044	28" x 28" opening		14	.571		104	19.15		123.15	145	
	4048	36" x 36" opening		12	.667		115	22.50		137.50	162	
	4050	48" x 24" opening	▼	12	.667	▼	171	22.50		193.50	223	
	4060	Eggcrate, lay-in, T-bar system										
	4070	48" x 24" sheet	1 Shee	40	.200	Ea.	76.50	6.70		83.20	95	
	5000	Transfer grille, vision proof, 8" x 4"		30	.267		19.85	8.95		28.80	36	
	5020	8" x 6"		28	.286		19.85	9.60		29.45	37	
	5040	8" x 8"		26	.308		22	10.30		32.30	40.50	
	5060	10" x 6"		24	.333		20.50	11.20		31.70	40	
	5080	10" x 10"		23	.348		26	11.65		37.65	47	
	5090	12" x 6"		23	.348		22	11.65		33.65	42.50	
	5100	12" x 10"		22	.364		28.50	12.20		40.70	50.50	
	5110	12" x 12"		22	.364		31	12.20		43.20	53	
	5120	14" x 10"		22	.364		31	12.20		43.20	53	
	5140	16" x 8"		21	.381		31	12.80		43.80	54	
	5160	18" x 12"		20	.400		40.50	13.40		53.90	65	
	5170	18" x 18"		21	.381		57	12.80		69.80	82.50	
	5180	20" x 12"		20	.400		43.50	13.40		56.90	68.50	
	5200	20" x 20"		19	.421		69	14.15		83.15	98	
	5220	24" x 12"		17	.471		49	15.80		64.80	78.50	
	5240	24" x 24"		16	.500		94.50	16.80		111.30	130	
	5260	30" x 6"		15	.533		43.50	17.90		61.40	75.50	
	5280	30" x 8"		15	.533		50.50	17.90		68.40	83.50	
	5300	30" x 12"		14	.571		63.50	19.15		82.65	99	
	5320	30" x 16"		13	.615		75	20.50		95.50	115	
	5340	30" x 20"		12	.667		95	22.50		117.50	140	
	5360	30" x 24"		11	.727		117	24.50		141.50	167	
	5380	30" x 30"	▼	10	.800		154	27		181	212	
	6000	For steel grilles instead of aluminum in above, deduct				▼	10%					
	6200	Plastic, eggcrate, lay-in, T-bar system										
	6210	48" x 24" sheet	1 Shee	50	.160	Ea.	14.95	5.35		20.30	25	
	6250	Steel door louver										
	6270	With fire link, steel only										
	6350	12" x 18"	1 Shee	22	.364	Ea.	135	12.20		147.20	168	
	6410	12" x 24"		21	.381		150	12.80		162.80	185	
	6470	18" x 18"		20	.400		150	13.40		163.40	186	
	6530	18" x 24"		19	.421		172	14.15		186.15	211	
	6600	18" x 36"		17	.471		195	15.80		210.80	240	
	6660	24" x 24"		15	.533		181	17.90		198.90	227	
	6720	24" x 30"		15	.533		195	17.90		212.90	243	
	6780	30" x 30"		13	.615		229	20.50		249.50	284	
	6840	30" x 36"		12	.667		265	22.50		287.50	325	
	6900	36" x 36"	▼	10	.800	▼	305	27		332	375	
600	0010	**LOUVERS**										600
	0100	Aluminum, extruded, with screen, mill finish										
	1000	Brick vent, (see also division 04090)										
	1100	Standard, 4" deep, 8" wide, 5" high	1 Shee	24	.333	Ea.	23	11.20		34.20	43	
	1200	Modular, 4" deep, 7-3/4" wide, 5" high		24	.333		24	11.20		35.20	44	
	1300	Speed brick, 4" deep, 11-5/8" wide, 3-7/8" high		24	.333		24	11.20		35.20	44	
	1400	Fuel oil brick, 4" deep, 8" wide, 5" high		24	.333	▼	42	11.20		53.20	63.50	
	2000	Cooling tower and mechanical equip., screens, light weight	▼	40	.200	S.F.	9.95	6.70		16.65	21.50	

15800 | Air Distribution

15850 | Air Outlets & Inlets

		CREW	DAILY OUTPUT	LABOR-HOURS	UNIT	MAT.	LABOR	EQUIP.	TOTAL	TOTAL INCL O&P
2020	Standard weight	1 Shee	35	.229	S.F.	27.50	7.65		35.15	42.50
2500	Dual combination, automatic, intake or exhaust		20	.400		38	13.40		51.40	62.50
2520	Manual operation		20	.400		28	13.40		41.40	51.50
2540	Electric or pneumatic operation		20	.400		28	13.40		41.40	51.50
2560	Motor, for electric or pneumatic		14	.571	Ea.	330	19.15		349.15	395
3000	Fixed blade, continuous line									
3100	Mullion type, stormproof	1 Shee	28	.286	S.F.	28	9.60		37.60	46
3200	Stormproof		28	.286		28	9.60		37.60	46
3300	Vertical line		28	.286		37	9.60		46.60	56
3500	For damper to use with above, add					50%	30%			
3520	Motor, for damper, electric or pneumatic	1 Shee	14	.571	Ea.	330	19.15		349.15	395
4000	Operating, 45°, manual, electric or pneumatic		24	.333	S.F.	28	11.20		39.20	48.50
4100	Motor, for electric or pneumatic		14	.571	Ea.	330	19.15		349.15	395
4200	Penthouse, roof		56	.143	S.F.	16.30	4.79		21.09	25.50
4300	Walls		40	.200		39.50	6.70		46.20	53.50
5000	Thinline, under 4" thick, fixed blade		40	.200		15.80	6.70		22.50	28
5010	Finishes, applied by mfr. at additional cost, available in colors									
5020	Prime coat only, add				S.F.	2.30			2.30	2.53
5040	Baked enamel finish coating, add					4.23			4.23	4.65
5060	Anodized finish, add					4.59			4.59	5.05
5080	Duranodic finish, add					8.30			8.30	9.15
5100	Fluoropolymer finish coating, add					13			13	14.30
9980	For small orders (under 10 pieces), add					25%				
0010	**REGISTERS**									
0980	Air supply									
1000	Ceiling/wall, O.B. damper, anodized aluminum									
1010	One or two way deflection, adj. curved face bars									
1014	6" x 6"	1 Shee	24	.333	Ea.	18.55	11.20		29.75	38
1020	8" x 4"		26	.308		18.55	10.30		28.85	36.50
1040	8" x 8"		24	.333		23	11.20		34.20	43
1060	10" x 6"		20	.400		21.50	13.40		34.90	44
1080	10" x 10"		19	.421		28.50	14.15		42.65	53.50
1100	12" x 6"		19	.421		21.50	14.15		35.65	45.50
1120	12" x 12"		18	.444		34	14.90		48.90	60.50
1140	14" x 8"		17	.471		29.50	15.80		45.30	57
1160	14" x 14"		18	.444		39.50	14.90		54.40	66.50
1170	16" x 16"		17	.471		44	15.80		59.80	73
1180	18" x 8"		18	.444		33	14.90		47.90	59.50
1200	18" x 18"		17	.471		60.50	15.80		76.30	91
1220	20" x 4"		19	.421		30.50	14.15		44.65	55.50
1240	20" x 6"		18	.444		30.50	14.90		45.40	56.50
1260	20" x 8"		18	.444		33.50	14.90		48.40	59.50
1280	20" x 20"		17	.471		73.50	15.80		89.30	105
1290	22" x 22"		15	.533		84.50	17.90		102.40	120
1300	24" x 4"		17	.471		37.50	15.80		53.30	65.50
1320	24" x 6"		16	.500		37.50	16.80		54.30	67
1340	24" x 8"		13	.615		41	20.50		61.50	77
1350	24" x 18"		12	.667		76	22.50		98.50	119
1360	24" x 24"		11	.727		103	24.50		127.50	152
1380	30" x 4"		16	.500		47	16.80		63.80	77.50
1400	30" x 6"		15	.533		47	17.90		64.90	79
1420	30" x 8"		14	.571		53.50	19.15		72.65	88.50
1440	30" x 24"		12	.667		134	22.50		156.50	182
1460	30" x 30"		10	.800		171	27		198	230
1504	4 way deflection, adjustable curved face bars									

15800 | Air Distribution

15850 | Air Outlets & Inlets

		CREW	DAILY OUTPUT	LABOR-HOURS	UNIT	MAT.	LABOR	EQUIP.	TOTAL	TOTAL INCL O&P
1510	6" x 6"	1 Shee	26	.308	Ea.	22.50	10.30		32.80	40.50
1514	8" x 8"		24	.333		28	11.20		39.20	48
1518	10" x 10"		19	.421		34.50	14.15		48.65	60
1522	12" x 6"		19	.421		26	14.15		40.15	50.50
1526	12" x 12"		18	.444		40.50	14.90		55.40	68
1530	14" x 14"		18	.444		47.50	14.90		62.40	75.50
1534	16" x 16"		17	.471		53	15.80		68.80	82.50
1538	18" x 18"		17	.471		72.50	15.80		88.30	104
1542	22" x 22"		16	.500		101	16.80		117.80	137
1980	One way deflection, adj. vert. or horiz. face bars									
1990	6" x 6"	1 Shee	26	.308	Ea.	19.45	10.30		29.75	37.50
2000	8" x 4"		26	.308		19.05	10.30		29.35	37
2020	8" x 8"		24	.333		22.50	11.20		33.70	42.50
2040	10" x 6"		20	.400		23	13.40		36.40	45.50
2060	10" x 10"		19	.421		26.50	14.15		40.65	51.50
2080	12" x 6"		19	.421		24.50	14.15		38.65	49
2100	12" x 12"		18	.444		32	14.90		46.90	58.50
2120	14" x 8"		17	.471		29.50	15.80		45.30	57
2140	14" x 12"		18	.444		35.50	14.90		50.40	62
2160	14" x 14"		18	.444		40	14.90		54.90	67
2180	16" x 6"		18	.444		29.50	14.90		44.40	55.50
2200	16" x 12"		18	.444		39	14.90		53.90	66
2220	16" x 16"		17	.471		49	15.80		64.80	78.50
2240	18" x 8"		18	.444		34.50	14.90		49.40	60.50
2260	18" x 12"		17	.471		41.50	15.80		57.30	70.50
2280	18" x 18"		16	.500		71	16.80		87.80	104
2300	20" x 10"		18	.444		40.50	14.90		55.40	67.50
2320	20" x 16"		16	.500		58.50	16.80		75.30	90.50
2340	20" x 20"		16	.500		74	16.80		90.80	108
2360	24" x 12"		15	.533		51	17.90		68.90	84
2380	24" x 16"		14	.571		68.50	19.15		87.65	105
2400	24" x 20"		12	.667		85.50	22.50		108	129
2420	24" x 24"		10	.800		104	27		131	157
2440	30" x 12"		13	.615		66.50	20.50		87	105
2460	30" x 16"		13	.615		86.50	20.50		107	127
2480	30" x 24"		11	.727		135	24.50		159.50	186
2500	30" x 30"		9	.889		172	30		202	236
2520	36" x 12"		11	.727		82.50	24.50		107	128
2540	36" x 24"		10	.800		166	27		193	225
2560	36" x 36"		8	1		278	33.50		311.50	355
2600	For 2 way deflect., adj. vert. or horiz. face bars, add					40%				
2700	Above registers in steel instead of aluminum, deduct					10%				
3000	Baseboard, hand adj. damper, enameled steel									
3012	8" x 6"	1 Shee	26	.308	Ea.	8.55	10.30		18.85	25.50
3020	10" x 6"		24	.333		9.60	11.20		20.80	28
3040	12" x 5"		23	.348		10.95	11.65		22.60	30
3060	12" x 6"		23	.348		10.10	11.65		21.75	29
3080	12" x 8"		22	.364		14.65	12.20		26.85	35
3100	14" x 6"		20	.400		10.95	13.40		24.35	32.50
4000	Floor, toe operated damper, enameled steel									
4020	4" x 8"	1 Shee	32	.250	Ea.	16.35	8.40		24.75	31
4040	4" x 12"		26	.308		18.65	10.30		28.95	36.50
4060	6" x 8"		28	.286		18.65	9.60		28.25	35.50
4080	6" x 14"		22	.364		21.50	12.20		33.70	43
4100	8" x 10"		22	.364		19.45	12.20		31.65	40.50
4120	8" x 16"		20	.400		32.50	13.40		45.90	56.50

15800 | Air Distribution

15850 | Air Outlets & Inlets

		CREW	DAILY OUTPUT	LABOR-HOURS	UNIT	MAT.	LABOR	EQUIP.	TOTAL	TOTAL INCL O&P
4140	10" x 10"	1 Shee	20	.400	Ea.	23.50	13.40		36.90	46
4160	10" x 16"		18	.444		35.50	14.90		50.40	62
4180	12" x 12"		18	.444		28.50	14.90		43.40	54.50
4200	12" x 24"		16	.500		94	16.80		110.80	129
4220	14" x 14"		16	.500		77.50	16.80		94.30	112
4240	14" x 20"		15	.533		103	17.90		120.90	142
4980	Air return									
5000	Ceiling or wall, fixed 45° face blades									
5010	Adjustable O.B. damper, anodized aluminum									
5020	4" x 8"	1 Shee	26	.308	Ea.	21.50	10.30		31.80	39.50
5040	6" x 8"		24	.333		21.50	11.20		32.70	41
5060	6" x 10"		19	.421		24	14.15		38.15	48.50
5070	6" x 12"		19	.421		26	14.15		40.15	50.50
5080	6" x 16"		18	.444		30.50	14.90		45.40	56.50
5100	8" x 10"		19	.421		26	14.15		40.15	50.50
5120	8" x 12"		16	.500		28.50	16.80		45.30	57.50
5140	10" x 10"		18	.444		28.50	14.90		43.40	54.50
5160	10" x 16"		17	.471		34.50	15.80		50.30	62.50
5170	12" x 12"		18	.444		36	14.90		50.90	62.50
5180	12" x 18"		18	.444		50	14.90		64.90	78
5184	12" x 24"		14	.571		60	19.15		79.15	95.50
5200	12" x 30"		12	.667		73.50	22.50		96	116
5210	12" x 36"		10	.800		92.50	27		119.50	144
5220	16" x 16"		17	.471		52	15.80		67.80	81.50
5240	18" x 18"		16	.500		62	16.80		78.80	94
5250	18" x 24"		13	.615		79.50	20.50		100	119
5254	18" x 30"		11	.727		97.50	24.50		122	145
5260	18" x 36"		10	.800		124	27		151	179
5270	20" x 20"		10	.800		73.50	27		100.50	123
5274	20" x 24"		12	.667		87	22.50		109.50	131
5280	24" x 24"		11	.727		103	24.50		127.50	151
5281	24" x 24"		11	.727		103	24.50		127.50	151
5290	24" x 30"		10	.800		131	27		158	186
5300	24" x 36"		8	1		159	33.50		192.50	227
5320	24" x 48"		6	1.333		233	44.50		277.50	325
5340	30" x 30"		6.50	1.231		166	41.50		207.50	247
5344	30" x 120"	Q-9	6	2.667		730	80.50		810.50	930
5360	36" x 36"	1 Shee	6	1.333		263	44.50		307.50	360
5370	40" x 32"	"	6	1.333		251	44.50		295.50	345
5400	Ceiling or wall, removable-reversible core, single deflection									
5402	Adjustable O.B. damper, radiused frame with border, aluminum									
5410	8" x 4"	1 Shee	26	.308	Ea.	59.50	10.30		69.80	81.50
5414	8" x 6"		24	.333		75	11.20		86.20	100
5418	10" x 6"		19	.421		79	14.15		93.15	109
5422	10" x 10"		18	.444		116	14.90		130.90	150
5426	12" x 6"		19	.421		84	14.15		98.15	115
5430	12" x 12"		18	.444		139	14.90		153.90	176
5434	16" x 16"		16	.500		214	16.80		230.80	262
5438	18" x 18"		17	.471		257	15.80		272.80	310
5442	20" x 20"		16	.500		297	16.80		313.80	350
5446	24" x 12"		15	.533		192	17.90		209.90	239
5450	24" x 18"		13	.615		295	20.50		315.50	355
5454	24" x 24"		11	.727		375	24.50		399.50	450
5458	30" x 12"		13	.615		239	20.50		259.50	295
5462	30" x 18"		12.50	.640		350	21.50		371.50	415
5466	30" x 24"		11	.727		440	24.50		464.50	525

15800 | Air Distribution

15850 | Air Outlets & Inlets

		CREW	DAILY OUTPUT	LABOR-HOURS	UNIT	MAT.	LABOR	EQUIP.	TOTAL	TOTAL INCL O&P	
700	5470	30" x 30"	1 Shee	9	.889	Ea.	640	30		670	750	700
	5474	36" x 12"		11	.727		285	24.50		309.50	355	
	5478	36" x 18"		10.50	.762		405	25.50		430.50	485	
	5482	36" x 24"		10	.800		510	27		537	600	
	5486	36" x 30"		9	.889		840	30		870	970	
	5490	36" x 36"	▼	8	1	▼	930	33.50		963.50	1,075	
	6000	For steel construction instead of aluminum, deduct					10%					

15854 | Ventilators

		CREW	DAILY OUTPUT	LABOR-HOURS	UNIT	MAT.	LABOR	EQUIP.	TOTAL	TOTAL INCL O&P	
300	0010	**VENTILATORS** Base, damper & bird screen, CFM in 5 MPH wind										300
	0500	Rotary syphon, galvanized, 6" neck diameter, 185 CFM	Q-9	16	1	Ea.	51	30		81	103	
	0520	8" neck diameter, 215 CFM		14	1.143		58	34.50		92.50	117	
	0540	10" neck diameters, 260 CFM		12	1.333		63	40.50		103.50	132	
	0560	12" neck diameter, 310 CFM		10	1.600		72	48.50		120.50	154	
	0580	14" neck diameter, 500 CFM		10	1.600		95	48.50		143.50	180	
	0600	16" neck diameter, 635 CFM		9	1.778		109	53.50		162.50	203	
	0620	18" neck diameter, 835 CFM		9	1.778		128	53.50		181.50	224	
	0640	20" neck diameter, 1080 CFM		8	2		148	60.50		208.50	256	
	0660	24" neck diameter, 1,530 CFM		8	2		293	60.50		353.50	415	
	0680	30" neck diameter, 2500 CFM		7	2.286		530	69		599	690	
	0700	36" neck diameter, 3,800 CFM		6	2.667		800	80.50		880.50	1,000	
	0720	42" neck diameter, 4,500 CFM	▼	4	4		1,300	121		1,421	1,600	
	0800	For aluminum, add					300%					
	0820	For stainless steel, add					600%					
	0830	For copper, add				▼	600%					
	1280	Spinner ventilators, wind driven, galvanized										
	1300	4" neck diameter, 180 CFM	Q-9	20	.800	Ea.	52	24		76	95	
	1320	5" neck diameter, 210 CFM		18	.889		58	27		85	106	
	1340	6" neck diameter, 250 CFM		16	1		58	30		88	111	
	1360	8" neck diameter, 360 CFM		14	1.143		60.50	34.50		95	120	
	1380	10" neck diameter, 540 CFM		12	1.333		77.50	40.50		118	148	
	1400	12" neck diameter, 770 CFM		10	1.600		84.50	48.50		133	167	
	1420	14" neck diameter, 830 CFM		10	1.600		117	48.50		165.50	204	
	1440	16" neck diameter, 1200 CFM		9	1.778		141	53.50		194.50	239	
	1460	18" neck diameter, 1700 CFM		9	1.778		164	53.50		217.50	264	
	1480	20" neck diameter, 2100 CFM		8	2		182	60.50		242.50	294	
	1500	24" neck diameter, 3,100 CFM		8	2		245	60.50		305.50	365	
	1520	30" neck diameter, 4500 CFM		7	2.286		510	69		579	665	
	1540	36" neck diameter, 5,500 CFM	▼	6	2.667		720	80.50		800.50	915	
	1600	For aluminum, add					300%					
	1620	For stainless steel, add					600%					
	1630	For copper, add				▼	600%					
	2000	Stationary, gravity, syphon, galvanized										
	2100	3" diameter, 40 CFM	Q-9	24	.667	Ea.	47	20		67	82.50	
	2120	4" neck diameter, 50 CFM		20	.800		51.50	24		75.50	94.50	
	2140	5" neck diameter, 58 CFM		18	.889		52	27		79	98.50	
	2160	6" neck diameter, 66 CFM		16	1		45	30		75	96	
	2180	7" neck diameter, 86 CFM		15	1.067		56.50	32		88.50	112	
	2200	8" neck diameter, 110 CFM		14	1.143		57	34.50		91.50	116	
	2220	10" neck diameter, 140 CFM		12	1.333		73.50	40.50		114	143	
	2240	12" neck diameter, 160 CFM		10	1.600		94.50	48.50		143	179	
	2260	14" neck diameter, 250 CFM		10	1.600		119	48.50		167.50	205	
	2280	16" neck diameter, 380 CFM		9	1.778		158	53.50		211.50	257	
	2300	18" neck diameter, 500 CFM		9	1.778		146	53.50		199.50	243	
	2320	20" neck diameter, 625 CFM	▼	8	2	▼	223	60.50		283.50	340	

15800 | Air Distribution

15854 | Ventilators

		CREW	DAILY OUTPUT	LABOR-HOURS	UNIT	MAT.	LABOR	EQUIP.	TOTAL	TOTAL INCL O&P
2340	24" neck diameter, 900 CFM	Q-9	8	2	Ea.	257	60.50		317.50	375
2360	30" neck diameter, 1375 CFM		7	2.286		585	69		654	750
2380	36" neck diameter, 2,000 CFM		6	2.667		805	80.50		885.50	1,000
2400	42" neck diameter, 3000 CFM		4	4		1,125	121		1,246	1,425
2410	48" neck diameter, 4000 CFM	▼	3	5.333		1,525	161		1,686	1,925
2500	For aluminum, add					300%				
2520	For stainless steel, add					600%				
2530	For copper, add					600%				
3000	Rotating chimney cap, galvanized, 4" neck diameter	Q-9	20	.800		40.50	24		64.50	82
3020	5" neck diameter		18	.889		41.50	27		68.50	87
3040	6" neck diameter		16	1		42	30		72	92.50
3060	7" neck diameter		15	1.067		45.50	32		77.50	99.50
3080	8" neck diameter		14	1.143		47	34.50		81.50	106
3100	10" neck diameter		12	1.333		61	40.50		101.50	129
3600	Stationary chimney rain cap, galvanized, 3" neck diameter		24	.667		35.50	20		55.50	70
3620	4" neck diameter		20	.800		36	24		60	77
3640	6" neck diameter		16	1		37.50	30		67.50	88
3680	8" neck diameter		14	1.143		40.50	34.50		75	98
3700	10" neck diameter		12	1.333		51.50	40.50		92	119
3720	12" neck diameter		10	1.600		85.50	48.50		134	169
3740	14" neck diameter		10	1.600		91.50	48.50		140	176
3760	16" neck diameter		9	1.778		112	53.50		165.50	206
3770	18" neck diameter		9	1.778		162	53.50		215.50	261
3780	20" neck diameter		8	2		216	60.50		276.50	330
3790	24" neck diameter		8	2		252	60.50		312.50	370
4200	Stationary mushroom, aluminum, 16" orifice diameter		10	1.600		246	48.50		294.50	345
4220	26" orifice diameter		6.15	2.602		365	78.50		443.50	520
4230	30" orifice diameter		5.71	2.802		530	84.50		614.50	715
4240	38" orifice diameter		5	3.200		760	96.50		856.50	985
4250	42" orifice diameter		4.70	3.404		1,000	103		1,103	1,250
4260	50" orifice diameter	▼	4.44	3.604	▼	1,200	109		1,309	1,500
5000	Relief vent									
5500	Rectangular, aluminum, galvanized curb									
5510	intake/exhaust, 0.05" SP									
5600	500 CFM, 12" x 16"	Q-9	8	2	Ea.	365	60.50		425.50	495
5620	750 CFM, 12" x 20"		7.20	2.222		400	67		467	545
5640	1000 CFM, 12" x 24"		6.60	2.424		435	73		508	590
5660	1500 CFM, 12" x 36"		5.80	2.759		520	83.50		603.50	705
5680	3000 CFM, 20" x 42"		4	4		750	121		871	1,000
5700	6000 CFM, 20" x 84"		2.60	6.154		1,150	186		1,336	1,525
5720	8000 CFM, 24" x 96"		2.30	6.957		1,300	210		1,510	1,750
5740	10,000 CFM, 48" x 60"		1.80	8.889		1,600	268		1,868	2,200
5760	12,500 CFM, 48" x 72"		1.60	10		1,875	300		2,175	2,550
5780	15,000 CFM, 48" x 96"		1.30	12.308		2,175	370		2,545	2,975
5800	20,000 CFM, 48" x 120"		1.20	13.333		2,525	405		2,930	3,400
5820	25,000 CFM, 60" x 120"		.90	17.778		3,325	535		3,860	4,475
5840	30,000 CFM, 72" x 120"		.70	22.857		4,375	690		5,065	5,900
5860	40,000 CFM, 96" x 120"		.60	26.667		5,625	805		6,430	7,450
5880	50,000 CFM, 96" x 144"	▼	.50	32	▼	5,975	965		6,940	8,075
7000	Note: sizes based on exhaust. Intake, with 0.125" SP									
7100	loss, approximately twice listed capacity.									
7500	For power open, spring return damper									
7600	motor for each S.F. of throat area	Q-9	100	.160	S.F.	40.50	4.83		45.33	52.50

15800 | Air Distribution

15860 | Air Cleaning Devices

			CREW	DAILY OUTPUT	LABOR-HOURS	UNIT	MAT.	LABOR	EQUIP.	TOTAL	TOTAL INCL O&P
100	0010	**AIR FILTERS**									100
	0050	Activated charcoal type, full flow				MCFM	600			600	660
	0500	Chemical media filtration type									
	1100	Industrial air fume & odor scrubber unit w/pump & motor									
	1110	corrosion resistant PVC construction									
	1120	Single pack filter, horizontal type									
	1130	500 CFM	Q-9	14	1.143	Ea.	3,700	34.50		3,734.50	4,100
	1140	1000 CFM		11	1.455		4,325	44		4,369	4,850
	1150	2000 CFM		8	2		5,250	60.50		5,310.50	5,875
	1160	3000 CFM		7	2.286		6,075	69		6,144	6,800
	1170	5000 CFM		5	3.200		7,925	96.50		8,021.50	8,850
	1180	8000 CFM		4	4		11,100	121		11,221	12,400
	1190	12,000 CFM		3	5.333		14,700	161		14,861	16,400
	1200	16,000 CFM		2.50	6.400		18,100	193		18,293	20,300
	1210	20,000 CFM		2	8		22,700	242		22,942	25,400
	1220	26,000 CFM		1.50	10.667		26,200	320		26,520	29,300
	1230	30,000 CFM	Q-10	2	12		29,800	375		30,175	33,300
	1240	40,000 CFM	"	1.50	16		39,000	500		39,500	43,700
	1250	50,000 CFM	Q-11	2	16		47,400	510		47,910	53,000
	1260	55,000 CFM		1.80	17.778		51,500	570		52,070	58,000
	1270	60,000 CFM		1.50	21.333		55,500	685		56,185	62,500
	1300	Double pack filter, horizontal type									
	1310	500 CFM	Q-9	10	1.600	Ea.	4,800	48.50		4,848.50	5,350
	1320	1000 CFM		8	2		5,750	60.50		5,810.50	6,425
	1330	2000 CFM		6	2.667		6,925	80.50		7,005.50	7,750
	1340	3000 CFM		5	3.200		8,300	96.50		8,396.50	9,275
	1350	5000 CFM		4	4		10,500	121		10,621	11,800
	1360	8000 CFM		3	5.333		13,900	161		14,061	15,500
	1370	12,000 CFM		2.50	6.400		18,900	193		19,093	21,100
	1380	16,000 CFM		2	8		24,700	242		24,942	27,500
	1390	20,000 CFM		1.50	10.667		29,900	320		30,220	33,400
	1400	26,000 CFM		1	16		34,600	485		35,085	38,800
	1410	30,000 CFM	Q-10	1.50	16		40,600	500		41,100	45,500
	1420	40,000 CFM	"	1.30	18.462		52,000	580		52,580	58,000
	1430	50,000 CFM	Q-11	1.50	21.333		65,000	685		65,685	72,500
	1440	55,000 CFM		1.30	24.615		71,000	790		71,790	79,000
	1450	60,000 CFM		1	32		77,000	1,025		78,025	86,000
	1500	Single pack filter, vertical type									
	1510	500 CFM	Q-9	24	.667	Ea.	3,525	20		3,545	3,900
	1520	1000 CFM		18	.889		4,175	27		4,202	4,625
	1530	2000 CFM		12	1.333		5,000	40.50		5,040.50	5,550
	1540	3000 CFM		9	1.778		5,825	53.50		5,878.50	6,500
	1550	5000 CFM		6	2.667		7,775	80.50		7,855.50	8,675
	1560	8000 CFM		4	4		11,100	121		11,221	12,400
	1570	12,000 CFM		3	5.333		15,100	161		15,261	16,800
	1580	16,000 CFM		2	8		17,600	242		17,842	19,800
	1590	20,000 CFM		1.80	8.889		22,200	268		22,468	24,800
	1600	24,000 CFM		1.60	10		24,800	300		25,100	27,800
	1650	Double pack filter, vertical type									
	1660	500 CFM	Q-9	22	.727	Ea.	4,575	22		4,597	5,050
	1670	1000 CFM		16	1		5,450	30		5,480	6,050
	1680	2000 CFM		10	1.600		6,550	48.50		6,598.50	7,275
	1690	3000 CFM		7	2.286		7,700	69		7,769	8,575
	1700	5000 CFM		5	3.200		10,100	96.50		10,196.50	11,200
	1710	8000 CFM		3	5.333		13,800	161		13,961	15,400
	1720	12,000 CFM		2.50	6.400		18,600	193		18,793	20,800

15800 | Air Distribution

15860 | Air Cleaning Devices

		CREW	DAILY OUTPUT	LABOR-HOURS	UNIT	2001 BARE COSTS MAT.	LABOR	EQUIP.	TOTAL	TOTAL INCL O&P
1730	16,000 CFM	Q-9	2	8	Ea.	23,400	242		23,642	26,200
1740	20,000 CFM		1.50	10.667		29,200	320		29,520	32,700
1750	24,000 CFM	↓	1	16	↓	32,600	485		33,085	36,600
1800	Inlet or outlet transition, horizontal									
1810	Single pack to 12,000 CFM, add					4%				
1820	Single pack to 30,000 CFM, add					6%				
1830	Single pack to 60,000 CFM, add					8%				
1840	Double pack to 12,000 CFM, add					3%				
1850	Double pack to 30,000 CFM, add					5%				
1860	Double pack to 60,000 CFM, add					6%				
1870	Inlet or outlet transition, vertical									
1880	Single pack to 5000 CFM, add					2%				
1890	Single pack to 24,000 CFM, add					3%				
1900	Double pack to 24,000 CFM, add					2%				
2000	Electronic air cleaner, duct mounted									
2150	400 - 1000 CFM	1 Shee	2.30	3.478	Ea.	545	117		662	775
2200	1000 - 1400 CFM		2.20	3.636		700	122		822	955
2250	1400 - 2000 CFM		2.10	3.810		760	128		888	1,025
2260	2000 - 2500 CFM	↓	2	4	↓	730	134		864	1,000
2950	Mechanical media filtration units									
3000	High efficiency type, with frame, non-supported				MCFM	45			45	49.50
3100	Supported type					55			55	60.50
4000	Medium efficiency, extended surface					5			5	5.50
4500	Permanent washable					20			20	22
5000	Renewable disposable roll				↓	120			120	132
5500	Throwaway glass or paper media type				Ea.	4.60			4.60	5.05
5800	Filter, bag type									
5810	90-95% efficiency									
5820	24" x 12" x 29", .75-1.25 MCFM	1 Shee	1.90	4.211	Ea.	30.50	141		171.50	252
5830	24" x 24" x 29", 1.5-2.5 MCFM	"	1.90	4.211	"	64.50	141		205.50	289
5850	80-85% efficiency									
5860	24" x 12" x 29", .75-1.25 MCFM	1 Shee	1.90	4.211	Ea.	27.50	141		168.50	249
5870	24" x 24" x 29", 1.5-2.5 MCFM	"	1.90	4.211	"	49	141		190	272
6000	HEPA filter complete w/particle board,									
6010	kraft paper frame, separator material									
6020	95% DOP efficiency									
6030	12" x 12" x 6", 150 CFM	1 Shee	3.70	2.162	Ea.	29	72.50		101.50	144
6034	24" x 12" x 6", 375 CFM		1.90	4.211		30.50	141		171.50	252
6038	24" x 18" x 6", 450 CFM		1.90	4.211		39	141		180	261
6042	24" x 24" x 6", 700 CFM		1.90	4.211		45.50	141		186.50	268
6046	12" x 12" x 12", 250 CFM		3.70	2.162		38.50	72.50		111	154
6050	24" x 12" x 12", 500 CFM		1.90	4.211		45.50	141		186.50	268
6054	24" x 18" x 12", 875 CFM		1.90	4.211		61.50	141		202.50	286
6058	24" x 24" x 12", 1000 CFM	↓	1.90	4.211	↓	64	141		205	289
6100	99% DOP efficiency									
6110	12" x 12" x 6", 150 CFM	1 Shee	3.70	2.162	Ea.	48.50	72.50		121	165
6114	24" x 12" x 6", 325 CFM		1.90	4.211		53.50	141		194.50	277
6118	24" x 18" x 6", 550 CFM		1.90	4.211		74.50	141		215.50	300
6122	24" x 24" x 6", 775 CFM		1.90	4.211		77	141		218	305
6126	12" x 12" x 12", 250 CFM		3.60	2.222		54.50	74.50		129	175
6130	24" x 12" x 12", 500 CFM		1.90	4.211		66.50	141		207.50	291
6134	24" x 18" x 12", 775 CFM		1.90	4.211		106	141		247	335
6138	24" x 24" x 12", 1100 CFM	↓	1.90	4.211	↓	110	141		251	340
6500	HEPA filter housing, 14 ga. galv. sheet metal									
6510	12" x 12" x 6"	1 Shee	2.50	3.200	Ea.	435	107		542	645
6514	24" x 12" x 6"	↓	2	4	↓	485	134		619	735

15800 | Air Distribution

15860 | Air Cleaning Devices

		CREW	DAILY OUTPUT	LABOR-HOURS	UNIT	MAT.	LABOR	EQUIP.	TOTAL	TOTAL INCL O&P	
100	6518	12" x 12" x 12"	1 Shee	2.40	3.333	Ea.	435	112		547	650	100
	6522	14" x 12" x 12"		2.30	3.478		485	117		602	710	
	6526	24" x 18" x 6"		1.90	4.211		560	141		701	840	
	6530	24" x 24" x 6"		1.80	4.444		560	149		709	850	
	6534	24" x 48" x 6"		1.70	4.706		750	158		908	1,075	
	6538	24" x 72" x 6"		1.60	5		965	168		1,133	1,300	
	6542	24" x 18" x 12"		1.80	4.444		560	149		709	850	
	6546	24" x 24" x 12"		1.70	4.706		560	158		718	865	
	6550	24" x 48" x 12"		1.60	5		750	168		918	1,075	
	6554	24" x 72" x 12"		1.50	5.333		965	179		1,144	1,325	
	6558	48" x 48" x 6"	Q-9	2.80	5.714		1,075	173		1,248	1,475	
	6562	48" x 72" x 6"		2.60	6.154		1,350	186		1,536	1,775	
	6566	48" x 96" x 6"		2.40	6.667		1,525	201		1,726	1,975	
	6570	48" x 48" x 12"		2.70	5.926		1,075	179		1,254	1,475	
	6574	48" x 72" x 12"		2.50	6.400		1,350	193		1,543	1,800	
	6578	48" x 96" x 12"		2.30	6.957		1,525	210		1,735	2,000	
	6582	114" x 72" x 12"		2	8		2,325	242		2,567	2,925	
500	0010	**EXHAUST SYSTEMS**										500
	0500	Engine exhaust, garage, in-floor system	A8.5-110									
	0510	Single tube outlet assemblies										
	0520	For transite pipe ducting, self-storing tube										
	0530	3" tubing adapter plate	1 Shee	16	.500	Ea.	162	16.80		178.80	204	
	0540	4" tubing adapter plate		16	.500		162	16.80		178.80	204	
	0550	5" tubing adapter plate		16	.500		162	16.80		178.80	204	
	0600	For vitrified tile ducting										
	0610	3" tubing adapter plate, self-storing tube	1 Shee	16	.500	Ea.	162	16.80		178.80	204	
	0620	4" tubing adapter plate, self-storing tube		16	.500		162	16.80		178.80	204	
	0660	5" tubing adapter plate, self-storing tube		16	.500		162	16.80		178.80	204	
	0800	Two tube outlet assemblies										
	0810	For transite pipe ducting, self-storing tube										
	0820	3" tubing, dual exhaust adapter plate	1 Shee	16	.500	Ea.	162	16.80		178.80	204	
	0850	For vitrified tile ducting										
	0860	3" tubing, dual exhaust, self-storing tube	1 Shee	16	.500	Ea.	162	16.80		178.80	204	
	0870	3" tubing, double outlet, non-storing tubes	"	16	.500	"	162	16.80		178.80	204	
	0900	Accessories for metal tubing, (overhead systems also)										
	0910	Adapters, for metal tubing end										
	0920	3" tail pipe type				Ea.	27			27	29.50	
	0930	4" tail pipe type					27			27	29.50	
	0940	5" tail pipe type					27			27	29.50	
	0990	5" diesel stack type					67.50			67.50	74.50	
	1000	6" diesel stack type					67.50			67.50	74.50	
	1100	Bullnose (guide) required for in-floor assemblies										
	1110	3" tubing size				Ea.	13.50			13.50	14.85	
	1120	4" tubing size					13.50			13.50	14.85	
	1130	5" tubing size					13.50			13.50	14.85	
	1150	Plain rings, for tubing end										
	1160	3" tubing size				Ea.	9			9	9.90	
	1170	4" tubing size				"	9			9	9.90	
	1200	Tubing, galvanized, flexible, (for overhead systems also)										
	1210	3" ID				L.F.	4.82			4.82	5.30	
	1220	4" ID					5.75			5.75	6.35	
	1230	5" ID					6.75			6.75	7.45	
	1240	6" ID					7.65			7.65	8.40	
	1250	Stainless steel, flexible, (for overhead sys, also)										
	1260	3" ID				L.F.	10			10	11	

15800 | Air Distribution

15860 | Air Cleaning Devices

		CREW	DAILY OUTPUT	LABOR-HOURS	UNIT	2001 BARE COSTS MAT.	LABOR	EQUIP.	TOTAL	TOTAL INCL O&P	
500	1270	4" ID	A8.5 -110			L.F.	13			13	14.30	500
	1280	5" ID					15.70			15.70	17.30	
	1290	6" ID					18.70			18.70	20.50	
	1500	Engine exhaust, garage, ovrhd. compon., for neoprene tubing										
	1510	Alternate metal tubing & accessories see above										
	1550	Adapters, for neoprene tubing end										
	1560	3" tail pipe, adjustable, neoprene				Ea.	27			27	29.50	
	1570	3" tail pipe, heavy wall neoprene					38			38	41.50	
	1580	4" tail pipe, heavy wall neoprene					39.50			39.50	43.50	
	1590	5" tail pipe, heavy wall neoprene					58.50			58.50	64.50	
	1650	Connectors, tubing										
	1660	3" interior, aluminum				Ea.	16.20			16.20	17.80	
	1670	4" interior, aluminum					27			27	29.50	
	1710	5" interior, neoprene					27			27	29.50	
	1750	3" spiralock, neoprene					9			9	9.90	
	1760	4" spiralock, neoprene					11.70			11.70	12.85	
	1780	Y for 3" ID tubing, neoprene, dual exhaust					19.80			19.80	22	
	1790	Y for 4" ID tubing, aluminum, dual exhaust					40.50			40.50	44.50	
	1850	Elbows, aluminum, splice into tubing for strap										
	1860	3" neoprene tubing size				Ea.	23.50			23.50	25.50	
	1870	4" neoprene tubing size					31.50			31.50	34.50	
	1900	Flange assemblies, connect tubing to overhead duct					29			29	31.50	
	2000	Hardware and accessories										
	2020	Cable, galvanized, 1/8" diameter				L.F.	.27			.27	.30	
	2040	Cleat, tie down cable or rope				Ea.	3.60			3.60	3.96	
	2060	Pulley					4.50			4.50	4.95	
	2080	Pulley hook, universal					3.60			3.60	3.96	
	2100	Rope, nylon, 1/4" diameter				L.F.	.26			.26	.29	
	2120	Winch, 1" diameter				Ea.	72			72	79	
	2150	Lifting strap, mounts on neoprene										
	2160	3" tubing size				Ea.	10.80			10.80	11.90	
	2170	4" tubing size					10.80			10.80	11.90	
	2180	5" tubing size					10.80			10.80	11.90	
	2190	6" tubing size					10.80			10.80	11.90	
	2200	Tubing, neoprene, 11' lengths										
	2210	3" ID				L.F.	6.10			6.10	6.70	
	2220	4" ID					8.90			8.90	9.80	
	2230	5" ID					12.95			12.95	14.20	
	2500	Engine exhaust, thru-door outlet										
	2510	3" tube size	1 Carp	16	.500	Ea.	31.50	14.60		46.10	57.50	
	2530	4" tube size	"	16	.500	"	31.50	14.60		46.10	57.50	
	3000	Tubing, exhaust, flex hose, with										
	3010	coupler, damper and tail pipe adapter										
	3020	Neoprene										
	3040	3" x 20'	1 Shee	6	1.333	Ea.	185	44.50		229.50	272	
	3050	4" x 15'		5.40	1.481		203	49.50		252.50	300	
	3060	4" x 20'		5	1.600		248	53.50		301.50	355	
	3070	5" x 15'		4.40	1.818		300	61		361	425	
	3100	Galvanized										
	3110	3" x 20'	1 Shee	6	1.333	Ea.	149	44.50		193.50	232	
	3120	4" x 17'		5.60	1.429		155	48		203	244	
	3130	4" x 20'		5	1.600		172	53.50		225.50	272	
	3140	5" x 17'		4.60	1.739		190	58.50		248.50	299	
	7500	Welding fume elimination accessories for garage exhaust systems										
	7600	Cut off (blast gate)										
	7610	3" tubing size, 3" x 6" opening	1 Shee	24	.333	Ea.	16.20	11.20		27.40	35	

15800 | Air Distribution

15860 | Air Cleaning Devices

			CREW	DAILY OUTPUT	LABOR-HOURS	UNIT	MAT.	LABOR	EQUIP.	TOTAL	TOTAL INCL O&P	
500	7620	4" tubing size, 4" x 8" opening	1 Shee A8.5-110	24	.333	Ea.	18	11.20		29.20	37	500
	7630	5" tubing size, 5" x 10" opening		24	.333		21.50	11.20		32.70	41.50	
	7640	6" tubing size		24	.333		25	11.20		36.20	45	
	7650	8" tubing size		24	.333		36	11.20		47.20	57	
	7700	Hoods, magnetic, with handle & screen										
	7710	3" tubing size, 3" x 6" opening	1 Shee	24	.333	Ea.	67.50	11.20		78.70	92	
	7720	4" tubing size, 4" x 8" opening		24	.333		67.50	11.20		78.70	92	
	7730	5" tubing size, 5" x 10" opening		24	.333		67.50	11.20		78.70	92	
	8000	Blower, for tailpipe exhaust system										
	8010	Direct drive										
	8012	495 CFM, 1/3 HP	Q-9	5	3.200	Ea.	1,000	96.50		1,096.50	1,250	
	8014	1445 CFM, 3/4 HP		4	4		1,100	121		1,221	1,375	
	8016	1840 CFM, 1-1/2 HP		3	5.333		1,500	161		1,661	1,900	
	8030	Beltdrive										
	8032	1400 CFM, 1 HP	Q-9	4	4	Ea.	1,200	121		1,321	1,500	
	8034	2023 CFM, 1-1/2 HP		3	5.333		1,800	161		1,961	2,225	
	8036	2750 CFM, 2 HP		2	8		2,200	242		2,442	2,800	
	8038	4400 CFM, 3 HP		1.80	8.889		3,400	268		3,668	4,175	
	8040	7060 CFM, 5 HP		1.40	11.429		6,000	345		6,345	7,125	

15900 | HVAC Instrumentation & Controls

15905 | HVAC Instrumentation

			CREW	DAILY OUTPUT	LABOR-HOURS	UNIT	MAT.	LABOR	EQUIP.	TOTAL	TOTAL INCL O&P	
960	0010	**WATER LEVEL CONTROLS**										960
	1000	Electric water feeder	1 Stpi	12	.667	Ea.	188	23		211	242	
	2000	Feeder cut-off combination										
	2100	Steam system up to 5000 sq. ft.	1 Stpi	12	.667	Ea.	365	23		388	435	
	2200	Steam system above 5000 sq. ft.		12	.667		445	23		468	525	
	2300	Steam and hot water, high pressure		10	.800		525	28		553	620	
	3000	Low water cut-off for hot water boiler, 50 psi maximum										
	3100	1" top & bottom equalizing pipes, manual reset	1 Stpi	14	.571	Ea.	198	19.90		217.90	248	
	3200	1" top & bottom equalizing pipes		14	.571		198	19.90		217.90	248	
	3300	2-1/2" side connection for nipple-to-boiler		14	.571		175	19.90		194.90	222	
	4000	Low water cut-off for low pressure steam with quick hook-up ftgs.										
	4100	For installation in gauge glass tappings	1 Stpi	16	.500	Ea.	146	17.45		163.45	188	
	4200	Built-in type, 2-1/2" tap - 3-1/8" insertion		16	.500		123	17.45		140.45	162	
	4300	Built-in type, 2-1/2" tap - 1-3/4" insertion		16	.500		131	17.45		148.45	171	
	4400	Side connection to 2-1/2" tapping		16	.500		139	17.45		156.45	180	
	5000	Pump control, low water cut-off and alarm switch		14	.571		360	19.90		379.90	430	
	9000	Water gauges, complete										
	9010	Rough brass, wheel type										
	9020	125 PSI at 350° F										
	9030	3/8" pipe size	1 Stpi	11	.727	Ea.	28.50	25.50		54	69.50	
	9040	1/2" pipe size	"	10	.800	"	31	28		59	76	
	9060	200 PSI at 400° F										
	9070	3/8" pipe size	1 Stpi	11	.727	Ea.	35.50	25.50		61	77.50	
	9080	1/2" pipe size		10	.800		35.50	28		63.50	81	
	9090	3/4" pipe size		9	.889		45	31		76	96.50	
	9130	Rough brass, chain lever type										

15900 | HVAC Instrumentation & Controls

15905 | HVAC Instrumentation

		CREW	DAILY OUTPUT	LABOR-HOURS	UNIT	2001 BARE COSTS				TOTAL INCL O&P
						MAT.	LABOR	EQUIP.	TOTAL	
9140	250 PSI at 400° F									
9200	Polished brass, wheel type									
9210	200 PSI at 400° F									
9220	3/8" pipe size	1 Stpi	11	.727	Ea.	68	25.50		93.50	113
9230	1/2" pipe size		10	.800		68.50	28		96.50	118
9240	3/4" pipe size		9	.889		72.50	31		103.50	127
9260	Polished brass, chain lever type									
9270	250 PSI at 400° F									
9280	1/2" pipe size	1 Stpi	10	.800	Ea.	115	28		143	168
9290	3/4" pipe size	"	9	.889	"	125	31		156	184
9400	Bronze, high pressure, ASME									
9410	1/2" pipe size	1 Stpi	10	.800	Ea.	123	28		151	177
9420	3/4" pipe size	"	9	.889	"	150	31		181	212
9460	Chain lever type									
9470	1/2" pipe size	1 Stpi	10	.800	Ea.	186	28		214	247
9480	3/4" pipe size	"	9	.889	"	196	31		227	263
9500	316 stainless steel, high pressure, ASME									
9510	500 PSI at 450° F									
9520	1/2" pipe size	1 Stpi	10	.800	Ea.	470	28		498	555
9530	3/4" pipe size	"	9	.889	"	455	31		486	545

15950 | Testing/Adjusting/Balancing

15955 | HVAC Test/Adjust/Balance

		CREW	DAILY OUTPUT	LABOR-HOURS	UNIT	2001 BARE COSTS				TOTAL INCL O&P
						MAT.	LABOR	EQUIP.	TOTAL	
0010	BALANCING, AIR (Subcontractor's quote incl. material & labor) R15050-710									
0900	Heating and ventilating equipment									
1000	Centrifugal fans, utility sets				Ea.					248.64
1100	Heating and ventilating unit									372.96
1200	In-line fan									372.96
1300	Propeller and wall fan									70.45
1400	Roof exhaust fan									165.76
2000	Air conditioning equipment, central station									538.72
2100	Built-up low pressure unit									497.28
2200	Built-up high pressure unit									580.16
2300	Built-up high pressure dual duct									911.68
2400	Built-up variable volume									1,077.44
2500	Multi-zone A.C. and heating unit									372.96
2600	For each zone over one, add									82.88
2700	Package A.C. unit									207.20
2800	Rooftop heating and cooling unit									290.08
3000	Supply, return, exhaust, registers & diffusers, avg. height ceiling									49.73
3100	High ceiling									74.59
3200	Floor height									41.44
3300	Off mixing box									33.15
3500	Induction unit									53.87
3600	Lab fume hood									248.64
3700	Linear supply									124.32
3800	Linear supply high									145.04
4000	Linear return									41.44
4100	Light troffers									49.73

15950 | Testing/Adjusting/Balancing

15955 | HVAC Test/Adjust/Balance

			CREW	DAILY OUTPUT	LABOR-HOURS	UNIT	MAT.	LABOR	EQUIP.	TOTAL	TOTAL INCL O&P	
100	4200	Moduline - master	R15050-710			Ea.					49.73	100
	4300	Moduline - slaves									24.86	
	4400	Regenerators									331.52	
	4500	Taps into ceiling plenums									62.16	
	4600	Variable volume boxes									49.73	
700	0010	**PIPING, TESTING**										700
	0100	Nondestructive testing										
	0110	Nondestructive hydraulic pressure test										
	0120	1" - 4" pipe										
	0140	0 - 250 L.F.	1 Stpi	.75	10.667	Ea.		370		370	560	
	0160	250 - 500 L.F.	"	.64	12.500			435		435	660	
	0180	500 - 1000 L.F.	Q-5	1	16			500		500	760	
	0200	1000 - 2000 L.F.	"	.53	30.189			945		945	1,425	
	0300	6" - 10" pipe										
	0320	0 - 250 L.F.	Q-5	1	16	Ea.		500		500	760	
	0340	250 - 500 L.F.		.33	48.485			1,525		1,525	2,300	
	0360	500 - 1000 L.F.		.21	76.190			2,400		2,400	3,600	
	0380	1000 - 2000 L.F.		.11	145			4,575		4,575	6,900	
	1000	Pneumatic pressure test, includes soaping joints										
	1120	1" - 4" pipe										
	1140	0 - 250 L.F.	Q-5	.90	17.778	Ea.	6.30	560		566.30	845	
	1160	250 - 500 L.F.		.55	29.091		12.60	915		927.60	1,400	
	1180	500 - 1000 L.F.		.45	35.556		18.90	1,125		1,143.90	1,700	
	1200	1000 - 2000 L.F.		.23	69.565		25	2,175		2,200	3,325	
	1300	6" - 10" pipe										
	1320	0 - 250 L.F.	Q-5	.45	35.556	Ea.	6.30	1,125		1,131.30	1,675	
	1340	250 - 500 L.F.		.29	55.172		12.60	1,725		1,737.60	2,650	
	1360	500 - 1000 L.F.		.18	88.889		25	2,800		2,825	4,225	
	1380	1000 - 2000 L.F.		.09	177		31.50	5,575		5,606.50	8,450	
	2000	X-Ray of welds										
	2110	2" diam.	1 Stpi	8	1	Ea.	3	35		38	56	
	2120	3" diam.		8	1		3	35		38	56	
	2130	4" diam.		8	1		4.50	35		39.50	57.50	
	2140	6" diam.		8	1		4.50	35		39.50	57.50	
	2150	8" diam.		6.60	1.212		4.50	42		46.50	69	
	2160	10" diam.		6	1.333		6	46.50		52.50	76.50	
	3000	Liquid penetration of welds										
	3110	2" diam.	1 Stpi	14	.571	Ea.	1	19.90		20.90	31	
	3120	3" diam.		13.60	.588		1	20.50		21.50	32	
	3130	4" diam.		13.40	.597		1	21		22	32.50	
	3140	6" diam.		13.20	.606		1	21		22	33	
	3150	8" diam.		13	.615		1.50	21.50		23	34	
	3160	10" diam.		12.80	.625		1.50	22		23.50	34.50	
900	0010	**BALANCING, WATER** (Subcontractor's quote incl. material & labor)										900
	0050	Air cooled condenser				Ea.					147.28	
	0100	Cabinet unit heater	R15050-710								50.50	
	0200	Chiller									357.68	
	0300	Convector									42.08	
	0400	Converter									210.40	
	0500	Cooling tower									273.52	
	0600	Fan coil unit, unit ventilator									75.74	
	0700	Fin tube and radiant panels									84.16	
	0800	Main and duct re-heat coils									77.85	
	0810	Heat exchanger									77.85	
	0900	Main balancing cocks									63.12	

15950 | Testing/Adjusting/Balancing

		15955	HVAC Test/Adjust/Balance	CREW	DAILY OUTPUT	LABOR-HOURS	UNIT	2001 BARE COSTS				TOTAL INCL O&P	
								MAT.	LABOR	EQUIP.	TOTAL		
900	1000		Pumps				Ea.					185.15	900
	1100		Unit heater	R15050 -710			↓					58.91	

For information about Means Estimating Seminars, see yellow pages 11 and 12 in back of book

Division 16 Electrical

Estimating Tips

16060 Grounding & Bonding
- When taking off grounding system, identify separately the type and size of wire and list each unique type of ground connection.

16100 Wiring Methods
- Conduit should be taken off in three main categories: power distribution, branch power, and branch lighting, so the estimator can concentrate on systems and components, therefore making it easier to ensure all items have been accounted for.
- For cost modifications for elevated conduit installation, add the percentages to labor according to the height of installation and only the quantities exceeding the different height levels, not to the total conduit quantities.
- Remember that aluminum wiring of equal ampacity is larger in diameter than copper and may require larger conduit.
- If more than three wires at a time are being pulled, deduct percentages from the labor hours of that grouping of wires.
- The estimator should take the weights of materials into consideration when completing a takeoff. Topics to consider include: How will the materials be supported? What methods of support are available? How high will the support structure have to reach? Will the final support structure be able to withstand the total burden? Is the support material included or separate from the fixture, equipment and material specified?

16200 Electrical Power
- Do not overlook the costs for equipment used in the installation. If scaffolding or highlifts are available in the field, contractors may use them in lieu of the proposed ladders and rolling staging.

16400 Low-Voltage Distribution
- Supports and concrete pads may be shown on drawings for the larger equipment, or the support system may be just a piece of plywood for the back of a panelboard. In either case, it must be included in the costs.

16500 Lighting
- Fixtures should be taken off room by room, using the fixture schedule, specifications, and the ceiling plan. For large concentrations of lighting fixtures in the same area deduct the percentages from labor hours.

16700 Communications
16800 Sound & Video
- When estimating material costs for special systems, it is always prudent to obtain manufacturers' quotations for equipment prices and special installation requirements which will affect the total costs.

Reference Numbers
Reference numbers are shown in bold squares at the beginning of some major classifications. These numbers refer to related items in the Reference Section. The reference information may be an estimating procedure, an alternate pricing method or technical information.

Note: Not all subdivisions listed here necessarily appear in this publication.

16100 | Wiring Methods

16132 | Conduit & Tubing

			Crew	Daily Output	Labor-Hours	Unit	Mat.	2001 Bare Costs Labor	Equip.	Total	Total Incl O&P
260	0010	**CUTTING AND DRILLING**										260
	0100	Hole drilling to 10' high, concrete wall									
	0110	8" thick, 1/2" pipe size	1 Elec	12	.667	Ea.		23		23	34
	0120	3/4" pipe size		12	.667			23		23	34
	0130	1" pipe size		9.50	.842			29		29	43
	0140	1-1/4" pipe size		9.50	.842			29		29	43
	0150	1-1/2" pipe size		9.50	.842			29		29	43
	0160	2" pipe size		4.40	1.818			62.50		62.50	93
	0170	2-1/2" pipe size		4.40	1.818			62.50		62.50	93
	0180	3" pipe size		4.40	1.818			62.50		62.50	93
	0190	3-1/2" pipe size		3.30	2.424			83		83	124
	0200	4" pipe size		3.30	2.424			83		83	124
	0500	12" thick, 1/2" pipe size		9.40	.851			29		29	43.50
	0520	3/4" pipe size		9.40	.851			29		29	43.50
	0540	1" pipe size		7.30	1.096			37.50		37.50	56
	0560	1-1/4" pipe size		7.30	1.096			37.50		37.50	56
	0570	1-1/2" pipe size		7.30	1.096			37.50		37.50	56
	0580	2" pipe size		3.60	2.222			76		76	114
	0590	2-1/2" pipe size		3.60	2.222			76		76	114
	0600	3" pipe size		3.60	2.222			76		76	114
	0610	3-1/2" pipe size		2.80	2.857			98		98	146
	0630	4" pipe size		2.50	3.200			110		110	164
	0650	16" thick, 1/2" pipe size		7.60	1.053			36		36	54
	0670	3/4" pipe size		7	1.143			39		39	58.50
	0690	1" pipe size		6	1.333			45.50		45.50	68.50
	0710	1-1/4" pipe size		5.50	1.455			50		50	74.50
	0730	1-1/2" pipe size		5.50	1.455			50		50	74.50
	0750	2" pipe size		3	2.667			91.50		91.50	137
	0770	2-1/2" pipe size		2.70	2.963			102		102	152
	0790	3" pipe size		2.50	3.200			110		110	164
	0810	3-1/2" pipe size		2.30	3.478			119		119	178
	0830	4" pipe size		2	4			137		137	205
	0850	20" thick, 1/2" pipe size		6.40	1.250			43		43	64
	0870	3/4" pipe size		6	1.333			45.50		45.50	68.50
	0890	1" pipe size		5	1.600			55		55	82
	0910	1-1/4" pipe size		4.80	1.667			57		57	85.50
	0930	1-1/2" pipe size		4.60	1.739			59.50		59.50	89
	0950	2" pipe size		2.70	2.963			102		102	152
	0970	2-1/2" pipe size		2.40	3.333			114		114	171
	0990	3" pipe size		2.20	3.636			125		125	186
	1010	3-1/2" pipe size		2	4			137		137	205
	1030	4" pipe size		1.70	4.706			161		161	241
	1050	24" thick, 1/2" pipe size		5.50	1.455			50		50	74.50
	1070	3/4" pipe size		5.10	1.569			54		54	80.50
	1090	1" pipe size		4.30	1.860			64		64	95.50
	1110	1-1/4" pipe size		4	2			68.50		68.50	102
	1130	1-1/2" pipe size		4	2			68.50		68.50	102
	1150	2" pipe size		2.40	3.333			114		114	171
	1170	2-1/2" pipe size		2.20	3.636			125		125	186
	1190	3" pipe size		2	4			137		137	205
	1210	3-1/2" pipe size		1.80	4.444			152		152	228
	1230	4" pipe size		1.50	5.333			183		183	273
	1500	Brick wall, 8" thick, 1/2" pipe size		18	.444			15.25		15.25	23
	1520	3/4" pipe size		18	.444			15.25		15.25	23
	1540	1" pipe size		13.30	.601			20.50		20.50	31
	1560	1-1/4" pipe size		13.30	.601			20.50		20.50	31

16100 | Wiring Methods

16132 | Conduit & Tubing

		CREW	DAILY OUTPUT	LABOR-HOURS	UNIT	MAT.	LABOR	EQUIP.	TOTAL	TOTAL INCL O&P
260 1580	1-1/2" pipe size	1 Elec	13.30	.601	Ea.		20.50		20.50	31
1600	2" pipe size		5.70	1.404			48		48	72
1620	2-1/2" pipe size		5.70	1.404			48		48	72
1640	3" pipe size		5.70	1.404			48		48	72
1660	3-1/2" pipe size		4.40	1.818			62.50		62.50	93
1680	4" pipe size		4	2			68.50		68.50	102
1700	12" thick, 1/2" pipe size		14.50	.552			18.90		18.90	28.50
1720	3/4" pipe size		14.50	.552			18.90		18.90	28.50
1740	1" pipe size		11	.727			25		25	37
1760	1-1/4" pipe size		11	.727			25		25	37
1780	1-1/2" pipe size		11	.727			25		25	37
1800	2" pipe size		5	1.600			55		55	82
1820	2-1/2" pipe size		5	1.600			55		55	82
1840	3" pipe size		5	1.600			55		55	82
1860	3-1/2" pipe size		3.80	2.105			72		72	108
1880	4" pipe size		3.30	2.424			83		83	124
1900	16" thick, 1/2" pipe size		12.30	.650			22.50		22.50	33.50
1920	3/4" pipe size		12.30	.650			22.50		22.50	33.50
1940	1" pipe size		9.30	.860			29.50		29.50	44
1960	1-1/4" pipe size		9.30	.860			29.50		29.50	44
1980	1-1/2" pipe size		9.30	.860			29.50		29.50	44
2000	2" pipe size		4.40	1.818			62.50		62.50	93
2010	2-1/2" pipe size		4.40	1.818			62.50		62.50	93
2030	3" pipe size		4.40	1.818			62.50		62.50	93
2050	3-1/2" pipe size		3.30	2.424			83		83	124
2070	4" pipe size		3	2.667			91.50		91.50	137
2090	20" thick, 1/2" pipe size		10.70	.748			25.50		25.50	38.50
2110	3/4" pipe size		10.70	.748			25.50		25.50	38.50
2130	1" pipe size		8	1			34.50		34.50	51
2150	1-1/4" pipe size		8	1			34.50		34.50	51
2170	1-1/2" pipe size		8	1			34.50		34.50	51
2190	2" pipe size		4	2			68.50		68.50	102
2210	2-1/2" pipe size		4	2			68.50		68.50	102
2230	3" pipe size		4	2			68.50		68.50	102
2250	3-1/2" pipe size		3	2.667			91.50		91.50	137
2270	4" pipe size		2.70	2.963			102		102	152
2290	24" thick, 1/2" pipe size		9.40	.851			29		29	43.50
2310	3/4" pipe size		9.40	.851			29		29	43.50
2330	1" pipe size		7.10	1.127			38.50		38.50	57.50
2350	1-1/4" pipe size		7.10	1.127			38.50		38.50	57.50
2370	1-1/2" pipe size		7.10	1.127			38.50		38.50	57.50
2390	2" pipe size		3.60	2.222			76		76	114
2410	2-1/2" pipe size		3.60	2.222			76		76	114
2430	3" pipe size		3.60	2.222			76		76	114
2450	3-1/2" pipe size		2.80	2.857			98		98	146
2470	4" pipe size		2.50	3.200			110		110	164
3000	Knockouts to 8' high, metal boxes & enclosures									
3020	With hole saw, 1/2" pipe size	1 Elec	53	.151	Ea.		5.20		5.20	7.75
3040	3/4" pipe size		47	.170			5.85		5.85	8.70
3050	1" pipe size		40	.200			6.85		6.85	10.25
3060	1-1/4" pipe size		36	.222			7.60		7.60	11.40
3070	1-1/2" pipe size		32	.250			8.60		8.60	12.80
3080	2" pipe size		27	.296			10.15		10.15	15.15
3090	2-1/2" pipe size		20	.400			13.70		13.70	20.50
4010	3" pipe size		16	.500			17.15		17.15	25.50
4030	3-1/2" pipe size		13	.615			21		21	31.50

For expanded coverage of these items see *Means Electrical Cost Data 2001*

16100 | Wiring Methods

16132 | Conduit & Tubing

		CREW	DAILY OUTPUT	LABOR-HOURS	UNIT	2001 BARE COSTS MAT.	LABOR	EQUIP.	TOTAL	TOTAL INCL O&P
260 4050	4" pipe size	1 Elec	11	.727	Ea.		25		25	37
4070	With hand punch set, 1/2" pipe size		40	.200			6.85		6.85	10.25
4090	3/4" pipe size		32	.250			8.60		8.60	12.80
4110	1" pipe size		30	.267			9.15		9.15	13.65
4130	1-1/4" pipe size		28	.286			9.80		9.80	14.65
4150	1-1/2" pipe size		26	.308			10.55		10.55	15.75
4170	2" pipe size		20	.400			13.70		13.70	20.50
4190	2-1/2" pipe size		17	.471			16.15		16.15	24
4200	3" pipe size		15	.533			18.30		18.30	27.50
4220	3-1/2" pipe size		12	.667			23		23	34
4240	4" pipe size		10	.800			27.50		27.50	41
4260	With hydraulic punch, 1/2" pipe size		44	.182			6.25		6.25	9.30
4280	3/4" pipe size		38	.211			7.20		7.20	10.80
4300	1" pipe size		38	.211			7.20		7.20	10.80
4320	1-1/4" pipe size		38	.211			7.20		7.20	10.80
4340	1-1/2" pipe size		38	.211			7.20		7.20	10.80
4360	2" pipe size		32	.250			8.60		8.60	12.80
4380	2-1/2" pipe size		27	.296			10.15		10.15	15.15
4400	3" pipe size		23	.348			11.95		11.95	17.80
4420	3-1/2" pipe size		20	.400			13.70		13.70	20.50
4440	4" pipe size		18	.444			15.25		15.25	23

16139 | Residential Wiring

		CREW	DAILY OUTPUT	LABOR-HOURS	UNIT	MAT.	LABOR	EQUIP.	TOTAL	TOTAL INCL O&P
700 0010	**RESIDENTIAL WIRING**									
0020	20' avg. runs and #14/2 wiring incl. unless otherwise noted									
1000	Service & panel, includes 24' SE-AL cable, service eye, meter,									
1010	Socket, panel board, main bkr., ground rod, 15 or 20 amp									
1020	1-pole circuit breakers, and misc. hardware									
1100	100 amp, with 10 branch breakers	1 Elec	1.19	6.723	Ea.	395	231		626	780
1110	With PVC conduit and wire		.92	8.696		430	298		728	915
1120	With RGS conduit and wire		.73	10.959		550	375		925	1,175
1150	150 amp, with 14 branch breakers		1.03	7.767		615	266		881	1,075
1170	With PVC conduit and wire		.82	9.756		680	335		1,015	1,250
1180	With RGS conduit and wire		.67	11.940		895	410		1,305	1,600
1200	200 amp, with 18 branch breakers	2 Elec	1.80	8.889		805	305		1,110	1,350
1220	With PVC conduit and wire		1.46	10.959		870	375		1,245	1,525
1230	With RGS conduit and wire		1.24	12.903		1,150	445		1,595	1,925
1800	Lightning surge suppressor for above services, add	1 Elec	32	.250		39.50	8.60		48.10	56.50
2000	Switch devices									
2100	Single pole, 15 amp, Ivory, with a 1-gang box, cover plate,									
2110	Type NM (Romex) cable	1 Elec	17.10	.468	Ea.	6.85	16.05		22.90	31.50
2120	Type MC (BX) cable		14.30	.559		16.95	19.20		36.15	47
2130	EMT & wire		5.71	1.401		17.65	48		65.65	91
2150	3-way, #14/3, type NM cable		14.55	.550		10.80	18.85		29.65	40
2170	Type MC cable		12.31	.650		21.50	22.50		44	57
2180	EMT & wire		5	1.600		20	55		75	104
2200	4-way, #14/3, type NM cable		14.55	.550		23.50	18.85		42.35	54
2220	Type MC cable		12.31	.650		34	22.50		56.50	71
2230	EMT & wire		5	1.600		32.50	55		87.50	118
2250	S.P., 20 amp, #12/2, type NM cable		13.33	.600		12.05	20.50		32.55	44
2270	Type MC cable		11.43	.700		21	24		45	59
2280	EMT & wire		4.85	1.649		23.50	56.50		80	111
2290	S.P. rotary dimmer, 600W, no wiring		17	.471		15.70	16.15		31.85	41.50
2300	S.P. rotary dimmer, 600W, type NM cable		14.55	.550		18.60	18.85		37.45	48.50
2320	Type MC cable		12.31	.650		28.50	22.50		51	65

16100 | Wiring Methods

16139 | Residential Wiring

		CREW	DAILY OUTPUT	LABOR-HOURS	UNIT	MAT.	LABOR	EQUIP.	TOTAL	TOTAL INCL O&P
700	2330	EMT & wire	1 Elec	5	1.600	Ea.	30	55		85	116
	2350	3-way rotary dimmer, type NM cable		13.33	.600		16	20.50		36.50	48
	2370	Type MC cable		11.43	.700		26	24		50	64.50
	2380	EMT & wire	↓	4.85	1.649	↓	27.50	56.50		84	115
	2400	Interval timer wall switch, 20 amp, 1-30 min., #12/2									
	2410	Type NM cable	1 Elec	14.55	.550	Ea.	29.50	18.85		48.35	60.50
	2420	Type MC cable		12.31	.650		36	22.50		58.50	73.50
	2430	EMT & wire	↓	5	1.600	↓	41	55		96	127
	2500	Decorator style									
	2510	S.P., 15 amp, type NM cable	1 Elec	17.10	.468	Ea.	10.40	16.05		26.45	35.50
	2520	Type MC cable		14.30	.559		20.50	19.20		39.70	51
	2530	EMT & wire		5.71	1.401		21	48		69	95
	2550	3-way, #14/3, type NM cable		14.55	.550		14.35	18.85		33.20	44
	2570	Type MC cable		12.31	.650		25	22.50		47.50	61
	2580	EMT & wire		5	1.600		23.50	55		78.50	108
	2600	4-way, #14/3, type NM cable		14.55	.550		27	18.85		45.85	57.50
	2620	Type MC cable		12.31	.650		37.50	22.50		60	75
	2630	EMT & wire		5	1.600		36	55		91	122
	2650	S.P., 20 amp, #12/2, type NM cable		13.33	.600		15.60	20.50		36.10	47.50
	2670	Type MC cable		11.43	.700		24.50	24		48.50	63
	2680	EMT & wire		4.85	1.649		27	56.50		83.50	114
	2700	S.P., slide dimmer, type NM cable		17.10	.468		25.50	16.05		41.55	52.50
	2720	Type MC cable		14.30	.559		36	19.20		55.20	68
	2730	EMT & wire		5.71	1.401		37.50	48		85.50	113
	2750	S.P., touch dimmer, type NM cable		17.10	.468		21.50	16.05		37.55	47.50
	2770	Type MC cable		14.30	.559		31.50	19.20		50.70	63.50
	2780	EMT & wire		5.71	1.401		33	48		81	108
	2800	3-way touch dimmer, type NM cable		13.33	.600		39	20.50		59.50	73.50
	2820	Type MC cable		11.43	.700		49	24		73	90
	2830	EMT & wire	↓	4.85	1.649	↓	50.50	56.50		107	141
	3000	Combination devices									
	3100	S.P. switch/15 amp recpt., Ivory, 1-gang box, plate									
	3110	Type NM cable	1 Elec	11.43	.700	Ea.	14.65	24		38.65	52
	3120	Type MC cable		10	.800		25	27.50		52.50	68
	3130	EMT & wire		4.40	1.818		26.50	62.50		89	122
	3150	S.P. switch/pilot light, type NM cable		11.43	.700		15.30	24		39.30	53
	3170	Type MC cable		10	.800		25.50	27.50		53	69
	3180	EMT & wire		4.43	1.806		27	62		89	122
	3190	2-S.P. switches, 2-#14/2, no wiring		14	.571		5.45	19.60		25.05	35.50
	3200	2-S.P. switches, 2-#14/2, type NM cables		10	.800		16.85	27.50		44.35	59.50
	3220	Type MC cable		8.89	.900		33.50	31		64.50	83
	3230	EMT & wire		4.10	1.951		28	67		95	131
	3250	3-way switch/15 amp recpt., #14/3, type NM cable		10	.800		21.50	27.50		49	64.50
	3270	Type MC cable		8.89	.900		32	31		63	81
	3280	EMT & wire		4.10	1.951		30.50	67		97.50	134
	3300	2-3 way switches, 2-#14/3, type NM cables		8.89	.900		28.50	31		59.50	77.50
	3320	Type MC cable		8	1		46.50	34.50		81	102
	3330	EMT & wire		4	2		35	68.50		103.50	141
	3350	S.P. switch/20 amp recpt., #12/2, type NM cable		10	.800		25.50	27.50		53	69
	3370	Type MC cable		8.89	.900		32	31		63	81.50
	3380	EMT & wire	↓	4.10	1.951	↓	37	67		104	141
	3400	Decorator style									
	3410	S.P. switch/15 amp recpt., type NM cable	1 Elec	11.43	.700	Ea.	18.20	24		42.20	56
	3420	Type MC cable		10	.800		28.50	27.50		56	72
	3430	EMT & wire		4.40	1.818		30	62.50		92.50	126
	3450	S.P. switch/pilot light, type NM cable	↓	11.43	.700	↓	18.85	24		42.85	56.50

For expanded coverage of these items see *Means Electrical Cost Data 2001*

16100 | Wiring Methods

16139 | Residential Wiring

		CREW	DAILY OUTPUT	LABOR-HOURS	UNIT	2001 BARE COSTS MAT.	LABOR	EQUIP.	TOTAL	TOTAL INCL O&P
3470	Type MC cable	1 Elec	10	.800	Ea.	29	27.50		56.50	73
3480	EMT & wire		4.40	1.818		30.50	62.50		93	127
3500	2-S.P. switches, 2-#14/2, type NM cables		10	.800		20.50	27.50		48	63.50
3520	Type MC cable		8.89	.900		37	31		68	87
3530	EMT & wire		4.10	1.951		31.50	67		98.50	135
3550	3-way/15 amp recpt., #14/3, type NM cable		10	.800		25	27.50		52.50	68.50
3570	Type MC cable		8.89	.900		35.50	31		66.50	85
3580	EMT & wire		4.10	1.951		34	67		101	138
3650	2-3 way switches, 2-#14/3, type NM cables		8.89	.900		32	31		63	81.50
3670	Type MC cable		8	1		50	34.50		84.50	106
3680	EMT & wire		4	2		38.50	68.50		107	145
3700	S.P. switch/20 amp recpt., #12/2, type NM cable		10	.800		29	27.50		56.50	73
3720	Type MC cable		8.89	.900		36	31		67	85.50
3730	EMT & wire		4.10	1.951		40.50	67		107.50	145
4000	Receptacle devices									
4010	Duplex outlet, 15 amp recpt., Ivory, 1-gang box, plate									
4015	Type NM cable	1 Elec	14.55	.550	Ea.	5.40	18.85		24.25	34
4020	Type MC cable		12.31	.650		15.50	22.50		38	50.50
4030	EMT & wire		5.33	1.501		16.20	51.50		67.70	95
4050	With #12/2, type NM cable		12.31	.650		6.70	22.50		29.20	41
4070	Type MC cable		10.67	.750		15.75	25.50		41.25	56
4080	EMT & wire		4.71	1.699		18.05	58.50		76.55	107
4100	20 amp recpt., #12/2, type NM cable		12.31	.650		9.95	22.50		32.45	44.50
4120	Type MC cable		10.67	.750		19	25.50		44.50	59.50
4130	EMT & wire		4.71	1.699		21.50	58.50		80	111
4140	For GFI see line 4300 below									
4150	Decorator style, 15 amp recpt., type NM cable	1 Elec	14.55	.550	Ea.	8.95	18.85		27.80	38
4170	Type MC cable		12.31	.650		19.05	22.50		41.55	54.50
4180	EMT & wire		5.33	1.501		19.75	51.50		71.25	98.50
4200	With #12/2, type NM cable		12.31	.650		10.25	22.50		32.75	45
4220	Type MC cable		10.67	.750		19.30	25.50		44.80	60
4230	EMT & wire		4.71	1.699		21.50	58.50		80	111
4250	20 amp recpt. #12/2, type NM cable		12.31	.650		13.50	22.50		36	48.50
4270	Type MC cable		10.67	.750		22.50	25.50		48	63.50
4280	EMT & wire		4.71	1.699		25	58.50		83.50	115
4300	GFI, 15 amp recpt., type NM cable		12.31	.650		33	22.50		55.50	70
4320	Type MC cable		10.67	.750		43	25.50		68.50	86
4330	EMT & wire		4.71	1.699		44	58.50		102.50	136
4350	GFI with #12/2, type NM cable		10.67	.750		34.50	25.50		60	76.50
4370	Type MC cable		9.20	.870		43.50	30		73.50	92.50
4380	EMT & wire		4.21	1.900		45.50	65		110.50	148
4400	20 amp recpt., #12/2 type NM cable		10.67	.750		36	25.50		61.50	78
4420	Type MC cable		9.20	.870		45	30		75	94
4430	EMT & wire		4.21	1.900		47.50	65		112.50	150
4500	Weather-proof cover for above receptacles, add		32	.250		4.40	8.60		13	17.65
4550	Air conditioner outlet, 20 amp-240 volt recpt.									
4560	30' of #12/2, 2 pole circuit breaker									
4570	Type NM cable	1 Elec	10	.800	Ea.	40.50	27.50		68	85.50
4580	Type MC cable		9	.889		52	30.50		82.50	103
4590	EMT & wire		4	2		51.50	68.50		120	159
4600	Decorator style, type NM cable		10	.800		44	27.50		71.50	89.50
4620	Type MC cable		9	.889		56	30.50		86.50	107
4630	EMT & wire		4	2		55	68.50		123.50	163
4650	Dryer outlet, 30 amp-240 volt recpt., 20' of #10/3									
4660	2 pole circuit breaker									
4670	Type NM cable	1 Elec	6.41	1.248	Ea.	47.50	43		90.50	117

16100 | Wiring Methods

16139 | Residential Wiring

			CREW	DAILY OUTPUT	LABOR-HOURS	UNIT	MAT.	LABOR	EQUIP.	TOTAL	TOTAL INCL O&P
700	4680	Type MC cable	1 Elec	5.71	1.401	Ea.	56	48		104	134	700
	4690	EMT & wire	↓	3.48	2.299	↓	53	79		132	176	
	4700	Range outlet, 50 amp-240 volt recpt., 30' of #8/3										
	4710	Type NM cable	1 Elec	4.21	1.900	Ea.	74	65		139	179	
	4720	Type MC cable		4	2		101	68.50		169.50	213	
	4730	EMT & wire		2.96	2.703		69	92.50		161.50	214	
	4750	Central vacuum outlet, Type NM cable		6.40	1.250		43.50	43		86.50	112	
	4770	Type MC cable		5.71	1.401		59	48		107	137	
	4780	EMT & wire	↓	3.48	2.299	↓	53.50	79		132.50	177	
	4800	30 amp-110 volt locking recpt., #10/2 circ. bkr.										
	4810	Type NM cable	1 Elec	6.20	1.290	Ea.	51.50	44.50		96	123	
	4830	EMT & wire	"	3.20	2.500	"	62.50	86		148.50	197	
	4900	Low voltage outlets										
	4910	Telephone recpt., 20' of 4/C phone wire	1 Elec	26	.308	Ea.	6.70	10.55		17.25	23	
	4920	TV recpt., 20' of RG59U coax wire, F type connector	"	16	.500	"	11.15	17.15		28.30	38	
	4950	Door bell chime, transformer, 2 buttons, 60' of bellwire										
	4970	Economy model	1 Elec	11.50	.696	Ea.	48.50	24		72.50	89	
	4980	Custom model		11.50	.696		83.50	24		107.50	128	
	4990	Luxury model, 3 buttons	↓	9.50	.842	↓	228	29		257	294	
	6000	Lighting outlets										
	6050	Wire only (for fixture), type NM cable	1 Elec	32	.250	Ea.	3.77	8.60		12.37	16.95	
	6070	Type MC cable		24	.333		11.60	11.45		23.05	30	
	6080	EMT & wire		10	.800		11.40	27.50		38.90	53.50	
	6100	Box (4"), and wire (for fixture), type NM cable		25	.320		7.65	11		18.65	25	
	6120	Type MC cable		20	.400		15.50	13.70		29.20	37.50	
	6130	EMT & wire		11	.727		15.30	25		40.30	54	
	6200	Fixtures (use with lines 6050 or 6100 above)										
	6210	Canopy style, economy grade	1 Elec	40	.200	Ea.	25.50	6.85		32.35	38.50	
	6220	Custom grade		40	.200		46	6.85		52.85	61	
	6250	Dining room chandelier, economy grade		19	.421		76	14.45		90.45	105	
	6260	Custom grade		19	.421		225	14.45		239.45	270	
	6270	Luxury grade		15	.533		495	18.30		513.30	575	
	6310	Kitchen fixture (fluorescent), economy grade		30	.267		51.50	9.15		60.65	70	
	6320	Custom grade		25	.320		161	11		172	193	
	6350	Outdoor, wall mounted, economy grade		30	.267		27	9.15		36.15	43	
	6360	Custom grade		30	.267		101	9.15		110.15	125	
	6370	Luxury grade		25	.320		227	11		238	266	
	6410	Outdoor PAR floodlights, 1 lamp, 150 watt		20	.400		20	13.70		33.70	42.50	
	6420	2 lamp, 150 watt each		20	.400		33.50	13.70		47.20	57.50	
	6430	For infrared security sensor, add		32	.250		87	8.60		95.60	109	
	6450	Outdoor, quartz-halogen, 300 watt flood		20	.400		38	13.70		51.70	62.50	
	6600	Recessed downlight, round, pre-wired, 50 or 75 watt trim		30	.267		35	9.15		44.15	52	
	6610	With shower light trim		30	.267		42	9.15		51.15	59.50	
	6620	With wall washer trim		28	.286		52.50	9.80		62.30	72.50	
	6630	With eye-ball trim	↓	28	.286		49	9.80		58.80	68.50	
	6640	For direct contact with insulation, add					1.60			1.60	1.76	
	6700	Porcelain lamp holder	1 Elec	40	.200		3.50	6.85		10.35	14.10	
	6710	With pull switch		40	.200		3.75	6.85		10.60	14.40	
	6750	Fluorescent strip, 1-20 watt tube, wrap around diffuser, 24"		24	.333		51.50	11.45		62.95	73.50	
	6770	2-40 watt tubes, 48"		20	.400		78	13.70		91.70	107	
	6780	With residential ballast		20	.400		89.50	13.70		103.20	119	
	6800	Bathroom heat lamp, 1-250 watt		28	.286		30.50	9.80		40.30	48	
	6810	2-250 watt lamps	↓	28	.286	↓	51	9.80		60.80	70.50	
	6820	For timer switch, see line 2400										
	6900	Outdoor post lamp, incl. post, fixture, 35' of #14/2										
	6910	Type NMC cable	1 Elec	3.50	2.286	Ea.	181	78.50		259.50	315	

For expanded coverage of these items see *Means Electrical Cost Data 2001*

16100 | Wiring Methods

16139 | Residential Wiring

			CREW	DAILY OUTPUT	LABOR-HOURS	UNIT	2001 BARE COSTS MAT.	LABOR	EQUIP.	TOTAL	TOTAL INCL O&P	
700	6920	Photo-eye, add	1 Elec	27	.296	Ea.	29	10.15		39.15	47	700
	6950	Clock dial time switch, 24 hr., w/enclosure, type NM cable	↓	11.43	.700		51	24		75	92	
	6970	Type MC cable		11	.727		61	25		86	104	
	6980	EMT & wire	↓	4.85	1.649	↓	61.50	56.50		118	152	
	7000	Alarm systems										
	7050	Smoke detectors, box, #14/3, type NM cable	1 Elec	14.55	.550	Ea.	28.50	18.85		47.35	59.50	
	7070	Type MC cable		12.31	.650		37	22.50		59.50	74	
	7080	EMT & wire	↓	5	1.600		35.50	55		90.50	121	
	7090	For relay output to security system, add				↓	11.75			11.75	12.95	
	8000	Residential equipment										
	8050	Disposal hook-up, incl. switch, outlet box, 3' of flex										
	8060	20 amp-1 pole circ. bkr., and 25' of #12/2										
	8070	Type NM cable	1 Elec	10	.800	Ea.	19.70	27.50		47.20	62.50	
	8080	Type MC cable		8	1		30	34.50		64.50	84	
	8090	EMT & wire	↓	5	1.600		32.50	55		87.50	118	
	8100	Trash compactor or dishwasher hook-up, incl. outlet box,										
	8110	3' of flex, 15 amp-1 pole circ. bkr., and 25' of #14/2										
	8120	Type NM cable	1 Elec	10	.800	Ea.	13.70	27.50		41.20	56	
	8130	Type MC cable		8	1		26	34.50		60.50	79.50	
	8140	EMT & wire	↓	5	1.600	↓	26.50	55		81.50	112	
	8150	Hot water sink dispensor hook-up, use line 8100										
	8200	Vent/exhaust fan hook-up, type NM cable	1 Elec	32	.250	Ea.	3.77	8.60		12.37	16.95	
	8220	Type MC cable		24	.333		11.60	11.45		23.05	30	
	8230	EMT & wire	↓	10	.800	↓	11.40	27.50		38.90	53.50	
	8250	Bathroom vent fan, 50 CFM (use with above hook-up)										
	8260	Economy model	1 Elec	15	.533	Ea.	21.50	18.30		39.80	51	
	8270	Low noise model		15	.533		30	18.30		48.30	60.50	
	8280	Custom model		12	.667		111	23		134	156	
	8300	Bathroom or kitchen vent fan, 110 CFM										
	8310	Economy model	1 Elec	15	.533	Ea.	56	18.30		74.30	89	
	8320	Low noise model	"	15	.533	"	75	18.30		93.30	110	
	8350	Paddle fan, variable speed (w/o lights)										
	8360	Economy model (AC motor)	1 Elec	10	.800	Ea.	100	27.50		127.50	151	
	8370	Custom model (AC motor)		10	.800		173	27.50		200.50	231	
	8380	Luxury model (DC motor)		8	1		340	34.50		374.50	425	
	8390	Remote speed switch for above, add	↓	12	.667	↓	23	23		46	59.50	
	8500	Whole house exhaust fan, ceiling mount, 36", variable speed										
	8510	Remote switch, incl. shutters, 20 amp-1 pole circ. bkr.										
	8520	30' of #12/2, type NM cable	1 Elec	4	2	Ea.	550	68.50		618.50	705	
	8530	Type MC cable		3.50	2.286		560	78.50		638.50	735	
	8540	EMT & wire	↓	3	2.667	↓	565	91.50		656.50	755	
	8600	Whirlpool tub hook-up, incl. timer switch, outlet box										
	8610	3' of flex, 20 amp-1 pole GFI circ. bkr.										
	8620	30' of #12/2, type NM cable	1 Elec	5	1.600	Ea.	75.50	55		130.50	165	
	8630	Type MC cable		4.20	1.905		82.50	65.50		148	188	
	8640	EMT & wire	↓	3.40	2.353	↓	84.50	80.50		165	213	
	8650	Hot water heater hook-up, incl. 1-2 pole circ. bkr., box;										
	8660	3' of flex, 20' of #10/2, type NM cable	1 Elec	5	1.600	Ea.	21	55		76	105	
	8670	Type MC cable		4.20	1.905		34	65.50		99.50	135	
	8680	EMT & wire	↓	3.40	2.353	↓	28.50	80.50		109	152	
	9000	Heating/air conditioning										
	9050	Furnace/boiler hook-up, incl. firestat, local on-off switch										
	9060	Emergency switch, and 40' of type NM cable	1 Elec	4	2	Ea.	43	68.50		111.50	150	
	9070	Type MC cable		3.50	2.286		59.50	78.50		138	183	
	9080	EMT & wire	↓	1.50	5.333	↓	61	183		244	340	
	9100	Air conditioner hook-up, incl. local 60 amp disc. switch										

16100 | Wiring Methods

16139 | Residential Wiring

			CREW	DAILY OUTPUT	LABOR-HOURS	UNIT	MAT.	LABOR	EQUIP.	TOTAL	TOTAL INCL O&P	
700	9110	3' sealtite, 40 amp, 2 pole circuit breaker										700
	9130	40' of #8/2, type NM cable	1 Elec	3.50	2.286	Ea.	140	78.50		218.50	271	
	9140	Type MC cable		3	2.667		185	91.50		276.50	340	
	9150	EMT & wire	↓	1.30	6.154	↓	148	211		359	480	
	9200	Heat pump hook-up, 1-40 & 1-100 amp 2 pole circ. bkr.										
	9210	Local disconnect switch, 3' sealtite										
	9220	40' of #8/2 & 30' of #3/2										
	9230	Type NM cable	1 Elec	1.30	6.154	Ea.	345	211		556	695	
	9240	Type MC cable		1.08	7.407		455	254		709	880	
	9250	EMT & wire	↓	.94	8.511	↓	345	292		637	815	
	9500	Thermostat hook-up, using low voltage wire										
	9520	Heating only	1 Elec	24	.333	Ea.	5.70	11.45		17.15	23.50	
	9530	Heating/cooling	"	20	.400	"	7.05	13.70		20.75	28.50	

16150 | Wiring Connections

			CREW	DAILY OUTPUT	LABOR-HOURS	UNIT	MAT.	LABOR	EQUIP.	TOTAL	TOTAL INCL O&P	
275	0010	**MOTOR CONNECTIONS**										275
	0020	Flexible conduit and fittings, 115 volt, 1 phase, up to 1 HP motor	1 Elec	8	1	Ea.	4.39	34.50		38.89	56	
	0050	2 HP motor		6.50	1.231		4.49	42		46.49	68	
	0100	3 HP motor		5.50	1.455		7.50	50		57.50	82.50	
	0120	230 volt, 10 HP motor, 3 phase		4.20	1.905		8.10	65.50		73.60	106	
	0150	15 HP motor		3.30	2.424		15.55	83		98.55	141	
	0200	25 HP motor		2.70	2.963		17.75	102		119.75	172	
	0400	50 HP motor		2.20	3.636		46.50	125		171.50	238	
	0600	100 HP motor		1.50	5.333		117	183		300	400	
	1500	460 volt, 5 HP motor, 3 phase		8	1		4.90	34.50		39.40	56.50	
	1520	10 HP motor		8	1		4.90	34.50		39.40	56.50	
	1530	25 HP motor		6	1.333		9.30	45.50		54.80	78.50	
	1540	30 HP motor		6	1.333		9.30	45.50		54.80	78.50	
	1550	40 HP motor		5	1.600		14.65	55		69.65	98	
	1560	50 HP motor		5	1.600		16.60	55		71.60	100	
	1570	60 HP motor		3.80	2.105		22	72		94	133	
	1580	75 HP motor		3.50	2.286		28.50	78.50		107	149	
	1590	100 HP motor		2.50	3.200		39.50	110		149.50	208	
	1600	125 HP motor		2	4		51.50	137		188.50	262	
	1610	150 HP motor		1.80	4.444		53	152		205	287	
	1620	200 HP motor		1.50	5.333		92.50	183		275.50	375	
	2005	460 Volt, 5 HP motor, 3 Phase, w/sealtite		8	1		8.95	34.50		43.45	61	
	2010	10 HP motor		8	1		8.95	34.50		43.45	61	
	2015	25 HP motor		6	1.333		16.50	45.50		62	86.50	
	2020	30 HP motor		6	1.333		16.50	45.50		62	86.50	
	2025	40 HP motor		5	1.600		29.50	55		84.50	115	
	2030	50 HP motor		5	1.600		28.50	55		83.50	114	
	2035	60 HP motor		3.80	2.105		42	72		114	155	
	2040	75 HP motor		3.50	2.286		45	78.50		123.50	167	
	2045	100 HP motor		2.50	3.200		60.50	110		170.50	231	
	2055	150 HP motor		1.80	4.444		64	152		216	299	
	2060	200 HP motor	↓	1.50	5.333	↓	395	183		578	705	

For expanded coverage of these items see Means Electrical Cost Data 2001

16200 | Electrical Power

16220 | Motors & Generators

		CREW	DAILY OUTPUT	LABOR-HOURS	UNIT	2001 BARE COSTS MAT.	LABOR	EQUIP.	TOTAL	TOTAL INCL O&P	
900	0010 **VARIABLE FREQUENCY DRIVES/ADJUSTABLE FREQUENCY DRIVES**										900
	0100 Enclosed (NEMA 1), 460 volt, for 3 HP motor size	1 Elec	.80	10	Ea.	1,500	345		1,845	2,150	
	0110 5 HP motor size		.80	10		1,725	345		2,070	2,400	
	0120 7.5 HP motor size		.67	11.940		2,450	410		2,860	3,275	
	0130 10 HP motor size		.67	11.940		2,450	410		2,860	3,275	
	0140 15 HP motor size	2 Elec	.89	17.978		2,650	615		3,265	3,825	
	0150 20 HP motor size		.89	17.978		3,900	615		4,515	5,200	
	0160 25 HP motor size		.67	23.881		4,300	820		5,120	5,950	
	0170 30 HP motor size		.67	23.881		4,950	820		5,770	6,675	
	0180 40 HP motor size		.67	23.881		7,275	820		8,095	9,225	
	0190 50 HP motor size		.53	30.189		7,925	1,025		8,950	10,300	
	0200 60 HP motor size	R-3	.56	35.714		8,875	1,200	254	10,329	11,800	
	0210 75 HP motor size		.56	35.714		12,000	1,200	254	13,454	15,300	
	0220 100 HP motor size		.50	40		13,800	1,350	284	15,434	17,500	
	0230 125 HP motor size		.50	40		16,300	1,350	284	17,934	20,200	
	0240 150 HP motor size		.50	40		18,700	1,350	284	20,334	22,900	
	0250 200 HP motor size		.42	47.619		21,000	1,625	340	22,965	25,900	
	1100 Custom-engineered, 460 volt, for 3 HP motor size	1 Elec	.56	14.286		2,425	490		2,915	3,400	
	1110 5 HP motor size		.56	14.286		2,650	490		3,140	3,650	
	1120 7.5 HP motor size		.47	17.021		3,575	585		4,160	4,800	
	1130 10 HP motor size		.47	17.021		3,775	585		4,360	5,025	
	1140 15 HP motor size	2 Elec	.62	25.806		4,650	885		5,535	6,450	
	1150 20 HP motor size		.62	25.806		5,475	885		6,360	7,350	
	1160 25 HP motor size		.47	34.043		6,325	1,175		7,500	8,700	
	1170 30 HP motor size		.47	34.043		7,575	1,175		8,750	10,100	
	1180 40 HP motor size		.47	34.043		8,075	1,175		9,250	10,600	
	1190 50 HP motor size		.37	43.243		10,200	1,475		11,675	13,400	
	1200 60 HP motor size	R-3	.39	51.282		14,100	1,725	365	16,190	18,500	
	1210 75 HP motor size		.39	51.282		15,300	1,725	365	17,390	19,800	
	1220 100 HP motor size		.35	57.143		16,600	1,925	405	18,930	21,500	
	1230 125 HP motor size		.35	57.143		19,000	1,925	405	21,330	24,200	
	1240 150 HP motor size		.35	57.143		21,100	1,925	405	23,430	26,500	
	1250 200 HP motor size		.29	68.966		24,600	2,325	490	27,415	31,100	
	2000 For complex & special design systems to meet specific										
	2010 requirements, obtain quote from vendor.										

16400 | Low-Voltage Distribution

16420 | Enclosed Controllers

		CREW	DAILY OUTPUT	LABOR-HOURS	UNIT	2001 BARE COSTS MAT.	LABOR	EQUIP.	TOTAL	TOTAL INCL O&P	
220	0010 **CONTROL STATIONS**										220
	0050 NEMA 1, heavy duty, stop/start	1 Elec	8	1	Ea.	113	34.50		147.50	175	
	0100 Stop/start, pilot light		6.20	1.290		153	44.50		197.50	234	
	0200 Hand/off/automatic		6.20	1.290		83.50	44.50		128	158	
	0400 Stop/start/reverse		5.30	1.509		153	52		205	246	
	0500 NEMA 7, heavy duty, stop/start		6	1.333		188	45.50		233.50	276	
	0600 Stop/start, pilot light		4	2		230	68.50		298.50	355	

16400 | Low-Voltage Distribution

16440 | Swbds, Panels & Cont Centers

		CREW	DAILY OUTPUT	LABOR-HOURS	UNIT	2001 BARE COSTS MAT.	LABOR	EQUIP.	TOTAL	TOTAL INCL O&P	
660	0010 **MOTOR STARTERS & CONTROLS**										660
	0050 Magnetic, FVNR, with enclosure and heaters, 480 volt										
	0080 2 HP, size 00	1 Elec	3.50	2.286	Ea.	167	78.50		245.50	300	
	0100 5 HP, size 0		2.30	3.478		202	119		321	400	
	0200 10 HP, size 1		1.60	5		227	172		399	505	
	0300 25 HP, size 2	2 Elec	2.20	7.273		425	249		674	840	
	0400 50 HP, size 3		1.80	8.889		695	305		1,000	1,225	
	0500 100 HP, size 4		1.20	13.333		1,550	455		2,005	2,375	
	0600 200 HP, size 5		.90	17.778		3,600	610		4,210	4,875	
	0610 400 HP, size 6		.80	20		10,000	685		10,685	12,000	
	0620 NEMA 7, 5 HP, size 0	1 Elec	1.60	5		865	172		1,037	1,200	
	0630 10 HP, size 1	"	1.10	7.273		905	249		1,154	1,375	
	0640 25 HP, size 2	2 Elec	1.80	8.889		1,450	305		1,755	2,050	
	0650 50 HP, size 3		1.20	13.333		2,150	455		2,605	3,050	
	0660 100 HP, size 4		.90	17.778		3,525	610		4,135	4,775	
	0670 200 HP, size 5		.50	32		8,250	1,100		9,350	10,700	
	0700 Combination, with motor circuit protectors, 5 HP, size 0	1 Elec	1.80	4.444		655	152		807	950	
	0800 10 HP, size 1	"	1.30	6.154		680	211		891	1,075	
	0900 25 HP, size 2	2 Elec	2	8		955	274		1,229	1,450	
	1000 50 HP, size 3		1.32	12.121		1,375	415		1,790	2,150	
	1200 100 HP, size 4		.80	20		3,000	685		3,685	4,325	
	1220 NEMA 7, 5 HP, size 0	1 Elec	1.30	6.154		1,525	211		1,736	2,000	
	1230 10 HP, size 1	"	1	8		1,575	274		1,849	2,125	
	1240 25 HP, size 2	2 Elec	1.32	12.121		2,075	415		2,490	2,925	
	1250 50 HP, size 3		.80	20		3,425	685		4,110	4,800	
	1260 100 HP, size 4		.60	26.667		5,325	915		6,240	7,225	
	1270 200 HP, size 5		.40	40		11,500	1,375		12,875	14,800	
	1400 Combination, with fused switch, 5 HP, size 0	1 Elec	1.80	4.444		500	152		652	780	
	1600 10 HP, size 1	"	1.30	6.154		535	211		746	905	
	1800 25 HP, size 2	2 Elec	2	8		870	274		1,144	1,375	
	2000 50 HP, size 3		1.32	12.121		1,475	415		1,890	2,250	
	2200 100 HP, size 4		.80	20		2,575	685		3,260	3,850	
	3500 Magnetic FVNR with NEMA 12, enclosure & heaters, 480 volt										
	3600 5 HP, size 0	1 Elec	2.20	3.636	Ea.	190	125		315	395	
	3700 10 HP, size 1	"	1.50	5.333		286	183		469	590	
	3800 25 HP, size 2	2 Elec	2	8		535	274		809	1,000	
	3900 50 HP, size 3		1.60	10		825	345		1,170	1,425	
	4000 100 HP, size 4		1	16		1,975	550		2,525	3,000	
	4100 200 HP, size 5		.80	20		4,725	685		5,410	6,225	
	4200 Combination, with motor circuit protectors, 5 HP, size 0	1 Elec	1.70	4.706		630	161		791	935	
	4300 10 HP, size 1	"	1.20	6.667		655	229		884	1,050	
	4400 25 HP, size 2	2 Elec	1.80	8.889		980	305		1,285	1,525	
	4500 50 HP, size 3		1.20	13.333		1,600	455		2,055	2,425	
	4600 100 HP, size 4		.74	21.622		3,600	740		4,340	5,075	
	4700 Combination, with fused switch, 5 HP, size 0	1 Elec	1.70	4.706		610	161		771	910	
	4800 10 HP, size 1	"	1.20	6.667		635	229		864	1,050	
	4900 25 HP, size 2	2 Elec	1.80	8.889		970	305		1,275	1,525	
	5000 50 HP, size 3		1.20	13.333		1,550	455		2,005	2,375	
	5100 100 HP, size 4		.74	21.622		3,150	740		3,890	4,550	
	5200 Factory installed controls, adders to size 0 thru 5										
	5300 Start-stop push button	1 Elec	32	.250	Ea.	40	8.60		48.60	57	
	5400 Hand-off-auto-selector switch		32	.250		40	8.60		48.60	57	
	5500 Pilot light		32	.250		74.50	8.60		83.10	95	
	5600 Start-stop-pilot		32	.250		115	8.60		123.60	139	
	5700 Auxiliary contact, NO or NC		32	.250		55	8.60		63.60	73.50	
	5800 NO-NC		32	.250		110	8.60		118.60	134	

For expanded coverage of these items see Means Electrical Cost Data 2001

Division Notes

	CREW	DAILY OUTPUT	LABOR-HOURS	UNIT	2001 BARE COSTS				TOTAL INCL O&P
					MAT.	LABOR	EQUIP.	TOTAL	

Division 17
Square Foot & Cubic Foot Costs

Estimating Tips

- The cost figures in Division 17 were derived from approximately 10,200 projects contained in the Means database of completed construction projects, and include the contractor's overhead and profit, but do not generally include architectural fees or land costs. The figures have been adjusted to January of the current year. New projects are added to our files each year, and outdated projects are discarded. For this reason, certain costs may not show a uniform annual progression. In no case are all subdivisions of a project listed.

- These projects were located throughout the U.S. and reflect a tremendous variation in square foot (S.F.) and cubic foot (C.F.) costs. This is due to differences, not only in labor and material costs, but also in individual owners' requirements. For instance, a bank in a large city would have different features than one in a rural area. This is true of all the different types of buildings analyzed. Therefore, caution should be exercised when using Division 17 costs. For example, for court houses, costs in the database are local court house costs and will not apply to the larger, more elaborate federal court houses. As a general rule, the projects in the 1/4 column do not include any site work or equipment, while the projects in the 3/4 column may include both equipment and site work. The median figures do not generally include site work.

- None of the figures "go with" any others. All individual cost items were computed and tabulated separately. Thus the sum of the median figures for Plumbing, HVAC and Electrical will not normally total up to the total Mechanical and Electrical costs arrived at by separate analysis and tabulation of the projects.

- Each building was analyzed as to total and component costs and percentages. The figures were arranged in ascending order with the results tabulated as shown. The 1/4 column shows that 25% of the projects had lower costs, 75% higher. The 3/4 column shows that 75% of the projects had lower costs, 25% had higher. The median column shows that 50% of the projects had lower costs, 50% had higher.

- There are two times when square foot costs are useful. The first is in the conceptual stage when no details are available. Then square foot costs make a useful starting point. The second is after the bids are in and the costs can be worked back into their appropriate units for information purposes. As soon as details become available in the project design, the square foot approach should be discontinued and the project priced as to its particular components. When more precision is required or for estimating the replacement cost of specific buildings, the current edition of *Means Square Foot Costs* should be used.

- In using the figures in Division 17, it is recommended that the median column be used for preliminary figures if no additional information is available. The median figures, when multiplied by the total city construction cost index figures (see City Cost Indexes) and then multiplied by the project size modifier in Reference Number R17100-100, should present a fairly accurate base figure, which would then have to be adjusted in view of the estimator's experience, local economic conditions, code requirements and the owner's particular requirements. There is no need to factor the percentage figures, as these should remain constant from city to city. All tabulations mentioning air conditioning had at least partial air conditioning.

- The editors of this book would greatly appreciate receiving cost figures on one or more of your recent projects which would then be included in the averages for next year. All cost figures received will be kept confidential except that they will be averaged with other similar projects to arrive at S.F. and C.F. cost figures for next year's book. See the last page of the book for details and the discount available for submitting one or more of your projects.

17100 | S.F., C.F. and % of Total Costs

17100 | S.F. & C.F. Costs

			UNIT	UNIT COSTS 1/4	UNIT COSTS MEDIAN	UNIT COSTS 3/4	% OF TOTAL 1/4	% OF TOTAL MEDIAN	% OF TOTAL 3/4	
010	0010	**APARTMENTS** Low Rise (1 to 3 story)	S.F.	43.95	55.05	73.60				010
	0020	Total project cost	C.F.	3.94	5.25	6.45				
	2720	Plumbing	S.F.	3.39	4.40	5.55	6.70%	9%	10.10%	
	2770	Heating, ventilating, air conditioning		2.18	2.69	3.95	4.20%	5.80%	7.70%	
	2900	Electrical		2.53	3.30	4.52	5.20%	6.70%	8.40%	
	3100	Total: Mechanical & Electrical	↓	8.75	11.15	14	16%	18.20%	23%	
	9000	Per apartment unit, total cost	Apt.	40,500	61,700	92,200				
	9500	Total: Mechanical & Electrical	"	7,700	12,200	15,900				
020	0010	**APARTMENTS** Mid Rise (4 to 7 story)	S.F.	57.15	69.50	85				020
	0020	Total project costs	C.F.	4.55	6.30	8.60				
	2720	Plumbing	S.F.	3.42	5.45	6.05	6.30%	7.40%	10%	
	2900	Electrical		4.02	5.50	6.40	6.70%	7.50%	8.90%	
	3100	Total: Mechanical & Electrical	↓	11.60	15.70	19.25	18.80%	21.60%	25.10%	
	9000	Per apartment unit, total cost	Apt.	65,900	77,800	128,700				
	9500	Total: Mechanical & Electrical	"	12,400	14,400	18,200				
030	0010	**APARTMENTS** High Rise (8 to 24 story)	S.F.	66.05	79.70	97.40				030
	0020	Total project costs	C.F.	5.62	7.85	9.55				
	2720	Plumbing	S.F.	4.87	5.75	7.05	7%	9.10%	10.60%	
	2900	Electrical		4.54	5.75	7.75	6.40%	7.70%	8.90%	
	3100	Total: Mechanical & Electrical	↓	13.60	17.35	20.90	18%	22.30%	24.40%	
	9000	Per apartment unit, total cost	Apt.	68,400	74,200	104,900				
	9500	Total: Mechanical & Electrical	"	14,800	16,900	18,000				
040	0010	**AUDITORIUMS**	S.F.	67.95	92.05	124				040
	0020	Total project costs	C.F.	4.30	6	9				
	2720	Plumbing	S.F.	4.35	6	7.60	6.30%	7.20%	9.05%	
	2900	Electrical		5.55	7.65	9.55	6.80%	9%	10.60%	
	3100	Total: Mechanical & Electrical	↓	10.60	14.25	24.80	14.40%	18.50%	23.60%	
050	0010	**AUTOMOTIVE SALES**	S.F.	49.50	56.95	83.85				050
	0020	Total project costs	C.F.	3.30	4.02	5.20				
	2720	Plumbing	S.F.	2.48	4.02	4.39	4.70%	6.50%	7%	
	2900	Electrical		3.94	6.13	6.95	7.30%	10%	12.40%	
	3100	Total: Mechanical & Electrical	↓	8.90	12.85	16.05	17%	20.40%	26%	
060	0010	**BANKS**	S.F.	98.60	123	156				060
	0020	Total project costs	C.F.	7.10	9.65	12.65				
	2720	Plumbing	S.F.	3.14	4.49	6.50	2.80%	3.90%	4.90%	
	2770	Heating, ventilating, air conditioning		6.10	7.95	10.55	5%	7.20%	8.50%	
	2900	Electrical		9.55	12.60	16.40	8.30%	10.30%	12.20%	
	3100	Total: Mechanical & Electrical	↓	22.65	30.50	37	16.60%	20.40%	24.40%	
130	0010	**CHURCHES**	S.F.	66.20	83.15	108				130
	0020	Total project costs	C.F.	4.16	5.25	6.90				
	2720	Plumbing	S.F.	2.61	3.65	5.40	3.60%	5%	6.30%	
	2770	Heating, ventilating, air conditioning		6.10	7.95	11.70	7.60%	10%	12.20%	
	2900	Electrical		5.50	7.60	9.85	7.30%	8.80%	10.90%	
	3100	Total: Mechanical & Electrical	↓	14.30	21.45	31.15	19.60%	22.90%	25.40%	
150	0010	**CLUBS, COUNTRY**	S.F.	69.85	82.85	109				150
	0020	Total project costs	C.F.	5.80	7.05	9.75				
	2720	Plumbing	S.F.	4.62	6.45	14.70	6.40%	9.60%	10%	
	2900	Electrical		5.65	8.10	10.65	7%	9.30%	11.40%	
	3100	Total: Mechanical & Electrical	↓	12.20	15.70	29.20	16.10%	22.30%	30.90%	
170	0010	**CLUBS, SOCIAL** Fraternal	S.F.	56.45	80.65	106				170
	0020	Total project costs	C.F.	3.39	5.45	6.30				
	2720	Plumbing	S.F.	3.25	4.50	5.45	5.20%	6.80%	8.30%	
	2770	Heating, ventilating, air conditioning		5.20	6.30	8.10	8.70%	11%	14.40%	
	2900	Electrical		4.55	7.10	8.60	7.30%	9.50%	11.40%	
	3100	Total: Mechanical & Electrical	↓	12.15	16.50	23.60	19.30%	23%	33.10%	

See R17100-100 and City Cost Indexes in the Reference Section

17100 | S.F., C.F. and % of Total Costs

17100 | S.F. & C.F. Costs

			UNIT	UNIT COSTS			% OF TOTAL			
				1/4	MEDIAN	3/4	1/4	MEDIAN	3/4	
180	0010	**CLUBS, Y.M.C.A.** R17100-100	S.F.	65.80	83.10	111				180
	0020	Total project costs	C.F.	3.32	5.55	8.30				
	2720	Plumbing	S.F.	4.79	9.05	10.15	6%	9.70%	16%	
	2900	Electrical		5.45	7.20	10.60	6.40%	9.20%	10.80%	
	3100	Total: Mechanical & Electrical	↓	11.95	17.10	25.50	16.20%	21.90%	28.50%	
190	0010	**COLLEGES** Classrooms & Administration	S.F.	82.30	109	146				190
	0020	Total project costs	C.F.	6	8.35	13.55				
	2720	Plumbing	S.F.	4.03	7.90	14.65	4.80%	8.90%	11.80%	
	2900	Electrical		6.75	10.40	12.80	8.40%	10%	12.10%	
	3100	Total: Mechanical & Electrical	↓	14.50	26.40	40.85	14.20%	24.70%	32.50%	
210	0010	**COLLEGES** Science, Engineering, Laboratories	S.F.	131	159	193				210
	0020	Total project costs	C.F.	7.75	11.35	12.85				
	2900	Electrical	S.F.	11.20	16.15	24.35	7.10%	9.60%	15.40%	
	3100	Total: Mechanical & Electrical	"	41.30	49.20	76.15	28%	31.60%	39.80%	
230	0010	**COLLEGES** Student Unions	S.F.	86.30	121	142				230
	0020	Total project costs	C.F.	4.82	6.30	8.15				
	3100	Total: Mechanical & Electrical	S.F.	22.70	32.45	35.15	17.40%	23.90%	28.90%	
250	0010	**COMMUNITY CENTERS**	"	64.50	86.60	115				250
	0020	Total project costs	C.F.	4.65	6.65	8.60				
	2720	Plumbing	S.F.	3.46	5.90	8.60	4.90%	7.10%	9.50%	
	2770	Heating, ventilating, air conditioning		5.70	8.20	11.30	7.70%	10.80%	13%	
	2900	Electrical		5.70	7.70	11.70	7.40%	9.40%	11.50%	
	3100	Total: Mechanical & Electrical	↓	21.15	25.25	36.40	25.10%	27.90%	34.40%	
280	0010	**COURT HOUSES**	S.F.	102	118	136				280
	0020	Total project costs	C.F.	7.80	9.45	13.10				
	2720	Plumbing	S.F.	4.89	6.82	9.80	5.90%	7.40%	9.60%	
	2900	Electrical		9	10.95	14.35	8.40%	9.80%	11.60%	
	3100	Total: Mechanical & Electrical	↓	20.20	27.70	35.55	20.10%	25.70%	28.70%	
300	0010	**DEPARTMENT STORES**	S.F.	38.05	51.45	64.95				300
	0020	Total project costs	C.F.	2	2.64	3.59				
	2720	Plumbing	S.F.	1.24	1.50	2.27	3.10%	4.20%	5.90%	
	2770	Heating, ventilating, air conditioning		3.46	5.35	8.05	8.30%	12.40%	14.80%	
	2900	Electrical		4.37	6	7.05	9.10%	12.20%	14.90%	
	3100	Total: Mechanical & Electrical	↓	7.70	9.95	16.10	20%	19%	26.50%	
310	0010	**DORMITORIES** Low Rise (1 to 3 story)	S.F.	62.20	88.75	107				310
	0020	Total project costs	C.F.	4.31	6.60	9.90				
	2720	Plumbing	S.F.	4.34	5.80	7.35	8%	9%	9.60%	
	2770	Heating, ventilating, air conditioning		4.59	5.50	7.30	4.60%	8%	10%	
	2900	Electrical		4.59	7	8.75	6.60%	8.90%	9.60%	
	3100	Total: Mechanical & Electrical	↓	24.10	25.25	26.30	22.50%	24.10%	29.10%	
	9000	Per bed, total cost	Bed	30,600	33,900	72,800				
320	0010	**DORMITORIES** Mid Rise (4 to 8 story)	S.F.	88.60	116	143				320
	0020	Total project costs	C.F.	9.75	10.70	12.85				
	2900	Electrical	S.F.	6.55	10.30	13.15	7.40%	8.90%	10.40%	
	3100	Total: Mechanical & Electrical	"	19.65	26.30	34.40	18.10%	19.50%	30.60%	
	9000	Per bed, total cost	Bed	12,600	28,800	59,200				
340	0010	**FACTORIES**	S.F.	33.80	49.85	77.15				340
	0020	Total project costs	C.F.	2.15	3.19	5.30				
	2720	Plumbing	S.F.	1.99	3.36	5.70	4.60%	6.20%	8.70%	
	2770	Heating, ventilating, air conditioning		3.52	5.05	6.80	5.20%	8.50%	11.40%	
	2900	Electrical		4.11	6.55	10.05	8.10%	10.50%	14.20%	
	3100	Total: Mechanical & Electrical	↓	9.90	15.90	24.15	20.40%	28.10%	34.50%	
360	0010	**FIRE STATIONS**	S.F.	64.80	87.05	112				360
	0020	Total project costs	C.F.	3.89	5.30	7.05				

For expanded coverage of these items see *Means Square Foot Costs 2001*

17100 | S.F., C.F. and % of Total Costs

17100 | S.F. & C.F. Costs

			UNIT	UNIT COSTS 1/4	MEDIAN	3/4	% OF TOTAL 1/4	MEDIAN	3/4	
360	2720	Plumbing	S.F.	4.21	6.20	9.15	6.20%	7.70%	9.70%	360
	2770	Heating, ventilating, air conditioning		3.63	5.95	9.10	4.90%	7.50%	9.30%	
	2900	Electrical		4.61	7.95	10.70	7%	9.20%	11%	
	3100	Total: Mechanical & Electrical		20.45	25.25	29.95	18.80%	23.80%	27.30%	
370	0010	**FRATERNITY HOUSES** and Sorority Houses	S.F.	66.50	85.65	121				370
	0020	Total project costs	C.F.	6.35	6.90	8.85				
	2720	Plumbing	S.F.	4.98	5.75	10.55	5.20%	8%	10.90%	
	2900	Electrical		4.39	9.50	11.60	6.60%	9.90%	10.70%	
	3100	Total: Mechanical & Electrical		12.15	17.25	20.75	14.60%	19.90%	20.70%	
380	0010	**FUNERAL HOMES**	S.F.	69.80	95.70	174				380
	0020	Total project costs	C.F.	7.15	7.95	15.30				
	2900	Electrical	S.F.	3.10	5.70	6.70	3.60%	4.40%	11%	
	3100	Total: Mechanical & Electrical	"	11.15	16.65	21.55	12.90%	17.80%	18.80%	
390	0010	**GARAGES, COMMERCIAL** (Service)	S.F.	38.30	60.80	83.60				390
	0020	Total project costs	C.F.	2.43	3.64	5.25				
	2720	Plumbing	S.F.	2.58	3.88	7.60	4.70%	7.40%	11%	
	2730	Heating & ventilating		3.61	4.92	6.80	5.30%	6.90%	9.60%	
	2900	Electrical		3.65	5.60	7.75	7.20%	9%	11.20%	
	3100	Total: Mechanical & Electrical		8.55	15.95	23.55	14.10%	22.50%	26.90%	
400	0010	**GARAGES, MUNICIPAL** (Repair)	S.F.	58.40	77.75	111				400
	0020	Total project costs	C.F.	3.65	4.63	7.55				
	2720	Plumbing	S.F.	2.62	5.05	9.50	3.60%	6.70%	8%	
	2730	Heating & ventilating		4.48	6.50	12.50	6.20%	7.40%	13.50%	
	2900	Electrical		4.31	6.35	9.80	6.30%	8%	11.70%	
	3100	Total: Mechanical & Electrical		11	19.75	32.55	15.20%	25.50%	32.70%	
410	0010	**GARAGES, PARKING**	S.F.	22	31.90	56.95				410
	0020	Total project costs	C.F.	2.13	2.90	4.22				
	2720	Plumbing	S.F.	.59	1	1.54	2.60%	3.40%	3.90%	
	2900	Electrical		.93	1.41	2.07	4.30%	5.20%	6.30%	
	3100	Total: Mechanical & Electrical		1.24	3.64	4.67	6.50%	9.40%	12.80%	
	9000	Per car, total cost	Car	9,600	12,000	15,600				
430	0010	**GYMNASIUMS**	S.F.	63	80.45	103				430
	0020	Total project costs	C.F.	3.15	4	5.25				
	2720	Plumbing	S.F.	4	4.93	5.90	5.40%	7.30%	7.90%	
	2770	Heating, ventilating, air conditioning	"	4.30	6.55	13.15	9%	11.10%	22.60%	
	2900	Electrical	S.F.	4.79	6	8.05	6.60%	8.30%	10.70%	
	3100	Total: Mechanical & Electrical		14.80	20.40	27.35	20.60%	26.20%	29.40%	
460	0010	**HOSPITALS**		130	152	225				460
	0020	Total project costs	C.F.	9.20	11.35	16.35				
	2720	Plumbing	S.F.	10.85	14.60	18.80	7.80%	9.40%	11.80%	
	2770	Heating, ventilating, air conditioning		15.35	20.35	27.40	8.40%	14.60%	17%	
	2900	Electrical		13.05	16.95	26.70	9.90%	12%	14.50%	
	3100	Total: Mechanical & Electrical		37.15	49.25	80.70	26.60%	33.10%	39%	
	9000	Per bed or person, total cost	Bed	46,800	99,300	156,600				
480	0010	**HOUSING** For the Elderly	S.F.	59.65	75.25	92.45				480
	0020	Total project costs	C.F.	4.23	5.90	7.55				
	2720	Plumbing	S.F.	4.43	5.80	7.75	8.10%	9.70%	10.90%	
	2730	Heating, ventilating, air conditioning		2.27	3.22	4.81	3.30%	5.60%	7.20%	
	2900	Electrical		4.44	6	7.70	7.30%	8.50%	10.50%	
	3100	Total: Mechanical & Electrical		15.30	17.95	23.75	18.10%	22%	29.10%	
	9000	Per rental unit, total cost	Unit	55,500	64,500	72,100				
	9500	Total: Mechanical & Electrical	"	12,200	14,200	16,600				
500	0010	**HOUSING** Public (Low Rise)	S.F.	51.85	69.70	90.60				500
	0020	Total project costs	C.F.	3.98	5.55	6.90				
	2720	Plumbing	S.F.	3.40	4.78	6.05	6.80%	9%	11.60%	
	2730	Heating, ventilating, air conditioning		1.82	3.53	3.86	4.20%	6%	6.40%	
	2900	Electrical		3.03	4.52	6.25	5.10%	6.60%	8.30%	
	3100	Total: Mechanical & Electrical		14.40	18.55	20.75	14.50%	17.60%	26.50%	

See R17100-100 and City Cost Indexes in the Reference Section

17100 | S.F., C.F. and % of Total Costs

17100 | S.F. & C.F. Costs

			UNIT	UNIT COSTS 1/4	UNIT COSTS MEDIAN	UNIT COSTS 3/4	% OF TOTAL 1/4	% OF TOTAL MEDIAN	% OF TOTAL 3/4	
500	9000	Per apartment, total cost	Apt.	55,000	62,600	78,600				500
	9500	Total: Mechanical & Electrical	"	10,000	13,000	15,000				
510	0010	ICE SKATING RINKS	S.F.	44.60	77.40	110				510
	0020	Total project costs	C.F.	3.15	3.22	3.71				
	2720	Plumbing	S.F.	1.60	3	3.07	3.10%	5.60%	6.70%	
	2900	Electrical		4.59	7.04	7.45	6.70%	15%	15.80%	
	3100	Total: Mechanical & Electrical		7.70	11.10	13.85	9.90%	25.90%	29.80%	
520	0010	JAILS	S.F.	132	168	217				520
	0020	Total project costs	C.F.	12.30	15.10	19.25				
	2720	Plumbing	S.F.	13.30	16.85	22.25	7%	8.90%	13.80%	
	2770	Heating, ventilating, air conditioning		11.75	15.71	30.35	7.50%	9.40%	17.70%	
	2900	Electrical		14.35	18.45	22.80	8.10%	11.40%	12.40%	
	3100	Total: Mechanical & Electrical		36.65	65	76.90	29.20%	31.10%	34.10%	
530	0010	LIBRARIES	S.F.	77.80	99.15	127				530
	0020	Total project costs	C.F.	5.45	6.80	8.85				
	2720	Plumbing	S.F.	3.15	4.45	6	3.60%	4.90%	5.70%	
	2770	Heating, ventilating, air conditioning		6.75	11.40	14.85	8%	11%	14.60%	
	2900	Electrical		8	10.25	12.85	8.30%	11%	12.10%	
	3100	Total: Mechanical & Electrical		23.05	31.55	39.50	18.90%	25.30%	27.60%	
550	0010	MEDICAL CLINICS	S.F.	76.55	94.80	119				550
	0020	Total project costs	C.F.	5.70	7.40	9.85				
	2720	Plumbing	S.F.	5.15	7.25	9.70	6.10%	8.40%	10%	
	2770	Heating, ventilating, air conditioning		6.25	8.05	11.85	6.70%	9%	11.30%	
	2900	Electrical		6.50	9.30	12.30	8.10%	10%	12.20%	
	3100	Total: Mechanical & Electrical		20.45	28.80	40.25	22%	27.60%	34.30%	
570	0010	MEDICAL OFFICES	S.F.	71.85	89.05	110				570
	0020	Total project costs	C.F.	5.35	7.35	10.05				
	2720	Plumbing	S.F.	4.03	6.20	8.45	5.70%	6.80%	8.60%	
	2770	Heating, ventilating, air conditioning		4.87	7.15	9.30	6.20%	8%	9.70%	
	2900	Electrical		5.70	8.30	11.60	7.60%	9.80%	11.40%	
	3100	Total: Mechanical & Electrical		14.20	20.35	30.25	18.50%	22%	24.90%	
590	0010	MOTELS	S.F.	46.10	68.25	88				590
	0020	Total project costs	C.F.	4.02	5.65	9.25				
	2720	Plumbing	S.F.	4.67	5.95	7.10	9.40%	10.60%	12.50%	
	2770	Heating, ventilating, air conditioning		2.85	4.24	7.60	5.60%	5.60%	10%	
	2900	Electrical		4.35	5.55	7.25	7.10%	8.20%	10.40%	
	3100	Total: Mechanical & Electrical		14.80	18.50	31.75	18.50%	21%	24.40%	
	9000	Per rental unit, total cost	Unit	23,500	44,600	48,200				
	9500	Total: Mechanical & Electrical	"	4,600	6,900	8,000				
600	0010	NURSING HOMES	S.F.	69.25	91.55	112				600
	0020	Total project costs	C.F.	5.55	7.10	9.70				
	2720	Plumbing	S.F.	6.50	8.30	11.50	9.40%	10.70%	14.20%	
	2770	Heating, ventilating, air conditioning		6.45	9.30	11.50	9.30%	11.40%	11.80%	
	2900	Electrical		7.15	8.95	12	9.70%	11%	13%	
	3100	Total: Mechanical & Electrical		17.05	23.85	34.95	26%	29.90%	30.50%	
	9000	Per bed or person, total cost	Bed	30,000	36,900	49,100				
610	0010	OFFICES Low Rise (1 to 4 story)	S.F.	58.35	74.30	98.95				610
	0020	Total project costs	C.F.	4.22	5.90	8				
	2720	Plumbing	S.F.	2.22	3.36	4.76	3.70%	4.50%	6.10%	
	2770	Heating, ventilating, air conditioning		4.80	6.65	9.80	7.20%	10.50%	11.90%	
	2900	Electrical		4.94	6.85	9.55	7.50%	9.60%	11.10%	
	3100	Total: Mechanical & Electrical		11.65	16.15	23.60	18%	21.80%	26.50%	
620	0010	OFFICES Mid Rise (5 to 10 story)	S.F.	64.35	78.05	106				620
	0020	Total project costs	C.F.	4.50	5.70	8.25				
	2720	Plumbing	S.F.	1.95	3.02	4.34	2.80%	3.70%	4.50%	
	2770	Heating, ventilating, air conditioning		4.90	7	11.15	7.60%	9.40%	11%	
	3100	Total: Mechanical & Electrical		12.20	15.60	30.95	17.90%	22.30%	29.90%	

For expanded coverage of these items see Means Square Foot Costs 2001

17100 | S.F., C.F. and % of Total Costs

17100 | S.F. & C.F. Costs

			UNIT	UNIT COSTS 1/4	UNIT COSTS MEDIAN	UNIT COSTS 3/4	% OF TOTAL 1/4	% OF TOTAL MEDIAN	% OF TOTAL 3/4	
630	0010	**OFFICES** High Rise (11 to 20 story) R17100-100	S.F.	77.90	99.85	123				630
	0020	Total project costs	C.F.	4.98	6.90	9.90				
	2900	Electrical	S.F.	3.98	5.75	8.70	5.80%	7%	10.50%	
	3100	Total: Mechanical & Electrical	"	14.55	18.60	34.10	15.40%	17%	34.10%	
640	0010	**POLICE STATIONS**	S.F.	93.30	125	157				640
	0020	Total project costs	C.F.	7.55	9.15	12.70				
	2720	Plumbing	S.F.	5.30	8.55	11.95	5.60%	6.90%	11.30%	
	2770	Heating, ventilating, air conditioning		8.30	11	14.95	7%	10.70%	12%	
	2900	Electrical		10.35	15.50	19.65	9.70%	11.90%	14.90%	
	3100	Total: Mechanical & Electrical		32.95	39.35	55.40	25.10%	31.30%	36.40%	
650	0010	**POST OFFICES**	S.F.	72.70	92	115				650
	0020	Total project costs	C.F.	4.44	5.70	6.75				
	2720	Plumbing	S.F.	3.37	4.33	5.25	4.40%	5.60%	5.60%	
	2770	Heating, ventilating, air conditioning		5.30	6.50	7.25	7%	8.40%	10.20%	
	2900	Electrical		6.15	8.55	10.10	7.20%	9.60%	11%	
	3100	Total: Mechanical & Electrical		10.85	21	26.50	16.40%	20.40%	22.30%	
660	0010	**POWER PLANTS**	S.F.	542	691	1,264				660
	0020	Total project costs	C.F.	14.31	28.95	66.70				
	2900	Electrical	S.F.	36.70	77.90	116	9.50%	12.80%	18.40%	
	8100	Total: Mechanical & Electrical	"	91.47	298	667	32.50%	32.60%	52.60%	
670	0010	**RELIGIOUS EDUCATION**	S.F.	58.75	76.25	88.90				670
	0020	Total project costs	C.F.	3.35	4.80	6.20				
	2720	Plumbing	S.F.	2.51	3.57	5.03	4.30%	4.90%	8.40%	
	2770	Heating, ventilating, air conditioning		5.90	6.65	7.35	9.80%	10%	11.80%	
	2900	Electrical		4.79	6.35	8.35	7.90%	9.10%	10.30%	
	3100	Total: Mechanical & Electrical		12.40	21.50	25.55	17.60%	20.50%	22%	
690	0010	**RESEARCH** Laboratories and facilities	S.F.	90.25	130	189				690
	0020	Total project costs	C.F.	6.75	11.30	15.20				
	2720	Plumbing	S.F.	9.05	11.85	18.65	7.30%	9.20%	13.30%	
	2770	Heating, ventilating, air conditioning		8.13	27.30	32.30	7.20%	16.50%	17.50%	
	2900	Electrical		10.40	17.45	29.70	9.60%	12%	15.80%	
	3100	Total: Mechanical & Electrical		24	45.55	86.75	23.90%	35.60%	41.70%	
700	0010	**RESTAURANTS**	S.F.	86.90	112	145				700
	0020	Total project costs	C.F.	7.25	9.60	12.65				
	2720	Plumbing	S.F.	6.90	8.40	12	6.10%	8.30%	9.10%	
	2770	Heating, ventilating, air conditioning		9.15	12.20	15.90	9.40%	12.20%	13%	
	2900	Electrical		9.25	11.45	14.85	8.50%	10.60%	11.60%	
	3100	Total: Mechanical & Electrical		28.95	30.20	38.20	21.10%	24%	29.50%	
	9000	Per seat unit, total cost	Seat	3,200	4,300	5,400				
	9500	Total: Mechanical & Electrical	"	675	960	1,263				
720	0010	**RETAIL STORES**	S.F.	40.55	54.65	71.85				720
	0020	Total project costs	C.F.	2.76	3.94	5.45				
	2720	Plumbing	S.F.	1.48	2.48	4.25	3.20%	4.60%	6.80%	
	2770	Heating, ventilating, air conditioning		3.20	4.39	6.60	6.70%	8.70%	10.10%	
	2900	Electrical		3.66	4.96	7.20	7.30%	9.90%	11.60%	
	3100	Total: Mechanical & Electrical		9.80	12.55	16.30	17.10%	21.40%	23.80%	
740	0010	**SCHOOLS** Elementary	S.F.	64.55	79.60	96				740
	0020	Total project costs	C.F.	4.37	5.60	7.25				
	2720	Plumbing	S.F.	3.74	5.40	7.25	5.70%	7.10%	9.40%	
	2730	Heating, ventilating, air conditioning		5.75	9.15	12.80	8.10%	10.80%	15.20%	
	2900	Electrical		6	7.65	9.80	8.40%	10%	11.60%	
	3100	Total: Mechanical & Electrical		20.80	23.05	32.55	22.20%	28%	30.40%	
	9000	Per pupil, total cost	Ea.	6,400	10,700	34,700				
	9500	Total: Mechanical & Electrical	"	2,175	2,750	11,000				
760	0010	**SCHOOLS** Junior High & Middle	S.F.	65.55	81.10	95				760
	0020	Total project costs	C.F.	4.31	5.65	6.35				
	2720	Plumbing	S.F.	4.35	4.84	6.40	5.60%	6.90%	8.10%	
	2770	Heating, ventilating, air conditioning		5.05	9.70	13.55	8.70%	12.70%	17.40%	
	2900	Electrical		6.35	7.95	9.60	7.80%	9.40%	10.60%	
	3100	Total: Mechanical & Electrical		18.30	23	30.55	23.30%	25.70%	27.30%	

17 SQUARE FOOT

See R17100-100 and City Cost Indexes in the Reference Section

17100 | S.F., C.F. and % of Total Costs

17100 | S.F. & C.F. Costs

			UNIT	UNIT COSTS			% OF TOTAL		
				1/4	MEDIAN	3/4	1/4	MEDIAN	3/4
760	9000	Per pupil, total cost R17100-100	Ea.	8,300	9,500	11,800			
780	0010	**SCHOOLS** Senior High	S.F.	70.80	81.10	114			
	0020	Total project costs	C.F.	4.54	5.90	8.30			
	2720	Plumbing	S.F.	3.66	6.70	11	5%	6.90%	12%
	2770	Heating, ventilating, air conditioning		8.20	9.30	17.90	8.90%	11%	15%
	2900	Electrical		6.95	9.15	15.25	8.30%	10.10%	12.60%
	3100	Total: Mechanical & Electrical		19	25.05	46.85	19.80%	23.10%	27.80%
	9000	Per pupil, total cost	Ea.	6,800	11,500	17,200			
800	0010	**SCHOOLS** Vocational	S.F.	57.80	80.75	103			
	0020	Total project costs	C.F.	3.54	5.10	7.15			
	2720	Plumbing	S.F.	3.71	5.65	8.15	5.40%	7%	8.50%
	2770	Heating, ventilating, air conditioning		5.20	9.65	16.15	8.80%	11.90%	14.60%
	2900	Electrical		5.95	8.25	11.65	8.40%	11.20%	13.80%
	3100	Total: Mechanical & Electrical		15.65	20.95	39.80	21.70%	27.30%	33.10%
	9000	Per pupil, total cost	Ea.	8,100	21,600	32,300			
830	0010	**SPORTS ARENAS**	S.F.	49.25	67.70	99.35			
	0020	Total project costs	C.F.	2.75	4.88	6.35			
	2720	Plumbing	S.F.	2.62	4.36	8.20	4.50%	6.30%	9.40%
	2770	Heating, ventilating, air conditioning		5.35	7.50	9.75	5.80%	10.20%	13.50%
	2900	Electrical		4.35	6.85	8.60	7.70%	9.80%	12.30%
	3100	Total: Mechanical & Electrical		12.80	22.75	29.35	13.40%	22.50%	30.80%
850	0010	**SUPERMARKETS**	S.F.	46.75	54.85	63.65			
	0020	Total project costs	C.F.	2.61	3.15	4.77			
	2720	Plumbing	S.F.	2.61	3.29	3.83	5.40%	6%	7.50%
	2770	Heating, ventilating, air conditioning		3.85	4.72	5.60	8.60%	8.60%	9.60%
	2900	Electrical		5.45	6.75	8	10.40%	12.40%	13.60%
	3100	Total: Mechanical & Electrical		15.30	16.50	22.85	17.40%	20.40%	28.40%
860	0010	**SWIMMING POOLS**	S.F.	82.65	127	177			
	0020	Total project costs	C.F.	6.10	7.55	8.25			
	2720	Plumbing	S.F.	7	8	11.15	9.70%	12.30%	12.30%
	2900	Electrical		5.70	8.85	11.50	7.50%	8%	8%
	3100	Total: Mechanical & Electrical		12.05	25.60	47.40	17.70%	24.90%	31.60%
870	0010	**TELEPHONE EXCHANGES**	S.F.	101	147	187			
	0020	Total project costs	C.F.	6.22	9.55	13.55			
	2720	Plumbing	S.F.	4.25	6.55	9.60	4.50%	5.80%	6.90%
	2770	Heating, ventilating, air conditioning		8.65	19.45	24.65	11.80%	11.80%	18.40%
	2900	Electrical		10.25	15.85	28.75	10.90%	14%	17.80%
	3100	Total: Mechanical & Electrical		22	30.25	75	27.30%	33.40%	44.40%
910	0010	**THEATERS**	S.F.	63.30	81.20	120			
	0020	Total project costs	C.F.	2.92	4.32	6.35			
	2720	Plumbing	S.F.	1.97	2.30	6.85	2.90%	4.70%	6.80%
	2770	Heating, ventilating, air conditioning		6.15	7.45	9.20	8%	12.20%	13.40%
	2900	Electrical		5.55	7.45	15.20	8%	10%	12.40%
	3100	Total: Mechanical & Electrical		14.25	21.30	43.70	22.90%	26.60%	27.50%
940	0010	**TOWN HALLS** City Halls & Municipal Buildings	S.F.	70.65	89.30	117			
	0020	Total project costs	C.F.	5.30	7.40	9.90			
	2720	Plumbing	S.F.	2.51	5.20	8.60	4.20%	6.10%	7.90%
	2770	Heating, ventilating, air conditioning		5.35	10.60	15.45	7%	9%	13.50%
	2900	Electrical		6.35	8.55	12.10	8.20%	9.50%	11.70%
	3100	Total: Mechanical & Electrical		22.40	23.25	29.50	22%	25.10%	30.40%
970	0010	**WAREHOUSES** And Storage Buildings	S.F.	26.35	36.80	56.70			
	0020	Total project costs	C.F.	1.33	2.16	3.58			
	2720	Plumbing	S.F.	.88	1.58	2.95	2.90%	4.80%	6.60%
	2730	Heating, ventilating, air conditioning		1	2.83	3.80	2.40%	5%	8.90%
	2900	Electrical		1.56	2.94	4.86	5%	7.30%	10%
	3100	Total: Mechanical & Electrical		4.35	6.40	14.60	12.80%	18.40%	26.20%

For expanded coverage of these items see *Means Square Foot Costs 2001*

17100 | S.F., C.F. and % of Total Costs

17100	S.F. & C.F. Costs		UNIT	UNIT COSTS			% OF TOTAL		
				1/4	MEDIAN	3/4	1/4	MEDIAN	3/4
990 0010	**WAREHOUSE & OFFICES** Combination	R17100 -100	S.F.	31.70	42.50	56.40				990
0020	Total project costs		C.F.	1.64	2.41	3.56			
2720	Plumbing		S.F.	1.23	2.11	3.33	3.60%	4.70%	6.30%
2770	Heating, ventilating, air conditioning			1.95	3.08	4.35	5%	5.60%	9.60%
2900	Electrical		S.F.	2.15	3.17	4.97	5.70%	7.70%	10%
3100	Total: Mechanical & Electrical		↓	5.05	8.40	10.50	13.80%	18.60%	23%

17 SQUARE FOOT

See R17100-100 and City Cost Indexes in the Reference Section

Assemblies Section

Table of Contents

Table No.		Page
Mechanical		411
A8.3-A8.4	Sample Worksheet	413
Heating—Hydronic Systems		
A8.3-110	Electric Boilers, Residential	414
A8.3-120	Electric Boilers, Commercial	415
General Space Heating		
A8.3-141	Unit Heaters	416
A8.3-142	Fin Tube Radiation	418
Apartment Heating		
A8.3-151	Fin Tube	420
Factory/Commercial Heating		
A8.3-161	Fin Tube	421
A8.3-162	Unit Heaters	422
A8.3-600	Solar Systems, Domestic Hot Water	423
A8.3-610	Closed Loop Hot Water, Immersed Exchanger	424
A8.3-620	Closed Loop Hot Water, External Exchanger	426
A8.3-630	Drainback Hot Water	428
A8.3-640	Draindown Hot Water	430
A8.3-650	Recirculation Hot Water	432
A8.3-660	Thermosyphon Hot Water	434

Table No.		Page
Solar Systems—Space Heat & Special		
A8.3-710	Closed Loop	436
A8.3-720	Swimming Pool Heater	438
A8.3-730	Air to Water Exchanger Hot Water	440
A8.3-740	Air to Water Exchanger Space/H.W.	442
Air Conditioning-Chilled Water Systems		
A8.4-110	Air Cooled Condenser	444
A8.4-120	Water Cooled Condenser	446
Direct Expansion Systems		
A8.4-210	Rooftop Single Zone	448
A8.4-220	Rooftop Multizone	450
A8.4-230	Self-contained Water Cooled	452
A8.4-240	Self-contained Air Cooled	454
A8.4-250	Split System Air Cooled	456
A8.4-810	Computer Cooling	458
Special Systems		
A8.5-110	Garage Exhaust	459
A8.5-120	Fume Hood Exhaust	460
A8.5-200	Dust Collection Vacuum	461
A8.5-400	Pumps	462
A8.5-800	Kitchen Ventilation	463

Table No.		Page
HVAC Systems		
A8.7-100	Unit Heaters	464
A8.7-200	Boiler Systems	466
A8.8-100	Chiller Systems	470
A8.8-150	Cooling Tower Systems	473
A8.8-200	Heat Exchangers	474
A8.8-310	Air Handling Unit Systems	475
A8.8-400	AC Unit, Packaged, Water Cooled Systems	481
A8.8-450	Heat Pump Systems	483
A8.8-500	AC Unit, Package, DX, Air Cooled Systems	484
A8.8-600	Fan Coil Air Conditioner Systems	486
A8.8-700	VAV Terminal Systems	488
A8.8-800	Thru-wall Unit Systems	490
A8.8-900	Fan Systems	491
Special Construction		493
A11.2-100	Fiberglass Tanks	494
A11.2-200	Steel Tanks	496
Site Construction		499
A12.3-110	Trenching	500
A12.3-310	Pipe Bedding	503
A12.3-510	Drainage & Sewage Piping	505
A12.3-520	Gas Service Piping	507
A12.3-530	Subdrainage Piping	508
A12.3-540	Water Distribution Piping	509
A12.3-710	Manholes & Catch Basins	510

How to Use the Assemblies Cost Tables

The following is a detailed explanation of a sample Assemblies Cost Table. Most Assembly Tables are separated into three parts: 1) an illustration of the system to be estimated; 2) the components and related costs of a typical system; and 3) the costs for similar systems with dimensional and/or size variations. For costs of the components that comprise these systems or "assemblies" refer to the Unit Price Section. Next to each bold number below is the item being described with the appropriate component of the sample entry following in parenthesis. In most cases, if the work is to be subcontracted, the general contractor will need to add an additional markup (R.S. Means suggests using 10%) to the "Total" figures.

System/Line Numbers (A8.3-151-1760)

Each Assemblies Cost Line has been assigned a unique identification number based on the UniFormat classification system.

```
         UniFormat Division
              ┌─┴─┐
              8.3  151  1760
              │     │     │
    Means Subdivision     │
        Means Major Classification
            Means Individual Line Number
```

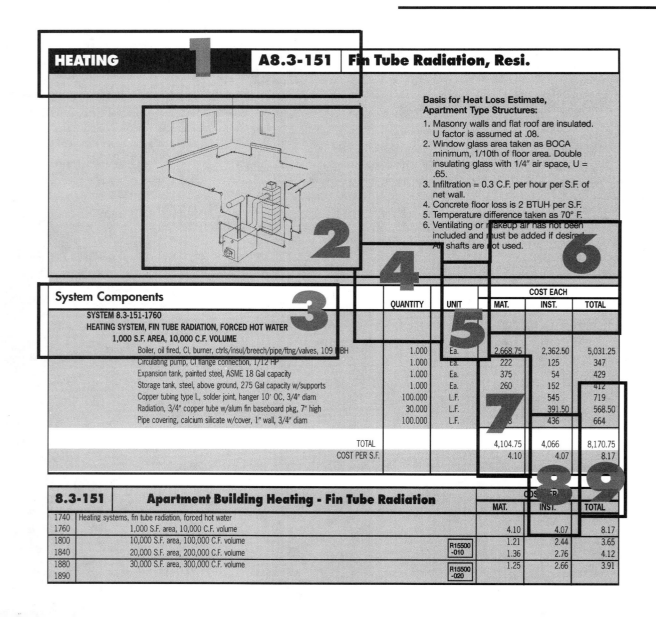

2 Illustration
At the top of most assembly pages is an illustration, a brief description, and the design criteria used to develop the cost.

3 System Components
The components of a typical system are listed separately to show what has been included in the development of the total system price. The table below contains prices for other similar systems with dimensional and/or size variations.

4 Quantity
This is the number of line item units required for one system unit. For example, we assume that it will take 30 linear feet of radiation for the 1000 S.F. area shown.

5 Unit of Measure for Each Item
The abbreviated designation indicates the unit of measure, as defined by industry standards, upon which the price of the component is based. For example, baseboard radiation is priced by the linear foot. For a complete listing of abbreviations, see the Reference Section.

6 Unit of Measure for Each System (Each)
Costs shown in the three right hand columns have been adjusted by the component quantity and unit of measure for the entire system. In this example, "Cost Each" is the unit of measure for this system or "assembly."

7 Materials (4,104.75)
This column contains the Materials Cost of each component. These cost figures are bare costs plus 10% for profit.

8 Installation (4,066.00)
Installation includes labor and equipment plus the installing contractor's overhead and profit. Equipment costs are the bare rental costs plus 10% for profit. The labor overhead and profit is defined on the inside back cover of this book.

9 Total (8,170.75)
The figure in this column is the sum of the material and installation costs.

Material Cost	+	Installation Cost	=	Total
$4,104.75	+	$4,066.00	=	$8,170.75

Division 8
Mechanical

MECHANICAL SYSTEMS — A8.3/A8.4 Sample Worksheet

Heating & Air Conditioning Estimate Worksheet

Equipment	Type	Table Number	Quantity	Unit Cost	Total Costs	Comments
Heat Source				$	$	
				$	$	
				$	$	
				$	$	
Pipe				$	$	
Duct				$	$	
Terminals				$	$	
Other				$	$	
				$	$	
				$	$	
				$	$	
Cold Source				$	$	
				$	$	
				$	$	
				$	$	
Pipe				$	$	
Duct				$	$	
Terminals				$	$	
Other				$	$	
SUBTOTAL				$	$	
Quality Complexity				%	$	
TOTAL					$	

413

HEATING | A8.3-110 | Hydronic, Electric Boilers

Boiler · Baseboard Radiation

Small Electric Boiler
System Considerations:
1. Terminal units are fin tube baseboard radiation rated at 720 BTU/hr with 200° water temperature or 820 BTU/hr steam.
2. Primary use being for residential or smaller supplementary areas, the floor levels are based on 7-1/2' ceiling heights.
3. All distribution piping is copper for boilers through 205 MBH. All piping for larger systems is steel pipe.

System Components	QUANTITY	UNIT	COST EACH MAT.	INST.	TOTAL
SYSTEM 8.3-110-1120					
SMALL HEATING SYSTEM, HYDRONIC, ELECTRIC BOILER					
1,480 S.F., 61 MBH, STEAM, 1 FLOOR					
Boiler, electric steam, std cntrls, trim, ftngs and valves, 18 KW, 61.4 MBH	1.000	Ea.	10,065	1,072.50	11,137.50
Copper tubing type L, solder joint, hanger 10'OC, 1-1/4" diam	160.000	L.F.	448	1,144	1,592
Radiation, 3/4" copper tube w/alum fin baseboard pkg 7" high	60.000	L.F.	354	783	1,137
Rough in baseboard panel or fin tube with valves & traps	10.000	Set	940	3,950	4,890
Pipe covering, calcium silicate w/cover 1" wall 1-1/4" diam	160.000	L.F.	348.80	718.40	1,067.20
Low water cut-off, quick hookup, in gage glass tappings	1.000	Ea.	161	26.50	187.50
TOTAL			12,316.80	7,694.40	20,011.20
COST PER S.F.			8.32	5.20	13.52

8.3-110	Small Heating Systems, Hydronic, Electric Boilers		COST PER S.F. MAT.	INST.	TOTAL
1100	Small heating systems, hydronic, electric boilers				
1120	Steam, 1 floor, 1480 S.F., 61 M.B.H.		8.30	5.20	13.50
1160	3,000 S.F., 123 M.B.H.	R15500 -010	4.80	4.55	9.35
1200	5,000 S.F., 205 M.B.H.		3.60	4.20	7.80
1240	2 floors, 12,400 S.F., 512 M.B.H.	R15500 -020	2.56	4.17	6.73
1280	3 floors, 24,800 S.F., 1023 M.B.H.		2.21	4.13	6.34
1320	34,750 S.F., 1,433 M.B.H.	R15500 -030	1.95	4	5.95
1360	Hot water, 1 floor, 1,000 S.F., 41 M.B.H.		7.75	2.91	10.66
1400	2,500 S.F., 103 M.B.H.		4.60	5.20	9.80
1440	2 floors, 4,850 S.F., 205 M.B.H.		4	6.20	10.20
1480	3 floors, 9,700 S.F., 410 M.B.H.		3.71	6.45	10.16

HEATING | A8.3-120 | Hydronic, Electric Boilers

Boiler

Unit Heater

Large Electric Boiler System Considerations:
1. Terminal units are all unit heaters of the same size. Quantities are varied to accommodate total requirements.
2. All air is circulated through the heaters a minimum of three times per hour.
3. As the capacities are adequate for commercial use, floor levels are based on 10' ceiling heights.
4. All distribution piping is black steel pipe.

System Components	QUANTITY	UNIT	COST EACH MAT.	COST EACH INST.	COST EACH TOTAL
SYSTEM 8.3-120-1240					
LARGE HEATING SYSTEM, HYDRONIC, ELECTRIC BOILER					
9,280 S.F., 150 KW, 510 MBH, 1 FLOOR					
Boiler, electric hot water, std ctrls, trim, ftngs, valves, 150 KW, 510 MBH	1.000	Ea.	14,137.50	2,325	16,462.50
Expansion tank, painted steel, 60 Gal capacity ASME	1.000	Ea.	2,175	126	2,301
Circulating pump, CI, close cpld, 50 GPM, 2 HP, 2" pipe conn	1.000	Ea.	1,200	250	1,450
Unit heater, 1 speed propeller, horizontal, 200° EWT, 72.7 MBH	7.000	Ea.	4,270	966	5,236
Unit heater piping hookup with controls	7.000	Set	2,415	6,265	8,680
Pipe, steel, black, schedule 40, welded, 2-1/2" diam	380.000	L.F.	2,109	6,957.80	9,066.80
Pipe covering, calcium silicate w/cover, 1" wall, 2-1/2" diam	380.000	L.F.	1,079.20	1,759.40	2,838.60
TOTAL			27,385.70	18,649.20	46,034.90
COST PER S.F.			2.95	2.01	4.96

8.3-120	Large Heating Systems, Hydronic, Electric Boilers		COST PER S.F. MAT.	COST PER S.F. INST.	COST PER S.F. TOTAL
1230	Large heating systems, hydronic, electric boilers				
1240	9,280 S.F., 150 K.W., 510 M.B.H., 1 floor		2.95	2.01	4.96
1280	14,900 S.F., 240 K.W., 820 M.B.H., 2 floors	R15500 -010	3.03	3.30	6.33
1320	18,600 S.F., 296 K.W., 1,010 M.B.H., 3 floors		2.93	3.60	6.53
1360	26,100 S.F., 420 K.W., 1,432 M.B.H., 4 floors	R15500 -020	3.06	3.55	6.61
1400	39,100 S.F., 666 K.W., 2,273 M.B.H., 4 floors		2.71	2.97	5.68
1440	57,700 S.F., 900 K.W., 3,071 M.B.H., 5 floors	R15500 -030	2.48	2.91	5.39
1480	111,700 S.F., 1,800 K.W., 6,148 M.B.H., 6 floors		2.19	2.48	4.67
1520	149,000 S.F., 2,400 K.W., 8,191 M.B.H., 8 floors		2.13	2.47	4.60
1560	223,300 S.F., 3,600 K.W., 12,283 M.B.H., 14 floors		2.14	2.82	4.96

MECHANICAL 8

HEATING — A8.3-141 | Hydronic, Fossil Fuel Boilers

Unit Heater

Fossil Fuel Boiler System Considerations:

1. Terminal units are horizontal unit heaters. Quantities are varied to accommodate total heat loss per building.
2. Unit heater selection was determined by their capacity to circulate the building volume a minimum of three times per hour in addition to the BTU output.
3. Systems shown are forced hot water. Steam boilers cost slightly more than hot water boilers. However, this is compensated for by the smaller size or fewer terminal units required with steam.
4. Floor levels are based on 10' story heights.
5. MBH requirements are gross boiler output.

System Components	QUANTITY	UNIT	COST EACH MAT.	INST.	TOTAL
SYSTEM 8.3-141-1280					
HEATING SYSTEM, HYDRONIC, FOSSIL FUEL, TERMINAL UNIT HEATERS					
CAST IRON BOILER, GAS, 80 MBH, 1,070 S.F. BUILDING					
Boiler, gas, hot water, CI, burner, controls, insulation, breeching, 80 MBH	1.000	Ea.	1,181.25	1,155	2,336.25
Pipe, steel, black, schedule 40, threaded, cplg & hngr 10'OC, 2" diam	200.000	L.F.	636	2,340	2,976
Unit heater, 1 speed propeller, horizontal, 200° EWT, 72.7 MBH	2.000	Ea.	1,220	276	1,496
Unit heater piping hookup with controls	2.000	Set	690	1,790	2,480
Expansion tank, painted steel, ASME, 18 Gal capacity	1.000	Ea.	1,400	63	1,463
Circulating pump, CI, flange connection, 1/12 HP	1.000	Ea.	222	125	347
Pipe covering, calcium silicate w/cover, 1" wall, 2" diam	200.000	L.F.	530	926	1,456
TOTAL			5,879.25	6,675	12,554.25
COST PER S.F.			5.49	6.24	11.73

8.3-141	Heating Systems, Unit Heaters		COST PER S.F. MAT.	INST.	TOTAL
1260	Heating systems, hydronic, fossil fuel, terminal unit heaters,				
1280	Cast iron boiler, gas, 80 M.B.H., 1,070 S.F. bldg.		5.50	6.25	11.75
1320	163 M.B.H., 2,140 S.F. bldg.		3.70	4.23	7.93
1360	544 M.B.H., 7,250 S.F. bldg.	R15500-010	2.57	3.08	5.65
1400	1,088 M.B.H., 14,500 S.F. bldg.		2.26	2.93	5.19
1440	3,264 M.B.H., 43,500 S.F. bldg.	R15500-020	1.86	2.22	4.08
1480	5,032 M.B.H., 67,100 S.F. bldg.		2.24	2.27	4.51
1520	Oil, 109 M.B.H., 1,420 S.F. bldg.	R15500-030	5.40	5.55	10.95
1560	235 M.B.H., 3,150 S.F. bldg.		4.02	4.09	8.11
1600	940 M.B.H., 12,500 S.F. bldg.		2.71	2.68	5.39
1640	1,600 M.B.H., 21,300 S.F. bldg.		2.62	2.56	5.18
1680	2,480 M.B.H., 33,100 S.F. bldg.		2.64	2.32	4.96
1720	3,350 M.B.H., 44,500 S.F. bldg.		2.28	2.36	4.64
1760	Coal, 148 M.B.H., 1,975 S.F. bldg.		4.63	3.63	8.26
1800	300 M.B.H., 4,000 S.F. bldg.		3.55	2.85	6.40
1840	2,360 M.B.H., 31,500 S.F. bldg.		2.50	2.36	4.86
1880	Steel boiler, gas, 72 M.B.H., 1,020 S.F. bldg.		5.50	4.61	10.11
1920	240 M.B.H., 3,200 S.F. bldg.		3.86	3.77	7.63
1960	480 M.B.H., 6,400 S.F. bldg.		3.15	2.78	5.93
2000	800 M.B.H., 10,700 S.F. bldg.		2.68	2.48	5.16
2040	1,960 M.B.H., 26,100 S.F. bldg.		2.38	2.24	4.62
2080	3,000 M.B.H., 40,000 S.F. bldg.		2.34	2.29	4.63

HEATING

A8.3-141 Hydronic, Fossil Fuel Boilers

8.3-141 Heating Systems, Unit Heaters

		COST PER S.F.		
		MAT.	INST.	TOTAL
2120	Oil, 97 M.B.H., 1,300 S.F. bldg.	4.90	5.25	10.15
2160	315 M.B.H., 4,550 S.F. bldg.	3.31	2.68	5.99
2200	525 M.B.H., 7,000 S.F. bldg.	3.50	2.78	6.28
2240	1,050 M.B.H., 14,000 S.F. bldg.	3.07	2.80	5.87
2280	2,310 M.B.H. 30,800 S.F. bldg.	2.66	2.41	5.07
2320	3,150 M.B.H., 42,000 S.F. bldg.	2.42	2.47	4.89

HEATING — A8.3-142 — Hydronic, Fossil Fuel Boilers

Fin Tube Radiator

Fossil Fuel Boiler System Considerations:

1. Terminal units are commercial steel fin tube radiation. Quantities are varied to accommodate total heat loss per building.
2. Systems shown are forced hot water. Steam boilers cost slightly more than hot water boilers. However, this is compensated for by the smaller size or fewer terminal units required with steam.
3. Floor levels are based on 10' story heights.
4. MBH requirements are gross boiler output.

System Components	QUANTITY	UNIT	COST EACH MAT.	COST EACH INST.	COST EACH TOTAL
SYSTEM 8.3-142-3240					
HEATING SYSTEM, HYDRONIC, FOSSIL FUEL, FIN TUBE RADIATION					
CAST IRON BOILER, GAS, 80 MBH, 1,070 S.F. BLDG.					
Boiler, gas, hot water, CI, burner, controls, insulation, breeching, 80 MBH	1.000	Ea.	1,181.25	1,155	2,336.25
Pipe, steel, black, schedule 40, threaded, 2" diam	200.000	L.F.	636	2,340	2,976
Radiation, steel 1-1/4" tube & 4-1/4" fin w/cover & damper, wall hung	80.000	L.F.	2,040	1,680	3,720
Rough-in, wall hung steel fin radiation with supply & balance valves	8.000	Set	1,168	3,680	4,848
Expansion tank, painted steel, ASME, 18 Gal capacity	1.000	Ea.	1,400	63	1,463
Circulating pump, CI flanged, 1/12 HP	1.000	Ea.	222	125	347
Pipe covering, calcium silicate w/cover, 1" wall, 2" diam	200.000	L.F.	530	926	1,456
TOTAL			7,177.25	9,969	17,146.25
COST PER S.F.			6.71	9.32	16.03

8.3-142	Heating System, Fin Tube Radiation		COST PER S.F. MAT.	COST PER S.F. INST.	COST PER S.F. TOTAL
3230	Heating systems, hydronic, fossil fuel, fin tube radiation				
3240	Cast iron boiler, gas, 80 MBH, 1,070 S.F. bldg.		6.70	9.30	16
3280	169 M.B.H., 2,140 S.F. bldg.		4.22	5.90	10.12
3320	544 M.B.H., 7,250 S.F. bldg.	R15500-010	3.47	5.15	8.62
3360	1,088 M.B.H., 14,500 S.F. bldg.		3.22	5.05	8.27
3400	3,264 M.B.H., 43,500 S.F. bldg.	R15500-020	2.91	4.43	7.34
3440	5,032 M.B.H., 67,100 S.F. bldg.		3.31	4.49	7.80
3480	Oil, 109 M.B.H., 1,420 S.F. bldg.	R15500-030	7.65	10.15	17.80
3520	235 M.B.H., 3,150 S.F. bldg.		4.92	6.10	11.02
3560	940 M.B.H., 12,500 S.F. bldg.		3.70	4.83	8.53
3600	1,600 M.B.H., 21,300 S.F. bldg.		3.67	4.79	8.46
3640	2,480 M.B.H., 33,100 S.F. bldg.		3.71	4.54	8.25
3680	3,350 M.B.H., 44,500 S.F. bldg.		3.34	4.59	7.93
3720	Coal, 148 M.B.H., 1,975 S.F. bldg.		5.50	5.65	11.15
3760	300 M.B.H., 4,000 S.F. bldg.		4.41	4.84	9.25
3800	2,360 M.B.H., 31,500 S.F. bldg.		3.54	4.55	8.09
3840	Steel boiler, gas, 72 M.B.H., 1,020 S.F. bldg.		7.70	8.85	16.55
3880	240 M.B.H., 3,200 S.F. bldg.		4.81	5.90	10.71
3920	480 M.B.H., 6,400 S.F. bldg.		4.08	4.84	8.92
3960	800 M.B.H., 10,700 S.F. bldg.		3.97	5.15	9.12
4000	1,960 M.B.H., 26,100 S.F. bldg.		3.41	4.43	7.84
4040	3,000 M.B.H., 40,000 S.F. bldg.		3.35	4.45	7.80
4080	Oil, 97 M.B.H., 1,300 S.F. bldg.		5.40	8.35	13.75
4120	315 M.B.H., 4,550 S.F. bldg.		4.15	4.57	8.72

HEATING — A8.3-142 Hydronic, Fossil Fuel Boilers

8.3-142	Heating System, Fin Tube Radiation	COST PER S.F.		
		MAT.	INST.	TOTAL
4160	525 M.B.H., 7,000 S.F. bldg.	4.51	4.94	9.45
4200	1,050 M.B.H., 14,000 S.F. bldg.	4.08	4.96	9.04
4240	2,310 M.B.H., 30,800 S.F. bldg.	3.68	4.57	8.25
4280	3,150 M.B.H., 42,000 S.F. bldg.	3.44	4.64	8.08

HEATING A8.3-151 Fin Tube Radiation, Resi.

Basis for Heat Loss Estimate, Apartment Type Structures:

1. Masonry walls and flat roof are insulated. U factor is assumed at .08.
2. Window glass area taken as BOCA minimum, 1/10th of floor area. Double insulating glass with 1/4" air space, U = .65.
3. Infiltration = 0.3 C.F. per hour per S.F. of net wall.
4. Concrete floor loss is 2 BTUH per S.F.
5. Temperature difference taken as 70° F.
6. Ventilating or makeup air has not been included and must be added if desired. Air shafts are not used.

System Components	QUANTITY	UNIT	COST EACH		
			MAT.	INST.	TOTAL
SYSTEM 8.3-151-1760					
HEATING SYSTEM, FIN TUBE RADIATION, FORCED HOT WATER					
1,000 S.F. AREA, 10,000 C.F. VOLUME					
Boiler, oil fired, CI, burner, ctrls/insul/breech/pipe/fttng/valves, 109 MBH	1.000	Ea.	2,668.75	2,362.50	5,031.25
Circulating pump, CI flange connection, 1/12 HP	1.000	Ea.	222	125	347
Expansion tank, painted steel, ASME 18 Gal capacity	1.000	Ea.	375	54	429
Storage tank, steel, above ground, 275 Gal capacity w/supports	1.000	Ea.	260	152	412
Copper tubing type L, solder joint, hanger 10' OC, 3/4" diam	100.000	L.F.	174	545	719
Radiation, 3/4" copper tube w/alum fin baseboard pkg, 7" high	30.000	L.F.	177	391.50	568.50
Pipe covering, calcium silicate w/cover, 1" wall, 3/4" diam	100.000	L.F.	228	436	664
TOTAL			4,104.75	4,066	8,170.75
COST PER S.F.			4.10	4.07	8.17

8.3-151	Apartment Building Heating - Fin Tube Radiation		COST PER S.F.		
			MAT.	INST.	TOTAL
1740	Heating systems, fin tube radiation, forced hot water				
1760	1,000 S.F. area, 10,000 C.F. volume		4.10	4.07	8.17
1800	10,000 S.F. area, 100,000 C.F. volume	R15500-010	1.21	2.44	3.65
1840	20,000 S.F. area, 200,000 C.F. volume		1.36	2.76	4.12
1880	30,000 S.F. area, 300,000 C.F. volume	R15500-020	1.25	2.66	3.91
1890					

HEATING A8.3-161 Fin Tube Radiation, Comm.

Fin Tube Radiator

Basis for Heat Loss Estimate, Factory or Commercial Type Structures:

1. Walls and flat roof are of lightly insulated concrete block or metal. U factor is assumed at .17.
2. Windows and doors are figured at 1-1/2 S.F. per linear foot of building perimeter. The U factor for flat single glass is 1.13.
3. Infiltration is approximately 5 C.F. per hour per S.F. of wall.
4. Concrete floor loss is 2 BTUH per S.F. of floor.
5. Temperature difference is assumed at 70° F.
6. Ventilation or makeup air has not been included and must be added if desired.

System Components	QUANTITY	UNIT	COST EACH		
			MAT.	INST.	TOTAL
SYSTEM 8.3-161-1960					
HEATING SYSTEM, FIN TUBE RADIATION, FORCED HOT WATER					
1,000 S.F. BLDG., ONE FLOOR					
Boiler, oil fired, CI, burner/ctrls/insul/breech/pipe/ftngs/valves, 109 MBH	1.000	Ea.	2,668.75	2,362.50	5,031.25
Expansion tank, painted steel, ASME 18 Gal capacity	1.000	Ea.	1,400	63	1,463
Storage tank, steel, above ground, 550 Gal capacity w/supports	1.000	Ea.	905	281	1,186
Circulating pump, CI flanged, 1/8 HP	1.000	Ea.	370	125	495
Pipe, steel, black, sch. 40, threaded, cplg & hngr 10'OC, 1-1/2" diam.	260.000	L.F.	652.60	2,431	3,083.60
Pipe covering, calcium silicate w/cover , 1" wall, 1-1/2" diam	260.000	L.F.	579.80	1,167.40	1,747.20
Radiation, steel 1-1/4" tube & 4-1/4" fin w/cover & damper, wall hung	42.000	L.F.	1,071	882	1,953
Rough in, steel fin tube radiation w/supply & balance valves	4.000	Set	584	1,840	2,424
TOTAL			8,231.15	9,151.90	17,383.05
COST PER S.F.			8.23	9.15	17.38

8.3-161	Commercial Building Heating - Fin Tube Radiation		COST PER S.F.		
			MAT.	INST.	TOTAL
1940	Heating systems, fin tube radiation, forced hot water				
1960	1,000 S.F. bldg, one floor		8.25	9.15	17.40
2000	10,000 S.F., 100,000 C.F., total two floors	R15500-010	2.30	3.51	5.81
2040	100,000 S.F., 1,000,000 C.F., total three floors		1.04	1.62	2.66
2080	1,000,000 S.F., 10,000,000 C.F., total five floors	R15500-020	.51	.85	1.36
2090					

MECHANICAL 8

HEATING — A8.3-162 Terminal Unit Heaters, Comm.

Unit Heater

Basis for Heat Loss Estimate, Factory or Commercial Type Structures:

1. Walls and flat roof are of lightly insulated concrete block or metal. U factor is assumed at .17.
2. Windows and doors are figured at 1-1/2 S.F. per linear foot of building perimeter. The U factor for flat single glass is 1.13.
3. Infiltration is approximately 5 C.F. per hour per S.F. of wall.
4. Concrete floor loss is 2 BTUH per S.F. of floor.
5. Temperature difference is assumed at 70° F.
6. Ventilation or makeup air has not been included and must be added if desired.

System Components	QUANTITY	UNIT	COST EACH MAT.	COST EACH INST.	COST EACH TOTAL
SYSTEM 8.3-162-1880					
HEATING SYSTEM, TERMINAL UNIT HEATERS, FORCED HOT WATER					
1,000 S.F. BLDG., ONE FLOOR					
Boiler oil fired, CI, burner/ctrls/insul/breech/pipe/ftngs/valves, 109 MBH	1.000	Ea.	2,668.75	2,362.50	5,031.25
Expansion tank, painted steel, ASME 18 Gal capacity	1.000	Ea.	1,400	63	1,463
Storage tank, steel, above ground, 550 Gal capacity w/supports	1.000	Ea.	905	281	1,186
Circulating pump, CI, flanged, 1/8 HP	1.000	Ea.	370	125	495
Pipe, steel, black, schedule 40, threaded, cplg & hngr 10'OC 1-1/2" diam	260.000	L.F.	652.60	2,431	3,083.60
Pipe covering, calcium silicate w/cover, 1" wall, 1-1/2" diam	260.000	L.F.	579.80	1,167.40	1,747.20
Unit heater, 1 speed propeller, horizontal, 200° EWT, 26.9 MBH	2.000	Ea.	850	190	1,040
Unit heater piping hookup with controls	2.000	Set	690	1,790	2,480
TOTAL			8,116.15	8,409.90	16,526.05
COST PER S.F.			8.12	8.41	16.53

8.3-162	Commercial Bldg. Heating - Terminal Unit Heaters		COST PER S.F. MAT.	COST PER S.F. INST.	COST PER S.F. TOTAL
1860	Heating systems, terminal unit heaters, forced hot water				
1880	1,000 S.F. bldg., one floor		8.10	8.40	16.50
1920	10,000 S.F. bldg., 100,000 C.F. total two floors		2.08	2.92	5
1960	100,000 S.F. bldg., 1,000,000 C.F. total three floors	R15500-010	1.07	1.49	2.56
2000	1,000,000 S.F. bldg., 10,000,000 C.F. total five floors		.68	.96	1.64
2010		R15500-020			

HEATING - SOLAR — A8.3-600 — General

Many styles of active solar energy systems exist. Those shown on the following page represent the majority of systems now being installed in different regions of the country.

The five active domestic hot water (DHW) systems which follow are typically specified as two or three panel systems with additional variations being the type of glazing and size of the storage tanks. Various combinations have been costed for the user's evaluation and comparison. The basic specifications from which the following systems were developed satisfy the construction detail requirements specified in the HUD Intermediate Minimum Property Standards (IMPS). If these standards are not complied with, the renewable energy system's costs could be significantly lower than shown.

To develop the system's specifications and costs it was necessary to make a number of assumptions about the systems. Certain systems are more appropriate to one climatic region than another or the systems may require modifications to be usable in particular locations. Specific instances in which the systems are not appropriate throughout the country as specified, or in which modification will be needed include the following:

- The freeze protection mechanisms provided in the DHW systems vary greatly. In harsh climates, where freeze is a major concern, a closed-loop indirect collection system may be more appropriate than a direct collection system.
- The thermosyphon water heater system described cannot be used when temperatures drop below 32° F.
- In warm climates it may be necessary to modify the systems installed to prevent overheating.

For each renewable resource (solar) system a schematic diagram and descriptive summary of the system is provided along with a list of all the components priced as part of the system. The costs were developed based on these specifications.

Considerations affecting costs which may increase or decrease beyond the estimates presented here include the following:

- Special structural qualities (allowance for earthquake, future expansion, high winds, and unusual spans or shapes);
- Isolated building site or rough terrain that would affect the transportation of personnel, material, or equipment;
- Unusual climatic conditions during the construction process;
- Substitution of other materials or system components for those used in the system specifications.

HEATING - SOLAR
A8.3-610 Closed Loop, Immersed Exch.

In this closed-loop indirect collection system, fluid with a low freezing temperature, such as propylene glycol, transports heat from the collectors to water storage. The transfer fluid is contained in a closed-loop consisting of collectors, supply and return piping, and a heat exchanger immersed in the storage tank. A typical two-or-three panel system contains 5 to 6 gallons of heat transfer fluid.

When the collectors become approximately 20° F warmer than the storage temperature, a controller activates the circulator. The circulator moves the fluid continuously through the collectors until the temperature difference between the collectors and storage is such that heat collection no longer occurs; at that point, the circulator shuts off. Since the heat transfer fluid has a very low freezing temperature, there is no need for it to be drained from the collectors between periods of collection.

HEATING - SOLAR

A8.3-610 Closed Loop, Immersed Exch.

System Components	QUANTITY	UNIT	COST EACH MAT.	COST EACH INST.	TOTAL
SYSTEM 8.3-610-2760					
SOLAR, CLOSED LOOP, HOT WATER SYSTEM, IMMERSED HEAT EXCHANGER					
3/4" TUBING, THREE 3' X 7' BLACK CHROME COLLECTORS					
A, B Differential controller, 2 sensors, thermostat, solar energy system	1.000	Ea.	85.50	34.50	120
C Thermometer 2" dial	3.000	Ea.	63	78	141
D, T Fill & drain valves, brass, 3/4" connection	3.000	Ea.	16.35	52.05	68.40
E Air vent, manual, 1/8" fitting	1.000	Ea.	2.13	13	15.13
F Air purger	1.000	Ea.	45	34.50	79.50
G Expansion tank	1.000	Ea.	64	13	77
I Valve, gate, bronze, NRS, soldered 3/4" diam	3.000	Ea.	55.65	63	118.65
J Neoprene vent flashing	2.000	Ea.	17.30	42	59.30
K Circulator, solar heated liquid, 1/25 HP	1.000	Ea.	153	53.50	206.50
N-1, N Relief valve, temp & press 150 psi 210°F self-closing 3/4" IPS	2.000	Ea.	17.30	27.70	45
O Pipe covering, urethane ultraviolet cover, 1" wall, 3/4" diam	20.000	L.F.	30.80	80.80	111.60
P Pipe covering, fiberglass, all service jacket, 1" wall, 3/4" diam	50.000	L.F.	66.50	161	227.50
Roof clamps for solar energy collector panel	3.000	Set	5.31	32.10	37.41
Q Collector panel solar blk chrome on copper, 1/8" temp glass, 3'x7'	3.000	Ea.	1,800	237	2,037
R-1 Valve, swing check, bronze, regrinding disc, 3/4" diam, soldered	1.000	Ea.	33	21	54
S Pressure gauge, 60 psi, 2-1/2" dial	1.000	Ea.	23	13	36
U Valve, water tempering, bronze, sweat connections, 3/4" diam	1.000	Ea.	54	21	75
W-2, W Tank, water storage immersed heat exchr elec elem 2"x2# insul 120 Gal	1.000	Ea.	925	297	1,222
X Valve, globe, bronze, rising stem, 3/4" diam, soldered	1.000	Ea.	47	21	68
Copper tubing type L, solder joint, hanger 10' OC 3/4" diam	20.000	L.F.	34.80	109	143.80
Copper tubing, type M, solder joint, hanger 10' OC 3/4" diam	70.000	L.F.	102.90	374.50	477.40
Sensor wire, #22-2 conductor multistranded	.500	C.L.F.	4.13	20.50	24.63
Solar energy heat transfer fluid, propylene glycol, anti-freeze	6.000	Gal.	58.20	89.10	147.30
Wrought copper fittings & solder, 3/4" diam	76.000	Ea.	40.28	1,672	1,712.28
TOTAL			**3,744.15**	**3,560.25**	**7,304.40**

8.3-610	Solar, Closed Loop, Hot Water Systems		COST EACH MAT.	COST EACH INST.	TOTAL
2550	Solar, closed loop, hot water system, immersed heat exchanger				
2560	3/8" tubing, 3 ea. 4' x 4'-4" vacuum tube collectors, 80 gal. tank		4,550	3,225	7,775
2580	1/2" tubing, 4 ea. 4' x 4'-4" vacuum tube collectors, 80 gal. tank		5,450	3,500	8,950
2600	120 gal. tank		5,550	3,525	9,075
2640	2 ea. 3'x7' black chrome collectors, 80 gal. tank	R13600 -610	2,925	3,325	6,250
2660	120 gal. tank		3,050	3,325	6,375
2700	2 ea. 3'x7' flat black collectors, 120 gal. tank		2,800	3,350	6,150
2720	3 ea. 3'x7' flat black collectors, 120 gal. tank		3,275	3,425	6,700
2760	3/4" tubing, 3 ea. 3'x7' black chrome collectors, 120 gal. tank		3,750	3,550	7,300
2780	3 ea. 3'x7' flat black collectors, 120 gal. tank		3,375	3,575	6,950
2800	2 ea. 4'x9' flat black w/plastic glazing collectors 120 gal. tank		3,025	3,575	6,600
2840	1" tubing, 4 ea. 2'x9' plastic absorber & glazing collectors 120 gal. tank		4,275	4,000	8,275
2860	4 ea. 3'x7' black chrome collectors, 120 gal. tank		4,500	4,000	8,500
2880	4 ea. 3'x7' flat black absorber collectors, 120 gal. tank		4,000	4,025	8,025

HEATING - SOLAR   A8.3-620   Closed Loop, External Exch.

In this closed-loop indirect collection system, fluid with a low freezing temperature, such as propylene glycol, transports heat from the collectors to water storage. The transfer fluid is contained in a closed-loop consisting of collectors, supply and return piping, and a remote heat exchanger. The heat exchanger transfers heat energy from the fluid in the collector loop to potable water circulated in a storage loop. A typical two-or-three panel system contains 5 to 6 gallons of heat transfer fluid.

When the collectors become approximately 20° F warmer than the storage temperature, a controller activates the circulator on the collector and storage loops. The circulators will move the fluid and potable water through the heat exchanger until heat collection no longer occurs. At that point, the system shuts down. Since the heat transfer medium is a fluid with a very low freezing temperature, there is no need for it to be drained from the system between periods of collection.

HEATING - SOLAR A8.3-620 Closed Loop, External Exch.

System Components

		QUANTITY	UNIT	COST EACH MAT.	COST EACH INST.	COST EACH TOTAL
	SYSTEM 8.3-620-2760 **SOLAR, CLOSED LOOP, ADD-ON HOT WATER SYS., EXTERNAL HEAT EXCHANGER** **3/4" TUBING, TWO 3'X7' BLACK CHROME COLLECTORS**					
A,B,G,L,K,M	Heat exchanger fluid-fluid pkg incl 2 circulators, expansion tank, Check valve, relief valve, controller, hi temp cutoff, & 2 sensors	1.000	Ea.	765	300	1,065
C	Thermometer, 2" dial	3.000	Ea.	63	78	141
D, T	Fill & drain valve, brass, 3/4" connection	1.000	Ea.	5.45	17.35	22.80
E	Air vent, manual, 1/8" fitting	2.000	Ea.	4.26	26	30.26
F	Air purger	1.000	Ea.	45	34.50	79.50
H	Strainer, Y type, bronze body, 3/4" IPS	1.000	Ea.	19.15	22	41.15
I	Valve, gate, bronze, NRS, soldered 3/4" diam	6.000	Ea.	111.30	126	237.30
J	Neoprene vent flashing	2.000	Ea.	17.30	42	59.30
N-1, N	Relief valve temp & press, 150 psi 210°F self-closing 3/4" IPS	1.000	Ea.	8.65	13.85	22.50
O	Pipe covering, urethane, ultraviolet cover, 1" wall 3/4" diam	20.000	L.F.	30.80	80.80	111.60
P	Pipe covering, fiberglass, all service jacket, 1" wall, 3/4" diam	50.000	L.F.	66.50	161	227.50
Q	Collector panel solar energy blk chrome on copper, 1/8" temp glass 3'x7'	2.000	Ea.	1,200	158	1,358
	Roof clamps for solar energy collector panels	2.000	Set	3.54	21.40	24.94
R	Valve, swing check, bronze, regrinding disc, 3/4" diam	2.000	Ea.	66	42	108
S	Pressure gauge, 60 psi, 2" dial	1.000	Ea.	23	13	36
U	Valve, water tempering, bronze, sweat connections, 3/4" diam	1.000	Ea.	54	21	75
W-2, V	Tank water storage w/heating element, drain, relief valve, existing	1.000	Ea.			
	Copper tubing type L, solder joint, hanger 10' OC 3/4" diam	20.000	L.F.	34.80	109	143.80
	Copper tubing, type M, solder joint, hanger 10' OC 3/4" diam	70.000	L.F.	102.90	374.50	477.40
	Sensor wire, #22-2 conductor multistranded	.500	C.L.F.	4.13	20.50	24.63
	Solar energy heat transfer fluid, propylene glycol anti-freeze	6.000	Gal.	58.20	89.10	147.30
	Wrought copper fittings & solder, 3/4" diam	76.000	Ea.	40.28	1,672	1,712.28
	Total			2,723.26	3,422	6,145.26

8.3-620	Solar, Closed Loop, Add-On Hot Water Systems		COST EACH MAT.	COST EACH INST.	COST EACH TOTAL
2550	Solar, closed loop, add-on hot water system, external heat exchanger				
2570	3/8" tubing, 3 ea. 4' x 4'-4" vacuum tube collectors		4,250	3,200	7,450
2580	1/2" tubing, 4 ea 4 x 4'-4" vacuum tube collectors		5,150	3,475	8,625
2600	2 ea 3'x7' black chrome collectors	R13600 -610	2,625	3,275	5,900
2620	3 ea 3'x7' black chrome collectors		3,225	3,350	6,575
2640	2 ea 3'x7' flat black collectors		2,375	3,275	5,650
2660	3 ea 3'x7' flat black collectors		2,850	3,375	6,225
2700	3/4" tubing, 3 ea 3'x7' black chrome collectors		3,325	3,500	6,825
2720	3 ea 3'x7' flat black absorber plate collectors		2,950	3,525	6,475
2740	2 ea 4'x9' flat black w/plastic glazing collectors		2,600	3,525	6,125
2760	2 ea 3'x7' black chrome collectors		2,725	3,425	6,150
2780	1" tubing, 4 ea 2'x9' plastic absorber & glazing collectors		3,875	3,975	7,850
2800	4 ea 3'x7' black chrome absorber collectors		4,100	3,975	8,075
2820	4 ea 3'x7' flat black absorber collectors		3,600	4,000	7,600

HEATING - SOLAR A8.3-630 Drain Back, Immersed Exch.

In the drainback indirect-collection system, the heat transfer fluid is distilled water contained in a loop consisting of collectors, supply and return piping, and an unpressurized holding tank. A large heat exchanger containing incoming potable water is immersed in the holding tank. When a controller activates solar collection, the distilled water is pumped through the collectors and heated and pumped back down to the holding tank. When the temperature differential between the water in the collectors and water in storage is such that collection no longer occurs, the pump turns off and gravity causes the distilled water in the collector loop to drain back to the holding tank. All the loop piping is pitched so that the water can drain out of the collectors and piping and not freeze there. As hot water is needed in the home, incoming water first flows through the holding tank with the immersed heat exchanger and is warmed and then flows through a conventional heater for any supplemental heating that is necessary.

HEATING - SOLAR | A8.3-630 | Drain Back, Immersed Exch.

System Components	QUANTITY	UNIT	COST EACH		
			MAT.	INST.	TOTAL
SYSTEM 8.3-630-2760					
SOLAR, DRAINBACK, ADD ON, HOT WATER, IMMERSED HEAT EXCHANGER					
3/4" TUBING, THREE EA 3'X7' BLACK CHROME COLLECTOR					
A, B Differential controller 2 sensors, thermostat, solar energy system	1.000	Ea.	85.50	34.50	120
C Thermometer 2" dial	3.000	Ea.	63	78	141
D, T Fill & drain valve, brass, 3/4" connection	1.000	Ea.	5.45	17.35	22.80
E-1 Automatic air vent 1/8" fitting	1.000	Ea.	9.70	13	22.70
H Strainer, Y type, bronze body, 3/4" IPS	1.000	Ea.	19.15	22	41.15
I Valve, gate, bronze, NRS, soldered 3/4" diam	2.000	Ea.	37.10	42	79.10
J Neoprene vent flashing	2.000	Ea.	17.30	42	59.30
L Circulator, solar heated liquid, 1/20 HP	1.000	Ea.	161	62.50	223.50
N Relief valve temp. & press. 150 psi 210°F self-closing 3/4" IPS	1.000	Ea.	8.65	13.85	22.50
O Pipe covering, urethane, ultraviolet cover, 1" wall, 3/4" diam	20.000	L.F.	30.80	80.80	111.60
P Pipe covering, fiberglass, all service jacket, 1" wall, 3/4" diam	50.000	L.F.	66.50	161	227.50
Q Collector panel solar energy blk chrome on copper, 1/8" temp glas 3'x7'	3.000	Ea.	1,800	237	2,037
Roof clamps for solar energy collector panels	3.000	Set	5.31	32.10	37.41
R Valve, swing check, bronze, regrinding disc, 3/4" diam	1.000	Ea.	33	21	54
U Valve, water tempering, bronze sweat connections, 3/4" diam	1.000	Ea.	54	21	75
W Tank, water storage immersed heat exchr elec 2"x1/2# insul 120 gal	1.000	Ea.	925	297	1,222
V Tank, water storage w/heating element, drain, relief valve, existing	1.000	Ea.			
X Valve, globe, bronze, rising stem, 3/4" diam, soldered	3.000	Ea.	141	63	204
Y Flow control valve	1.000	Ea.	54.50	18.90	73.40
Z Valve, ball, bronze, solder 3/4" diam, solar loop flow control	1.000	Ea.	11.65	21	32.65
Copper tubing, type L, solder joint, hanger 10' OC 3/4" diam	20.000	L.F.	34.80	109	143.80
Copper tubing, type M, solder joint, hanger 10' OC 3/4" diam	70.000	L.F.	102.90	374.50	477.40
Sensor wire, #22-2 conductor, multistranded	.500	C.L.F.	4.13	20.50	24.63
Wrought copper fittings & solder, 3/4" diam	76.000	Ea.	40.28	1,672	1,712.28
Total			3,710.72	3,454	7,164.72

8.3-630	Solar, Drainback, Hot Water Systems	COST EACH		
		MAT.	INST.	TOTAL
2550	Solar, drainback, hot water, immersed heat exchanger R13600-610			
2560	3/8" tubing, 3 ea. 4' x 4'-4" vacuum tube collectors	4,500	3,100	7,600
2580	1/2" tubing, 4 ea 4' x 4'-4" vacuum tube collectors, 80 gal tank	5,400	3,375	8,775
2600	120 gal tank	5,500	3,425	8,925
2640	2 ea. 3'x7' blk chrome collectors, 80 gal tank	2,875	3,175	6,050
2660	3 ea. 3'x7' blk chrome collectors, 120 gal tank	3,575	3,300	6,875
2700	2 ea. 3'x7' flat blk collectors, 120 gal tank	2,725	3,225	5,950
2720	3 ea. 3'x7' flat blk collectors, 120 gal tank	3,200	3,325	6,525
2760	3/4" tubing, 3 ea 3'x7' black chrome collectors, 120 gal tank	3,700	3,450	7,150
2780	3 ea. 3'x7' flat black absorber collectors, 120 gal tank	3,325	3,475	6,800
2800	2 ea. 4'x9' flat blk w/plastic glazing collectors 120 gal tank	3,000	3,475	6,475
2840	1" tubing, 4 ea. 2'x9' plastic absorber & glazing collectors, 120 gal tank	4,300	3,900	8,200
2860	4 ea. 3'x7' black chrome absorber collectors, 120 gal tank	4,525	3,925	8,450
2880	4 ea. 3'x7' flat black absorber collectors, 120 gal tank	4,025	3,950	7,975

HEATING - SOLAR

A8.3-640 Draindown, Direct Collection

In the draindown direct-collection system, incoming domestic water is heated in the collectors. When the controller activates solar collection, domestic water is first heated as it flows through the collectors and is then pumped to storage. When conditions are no longer suitable for heat collection, the pump shuts off and the water in the loop drains down and out of the system by means of solenoid valves and properly pitched piping.

HEATING - SOLAR A8.3-640 | Draindown, Direct Collection

System Components

		QUANTITY	UNIT	COST EACH MAT.	COST EACH INST.	TOTAL
	SYSTEM 8.3-640-2760					
	SOLAR, DRAINDOWN, HOT WATER, DIRECT COLLECTION					
	3/4" TUBING, THREE 3'X7' BLACK CHROME COLLECTORS					
A, B	Differential controller, 2 sensors, thermostat, solar energy system	1.000	Ea.	85.50	34.50	120
A-1	Solenoid valve, solar heating loop, brass, 3/4" diam, 24 volts	3.000	Ea.	699	138	837
B-1	Solar energy sensor, freeze prevention	1.000	Ea.	22	13	35
C	Thermometer, 2" dial	3.000	Ea.	63	78	141
E-1	Vacuum relief valve, 3/4" diam	1.000	Ea.	27.50	13	40.50
F-1	Air vent, automatic, 1/8" fitting	1.000	Ea.	9.70	13	22.70
H	Strainer, Y type, bronze body, 3/4" IPS	1.000	Ea.	19.15	22	41.15
I	Valve, gate, bronze, NRS, soldered, 3/4" diam	2.000	Ea.	37.10	42	79.10
J	Vent flashing neoprene	2.000	Ea.	17.30	42	59.30
K	Circulator, solar heated liquid, 1/25 HP	1.000	Ea.	153	53.50	206.50
N	Relief valve temp & press 150 psi 210°F self-closing 3/4" IPS	1.000	Ea.	8.65	13.85	22.50
O	Pipe covering, urethane, ultraviolet cover, 1" wall, 3/4" diam	20.000	L.F.	30.80	80.80	111.60
P	Pipe covering, fiberglass, all service jacket, 1" wall, 3/4" diam	50.000	L.F.	66.50	161	227.50
	Roof clamps for solar energy collector panels	3.000	Set	5.31	32.10	37.41
Q	Collector panel solar energy blk chrome on copper, 1/8" temp glass 3'x7'	3.000	Ea.	1,800	237	2,037
R	Valve, swing check, bronze, regrinding disc, 3/4" diam, soldered	2.000	Ea.	66	42	108
T	Drain valve, brass, 3/4" connection	2.000	Ea.	10.90	34.70	45.60
U	Valve, water tempering, bronze, sweat connections, 3/4" diam	1.000	Ea.	54	21	75
W-2, W	Tank, water storage elec elem 2"x1/2# insul 120 gal	1.000	Ea.	925	297	1,222
X	Valve, globe, bronze, rising stem, 3/4" diam, soldered	1.000	Ea.	47	21	68
	Copper tubing, type L, solder joints, hangers 10' OC 3/4" diam	20.000	L.F.	34.80	109	143.80
	Copper tubing, type M, solder joints, hangers 10' OC 3/4" diam	70.000	L.F.	102.90	374.50	477.40
	Sensor wire, #22-2 conductor, multistranded	.500	C.L.F.	4.13	20.50	24.63
	Wrought copper fittings & solder, 3/4" diam	76.000	Ea.	40.28	1,672	1,712.28
	Total			**4,329.52**	**3,565.45**	**7,894.97**

8.3-640	Solar, Draindown, Hot Water Systems	MAT.	INST.	TOTAL
2550	Solar, draindown, hot water			
2560	3/8" tubing, 3 ea. 4' x 4'-4" vacuum tube collectors, 80 gal tank	5,100	3,400	8,500
2580	1/2" tubing, 4 ea. 4' x 4'-4" vacuum tube collectors, 80 gal tank	6,025	3,500	9,525
2600	120 gal tank	6,150	3,525	9,675
2640	2 ea 3'x7' black chrome collectors, 80 gal tank	3,500	3,300	6,800
2660	3 ea 3'x7' black chrome collectors, 120 gal tank	4,225	3,425	7,650
2700	2 ea 3'x7' flat black collectors, 120 gal tank	3,375	3,350	6,725
2720	3 ea 3'x7' flat black collectors, 120 gal tank	3,825	3,425	7,250
2760	3/4" tubing, 3 ea 3'x7' black chrome collectors, 120 gal tank	4,325	3,575	7,900
2780	3 ea 3'x7' flat collectors, 120 gal tank	3,950	3,575	7,525
2800	2 ea. 4'x9' flat black & plastic glazing collectors, 120 gal tank	3,600	3,600	7,200
2840	1" tubing, 4 ea. 2'x9' plastic absorber & glazing collectors, 120 gal tank	4,875	4,025	8,900
2860	4 ea 3'x7' black chrome absorber collectors, 120 gal tank	5,100	4,025	9,125
2880	4 ea 3'x7' flat black absorber collectors, 120 gal tank	4,600	4,050	8,650

Note at row 2580/2600: R13600-610

HEATING - SOLAR | A8.3-650 | Recirculation, Hot Water

In the recirculation system, (a direct-collection system), incoming domestic water is heated in the collectors. When the controller activates solar collection, domestic water is heated as it flows through the collectors and then it flows back to storage. When conditions are not suitable for heat collection, the pump shuts off and the flow of the water stops. In this type of system, water remains in the collector loop at all times. A "frost sensor" at the collector activates the circulation of warm water from storage through the collectors when protection from freezing is required.

HEATING - SOLAR   A8.3-650   Recirculation, Hot Water

System Components

	System Components	QUANTITY	UNIT	MAT.	INST.	TOTAL
	SYSTEM 8.3-650-2820					
	SOLAR, RECIRCULATION, HOT WATER					
	3/4" TUBING, TWO 3'X7' BLACK CHROME COLLECTORS					
A, B	Differential controller 2 sensors, thermostat, for solar energy system	1.000	Ea.	85.50	34.50	120
A-1	Solenoid valve, solar heating loop, brass, 3/4" IPS, 24 volts	2.000	Ea.	466	92	558
B-1	Solar energy sensor freeze prevention	1.000	Ea.	22	13	35
C	Thermometer, 2" dial	3.000	Ea.	63	78	141
D	Drain valve, brass, 3/4" connection	1.000	Ea.	5.45	17.35	22.80
F-1	Air vent, automatic, 1/8" fitting	1.000	Ea.	9.70	13	22.70
H	Strainer, Y type, bronze body, 3/4" IPS	1.000	Ea.	19.15	22	41.15
I	Valve, gate, bronze, 125 lb, soldered 3/4" diam	3.000	Ea.	55.65	63	118.65
J	Vent flashing, neoprene	2.000	Ea.	17.30	42	59.30
L	Circulator, solar heated liquid, 1/20 HP	1.000	Ea.	161	62.50	223.50
N	Relief valve, temp & press 150 psi 210°F self-closing 3/4" IPS	2.000	Ea.	17.30	27.70	45
O	Pipe covering, urethane, ultraviolet cover, 1" wall, 3/4" diam	20.000	L.F.	30.80	80.80	111.60
P	Pipe covering, fiberglass, all service jacket, 1" wall 3/4" diam	50.000	L.F.	66.50	161	227.50
Q	Collector panel solar energy blk chrome on copper, 1/8" temp glass 3'x7'	2.000	Ea.	1,200	158	1,358
	Roof clamps for solar energy collector panels	2.000	Set	3.54	21.40	24.94
R	Valve, swing check, bronze, 125 lb, regrinding disc, soldered 3/4" diam	2.000	Ea.	66	42	108
U	Valve, water tempering, bronze, sweat connections, 3/4" diam	1.000	Ea.	54	21	75
V	Tank, water storage, w/heating element, drain, relief valve, existing	1.000	Ea.			
X	Valve, globe, bronze, 125 lb, 3/4" diam	2.000	Ea.	94	42	136
	Copper tubing, type L, solder joints, hangers 10' OC 3/4" diam	20.000	L.F.	34.80	109	143.80
	Copper tubing, type M, solder joints, hangers 10' OC 3/4" diam	70.000	L.F.	102.90	374.50	477.40
	Wrought copper fittings & solder, 3/4" diam	76.000	Ea.	40.28	1,672	1,712.28
	Sensor wire, #22-2 conductor, multistranded	.500	C.L.F.	4.13	20.50	24.63
	Total			**2,619**	**3,167.25**	**5,786.25**

8.3-650 Solar Recirculation, Domestic Hot Water Systems

		MAT.	INST.	TOTAL
2550	Solar recirculation, hot water			
2560	3/8" tubing, 3 ea. 4' x 4'-4" vacuum tube collectors	4,125	2,950	7,075
2580	1/2" tubing, 4 ea. 4' x 4'-4" vacuum tube collectors	5,025	3,225	8,250
2640	2 ea. 3'x7' black chrome collectors [R13600-610]	2,500	3,025	5,525
2660	3 ea. 3'x7' black chrome collectors	3,100	3,100	6,200
2700	2 ea. 3'x7' flat black collectors	2,250	3,025	5,275
2720	3 ea. 3'x7' flat black collectors	2,725	3,125	5,850
2760	3/4" tubing, 3 ea. 3'x7' black chrome collectors	3,225	3,250	6,475
2780	3 ea. 3'x7' flat black absorber plate collectors	2,850	3,275	6,125
2800	2 ea. 4'x9' flat black w/plastic glazing collectors	2,500	3,275	5,775
2820	2 ea. 3'x7' black chrome collectors	2,625	3,175	5,800
2840	1" tubing, 4 ea. 2'x9' black plastic absorber & glazing collectors	3,800	3,700	7,500
2860	4 ea. 3'x7' black chrome absorber collectors	4,025	3,725	7,750
2880	4 ea. 3'x7' flat black absorber collectors	3,525	3,750	7,275

HEATING - SOLAR A8.3-660 Thermosyphon, Hot Water

The thermosyphon domestic hot water system, a direct collection system, operates under city water pressure and does not require pumps for system operation. An insulated water storage tank is located above the collectors. As the sun heats the collectors, warm water in them rises by means of natural convection; the colder water in the storage tank flows into the collectors by means of gravity. As long as the sun is shining the water continues to flow through the collectors and to become warmer.

To prevent freezing, the system must be drained or the collectors covered with an insulated lid when the temperature drops below 32°F.

HEATING - SOLAR

A8.3-660 Thermosyphon, Hot Water

System Components	QUANTITY	UNIT	COST EACH		
			MAT.	INST.	TOTAL
SYSTEM 8.3-660-0940					
SOLAR, THERMOSYPHON, WATER HEATER					
3/4" TUBING, TWO 3'X7' BLACK CHROME COLLECTORS					
D-1 Framing lumber, fir, 2" x 6" x 8', tank cradle	.008	M.B.F.	5.20	4.68	9.88
F-1 Framing lumber, fir, 2" x 4" x 24', sleepers	.016	M.B.F.	10.56	14.08	24.64
I Valve, gate, bronze, 125 lb, soldered 1/2" diam	2.000	Ea.	32.50	34.70	67.20
J Vent flashing, neoprene	4.000	Ea.	34.60	84	118.60
O Pipe covering, urethane, ultraviolet cover, 1" wall, 1/2" diam	40.000	L.F.	47.20	158.40	205.60
P Pipe covering fiberglass all service jacket 1" wall 1/2" diam	160.000	L.F.	187.20	494.40	681.60
Q Collector panel solar, blk chrome on copper, 3/16" temp glass 3'-8" x 6'	2.000	Ea.	950	166	1,116
Y Flow control valve, globe, bronze, 125#, soldered, 1/2" diam	1.000	Ea.	34.50	17.35	51.85
U Valve, water tempering, bronze, sweat connections, 1/2" diam	1.000	Ea.	44	17.35	61.35
W Tank, water storage, solar energy system, 80 Gal, 2" x 1/2 lb insul	1.000	Ea.	820	260	1,080
Copper tubing type L, solder joints, hangers 10' OC 1/2" diam	150.000	L.F.	207	772.50	979.50
Copper tubing type M, solder joints, hangers 10' OC 1/2" diam	50.000	L.F.	58	247.50	305.50
Sensor wire, #22-2 gauge multistranded	.500	C.L.F.	4.13	20.50	24.63
Wrought copper fittings & solder, 1/2" diam	75.000	Ea.	18	1,575	1,593
Total			2,452.89	3,866.46	6,319.35

8.3-660	Thermosyphon, Hot Water		MAT.	INST.	TOTAL
0960 0970	Solar, thermosyphon, hot water, two collector system	R13600 -610			

HEATING - SOLAR A8.3-710 Closed Loop, Space/Hot Water

In this closed-loop indirect collection system, fluid with a low freezing temperature, propylene glycol, transports heat from the collectors to water storage. The transfer fluid is contained in a closed loop consisting of collectors, supply and return piping, and a heat exchanger immersed in the storage tank.

When the collectors become approximately 20° F warmer than the storage temperature, the controller activates the circulator. The circulator moves the fluid continuously until the temperature difference between fluid in the collectors and storage is such that the collection will no longer occur and then the circulator turns off. Since the heat transfer fluid has a very low freezing temperature, there is no need for it to be drained from the collectors between periods of collection.

HEATING - SOLAR — A8.3-710 Closed Loop, Space/Hot Water

System Components

		QUANTITY	UNIT	COST EACH MAT.	COST EACH INST.	COST EACH TOTAL
	SYSTEM 8.3-710-2750					
	SOLAR, CLOSED LOOP, SPACE/HOT WATER					
	1" TUBING, TEN 3'X7' BLK CHROME ON COPPER ABSORBER COLLECTORS					
A, B	Differential controller 2 sensors, thermostat, solar energy system	2.000	Ea.	171	69	240
C	Thermometer, 2" dial	10.000	Ea.	210	260	470
C-1	Heat exchanger, solar energy system, fluid to air, up flow, 80 MBH	1.000	Ea.	460	250	710
D, T	Fill & drain valves, brass, 3/4" connection	5.000	Ea.	27.25	86.75	114
D-1	Fan center	1.000	Ea.	119	91	210
E	Air vent, manual, 1/8" fitting	1.000	Ea.	2.13	13	15.13
E-2	Thermostat, 2 stage for sensing room temperature	1.000	Ea.	119	52	171
F	Air purger	2.000	Ea.	90	69	159
F-1	Controller, liquid temperature, solar energy system	1.000	Ea.	101	83	184
G	Expansion tank	2.000	Ea.	128	26	154
I	Valve, gate, bronze, 125 lb, soldered, 1" diam	5.000	Ea.	132.50	110	242.50
J	Vent flashing, neoprene	2.000	Ea.	17.30	42	59.30
K	Circulator, solar heated liquid, 1/25 HP	2.000	Ea.	306	107	413
L	Circulator, solar heated liquid, 1/20 HP	1.000	Ea.	161	62.50	223.50
N	Relief valve, temp & pressure 150 psi 210°F self-closing	4.000	Ea.	34.60	55.40	90
N-1	Relief valve, pressure poppet, bronze, 30 psi, 3/4" IPS	3.000	Ea.	145.50	44.55	190.05
O	Pipe covering, urethane, ultraviolet cover, 1" wall, 1" diam	50.000	L.F.	77	202	279
P	Pipe covering, fiberglass, all service jacket, 1" wall, 1" diam	60.000	L.F.	79.80	193.20	273
Q	Collector panel solar energy blk chrome on copper 1/8" temp glass 3'x7'	10.000	Ea.	6,000	790	6,790
	Roof clamps for solar energy collector panel	10.000	Set	17.70	107	124.70
R	Valve, swing check, bronze, 125 lb, regrinding disc, 3/4" & 1" diam	4.000	Ea.	198	88	286
S	Pressure gage, 0-60 psi, for solar energy system	2.000	Ea.	46	26	72
U	Valve, water tempering, bronze, sweat connections, 3/4" diam	1.000	Ea.	54	21	75
W-2, W-1, W	Tank, water storage immersed heat xchr elec elem 2"x1/2# ins 120 gal	4.000	Ea.	3,700	1,188	4,888
X	Valve, globe, bronze, 125 lb, rising stem, 1" diam	3.000	Ea.	220.50	66	286.50
Y	Valve, flow control	1.000	Ea.	54.50	18.90	73.40
	Copper tubing type M, solder joint, hanger 10' OC, 1" diam	110.000	L.F.	205.70	654.50	860.20
	Copper tubing, type L, solder joint, hanger 10' OC 3/4" diam	20.000	L.F.	34.80	109	143.80
	Wrought copper fittings & solder, 3/4" & 1" diam	121.000	Ea.	148.83	3,146	3,294.83
	Sensor, wire, #22-2 conductor, multistranded	.700	C.L.F.	5.78	28.70	34.48
	Ductwork, galvanized steel, for heat exchanger	8.000	Lb.	11.60	39.52	51.12
	Solar energy heat transfer fluid propylene glycol, anti-freeze	25.000	Gal.	242.50	371.25	613.75
	TOTAL			13,320.99	8,470.27	21,791.26

8.3-710 Solar, Closed Loop, Space/Hot Water Systems

		COST EACH MAT.	COST EACH INST.	COST EACH TOTAL
2540	Solar, closed loop, space/hot water			
2550	1/2" tubing, 12 ea. 4'x4'4" vacuum tube collectors	17,900	7,925	25,825
2600	3/4" tubing, 12 ea. 4'x4'4" vacuum tube collectors	18,200	8,150	26,350
2650	10 ea. 3' x 7' black chrome absorber collectors	13,100	7,925	21,025
2700	10 ea. 3' x 7' flat black absorber collectors	11,800	7,950	19,750
2750	1" tubing, 10 ea. 3' x 7' black chrome absorber collectors	13,300	8,475	21,775
2800	10 ea. 3' x 7' flat black absorber collectors	12,100	8,525	20,625
2850	6 ea. 4' x 9' flat black w/plastic glazing collectors	10,600	8,475	19,075
2900	12 ea. 2' x 9' plastic absorber and glazing collectors	13,900	8,625	22,525

(Row 2650 reference: R13600-610)

MECHANICAL 8

HEATING - SOLAR A8.3-720 Swimming Pool Heater

This draindown pool system uses a differential thermostat similar to those used in solar domestic hot water and space heating applications. To heat the pool, the pool water passes through the conventional pump-filter loop and then flows through the collectors. When collection is not possible, or when the pool temperature is reached, all water drains from the solar loop back to the pool through the existing piping. The modes are controlled by solenoid valves or other automatic valves in conjunction with a vacuum breaker relief valve, which facilitates draindown.

HEATING - SOLAR   A8.3-720 Swimming Pool Heater

System Components

		QUANTITY	UNIT	COST EACH MAT.	COST EACH INST.	COST EACH TOTAL
	SYSTEM 8.3-720-2640					
	SOLAR SWIMMING POOL HEATER, ROOF MOUNTED COLLECTORS					
	TEN 4' X 10' FULLY WETTED UNGLAZED PLASTIC ABSORBERS					
A	Differential thermostat/controller, 110V, adj pool pump system	1.000	Ea.	282	208	490
A-1	Solenoid valve, PVC, normally 1 open 1 closed (included)	2.000	Ea.			
B	Sensor, thermistor type (included)	2.000	Ea.			
E-1	Valve, vacuum relief	1.000	Ea.	27.50	13	40.50
Q	Collector panel, solar energy, plastic, liquid full wetted, 4' x 10'	10.000	Ea.	2,150	1,500	3,650
R	Valve, ball check, PVC, socket, 1-1/2" diam	1.000	Ea.	75	21	96
Z	Valve, ball, PVC, socket, 1-1/2" diam	3.000	Ea.	187.50	63	250.50
	Pipe, PVC, sch 40, 1-1/2" diam	80.000	L.F.	251.20	924	1,175.20
	Pipe fittings, PVC sch 40, socket joint, 1-1/2" diam	10.000	Ea.	11	210	221
	Sensor wire, #22-2 conductor, multistranded	.500	C.L.F.	4.13	20.50	24.63
	Roof clamps for solar energy collector panels	10.000	Set	17.70	107	124.70
	Roof strap, teflon for solar energy collector panels	26.000	L.F.	271.70	52.78	324.48
	TOTAL			3,277.73	3,119.28	6,397.01

8.3-720	Solar Swimming Pool Heater Systems		COST EACH MAT.	COST EACH INST.	COST EACH TOTAL
2530	Solar swimming pool heater systems, roof mounted collectors				
2540	10 ea. 3'x7' black chrome absorber, 1/8" temp. glass		7,125	2,400	9,525
2560	10 ea. 4'x8' black chrome absorber, 3/16" temp. glass	R13600 -610	8,225	2,875	11,100
2580	10 ea. 3'8"x6' flat black absorber, 3/16" temp. glass		5,875	2,450	8,325
2600	10 ea. 4'x9' flat black absorber, plastic glazing		6,525	2,975	9,500
2620	10 ea. 2'x9' rubber absorber, plastic glazing		12,000	3,225	15,225
2640	10 ea. 4'x10' fully wetted unglazed plastic absorber		3,275	3,125	6,400
2660	Ground mounted collectors				
2680	10 ea. 3'x7' black chrome absorber, 1/8" temp. glass		7,225	2,775	10,000
2700	10 ea. 4'x8' black chrome absorber, 3/16" temp glass		8,325	3,250	11,575
2720	10 ea. 3'8"x6' flat blk absorber, 3/16" temp. glass		5,975	2,825	8,800
2740	10 ea. 4'x9' flat blk absorber, plastic glazing		6,625	3,350	9,975
2760	10 ea. 2'x9' rubber absorber, plastic glazing		12,100	3,500	15,600
2780	10 ea. 4'x10' fully wetted unglazed plastic absorber		3,375	3,500	6,875

HEATING - SOLAR — A8.3-730 — Air to Water Heat Exchange

This domestic hot water pre-heat system includes heat exchanger with a circulating pump, blower, air-to-water, coil and controls, mounted in the upper collector manifold. Heat from the hot air coming out of the collectors is transferred through the heat exchanger. For each degree of DHW preheating gained, one degree less heating is needed from the fuel fired water heater. The system is simple, inexpensive to operate and can provide a substantial portion of DHW requirements for modest additional cost.

HEATING - SOLAR   A8.3-730   Air to Water Heat Exchange

System Components

		QUANTITY	UNIT	COST EACH		
				MAT.	INST.	TOTAL
	SYSTEM 8.3-730-2560					
	SOLAR, HOT WATER, AIR TO WATER HEAT EXCHANGE					
	THREE COLLECTORS, OPTICAL BLACK ON ALUMINUM, 10'X2', 80GAL TANK					
A, B	Differential controller, 2 sensors, thermostat, solar energy system	1.000	Ea.	85.50	34.50	120
C	Thermometer, 2" dial	2.000	Ea.	42	52	94
C-1	Heat exchanger, air to fluid, up flow 70 MBH	1.000	Ea.	345	214	559
E	Air vent, manual, for solar energy system 1/8" fitting	1.000	Ea.	2.13	13	15.13
F	Air purger	1.000	Ea.	45	34.50	79.50
G	Expansion tank	1.000	Ea.	64	13	77
H	Strainer, Y type, bronze body, 1/2" IPS	1.000	Ea.	17.50	21	38.50
I	Valve, gate, bronze, 125 lb, NRS, soldered 1/2" diam	5.000	Ea.	81.25	86.75	168
N	Relief valve, temp & pressure solar 150 psi 210°F self-closing	1.000	Ea.	8.65	13.85	22.50
P	Pipe covering, fiberglass, all service jacket, 1" wall, 1/2" diam	60.000	L.F.	70.20	185.40	255.60
Q	Collector panel solar energy, air, black on alum plate, flush mount, 10'x2'	30.000	L.F.	4,830	232.50	5,062.50
B-1, R-1	Shutter damper with shutter motor	1.000	Ea.	156.50	108.50	265
R	Backflow preventer for solar energy system, 1/2" diam	2.000	Ea.	128	52	180
T	Drain valve, brass, 3/4" connection	1.000	Ea.	5.45	17.35	22.80
U	Valve, water tempering, bronze, sweat connections, 1/2" diam	1.000	Ea.	44	17.35	61.35
W	Tank, water storage, solar energy system, 80 Gal, 2" x 2 lb insul	1.000	Ea.	820	260	1,080
V	Tank, water storage, w/heating element, drain, relief valve, existing	1.000	System			
X	Valve, globe, bronze, 125 lb, rising stem, 1/2" diam	1.000	Ea.	34.50	17.35	51.85
	Copper tubing type M, solder joints, hangers 10' OC 1/2" diam	50.000	L.F.	58	247.50	305.50
	Copper tubing type L, solder joints, hangers 10' OC 1/2" diam	10.000	L.F.	13.80	51.50	65.30
	Wrought copper fittings & solder, 1/2" diam	10.000	Ea.	2.40	210	212.40
	Sensor wire, #22-2 conductor multistranded	.500	C.L.F.	4.13	20.50	24.63
Q-1, Q-2	Ductwork, fiberglass, aluminzed jacket, 1-1/2" thick, 8" diam	32.000	S.F.	143.36	132.48	275.84
Q-3	Manifold for flush mount solar energy collector panels, air	6.000	L.F.	177	27.96	204.96
	TOTAL			7,178.37	2,062.99	9,241.36

8.3-730	Solar Hot Water, Air To Water Heat Exchange		COST EACH		
			MAT.	INST.	TOTAL
2550	Solar hot water, air to water heat exchange				
2560	Three collectors, optical black on aluminum, 10' x 2', 80 Gal tank		7,175	2,075	9,250
2580	Four collectors, optical black on aluminum, 10' x 2', 80 Gal tank	R13600-610	8,875	2,175	11,050
2600	Four collectors, optical black on aluminum, 10' x 2', 120 Gal tank		9,000	2,225	11,225

HEATING - SOLAR

A8.3-740 Air to Water Heat Exchange

The complete Solar Air Heating System provides maximum savings of conventional fuel with both space heating and year-round domestic hot water heating. It allows for the home air conditioning to operate simultaneously and independently from the solar domestic water heating in summer. The system's modes of operation are:

Mode 1: The building is heated directly from the collectors with air circulated by the Solar Air Mover.

Mode 2: When heat is not needed in the building, the dampers change within the air mover to circulate the air from the collectors to the rock storage bin.

Mode 3: When heat is not available from the collector array and is available in rock storage, the air mover draws heated air from rock storage and directs it into the building. When heat is not available from the collectors or the rock storage bin, the auxiliary heating unit will provide heat for the building. The size of the collector array is typically 25% the size of the main floor area.

HEATING - SOLAR   A8.3-740   Air to Water Heat Exchange

System Components	QUANTITY	UNIT	COST EACH MAT.	COST EACH INST.	COST EACH TOTAL
SYSTEM 8.3-740-0510					
SOLAR, SPACE/HOT WATER, AIR TO WATER HEAT EXCHANGE					
A, B Differential controller 2 sensors thermos., solar energy sys liquid loop	1.000	Ea.	85.50	34.50	120
A-1, B Differential controller 2 sensors 6 station solar energy sys air loop	1.000	Ea.	222	139	361
B-1 Solar energy sensor, freeze prevention	1.000	Ea.	22	13	35
C Thermometer for solar energy system, 2" dial	2.000	Ea.	42	52	94
C-1 Heat exchanger, solar energy system, air to fluid, up flow, 70 MBH	1.000	Ea.	345	214	559
D Drain valve, brass, 3/4" connection	2.000	Ea.	10.90	34.70	45.60
E Air vent, manual, for solar energy system 1/8" fitting	1.000	Ea.	2.13	13	15.13
E-1 Thermostat, 2 stage for sensing room temperature	1.000	Ea.	119	52	171
F Air purger	1.000	Ea.	45	34.50	79.50
G Expansion tank, for solar energy system	1.000	Ea.	64	13	77
I Valve, gate, bronze, 125 lb, NRS, soldered 3/4" diam	2.000	Ea.	37.10	42	79.10
K Circulator, solar heated liquid, 1/25 HP	1.000	Ea.	153	53.50	206.50
N Relief valve temp & press 150 psi 210°F self-closing, 3/4" IPS	1.000	Ea.	8.65	13.85	22.50
N-1 Relief valve, pressure, poppet, bronze, 30 psi, 3/4" IPS	1.000	Ea.	48.50	14.85	63.35
P Pipe covering, fiberglass, all service jacket, 1" wall, 3/4" diam	60.000	L.F.	79.80	193.20	273
Q-3 Manifold for flush mount solar energy collector panels	20.000	L.F.	590	93.20	683.20
Q Collector panel solar energy, air, black on alum. plate, flush mount 10'x2'	100.000	L.F.	16,100	775	16,875
R Valve, swing check, bronze, 125 lb, regrinding disc, 3/4" diam	2.000	Ea.	66	42	108
S Pressure gage, 2" dial, for solar energy system	1.000	Ea.	23	13	36
U Valve, water tempering, bronze, sweat connections, 3/4" diam	1.000	Ea.	54	21	75
W-2, W Tank, water storage, solar, elec element 2"x1/2# insul, 80 Gal	1.000	Ea.	820	260	1,080
X Valve, globe, bronze, 125 lb, soldered, 3/4" diam	1.000	Ea.	47	21	68
Copper tubing type L, solder joints, hangers 10' OC 3/4" diam	10.000	L.F.	17.40	54.50	71.90
Copper tubing type M, solder joints, hangers 10' OC 3/4" diam	60.000	L.F.	88.20	321	409.20
Wrought copper fittings & solder, 3/4" diam	26.000	Ea.	13.78	572	585.78
Sensor wire, #22-2 conductor multistranded	1.200	C.L.F.	9.90	49.20	59.10
Q-1 Duct work, rigid fiberglass, rectangular	400.000	S.F.	364	1,328	1,692
Ductwork, transition pieces	8.000	Ea.	118.40	248	366.40
R-2 Shutter/damper for solar heater circulator	9.000	Ea.	445.50	558	1,003.50
K-1 Shutter motor for solar heater circulator blower	9.000	Ea.	963	418.50	1,381.50
K-1 Fan, solar energy heated air circulator, space & DHW system	1.000	Ea.	1,550	1,500	3,050
Z-1 Tank, solar energy air storage, 6'-3"H 7'x7' = 306 CF/2000 Gal	1.000	Ea.	8,950	965	9,915
Z-2 Crushed stone 1-1/2"	11.000	C.Y.	225.50	54.45	279.95
C-2 Themometer, remote probe, 2" dial	1.000	Ea.	32	52	84
R-1 Solenoid valve	1.000	Ea.	233	46	279
V, R-4, R-3 Furnace, supply diffusers, return grilles, existing	1.000	Ea.	8,950	965	9,915
Total			**31,769.76**	**8,254.50**	**40,024.26**

8.3-740	Air To Water Heat Exchange		COST EACH MAT.	COST EACH INST.	COST EACH TOTAL
1210	Solar, air to water heat exchange for space / hot water heating	R13600 -610			
1220					

AIR CONDITIONING — A8.4-110 Chilled Water, Air Cooled

Design Assumptions: The chilled water, air cooled systems priced, utilize reciprocating hermetic compressors and propeller-type condenser fans. Piping with pumps and expansion tanks is included based on a two pipe system. No ducting is included and the fan-coil units are cooling only. Water treatment and balancing are not included. Chilled water piping is insulated. Area distribution is through the use of multiple fan coil units. Fewer but larger fan coil units with duct distribution would be approximately the same S.F. cost.

System Components	QUANTITY	UNIT	COST EACH MAT.	COST EACH INST.	COST EACH TOTAL
SYSTEM 8.4-110-1200					
PACKAGED CHILLER, AIR COOLED, WITH FAN COIL UNIT					
APARTMENT CORRIDORS, 3,000 S.F., 5.50 TON					
Fan coil air conditioning unit, cabinet mounted & filters chilled water	1.000	Ea.	3,345.23	348.27	3,693.50
Water chiller, air conditioning unit, reciprocating, air cooled,	1.000	Ea.	4,702.50	1,313.13	6,015.63
Chilled water unit coil connections	1.000	Ea.	515	950	1,465
Chilled water distribution piping	440.000	L.F.	4,290	12,760	17,050
TOTAL			12,852.73	15,371.40	28,224.13
COST PER S.F.			4.28	5.12	9.40

*Cooling requirements would lead to choosing a water cooled unit

8.4-110	Chilled Water, Air Cooled Condenser Systems	COST PER S.F. MAT.	COST PER S.F. INST.	COST PER S.F. TOTAL
1180	Packaged chiller, air cooled, with fan coil unit			
1200	Apartment corridors, 3,000 S.F., 5.50 ton	4.28	5.10	9.38
1240	6,000 S.F., 11.00 ton	3.39	4.12	7.51
1280	10,000 S.F., 18.33 ton	3	3.13	6.13
1320	20,000 S.F., 36.66 ton	2.45	2.27	4.72
1360	40,000 S.F., 73.33 ton	2.93	2.29	5.22
1440	Banks and libraries, 3,000 S.F., 12.50 ton	6.50	6.05	12.55
1480	6,000 S.F., 25.00 ton	6.10	4.91	11.01
1520	10,000 S.F., 41.66 ton	5	3.43	8.43
1560	20,000 S.F., 83.33 ton	5.40	3.12	8.52
1600	40,000 S.F., 167 ton*			
1680	Bars and taverns, 3,000 S.F., 33.25 ton	12.55	7.15	19.70
1720	6,000 S.F., 66.50 ton	11.85	5.55	17.40
1760	10,000 S.F., 110.83 ton	9.80	1.78	11.58
1800	20,000 S.F., 220 ton*			
1840	40,000 S.F., 440 ton*			
1920	Bowling alleys, 3,000 S.F., 17.00 ton	8.35	6.80	15.15
1960	6,000 S.F., 34.00 ton	6.80	4.77	11.57
2000	10,000 S.F., 56.66 ton	6.25	3.47	9.72
2040	20,000 S.F., 113.33 ton	6.05	3.15	9.20
2080	40,000 S.F., 227 ton*			
2160	Department stores, 3,000 S.F., 8.75 ton	6.20	5.70	11.90
2200	6,000 S.F., 17.50 ton	4.57	4.43	9
2240	10,000 S.F., 29.17 ton	3.84	3.20	7.04
2280	20,000 S.F., 58.33 ton	3.45	2.36	5.81
2320	40,000 S.F., 116.66 ton	3.77	2.34	6.11

AIR CONDITIONING | A8.4-110 | Chilled Water, Air Cooled

8.4-110	Chilled Water, Air Cooled Condenser Systems	COST PER S.F.		
		MAT.	INST.	TOTAL
2400	Drug stores, 3,000 S.F., 20.00 ton	9.60	7	16.60
2440	6,000 S.F., 40.00 ton	8.05	5.20	13.25
2480	10,000 S.F., 66.66 ton	8.40	4.48	12.88
2520	20,000 S.F., 133.33 ton	7.40	3.33	10.73
2560	40,000 S.F., 267 ton*			
2640	Factories, 2,000 S.F., 10.00 ton	5.60	5.80	11.40
2680	6,000 S.F., 20.00 ton	5.20	4.67	9.87
2720	10,000 S.F., 33.33 ton	4.23	3.27	7.50
2760	20,000 S.F., 66.66 ton	4.62	2.95	7.57
2800	40,000 S.F., 133.33 ton	4.15	2.41	6.56
2880	Food supermarkets, 3,000 S.F., 8.50 ton	6.05	5.70	11.75
2920	6,000 S.F., 17.00 ton	4.48	4.41	8.89
2960	10,000 S.F., 28.33 ton	3.66	3.10	6.76
3000	20,000 S.F., 56.66 ton	3.32	2.31	5.63
3040	40,000 S.F., 113.33 ton	3.64	2.35	5.99
3120	Medical centers, 3,000 S.F., 7.00 ton	5.35	5.55	10.90
3160	6,000 S.F., 14.00 ton	3.94	4.26	8.20
3200	10,000 S.F., 23.33 ton	3.59	3.25	6.84
3240	20,000 S.F., 46.66 ton	2.99	2.38	5.37
3280	40,000 S.F., 93.33 ton	3.23	2.30	5.53
3360	Offices, 3,000 S.F., 9.50 ton	5.20	5.65	10.85
3400	6,000 S.F., 19.00 ton	5	4.63	9.63
3440	10,000 S.F., 31.66 ton	4.08	3.25	7.33
3480	20,000 S.F., 63.33 ton	4.57	3.02	7.59
3520	40,000 S.F., 126.66 ton	4.06	2.44	6.50
3600	Restaurants, 3,000 S.F., 15.00 ton	7.45	6.25	13.70
3640	6,000 S.F., 30.00 ton	6.35	4.77	11.12
3680	10,000 S.F., 50.00 ton	5.85	3.52	9.37
3720	20,000 S.F., 100.00 ton	6.05	3.29	9.34
3760	40,000 S.F., 200 ton*			
3840	Schools and colleges, 3,000 S.F., 11.50 ton	6.15	5.95	12.10
3880	6,000 S.F., 23.00 ton	5.75	4.82	10.57
3920	10,000 S.F., 38.33 ton	4.69	3.37	8.06
3960	20,000 S.F., 76.66 ton	5.20	3.15	8.35
4000	40,000 S.F., 153 ton*			

AIR CONDITIONING — A8.4-120 — Chilled Water, Water Cooled

General: Water cooled chillers are available in the same sizes as air cooled units. They are also available in larger capacities.

Design Assumptions: The chilled water systems with water cooled condenser include reciprocating hermetic compressors, water cooling tower, pumps, piping and expansion tanks and are based on a two pipe system. Chilled water piping is insulated. No ducts are included and fan-coil units are cooling only. Area distribution is through use of multiple fan coil units. Fewer but larger fan coil units with duct distribution would be approximately the same S.F. cost. Water treatment and balancing are not included.

System Components	QUANTITY	UNIT	COST EACH MAT.	COST EACH INST.	COST EACH TOTAL
SYSTEM 8.4-120-1320					
PACKAGED CHILLER, WATER COOLED, WITH FAN COIL UNIT					
APARTMENT CORRIDORS, 4,000 S.F., 7.33 TON					
Fan coil air conditioner unit, cabinet mounted & filters, chilled water	2.000	Ea.	4,458.48	464.17	4,922.65
Water chiller, reciprocating, water cooled, 1 compressor semihermetic	1.000	Ea.	9,892.80	2,267.10	12,159.90
Cooling tower, draw thru single flow, belt drive	1.000	Ea.	630.38	96.02	726.40
Cooling tower pumps & piping	1.000	System	315.19	227.23	542.42
Chilled water unit coil connections	2.000	Ea.	1,030	1,900	2,930
Chilled water distribution piping	520.000	L.F.	5,070	15,080	20,150
TOTAL			21,396.85	20,034.52	41,431.37
COST PER S.F.			5.35	5.01	10.36

*Cooling requirements would lead to choosing a water cooled unit

8.4-120	Chilled Water, Cooling Tower Systems	COST PER S.F. MAT.	INST.	TOTAL
1300	Packaged chiller, water cooled, with fan coil unit			
1320	Apartment corridors, 4,000 S.F., 7.33 ton	5.35	5	10.35
1360	6,000 S.F., 11.00 ton	4.22	4.33	8.55
1400	10,000 S.F., 18.33 ton	3.53	3.20	6.73
1440	20,000 S.F., 26.66 ton	2.80	2.39	5.19
1480	40,000 S.F., 73.33 ton	3.14	2.46	5.60
1520	60,000 S.F., 110.00 ton	3.07	2.53	5.60
1600	Banks and libraries, 4,000 S.F., 16.66 ton	8.05	5.50	13.55
1640	6,000 S.F., 25.00 ton	6.50	4.85	11.35
1680	10,000 S.F., 41.66 ton	5.20	3.62	8.82
1720	20,000 S.F., 83.33 ton	5.85	3.46	9.31
1760	40,000 S.F., 166.66 ton	5.65	4.32	9.97
1800	60,000 S.F., 250.00 ton	5.55	4.57	10.12
1880	Bars and taverns, 4,000 S.F., 44.33 ton	12.65	6.95	19.60
1920	6,000 S.F., 66.50 ton	12.95	7.20	20.15
1960	10,000 S.F., 110.83 ton	12.75	5.70	18.45
2000	20,000 S.F., 221.66 ton	11.90	6.20	18.10
2040	40,000 S.F., 440 ton*			
2080	60,000 S.F., 660 ton*			
2160	Bowling alleys, 4,000 S.F., 22.66 ton	9.45	6.05	15.50
2200	6,000 S.F., 34.00 ton	7.95	5.35	13.30
2240	10,000 S.F., 56.66 ton	6.50	3.92	10.42
2280	20,000 S.F., 113.33 ton	6.90	3.66	10.56
2320	40,000 S.F., 226.66 ton	6.65	4.45	11.10

Reference: R15700-020

AIR CONDITIONING

A8.4-120 Chilled Water, Water Cooled

8.4-120 Chilled Water, Cooling Tower Systems

		COST PER S.F.		
		MAT.	INST.	TOTAL
2360	60,000 S.F., 340 ton			
2440	Department stores, 4,000 S.F., 11.66 ton	6.05	5.45	11.50
2480	6,000 S.F., 17.50 ton	5.85	4.49	10.34
2520	10,000 S.F., 29.17 ton	4.25	3.35	7.60
2560	20,000 S.F., 58.33 ton	3.29	2.49	5.78
2600	40,000 S.F., 116.66 ton	3.96	2.58	6.54
2640	60,000 S.F., 175.00 ton	4.49	4.28	8.77
2720	Drug stores, 4,000 S.F., 26.66 ton	9.90	6.25	16.15
2760	6,000 S.F., 40.00 ton	8.25	5.35	13.60
2800	10,000 S.F., 66.66 ton	8.15	4.87	13.02
2840	20,000 S.F., 133.33 ton	7.95	3.83	11.78
2880	40,000 S.F., 266.67 ton	7.75	4.90	12.65
2920	60,000 S.F., 400 ton*			
3000	Factories, 4,000 S.F., 13.33 ton	6.80	5.25	12.05
3040	6,000 S.F., 20.00 ton	5.90	4.50	10.40
3080	10,000 S.F., 33.33 ton	4.71	3.45	8.16
3120	20,000 S.F., 66.66 ton	4.44	3.10	7.54
3160	40,000 S.F., 133.33 ton	4.42	2.65	7.07
3200	60,000 S.F., 200.00 ton	4.85	4.38	9.23
3280	Food supermarkets, 4,000 S.F., 11.33 ton	5.95	5.40	11.35
3320	6,000 S.F., 17.00 ton	5.15	4.42	9.57
3360	10,000 S.F., 28.33 ton	4.15	3.33	7.48
3400	20,000 S.F., 56.66 ton	3.38	2.50	5.88
3440	40,000 S.F., 113.33 ton	3.91	2.58	6.49
3480	60,000 S.F., 170.00 ton	4.42	4.27	8.69
3560	Medical centers, 4,000 S.F., 9.33 ton	5.05	4.95	10
3600	6,000 S.F., 14.00 ton	4.96	4.31	9.27
3640	10,000 S.F., 23.33 ton	3.82	3.21	7.03
3680	20,000 S.F., 46.66 ton	2.98	2.43	5.41
3720	40,000 S.F., 93.33 ton	3.57	2.52	6.09
3760	60,000 S.F., 140.00 ton	4.06	4.21	8.27
3840	Offices, 4,000 S.F., 12.66 ton	6.55	5.20	11.75
3880	6,000 S.F., 19.00 ton	5.75	4.63	10.38
3920	10,000 S.F., 31.66 ton	4.63	3.50	8.13
3960	20,000 S.F., 63.33 ton	4.35	3.12	7.47
4000	40,000 S.F., 126.66 ton	4.70	4.14	8.84
4040	60,000 S.F., 190.00 ton	4.68	4.34	9.02
4120	Restaurants, 4,000 S.F., 20.00 ton	8.45	5.60	14.05
4160	6,000 S.F., 30.00 ton	7.15	5	12.15
4200	10,000 S.F., 50.00 ton	5.90	3.76	9.66
4240	20,000 S.F., 100.00 ton	6.65	3.65	10.30
4280	40,000 S.F., 200.00 ton	5.90	4.19	10.09
4320	60,000 S.F., 300.00 ton	6.25	4.73	10.98
4400	Schools and colleges, 4,000 S.F., 15.33 ton	7.55	5.40	12.95
4440	6,000 S.F., 23.00 ton	6.10	4.76	10.86
4480	10,000 S.F., 38.33 ton	4.85	3.54	8.39
4520	20,000 S.F., 76.66 ton	5.55	3.43	8.98
4560	40,000 S.F., 153.33 ton	5.25	4.22	9.47
4600	60,000 S.F., 230.00 ton	5.10	4.40	9.50

AIR CONDITIONING — A8.4-210 Rooftop, Single Zone

System Description: Rooftop single zone units are electric cooling and gas heat. Duct systems are low velocity, galvanized steel supply and return. Price variations between sizes are due to several factors. Jumps in the cost of the rooftop unit occur when the manufacturer shifts from the largest capacity unit on a small frame to the smallest capacity on the next larger frame, or changes from one compressor to two. As the unit capacity increases for larger areas the duct distribution grows in proportion. For most applications there is a tradeoff point where it is less expensive and more efficient to utilize smaller units with short simple distribution systems. Larger units also require larger initial supply and return ducts which can create a space problem. Supplemental heat may be desired in colder locations. The table below is based on one unit supplying the area listed. The 10,000 S.F. unit for bars and taverns is not listed because a nominal 110 ton unit would be required and this is above the normal single zone rooftop capacity.

System Components	QUANTITY	UNIT	COST EACH		
			MAT.	INST.	TOTAL
SYSTEM 8.4-210-1280					
ROOFTOP, SINGLE ZONE, AIR CONDITIONER					
APARTMENT CORRIDORS, 500 S.F., .92 TON					
Rooftop air-conditioner, 1 zone, electric cool, standard controls, curb	1.000	Ea.	2,127.50	494.50	2,622
Ductwork package for rooftop single zone units	1.000	System	265.88	713	978.88
TOTAL			2,393.38	1,207.50	3,600.88
COST PER S.F.			4.79	2.42	7.21

*Size would suggest multiple units

8.4-210	Rooftop Single Zone Unit Systems	COST PER S.F.		
		MAT.	INST.	TOTAL
1260	Rooftop, single zone, air conditioner			
1280	Apartment corridors, 500 S.F., .92 ton	4.79	2.42	7.21
1320	1,000 S.F., 1.83 ton	4.76	2.40	7.16
1360	1500 S.F., 2.75 ton	2.80	1.99	4.79
1400	3,000 S.F., 5.50 ton	2.42	1.92	4.34
1440	5,000 S.F., 9.17 ton	2.44	1.75	4.19
1480	10,000 S.F., 18.33 ton	2.60	1.64	4.24
1560	Banks or libraries, 500 S.F., 2.08 ton	10.80	5.45	16.25
1600	1,000 S.F., 4.17 ton	6.35	4.53	10.88
1640	1,500 S.F., 6.25 ton	5.50	4.36	9.86
1680	3,000 S.F., 12.50 ton	5.55	3.97	9.52
1720	5,000 S.F., 20.80 ton	5.90	3.72	9.62
1760	10,000 S.F., 41.67 ton	5.55	3.71	9.26
1840	Bars and taverns, 500 S.F. 5.54 ton	13	7.30	20.30
1880	1,000 S.F., 11.08 ton	13.10	6.25	19.35
1920	1,500 S.F., 16.62 ton	11.80	5.80	17.60
1960	3,000 S.F., 33.25 ton	13.40	5.55	18.95
2000	5,000 S.F., 55.42 ton	12.80	5.55	18.35
2040	10,000 S.F., 110.83 ton*			
2080	Bowling alleys, 500 S.F., 2.83 ton	8.65	6.15	14.80
2120	1,000 S.F., 5.67 ton	7.45	5.90	13.35
2160	1,500 S.F., 8.50 ton	7.55	5.40	12.95
2200	3,000 S.F., 17.00 ton	6.85	5.15	12
2240	5,000 S.F., 28.33 ton	7.70	5.05	12.75
2280	10,000 S.F., 56.67 ton	7.35	5.05	12.40
2360	Department stores, 500 S.F., 1.46 ton	7.60	3.83	11.43
2400	1,000 S.F., 2.92 ton	4.45	3.17	7.62
2440	1,500 S.F., 4.37 ton	3.84	3.05	6.89

AIR CONDITIONING A8.4-210 Rooftop, Single Zone

8.4-210 Rooftop Single Zone Unit Systems

		COST PER S.F.		
		MAT.	INST.	TOTAL
2480	3,000 S.F., 8.75 ton	3.87	2.78	6.65
2520	5,000 S.F., 14.58 ton	3.52	2.66	6.18
2560	10,000 S.F., 29.17 ton	3.95	2.59	6.54
2640	Drug stores, 500 S.F., 3.33 ton	10.15	7.25	17.40
2680	1,000 S.F., 6.67 ton	8.80	6.95	15.75
2720	1,500 S.F., 10.00 ton	8.85	6.35	15.20
2760	3,000 S.F., 20.00 ton	9.45	5.95	15.40
2800	5,000 S.F., 33.33 ton	9.05	5.95	15
2840	10,000 S.F., 66.67 ton	8.65	5.95	14.60
2920	Factories, 500 S.F., 1.67 ton	8.65	4.37	13.02
2960	1,000 S.F., 3.33 ton	5.10	3.62	8.72
3000	1,500 S.F., 5.00 ton	4.39	3.48	7.87
3040	3,000 S.F., 10.00 ton	4.43	3.17	7.60
3080	5,000 S.F., 16.67 ton	4.03	3.04	7.07
3120	10,000 S.F., 33.33 ton	4.52	2.96	7.48
3200	Food supermarkets, 500 S.F., 1.42 ton	7.35	3.72	11.07
3240	1,000 S.F., 2.83 ton	4.29	3.07	7.36
3280	1,500 S.F., 4.25 ton	3.74	2.97	6.71
3320	3,000 S.F., 8.50 ton	3.77	2.70	6.47
3360	5,000 S.F., 14.17 ton	3.43	2.59	6.02
3400	10,000 S.F., 28.33 ton	3.84	2.52	6.36
3480	Medical centers, 500 S.F., 1.17 ton	6.05	3.06	9.11
3520	1,000 S.F., 2.33 ton	6.05	3.06	9.11
3560	1,500 S.F., 3.50 ton	3.56	2.54	6.10
3600	3,000 S.F., 7.00 ton	3.07	2.44	5.51
3640	5,000 S.F., 11.67 ton	3.10	2.22	5.32
3680	10,000 S.F., 23.33 ton	3.31	2.09	5.40
3760	Offices, 500 S.F., 1.58 ton	8.25	4.15	12.40
3800	1,000 S.F., 3.17 ton	4.84	3.45	8.29
3840	1,500 S.F., 4.75 ton	4.18	3.31	7.49
3880	3,000 S.F., 9.50 ton	4.21	3.01	7.22
3920	5,000 S.F., 15.83 ton	3.82	2.89	6.71
3960	10,000 S.F., 31.67 ton	4.30	2.81	7.11
4000	Restaurants, 500 S.F., 2.50 ton	13	6.55	19.55
4040	1,000 S.F., 5.00 ton	6.60	5.25	11.85
4080	1,500 S.F., 7.50 ton	6.65	4.77	11.42
4120	3,000 S.F., 15.00 ton	6.05	4.57	10.62
4160	5,000 S.F., 25.00 ton	7.10	4.48	11.58
4200	10,000 S.F., 50.00 ton	6.50	4.45	10.95
4240	Schools and colleges, 500 S.F., 1.92 ton	9.95	5.05	15
4280	1,000 S.F., 3.83 ton	5.85	4.17	10.02
4320	1,500 S.F., 5.75 ton	5.05	4.01	9.06
4360	3,000 S.F., 11.50 ton	5.10	3.65	8.75
4400	5,000 S.F., 19.17 ton	5.45	3.43	8.88
4440	10,000 S.F., 38.33 ton	5.10	3.41	8.51
5000				
5100				
5200				
9000	Components of ductwork packages for above systems, per ton of cooling:			
9010	Ductwork; galvanized steel, 120 pounds			
9020	Insulation; fiberglass 2" thick, FRK faced, 52 SF			
9030	Diffusers; aluminum, 24" x 12", one			
9040	Registers; aluminum, return with dampers, one			

AIR CONDITIONING — A8.4-220 Rooftop Multizone

System Description: Rooftop units are multizone with up to 12 zones, and include electric cooling, gas heat, thermostats, filters, supply and return fans complete. Duct systems are low velocity, galvanized steel supply and return with insulated supplies.

Multizone units cost more per ton of cooling than single zone. However, they offer flexibility where load conditions are varied due to heat generating areas or exposure to radiational heating. For example, perimeter offices on the "sunny side" may require cooling at the same time "shady side" or central offices may require heating. It is possible to accomplish similar results using duct heaters in branches of the single zone unit. However, heater location could be a problem and total system operating energy efficiency could be lower.

System Components	QUANTITY	UNIT	COST EACH MAT.	INST.	TOTAL
SYSTEM 8.4-220-1280					
ROOFTOP, MULTIZONE, AIR CONDITIONER					
APARTMENT CORRIDORS, 3,000 S.F., 5.50 TON					
Rooftop multizone unit, standard controls, curb	1.000	Ea.	19,316	1,155	20,471
Ductwork package for rooftop multizone units	1.000	System	2,282.50	8,250	10,532.50
TOTAL			21,598.50	9,405	31,003.50
COST PER S.F.			7.20	3.14	10.34

Note A: Small single zone unit recommended
Note B: A combination of multizone units recommended

8.4-220	Rooftop Multizone Unit Systems		COST PER S.F. MAT.	INST.	TOTAL
1240	Rooftop, multizone, air conditioner				
1260	Apartment corridors, 1,500 S.F., 2.75 ton. See Note A.				
1280	3,000 S.F., 5.50 ton	R15700 -020	7.20	3.14	10.34
1320	10,000 S.F., 18.30 ton		5.35	3.03	8.38
1360	15,000 S.F., 27.50 ton		4.49	3.01	7.50
1400	20,000 S.F., 36.70 ton		4.61	3.01	7.62
1440	25,000 S.F., 45.80 ton		4.40	3.03	7.43
1520	Banks or libraries, 1,500 S.F., 6.25 ton		16.35	7.15	23.50
1560	3,000 S.F., 12.50 ton		13.90	7	20.90
1600	10,000 S.F., 41.67 ton		10	6.90	16.90
1640	15,000 S.F., 62.50 ton		8.60	6.85	15.45
1680	20,000 S.F., 83.33 ton		8.60	6.85	15.45
1720	25,000 S.F., 104.00 ton		8.15	6.80	14.95
1800	Bars and taverns, 1,500 S.F., 16.62 ton		34.50	10.25	44.75
1840	3,000 S.F., 33.24 ton		25.50	9.90	35.40
1880	10,000 S.F., 110.83 ton		19.35	9.85	29.20
1920	15,000 S.F., 165 ton, See Note B				
1960	20,000 S.F., 220 ton, See Note B				
2000	25,000 S.F., 275 ton, See Note B				
2080	Bowling alleys, 1,500 S.F., 8.50 ton		22.50	9.70	32.20
2120	3,000 S.F., 17.00 ton		18.95	9.50	28.45
2160	10,000 S.F., 56.70 ton		13.60	9.40	23
2200	15,000 S.F., 85.00 ton		11.70	9.30	21
2240	20,000 S.F., 113.00 ton		11.05	9.25	20.30

AIR CONDITIONING | A8.4-220 | Rooftop Multizone

8.4-220 Rooftop Multizone Unit Systems

Line	Description	MAT.	INST.	TOTAL
2280	25,000 S.F., 140.00 ton see Note B			
2360	Department stores, 1,500 S.F., 4.37 ton, See Note A.			
2400	3,000 S.F., 8.75 ton	11.45	4.99	16.44
2440	10,000 S.F., 29.17 ton	7.35	4.80	12.15
2480	15,000 S.F., 43.75 ton	7	4.83	11.83
2520	20,000 S.F., 58.33 ton	6	4.79	10.79
2560	25,000 S.F., 72.92 ton	6	4.80	10.80
2640	Drug stores, 1,500 S.F., 10.00 ton	26	11.40	37.40
2680	3,000 S.F., 20.00 ton	19.45	11	30.45
2720	10,000 S.F., 66.66 ton	13.75	10.95	24.70
2760	15,000 S.F., 100.00 ton	13	10.95	23.95
2800	20,000 S.F., 135 ton, See Note B			
2840	25,000 S.F., 165 ton, See Note B			
2920	Factories, 1,500 S.F., 5 ton, See Note A			
2960	3,000 S.F., 10.00 ton	13.10	5.70	18.80
3000	10,000 S.F., 33.33 ton	8.40	5.50	13.90
3040	15,000 S.F., 50.00 ton	8	5.50	13.50
3080	20,000 S.F., 66.66 ton	6.90	5.45	12.35
3120	25,000 S.F., 83.33 ton	6.90	5.50	12.40
3200	Food supermarkets, 1,500 S.F., 4.25 ton, See Note A			
3240	3,000 S.F., 8.50 ton	11.15	4.85	16
3280	10,000 S.F., 28.33 ton	6.95	4.65	11.60
3320	15,000 S.F., 42.50 ton	6.80	4.69	11.49
3360	20,000 S.F., 56.67 ton	5.85	4.65	10.50
3400	25,000 S.F., 70.83 ton	5.85	4.65	10.50
3480	Medical centers, 1,500 S.F., 3.5 ton, See Note A			
3520	3,000 S.F., 7.00 ton	9.15	3.99	13.14
3560	10,000 S.F., 23.33 ton	5.95	3.84	9.79
3600	15,000 S.F., 35.00 ton	5.85	3.83	9.68
3640	20,000 S.F., 46.66 ton	5.60	3.86	9.46
3680	25,000 S.F., 58.33 ton	4.82	3.83	8.65
3760	Offices, 1,500 S.F., 4.75 ton, See Note A			
3800	3,000 S.F., 9.50 ton	12.45	5.40	17.85
3840	10,000 S.F., 31.66 ton	7.95	5.20	13.15
3880	15,000 S.F., 47.50 ton	7.60	5.25	12.85
3920	20,000 S.F., 63.33 ton	6.55	5.20	11.75
3960	25,000 S.F., 79.16 ton	6.55	5.20	11.75
4000	Restaurants, 1,500 S.F., 7.50 ton	19.65	8.55	28.20
4040	3,000 S.F., 15.00 ton	16.70	8.40	25.10
4080	10,000 S.F., 50.00 ton	12	8.25	20.25
4120	15,000 S.F., 75.00 ton	10.35	8.20	18.55
4160	20,000 S.F., 100.00 ton	9.75	8.20	17.95
4200	25,000 S.F., 125 ton, See Note B			
4240	Schools and colleges, 1,500 S.F., 5.75 ton	15.05	6.55	21.60
4280	3,000 S.F., 11.50 ton	12.80	6.40	19.20
4320	10,000 S.F., 38.33 ton	9.20	6.35	15.55
4360	15,000 S.F., 57.50 ton	7.90	6.30	14.20
4400	20,000 S.F., 76.66 ton	7.90	6.30	14.20
4440	25,000 S.F., 95.83 ton	7.50	6.30	13.80
4450				
9000	Components of ductwork packages for above systems, per ton of cooling:			
9010	Ductwork; galvanized steel, 240 pounds			
9020	Insulation; fiberglass, 2" thick, FRK faced, 104 SF			
9030	Diffusers; aluminum, 24" x 12", two			
9040	Registers; aluminum, return with dampers, one			

AIR CONDITIONING — A8.4-230 Self-contained, Water Cooled

System Description: Self-contained, single package water cooled units include cooling tower, pump, piping allowance. Systems for 1000 S.F. and up include duct and diffusers to provide for even distribution of air. Smaller units distribute air through a supply air plenum, which is integral with the unit.

Returns are not ducted and supplies are not insulated.

Hot water or steam heating coils are included but piping to boiler and the boiler itself is not included.

Where local codes or conditions permit single pass cooling for the smaller units, deduct 10%.

System Components	QUANTITY	UNIT	COST EACH MAT.	INST.	TOTAL
SYSTEM 8.4-230-1300 SELF-CONTAINED, WATER COOLED UNIT APARTMENT CORRIDORS, 500 S.F., .92 TON					
Self contained, water cooled, single package air conditioner unit	1.000	Ea.	1,134	378	1,512
Ductwork package for water or air cooled packaged units	1.000	System	89.70	542.80	632.50
Cooling tower, draw thru single flow, belt drive	1.000	Ea.	79.12	12.05	91.17
Cooling tower pumps & piping	1.000	System	39.56	28.52	68.08
TOTAL			1,342.38	961.37	2,303.75
COST PER S.F.			2.68	1.92	4.60

8.4-230	Self-contained, Water Cooled Unit Systems	COST PER S.F. MAT.	INST.	TOTAL
1280	Self-contained, water cooled unit	2.69	1.93	4.62
1300	Apartment corridors, 500 S.F., .92 ton	2.68	1.92	4.60
1320	1,000 S.F., 1.83 ton	2.68	1.91	4.59
1360	3,000 S.F., 5.50 ton	2.04	1.63	3.67
1400	5,000 S.F., 9.17 ton	2.07	1.52	3.59
1440	10,000 S.F., 18.33 ton	1.89	1.35	3.24
1520	Banks or libraries, 500 S.F., 2.08 ton	5.70	1.89	7.59
1560	1,000 S.F., 4.17 ton	4.64	3.72	8.36
1600	3,000 S.F., 12.50 ton	4.70	3.45	8.15
1640	5,000 S.F., 20.80 ton	4.29	3.07	7.36
1680	10,000 S.F., 41.66 ton	3.75	3.08	6.83
1760	Bars and taverns, 500 S.F., 5.54 ton	11.15	3.33	14.48
1800	1,000 S.F., 11.08 ton	11.85	5.90	17.75
1840	3,000 S.F., 33.25 ton	9.80	4.65	14.45
1880	5,000 S.F., 55.42 ton	9.20	4.85	14.05
1920	10,000 S.F., 110.00 ton	9	4.79	13.79
2000	Bowling alleys, 500 S.F., 2.83 ton	7.75	2.59	10.34
2040	1,000 S.F., 5.66 ton	6.30	5.05	11.35
2080	3,000 S.F., 17.00 ton	5.85	4.19	10.04
2120	5,000 S.F., 28.33 ton	5.30	4.05	9.35
2160	10,000 S.F., 56.66 ton	4.98	4.14	9.12
2200	Department stores, 500 S.F., 1.46 ton	3.99	1.33	5.32
2240	1,000 S.F., 2.92 ton	3.24	2.60	5.84
2280	3,000 S.F., 8.75 ton	3.28	2.42	5.70
2320	5,000 S.F., 14.58 ton	3	2.16	5.16
2360	10,000 S.F., 29.17 ton	2.73	2.09	4.82

Reference: R15700-020 (at row 1360)

AIR CONDITIONING — A8.4-230 Self-contained, Water Cooled

8.4-230 Self-contained, Water Cooled Unit Systems

		COST PER S.F.		
		MAT.	INST.	TOTAL
2440	Drug stores, 500 S.F., 3.33 ton	9.10	3.05	12.15
2480	1,000 S.F., 6.66 ton	7.40	5.95	13.35
2520	3,000 S.F., 20.00 ton	6.85	4.94	11.79
2560	5,000 S.F., 33.33 ton	7.20	4.88	12.08
2600	10,000 S.F., 66.66 ton	5.85	4.88	10.73
2680	Factories, 500 S.F., 1.66 ton	4.55	1.51	6.06
2720	1,000 S.F. 3.37 ton	3.74	3	6.74
2760	3,000 S.F., 10.00 ton	3.76	2.76	6.52
2800	5,000 S.F., 16.66 ton	3.43	2.46	5.89
2840	10,000 S.F., 33.33 ton	3.12	2.38	5.50
2920	Food supermarkets, 500 S.F., 1.42 ton	3.86	1.30	5.16
2960	1,000 S.F., 2.83 ton	4.14	2.97	7.11
3000	3,000 S.F., 8.50 ton	3.15	2.53	5.68
3040	5,000 S.F., 14.17 ton	3.19	2.35	5.54
3080	10,000 S.F., 28.33 ton	2.66	2.03	4.69
3160	Medical centers, 500 S.F., 1.17 ton	3.18	1.06	4.24
3200	1,000 S.F., 2.33 ton	3.42	2.44	5.86
3240	3,000 S.F., 7.00 ton	2.60	2.08	4.68
3280	5,000 S.F., 11.66 ton	2.63	1.94	4.57
3320	10,000 S.F., 23.33 ton	2.40	1.73	4.13
3400	Offices, 500 S.F., 1.58 ton	4.33	1.45	5.78
3440	1,000 S.F., 3.17 ton	4.64	3.32	7.96
3480	3,000 S.F., 9.50 ton	3.57	2.63	6.20
3520	5,000 S.F., 15.83 ton	3.26	2.34	5.60
3560	10,000 S.F., 31.67 ton	2.97	2.27	5.24
3640	Restaurants, 500 S.F., 2.50 ton	6.85	2.29	9.14
3680	1,000 S.F., 5.00 ton	5.55	4.47	10.02
3720	3,000 S.F., 15.00 ton	5.65	4.16	9.81
3760	5,000 S.F., 25.00 ton	5.15	3.71	8.86
3800	10,000 S.F., 50.00 ton	3.27	3.42	6.69
3880	Schools and colleges, 500 S.F., 1.92 ton	5.25	1.75	7
3920	1,000 S.F., 3.83 ton	4.25	3.42	7.67
3960	3,000 S.F., 11.50 ton	4.31	3.18	7.49
4000	5,000 S.F., 19.17 ton	3.94	2.84	6.78
4040	10,000 S.F., 38.33 ton	3.44	2.84	6.28
4050				
9000	Components of ductwork packages for above systems, per ton of cooling:			
9010	Ductwork; galvanized steel, 108 pounds			
9020	Diffusers, aluminum, 24" x 12", two			

AIR CONDITIONING | A8.4-240 | Self-contained, Air Cooled

System Description: Self-contained air cooled units with remote air cooled condenser and interconnecting tubing. Systems for 1000 S.F. and up include duct and diffusers. Smaller units distribute air directly.

Returns are not ducted and supplies are not insulated.

Potential savings may be realized by using a single zone rooftop system or through-the-wall unit, especially in the smaller capacities, if the application permits.

Hot water or steam heating coils are included but piping to boiler and the boiler itself is not included.

Condenserless models are available for 15% less where remote refrigerant source is available.

System Components	QUANTITY	UNIT	COST EACH		
			MAT.	INST.	TOTAL
SYSTEM 8.4-240-1320					
SELF-CONTAINED, AIR COOLED UNIT					
APARTMENT CORRIDORS, 500 S.F., .92 TON					
Air cooled, package unit	1.000	Ea.	1,640.50	248.20	1,888.70
Ductwork package for water or air cooled packaged units	1.000	System	89.70	542.80	632.50
Refrigerant piping	1.000	System	90.62	349.60	440.22
Air cooled condenser, direct drive, propeller fan	1.000	Ea.	272.80	97.65	370.45
TOTAL			2,093.62	1,238.25	3,331.87
COST PER S.F.			4.19	2.48	6.67

8.4-240	Self-contained, Air Cooled Unit Systems	COST PER S.F.		
		MAT.	INST.	TOTAL
1300	Self-contained, air cooled unit			
1320	Apartment corridors, 500 S.F., .92 ton	4.19	2.48	6.67
1360	1,000 S.F., 1.83 ton	4.13	2.46	6.59
1400	3,000 S.F., 5.50 ton	3.40	2.32	5.72
1440	5,000 S.F., 9.17 ton	2.89	2.20	5.09
1480	10,000 S.F., 18.33 ton	2.41	2.03	4.44
1560	Banks or libraries, 500 S.F., 2.08 ton	8.95	3.12	12.07
1600	1,000 S.F., 4.17 ton	7.70	5.25	12.95
1640	3,000 S.F., 12.50 ton	6.60	5	11.60
1680	5,000 S.F., 20.80 ton	5.50	4.59	10.09
1720	10,000 S.F., 41.66 ton	5.05	4.50	9.55
1800	Bars and taverns, 500 S.F., 5.54 ton	16.40	6.85	23.25
1840	1,000 S.F., 11.08 ton	16.95	10.05	27
1880	3,000 S.F., 33.25 ton	13.05	8.70	21.75
1920	5,000 S.F., 55.42 ton	12.50	8.75	21.25
1960	10,000 S.F., 110.00 ton	12.65	8.70	21.35
2040	Bowling alleys, 500 S.F., 2.83 ton	12.25	4.26	16.51
2080	1,000 S.F., 5.66 ton	10.50	7.15	17.65
2120	3,000 S.F., 17.00 ton	7.45	6.25	13.70
2160	5,000 S.F., 28.33 ton	6.95	6.10	13.05
2200	10,000 S.F., 56.66 ton	6.65	6.15	12.80
2240	Department stores, 500 S.F., 1.46 ton	6.30	2.20	8.50
2280	1,000 S.F., 2.92 ton	5.40	3.68	9.08
2320	3,000 S.F., 8.75 ton	4.59	3.50	8.09

AIR CONDITIONING | A8.4-240 Self-contained, Air Cooled

8.4-240 Self-contained, Air Cooled Unit Systems

		COST PER S.F.		
		MAT.	INST.	TOTAL
2360	5,000 S.F., 14.58 ton	4.59	3.50	8.09
2400	10,000 S.F., 29.17 ton	3.57	3.14	6.71
2480	Drug stores, 500 S.F., 3.33 ton	14.40	5	19.40
2520	1,000 S.F., 6.66 ton	12.30	8.40	20.70
2560	3,000 S.F., 20.00 ton	8.80	7.35	16.15
2600	5,000 S.F., 33.33 ton	8.15	7.20	15.35
2640	10,000 S.F., 66.66 ton	7.90	7.25	15.15
2720	Factories, 500 S.F., 1.66 ton	7.30	2.54	9.84
2760	1,000 S.F., 3.33 ton	6.20	4.22	10.42
2800	3,000 S.F., 10.00 ton	5.25	4.01	9.26
2840	5,000 S.F., 16.66 ton	4.35	3.68	8.03
2880	10,000 S.F., 33.33 ton	4.07	3.60	7.67
2960	Food supermarkets, 500 S.F., 1.42 ton	6.15	2.14	8.29
3000	1,000 S.F., 2.83 ton	6.35	3.80	10.15
3040	3,000 S.F., 8.50 ton	5.25	3.58	8.83
3080	5,000 S.F., 14.17 ton	4.47	3.40	7.87
3120	10,000 S.F., 28.33 ton	3.47	3.06	6.53
3200	Medical centers, 500 S.F., 1.17 ton	5.05	1.77	6.82
3240	1,000 S.F., 2.33 ton	5.30	3.15	8.45
3280	3,000 S.F., 7.00 ton	4.33	2.95	7.28
3320	5,000 S.F., 16.66 ton	3.70	2.81	6.51
3360	10,000 S.F., 23.33 ton	3.08	2.58	5.66
3440	Offices, 500 S.F., 1.58 ton	6.85	2.38	9.23
3480	1,000 S.F., 3.16 ton	7.20	4.25	11.45
3520	3,000 S.F., 9.50 ton	5	3.80	8.80
3560	5,000 S.F., 15.83 ton	4.18	3.50	7.68
3600	10,000 S.F., 31.66 ton	3.88	3.41	7.29
3680	Restaurants, 500 S.F., 2.50 ton	10.75	3.75	14.50
3720	1,000 S.F., 5.00 ton	9.25	6.30	15.55
3760	3,000 S.F., 15.00 ton	7.90	6	13.90
3800	5,000 S.F., 25.00 ton	6.10	5.40	11.50
3840	10,000 S.F., 50.00 ton	6.10	5.35	11.45
3920	Schools and colleges, 500 S.F., 1.92 ton	8.25	2.88	11.13
3960	1,000 S.F., 3.83 ton	7.15	4.84	11.99
4000	3,000 S.F., 11.50 ton	6.05	4.61	10.66
4040	5,000 S.F., 19.17 ton	5.05	4.24	9.29
4080	10,000 S.F., 38.33 ton	4.64	4.14	8.78
4090				
9000	Components of ductwork packages for above systems, per ton of cooling:			
9010	Ductwork; galvanized steel, 102 pounds			
9020	Diffusers; aluminum, 24" x 12", two			

AIR CONDITIONING — A8.4-250 Split System, Air Cooled

General: Split systems offer several important advantages which should be evaluated when a selection is to be made. They provide a greater degree of flexibility in component selection which permits an accurate match-up of the proper equipment size and type with the particular needs of the building. This allows for maximum use of modern energy saving concepts in heating and cooling. Outdoor installation of the air cooled condensing unit allows space savings in the building and also isolates the equipment operating sounds from building occupants.

Design Assumptions: The systems below are comprised of a direct expansion air handling unit and air cooled condensing unit with interconnecting copper tubing. Ducts and diffusers are also included for distribution of air. Systems are priced for cooling only. Heat can be added as desired either by putting hot water/steam coils into the air unit or into the duct supplying the particular area of need. Gas fired duct furnaces are also available. Refrigerant liquid line is insulated.

System Components	QUANTITY	UNIT	COST EACH MAT.	COST EACH INST.	COST EACH TOTAL
SYSTEM 8.4-250-1280					
SPLIT SYSTEM, AIR COOLED CONDENSING UNIT					
APARTMENT CORRIDORS, 1,000 S.F., 1.80 TON					
Fan coil AC unit, cabinet mntd & filters direct expansion air cool	1.000	Ea.	402.60	92.60	495.20
Ductwork package, for split system, remote condensing unit	1.000	System	83.72	516.98	600.70
Refrigeration piping	1.000	System	105.23	576.45	681.68
Condensing unit, air cooled, incls compressor & standard controls	1.000	Ea.	1,030.90	372.10	1,403
TOTAL			1,622.45	1,558.13	3,180.58
COST PER S.F.			1.62	1.56	3.18

*Cooling requirements would lead to choosing a water cooled unit

8.4-250	Split Systems With Air Cooled Condensing Units	COST PER S.F. MAT.	COST PER S.F. INST.	COST PER S.F. TOTAL
1260	Split system, air cooled condensing unit			
1280	Apartment corridors, 1,000 S.F., 1.83 ton	1.62	1.56	3.18
1320	2,000 S.F., 3.66 ton	1.32	1.57	2.89
1360	5,000 S.F., 9.17 ton	1.62	1.98	3.60
1400	10,000 S.F., 18.33 ton	1.86	2.16	4.02
1440	20,000 S.F., 36.66 ton	1.68	2.19	3.87
1520	Banks and libraries, 1,000 S.F., 4.17 ton	3.02	3.58	6.60
1560	2,000 S.F., 8.33 ton	3.67	4.49	8.16
1600	5,000 S.F., 20.80 ton	4.21	4.89	9.10
1640	10,000 S.F., 41.66 ton	3.81	4.98	8.79
1680	20,000 S.F., 83.32 ton	3.71	5.15	8.86
1760	Bars and taverns, 1,000 S.F., 11.08 ton	8.95	7.05	16
1800	2,000 S.F., 22.16 ton	12.65	8.45	21.10
1840	5,000 S.F., 55.42 ton	9.10	7.90	17
1880	10,000 S.F., 110.84 ton	9.25	8.25	17.50
1920	20,000 S.F., 220 ton*			
2000	Bowling alleys, 1,000 S.F., 5.66 ton	4.22	6.70	10.92
2040	2,000 S.F., 11.33 ton	5	6.10	11.10
2080	5,000 S.F., 28.33 ton	5.75	6.65	12.40
2120	10,000 S.F., 56.66 ton	5.15	6.80	11.95
2160	20,000 S.F., 113.32 ton	5.45	7.30	12.75
2320	Department stores, 1,000 S.F., 2.92 ton	2.13	2.46	4.59
2360	2,000 S.F., 5.83 ton	2.17	3.46	5.63
2400	5,000 S.F., 14.58 ton	2.57	3.14	5.71

R15700-020

AIR CONDITIONING

A8.4-250 Split System, Air Cooled

8.4-250 Split Systems With Air Cooled Condensing Units

		COST PER S.F.		
		MAT.	INST.	TOTAL
2440	10,000 S.F., 29.17 ton	2.95	3.43	6.38
2480	20,000 S.F., 58.33 ton	2.67	3.49	6.16
2560	Drug stores, 1,000 S.F., 6.66 ton	4.96	7.90	12.86
2600	2,000 S.F., 13.32 ton	5.85	7.20	13.05
2640	5,000 S.F., 33.33 ton	6.10	8	14.10
2680	10,000 S.F., 66.66 ton	5.85	8.30	14.15
2720	20,000 S.F., 133.32 ton*			
2800	Factories, 1,000 S.F., 3.33 ton	2.43	2.81	5.24
2840	2,000 S.F., 6.66 ton	2.47	3.95	6.42
2880	5,000 S.F., 16.66 ton	3.37	3.92	7.29
2920	10,000 S.F., 33.33 ton	3.05	3.98	7.03
2960	20,000 S.F., 66.66 ton	2.91	4.15	7.06
3040	Food supermarkets, 1,000 S.F., 2.83 ton	2.07	2.38	4.45
3080	2,000 S.F., 5.66 ton	2.10	3.35	5.45
3120	5,000 S.F., 14.66 ton	2.49	3.05	5.54
3160	10,000 S.F., 28.33 ton	2.86	3.34	6.20
3200	20,000 S.F., 56.66 ton	2.59	3.39	5.98
3280	Medical centers, 1,000 S.F., 2.33 ton	1.73	1.93	3.66
3320	2,000 S.F., 4.66 ton	1.72	2.76	4.48
3360	5,000 S.F., 11.66 ton	2.05	2.52	4.57
3400	10,000 S.F., 23.33 ton	2.36	2.75	5.11
3440	20,000 S.F., 46.66 ton	2.14	2.79	4.93
3520	Offices, 1,000 S.F., 3.17 ton	2.31	2.67	4.98
3560	2,000 S.F., 6.33 ton	2.36	3.76	6.12
3600	5,000 S.F., 15.83 ton	2.79	3.41	6.20
3640	10,000 S.F., 31.66 ton	3.20	3.73	6.93
3680	20,000 S.F., 63.32 ton	2.77	3.96	6.73
3760	Restaurants, 1,000 S.F., 5.00 ton	3.73	5.95	9.68
3800	2,000 S.F., 10.00 ton	4.42	5.40	9.82
3840	5,000 S.F., 25.00 ton	5.05	5.90	10.95
3880	10,000 S.F., 50.00 ton	4.57	6	10.57
3920	20,000 S.F., 100.00 ton	4.82	6.50	11.32
4000	Schools and colleges, 1,000 S.F., 3.83 ton	2.77	3.28	6.05
4040	2,000 S.F., 7.66 ton	3.38	4.13	7.51
4080	5,000 S.F., 19.17 ton	3.87	4.52	8.39
4120	10,000 S.F., 38.33 ton	3.50	4.59	8.09
4160	20,000 S.F., 76.66 ton	3.37	4.77	8.14
5000				
5100				
5200				
9000	Components of ductwork packages for above systems, per ton of cooling:			
9010	Ductwork; galvanized steel, 102 pounds			
9020	Diffusers; aluminum, 24" x 12", two			

AIR CONDITIONING | A8.4-810 | Computer Cooling

Computer rooms impose special requirements on air conditioning systems. A prime requirement is reliability, due to the potential monetary loss that could be incurred by a system failure. A second basic requirement is the tolerance of control with which temperature and humidity are regulated, and dust eliminated. As the air conditioning system reliability is so vital, the additional cost of reserve capacity and redundant components is often justified.

System Descriptions: Computer areas may be environmentally controlled by one of three methods as follows:

1. Self-contained Units
These are units built to higher standards of performance and reliability. They usually contain alarms and controls to indicate component operation failure, filter change, etc. It should be remembered that these units in the room will occupy space that is relatively expensive to build and that all alterations and service of the equipment will also have to be accomplished within the computer area.

2. Decentralized Air Handling Units In operation these are similar to the self-contained units except that their cooling capability comes from remotely located refrigeration equipment as refrigerant or chilled water. As no compressors or refrigerating equipment are required in the air units, they are smaller and require less service than self-contained units. An added plus for this type of system occurs if some of the computer components themselves also require chilled water for cooling.

3. Central System Supply Cooling is obtained from a central source which, since it is not located within the computer room, may have excess capacity and permit greater flexibility without interfering with the computer components. System performance criteria must still be met.

Note: The costs shown below do not include an allowance for ductwork or piping.

8.4-810	Computer Room Cooling Units	MAT.	INST.	TOTAL
0560	Computer room unit, air cooled, includes remote condenser			
0580	3 ton	7,725	1,525	9,250
0600	5 ton	10,300	1,675	11,975
0620	8 ton	21,600	2,800	24,400
0640	10 ton	22,400	3,025	25,425
0660	15 ton	25,000	3,450	28,450
0680	20 ton	29,900	4,075	33,975
0700	23 ton	31,400	4,200	35,600
0800	Chilled water, for connection to existing chiller system			
0820	5 ton	8,725	1,025	9,750
0840	8 ton	12,500	1,525	14,025
0860	10 ton	13,800	1,550	15,350
0880	15 ton	14,800	1,575	16,375
0900	20 ton	15,200	1,650	16,850
0920	23 ton	16,000	1,800	17,800
1000	Glycol system, complete except for interconnecting tubing			
1020	3 ton	9,475	1,900	11,375
1040	5 ton	12,300	2,000	14,300
1060	8 ton	26,700	3,300	30,000
1080	10 ton	28,900	3,600	32,500
1100	15 ton	35,000	4,525	39,525
1120	20 ton	37,500	4,900	42,400
1140	23 ton	40,900	5,350	46,250
1240	Water cooled, not including condenser water supply or cooling tower			
1260	3 ton	6,975	1,225	8,200
1280	5 ton	9,850	1,400	11,250
1300	8 ton	20,300	2,300	22,600
1320	15 ton	24,800	2,800	27,600
1340	20 ton	29,300	3,100	32,400
1360	23 ton	31,000	3,475	34,475

SPECIAL | A8.5-110 | Garage Exhaust

Vitrified Clay Garage Exhaust System | Dual Exhaust System

System Components	QUANTITY	UNIT	COST EACH MAT.	COST EACH INST.	COST EACH TOTAL
SYSTEM 8.5-110-1040					
GARAGE, EXHAUST, SINGLE OUTLET, 3" EXHAUST, CARS & LIGHT TRUCKS					
A Outlet top assy. for engine exhaust system with adapters and ftngs, 3" diam	1.000	Ea.	178	26	204
F Bullnose (guide) for engine exhaust system, 3" diam	1.000	Ea.	14.85		14.85
G Galvanized flexible tubing for engine exhaust system, 3" diam	8.000	L.F.	42.40		42.40
H Adapter for metal tubing end of engine exhaust system, 3" tail pipe	1.000	Ea.	29.50		29.50
J Pipe, sewer, vitrified clay, premium joint, 8" diam.	18.000	L.F.	73.26	118.44	191.70
Excavate utility trench w/chain trencher, 12 H.P., oper. walking	18.000	L.F.		23.40	23.40
Backfill utility trench by hand, incl. compaction, 8" wide 24" deep	18.000	L.F.		32.22	32.22
Stand for blower, concrete over polystyrene core, 6" high	1.000	Ea.	2.38	12.95	15.33
AC&V duct spiral reducer 10"x8"	1.000	Ea.	7.80	23.50	31.30
AC&V duct spiral reducer 12"x10"	1.000	Ea.	11.10	31	42.10
AC&V duct spiral preformed 45° elbow 8" diam	1.000	Ea.	3.64	26.50	30.14
AC&V utility fan, belt drive, 3 phase, 2000 CFM, 1 HP	2.000	Ea.	1,130	414	1,544
Safety switch, heavy duty fused, 240V, 3 pole, 30 amp	1.000	Ea.	120	128	248
Total			1,612.93	836.01	2,448.94

8.5-110	Garage Exhaust Systems	COST PER BAY MAT.	COST PER BAY INST.	COST PER BAY TOTAL
1040	Garage, single exhaust, 3" outlet, cars & light trucks, one bay	1,625	835	2,460
1060	Additional bays up to seven bays	305	122	427
1500	4" outlet, trucks, one bay	1,625	835	2,460
1520	Additional bays up to six bays	315	122	437
1600	5" outlet, diesel trucks, one bay	1,675	835	2,510
1650	Additional single bays up to six	375	142	517
1700	Two adjoining bays	1,675	835	2,510
2000	Dual exhaust, 3" outlets, pair of adjoining bays	1,900	930	2,830
2100	Additional pairs of adjoining bays	570	142	712

SPECIAL | A8.5-120 | Fume Hood Exhaust

System Components	QUANTITY	UNIT	COST EACH		
			MAT.	INST.	TOTAL
SYSTEM 8.5-120-1010					
FUME HOOD EXHAUST SYSTEM, 3' LONG, 1000 CFM					
Fume hood	3.000	L.F.	3,900	915	4,815
Ductwork, plastic, FM approved for acid fume & smoke, straight, 10" dia	60.000	L.F.	1,350	528	1,878
Ductwork, plastic, FM approved for acid fume & smoke, elbow 90°, 10" dia	3.000	Ea.	348	286.50	634.50
Fan, corrosive fume resistant plastic, 1630 CFM, 1/2 HP	1.000	Ea.	3,300	238	3,538
Fan control switch & mounting base	1.000	Ea.	69	26.50	95.50
Roof fan balancing	1.000	Ea.		166	166
TOTAL			8,967	2,160	11,127

8.5-120	Fume Hood Exhaust Systems	COST EACH		
		MAT.	INST.	TOTAL
1010	Fume hood exhaust system, 3' long, 1000 CFM	8,975	2,150	11,125
1020	4' long, 2000 CFM	10,700	2,725	13,425
1030	6' long, 3500 CFM	14,900	4,075	18,975
1040	6' long, 5000 CFM	20,400	4,800	25,200
1050	10' long, 8000 CFM	27,400	6,575	33,975

SPECIAL | A8.5-200 | Dust Collecting Vacuum Systems

Commercial/Industrial Dust Collection System

System Components	QUANTITY	UNIT	COST EACH		
			MAT.	INST.	TOTAL
SYSTEM 8.5-210-1010					
COMMERCIAL/INDUSTRIAL VACUUM DUST COLLECTION SYSTEM, 500 CFM					
Dust collection central vac unit inc. stand, filter & shaker, 500 CFM 2 HP	1.000	Ea.	3,400	395	3,795
Galvanized tubing, 16 ga., 4" OD	80.000	L.F.	331.20	165.60	496.80
Galvanized tubing, 90° ell, 4" dia.	3.000	Ea.	70.50	44.70	115.20
Galvanized tubing, 45° ell, 4" dia.	4.000	Ea.	82	59.60	141.60
Galvanized tubing, TY slip fit, 4" dia.	4.000	Ea.	200	100	300
Flexible rubber hose, 4" dia.	32.000	L.F.	249.60	66.24	315.84
Galvanized air gate valve, 4" dia.	4.000	Ea.	600	130	730
Galvanized compression coupling, 4" dia.	36.000	Ea.	707.40	810	1,517.40
Pipe hangers, steel, 4" pipe size	16.000	Ea.	38.24	108.80	147.04
TOTAL			5,678.94	1,879.94	7,558.88

8.5-210	Commercial/Industrial Vacuum Dust Collection	COST EACH		
		MAT.	INST.	TOTAL
1010	Commercial / industrial, vacuum dust collection system, 500 CFM	5,675	1,875	7,550
1020	1000 CFM	7,275	2,125	9,400
1030	1500 CFM	8,975	2,375	11,350
1040	3000 CFM	18,300	3,325	21,625
1050	5000 CFM	27,400	4,975	32,375

SPECIAL A8.5-400 | Pumps

Base Mounted End-Suction Pump

System Components	QUANTITY	UNIT	MAT.	INST.	TOTAL
SYSTEM 8.5-414-1010					
PUMP, BASE MTD. WITH MOTOR, END-SUCTION, 2-1/2" SIZE, 3 HP, TO 150 GPM					
Pump, circulating, CI, base mounted, 2-1/2" size, 3 HP, to 150 GPM	1.000	Ea.	1,975	415	2,390
Pipe, black steel, Sch. 40, on yoke & roll hangers, 10' O.C., 2-1/2" dia	12.000	L.F.	66.60	219.72	286.32
Elbow, 90°, weld joint, steel, 2-1/2" pipe size	1.000	Ea.	9.50	107.35	116.85
Flange, weld neck, 150 LB, 2-1/2" pipe size	8.000	Ea.	164	431.20	595.20
Valve, iron body, gate, 125 lb., N.R.S., flanged, 2-1/2" size	1.000	Ea.	310	150	460
Strainer, Y type, iron body, flanged, 125 lb., 2-1/2" pipe size	1.000	Ea.	80	152	232
Multipurpose valve, CI body, 2" size	1.000	Ea.	370	52.50	422.50
Expansion joint, flanged spool, 6" F to F, 2-1/2" dia	2.000	Ea.	374	122	496
T-O-L, weld joint, socket, 1/4" pipe size, nozzle	2.000	Ea.	9.48	74.62	84.10
T-O-L, weld joint, socket, 1/2" pipe size, nozzle	2.000	Ea.	9.48	78.10	87.58
Control gauges, pressure or vacuum, 3-1/2" diameter dial	2.000	Ea.	45	26.30	71.30
Pressure/temperature relief plug, 316 SS, 3/4" OD, 7-1/2" insertion	2.000	Ea.	92	26.30	118.30
Insulation, fiberglass pipe covering, 1-1/2" wall, 2-1/2" IPS	12.000	L.F.	42.36	49.44	91.80
Pump control system	1.000	Ea.	925	390	1,315
Pump balancing	1.000	Ea.		185	185
TOTAL			4,472.42	2,479.53	6,951.95

8.5-414	Circulating Pump Systems, End Suction	MAT.	INST.	TOTAL
1010	Pump, base mtd with motor, end-suction, 2-1/2" size, 3 HP, to 150 GPM	4,475	2,475	6,950
1020	3" size, 5 HP, to 225 GPM	5,000	2,825	7,825
1030	4" size, 7-1/2 HP, to 350 GPM	6,075	3,425	9,500
1040	5" size, 15 HP, to 1000 GPM	8,550	5,175	13,725
1050	6" size, 25 HP, to 1550 GPM	9,900	6,250	16,150

8.5-420	Circulating Pump Systems, Double Suction	MAT.	INST.	TOTAL
1010	Pump, base mtd w/ motor, double suction, 6" size, 50 HP, to 1200 GPM	15,300	8,950	24,250
1020	8" size, 75 HP, to 2500 GPM	21,900	12,100	34,000
1030	8" size, 100 HP, to 3000 GPM	32,400	13,500	45,900
1040	10" size, 150 HP, to 4000 GPM	45,900	15,400	61,300

SPECIAL | A8.5-800 | Kitchen Ventilation, Comm'l

Commercial Kitchen Exhaust/Make-up Air Rooftop System

System Components	QUANTITY	UNIT	COST EACH		
			MAT.	INST.	TOTAL
SYSTEM 8.5-810-1010					
COMMERCIAL KITCHEN EXHAUST/MAKE-UP AIR SYSTEM, ROOFTOP, GAS, 2000 CFM					
Make-up air rooftop unit, gas, 2090 CFM, 159 MBH	1.000	Ea.	4,293	426.60	4,719.60
Pipe, black steel, Sch 40, thrded, nipples, W/cplngs/hngrs, 10' OC, 1" dia.	20.000	L.F.	42.02	172.70	214.72
Elbow, 90°, black steel, straight, 3/4" dia.	2.000	Ea.	3.24	59	62.24
Elbow, 90°, black steel, straight, 1" dia.	3.000	Ea.	8.46	96	104.46
Tee, black steel, reducing, 1" dia.	1.000	Ea.	5	52	57
Union, black with brass seat, 1" dia.	1.000	Ea.	8.85	34.50	43.35
Pipe nipples, black, 3/4" dia	2.000	Ea.	3	13.60	16.60
Pipe cap, malleable iron, black, 1" dia.	1.000	Ea.	2.21	13.85	16.06
Gas cock, brass, 1" size	1.000	Ea.	13.85	22	35.85
Ductwork, 20" x 14" fabricated, galv. steel, 30 LF	220.000	Lb.	138.60	1,001	1,139.60
Ductwork, 18" x 10" fabricated, galv. steel, 12 LF	72.000	Lb.	45.36	327.60	372.96
Ductwork, 30" x 10" fabricated, galv. steel, 30 LF	260.000	Lb.	163.80	1,183	1,346.80
Ductwork, 16" x 10" fabricated, galv. steel, 12 LF	70.000	Lb.	44.10	318.50	362.60
Kitchen ventilation, island style, water wash, Stainless Steel	5.000	L.F.	8,250	815	9,065
Control system, Make-up Air Unit / Exhauster	1.000	Ea.	2,000	1,475	3,475
Rooftop unit heat/cool balancing	1.000	Ea.		290	290
Roof fan balancing	1.000	Ea.		166	166
TOTAL			15,021.49	6,466.35	21,487.84

8.5-810	Kitchen Exhaust/Make-Up Air	COST EACH		
		MAT.	INST.	TOTAL
1010	Commercial kitchen exhaust/make-up air system, rooftop, gas, 2000 CFM	15,000	6,475	21,475
1020	3000 CFM	15,900	8,850	24,750
1030	5000 CFM	19,800	10,700	30,500
1040	8000 CFM	31,800	15,900	47,700
1050	12,000 CFM	45,900	19,300	65,200
1060	16,000 CFM	56,500	30,700	87,200

HVAC   A8.7-100 Unit Heaters

Cabinet Unit Heater   Gas Unit Heater   Hydronic Unit Heater

System Components	QUANTITY	UNIT	COST EACH MAT.	COST EACH INST.	COST EACH TOTAL
SYSTEM 8.7-110-1010					
SPACE HEATER, SUSPENDED, GAS FIRED, PROPELLER FAN, 20 MBH					
Space heater, propeller fan, 20 MBH output	1.000	Ea.	450	89	539
Pipe, black steel, Sch 40, threaded, W/coupling & hangers, 10' OC, 3/4" dia	20.000	L.F.	31.50	142.80	174.30
Elbow, 90°, black steel, straight, 1/2" dia.	3.000	Ea.	3.99	82.50	86.49
Elbow, 90°, black steel, straight, 3/4" dia.	3.000	Ea.	4.86	88.50	93.36
Tee, black steel, reducing, 3/4" dia.	1.000	Ea.	3.97	46	49.97
Union, black with brass seat, 3/4" dia.	1.000	Ea.	6.80	32	38.80
Pipe nipples, black, 1/2" dia	2.000	Ea.	2.70	13.20	15.90
Pipe nipples, black, 3/4" dia	2.000	Ea.	1.50	6.80	8.30
Pipe cap, black, 3/4" dia.	1.000	Ea.	1.84	13	14.84
Gas cock, brass, 3/4" size	1.000	Ea.	11.35	18.90	30.25
Thermostat, 1 set back, electric, timed	1.000	Ea.	87	52	139
Wiring, thermostat hook-up	1.000	Ea.	6.25	17.05	23.30
Vent chimney, prefab metal, U.L. listed, gas, double wall, galv. st, 4" dia	12.000	L.F.	57.48	131.40	188.88
Vent chimney, gas, double wall, galv. steel, elbow 90°, 4" dia.	2.000	Ea.	38.50	44	82.50
Vent chimney, gas, double wall, galv. steel, Tee, 4" dia.	1.000	Ea.	24.50	28.50	53
Vent chimney, gas, double wall, galv. steel, T cap, 4" dia.	1.000	Ea.	1.86	17.75	19.61
Vent chimney, gas, double wall, galv. steel, roof flashing, 4" dia.	1.000	Ea.	6.45	22	28.45
Vent chimney, gas, double wall, galv. steel, top, 4" dia.	1.000	Ea.	9.25	16.95	26.20
TOTAL			749.80	862.35	1,612.15

8.7-110	Unit Heaters, Gas	COST EACH MAT.	INST.	TOTAL
1010	Space heater, suspended, gas fired, propeller fan, 20 MBH	750	860	1,610
1020	60 MBH	880	895	1,775
1030	100 MBH	1,100	950	2,050
1040	160 MBH	1,300	1,050	2,350
1050	200 MBH	1,575	1,150	2,725
1060	280 MBH	1,975	1,225	3,200
1070	320 MBH	2,725	1,400	4,125

8.7-120	Unit Heaters, Hydronic	COST EACH MAT.	INST.	TOTAL
1010	Space heater, suspended, horiz. mount, HW, prop. fan, 20 MBH	1,300	925	2,225
1020	60 MBH	1,600	1,050	2,650
1030	100 MBH	1,900	1,150	3,050
1040	150 MBH	2,400	1,250	3,650

HVAC | A8.7-100 Unit Heaters

8.7-120 Unit Heaters, Hydronic

		MAT.	INST.	TOTAL
1050	200 MBH	2,600	1,400	4,000
1060	300 MBH	2,875	1,650	4,525

8.7-130 Cabinet Unit Heaters, Hydronic

		MAT.	INST.	TOTAL
1010	Unit heater, cabinet type, horizontal blower, hot water, 20 MBH	1,675	905	2,580
1020	60 MBH	2,375	1,000	3,375
1030	100 MBH	2,550	1,100	3,650
1040	120 MBH	2,625	1,150	3,775

HVAC — A8.7-210 Boilers, Electric

Electric Boiler, Hot Water

System Components

SYSTEM 8.7-210-1010
BOILER, ELECTRIC, HOT WATER, 15 kW, 52 MBH

System Components	QUANTITY	UNIT	MAT.	INST.	TOTAL
Boilers, electric, ASME, hot water, 15 KW, 52 MBH	1.000	Ea.	3,750	900	4,650
Pipe, black steel, Sch 40, threaded, W/coupling & hangers, 10' OC, 3/4" dia	10.000	L.F.	17.25	78.20	95.45
Pipe, black steel, Sch 40, threaded, W/coupling & hangers, 10' OC, 1" dia	20.000	L.F.	41.07	168.78	209.85
Elbow, 90°, black, straight, 3/4" dia.	6.000	Ea.	9.72	177	186.72
Elbow, 90°, black, straight, 1" dia.	6.000	Ea.	16.92	192	208.92
Tee, black, straight, 3/4" dia.	1.000	Ea.	2.57	46	48.57
Tee, black, straight, 1" dia.	2.000	Ea.	8.78	104	112.78
Tee, black, reducing, 1" dia.	2.000	Ea.	10	104	114
Union, black with brass seat, 3/4" dia.	1.000	Ea.	6.80	32	38.80
Union, black with brass seat, 1" dia.	2.000	Ea.	17.70	69	86.70
Pipe nipples, 3/4" diam.	3.000	Ea.	2.25	10.20	12.45
Pipe nipples, 1" diam.	3.000	Ea.	2.87	11.78	14.65
Valves, bronze, gate, N.R.S., threaded, class 150, 3/4" size	2.000	Ea.	56	42	98
Valves, bronze, gate, N.R.S., threaded, class 150, 1" size	2.000	Ea.	73	44	117
Thermometer, stem type, 9" case, 8" stem, 3/4" NPT	1.000	Ea.	260	30.10	290.10
Tank, steel, liquid expansion, ASME, painted, 15 gallon capacity	1.000	Ea.	370	44.50	414.50
Pump, circulating, bronze, flange connection, 3/4" to 1-1/2" size, 1/8 HP	1.000	Ea.	585	125	710
Insulation, fiberglass pipe covering, 1" wall, 1" IPS	20.000	L.F.	27.20	67.40	94.60
TOTAL			5,257.13	2,245.96	7,503.09

8.7-210 Electric Boiler, Hot Water

		MAT.	INST.	TOTAL
1010	Boiler, electric, hot water, 15 kW, 52 MBH	5,250	2,250	7,500
1020	30 kW, 103 MBH	5,675	2,375	8,050
1030	60 kW, 205 MBH	5,725	2,450	8,175
1040	120 kW, 410 MBH	8,625	2,950	11,575
1050	150 kW, 510 MBH	14,600	5,825	20,425
1060	210 kW, 716 MBH	17,400	6,400	23,800
1070	296 kW, 1010 MBH	18,300	6,575	24,875
1080	444 kW, 1515 MBH	25,500	8,875	34,375
1090	666 kW, 2273 MBH	32,800	9,575	42,375
1100	900 kW, 3071 MBH	42,200	11,100	53,300
1110	1320 kW, 4505 MBH	53,000	11,700	64,700
1120	2100 kW, 7167 MBH	69,500	14,000	83,500
1130	2610 kW, 8905 MBH	82,500	17,000	99,500
1140	3600 kW, 12,283 MBH	95,500	20,300	115,800

HVAC — A8.7-210 Boilers, Electric

8.7-215	Electric Boiler, Steam	COST EACH		
		MAT.	INST.	TOTAL
1010	Boiler, electric, steam, 18 kW, 61.4 MBH	9,750	2,150	11,900
1020	36 kW, 123 MBH	10,100	2,275	12,375
1030	60 kW, 205 MBH	13,700	4,650	18,350
1040	120 kW, 409 MBH	15,300	5,100	20,400
1050	150 kW, 512 MBH	18,800	7,900	26,700
1060	210 kW, 716 MBH	22,000	8,450	30,450
1070	300 kW, 1023 MBH	24,600	9,250	33,850
1080	510 kW, 1740 MBH	28,800	12,000	40,800
1090	720 kW, 2456 MBH	34,800	12,600	47,400
1100	1080 kW, 3685 MBH	42,300	15,000	57,300
1110	1260 kW, 4300 MBH	52,000	15,900	67,900
1120	1620 kW, 5527 MBH	59,000	17,900	76,900
1130	2070 kW, 7063 MBH	71,000	18,800	89,800
1140	2340 kW, 7984 MBH	78,000	21,300	99,300

HVAC A8.7-220 Boilers, Gas or Gas / Oil

Cast Iron Boiler, Hot Water, Gas Fired

System Components	QUANTITY	UNIT	COST EACH MAT.	COST EACH INST.	COST EACH TOTAL
SYSTEM 8.7-220-1010					
BOILER, CAST IRON, GAS, HOT WATER, 100 MBH					
Boilers, gas fired, std controls, CI, insulated, HW, gross output 100 MBH	1.000	Ea.	1,300	1,200	2,500
Pipe, black steel, Sch 40, threaded, W/coupling & hangers, 10' OC, 3/4" dia	20.000	L.F.	33.75	153	186.75
Pipe, black steel, Sch 40, threaded, W/coupling & hangers, 10' OC, 1" dia.	20.000	L.F.	41.07	168.78	209.85
Elbow, 90°, black, straight, 3/4" dia.	9.000	Ea.	14.58	265.50	280.08
Elbow, 90°, black, straight, 1" dia.	6.000	Ea.	16.92	192	208.92
Tee, black, straight, 3/4" dia.	2.000	Ea.	5.14	92	97.14
Tee, black, straight, 1" dia.	2.000	Ea.	8.78	104	112.78
Tee, black, reducing, 1" dia.	2.000	Ea.	10	104	114
Pipe cap, black, 3/4" dia.	1.000	Ea.	1.84	13	14.84
Union, black with brass seat, 3/4" dia.	2.000	Ea.	13.60	64	77.60
Union, black with brass seat, 1" dia.	2.000	Ea.	17.70	69	86.70
Pipe nipples, black, 3/4" dia	5.000	Ea.	3.75	17	20.75
Pipe nipples, black, 1" dia	3.000	Ea.	2.87	11.78	14.65
Valves, bronze, gate, N.R.S., threaded, class 150, 3/4" size	2.000	Ea.	56	42	98
Valves, bronze, gate, N.R.S., threaded, class 150, 1" size	2.000	Ea.	73	44	117
Gas cock, brass, 3/4" size	1.000	Ea.	11.35	18.90	30.25
Thermometer, stem type, 9" case, 8" stem, 3/4" NPT	2.000	Ea.	260	30.10	290.10
Tank, steel, liquid expansion, ASME, painted, 15 gallon capacity	1.000	Ea.	370	44.50	414.50
Pump, circulating, bronze, flange connection, 3/4" to 1-1/2" size, 1/8 HP	1.000	Ea.	585	125	710
Vent chimney, all fuel, pressure tight, double wall, SS, 6" dia	20.000	L.F.	780	249	1,029
Vent chimney, elbow, 90° fixed, 6" dia	2.000	Ea.	440	50	490
Vent chimney, Tee, 6" dia	2.000	Ea.	290	62	352
Vent chimney, ventilated roof thimble, 6" dia	1.000	Ea.	210	28.50	238.50
Vent chimney, adjustable roof flashing, 6" dia	1.000	Ea.	53.50	25	78.50
Vent chimney, stack cap, 6" diameter	1.000	Ea.	154	16.20	170.20
Insulation, fiberglass pipe covering, 1" wall, 1" IPS	20.000	L.F.	27.20	67.40	94.60
TOTAL			4,780.05	3,256.66	8,036.71

8.7-220	Boiler, Cast Iron, Hot Water, Gas		COST EACH MAT.	COST EACH INST.	COST EACH TOTAL
1010	Boiler, cast iron, gas, hot water, 100 MBH		4,775	3,250	8,025
1020	200 MBH		5,525	3,700	9,225
1030	320 MBH		5,875	4,025	9,900
1040	440 MBH		7,875	5,200	13,075
1050	544 MBH		11,800	8,350	20,150
1060	765 MBH		14,600	8,950	23,550

HVAC — A8.7-220 Boilers, Gas or Gas / Oil

8.7-220 Boiler, Cast Iron, Hot Water, Gas

		MAT.	INST.	TOTAL
1070	1088 MBH	16,400	9,550	25,950
1080	1530 MBH	20,100	12,800	32,900
1090	2312 MBH	24,300	14,900	39,200
1100	2856 MBH	30,200	16,000	46,200
1110	3808 MBH	33,900	18,800	52,700
1120	4720 MBH	63,500	21,000	84,500
1130	6100 MBH	70,500	23,300	93,800
1140	6970 MBH	77,000	28,700	105,700

8.7-225 Boiler, Cast Iron, Steam, Gas

		MAT.	INST.	TOTAL
1010	Boiler, cast iron, gas, steam, 100 MBH	4,000	3,125	7,125
1020	200 MBH	7,000	5,900	12,900
1030	320 MBH	8,200	6,650	14,850
1040	544 MBH	11,600	10,600	22,200
1050	765 MBH	13,400	11,000	24,400
1060	1275 MBH	16,400	12,100	28,500
1070	2675 MBH	24,900	17,700	42,600
1080	3570 MBH	30,000	19,900	49,900
1090	4720 MBH	34,000	21,100	55,100
1100	6100 MBH	67,000	24,500	91,500
1110	6970 MBH	73,500	28,700	102,200

8.7-230 Boiler, Cast Iron, Hot Water, Gas/Oil

		MAT.	INST.	TOTAL
1010	Boiler, cast iron, gas & oil, hot water, 584 MBH	14,100	9,250	23,350
1020	876 MBH	19,700	9,900	29,600
1030	1168 MBH	23,200	11,100	34,300
1040	1460 MBH	27,100	13,600	40,700
1050	2044 MBH	33,300	14,000	47,300
1060	2628 MBH	53,500	16,900	70,400
1070	3210 MBH	62,500	18,200	80,700
1080	3796 MBH	71,500	18,700	90,200
1090	4672 MBH	84,500	19,600	104,100
1100	5256 MBH	101,000	23,700	124,700
1110	6000 MBH	117,000	24,100	141,100
1120	9800 MBH	143,500	41,000	184,500
1130	12,200 MBH	149,000	47,700	196,700
1140	13,500 MBH	150,500	50,500	201,000

8.7-235 Boiler, Cast Iron, Steam, Gas/Oil

		MAT.	INST.	TOTAL
1010	Boiler, cast iron, gas & oil, steam, 810 MBH	15,500	11,900	27,400
1020	1360 MBH	18,200	12,700	30,900
1030	2040 MBH	23,300	16,500	39,800
1040	2700 MBH	25,700	18,300	44,000
1050	3270 MBH	30,300	20,400	50,700
1060	3770 MBH	54,000	20,800	74,800
1070	4650 MBH	63,000	23,600	86,600
1080	5520 MBH	70,000	24,800	94,800
1090	6100 MBH	71,000	26,100	97,100
1100	6970 MBH	78,000	31,700	109,700

HVAC — A8.8-100 Chillers

Chilled Water Piping

Reciprocating Chiller, Water Cooled

Condenser Water Piping

System Components	QUANTITY	UNIT	MAT.	INST.	TOTAL
SYSTEM 8.8-110-1010					
CHILLER, RECIPROCATING, WATER COOLED, STD. CONTROLS, 60 TON					
Water chiller, reciprocating, water cooled, 60 ton cooling	1.000	Ea.	29,400	6,325	35,725
Pipe, black steel, welded, Sch. 40, on yoke & roll hgrs, 10' O.C., 4" dia	40.000	L.F.	346	919.60	1,265.60
Elbow, 90°, weld joint, butt, 4" pipe size	8.000	Ea.	149.20	1,376	1,525.20
Flange, welding neck, 150 lb., 4" pipe size	16.000	Ea.	432	1,376.80	1,808.80
Valves, iron body, butterfly, lug type, lever actuator, 4" size	4.000	Ea.	448	600	1,048
Expansion joints, bellows type, flange spool, 6" F to F, 4" dia	4.000	Ea.	908	362	1,270
T-O-L socket, weld joint, 1/4" pipe size, nozzle	4.000	Ea.	18.96	149.24	168.20
T-O-L socket, weld joint, 3/4" pipe size, nozzle	4.000	Ea.	22.20	163	185.20
Thermometers, stem type, 9" case, 8" stem, 3/4" NPT	4.000	Ea.	520	60.20	580.20
Gauges, pressure or vacuum, 3-1/2" diameter dial	4.000	Ea.	90	52.60	142.60
Insulation, fiberglass pipe covering, 1-1/2" wall, 4" IPS	20.000	L.F.	83.20	106	189.20
Chiller balancing	1.000	Ea.		360	360
TOTAL			32,417.56	11,850.44	44,268

8.8-110	Reciprocating Chiller, Water Cooled	MAT.	INST.	TOTAL
1010	Chiller, reciprocating, water cooled, std. controls, 60 Ton	32,400	11,800	44,200
1020	100 Ton	63,000	17,800	80,800
1030	150 Ton	84,000	19,500	103,500
1040	200 Ton	101,500	20,400	121,900

8.8-115	Reciprocating Chiller, Air Cooled	MAT.	INST.	TOTAL
1010	Chiller, reciprocating, air cooled, std. controls, 20 Ton	18,400	6,950	25,350
1020	30 Ton	22,200	7,300	29,500
1030	40 Ton	28,700	7,800	36,500
1040	60 Ton	40,600	8,775	49,375
1050	70 Ton	46,400	9,025	55,425
1060	80 Ton	49,600	9,175	58,775
1070	90 Ton	53,500	9,325	62,825
1080	100 Ton	59,500	9,475	68,975
1090	110 Ton	65,500	10,200	75,700
1100	130 Ton	74,000	10,400	84,400
1110	150 Ton	85,500	10,500	96,000
1120	175 Ton	90,500	10,700	101,200
1130	210 Ton	112,500	11,600	124,100

HVAC — A8.8-120 | Chillers

Chilled Water Piping

Centrifugal Chiller, Water Cooled, Hermetic

Condenser Water Piping

System Components	QUANTITY	UNIT	COST EACH MAT.	COST EACH INST.	COST EACH TOTAL
SYSTEM 8.8-120-1010					
CHILLER, CENTRIF., WATER COOLED, PKGD. HERMETIC, STD. CONTROLS, 200 TON					
Water chiller, centrifugal, water cooled, not incl tower, 200 ton	1.000	Ea.	94,000	12,600	106,600
Pipe, black steel, welded, Sch. 40, on yoke & roll hgrs, 10' o.c., 6" dia	40.000	L.F.	740	1,422.80	2,162.80
Elbow, 90°, weld joint, 6" pipe size	8.000	Ea.	356	2,040	2,396
Flange, welding neck, 150 lb., 6" pipe size	16.000	Ea.	656	2,032.80	2,688.80
Valves, iron body, butterfly, lug type, lever actuator, 6" size	4.000	Ea.	744	932	1,676
Expansion joints, bellows type, flange spool, 6" F to F, 6" dia	4.000	Ea.	1,148	444	1,592
T-O-L socket, weld jiont, 1/4" pipe size, nozzle	4.000	Ea.	18.96	149.24	168.20
T-O-L socket, weld jiont, 3/4" pipe size, nozzle	4.000	Ea.	22.20	163	185.20
Thermometers, stem type, 9" case, 8" stem, 3/4" NPT	4.000	Ea.	520	60.20	580.20
Gauges, pressure or vacuum, 3-1/2" diameter dial	4.000	Ea.	90	52.60	142.60
Insulation, fiberglass pipe covering, 1-1/2" wall, 6" IPS	40.000	L.F.	190.80	270	460.80
Chiller balancing	1.000	Ea.		360	360
TOTAL			98,485.96	20,526.64	119,012.60

8.8-120	Centrifugal Chiller, Water Cooled	COST EACH MAT.	COST EACH INST.	COST EACH TOTAL
1010	Chiller, centrif., water cooled, pkgd. hermetic, std controls, 200 Ton	98,500	20,500	119,000
1020	400 Ton	165,000	28,400	193,400
1030	1,000 Ton	401,500	42,700	444,200
1040	1,500 Ton	585,500	52,000	637,500

HVAC A8.8-130 | Chillers

Chilled Water Piping (Condenser Water Piping Similar) | Gas Absorption Chiller, Water Cooled | Gas Piping

System Components	QUANTITY	UNIT	MAT.	INST.	TOTAL
SYSTEM 8.8-130-1010					
CHILLER, GAS FIRED ABSORPTION, WATER COOLED, STD. CONTROLS, 800 TON					
Absorption water chiller, gas fired, water cooled, 800 ton	1.000	Ea.	549,500	19,100	568,600
Pipe, black steel, weld, Sch. 40, on yoke & roll hangers, 10' O.C., 10" dia	20.000	L.F.	950	1,062.20	2,012.20
Pipe, black steel, welded, Sch. 40, (no hgrs incl), 14" dia.	44.000	L.F.	1,650	3,735.60	5,385.60
Pipe, blacksteel, welded, Sch. 40, on yoke & roll hangers, 10' O.C., 4" dia	20.000	L.F.	173	459.80	632.80
Elbow, 90°, weld joint, 10" pipe size	4.000	Ea.	608	1,708	2,316
Elbow, 90°, weld joint, 14" pipe size	6.000	Ea.	1,830	3,813	5,643
Elbow, 90°, weld joint, 4" pipe size	3.000	Ea.	55.95	516	571.95
Flange, welding neck, 150 lb., 10" pipe size	8.000	Ea.	928	1,699.60	2,627.60
Flange, welding neck, 150 lb., 14" pipe size	14.000	Ea.	3,402	3,969	7,371
Flange, welding neck, 150 lb., 4" pipe size	3.000	Ea.	81	258.15	339.15
Valves, iron body, butterfly, lug type, gear operated, 10" size	2.000	Ea.	1,050	582	1,632
Valves, iron body, butterfly, lug type, gear operated, 14" size	5.000	Ea.	4,475	2,525	7,000
Valves, semi-steel, lubricated plug valve, flanged, 4" pipe size	1.000	Ea.	365	250	615
Expansion joints, bellows type, flange spool, 6" F to F, 10" dia	2.000	Ea.	880	304	1,184
Expansion joints, bellows type, flange spool, 10" F to F, 14" dia	2.000	Ea.	1,430	398	1,828
T-O-L socket, weld joint, 1/4" pipe size, nozzle	4.000	Ea.	18.96	149.24	168.20
T-O-L socket, weld joint, 3/4" pipe size, nozzle	4.000	Ea.	22.20	163	185.20
Thermometers, stem type, 9" case, 8" stem, 3/4" NPT	4.000	Ea.	520	60.20	580.20
Gauges, pressure or vacuum, 3-1/2" diameter dial	4.000	Ea.	90	52.60	142.60
Insulation, fiberglass pipe covering, 2" wall, 10" IPS	20.000	L.F.	217	212	429
Vent chimney, all fuel, pressure tight, double wall, SS, 12" dia	20.000	L.F.	1,220	339	1,559
Chiller balancing	1.000	Ea.		360	360
TOTAL			569,466.11	41,716.39	611,182.50

8.8-130	Gas Absorption Chiller, Water Cooled	MAT.	INST.	TOTAL
1010	Chiller, gas fired absorption, water cooled, std. ctrls, 800 Ton	569,500	41,700	611,200
1020	1000 Ton	720,500	54,000	774,500

8.8-140	Steam Absorption Chiller, Water Cooled	MAT.	INST.	TOTAL
1010	Chiller, steam absorption, water cooled, std. ctrls, 750 Ton	372,000	49,700	421,700
1020	955 Ton	433,000	59,000	492,000
1030	1465 Ton	624,000	68,000	692,000
1040	1660 Ton	744,500	77,500	822,000

HVAC A8.8-150 Cooling Towers

Cold Water Fill

Drain To Waste Piping

Cooling Tower

Condenser Water Piping

System Components	QUANTITY	UNIT	COST EACH MAT.	INST.	TOTAL
SYSTEM 8.8-160-1010					
COOLING TOWER, GALVANIZED STEEL, PACKAGED UNIT, DRAW THRU, 60 TON					
Cooling towers, draw thru, single flow, belt drive, 60 tons	1.000	Ea.	5,160	786	5,946
Pipe, black steel, welded, Sch. 40, on yoke & roll hangers, 10' OC, 4" dia.	40.000	L.F.	346	919.60	1,265.60
Tubing, copper, Type L, couplings & hangers 10' OC, 1-1/2" dia.	20.000	L.F.	67.40	160	227.40
Elbow, 90°, copper, cu x cu, 1-1/2" dia.	3.000	Ea.	8.70	96	104.70
Elbow, 90°, weld joint, 4" pipe size	6.000	Ea.	111.90	1,032	1,143.90
Flange, welding neck, 150 lb., 4" pipe size	6.000	Ea.	162	516.30	678.30
Valves, bronze, gate, N.R.S., soldered, 125 psi, 1-1/2" size	1.000	Ea.	49.50	32	81.50
Valves, iron body, butterfly, lug type, lever actuator, 4" size	2.000	Ea.	224	300	524
Electric heat trace system, 400 degree, 115v, 5 watts per L.F.	160.000	L.F.	1,024	123.20	1,147.20
Insulation, fiberglass pipe covering, , 1-1/2" wall, 4" IPS	40.000	L.F.	166.40	212	378.40
Insulation, fiberglass pipe covering, 1" wall, 1-1/2" IPS	20.000	L.F.	33.20	70.60	103.80
Cooling tower condenser control system	1.000	Ea.	4,000	1,750	5,750
Cooling tower balancing	1.000	Ea.		274	274
TOTAL			11,353.10	6,271.70	17,624.80

8.8-160	Galvanized Draw Through Cooling Tower	COST EACH MAT.	INST.	TOTAL
1010	Cooling tower, galvanized steel, packaged unit, draw thru, 60 Ton	11,400	6,275	17,675
1020	110 Ton	14,900	7,375	22,275
1030	300 Ton	28,500	11,000	39,500
1040	600 Ton	50,500	16,400	66,900
1050	1000 Ton	63,000	24,300	87,300

8.8-165	Fiberglass Draw Through Cooling Tower	COST EACH MAT.	INST.	TOTAL
1010	Cooling tower, fiberglass, packaged unit, draw thru, 60 Ton	9,250	6,275	15,525
1020	125 Ton	13,000	7,400	20,400
1030	300 Ton	22,700	11,000	33,700
1040	600 Ton	38,700	16,400	55,100
1050	1000 Ton	61,000	24,300	85,300

8.8-170	Stainless Steel Draw Through Cooling Tower	COST EACH MAT.	INST.	TOTAL
1010	Cooling tower, stainless steel, packaged unit, draw thru, 60 Ton	14,500	6,275	20,775
1020	110 Ton	20,000	7,400	27,400
1030	300 Ton	45,000	11,000	56,000
1040	600 Ton	69,000	16,400	85,400
1050	1000 Ton	108,000	24,300	132,300

HVAC — A8.8-200 Heat Exchanger

Plate Heat Exchanger

Shell and Tube Heat Exchanger

System Components	QUANTITY	UNIT	COST EACH MAT.	COST EACH INST.	COST EACH TOTAL
SYSTEM 8.8-220-1010					
SHELL & TUBE HEAT EXCHANGER, 40 GPM					
Heat exchanger, 4 pass, 3/4" O.D. copper tubes, by steam at 10 psi, 40 GPM	1.000	Ea.	2,300	190	2,490
Pipe, black steel, Sch 40, threaded, W/couplings & hangers, 10' OC, 2" dia.	40.000	L.F.	160.59	590.85	751.44
Elbow, 90°, straight, 2" dia.	16.000	Ea.	168	664	832
Reducer, black steel, concentric, 2" dia.	2.000	Ea.	23.70	71	94.70
Valves, bronze, gate, rising stem, threaded, class 150, 2" size	6.000	Ea.	486	228	714
Valves, bronze, globe, class 150, rising stem, threaded, 2" size	2.000	Ea.	420	76	496
Strainers, Y type, bronze body, screwed, 150 lb., 2" pipe size	2.000	Ea.	118	65	183
Valve, electric motor actuated, brass, 2 way, screwed, 1-1/2" pipe size	1.000	Ea.	375	32.50	407.50
Union, black with brass seat, 2" dia.	8.000	Ea.	147.60	352	499.60
Tee, black, straight, 2" dia.	9.000	Ea.	112.05	612	724.05
Tee, black, reducing run and outlet, 2" dia.	4.000	Ea.	77.20	272	349.20
Pipe nipple, black, 2" dia.	21.000	Ea.	1.59	5.85	7.44
Thermometers, stem type, 9" case, 8" stem, 3/4" NPT	2.000	Ea.	260	30.10	290.10
Gauges, pressure or vacuum, 3-1/2" diameter dial	2.000	Ea.	45	26.30	71.30
Insulation, fiberglass pipe covering, 1" wall, 2" IPS	40.000	L.F.	72.80	148.40	221.20
Heat exchanger control system	1.000	Ea.	1,950	1,350	3,300
Coil balancing	1.000	Ea.		78	78
TOTAL			6,717.53	4,792	11,509.53

8.8-210	Heat Exchanger, Plate Type	COST EACH MAT.	COST EACH INST.	COST EACH TOTAL
1010	Plate heat exchanger, 400 GPM	34,900	10,000	44,900
1020	800 GPM	57,500	12,700	70,200
1030	1200 GPM	112,500	17,200	129,700
1040	1800 GPM	167,000	21,300	188,300

8.8-220	Heat Exchanger, Shell & Tube	COST EACH MAT.	COST EACH INST.	COST EACH TOTAL
1010	Shell & tube heat exchanger, 40 GPM	6,725	4,800	11,525
1020	96 GPM	11,700	8,900	20,600
1030	240 GPM	20,900	12,000	32,900
1040	600 GPM	39,400	16,700	56,100

HVAC | A8.8-310 | Air Handling Unit, Field Fabricated

Built-Up Air Handling Unit

System Components	QUANTITY	UNIT	COST EACH		
			MAT.	INST.	TOTAL
SYSTEM 8.8-310-1010					
AHU, FIELD FAB., BLT. UP, COOL/HEAT COILS, FLTRS., CONST. VOL., 40,000 CFM					
AHU, Built-up, cool/heat coils, filters, mix box, constant volume	1.000	Ea.	11,400	3,925	15,325
Pipe, black steel, welded, Sch. 40, on yoke & roll hgrs, 10' OC, 3" dia.	26.000	L.F.	156	519.22	675.22
Pipe, black steel, welded, Sch. 40, on yoke & roll hgrs, 10' OC, 4" dia	20.000	L.F.	173	459.80	632.80
Elbow, 90°, weld joint, 3" pipe size	5.000	Ea.	54.50	614	668.50
Elbow, 90°, weld joint, 4" pipe size	4.000	Ea.	74.60	688	762.60
Tee, welded, reducing on outlet, 3" by 2-1/2" pipe	3.000	Ea.	100.50	643.50	744
Tee, welded, reducing on outlet, 4" by 3" pipe	2.000	Ea.	85	574	659
Reducer, eccentric, weld joint, 3" pipe size	4.000	Ea.	73.20	429.40	502.60
Reducer, eccentric, weld joint, 4" pipe size	4.000	Ea.	78.80	573.80	652.60
Flange, weld neck, 150 LB, 2-1/2" pipe size	3.000	Ea.	61.50	161.70	223.20
Flange, weld neck, 150 lb., 3" pipe size	14.000	Ea.	315	859.60	1,174.60
Flange, weld neck, 150 lb., 4" pipe size	10.000	Ea.	270	860.50	1,130.50
Valves, iron body, butterfly, lug type, lever actuator, 3" size	2.000	Ea.	183	187	370
Valves, iron body, butterfly, lug type, lever actuator, 4" size	2.000	Ea.	224	300	524
Strainer, Y type, iron body, flanged, 125 lb., 3" pipe size	1.000	Ea.	93	168	261
Strainer, Y type, iron body, flanged, 125 lb., 4" pipe size	1.000	Ea.	170	253	423
Valve, electric motor actuated, iron body, 3 way, flanged, 2-1/2" pipe	1.000	Ea.	920	207	1,127
Valve, electric motor actuated, iron body, two way, flanged, 3" pipe size	1.000	Ea.	495	207	702
Expansion joints, bellows type, flange spool, 6" F to F, 3" dia	2.000	Ea.	424	133	557
Expansion joints, bellows type, flange spool, 6" F to F, 4" dia	2.000	Ea.	454	181	635
T-O-L socket, weld joint, 1/4" pipe size, nozzle	4.000	Ea.	18.96	149.24	168.20
T-O-L socket, weld joint, 1/2" pipe size, nozzle	12.000	Ea.	56.88	468.60	525.48
Thermometers, stem type, 9" case, 8" stem, 3/4" NPT	4.000	Ea.	520	60.20	580.20
Pressure/temperature relief plug, 316 SS, 3/4"OD, 7-1/2" insertion	4.000	Ea.	184	52.60	236.60
Insulation, fiberglass pipe covering, 1-1/2" wall, 3" IPS	26.000	L.F.	94.64	113.36	208
Insulation, fiberglass pipe covering, 1-1/2" wall, 4" IPS	20.000	L.F.	83.20	106	189.20
Fan coil control system	1.000	Ea.	900	390	1,290
Re-heat coil balancing	2.000	Ea.		156	156
Air conditioner balancing	1.000	Ea.		495	495
TOTAL			17,662.78	13,935.52	31,598.30

8.8-310	Field Fabricated Air Handling Unit	COST EACH		
		MAT.	INST.	TOTAL
1010	AHU, Field fab, blt up, cool/heat coils, fltrs, const vol, 40,000 CFM	17,700	13,900	31,600
1020	60,000 CFM	57,500	21,400	78,900
1030	75,000 CFM	69,500	22,500	92,000

MECHANICAL 8

HVAC | A8.8-310 | Air Handling Unit, Field Fabricated

8.8-312		Field Fabricated VAV Air Handling Unit	COST EACH		
			MAT.	INST.	TOTAL
1010	AHU, Field fab, blt up, cool/heat coils, fltrs, VAV, 75,000 CFM		71,500	25,700	97,200
1020	100,000 CFM		117,000	34,000	151,000
1030	150,000 CFM		190,000	54,500	244,500

HVAC A8.8-320 | Air Handling Unit, Central Station

Chilled Water Piping | Central Station Air Handling Unit | Hot Water Piping

System Components	QUANTITY	UNIT	COST EACH MAT.	COST EACH INST.	COST EACH TOTAL
SYSTEM 8.8-320-1010					
AHU, CENTRAL STATION, COOL/HEAT COILS, CONST. VOL., FLTRS., 2,000 CFM					
Central station air handling unit, chilled water, 1900 CFM	1.000	Ea.	3,277.50	793.50	4,071
Pipe, black steel, Sch 40, threaded, W/cplgs & hangers, 10' OC, 1" dia.	26.000	L.F.	62.08	255.13	317.21
Pipe, black steel, Sch 40, threaded, W/cplg & hangers, 10' OC, 1-1/4" dia.	20.000	L.F.	58.24	218.40	276.64
Elbow, 90°, black, straight, 1" dia.	4.000	Ea.	11.28	128	139.28
Elbow, 90°, black, straight, 1-1/4" dia.	4.000	Ea.	18.48	136	154.48
Tee, black, reducing, 1" dia.	8.000	Ea.	40	416	456
Tee, black, reducing, 1-1/4" dia.	8.000	Ea.	69.60	428	497.60
Reducer, concentric, black, 1" dia.	4.000	Ea.	19.08	110	129.08
Reducer, concentric, black, 1-1/4" dia.	4.000	Ea.	22.80	116	138.80
Union, black with brass seat, 1" dia.	4.000	Ea.	35.40	138	173.40
Union, black with brass seat, 1-1/4" dia.	2.000	Ea.	25.40	71	96.40
Pipe nipple, black, 1" dia	13.000	Ea.	12.42	51.03	63.45
Pipe nipple, black, 1-1/4" dia	12.000	Ea.	13.44	50.40	63.84
Valves, bronze, gate, N.R.S., threaded, class 150, 1" size	2.000	Ea.	73	44	117
Valves, bronze, gate, N.R.S., threaded, class 150, 1-1/4" size	2.000	Ea.	98	55	153
Strainer, Y type, bronze body, screwed, 150 lb., 1" pipe size	1.000	Ea.	20.50	25	45.50
Strainer, Y type, bronze body, screwed, 150 lb., 1-1/4" pipe size	1.000	Ea.	46	28	74
Valve, electric motor actuated, brass, 3 way, screwed, 3/4" pipe	1.000	Ea.	266	23	289
Valve, electric motor actuated, brass, 2 way, screwed, 1" pipe size	1.000	Ea.	269	22	291
Thermometers, stem type, 9" case, 8" stem, 3/4" NPT	4.000	Ea.	520	60.20	580.20
Pressure/temperature relief plug, 316 SS, 3/4"OD, 7-1/2" insertion	4.000	Ea.	184	52.60	236.60
Insulation, fiberglass pipe covering, 1" wall, 1" IPS	26.000	L.F.	35.36	87.62	122.98
Insulation, fiberglass pipe covering, 1" wall, 1-1/4" IPS	20.000	L.F.	30.80	70.60	101.40
Fan coil control system	1.000	Ea.	900	390	1,290
Re-heat coil balancing	2.000	Ea.		156	156
Multizone A/C balancing	1.000	Ea.		249	249
TOTAL			6,108.38	4,174.48	10,282.86

8.8-320	Central Station Air Handling Unit	COST EACH MAT.	COST EACH INST.	COST EACH TOTAL
1010	AHU, Central station, cool/heat coils, constant vol, fltrs, 2,000 CFM	6,100	4,175	10,275
1020	5,000 CFM	9,275	5,500	14,775
1030	10,000 CFM	12,400	7,900	20,300
1040	15,000 CFM	17,400	10,900	28,300
1050	20,000 CFM	20,700	12,800	33,500

HVAC — A8.8-320 Air Handling Unit, Central Station

8.8-322	Central Station Air Handling Unit, VAV	COST EACH		
		MAT.	INST.	TOTAL
1010	AHU, Central station, cool/heat coils, VAV, fltrs, 5,000 CFM	11,400	5,475	16,875
1020	10,000 CFM	15,500	7,750	23,250
1030	15,000 CFM	20,600	10,300	30,900
1040	20,000 CFM	25,600	12,600	38,200
1050	30,000 CFM	37,500	14,400	51,900

HVAC A8.8-330 Air Handling Unit, Rooftop

Rooftop Air Handling Unit

System Components	QUANTITY	UNIT	COST EACH MAT.	COST EACH INST.	COST EACH TOTAL
SYSTEM 8.8-330-1010					
AHU, ROOFTOP, COOL/HEAT COILS, FLTRS., CONST. VOL., 2,000 CFM					
AHU, Built-up, cool/heat coils, filter, mix box, rooftop, constant volume	1.000	Ea.	3,700	760	4,460
Pipe, black steel, Sch 40, threaded, W/cplgs & hangers, 10' OC, 1" dia.	26.000	L.F.	62.08	255.13	317.21
Pipe, black steel, Sch 40, threaded, W/cplg & hangers, 10' OC, 1-1/4" dia.	20.000	L.F.	58.24	218.40	276.64
Elbow, 90°, black, straight, 1" dia.	4.000	Ea.	11.28	128	139.28
Elbow, 90°, black, straight, 1-1/4" dia.	4.000	Ea.	18.48	136	154.48
Tee, black, reducing, 1" dia.	8.000	Ea.	40	416	456
Tee, black, reducing, 1-1/4" dia.	8.000	Ea.	69.60	428	497.60
Reducer, black, concentric, 1" dia.	4.000	Ea.	19.08	110	129.08
Reducer, black, concentric, 1-1/4" dia.	4.000	Ea.	22.80	116	138.80
Union, black with brass seat, 1" dia.	4.000	Ea.	35.40	138	173.40
Union, black with brass seat, 1-1/4" dia.	2.000	Ea.	25.40	71	96.40
Pipe nipples, 1" dia	13.000	Ea.	12.42	51.03	63.45
Pipe nipples, 1-1/4" dia	12.000	Ea.	13.44	50.40	63.84
Valves, bronze, gate, N.R.S., threaded, class 150, 1" size	2.000	Ea.	73	44	117
Valves, bronze, gate, N.R.S., threaded, class 150, 1-1/4" size	2.000	Ea.	98	55	153
Strainer, Y type, bronze body, screwed, 150 lb., 1" pipe size	1.000	Ea.	20.50	25	45.50
Strainer, Y type, bronze body, screwed, 150 lb., 1-1/4" pipe size	1.000	Ea.	46	28	74
Valve, electric motor actuated, brass, 3 way, screwed, 3/4" pipe	1.000	Ea.	266	23	289
Valve, electric motor actuated, brass, 2 way, screwed, 1" pipe size	1.000	Ea.	269	22	291
Thermometers, stem type, 9" case, 8" stem, 3/4" NPT	4.000	Ea.	520	60.20	580.20
Pressure/temperature relief plug, 316 SS, 3/4"OD, 7-1/2" insertion	4.000	Ea.	184	52.60	236.60
Insulation, fiberglass pipe covering, 1" wall, 1" IPS	26.000	L.F.	35.36	87.62	122.98
Insulation, fiberglass pipe covering, 1" wall, 1-1/4" IPS	20.000	L.F.	30.80	70.60	101.40
Insulation pipe covering .016" aluminum jacket finish, add	45.000	S.F.	19.80	166.95	186.75
Fan coil control system	1.000	Ea.	900	390	1,290
Re-heat coil balancing	2.000	Ea.		156	156
Rooftop unit heat/cool balancing	1.000	Ea.		290	290
TOTAL			6,550.68	4,348.93	10,899.61

8.8-330	Rooftop Air Handling Unit	COST EACH MAT.	COST EACH INST.	COST EACH TOTAL
1010	AHU, Rooftop, cool/heat coils, constant vol., fltrs, 2,000 CFM	6,550	4,350	10,900
1020	5,000 CFM	10,100	5,550	15,650
1030	10,000 CFM	13,400	7,875	21,275
1040	15,000 CFM	18,900	10,700	29,600
1050	20,000 CFM	22,600	12,800	35,400

HVAC | A8.8-330 | Air Handling Unit, Rooftop

8.8-332	Rooftop Air Handling Unit, VAV	COST EACH		
		MAT.	INST.	TOTAL
1010	AHU, Rooftop, cool/heat coils, VAV, fltrs, 5,000 CFM	12,000	5,725	17,725
1020	10,000 CFM	16,100	7,950	24,050
1030	15,000 CFM	21,600	10,600	32,200
1040	20,000 CFM	26,900	12,900	39,800
1050	30,000 CFM	39,800	14,800	54,600

HVAC | A8.8-400 | AC Unit, Package, Water Cooled

Condenser Supply Water Piping

Condenser Return Water Piping

Self-Contained Air Conditioner Unit, Water Cooled

System Components	QUANTITY	UNIT	COST EACH		
			MAT.	INST.	TOTAL
SYSTEM 8.8-410-1010					
A/C, SELF CONTAINED, SINGLE PKG., WATER COOLED, ELECT. HEAT, 5 TON					
Self-contained, water cooled, elect. heat, not inc tower, 5 ton, const vol	1.000	Ea.	6,575	980	7,555
Pipe, black steel, Sch 40, threaded, W/cplg & hangers, 10' OC, 1-1/4" dia.	20.000	L.F.	48.16	180.60	228.76
Elbow, 90°, black, straight, 1-1/4" dia.	6.000	Ea.	27.72	204	231.72
Tee, black, straight, 1-1/4" dia.	2.000	Ea.	14.20	107	121.20
Tee, black, reducing, 1-1/4" dia.	2.000	Ea.	17.40	107	124.40
Thermometers, stem type, 9" case, 8" stem, 3/4" NPT	2.000	Ea.	260	30.10	290.10
Union, black with brass seat, 1-1/4" dia.	2.000	Ea.	25.40	71	96.40
Pipe nipple, black, 1-1/4" dia	3.000	Ea.	3.36	12.60	15.96
Valves, bronze, gate, N.R.S., threaded, class 150, 1-1/4" size	2.000	Ea.	98	55	153
Circuit setter, bal valve, bronze body, threaded, 1-1/4" pipe size	1.000	Ea.	96.50	28	124.50
Insulation, fiberglass pipe covering, 1" wall, 1-1/4" IPS	20.000	L.F.	30.80	70.60	101.40
Control system, pneumatic, A/C Unit with heat	1.000	Ea.	2,000	1,475	3,475
Re-heat coil balancing	1.000	Ea.		78	78
Rooftop unit heat/cool balancing	1.000	Ea.		290	290
TOTAL			9,196.54	3,688.90	12,885.44

8.8-410	AC Unit, Package, Elec. Ht., Water Cooled	COST EACH		
		MAT.	INST.	TOTAL
1010	A/C, Self contained, single pkg., water cooled, elect. heat, 5 Ton	9,200	3,700	12,900
1020	10 Ton	15,000	4,550	19,550
1030	20 Ton	19,200	5,925	25,125
1040	30 Ton	33,800	6,450	40,250
1050	40 Ton	42,700	7,125	49,825
1060	50 Ton	53,500	8,550	62,050

8.8-412	AC Unit, Package, Elec. Ht., Water Cooled, VAV	COST EACH		
		MAT.	INST.	TOTAL
1010	A/C, Self contained, single pkg., water cooled, elect. ht, VAV, 10 Ton	17,700	4,550	22,250
1020	20 Ton	23,500	5,925	29,425
1030	30 Ton	39,300	6,450	45,750
1040	40 Ton	50,500	7,125	57,625
1050	50 Ton	62,500	8,550	71,050
1060	60 Ton	72,000	10,600	82,600

HVAC | A8.8-400 | AC Unit, Package, Water Cooled

8.8-420 | AC Unit, Package, Hot Water Coil, Water Cooled

		COST EACH		
		MAT.	INST.	TOTAL
1010	A/C, Self contn'd, single pkg., water cool, H/W ht, const. vol, 5 Ton	7,975	5,125	13,100
1020	10 Ton	12,600	6,100	18,700
1030	20 Ton	20,800	7,700	28,500
1040	30 Ton	27,000	9,975	36,975
1050	40 Ton	33,000	12,000	45,000
1060	50 Ton	40,500	14,700	55,200

8.8-422 | AC Unit, Package, HW Coil, Water Cooled, VAV

		COST EACH		
		MAT.	INST.	TOTAL
1010	A/C, Self contn'd, single pkg., water cool, H/W ht, VAV, 10 Ton	15,300	5,750	21,050
1020	20 Ton	20,400	7,575	27,975
1030	30 Ton	34,400	9,675	44,075
1040	40 Ton	46,300	10,500	56,800
1050	50 Ton	55,500	12,400	67,900
1060	60 Ton	66,500	13,500	80,000

HVAC | A8.8-450 | Heat Pump

Heat Pump, Air/Air, Rooftop w/ Economizer

System Components	QUANTITY	UNIT	COST EACH		
			MAT.	INST.	TOTAL
SYSTEM 8.8-470-1010 HEAT PUMP, ROOF TOP, AIR/AIR, CURB., ECONOMIZER., SUP. ELECT. HEAT, 10 TON					
Heat pump, air to air, rooftop, curb, economizer, elect ht, 10 ton cooling	1.000	Ea.	38,600	2,950	41,550
Ductwork, 32" x 12" fabricated, galvanized steel, 16 LF	176.000	Lb.	255.20	869.44	1,124.64
Insulation, ductwork, blanket type, fiberglass, 1" thk, 1-1/2 LB density	117.000	S.F.	50.31	248.04	298.35
Rooftop unit heat/cool balancing	1.000	Ea.		290	290
TOTAL			38,905.51	4,357.48	43,262.99

8.8-460	Heat Pump, Water Source, Central Station	COST EACH		
		MAT.	INST.	TOTAL
1010	Heat pump, central station, water source, constant vol., 5 Ton	2,975	2,075	5,050
1020	10 Ton	9,225	2,900	12,125
1030	20 Ton	10,200	5,550	15,750
1040	30 Ton	18,800	8,000	26,800
1050	40 Ton	19,300	8,725	28,025
1060	50 Ton	28,200	11,800	40,000

8.8-466	Heat Pump, Water Source, Console	COST EACH		
		MAT.	INST.	TOTAL
1020	Heat pump, console, water source, 1 Ton	1,325	1,150	2,475
1030	1-1/2 Ton	1,525	1,175	2,700
1040	2 Ton	1,950	1,400	3,350
1050	3 Ton	2,100	1,475	3,575
1060	3-1/2 Ton	2,200	1,525	3,725

8.8-468	Heat Pump, Horizontal, Duct Mounted	COST EACH		
		MAT.	INST.	TOTAL
1020	Heat pump, horizontal, ducted, water source, 1 Ton	1,850	2,450	4,300
1030	1-1/2 Ton	2,300	3,100	5,400
1040	2 Ton	2,675	3,700	6,375
1050	3 Ton	3,275	4,850	8,125
1060	3-1/2 Ton	3,375	5,025	8,400

8.8-470	Heat Pump, Roof Top, Air/Air	COST EACH		
		MAT.	INST.	TOTAL
1010	Heat pump, roof top, air/air, curb, economizer, sup. elect heat, 10 Ton	38,900	4,350	43,250
1020	20 Ton	55,000	7,425	62,425
1030	30 Ton	75,000	10,300	85,300
1040	40 Ton	83,000	11,900	94,900
1050	50 Ton	97,000	13,400	110,400
1060	60 Ton	102,500	15,200	117,700

HVAC — A8.8-500 — AC Unit, Package, DX, Air Cooled

Rooftop Air Conditioner Unit

System Components	QUANTITY	UNIT	COST EACH MAT.	COST EACH INST.	COST EACH TOTAL
SYSTEM 8.8-510-1010					
A/C, ROOFTOP, DX COOL, GAS HEAT, CURB, ECONOMIZER, FILTERS., 5 TON					
Roof top A/C, curb, economizer, sgl zone, elec cool, gas ht, 5 ton, 112 MBH	1.000	Ea.	5,150	1,350	6,500
Pipe, black steel, Sch 40, threaded, W/cplgs & hangers, 10' OC, 1" dia	20.000	L.F.	40.11	164.85	204.96
Elbow, 90°, black, straight, 3/4" dia.	3.000	Ea.	4.86	88.50	93.36
Elbow, 90°, black, straight, 1" dia.	3.000	Ea.	8.46	96	104.46
Tee, black, reducing, 1" dia.	1.000	Ea.	5	52	57
Union, black with brass seat, 1" dia.	1.000	Ea.	8.85	34.50	43.35
Pipe nipple, black, 3/4" dia	2.000	Ea.	3	13.60	16.60
Pipe nipple, black, 1" dia	2.000	Ea.	1.91	7.85	9.76
Cap, black, 1" dia.	1.000	Ea.	2.21	13.85	16.06
Gas cock, brass, 1" size	1.000	Ea.	13.85	22	35.85
Control system, pneumatic, Rooftop A/C unit	1.000	Ea.	2,000	1,475	3,475
Rooftop unit heat/cool balancing	1.000	Ea.		290	290
TOTAL			7,238.25	3,608.15	10,846.40

8.8-510 Rooftop Air Conditioner, Const. Volume

		MAT.	INST.	TOTAL
1010	A/C, Rooftop, DX cool, gas heat, curb, economizer, fltrs, 5 Ton	7,250	3,600	10,850
1020	7-1/2 Ton	10,100	3,750	13,850
1030	12-1/2 Ton	14,000	4,125	18,125
1040	18 Ton	18,700	4,500	23,200
1050	25 Ton	29,200	5,150	34,350
1060	40 Ton	43,900	6,875	50,775

8.8-512 Rooftop Air Conditioner, Variable Air Volume

		MAT.	INST.	TOTAL
1010	A/C, Rooftop, DX cool, gas heat, curb, ecmizr, fltrs, VAV, 12-1/2 Ton	16,700	4,400	21,100
1020	18 Ton	21,600	4,850	26,450
1030	25 Ton	33,500	5,575	39,075
1040	40 Ton	52,000	7,575	59,575
1050	60 Ton	72,500	10,300	82,800
1060	80 Ton	112,500	12,900	125,400

8.8-520 Packaged A/C, Elec. Ht., Const. Volume

		MAT.	INST.	TOTAL
1010	A/C Packaged, DX, air cooled, electric heat, constant vol., 5 Ton	8,875	2,625	11,500
1020	10 Ton	14,600	3,625	18,225
1030	20 Ton	22,900	4,750	27,650
1040	30 Ton	31,200	5,275	36,475
1050	40 Ton	40,000	6,050	46,050
1060	50 Ton	50,500	6,775	57,275

HVAC — A8.8-500 | AC Unit, Package, DX, Air Cooled

8.8-522 Packaged Air Conditioner, Elect. Heat, VAV

			COST EACH		
			MAT.	INST.	TOTAL
1010	A/C Packaged, DX, air cooled, electric heat, VAV, 10 Ton		17,500	4,550	22,050
1020		20 Ton	27,600	6,000	33,600
1030		30 Ton	37,000	6,550	43,550
1040		40 Ton	48,500	7,500	56,000
1050		50 Ton	60,000	8,325	68,325
1060		60 Ton	69,000	9,900	78,900

8.8-530 Packaged A/C, Hot Wtr. Heat, Const. Volume

			COST EACH		
			MAT.	INST.	TOTAL
1010	A/C Packaged, DX, air cooled, H/W heat, constant vol., 5 Ton		8,100	3,850	11,950
1020		10 Ton	12,700	5,050	17,750
1030		20 Ton	20,500	6,800	27,300
1040		30 Ton	26,200	8,625	34,825
1050		40 Ton	35,500	9,675	45,175
1060		50 Ton	43,200	10,500	53,700

8.8-532 Packaged A/C, Hot Wtr. Heat, VAV

			COST EACH		
			MAT.	INST.	TOTAL
1010	A/C Packaged, DX, air cooled, H/W heat, VAV, 10 Ton		15,800	5,900	21,700
1020		20 Ton	25,300	7,775	33,075
1030		30 Ton	32,000	9,825	41,825
1040		40 Ton	43,900	10,600	54,500
1050		50 Ton	52,500	12,300	64,800
1060		60 Ton	63,500	13,900	77,400

HVAC | A8.8-600 | Fan Coil Air Conditioner

Horizontal Fan Air Conditioning System

Labels: Galvanized Steel Duct, Fiberglass Duct Insulation, Flexible Fiberglass Duct, Supply Piping, Return Piping

System Components	QUANTITY	UNIT	COST EACH MAT.	COST EACH INST.	COST EACH TOTAL
SYSTEM 8.8-625-1010					
FAN COIL A/C SYSTEM, HORIZONTAL, W/HOUSING, CONTROLS, 2 PIPE, 1/2 TON					
Fan coil A/C, horizontal housing, filters, chilled water, 1/2 ton cooling	1.000	Ea.	930	95	1,025
Pipe, black steel, Sch 40, threaded, W/cplgs & hangers, 10' OC, 3/4" dia.	20.000	L.F.	32.25	146.20	178.45
Elbow, 90°, black, straight, 3/4" dia.	6.000	Ea.	9.72	177	186.72
Tee, black, straight, 3/4" dia.	2.000	Ea.	5.14	92	97.14
Union, black with brass seat, 3/4" dia.	2.000	Ea.	13.60	64	77.60
Pipe nipples, 3/4" diam	3.000	Ea.	2.25	10.20	12.45
Valves, bronze, gate, N.R.S., threaded, class 150, 3/4" size	2.000	Ea.	56	42	98
Circuit setter, balance valve, bronze body, threaded, 3/4" pipe size	1.000	Ea.	43	21	64
Insulation, fiberglass pipe covering, 1" wall, 3/4" IPS	20.000	L.F.	26.60	64.40	91
Ductwork, 12" x 8" fabricated, galvanized steel, 12 LF	55.000	Lb.	79.75	271.70	351.45
Insulation, ductwork, blanket type, fiberglass, 1" thk, 1-1/2 LB density	40.000	S.F.	17.20	84.80	102
Diffusers, aluminum, OB damper, ceiling, perf, 24"x24" panel size, 6"x6"	2.000	Ea.	174	52	226
Ductwork, flexible, fiberglass fabric, insulated, 1"thk, PE jacket, 6" dia	16.000	L.F.	37.92	45.92	83.84
Round volume control damper 6" dia.	2.000	Ea.	48	37.70	85.70
Fan coil unit balancing	1.000	Ea.		54	54
Re-heat coil balancing	1.000	Ea.		78	78
Diffuser/register, high, balancing	2.000	Ea.		149	149
TOTAL			1,475.43	1,484.92	2,960.35

8.8-610	Fan Coil A/C Unit, Two Pipe	COST EACH MAT.	COST EACH INST.	COST EACH TOTAL
1010	Fan coil A/C system, cabinet mounted, controls, 2 pipe, 1/2 Ton	995	790	1,785
1020	1 Ton	1,175	880	2,055
1030	1-1/2 Ton	1,250	895	2,145
1040	2 Ton	1,900	1,075	2,975
1050	3 Ton	2,450	1,125	3,575

8.8-615	Fan Coil A/C Unit, Two Pipe, Electric Heat	COST EACH MAT.	COST EACH INST.	COST EACH TOTAL
1010	Fan coil A/C system, cabinet mntd, elect. ht, controls, 2 pipe, 1/2 Ton	1,175	790	1,965
1020	1 Ton	1,425	880	2,305
1030	1-1/2 Ton	1,700	895	2,595
1040	2 Ton	2,550	1,100	3,650
1050	3 Ton	4,300	1,125	5,425

Important: See the Reference Section for critical supporting data - Reference Nos., Crews, & City Cost Indexes

HVAC — A8.8-600 Fan Coil Air Conditioner

8.8-620 Fan Coil A/C Unit, Four Pipe

		MAT.	INST.	TOTAL
1010	Fan coil A/C system, cabinet mounted, controls, 4 pipe, 1/2 Ton	1,500	1,525	3,025
1020	1 Ton	1,725	1,625	3,350
1030	1-1/2 Ton	1,850	1,650	3,500
1040	2 Ton	2,325	1,700	4,025
1050	3 Ton	3,150	1,825	4,975

8.8-625 Fan Coil A/C, Horizontal, Duct Mount, 2 Pipe

		MAT.	INST.	TOTAL
1010	Fan coil A/C system, horizontal w/housing, controls, 2 pipe, 1/2 Ton	1,475	1,475	2,950
1020	1 Ton	1,975	2,175	4,150
1030	1-1/2 Ton	2,500	2,825	5,325
1040	2 Ton	3,550	3,400	6,950
1050	3 Ton	3,800	4,500	8,300
1060	3-1/2 Ton	4,000	4,650	8,650
1070	4 Ton	4,100	5,275	9,375
1080	5 Ton	4,750	6,700	11,450
1090	6 Ton	4,825	7,200	12,025
1100	7 Ton	4,850	7,475	12,325
1110	8 Ton	4,850	7,625	12,475
1120	10 Ton	5,325	7,925	13,250

8.8-630 Fan Coil A/C, Horiz., Duct Mount, 2 Pipe, Elec. Ht.

		MAT.	INST.	TOTAL
1010	Fan coil A/C system, horiz. hsng, elect. ht, ctrls, 2 pipe, 1/2 Ton	1,675	1,475	3,150
1020	1 Ton	2,225	2,175	4,400
1030	1-1/2 Ton	2,675	2,825	5,500
1040	2 Ton	3,425	3,400	6,825
1050	3 Ton	5,475	4,650	10,125
1060	3-1/2 Ton	6,150	4,650	10,800
1070	4 Ton	6,900	5,300	12,200
1080	5 Ton	7,800	6,700	14,500
1090	6 Ton	7,875	7,225	15,100
1100	7 Ton	8,100	7,175	15,275
1110	8 Ton	8,700	7,250	15,950
1120	10 Ton	9,225	7,475	16,700

8.8-635 Fan Coil A/C, Horizontal, Duct Mount, 4 Pipe

		MAT.	INST.	TOTAL
1010	Fan coil A/C system, horiz. w/cabinet, controls, 4 pipe, 1/2 Ton	2,025	2,225	4,250
1020	1 Ton	2,600	2,975	5,575
1030	1-1/2 Ton	3,225	3,575	6,800
1040	2 Ton	4,275	4,000	8,275
1050	3 Ton	4,500	5,350	9,850
1060	3.5 Ton	5,100	6,525	11,625
1070	4 Ton	5,100	6,600	11,700
1080	5 Ton	6,175	9,100	15,275
1090	6 Ton	6,475	10,800	17,275
1100	7 Ton	7,350	12,100	19,450
1110	8 Ton	7,425	12,800	20,225
1120	10 Ton	9,075	15,700	24,775

HVAC A8.8-700 VAV Terminals

VAV Terminal   Fan Powered VAV Terminal

System Components	QUANTITY	UNIT	COST EACH MAT.	COST EACH INST.	COST EACH TOTAL
SYSTEM 8.8-710-1010					
VAV TERMINAL, COOLING ONLY, WITH ACTUATOR / CONTROLS, 200 CFM					
Mixing box variable volume 300 to 600CFM, cool only	1.000	Ea.	385	74.80	459.80
Ductwork, 12" x 8" fabricated, galvanized steel, 12 LF	55.000	Lb.	79.75	271.70	351.45
Insulation, ductwork, blanket type, fiberglass, 1" thk, 1-1/2 LB density	40.000	S.F.	17.20	84.80	102
Diffusers, aluminum, OB damper, ceiling, perf, 24"x24" panel size, 6"x6"	2.000	Ea.	174	52	226
Ductwork, flexible, fiberglass fabric, insulated, 1"thk, PE jacket, 6" dia	16.000	L.F.	37.92	45.92	83.84
Round volume control damper 6" dia.	2.000	Ea.	48	37.70	85.70
VAV box balancing	1.000	Ea.		49.50	49.50
Diffuser/register, high, balancing	2.000	Ea.		149	149
TOTAL			741.87	765.42	1,507.29

8.8-710	VAV Terminal, Cooling Only	MAT.	INST.	TOTAL
1010	VAV Terminal, cooling only, with actuator / controls, 200 CFM	740	765	1,505
1020	400 CFM	975	1,350	2,325
1030	600 CFM	1,300	2,000	3,300
1040	800 CFM	1,400	2,450	3,850
1050	1000 CFM	1,625	2,825	4,450
1060	1250 CFM	1,725	3,600	5,325
1070	1500 CFM	1,875	4,250	6,125
1080	2000 CFM	2,300	6,025	8,325

8.8-715	VAV Terminal, Hot Water Reheat	MAT.	INST.	TOTAL
1010	VAV Terminal, cooling, HW reheat, with actuator / controls, 200 CFM	1,425	1,425	2,850
1020	400 CFM	1,650	1,925	3,575
1030	600 CFM	1,975	2,650	4,625
1040	800 CFM	2,100	3,100	5,200
1050	1000 CFM	2,300	3,475	5,775
1060	1250 CFM	2,475	4,275	6,750
1070	1500 CFM	2,700	5,075	7,775
1080	2000 CFM	3,150	6,900	10,050

8.8-720	Fan Powered VAV Terminal, Cooling Only	MAT.	INST.	TOTAL
1010	VAV Terminal, cooling, fan powrd, with actuator / controls, 200 CFM	1,200	765	1,965
1020	400 CFM	1,375	1,250	2,625
1030	600 CFM	1,750	2,000	3,750
1040	800 CFM	1,800	2,400	4,200

HVAC — A8.8-700 VAV Terminals

8.8-720 Fan Powered VAV Terminal, Cooling Only

		COST EACH		
		MAT.	INST.	TOTAL
1050	1000 CFM	1,975	2,775	4,750
1060	1250 CFM	2,325	3,650	5,975
1070	1500 CFM	2,450	4,275	6,725
1080	2000 CFM	2,925	6,200	9,125

8.8-725 Fan Powered VAV Terminal, Hot Water Reheat

		COST EACH		
		MAT.	INST.	TOTAL
1010	VAV Terminal, cool, HW reht, fan powrd, with actuator/ctrls, 200 CFM	1,800	1,400	3,200
1020	400 CFM	2,075	1,950	4,025
1030	600 CFM	2,375	2,650	5,025
1040	800 CFM	2,425	3,050	5,475
1050	1000 CFM	2,600	3,450	6,050
1060	1250 CFM	3,025	4,300	7,325
1070	1500 CFM	3,250	5,125	8,375
1080	2000 CFM	3,750	7,075	10,825

HVAC — A8.8-800 Thru-Wall Units

Thru-Wall Heat Pump   Thru-Wall Air Conditioner

System Components	QUANTITY	UNIT	COST EACH MAT.	COST EACH INST.	COST EACH TOTAL
SYSTEM 8.8-810-1010					
A/C UNIT, THRU-THE-WALL, SUP. ELECT. HEAT, CABINET, LOUVER, 1/2 TON					
A/C unit, thru-wall, electric heat., cabinet, louver, 1/2 ton	1.000	Ea.	1,000	126	1,126
TOTAL			1,000	126	1,126

8.8-810	Thru-Wall A/C Unit	MAT.	INST.	TOTAL
1010	A/C Unit, thru-the-wall, sup. elect. heat, cabinet, louver, 1/2 Ton	1,000	126	1,126
1020	3/4 Ton	1,025	152	1,177
1030	1 Ton	1,075	190	1,265
1040	1-1/2 Ton	1,500	315	1,815
1050	2 Ton	1,500	400	1,900

8.8-820	Thru-Wall Heat Pump	MAT.	INST.	TOTAL
1010	Heat pump, thru-the-wall, cabinet, louver, 1/2 Ton	1,325	95	1,420
1020	3/4 Ton	1,350	126	1,476
1030	1 Ton	1,450	190	1,640
1040	Sup. elect. heat, 1-1/2 Ton	2,525	490	3,015
1050	2 Ton	2,775	505	3,280

HVAC A8.8-900 | Fan Systems

Axial Flow Centrifugal Fan Belt Drive | Belt Drive Utility Set | Centrifugal Roof Exhaust Fan Direct Drive

System Components	QUANTITY	UNIT	MAT.	INST.	TOTAL
SYSTEM 8.8-910-1010					
FAN SYSTEM, IN-LINE CENTRIFUGAL, 500 CFM					
Fans, in-line centrifugal, supply/exhaust, 500 CFM, 10" dia conn	1.000	Ea.	770	315	1,085
Louver, alum., W/screen, damper, mill fin, fxd blade, cont line, stormproof	4.050	S.F.	125.55	59.94	185.49
Motor for louver damper, electric or pneumatic	1.000	Ea.	365	29.50	394.50
Ductwork, 12" x 8" fabricated, galvanized steel, 60 LF	200.000	Lb.	234	1,232.40	1,466.40
Grilles, aluminum, air supply, single deflection, adjustable, 12" x 6"	1.000	Ea.	46.35	54	100.35
Fire damper, curtain type, vertical, 12" x 8"	1.000	Ea.	21.50	18.85	40.35
Duct access door, insulated, 10" x 10"	1.000	Ea.	17.90	37.50	55.40
Duct access door, insulated, 16" x 12"	1.000	Ea.	27	46	73
Pressure controller/switch, air flow controller	1.000	Ea.	118	35	153
Fan control switch and mounting base	1.000	Ea.	69	26.50	95.50
Fan balancing	1.000	Ea.		249	249
Diffuser/register balancing	3.000	Ea.		148.50	148.50
TOTAL			1,794.30	2,252.19	4,046.49

8.8-910	Centrifugal In Line Fan Systems	MAT.	INST.	TOTAL
1010	Fan system, in-line centrifugal, 500 CFM	1,800	2,250	4,050
1020	1300 CFM	2,175	3,900	6,075
1030	1500 CFM	2,500	4,500	7,000
1040	2500 CFM	3,625	13,800	17,425
1050	3500 CFM	4,575	18,600	23,175
1060	5000 CFM	5,625	27,500	33,125
1070	7500 CFM	6,800	28,100	34,900
1080	10,000 CFM	8,175	33,000	41,175

8.8-915	Utility Set Fan Systems	MAT.	INST.	TOTAL
1010	Utility fan set system, belt drive, 2000 CFM	1,550	7,300	8,850
1020	3500 CFM	2,525	12,900	15,425
1030	5000 CFM	3,200	17,900	21,100
1040	7500 CFM	3,900	20,800	24,700
1050	10,000 CFM	5,525	27,300	32,825
1060	15,000 CFM	6,125	27,800	33,925
1070	20,000 CFM	6,700	31,500	38,200

HVAC | A8.8-900 | Fan Systems

8.8-940	Roof Exhaust Fan Systems	COST EACH		
		MAT.	INST.	TOTAL
1010	Roof vent. system, centrifugal, alum., galv curb, BDD, 500 CFM	755	1,475	2,230
1020	800 CFM	975	2,850	3,825
1030	1500 CFM	1,150	4,225	5,375
1040	2750 CFM	2,125	9,550	11,675
1050	3500 CFM	2,500	12,300	14,800
1060	5000 CFM	3,550	20,300	23,850
1070	8500 CFM	4,225	26,100	30,325
1080	13,800 CFM	5,950	37,900	43,850

Division 11
Special Construction

SPECIAL CONSTR. | A11.2-100 | Tanks, Fiberglass

Fiberglass Underground Fuel Storage Tank, Single Wall

System Components	QUANTITY	UNIT	COST EACH		
			MAT.	INST.	TOTAL
SYSTEM 11.2-105-1010					
STORAGE TANK, FUEL, UNDERGROUND, SINGLE WALL FIBERGLASS, 550 GAL.					
Tank, fiberglass, underground, single wall, U.L. listed, 550 gal	1.000	Ea.	1,600	284	1,884
Tank, fiberglass, manways, add	1.000	Ea.	925		925
Tank, for hold-downs 500-4000 gal, add	1.000	Ea.	178	100	278
Foot valve, single poppet, 3/4" dia	1.000	Ea.	42	23.50	65.50
Tubing, copper, Type L, 3/4" dia	120.000	L.F.	208.80	654	862.80
Elbow 90°, copper, wrought, cu x cu, 3/4" dia.	8.000	Ea.	4.24	176	180.24
Fuel oil specialties, valve, ball chk, globe type, fusible, 3/4" dia	2.000	Ea.	59	42	101
Elbow 90°, black steel, straight, 2" dia.	3.000	Ea.	31.50	124.50	156
Union, black steel, with brass seat, 2" dia.	1.000	Ea.	18.45	44	62.45
Pipe and nipples, steel, Sch. 40, threaded, black, 2" dia	2.000	Ea.	120.84	444.60	565.44
Vent protector / breather, 2" dia.	1.000	Ea.	33	13.15	46.15
Pipe, black steel, welded, Sch. 40, 3" dia.	30.000	L.F.	180	599.10	779.10
Elbow 90°, steel, weld joint, butt, 3" dia.	3.000	Ea.	32.70	368.40	401.10
Flange, weld neck, 150 lb., 3" pipe size	1.000	Ea.	22.50	61.40	83.90
Fuel fill box, locking inner cover, 3" dia.	1.000	Ea.	106	84	190
Remote tank gauging system, 30', 5" pointer travel	1.000	Ea.	1,325	168	1,493
Tank leak detector system, 8 channel, external monitoring	1.000	Ea.	1,750		1,750
Tank leak detection system, probes, well monitoring, liquid phase detection	2.000	Ea.	1,570		1,570
Excavating, trench, NO sheeting or dewatering, 6'-10' D,3/4 CY hyd backhoe	18.000	C.Y.		91.44	91.44
Concrete hold down pad, 6' x 5' x 8" thick	30.000	C.Y.	56.70	29.10	85.80
Reinforcing in hold down pad, #3 to #7	120.000	Lb.	36	50.40	86.40
Stone back-fill, hand spread, pea gravel	9.000	C.Y.	292.50	283.50	576
Corrosion resistance, wrap & coat, small diam pipe, 1" diam, add	120.000	L.F.	134.40		134.40
Corrosion resistance, wrap & coat, small diameter pipe, 2" diam, add	36.000	L.F.	44.28		44.28
Corrosion resistance wrap & coat, 4" diam, add	30.000	L.F.	44.70		44.70
TOTAL			**8,815.61**	**3,641.09**	**12,456.70**

11.2-105	Fiberglass Fuel Tank, Single Wall	COST EACH		
		MAT.	INST.	TOTAL
1010	Storage tank, fuel, underground, single wall fiberglass, 550 Gal.	8,825	3,650	12,475
1020	2000 Gal.	11,000	5,000	16,000
1030	4000 Gal.	15,000	6,800	21,800
1040	6000 Gal.	16,100	7,600	23,700
1050	8000 Gal.	20,200	9,325	29,525
1060	10,000 Gal.	24,000	10,200	34,200

SPECIAL CONSTR. | A11.2-100 Tanks, Fiberglass

11.2-105 Fiberglass Fuel Tank, Single Wall

		\multicolumn{3}{c}{COST EACH}		
		MAT.	INST.	TOTAL
1070	15,000 Gal.	27,700	12,600	40,300
1080	20,000 Gal.	35,100	15,300	50,400
1090	25,000 Gal.	44,900	17,900	62,800
1100	30,000 Gal.	52,000	19,300	71,300
1110	40,000 Gal.	65,500	24,400	89,900
1120	48,000 Gal.	91,000	28,000	119,000

11.2-110 Fiberglass Fuel Tank, Double Wall

		COST EACH		
		MAT.	INST.	TOTAL
1010	Storage tank, fuel, underground, double wall fiberglass, 550 Gal.	11,000	4,175	15,175
1020	2500 Gal.	15,500	5,800	21,300
1030	4000 Gal.	19,600	7,200	26,800
1040	6000 Gal.	21,300	8,150	29,450
1050	8000 Gal.	27,100	9,850	36,950
1060	10,000 Gal.	30,900	11,400	42,300
1070	15,000 Gal.	40,600	13,100	53,700
1080	20,000 Gal.	49,200	16,000	65,200
1090	25,000 Gal.	59,000	18,500	77,500
1100	30,000 Gal.	67,500	19,700	87,200

SPECIAL CONSTR. | A11.2-200 Tanks, Steel

Steel Underground Fuel Storage Tank, Single Wall

System Components	QUANTITY	UNIT	COST EACH MAT.	COST EACH INST.	COST EACH TOTAL
SYSTEM 11.2-205-1010					
STORAGE TANK, FUEL, UNDERGROUND, SINGLE WALL STEEL, 550 GAL.					
Tank, steel, underground, sti-P3, set in place, 500 gal, 7 Ga. shell	1.000	Ea.	905	281	1,186
Tank, for manways, add	1.000	Ea.	610		610
Tanks, for hold-downs, 500-2000 gal, add	1.000	Ea.	295	100	395
Foot valve, single poppet, 3/4" dia	1.000	Ea.	42	23.50	65.50
Tubing, copper, Type L, 3/4" dia.	120.000	L.F.	208.80	654	862.80
Elbow 90°, copper, wrought, cu x cu, 3/4" dia.	8.000	Ea.	4.24	176	180.24
Fuel oil valve, ball chk, globe type, fusible, 3/4" dia.	2.000	Ea.	59	42	101
Elbow 90°, black steel, straight, 2" dia.	3.000	Ea.	31.50	124.50	156
Union, black steel, with brass seat, 3/4" dia.	1.000	Ea.	6.80	32	38.80
Pipe and nipples, steel, Sch. 40, threaded, black, 2" dia	2.000	Ea.	120.84	444.60	565.44
Vent protector / breather, 2" dia.	1.000	Ea.	33	13.15	46.15
Pipe, black steel, welded, Sch. 40, 3" dia.	30.000	L.F.	180	599.10	779.10
Elbow 90°, steel, weld joint, butt, 90° elbow, 3" dia.	3.000	Ea.	32.70	368.40	401.10
Flange, steel, weld neck, 150 lb., 3" pipe size	1.000	Ea.	22.50	61.40	83.90
Fuel fill box, locking inner cover, 3" dia.	1.000	Ea.	106	84	190
Remote tank gauging system, 30', 5" pointer travel	1.000	Ea.	1,325	168	1,493
Tank leak detector system, 8 channel, external monitoring	1.000	Ea.	1,750		1,750
Tank leak detection system, probes, well monitoring, liquid phase detection	2.000	Ea.	1,570		1,570
Excavation, NO sheeting or dewatering, 6'-10' D, 3/4 CY hyd backhoe	18.000	C.Y.		91.44	91.44
Concrete hold down pad, 6' x 5' x 8" thick	30.000	C.Y.	56.70	29.10	85.80
Reinforcing in hold down pad, #3 to #7	120.000	Lb.	36	50.40	86.40
Stone back-fill, hand spread, pea gravel	9.000	C.Y.	292.50	283.50	576
Corrosion resistance, wrap & coat, small diam pipe, 1" dia., add	120.000	L.F.	134.40		134.40
Corrosion resistance, wrap & coat, small diameter pipe, 2" dia., add	36.000	L.F.	44.28		44.28
Corrosion resistance wrap & coat, 4" dia., add	30.000	L.F.	44.70		44.70
TOTAL			**7,910.96**	**3,626.09**	**11,537.05**

11.2-205	Steel Fuel Tank, Single Wall	COST EACH MAT.	COST EACH INST.	COST EACH TOTAL
1010	Storage tank, fuel, underground, single wall steel, 550 Gal.	7,900	3,625	11,525
1020	2000 Gal.	10,600	5,000	15,600
1030	5000 Gal.	16,500	7,075	23,575
1040	10,000 Gal.	26,900	10,800	37,700
1050	15,000 Gal.	31,200	12,600	43,800
1060	20,000 Gal.	40,200	15,200	55,400

SPECIAL CONSTR. A11.2-200 Tanks, Steel

11.2-205 Steel Fuel Tank, Single Wall

		MAT.	INST.	TOTAL
1070	25,000 Gal.	46,900	17,800	64,700
1080	30,000 Gal.	53,000	19,000	72,000
1090	40,000 Gal.	67,000	24,000	91,000
1100	50,000 Gal.	81,000	29,000	110,000

11.2-210 Steel Fuel Tank, Double Wall

		MAT.	INST.	TOTAL
1010	Storage tank, fuel, underground, double wall steel, 500 Gal.	10,500	4,250	14,750
1020	2000 Gal.	13,900	5,600	19,500
1030	4000 Gal.	19,500	7,300	26,800
1040	6000 Gal.	23,500	8,100	31,600
1050	8000 Gal.	28,000	9,975	37,975
1060	10,000 Gal.	32,700	11,500	44,200
1070	15,000 Gal.	39,600	13,500	53,100
1080	20,000 Gal.	47,800	16,200	64,000
1090	25,000 Gal.	62,000	18,600	80,600
1100	30,000 Gal.	71,000	19,800	90,800
1110	40,000 Gal.	86,500	24,900	111,400
1120	50,000 Gal.	102,500	29,800	132,300

11.2-220 Steel Tank, Above Ground, Single Wall

		MAT.	INST.	TOTAL
1010	Storage tank, fuel, above ground, single wall steel, 550 Gal.	5,075	3,075	8,150
1020	2000 Gal.	6,375	3,150	9,525
1030	5000 Gal.	9,800	4,850	14,650
1040	10,000 Gal.	12,800	5,150	17,950
1050	15,000 Gal.	15,800	5,300	21,100
1060	20,000 Gal.	17,600	5,450	23,050
1070	25,000 Gal.	20,700	5,575	26,275
1080	30,000 Gal.	23,900	5,800	29,700

11.2-230 Steel Tank, Above Ground, Double Wall

		MAT.	INST.	TOTAL
1010	Storage tank, fuel, above ground, double wall steel, 500 Gal.	9,125	3,125	12,250
1020	2000 Gal.	11,500	3,200	14,700
1030	4000 Gal.	15,900	3,250	19,150
1040	6000 Gal.	18,700	5,025	23,725
1050	8000 Gal.	20,700	5,150	25,850
1060	10,000 Gal.	22,000	5,250	27,250
1070	15,000 Gal.	28,900	5,425	34,325
1080	20,000 Gal.	31,800	5,575	37,375
1090	25,000 Gal.	36,900	5,750	42,650
1100	30,000 Gal.	39,700	5,950	45,650

Division 12
Site Work

SITE WORK — A12.3-110 Trenching

Trenching Systems are shown on a cost per linear foot basis. The systems include: excavation; backfill and removal of spoil; and compaction for various depths and trench bottom widths. The backfill has been reduced to accommodate a pipe of suitable diameter and bedding.

The slope for trench sides varies from 0:1 to 2:1.

The Expanded System Listing shows Trenching Systems that range from 2' to 12' in width. Depths range from 2' to 25'.

System Components	QUANTITY	UNIT	COST PER L.F.		
			EQUIP.	LABOR	TOTAL
SYSTEM 12.3-110-1310					
TRENCHING, BACKHOE, 0 TO 1 SLOPE, 2' WIDE, 2' DP, 3/8 C.Y. BUCKET					
Excavation, trench, hyd. backhoe, track mtd., 3/8 C.Y. bucket	.174	C.Y.	.23	.76	.99
Backfill and load spoil, from stockpile	.174	C.Y.	.11	.22	.33
Compaction by rammer tamper, 8" lifts, 4 passes	.014	C.Y.	.01	.03	.04
Remove excess spoil, 6 C.Y. dump truck, 2 mile roundtrip	.160	C.Y.	.65	.47	1.12
Total			1	1.48	2.48

12.3-110	Trenching	COST PER L.F.		
		EQUIP.	LABOR	TOTAL
1310	Trenching, backhoe, 0 to 1 slope, 2' wide, 2' deep, 3/8 C.Y. bucket	1	1.48	2.48
1320	3' deep, 3/8 C.Y. bucket	1.22	2.16	3.38
1330	4' deep, 3/8 C.Y. bucket	1.43	2.86	4.29
1340	6' deep, 3/8 C.Y. bucket	1.84	3.64	5.48
1350	8' deep, 1/2 C.Y. bucket	2.09	4.62	6.71
1360	10' deep, 1 C.Y. bucket	3.13	5.55	8.68
1400	4' wide, 2' deep, 3/8 C.Y. bucket	2.10	2.99	5.09
1410	3' deep, 3/8 C.Y. bucket	2.96	4.44	7.40
1420	4' deep, 1/2 C.Y. bucket	2.84	4.45	7.29
1430	6' deep, 1/2 C.Y. bucket	3.67	6.75	10.42
1440	8' deep, 1/2 C.Y. bucket	5.60	8.85	14.45
1450	10' deep, 1 C.Y. bucket	6.60	10.95	17.55
1460	12' deep, 1 C.Y. bucket	8.25	14.05	22.30
1470	15' deep, 1-1/2 C.Y. bucket	7.25	12	19.25
1480	18' deep, 2-1/2 C.Y. bucket	9.85	17.15	27
1520	6' wide, 6' deep, 5/8 C.Y. bucket	7.40	9.50	16.90
1530	8' deep, 3/4 C.Y. bucket	8.55	11.45	20
1540	10' deep, 1 C.Y. bucket	9.80	13.80	23.60
1550	12' deep, 1-1/4 C.Y. bucket	8.70	10.85	19.55
1560	16' deep, 2 C.Y. bucket	14.20	12.35	26.55
1570	20' deep, 3-1/2 C.Y. bucket	15.15	22	37.15
1580	24' deep, 3-1/2 C.Y. bucket	24	37	61
1640	8' wide, 12' deep, 1-1/4 C.Y. bucket	13.30	14.85	28.15
1650	15' deep, 1-1/2 C.Y. bucket	15.40	18.25	33.65
1660	18' deep, 2-1/2 C.Y. bucket	22	18.45	40.45
1680	24' deep, 3-1/2 C.Y. bucket	33.50	49.50	83
1730	10' wide, 20' deep, 3-1/2 C.Y. bucket	28	37.50	65.50
1740	24' deep, 3-1/2 C.Y. bucket	42.50	62	104.50
1780	12' wide, 20' deep, 3-1/2 C.Y. bucket	34	45	79
1790	25' deep, bucket	57	83	140
1800	1/2 to 1 slope, 2' wide, 2' deep, 3/8 C.Y. bucket	1.34	1.90	3.24
1810	3' deep, 3/8 C.Y. bucket	1.84	3.14	4.98
1820	4' deep, 3/8 C.Y. bucket	2.31	4.58	6.89
1840	6' deep, 3/8 C.Y. bucket	3.44	7.05	10.49
1860	8' deep, 1/2 C.Y. bucket	4.67	10.75	15.42
1880	10' deep, 1 C.Y. bucket	8.40	15.05	23.45

SITE WORK — A12.3-110 Trenching

12.3-110	Trenching	COST PER L.F.		
		EQUIP.	LABOR	TOTAL
2300	4' wide, 2' deep, 3/8 C.Y. bucket	2.23	3.16	5.39
2310	3' deep, 3/8 C.Y. bucket	3.67	5.25	8.92
2320	4' deep, 1/2 C.Y. bucket	3.69	5.45	9.14
2340	6' deep, 1/2 C.Y. bucket	5.20	9.45	14.65
2360	8' deep, 1/2 C.Y. bucket	8.45	13.05	21.50
2380	10' deep, 1 C.Y. bucket	11.60	19.25	30.85
2400	12' deep, 1 C.Y. bucket	12.65	20.50	33.15
2430	15' deep, 1-1/2 C.Y. bucket	15.65	25.50	41.15
2460	18' deep, 2-1/2 C.Y. bucket	27.50	30.50	58
2840	6' wide, 6' deep, 5/8 C.Y. bucket	8.25	10.50	18.75
2860	8' deep, 3/4 C.Y. bucket	11.80	15.75	27.55
2880	10' deep, 1 C.Y. bucket	14.55	20.50	35.05
2900	12' deep, 1-1/4 C.Y. bucket	13.75	17.30	31.05
2940	16' deep, 2 C.Y. bucket	21	27.50	48.50
2980	20' deep, 3-1/2 C.Y. bucket	32	47.50	79.50
3020	24' deep, 3-1/2 C.Y. bucket	57.50	88.50	146
3100	8' wide, 12' deep, 1-1/4 C.Y. bucket	19.20	21	40.20
3120	15' deep, 1-1/2 C.Y. bucket	24	28.50	52.50
3140	18' deep, 2-1/2 C.Y. bucket	37.50	31	68.50
3180	24' deep, 3-1/2 C.Y. bucket	67	99.50	166.50
3270	10' wide, 20' deep, 3-1/2 C.Y. bucket	42	59.50	101.50
3280	24' deep, 3-1/2 C.Y. bucket	77	112	189
3370	12' wide, 20' deep, 3-1/2 C.Y. bucket	48	66	114
3380	25' deep, 3-1/2 C.Y. bucket	90	130	220
3500	1 to 1 slope, 2' wide, 2' deep, 3/8 C.Y. bucket	1.56	2.22	3.78
3520	3' deep, 3/8 C.Y. bucket	2.38	3.88	6.26
3540	4' deep, 3/8 C.Y. bucket	3.01	5.85	8.86
3560	6' deep, 3/8 C.Y. bucket	4.63	9.30	13.93
3580	8' deep, 1/2 C.Y. bucket	6.45	14.65	21.10
3600	10' deep, 1 C.Y. bucket	13.75	24	37.75
3800	4' wide, 2' deep, 3/8 C.Y. bucket	2.66	3.80	6.46
3820	3' deep, 3/8 C.Y. bucket	4.69	6.65	11.34
3840	4' deep, 1/2 C.Y. bucket	4.45	6.10	10.55
3860	6' deep, 1/2 C.Y. bucket	6.45	11.35	17.80
3880	8' deep, 1/2 C.Y. bucket	11.30	17.20	28.50
3900	10' deep, 1 C.Y. bucket	15.25	24.50	39.75
3920	12' deep, 1 C.Y. bucket	21.50	36	57.50
3940	15' deep, 1-1/2 C.Y. bucket	21.50	33.50	55
3960	18' deep, 2-1/2 C.Y. bucket	39	40	79
4030	6' wide, 6' deep, 5/8 C.Y. bucket	10.05	12.80	22.85
4040	8' deep, 3/4 C.Y. bucket	14.85	20.50	35.35
4050	10' deep, 1 C.Y. bucket	19	28.50	47.50
4060	12' deep, 1-1/4 C.Y. bucket	18.55	25.50	44.05
4070	16' deep, 2 C.Y. bucket	37	36	73
4080	20' deep, 3-1/2 C.Y. bucket	46.50	75	121.50
4090	24' deep, 3-1/2 C.Y. bucket	87.50	146	233.50
4500	8' wide, 12' deep, 1-1/4 C.Y. bucket	25	29	54
4550	15' deep, 1-1/2 C.Y. bucket	32	41	73
4600	18' deep, 2-1/2 C.Y. bucket	51	47.50	98.50
4650	24' deep, 3-1/2 C.Y. bucket	97	156	253
4800	10' wide, 20' deep, 3-1/2 C.Y. bucket	60	87	147
4850	24' deep, 3-1/2 C.Y. bucket	105	165	270
4950	12' wide, 20' deep, 3-1/2 C.Y. bucket	66	93	159
4980	25' deep, 3-1/2 C.Y. bucket	128	201	329
5000	1-1/2 to 1 slope, 2' wide, 2' deep, 3/8 C.Y. bucket	1.81	2.57	4.38
5020	3' deep, 3/8 C.Y. bucket	2.90	4.60	7.50
5040	4' deep, 3/8 C.Y. bucket	3.68	6.85	10.53
5060	6' deep, 3/8 C.Y. bucket	5.65	10.95	16.60
5080	8' deep, 1/2 C.Y. bucket	7.90	17.30	25.20
5100	10' deep, 1 C.Y. bucket	15.10	25	40.10

For expanded coverage of these items see Means Site Work & Landscape Cost Data 2001

SITE WORK

A12.3-110 Trenching

12.3-110	Trenching	COST PER L.F.		
		EQUIP.	LABOR	TOTAL
5300	4' wide, 2' deep, 3/8 C.Y. bucket	2.52	3.57	6.09
5320	3' deep, 3/8 C.Y. bucket	4.57	6.50	11.07
5340	4' deep, 1/2 C.Y. bucket	5.30	7.15	12.45
5360	6' deep, 1/2 C.Y. bucket	7.60	12.85	20.45
5380	8' deep, 1/2 C.Y. bucket	13.45	19.55	33
5400	10' deep, 1 C.Y. bucket	18.30	28	46.30
5420	12' deep, 1 C.Y. bucket	26.50	42	68.50
5450	15' deep, 1-1/2 C.Y. bucket	26	37	63
5480	18' deep, 2-1/2 C.Y. bucket	48	43.50	91.50
5660	6' wide, 6' deep, 5/8 C.Y. bucket	11.65	14.30	25.95
5680	8' deep, 3/4 C.Y. bucket	17.30	23	40.30
5700	10' deep, 1 C.Y. bucket	22	31.50	53.50
5720	12' deep, 1-1/4 C.Y. bucket	21.50	27	48.50
5760	16' deep, 2 C.Y. bucket	44.50	38.50	83
5800	20' deep, 3-1/2 C.Y. bucket	56.50	85.50	142
5840	24' deep, 3-1/2 C.Y. bucket	107	166	273
6020	8' wide, 12' deep, 1-1/4 C.Y. bucket	29	31.50	60.50
6050	15' deep, 1-1/2 C.Y. bucket	37.50	45	82.50
6080	18' deep, 2-1/2 C.Y. bucket	61	51	112
6140	24' deep, 3-1/2 C.Y. bucket	117	175	292
6300	10' wide, 20' deep, 3-1/2 C.Y. bucket	71	97	168
6350	24' deep, 3-1/2 C.Y. bucket	125	184	309
6450	12' wide, 20' deep, 3-1/2 C.Y. bucket	78.50	103	181.50
6480	25' deep, 3-1/2 C.Y. bucket	152	222	374
6600	2 to 1 slope, 2' wide, 2' deep, 3/8 C.Y. bucket	2.57	2.38	4.95
6620	3' deep, 3/8 C.Y. bucket	3.35	5.10	8.45
6640	4' deep, 3/8 C.Y. bucket	4.22	7.65	11.87
6660	6' deep, 3/8 C.Y. bucket	5.85	11.60	17.45
6680	8' deep, 1/2 C.Y. bucket	9	19.50	28.50
6700	10' deep, 1 C.Y. bucket	17.20	28	45.20
6900	4' wide, 2' deep, 3/8 C.Y. bucket	2.52	3.57	6.09
6920	3' deep, 3/8 C.Y. bucket	4.69	6.65	11.34
6940	4' deep, 1/2 C.Y. bucket	5.90	7.65	13.55
6960	6' deep, 1/2 C.Y. bucket	8.45	13.95	22.40
6980	8' deep, 1/2 C.Y. bucket	14.95	21.50	36.45
7000	10' deep, 1 C.Y. bucket	20.50	31	51.50
7020	12' deep, 1 C.Y. bucket	29.50	46.50	76
7050	15' deep, 1-1/2 C.Y. bucket	29	41.50	70.50
7080	18' deep, 2-1/2 C.Y. bucket	54.50	49.50	104
7260	6' wide, 6' deep, 5/8 C.Y. bucket	12.85	15.25	28.10
7280	8' deep, 3/4 C.Y. bucket	19	24.50	43.50
7300	10' deep, 3/4 C.Y. bucket	25	35	60
7320	12' deep, 1-1/4 C.Y. bucket	24	30.50	54.50
7360	16' deep, 2 C.Y. bucket	49.50	43	92.50
7400	20' deep, 3-1/2 C.Y. bucket	63.50	95.50	159
7440	24' deep, 3-1/2 C.Y. bucket	120	187	307
7620	8' wide, 12' deep, 1-1/4 C.Y. bucket	32.50	34.50	67
7650	15' deep, 1-1/2 C.Y. bucket	42	49.50	91.50
7680	18' deep, 2-1/2 C.Y. bucket	68	56.50	124.50
7740	24' deep, 3-1/2 C.Y. bucket	131	195	326
7920	10' wide, 20' deep, 3-1/2 C.Y. bucket	79	106	185
7940	24' deep, 3-1/2 C.Y. bucket	139	203	342
8060	12' wide, 20' deep, 3-1/2 C.Y. bucket	87	112	199
8080	25' deep, 3-1/2 C.Y. bucket	168	243	411

Important: See the Reference Section for critical supporting data - Reference Nos., Crews, & City Cost Indexes

SITE WORK — A12.3-310 Pipe Bedding

The Pipe Bedding System is shown for various pipe diameters. Compacted bank sand is used for pipe bedding and to fill 12" over the pipe. No backfill is included. Various side slopes are shown to accommodate different soil conditions. Pipe sizes vary from 6" to 84" diameter.

System Components	QUANTITY	UNIT	COST PER L.F. MAT.	COST PER L.F. INST.	COST PER L.F. TOTAL
SYSTEM 12.3-310-1440					
PIPE BEDDING, SIDE SLOPE 0 TO 1, 1' WIDE, PIPE SIZE 6" DIAMETER					
Borrow, bank sand, 2 mile haul, machine spread	.067	C.Y.	.26	.33	.59
Compaction, vibrating plate	.067	C.Y.		.12	.12
TOTAL			.26	.45	.71

12.3-310	Pipe Bedding	MAT.	INST.	TOTAL
1440	Pipe bedding, side slope 0 to 1, 1' wide, pipe size 6" diameter	.26	.45	.71
1460	2' wide, pipe size 8" diameter	.58	.99	1.57
1480	Pipe size 10" diameter	.59	1.02	1.61
1500	Pipe size 12" diameter	.61	1.06	1.67
1520	3' wide, pipe size 14" diameter	1	1.73	2.73
1540	Pipe size 15" diameter	1.02	1.75	2.77
1560	Pipe size 16" diameter	1.03	1.77	2.80
1580	Pipe size 18" diameter	1.05	1.81	2.86
1600	4' wide, pipe size 20" diameter	1.52	2.61	4.13
1620	Pipe size 21" diameter	1.53	2.64	4.17
1640	Pipe size 24" diameter	1.57	2.70	4.27
1660	Pipe size 30" diameter	1.61	2.77	4.38
1680	6' wide, pipe size 32" diameter	2.81	4.85	7.66
1700	Pipe size 36" diameter	2.89	4.98	7.87
1720	7' wide, pipe size 48" diameter	3.76	6.50	10.26
1740	8' wide, pipe size 60" diameter	4.69	8.05	12.74
1760	10' wide, pipe size 72" diameter	6.75	11.65	18.40
1780	12' wide, pipe size 84" diameter	9.20	15.85	25.05
2140	Side slope 1/2 to 1, 1' wide, pipe size 6" diameter	.55	.95	1.50
2160	2' wide, pipe size 8" diameter	.91	1.57	2.48
2180	Pipe size 10" diameter	.99	1.70	2.69
2200	Pipe size 12" diameter	1.06	1.84	2.90
2220	3' wide, pipe size 14" diameter	1.52	2.62	4.14
2240	Pipe size 15" diameter	1.56	2.68	4.24
2260	Pipe size 16" diameter	1.61	2.77	4.38
2280	Pipe size 18" diameter	1.70	2.94	4.64
2300	4' wide, pipe size 20" diameter	2.24	3.87	6.11
2320	Pipe size 21" diameter	2.30	3.96	6.26
2340	Pipe size 24" diameter	2.46	4.25	6.71
2360	Pipe size 30" diameter	2.76	4.76	7.52
2380	6' wide, pipe size 32" diameter	4.07	7	11.07
2400	Pipe size 36" diameter	4.36	7.50	11.86
2420	7' wide, pipe size 48" diameter	5.95	10.25	16.20
2440	8' wide, pipe size 60" diameter	7.75	13.35	21.10
2460	10' wide, pipe size 72" diameter	10.85	18.70	29.55
2480	12' wide, pipe size 84" diameter	14.45	25	39.45
2620	Side slope 1 to 1, 1' wide, pipe size 6" diameter	.84	1.45	2.29
2640	2' wide, pipe size 8" diameter	1.26	2.18	3.44
2660	Pipe size 10" diameter	1.38	2.39	3.77
2680	Pipe size 12" diameter	1.52	2.62	4.14

For expanded coverage of these items see *Means Site Work & Landscape Cost Data 2001*

SITE WORK — A12.3-310 Pipe Bedding

12.3-310 Pipe Bedding

		COST PER L.F.		
		MAT.	INST.	TOTAL
2700	3' wide, pipe size 14" diameter	2.04	3.52	5.56
2720	Pipe size 15" diameter	2.11	3.63	5.74
2740	Pipe size 16" diameter	2.19	3.76	5.95
2760	Pipe size 18" diameter	2.35	4.06	6.41
2780	4' wide, pipe size 20" diameter	2.97	5.10	8.07
2800	Pipe size 21" diameter	3.06	5.25	8.31
2820	Pipe size 24" diameter	3.35	5.75	9.10
2840	Pipe size 30" diameter	3.91	6.75	10.66
2860	6' wide, pipe size 32" diameter	5.35	9.20	14.55
2880	Pipe size 36" diameter	5.80	10.05	15.85
2900	7' wide, pipe size 48" diameter	8.15	14.05	22.20
2920	8' wide, pipe size 60" diameter	10.80	18.65	29.45
2940	10' wide, pipe size 72" diameter	14.90	25.50	40.40
2960	12' wide, pipe size 84" diameter	19.65	34	53.65
3000	Side slope 1-1/2 to 1, 1' wide, pipe size 6" diameter	1.13	1.94	3.07
3020	2' wide, pipe size 8" diameter	1.60	2.77	4.37
3040	Pipe size 10" diameter	1.78	3.06	4.84
3060	Pipe size 12" diameter	1.97	3.39	5.36
3080	3' wide, pipe size 14" diameter	2.55	4.38	6.93
3100	Pipe size 15" diameter	2.65	4.57	7.22
3120	Pipe size 16" diameter	2.76	4.76	7.52
3140	Pipe size 18" diameter	3.01	5.20	8.21
3160	4' wide, pipe size 20" diameter	3.71	6.40	10.11
3180	Pipe size 21" diameter	3.82	6.60	10.42
3200	Pipe size 24" diameter	4.23	7.30	11.53
3220	Pipe size 30" diameter	5.10	8.75	13.85
3240	6' wide, pipe size 32" diameter	6.60	11.35	17.95
3260	Pipe size 36" diameter	7.30	12.55	19.85
3280	7' wide, pipe size 48" diameter	10.35	17.80	28.15
3300	8' wide, pipe size 60" diameter	13.85	24	37.85
3320	10' wide, pipe size 72" diameter	19	33	52
3340	12' wide, pipe size 84" diameter	25	43	68
3400	Side slope 2 to 1, 1' wide, pipe size 6" diameter	1.45	2.49	3.94
3420	2' wide, pipe size 8" diameter	1.94	3.35	5.29
3440	Pipe size 10" diameter	2.17	3.74	5.91
3460	Pipe size 12" diameter	2.42	4.17	6.59
3480	3' wide, pipe size 14" diameter	3.06	5.25	8.31
3500	Pipe size 15" diameter	3.19	5.50	8.69
3520	Pipe size 16" diameter	3.34	5.75	9.09
3540	Pipe size 18" diameter	3.66	6.30	9.96
3560	4' wide, pipe size 20" diameter	4.43	7.65	12.08
3580	Pipe size 21" diameter	4.59	7.90	12.49
3600	Pipe size 24" diameter	5.10	8.80	13.90
3620	Pipe size 30" diameter	6.25	10.75	17
3640	6' wide, pipe size 32" diameter	7.85	13.55	21.40
3660	Pipe size 36" diameter	8.75	15.10	23.85
3680	7' wide, pipe size 48" diameter	12.50	21.50	34
3700	8' wide, pipe size 60" diameter	16.95	29	45.95
3720	10' wide, pipe size 72" diameter	23	39.50	62.50
3740	12' wide, pipe size 84" diameter	30	52	82

SITE WORK

A12.3-510 Drainage & Sewage Piping

12.3-510 Drainage & Sewage Piping

		COST PER L.F.		
		MAT.	INST.	TOTAL
2000	Piping, excavation & backfill excluded, PVC, plain			
2130	4" diameter	1.55	2.59	4.14
2150	6" diameter	2.65	2.78	5.43
2160	8" diameter	4.84	2.90	7.74
2900	Box culvert, precast, 8' long			
3000	6' x 3'	164	23.50	187.50
3020	6' x 7'	253	26.50	279.50
3040	8' x 3'	186	25	211
3060	8' x 8'	310	33	343
3080	10' x 3'	400	30	430
3100	10' x 8'	385	41	426
3120	12' x 3'	315	33	348
3140	12' x 8'	550	49.50	599.50
4000	Concrete, nonreinforced			
4150	6" diameter	3.83	7.55	11.38
4160	8" diameter	4.21	8.95	13.16
4170	10" diameter	4.67	9.25	13.92
4180	12" diameter	5.75	10	15.75
4200	15" diameter	6.70	11.10	17.80
4220	18" diameter	8.25	13.95	22.20
4250	24" diameter	12.45	20	32.45
4400	Reinforced, no gasket			
4580	12" diameter	10.75	9.55	20.30
4600	15" diameter	13.30	11.45	24.75
4620	18" diameter	13.40	15.45	28.85
4650	24" diameter	19.60	20	39.60
4670	30" diameter	27	32.50	59.50
4680	36" diameter	38.50	39.50	78
4690	42" diameter	52.50	44	96.50
4700	48" diameter	65.50	49	114.50
4720	60" diameter	105	65	170
4730	72" diameter	148	78.50	226.50
4740	84" diameter	269	98	367
4800	With gasket			
4980	12" diameter	11.90	7.85	19.75
5000	15" diameter	14.25	8.25	22.50
5020	18" diameter	17.85	8.70	26.55
5050	24" diameter	27	9.70	36.70
5070	30" diameter	35.50	32.50	68
5080	36" diameter	53.50	39.50	93
5090	42" diameter	70	39.50	109.50
5100	48" diameter	87.50	49	136.50
5120	60" diameter	118	63	181
5130	72" diameter	226	78.50	304.50
5140	84" diameter	325	131	456
5700	Corrugated metal, alum. or galv. bit. coated			
5760	8" diameter	6.45	6.05	12.50
5770	10" diameter	7.70	7.70	15.40
5780	12" diameter	9.30	9.55	18.85
5800	15" diameter	11.25	10	21.25
5820	18" diameter	14.50	10.55	25.05
5850	24" diameter	17.75	12.50	30.25
5870	30" diameter	23	23.50	46.50
5880	36" diameter	34	23.50	57.50
5900	48" diameter	52	28.50	80.50
5920	60" diameter	67	41.50	108.50
5930	72" diameter	102	69.50	171.50
6000	Plain			
6060	8" diameter	5.05	5.65	10.70
6070	10" diameter	5.60	7.15	12.75

For expanded coverage of these items see *Means Site Work & Landscape Cost Data 2001*

SITE WORK — A12.3-510 Drainage & Sewage Piping

12.3-510 Drainage & Sewage Piping

			COST PER L.F.		
			MAT.	INST.	TOTAL
6080		12" diameter	6.35	9.10	15.45
6100		15" diameter	8.40	9.10	17.50
6120		18" diameter	11.30	9.75	21.05
6140		24" diameter	15.50	11.45	26.95
6170		30" diameter	19.70	22	41.70
6180		36" diameter	32.50	22	54.50
6200		48" diameter	43.50	26	69.50
6220		60" diameter	68	40	108
6230		72" diameter	101	93.50	194.50
6300	Steel or alum. oval arch, coated & paved invert				
6400		15" equivalent diam	19.10	10	29.10
6420		18" equivalent diam	24.50	13.40	37.90
6450		24" equivalent diam	35.50	16.05	51.55
6470		30" equivalent diam	43.50	20	63.50
6480		36" equivalent diam	64	28.50	92.50
6490		42" equivalent diam	77.50	31.50	109
6500		48" equivalent diam	87.50	38	125.50
6600	Plain				
6700		15" equivalent diam	10.50	8.90	19.40
6720		18" equivalent diam	12.45	11.45	23.90
6750		24" equivalent diam	20	13.40	33.40
6770		30" equivalent diam	25	26.50	51.50
6780		36" equivalent diam	41.50	26.50	68
6790		42" equivalent diam	49	31	80
6800		48" equivalent diam	43	38	81
8000	Polyvinyl chloride SDR 35				
8130		4" diameter	1.55	2.59	4.14
8150		6" diameter	2.65	2.78	5.43
8160		8" diameter	4.84	2.90	7.74
8170		10" diameter	7.20	3.99	11.19
8180		12" diameter	8.10	4.12	12.22
8200		15" diameter	12.30	6.90	19.20
9000	Vitrified clay, C-2000, premium joint				
9030		4" diameter	1.69	3.67	5.36
9050		6" diameter	2.87	4.86	7.73
9060		8" diameter	4.07	6.60	10.67
9070		10" diameter	6.80	6.90	13.70
9080		12" diameter	9.10	8.80	17.90
9100		15" diameter	15.90	11.95	27.85
9120		18" diameter	23.50	15	38.50
9150		24" diameter	43	29.50	72.50
9170		30" diameter	80	48	128
9180		36" diameter	125	74.50	199.50

Important: See the Reference Section for critical supporting data - Reference Nos., Crews, & City Cost Indexes

SITE WORK — A12.3-520 Gas Service Piping

12.3-520 Gas Service Piping

		COST PER L.F.		
		MAT.	INST.	TOTAL
2000	Piping, excavation & backfill excluded, polyethylene			
2070	1-1/4" diam, SDR 10	.71	2.47	3.18
2090	2" diam, SDR 11	.87	2.76	3.63
2110	3" diam, SDR 11.5	1.85	3.30	5.15
2130	4" diam, SDR 11	4.22	6.25	10.47
2150	6" diam, SDR 21	13.15	6.70	19.85
2160	8" diam, SDR 21	17.95	8.05	26
3000	Steel, schedule 40, plain end, tarred & wrapped			
3060	1" diameter	2.33	5.65	7.98
3090	2" diameter	3.66	6.05	9.71
3110	3" diameter	6.05	6.55	12.60
3130	4" diameter	7.90	10.35	18.25
3140	5" diameter	11.45	12	23.45
3150	6" diameter	14	14.70	28.70
3160	8" diameter	22	18.90	40.90

SITE WORK

A12.3-530 Subdrainage Piping

12.3-530	Subdrainage Piping	COST PER L.F.		
		MAT.	INST.	TOTAL
2000	Piping, excavation & backfill excluded, PVC, perforated			
2110	3" diameter	1.55	2.59	4.14
2130	4" diameter	1.55	2.59	4.14
2140	5" diameter	2.65	2.78	5.43
2150	6" diameter	2.65	2.78	5.43
3000	Metal alum. or steel, perforated asphalt coated			
3150	6" diameter	2.88	5.25	8.13
3160	8" diameter	4.19	5.40	9.59
3170	10" diameter	5.25	5.55	10.80
3180	12" diameter	5.85	7.05	12.90
3220	18" diameter	8.80	9.75	18.55
4000	Porous wall concrete			
4130	4" diameter	1.96	2.90	4.86
4150	6" diameter	2.55	3.09	5.64
4160	8" diameter	3.14	4.24	7.38
4180	12" diameter	6.65	4.62	11.27
4200	15" diameter	7.65	5.75	13.40
4220	18" diameter	10.10	8	18.10
5000	Vitrified clay C-211, perforated			
5130	4" diameter	2.10	5	7.10
5150	6" diameter	3.58	6.40	9.98
5160	8" diameter	5.10	6.90	12
5180	12" diameter	10.55	7.25	17.80

SITE WORK

A12.3-540 Water Distribution Piping

12.3-540 Water Distribution Piping

		COST PER L.F.		
		MAT.	INST.	TOTAL
2000	Piping. excav. & backfill excl., ductile iron class 250, mech. joint			
2130	4" diameter	9.65	6.75	16.40
2150	6" diameter	10	7.70	17.70
2160	8" diameter	12.85	9	21.85
2170	10" diameter	16.65	10.80	27.45
2180	12" diameter	21.50	18.25	39.75
2210	16" diameter	28.50	28.50	57
2220	18" diameter	37.50	31	68.50
3000	Tyton joint			
3130	4" diameter	7.10	6.25	13.35
3150	6" diameter	8.15	7.20	15.35
3160	8" diameter	11.55	8.45	20
3170	10" diameter	17.55	9.90	27.45
3180	12" diameter	18.50	12.45	30.95
3210	16" diameter	28.50	25.50	54
3220	18" diameter	32	30.50	62.50
3230	20" diameter	35	32.50	67.50
3250	24" diameter	45	33	78
4000	Copper tubing, type K			
4050	3/4" diameter	1.45	1.87	3.32
4060	1" diameter	1.85	2.34	4.19
4080	1-1/2" diameter	2.97	2.83	5.80
4090	2" diameter	4.74	3.26	8
4110	3" diameter	9.50	5.60	15.10
4130	4" diameter	14.80	7.90	22.70
4150	6" diameter	46.50	12.25	58.75
5000	Polyvinyl chloride class 160, S.D.R. 26			
5130	1-1/2" diameter	.65	3.24	3.89
5150	2" diameter	.98	3.89	4.87
5160	4" diameter	3.40	4.86	8.26
5170	6" diameter	7.35	5.40	12.75
5180	8" diameter	12.40	8.25	20.65
6000	Polyethylene 160 psi, S.D.R. 7			
6050	3/4" diameter	.23	1.85	2.08
6060	1" diameter	.39	2	2.39
6080	1-1/2" diameter	.89	2.16	3.05
6090	2" diameter	1.52	2.66	4.18

SITE WORK — A12.3-710 Manholes & Catch Basins

Manhole | Catch Basin

The Manhole and Catch Basin System includes: excavation with a backhoe; a formed concrete footing; frame and cover; cast iron steps and compacted backfill.

The Expanded System Listing shows manholes that have a 4', 5' and 6' inside diameter riser. Depths range from 4' to 14'. Construction material shown is either concrete, concrete block, precast concrete, or brick.

System Components	QUANTITY	UNIT	COST PER EACH		
			MAT.	INST.	TOTAL
SYSTEM 12.3-710-1920					
MANHOLE/CATCH BASIN, BRICK, 4' I.D. RISER, 4' DEEP					
Excavation, hydraulic backhoe, 3/8 C.Y. bucket	14.815	C.Y.		75.26	75.26
Trim sides and bottom of excavation	64.000	S.F.		40.32	40.32
Forms in place, manhole base, 4 uses	20.000	SFCA	13.20	66.60	79.80
Reinforcing in place footings, #4 to #7	.019	Ton	11.02	16.06	27.08
Concrete, 3000 psi, incl. place and vibrate under 1 CY, direct chute	.925	C.Y.	65.68	31.47	97.15
Catch basin or MH, brick, 4' ID, 4' deep	1.000	Ea.	298	650	948
Catch basin or MH steps; heavy galvanized cast iron	1.000	Ea.	10.40	9.15	19.55
Catch basin or MH frame and cover	1.000	Ea.	174	143.50	317.50
Backfill with wheeled front end loader	12.954	C.Y.	108.17	19.04	127.21
Backfill compaction, 12" lifts, air tamp	12.954	C.Y.		74.88	74.88
TOTAL			680.47	1,126.28	1,806.75

12.3-710	Manholes & Catch Basins	COST PER EACH		
		MAT.	INST.	TOTAL
1920	Manhole/catch basin, brick, 4' I.D. riser, 4' deep	680	1,125	1,805
1940	6' deep	915	1,575	2,490
1960	8' deep	1,175	2,150	3,325
1980	10' deep	1,500	2,650	4,150
3000	12' deep	1,925	2,900	4,825
3020	14' deep	2,425	4,050	6,475
3200	Block, 4' I.D. riser, 4' deep	630	910	1,540
3220	6' deep	825	1,300	2,125
3240	8' deep	1,050	1,775	2,825
3260	10' deep	1,225	2,175	3,400
3280	12' deep	1,525	2,800	4,325
3300	14' deep	1,850	3,400	5,250
4620	Concrete, cast-in-place, 4' I.D. riser, 4' deep	760	1,600	2,360
4640	6' deep	1,050	2,125	3,175
4660	8' deep	1,425	3,075	4,500
4680	10' deep	1,725	3,800	5,525
4700	12' deep	2,100	4,725	6,825
4720	14' deep	2,550	5,650	8,200
5820	Concrete, precast, 4' I.D. riser, 4' deep	715	835	1,550
5840	6' deep	960	1,125	2,085
5860	8' deep	1,150	1,600	2,750
5880	10' deep	1,425	1,950	3,375
5900	12' deep	1,800	2,400	4,200
5920	14' deep	2,200	3,075	5,275

SITE WORK | A12.3-710 Manholes & Catch Basins

12.3-710	Manholes & Catch Basins	COST PER EACH		
		MAT.	INST.	TOTAL
6000	5' I.D. riser, 4' deep	910	910	1,820
6020	6' deep	1,200	1,275	2,475
6040	8' deep	1,525	1,675	3,200
6060	10' deep	1,925	2,150	4,075
6080	12' deep	2,375	2,750	5,125
6100	14' deep	2,875	3,350	6,225
6200	6' I.D. riser, 4' deep	1,300	1,200	2,500
6220	6' deep	1,675	1,600	3,275
6240	8' deep	2,075	2,200	4,275
6260	10' deep	2,625	2,825	5,450
6280	12' deep	3,225	3,525	6,750
6300	14' deep	3,875	4,275	8,150

For information about Means Estimating Seminars, see yellow pages 11 and 12 in back of book

For expanded coverage of these items see *Means Site Work & Landscape Cost Data 2001*

Reference Section

All the reference information is in one section, making it easy to find what you need to know... and easy to use the book on a daily basis. This section is visually identified by a vertical gray bar on the edge of pages.

In the reference number information that follows, you'll see the background that relates to the "reference numbers" that appeared in the Unit Price Section. You'll find reference tables, explanations and estimating information that support how we arrived at the unit price data. Also included are alternate pricing methods, technical data and estimating procedures along with information on design and economy in construction.

Also in this Reference Section, we've included Change Orders, information on pricing changes to contract documents; Crew Listings, a full listing of all the crews, equipment and their costs; Historical Cost Indexes for cost comparisons over time; City Cost Indexes and Location Factors for adjusting costs to the region you are in, and an explanation of all abbreviations used in the book.

Table of Contents

Reference Numbers

R011 Overhead & Misc. Data
- R01100-005 Tips for Acurate Estimating...... 514
- R01100-040 Builder's Risk Insurance........... 514
- R01100-060 Workers' Compensation........... 515
- R01100-070 Contractor's Overhead & Profit..................................... 518
- R01100-080 Performance Bond 519
- R01100-090 Sales Tax by State 519
- R01100-100 Unemployment & Social Security Taxes 519
- R01100-110 Overtime................................... 520
- R01100-710 Unit Gross Area Require. 520
- R01100-720 Floor Area Ratios 521
- R01100-730 Occupancy Determinations..... 521
- R01107-030 Engineering Fees 522

R012 Price & Payment Procedures
- R01250-010 Repair & Remodeling................ 523

R015 Construction Aids
- R01540-100 Steel Tubular Scaffolding.......... 524
- R01559-100 Contractor Equipment 525

R021 Subsurface Invest. & Demolition
- R02115-200 Underground Storage Tank Removal 526

R022 Site Prep. & Excav. Support
- R02240-900 Wellpoints................................. 526
- R02250-400 Wood Sheet Piling.................... 527
- R02250-450 Steel Sheet Piling...................... 528

R132 Hazardous Material Remediation
- R13281-120 Asbestos Removal Process........ 529

R136 Solar & Wind Energy Equip.
- R13600-610 Solar Heating (Space & H.W.).. 529

R150 Basic Materials & Methods
- R15050-710 Subcontractors 530
- R15050-720 Demolition (Selective vs. Removal for Replacement) 530

R151 Plumbing
- R15100-050 Pipe Material Considerations... 531
- R15100-070 Piping to 10' High 532
- R15100-080 Valve Materials.......................... 533
- R15100-090 Valve Select. Considerations.... 533

Reference Numbers (cont.)

R155 Heating
- R15500-010 Heating Systems & Heat Transmission............................. 536
- R15500-020 Heating Approximations for Quick Estimating...................... 536
- R15500-030 Basics of a Heating System 537
- R15500-050 Factor for Determining Heat Loss 538
- R15500-070 Transmission of Heat............... 539
- R15500-080 Transmission of Heat (Low Rate).............................. 539

R157 Air Conditioning & Ventilation
- R15700-010 Air Conditioning...................... 540
- R15700-020 Air Conditioning Requirements 540
- R15700-030 Psychrometric Table................ 541
- R15700-040 Recommended Ventilation Air Changes 542
- R15700-090 Quality/Complexity Adjust. for A/C Systems........... 542

R158 Air Distribution
- R15800-080 Loudness Levels for Moving Air thru Fans, Diffusers, Registers, etc..................... 543
- R15810-050 Ductwork (Duct Weight in Pounds per L.F. Straight Runs)......................... 544
- R15810-070 Duct Fabrication/Installation ... 544
- R15810-100 Sheet Metal Calculator (Weight in Lb./Ft. of Length)... 545
- R15810-110 Ductwork Packages (per Ton of Cooling) 546
- R15850-060 Diffuser Evaluation.................. 546

R171 Information
- R17100-100 Square Foot Project Size Modifier 547

Change Orders	548
Crew Listings	552
Historical Cost Indexes	577
City Cost Indexes	578
Location Factors	598
Abbreviations	604

General Requirements — R011 — Overhead & Miscellaneous Data

R01100-005 Tips for Accurate Estimating

1. Use pre-printed or columnar forms for orderly sequence of dimensions and locations and for recording telephone quotations.
2. Use only the front side of each paper or form except for certain pre-printed summary forms.
3. Be consistent in listing dimensions: For example, length x width x height. This helps in rechecking to ensure that, the total length of partitions is appropriate for the building area.
4. Use printed (rather than measured) dimensions where given.
5. Add up multiple printed dimensions for a single entry where possible.
6. Measure all other dimensions carefully.
7. Use each set of dimensions to calculate multiple related quantities.
8. Convert foot and inch measurements to decimal feet when listing. Memorize decimal equivalents to .01 parts of a foot (1/8" equals approximately .01').
9. Do not "round off" quantities until the final summary.
10. Mark drawings with different colors as items are taken off.
11. Keep similar items together, different items separate.
12. Identify location and drawing numbers to aid in future checking for completeness.
13. Measure or list everything on the drawings or mentioned in the specifications.
14. It may be necessary to list items not called for to make the job complete.
15. Be alert for: Notes on plans such as N.T.S. (not to scale); changes in scale throughout the drawings; reduced size drawings; discrepancies between the specifications and the drawings.
16. Develop a consistent pattern of performing an estimate. For example:
    a. Start the quantity takeoff at the lower floor and move to the next higher floor.
    b. Proceed from the main section of the building to the wings.
    c. Proceed from south to north or vice versa, clockwise or counterclockwise.
    d. Take off floor plan quantities first, elevations next, then detail drawings.
17. List all gross dimensions that can be either used again for different quantities, or used as a rough check of other quantities for verification (exterior perimeter, gross floor area, individual floor areas, etc.).
18. Utilize design symmetry or repetition (repetitive floors, repetitive wings, symmetrical design around a center line, similar room layouts, etc.). Note: Extreme caution is needed here so as not to omit or duplicate an area.
19. Do not convert units until the final total is obtained. For instance, when estimating concrete work, keep all units to the nearest cubic foot, then summarize and convert to cubic yards.
20. When figuring alternatives, it is best to total all items involved in the basic system, then total all items involved in the alternates. Therefore you work with positive numbers in all cases. When adds and deducts are used, it is often confusing whether to add or subtract a portion of an item; especially on a complicated or involved alternate.

R01100-040 Builder's Risk Insurance

Builder's Risk Insurance is insurance on a building during construction. Premiums are paid by the owner or the contractor. Blasting, collapse and underground insurance would raise total insurance costs above those listed. Floater policy for materials delivered to the job runs $.75 to $1.25 per $100 value. Contractor equipment insurance runs $.50 to $1.50 per $100 value. Insurance for miscellaneous tools to $1,500 value runs from $3.00 to $7.50 per $100 value.

Tabulated below are New England Builder's Risk insurance rates in dollars per $100 value for $1,000 deductible. For $25,000 deductible, rates can be reduced 13% to 34%. On contracts over $1,000,000, rates may be lower than those tabulated. Policies are written annually for the total completed value in place. For "all risk" insurance (excluding flood, earthquake and certain other perils) add $.025 to total rates below.

Coverage	Frame Construction (Class 1) Range	Frame Construction (Class 1) Average	Brick Construction (Class 4) Range	Brick Construction (Class 4) Average	Fire Resistive (Class 6) Range	Fire Resistive (Class 6) Average
Fire Insurance	$.300 to $.420	$.394	$.132 to $.189	$.174	$.052 to $.080	$.070
Extended Coverage	.115 to .150	.144	.080 to .105	.101	.081 to .105	.100
Vandalism	.012 to .016	.015	.008 to .011	.011	.008 to .011	.010
Total Annual Rate	$.427 to $.586	$.553	$.220 to $.305	$.286	$.141 to $.196	$.180

General Requirements — R011 Overhead & Miscellaneous Data

R01100-060 Workers' Compensation Insurance Rates by Trade

The table below tabulates the national averages for Workers' Compensation insurance rates by trade and type of building. The average "Insurance Rate" is multiplied by the "% of Building Cost" for each trade. This produces the "Workers' Compensation Cost" by % of total labor cost, to be added for each trade by building type to determine the weighted average Workers' Compensation rate for the building types analyzed.

Trade	Insurance Rate (% Labor Cost) Range	Insurance Rate (% Labor Cost) Average	% of Building Cost Office Bldgs.	% of Building Cost Schools & Apts.	% of Building Cost Mfg.	Workers' Compensation Office Bldgs.	Workers' Compensation Schools & Apts.	Workers' Compensation Mfg.
Excavation, Grading, etc.	3.8% to 26.6%	10.9%	4.8%	4.9%	4.5%	.52%	.53%	.49%
Piles & Foundations	5.9 to 80.1	27.7	7.1	5.2	8.7	1.97	1.44	2.41
Concrete	5.9 to 35.5	17.8	5.0	14.8	3.7	.89	2.63	.66
Masonry	5.3 to 48.6	17.0	6.9	7.5	1.9	1.17	1.28	.32
Structural Steel	5.9 to 132.9	40.9	10.7	3.9	17.6	4.38	1.60	7.20
Miscellaneous & Ornamental Metals	4.9 to 34.0	13.2	2.8	4.0	3.6	.37	.53	.48
Carpentry & Millwork	5.9 to 49.9	19.0	3.7	4.0	0.5	.70	.76	.10
Metal or Composition Siding	5.9 to 36.7	17.6	2.3	0.3	4.3	.40	.05	.76
Roofing	5.9 to 88.8	34.3	2.3	2.6	3.1	.79	.89	1.06
Doors & Hardware	4.1 to 28.8	11.4	0.9	1.4	0.4	.10	.16	.05
Sash & Glazing	5.7 to 32.4	14.1	3.5	4.0	1.0	.49	.56	.14
Lath & Plaster	5.2 to 44.6	15.7	3.3	6.9	0.8	.52	1.08	.13
Tile, Marble & Floors	3.7 to 29.5	10.3	2.6	3.0	0.5	.27	.31	.05
Acoustical Ceilings	3.5 to 26.7	11.8	2.4	0.2	0.3	.28	.02	.04
Painting	5.9 to 41.1	14.8	1.5	1.6	1.6	.22	.24	.24
Interior Partitions	5.9 to 49.9	19.0	3.9	4.3	4.4	.74	.82	.84
Miscellaneous Items	2.6 to 128.0	18.5	5.2	3.7	9.7	.96	.68	1.79
Elevators	2.0 to 18.9	8.4	2.1	1.1	2.2	.18	.09	.18
Sprinklers	3.0 to 20.6	8.8	0.5	—	2.0	.04	—	.18
Plumbing	3.0 to 15.9	8.5	4.9	7.2	5.2	.42	.61	.44
Heat., Vent., Air Conditioning	3.5 to 32.5	11.9	13.5	11.0	12.9	1.61	1.31	1.54
Electrical	2.7 to 11.0	6.8	10.1	8.4	11.1	.69	.57	.75
Total	2.0% to 132.9%	—	100.0%	100.0%	100.0%	17.71%	16.16%	19.85%

Overall Weighted Average 17.91%

Workers' Compensation Insurance Rates by States

The table below lists the weighted average Workers' Compensation base rate for each state with a factor comparing this with the national average of 17.5%.

State	Weighted Average	Factor	State	Weighted Average	Factor	State	Weighted Average	Factor
Alabama	30.5%	174	Kentucky	19.8%	113	North Dakota	15.0%	86
Alaska	11.1	63	Louisiana	27.3	156	Ohio	19.1	109
Arizona	12.5	71	Maine	21.8	125	Oklahoma	21.9	125
Arkansas	14.2	81	Maryland	11.9	68	Oregon	19.2	110
California	18.6	106	Massachusetts	17.5	100	Pennsylvania	22.9	131
Colorado	26.5	151	Michigan	20.6	118	Rhode Island	20.9	119
Connecticut	19.4	111	Minnesota	37.7	215	South Carolina	13.8	79
Delaware	12.3	70	Mississippi	22.8	130	South Dakota	14.6	83
District of Columbia	21.4	122	Missouri	15.8	90	Tennessee	13.0	74
Florida	28.0	160	Montana	37.3	213	Texas	16.4	94
Georgia	24.9	142	Nebraska	15.7	90	Utah	12.9	74
Hawaii	16.7	95	Nevada	19.4	111	Vermont	14.9	85
Idaho	10.8	62	New Hampshire	23.2	133	Virginia	10.9	62
Illinois	23.3	133	New Jersey	11.0	63	Washington	10.0	57
Indiana	7.2	41	New Mexico	23.1	132	West Virginia	14.3	82
Iowa	12.1	69	New York	16.4	94	Wisconsin	13.7	78
Kansas	9.6	55	North Carolina	13.3	76	Wyoming	6.5	37

Weighted Average for U.S. is 17.9% of payroll = 100%

Rates in the following table are the base or manual costs per $100 of payroll for Workers' Compensation in each state. Rates are usually applied to straight time wages only and not to premium time wages and bonuses.

The weighted average skilled worker rate for 35 trades is 17.5%. For bidding purposes, apply the full value of Workers' Compensation directly to total labor costs, or if labor is 38%, materials 42% and overhead and profit 20% of total cost, carry 38/80 x 17.5% =8.3% of cost (before overhead and profit) into overhead. Rates vary not only from state to state but also with the experience rating of the contractor.

Rates are the most current available at the time of publication.

General Requirements — R011 Overhead & Miscellaneous Data

R01100-060 Workers' Compensation Insurance Rates by Trade and State (cont.)

State	Carpentry — 3 stories or less	Carpentry — interior cab. work	Carpentry — general	Concrete Work — NOC	Concrete Work — flat (flr., sdwk.)	Electrical Wiring — inside	Excavation — earth NOC	Excavation — rock	Glaziers	Insulation Work	Lathing	Masonry	Painting & Decorating	Pile Driving	Plastering	Plumbing	Roofing	Sheet Metal Work (HVAC)	Steel Erection — door & sash	Steel Erection — inter., ornam.	Steel Erection — structure	Steel Erection — NOC	Tile Work — (interior ceramic)	Waterproofing	Wrecking
	5651	5437	5403	5213	5221	5190	6217	6217	5462	5479	5443	5022	5474	6003	5480	5183	5551	5538	5102	5102	5040	5057	5348	9014	5701
AL	29.00	14.42	30.31	23.87	18.39	10.36	19.14	19.14	32.36	21.66	14.63	25.12	19.63	49.29	44.55	12.07	75.95	30.15	17.28	17.28	52.35	65.34	20.31	7.12	54.68
AK	8.81	8.42	9.30	9.30	5.59	5.67	8.82	8.82	9.10	10.01	5.64	10.99	8.49	29.93	9.97	5.25	17.33	6.01	9.40	9.40	17.20	17.20	6.23	6.38	25.07
AZ	12.94	7.31	23.11	11.90	7.03	5.46	7.53	7.53	11.21	13.57	6.80	11.90	8.89	17.26	10.66	6.75	17.09	8.80	8.71	8.71	31.81	20.24	6.29	4.29	50.31
AR	16.12	10.60	15.38	16.45	5.97	7.93	9.21	9.21	12.58	15.29	12.66	9.32	11.10	19.14	13.01	6.58	30.74	10.28	8.19	8.19	27.11	25.39	7.80	4.58	27.11
CA	28.28	9.07	28.28	13.18	13.18	9.93	7.88	7.88	16.37	23.34	10.90	14.55	19.45	21.18	18.12	11.59	39.79	15.33	13.55	13.55	23.93	21.91	7.74	19.45	21.91
CO	32.22	13.54	20.95	20.96	15.13	9.20	15.99	15.99	15.98	21.31	14.68	31.06	21.85	41.53	37.40	14.50	59.10	12.54	11.83	11.83	73.46	43.49	15.04	12.26	73.46
CT	22.88	13.79	21.35	15.44	13.85	5.94	8.19	8.19	21.67	26.13	8.60	23.82	16.74	20.45	16.66	9.16	29.83	9.89	11.52	11.52	66.97	27.81	11.14	5.71	50.86
DE	13.83	13.83	11.56	9.28	6.29	5.82	8.54	8.54	11.18	11.56	10.51	9.85	13.41	11.64	10.51	5.41	22.89	10.29	10.21	10.21	26.69	10.21	7.54	9.85	25.69
DC	16.64	7.93	16.96	29.16	16.85	7.03	12.07	12.07	21.57	18.23	11.28	25.62	14.04	32.48	18.37	10.51	23.16	11.62	24.74	24.74	60.38	32.31	17.92	4.39	66.43
FL	36.73	23.28	29.09	29.92	16.15	10.51	15.27	15.27	17.81	25.18	24.27	28.72	28.92	49.35	28.79	13.37	53.20	17.12	18.67	18.67	43.71	50.61	12.56	7.89	43.71
GA	31.04	16.02	32.71	20.67	15.12	9.82	19.55	19.55	21.97	19.15	26.66	23.03	21.20	40.88	19.50	11.14	40.72	18.65	13.74	13.74	36.03	53.09	13.08	10.16	36.03
HI	17.38	11.62	28.20	13.74	12.27	7.10	7.73	7.73	20.55	21.19	10.94	17.87	11.33	20.15	15.61	6.06	33.24	8.02	11.11	11.11	31.71	21.70	9.78	11.54	31.71
ID	8.79	6.95	14.19	10.28	10.45	4.56	6.41	6.41	6.96	10.12	7.12	7.92	7.51	14.30	8.97	3.95	20.55	5.87	8.05	8.05	27.60	20.85	4.78	6.59	26.68
IL	20.76	12.31	21.64	35.48	12.83	8.86	8.69	8.69	23.09	24.90	14.24	20.34	13.15	37.63	13.78	11.68	35.88	14.58	21.53	21.53	58.63	52.47	13.87	6.26	58.63
IN	7.08	4.06	7.07	6.68	3.23	3.00	3.79	3.79	5.72	7.01	3.48	5.32	5.94	9.68	5.16	3.00	11.87	4.29	4.92	4.92	28.84	13.32	3.69	3.26	27.19
IA	10.22	5.19	12.01	9.18	6.42	3.84	5.56	5.56	11.62	16.49	5.95	10.93	7.58	16.30	9.26	5.01	17.38	6.09	9.80	9.80	45.02	27.20	5.60	4.16	45.02
KS	12.10	7.50	9.17	10.12	6.55	4.01	5.02	5.02	7.59	13.37	5.73	9.94	7.28	12.42	8.78	4.96	20.54	5.93	5.93	5.93	19.14	17.05	4.16	3.66	16.49
KY	16.47	15.26	24.34	22.12	9.53	8.95	13.67	13.67	15.96	14.68	10.62	22.00	19.83	30.44	13.94	10.38	32.83	15.92	13.21	13.21	45.21	33.73	10.60	8.85	45.21
LA	29.63	22.95	26.01	20.86	14.80	10.48	24.73	24.73	23.88	20.82	14.00	21.27	27.63	54.39	21.42	13.78	49.11	21.35	15.51	15.51	66.41	36.94	12.82	10.26	66.41
ME	13.43	10.42	43.59	24.29	11.12	10.02	14.75	14.75	13.35	15.39	14.03	17.26	16.07	31.44	17.95	11.12	32.07	17.70	15.08	15.08	37.84	61.81	11.66	9.37	37.84
MD	10.55	5.95	10.55	11.35	5.05	5.15	9.25	9.25	13.20	13.05	6.45	11.35	6.75	27.45	6.65	5.55	22.80	7.00	9.15	9.15	26.80	18.50	6.35	3.50	26.80
MA	11.05	7.61	13.61	22.12	9.81	3.69	5.91	5.91	9.92	18.24	8.84	17.55	9.04	17.59	7.39	5.69	67.48	7.65	11.96	11.96	50.78	46.56	10.48	3.89	34.86
MI	16.08	9.61	17.85	24.50	12.37	5.71	10.99	10.99	9.62	20.95	11.61	19.18	17.86	64.20	18.33	8.19	43.08	10.91	14.52	14.52	36.41	31.20	13.08	10.42	41.58
MN	27.66	28.76	49.94	30.20	22.59	8.28	19.73	19.73	30.01	37.20	23.93	29.89	19.91	80.08	23.93	15.88	88.83	13.91	27.21	27.21	132.92	31.83	29.54	8.71	132.92
MS	24.99	20.10	26.38	17.69	10.13	9.84	15.53	15.53	18.15	21.31	11.97	19.52	16.75	55.90	22.89	10.23	33.43	21.81	12.10	12.10	46.09	36.57	12.26	9.38	37.42
MO	18.72	7.52	11.84	13.28	10.66	5.79	11.05	11.05	8.97	19.99	9.67	13.57	11.77	16.88	14.22	7.10	25.74	9.76	13.82	13.82	47.98	31.96	5.83	5.71	47.98
MT	30.63	15.19	47.53	32.78	24.74	11.01	26.63	26.63	25.04	38.54	19.54	48.55	41.06	53.68	27.00	15.03	85.06	16.93	33.95	33.95	86.98	55.25	13.30	12.10	86.98
NE	16.19	9.23	15.94	17.31	12.73	5.43	10.19	10.19	12.36	16.79	9.34	18.07	11.58	22.01	14.62	9.06	33.91	13.86	10.75	10.75	25.03	26.96	7.88	5.47	25.03
NV	19.86	10.88	14.94	14.89	11.46	9.64	9.26	9.26	13.64	22.78	14.60	14.48	16.01	27.25	17.93	10.76	33.53	18.03	16.53	16.53	45.36	32.56	18.03	7.43	45.36
NH	16.38	9.82	19.68	28.99	9.29	7.10	13.87	13.87	12.87	43.98	10.90	25.97	17.94	28.09	17.33	9.74	58.76	12.48	13.27	13.27	55.90	49.67	12.82	7.37	55.90
NJ	10.40	8.85	10.40	10.83	8.37	4.23	7.95	7.95	7.36	12.78	7.88	10.77	11.49	11.32	7.88	5.48	29.41	6.20	8.03	8.03	22.52	12.96	5.74	5.37	29.92
NM	25.35	13.09	20.45	19.66	13.39	6.34	10.95	10.95	14.58	31.49	16.25	26.97	16.87	33.27	15.64	12.10	43.41	16.46	21.26	21.26	67.82	29.91	9.67	10.65	67.82
NY	15.13	6.82	14.34	22.04	15.14	6.77	8.75	8.75	16.13	11.09	14.58	21.00	13.37	29.90	11.57	8.49	32.78	15.10	14.46	14.46	21.93	20.26	12.11	5.85	32.53
NC	13.08	11.23	15.08	16.19	6.34	7.36	9.18	9.18	8.25	12.80	6.49	11.00	10.42	18.01	15.63	6.06	29.08	10.36	10.19	10.19	29.07	15.79	7.44	3.74	39.03
ND	12.70	12.70	12.70	9.80	9.80	5.57	9.09	9.06	8.83	12.70	11.94	10.99	9.02	25.96	11.94	7.26	28.69	7.26	12.70	12.70	25.96	25.96	8.80	28.69	20.13
OH	17.06	17.06	17.06	15.16	15.16	9.12	15.16	15.16	22.67	17.90	17.90	20.44	22.67	15.16	17.90	10.47	33.45	32.47	15.16	15.16	15.16	15.16	13.22	15.16	15.16
OK	26.16	11.79	19.33	15.51	11.50	7.41	18.45	18.45	13.55	18.00	12.05	16.80	17.10	36.26	18.29	8.72	42.26	12.45	11.17	11.17	69.45	46.76	11.42	8.20	69.45
OR	29.51	9.28	20.03	16.50	10.61	6.63	12.94	12.94	14.95	19.68	12.27	12.43	19.83	22.05	13.85	8.38	38.11	11.85	11.01	11.01	61.69	25.54	9.92	12.10	61.69
PA	15.80	15.80	19.85	28.18	11.84	7.80	12.74	12.74	15.69	19.85	18.94	18.96	22.05	32.12	18.94	11.57	43.30	12.58	24.71	24.71	59.05	24.71	11.83	18.96	81.34
RI	19.53	11.65	18.07	18.23	16.24	4.43	10.38	10.38	12.85	22.78	11.97	25.11	24.13	37.66	17.25	8.52	33.92	10.29	14.07	14.07	59.49	37.50	14.36	7.70	78.79
SC	19.10	13.03	19.24	12.88	6.19	7.02	8.32	8.32	12.22	10.69	7.27	9.55	11.19	16.64	18.89	6.88	28.98	11.42	9.49	9.49	21.80	23.66	6.23	4.65	21.80
SD	13.70	6.67	17.59	16.61	6.07	5.60	9.23	9.23	11.50	15.30	7.77	10.02	12.97	31.49	12.79	8.27	20.70	7.55	9.64	9.64	45.63	21.23	6.90	4.75	46.57
TN	13.22	9.52	11.99	12.39	7.53	5.00	8.68	8.68	8.96	15.04	8.51	12.73	10.88	15.62	12.19	5.91	24.81	8.76	7.36	7.36	45.06	13.95	5.89	4.98	45.06
TX	16.89	12.49	16.89	18.90	10.64	8.69	11.28	11.28	11.91	14.50	12.98	16.51	11.74	22.91	20.99	8.82	27.56	15.13	11.85	11.85	27.79	21.44	9.25	7.91	28.44
UT	11.17	11.17	11.17	17.68	7.99	7.05	6.38	6.38	9.55	10.95	12.32	13.75	15.29	17.24	10.60	6.41	26.27	6.30	8.99	8.99	23.51	23.51	6.06	5.83	26.53
VT	11.93	7.61	14.88	34.92	9.35	4.51	7.91	7.91	11.95	11.32	8.84	13.76	7.82	19.81	10.86	6.71	24.32	10.29	10.29	10.29	32.12	35.79	6.45	8.56	32.12
VA	10.00	6.46	8.65	9.29	5.47	3.45	6.13	6.13	10.17	8.77	7.59	8.07	8.11	24.38	8.20	5.60	17.80	6.74	10.42	10.42	19.28	30.25	6.78	2.57	19.28
WA	7.57	7.57	7.57	6.22	6.22	2.71	7.32	7.32	9.42	7.30	7.57	9.87	8.81	14.33	10.96	4.47	19.48	3.47	12.96	12.96	12.96	12.96	8.93	10.23	12.96
WV	14.98	14.98	14.98	26.36	26.36	5.02	9.10	9.10	6.96	6.96	16.88	14.51	16.81	10.03	16.81	6.65	12.47	6.96	14.98	14.98	11.89	15.59	16.02	4.67	13.27
WI	10.31	8.30	19.64	9.42	8.44	5.53	6.39	6.39	10.37	12.12	13.99	15.11	10.03	18.39	11.59	5.53	32.21	8.32	12.89	12.89	32.11	16.18	10.39	3.63	36.51
WY	5.94	5.94	5.94	5.94	5.94	5.94	5.94	5.94	5.94	5.94	5.94	5.94	5.94	5.94	5.94	5.94	5.94	5.94	5.94	5.94	5.94	5.94	5.94	5.94	5.94
AVG.	17.57	11.40	19.01	17.82	11.14	6.79	10.93	10.93	14.08	17.65	11.76	17.04	14.81	27.68	15.71	8.47	34.33	11.93	13.21	13.21	40.87	29.19	10.30	8.04	42.15

General Requirements — R011 Overhead & Miscellaneous Data

R01100-060 Workers' Compensation (cont.) (Canada in Canadian dollars)

Province		Alberta	British Columbia	Manitoba	Ontario	New Brunswick	Newfndld. & Labrador	Northwest Territories	Nova Scotia	Prince Edward Island	Quebec	Saskatchewan	Yukon
Carpentry—3 stories or less	Rate	2.53	5.20	5.35	6.66	3.76	6.00	2.50	6.96	5.96	13.01	5.38	2.35
	Code	40400	721028	40102	723	422	403	4-41	4022	401	80110	B12-02	4-042
Carpentry—interior cab. work	Rate	1.70	5.44	5.35	6.66	3.85	6.00	2.50	6.96	3.64	13.01	3.86	2.35
	Code	42133	721021	40102	723	427	403	4-41	4022	402	80110	B11-25	4-042
CARPENTRY—general	Rate	2.53	5.20	5.35	6.66	3.76	6.00	2.50	6.96	5.96	13.01	5.38	2.35
	Code	40400	721028	40102	723	422	403	4-41	4022	401	80110	B12-02	4-042
CONCRETE WORK—NOC	Rate	3.63	7.33	6.72	15.22	3.76	6.00	2.50	6.31	5.96	15.31	7.99	3.25
	Code	42104	721010	40110	745	422	403	4-41	4224	401	80100	B14-04	2-032
CONCRETE WORK—flat (flr. sidewalk)	Rate	3.63	7.33	6.72	15.22	3.76	6.00	2.50	6.31	5.96	15.31	7.99	3.25
	Code	42104	721010	40110	745	422	403	4-41	4224	401	80100	B14-04	2-032
ELECTRICAL Wiring—inside	Rate	1.73	3.93	3.50	3.20	1.49	4.83	2.00	2.70	3.64	6.97	3.86	2.35
	Code	42124	721019	40203	704	426	400	4-46	4261	402	80170	B11-05	4-041
EXCAVATION—earth NOC	Rate	2.19	4.03	5.35	4.54	2.72	6.00	2.50	3.64	3.76	7.40	4.69	3.25
	Code	40604	721031	40706	711	421	403	4-43	4214	404	80030	R11-06	2-016
EXCAVATION—rock	Rate	2.19	4.03	5.35	4.54	2.72	6.00	2.50	3.64	3.76	7.40	4.69	3.25
	Code	40604	721031	40706	711	421	403	4-43	4214	404	80030	R11-06	2-016
GLAZIERS	Rate	2.14	3.37	3.50	8.82	3.85	4.55	2.50	9.30	3.64	16.25	7.99	2.35
	Code	42121	715020	40109	751	423	402	4-41	4233	402	80150	B13-04	4-042
INSULATION WORK	Rate	2.53	6.46	5.35	8.82	3.85	4.55	2.50	9.30	5.96	14.15	5.38	3.25
	Code	42184	721029	40102	751	423	402	4-41	4234	401	80120	B12-07	2-035
LATHING	Rate	4.58	8.12	5.35	6.66	3.85	4.55	2.50	6.34	3.64	14.15	7.99	3.25
	Code	42135	721033	40102	723	427	402	4-41	4271	402	80120	B13-02	2-036
MASONRY	Rate	3.63	8.12	5.35	14.22	3.85	6.00	2.50	9.30	5.96	15.31	7.99	3.25
	Code	42102	721037	40102	741	423	403	4-41	4231	401	80100	B15-01	2-032
PAINTING & DECORATING	Rate	2.93	6.46	6.72	8.87	3.85	4.55	2.50	6.34	3.64	14.15	5.38	3.25
	Code	42111	721041	40105	719	427	402	4-41	4275	402	80120	B12-01	2-036
PILE DRIVING	Rate	3.63	14.60	5.35	5.27	3.76	9.60	2.50	6.31	5.96	7.40	5.38	3.25
	Code	42159	722004	40706	732	422	404	4-43	4221	401	80030	B12-10	2-030
PLASTERING	Rate	4.58	8.12	6.72	8.87	3.85	4.55	2.50	6.34	3.64	14.15	7.99	3.25
	Code	42135	721042	40108	719	427	402	4-41	4271	402	80120	B13-02	2-036
PLUMBING	Rate	1.73	4.30	3.50	4.04	1.88	3.88	2.00	3.64	3.64	7.55	3.86	2.35
	Code	42122	721043	40204	707	424	401	4-46	4241	402	80160	B11-01	4-039
ROOFING	Rate	7.69	10.43	6.72	12.45	7.72	6.00	2.50	9.30	5.96	21.16	7.99	3.25
	Code	42118	721036	40403	728	430	403	4-41	4236	401	80130	B15-02	2-031
SHEET METAL WORK (HVAC)	Rate	1.73	4.30	5.35	4.04	1.88	3.88	2.00	3.64	3.64	7.55	3.86	2.35
	Code	42117	721043	40402	707	424	401	4-46	4244	402	80160	B11-07	4-040
STEEL ERECTION—door & sash	Rate	2.53	14.60	5.35	18.61	3.76	9.60	2.50	9.33	5.96	30.77	7.99	3.25
	Code	42106	722005	40502	748	422	404	4-41	4227	401	80080	B15-03	2-012
STEEL ERECTION—inter., ornam.	Rate	2.53	14.60	5.35	18.61	3.76	6.00	2.50	9.33	5.96	30.77	7.99	3.25
	Code	42106	722005	40502	748	422	403	4-41	4227	401	80080	B15-03	2-012
STEEL ERECTION—structure	Rate	2.53	14.60	5.35	18.61	3.76	9.60	2.50	9.33	5.96	30.77	7.99	3.25
	Code	42106	722005	40502	748	422	404	4-41	4227	401	80080	B15-04	2-012
STEEL ERECTION—NOC	Rate	2.53	14.60	5.35	18.61	3.76	9.60	2.50	9.33	5.96	30.77	7.99	3.25
	Code	42106	722005	40502	748	422	404	4-41	4227	401	80080	B15-03	2-012
TILE WORK—inter. (ceramic)	Rate	2.56	3.57	2.31	8.87	3.85	4.55	2.50	6.34	3.64	14.15	7.99	3.25
	Code	42113	721054	40103	719	427	402	4-41	4276	402	80120	B13-01	2-034
WATERPROOFING	Rate	2.93	6.46	5.35	6.66	3.85	6.00	2.50	9.30	3.64	21.16	3.86	3.25
	Code	42139	721016	40102	723	423	403	4-41	4239	402	80130	B11-17	2-030
WRECKING	Rate	12.86	6.75	6.72	18.61	2.72	6.00	2.50	3.64	5.96	25.35	7.99	3.25
	Code	42108	721005	40106	748	421	403	4-43	4211	401	80220	B14-07	2-030

General Requirements | R011 | Overhead & Miscellaneous Data

R01100-070 Contractor's Overhead & Profit

Below are the **average** installing contractor's percentage mark-ups applied to base labor rates to arrive at typical billing rates.

Column A: Labor rates are based on union wages averaged for 30 major U.S. cities. Base rates including fringe benefits are listed hourly and daily. These figures are the sum of the wage rate and employer-paid fringe benefits such as vacation pay, employer-paid health and welfare costs, pension costs, plus appropriate training and industry advancement funds costs.

Column B: Workers' Compensation rates are the national average of state rates established for each trade.

Column C: Column C lists average fixed overhead figures for all trades. Included are Federal and State Unemployment costs set at 7.0%; Social Security Taxes (FICA) set at 7.65%; Builder's Risk Insurance costs set at 0.34%; and Public Liability costs set at 1.55%. All the percentages except those for Social Security Taxes vary from state to state as well as from company to company.

Columns D and E: Percentages in Columns D and E are based on the presumption that the installing contractor has annual billing of $1,500,000 and up. Overhead percentages may increase with smaller annual billing. The overhead percentages for any given contractor may vary greatly and depend on a number of factors, such as the contractor's annual volume, engineering and logistical support costs, and staff requirements. The figures for overhead and profit will also vary depending on the type of job, the job location, and the prevailing economic conditions. All factors should be examined very carefully for each job.

Column F: Column F lists the total of Columns B, C, D, and E.

Column G: Column G is Column A (hourly base labor rate) multiplied by the percentage in Column F (O&P percentage).

Column H: Column H is the total of Column A (hourly base labor rate) plus Column G (Total O&P).

Column I: Column I is Column H multiplied by eight hours.

		A		B	C	D	E	F	G	H	I
		Base Rate Incl. Fringes		Workers' Comp. Ins.	Average Fixed Overhead	Overhead	Profit	Total Overhead & Profit		Rate with O & P	
Abbr.	Trade	Hourly	Daily					%	Amount	Hourly	Daily
Skwk	Skilled Workers Average (35 trades)	$29.85	$238.80	17.5%	16.5%	13.0%	10.0%	57.0%	$17.00	$46.85	$374.80
	Helpers Average (5 trades)	22.15	177.20	19.3		11.0		56.8	12.55	34.70	277.60
	Foreman Average, Inside ($0.50 over trade)	30.35	242.80	17.5		13.0		57.0	17.30	47.65	381.20
	Foreman Average, Outside ($2.00 over trade)	31.85	254.80	17.5		13.0		57.0	18.15	50.00	400.00
Clab	Common Building Laborers	22.85	182.80	19.0		11.0		56.5	12.90	35.75	286.00
Asbe	Asbestos/Insulation Workers/Pipe Coverers	32.15	257.20	17.7		16.0		60.2	19.35	51.50	412.00
Boil	Boilermakers	34.55	276.40	15.9		16.0		58.4	20.20	54.75	438.00
Bric	Bricklayers	29.60	236.80	17.0		11.0		54.5	16.15	45.75	366.00
Brhe	Bricklayer Helpers	22.95	183.60	17.0		11.0		54.5	12.50	35.45	283.60
Carp	Carpenters	29.15	233.20	19.0		11.0		56.5	16.45	45.60	364.80
Cefi	Cement Finishers	27.95	223.60	11.1		11.0		48.6	13.60	41.55	332.40
Elec	Electricians	34.30	274.40	6.8		16.0		49.3	16.90	51.20	409.60
Elev	Elevator Constructors	35.65	285.20	8.4		16.0		50.9	18.15	53.80	430.40
Eqhv	Equipment Operators, Crane or Shovel	31.15	249.20	10.9		14.0		51.4	16.00	47.15	377.20
Eqmd	Equipment Operators, Medium Equipment	30.35	242.80	10.9		14.0		51.4	15.60	45.95	367.60
Eqlt	Equipment Operators, Light Equipment	28.75	230.00	10.9		14.0		51.4	14.80	43.55	348.40
Eqol	Equipment Operators, Oilers	25.75	206.00	10.9		14.0		51.4	13.25	39.00	312.00
Eqmm	Equipment Operators, Master Mechanics	31.80	254.40	10.9		14.0		51.4	16.35	48.15	385.20
Glaz	Glaziers	28.55	228.40	14.1		11.0		51.6	14.75	43.30	346.40
Lath	Lathers	28.20	225.60	11.8		11.0		49.3	13.90	42.10	336.80
Marb	Marble Setters	29.20	233.60	17.0		11.0		54.5	15.90	45.10	360.80
Mill	Millwrights	30.55	244.40	11.2		11.0		48.7	14.90	45.45	363.60
Mstz	Mosaic and Terrazzo Workers	28.30	226.40	10.3		11.0		47.8	13.55	41.85	334.80
Pord	Painters, Ordinary	26.10	208.80	14.8		11.0		52.3	13.65	39.75	318.00
Psst	Painters, Structural Steel	27.15	217.20	50.0		11.0		87.5	23.75	50.90	407.20
Pape	Paper Hangers	26.25	210.00	14.8		11.0		52.3	13.75	40.00	320.00
Pile	Pile Drivers	28.80	230.40	27.7		16.0		70.2	20.20	49.00	392.00
Plas	Plasterers	27.30	218.40	15.7		11.0		53.2	14.50	41.80	334.40
Plah	Plasterer Helpers	23.25	186.00	15.7		11.0		53.2	12.35	35.60	284.80
Plum	Plumbers	34.45	275.60	8.5		16.0		51.0	17.55	52.00	416.00
Rodm	Rodmen (Reinforcing)	32.75	262.00	29.2		14.0		69.7	22.85	55.60	444.80
Rofc	Roofers, Composition	25.70	205.60	34.3		11.0		71.8	18.45	44.15	353.20
Rots	Roofers, Tile and Slate	26.00	208.00	34.3		11.0		71.8	18.65	44.65	357.20
Rohe	Roofer Helpers (Composition)	19.10	152.80	34.3		11.0		71.8	13.70	32.80	262.40
Shee	Sheet Metal Workers	33.55	268.40	11.9		16.0		54.4	18.25	51.80	414.40
Spri	Sprinkler Installers	34.45	275.60	8.8		16.0		51.3	17.65	52.10	416.80
Stpi	Steamfitters or Pipefitters	34.85	278.80	8.5		16.0		51.0	17.75	52.60	420.80
Ston	Stone Masons	29.75	238.00	17.0		11.0		54.5	16.20	45.95	367.60
Sswk	Structural Steel Workers	32.85	262.80	40.9		14.0		81.4	26.75	59.60	476.80
Tilf	Tile Layers	28.15	225.20	10.3		11.0		47.8	13.45	41.60	332.80
Tilh	Tile Layer Helpers	22.70	181.60	10.3		11.0		47.8	10.85	33.55	268.40
Trlt	Truck Drivers, Light	23.40	187.20	15.3		11.0		52.8	12.35	35.75	286.00
Trhv	Truck Drivers, Heavy	23.95	191.60	15.3		11.0		52.8	12.65	36.60	292.80
Sswl	Welders, Structural Steel	32.85	262.80	40.9		14.0		81.4	26.75	59.60	476.80
Wrck	*Wrecking	22.85	182.80	42.2	▼	11.0	▼	79.7	18.20	41.05	328.40

*Not included in Averages.

General Requirements — R011 Overhead & Miscellaneous Data

R01100-080 Performance Bond

This table shows the cost of a Performance Bond for a construction job scheduled to be completed in 12 months. Add 1% of the premium cost per month for jobs requiring more than 12 months to complete. The rates are "standard" rates offered to contractors that the bonding company considers financially sound and capable of doing the work. Preferred rates are offered by some bonding companies based upon financial strength of the contractor. Actual rates vary from contractor to contractor and from bonding company to bonding company. Contractors should prequalify through a bonding agency before submitting a bid on a contract that requires a bond.

Contract Amount	Building Construction Class B Projects	Highways & Bridges Class A New Construction	Highways & Bridges Class A-1 Highway Resurfacing
First $ 100,000 bid	$25.00 per M	$15.00 per M	$9.40 per M
Next 400,000 bid	$ 2,500 plus $15.00 per M	$ 1,500 plus $10.00 per M	$ 940 plus $7.20 per M
Next 2,000,000 bid	8,500 plus 10.00 per M	5,500 plus 7.00 per M	3,820 plus 5.00 per M
Next 2,500,000 bid	28,500 plus 7.50 per M	19,500 plus 5.50 per M	15,820 plus 4.50 per M
Next 2,500,000 bid	47,250 plus 7.00 per M	33,250 plus 5.00 per M	28,320 plus 4.50 per M
Over 7,500,000 bid	64,750 plus 6.00 per M	45,750 plus 4.50 per M	39,570 plus 4.00 per M

R01100-090 Sales Tax by State

State sales tax on materials is tabulated below (5 states have no sales tax). Many states allow local jurisdictions, such as a county or city, to levy additional sales tax.

Some projects may be sales tax exempt, particularly those constructed with public funds.

State	Tax (%)	State	Tax (%)	State	Tax (%)	State	Tax (%)
Alabama	4	Illinois	6.25	Montana	0	Rhode Island	7
Alaska	0	Indiana	5	Nebraska	4.5	South Carolina	5
Arizona	5	Iowa	5	Nevada	6.875	South Dakota	4
Arkansas	4.625	Kansas	4.9	New Hampshire	0	Tennessee	6
California	7.25	Kentucky	6	New Jersey	6	Texas	6.25
Colorado	3	Louisiana	4	New Mexico	5	Utah	4.75
Connecticut	6	Maine	6	New York	4	Vermont	5
Delaware	0	Maryland	5	North Carolina	4	Virginia	4.5
District of Columbia	5.75	Massachusetts	5	North Dakota	5	Washington	6.5
Florida	6	Michigan	6	Ohio	5	West Virginia	6
Georgia	4	Minnesota	6.5	Oklahoma	4.5	Wisconsin	5
Hawaii	4	Mississippi	7	Oregon	0	Wyoming	4
Idaho	5	Missouri	4.225	Pennsylvania	6	Average	4.71 %

Sales Tax by Province (Canada)

GST - a value-added tax, which the federal government imposes on most goods and services provided in or imported into Canada.
PST - a retail sales tax, which five of the provinces impose on the price of most goods and some services.

QST - a value-added tax, similar to the federal GST, which Quebec imposes.
HST - Three provinces have combined their retail sales tax with the federal GST into one harmonized tax.

Province	PST (%)	QST (%)	GST (%)	HST (%)
Alberta	0	0	7	0
British Columbia	7	0	7	0
Manitoba	7	0	7	0
New Brunswick	0	0	0	15
Newfoundland	0	0	0	15
Northwest Territories	0	0	7	0
Nova Scotia	0	0	0	15
Ontario	8	0	7	0
Prince Edward Island	10.7	0	7	0
Quebec	0	8.02	7	0
Saskatchewan	6	0	7	0
Yukon	0	0	7	0

R01100-100 Unemployment Taxes and Social Security Taxes

Mass. State Unemployment tax ranges from 1.325% to 7.225% plus an experience rating assessment the following year, on the first $10,800 of wages. Federal Unemployment tax is 6.2% of the first $7,000 of wages. This is reduced by a credit for payment to the state. The minimum Federal Unemployment tax is .8% after all credits.

Combined rates in Mass. thus vary from 2.125% to 8.025% of the first $10,800 of wages. Combined average U.S. rate is about 7.0% of the first $7,000. Contractors with permanent workers will pay less since the average annual wages for skilled workers is $29.85 x 2,000 hours or about $59,700 per year. The average combined rate for U.S. would thus be 7.0% x $7,000 ÷ $59,700 = 0.8% of total wages for permanent employees.

Rates vary not only from state to state but also with the experience rating of the contractor.

Social Security (FICA) for 2001 is estimated at time of publication to be 7.65% of wages up to $76,200.

General Requirements | R011 | Overhead & Miscellaneous Data

R01100-110 Overtime

One way to improve the completion date of a project or eliminate negative float from a schedule is to compress activity duration times. This can be achieved by increasing the crew size or working overtime with the proposed crew.

To determine the costs of working overtime to compress activity duration times, consider the following examples. Below is an overtime efficiency and cost chart based on a five, six, or seven day week with an eight through twelve hour day. Payroll percentage increases for time and one half and double time are shown for the various working days.

Days per Week	Hours per Day	Production Efficiency					Payroll Cost Factors	
		1 Week	2 Weeks	3 Weeks	4 Weeks	Average 4 Weeks	@ 1-1/2 Times	@ 2 Times
5	8	100%	100%	100%	100%	100 %	100 %	100 %
	9	100	100	95	90	96.25	105.6	111.1
	10	100	95	90	85	91.25	110.0	120.0
	11	95	90	75	65	81.25	113.6	127.3
	12	90	85	70	60	76.25	116.7	133.3
6	8	100	100	95	90	96.25	108.3	116.7
	9	100	95	90	85	92.50	113.0	125.9
	10	95	90	85	80	87.50	116.7	133.3
	11	95	85	70	65	78.75	119.7	139.4
	12	90	80	65	60	73.75	122.2	144.4
7	8	100	95	85	75	88.75	114.3	128.6
	9	95	90	80	70	83.75	118.3	136.5
	10	90	85	75	65	78.75	121.4	142.9
	11	85	80	65	60	72.50	124.0	148.1
	12	85	75	60	55	68.75	126.2	152.4

R01100-710 Unit Gross Area Requirements

The figures in the table below indicate typical ranges in square feet as a function of the "occupant" unit. This table is best used in the preliminary design stages to help determine the probable size requirement for the total project. See R17100-100 for the typical total size ranges for various types of buildings.

Building Type	Unit	Gross Area in S.F.		
		1/4	Median	3/4
Apartments	Unit	660	860	1,100
Auditorium & Play Theaters	Seat	18	25	38
Bowling Alleys	Lane		940	
Churches & Synagogues	Seat	20	28	39
Dormitories	Bed	200	230	275
Fraternity & Sorority Houses	Bed	220	315	370
Garages, Parking	Car	325	355	385
Hospitals	Bed	685	850	1,075
Hotels	Rental Unit	475	600	710
Housing for the elderly	Unit	515	635	755
Housing, Public	Unit	700	875	1,030
Ice Skating Rinks	Total	27,000	30,000	36,000
Motels	Rental Unit	360	465	620
Nursing Homes	Bed	290	350	450
Restaurants	Seat	23	29	39
Schools, Elementary	Pupil	65	77	90
Junior High & Middle		85	110	129
Senior High		102	130	145
Vocational		110	135	195
Shooting Ranges	Point		450	
Theaters & Movies	Seat		15	

General Requirements | R011 | Overhead & Miscellaneous Data

R01100-720 Floor Area Ratios

Table below lists commonly used gross to net area and net to gross area ratios expressed in % for various building types.

Building Type	Gross to Net Ratio	Net to Gross Ratio	Building Type	Gross to Net Ratio	Net to Gross Ratio
Apartment	156	64	School Buildings (campus type)		
Bank	140	72	Administrative	150	67
Church	142	70	Auditorium	142	70
Courthouse	162	61	Biology	161	62
Department Store	123	81	Chemistry	170	59
Garage	118	85	Classroom	152	66
Hospital	183	55	Dining Hall	138	72
Hotel	158	63	Dormitory	154	65
Laboratory	171	58	Engineering	164	61
Library	132	76	Fraternity	160	63
Office	135	75	Gymnasium	142	70
Restaurant	141	70	Science	167	60
Warehouse	108	93	Service	120	83
			Student Union	172	59

The gross area of a building is the total floor area based on outside dimensions.

The net area of a building is the usable floor area for the function intended and excludes such items as stairways, corridors and mechanical rooms. In the case of a commercial building, it might be considered as the "leasable area."

R01100-730 Occupancy Determinations

		S.F. Required per Person		
	Description	BOCA	SBC	UBC
Assembly Areas	Fixed Seats	**	6	7
	Movable Seats		15	15
	Concentrated	7		
	Unconcentrated	15		
	Standing Space	3		
Educational	Unclassified			
	Classrooms	20	40	20
	Shop Areas	50	100	50
Institutional	Unclassified		125	
	In-Patient Areas	240		
	Sleeping Areas	120		
Mercantile	Basement	30	30	20
	Ground Floor	30	30	30
	Upper Floors	60	60	50
Office		100	100	100

BOCA=Building Officials & Code Administrators
SBC=Southern Building Code
UBC=Uniform Building Code

** The occupancy load for assembly area with fixed seats shall be determined by the number of fixed seats installed.

General Requirements R011 Overhead & Miscellaneous Data

R01107-030 Engineering Fees

Typical **Structural Engineering Fees** based on type of construction and total project size. These fees are included in Architectural Fees.

Type of Construction	Total Project Size (in thousands of dollars)			
	$500	$500-$1,000	$1,000-$5,000	Over $5000
Industrial buildings, factories & warehouses	Technical payroll times 2.0 to 2.5	1.60%	1.25%	1.00%
Hotels, apartments, offices, dormitories, hospitals, public buildings, food stores		2.00%	1.70%	1.20%
Museums, banks, churches and cathedrals		2.00%	1.75%	1.25%
Thin shells, prestressed concrete, earthquake resistive		2.00%	1.75%	1.50%
Parking ramps, auditoriums, stadiums, convention halls, hangars & boiler houses		2.50%	2.00%	1.75%
Special buildings, major alterations, underpinning & future expansion		Add to above 0.5%	Add to above 0.5%	Add to above 0.5%

For complex reinforced concrete or unusually complicated structures, add 20% to 50%.

Typical **Mechanical and Electrical Engineering Fees** are based on the size of the subcontract. The fee structure for Mechanical Engineering is shown below; Electrical Engineering fees range from 2 to 5%. These fees are included in Architectural Fees.

Type of Construction	Subcontract Size							
	$25,000	$50,000	$100,000	$225,000	$350,000	$500,000	$750,000	$1,000,000
Simple structures	6.4%	5.7%	4.8%	4.5%	4.4%	4.3%	4.2%	4.1%
Intermediate structures	8.0	7.3	6.5	5.6	5.1	5.0	4.9	4.8
Complex structures	12.0	9.0	9.0	8.0	7.5	7.5	7.0	7.0

For renovations, add 15% to 25% to applicable fee.

General Requirements — R012 Special Project Procedures

R01250-010 Repair and Remodeling

Cost figures are based on new construction utilizing the most cost-effective combination of labor, equipment and material with the work scheduled in proper sequence to allow the various trades to accomplish their work in an efficient manner.

The costs for repair and remodeling work must be modified due to the following factors that may be present in any given repair and remodeling project.

1. Equipment usage curtailment due to the physical limitations of the project, with only hand-operated equipment being used.
2. Increased requirement for shoring and bracing to hold up the building while structural changes are being made and to allow for temporary storage of construction materials on above-grade floors.
3. Material handling becomes more costly due to having to move within the confines of an enclosed building. For multi-story construction, low capacity elevators and stairwells may be the only access to the upper floors.
4. Large amount of cutting and patching and attempting to match the existing construction is required. It is often more economical to remove entire walls rather than create many new door and window openings. This sort of trade-off has to be carefully analyzed.
5. Cost of protection of completed work is increased since the usual sequence of construction usually cannot be accomplished.
6. Economies of scale usually associated with new construction may not be present. If small quantities of components must be custom fabricated due to job requirements, unit costs will naturally increase. Also, if only small work areas are available at a given time, job scheduling between trades becomes difficult and subcontractor quotations may reflect the excessive start-up and shut-down phases of the job.
7. Work may have to be done on other than normal shifts and may have to be done around an existing production facility which has to stay in production during the course of the repair and remodeling.
8. Dust and noise protection of adjoining non-construction areas can involve substantial special protection and alter usual construction methods.
9. Job may be delayed due to unexpected conditions discovered during demolition or removal. These delays ultimately increase construction costs.
10. Piping and ductwork runs may not be as simple as for new construction. Wiring may have to be snaked through walls and floors.
11. Matching "existing construction" may be impossible because materials may no longer be manufactured. Substitutions may be expensive.
12. Weather protection of existing structure requires additional temporary structures to protect building at openings.
13. On small projects, because of local conditions, it may be necessary to pay a tradesman for a minimum of four hours for a task that is completed in one hour.

All of the above areas can contribute to increased costs for a repair and remodeling project. Each of the above factors should be considered in the planning, bidding and construction stage in order to minimize the increased costs associated with repair and remodeling jobs.

General Requirements | R015 | Construction Aids

R01540-100 Steel Tubular Scaffolding

On new construction, tubular scaffolding is efficient up to 60' high or five stories. Above this it is usually better to use a hung scaffolding if construction permits. Swing scaffolding operations may interfere with tenants. In this case, the tubular is more practical at all heights.

In repairing or cleaning the front of an existing building the cost of tubular scaffolding per S.F. of building front increases as the height increases above the first tier. The first tier cost is relatively high due to leveling and alignment.

The minimum efficient crew for erection is three workers. For heights over 50', a crew of four is more efficient. Use two or more on top and two at the bottom for handing up or hoisting. Four workers can erect and dismantle about nine frames per hour up to five stories. From five to eight stories they will average six frames per hour. With 7' horizontal spacing this will run about 400 S.F. and 265 S.F. of wall surface, respectively. Time for placing planks must be added to the above. On heights above 50', five planks can be placed per labor-hour.

The cost per 1,000 S.F. of building front in the table below was developed by pricing the materials required for a typical tubular scaffolding system eleven frames long and two frames high. Planks were figured five wide for standing plus two wide for materials.

Frames are 5' wide and usually spaced 7' O.C. horizontally. Sidewalk frames are 6' wide. Rental rates will be lower for jobs over three months duration.

For jobs under twenty-five frames, add 50% to rental cost. These figures do not include accessories which are listed separately below. Large quantities for long periods can reduce rental rates by 20%.

Item	Unit	Monthly Rent	Per 1,000 S.F. of Building Front	
			No. of Pieces	Rental per Month
5' Wide Standard Frame, 6'-4" High	Ea.	$ 3.75	24	$ 90.00
Leveling Jack & Plate		1.50	24	36.00
Cross Brace		.60	44	26.40
Side Arm Bracket, 21"		1.50	12	18.00
Guardrail Post		1.00	12	12.00
Guardrail, 7' section		.75	22	16.50
Stairway Section		10.00	2	20.00
Stairway Starter Bar		.10	1	.10
Stairway Inside Handrail		5.00	2	10.00
Stairway Outside Handrail		5.00	2	10.00
Walk-Thru Frame Guardrail		2.00	2	4.00
			Total	$243.00
			Per C.S.F., 1 Use/Mo.	$ 24.30

Scaffolding is often used as falsework over 15' high during construction of cast-in-place concrete beams and slabs. Two foot wide scaffolding is generally used for heavy beam construction. The span between frames depends upon the load to be carried with a maximum span of 5'.

Heavy duty shoring frames with a capacity of 10,000#/leg can be spaced up to 10' O. C. depending upon form support design and loading.

Scaffolding used as horizontal shoring requires less than half the material required with conventional shoring.

On new construction, erection is done by carpenters.

Rolling towers supporting horizontal shores can reduce labor and speed the job. For maintenance work, catwalks with spans up to 70' can be supported by the rolling towers.

General Requirements — R015 Construction Aids

R01590-100 Contractor Equipment

Rental Rates shown in the front of the book pertain to late model high quality machines in excellent working condition, rented from equipment dealers. Rental rates from contractors may be substantially lower than the rental rates from equipment dealers depending upon economic conditions. For older, less productive machines, reduce rates by a maximum of 15%. Any overtime must be added to the base rates. For shift work, rates are lower. Usual rule of thumb is 150% of one shift rate for two shifts; 200% for three shifts.

For periods of less than one week, operated equipment is usually more economical to rent than renting bare equipment and hiring an operator.

Equipment moving and mobilization costs must be added to rental rates where applicable. A large crane, for instance, may take two days to erect and two days to dismantle.

Rental rates vary throughout the country with larger cities generally having lower rates. Lease plans for new equipment are available for periods in excess of six months with a percentage of payments applying toward purchase.

Monthly rental rates vary from 2% to 5% of the cost of the equipment depending on the anticipated life of the equipment and its wearing parts. Weekly rates are about 1/3 the monthly rates and daily rental rates about 1/3 the weekly rate.

The hourly operating costs for each piece of equipment include costs to the user such as fuel, oil, lubrication, normal expendables for the equipment, and a percentage of mechanic's wages chargeable to maintenance. The hourly operating costs listed do not include the operator's wages.

The daily cost for equipment used in the standard crews is figured by dividing the weekly rate by five, then adding eight times the hourly operating cost to give the total daily equipment cost, not including the operator. This figure is in the right hand column of Division 01590 under Crew Equipment Cost/Day.

Pile Driving rates shown for pile hammer and extractor do not include leads, crane, boiler or compressor. Vibratory pile driving requires an added field specialist during set-up and pile driving operation for the electric model. The hydraulic model requires a field specialist for set-up only. Up to 125 reuses of sheet piling are possible using vibratory drivers. For normal conditions, crane capacity for hammer type and size are as follows.

Crane Capacity	Hammer Type and Size		
	Air or Steam	Diesel	Vibratory
25 ton	to 8,750 ft.-lb.		70 H.P.
40 ton	15,000 ft.-lb.	to 32,000 ft.-lb.	170 H.P.
60 ton	25,000 ft.-lb.		300 H.P.
100 ton		112,000 ft.-lb.	

Cranes should be specified for the job by size, building and site characteristics, availability, performance characteristics, and duration of time required.

Backhoes & Shovels rent for about the same as equivalent size cranes but maintenance and operating expense is higher. Crane operators rate must be adjusted for high boom heights. Average adjustments: for 150' boom add $.80 per hour; over 185', add $1.40 per hour; over 210', add $1.85 per hour; over 250', add $2.65 per hour and over 295', add $4.20 per hour.

Tower Cranes

Capacity In Kip-Feet	Typical Jib Length in Feet	Speed at Maximum Reach and Load	Purchase Price (New)		Monthly Rental, to 6 mo.	
			Crane & 80' Mast	Mast Sections	Crane & 80' Mast	Mast Sections
725	100	350 FPM	$ 256,000	$ 670 /L.F.	$ 6,400	$16.80 /L.F.
900	100	500	303,000	790	7,600	19.60
*1100	130	1000	418,000	910	10,400	22.70
1450	150	1000	580,000	1,300	14,500	32.40
2150	200	1000	729,000	1,480	18,200	37.00
3000	200	1000	1,018,000	1,510	25,500	37.70

*Most widely used.

Tower Cranes of the climbing or static type have jibs from 50' to 200' and capacities at maximum reach range from 4,000 to 14,000 pounds. Lifting capacities increase up to maximum load as the hook radius decreases.

Typical rental rates, based on purchase price are about 2% to 3% per month.

Erection and dismantling runs between $13,000 and $82,000. Climbing operation takes ten labor hours per 20' climb. Crane dead time is about five hours per 40' climb. If crane is bolted to side of the building add cost of ties and extra mast sections. Mast sections cost $600 to $1,500 per vertical foot or can be rented at 2% to 3% of purchase price per month. Contractors using climbers claim savings of $1.50 per C.Y. of concrete placed, plus $.21 per S.F. of formwork. Climbing cranes have from 80' to 180' of mast while static cranes have 80' to 800' of mast.

Truck Cranes can be converted to tower cranes by using tower attachments. Mast heights over 400' have been used. See Division 01590-600 for rental rates of high boom cranes.

A single 100' high material **Hoist and Tower** can be erected and dismantled for about $15,600; a double 100' high hoist and tower for about $20,700. Erection costs for additional heights are $100 and $130 per vertical foot respectively up to 150' and $100 to $165 per vertical foot over 150' high. A 40' high portable Buck hoist costs about $5,300 to erect and dismantle. Additional heights run $80 per vertical foot to 80' and $100 per vertical foot for the next 100'. Most material hoists do not meet local code requirements for carrying personnel.

A 150' high **Personnel Hoist** requires about 500 to 800 labor hours to erect and dismantle with costs ranging from $16,400 to $26,300. Budget erection cost is $165 per vertical foot for all trades. Local code requirements or labor scarcity requiring overtime can add up to 50% to any of the above erection costs.

Earthmoving Equipment: The selection of earthmoving equipment depends upon the type and quantity of material, moisture content, haul distance, haul road, time available, and equipment available. Short haul cut and fill operations may require dozers only, while another operation may require excavators, a fleet of trucks, and spreading and compaction equipment. Stockpiled material and granular material are easily excavated with front end loaders. Scrapers are most economically used with hauls between 300' and 1-1/2 miles if adequate haul roads can be maintained. Shovels are often used for blasted rock and any material where a vertical face of 8' or more can be excavated. Special conditions may dictate the use of draglines, clamshells, or backhoes. Spreading and compaction equipment must be matched to the soil characteristics, the compaction required and the rate the fill is being supplied.

Site Construction — R021 | Subsurface Invest. & Demol.

R02115-200 Underground Storage Tank Removal

Underground Storage Tank Removal can be divided into two categories: Non-Leaking and Leaking. Prior to removing an underground storage tank, tests should be made, with the proper authorities present, to determine whether a tank has been leaking or the surrounding soil has been contaminated.

To safely remove Liquid Underground Storage Tanks:
1. Excavate to the top of the tank.
2. Disconnect all piping.
3. Open all tank vents and access ports.
4. Remove all liquids and/or sludge.
5. Purge the tank with an inert gas.
6. Provide access to the inside of the tank and clean out the interior using proper personal protective equipment (PPE).
7. Excavate soil surrounding the tank using proper PPE for on-site personnel.
8. Pull and properly dispose of the tank.
9. Clean up the site of all contaminated material.
10. Install new tanks or close the excavation.

Site Construction — R022 | Site Prep. & Excav. Support

R02240-900 Wellpoints

A single stage wellpoint system is usually limited to dewatering an average 15' depth below normal ground water level. Multi-stage systems are employed for greater depth with the pumping equipment installed only at the lowest header level. Ejectors, with unlimited lift capacity, can be economical when two or more stages of wellpoints can be replaced or when horizontal clearance is restricted, such as in deep trenches or tunneling projects, and where low water flows are expected. Wellpoints are usually spaced on 2-1/2' to 10' centers along a header pipe. Wellpoint spacing, header size, and pump size are all determined by the expected flow, as dictated by soil conditions.

In almost all soils encountered in wellpoint dewatering, the wellpoints may be jetted into place. Cemented soils and stiff clays may require sand wicks about 12" in diameter around each wellpoint to increase efficiency and eliminate weeping into the excavation. These sand wicks require 1/2 to 3 C.Y. of washed filter sand and are installed by using a 12" diameter steel casing and hole puncher jetted into the ground 2' deeper than the wellpoint. Rock may require predrilled holes.

Labor required for the complete installation and removal of a single stage wellpoint system is in the range of 3/4 to 2 labor-hours per linear foot of header, depending upon jetting conditions, wellpoint spacing, etc.

Continuous pumping is necessary except in some free draining soil where temporary flooding is permissible (as in trenches which are backfilled after each day's work). Good practice requires provision of a stand-by pump during the continuous pumping operation.

Systems for continuous trenching below the water table should be installed three to four times the length of expected daily progress to insure uninterrupted digging, and header pipe size should not be changed during the job.

For pervious free draining soils, deep wells in place of wellpoints may be economical because of lower installation and maintenance costs. Daily production ranges between two to three wells per day, for 25' to 40' depths, to one well per day for depths over 50'.

Detailed analysis and estimating for any dewatering problem is available at no cost from wellpoint manufacturers. Major firms will quote "sufficient equipment" quotes or their affiliates offer lump sum proposals to cover complete dewatering responsibility.

	Description for 200' System with 8" Header	Quantities		1st Month Unit	1st Month Total	Thereafter Unit	Thereafter Total
Equipment & Material	Wellpoints 25' long, 2" diameter @ 5' O.C.	40	Each	$ 28.50	$ 1,140.00	$ 21.38	$ 855.00
	Header pipe, 8" diameter	200	L.F.	5.40	1,080.00	2.70	540.00
	Discharge pipe, 8" diameter	100	L.F.	3.66	366.00	2.38	238.00
	8" valves	3	Each	140.00	420.00	91.00	273.00
	Combination Jetting & Wellpoint pump (standby)	1	Each	2,275.00	2,275.00	1,592.50	1,592.50
	Wellpoint pump, 8" diameter	1	Each	2,225.00	2,225.00	1,557.50	1,557.50
	Transportation to and from site	1	Day	674.40	674.40	—	—
	Fuel 30 days x 60 gal./day	1800	Gallons	1.05	1,890.00	1.05	1,890.00
	Lubricants for 30 days x 16 lbs./day	480	Lbs.	1.10	528.00	1.10	528.00
	Sand for points	40	C.Y.	16.60	664.00	—	—
	Equipment & Materials Subtotal				$11,262.40		$ 7,474.00
Labor	Technician to supervise installation	1	Week	$ 31.80	$ 1,272.00	—	—
	Labor for installation and removal of system	300	Labor-hours	22.85	6,855.00	—	—
	4 Operators straight time 40 hrs./wk. for 4.33 wks.	693	Hrs.	28.75	19,923.75	$ 28.75	$19,923.75
	4 Operators overtime 2 hrs./wk. for 4.33 wks.	35	Hrs.	57.50	2,012.50	57.50	2,012.50
	Labor Subtotal				$30,063.25		$21,936.25
	Total Cost				$41,325.65		$29,410.25
	Monthly Cost/L.F. Header				$ 206.63		$ 147.05

Site Construction | R022 | Site Prep. & Excav. Support

R02250-400 Wood Sheet Piling

Wood sheet piling may be used for depths to 20' where there is no ground water. If moderate ground water is encountered Tongue & Groove sheeting will help to keep it out. When considerable ground water is present, steel sheeting must be used.

For estimating purposes on trench excavation, sizes are as follows:

Depth	Sheeting	Wales	Braces	B.F. per S.F.
To 8'	3 x 12's	6 x 8's, 2 line	6 x 8's, @ 10'	4.0 @ 8'
8' x 12'	3 x 12's	10 x 10's, 2 line	10 x 10's, @ 9'	5.0 average
12' to 20'	3 x 12's	12 x 12's, 3 line	12 x 12's, @ 8'	7.0 average

Sheeting to be toed in at least 2' depending upon soil conditions. A five person crew with an air compressor and sheeting driver can drive and brace 440 SF/day at 8' deep, 360 SF/day at 12' deep, and 320 SF/day at 16' deep. For normal soils, piling can be pulled in 1/3 the time to install. Pulling difficulty increases with the time in the ground. Production can be increased by high pressure jetting. Figures below assume 50% of lumber is salvaged and includes pulling costs. Some jurisdictions require an equipment operator in addition to Crew B-31.

Sheeting Pulled

Daily Cost Crew B-31	L.H./Day	Hourly Cost	Daily Cost	8' Depth, 440 S.F./Day		16' Depth, 320 S.F./Day	
				To Drive (1 Day)	To Pull (1/3 Day)	To Drive (1 Day)	To Pull (1/3 Day)
1 Foreman	8	$24.85	$ 198.80	$ 198.80	$ 66.20	$ 198.80	$ 66.20
3 Laborers	24	22.85	548.40	548.40	182.62	548.40	182.62
1 Carpenter	8	29.15	233.20	233.20	77.66	233.20	77.66
1 Air Compressor			125.20	125.20	41.69	125.20	41.69
1 Sheeting Driver			12.50	12.50	4.16	12.50	4.16
2 -50 Ft. Air Hoses, 1-1/2" Diam.			11.80	11.80	3.93	11.80	3.93
Lumber (50% salvage)				1.76 MBF 559.68		2.24 MBF 712.32	
Total			$1,129.90	$1,689.58	$376.26	$1,842.22	$376.26
Total/S.F.				$ 3.84	$ .86	$ 5.76	$ 1.18
Total (Drive and Pull)/S.F.				$ 4.70		$ 6.93	

Sheeting Left in Place

Daily Cost	8' Depth 440 S.F./Day		10' Depth 400 S.F./Day		12' Depth 360 S.F./Day		16' Depth 320 S.F./Day		18' Depth 305 S.F./Day		20' Depth 280 S.F./Day	
Crew B-31		$1,129.90		$1,129.90		$1,129.90		$1,129.90		$1,129.90		$1,129.90
Lumber	1.76 M	1,119.36	1.8 M	1,144.80	1.8 M	1,144.80	2.24 M	1,424.64	2.1 M	1,335.60	1.9 M	1,208.40
Total in Place		$2,249.26		$2,274.70		$2,274.70		$2,554.54		$2,465.50		$2,338.30
Total/S.F.		$ 5.11		$ 5.69		$ 6.32		$ 7.98		$ 8.08		$ 8.35
Total/M.B.F.		$1,277.99		$1,263.72		$1,263.72		$1,140.42		$1,174.05		$1,230.68

Site Construction — R022 Site Prep. & Excav. Support

R02250-450 Steel Sheet Piling

Limiting weights are 22 to 38#/S.F. of wall surface with 27#/S.F. average for usual types and sizes. (Weights of piles themselves are from 30.7#/L.F. to 57#/L.F. but they are 15" to 21" wide.) Lightweight sections 12" to 28" wide from 3 ga. to 12 ga. thick are also available for shallow excavations. Piles may be driven two at a time with an impact or vibratory hammer (use vibratory to pull) hung from a crane without leads. A reasonable estimate of the life of steel sheet piling is 10 uses with up to 125 uses possible if a vibratory hammer is used. Used piling costs from 50% to 80% of new piling depending on location and market conditions. Sheet piling and H piles can be rented for about 30% of the delivered mill price for the first month and 5% per month thereafter. Allow 1 labor-hour per pile for cleaning and trimming after driving. These costs increase with depth and hydrostatic head. Vibratory drivers are faster in wet granular soils and are excellent for pile extraction. Pulling difficulty increases with the time in the ground and may cost more than driving. It is often economical to abandon the sheet piling, especially if it can be used as the outer wall form. Allow about 1/3 additional length or more, for toeing into ground. Add bracing, waler and strut costs. Waler costs can equal the cost per ton of sheeting.

Cost of Sheet Piling & Production Rate by Ton & S.F.

Depth of Excavation	15' Depth		20' Depth		25' Depth	
Description of Pile	22 psf = 90.9 S.F./Ton		27 psf = 74 S.F./Ton		38 psf = 52.6 S.F./Ton	
Type of operation: Left in Place or Removed	Drive & Left	Drive & Extract*	Drive & Left	Drive & Extract*	Drive & Left	Drive & Extract*
Labor & Equip. to Drive 1 ton	$358.13	$536.20	$298.72	$448.08	$203.76	$304.12
Piling (75% Salvage)	812.00	203.00	812.00	203.00	812.00	203.00
Cost/Ton in Place	$1,170.13	$739.20	$1,110.72	$651.08	$1,015.76	$507.12
Production Rate, Tons/Day	10.81	7.22	12.96	8.64	19.00	12.73
Cost, S.F. in Place (incl. 33% toe in)	$12.87	$8.13	$15.01	$8.80	$19.31	$9.64
Production Rate, S.F./Day (incl. 33% toe in)	983	656	960	640	1,000	670

Crew B-40	Bare Costs Hr.	Bare Costs Daily	Inc. Subs O&P Hr.	Inc. Subs O&P Daily
1 Pile Driver Foreman	$30.80	$246.40	$52.40	$419.20
4 Pile Drivers	28.80	921.60	49.00	1,568.00
1 Equip. Oper. Oiler	25.75	206.00	39.00	312.00
2 Equip. Oper. (crane)	31.15	498.40	47.15	754.40
1 Crane, 40 Ton		716.00		787.60
Vibratory Hammer & Gen.		1,283.00		1,411.30
64 L.H., Daily Totals		$3,871.40		$5,252.50

Installation Cost per Ton
for 15' Deep Excavation plus 33% toe-in
$3,871.40 per day/10.81 tons = $358.13 per ton

Installation Cost per S.F.

$$\frac{2000\#}{22\#/S.F.} = 90.9 \text{ S.F.} \times 10.81 \text{ tons} = 983 \text{ S.F. per day}$$

$3,871.40 / 983 S.F. per day = $3.94 per S.F.

*For driving & extracting, two mobilizations & demobilizations will be necessary

Special Construction | R132 | Hazardous Material Remediation

R13281-120 Asbestos Removal Process

Asbestos removal is accomplished by a specialty contractor who understands the federal and state regulations regarding the handling and disposal of the material. The process of asbestos removal is divided into many individual steps. An accurate estimate can be calculated only after all the steps have been priced.

The steps are generally as follows:

1. Obtain an asbestos abatement plan from an industrial hygienist.
2. Monitor the air quality in and around the removal area and along the path of travel between the removal area and transport area. This establishes the background contamination.
3. Construct a two part decontamination chamber at entrance to removal area.
4. Install a HEPA filter to create a negative pressure in the removal area.
5. Install wall, floor and ceiling protection as required by the plan, usually 2 layers of fireproof 6 mil polyethylene.
6. Industrial hygienist visually inspects work area to verify compliance with plan.
7. Provide temporary supports for conduit and piping affected by the removal process.
8. Proceed with asbestos removal and bagging process. Monitor air quality as described in Step #2. Discontinue operations when contaminate levels exceed applicable standards.
9. Document the legal disposal of materials in accordance with EPA standards.
10. Thoroughly clean removal area including all ledges, crevices and surfaces.
11. Post abatement inspection by industrial hygienist to verify plan compliance.
12. Provide a certificate from a licensed industrial hygienist attesting that contaminate levels are within acceptable standards before returning area to regular use.

Special Construction | R136 | Solar & Wind Energy Equip.

R13600-610 Solar Heating (Space and Hot Water)

Collectors should face as close to due South as possible, however, variations of up to 20 degrees on either side of true South are acceptable. Local climate and collector type may influence the choice between east or west deviations. Obviously they should be located so they are not shaded from the sun's rays. Incline collectors at a slope of latitude minus 5 degrees for domestic hot water and latitude plus 15 degrees for space heating.

Flat plate collectors consist of a number of components as follows: Insulation to reduce heat loss through the bottom and sides of the collector. The enclosure which contains all the components in this assembly is usually weatherproof and prevents dust, wind and water from coming in contact with the absorber plate. The cover plate usually consists of one or more layers of a variety of glass or plastic and reduces the reradiation by creating an air space which traps the heat between the cover and the absorber plates.

The absorber plate must have a good thermal bond with the fluid passages. The absorber plate is usually metallic and treated with a surface coating which improves absorptivity. Black or dark paints or selective coatings are used for this purpose, and the design of this passage and plate combination helps determine a solar system's effectiveness.

Heat transfer fluid passage tubes are attached above and below or integral with an absorber plate for the purpose of transferring thermal energy from the absorber plate to a heat transfer medium. The heat exchanger is a device for transferring thermal energy from one fluid to another.

Piping and storage tanks should be well insulated to minimize heat losses.

Size domestic water heating storage tanks to hold 20 gallons of water per user, minimum, plus 10 gallons per dishwasher or washing machine. For domestic water heating an optimum collector size is approximately 3/4 square foot of area per gallon of water storage. For space heating of residences and small commercial applications the collector is commonly sized between 30% and 50% of the internal floor area. For space heating of large commercial applications, collector areas less than 30% of the internal floor area can still provide significant heat reductions.

A supplementary heat source is recommended for Northern states for December through February.

The solar energy transmission per square foot of collector surface varies greatly with the material used. Initial cost, heat transmittance and useful life are obviously interrelated.

Mechanical — R150 Basic Materials & Methods

R15050-710 Subcontractors

On the unit cost pages of the R.S. Means Cost Data books, the last column is entitled "Total Incl. O&P". This is normally the cost of the installing contractor. In Division 15, this is the cost of the mechanical contractor. If the particular work being estimated is to be performed by a sub to the mechanical contractor, the mechanical's profit and handling charge (usually 10%) is added to the total of the last column.

R15050-720 Demolition (Selective vs. Removal for Replacement)

Demolition can be divided into two basic categories.

One type of demolition involves the removal of material with no concern for its replacement. The labor-hours to estimate this work are found in Div. 15055 under "Selective Demolition". It is selective in that individual items or all the material installed as a system or trade grouping such as plumbing or heating systems are removed. This may be accomplished by the easiest way possible, such as sawing, torch cutting, or sledge hammer as well as simple unbolting.

The second type of demolition is the removal of some item for repair or replacement. This removal may involve careful draining, opening of unions, disconnecting and tagging of electrical connections, capping of pipes/ducts to prevent entry of debris or leakage of the material contained as well as transport of the item away from its in-place location to a truck/dumpster. An approximation of the time required to accomplish this type of demolition is to use half of the time indicated as necessary to install a new unit. For example; installation of a new pump might be listed as requiring 6 labor-hours so if we had to estimate the removal of the old pump we would allow an additional 3 hours for a total of 9 hours. That is, the complete replacement of a defective pump with a new pump would be estimated to take 9 labor-hours.

Mechanical — R151 Plumbing

R15100-050 Pipe Material Considerations

1. Malleable fittings should be used for gas service.
2. Malleable fittings are used where there are stresses/strains due to expansion and vibration.
3. Cast fittings may be broken as an aid to disassembling of heating lines frozen by long use, temperature and minerals.
4. Cast iron pipe is extensively used for underground and submerged service.
5. Type M (light wall) copper tubing is available in hard temper only and is used for nonpressure and less severe applications than K and L.
6. Type L (medium wall) copper tubing, available hard or soft for interior service.
7. Type K (heavy wall) copper tubing, available in hard or soft temper for use where conditions are severe. For underground and interior service.
8. Hard drawn tubing requires fewer hangers or supports but should not be bent. Silver brazed fittings are recommended, however soft solder is normally used.
9. Type DMV (very light wall) copper tubing designed for drainage, waste and vent plus other non-critical pressure services.

Domestic/Imported Pipe and Fittings Cost

The prices shown in this publication for steel/cast iron pipe and steel, cast iron, malleable iron fittings are based on domestic production sold at the normal trade discounts. The above listed items of foreign manufacture may be available at prices of 1/3 to 1/2 those shown. Some imported items after minor machining or finishing operations are being sold as domestic to further complicate the system.

Caution: Most pipe prices in this book also include a coupling and pipe hangers which for the larger sizes can add significantly to the per meter cost and should be taken into account when comparing "book cost" with quoted supplier's cost.

Mechanical — R151 Plumbing

R15100-070 Piping to 10' High

When taking off pipe, it is important to identify the different material types and joining procedures, as well as distances between supports and components required for proper support.

During the takeoff, measure through all fittings. Do not subtract the lengths of the fittings, valves, or strainers, etc. This added length plus the final rounding of the totals will compensate for nipples and waste.

When rounding off totals always increase the actual amount to correspond with manufacturer's shipping lengths.

A. Both red brass and yellow brass pipe are normally furnished in 12' lengths, plain end. Division 15107-220 includes in the linear foot costs two field threads and one coupling per 10' length. A carbon steel clevis type hanger assembly every 10' is also prorated into the linear foot costs, including both material and labor.

B. Cast iron soil pipe is furnished in either 5' or 10' lengths. For pricing purposes, Division 15107-320 in *Means Plumbing Cost Data* features 10' lengths with a joint and a carbon steel clevis hanger assembly every 5' prorated into the per foot costs of both material and labor.

Three methods of joining are considered in Division 15107-320, lead and oakum poured joints, or push-on gasket type joints for the bell and spigot pipe, and a joint clamp for the no-hub soil pipe. The labor and material costs for each of these individual joining procedures are also prorated into the linear costs per foot.

C. Copper tubing in Division 15107-420 covers types K, L, M, and DWV which are furnished in 20' lengths. Means pricing data is based on a tubing cut each length and a coupling and two soft soldered joints every 10'. A carbon steel, clevis type hanger assembly every 10' is also prorated into the per foot costs. The prices for refrigeration tubing are for materials only. Labor for full lengths may be based on the type L labor but short cut measures in tight areas can increase the installation labor-hours from 20 to 40%.

D. Corrosion-resistant piping does not lend itself to one particular standard of hanging or support assembly due to its diversity of application and placement. The several varieties of corrosion-resistant piping do not include any material or labor costs for hanger assemblies (See Division 15060-300 for appropriate selection). The estimator should refer to Division 15107-560 in *Means Plumbing Cost Data* for the compatible corrosion-resistant fittings and piping.

E. Glass pipe, as shown in *Means Plumbing Cost Data* Division 15106-120, is furnished in standard lengths either 5' or 10' long, beaded on one end. Special orders for diverse lengths beaded on both ends are also available. For pricing purposes, R.S. Means features 10' lengths with a coupling and a carbon steel band hanger assembly every 10' prorated into the per foot linear costs.

Glass pipe is also available with conical ends and standard lengths ranging from 6" through 3' in 6" increments, then up to 10' in 12" increments. Special lengths can be customized for particular installation requirements.

For pricing purposes, Means has based the labor and material pricing on 10' lengths. Included in these costs per linear foot are the prorated costs for a flanged assembly every 10' consisting of two flanges, a gasket, two insertable seals, and the required number of bolts and nuts. A carbon steel band hanger assembly based on 10' center lines has also been prorated into the costs per foot for labor and materials.

F. Plastic pipe of several compositions and joining methods are considered in Division 15108-520. Fiberglass reinforced pipe (FRP) is priced, based on 10' lengths (20' lengths are also available), with coupling and epoxy joints every 10'. FRP is furnished in both "General Service" and "High Strength." A carbon steel clevis hanger assembly, 3 for every 10', is built into the prorated labor and material costs on a per foot basis.

The PVC and CPVC pipe schedules 40, 80 and 120, plus SDR ratings are all based on 20' lengths with a coupling installed every 10', as well as a carbon steel clevis hanger assembly every 3'. The PVC and ABS type DWV piping is based on 10' lengths with solvent weld couplings every 10', and with carbon steel clevis hanger assemblies, 3 for every 10'. The rest of the plastic piping in this section is based on flexible 100' coils and does not include any coupling or supports.

This section ends with PVC drain and sewer piping based on 10' lengths with bell and spigot ends and 0-ring type, push-on joints. Material costs only are given, but the complete installation costs should be similar to those shown in line number 02620-270.

G. Stainless steel piping, Division 15107-900, includes both weld end and threaded piping, both in the type 304 and 316 specification and in the following schedules, 5, 10, 40, 80, and 160. Although this piping is usually furnished in 20' lengths, this cost grouping has a joint (either heli-arc butt-welded or threads and coupling) every 10'. A carbon steel clevis type hanger assembly is also included at 10' intervals and prorated into the linear foot costs.

H. Division 15107-620 covers carbon steel pipe, both black and galvanized. This section encompasses schedules 40 (standard) and 80 (extra heavy).

Several common methods of joining steel pipe — such as thread and coupled, butt welded, and flanged (150 lb. weld neck flanges) are also included.

For estimating purposes, it is assumed that the piping is purchased in 20' lengths and that a compatible joint is made up every 10'. These joints are prorated into the labor and material costs per linear foot. The following hanger and support assemblies every 10' are also included: carbon steel clevis for the T & C pipe, and single rod roll type for both the welded and flanged piping. All of these hangers are oversized to accommodate pipe insulation 3/4" thick through 5" pipe size and 1-1/2" thick from 6" through 12" pipe size.

I. Division 15107-690 prices grooved joint steel pipe, both black and galvanized, in schedules 10, 40, and 80, furnished in 20' lengths. This section describes two joining methods: cut groove and roll groove. The schedule 10 piping is roll-grooved, while the heavier schedules are cut-grooved. The labor and material costs are prorated into per linear foot prices, including a coupled joint every 10', as well as a carbon steel clevis hanger assembly.

Notes:

The pipe hanger assemblies mentioned in the preceding paragraphs include the described hanger; appropriately sized steel, box-type insert and nut; plus a threaded hanger rod. On average, the distance from the pipe center line to the insert face is 2'.

C clamps are used when the pipe is to be supported from steel shapes rather than anchored in the slab. C clamps are slightly less costly than inserts. However, to save time in estimating, it is advisable to use the given line number cost, rather than substituting a C clamp for the insert.

Add to piping labor for elevated installation:

10' to 15' high	10%	30' to 35' high	40%
15' to 20' high	20%	35' to 40' high	50%
20' to 25' high	25%	Over 40' high	55%
25' to 30' high	35%		

When using the percentage adds for elevated piping installations as shown above, bear in mind that the given heights are for the pipe supports, even though the insert, anchor, or clamp may be several feet higher than the pipe itself.

An allowance has been included in the piping installation time for testing and minor tightening of leaking joints, fittings, stuffing boxes, packing glands, etc. For extraordinary test requirements such as x-rays, prolonged pressure or demonstration tests, a percentage of the piping labor, based on the estimator's experience, must be added to the labor total. A testing service specializing in weld x-rays should be consulted for pricing if it is an estimate requirement. Equipment installation time includes start-up with associated adjustments.

Mechanical | R151 | Plumbing

R15100-080 Valve Materials

VALVE MATERIALS

Bronze:
Bronze is one of the oldest materials used to make valves. It is most commonly used in hot and cold water systems and other non-corrosive services. It is often used as a seating surface in larger iron body valves to ensure tight closure.

Carbon Steel:
Carbon steel is a high strength material. Therefore, valves made from this metal are used in higher pressure services, such as steam lines up to 600 psi at 850°F. Many steel valves are available with butt-weld ends for economy and are generally used in high pressure steam service as well as other higher pressure non-corrosive services.

Forged Steel:
Valves from tough carbon steel are used in service up to 2000 psi and temperatures up to 1000°F in Gate, Globe and Check valves.

Iron:
Valves are normally used in medium to large pipe lines to control non-corrosive fluid and gases, where pressures do not exceed 250 psi at 450° or 500 psi cold water, oil or gas.

Stainless Steel:
Developed steel alloys can be used in over 90% corrosive services.

Plastic PVC:
This is used in a great variety of valves generally in high corrosive service with lower temperatures and pressures.

VALVE SERVICE PRESSURES

Pressure ratings on valves provide an indication of the safe operating pressure for a valve at some elevated temperature. This temperature is dependent upon the materials used and the fabrication of the valve. When specific data is not available, a good "rule-of-thumb" to follow is the temperature of saturated steam on the primary rating indicated on the valve body. Example: The valve has the number 150S printed on the side indicating 150 psi and hence, a maximum operating temperature of 367°F (temperature of saturated steam and 150 psi).

DEFINITIONS

1. "WOG" – Water, oil, gas (cold working pressures).
2. "SWP" – Steam working pressure.
3. 100% area (full port) – means the area through the valve is equal to or greater than the area of standard pipe.
4. "Standard Opening" – means that the area through the valve is less than the area of standard pipe and therefore these valves should be used only where restriction of flow is unimportant.
5. "Round Port" – means the valve has a full round opening through the plug and body, of the same size and area as standard pipe.
6. "Rectangular Port" – valves have rectangular shaped ports through the plug body. The area of the port is either equal to 100% of the area of standard pipe, or restricted (standard opening). In either case it is clearly marked.
7. "ANSI" – American National Standards Institute.

R15100-090 Valve Selection Considerations

INTRODUCTION: In any piping application, valve performance is critical. Valves should be selected to give the best performance at the lowest cost.

The following is a list of performance characteristics generally expected of valves.

1. Stopping flow or starting it.
2. Throttling flow (Modulation).
3. Flow direction changing.
4. Checking backflow (Permitting flow in only one direction).
5. Relieving or regulating pressure.

In order to properly select the right valve, some facts must be determined.

A. What liquid or gas will flow through the valve?
B. Does the fluid contain suspended particles?
C. Does the fluid remain in liquid form at all times?
D. Which metals does fluid corrode?
E. What are the pressure and temperature limits? (As temperature and pressure rise, so will the price of the valve.)
F. Is there constant line pressure?
G. Is the valve merely an on-off valve?
H. Will checking of backflow be required?
I. Will the valve operate frequently or infrequently?

Valves are classified by design type into such classifications as Gate, Globe, Angle, Check, Ball, Butterfly and Plug. They are also classified by end connection, stem, pressure restrictions and material such as bronze, cast iron, etc. Each valve has a specific use. A quality valve used correctly will provide a lifetime of trouble-free service, but a high quality valve installed in the wrong service may require frequent attention.

Mechanical — R151 Plumbing

STEM TYPES

(OS & Y)—Rising Stem-Outside Screw and Yoke

Offers a visual indication of whether the valve is open or closed. Recommended where high temperatures, corrosives, and solids in the line might cause damage to inside-valve stem threads. The stem threads are engaged by the yoke bushing so the stem rises through the hand wheel as it is turned.

(R.S.)—Rising Stem-Inside Screw

Adequate clearance for operation must be provided because both the hand wheel and the stem rise.

The valve wedge position is indicated by the position of the stem and hand wheel.

(N.R.S.)—Non-Rising Stem-Inside Screw

A minimum clearance is required for operating this type of valve. Excessive wear or damage to stem threads inside the valve may be caused by heat, corrosion, and solids. Because the hand wheel and stem do not rise, wedge position cannot be visually determined.

VALVE TYPES

Gate Valves

Provide full flow, minute pressure drop, minimum turbulence and minimum fluid trapped in the line.
They are normally used where operation is infrequent.

Globe Valves

Globe valves are designed for throttling and/or frequent operation with positive shut-off. Particular attention must be paid to the several types of seating materials available to avoid unnecessary wear. The seats must be compatible with the fluid in service and may be composition or metal. The configuration of the Globe valve opening causes turbulence which results in increased resistance. Most bronze Globe valves are rising stem-inside screw, but they are also available on O.S. & Y.

Angle Valves

The fundamental difference between the Angle valve and the Globe valve is the fluid flow through the Angle valve. It makes a 90° turn and offers less resistance to flow than the Globe valve while replacing an elbow. An Angle valve thus reduces the number of joints and installation time.

Check Valves

Check valves are designed to prevent backflow by automatically seating when the direction of fluid is reversed.
Swing Check valves are generally installed with Gate-valves, as they provide comparable full flow. Usually recommended for lines where flow velocities are low and should not be used on lines with pulsating flow. Recommended for horizontal installation, or in vertical lines only where flow is upward.

Lift Check Valves

These are commonly used with Globe and Angle valves since they have similar diaphragm seating arrangements and are recommended for preventing backflow of steam, air, gas and water, and on vapor lines with high flow velocities. For horizontal lines, horizontal lift checks should be used and vertical lift checks for vertical lines.

Ball Valves

Ball valves are light and easily installed, yet because of modern elastomeric seats, provide tight closure. Flow is controlled by rotating up to 90° a drilled ball which fits tightly against resilient seals. This ball seats with flow in either direction, and valve handle indicates the degree of opening. Recommended for frequent operation readily adaptable to automation, ideal for installation where space is limited.

Butterfly Valves

Butterfly valves provide bubble-tight closure with excellent throttling characteristics. They can be used for full-open, closed and for throttling applications.

The Butterfly valve consists of a disc within the valve body which is controlled by a shaft. In its closed position, the valve disc seals against a resilient seat. The disc position throughout the full 90° rotation is visually indicated by the position of the operator.

A Butterfly valve is only a fraction of the weight of a Gate valve and requires no gaskets between flanges in most cases. Recommended for frequent operation and adaptable to automation where space is limited.

Wafer and Lug type bodies when installed between two pipe flanges, can be easily removed from the line. The pressure of the bolted flanges holds the valve in place.
Locating lugs makes installation easier.

Plug Valves

Lubricated plug valves, because of the wide range of service to which they are adapted, may be classified as all purpose valves. They can be safely used at all pressure and vacuums, and at all temperatures up to the limits of available lubricants. They are the most satisfactory valves for the handling of gritty suspensions and many other destructive, erosive, corrosive and chemical solutions.

Mechanical — R155 Heating

R15500-010 Heating Systems

Heating Systems

The basic function of a heating system is to bring an enclosed volume up to a desired temperature and then maintain that temperature within a reasonable range. To accomplish this, the selected system must have sufficient capacity to offset transmission losses resulting from the temperature difference on the interior and exterior of the enclosing walls in addition to losses due to cold air infiltration through cracks, crevices and around doors and windows. The amount of heat to be furnished is dependent upon the building size, construction, temperature difference, air leakage, use, shape, orientation and exposure. Air circulation is also an important consideration. Circulation will prevent stratification which could result in heat losses through uneven temperatures at various levels. For example, the most efficient use of unit heaters can usually be achieved by circulating the space volume through the total number of units once every 20 minutes or 3 times an hour. This general rule must, of course, be adapted for special cases such as large buildings with low ratios of heat transmitting surface to cubical volume. The type of occupancy of a building will have considerable bearing on the number of heat transmitting units and the location selected. It is axiomatic, however, that the basis of any successful heating system is to provide the maximum amount of heat at the points of maximum heat loss such as exposed walls, windows, and doors. Large roof areas, wind direction, and wide doorways create problems of excessive heat loss and require special consideration and treatment.

Heat Transmission

Heat transfer is an important parameter to be considered during selection of the exterior wall style, material and window area. A high rate of transfer will permit greater heat loss during the wintertime with the resultant increase in heating energy costs and a greater rate of heat gain in the summer with proportionally greater cooling cost. Several terms are used to describe various aspects of heat transfer. However, for general estimating purposes this book lists U values for systems of construction materials. U is the "overall heat transfer coefficient." It is defined as the heat flow per hour through one square foot when the temperature difference in the air on either side of the structure wall, roof, ceiling or floor is one degree Fahrenheit. The structural segment may be a single homogeneous material or a composite.

Total heat transfer is found using the following equation:

$Q = AU(T_2 - T_1)$ where
- Q = Heat flow, BTU per hour
- A = Area, square feet
- U = Overall heat transfer coefficient
- $(T_2 - T_1)$ = Difference in temperature of air on each side of the construction component. (Also abbreviated TD)

Note that heat can flow through all surfaces of any building and this flow is in addition to heat gain or loss due to ventilation, infiltration and generation (appliances, machinery, people).

R15500-020 Heating Approximations for Quick Estimating

Oil Piping & Boiler Room Piping
- Small System .... 20 to 30% of Boiler
- Complex System with Pumps, Headers, Etc. .... 80 to 110% of Boiler

Breeching With Insulation:
- Small .... 10 to 15% of Boiler
- Large .... 15 to 25% of Boiler

Coils: .... 15 to 30% of Containing Unit

Balancing (Independent) .... 1/2% of H.V.A.C. Estimate

Quality/Complexity Adjustment: For all heating installations add these adjustments to the estimate to more closely allow for the equipment and conditions of the particular job under consideration.
- Economy installation, add .... 0 to 5% of System
- Good quality, medium complexity, add .... 5 to 15% of System
- Above average quality and complexity, add .... 15 to 25% of System

Mechanical — R155 Heating

R15500-030 The Basics of a Heating System

The function of a heating system is to achieve and maintain a desired temperature in a room or building by replacing the amount of heat being dissipated. There are four kinds of heating systems: hot-water, steam, warm-air and electric resistance. Each has certain essential and similar elements with the exception of electric resistance heating which is covered in Section 16.

The basic elements of a heating system are:

A. A **combustion chamber** in which fuel is burned and heat transferred to a conveying medium.

B. The **"fluid"** used for conveying the heat (water, steam or air).

C. **Conductors** or pipes for transporting the fluid to specific desired locations.

D. A means of disseminating the heat, sometimes called **terminal units**.

A. The **combustion chamber** in a furnace heats air which is then distributed. This is called a warm-air system.

The combustion chamber in a boiler heats water which is either distributed as hot water or steam and this is termed a hydronic system.

The maximum allowable working pressures are limited by ASME "Code for Heating Boilers" to 15 PSI for steam and 160 PSI for hot water heating boilers, with a maximum temperature limitation of 250°F. Hot water boilers are generally rated for a working pressure of 30 PSI. High pressure boilers are governed by the ASME "Code for Power Boilers" which is used almost universally for boilers operating over 15 PSIG. High pressure boilers used for a combination of heating/process loads are usually designed for 150 PSIG.

Boiler ratings are usually indicated as either Gross or Net Output. The Gross Load is equal to the Net Load plus a piping and pickup allowance. When this allowance cannot be determined, divide the gross output rating by 1.25 for a value equal to or greater than the net heat loss requirement of the building.

B. Of the three **fluids** used, steam carries the greatest amount of heat per unit volume. This is due to the fact that it gives up its latent heat of vaporization at a temperature considerably above room temperature. Another advantage is that the pressure to produce a positive circulation is readily available. Piping conducts the steam to terminal units and returns condensate to the boiler.

The **steam system** is well adapted to large buildings because of its positive circulation, its comparatively economical installation and its ability to deliver large quantities of heat. Nearly all large office buildings, stores, hotels, and industrial buildings are so heated, in addition to many residences.

Hot water, when used as the heat carrying fluid, gives up a portion of its sensible heat and then returns to the boiler or heating apparatus for reheating. As the heat conveyed by each pound of water is about one-fiftieth of the heat conveyed by a pound of steam, it is necessary to circulate about fifty times as much water as steam by weight (although only one-thirtieth as much by volume). The hot water system is usually, although not necessarily, designed to operate at temperatures below that of the ordinary steam system and so the amount of heat transfer surface must be correspondingly greater. A temperature of 190°F to 200°F is normally the maximum. Circulation in small buildings may depend on the difference in density between hot water and the cool water returning to the boiler; circulating pumps are normally used to maintain a desired rate of flow. Pumps permit a greater degree of flexibility and better control.

In **warm-air** furnace systems, cool air is taken from one or more points in the building, passed over the combustion chamber and flue gas passages and then distributed through a duct system. A disadvantage of this system is that the ducts take up much more building volume than steam or hot water pipes. Advantages of this system are the relative ease with which humidification can be accomplished by the evaporation of water as the air circulates through the heater, and the lack of need for expensive disseminating units as the warm air simply becomes part of the interior atmosphere of the building.

C. **Conductors** (pipes and ducts) have been lightly treated in the discussion of conveying fluids. For more detailed information such as sizing and distribution methods, the reader is referred to technical publications such as the American Society of Heating, Refrigerating and Air-Conditioning Engineers "Handbook of Fundamentals."

D. **Terminal units** come in an almost infinite variety of sizes and styles, but the basic principles of operation are very limited. As previously mentioned, warm-air systems require only a simple register or diffuser to mix heated air with that present in the room. Special application items such as radiant coils and infrared heaters are available to meet particular conditions but are not usually considered for general heating needs. Most heating is accomplished by having air flow over coils or pipes containing the heat transporting medium (steam, hot-water, electricity). These units, while varied, may be separated into two general types, (1) radiator/convectors and (2) unit heaters.

Radiator/convectors may be cast, fin-tube or pipe assemblies. They may be direct, indirect, exposed, concealed or mounted within a cabinet enclosure, upright or baseboard style. These units are often collectively referred to as "radiatiors" or "radiation" although none gives off heat either entirely by radiation or by convection but rather a combination of both. The air flows over the units as a gravity "current." It is necessary to have one or more heat-emitting units in each room. The most efficient placement is low along an outside wall or under a window to counteract the cold coming into the room and achieve an even distribution.

In contrast to radiator/convectors which operate most effectively against the walls of smaller rooms, **unit heaters** utilize a fan to move air over heating coils and are very effective in locations of relatively large volume. Unit heaters, while usually suspended overhead, may be floor mounted. They also may take in fresh outside air for ventilation. The heat distributed by unit heaters may be from a remote source and conveyed by a fluid or it may be from the combustion of fuel in each individual heater. In the latter case the only piping required would be for fuel, however, a vent for the products of combustion would be necessary.

The following list gives may of the advantages of unit heaters for applications other than office or residential:

a. Large capacity so smaller number of units are required, **b.** Piping system simplified, **c.** Space saved where they are located overhead out of the way, **d.** Rapid heating directed where needed with effective wide distribution, **e.** Difference between floor and ceiling temperature reduced, **f.** Circulation of air obtained, and ventilation with introduction of fresh air possible, **g.** Heat output flexible and easily controlled.

Mechanical | R155 | Heating

R15500-050 Factor for Determining Heat Loss for Various Types of Buildings

General: While the most accurate estimates of heating requirements would naturally be based on detailed information about the building being considered, it is possible to arrive at a reasonable approximation using the following procedure:

1. Calculate the cubic volume of the room or building.
2. Select the appropriate factor from Table 1 below. Note that the factors apply only to inside temperatures listed in the first column and to 0°F outside temperature.
3. If the building has bad north and west exposures, multiply the heat loss factor by 1.1.
4. If the outside design temperature is other than 0°F, multiply the factor from Table 1 by the factor from Table 2.
5. Multiply the cubic volume by the factor selected from Table 1. This will give the estimated BTUH heat loss which must be made up to maintain inside temperature.

Table 1 — Building Type	Conditions	Qualifications	Loss Factor*
Factories & Industrial Plants General Office Areas at 70°F	One Story	Skylight in Roof	6.2
		No Skylight in Roof	5.7
	Multiple Story	Two Story	4.6
		Three Story	4.3
		Four Story	4.1
		Five Story	3.9
		Six Story	3.6
	All Walls Exposed	Flat Roof	6.9
		Heated Space Above	5.2
	One Long Warm Common Wall	Flat Roof	6.3
		Heated Space Above	4.7
	Warm Common Walls on Both Long Sides	Flat Roof	5.8
		Heated Space Above	4.1
Warehouses at 60°F	All Walls Exposed	Skylights in Roof	5.5
		No Skylight in Roof	5.1
		Heated Space Above	4.0
	One Long Warm Common Wall	Skylight in Roof	5.0
		No Skylight in Roof	4.9
		Heated Space Above	3.4
	Warm Common Walls on Both Long Sides	Skylight in Roof	4.7
		No Skylight in Roof	4.4
		Heated Space Above	3.0

*Note: This table tends to be conservative particularly for new buildings designed for minimum energy consumption.

Table 2 — Outside Design Temperature Correction Factor (for Degrees Fahrenheit)									
Outside Design Temperature	50	40	30	20	10	0	-10	-20	-30
Correction Factor	.29	.43	.57	.72	.86	1.00	1.14	1.28	1.43

R15500-070 and R15500-080 Transmission of Heat

R15500-070 and R15500-080 provide a way to calculate heat transmission of various construction materials from their U values and the TD (Temperature Difference).

1. From the Exterior Enclosure Division or elsewhere, determine U values for the construction desired.
2. Determine the coldest design temperature. The difference between this temperature and the desired interior temperature is the TD (temperature difference).
3. Enter R15500-070 or R15500-080 at correct U Value. Cross horizontally to the intersection with appropriate TD. Read transmission per square foot from bottom of figure.
4. Multiply this value of BTU per hour transmission per square foot of area by the total surface area of that type of construction.

Mechanical | R155 Heating

R15500-070 Transmission of Heat

R15500-080 Transmission of Heat (Low Rate)

Mechanical — R157 Air Conditioning & Ventilation

R15700-010 Air Conditioning

General: The purpose of air conditioning is to control the environment of a space so that comfort is provided for the occupants and/or conditions are suitable for the processes or equipment contained therein. The several items which should be evaluated to define system objectives are:

Temperature Control
Humidity Control
Cleanliness
Odor, smoke and fumes
Ventilation

Efforts to control the above parameters must also include consideration of the degree or tolerance of variation, the noise level introduced, the velocity of air motion and the energy requirements to accomplish the desired results.

The variation in **temperature** and **humidity** is a function of the sensor and the controller. The controller reacts to a signal from the sensor and produces the appropriate suitable response in either the terminal unit, the conductor of the transporting medium (air, steam, chilled water, etc.), or the source (boiler, evaporating coils, etc.).

The **noise level** is a by-product of the energy supplied to moving components of the system. Those items which usually contribute the most noise are pumps, blowers, fans, compressors and diffusers. The level of noise can be partially controlled through use of vibration pads, isolators, proper sizing, shields, baffles and sound absorbing liners.

Some **air motion** is necessary to prevent stagnation and stratification. The maximum acceptable velocity varies with the degree of heating or cooling which is taking place. Most people feel air moving past them at velocities in excess of 25 FPM as an annoying draft. However, velocities up to 45 FPM may be acceptable in certain cases. Ventilation, expressed as air changes per hour and percentage of fresh air, is usually an item regulated by local codes.

Selection of the system to be used for a particular application is usually a trade-off. In some cases the building size, style, or room available for mechanical use limits the range of possibilities. Prime factors influencing the decision are first cost and total life (operating, maintenance and replacement costs). The accuracy with which each parameter is determined will be an important measure of the reliability of the decision and subsequent satisfactory operation of the installed system.

Heat delivery may be desired from an air conditioning system. Heating capability usually is added as follows: A gas fired burner or hot water/steam/electric coils may be added to the air handling unit directly and heat all air equally. For limited or localized heat requirements the water/steam/electric coils may be inserted into the duct branch supplying the cold areas. Gas fired duct furnaces are also available.

Note: When water or steam coils are used the cost of the piping and boiler must also be added. For a rough estimate use the cost per square foot of the appropriate sized hydronic system with unit heaters. This will provide a cost for the boiler and piping, and the unit heaters of the system would equate to the approximate cost of the heating coils.

R15700-020 Air Conditioning Requirements

BTU's per hour per S.F. of floor area and S.F. per ton of air conditioning.

Type of Building	BTU per S.F.	S.F. per Ton	Type of Building	BTU per S.F.	S.F. per Ton	Type of Building	BTU per S.F.	S.F. per Ton
Apartments, Individual	26	450	Dormitory, Rooms	40	300	Libraries	50	240
Corridors	22	550	Corridors	30	400	Low Rise Office, Exterior	38	320
Auditoriums & Theaters	40	300/18*	Dress Shops	43	280	Interior	33	360
Banks	50	240	Drug Stores	80	150	Medical Centers	28	425
Barber Shops	48	250	Factories	40	300	Motels	28	425
Bars & Taverns	133	90	High Rise Office—Ext. Rms.	46	263	Office (small suite)	43	280
Beauty Parlors	66	180	Interior Rooms	37	325	Post Office, Individual Office	42	285
Bowling Alleys	68	175	Hospitals, Core	43	280	Central Area	46	260
Churches	36	330/20*	Perimeter	46	260	Residences	20	600
Cocktail Lounges	68	175	Hotel, Guest Rooms	44	275	Restaurants	60	200
Computer Rooms	141	85	Corridors	30	400	Schools & Colleges	46	260
Dental Offices	52	230	Public Spaces	55	220	Shoe Stores	55	220
Dept. Stores, Basement	34	350	Industrial Plants, Offices	38	320	Shop'g. Ctrs., Supermarkets	34	350
Main Floor	40	300	General Offices	34	350	Retail Stores	48	250
Upper Floor	30	400	Plant Areas	40	300	Specialty	60	200

*Persons per ton
12,000 BTU = 1 ton of air conditioning

Mechanical — R157 Air Conditioning & Ventilation

R15700-030 Psychrometric Table

Dewpoint or Saturation Temperature (F)

Relative humidity (%)	32	35	40	45	50	55	60	65	70	75	80	85	90	95	100
100	32	35	40	45	50	55	60	65	70	75	80	85	90	95	100
90	30	33	37	42	47	52	57	62	67	72	77	82	87	92	97
80	27	30	34	39	44	49	54	58	64	68	73	78	83	88	93
70	24	27	31	36	40	45	50	55	60	64	69	74	79	84	88
60	20	24	28	32	36	41	46	51	55	60	65	69	74	79	83
50	16	20	24	28	33	36	41	46	50	55	60	64	69	73	78
40	12	15	18	23	27	31	35	40	45	49	53	58	62	67	71
30	8	10	14	18	21	25	29	33	37	42	46	50	54	59	62
20	6	7	8	9	13	16	20	24	28	31	35	40	43	48	52
10	4	4	5	5	6	8	9	10	13	17	20	24	27	30	34
	32	35	40	45	50	55	60	65	70	75	80	85	90	95	100

Dry bulb temperature (F)

This table shows the relationship between RELATIVE HUMIDITY, DRY BULB TEMPERATURE AND DEWPOINT.

As an example, assume that the thermometer in a room reads 75°F, and we know that the relative humidity is 50%. The chart shows the dewpoint temperature to be 55°. That is, any surface colder than 55°F will "sweat" or collect condensing moisture. This surface could be the outside of an uninsulated chilled water pipe in the summertime, or the inside surface of a wall or deck in the wintertime. After determining the extreme ambient parameters, the table at the left is useful in determining which surfaces need insulation or vapor barrier protection.

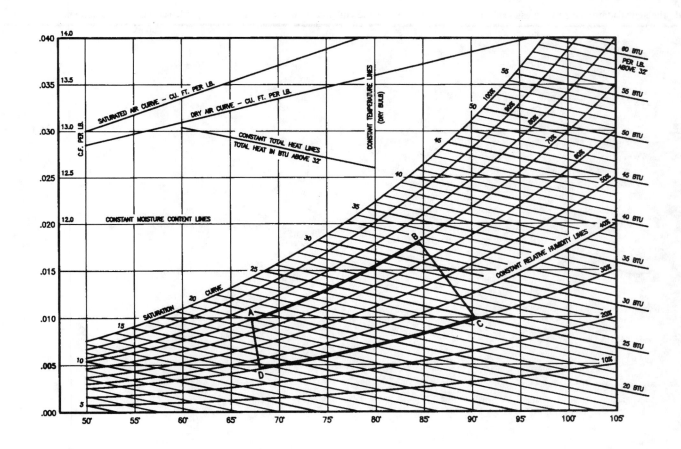

TEMPERATURE DEGREES FAHRENHEIT
TOTAL PRESSURE = 14.696 LB. PER SQ. IN. ABS.

Psychrometric chart showing different variables based on one pound of dry air. Space marked A B C D is temperature-humidity range which is most comfortable for majority of people.

Mechanical | R157 | Air Conditioning & Ventilation

R15700-040 Recommended Ventilation Air Changes

Table below lists range of time in minutes per change for various types of facilities.

Facility	Min/Change	Facility	Min/Change	Facility	Min/Change
Assembly Halls	2-10	Dance Halls	2-10	Laundries	1-3
Auditoriums	2-10	Dining Rooms	3-10	Markets	2-10
Bakeries	2-3	Dry Cleaners	1-5	Offices	2-10
Banks	3-10	Factories	2-5	Pool Rooms	2-5
Bars	2-5	Garages	2-10	Recreation Rooms	2-10
Beauty Parlors	2-5	Generator Rooms	2-5	Sales Rooms	2-10
Boiler Rooms	1-5	Gymnasiums	2-10	Theaters	2-8
Bowling Alleys	2-10	Kitchens-Hospitals	2-5	Toilets	2-5
Churches	5-10	Kitchens-Restaurant	1-3	Transformer Rooms	1-5

CFM air required for changes = Volume of room in cubic feet ÷ Minutes per change.

R15700-090 Quality/Complexity Adjustment for Air Conditioning Systems

Economy installation, add	0 to 5%
Good quality, medium complexity, add	5 to 15%
Above average quality and complexity, add	15 to 25%

Add the above adjustments to the estimate to more closely allow for the equipment and conditions of the particular job under consideration.

Mechanical — R158 Air Distribution

R15800-080 Loudness Levels for Moving Air thru Fans, Diffusers, Register, Etc. (Measured in Sones)

Recommended Loudness Levels (Sones)

Area	Very Quiet	Quiet	Noisy	Area	Very Quiet	Quiet	Noisy	Area	Very Quiet	Quiet	Noisy
Auditoriums				**Hotels**				**Offices**			
Auditorium lobbies	3	4	6	Banquet Rooms	1.5	3	6	Conference rooms	1	1.7	3
Concert and opera halls	0.8	1	1.5	Individual rooms, suites	1	2	4	Drafting	2	4	8
Courtrooms	2	3	4	Kitchens and laundries	4	7	10	General open offices	2	4	8
Lecture halls	1.5	2	3	Lobbies	2	4	8	Halls and corridors	2.5	5	10
Movie theaters	1.5	2	3	**Indoor Sports**				Professional offices	1.5	3	6
Churches and Schools				Bowling alleys	3	4	6	Reception room	1.5	3	6
Kitchens	4	6	8	Gymnasiums	2	4	6	Tabulation & computation	3	6	12
Laboratories	2	4	6	Swimming pools	4	7	10	**Public Buildings**			
Libraries	1.5	2	3	**Manufacturing Areas**				Banks	2	4	6
Recreation halls	2	4	8	Assembly lines	5	12	30	Court houses	2.5	4	6
Sanctuaries	1	1.7	3	Foreman's office	3	5	8	Museums	2	3	4
Schools and classrooms	1.5	2.5	4	Foundries	10	20	40	Planetariums	1.5	2	3
Hospital and Clinics				General storage	5	10	20	Post offices	2.5	4	6
Laboratories	2	4	6	Heavy machinery	10	25	60	Public libraries	1.5	2	4
Lobbies, waiting rooms	2	4	6	Light machinery	5	12	30	Waiting rooms	3	5	8
Halls and corridors	2	4	6	Tool maintenance	4	7	10	**Restaurants**			
Operating rooms	1.5	2.5	4	**Stores**				Cafeterias	3	6	10
Private rooms	1	1.7	3	Department stores	3	6	8	Night clubs	2.5	4	6
Wards	1.5	2.5	4	Supermarkets	4	7	10	Restaurants	2	4	8

Mechanical | R158 | Air Distribution

R15810-050 Ductwork

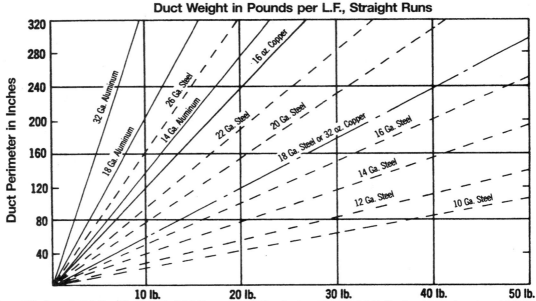

Duct Weight in Pounds per L.F., Straight Runs

Add to the above for fittings; 90° elbow is 3 L.F.; 45° elbow is 2.5 L.F.; offset is 4 L.F.; transition offset is 6 L.F.; square-to-round transition is 4 L.F.; 90° reducing elbow is 5 L.F. For bracing and waste, add 20% to aluminum and copper, 15% to steel.

R15810-070 Duct Fabrication/Installation

The labor cost for sheet metal duct using lines 15810-600-0070 thru 1060 includes both the cost of fabrication and installation of the duct. The split is approximately 40% for fabrication, 60% for installation. It is for this reason that the percentage add for elevated installation is less than the percentage add for prefabricated duct as found starting on line 15810-600-1283.

Example: assume a piece of duct cost $100 installed (labor only)

Sheet Metal Fabrication = 40% = $40
Installation = 60% = $60

The add for elevated installation is:

$$\frac{\text{Based on total labor}}{\text{(fabrication \& installation)}} = \$100 \times 6\% = \$6.00$$

$$\frac{\text{Based on installation cost only}}{\text{(Material purchased prefabricated)}} = \$60 \times 10\% = \$6.00$$

The $6.00 markup (10' to 15' high) is the same.

Mechanical

R158 Air Distribution

R15810-100 Sheet Metal Calculator (Weight in Lb./Ft. of Length)

Gauge	26	24	22	20	18	16	Gauge	26	24	22	20	18	16
Wt.-Lb./S.F.	.906	1.156	1.406	1.656	2.156	2.656	Wt.-Lb./S.F.	.906	1.156	1.406	1.656	2.156	2.656
SMACNA Max. Dimension – Long Side		30"	54"	84"	85" Up		SMACNA Max. Dimension – Long Side		30"	54"	84"	85" Up	
Sum-2 sides							Sum-2 Sides						
2	.3	.40	.50	.60	.80	.90	56	9.3	12.0	14.0	16.2	21.3	25.2
3	.5	.65	.80	.90	1.1	1.4	57	9.5	12.3	14.3	16.5	21.7	25.7
4	.7	.85	1.0	1.2	1.5	1.8	58	9.7	12.5	14.5	16.8	22.0	26.1
5	.8	1.1	1.3	1.5	1.9	2.3	59	9.8	12.7	14.8	17.1	22.4	26.6
6	1.0	1.3	1.5	1.7	2.3	2.7	60	10.0	12.9	15.0	17.4	22.8	27.0
7	1.2	1.5	1.8	2.0	2.7	3.2	61	10.2	13.1	15.3	17.7	23.2	27.5
8	1.3	1.7	2.0	2.3	3.0	3.6	62	10.3	13.3	15.5	18.0	23.6	27.9
9	1.5	1.9	2.3	2.6	3.4	4.1	63	10.5	13.5	15.8	18.3	24.0	28.4
10	1.7	2.2	2.5	2.9	3.8	4.5	64	10.7	13.7	16.0	18.6	24.3	28.8
11	1.8	2.4	2.8	3.2	4.2	5.0	65	10.8	13.9	16.3	18.9	24.7	29.3
12	2.0	2.6	3.0	3.5	4.6	5.4	66	11.0	14.1	16.5	19.1	25.1	29.7
13	2.2	2.8	3.3	3.8	4.9	5.9	67	11.2	14.3	16.8	19.4	25.5	30.2
14	2.3	3.0	3.5	4.1	5.3	6.3	68	11.3	14.6	17.0	19.7	25.8	30.6
15	2.5	3.2	3.8	4.4	5.7	6.8	69	11.5	14.8	17.3	20.0	26.2	31.1
16	2.7	3.4	4.0	4.6	6.1	7.2	70	11.7	15.0	17.5	20.3	26.6	31.5
17	2.8	3.7	4.3	4.9	6.5	7.7	71	11.8	15.2	17.8	20.6	27.0	32.0
18	3.0	3.9	4.5	5.2	6.8	8.1	72	12.0	15.4	18.0	20.9	27.4	32.4
19	3.2	4.1	4.8	5.5	7.2	8.6	73	12.2	15.6	18.3	21.2	27.7	32.9
20	3.3	4.3	5.0	5.8	7.6	9.0	74	12.3	15.8	18.5	21.5	28.1	33.3
21	3.5	4.5	5.3	6.1	8.0	9.5	75	12.5	16.1	18.8	21.8	28.5	33.8
22	3.7	4.7	5.5	6.4	8.4	9.9	76	12.7	16.3	19.0	22.0	28.9	34.2
23	3.8	5.0	5.8	6.7	8.7	10.4	77	12.8	16.5	19.3	22.3	29.3	34.7
24	4.0	5.2	6.0	7.0	9.1	10.8	78	13.0	16.7	19.5	22.6	29.6	35.1
25	4.2	5.4	6.3	7.3	9.5	11.3	79	13.2	16.9	19.8	22.9	30.0	35.6
26	4.3	5.6	6.5	7.5	9.9	11.7	80	13.3	17.1	20.0	23.2	30.4	36.0
27	4.5	5.8	6.8	7.8	10.3	12.2	81	13.5	17.3	20.3	23.5	30.8	36.5
28	4.7	6.0	7.0	8.1	10.6	12.6	82	13.7	17.5	20.5	23.8	31.2	36.9
29	4.8	6.2	7.3	8.4	11.0	13.1	83	13.8	17.8	20.8	24.1	31.5	37.4
30	5.0	6.5	7.5	8.7	11.4	13.5	84	14.0	18.0	21.0	24.4	31.9	37.8
31	5.2	6.7	7.8	9.0	11.8	14.0	85	14.2	18.2	21.3	24.7	32.3	38.3
32	5.3	6.9	8.0	9.3	12.2	14.4	86	14.3	18.4	21.5	24.9	32.7	38.7
33	5.5	7.1	8.3	9.6	12.5	14.9	87	14.5	18.6	21.8	25.2	33.1	39.2
34	5.7	7.3	8.5	9.9	12.9	15.3	88	14.7	18.8	22.0	25.5	33.4	39.6
35	5.8	7.5	8.8	10.2	13.3	15.8	89	14.8	19.0	22.3	25.8	33.8	40.1
36	6.0	7.8	9.0	10.4	13.7	16.2	90	15.0	19.3	22.5	26.1	34.2	40.5
37	6.2	8.0	9.3	10.7	14.1	16.7	91	15.2	19.5	22.8	26.4	34.6	41.0
38	6.3	8.2	9.5	11.0	14.4	17.1	92	15.3	19.7	23.0	26.7	35.0	41.4
39	6.5	8.4	9.8	11.3	14.8	17.6	93	15.5	19.9	23.3	27.0	35.3	41.9
40	6.7	8.6	10.0	11.6	15.2	18.0	94	15.7	20.1	23.5	27.3	35.7	42.3
41	6.8	8.8	10.3	11.9	15.6	18.5	95	15.8	20.3	23.8	27.6	36.1	42.8
42	7.0	9.0	10.5	12.2	16.0	18.9	96	16.0	20.5	24.0	27.8	36.5	43.2
43	7.2	9.2	10.8	12.5	16.3	19.4	97	16.2	20.8	24.3	28.1	36.9	43.7
44	7.3	9.5	11.0	12.8	16.7	19.8	98	16.3	21.0	24.5	28.4	37.2	44.1
45	7.5	9.7	11.3	13.1	17.1	20.3	99	16.5	21.2	24.8	28.7	37.6	44.6
46	7.7	9.9	11.5	13.3	17.5	20.7	100	16.7	21.4	25.0	29.0	38.0	45.0
47	7.8	10.1	11.8	13.6	17.9	21.2	101	16.8	21.6	25.3	29.3	38.4	45.5
48	8.0	10.3	12.0	13.9	18.2	21.6	102	17.0	21.8	25.5	29.6	38.8	45.9
49	8.2	10.5	12.3	14.2	18.6	22.1	103	17.2	22.0	25.8	29.9	39.1	46.4
50	8.3	10.7	12.5	14.5	19.0	22.5	104	17.3	22.3	26.0	30.2	39.5	46.8
51	8.5	11.0	12.8	14.8	19.4	23.0	105	17.5	22.5	26.3	30.5	39.9	47.3
52	8.7	11.2	13.0	15.1	19.8	23.4	106	17.7	22.7	26.5	30.7	40.3	47.7
53	8.8	11.4	13.3	15.4	20.1	23.9	107	17.8	22.9	26.8	31.0	40.7	48.2
54	9.0	11.6	13.5	15.7	20.5	24.3	108	18.0	23.1	27.0	31.3	41.0	48.6
55	9.2	11.8	13.8	16.0	20.9	24.8	109	18.2	23.3	27.3	31.6	41.4	49.1
							110	18.3	23.5	27.5	31.9	41.8	49.5

Example: If duct is 34" x 20" x 15' long, 34" is greater than 30" maximum, for 24 ga. so must be 22 ga. 34" + 20" = 54" going across from 54" find 13.5 lb. per foot. 13.5 x 15' = 202.5 lbs. For S.F. of surface area 202.5 ÷ 1.406 = 144 S.F.

Note: Figures include an allowance for scrap.

Mechanical — R158 Air Distribution

R15810-110 Ductwork Packages (per Ton of Cooling)

System	Sheet Metal	Insulation	Diffusers	Return Register
Roof Top Unit Single Zone	120 Lbs.	52 S.F.	1	1
Roof Top Unit Multizone	240 Lbs.	104 S.F.	2	1
Self-contained Air or Water Cooled	108 Lbs.	—	2	—
Split System Air Cooled	102 Lbs.	—	2	—

Systems reflect most common usage.
Refer to system graphics for duct layout.

R15850-060 Diffuser Evaluation

CFM = $V \times A_n \times K$ where V = Outlet velocity in feet per minute. A_n = Neck area in square feet and K = Diffuser delivery factor. An undersized diffuser for a desired CFM will produce a high velocity and noise level. When air moves past people at a velocity in excess of 25 FPM, an annoying draft is felt. An oversized diffuser will result in low velocity with poor mixing. Consideration must be given to avoid vertical stratification or horizontal areas of stagnation.

S.F. & C.F. Costs R171 | Information

R17100-100 Square Foot Project Size Modifier

One factor that affects the S.F. cost of a particular building is the size. In general, for buildings built to the same specifications in the same locality, the larger building will have the lower S.F. cost. This is due mainly to the decreasing contribution of the exterior walls plus the economy of scale usually achievable in larger buildings. The Area Conversion Scale shown below will give a factor to convert costs for the typical size building to an adjusted cost for the particular project.

The Square Foot Base Size lists the median costs, most typical project size in our accumulated data and the range in size of the projects.

The Size Factor for your project is determined by dividing your project area in S.F. by the typical project size for the particular Building Type. With this factor, enter the Area Conversion Scale at the appropriate Size Factor and determine the appropriate cost multiplier for your building size.

Example: Determine the cost per S.F. for a 100,000 S.F. Mid-rise apartment building.

$$\frac{\text{Proposed building area} = 100{,}000 \text{ S.F.}}{\text{Typical size from below} = 50{,}000 \text{ S.F.}} = 2.00$$

Enter Area Conversion scale at 2.0, intersect curve, read horizontally the appropriate cost multiplier of .94. Size adjusted cost becomes .94 x $69.50 = $65.35 based on national average costs.

Note: For Size Factors less than .50, the Cost Multiplier is 1.1
      For Size Factors greater than 3.5, the Cost Multiplier is .90

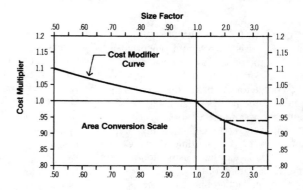

Square Foot Base Size

Building Type	Median Cost per S.F.	Typical Size Gross S.F.	Typical Range Gross S.F.	Building Type	Median Cost per S.F.	Typical Size Gross S.F.	Typical Range Gross S.F.
Apartments, Low Rise	$ 55.05	21,000	9,700 - 37,200	Jails	$168.00	40,000	5,500 - 145,000
Apartments, Mid Rise	69.50	50,000	32,000 - 100,000	Libraries	99.15	12,000	7,000 - 31,000
Apartments, High Rise	79.70	145,000	95,000 - 600,000	Medical Clinics	94.80	7,200	4,200 - 15,700
Auditoriums	92.05	25,000	7,600 - 39,000	Medical Offices	89.05	6,000	4,000 - 15,000
Auto Sales	56.95	20,000	10,800 - 28,600	Motels	68.25	40,000	15,800 - 120,000
Banks	123.00	4,200	2,500 - 7,500	Nursing Homes	91.55	23,000	15,000 - 37,000
Churches	83.15	17,000	2,000 - 42,000	Offices, Low Rise	74.30	20,000	5,000 - 80,000
Clubs, Country	82.85	6,500	4,500 - 15,000	Offices, Mid Rise	78.05	120,000	20,000 - 300,000
Clubs, Social	80.65	10,000	6,000 - 13,500	Offices, High Rise	99.85	260,000	120,000 - 800,000
Clubs, YMCA	83.10	28,300	12,800 - 39,400	Police Stations	125.00	10,500	4,000 - 19,000
Colleges (Class)	109.00	50,000	15,000 - 150,000	Post Offices	92.00	12,400	6,800 - 30,000
Colleges (Science Lab)	159.00	45,600	16,600 - 80,000	Power Plants	691.00	7,500	1,000 - 20,000
College (Student Union)	121.00	33,400	16,000 - 85,000	Religious Education	76.25	9,000	6,000 - 12,000
Community Center	86.60	9,400	5,300 - 16,700	Research	130.00	19,000	6,300 - 45,000
Court Houses	118.00	32,400	17,800 - 106,000	Restaurants	112.00	4,400	2,800 - 6,000
Dept. Stores	51.45	90,000	44,000 - 122,000	Retail Stores	54.65	7,200	4,000 - 17,600
Dormitories, Low Rise	88.75	25,000	10,000 - 95,000	Schools, Elementary	79.60	41,000	24,500 - 55,000
Dormitories, Mid Rise	116.00	85,000	20,000 - 200,000	Schools, Jr. High	81.10	92,000	52,000 - 119,000
Factories	49.85	26,400	12,900 - 50,000	Schools, Sr. High	81.10	101,000	50,500 - 175,000
Fire Stations	87.05	5,800	4,000 - 8,700	Schools, Vocational	80.75	37,000	20,500 - 82,000
Fraternity Houses	85.65	12,500	8,200 - 14,800	Sports Arenas	67.70	15,000	5,000 - 40,000
Funeral Homes	95.70	10,000	4,000 - 20,000	Supermarkets	54.85	44,000	12,000 - 60,000
Garages, Commercial	60.80	9,300	5,000 - 13,600	Swimming Pools	127.00	20,000	10,000 - 32,000
Garages, Municipal	77.75	8,300	4,500 - 12,600	Telephone Exchange	147.00	4,500	1,200 - 10,600
Garages, Parking	31.90	163,000	76,400 - 225,300	Theaters	81.20	10,500	8,800 - 17,500
Gymnasiums	80.45	19,200	11,600 - 41,000	Town Halls	89.30	10,800	4,800 - 23,400
Hospitals	152.00	55,000	27,200 - 125,000	Warehouses	36.80	25,000	8,000 - 72,000
House (Elderly)	75.25	37,000	21,000 - 66,000	Warehouse & Office	42.50	25,000	8,000 - 72,000
Housing (Public)	69.70	36,000	14,400 - 74,400				
Ice Rinks	77.40	29,000	27,200 - 33,600				

Change Orders

Change Order Considerations

A Change Order is a written document, usually prepared by the design professional, and signed by the owner, the architect/engineer and the contractor. A change order states the agreement of the parties to: an addition, deletion, or revision in the work; an adjustment in the contract sum, if any; or an adjustment in the contract time, if any. Change orders, or "extras" in the construction process occur after execution of the construction contract and impact architects/engineers, contractors and owners.

Change orders that are properly recognized and managed can ensure orderly, professional and profitable progress for all who are involved in the project. There are many causes for change orders and change order requests. In all cases, change orders or change order requests should be addressed promptly and in a precise and prescribed manner. The following paragraphs include information regarding change order pricing and procedures.

The Causes of Change Orders

Reasons for issuing change orders include:

- Unforeseen field conditions that require a change in the work
- Correction of design discrepancies, errors or omissions in the contract documents
- Owner-requested changes, either by design criteria, scope of work, or project objectives
- Completion date changes for reasons unrelated to the construction process
- Changes in building code interpretations, or other public authority requirements that require a change in the work
- Changes in availability of existing or new materials and products

Procedures

Properly written contract documents must include the correct change order procedures for all parties—owners, design professionals and contractors—to follow in order to avoid costly delays and litigation.

Being "in the right" is not always a sufficient or acceptable defense. The contract provisions requiring notification and documentation must be adhered to within a defined or reasonable time frame.

The appropriate method of handling change orders is by a written proposal and acceptance by all parties involved. Prior to starting work on a project, all parties should identify their authorized agents who may sign and accept change orders, as well as any limits placed on their authority.

Time may be a critical factor when the need for a change arises. For such cases, the contractor might be directed to proceed on a "time and materials" basis, rather than wait for all paperwork to be processed—a delay that could impede progress. In this situation, the contractor must still follow the prescribed change order procedures, including but not limited to, notification and documentation.

All forms used for change orders should be dated and signed by the proper authority. Lack of documentation can be very costly, especially if legal judgments are to be made and if certain field personnel are no longer available. For time and material change orders, the contractor should keep accurate daily records of all labor and material allocated to the change. Forms that can be used to document change order work are available in *Means Forms for Building Construction Professionals*.

Owners or awarding authorities who do considerable and continual building construction (such as the federal government) realize the inevitability of change orders for numerous reasons, both predictable and unpredictable. As a result, the federal government, the American Institute of Architects (AIA), the Engineers Joint Contract Documents Committee (EJCDC) and other contractor, legal and technical organizations have developed standards and procedures to be followed by all parties to achieve contract continuance and timely completion, while being financially fair to all concerned.

In addition to the change order standards put forth by industry associations, there are also many books available on the subject.

Pricing Change Orders

When pricing change orders, regardless of their cause, the most significant factor is *when* the change occurs. The need for a change may be perceived in the field or requested by the architect/engineer *before* any of the actual installation has begun, or may evolve or appear *during* construction when the item of work in question is partially installed. In the latter cases, the original sequence of construction is disrupted, along with all contiguous and supporting systems. Change orders cause the greatest impact when they occur *after* the installation has been completed and must be uncovered, or even replaced. Post-completion changes may be caused by necessary design changes, product failure, or changes in the owner's requirements that are not discovered until the building or the systems begin to function.

Specified procedures of notification and record keeping must be adhered to and enforced regardless of the stage of construction: *before*, *during*, or *after* installation. Some bidding documents anticipate change orders by requiring that unit prices including overhead and profit percentages—for additional as well as deductible changes—be listed. Generally these unit prices do not fully take into account the ripple effect, or impact on other trades, and should be used for general guidance only.

When pricing change orders, it is important to classify the time frame in which the change occurs. There are two basic time frames for change orders: *pre-installation change orders*, which occur before the start of construction, and *post-installation change orders*, which involve reworking after the original installation. Change orders that occur between these stages may be priced according to the extent of work completed using a combination of techniques developed for pricing *pre-* and *post-installation* changes.

The following factors are the basis for a check list to use when preparing a change order estimate.

Factors To Consider When Pricing Change Orders

As an estimator begins to prepare a change order, the following questions should be reviewed to determine their impact on the final price.

General

- Is the change order work *pre-installation* or *post-installation*?

  Change order work costs vary according to how much of the installation has been completed. Once workers have the project scoped in their mind, even though they have not started, it can be difficult to refocus. Consequently they may spend more than the normal amount of time understanding the change. Also, modifications to work in place such as trimming or refitting usually take more time than was initially estimated. The greater the amount of work in place, the more reluctant workers are to change it. Psychologically they may resent the change and as a result the rework takes longer than normal. Post-installation change order estimates must include demolition of existing work as required to accomplish the change. If the work is performed at a later time, additional obstacles such as building finishes may be present which must be protected. Regardless of whether the change occurs pre-installation or post-installation, attempt to isolate the identifiable factors and price them separately. For example, add shipping costs that may be required pre-installation or any demolition required post-installation. Then analyze the potential impact on productivity of psychological and/or learning curve factors and adjust the output rates accordingly. One approach is to break down the typical workday into segments and quantify the impact on each segment. The following chart may be useful as a guide:

	Activities (Productivity) Expressed as Percentages of a Workday		
Task	Means Mechanical Cost Data (for New Construction)	Pre-Installation Change Orders	Post-Installation Change Orders
1. Study plans	3%	6%	6%
2. Material procurement	3%	3%	3%
3. Receiving and storing	3%	3%	3%
4. Mobilization	5%	5%	5%
5. Site movement	5%	5%	8%
6. Layout and marking	8%	10%	12%
7. Actual installation	64%	59%	54%
8. Clean-up	3%	3%	3%
9. Breaks—non-productive	6%	6%	6%
Total	100%	100%	100%

Change Order Installation Efficiency

The labor-hours expressed (for new construction) are based on average installation time, using an efficiency level of approximately 60-65%. For change order situations, adjustments to this efficiency level should reflect the daily labor-hour allocation for that particular occurrence.

If any of the specific percentages expressed in the above chart do not apply to a particular project situation, then those percentage points should be reallocated to the appropriate task(s). Example: Using data for new construction, assume there is no new material being utilized. The percentages for Tasks 2 and 3 would therefore be reallocated to other tasks. If the time required for Tasks 2 and 3 can now be applied to installation, we can add the time allocated for *Material Procurement* and *Receiving and Storing* to the *Actual Installation* time for new construction, thereby increasing the Actual Installation percentage.

This chart shows that, due to reduced productivity, labor costs will be higher than those for new construction by 5% to 15% for pre-installation change orders and by 15% to 25% for post-installation change orders. Each job and change order is unique and must be examined individually. Many factors, covered elsewhere in this section, can each have a significant impact on productivity and change order costs. All such factors should be considered in every case.

- Will the change substantially delay the original completion date?

    A significant change in the project may cause the original completion date to be extended. The extended schedule may subject the contractor to new wage rates dictated by relevant labor contracts. Project supervision and other project overhead must also be extended beyond the original completion date. The schedule extension may also put installation into a new weather season. For example, underground piping scheduled for October installation was delayed until January. As a result, frost penetrated the trench area, thereby changing the degree of difficulty of the task. Changes and delays may have a ripple effect throughout the project. This effect must be analyzed and negotiated with the owner.

- What is the net effect of a deduct change order?

    In most cases, change orders resulting in a deduction or credit reflect only bare costs. The contractor may retain the overhead and profit based on the original bid.

Materials

- Will you have to pay more or less for the new material, required by the change order, than you paid for the original purchase?

    The same material prices or discounts will usually apply to materials purchased for change orders as new construction. In some instances, however, the contractor may forfeit the advantages of competitive pricing for change orders. Consider the following example:

    A contractor purchased over $20,000 worth of fan coil units for an installation, and obtained the maximum discount. Some time later it was determined the project required an additional matching unit. The contractor has to purchase this unit from the original supplier to ensure a match. The supplier at this time may not discount the unit because of the small quantity, and the fact that he is no longer in a competitive situation. The impact of quantity on purchase can add between 0% and 25% to material prices and/or subcontractor quotes.

- If materials have been ordered or delivered to the job site, will they be subject to a cancellation charge or restocking fee?

    Check with the supplier to determine if ordered materials are subject to a cancellation charge. Delivered materials not used as result of a change order may be subject to a restocking fee if returned to the supplier. Common restocking charges run between 20% and 40%. Also, delivery charges to return the goods to the supplier must be added.

Labor

- How efficient is the existing crew at the actual installation?

    Is the same crew that performed the initial work going to do the change order? Possibly the change consists of the installation of a unit identical to one already installed; therefore the change should take less time. Be sure to consider this potential productivity increase and modify the productivity rates accordingly.

- If the crew size is increased, what impact will that have on supervision requirements?

    Under most bargaining agreements or management practices, there is a point at which a working foreman is replaced by a nonworking foreman. This replacement increases project overhead by adding a nonproductive worker. If additional workers are added to accelerate the project or to perform changes while maintaining the schedule, be sure to add additional supervision time if warranted. Calculate the hours involved and the additional cost directly if possible.

- What are the other impacts of increased crew size?

    The larger the crew, the greater the potential for productivity to decrease. Some of the factors that cause this productivity loss are: overcrowding (producing restrictive conditions in the working space), and possibly a shortage of any special tools and equipment required. Such factors affect not only the crew working on the elements directly involved in the change order, but other crews whose movement may also be hampered.

    As the crew increases, check its basic composition for changes by the addition or deletion of apprentices or nonworking foreman and quantify the potential effects of equipment shortages or other logistical factors.

- As new crews, unfamiliar with the project, are brought onto the site, how long will it take them to become oriented to the project requirements?

    The orientation time for a new crew to become 100% effective varies with the site and type of project. Orientation is easiest at a new construction site, and most difficult at existing, very restrictive renovation sites. The type of work also affects orientation time. When all elements of the work are exposed, such as concrete or masonry work, orientation is decreased. When the work is concealed or less visible, such as existing electrical systems, orientation takes longer. Usually orientation can be accomplished in one day or less. Costs for added orientation should be itemized and added to the total estimated cost.

- How much actual production can be gained by working overtime?

    Short term overtime can be used effectively to accomplish more work in a day. However, as overtime is scheduled to run beyond several weeks, studies have shown marked decreases in output. The following chart shows the effect of long term overtime on worker efficiency. If the anticipated change requires extended overtime to keep the job on schedule, these factors can be used as a guide to predict the impact on time and cost. Add project overhead, particularly supervision, that may also be incurred.

Days per Week	Hours per Day	Production Efficiency					Payroll Cost Factors	
		1 Week	2 Weeks	3 Weeks	4 Weeks	Average 4 Weeks	@ 1-1/2 Times	@ 2 Times
5	8	100%	100%	100%	100%	100%	100%	100%
	9	100	100	95	90	96.25	105.6	111.1
	10	100	95	90	85	91.25	110.0	120.0
	11	95	90	75	65	81.25	113.6	127.3
	12	90	85	70	60	76.25	116.7	133.3
6	8	100	100	95	90	96.25	108.3	116.7
	9	100	95	90	85	92.50	113.0	125.9
	10	95	90	85	80	87.50	116.7	133.3
	11	95	85	70	65	78.75	119.7	139.4
	12	90	80	65	60	73.75	122.2	144.4
7	8	100	95	85	75	88.75	114.3	128.6
	9	95	90	80	70	83.75	118.3	136.5
	10	90	85	75	65	78.75	121.4	142.9
	11	85	80	65	60	72.50	124.0	148.1
	12	85	75	60	55	68.75	126.2	152.4

Effects of Overtime

Caution: Under many labor agreements, Sundays and holidays are paid at a higher premium than the normal overtime rate.

The use of long-term overtime is counterproductive on almost any construction job; that is, the longer the period of overtime, the lower the actual production rate. Numerous studies have been conducted, and while they have resulted in slightly different numbers, all reach the same conclusion. The figure above tabulates the effects of overtime work on efficiency.

As illustrated, there can be a difference between the *actual* payroll cost per hour and the *effective* cost per hour for overtime work. This is due to the reduced production efficiency with the increase in weekly hours beyond 40. This difference between actual and effective cost results from overtime work over a prolonged period. Short-term overtime work does not result in as great a reduction in efficiency, and in such cases, effective cost may not vary significantly from the actual payroll cost. As the total hours per week are increased on a regular basis, more time is lost because of fatigue, lowered morale, and an increased accident rate.

As an example, assume a project where workers are working 6 days a week, 10 hours per day. From the figure above (based on productivity studies), the average effective productive hours over a four-week period are:

$$0.875 \times 60 = 52.5$$

Depending upon the locale and day of week, overtime hours may be paid at time and a half or double time. For time and a half, the overall (average) *actual* payroll cost (including regular and overtime hours) is determined as follows:

$$\frac{40 \text{ reg. hrs.} + (20 \text{ overtime hrs.} \times 1.5)}{60 \text{ hrs.}} = 1.167$$

Based on 60 hours, the payroll cost per hour will be 116.7% of the normal rate at 40 hours per week. However, because the effective production (efficiency) for 60 hours is reduced to the equivalent of 52.5 hours, the effective cost of overtime is calculated as follows:

For time and a half:

$$\frac{40 \text{ reg. hrs.} + (20 \text{ overtime hrs.} \times 1.5)}{52.5 \text{ hrs.}} = 1.33$$

Installed cost will be 133% of the normal rate (for labor).

Thus, when figuring overtime, the actual cost per unit of work will be higher than the apparent overtime payroll dollar increase, due to the reduced productivity of the longer workweek. These efficiency calculations are true only for those cost factors determined by hours worked. Costs that are applied weekly or monthly, such as equipment rentals, will not be similarly affected.

Equipment

- What equipment is required to complete the change order?

Change orders may require extending the rental period of equipment already on the job site, or the addition of special equipment brought in to accomplish the change work. In either case, the additional rental charges and operator labor charges must be added.

Summary

The preceding considerations and others you deem appropriate should be analyzed and applied to a change order estimate. The impact of each should be quantified and listed on the estimate to form an audit trail.

Change orders that are properly identified, documented, and managed help to ensure the orderly, professional and profitable progress of the work. They also minimize potential claims or disputes at the end of the project.

Crews

Crew No.	Bare Costs		Incl. Subs O & P		Cost Per Labor-Hour	
Crew A-1	Hr.	Daily	Hr.	Daily	Bare Costs	Incl. O&P
1 Building Laborer	$22.85	$182.80	$35.75	$286.00	$22.85	$35.75
1 Gas Eng. Power Tool		64.80		71.30	8.10	8.91
8 L.H., Daily Totals		$247.60		$357.30	$30.95	$44.66
Crew A-1A	Hr.	Daily	Hr.	Daily	Bare Costs	Incl. O&P
1 Skilled Worker	$29.85	$238.80	$46.85	$374.80	$29.85	$46.85
1 Shot Blaster, 20"		146.40		161.05	18.30	20.13
8 L.H., Daily Totals		$385.20		$535.85	$48.15	$66.98
Crew A-2	Hr.	Daily	Hr.	Daily	Bare Costs	Incl. O&P
2 Laborers	$22.85	$365.60	$35.75	$572.00	$23.03	$35.75
1 Truck Driver (light)	23.40	187.20	35.75	286.00		
1 Light Truck, 1.5 Ton		158.00		173.80	6.58	7.24
24 L.H., Daily Totals		$710.80		$1031.80	$29.61	$42.99
Crew A-2A	Hr.	Daily	Hr.	Daily	Bare Costs	Incl. O&P
2 Laborers	$22.85	$365.60	$35.75	$572.00	$23.03	$35.75
1 Truck Driver (light)	23.40	187.20	35.75	286.00		
1 Light Truck, 1.5 Ton		158.00		173.80		
1 Concrete Saw		98.40		108.25	10.68	11.75
24 L.H., Daily Totals		$809.20		$1140.05	$33.71	$47.50
Crew A-3	Hr.	Daily	Hr.	Daily	Bare Costs	Incl. O&P
1 Truck Driver (heavy)	$23.95	$191.60	$36.60	$292.80	$23.95	$36.60
1 Dump Truck, 12 Ton		370.40		407.45	46.30	50.93
8 L.H., Daily Totals		$562.00		$700.25	$70.25	$87.53
Crew A-3A	Hr.	Daily	Hr.	Daily	Bare Costs	Incl. O&P
1 Truck Driver (light)	$23.40	$187.20	$35.75	$286.00	$23.40	$35.75
1 Pickup Truck (4x4)		89.20		98.10	11.15	12.27
8 L.H., Daily Totals		$276.40		$384.10	$34.55	$48.02
Crew A-3B	Hr.	Daily	Hr.	Daily	Bare Costs	Incl. O&P
1 Equip. Oper. (medium)	$30.35	$242.80	$45.95	$367.60	$27.15	$41.28
1 Truck Driver (heavy)	23.95	191.60	36.60	292.80		
1 Dump Truck, 16 Ton		482.80		531.10		
1 F.E. Loader, 3 C.Y.		391.80		431.00	54.66	60.13
16 L.H., Daily Totals		$1309.00		$1622.50	$81.81	$101.41
Crew A-3C	Hr.	Daily	Hr.	Daily	Bare Costs	Incl. O&P
1 Equip. Oper. (light)	$28.75	$230.00	$43.55	$348.40	$28.75	$43.55
1 Wheeled Skid Steer Loader		220.00		242.00	27.50	30.25
8 L.H., Daily Totals		$450.00		$590.40	$56.25	$73.80
Crew A-4	Hr.	Daily	Hr.	Daily	Bare Costs	Incl. O&P
2 Carpenters	$29.15	$466.40	$45.60	$729.60	$28.13	$43.65
1 Painter, Ordinary	26.10	208.80	39.75	318.00		
24 L.H., Daily Totals		$675.20		$1047.60	$28.13	$43.65
Crew A-5	Hr.	Daily	Hr.	Daily	Bare Costs	Incl. O&P
2 Laborers	$22.85	$365.60	$35.75	$572.00	$22.91	$35.75
.25 Truck Driver (light)	23.40	46.80	35.75	71.50		
.25 Light Truck, 1.5 Ton		39.50		43.45	2.19	2.41
18 L.H., Daily Totals		$451.90		$686.95	$25.10	$38.16
Crew A-6	Hr.	Daily	Hr.	Daily	Bare Costs	Incl. O&P
1 Chief Of Party	$34.55	$276.40	$54.75	$438.00	$32.20	$50.80
1 Instrument Man	29.85	238.80	46.85	374.80		
16 L.H., Daily Totals		$515.20		$812.80	$32.20	$50.80
Crew A-7	Hr.	Daily	Hr.	Daily	Bare Costs	Incl. O&P
1 Chief Of Party	$34.55	$276.40	$54.75	$438.00	$30.98	$48.30
1 Instrument Man	29.85	238.80	46.85	374.80		
1 Rodman/Chainman	28.55	228.40	43.30	346.40		
24 L.H., Daily Totals		$743.60		$1159.20	$30.98	$48.30
Crew A-8	Hr.	Daily	Hr.	Daily	Bare Costs	Incl. O&P
1 Chief Of Party	$34.55	$276.40	$54.75	$438.00	$30.38	$47.05
1 Instrument Man	29.85	238.80	46.85	374.80		
2 Rodmen/Chainmen	28.55	456.80	43.30	692.80		
32 L.H., Daily Totals		$972.00		$1505.60	$30.38	$47.05
Crew A-9	Hr.	Daily	Hr.	Daily	Bare Costs	Incl. O&P
1 Asbestos Foreman	$32.65	$261.20	$52.30	$418.40	$32.21	$51.60
7 Asbestos Workers	32.15	1800.40	51.50	2884.00		
64 L.H., Daily Totals		$2061.60		$3302.40	$32.21	$51.60
Crew A-10	Hr.	Daily	Hr.	Daily	Bare Costs	Incl. O&P
1 Asbestos Foreman	$32.65	$261.20	$52.30	$418.40	$32.21	$51.60
7 Asbestos Workers	32.15	1800.40	51.50	2884.00		
64 L.H., Daily Totals		$2061.60		$3302.40	$32.21	$51.60
Crew A-10A	Hr.	Daily	Hr.	Daily	Bare Costs	Incl. O&P
1 Asbestos Foreman	$32.65	$261.20	$52.30	$418.40	$32.32	$51.77
2 Asbestos Workers	32.15	514.40	51.50	824.00		
24 L.H., Daily Totals		$775.60		$1242.40	$32.32	$51.77
Crew A-10B	Hr.	Daily	Hr.	Daily	Bare Costs	Incl. O&P
1 Asbestos Foreman	$32.65	$261.20	$52.30	$418.40	$32.28	$51.70
3 Asbestos Workers	32.15	771.60	51.50	1236.00		
32 L.H., Daily Totals		$1032.80		$1654.40	$32.28	$51.70
Crew A-10C	Hr.	Daily	Hr.	Daily	Bare Costs	Incl. O&P
3 Asbestos Workers	$32.15	$771.60	$51.50	$1236.00	$32.15	$51.50
1 Flatbed Truck		158.00		173.80	6.58	7.24
24 L.H., Daily Totals		$929.60		$1409.80	$38.73	$58.74
Crew A-10D	Hr.	Daily	Hr.	Daily	Bare Costs	Incl. O&P
2 Asbestos Workers	$32.15	$514.40	$51.50	$824.00	$30.30	$47.29
1 Equip. Oper. (crane)	31.15	249.20	47.15	377.20		
1 Equip. Oper. Oiler	25.75	206.00	39.00	312.00		
1 Hydraulic Crane, 33 Ton		804.80		885.30	25.15	27.67
32 L.H., Daily Totals		$1774.40		$2398.50	$55.45	$74.96
Crew A-11	Hr.	Daily	Hr.	Daily	Bare Costs	Incl. O&P
1 Asbestos Foreman	$32.65	$261.20	$52.30	$418.40	$32.21	$51.60
7 Asbestos Workers	32.15	1800.40	51.50	2884.00		
2 Chipping Hammers		35.20		38.70	.55	.61
64 L.H., Daily Totals		$2096.80		$3341.10	$32.76	$52.21
Crew A-12	Hr.	Daily	Hr.	Daily	Bare Costs	Incl. O&P
1 Asbestos Foreman	$32.65	$261.20	$52.30	$418.40	$32.21	$51.60
7 Asbestos Workers	32.15	1800.40	51.50	2884.00		
1 Large Prod. Vac. Loader		520.80		572.90	8.14	8.95
64 L.H., Daily Totals		$2582.40		$3875.30	$40.35	$60.55
Crew A-13	Hr.	Daily	Hr.	Daily	Bare Costs	Incl. O&P
1 Equip. Oper. (light)	$28.75	$230.00	$43.55	$348.40	$28.75	$43.55
1 Large Prod. Vac. Loader		520.80		572.90	65.10	71.61
8 L.H., Daily Totals		$750.80		$921.30	$93.85	$115.16

Crews

Crew No.	Bare Costs		Incl. Subs O & P		Cost Per Labor-Hour	
Crew B-1	Hr.	Daily	Hr.	Daily	Bare Costs	Incl. O&P
1 Labor Foreman (outside)	$24.85	$198.80	$38.90	$311.20	$23.52	$36.80
2 Laborers	22.85	365.60	35.75	572.00		
24 L.H., Daily Totals		$564.40		$883.20	$23.52	$36.80
Crew B-2	Hr.	Daily	Hr.	Daily	Bare Costs	Incl. O&P
1 Labor Foreman (outside)	$24.85	$198.80	$38.90	$311.20	$23.25	$36.38
4 Laborers	22.85	731.20	35.75	1144.00		
40 L.H., Daily Totals		$930.00		$1455.20	$23.25	$36.38
Crew B-3	Hr.	Daily	Hr.	Daily	Bare Costs	Incl. O&P
1 Labor Foreman (outside)	$24.85	$198.80	$38.90	$311.20	$24.80	$38.26
2 Laborers	22.85	365.60	35.75	572.00		
1 Equip. Oper. (med.)	30.35	242.80	45.95	367.60		
2 Truck Drivers (heavy)	23.95	383.20	36.60	585.60		
1 F.E. Loader, T.M., 2.5 C.Y.		770.00		847.00		
2 Dump Trucks, 16 Ton		965.60		1062.15	36.16	39.77
48 L.H., Daily Totals		$2926.00		$3745.55	$60.96	$78.03
Crew B-3A	Hr.	Daily	Hr.	Daily	Bare Costs	Incl. O&P
4 Laborers	$22.85	$731.20	$35.75	$1144.00	$24.35	$37.79
1 Equip. Oper. (med.)	30.35	242.80	45.95	367.60		
1 Hyd. Excavator, 1.5 C.Y.		593.20		652.50	14.83	16.31
40 L.H., Daily Totals		$1567.20		$2164.10	$39.18	$54.10
Crew B-3B	Hr.	Daily	Hr.	Daily	Bare Costs	Incl. O&P
2 Laborers	$22.85	$365.60	$35.75	$572.00	$25.00	$38.51
1 Equip. Oper. (med.)	30.35	242.80	45.95	367.60		
1 Truck Drivers (heavy)	23.95	191.60	36.60	292.80		
1 Backhoe Loader, 80 H.P.		228.40		251.25		
1 Dump Trucks, 16 Ton		482.80		531.10	22.23	24.45
32 L.H., Daily Totals		$1511.20		$2014.75	$47.23	$62.96
Crew B-3C	Hr.	Daily	Hr.	Daily	Bare Costs	Incl. O&P
3 Laborers	$22.85	$548.40	$35.75	$858.00	$24.73	$38.30
1 Equip. Oper. (med.)	30.35	242.80	45.95	367.60		
1 F.E. Loader, 3.75 C.Y.		518.80		570.70	16.21	17.83
32 L.H., Daily Totals		$1310.00		$1796.30	$40.94	$56.13
Crew B-4	Hr.	Daily	Hr.	Daily	Bare Costs	Incl. O&P
1 Labor Foreman (outside)	$24.85	$198.80	$38.90	$311.20	$23.37	$36.42
4 Laborers	22.85	731.20	35.75	1144.00		
1 Truck Driver (heavy)	23.95	191.60	36.60	292.80		
1 Tractor, 4 x 2, 195 H.P.		320.40		352.45		
1 Platform Trailer		170.40		187.45	10.23	11.25
48 L.H., Daily Totals		$1612.40		$2287.90	$33.60	$47.67
Crew B-5	Hr.	Daily	Hr.	Daily	Bare Costs	Incl. O&P
1 Labor Foreman (outside)	$24.85	$198.80	$38.90	$311.20	$25.28	$39.11
4 Laborers	22.85	731.20	35.75	1144.00		
2 Equip. Oper. (med.)	30.35	485.60	45.95	735.20		
1 Air Compr., 250 C.F.M.		125.20		137.70		
2 Air Tools & Accessories		39.20		43.10		
2-50 Ft. Air Hoses, 1.5" Dia.		11.80		13.00		
1 F.E. Loader, T.M., 2.5 C.Y.		770.00		847.00	16.90	18.59
56 L.H., Daily Totals		$2361.80		$3231.20	$42.18	$57.70

Crew No.	Bare Costs		Incl. Subs O & P		Cost Per Labor-Hour	
Crew B-5A	Hr.	Daily	Hr.	Daily	Bare Costs	Incl. O&P
1 Foreman	$24.85	$198.80	$38.90	$311.20	$24.94	$38.50
6 Laborers	22.85	1096.80	35.75	1716.00		
2 Equip. Oper. (med.)	30.35	485.60	45.95	735.20		
1 Equip. Oper. (light)	28.75	230.00	43.55	348.40		
2 Truck Drivers (heavy)	23.95	383.20	36.60	585.60		
1 Air Compr. 365 C.F.M.		154.80		170.30		
2 Pavement Breakers		39.20		43.10		
8 Air Hoses w/Coup.,1"		29.20		32.10		
2 Dump Trucks, 12 Ton		740.80		814.90	10.04	11.05
96 L.H., Daily Totals		$3358.40		$4756.80	$34.98	$49.55
Crew B-5B	Hr.	Daily	Hr.	Daily	Bare Costs	Incl. O&P
1 Powderman	$29.85	$238.80	$46.85	$374.80	$27.07	$41.42
2 Equip. Oper. (med.)	30.35	485.60	45.95	735.20		
3 Truck Drivers (heavy)	23.95	574.80	36.60	878.40		
1 F.E. Ldr. 2-1/2 CY		391.80		431.00		
3 Dump Trucks, 16 Ton		1448.40		1593.25		
1 Air Compr. 365 C.F.M.		154.80		170.30	41.56	45.72
48 L.H., Daily Totals		$3294.20		$4182.95	$68.63	$87.14
Crew B-5C	Hr.	Daily	Hr.	Daily	Bare Costs	Incl. O&P
3 Laborers	$22.85	$548.40	$35.75	$858.00	$25.46	$39.07
1 Equip. Oper. (medium)	30.35	242.80	45.95	367.60		
2 Truck Drivers (heavy)	23.95	383.20	36.60	585.60		
1 Equip. Oper. (crane)	31.15	249.20	47.15	377.20		
1 Equip. Oper. Oiler	25.75	206.00	39.00	312.00		
2 Dump Truck, 16 Ton		965.60		1062.15		
1 F.E. Lder, 3.75 C.Y.		990.80		1089.90		
1 Hyd. Crane, 25 Ton		602.00		662.20	39.98	43.97
64 L.H., Daily Totals		$4188.00		$5314.65	$65.44	$83.04
Crew B-6	Hr.	Daily	Hr.	Daily	Bare Costs	Incl. O&P
2 Laborers	$22.85	$365.60	$35.75	$572.00	$24.82	$38.35
1 Equip. Oper. (light)	28.75	230.00	43.55	348.40		
1 Backhoe Loader, 48 H.P.		181.00		199.10	7.54	8.30
24 L.H., Daily Totals		$776.60		$1119.50	$32.36	$46.65
Crew B-6A	Hr.	Daily	Hr.	Daily	Bare Costs	Incl. O&P
.5 Labor Foreman (outside)	$24.85	$99.40	$38.90	$155.60	$26.25	$40.46
1 Laborers	22.85	182.80	35.75	286.00		
1 Equip. Oper. (med.)	30.35	242.80	45.95	367.60		
1 Vacuum Trk.,5000 Gal.		302.20		332.40	15.11	16.62
20 L.H., Daily Totals		$827.20		$1141.60	$41.36	$57.08
Crew B-7	Hr.	Daily	Hr.	Daily	Bare Costs	Incl. O&P
1 Labor Foreman (outside)	$24.85	$198.80	$38.90	$311.20	$24.43	$37.98
4 Laborers	22.85	731.20	35.75	1144.00		
1 Equip. Oper. (med.)	30.35	242.80	45.95	367.60		
1 Chipping Machine		190.60		209.65		
1 F.E. Loader, T.M., 2.5 C.Y.		770.00		847.00		
2 Chain Saws, 36"		86.40		95.05	21.81	23.99
48 L.H., Daily Totals		$2219.80		$2974.50	$46.24	$61.97
Crew B-7A	Hr.	Daily	Hr.	Daily	Bare Costs	Incl. O&P
2 Laborers	$22.85	$365.60	$35.75	$572.00	$24.82	$38.35
1 Equip. Oper. (light)	28.75	230.00	43.55	348.40		
1 Rake w/Tractor		139.70		153.65		
2 Chain Saws, 18"		52.40		57.65	8.00	8.80
24 L.H., Daily Totals		$787.70		$1131.70	$32.82	$47.15

Crews

Crew No.	Bare Costs		Incl. Subs O & P		Cost Per Labor-Hour	
Crew B-8	Hr.	Daily	Hr.	Daily	Bare Costs	Incl. O&P
1 Labor Foreman (outside)	$24.85	$198.80	$38.90	$311.20	$25.61	$39.31
2 Laborers	22.85	365.60	35.75	572.00		
2 Equip. Oper. (med.)	30.35	485.60	45.95	735.20		
1 Equip. Oper. Oiler	25.75	206.00	39.00	312.00		
2 Truck Drivers (heavy)	23.95	383.20	36.60	585.60		
1 Hyd. Crane, 25 Ton		629.60		692.55		
1 F.E. Loader, T.M., 2.5 C.Y.		770.00		847.00		
2 Dump Trucks, 16 Ton		965.60		1062.15	36.96	40.65
64 L.H., Daily Totals		$4004.40		$5117.70	$62.57	$79.96
Crew B-9	Hr.	Daily	Hr.	Daily	Bare Costs	Incl. O&P
1 Labor Foreman (outside)	$24.85	$198.80	$38.90	$311.20	$23.25	$36.38
4 Laborers	22.85	731.20	35.75	1144.00		
1 Air Compr., 250 C.F.M.		125.20		137.70		
2 Air Tools & Accessories		39.20		43.10		
2-50 Ft. Air Hoses, 1.5" Dia.		11.80		13.00	4.41	4.85
40 L.H., Daily Totals		$1106.20		$1649.00	$27.66	$41.23
Crew B-9A	Hr.	Daily	Hr.	Daily	Bare Costs	Incl. O&P
2 Laborers	$22.85	$365.60	$35.75	$572.00	$23.22	$36.03
1 Truck Driver (heavy)	23.95	191.60	36.60	292.80		
1 Water Tanker		184.00		202.40		
1 Tractor		320.40		352.45		
2-50 Ft. Disch. Hoses		12.80		14.10	21.55	23.71
24 L.H., Daily Totals		$1074.40		$1433.75	$44.77	$59.74
Crew B-9B	Hr.	Daily	Hr.	Daily	Bare Costs	Incl. O&P
2 Laborers	$22.85	$365.60	$35.75	$572.00	$23.22	$36.03
1 Truck Driver (heavy)	23.95	191.60	36.60	292.80		
2-50 Ft. Disch. Hoses		12.80		14.10		
1 Water Tanker		184.00		202.40		
1 Tractor		320.40		352.45		
1 Pressure Washer		57.40		63.15	23.94	26.34
24 L.H., Daily Totals		$1131.80		$1496.90	$47.16	$62.37
Crew B-9C	Hr.	Daily	Hr.	Daily	Bare Costs	Incl. O&P
1 Labor Foreman (outside)	$24.85	$198.80	$38.90	$311.20	$23.25	$36.38
4 Laborers	22.85	731.20	35.75	1144.00		
1 Air Compr., 250 C.F.M.		125.20		137.70		
2-50 Ft. Air Hoses, 1.5" Dia.		11.80		13.00		
2 Breaker, Pavement, 60 lb.		39.20		43.10	4.41	4.85
40 L.H., Daily Totals		$1106.20		$1649.00	$27.66	$41.23
Crew B-10	Hr.	Daily	Hr.	Daily	Bare Costs	Incl. O&P
1 Equip. Oper. (med.)	$30.35	$242.80	$45.95	$367.60	$27.85	$42.55
.5 Laborer	22.85	91.40	35.75	143.00		
12 L.H., Daily Totals		$334.20		$510.60	$27.85	$42.55
Crew B-10A	Hr.	Daily	Hr.	Daily	Bare Costs	Incl. O&P
1 Equip. Oper. (med.)	$30.35	$242.80	$45.95	$367.60	$27.85	$42.55
.5 Laborer	22.85	91.40	35.75	143.00		
1 Roll. Compact., 2K Lbs.		111.00		122.10	9.25	10.18
12 L.H., Daily Totals		$445.20		$632.70	$37.10	$52.73
Crew B-10B	Hr.	Daily	Hr.	Daily	Bare Costs	Incl. O&P
1 Equip. Oper. (med.)	$30.35	$242.80	$45.95	$367.60	$27.85	$42.55
.5 Laborer	22.85	91.40	35.75	143.00		
1 Dozer, 200 H.P.		735.00		808.50	61.25	67.38
12 L.H., Daily Totals		$1069.20		$1319.10	$89.10	$109.93

Crew No.	Bare Costs		Incl. Subs O & P		Cost Per Labor-Hour	
Crew B-10C	Hr.	Daily	Hr.	Daily	Bare Costs	Incl. O&P
1 Equip. Oper. (med.)	$30.35	$242.80	$45.95	$367.60	$27.85	$42.55
.5 Laborer	22.85	91.40	35.75	143.00		
1 Dozer, 200 H.P.		735.00		808.50		
1 Vibratory Roller, Towed		168.40		185.25	75.28	82.81
12 L.H., Daily Totals		$1237.60		$1504.35	$103.13	$125.36
Crew B-10D	Hr.	Daily	Hr.	Daily	Bare Costs	Incl. O&P
1 Equip. Oper. (med.)	$30.35	$242.80	$45.95	$367.60	$27.85	$42.55
.5 Laborer	22.85	91.40	35.75	143.00		
1 Dozer, 200 H.P.		735.00		808.50		
1 Sheepsft. Roller, Towed		69.80		76.80	67.07	73.77
12 L.H., Daily Totals		$1139.00		$1395.90	$94.92	$116.32
Crew B-10E	Hr.	Daily	Hr.	Daily	Bare Costs	Incl. O&P
1 Equip. Oper. (med.)	$30.35	$242.80	$45.95	$367.60	$27.85	$42.55
.5 Laborer	22.85	91.40	35.75	143.00		
1 Tandem Roller, 5 Ton		133.20		146.50	11.10	12.21
12 L.H., Daily Totals		$467.40		$657.10	$38.95	$54.76
Crew B-10F	Hr.	Daily	Hr.	Daily	Bare Costs	Incl. O&P
1 Equip. Oper. (med.)	$30.35	$242.80	$45.95	$367.60	$27.85	$42.55
.5 Laborer	22.85	91.40	35.75	143.00		
1 Tandem Roller, 10 Ton		210.40		231.45	17.53	19.29
12 L.H., Daily Totals		$544.60		$742.05	$45.38	$61.84
Crew B-10G	Hr.	Daily	Hr.	Daily	Bare Costs	Incl. O&P
1 Equip. Oper. (med.)	$30.35	$242.80	$45.95	$367.60	$27.85	$42.55
.5 Laborer	22.85	91.40	35.75	143.00		
1 Sheepsft. Roll., 130 H.P.		616.00		677.60	51.33	56.47
12 L.H., Daily Totals		$950.20		$1188.20	$79.18	$99.02
Crew B-10H	Hr.	Daily	Hr.	Daily	Bare Costs	Incl. O&P
1 Equip. Oper. (med.)	$30.35	$242.80	$45.95	$367.60	$27.85	$42.55
.5 Laborer	22.85	91.40	35.75	143.00		
1 Diaphr. Water Pump, 2"		30.80		33.90		
1-20 Ft. Suction Hose, 2"		6.50		7.15		
2-50 Ft. Disch. Hoses, 2"		10.80		11.90	4.01	4.41
12 L.H., Daily Totals		$382.30		$563.55	$31.86	$46.96
Crew B-10I	Hr.	Daily	Hr.	Daily	Bare Costs	Incl. O&P
1 Equip. Oper. (med.)	$30.35	$242.80	$45.95	$367.60	$27.85	$42.55
.5 Laborer	22.85	91.40	35.75	143.00		
1 Diaphr. Water Pump, 4"		62.00		68.20		
1-20 Ft. Suction Hose, 4"		12.50		13.75		
2-50 Ft. Disch. Hoses, 4"		17.00		18.70	7.63	8.39
12 L.H., Daily Totals		$425.70		$611.25	$35.48	$50.94
Crew B-10J	Hr.	Daily	Hr.	Daily	Bare Costs	Incl. O&P
1 Equip. Oper. (med.)	$30.35	$242.80	$45.95	$367.60	$27.85	$42.55
.5 Laborer	22.85	91.40	35.75	143.00		
1 Centr. Water Pump, 3"		39.20		43.10		
1-20 Ft. Suction Hose, 3"		9.50		10.45		
2-50 Ft. Disch. Hoses, 3"		12.80		14.10	5.13	5.64
12 L.H., Daily Totals		$395.70		$578.25	$32.98	$48.19

Crews

Crew No.		Bare Costs		Incl. Subs O & P		Cost Per Labor-Hour	
Crew B-10K	Hr.	Daily	Hr.	Daily	Bare Costs	Incl. O&P
1 Equip. Oper. (med.)	$30.35	$242.80	$45.95	$367.60	$27.85	$42.55
.5 Laborer	22.85	91.40	35.75	143.00		
1 Centr. Water Pump, 6"		164.40		180.85		
1-20 Ft. Suction Hose, 6"		22.50		24.75		
2-50 Ft. Disch. Hoses, 6"		41.10		45.20	19.00	20.90
12 L.H., Daily Totals		$562.20		$761.40	$46.85	$63.45
Crew B-10L	Hr.	Daily	Hr.	Daily	Bare Costs	Incl. O&P
1 Equip. Oper. (med.)	$30.35	$242.80	$45.95	$367.60	$27.85	$42.55
.5 Laborer	22.85	91.40	35.75	143.00		
1 Dozer, 75 H.P.		266.00		292.60	22.17	24.38
12 L.H., Daily Totals		$600.20		$803.20	$50.02	$66.93
Crew B-10M	Hr.	Daily	Hr.	Daily	Bare Costs	Incl. O&P
1 Equip. Oper. (med.)	$30.35	$242.80	$45.95	$367.60	$27.85	$42.55
.5 Laborer	22.85	91.40	35.75	143.00		
1 Dozer, 300 H.P.		1027.00		1129.70	85.58	94.14
12 L.H., Daily Totals		$1361.20		$1640.30	$113.43	$136.69
Crew B-10N	Hr.	Daily	Hr.	Daily	Bare Costs	Incl. O&P
1 Equip. Oper. (med.)	$30.35	$242.80	$45.95	$367.60	$27.85	$42.55
.5 Laborer	22.85	91.40	35.75	143.00		
1 F.E. Loader, T.M., 1.5 C.Y		307.20		337.90	25.60	28.16
12 L.H., Daily Totals		$641.40		$848.50	$53.45	$70.71
Crew B-10O	Hr.	Daily	Hr.	Daily	Bare Costs	Incl. O&P
1 Equip. Oper. (med.)	$30.35	$242.80	$45.95	$367.60	$27.85	$42.55
.5 Laborer	22.85	91.40	35.75	143.00		
1 F.E. Loader, T.M., 2.25 C.Y.		457.00		502.70	38.08	41.89
12 L.H., Daily Totals		$791.20		$1013.30	$65.93	$84.44
Crew B-10P	Hr.	Daily	Hr.	Daily	Bare Costs	Incl. O&P
1 Equip. Oper. (med.)	$30.35	$242.80	$45.95	$367.60	$27.85	$42.55
.5 Laborer	22.85	91.40	35.75	143.00		
1 F.E. Loader, T.M., 2.5 C.Y.		770.00		847.00	64.17	70.58
12 L.H., Daily Totals		$1104.20		$1357.60	$92.02	$113.13
Crew B-10Q	Hr.	Daily	Hr.	Daily	Bare Costs	Incl. O&P
1 Equip. Oper. (med.)	$30.35	$242.80	$45.95	$367.60	$27.85	$42.55
.5 Laborer	22.85	91.40	35.75	143.00		
1 F.E. Loader, T.M., 5 C.Y.		990.80		1089.90	82.57	90.82
12 L.H., Daily Totals		$1325.00		$1600.50	$110.42	$133.37
Crew B-10R	Hr.	Daily	Hr.	Daily	Bare Costs	Incl. O&P
1 Equip. Oper. (med.)	$30.35	$242.80	$45.95	$367.60	$27.85	$42.55
.5 Laborer	22.85	91.40	35.75	143.00		
1 F.E. Loader, W.M., 1 C.Y.		227.60		250.35	18.97	20.86
12 L.H., Daily Totals		$561.80		$760.95	$46.82	$63.41
Crew B-10S	Hr.	Daily	Hr.	Daily	Bare Costs	Incl. O&P
1 Equip. Oper. (med.)	$30.35	$242.80	$45.95	$367.60	$27.85	$42.55
.5 Laborer	22.85	91.40	35.75	143.00		
1 F.E. Loader, W.M., 1.5 C.Y.		284.40		312.85	23.70	26.07
12 L.H., Daily Totals		$618.60		$823.45	$51.55	$68.62
Crew B-10T	Hr.	Daily	Hr.	Daily	Bare Costs	Incl. O&P
1 Equip. Oper. (med.)	$30.35	$242.80	$45.95	$367.60	$27.85	$42.55
.5 Laborer	22.85	91.40	35.75	143.00		
1 F.E. Loader, W.M., 2.5 C.Y.		391.80		431.00	32.65	35.92
12 L.H., Daily Totals		$726.00		$941.60	$60.50	$78.47
Crew B-10U	Hr.	Daily	Hr.	Daily	Bare Costs	Incl. O&P
1 Equip. Oper. (med.)	$30.35	$242.80	$45.95	$367.60	$27.85	$42.55
.5 Laborer	22.85	91.40	35.75	143.00		
1 F.E. Loader, W.M., 5.5 C.Y.		814.60		896.05	67.88	74.67
12 L.H., Daily Totals		$1148.80		$1406.65	$95.73	$117.22
Crew B-10V	Hr.	Daily	Hr.	Daily	Bare Costs	Incl. O&P
1 Equip. Oper. (med.)	$30.35	$242.80	$45.95	$367.60	$27.85	$42.55
.5 Laborer	22.85	91.40	35.75	143.00		
1 Dozer, 700 H.P.		2629.00		2891.90	219.08	240.99
12 L.H., Daily Totals		$2963.20		$3402.50	$246.93	$283.54
Crew B-10W	Hr.	Daily	Hr.	Daily	Bare Costs	Incl. O&P
1 Equip. Oper. (med.)	$30.35	$242.80	$45.95	$367.60	$27.85	$42.55
.5 Laborer	22.85	91.40	35.75	143.00		
1 Dozer, 105 H.P.		399.20		439.10	33.27	36.59
12 L.H., Daily Totals		$733.40		$949.70	$61.12	$79.14
Crew B-10X	Hr.	Daily	Hr.	Daily	Bare Costs	Incl. O&P
1 Equip. Oper. (med.)	$30.35	$242.80	$45.95	$367.60	$27.85	$42.55
.5 Laborer	22.85	91.40	35.75	143.00		
1 Dozer, 410 H.P.		1323.00		1455.30	110.25	121.28
12 L.H., Daily Totals		$1657.20		$1965.90	$138.10	$163.83
Crew B-10Y	Hr.	Daily	Hr.	Daily	Bare Costs	Incl. O&P
1 Equip. Oper. (med.)	$30.35	$242.80	$45.95	$367.60	$27.85	$42.55
.5 Laborer	22.85	91.40	35.75	143.00		
1 Vibratory Drum Roller		357.20		392.90	29.77	32.74
12 L.H., Daily Totals		$691.40		$903.50	$57.62	$75.29
Crew B-11	Hr.	Daily	Hr.	Daily	Bare Costs	Incl. O&P
1 Equipment Oper. (med.)	$30.35	$242.80	$45.95	$367.60	$26.60	$40.85
1 Laborer	22.85	182.80	35.75	286.00		
16 L.H., Daily Totals		$425.60		$653.60	$26.60	$40.85
Crew B-11A	Hr.	Daily	Hr.	Daily	Bare Costs	Incl. O&P
1 Equipment Oper. (med.)	$30.35	$242.80	$45.95	$367.60	$26.60	$40.85
1 Laborer	22.85	182.80	35.75	286.00		
1 Dozer, 200 H.P.		735.00		808.50	45.94	50.53
16 L.H., Daily Totals		$1160.60		$1462.10	$72.54	$91.38
Crew B-11B	Hr.	Daily	Hr.	Daily	Bare Costs	Incl. O&P
1 Equipment Oper. (med.)	$30.35	$242.80	$45.95	$367.60	$26.60	$40.85
1 Laborer	22.85	182.80	35.75	286.00		
1 Dozer, 200 H.P.		735.00		808.50		
1 Air Powered Tamper		21.40		23.55		
1 Air Compr. 365 C.F.M.		154.80		170.30		
2-50 Ft. Air Hoses, 1.5" Dia.		11.80		13.00	57.69	63.46
16 L.H., Daily Totals		$1348.60		$1668.95	$84.29	$104.31
Crew B-11C	Hr.	Daily	Hr.	Daily	Bare Costs	Incl. O&P
1 Equipment Oper. (med.)	$30.35	$242.80	$45.95	$367.60	$26.60	$40.85
1 Laborer	22.85	182.80	35.75	286.00		
1 Backhoe Loader, 48 H.P.		181.00		199.10	11.31	12.44
16 L.H., Daily Totals		$606.60		$852.70	$37.91	$53.29
Crew B-11K	Hr.	Daily	Hr.	Daily	Bare Costs	Incl. O&P
1 Equipment Oper. (med.)	$30.35	$242.80	$45.95	$367.60	$26.60	$40.85
1 Laborer	22.85	182.80	35.75	286.00		
1 Trencher, 8' D., 16" W.		711.40		782.55	44.46	48.91
16 L.H., Daily Totals		$1137.00		$1436.15	$71.06	$89.76

Crews

Crew No.	Bare Costs		Incl. Subs O & P		Cost Per Labor-Hour	
Crew B-11L	Hr.	Daily	Hr.	Daily	Bare Costs	Incl. O&P
1 Equipment Oper. (med.)	$30.35	$242.80	$45.95	$367.60	$26.60	$40.85
1 Laborer	22.85	182.80	35.75	286.00		
1 Grader, 30,000 Lbs.		459.40		505.35	28.71	31.58
16 L.H., Daily Totals		$885.00		$1158.95	$55.31	$72.43
Crew B-11M	Hr.	Daily	Hr.	Daily	Bare Costs	Incl. O&P
1 Equipment Oper. (med.)	$30.35	$242.80	$45.95	$367.60	$26.60	$40.85
1 Laborer	22.85	182.80	35.75	286.00		
1 Backhoe Loader, 80 H.P.		228.40		251.25	14.28	15.70
16 L.H., Daily Totals		$654.00		$904.85	$40.88	$56.55
Crew B-11N	Hr.	Daily	Hr.	Daily	Bare Costs	Incl. O&P
1 Labor Foreman	$24.85	$198.80	$38.90	$311.20	$25.47	$38.93
2 Equipment Operators (med.)	30.35	485.60	45.95	735.20		
6 Truck Drivers (hvy.)	23.95	1149.60	36.60	1756.80		
1 F.E. Loader, 5.5 C.Y.		814.60		896.05		
1 Dozer, 400 H.P.		1323.00		1455.30		
6 Off Hwy. Tks. 50 Ton		8796.00		9675.60	151.86	167.04
72 L.H., Daily Totals		$12767.60		$14830.15	$177.33	$205.97
Crew B-11Q	Hr.	Daily	Hr.	Daily	Bare Costs	Incl. O&P
1 Equipment Operator (med.)	$30.35	$242.80	$45.95	$367.60	$27.85	$42.55
.5 Laborer	22.85	91.40	35.75	143.00		
1 Dozer, 140 H.P.		493.20		542.50	41.10	45.21
12 L.H., Daily Totals		$827.40		$1053.10	$68.95	$87.76
Crew B-11R	Hr.	Daily	Hr.	Daily	Bare Costs	Incl. O&P
1 Equipment Operator (med.)	$30.35	$242.80	$45.95	$367.60	$27.85	$42.55
.5 Laborer	22.85	91.40	35.75	143.00		
1 Dozer, 215 H.P.		735.00		808.50	61.25	67.38
12 L.H., Daily Totals		$1069.20		$1319.10	$89.10	$109.93
Crew B-11S	Hr.	Daily	Hr.	Daily	Bare Costs	Incl. O&P
1 Equipment Operator	$30.35	$242.80	$45.95	$367.60	$27.85	$42.55
.5 Laborer	22.85	91.40	35.75	143.00		
1 Dozer, 285 H.P.		1027.00		1129.70	85.58	94.14
12 L.H., Daily Totals		$1361.20		$1640.30	$113.43	$136.69
Crew B-11T	Hr.	Daily	Hr.	Daily	Bare Costs	Incl. O&P
1 Equipment Operator (med.)	$30.35	$242.80	$45.95	$367.60	$27.85	$42.55
.5 Laborer	22.85	91.40	35.75	143.00		
1 Dozer, 370 H.P.		1323.00		1455.30	110.25	121.28
12 L.H., Daily Totals		$1657.20		$1965.90	$138.10	$163.83
Crew B-11U	Hr.	Daily	Hr.	Daily	Bare Costs	Incl. O&P
1 Equipment Operator (med.)	$30.35	$242.80	$45.95	$367.60	$27.85	$42.55
.5 Laborer	22.85	91.40	35.75	143.00		
1 Dozer, 520 H.P.		1634.00		1797.40	136.17	149.78
12 L.H., Daily Totals		$1968.20		$2308.00	$164.02	$192.33
Crew B-11V	Hr.	Daily	Hr.	Daily	Bare Costs	Incl. O&P
3 Laborer	$22.85	$548.40	$35.75	$858.00	$22.85	$35.75
1 Roll. Compact., 2K Lbs.		111.00		122.10	4.63	5.09
24 L.H., Daily Totals		$659.40		$980.10	$27.48	$40.84
Crew B-12	Hr.	Daily	Hr.	Daily	Bare Costs	Incl. O&P
1 Equip. Oper. (crane)	$31.15	$249.20	$47.15	$377.20	$28.45	$43.08
1 Equip. Oper. Oiler	25.75	206.00	39.00	312.00		
16 L.H., Daily Totals		$455.20		$689.20	$28.45	$43.08

Crew No.	Bare Costs		Incl. Subs O & P		Cost Per Labor-Hour	
Crew B-12A	Hr.	Daily	Hr.	Daily	Bare Costs	Incl. O&P
1 Equip. Oper. (crane)	$31.15	$249.20	$47.15	$377.20	$28.45	$43.08
1 Equip. Oper. Oiler	25.75	206.00	39.00	312.00		
1 Hyd. Excavator, 1 C.Y.		490.80		539.90	30.68	33.74
16 L.H., Daily Totals		$946.00		$1229.10	$59.13	$76.82
Crew B-12B	Hr.	Daily	Hr.	Daily	Bare Costs	Incl. O&P
1 Equip. Oper. (crane)	$31.15	$249.20	$47.15	$377.20	$28.45	$43.08
1 Equip. Oper. Oiler	25.75	206.00	39.00	312.00		
1 Hyd. Excavator, 1.5 C.Y.		593.20		652.50	37.08	40.78
16 L.H., Daily Totals		$1048.40		$1341.70	$65.53	$83.86
Crew B-12C	Hr.	Daily	Hr.	Daily	Bare Costs	Incl. O&P
1 Equip. Oper. (crane)	$31.15	$249.20	$47.15	$377.20	$28.45	$43.08
1 Equip. Oper. Oiler	25.75	206.00	39.00	312.00		
1 Hyd. Excavator, 2 C.Y.		854.80		940.30	53.43	58.77
16 L.H., Daily Totals		$1310.00		$1629.50	$81.88	$101.85
Crew B-12D	Hr.	Daily	Hr.	Daily	Bare Costs	Incl. O&P
1 Equip. Oper. (crane)	$31.15	$249.20	$47.15	$377.20	$28.45	$43.08
1 Equip. Oper. Oiler	25.75	206.00	39.00	312.00		
1 Hyd. Excavator, 3.5 C.Y.		1974.00		2171.40	123.38	135.71
16 L.H., Daily Totals		$2429.20		$2860.60	$151.83	$178.79
Crew B-12E	Hr.	Daily	Hr.	Daily	Bare Costs	Incl. O&P
1 Equip. Oper. (crane)	$31.15	$249.20	$47.15	$377.20	$28.45	$43.08
1 Equip. Oper. Oiler	25.75	206.00	39.00	312.00		
1 Hyd. Excavator, .5 C.Y.		318.40		350.25	19.90	21.89
16 L.H., Daily Totals		$773.60		$1039.45	$48.35	$64.97
Crew B-12F	Hr.	Daily	Hr.	Daily	Bare Costs	Incl. O&P
1 Equip. Oper. (crane)	$31.15	$249.20	$47.15	$377.20	$28.45	$43.08
1 Equip. Oper. Oiler	25.75	206.00	39.00	312.00		
1 Hyd. Excavator, .75 C.Y.		413.60		454.95	25.85	28.44
16 L.H., Daily Totals		$868.80		$1144.15	$54.30	$71.52
Crew B-12G	Hr.	Daily	Hr.	Daily	Bare Costs	Incl. O&P
1 Equip. Oper. (crane)	$31.15	$249.20	$47.15	$377.20	$28.45	$43.08
1 Equip. Oper. Oiler	25.75	206.00	39.00	312.00		
1 Power Shovel, .5 C.Y.		357.75		393.50		
1 Clamshell Bucket, .5 C.Y.		48.00		52.80	24.23	26.66
16 L.H., Daily Totals		$860.95		$1135.50	$52.68	$69.74
Crew B-12H	Hr.	Daily	Hr.	Daily	Bare Costs	Incl. O&P
1 Equip. Oper. (crane)	$31.15	$249.20	$47.15	$377.20	$28.45	$43.08
1 Equip. Oper. Oiler	25.75	206.00	39.00	312.00		
1 Power Shovel, 1 C.Y.		547.20		601.90		
1 Clamshell Bucket, 1 C.Y.		71.00		78.10	38.64	42.50
16 L.H., Daily Totals		$1073.40		$1369.20	$67.09	$85.58
Crew B-12I	Hr.	Daily	Hr.	Daily	Bare Costs	Incl. O&P
1 Equip. Oper. (crane)	$31.15	$249.20	$47.15	$377.20	$28.45	$43.08
1 Equip. Oper. Oiler	25.75	206.00	39.00	312.00		
1 Power Shovel, .75 C.Y.		473.40		520.75		
1 Dragline Bucket, .75 C.Y.		21.20		23.30	30.91	34.00
16 L.H., Daily Totals		$949.80		$1233.25	$59.36	$77.08

Crews

Crew No.	Bare Costs		Incl. Subs O & P		Cost Per Labor-Hour	
Crew B-12J	Hr.	Daily	Hr.	Daily	Bare Costs	Incl. O&P
1 Equip. Oper. (crane)	$31.15	$249.20	$47.15	$377.20	$28.45	$43.08
1 Equip. Oper. Oiler	25.75	206.00	39.00	312.00		
1 Gradall, 3 Ton, .5 C.Y.		648.40		713.25	40.53	44.58
16 L.H., Daily Totals		$1103.60		$1402.45	$68.98	$87.66
Crew B-12K	Hr.	Daily	Hr.	Daily	Bare Costs	Incl. O&P
1 Equip. Oper. (crane)	$31.15	$249.20	$47.15	$377.20	$28.45	$43.08
1 Equip. Oper. Oiler	25.75	206.00	39.00	312.00		
1 Gradall, 3 Ton, 1 C.Y.		774.80		852.30	48.43	53.27
16 L.H., Daily Totals		$1230.00		$1541.50	$76.88	$96.35
Crew B-12L	Hr.	Daily	Hr.	Daily	Bare Costs	Incl. O&P
1 Equip. Oper. (crane)	$31.15	$249.20	$47.15	$377.20	$28.45	$43.08
1 Equip. Oper. Oiler	25.75	206.00	39.00	312.00		
1 Power Shovel, .5 C.Y.		357.75		393.50		
1 F.E. Attachment, .5 C.Y.		37.00		40.70	23.55	25.90
16 L.H., Daily Totals		$849.95		$1123.40	$52.00	$68.98
Crew B-12M	Hr.	Daily	Hr.	Daily	Bare Costs	Incl. O&P
1 Equip. Oper. (crane)	$31.15	$249.20	$47.15	$377.20	$28.45	$43.08
1 Equip. Oper. Oiler	25.75	206.00	39.00	312.00		
1 Power Shovel, .75 C.Y.		473.40		520.75		
1 F.E. Attachment, .75 C.Y.		40.00		44.00	32.09	35.30
16 L.H., Daily Totals		$968.60		$1253.95	$60.54	$78.38
Crew B-12N	Hr.	Daily	Hr.	Daily	Bare Costs	Incl. O&P
1 Equip. Oper. (crane)	$31.15	$249.20	$47.15	$377.20	$28.45	$43.08
1 Equip. Oper. Oiler	25.75	206.00	39.00	312.00		
1 Power Shovel, 1 C.Y.		547.20		601.90		
1 F.E. Attachment, 1 C.Y.		44.40		48.85	36.98	40.67
16 L.H., Daily Totals		$1046.80		$1339.95	$65.43	$83.75
Crew B-12O	Hr.	Daily	Hr.	Daily	Bare Costs	Incl. O&P
1 Equip. Oper. (crane)	$31.15	$249.20	$47.15	$377.20	$28.45	$43.08
1 Equip. Oper. Oiler	25.75	206.00	39.00	312.00		
1 Power Shovel, 1.5 C.Y.		716.00		787.60		
1 F.E. Attachment, 1.5 C.Y.		134.20		147.60	53.14	58.45
16 L.H., Daily Totals		$1305.40		$1624.40	$81.59	$101.53
Crew B-12P	Hr.	Daily	Hr.	Daily	Bare Costs	Incl. O&P
1 Equip. Oper. (crane)	$31.15	$249.20	$47.15	$377.20	$28.45	$43.08
1 Equip. Oper. Oiler	25.75	206.00	39.00	312.00		
1 Crawler Crane, 40 Ton		716.00		787.60		
1 Dragline Bucket, 1.5 C.Y.		35.20		38.70	46.95	51.65
16 L.H., Daily Totals		$1206.40		$1515.50	$75.40	$94.73
Crew B-12Q	Hr.	Daily	Hr.	Daily	Bare Costs	Incl. O&P
1 Equip. Oper. (crane)	$31.15	$249.20	$47.15	$377.20	$28.45	$43.08
1 Equip. Oper. Oiler	25.75	206.00	39.00	312.00		
1 Hyd. Excavator, 5/8 C.Y.		360.00		396.00	22.50	24.75
16 L.H., Daily Totals		$815.20		$1085.20	$50.95	$67.83
Crew B-12R	Hr.	Daily	Hr.	Daily	Bare Costs	Incl. O&P
1 Equip. Oper. (crane)	$31.15	$249.20	$47.15	$377.20	$28.45	$43.08
1 Equip. Oper. Oiler	25.75	206.00	39.00	312.00		
1 Hyd. Excavator, 1.5 C.Y.		593.20		652.50	37.08	40.78
16 L.H., Daily Totals		$1048.40		$1341.70	$65.53	$83.86

Crew No.	Bare Costs		Incl. Subs O & P		Cost Per Labor-Hour	
Crew B-12S	Hr.	Daily	Hr.	Daily	Bare Costs	Incl. O&P
1 Equip. Oper. (crane)	$31.15	$249.20	$47.15	$377.20	$28.45	$43.08
1 Equip. Oper. Oiler	25.75	206.00	39.00	312.00		
1 Hyd. Excavator, 2.5 C.Y.		1557.00		1712.70	97.31	107.04
16 L.H., Daily Totals		$2012.20		$2401.90	$125.76	$150.12
Crew B-12T	Hr.	Daily	Hr.	Daily	Bare Costs	Incl. O&P
1 Equip. Oper. (crane)	$31.15	$249.20	$47.15	$377.20	$28.45	$43.08
1 Equip. Oper. Oiler	25.75	206.00	39.00	312.00		
1 Crawler Crane, 75 Ton		933.20		1026.50		
1 F.E. Attachment, 3 C.Y.		248.40		273.25	73.85	81.24
16 L.H., Daily Totals		$1636.80		$1988.95	$102.30	$124.32
Crew B-12V	Hr.	Daily	Hr.	Daily	Bare Costs	Incl. O&P
1 Equip. Oper. (crane)	$31.15	$249.20	$47.15	$377.20	$28.45	$43.08
1 Equip. Oper. Oiler	25.75	206.00	39.00	312.00		
1 Crawler Crane, 75 Ton		933.20		1026.50		
1 Dragline Bucket, 3 C.Y.		74.00		81.40	62.95	69.25
16 L.H., Daily Totals		$1462.40		$1797.10	$91.40	$112.33
Crew B-13	Hr.	Daily	Hr.	Daily	Bare Costs	Incl. O&P
1 Labor Foreman (outside)	$24.85	$198.80	$38.90	$311.20	$24.74	$38.29
4 Laborers	22.85	731.20	35.75	1144.00		
1 Equip. Oper. (crane)	31.15	249.20	47.15	377.20		
1 Equip. Oper. Oiler	25.75	206.00	39.00	312.00		
1 Hyd. Crane, 25 Ton		629.60		692.55	11.24	12.37
56 L.H., Daily Totals		$2014.80		$2836.95	$35.98	$50.66
Crew B-13A	Hr.	Daily	Hr.	Daily	Bare Costs	Incl. O&P
1 Foreman	$24.85	$198.80	$38.90	$311.20	$25.59	$39.36
2 Laborers	22.85	365.60	35.75	572.00		
2 Equipment Operator	30.35	485.60	45.95	735.20		
2 Truck Drivers (heavy)	23.95	383.20	36.60	585.60		
1 Crane, 75 Ton		933.20		1026.50		
1 F.E. Lder, 3.75 C.Y.		990.80		1089.90		
2 Dump Trucks, 12 Ton		740.80		814.90	47.59	52.34
56 L.H., Daily Totals		$4098.00		$5135.30	$73.18	$91.70
Crew B-13B	Hr.	Daily	Hr.	Daily	Bare Costs	Incl. O&P
1 Labor Foreman (outside)	$24.85	$198.80	$38.90	$311.20	$24.74	$38.29
4 Laborers	22.85	731.20	35.75	1144.00		
1 Equip. Oper. (crane)	31.15	249.20	47.15	377.20		
1 Equip. Oper. Oiler	25.75	206.00	39.00	312.00		
1 Hyd. Crane, 55 Ton		901.60		991.75	16.10	17.71
56 L.H., Daily Totals		$2286.80		$3136.15	$40.84	$56.00
Crew B-13C	Hr.	Daily	Hr.	Daily	Bare Costs	Incl. O&P
1 Labor Foreman (outside)	$24.85	$198.80	$38.90	$311.20	$24.74	$38.29
4 Laborers	22.85	731.20	35.75	1144.00		
1 Equip. Oper. (crane)	31.15	249.20	47.15	377.20		
1 Equip. Oper. Oiler	25.75	206.00	39.00	312.00		
1 Crawler Crane, 100 Ton		1209.00		1329.90	21.59	23.75
56 L.H., Daily Totals		$2594.20		$3474.30	$46.33	$62.04
Crew B-14	Hr.	Daily	Hr.	Daily	Bare Costs	Incl. O&P
1 Labor Foreman (outside)	$24.85	$198.80	$38.90	$311.20	$24.17	$37.58
4 Laborers	22.85	731.20	35.75	1144.00		
1 Equip. Oper. (light)	28.75	230.00	43.55	348.40		
1 Backhoe Loader, 48 H.P.		181.00		199.10	3.77	4.15
48 L.H., Daily Totals		$1341.00		$2002.70	$27.94	$41.73

Crews

Crew No.	Bare Costs		Incl. Subs O & P		Cost Per Labor-Hour	
Crew B-15	**Hr.**	**Daily**	**Hr.**	**Daily**	**Bare Costs**	**Incl. O&P**
1 Equipment Oper. (med)	$30.35	$242.80	$45.95	$367.60	$25.62	$39.15
.5 Laborer	22.85	91.40	35.75	143.00		
2 Truck Drivers (heavy)	23.95	383.20	36.60	585.60		
2 Dump Trucks, 16 Ton		965.60		1062.15		
1 Dozer, 200 H.P.		735.00		808.50	60.74	66.81
28 L.H., Daily Totals		$2418.00		$2966.85	$86.36	$105.96
Crew B-16	**Hr.**	**Daily**	**Hr.**	**Daily**	**Bare Costs**	**Incl. O&P**
1 Labor Foreman (outside)	$24.85	$198.80	$38.90	$311.20	$23.63	$36.75
2 Laborers	22.85	365.60	35.75	572.00		
1 Truck Driver (heavy)	23.95	191.60	36.60	292.80		
1 Dump Truck, 16 Ton		482.80		531.10	15.09	16.60
32 L.H., Daily Totals		$1238.80		$1707.10	$38.72	$53.35
Crew B-17	**Hr.**	**Daily**	**Hr.**	**Daily**	**Bare Costs**	**Incl. O&P**
2 Laborers	$22.85	$365.60	$35.75	$572.00	$24.60	$37.91
1 Equip. Oper. (light)	28.75	230.00	43.55	348.40		
1 Truck Driver (heavy)	23.95	191.60	36.60	292.80		
1 Backhoe Loader, 48 H.P.		181.00		199.10		
1 Dump Truck, 12 Ton		370.40		407.45	17.23	18.95
32 L.H., Daily Totals		$1338.60		$1819.75	$41.83	$56.86
Crew B-18	**Hr.**	**Daily**	**Hr.**	**Daily**	**Bare Costs**	**Incl. O&P**
1 Labor Foreman (outside)	$24.85	$198.80	$38.90	$311.20	$23.52	$36.80
2 Laborers	22.85	365.60	35.75	572.00		
1 Vibrating Compactor		60.60		66.65	2.53	2.78
24 L.H., Daily Totals		$625.00		$949.85	$26.05	$39.58
Crew B-19	**Hr.**	**Daily**	**Hr.**	**Daily**	**Bare Costs**	**Incl. O&P**
1 Pile Driver Foreman	$30.80	$246.40	$52.40	$419.20	$29.26	$47.71
4 Pile Drivers	28.80	921.60	49.00	1568.00		
2 Equip. Oper. (crane)	31.15	498.40	47.15	754.40		
1 Equip. Oper. Oiler	25.75	206.00	39.00	312.00		
1 Crane, 40 Ton & Access.		716.00		787.60		
60 L.F. Leads, 15K Ft. Lbs.		312.00		343.20		
1 Hammer, 15K Ft. Lbs.		313.40		344.75		
1 Air Compr., 600 C.F.M.		251.00		276.10		
2-50 Ft. Air Hoses, 3" Dia.		34.60		38.05	25.42	27.96
64 L.H., Daily Totals		$3499.40		$4843.30	$54.68	$75.67
Crew B-19A	**Hr.**	**Daily**	**Hr.**	**Daily**	**Bare Costs**	**Incl. O&P**
1 Pile Driver Foreman	$30.80	$246.40	$52.40	$419.20	$29.26	$47.71
4 Pile Drivers	28.80	921.60	49.00	1568.00		
2 Equip. Oper. (crane)	31.15	498.40	47.15	754.40		
1 Equip. Oper. Oiler	25.75	206.00	39.00	312.00		
1 Crawler Crane, 75 Ton		933.20		1026.50		
60 L.F. Leads, 25K Ft. Lbs.		432.00		475.20		
1 Air Compressor, 750 CFM		294.20		323.60		
4-50 Ft. Air Hose, 3" Dia.		69.20		76.10	27.01	29.71
64 L.H., Daily Totals		$3601.00		$4955.00	$56.27	$77.42
Crew B-20	**Hr.**	**Daily**	**Hr.**	**Daily**	**Bare Costs**	**Incl. O&P**
1 Labor Foreman (out)	$24.85	$198.80	$38.90	$311.20	$25.85	$40.50
1 Skilled Worker	29.85	238.80	46.85	374.80		
1 Laborer	22.85	182.80	35.75	286.00		
24 L.H., Daily Totals		$620.40		$972.00	$25.85	$40.50

Crew No.	Bare Costs		Incl. Subs O & P		Cost Per Labor-Hour	
Crew B-20A	**Hr.**	**Daily**	**Hr.**	**Daily**	**Bare Costs**	**Incl. O&P**
1 Labor Foreman	$24.85	$198.80	$38.90	$311.20	$27.43	$42.06
1 Laborer	22.85	182.80	35.75	286.00		
1 Plumber	34.45	275.60	52.00	416.00		
1 Plumber Apprentice	27.55	220.40	41.60	332.80		
32 L.H., Daily Totals		$877.60		$1346.00	$27.43	$42.06
Crew B-21	**Hr.**	**Daily**	**Hr.**	**Daily**	**Bare Costs**	**Incl. O&P**
1 Labor Foreman (out)	$24.85	$198.80	$38.90	$311.20	$26.61	$41.45
1 Skilled Worker	29.85	238.80	46.85	374.80		
1 Laborer	22.85	182.80	35.75	286.00		
.5 Equip. Oper. (crane)	31.15	124.60	47.15	188.60		
.5 S.P. Crane, 5 Ton		142.00		156.20	5.07	5.58
28 L.H., Daily Totals		$887.00		$1316.80	$31.68	$47.03
Crew B-21A	**Hr.**	**Daily**	**Hr.**	**Daily**	**Bare Costs**	**Incl. O&P**
1 Labor Foreman	$24.85	$198.80	$38.90	$311.20	$28.17	$43.08
1 Laborer	22.85	182.80	35.75	286.00		
1 Plumber	34.45	275.60	52.00	416.00		
1 Plumber Apprentice	27.55	220.40	41.60	332.80		
1 Equip. Oper. (crane)	31.15	249.20	47.15	377.20		
1 S.P. Crane, 12 Ton		425.20		467.70	10.63	11.69
40 L.H., Daily Totals		$1552.00		$2190.90	$38.80	$54.77
Crew B-22	**Hr.**	**Daily**	**Hr.**	**Daily**	**Bare Costs**	**Incl. O&P**
1 Labor Foreman (out)	$24.85	$198.80	$38.90	$311.20	$26.91	$41.83
1 Skilled Worker	29.85	238.80	46.85	374.80		
1 Laborer	22.85	182.80	35.75	286.00		
.75 Equip. Oper. (crane)	31.15	186.90	47.15	282.90		
.75 S.P. Crane, 5 Ton		213.00		234.30	7.10	7.81
30 L.H., Daily Totals		$1020.30		$1489.20	$34.01	$49.64
Crew B-22A	**Hr.**	**Daily**	**Hr.**	**Daily**	**Bare Costs**	**Incl. O&P**
1 Labor Foreman (out)	$24.85	$198.80	$38.90	$311.20	$26.06	$40.55
1 Skilled Worker	29.85	238.80	46.85	374.80		
2 Laborers	22.85	365.60	35.75	572.00		
.75 Equipment Oper. (crane)	31.15	186.90	47.15	282.90		
.75 Crane, 5 Ton		213.00		234.30		
1 Generator, 5 KW		46.90		51.60		
1 Butt Fusion Machine		302.20		332.40	14.79	16.27
38 L.H., Daily Totals		$1552.20		$2159.20	$40.85	$56.82
Crew B-22B	**Hr.**	**Daily**	**Hr.**	**Daily**	**Bare Costs**	**Incl. O&P**
1 Skilled Worker	$29.85	$238.80	$46.85	$374.80	$26.35	$41.30
1 Laborer	22.85	182.80	35.75	286.00		
1 Electro Fusion Machine		118.60		130.45	7.41	8.15
16 L.H., Daily Totals		$540.20		$791.25	$33.76	$49.45
Crew B-23	**Hr.**	**Daily**	**Hr.**	**Daily**	**Bare Costs**	**Incl. O&P**
1 Labor Foreman (outside)	$24.85	$198.80	$38.90	$311.20	$23.25	$36.38
4 Laborers	22.85	731.20	35.75	1144.00		
1 Drill Rig, Wells		1870.00		2057.00		
1 Light Truck, 3 Ton		184.20		202.60	51.36	56.49
40 L.H., Daily Totals		$2984.20		$3714.80	$74.61	$92.87
Crew B-23A	**Hr.**	**Daily**	**Hr.**	**Daily**	**Bare Costs**	**Incl. O&P**
1 Labor Foreman (outside)	$24.85	$198.80	$38.90	$311.20	$26.02	$40.20
1 Laborers	22.85	182.80	35.75	286.00		
1 Equip. Operator (medium)	30.35	242.80	45.95	367.60		
1 Drill Rig, Wells		1870.00		2057.00		
1 Pickup Truck, 3/4 Ton		83.00		91.30	81.38	89.51
24 L.H., Daily Totals		$2577.40		$3113.10	$107.40	$129.71

Crews

Crew No.	Bare Costs		Incl. Subs O & P		Cost Per Labor-Hour	
Crew B-23B	Hr.	Daily	Hr.	Daily	Bare Costs	Incl. O&P
1 Labor Foreman (outside)	$24.85	$198.80	$38.90	$311.20	$26.02	$40.20
1 Laborer	22.85	182.80	35.75	286.00		
1 Equip. Operator (medium)	30.35	242.80	45.95	367.60		
1 Drill Rig, Wells		1870.00		2057.00		
1 Pickup Truck, 3/4 Ton		83.00		91.30		
1 Pump, Cntfgl, 6"		164.40		180.85	88.23	97.05
24 L.H., Daily Totals		$2741.80		$3293.95	$114.25	$137.25
Crew B-24	Hr.	Daily	Hr.	Daily	Bare Costs	Incl. O&P
1 Cement Finisher	$27.95	$223.60	$41.55	$332.40	$26.65	$40.97
1 Laborer	22.85	182.80	35.75	286.00		
1 Carpenter	29.15	233.20	45.60	364.80		
24 L.H., Daily Totals		$639.60		$983.20	$26.65	$40.97
Crew B-25	Hr.	Daily	Hr.	Daily	Bare Costs	Incl. O&P
1 Labor Foreman	$24.85	$198.80	$38.90	$311.20	$25.08	$38.82
7 Laborers	22.85	1279.60	35.75	2002.00		
3 Equip. Oper. (med.)	30.35	728.40	45.95	1102.80		
1 Asphalt Paver, 130 H.P		1281.00		1409.10		
1 Tandem Roller, 10 Ton		210.40		231.45		
1 Roller, Pneumatic Wheel		215.60		237.15	19.40	21.34
88 L.H., Daily Totals		$3913.80		$5293.70	$44.48	$60.16
Crew B-25B	Hr.	Daily	Hr.	Daily	Bare Costs	Incl. O&P
1 Labor Foreman	$24.85	$198.80	$38.90	$311.20	$25.52	$39.41
7 Laborers	22.85	1279.60	35.75	2002.00		
4 Equip. Oper. (medium)	30.35	971.20	45.95	1470.40		
1 Asphalt Paver, 130 H.P.		1281.00		1409.10		
2 Rollers, Steel Wheel		420.80		462.90		
1 Roller, Pneumatic Wheel		215.60		237.15	19.97	21.97
96 L.H., Daily Totals		$4367.00		$5892.75	$45.49	$61.38
Crew B-25C	Hr.	Daily	Hr.	Daily	Bare Costs	Incl. O&P
1 Labor Foreman	$24.85	$198.80	$38.90	$311.20	$25.68	$39.67
3 Laborers	22.85	548.40	35.75	858.00		
2 Equip. Oper. (medium)	30.35	485.60	45.95	735.20		
1 Asphalt Paver, 130 H.P.		1281.00		1409.10		
1 Rollers, Steel Wheel		210.40		231.45	31.07	34.18
48 L.H., Daily Totals		$2724.20		$3544.95	$56.75	$73.85
Crew B-26	Hr.	Daily	Hr.	Daily	Bare Costs	Incl. O&P
1 Labor Foreman (outside)	$24.85	$198.80	$38.90	$311.20	$25.76	$40.22
6 Laborers	22.85	1096.80	35.75	1716.00		
2 Equip. Oper. (med.)	30.35	485.60	45.95	735.20		
1 Rodman (reinf.)	32.75	262.00	55.60	444.80		
1 Cement Finisher	27.95	223.60	41.55	332.40		
1 Grader, 30,000 Lbs.		459.40		505.35		
1 Paving Mach. & Equip.		1371.00		1508.10	20.80	22.88
88 L.H., Daily Totals		$4097.20		$5553.05	$46.56	$63.10
Crew B-27	Hr.	Daily	Hr.	Daily	Bare Costs	Incl. O&P
1 Labor Foreman (outside)	$24.85	$198.80	$38.90	$311.20	$23.35	$36.54
3 Laborers	22.85	548.40	35.75	858.00		
1 Berm Machine		144.20		158.60	4.51	4.96
32 L.H., Daily Totals		$891.40		$1327.80	$27.86	$41.50
Crew B-28	Hr.	Daily	Hr.	Daily	Bare Costs	Incl. O&P
2 Carpenters	$29.15	$466.40	$45.60	$729.60	$27.05	$42.32
1 Laborer	22.85	182.80	35.75	286.00		
24 L.H., Daily Totals		$649.20		$1015.60	$27.05	$42.32

Crew No.	Bare Costs		Incl. Subs O & P		Cost Per Labor-Hour	
Crew B-29	Hr.	Daily	Hr.	Daily	Bare Costs	Incl. O&P
1 Labor Foreman (outside)	$24.85	$198.80	$38.90	$311.20	$24.74	$38.29
4 Laborers	22.85	731.20	35.75	1144.00		
1 Equip. Oper. (crane)	31.15	249.20	47.15	377.20		
1 Equip. Oper. Oiler	25.75	206.00	39.00	312.00		
1 Gradall, 3 Ton, 1/2 C.Y.		648.40		713.25	11.58	12.74
56 L.H., Daily Totals		$2033.60		$2857.65	$36.32	$51.03
Crew B-30	Hr.	Daily	Hr.	Daily	Bare Costs	Incl. O&P
1 Equip. Oper. (med.)	$30.35	$242.80	$45.95	$367.60	$26.08	$39.72
2 Truck Drivers (heavy)	23.95	383.20	36.60	585.60		
1 Hyd. Excavator, 1.5 C.Y.		593.20		652.50		
2 Dump Trucks, 16 Ton		965.60		1062.15	64.95	71.45
24 L.H., Daily Totals		$2184.80		$2667.85	$91.03	$111.17
Crew B-31	Hr.	Daily	Hr.	Daily	Bare Costs	Incl. O&P
1 Labor Foreman (outside)	$24.85	$198.80	$38.90	$311.20	$24.51	$38.35
3 Laborers	22.85	548.40	35.75	858.00		
1 Carpenter	29.15	233.20	45.60	364.80		
1 Air Compr., 250 C.F.M.		125.20		137.70		
1 Sheeting Driver		12.50		13.75		
2-50 Ft. Air Hoses, 1.5" Dia.		11.80		13.00	3.74	4.11
40 L.H., Daily Totals		$1129.90		$1698.45	$28.25	$42.46
Crew B-32	Hr.	Daily	Hr.	Daily	Bare Costs	Incl. O&P
1 Laborer	$22.85	$182.80	$35.75	$286.00	$28.48	$43.40
3 Equip. Oper. (med.)	30.35	728.40	45.95	1102.80		
1 Grader, 30,000 Lbs.		459.40		505.35		
1 Tandem Roller, 10 Ton		210.40		231.45		
1 Dozer, 200 H.P.		735.00		808.50	43.90	48.29
32 L.H., Daily Totals		$2316.00		$2934.10	$72.38	$91.69
Crew B-32A	Hr.	Daily	Hr.	Daily	Bare Costs	Incl. O&P
1 Laborer	$22.85	$182.80	$35.75	$286.00	$27.85	$42.55
2 Equip. Oper. (medium)	30.35	485.60	45.95	735.20		
1 Grader, 30,000 Lbs.		459.40		505.35		
1 Roller, Vibratory, 29,000 Lbs.		451.20		496.30	37.94	41.74
24 L.H., Daily Totals		$1579.00		$2022.85	$65.79	$84.29
Crew B-32B	Hr.	Daily	Hr.	Daily	Bare Costs	Incl. O&P
1 Laborer	$22.85	$182.80	$35.75	$286.00	$27.85	$42.55
2 Equip. Oper. (medium)	30.35	485.60	45.95	735.20		
1 Dozer, 200 H.P.		735.00		808.50		
1 Roller, Vibratory, 29,000 Lbs.		451.20		496.30	49.43	54.37
24 L.H., Daily Totals		$1854.60		$2326.00	$77.28	$96.92
Crew B-32C	Hr.	Daily	Hr.	Daily	Bare Costs	Incl. O&P
1 Labor Foreman	$24.85	$198.80	$38.90	$311.20	$26.93	$41.38
2 Laborers	22.85	365.60	35.75	572.00		
3 Equip. Oper. (medium)	30.35	728.40	45.95	1102.80		
1 Grader, 30,000 Lbs.		459.40		505.35		
1 Roller, Steel Wheel		210.40		231.45		
1 Dozer, 200 H.P.		735.00		808.50	29.27	32.19
48 L.H., Daily Totals		$2697.60		$3531.30	$56.20	$73.57
Crew B-33	Hr.	Daily	Hr.	Daily	Bare Costs	Incl. O&P
1 Equip. Oper. (med.)	$30.35	$242.80	$45.95	$367.60	$28.21	$43.04
.5 Laborer	22.85	91.40	35.75	143.00		
.25 Equip. Oper. (med.)	30.35	60.70	45.95	91.90		
14 L.H., Daily Totals		$394.90		$602.50	$28.21	$43.04

Crews

Crew No.	Bare Costs		Incl. Subs O & P		Cost Per Labor-Hour	
Crew B-33A	Hr.	Daily	Hr.	Daily	Bare Costs	Incl. O&P
1 Equip. Oper. (med.)	$30.35	$242.80	$45.95	$367.60	$28.21	$43.04
.5 Laborer	22.85	91.40	35.75	143.00		
.25 Equip. Oper. (med.)	30.35	60.70	45.95	91.90		
1 Scraper, Towed, 7 C.Y.		123.35		135.70		
1.25 Dozer, 300 H.P.		1283.75		1412.15	100.51	110.56
14 L.H., Daily Totals		$1802.00		$2150.35	$128.72	$153.60
Crew B-33B	Hr.	Daily	Hr.	Daily	Bare Costs	Incl. O&P
1 Equip. Oper. (med.)	$30.35	$242.80	$45.95	$367.60	$28.21	$43.04
.5 Laborer	22.85	91.40	35.75	143.00		
.25 Equip. Oper. (med.)	30.35	60.70	45.95	91.90		
1 Scraper, Towed, 10 C.Y.		157.25		173.00		
1.25 Dozer, 300 H.P.		1283.75		1412.15	102.93	113.22
14 L.H., Daily Totals		$1835.90		$2187.65	$131.14	$156.26
Crew B-33C	Hr.	Daily	Hr.	Daily	Bare Costs	Incl. O&P
1 Equip. Oper. (med.)	$30.35	$242.80	$45.95	$367.60	$28.21	$43.04
.5 Laborer	22.85	91.40	35.75	143.00		
.25 Equip. Oper. (med.)	30.35	60.70	45.95	91.90		
1 Scraper, Towed, 12 C.Y.		157.25		173.00		
1.25 Dozer, 300 H.P.		1283.75		1412.15	102.93	113.22
14 L.H., Daily Totals		$1835.90		$2187.65	$131.14	$156.26
Crew B-33D	Hr.	Daily	Hr.	Daily	Bare Costs	Incl. O&P
1 Equip. Oper. (med.)	$30.35	$242.80	$45.95	$367.60	$28.21	$43.04
.5 Laborer	22.85	91.40	35.75	143.00		
.25 Equip. Oper. (med.)	30.35	60.70	45.95	91.90		
1 S.P. Scraper, 14 C.Y.		1597.00		1756.70		
.25 Dozer, 300 H.P.		256.75		282.45	132.41	145.65
14 L.H., Daily Totals		$2248.65		$2641.65	$160.62	$188.69
Crew B-33E	Hr.	Daily	Hr.	Daily	Bare Costs	Incl. O&P
1 Equip. Oper. (med.)	$30.35	$242.80	$45.95	$367.60	$28.21	$43.04
.5 Laborer	22.85	91.40	35.75	143.00		
.25 Equip. Oper. (med.)	30.35	60.70	45.95	91.90		
1 S.P. Scraper, 24 C.Y.		1932.00		2125.20		
.25 Dozer, 300 H.P.		256.75		282.45	156.34	171.97
14 L.H., Daily Totals		$2583.65		$3010.15	$184.55	$215.01
Crew B-33F	Hr.	Daily	Hr.	Daily	Bare Costs	Incl. O&P
1 Equip. Oper. (med.)	$30.35	$242.80	$45.95	$367.60	$28.21	$43.04
.5 Laborer	22.85	91.40	35.75	143.00		
.25 Equip. Oper. (med.)	30.35	60.70	45.95	91.90		
1 Elev. Scraper, 11 C.Y.		668.60		735.45		
.25 Dozer, 300 H.P.		256.75		282.45	66.10	72.71
14 L.H., Daily Totals		$1320.25		$1620.40	$94.31	$115.75
Crew B-33G	Hr.	Daily	Hr.	Daily	Bare Costs	Incl. O&P
1 Equip. Oper. (med.)	$30.35	$242.80	$45.95	$367.60	$28.21	$43.04
.5 Laborer	22.85	91.40	35.75	143.00		
.25 Equip. Oper. (med.)	30.35	60.70	45.95	91.90		
1 Elev. Scraper, 20 C.Y.		1178.00		1295.80		
.25 Dozer, 300 H.P.		256.75		282.45	102.48	112.73
14 L.H., Daily Totals		$1829.65		$2180.75	$130.69	$155.77
Crew B-33H	Hr.	Daily	Hr.	Daily	Bare Costs	Incl. O&P
.25 Laborer	$22.85	$45.70	$35.75	$71.50	$29.06	$44.19
1 Equipment Operator (med.)	30.35	242.80	45.95	367.60		
.2 Equipment Operator (med.)	30.35	48.56	45.95	73.52		
1 Scraper, 32-44 C.Y.		2385.00		2623.50		
.2 Dozer, 298 kw		264.60		291.05	228.41	251.26
11. L.H., Daily Totals		$2986.66		$3427.17	$257.47	$295.45
Crew B-33J	Hr.	Daily	Hr.	Daily	Bare Costs	Incl. O&P
1 Equipment Operator (med.)	$30.35	$242.80	$45.95	$367.60	$30.35	$45.95
1 Scraper 17 C.Y.		1597.00		1756.70	199.63	219.59
8 L.H., Daily Totals		$1839.80		$2124.30	$229.98	$265.54
Crew B-34A	Hr.	Daily	Hr.	Daily	Bare Costs	Incl. O&P
1 Truck Driver (heavy)	$23.95	$191.60	$36.60	$292.80	$23.95	$36.60
1 Dump Truck, 12 Ton		370.40		407.45	46.30	50.93
8 L.H., Daily Totals		$562.00		$700.25	$70.25	$87.53
Crew B-34B	Hr.	Daily	Hr.	Daily	Bare Costs	Incl. O&P
1 Truck Driver (heavy)	$23.95	$191.60	$36.60	$292.80	$23.95	$36.60
1 Dump Truck, 16 Ton		482.80		531.10	60.35	66.39
8 L.H., Daily Totals		$674.40		$823.90	$84.30	$102.99
Crew B-34C	Hr.	Daily	Hr.	Daily	Bare Costs	Incl. O&P
1 Truck Driver (heavy)	$23.95	$191.60	$36.60	$292.80	$23.95	$36.60
1 Truck Tractor, 40 Ton		413.00		454.30		
1 Dump Trailer, 16.5 C.Y.		140.00		154.00	69.13	76.04
8 L.H., Daily Totals		$744.60		$901.10	$93.08	$112.64
Crew B-34D	Hr.	Daily	Hr.	Daily	Bare Costs	Incl. O&P
1 Truck Driver (heavy)	$23.95	$191.60	$36.60	$292.80	$23.95	$36.60
1 Truck Tractor, 40 Ton		413.00		454.30		
1 Dump Trailer, 20 C.Y.		145.60		160.15	69.83	76.81
8 L.H., Daily Totals		$750.20		$907.25	$93.78	$113.41
Crew B-34E	Hr.	Daily	Hr.	Daily	Bare Costs	Incl. O&P
1 Truck Driver (heavy)	$23.95	$191.60	$36.60	$292.80	$23.95	$36.60
1 Truck, Off Hwy., 25 Ton		901.40		991.55	112.68	123.94
8 L.H., Daily Totals		$1093.00		$1284.35	$136.63	$160.54
Crew B-34F	Hr.	Daily	Hr.	Daily	Bare Costs	Incl. O&P
1 Truck Driver (heavy)	$23.95	$191.60	$36.60	$292.80	$23.95	$36.60
1 Truck, Off Hwy., 22 C.Y.		1100.00		1210.00	137.50	151.25
8 L.H., Daily Totals		$1291.60		$1502.80	$161.45	$187.85
Crew B-34G	Hr.	Daily	Hr.	Daily	Bare Costs	Incl. O&P
1 Truck Driver (heavy)	$23.95	$191.60	$36.60	$292.80	$23.95	$36.60
1 Truck, Off Hwy., 34 C.Y.		1466.00		1612.60	183.25	201.58
8 L.H., Daily Totals		$1657.60		$1905.40	$207.20	$238.18
Crew B-34H	Hr.	Daily	Hr.	Daily	Bare Costs	Incl. O&P
1 Truck Driver (heavy)	$23.95	$191.60	$36.60	$292.80	$23.95	$36.60
1 Truck, Off Hwy., 42 C.Y.		1561.00		1717.10	195.13	214.64
8 L.H., Daily Totals		$1752.60		$2009.90	$219.08	$251.24
Crew B-34J	Hr.	Daily	Hr.	Daily	Bare Costs	Incl. O&P
1 Truck Driver (heavy)	$23.95	$191.60	$36.60	$292.80	$23.95	$36.60
1 Truck, Off Hwy., 60 C.Y.		2259.00		2484.90	282.38	310.61
8 L.H., Daily Totals		$2450.60		$2777.70	$306.33	$347.21

Crews

Crew No.		Bare Costs		Incl. Subs O & P		Cost Per Labor-Hour	
Crew B-34K	Hr.	Daily	Hr.	Daily	Bare Costs	Incl. O&P
1 Truck Driver (heavy)	$23.95	$191.60	$36.60	$292.80	$23.95	$36.60
1 Truck Tractor, 240 H.P.		503.60		553.95		
1 Low Bed Trailer		286.60		315.25	98.78	108.65
8 L.H., Daily Totals		$981.80		$1162.00	$122.73	$145.25
Crew B-35	Hr.	Daily	Hr.	Daily	Bare Costs	Incl. O&P
1 Laborer Foreman (out)	$24.85	$198.80	$38.90	$311.20	$28.15	$43.27
1 Skilled Worker	29.85	238.80	46.85	374.80		
1 Welder (plumber)	34.45	275.60	52.00	416.00		
1 Laborer	22.85	182.80	35.75	286.00		
1 Equip. Oper. (crane)	31.15	249.20	47.15	377.20		
1 Equip. Oper. Oiler	25.75	206.00	39.00	312.00		
1 Electric Welding Mach.		100.65		110.70		
1 Hyd. Excavator, .75 C.Y.		413.60		454.95	10.71	11.78
48 L.H., Daily Totals		$1865.45		$2642.85	$38.86	$55.05
Crew B-35A	Hr.	Daily	Hr.	Daily	Bare Costs	Incl. O&P
1 Laborer Foreman (out)	$24.85	$198.80	$38.90	$311.20	$27.39	$42.20
2 Laborers	22.85	365.60	35.75	572.00		
1 Skilled Worker	29.85	238.80	46.85	374.80		
1 Welder (plumber)	34.45	275.60	52.00	416.00		
1 Equip. Oper. (crane)	31.15	249.20	47.15	377.20		
1 Equip. Oper. Oiler	25.75	206.00	39.00	312.00		
1 Welder, 300 amp		84.00		92.40		
1 Crane, 75 Ton		933.20		1026.50	18.16	19.98
56 L.H., Daily Totals		$2551.20		$3482.10	$45.55	$62.18
Crew B-36	Hr.	Daily	Hr.	Daily	Bare Costs	Incl. O&P
1 Labor Foreman (outside)	$24.85	$198.80	$38.90	$311.20	$26.25	$40.46
2 Laborers	22.85	365.60	35.75	572.00		
2 Equip. Oper. (med.)	30.35	485.60	45.95	735.20		
1 Dozer, 200 H.P.		735.00		808.50		
1 Aggregate Spreader		73.00		80.30		
1 Tandem Roller, 10 Ton		210.40		231.45	25.46	28.01
40 L.H., Daily Totals		$2068.40		$2738.65	$51.71	$68.47
Crew B-36A	Hr.	Daily	Hr.	Daily	Bare Costs	Incl. O&P
1 Labor Foreman (outside)	$24.85	$198.80	$38.90	$311.20	$27.42	$42.03
2 Laborers	22.85	365.60	35.75	572.00		
4 Equip. Oper. (med.)	30.35	971.20	45.95	1470.40		
1 Dozer, 200 H.P.		735.00		808.50		
1 Aggregate Spreader		73.00		80.30		
1 Roller, Steel Wheel		210.40		231.45		
1 Roller, Pneumatic Wheel		215.60		237.15	22.04	24.24
56 L.H., Daily Totals		$2769.60		$3711.00	$49.46	$66.27
Crew B-36B	Hr.	Daily	Hr.	Daily	Bare Costs	Incl. O&P
1 Labor Foreman (outside)	$24.85	$198.80	$38.90	$311.20	$26.99	$41.35
2 Laborers	22.85	365.60	35.75	572.00		
4 Equip. Oper. (medium)	30.35	971.20	45.95	1470.40		
1 Truck Driver, Heavy	23.95	191.60	36.60	292.80		
1 Grader, 30,000 Lbs.		459.40		505.35		
1 F.E. Loader, crl, 1.5 C.Y.		358.80		394.70		
1 Dozer, 300 H.P.		1027.00		1129.70		
1 Roller, Vibratory		451.20		496.30		
1 Truck, Tractor, 240 H.P.		503.60		553.95		
1 Water Tanker, 5000 Gal.		184.00		202.40	46.63	51.29
64 L.H., Daily Totals		$4711.20		$5928.80	$73.62	$92.64
Crew B-37	Hr.	Daily	Hr.	Daily	Bare Costs	Incl. O&P
1 Labor Foreman (outside)	$24.85	$198.80	$38.90	$311.20	$24.17	$37.58
4 Laborers	22.85	731.20	35.75	1144.00		
1 Equip. Oper. (light)	28.75	230.00	43.55	348.40		
1 Tandem Roller, 5 Ton		133.20		146.50	2.78	3.05
48 L.H., Daily Totals		$1293.20		$1950.10	$26.95	$40.63
Crew B-38	Hr.	Daily	Hr.	Daily	Bare Costs	Incl. O&P
1 Labor Foreman (outside)	$24.85	$198.80	$38.90	$311.20	$25.93	$39.98
2 Laborers	22.85	365.60	35.75	572.00		
1 Equip. Oper. (light)	28.75	230.00	43.55	348.40		
1 Equip. Oper. (medium)	30.35	242.80	45.95	367.60		
1 Backhoe Loader, 48 H.P.		181.00		199.10		
1 Hyd.Hammer, (1200 lb)		190.40		209.45		
1 F.E. Loader (170 H.P.)		518.80		570.70		
1 Pavt. Rem. Bucket		45.60		50.15	23.40	25.73
40 L.H., Daily Totals		$1973.00		$2628.60	$49.33	$65.71
Crew B-39	Hr.	Daily	Hr.	Daily	Bare Costs	Incl. O&P
1 Labor Foreman (outside)	$24.85	$198.80	$38.90	$311.20	$24.17	$37.58
4 Laborers	22.85	731.20	35.75	1144.00		
1 Equip. Oper. (light)	28.75	230.00	43.55	348.40		
1 Air Compr., 250 C.F.M.		125.20		137.70		
2 Air Tools & Accessories		39.20		43.10		
2-50 Ft. Air Hoses, 1.5" Dia.		11.80		13.00	3.67	4.04
48 L.H., Daily Totals		$1336.20		$1997.40	$27.84	$41.62
Crew B-40	Hr.	Daily	Hr.	Daily	Bare Costs	Incl. O&P
1 Pile Driver Foreman (out)	$30.80	$246.40	$52.40	$419.20	$29.26	$47.71
4 Pile Drivers	28.80	921.60	49.00	1568.00		
2 Equip. Oper. (crane)	31.15	498.40	47.15	754.40		
1 Equip. Oper. Oiler	25.75	206.00	39.00	312.00		
1 Crane, 40 Ton		716.00		787.60		
1 Vibratory Hammer & Gen.		1283.00		1411.30	31.23	34.36
64 L.H., Daily Totals		$3871.40		$5252.50	$60.49	$82.07
Crew B-41	Hr.	Daily	Hr.	Daily	Bare Costs	Incl. O&P
1 Labor Foreman (outside)	$24.85	$198.80	$38.90	$311.20	$23.72	$36.99
4 Laborers	22.85	731.20	35.75	1144.00		
.25 Equip. Oper. (crane)	31.15	62.30	47.15	94.30		
.25 Equip. Oper. Oiler	25.75	51.50	39.00	78.00		
.25 Crawler Crane, 40 Ton		179.00		196.90	4.07	4.48
44 L.H., Daily Totals		$1222.80		$1824.40	$27.79	$41.47
Crew B-42	Hr.	Daily	Hr.	Daily	Bare Costs	Incl. O&P
1 Labor Foreman (outside)	$24.85	$198.80	$38.90	$311.20	$25.75	$40.96
4 Laborers	22.85	731.20	35.75	1144.00		
1 Equip. Oper. (crane)	31.15	249.20	47.15	377.20		
1 Equip. Oper. Oiler	25.75	206.00	39.00	312.00		
1 Welder	32.85	262.80	59.60	476.80		
1 Hyd. Crane, 25 Ton		629.60		692.55		
1 Gas Welding Machine		84.00		92.40		
1 Horz. Boring Csg. Mch.		523.20		575.50	19.33	21.26
64 L.H., Daily Totals		$2884.80		$3981.65	$45.08	$62.22
Crew B-43	Hr.	Daily	Hr.	Daily	Bare Costs	Incl. O&P
1 Labor Foreman (outside)	$24.85	$198.80	$38.90	$311.20	$25.05	$38.72
3 Laborers	22.85	548.40	35.75	858.00		
1 Equip. Oper. (crane)	31.15	249.20	47.15	377.20		
1 Equip. Oper. Oiler	25.75	206.00	39.00	312.00		
1 Drill Rig & Augers		1870.00		2057.00	38.96	42.85
48 L.H., Daily Totals		$3072.40		$3915.40	$64.01	$81.57

Crews

Crew B-44

Crew B-44	Hr.	Daily	Hr.	Daily	Bare Costs	Incl. O&P
1 Pile Driver Foreman	$30.80	$246.40	$52.40	$419.20	$28.89	$47.31
4 Pile Drivers	28.80	921.60	49.00	1568.00		
2 Equip. Oper. (crane)	31.15	498.40	47.15	754.40		
1 Laborer	22.85	182.80	35.75	286.00		
1 Crane, 40 Ton, & Access.		716.00		787.60		
45 L.F. Leads, 15K Ft. Lbs.		234.00		257.40	14.84	16.33
64 L.H., Daily Totals		$2799.20		$4072.60	$43.73	$63.64

Crew B-45

Crew B-45	Hr.	Daily	Hr.	Daily	Bare Costs	Incl. O&P
1 Equip. Oper. (med.)	$30.35	$242.80	$45.95	$367.60	$27.15	$41.28
1 Truck Driver (heavy)	23.95	191.60	36.60	292.80		
1 Dist. Tank Truck, 3K Gal.		328.80		361.70		
1 Tractor, 4 x 2, 250 H.P.		397.60		437.35	45.40	49.94
16 L.H., Daily Totals		$1160.80		$1459.45	$72.55	$91.22

Crew B-46

Crew B-46	Hr.	Daily	Hr.	Daily	Bare Costs	Incl. O&P
1 Pile Driver Foreman	$30.80	$246.40	$52.40	$419.20	$26.16	$42.94
2 Pile Drivers	28.80	460.80	49.00	784.00		
3 Laborers	22.85	548.40	35.75	858.00		
1 Chain Saw, 36" Long		43.20		47.50	.90	.99
48 L.H., Daily Totals		$1298.80		$2108.70	$27.06	$43.93

Crew B-47

Crew B-47	Hr.	Daily	Hr.	Daily	Bare Costs	Incl. O&P
1 Blast Foreman	$24.85	$198.80	$38.90	$311.20	$25.48	$39.40
1 Driller	22.85	182.80	35.75	286.00		
1 Equip. Oper. (light)	28.75	230.00	43.55	348.40		
1 Crawler Type Drill, 4"		455.60		501.15		
1 Air Compr., 600 C.F.M.		251.00		276.10		
2-50 Ft. Air Hoses, 3" Dia.		34.60		38.05	30.88	33.97
24 L.H., Daily Totals		$1352.80		$1760.90	$56.36	$73.37

Crew B-47A

Crew B-47A	Hr.	Daily	Hr.	Daily	Bare Costs	Incl. O&P
1 Drilling Foreman	$24.85	$198.80	$38.90	$311.20	$27.25	$41.68
1 Equip. Oper. (heavy)	31.15	249.20	47.15	377.20		
1 Oiler	25.75	206.00	39.00	312.00		
1 Quarry Drill		721.20		793.30	30.05	33.06
24 L.H., Daily Totals		$1375.20		$1793.70	$57.30	$74.74

Crew B-47C

Crew B-47C	Hr.	Daily	Hr.	Daily	Bare Costs	Incl. O&P
1 Laborer	$22.85	$182.80	$35.75	$286.00	$25.80	$39.65
1 Equip. Oper. (light)	28.75	230.00	43.55	348.40		
1 Air Compressor, 750 CFM		294.20		323.60		
2-50' Air Hose, 3"		34.60		38.05		
1 Air Track Drill, 4"		455.60		501.15	49.03	53.93
16 L.H., Daily Totals		$1197.20		$1497.20	$74.83	$93.58

Crew B-47E

Crew B-47E	Hr.	Daily	Hr.	Daily	Bare Costs	Incl. O&P
1 Laborer Foreman	$24.85	$198.80	$38.90	$311.20	$23.35	$36.54
3 Laborers	22.85	548.40	35.75	858.00		
1 Truck, Flatbed, 3 Ton		184.20		202.60	5.76	6.33
32 L.H., Daily Totals		$931.40		$1371.80	$29.11	$42.87

Crew B-48

Crew B-48	Hr.	Daily	Hr.	Daily	Bare Costs	Incl. O&P
1 Labor Foreman (outside)	$24.85	$198.80	$38.90	$311.20	$25.58	$39.41
3 Laborers	22.85	548.40	35.75	858.00		
1 Equip. Oper. (crane)	31.15	249.20	47.15	377.20		
1 Equip. Oper. Oiler	25.75	206.00	39.00	312.00		
1 Equip. Oper. (light)	28.75	230.00	43.55	348.40		
1 Centr. Water Pump, 6"		164.40		180.85		
1-20 Ft. Suction Hose, 6"		22.50		24.75		
1-50 Ft. Disch. Hose, 6"		20.55		22.60		
1 Drill Rig & Augers		1870.00		2057.00	37.10	40.81
56 L.H., Daily Totals		$3509.85		$4492.00	$62.68	$80.22

Crew B-49

Crew B-49	Hr.	Daily	Hr.	Daily	Bare Costs	Incl. O&P
1 Labor Foreman (outside)	$24.85	$198.80	$38.90	$311.20	$26.69	$41.82
3 Laborers	22.85	548.40	35.75	858.00		
2 Equip. Oper. (crane)	31.15	498.40	47.15	754.40		
2 Equip. Oper. Oilers	25.75	412.00	39.00	624.00		
1 Equip. Oper. (light)	28.75	230.00	43.55	348.40		
2 Pile Drivers	28.80	460.80	49.00	784.00		
1 Hyd. Crane, 25 Ton		629.60		692.55		
1 Centr. Water Pump, 6"		164.40		180.85		
1-20 Ft. Suction Hose, 6"		22.50		24.75		
1-50 Ft. Disch. Hose, 6"		20.55		22.60		
1 Drill Rig & Augers		1870.00		2057.00	30.76	33.84
88 L.H., Daily Totals		$5055.45		$6657.75	$57.45	$75.66

Crew B-50

Crew B-50	Hr.	Daily	Hr.	Daily	Bare Costs	Incl. O&P
2 Pile Driver Foremen	$30.80	$492.80	$52.40	$838.40	$27.93	$45.67
6 Pile Drivers	28.80	1382.40	49.00	2352.00		
2 Equip. Oper. (crane)	31.15	498.40	47.15	754.40		
1 Equip. Oper. Oiler	25.75	206.00	39.00	312.00		
3 Laborers	22.85	548.40	35.75	858.00		
1 Crane, 40 Ton		716.00		787.60		
60 L.F. Leads, 15K Ft. Lbs.		312.00		343.20		
1 Hammer, 15K Ft. Lbs.		313.40		344.75		
1 Air Compr., 600 C.F.M.		251.00		276.10		
2-50 Ft. Air Hoses, 3" Dia.		34.60		38.05		
1 Chain Saw, 36" Long		43.20		47.50	14.91	16.40
112 L.H., Daily Totals		$4798.20		$6952.00	$42.84	$62.07

Crew B-51

Crew B-51	Hr.	Daily	Hr.	Daily	Bare Costs	Incl. O&P
1 Labor Foreman (outside)	$24.85	$198.80	$38.90	$311.20	$23.28	$36.28
4 Laborers	22.85	731.20	35.75	1144.00		
1 Truck Driver (light)	23.40	187.20	35.75	286.00		
1 Light Truck, 1.5 Ton		158.00		173.80	3.29	3.62
48 L.H., Daily Totals		$1275.20		$1915.00	$26.57	$39.90

Crew B-52

Crew B-52	Hr.	Daily	Hr.	Daily	Bare Costs	Incl. O&P
1 Carpenter Foreman	$31.15	$249.20	$48.75	$390.00	$26.91	$41.99
1 Carpenter	29.15	233.20	45.60	364.80		
3 Laborers	22.85	548.40	35.75	858.00		
1 Cement Finisher	27.95	223.60	41.55	332.40		
.5 Rodman (reinf.)	32.75	131.00	55.60	222.40		
.5 Equip. Oper. (med.)	30.35	121.40	45.95	183.80		
.5 F.E. Ldr., T.M., 2.5 C.Y.		385.00		423.50	6.88	7.56
56 L.H., Daily Totals		$1891.80		$2774.90	$33.79	$49.55

Crew B-53

Crew B-53	Hr.	Daily	Hr.	Daily	Bare Costs	Incl. O&P
1 Equip. Oper. (light)	$28.75	$230.00	$43.55	$348.40	$28.75	$43.55
1 Trencher, Chain, 12 H.P.		96.80		106.50	12.10	13.31
8 L.H., Daily Totals		$326.80		$454.90	$40.85	$56.86

Crews

Crew No.	Bare Costs		Incl. Subs O & P		Cost Per Labor-Hour	
Crew B-54	Hr.	Daily	Hr.	Daily	Bare Costs	Incl. O&P
1 Equip. Oper. (light)	$28.75	$230.00	$43.55	$348.40	$28.75	$43.55
1 Trencher, Chain, 40 H.P.		196.40		216.05	24.55	27.01
8 L.H., Daily Totals		$426.40		$564.45	$53.30	$70.56
Crew B-54A	Hr.	Daily	Hr.	Daily	Bare Costs	Incl. O&P
.17 Labor Foreman (outside)	$24.85	$33.80	$38.90	$52.90	$29.55	$44.93
1 Equipment Operator (med.)	30.35	242.80	45.95	367.60		
1 Wheel Trencher, 67 H.P.		424.80		467.30	45.38	49.92
9.36 L.H., Daily Totals		$701.40		$887.80	$74.93	$94.85
Crew B-54B	Hr.	Daily	Hr.	Daily	Bare Costs	Incl. O&P
.25 Labor Foreman (outside)	$24.85	$49.70	$38.90	$77.80	$29.25	$44.54
1 Equipment Operator (med.)	30.35	242.80	45.95	367.60		
1 Wheel Trencher, 150 H.P.		636.80		700.50	63.68	70.05
10 L.H., Daily Totals		$929.30		$1145.90	$92.93	$114.59
Crew B-55	Hr.	Daily	Hr.	Daily	Bare Costs	Incl. O&P
2 Laborers	$22.85	$365.60	$35.75	$572.00	$23.03	$35.75
1 Truck Driver (light)	23.40	187.20	35.75	286.00		
1 Auger, 4" to 36" Dia		519.40		571.35		
1 Flatbed 3 Ton Truck		184.20		202.60	29.32	32.25
24 L.H., Daily Totals		$1256.40		$1631.95	$52.35	$68.00
Crew B-56	Hr.	Daily	Hr.	Daily	Bare Costs	Incl. O&P
1 Laborer	$22.85	$182.80	$35.75	$286.00	$25.80	$39.65
1 Equip. Oper. (light)	28.75	230.00	43.55	348.40		
1 Crawler Type Drill, 4"		455.60		501.15		
1 Air Compr., 600 C.F.M.		251.00		276.10		
1-50 Ft. Air Hose, 3" Dia.		17.30		19.05	45.24	49.77
16 L.H., Daily Totals		$1136.70		$1430.70	$71.04	$89.42
Crew B-57	Hr.	Daily	Hr.	Daily	Bare Costs	Incl. O&P
1 Labor Foreman (outside)	$24.85	$198.80	$38.90	$311.20	$26.03	$40.02
2 Laborers	22.85	365.60	35.75	572.00		
1 Equip. Oper. (crane)	31.15	249.20	47.15	377.20		
1 Equip. Oper. (light)	28.75	230.00	43.55	348.40		
1 Equip. Oper. Oiler	25.75	206.00	39.00	312.00		
1 Power Shovel, 1 C.Y.		547.20		601.90		
1 Clamshell Bucket, 1 C.Y.		71.00		78.10		
1 Centr. Water Pump, 6"		164.40		180.85		
1-20 Ft. Suction Hose, 6"		22.50		24.75		
20-50 Ft. Disch. Hoses, 6"		411.00		452.10	25.34	27.87
48 L.H., Daily Totals		$2465.70		$3258.50	$51.37	$67.89
Crew B-58	Hr.	Daily	Hr.	Daily	Bare Costs	Incl. O&P
2 Laborers	$22.85	$365.60	$35.75	$572.00	$24.82	$38.35
1 Equip. Oper. (light)	28.75	230.00	43.55	348.40		
1 Backhoe Loader, 48 H.P.		181.00		199.10		
1 Small Helicopter, w/pilot		1986.00		2184.60	90.29	99.32
24 L.H., Daily Totals		$2762.60		$3304.10	$115.11	$137.67
Crew B-59	Hr.	Daily	Hr.	Daily	Bare Costs	Incl. O&P
1 Truck Driver (heavy)	$23.95	$191.60	$36.60	$292.80	$23.95	$36.60
1 Truck, 30 Ton		320.40		352.45		
1 Water tank, 5000 Gal.		184.00		202.40	63.05	69.36
8 L.H., Daily Totals		$696.00		$847.65	$87.00	$105.96

Crew No.	Bare Costs		Incl. Subs O & P		Cost Per Labor-Hour	
Crew B-59A	Hr.	Daily	Hr.	Daily	Bare Costs	Incl. O&P
2 Laborers	$22.85	$365.60	$35.75	$572.00	$23.22	$36.03
1 Truck Driver (heavy)	23.95	191.60	36.60	292.80		
1 Water tank, 5K w/pum		184.00		202.40		
1 Truck, 30 Ton		320.40		352.45	21.02	23.12
24 L.H., Daily Totals		$1061.60		$1419.65	$44.24	$59.15
Crew B-60	Hr.	Daily	Hr.	Daily	Bare Costs	Incl. O&P
1 Labor Foreman (outside)	$24.85	$198.80	$38.90	$311.20	$26.42	$40.52
2 Laborers	22.85	365.60	35.75	572.00		
1 Equip. Oper. (crane)	31.15	249.20	47.15	377.20		
2 Equip. Oper. (light)	28.75	460.00	43.55	696.80		
1 Equip. Oper. Oiler	25.75	206.00	39.00	312.00		
1 Crawler Crane, 40 Ton		716.00		787.60		
45 L.F. Leads, 15K Ft. Lbs.		234.00		257.40		
1 Backhoe Loader, 48 H.P.		181.00		199.10	20.20	22.22
56 L.H., Daily Totals		$2610.60		$3513.30	$46.62	$62.74
Crew B-61	Hr.	Daily	Hr.	Daily	Bare Costs	Incl. O&P
1 Labor Foreman (outside)	$24.85	$198.80	$38.90	$311.20	$24.43	$37.94
3 Laborers	22.85	548.40	35.75	858.00		
1 Equip. Oper. (light)	28.75	230.00	43.55	348.40		
1 Cement Mixer, 2 C.Y.		222.00		244.20		
1 Air Compr., 160 C.F.M.		98.40		108.25	8.01	8.81
40 L.H., Daily Totals		$1297.60		$1870.05	$32.44	$46.75
Crew B-62	Hr.	Daily	Hr.	Daily	Bare Costs	Incl. O&P
2 Laborers	$22.85	$365.60	$35.75	$572.00	$24.82	$38.35
1 Equip. Oper. (light)	28.75	230.00	43.55	348.40		
1 Loader, Skid Steer		114.00		125.40	4.75	5.23
24 L.H., Daily Totals		$709.60		$1045.80	$29.57	$43.58
Crew B-63	Hr.	Daily	Hr.	Daily	Bare Costs	Incl. O&P
4 Laborers	$22.85	$731.20	$35.75	$1144.00	$24.03	$37.31
1 Equip. Oper. (light)	28.75	230.00	43.55	348.40		
1 Loader, Skid Steer		114.00		125.40	2.85	3.14
40 L.H., Daily Totals		$1075.20		$1617.80	$26.88	$40.45
Crew B-64	Hr.	Daily	Hr.	Daily	Bare Costs	Incl. O&P
1 Laborer	$22.85	$182.80	$35.75	$286.00	$23.13	$35.75
1 Truck Driver (light)	23.40	187.20	35.75	286.00		
1 Power Mulcher (small)		124.80		137.30		
1 Light Truck, 1.5 Ton		158.00		173.80	17.68	19.44
16 L.H., Daily Totals		$652.80		$883.10	$40.81	$55.19
Crew B-65	Hr.	Daily	Hr.	Daily	Bare Costs	Incl. O&P
1 Laborer	$22.85	$182.80	$35.75	$286.00	$23.13	$35.75
1 Truck Driver (light)	23.40	187.20	35.75	286.00		
1 Power Mulcher (large)		232.80		256.10		
1 Light Truck, 1.5 Ton		158.00		173.80	24.43	26.87
16 L.H., Daily Totals		$760.80		$1001.90	$47.56	$62.62
Crew B-66	Hr.	Daily	Hr.	Daily	Bare Costs	Incl. O&P
1 Equip. Oper. (light)	$28.75	$230.00	$43.55	$348.40	$28.75	$43.55
1 Backhoe Ldr. w/Attchmt.		168.80		185.70	21.10	23.21
8 L.H., Daily Totals		$398.80		$534.10	$49.85	$66.76

Crews

Crew No.		Bare Costs		Incl. Subs O & P		Cost Per Labor-Hour	
Crew B-67	Hr.	Daily	Hr.	Daily	Bare Costs	Incl. O&P
1 Millwright	$30.55	$244.40	$45.45	$363.60	$29.65	$44.50
1 Equip. Oper. (light)	28.75	230.00	43.55	348.40		
1 Forklift		198.20		218.00	12.39	13.63
16 L.H., Daily Totals		$672.60		$930.00	$42.04	$58.13
Crew B-68	Hr.	Daily	Hr.	Daily	Bare Costs	Incl. O&P
2 Millwrights	$30.55	$488.80	$45.45	$727.20	$29.95	$44.82
1 Equip. Oper. (light)	28.75	230.00	43.55	348.40		
1 Forklift		198.20		218.00	8.26	9.08
24 L.H., Daily Totals		$917.00		$1293.60	$38.21	$53.90
Crew B-69	Hr.	Daily	Hr.	Daily	Bare Costs	Incl. O&P
1 Labor Foreman (outside)	$24.85	$198.80	$38.90	$311.20	$25.05	$38.72
3 Laborers	22.85	548.40	35.75	858.00		
1 Equip Oper. (crane)	31.15	249.20	47.15	377.20		
1 Equip Oper. Oiler	25.75	206.00	39.00	312.00		
1 Truck Crane, 80 Ton		1325.00		1457.50	27.60	30.36
48 L.H., Daily Totals		$2527.40		$3315.90	$52.65	$69.08
Crew B-69A	Hr.	Daily	Hr.	Daily	Bare Costs	Incl. O&P
1 Labor Foreman	$24.85	$198.80	$38.90	$311.20	$25.28	$38.94
3 Laborers	22.85	548.40	35.75	858.00		
1 Equip. Oper. (medium)	30.35	242.80	45.95	367.60		
1 Concrete Finisher	27.95	223.60	41.55	332.40		
1 Curb Paver		444.60		489.05	9.26	10.19
48 L.H., Daily Totals		$1658.20		$2358.25	$34.54	$49.13
Crew B-69B	Hr.	Daily	Hr.	Daily	Bare Costs	Incl. O&P
1 Labor Foreman	$24.85	$198.80	$38.90	$311.20	$25.28	$38.94
3 Laborers	22.85	548.40	35.75	858.00		
1 Equip. Oper. (medium)	30.35	242.80	45.95	367.60		
1 Cement Finisher	27.95	223.60	41.55	332.40		
1 Curb/Gutter Paver		897.60		987.35	18.70	20.57
48 L.H., Daily Totals		$2111.20		$2856.55	$43.98	$59.51
Crew B-70	Hr.	Daily	Hr.	Daily	Bare Costs	Incl. O&P
1 Labor Foreman (outside)	$24.85	$198.80	$38.90	$311.20	$26.35	$40.57
3 Laborers	22.85	548.40	35.75	858.00		
3 Equip. Oper. (med.)	30.35	728.40	45.95	1102.80		
1 Motor Grader, 30,000 Lb.		459.40		505.35		
1 Grader Attach., Ripper		67.20		73.90		
1 Road Sweeper, S.P.		258.00		283.80		
1 F.E. Loader, 1-3/4 C.Y.		284.40		312.85	19.09	21.00
56 L.H., Daily Totals		$2544.60		$3447.90	$45.44	$61.57
Crew B-71	Hr.	Daily	Hr.	Daily	Bare Costs	Incl. O&P
1 Labor Foreman (outside)	$24.85	$198.80	$38.90	$311.20	$26.35	$40.57
3 Laborers	22.85	548.40	35.75	858.00		
3 Equip. Oper. (med.)	30.35	728.40	45.95	1102.80		
1 Pvmt. Profiler, 450 H.P.		3506.00		3856.60		
1 Road Sweeper, S.P.		258.00		283.80		
1 F.E. Loader, 1-3/4 C.Y.		284.40		312.85	72.29	79.52
56 L.H., Daily Totals		$5524.00		$6725.25	$98.64	$120.09

Crew No.		Bare Costs		Incl. Subs O & P		Cost Per Labor-Hour	
Crew B-72	Hr.	Daily	Hr.	Daily	Bare Costs	Incl. O&P
1 Labor Foreman (outside)	$24.85	$198.80	$38.90	$311.20	$26.85	$41.24
3 Laborers	22.85	548.40	35.75	858.00		
4 Equip. Oper. (med.)	30.35	971.20	45.95	1470.40		
1 Pvmt. Profiler, 450 H.P.		3506.00		3856.60		
1 Hammermill, 250 H.P.		1234.00		1357.40		
1 Windrow Loader		884.00		972.40		
1 Mix Paver 165 H.P.		1428.00		1570.80		
1 Roller, Pneu. Tire, 12 T.		215.60		237.15	113.56	124.91
64 L.H., Daily Totals		$8986.00		$10633.95	$140.41	$166.15
Crew B-73	Hr.	Daily	Hr.	Daily	Bare Costs	Incl. O&P
1 Labor Foreman (outside)	$24.85	$198.80	$38.90	$311.20	$27.79	$42.52
2 Laborers	22.85	365.60	35.75	572.00		
5 Equip. Oper. (med.)	30.35	1214.00	45.95	1838.00		
1 Road Mixer, 310 H.P.		1112.00		1223.20		
1 Roller, Tandem, 12 Ton		210.40		231.45		
1 Hammermill, 250 H.P.		1234.00		1357.40		
1 Motor Grader, 30,000 Lb.		459.40		505.35		
.5 F.E. Loader, 1-3/4 C.Y.		142.20		156.40		
.5 Truck, 30 Ton		160.20		176.20		
.5 Water Tank 5000 Gal.		92.00		101.20	53.28	58.61
64 L.H., Daily Totals		$5188.60		$6472.40	$81.07	$101.13
Crew B-74	Hr.	Daily	Hr.	Daily	Bare Costs	Incl. O&P
1 Labor Foreman (outside)	$24.85	$198.80	$38.90	$311.20	$27.13	$41.46
1 Laborer	22.85	182.80	35.75	286.00		
4 Equip. Oper. (med.)	30.35	971.20	45.95	1470.40		
2 Truck Drivers (heavy)	23.95	383.20	36.60	585.60		
1 Motor Grader, 30,000 Lb.		459.40		505.35		
1 Grader Attach., Ripper		67.20		73.90		
2 Stabilizers, 310 H.P.		1601.60		1761.75		
1 Flatbed Truck, 3 Ton		184.20		202.60		
1 Chem. Spreader, Towed		92.40		101.65		
1 Vibr. Roller, 29,000 Lb.		451.20		496.30		
1 Water Tank 5000 Gal.		184.20		202.40		
1 Truck, 30 Ton		320.40		352.45	52.51	57.76
64 L.H., Daily Totals		$5096.40		$6349.60	$79.64	$99.22
Crew B-75	Hr.	Daily	Hr.	Daily	Bare Costs	Incl. O&P
1 Labor Foreman (outside)	$24.85	$198.80	$38.90	$311.20	$27.58	$42.15
1 Laborer	22.85	182.80	35.75	286.00		
4 Equip. Oper. (med.)	30.35	971.20	45.95	1470.40		
1 Truck Driver (heavy)	23.95	191.60	36.60	292.80		
1 Motor Grader, 30,000 Lb.		459.40		505.35		
1 Grader Attach., Ripper		67.20		73.90		
2 Stabilizers, 310 H.P.		1601.60		1761.75		
1 Dist. Truck, 3000 Gal.		328.80		361.70		
1 Vibr. Roller, 29,000 Lb.		451.20		496.30	51.93	57.13
56 L.H., Daily Totals		$4452.60		$5559.40	$79.51	$99.28
Crew B-76	Hr.	Daily	Hr.	Daily	Bare Costs	Incl. O&P
1 Dock Builder Foreman	$30.80	$246.40	$52.40	$419.20	$29.21	$47.86
5 Dock Builders	28.80	1152.00	49.00	1960.00		
2 Equip. Oper. (crane)	31.15	498.40	47.15	754.40		
1 Equip. Oper. Oiler	25.75	206.00	39.00	312.00		
1 Crawler Crane, 50 Ton		885.80		974.40		
1 Barge, 400 Ton		428.40		471.25		
1 Hammer, 15K Ft. Lbs.		313.40		344.75		
60 L.F. Leads, 15K Ft. Lbs.		312.00		343.20		
1 Air Compr., 600 C.F.M.		251.00		276.10		
2-50 Ft. Air Hoses, 3" Dia.		34.60		38.05	30.91	34.00
72 L.H., Daily Totals		$4328.00		$5893.35	$60.12	$81.86

Crews

Crew No.	Bare Costs		Incl. Subs O & P		Cost Per Labor-Hour	
	Hr.	Daily	Hr.	Daily	Bare Costs	Incl. O&P
Crew B-77						
1 Labor Foreman	$24.85	$198.80	$38.90	$311.20	$23.36	$36.38
3 Laborers	22.85	548.40	35.75	858.00		
1 Truck Driver (light)	23.40	187.20	35.75	286.00		
1 Crack Cleaner, 25 H.P.		71.20		78.30		
1 Crack Filler, Trailer Mtd.		142.40		156.65		
1 Flatbed Truck, 3 Ton		184.20		202.60	9.95	10.94
40 L.H., Daily Totals		$1332.20		$1892.75	$33.31	$47.32
Crew B-78						
1 Labor Foreman	$24.85	$198.80	$38.90	$311.20	$23.28	$36.28
4 Laborers	22.85	731.20	35.75	1144.00		
1 Truck Driver (light)	23.40	187.20	35.75	286.00		
1 Paint Striper, S.P.		201.80		222.00		
1 Flatbed Truck, 3 Ton		184.20		202.60		
1 Pickup Truck, 3/4 Ton		83.00		91.30	9.77	10.75
48 L.H., Daily Totals		$1586.20		$2257.10	$33.05	$47.03
Crew B-79						
1 Labor Foreman	$24.85	$198.80	$38.90	$311.20	$23.36	$36.38
3 Laborers	22.85	548.40	35.75	858.00		
1 Truck Driver (light)	23.40	187.20	35.75	286.00		
1 Thermo. Striper, T.M.		390.00		429.00		
1 Flatbed Truck, 3 Ton		184.20		202.60		
2 Pickup Truck, 3/4 Ton		166.00		182.60	18.51	20.36
40 L.H., Daily Totals		$1674.60		$2269.40	$41.87	$56.74
Crew B-80						
1 Labor Foreman	$24.85	$198.80	$38.90	$311.20	$24.96	$38.49
1 Laborer	22.85	182.80	35.75	286.00		
1 Truck Driver (light)	23.40	187.20	35.75	286.00		
1 Equip. Oper. (light)	28.75	230.00	43.55	348.40		
1 Flatbed Truck, 3 Ton		184.20		202.60		
1 Fence Post Auger, T.M.		363.00		399.30	17.10	18.81
32 L.H., Daily Totals		$1346.00		$1833.50	$42.06	$57.30
Crew B-80A						
3 Laborers	$22.85	$548.40	$35.75	$858.00	$22.85	$35.75
1 Flatbed Truck, 3 Ton		184.20		202.60	7.68	8.44
24 L.H., Daily Totals		$732.60		$1060.60	$30.53	$44.19
Crew B-80B						
3 Laborers	$22.85	$548.40	$35.75	$858.00	$24.33	$37.70
1 Equip. Oper. (light)	28.75	230.00	43.55	348.40		
1 Crane, Flatbed Mnt.		222.40		244.65	6.95	7.65
32 L.H., Daily Totals		$1000.80		$1451.05	$31.28	$45.35
Crew B-81						
1 Laborer	$22.85	$182.80	$35.75	$286.00	$25.72	$39.43
1 Equip. Oper. (med.)	30.35	242.80	45.95	367.60		
1 Truck Driver (heavy)	23.95	191.60	36.60	292.80		
1 Hydromulcher, T.M.		263.20		289.50		
1 Tractor Truck, 4x2		320.40		352.45	24.32	26.75
24 L.H., Daily Totals		$1200.80		$1588.35	$50.04	$66.18
Crew B-82						
1 Laborer	$22.85	$182.80	$35.75	$286.00	$25.80	$39.65
1 Equip. Oper. (light)	28.75	230.00	43.55	348.40		
1 Horiz. Borer, 6 H.P.		46.40		51.05	2.90	3.19
16 L.H., Daily Totals		$459.20		$685.45	$28.70	$42.84

Crew No.	Bare Costs		Incl. Subs O & P		Cost Per Labor-Hour	
	Hr.	Daily	Hr.	Daily	Bare Costs	Incl. O&P
Crew B-83						
1 Tugboat Captain	$30.35	$242.80	$45.95	$367.60	$26.60	$40.85
1 Tugboat Hand	22.85	182.80	35.75	286.00		
1 Tugboat, 250 H.P.		466.00		512.60	29.13	32.04
16 L.H., Daily Totals		$891.60		$1166.20	$55.73	$72.89
Crew B-84						
1 Equip. Oper. (med.)	$30.35	$242.80	$45.95	$367.60	$30.35	$45.95
1 Rotary Mower/Tractor		179.20		197.10	22.40	24.64
8 L.H., Daily Totals		$422.00		$564.70	$52.75	$70.59
Crew B-85						
3 Laborers	$22.85	$548.40	$35.75	$858.00	$24.57	$37.96
1 Equip. Oper. (med.)	30.35	242.80	45.95	367.60		
1 Truck Driver (heavy)	23.95	191.60	36.60	292.80		
1 Aerial Lift Truck, 80'		530.80		583.90		
1 Brush Chipper, 130 H.P.		190.60		209.65		
1 Pruning Saw, Rotary		22.00		24.20	18.59	20.44
40 L.H., Daily Totals		$1726.20		$2336.15	$43.16	$58.40
Crew B-86						
1 Equip. Oper. (med.)	$30.35	$242.80	$45.95	$367.60	$30.35	$45.95
1 Stump Chipper, S.P.		166.65		183.30	20.83	22.91
8 L.H., Daily Totals		$409.45		$550.90	$51.18	$68.86
Crew B-86A						
1 Equip. Oper. (medium)	$30.35	$242.80	$45.95	$367.60	$30.35	$45.95
1 Grader, 30,000 Lbs.		459.40		505.35	57.43	63.17
8 L.H., Daily Totals		$702.20		$872.95	$87.78	$109.12
Crew B-86B						
1 Equip. Oper. (medium)	$30.35	$242.80	$45.95	$367.60	$30.35	$45.95
1 Dozer, 200 H.P.		735.00		808.50	91.88	101.06
8 L.H., Daily Totals		$977.80		$1176.10	$122.23	$147.01
Crew B-87						
1 Laborer	$22.85	$182.80	$35.75	$286.00	$28.85	$43.91
4 Equip. Oper. (med.)	30.35	971.20	45.95	1470.40		
2 Feller Bunchers, 50 H.P.		746.00		820.60		
1 Log Chipper, 22" Tree		1789.00		1967.90		
1 Dozer, 105 H.P.		266.00		292.60		
1 Chainsaw, Gas, 36" Long		43.20		47.50	71.11	78.22
40 L.H., Daily Totals		$3998.20		$4885.00	$99.96	$122.13
Crew B-88						
1 Laborer	$22.85	$182.80	$35.75	$286.00	$29.28	$44.49
6 Equip. Oper. (med.)	30.35	1456.80	45.95	2205.60		
2 Feller Bunchers, 50 H.P.		746.00		820.60		
1 Log Chipper, 22" Tree		1789.00		1967.90		
2 Log Skidders, 50 H.P.		1026.80		1129.50		
1 Dozer, 105 H.P.		266.00		292.60		
1 Chainsaw, Gas, 36" Long		43.20		47.50	69.13	76.04
56 L.H., Daily Totals		$5510.60		$6749.70	$98.41	$120.53
Crew B-89						
1 Equip. Oper. (light)	$28.75	$230.00	$43.55	$348.40	$26.08	$39.65
1 Truck Driver (light)	23.40	187.20	35.75	286.00		
1 Truck, Stake Body, 3 Ton		184.20		202.60		
1 Concrete Saw		98.40		108.25		
1 Water Tank, 65 Gal.		13.50		14.85	18.51	20.36
16 L.H., Daily Totals		$713.30		$960.10	$44.59	$60.01

Crews

Crew No.	Bare Costs		Incl. Subs O & P		Cost Per Labor-Hour	
	Hr.	Daily	Hr.	Daily	Bare Costs	Incl. O&P
Crew B-89A						
1 Skilled Worker	$29.85	$238.80	$46.85	$374.80	$26.35	$41.30
1 Laborer	22.85	182.80	35.75	286.00		
1 Core Drill (large)		101.45		111.60	6.34	6.97
16 L.H., Daily Totals		$523.05		$772.40	$32.69	$48.27
Crew B-89B	Hr.	Daily	Hr.	Daily	Bare Costs	Incl. O&P
1 Equip. Oper. (light)	$28.75	$230.00	$43.55	$348.40	$26.08	$39.65
1 Truck Driver, Light	23.40	187.20	35.75	286.00		
1 Wall Saw, Hydraulic, 10 H.P.		106.05		116.65		
1 Generator, Diesel, 100 KW		211.05		232.15		
1 Water Tank, 65 Gal.		13.50		14.85		
1 Flatbed Truck, 3 Ton		184.20		202.60	32.18	35.39
16 L.H., Daily Totals		$932.00		$1200.65	$58.26	$75.04
Crew B-90	Hr.	Daily	Hr.	Daily	Bare Costs	Incl. O&P
1 Labor Foreman (outside)	$24.85	$198.80	$38.90	$311.20	$24.85	$38.31
3 Laborers	22.85	548.40	35.75	858.00		
2 Equip. Oper. (light)	28.75	460.00	43.55	696.80		
2 Truck Drivers (heavy)	23.95	383.20	36.60	585.60		
1 Road Mixer, 310 H.P.		1112.00		1223.20		
1 Dist. Truck, 2000 Gal.		314.80		346.30	22.29	24.52
64 L.H., Daily Totals		$3017.20		$4021.10	$47.14	$62.83
Crew B-90A	Hr.	Daily	Hr.	Daily	Bare Costs	Incl. O&P
1 Labor Foreman	$24.85	$198.80	$38.90	$311.20	$27.42	$42.03
2 Laborers	22.85	365.60	35.75	572.00		
4 Equip. Oper. (medium)	30.35	971.20	45.95	1470.40		
2 Graders, 30,000 Lbs.		918.80		1010.70		
1 Roller, Steel Wheel		210.40		231.45		
1 Roller, Pneumatic Wheel		215.60		237.15	24.01	26.42
56 L.H., Daily Totals		$2880.40		$3832.90	$51.43	$68.45
Crew B-90B	Hr.	Daily	Hr.	Daily	Bare Costs	Incl. O&P
1 Labor Foreman	$24.85	$198.80	$38.90	$311.20	$26.93	$41.38
2 Laborers	22.85	365.60	35.75	572.00		
3 Equip. Oper. (medium)	30.35	728.40	45.95	1102.80		
1 Roller, Steel Wheel		210.40		231.45		
1 Roller, Pneumatic Wheel		215.60		237.15		
1 Road Mixer, 310 H.P.		1112.00		1223.20	32.04	35.25
48 L.H., Daily Totals		$2830.80		$3677.80	$58.97	$76.63
Crew B-91	Hr.	Daily	Hr.	Daily	Bare Costs	Incl. O&P
1 Labor Foreman (outside)	$24.85	$198.80	$38.90	$311.20	$26.99	$41.35
2 Laborers	22.85	365.60	35.75	572.00		
4 Equip. Oper. (med.)	30.35	971.20	45.95	1470.40		
1 Truck Driver (heavy)	23.95	191.60	36.60	292.80		
1 Dist. Truck, 3000 Gal.		328.80		361.70		
1 Aggreg. Spreader, S.P.		628.60		691.45		
1 Roller, Pneu. Tire, 12 Ton		215.60		237.15		
1 Roller, Steel, 10 Ton		210.40		231.45	21.62	23.78
64 L.H., Daily Totals		$3110.60		$4168.15	$48.61	$65.13
Crew B-92	Hr.	Daily	Hr.	Daily	Bare Costs	Incl. O&P
1 Labor Foreman (outside)	$24.85	$198.80	$38.90	$311.20	$23.35	$36.54
3 Laborers	22.85	548.40	35.75	858.00		
1 Crack Cleaner, 25 H.P.		71.20		78.30		
1 Air Compressor		78.30		86.15		
1 Tar Kettle, T.M.		37.50		41.25		
1 Flatbed Truck, 3 Ton		184.20		202.60	11.60	12.76
32 L.H., Daily Totals		$1118.40		$1577.50	$34.95	$49.30

Crew No.	Bare Costs		Incl. Subs O & P		Cost Per Labor-Hour	
	Hr.	Daily	Hr.	Daily	Bare Costs	Incl. O&P
Crew B-93						
1 Equip. Oper. (med.)	$30.35	$242.80	$45.95	$367.60	$30.35	$45.95
1 Feller Buncher, 50 H.P.		373.00		410.30	46.63	51.29
8 L.H., Daily Totals		$615.80		$777.90	$76.98	$97.24
Crew B-94A	Hr.	Daily	Hr.	Daily	Bare Costs	Incl. O&P
1 Laborer	$22.85	$182.80	$35.75	$286.00	$22.85	$35.75
1 Diaph. Water Pump, 2"		30.80		33.90		
1-20 Ft. Suction Hose, 2"		6.50		7.15		
2-50 Ft. Disch. Hoses, 2"		10.80		11.90	6.01	6.61
8 L.H., Daily Totals		$230.90		$338.95	$28.86	$42.36
Crew B-94B	Hr.	Daily	Hr.	Daily	Bare Costs	Incl. O&P
1 Laborer	$22.85	$182.80	$35.75	$286.00	$22.85	$35.75
1 Diaph. Water Pump, 4"		62.00		68.20		
1-20 Ft. Suction Hose, 4"		12.50		13.75		
2-50 Ft. Disch. Hoses, 4"		17.00		18.70	11.44	12.58
8 L.H., Daily Totals		$274.30		$386.65	$34.29	$48.33
Crew B-94C	Hr.	Daily	Hr.	Daily	Bare Costs	Incl. O&P
1 Laborer	$22.85	$182.80	$35.75	$286.00	$22.85	$35.75
1 Centr. Water Pump, 3"		39.20		43.10		
1-20 Ft. Suction Hose, 3"		9.50		10.45		
2-50 Ft. Disch. Hoses, 3"		12.80		14.10	7.69	8.46
8 L.H., Daily Totals		$244.30		$353.65	$30.54	$44.21
Crew B-94D	Hr.	Daily	Hr.	Daily	Bare Costs	Incl. O&P
1 Laborer	$22.85	$182.80	$35.75	$286.00	$22.85	$35.75
1 Centr. Water Pump, 6"		164.40		180.85		
1-20 Ft. Suction Hose, 6"		22.50		24.75		
2-50 Ft. Disch. Hoses, 6"		41.10		45.20	28.50	31.35
8 L.H., Daily Totals		$410.80		$536.80	$51.35	$67.10
Crew B-95	Hr.	Daily	Hr.	Daily	Bare Costs	Incl. O&P
1 Equip. Oper. (crane)	$31.15	$249.20	$47.15	$377.20	$27.00	$41.45
1 Laborer	22.85	182.80	35.75	286.00		
16 L.H., Daily Totals		$432.00		$663.20	$27.00	$41.45
Crew B-95A	Hr.	Daily	Hr.	Daily	Bare Costs	Incl. O&P
1 Equip. Oper. (crane)	$31.15	$249.20	$47.15	$377.20	$27.00	$41.45
1 Laborer	22.85	182.80	35.75	286.00		
1 Hyd. Excavator, 5/8 C.Y.		360.00		396.00	22.50	24.75
16 L.H., Daily Totals		$792.00		$1059.20	$49.50	$66.20
Crew B-95B	Hr.	Daily	Hr.	Daily	Bare Costs	Incl. O&P
1 Equip. Oper. (crane)	$31.15	$249.20	$47.15	$377.20	$27.00	$41.45
1 Laborer	22.85	182.80	35.75	286.00		
1 Hyd. Excavator, 1.5 C.Y.		593.20		652.50	37.08	40.78
16 L.H., Daily Totals		$1025.20		$1315.70	$64.08	$82.23
Crew B-95C	Hr.	Daily	Hr.	Daily	Bare Costs	Incl. O&P
1 Equip. Oper. (crane)	$31.15	$249.20	$47.15	$377.20	$27.00	$41.45
1 Laborer	22.85	182.80	35.75	286.00		
1 Hyd. Excavator, 2.5 C.Y.		1557.00		1712.70	97.31	107.04
16 L.H., Daily Totals		$1989.00		$2375.90	$124.31	$148.49
Crew C-1	Hr.	Daily	Hr.	Daily	Bare Costs	Incl. O&P
3 Carpenters	$29.15	$699.60	$45.60	$1094.40	$27.57	$43.14
1 Laborer	22.85	182.80	35.75	286.00		
32 L.H., Daily Totals		$882.40		$1380.40	$27.57	$43.14

Crews

Crew No.		Bare Costs		Incl. Subs O & P		Cost Per Labor-Hour	
Crew C-2	Hr.	Daily	Hr.	Daily	Bare Costs	Incl. O&P	
1 Carpenter Foreman (out)	$31.15	$249.20	$48.75	$390.00	$28.43	$44.48	
4 Carpenters	29.15	932.80	45.60	1459.20			
1 Laborer	22.85	182.80	35.75	286.00			
48 L.H., Daily Totals		$1364.80		$2135.20	$28.43	$44.48	
Crew C-2A	Hr.	Daily	Hr.	Daily	Bare Costs	Incl. O&P	
1 Carpenter Foreman (out)	$31.15	$249.20	$48.75	$390.00	$28.23	$43.81	
3 Carpenters	29.15	699.60	45.60	1094.40			
1 Cement Finisher	27.95	223.60	41.55	332.40			
1 Laborer	22.85	182.80	35.75	286.00			
48 L.H., Daily Totals		$1355.20		$2102.80	$28.23	$43.81	
Crew C-3	Hr.	Daily	Hr.	Daily	Bare Costs	Incl. O&P	
1 Rodman Foreman	$34.75	$278.00	$58.95	$471.60	$30.02	$49.55	
4 Rodmen (reinf.)	32.75	1048.00	55.60	1779.20			
1 Equip. Oper. (light)	28.75	230.00	43.55	348.40			
2 Laborers	22.85	365.60	35.75	572.00			
3 Stressing Equipment		78.00		85.80			
.5 Grouting Equipment		116.60		128.25	3.04	3.34	
64 L.H., Daily Totals		$2116.20		$3385.25	$33.06	$52.89	
Crew C-4	Hr.	Daily	Hr.	Daily	Bare Costs	Incl. O&P	
1 Rodman Foreman	$34.75	$278.00	$58.95	$471.60	$33.25	$56.44	
3 Rodmen (reinf.)	32.75	786.00	55.60	1334.40			
3 Stressing Equipment		78.00		85.80	2.44	2.68	
32 L.H., Daily Totals		$1142.00		$1891.80	$35.69	$59.12	
Crew C-5	Hr.	Daily	Hr.	Daily	Bare Costs	Incl. O&P	
1 Rodman Foreman	$34.75	$278.00	$58.95	$471.60	$31.81	$52.50	
4 Rodmen (reinf.)	32.75	1048.00	55.60	1779.20			
1 Equip. Oper. (crane)	31.15	249.20	47.15	377.20			
1 Equip. Oper. Oiler	25.75	206.00	39.00	312.00			
1 Hyd. Crane, 25 Ton		629.60		692.55	11.24	12.37	
56 L.H., Daily Totals		$2410.80		$3632.55	$43.05	$64.87	
Crew C-6	Hr.	Daily	Hr.	Daily	Bare Costs	Incl. O&P	
1 Labor Foreman (outside)	$24.85	$198.80	$38.90	$311.20	$24.03	$37.24	
4 Laborers	22.85	731.20	35.75	1144.00			
1 Cement Finisher	27.95	223.60	41.55	332.40			
2 Gas Engine Vibrators		76.00		83.60	1.58	1.74	
48 L.H., Daily Totals		$1229.60		$1871.20	$25.61	$38.98	
Crew C-7	Hr.	Daily	Hr.	Daily	Bare Costs	Incl. O&P	
1 Labor Foreman (outside)	$24.85	$198.80	$38.90	$311.20	$24.79	$38.24	
5 Laborers	22.85	914.00	35.75	1430.00			
1 Cement Finisher	27.95	223.60	41.55	332.40			
1 Equip. Oper. (med.)	30.35	242.80	45.95	367.60			
1 Equip. Oper. (oiler)	25.75	206.00	39.00	312.00			
2 Gas Engine Vibrators		76.00		83.60			
1 Concrete Bucket, 1 C.Y.		27.60		30.35			
1 Hyd. Crane, 55 Ton		901.60		991.75	13.96	15.36	
72 L.H., Daily Totals		$2790.40		$3858.90	$38.75	$53.60	
Crew C-7A	Hr.	Daily	Hr.	Daily	Bare Costs	Incl. O&P	
1 Labor Foreman (outside)	$24.85	$198.80	$38.90	$311.20	$23.38	$36.36	
5 Laborers	22.85	914.00	35.75	1430.00			
2 Truck Driver (Heavy)	23.95	383.20	36.60	585.60			
2 Conc. Transit Mixers		1684.00		1852.40	26.31	28.94	
64 L.H., Daily Totals		$3180.00		$4179.20	$49.69	$65.30	

Crew No.		Bare Costs		Incl. Subs O & P		Cost Per Labor-Hour	
Crew C-7B	Hr.	Daily	Hr.	Daily	Bare Costs	Incl. O&P	
1 Labor Foreman (outside)	$24.85	$198.80	$38.90	$311.20	$24.50	$37.98	
5 Laborers	22.85	914.00	35.75	1430.00			
1 Equipment Operator (heavy)	31.15	249.20	47.15	377.20			
1 Equipment Oiler	25.75	206.00	39.00	312.00			
1 Conc. Bucket, 2 C.Y.		37.00		40.70			
1 Truck Crane, 165 Ton		1714.00		1885.40	27.36	30.10	
64 L.H., Daily Totals		$3319.00		$4356.50	$51.86	$68.08	
Crew C-7C	Hr.	Daily	Hr.	Daily	Bare Costs	Incl. O&P	
1 Labor Foreman (outside)	$24.85	$198.80	$38.90	$311.20	$24.98	$38.69	
5 Laborers	22.85	914.00	35.75	1430.00			
2 Equipment Operator (medium)	30.35	485.60	45.95	735.20			
2 Wheel Loader, 4 C.Y.		1037.60		1141.35	16.21	17.83	
64 L.H., Daily Totals		$2636.00		$3617.75	$41.19	$56.52	
Crew C-7D	Hr.	Daily	Hr.	Daily	Bare Costs	Incl. O&P	
1 Labor Foreman (outside)	$24.85	$198.80	$38.90	$311.20	$24.21	$37.66	
5 Laborers	22.85	914.00	35.75	1430.00			
1 Equipment Operator (med.)	30.35	242.80	45.95	367.60			
1 Concrete Conveyer		190.00		209.00	3.39	3.73	
56 L.H., Daily Totals		$1545.60		$2317.80	$27.60	$41.39	
Crew C-8	Hr.	Daily	Hr.	Daily	Bare Costs	Incl. O&P	
1 Labor Foreman (outside)	$24.85	$198.80	$38.90	$311.20	$25.66	$39.31	
3 Laborers	22.85	548.40	35.75	858.00			
2 Cement Finishers	27.95	447.20	41.55	664.80			
1 Equip. Oper. (med.)	30.35	242.80	45.95	367.60			
1 Concrete Pump (small)		580.00		638.00	10.36	11.39	
56 L.H., Daily Totals		$2017.20		$2839.60	$36.02	$50.70	
Crew C-8A	Hr.	Daily	Hr.	Daily	Bare Costs	Incl. O&P	
1 Labor Foreman (outside)	$24.85	$198.80	$38.90	$311.20	$24.88	$38.21	
3 Laborers	22.85	548.40	35.75	858.00			
2 Cement Finishers	27.95	447.20	41.55	664.80			
48 L.H., Daily Totals		$1194.40		$1834.00	$24.88	$38.21	
Crew C-8B	Hr.	Daily	Hr.	Daily	Bare Costs	Incl. O&P	
1 Labor Foreman (outside)	$24.85	$198.80	$38.90	$311.20	$24.75	$38.42	
3 Laborers	22.85	548.40	35.75	858.00			
1 Equipment Operator	30.35	242.80	45.95	367.60			
1 Vibrating Screed		53.90		59.30			
1 Vibratory Roller		451.20		496.30			
1 Dozer, 200 H.P.		735.00		808.50	31.00	34.10	
40 L.H., Daily Totals		$2230.10		$2900.90	$55.75	$72.52	
Crew C-8C	Hr.	Daily	Hr.	Daily	Bare Costs	Incl. O&P	
1 Labor Foreman (outside)	$24.85	$198.80	$38.90	$311.20	$25.28	$38.94	
3 Laborers	22.85	548.40	35.75	858.00			
1 Cement Finisher	27.95	223.60	41.55	332.40			
1 Equipment Operator (med.)	30.35	242.80	45.95	367.60			
1 Shotcrete Rig, 12 CY/hr		341.20		375.30	7.11	7.82	
48 L.H., Daily Totals		$1554.80		$2244.50	$32.39	$46.76	
Crew C-8D	Hr.	Daily	Hr.	Daily	Bare Costs	Incl. O&P	
1 Labor Foreman (outside)	$24.85	$198.80	$38.90	$311.20	$26.10	$39.94	
1 Laborers	22.85	182.80	35.75	286.00			
1 Cement Finisher	27.95	223.60	41.55	332.40			
1 Equipment Operator (light)	28.75	230.00	43.55	348.40			
1 Compressor, 250 CFM		125.20		137.70			
2 Hoses, 1", 50'		7.30		8.05	4.14	4.55	
32 L.H., Daily Totals		$967.70		$1423.75	$30.24	$44.49	

Crews

Crew C-8E

Crew C-8E	Hr.	Daily	Hr.	Daily	Bare Costs	Incl. O&P
1 Labor Foreman (outside)	$24.85	$198.80	$38.90	$311.20	$26.10	$39.94
1 Laborers	22.85	182.80	35.75	286.00		
1 Cement Finisher	27.95	223.60	41.55	332.40		
1 Equipment Operator (light)	28.75	230.00	43.55	348.40		
1 Compressor, 250 CFM		125.20		137.70		
2 Hoses, 1", 50'		7.30		8.05		
1 Concrete Pump (small)		580.00		638.00	22.27	24.49
32 L.H., Daily Totals		$1547.70		$2061.75	$48.37	$64.43

Crew C-10

Crew C-10	Hr.	Daily	Hr.	Daily	Bare Costs	Incl. O&P
1 Laborer	$22.85	$182.80	$35.75	$286.00	$26.25	$39.62
2 Cement Finishers	27.95	447.20	41.55	664.80		
24 L.H., Daily Totals		$630.00		$950.80	$26.25	$39.62

Crew C-11

Crew C-11	Hr.	Daily	Hr.	Daily	Bare Costs	Incl. O&P
1 Struc. Steel Foreman	$34.85	$278.80	$63.20	$505.60	$32.09	$56.33
6 Struc. Steel Workers	32.85	1576.80	59.60	2860.80		
1 Equip. Oper. (crane)	31.15	249.20	47.15	377.20		
1 Equip. Oper. Oiler	25.75	206.00	39.00	312.00		
1 Truck Crane, 150 Ton		1478.00		1625.80	20.53	22.58
72 L.H., Daily Totals		$3788.80		$5681.40	$52.62	$78.91

Crew C-12

Crew C-12	Hr.	Daily	Hr.	Daily	Bare Costs	Incl. O&P
1 Carpenter Foreman (out)	$31.15	$249.20	$48.75	$390.00	$28.77	$44.74
3 Carpenters	29.15	699.60	45.60	1094.40		
1 Laborer	22.85	182.80	35.75	286.00		
1 Equip. Oper. (crane)	31.15	249.20	47.15	377.20		
1 Hyd. Crane, 12 Ton		514.80		566.30	10.73	11.80
48 L.H., Daily Totals		$1895.60		$2713.90	$39.50	$56.54

Crew C-13

Crew C-13	Hr.	Daily	Hr.	Daily	Bare Costs	Incl. O&P
1 Struc. Steel Worker	$32.85	$262.80	$59.60	$476.80	$31.62	$54.93
1 Welder	32.85	262.80	59.60	476.80		
1 Carpenter	29.15	233.20	45.60	364.80		
1 Gas Welding Machine		84.00		92.40	3.50	3.85
24 L.H., Daily Totals		$842.80		$1410.80	$35.12	$58.78

Crew C-14

Crew C-14	Hr.	Daily	Hr.	Daily	Bare Costs	Incl. O&P
1 Carpenter Foreman (out)	$31.15	$249.20	$48.75	$390.00	$28.45	$45.08
5 Carpenters	29.15	1166.00	45.60	1824.00		
4 Laborers	22.85	731.20	35.75	1144.00		
4 Rodmen (reinf.)	32.75	1048.00	55.60	1779.20		
2 Cement Finishers	27.95	447.20	41.55	664.80		
1 Equip. Oper. (crane)	31.15	249.20	47.15	377.20		
1 Equip. Oper. Oiler	25.75	206.00	39.00	312.00		
1 Crane, 80 Ton, & Tools		1325.00		1457.50	9.20	10.12
144 L.H., Daily Totals		$5421.80		$7948.70	$37.65	$55.20

Crew C-14A

Crew C-14A	Hr.	Daily	Hr.	Daily	Bare Costs	Incl. O&P
1 Carpenter Foreman (out)	$31.15	$249.20	$48.75	$390.00	$29.30	$46.39
16 Carpenters	29.15	3731.20	45.60	5836.80		
4 Rodmen (reinf.)	32.75	1048.00	55.60	1779.20		
2 Laborers	22.85	365.60	35.75	572.00		
1 Cement Finisher	27.95	223.60	41.55	332.40		
1 Equip. Oper. (med.)	30.35	242.80	45.95	367.60		
1 Gas Engine Vibrator		38.00		41.80		
1 Concrete Pump (small)		580.00		638.00	3.09	3.40
200 L.H., Daily Totals		$6478.40		$9957.80	$32.39	$49.79

Crew C-14B

Crew C-14B	Hr.	Daily	Hr.	Daily	Bare Costs	Incl. O&P
1 Carpenter Foreman (out)	$31.15	$249.20	$48.75	$390.00	$29.25	$46.20
16 Carpenters	29.15	3731.20	45.60	5836.80		
4 Rodmen (reinf.)	32.75	1048.00	55.60	1779.20		
2 Laborers	22.85	365.60	35.75	572.00		
2 Cement Finishers	27.95	447.20	41.55	664.80		
1 Equip. Oper. (med.)	30.35	242.80	45.95	367.60		
1 Gas Engine Vibrator		38.00		41.80		
1 Concrete Pump (small)		580.00		638.00	2.97	3.27
208 L.H., Daily Totals		$6702.00		$10290.20	$32.22	$49.47

Crew C-14C

Crew C-14C	Hr.	Daily	Hr.	Daily	Bare Costs	Incl. O&P
1 Carpenter Foreman (out)	$31.15	$249.20	$48.75	$390.00	$27.92	$44.15
6 Carpenters	29.15	1399.20	45.60	2188.80		
2 Rodmen (reinf.)	32.75	524.00	55.60	889.60		
4 Laborers	22.85	731.20	35.75	1144.00		
1 Cement Finisher	27.95	223.60	41.55	332.40		
1 Gas Engine Vibrator		38.00		41.80	.34	.37
112 L.H., Daily Totals		$3165.20		$4986.60	$28.26	$44.52

Crew C-14D

Crew C-14D	Hr.	Daily	Hr.	Daily	Bare Costs	Incl. O&P
1 Carpenter Foreman (out)	$31.15	$249.20	$48.75	$390.00	$29.01	$45.59
18 Carpenters	29.15	4197.60	45.60	6566.40		
2 Rodmen (reinf.)	32.75	524.00	55.60	889.60		
2 Laborers	22.85	365.60	35.75	572.00		
1 Cement Finisher	27.95	223.60	41.55	332.40		
1 Equip. Oper. (med.)	30.35	242.80	45.95	367.60		
1 Gas Engine Vibrator		38.00		41.80		
1 Concrete Pump (small)		580.00		638.00	3.09	3.40
200 L.H., Daily Totals		$6420.80		$9797.80	$32.10	$48.99

Crew C-14E

Crew C-14E	Hr.	Daily	Hr.	Daily	Bare Costs	Incl. O&P
1 Carpenter Foreman (out)	$31.15	$249.20	$48.75	$390.00	$28.81	$46.47
2 Carpenters	29.15	466.40	45.60	729.60		
4 Rodmen (reinf.)	32.75	1048.00	55.60	1779.20		
3 Laborers	22.85	548.40	35.75	858.00		
1 Cement Finisher	27.95	223.60	41.55	332.40		
1 Gas Engine Vibrator		38.00		41.80	.43	.48
88 L.H., Daily Totals		$2573.60		$4131.00	$29.24	$46.95

Crew C-14F

Crew C-14F	Hr.	Daily	Hr.	Daily	Bare Costs	Incl. O&P
1 Laborer Foreman (out)	$24.85	$198.80	$38.90	$311.20	$26.47	$39.97
2 Laborers	22.85	365.60	35.75	572.00		
6 Cement Finishers	27.95	1341.60	41.55	1994.40		
1 Gas Engine Vibrator		38.00		41.80	.53	.58
72 L.H., Daily Totals		$1944.00		$2919.40	$27.00	$40.55

Crew C-14G

Crew C-14G	Hr.	Daily	Hr.	Daily	Bare Costs	Incl. O&P
1 Laborer Foreman (out)	$24.85	$198.80	$38.90	$311.20	$26.05	$39.51
2 Laborers	22.85	365.60	35.75	572.00		
4 Cement Finishers	27.95	894.40	41.55	1329.60		
1 Gas Engine Vibrator		38.00		41.80	.68	.75
56 L.H., Daily Totals		$1496.80		$2254.60	$26.73	$40.26

Crew C-14H

Crew C-14H	Hr.	Daily	Hr.	Daily	Bare Costs	Incl. O&P
1 Carpenter Foreman (out)	$31.15	$249.20	$48.75	$390.00	$28.83	$45.48
2 Carpenters	29.15	466.40	45.60	729.60		
1 Rodman (reinf.)	32.75	262.00	55.60	444.80		
1 Laborer	22.85	182.80	35.75	286.00		
1 Cement Finisher	27.95	223.60	41.55	332.40		
1 Gas Engine Vibrator		38.00		41.80	.79	.87
48 L.H., Daily Totals		$1422.00		$2224.60	$29.62	$46.35

Crews

Crew No.		Bare Costs		Incl. Subs O & P		Cost Per Labor-Hour	
Crew C-15	Hr.	Daily	Hr.	Daily	Bare Costs	Incl. O&P
1 Carpenter Foreman (out)	$31.15	$249.20	$48.75	$390.00	$27.41	$42.88
2 Carpenters	29.15	466.40	45.60	729.60		
3 Laborers	22.85	548.40	35.75	858.00		
2 Cement Finishers	27.95	447.20	41.55	664.80		
1 Rodman (reinf.)	32.75	262.00	55.60	444.80		
72 L.H., Daily Totals		$1973.20		$3087.20	$27.41	$42.88
Crew C-16	Hr.	Daily	Hr.	Daily	Bare Costs	Incl. O&P
1 Labor Foreman (outside)	$24.85	$198.80	$38.90	$311.20	$27.24	$42.93
3 Laborers	22.85	548.40	35.75	858.00		
2 Cement Finishers	27.95	447.20	41.55	664.80		
1 Equip. Oper. (med.)	30.35	242.80	45.95	367.60		
2 Rodmen (reinf.)	32.75	524.00	55.60	889.60		
1 Concrete Pump (small)		580.00		638.00	8.06	8.86
72 L.H., Daily Totals		$2541.20		$3729.20	$35.30	$51.79
Crew C-17	Hr.	Daily	Hr.	Daily	Bare Costs	Incl. O&P
2 Skilled Worker Foremen	$31.85	$509.60	$50.00	$800.00	$30.25	$47.48
8 Skilled Workers	29.85	1910.40	46.85	2998.40		
80 L.H., Daily Totals		$2420.00		$3798.40	$30.25	$47.48
Crew C-17A	Hr.	Daily	Hr.	Daily	Bare Costs	Incl. O&P
2 Skilled Worker Foremen	$31.85	$509.60	$50.00	$800.00	$30.26	$47.48
8 Skilled Workers	29.85	1910.40	46.85	2998.40		
.125 Equip. Oper. (crane)	31.15	31.15	47.15	47.15		
.125 Crane, 80 Ton, & Tools		165.63		182.20	2.04	2.25
81 L.H., Daily Totals		$2616.78		$4027.75	$32.30	$49.73
Crew C-17B	Hr.	Daily	Hr.	Daily	Bare Costs	Incl. O&P
2 Skilled Worker Foremen	$31.85	$509.60	$50.00	$800.00	$30.27	$47.47
8 Skilled Workers	29.85	1910.40	46.85	2998.40		
.25 Equip. Oper. (crane)	31.15	62.30	47.15	94.30		
.25 Crane, 80 Ton, & Tools		331.25		364.40		
.25 Walk Behind Power Tools		13.25		14.60	4.20	4.62
82 L.H., Daily Totals		$2826.80		$4271.70	$34.47	$52.09
Crew C-17C	Hr.	Daily	Hr.	Daily	Bare Costs	Incl. O&P
2 Skilled Worker Foremen	$31.85	$509.60	$50.00	$800.00	$30.28	$47.47
8 Skilled Workers	29.85	1910.40	46.85	2998.40		
.375 Equip. Oper. (crane)	31.15	93.45	47.15	141.45		
.375 Crane, 80 Ton & Tools		496.88		546.55	5.99	6.59
83 L.H., Daily Totals		$3010.33		$4486.40	$36.27	$54.06
Crew C-17D	Hr.	Daily	Hr.	Daily	Bare Costs	Incl. O&P
2 Skilled Worker Foremen	$31.85	$509.60	$50.00	$800.00	$30.29	$47.46
8 Skilled Workers	29.85	1910.40	46.85	2998.40		
.5 Equip. Oper. (crane)	31.15	124.60	47.15	188.60		
.5 Crane, 80 Ton & Tools		662.50		728.75	7.89	8.68
84 L.H., Daily Totals		$3207.10		$4715.75	$38.18	$56.14
Crew C-17E	Hr.	Daily	Hr.	Daily	Bare Costs	Incl. O&P
2 Skilled Worker Foremen	$31.85	$509.60	$50.00	$800.00	$30.25	$47.48
8 Skilled Workers	29.85	1910.40	46.85	2998.40		
1 Hyd. Jack with Rods		73.25		80.60	.92	1.01
80 L.H., Daily Totals		$2493.25		$3879.00	$31.17	$48.49
Crew C-18	Hr.	Daily	Hr.	Daily	Bare Costs	Incl. O&P
.125 Labor Foreman (out)	$24.85	$24.85	$38.90	$38.90	$23.07	$36.10
1 Laborer	22.85	182.80	35.75	286.00		
1 Concrete Cart, 10 C.F.		50.60		55.65	5.62	6.18
9 L.H., Daily Totals		$258.25		$380.55	$28.69	$42.28
Crew C-19	Hr.	Daily	Hr.	Daily	Bare Costs	Incl. O&P
.125 Labor Foreman (out)	$24.85	$24.85	$38.90	$38.90	$23.07	$36.10
1 Laborer	22.85	182.80	35.75	286.00		
1 Concrete Cart, 18 C.F.		76.00		83.60	8.44	9.29
9 L.H., Daily Totals		$283.65		$408.50	$31.51	$45.39
Crew C-20	Hr.	Daily	Hr.	Daily	Bare Costs	Incl. O&P
1 Labor Foreman (outside)	$24.85	$198.80	$38.90	$311.20	$24.68	$38.14
5 Laborers	22.85	914.00	35.75	1430.00		
1 Cement Finisher	27.95	223.60	41.55	332.40		
1 Equip. Oper. (med.)	30.35	242.80	45.95	367.60		
2 Gas Engine Vibrators		76.00		83.60		
1 Concrete Pump (small)		580.00		638.00	10.25	11.28
64 L.H., Daily Totals		$2235.20		$3162.80	$34.93	$49.42
Crew C-21	Hr.	Daily	Hr.	Daily	Bare Costs	Incl. O&P
1 Labor Foreman (outside)	$24.85	$198.80	$38.90	$311.20	$24.68	$38.14
5 Laborers	22.85	914.00	35.75	1430.00		
1 Cement Finisher	27.95	223.60	41.55	332.40		
1 Equip. Oper. (med.)	30.35	242.80	45.95	367.60		
2 Gas Engine Vibrators		76.00		83.60		
1 Concrete Conveyer		190.00		209.00	4.16	4.57
64 L.H., Daily Totals		$1845.20		$2733.80	$28.84	$42.71
Crew C-22	Hr.	Daily	Hr.	Daily	Bare Costs	Incl. O&P
1 Rodman Foreman	$34.75	$278.00	$58.95	$471.60	$32.93	$55.64
4 Rodmen (reinf.)	32.75	1048.00	55.60	1779.20		
.125 Equip. Oper. (crane)	31.15	31.15	47.15	47.15		
.125 Equip. Oper. Oiler	25.75	25.75	39.00	39.00		
.125 Hyd. Crane, 25 Ton		78.70		86.55	1.87	2.06
42 L.H., Daily Totals		$1461.60		$2423.50	$34.80	$57.70
Crew C-23	Hr.	Daily	Hr.	Daily	Bare Costs	Incl. O&P
2 Skilled Worker Foremen	$31.85	$509.60	$50.00	$800.00	$29.97	$46.72
6 Skilled Workers	29.85	1432.80	46.85	2248.80		
1 Equip. Oper. (crane)	31.15	249.20	47.15	377.20		
1 Equip. Oper. Oiler	25.75	206.00	39.00	312.00		
1 Crane, 90 Ton		1290.00		1419.00	16.13	17.74
80 L.H., Daily Totals		$3687.60		$5157.00	$46.10	$64.46
Crew C-24	Hr.	Daily	Hr.	Daily	Bare Costs	Incl. O&P
2 Skilled Worker Foremen	$31.85	$509.60	$50.00	$800.00	$29.97	$46.72
6 Skilled Workers	29.85	1432.80	46.85	2248.80		
1 Equip. Oper. (crane)	31.15	249.20	47.15	377.20		
1 Equip. Oper. Oiler	25.75	206.00	39.00	312.00		
1 Truck Crane, 150 Ton		1478.00		1625.80	18.48	20.32
80 L.H., Daily Totals		$3875.60		$5363.80	$48.45	$67.04
Crew C-25	Hr.	Daily	Hr.	Daily	Bare Costs	Incl. O&P
2 Rodmen (reinf.)	$32.75	$524.00	$55.60	$889.60	$25.93	$44.20
2 Rodmen Helpers	19.10	305.60	32.80	524.80		
32 L.H., Daily Totals		$829.60		$1414.40	$25.93	$44.20

Crews

Crew No.	Bare Costs		Incl. Subs O & P		Cost Per Labor-Hour	
Crew D-1	Hr.	Daily	Hr.	Daily	Bare Costs	Incl. O&P
1 Bricklayer	$29.60	$236.80	$45.75	$366.00	$26.28	$40.60
1 Bricklayer Helper	22.95	183.60	35.45	283.60		
16 L.H., Daily Totals		$420.40		$649.60	$26.28	$40.60
Crew D-2	Hr.	Daily	Hr.	Daily	Bare Costs	Incl. O&P
3 Bricklayers	$29.60	$710.40	$45.75	$1098.00	$27.14	$41.99
2 Bricklayer Helpers	22.95	367.20	35.45	567.20		
.5 Carpenter	29.15	116.60	45.60	182.40		
44 L.H., Daily Totals		$1194.20		$1847.60	$27.14	$41.99
Crew D-3	Hr.	Daily	Hr.	Daily	Bare Costs	Incl. O&P
3 Bricklayers	$29.60	$710.40	$45.75	$1098.00	$27.05	$41.82
2 Bricklayer Helpers	22.95	367.20	35.45	567.20		
.25 Carpenter	29.15	58.30	45.60	91.20		
42 L.H., Daily Totals		$1135.90		$1756.40	$27.05	$41.82
Crew D-4	Hr.	Daily	Hr.	Daily	Bare Costs	Incl. O&P
1 Bricklayer	$29.60	$236.80	$45.75	$366.00	$26.06	$40.05
2 Bricklayer Helpers	22.95	367.20	35.45	567.20		
1 Equip. Oper. (light)	28.75	230.00	43.55	348.40		
1 Grout Pump, 50 C.F./hr		83.70		92.05		
1 Hoses & Hopper		33.20		36.50		
1 Accessories		11.00		12.10	4.00	4.40
32 L.H., Daily Totals		$961.90		$1422.25	$30.06	$44.45
Crew D-5	Hr.	Daily	Hr.	Daily	Bare Costs	Incl. O&P
1 Bricklayer	$29.60	$236.80	$45.75	$366.00	$29.60	$45.75
8 L.H., Daily Totals		$236.80		$366.00	$29.60	$45.75
Crew D-6	Hr.	Daily	Hr.	Daily	Bare Costs	Incl. O&P
3 Bricklayers	$29.60	$710.40	$45.75	$1098.00	$26.39	$40.80
3 Bricklayer Helpers	22.95	550.80	35.45	850.80		
.25 Carpenter	29.15	58.30	45.60	91.20		
50 L.H., Daily Totals		$1319.50		$2040.00	$26.39	$40.80
Crew D-7	Hr.	Daily	Hr.	Daily	Bare Costs	Incl. O&P
1 Tile Layer	$28.15	$225.20	$41.60	$332.80	$25.43	$37.58
1 Tile Layer Helper	22.70	181.60	33.55	268.40		
16 L.H., Daily Totals		$406.80		$601.20	$25.43	$37.58
Crew D-8	Hr.	Daily	Hr.	Daily	Bare Costs	Incl. O&P
3 Bricklayers	$29.60	$710.40	$45.75	$1098.00	$26.94	$41.63
2 Bricklayer Helpers	22.95	367.20	35.45	567.20		
40 L.H., Daily Totals		$1077.60		$1665.20	$26.94	$41.63
Crew D-9	Hr.	Daily	Hr.	Daily	Bare Costs	Incl. O&P
3 Bricklayers	$29.60	$710.40	$45.75	$1098.00	$26.28	$40.60
3 Bricklayer Helpers	22.95	550.80	35.45	850.80		
48 L.H., Daily Totals		$1261.20		$1948.80	$26.28	$40.60
Crew D-10	Hr.	Daily	Hr.	Daily	Bare Costs	Incl. O&P
1 Bricklayer Foreman	$31.60	$252.80	$48.80	$390.40	$27.65	$42.52
1 Bricklayer	29.60	236.80	45.75	366.00		
2 Bricklayer Helpers	22.95	367.20	35.45	567.20		
1 Equip. Oper. (crane)	31.15	249.20	47.15	377.20		
1 Truck Crane, 12.5 Ton		425.20		467.70	10.63	11.69
40 L.H., Daily Totals		$1531.20		$2168.50	$38.28	$54.21

Crew No.	Bare Costs		Incl. Subs O & P		Cost Per Labor-Hour	
Crew D-11	Hr.	Daily	Hr.	Daily	Bare Costs	Incl. O&P
1 Bricklayer Foreman	$31.60	$252.80	$48.80	$390.40	$28.05	$43.33
1 Bricklayer	29.60	236.80	45.75	366.00		
1 Bricklayer Helper	22.95	183.60	35.45	283.60		
24 L.H., Daily Totals		$673.20		$1040.00	$28.05	$43.33
Crew D-12	Hr.	Daily	Hr.	Daily	Bare Costs	Incl. O&P
1 Bricklayer Foreman	$31.60	$252.80	$48.80	$390.40	$26.78	$41.36
1 Bricklayer	29.60	236.80	45.75	366.00		
2 Bricklayer Helpers	22.95	367.20	35.45	567.20		
32 L.H., Daily Totals		$856.80		$1323.60	$26.78	$41.36
Crew D-13	Hr.	Daily	Hr.	Daily	Bare Costs	Incl. O&P
1 Bricklayer Foreman	$31.60	$252.80	$48.80	$390.40	$27.90	$43.03
1 Bricklayer	29.60	236.80	45.75	366.00		
2 Bricklayer Helpers	22.95	367.20	35.45	567.20		
1 Carpenter	29.15	233.20	45.60	364.80		
1 Equip. Oper. (crane)	31.15	249.20	47.15	377.20		
1 Truck Crane, 12.5 Ton		425.20		467.70	8.86	9.74
48 L.H., Daily Totals		$1764.40		$2533.30	$36.76	$52.77
Crew E-1	Hr.	Daily	Hr.	Daily	Bare Costs	Incl. O&P
1 Welder Foreman	$34.85	$278.80	$63.20	$505.60	$32.15	$55.45
1 Welder	32.85	262.80	59.60	476.80		
1 Equip. Oper. (light)	28.75	230.00	43.55	348.40		
1 Gas Welding Machine		84.00		92.40	3.50	3.85
24 L.H., Daily Totals		$855.60		$1423.20	$35.65	$59.30
Crew E-2	Hr.	Daily	Hr.	Daily	Bare Costs	Incl. O&P
1 Struc. Steel Foreman	$34.85	$278.80	$63.20	$505.60	$31.88	$55.39
4 Struc. Steel Workers	32.85	1051.20	59.60	1907.20		
1 Equip. Oper. (crane)	31.15	249.20	47.15	377.20		
1 Equip. Oper. Oiler	25.75	206.00	39.00	312.00		
1 Crane, 90 Ton		1290.00		1419.00	23.04	25.34
56 L.H., Daily Totals		$3075.20		$4521.00	$54.92	$80.73
Crew E-3	Hr.	Daily	Hr.	Daily	Bare Costs	Incl. O&P
1 Struc. Steel Foreman	$34.85	$278.80	$63.20	$505.60	$33.52	$60.80
1 Struc. Steel Worker	32.85	262.80	59.60	476.80		
1 Welder	32.85	262.80	59.60	476.80		
1 Gas Welding Machine		84.00		92.40	3.50	3.85
24 L.H., Daily Totals		$888.40		$1551.60	$37.02	$64.65
Crew E-4	Hr.	Daily	Hr.	Daily	Bare Costs	Incl. O&P
1 Struc. Steel Foreman	$34.85	$278.80	$63.20	$505.60	$33.35	$60.50
3 Struc. Steel Workers	32.85	788.40	59.60	1430.40		
1 Gas Welding Machine		84.00		92.40	2.63	2.89
32 L.H., Daily Totals		$1151.20		$2028.40	$35.98	$63.39
Crew E-5	Hr.	Daily	Hr.	Daily	Bare Costs	Incl. O&P
2 Struc. Steel Foremen	$34.85	$557.60	$63.20	$1011.20	$32.37	$57.02
5 Struc. Steel Workers	32.85	1314.00	59.60	2384.00		
1 Equip. Oper. (crane)	31.15	249.20	47.15	377.20		
1 Welder	32.85	262.80	59.60	476.80		
1 Equip. Oper. Oiler	25.75	206.00	39.00	312.00		
1 Crane, 90 Ton		1290.00		1419.00		
1 Gas Welding Machine		84.00		92.40	17.18	18.89
80 L.H., Daily Totals		$3963.60		$6072.60	$49.55	$75.91

Crews

Crew No.		Bare Costs		Incl. Subs O & P		Cost Per Labor-Hour	
Crew E-6	Hr.	Daily	Hr.	Daily	Bare Costs	Incl. O&P
3 Struc. Steel Foremen	$34.85	$836.40	$63.20	$1516.80	$32.42	$57.21
9 Struc. Steel Workers	32.85	2365.20	59.60	4291.20		
1 Equip. Oper. (crane)	31.15	249.20	47.15	377.20		
1 Welder	32.85	262.80	59.60	476.80		
1 Equip. Oper. Oiler	25.75	206.00	39.00	312.00		
1 Equip. Oper. (light)	28.75	230.00	43.55	348.40		
1 Crane, 90 Ton		1290.00		1419.00		
1 Gas Welding Machine		84.00		92.40		
1 Air Compr., 160 C.F.M.		98.40		108.25		
2 Impact Wrenches		50.40		55.45	11.90	13.09
128 L.H., Daily Totals		$5672.40		$8997.50	$44.32	$70.30
Crew E-7	Hr.	Daily	Hr.	Daily	Bare Costs	Incl. O&P
1 Struc. Steel Foreman	$34.85	$278.80	$63.20	$505.60	$32.37	$57.02
4 Struc. Steel Workers	32.85	1051.20	59.60	1907.20		
1 Equip. Oper. (crane)	31.15	249.20	47.15	377.20		
1 Equip. Oper. Oiler	25.75	206.00	39.00	312.00		
1 Welder Foreman	34.85	278.80	63.20	505.60		
2 Welders	32.85	525.60	59.60	953.60		
1 Crane, 90 Ton		1290.00		1419.00		
2 Gas Welding Machines		168.00		184.80	18.23	20.05
80 L.H., Daily Totals		$4047.60		$6165.00	$50.60	$77.07
Crew E-8	Hr.	Daily	Hr.	Daily	Bare Costs	Incl. O&P
1 Struc. Steel Foreman	$34.85	$278.80	$63.20	$505.60	$32.17	$56.38
4 Struc. Steel Workers	32.85	1051.20	59.60	1907.20		
1 Welder Foreman	34.85	278.80	63.20	505.60		
4 Welders	32.85	1051.20	59.60	1907.20		
1 Equip. Oper. (crane)	31.15	249.20	47.15	377.20		
1 Equip. Oper. Oiler	25.75	206.00	39.00	312.00		
1 Equip. Oper. (light)	28.75	230.00	43.55	348.40		
1 Crane, 90 Ton		1290.00		1419.00		
4 Gas Welding Machines		336.00		369.60	15.63	17.20
104 L.H., Daily Totals		$4971.20		$7651.80	$47.80	$73.58
Crew E-9	Hr.	Daily	Hr.	Daily	Bare Costs	Incl. O&P
2 Struc. Steel Foremen	$34.85	$557.60	$63.20	$1011.20	$32.42	$57.21
5 Struc. Steel Workers	32.85	1314.00	59.60	2384.00		
1 Welder Foreman	34.85	278.80	63.20	505.60		
5 Welders	32.85	1314.00	59.60	2384.00		
1 Equip. Oper. (crane)	31.15	249.20	47.15	377.20		
1 Equip. Oper. Oiler	25.75	206.00	39.00	312.00		
1 Equip. Oper. (light)	28.75	230.00	43.55	348.40		
1 Crane, 90 Ton		1290.00		1419.00		
5 Gas Welding Machines		420.00		462.00	13.36	14.70
128 L.H., Daily Totals		$5859.60		$9203.40	$45.78	$71.91
Crew E-10	Hr.	Daily	Hr.	Daily	Bare Costs	Incl. O&P
1 Welder Foreman	$34.85	$278.80	$63.20	$505.60	$33.85	$61.40
1 Welder	32.85	262.80	59.60	476.80		
4 Gas Welding Machines		336.00		369.60		
1 Truck, 3 Ton		184.20		202.60	32.51	35.76
16 L.H., Daily Totals		$1061.80		$1554.60	$66.36	$97.16
Crew E-11	Hr.	Daily	Hr.	Daily	Bare Costs	Incl. O&P
2 Painters, Struc. Steel	$27.15	$434.40	$50.90	$814.40	$26.48	$45.28
1 Building Laborer	22.85	182.80	35.75	286.00		
1 Equip. Oper. (light)	28.75	230.00	43.55	348.40		
1 Air Compressor 250 C.F.M.		125.20		137.70		
1 Sand Blaster		33.20		36.50		
1 Sand Blasting Accessories		11.00		12.10	5.29	5.82
32 L.H., Daily Totals		$1016.60		$1635.10	$31.77	$51.10

Crew No.		Bare Costs		Incl. Subs O & P		Cost Per Labor-Hour	
Crew E-12	Hr.	Daily	Hr.	Daily	Bare Costs	Incl. O&P
1 Welder Foreman	$34.85	$278.80	$63.20	$505.60	$31.80	$53.38
1 Equip. Oper. (light)	28.75	230.00	43.55	348.40		
1 Gas Welding Machine		84.00		92.40	5.25	5.78
16 L.H., Daily Totals		$592.80		$946.40	$37.05	$59.16
Crew E-13	Hr.	Daily	Hr.	Daily	Bare Costs	Incl. O&P
1 Welder Foreman	$34.85	$278.80	$63.20	$505.60	$32.82	$56.65
.5 Equip. Oper. (light)	28.75	115.00	43.55	174.20		
1 Gas Welding Machine		84.00		92.40	7.00	7.70
12 L.H., Daily Totals		$477.80		$772.20	$39.82	$64.35
Crew E-14	Hr.	Daily	Hr.	Daily	Bare Costs	Incl. O&P
1 Welder Foreman	$34.85	$278.80	$63.20	$505.60	$34.85	$63.20
1 Gas Welding Machine		84.00		92.40	10.50	11.55
8 L.H., Daily Totals		$362.80		$598.00	$45.35	$74.75
Crew E-16	Hr.	Daily	Hr.	Daily	Bare Costs	Incl. O&P
1 Welder Foreman	$34.85	$278.80	$63.20	$505.60	$33.85	$61.40
1 Welder	32.85	262.80	59.60	476.80		
1 Gas Welding Machine		84.00		92.40	5.25	5.78
16 L.H., Daily Totals		$625.60		$1074.80	$39.10	$67.18
Crew E-17	Hr.	Daily	Hr.	Daily	Bare Costs	Incl. O&P
1 Structural Steel Foreman	$34.85	$278.80	$63.20	$505.60	$33.85	$61.40
1 Structural Steel Worker	32.85	262.80	59.60	476.80		
1 Power Tool		2.90		3.20	.18	.20
16 L.H., Daily Totals		$544.50		$985.60	$34.03	$61.60
Crew E-18	Hr.	Daily	Hr.	Daily	Bare Costs	Incl. O&P
1 Structural Steel Foreman	$34.85	$278.80	$63.20	$505.60	$32.75	$57.59
3 Structural Steel Workers	32.85	788.40	59.60	1430.40		
1 Equipment Operator (med.)	30.35	242.80	45.95	367.60		
1 Crane, 20 Ton		554.80		610.30	13.87	15.26
40 L.H., Daily Totals		$1864.80		$2913.90	$46.62	$72.85
Crew E-19	Hr.	Daily	Hr.	Daily	Bare Costs	Incl. O&P
1 Structural Steel Worker	$32.85	$262.80	$59.60	$476.80	$32.15	$55.45
1 Structural Steel Foreman	34.85	278.80	63.20	505.60		
1 Equip. Oper. (light)	28.75	230.00	43.55	348.40		
1 Power Tool		2.90		3.20		
1 Crane, 20 Ton		554.80		610.30	23.24	25.56
24 L.H., Daily Totals		$1329.30		$1944.30	$55.39	$81.01
Crew E-20	Hr.	Daily	Hr.	Daily	Bare Costs	Incl. O&P
1 Structural Steel Foreman	$34.85	$278.80	$63.20	$505.60	$32.00	$55.92
5 Structural Steel Workers	32.85	1314.00	59.60	2384.00		
1 Equip. Oper. (crane)	31.15	249.20	47.15	377.20		
1 Oiler	25.75	206.00	39.00	312.00		
1 Power Tool		2.90		3.20		
1 Crane, 40 Ton		751.60		826.75	11.79	12.97
64 L.H., Daily Totals		$2802.50		$4408.75	$43.79	$68.89
Crew E-22	Hr.	Daily	Hr.	Daily	Bare Costs	Incl. O&P
1 Skilled Worker Foreman	$31.85	$254.80	$50.00	$400.00	$30.52	$47.90
2 Skilled Worker	29.85	477.60	46.85	749.60		
24 L.H., Daily Totals		$732.40		$1149.60	$30.52	$47.90

Crews

Crew No.	Bare Costs		Incl. Subs O & P		Cost Per Labor-Hour	
Crew E-24	Hr.	Daily	Hr.	Daily	Bare Costs	Incl. O&P
3 Structural Steel Worker	$32.85	$788.40	$59.60	$1430.40	$32.22	$56.19
1 Equipment Operator (medium)	30.35	242.80	45.95	367.60		
1-25 Ton Crane		629.60		692.55	19.68	21.64
32 L.H., Daily Totals		$1660.80		$2490.55	$51.90	$77.83
Crew F-3	Hr.	Daily	Hr.	Daily	Bare Costs	Incl. O&P
4 Carpenters	$29.15	$932.80	$45.60	$1459.20	$29.55	$45.91
1 Equip. Oper. (crane)	31.15	249.20	47.15	377.20		
1 Hyd. Crane, 12 Ton		514.80		566.30	12.87	14.16
40 L.H., Daily Totals		$1696.80		$2402.70	$42.42	$60.07
Crew F-4	Hr.	Daily	Hr.	Daily	Bare Costs	Incl. O&P
4 Carpenters	$29.15	$932.80	$45.60	$1459.20	$28.92	$44.76
1 Equip. Oper. (crane)	31.15	249.20	47.15	377.20		
1 Equip. Oper. Oiler	25.75	206.00	39.00	312.00		
1 Hyd. Crane, 55 Ton		901.60		991.75	18.78	20.66
48 L.H., Daily Totals		$2289.60		$3140.15	$47.70	$65.42
Crew F-5	Hr.	Daily	Hr.	Daily	Bare Costs	Incl. O&P
1 Carpenter Foreman	$31.15	$249.20	$48.75	$390.00	$29.65	$46.39
3 Carpenters	29.15	699.60	45.60	1094.40		
32 L.H., Daily Totals		$948.80		$1484.40	$29.65	$46.39
Crew F-6	Hr.	Daily	Hr.	Daily	Bare Costs	Incl. O&P
2 Carpenters	$29.15	$466.40	$45.60	$729.60	$27.03	$41.97
2 Building Laborers	22.85	365.60	35.75	572.00		
1 Equip. Oper. (crane)	31.15	249.20	47.15	377.20		
1 Hyd. Crane, 12 Ton		514.80		566.30	12.87	14.16
40 L.H., Daily Totals		$1596.00		$2245.10	$39.90	$56.13
Crew F-7	Hr.	Daily	Hr.	Daily	Bare Costs	Incl. O&P
2 Carpenters	$29.15	$466.40	$45.60	$729.60	$26.00	$40.67
2 Building Laborers	22.85	365.60	35.75	572.00		
32 L.H., Daily Totals		$832.00		$1301.60	$26.00	$40.67
Crew G-1	Hr.	Daily	Hr.	Daily	Bare Costs	Incl. O&P
1 Roofer Foreman	$27.70	$221.60	$47.60	$380.80	$24.10	$41.40
4 Roofers, Composition	25.70	822.40	44.15	1412.80		
2 Roofer Helpers	19.10	305.60	32.80	524.80		
1 Application Equipment		181.60		199.75		
1 Tar Kettle/Pot		60.95		67.05		
1 Crew Truck		90.40		99.45	5.95	6.54
56 L.H., Daily Totals		$1682.55		$2684.65	$30.05	$47.94
Crew G-2	Hr.	Daily	Hr.	Daily	Bare Costs	Incl. O&P
1 Plasterer	$27.30	$218.40	$41.80	$334.40	$24.47	$37.72
1 Plasterer Helper	23.25	186.00	35.60	284.80		
1 Building Laborer	22.85	182.80	35.75	286.00		
1 Grouting Equipment		233.20		256.50	9.72	10.69
24 L.H., Daily Totals		$820.40		$1161.70	$34.19	$48.41
Crew G-3	Hr.	Daily	Hr.	Daily	Bare Costs	Incl. O&P
2 Sheet Metal Workers	$33.55	$536.80	$51.80	$828.80	$28.20	$43.78
2 Building Laborers	22.85	365.60	35.75	572.00		
32 L.H., Daily Totals		$902.40		$1400.80	$28.20	$43.78

Crew No.	Bare Costs		Incl. Subs O & P		Cost Per Labor-Hour	
Crew G-4	Hr.	Daily	Hr.	Daily	Bare Costs	Incl. O&P
1 Labor Foreman (outside)	$24.85	$198.80	$38.90	$311.20	$23.52	$36.80
2 Building Laborers	22.85	365.60	35.75	572.00		
1 Light Truck, 1.5 Ton		158.00		173.80		
1 Air Compr., 160 C.F.M.		98.40		108.25	10.68	11.75
24 L.H., Daily Totals		$820.80		$1165.25	$34.20	$48.55
Crew G-5	Hr.	Daily	Hr.	Daily	Bare Costs	Incl. O&P
1 Roofer Foreman	$27.70	$221.60	$47.60	$380.80	$23.46	$40.30
2 Roofers, Composition	25.70	411.20	44.15	706.40		
2 Roofer Helpers	19.10	305.60	32.80	524.80		
1 Application Equipment		181.60		199.75	4.54	4.99
40 L.H., Daily Totals		$1120.00		$1811.75	$28.00	$45.29
Crew G-6A	Hr.	Daily	Hr.	Daily	Bare Costs	Incl. O&P
2 Roofers Composition	$25.70	$411.20	$44.15	$706.40	$25.70	$44.15
1 Small Compressor		25.60		28.15		
2 Pneumatic Nailers		40.30		44.35	4.12	4.53
16 L.H., Daily Totals		$477.10		$778.90	$29.82	$48.68
Crew G-7	Hr.	Daily	Hr.	Daily	Bare Costs	Incl. O&P
1 Carpenter	$29.15	$233.20	$45.60	$364.80	$29.15	$45.60
1 Small Compressor		25.60		28.15		
1 Pneumatic Nailer		20.15		22.15	5.72	6.29
8 L.H., Daily Totals		$278.95		$415.10	$34.87	$51.89
Crew H-1	Hr.	Daily	Hr.	Daily	Bare Costs	Incl. O&P
2 Glaziers	$28.55	$456.80	$43.30	$692.80	$30.70	$51.45
2 Struc. Steel Workers	32.85	525.60	59.60	953.60		
32 L.H., Daily Totals		$982.40		$1646.40	$30.70	$51.45
Crew H-2	Hr.	Daily	Hr.	Daily	Bare Costs	Incl. O&P
2 Glaziers	$28.55	$456.80	$43.30	$692.80	$26.65	$40.78
1 Building Laborer	22.85	182.80	35.75	286.00		
24 L.H., Daily Totals		$639.60		$978.80	$26.65	$40.78
Crew H-3	Hr.	Daily	Hr.	Daily	Bare Costs	Incl. O&P
1 Glazier	$28.55	$228.40	$43.30	$346.40	$25.35	$39.00
1 Helper	22.15	177.20	34.70	277.60		
16 L.H., Daily Totals		$405.60		$624.00	$25.35	$39.00
Crew J-1	Hr.	Daily	Hr.	Daily	Bare Costs	Incl. O&P
3 Plasterers	$27.30	$655.20	$41.80	$1003.20	$25.68	$39.32
2 Plasterer Helpers	23.25	372.00	35.60	569.60		
1 Mixing Machine, 6 C.F.		59.00		64.90	1.48	1.62
40 L.H., Daily Totals		$1086.20		$1637.70	$27.16	$40.94
Crew J-2	Hr.	Daily	Hr.	Daily	Bare Costs	Incl. O&P
3 Plasterers	$27.30	$655.20	$41.80	$1003.20	$26.10	$39.78
2 Plasterer Helpers	23.25	372.00	35.60	569.60		
1 Lather	28.20	225.60	42.10	336.80		
1 Mixing Machine, 6 C.F.		59.00		64.90	1.23	1.35
48 L.H., Daily Totals		$1311.80		$1974.50	$27.33	$41.13
Crew J-3	Hr.	Daily	Hr.	Daily	Bare Costs	Incl. O&P
1 Terrazzo Worker	$28.30	$226.40	$41.85	$334.80	$25.70	$38.00
1 Terrazzo Helper	23.10	184.80	34.15	273.20		
1 Terrazzo Grinder, Electric		60.50		66.55		
1 Terrazzo Mixer		80.20		88.20	8.79	9.67
16 L.H., Daily Totals		$551.90		$762.75	$34.49	$47.67

Crews

Crew No.	Bare Costs		Incl. Subs O & P		Cost Per Labor-Hour	
Crew J-4	Hr.	Daily	Hr.	Daily	Bare Costs	Incl. O&P
1 Tile Layer	$28.15	$225.20	$41.60	$332.80	$25.43	$37.58
1 Tile Layer Helper	22.70	181.60	33.55	268.40		
16 L.H., Daily Totals		$406.80		$601.20	$25.43	$37.58
Crew K-1	Hr.	Daily	Hr.	Daily	Bare Costs	Incl. O&P
1 Carpenter	$29.15	$233.20	$45.60	$364.80	$26.27	$40.67
1 Truck Driver (light)	23.40	187.20	35.75	286.00		
1 Truck w/Power Equip.		184.20		202.60	11.51	12.66
16 L.H., Daily Totals		$604.60		$853.40	$37.78	$53.33
Crew K-2	Hr.	Daily	Hr.	Daily	Bare Costs	Incl. O&P
1 Struc. Steel Foreman	$34.85	$278.80	$63.20	$505.60	$30.37	$52.85
1 Struc. Steel Worker	32.85	262.80	59.60	476.80		
1 Truck Driver (light)	23.40	187.20	35.75	286.00		
1 Truck w/Power Equip.		184.20		202.60	7.68	8.44
24 L.H., Daily Totals		$913.00		$1471.00	$38.05	$61.29
Crew L-1	Hr.	Daily	Hr.	Daily	Bare Costs	Incl. O&P
1 Electrician	$34.30	$274.40	$51.20	$409.60	$34.38	$51.60
1 Plumber	34.45	275.60	52.00	416.00		
16 L.H., Daily Totals		$550.00		$825.60	$34.38	$51.60
Crew L-2	Hr.	Daily	Hr.	Daily	Bare Costs	Incl. O&P
1 Carpenter	$29.15	$233.20	$45.60	$364.80	$25.65	$40.15
1 Carpenter Helper	22.15	177.20	34.70	277.60		
16 L.H., Daily Totals		$410.40		$642.40	$25.65	$40.15
Crew L-3	Hr.	Daily	Hr.	Daily	Bare Costs	Incl. O&P
1 Carpenter	$29.15	$233.20	$45.60	$364.80	$31.54	$48.55
.5 Electrician	34.30	137.20	51.20	204.80		
.5 Sheet Metal Worker	33.55	134.20	51.80	207.20		
16 L.H., Daily Totals		$504.60		$776.80	$31.54	$48.55
Crew L-3A	Hr.	Daily	Hr.	Daily	Bare Costs	Incl. O&P
1 Carpenter Foreman (outside)	$31.15	$249.20	$48.75	$390.00	$31.95	$49.77
.5 Sheet Metal Worker	33.55	134.20	51.80	207.20		
12 L.H., Daily Totals		$383.40		$597.20	$31.95	$49.77
Crew L-4	Hr.	Daily	Hr.	Daily	Bare Costs	Incl. O&P
2 Skilled Workers	$29.85	$477.60	$46.85	$749.60	$27.28	$42.80
1 Helper	22.15	177.20	34.70	277.60		
24 L.H., Daily Totals		$654.80		$1027.20	$27.28	$42.80
Crew L-5	Hr.	Daily	Hr.	Daily	Bare Costs	Incl. O&P
1 Struc. Steel Foreman	$34.85	$278.80	$63.20	$505.60	$32.89	$58.34
5 Struc. Steel Workers	32.85	1314.00	59.60	2384.00		
1 Equip. Oper. (crane)	31.15	249.20	47.15	377.20		
1 Hyd. Crane, 25 Ton		629.60		692.55	11.24	12.37
56 L.H., Daily Totals		$2471.60		$3959.35	$44.13	$70.71
Crew L-5A	Hr.	Daily	Hr.	Daily	Bare Costs	Incl. O&P
1 Structural Steel Foreman	$34.85	$278.80	$63.20	$505.60	$32.92	$57.39
2 Structural Steel Workers	32.85	525.60	59.60	953.60		
1 Equip. Oper. (crane)	31.15	249.20	47.15	377.20		
1 Crane, SP, 25 Ton		602.00		662.20	18.81	20.69
32 L.H., Daily Totals		$1655.60		$2498.60	$51.73	$78.08

Crew No.	Bare Costs		Incl. Subs O & P		Cost Per Labor-Hour	
Crew L-6	Hr.	Daily	Hr.	Daily	Bare Costs	Incl. O&P
1 Plumber	$34.45	$275.60	$52.00	$416.00	$34.40	$51.73
.5 Electrician	34.30	137.20	51.20	204.80		
12 L.H., Daily Totals		$412.80		$620.80	$34.40	$51.73
Crew L-7	Hr.	Daily	Hr.	Daily	Bare Costs	Incl. O&P
2 Carpenters	$29.15	$466.40	$45.60	$729.60	$28.09	$43.59
1 Building Laborer	22.85	182.80	35.75	286.00		
.5 Electrician	34.30	137.20	51.20	204.80		
28 L.H., Daily Totals		$786.40		$1220.40	$28.09	$43.59
Crew L-8	Hr.	Daily	Hr.	Daily	Bare Costs	Incl. O&P
2 Carpenters	$29.15	$466.40	$45.60	$729.60	$30.21	$46.88
.5 Plumber	34.45	137.80	52.00	208.00		
20 L.H., Daily Totals		$604.20		$937.60	$30.21	$46.88
Crew L-9	Hr.	Daily	Hr.	Daily	Bare Costs	Incl. O&P
1 Labor Foreman (inside)	$23.35	$186.80	$36.55	$292.40	$26.46	$42.94
2 Building Laborers	22.85	365.60	35.75	572.00		
1 Struc. Steel Worker	32.85	262.80	59.60	476.80		
.5 Electrician	34.30	137.20	51.20	204.80		
36 L.H., Daily Totals		$952.40		$1546.00	$26.46	$42.94
Crew L-10	Hr.	Daily	Hr.	Daily	Bare Costs	Incl. O&P
1 Structural Steel Foreman	$34.85	$278.80	$63.20	$505.60	$32.95	$56.65
1 Structural Steel Worker	32.85	262.80	59.60	476.80		
1 Equip. Oper. (crane)	31.15	249.20	47.15	377.20		
1 Hyd. Crane, 12 Ton		514.80		566.30	21.45	23.60
24 L.H., Daily Totals		$1305.60		$1925.90	$54.40	$80.25
Crew M-1	Hr.	Daily	Hr.	Daily	Bare Costs	Incl. O&P
3 Elevator Constructors	$35.65	$855.60	$53.80	$1291.20	$33.86	$51.10
1 Elevator Apprentice	28.50	228.00	43.00	344.00		
5 Hand Tools		95.00		104.50	2.97	3.27
32 L.H., Daily Totals		$1178.60		$1739.70	$36.83	$54.37
Crew M-3	Hr.	Daily	Hr.	Daily	Bare Costs	Incl. O&P
1 Electrician Foreman (out)	$36.30	$290.40	$54.20	$433.60	$30.80	$46.64
1 Common Laborer	22.85	182.80	35.75	286.00		
.25 Equipment Operator, Medium	30.35	60.70	45.95	91.90		
1 Elevator Constructor	35.65	285.20	53.80	430.40		
1 Elevator Apprentice	28.50	228.00	43.00	344.00		
.25 Crane, SP, 4 x 4, 20 ton		134.35		147.80	3.95	4.35
34 L.H., Daily Totals		$1181.45		$1733.70	$34.75	$50.99
Crew M-4	Hr.	Daily	Hr.	Daily	Bare Costs	Incl. O&P
1 Electrician Foreman (out)	$36.30	$290.40	$54.20	$433.60	$30.56	$46.29
1 Common Laborer	22.85	182.80	35.75	286.00		
.25 Equipment Operator, Crane	31.15	62.30	47.15	94.30		
.25 Equipment Operator, Oiler	25.75	51.50	39.00	78.00		
1 Elevator Constructor	35.65	285.20	53.80	430.40		
1 Elevator Apprentice	28.50	228.00	43.00	344.00		
.25 Crane, Hyd, SP, 4WD, 40 Ton		218.00		239.80	6.06	6.66
36 L.H., Daily Totals		$1318.20		$1906.10	$36.62	$52.95
Crew Q-1	Hr.	Daily	Hr.	Daily	Bare Costs	Incl. O&P
1 Plumber	$34.45	$275.60	$52.00	$416.00	$31.00	$46.80
1 Plumber Apprentice	27.55	220.40	41.60	332.80		
16 L.H., Daily Totals		$496.00		$748.80	$31.00	$46.80

Crews

Crew No.	Bare Costs		Incl. Subs O & P		Cost Per Labor-Hour	
	Hr.	Daily	Hr.	Daily	Bare Costs	Incl. O&P
Crew Q-1C						
1 Plumber	$34.45	$275.60	$52.00	$416.00	$30.78	$46.52
1 Plumber Apprentice	27.55	220.40	41.60	332.80		
1 Equip. Oper. (medium)	30.35	242.80	45.95	367.60		
1 Trencher, Chain		711.40		782.55	29.64	32.61
24 L.H., Daily Totals		$1450.20		$1898.95	$60.42	$79.13
Crew Q-2	Hr.	Daily	Hr.	Daily	Bare Costs	Incl. O&P
2 Plumbers	$34.45	$551.20	$52.00	$832.00	$32.15	$48.53
1 Plumber Apprentice	27.55	220.40	41.60	332.80		
24 L.H., Daily Totals		$771.60		$1164.80	$32.15	$48.53
Crew Q-3	Hr.	Daily	Hr.	Daily	Bare Costs	Incl. O&P
1 Plumber Foreman (inside)	$34.95	$279.60	$52.75	$422.00	$32.85	$49.59
2 Plumbers	34.45	551.20	52.00	832.00		
1 Plumber Apprentice	27.55	220.40	41.60	332.80		
32 L.H., Daily Totals		$1051.20		$1586.80	$32.85	$49.59
Crew Q-4	Hr.	Daily	Hr.	Daily	Bare Costs	Incl. O&P
1 Plumber Foreman (inside)	$34.95	$279.60	$52.75	$422.00	$32.85	$49.59
1 Plumber	34.45	275.60	52.00	416.00		
1 Welder (plumber)	34.45	275.60	52.00	416.00		
1 Plumber Apprentice	27.55	220.40	41.60	332.80		
1 Electric Welding Mach.		100.65		110.70	3.15	3.46
32 L.H., Daily Totals		$1151.85		$1697.50	$36.00	$53.05
Crew Q-5	Hr.	Daily	Hr.	Daily	Bare Costs	Incl. O&P
1 Steamfitter	$34.85	$278.80	$52.60	$420.80	$31.38	$47.38
1 Steamfitter Apprentice	27.90	223.20	42.15	337.20		
16 L.H., Daily Totals		$502.00		$758.00	$31.38	$47.38
Crew Q-6	Hr.	Daily	Hr.	Daily	Bare Costs	Incl. O&P
2 Steamfitters	$34.85	$557.60	$52.60	$841.60	$32.53	$49.12
1 Steamfitter Apprentice	27.90	223.20	42.15	337.20		
24 L.H., Daily Totals		$780.80		$1178.80	$32.53	$49.12
Crew Q-7	Hr.	Daily	Hr.	Daily	Bare Costs	Incl. O&P
1 Steamfitter Foreman (inside)	$35.35	$282.80	$53.40	$427.20	$33.24	$50.19
2 Steamfitters	34.85	557.60	52.60	841.60		
1 Steamfitter Apprentice	27.90	223.20	42.15	337.20		
32 L.H., Daily Totals		$1063.60		$1606.00	$33.24	$50.19
Crew Q-8	Hr.	Daily	Hr.	Daily	Bare Costs	Incl. O&P
1 Steamfitter Foreman (inside)	$35.35	$282.80	$53.40	$427.20	$33.24	$50.19
1 Steamfitter	34.85	278.80	52.60	420.80		
1 Welder (steamfitter)	34.85	278.80	52.60	420.80		
1 Steamfitter Apprentice	27.90	223.20	42.15	337.20		
1 Electric Welding Mach.		100.65		110.70	3.15	3.46
32 L.H., Daily Totals		$1164.25		$1716.70	$36.39	$53.65
Crew Q-9	Hr.	Daily	Hr.	Daily	Bare Costs	Incl. O&P
1 Sheet Metal Worker	$33.55	$268.40	$51.80	$414.40	$30.20	$46.63
1 Sheet Metal Apprentice	26.85	214.80	41.45	331.60		
16 L.H., Daily Totals		$483.20		$746.00	$30.20	$46.63
Crew Q-10	Hr.	Daily	Hr.	Daily	Bare Costs	Incl. O&P
2 Sheet Metal Workers	$33.55	$536.80	$51.80	$828.80	$31.32	$48.35
1 Sheet Metal Apprentice	26.85	214.80	41.45	331.60		
24 L.H., Daily Totals		$751.60		$1160.40	$31.32	$48.35
Crew Q-11	Hr.	Daily	Hr.	Daily	Bare Costs	Incl. O&P
1 Sheet Metal Foreman (inside)	$34.05	$272.40	$52.55	$420.40	$32.00	$49.40
2 Sheet Metal Workers	33.55	536.80	51.80	828.80		
1 Sheet Metal Apprentice	26.85	214.80	41.45	331.60		
32 L.H., Daily Totals		$1024.00		$1580.80	$32.00	$49.40
Crew Q-12	Hr.	Daily	Hr.	Daily	Bare Costs	Incl. O&P
1 Sprinkler Installer	$34.45	$275.60	$52.10	$416.80	$31.00	$46.90
1 Sprinkler Apprentice	27.55	220.40	41.70	333.60		
16 L.H., Daily Totals		$496.00		$750.40	$31.00	$46.90
Crew Q-13	Hr.	Daily	Hr.	Daily	Bare Costs	Incl. O&P
1 Sprinkler Foreman (inside)	$34.95	$279.60	$52.90	$423.20	$32.85	$49.70
2 Sprinkler Installers	34.45	551.20	52.10	833.60		
1 Sprinkler Apprentice	27.55	220.40	41.70	333.60		
32 L.H., Daily Totals		$1051.20		$1590.40	$32.85	$49.70
Crew Q-14	Hr.	Daily	Hr.	Daily	Bare Costs	Incl. O&P
1 Asbestos Worker	$32.15	$257.20	$51.50	$412.00	$28.93	$46.33
1 Asbestos Apprentice	25.70	205.60	41.15	329.20		
16 L.H., Daily Totals		$462.80		$741.20	$28.93	$46.33
Crew Q-15	Hr.	Daily	Hr.	Daily	Bare Costs	Incl. O&P
1 Plumber	$34.45	$275.60	$52.00	$416.00	$31.00	$46.80
1 Plumber Apprentice	27.55	220.40	41.60	332.80		
1 Electric Welding Mach.		100.65		110.70	6.29	6.92
16 L.H., Daily Totals		$596.65		$859.50	$37.29	$53.72
Crew Q-16	Hr.	Daily	Hr.	Daily	Bare Costs	Incl. O&P
2 Plumbers	$34.45	$551.20	$52.00	$832.00	$32.15	$48.53
1 Plumber Apprentice	27.55	220.40	41.60	332.80		
1 Electric Welding Mach.		100.65		110.70	4.19	4.61
24 L.H., Daily Totals		$872.25		$1275.50	$36.34	$53.14
Crew Q-17	Hr.	Daily	Hr.	Daily	Bare Costs	Incl. O&P
1 Steamfitter	$34.85	$278.80	$52.60	$420.80	$31.38	$47.38
1 Steamfitter Apprentice	27.90	223.20	42.15	337.20		
1 Electric Welding Mach.		100.65		110.70	6.29	6.92
16 L.H., Daily Totals		$602.65		$868.70	$37.67	$54.30
Crew Q-17A	Hr.	Daily	Hr.	Daily	Bare Costs	Incl. O&P
1 Steamfitter	$34.85	$278.80	$52.60	$420.80	$31.30	$47.30
1 Steamfitter Apprentice	27.90	223.20	42.15	337.20		
1 Equip. Oper. (crane)	31.15	249.20	47.15	377.20		
1 Truck Crane, 12 Ton		514.80		566.30		
1 Electric Welding Mach.		100.65		110.70	25.64	28.21
24 L.H., Daily Totals		$1366.65		$1812.20	$56.94	$75.51
Crew Q-18	Hr.	Daily	Hr.	Daily	Bare Costs	Incl. O&P
2 Steamfitters	$34.85	$557.60	$52.60	$841.60	$32.53	$49.12
1 Steamfitter Apprentice	27.90	223.20	42.15	337.20		
1 Electric Welding Mach.		100.65		110.70	4.19	4.61
24 L.H., Daily Totals		$881.45		$1289.50	$36.72	$53.73
Crew Q-19	Hr.	Daily	Hr.	Daily	Bare Costs	Incl. O&P
1 Steamfitter	$34.85	$278.80	$52.60	$420.80	$32.35	$48.65
1 Steamfitter Apprentice	27.90	223.20	42.15	337.20		
1 Electrician	34.30	274.40	51.20	409.60		
24 L.H., Daily Totals		$776.40		$1167.60	$32.35	$48.65

Crews

Crew No.	Bare Costs		Incl. Subs O & P		Cost Per Labor-Hour	
Crew Q-20	Hr.	Daily	Hr.	Daily	Bare Costs	Incl. O&P
1 Sheet Metal Worker	$33.55	$268.40	$51.80	$414.40	$31.02	$47.54
1 Sheet Metal Apprentice	26.85	214.80	41.45	331.60		
.5 Electrician	34.30	137.20	51.20	204.80		
20 L.H., Daily Totals		$620.40		$950.80	$31.02	$47.54
Crew Q-21	Hr.	Daily	Hr.	Daily	Bare Costs	Incl. O&P
2 Steamfitters	$34.85	$557.60	$52.60	$841.60	$32.97	$49.64
1 Steamfitter Apprentice	27.90	223.20	42.15	337.20		
1 Electrician	34.30	274.40	51.20	409.60		
32 L.H., Daily Totals		$1055.20		$1588.40	$32.97	$49.64
Crew Q-22	Hr.	Daily	Hr.	Daily	Bare Costs	Incl. O&P
1 Plumber	$34.45	$275.60	$52.00	$416.00	$31.00	$46.80
1 Plumber Apprentice	27.55	220.40	41.60	332.80		
1 Truck Crane, 12 Ton		514.80		566.30	32.18	35.39
16 L.H., Daily Totals		$1010.80		$1315.10	$63.18	$82.19
Crew Q-22A	Hr.	Daily	Hr.	Daily	Bare Costs	Incl. O&P
1 Plumber	$34.45	$275.60	$52.00	$416.00	$29.00	$44.13
1 Plumber Apprentice	27.55	220.40	41.60	332.80		
1 Laborer	22.85	182.80	35.75	286.00		
1 Equip. Oper. (crane)	31.15	249.20	47.15	377.20		
1 Truck Crane, 12 Ton		514.80		566.30	16.09	17.70
32 L.H., Daily Totals		$1442.80		$1978.30	$45.09	$61.83
Crew Q-23	Hr.	Daily	Hr.	Daily	Bare Costs	Incl. O&P
1 Plumber Foreman	$36.45	$291.60	$55.05	$440.40	$33.75	$51.00
1 Plumber	34.45	275.60	52.00	416.00		
1 Equip. Oper. (medium)	30.35	242.80	45.95	367.60		
1 Power Tools		2.90		3.20		
1 Crane, 20 Ton		554.80		610.30	23.24	25.56
24 L.H., Daily Totals		$1367.70		$1837.50	$56.99	$76.56
Crew R-1	Hr.	Daily	Hr.	Daily	Bare Costs	Incl. O&P
1 Electrician Foreman	$34.80	$278.40	$51.95	$415.60	$30.33	$45.83
3 Electricians	34.30	823.20	51.20	1228.80		
2 Helpers	22.15	354.40	34.70	555.20		
48 L.H., Daily Totals		$1456.00		$2199.60	$30.33	$45.83
Crew R-1A	Hr.	Daily	Hr.	Daily	Bare Costs	Incl. O&P
1 Electrician	$34.30	$274.40	$51.20	$409.60	$28.23	$42.95
1 Helper	22.15	177.20	34.70	277.60		
16 L.H., Daily Totals		$451.60		$687.20	$28.23	$42.95
Crew R-2	Hr.	Daily	Hr.	Daily	Bare Costs	Incl. O&P
1 Electrician Foreman	$34.80	$278.40	$51.95	$415.60	$30.45	$46.01
3 Electricians	34.30	823.20	51.20	1228.80		
2 Helpers	22.15	354.40	34.70	555.20		
1 Equip. Oper. (crane)	31.15	249.20	47.15	377.20		
1 S.P. Crane, 5 Ton		284.00		312.40	5.07	5.58
56 L.H., Daily Totals		$1989.20		$2889.20	$35.52	$51.59
Crew R-3	Hr.	Daily	Hr.	Daily	Bare Costs	Incl. O&P
1 Electrician Foreman	$34.80	$278.40	$51.95	$415.60	$33.87	$50.69
1 Electrician	34.30	274.40	51.20	409.60		
.5 Equip. Oper. (crane)	31.15	124.60	47.15	188.60		
.5 S.P. Crane, 5 Ton		142.00		156.20	7.10	7.81
20 L.H., Daily Totals		$819.40		$1170.00	$40.97	$58.50

Crew No.	Bare Costs		Incl. Subs O & P		Cost Per Labor-Hour	
Crew R-4	Hr.	Daily	Hr.	Daily	Bare Costs	Incl. O&P
1 Struc. Steel Foreman	$34.85	$278.80	$63.20	$505.60	$33.54	$58.64
3 Struc. Steel Workers	32.85	788.40	59.60	1430.40		
1 Electrician	34.30	274.40	51.20	409.60		
1 Gas Welding Machine		84.00		92.40	2.10	2.31
40 L.H., Daily Totals		$1425.60		$2438.00	$35.64	$60.95
Crew R-5	Hr.	Daily	Hr.	Daily	Bare Costs	Incl. O&P
1 Electrician Foreman	$34.80	$278.40	$51.95	$415.60	$29.93	$45.27
4 Electrician Linemen	34.30	1097.60	51.20	1638.40		
2 Electrician Operators	34.30	548.80	51.20	819.20		
4 Electrician Groundmen	22.15	708.80	34.70	1110.40		
1 Crew Truck		90.40		99.45		
1 Tool Van		113.90		125.30		
1 Pickup Truck, 3/4 Ton		83.00		91.30		
.2 Crane, 55 Ton		180.32		198.35		
.2 Crane, 12 Ton		102.96		113.25		
.2 Auger, Truck Mtd.		374.00		411.40		
1 Tractor w/Winch		263.00		289.30	13.72	15.09
88 L.H., Daily Totals		$3841.18		$5311.95	$43.65	$60.36
Crew R-6	Hr.	Daily	Hr.	Daily	Bare Costs	Incl. O&P
1 Electrician Foreman	$34.80	$278.40	$51.95	$415.60	$29.93	$45.27
4 Electrician Linemen	34.30	1097.60	51.20	1638.40		
2 Electrician Operators	34.30	548.80	51.20	819.20		
4 Electrician Groundmen	22.15	708.80	34.70	1110.40		
1 Crew Truck		90.40		99.45		
1 Tool Van		113.90		125.30		
1 Pickup Truck, 3/4 Ton		83.00		91.30		
.2 Crane, 55 Ton		180.32		198.35		
.2 Crane, 12 Ton		102.96		113.25		
.2 Auger, Truck Mtd.		374.00		411.40		
1 Tractor w/Winch		263.00		289.30		
3 Cable Trailers		453.15		498.45		
.5 Tensioning Rig		149.75		164.75		
.5 Cable Pulling Rig		882.00		970.20	30.60	33.66
88 L.H., Daily Totals		$5326.08		$6945.35	$60.53	$78.93
Crew R-7	Hr.	Daily	Hr.	Daily	Bare Costs	Incl. O&P
1 Electrician Foreman	$34.80	$278.40	$51.95	$415.60	$24.26	$37.58
5 Electrician Groundmen	22.15	886.00	34.70	1388.00		
1 Crew Truck		90.40		99.45	1.88	2.07
48 L.H., Daily Totals		$1254.80		$1903.05	$26.14	$39.65
Crew R-8	Hr.	Daily	Hr.	Daily	Bare Costs	Incl. O&P
1 Electrician Foreman	$34.80	$278.40	$51.95	$415.60	$30.33	$45.83
3 Electrician Linemen	34.30	823.20	51.20	1228.80		
2 Electrician Groundmen	22.15	354.40	34.70	555.20		
1 Pickup Truck, 3/4 Ton		83.00		91.30		
1 Crew Truck		90.40		99.45	3.61	3.97
48 L.H., Daily Totals		$1629.40		$2390.35	$33.94	$49.80
Crew R-9	Hr.	Daily	Hr.	Daily	Bare Costs	Incl. O&P
1 Electrician Foreman	$34.80	$278.40	$51.95	$415.60	$28.29	$43.04
1 Electrician Lineman	34.30	274.40	51.20	409.60		
2 Electrician Operators	34.30	548.80	51.20	819.20		
4 Electrician Groundmen	22.15	708.80	34.70	1110.40		
1 Pickup Truck, 3/4 Ton		83.00		91.30		
1 Crew Truck		90.40		99.45	2.71	2.98
64 L.H., Daily Totals		$1983.80		$2945.55	$31.00	$46.02

Crews

Crew No.	Bare Costs		Incl. Subs O & P		Cost Per Labor-Hour	
Crew R-10	**Hr.**	**Daily**	**Hr.**	**Daily**	**Bare Costs**	**Incl. O&P**
1 Electrician Foreman	$34.80	$278.40	$51.95	$415.60	$32.36	$48.58
4 Electrician Linemen	34.30	1097.60	51.20	1638.40		
1 Electrician Groundman	22.15	177.20	34.70	277.60		
1 Crew Truck		90.40		99.45		
3 Tram Cars		485.70		534.25	12.00	13.20
48 L.H., Daily Totals		$2129.30		$2965.30	$44.36	$61.78
Crew R-11	**Hr.**	**Daily**	**Hr.**	**Daily**	**Bare Costs**	**Incl. O&P**
1 Electrician Foreman	$34.80	$278.40	$51.95	$415.60	$31.00	$46.74
4 Electricians	34.30	1097.60	51.20	1638.40		
1 Helper	22.15	177.20	34.70	277.60		
1 Common Laborer	22.85	182.80	35.75	286.00		
1 Crew Truck		90.40		99.45		
1 Crane, 12 Ton		514.80		566.30	10.81	11.89
56 L.H., Daily Totals		$2341.20		$3283.35	$41.81	$58.63
Crew R-12	**Hr.**	**Daily**	**Hr.**	**Daily**	**Bare Costs**	**Incl. O&P**
1 Carpenter Foreman	$29.65	$237.20	$46.40	$371.20	$27.35	$43.40
4 Carpenters	29.15	932.80	45.60	1459.20		
4 Common Laborers	22.85	731.20	35.75	1144.00		
1 Equip. Oper. (med.)	30.35	242.80	45.95	367.60		
1 Steel Worker	32.85	262.80	59.60	476.80		
1 Dozer, 200 H.P.		735.00		808.50		
1 Pickup Truck, 3/4 Ton		83.00		91.30	9.30	10.23
88 L.H., Daily Totals		$3224.80		$4718.60	$36.65	$53.63
Crew R-13	**Hr.**	**Daily**	**Hr.**	**Daily**	**Bare Costs**	**Incl. O&P**
1 Electrician Foreman	$34.80	$278.40	$51.95	$415.60	$32.43	$48.62
3 Electricians	34.30	823.20	51.20	1228.80		
1 Equipment Operator	31.15	249.20	47.15	377.20		
1 Equipment Oiler	25.75	206.00	39.00	312.00		
.25-1 Hyd. Crane, 33 Ton		201.20		221.30	4.19	4.61
48 L.H., Daily Totals		$1758.00		$2554.90	$36.62	$53.23
Crew R-15	**Hr.**	**Daily**	**Hr.**	**Daily**	**Bare Costs**	**Incl. O&P**
1 Electrician Foreman	$34.80	$278.40	$51.95	$415.60	$33.46	$50.05
4 Electricians	34.30	1097.60	51.20	1638.40		
1 Equipment Operator	28.75	230.00	43.55	348.40		
1 Aerial Lift Truck		211.00		232.10	4.40	4.84
48 L.H., Daily Totals		$1817.00		$2634.50	$37.86	$54.89
Crew R-18	**Hr.**	**Daily**	**Hr.**	**Daily**	**Bare Costs**	**Incl. O&P**
.25 Electrician Foreman	$34.80	$69.60	$51.95	$103.90	$26.86	$41.10
1 Electrician	34.30	274.40	51.20	409.60		
2 Helpers	22.15	354.40	34.70	555.20		
26 L.H., Daily Totals		$698.40		$1068.70	$26.86	$41.10
Crew R-19	**Hr.**	**Daily**	**Hr.**	**Daily**	**Bare Costs**	**Incl. O&P**
.5 Electrician Foreman	$34.80	$139.20	$51.95	$207.80	$34.40	$51.35
2 Electricians	34.30	548.80	51.20	819.20		
20 L.H., Daily Totals		$688.00		$1027.00	$34.40	$51.35
Crew R-21	**Hr.**	**Daily**	**Hr.**	**Daily**	**Bare Costs**	**Incl. O&P**
1 Electrician Foreman	$34.80	$278.40	$51.95	$415.60	$34.33	$51.25
3 Electricians	34.30	823.20	51.20	1228.80		
.1 Equip. Oper. (med.)	30.35	24.28	45.95	36.76		
.1 Hyd. Crane 25 Ton		60.20		66.20	1.84	2.02
32. L.H., Daily Totals		$1186.08		$1747.36	$36.17	$53.27

Crew No.	Bare Costs		Incl. Subs O & P		Cost Per Labor-Hour	
Crew R-22	**Hr.**	**Daily**	**Hr.**	**Daily**	**Bare Costs**	**Incl. O&P**
.66 Electrician Foreman	$34.80	$183.74	$51.95	$274.30	$29.16	$44.22
2 Helpers	22.15	354.40	34.70	555.20		
2 Electricians	34.30	548.80	51.20	819.20		
37.28 L.H., Daily Totals		$1086.94		$1648.70	$29.16	$44.22
Crew R-30	**Hr.**	**Daily**	**Hr.**	**Daily**	**Bare Costs**	**Incl. O&P**
.25 Electricians	$36.30	$72.60	$54.20	$108.40	$27.41	$41.92
1 Electricians	34.30	274.40	51.20	409.60		
2 Laborers, (Semi-Skilled)	22.85	365.60	35.75	572.00		
26 L.H., Daily Totals		$712.60		$1090.00	$27.41	$41.92

Historical Cost Indexes

The table below lists both the Means Historical Cost Index based on Jan. 1, 1993 = 100 as well as the computed value of an index based on Jan. 1, 2001 costs. Since the Jan. 1, 2001 figure is estimated, space is left to write in the actual index figures as they become available through either the quarterly "Means Construction Cost Indexes" or as printed in the "Engineering News-Record." To compute the actual index based on Jan. 1, 2001 = 100, divide the Historical Cost Index for a particular year by the actual Jan. 1, 2001 Construction Cost Index. Space has been left to advance the index figures as the year progresses.

Year	Historical Cost Index Jan. 1, 1993 = 100		Current Index Based on Jan. 1, 2001 = 100		Year	Historical Cost Index Jan. 1, 1993 = 100	Current Index Based on Jan. 1, 2001 = 100		Year	Historical Cost Index Jan. 1, 1993 = 100	Current Index Based on Jan. 1, 2001 = 100	
	Est.	Actual	Est.	Actual		Actual	Est.	Actual		Actual	Est.	Actual
Oct 2001					July 1986	84.2	69.2		July 1968	24.9	20.4	
July 2001					1985	82.6	67.8		1967	23.5	19.3	
April 2001					1984	82.0	67.3		1966	22.7	18.6	
Jan 2001	121.8		100.0	100.0	1983	80.2	65.8		1965	21.7	17.8	
July 2000		120.9	99.3		1982	76.1	62.5		1964	21.2	17.4	
1999		117.6	96.6		1981	70.0	57.5		1963	20.7	17.0	
1998		115.1	94.5		1980	62.9	51.6		1962	20.2	16.6	
1997		112.8	92.6		1979	57.8	47.5		1961	19.8	16.3	
1996		110.2	90.5		1978	53.5	43.9		1960	19.7	16.2	
1995		107.6	88.3		1977	49.5	40.6		1959	19.3	15.8	
1994		104.4	85.7		1976	46.9	38.5		1958	18.8	15.4	
1993		101.7	83.5		1975	44.8	36.8		1957	18.4	15.1	
1992		99.4	81.6		1974	41.4	34.0		1956	17.6	14.4	
1991		96.8	79.5		1973	37.7	31.0		1955	16.6	13.6	
1990		94.3	77.4		1972	34.8	28.6		1954	16.0	13.1	
1989		92.1	75.6		1971	32.1	26.4		1953	15.8	13.0	
1988		89.9	73.8		1970	28.7	23.6		1952	15.4	12.6	
1987		87.7	72.0		1969	26.9	22.1		1951	15.0	12.3	

Adjustments to Costs

The Historical Cost Index can be used to convert National Average building costs at a particular time to the approximate building costs for some other time.

Example:

Estimate and compare construction costs for different years in the same city.

To estimate the National Average construction cost of a building in 1970, knowing that it cost $900,000 in 2001:

INDEX in 1970 = 28.7

INDEX in 2001 = 121.8

Note: The City Cost Indexes for Canada can be used to convert U.S. National averages to local costs in Canadian dollars.

Time Adjustment using the Historical Cost Indexes:

$$\frac{\text{Index for Year A}}{\text{Index for Year B}} \times \text{Cost in Year B} = \text{Cost in Year A}$$

$$\frac{\text{INDEX 1970}}{\text{INDEX 2001}} \times \text{Cost 2001} = \text{Cost 1970}$$

$$\frac{28.7}{121.8} \times \$900,000 = .236 \times \$900,000 = \$212,400$$

The construction cost of the building in 1970 is $212,400.

How to Use the City Cost Indexes

What you should know before you begin

Means City Cost Indexes (CCI) are an extremely useful tool to use when you want to compare costs from city to city and region to region.

This publication contains average construction cost indexes for 305 major U.S. and Canadian cities and Location Factors covering over 930 three-digit zip code locations.

Keep in mind that a City Cost Index number is a *percentage ratio* of a specific city's cost to the national average cost of the same item at a stated time period.

In other words, these index figures represent relative construction *factors* (or, if you prefer, multipliers) for Material and Installation costs, as well as the weighted average for Total In Place costs for each CSI MasterFormat division. Installation costs include both labor and equipment rental costs.

The 30 City Average Index is the average of 30 major U.S. cities and serves as a National Average.

Index figures for both material and installation are based on the 30 major city average of 100 and represent the cost relationship as of July 1, 2000. The index for each division is computed from representative material and labor quantities for that division. The weighted average for each city is a weighted total of the components listed above it, but does not include relative productivity between trades or cities.

As changes occur in local material prices, labor rates and equipment rental rates, the impact of these changes should be accurately measured by the change in the City Cost Index for each particular city (as compared to the 30 City Average).

Therefore, if you know (or have estimated) building costs in one city today, you can easily convert those costs to expected building costs in another city.

In addition, by using the Historical Cost Index, you can easily convert National Average building costs at a particular time to the approximate building costs for some other time. The City Cost Indexes can then be applied to calculate the costs for a particular city.

Quick Calculations

Location Adjustment Using the City Cost Indexes:

$$\frac{\text{Index for City A}}{\text{Index for City B}} \times \text{Cost in City B} = \text{Cost in City A}$$

Time Adjustment for the National Average Using the Historical Cost Index:

$$\frac{\text{Index for Year A}}{\text{Index for Year B}} \times \text{Cost in Year B} = \text{Cost in Year A}$$

Adjustment from the National Average:

$$\frac{\text{Index for City A}}{100} \times \text{National Average Cost} = \text{Cost in City A}$$

Since each of the other R.S. Means publications contains many different items, any *one* item multiplied by the particular city index may give incorrect results. However, the larger the number of items compiled, the closer the results should be to actual costs for that particular city.

The City Cost Indexes for Canadian cities are calculated using Canadian material and equipment prices and labor rates, in Canadian dollars. Therefore, indexes for Canadian cities can be used to convert U.S. National Average prices to local costs in Canadian dollars.

How to use this section

1. Compare costs from city to city.

In using the Means Indexes, remember that an index number is not a fixed number but a *ratio:* It's a percentage ratio of a building component's cost at any stated time to the National Average cost of that same component at the same time period. Put in the form of an equation:

$$\frac{\text{Specific City Cost}}{\text{National Average Cost}} \times 100 = \text{City Index Number}$$

Therefore, when making cost comparisons between cities, do not subtract one city's index number from the index number of another city and read the result as a percentage difference. Instead, divide one city's index number by that of the other city. The resulting number may then be used as a multiplier to calculate cost differences from city to city.

The formula used to find cost differences between cities for the purpose of comparison is as follows:

$$\frac{\text{City A Index}}{\text{City B Index}} \times \text{City B Cost (Known)} = \text{City A Cost (Unknown)}$$

In addition, you can use *Means CCI* to calculate and compare costs division by division between cities using the same basic formula. (Just be sure that you're comparing similar divisions.)

2. Compare a specific city's construction costs with the National Average.

When you're studying construction location feasibility, it's advisable to compare a prospective project's cost index with an index of the National Average cost.

For example, divide the weighted average index of construction costs of a specific city by that of the 30 City Average, which = 100.

$$\frac{\text{City Index}}{100} = \% \text{ of National Average}$$

As a result, you get a ratio that indicates the relative cost of construction in that city in comparison with the National Average.

3. Convert U.S. National Average to actual costs in Canadian City

$$\frac{\text{Index for Canadian City}}{100} \times \text{National Average Cost} = \text{Cost in Canadian City in \$ CAN}$$

4. Adjust construction cost data based on a National Average.

When you use a source of construction cost data which is based on a National Average (such as *Means cost data publications*), it is necessary to adjust those costs to a specific location.

$$\frac{\text{City Index}}{100} \times \text{"Book" Cost Based on National Average Costs} = \text{City Cost (Unknown)}$$

5. When applying the City Cost Indexes to demolition projects, use the appropriate division index. For example, for removal of existing doors and windows, use the Division 8 index.

What you might like to know about how we developed the Indexes

To create a reliable index, R.S. Means researched the building type most often constructed in the United States and Canada. Because it was concluded that no one type of building completely represented the building construction industry, nine different types of buildings were combined to create a composite model.

The exact material, labor and equipment quantities are based on detailed analysis of these nine building types, then each quantity is weighted in proportion to expected usage. These various material items, labor hours, and equipment rental rates are thus combined to form a composite building representing as closely as possible the actual usage of materials, labor and equipment used in the North American Building Construction Industry.

The following structures were chosen to make up that composite model:

1. Factory, 1 story
2. Office, 2–4 story
3. Store, Retail
4. Town Hall, 2–3 story
5. High School, 2–3 story
6. Hospital, 4–8 story
7. Garage, Parking
8. Apartment, 1–3 story
9. Hotel/Motel, 2–3 story

For the purposes of ensuring the timeliness of the data, the components of the index for the composite model have been streamlined. They currently consist of:

- specific quantities of 66 commonly used construction materials;
- specific labor-hours for 21 building construction trades; and
- specific days of equipment rental for 6 types of construction equipment (normally used to install the 66 material items by the 21 trades.)

A sophisticated computer program handles the updating of all costs for each city on a quarterly basis. Material and equipment price quotations are gathered quarterly from over 305 cities in the United States and Canada. These prices and the latest negotiated labor wage rates for 21 different building trades are used to compile the quarterly update of the City Cost Index.

The 30 major U.S. cities used to calculate the National Average are:

Atlanta, GA
Baltimore, MD
Boston, MA
Buffalo, NY
Chicago, IL
Cincinnati, OH
Cleveland, OH
Columbus, OH
Dallas, TX
Denver, CO
Detroit, MI
Houston, TX
Indianapolis, IN
Kansas City, MO
Los Angeles, CA

Memphis, TN
Milwaukee, WI
Minneapolis, MN
Nashville, TN
New Orleans, LA
New York, NY
Philadelphia, PA
Phoenix, AZ
Pittsburgh, PA
St. Louis, MO
San Antonio, TX
San Diego, CA
San Francisco, CA
Seattle, WA
Washington, DC

F.Y.I.: The CSI MasterFormat Divisions

1. General Requirements
2. Site Work
3. Concrete
4. Masonry
5. Metals
6. Wood & Plastics
7. Thermal & Moisture Protection
8. Doors & Windows
9. Finishes
10. Specialties
11. Equipment
12. Furnishings
13. Special Construction
14. Conveying Systems
15. Mechanical
16. Electrical

The information presented in the CCI is organized according to the Construction Specifications Institute (CSI) MasterFormat.

What the CCI does not indicate

The weighted average for each city is a total of the components listed above weighted to reflect typical usage, but it does *not* include the productivity variations between trades or cities.

In addition, the CCI does not take into consideration factors such as the following:

- managerial efficiency
- competitive conditions
- automation
- restrictive union practices
- Unique local requirements
- regional variations due to specific building codes

City Cost Indexes

DIVISION		UNITED STATES 30 CITY AVERAGE			ALABAMA												
					BIRMINGHAM			HUNTSVILLE			MOBILE			MONTGOMERY			TUSCALOOSA		
		MAT.	INST.	TOTAL	MAT.	INST.	TOTAL	MAT.	INST.	TOTAL	MAT.	INST.	TOTAL	MAT.	INST.	TOTAL	MAT.	INST.	TOTAL
02	SITE CONSTRUCTION	100.0	100.0	100.0	88.6	93.8	92.6	86.6	93.0	91.5	97.9	86.8	89.4	98.4	87.3	89.9	87.1	92.7	91.4
03100	CONCRETE FORMS & ACCESSORIES	100.0	100.0	100.0	94.4	77.4	79.7	96.0	57.6	62.9	96.0	60.9	65.7	94.2	55.0	60.4	96.0	45.2	52.2
03200	CONCRETE REINFORCEMENT	100.0	100.0	100.0	92.8	91.0	91.8	92.8	89.2	90.8	95.8	67.2	79.5	95.8	90.0	92.5	92.8	90.3	91.4
03300	CAST-IN-PLACE CONCRETE	100.0	100.0	100.0	93.9	68.1	83.3	88.8	65.3	79.1	93.8	63.3	81.2	95.5	57.0	79.6	92.5	54.5	76.8
03	CONCRETE	100.0	100.0	100.0	89.8	77.7	83.8	87.4	67.7	77.6	90.2	64.4	77.3	90.9	63.9	77.4	89.2	58.8	74.0
04	MASONRY	100.0	100.0	100.0	85.8	72.4	77.5	85.7	58.1	68.6	86.3	58.5	69.2	86.8	40.7	58.5	86.0	43.0	59.5
05	METALS	100.0	100.0	100.0	98.6	95.9	97.6	98.2	93.9	96.7	96.8	86.2	93.0	96.8	93.7	95.7	97.3	94.3	96.2
06	WOOD & PLASTICS	100.0	100.0	100.0	95.3	77.6	86.2	94.4	56.1	74.6	94.4	61.2	77.2	92.3	56.1	73.5	94.4	44.5	68.6
07	THERMAL & MOISTURE PROTECTION	100.0	100.0	100.0	96.8	80.1	88.9	96.6	72.7	85.3	96.6	73.9	85.8	96.3	66.9	82.3	96.5	65.4	81.8
08	DOORS & WINDOWS	100.0	100.0	100.0	98.3	77.9	93.4	98.3	60.4	89.1	98.3	60.3	89.1	98.3	61.9	89.5	98.3	59.9	89.0
09200	PLASTER & GYPSUM BOARD	100.0	100.0	100.0	110.0	77.5	90.1	107.2	55.2	75.4	107.2	60.5	78.6	107.2	55.2	75.4	107.2	43.3	68.1
095,098	CEILINGS & ACOUSTICAL TREATMENT	100.0	100.0	100.0	95.1	77.5	83.5	95.1	55.2	68.8	95.1	60.5	72.3	95.1	55.2	68.8	95.1	43.3	60.9
09600	FLOORING	100.0	100.0	100.0	101.6	60.9	91.7	101.6	52.7	89.8	110.2	66.0	99.5	110.2	34.1	91.8	101.6	50.1	89.1
097,099	WALL FINISHES, PAINTS & COATINGS	100.0	100.0	100.0	92.9	54.9	70.8	92.9	53.8	70.2	92.9	62.3	75.1	92.9	61.2	74.5	92.9	52.7	69.5
09	FINISHES	100.0	100.0	100.0	98.8	71.7	85.0	98.2	55.2	76.3	102.1	61.6	81.5	102.2	50.7	75.9	98.2	45.5	71.3
10 - 14	TOTAL DIV. 10000 - 14000	100.0	100.0	100.0	100.0	84.7	96.8	100.0	78.4	95.4	100.0	74.5	94.6	100.0	76.0	94.9	100.0	73.0	94.3
15	MECHANICAL	100.0	100.0	100.0	99.9	72.2	87.1	99.9	54.2	78.8	99.9	67.0	84.7	99.9	48.8	76.3	99.9	43.0	73.6
16	ELECTRICAL	100.0	100.0	100.0	96.3	67.1	76.2	96.7	73.9	81.0	96.7	61.5	72.5	96.3	72.4	79.9	96.7	67.1	76.3
01 - 16	WEIGHTED AVERAGE	100.0	100.0	100.0	96.6	77.2	87.2	96.2	68.6	82.8	97.1	68.0	83.0	97.2	63.9	81.0	96.3	61.1	79.1

DIVISION		ALASKA									ARIZONA								
		ANCHORAGE			FAIRBANKS			JUNEAU			FLAGSTAFF			MESA/TEMPE			PHOENIX		
		MAT.	INST.	TOTAL	MAT.	INST.	TOTAL	MAT.	INST.	TOTAL	MAT.	INST.	TOTAL	MAT.	INST.	TOTAL	MAT.	INST.	TOTAL
02	SITE CONSTRUCTION	132.6	135.4	134.8	117.3	135.4	131.2	128.6	135.4	133.8	86.9	99.6	96.7	81.6	99.3	95.1	82.0	100.0	95.8
03100	CONCRETE FORMS & ACCESSORIES	132.5	115.8	118.1	134.1	119.6	121.6	133.8	115.8	118.3	108.9	74.3	79.0	103.6	72.2	76.5	105.0	79.5	83.0
03200	CONCRETE REINFORCEMENT	141.4	106.5	121.5	118.8	106.6	111.8	105.2	106.5	105.9	103.6	80.9	90.7	103.9	77.9	89.0	102.0	81.4	90.3
03300	CAST-IN-PLACE CONCRETE	186.6	115.3	157.2	158.0	115.9	140.6	187.4	115.4	157.7	96.8	84.9	91.9	96.8	76.6	88.4	96.9	84.9	91.9
03	CONCRETE	152.2	113.2	132.7	130.0	115.0	122.5	148.1	113.2	130.6	122.5	78.9	100.7	99.6	74.6	87.1	99.2	81.4	90.3
04	MASONRY	195.5	123.6	151.3	182.8	123.6	146.4	184.8	123.6	147.1	103.1	65.7	80.2	110.7	57.8	78.2	97.5	73.6	82.8
05	METALS	130.0	99.3	119.0	130.1	99.7	119.1	130.3	99.4	119.2	98.1	72.7	88.9	98.3	72.5	89.0	99.6	75.1	90.8
06	WOOD & PLASTICS	123.8	113.8	118.6	124.1	118.5	121.2	123.8	113.8	118.6	109.7	74.1	91.3	101.4	79.1	89.9	102.7	80.9	91.4
07	THERMAL & MOISTURE PROTECTION	199.7	114.1	159.1	195.7	116.4	158.1	196.4	114.1	157.3	111.5	74.9	94.1	106.5	70.1	89.2	106.4	76.8	92.4
08	DOORS & WINDOWS	131.3	107.3	125.6	128.4	110.2	124.0	128.4	107.3	123.3	103.0	74.0	96.0	99.8	74.3	93.6	98.9	77.7	93.8
09200	PLASTER & GYPSUM BOARD	129.9	114.2	120.3	129.9	119.0	123.3	129.9	114.2	120.3	93.2	73.4	81.1	93.8	78.3	84.4	94.3	80.5	85.7
095,098	CEILINGS & ACOUSTICAL TREATMENT	131.7	114.2	120.1	131.7	119.0	123.3	131.7	114.2	120.1	99.2	73.4	82.2	103.3	78.4	86.9	103.3	80.2	88.1
09600	FLOORING	126.3	128.2	126.8	126.4	128.2	126.9	126.3	128.2	126.8	99.0	61.6	89.9	101.4	69.9	93.8	101.8	69.9	94.1
097,099	WALL FINISHES, PAINTS & COATINGS	117.6	110.6	113.5	117.6	116.0	116.7	117.6	110.6	113.5	99.1	58.5	75.5	109.9	62.8	82.5	109.9	70.5	87.0
09	FINISHES	140.1	117.8	128.7	138.0	121.2	129.4	138.7	117.8	128.0	96.8	69.9	83.1	98.2	70.9	84.3	98.4	76.9	87.4
10 - 14	TOTAL DIV. 10000 - 14000	100.0	114.8	103.1	100.0	115.5	103.3	100.0	114.8	103.1	100.0	83.9	96.6	100.0	78.7	95.5	100.0	85.0	96.8
15	MECHANICAL	106.4	108.0	107.1	106.4	118.5	112.0	106.4	108.6	107.4	100.3	82.2	91.9	100.1	73.8	87.9	100.1	82.3	91.8
16	ELECTRICAL	161.7	110.5	126.5	164.6	110.5	127.5	164.6	110.5	127.5	100.4	57.4	70.9	98.1	38.8	57.3	107.2	63.5	77.2
01 - 16	WEIGHTED AVERAGE	134.5	114.1	124.5	130.1	117.0	123.7	132.9	114.2	123.8	102.8	74.8	89.2	99.6	68.7	84.6	99.6	78.4	89.3

DIVISION		ARIZONA						ARKANSAS											
		PRESCOTT			TUCSON			FORT SMITH			JONESBORO			LITTLE ROCK			PINE BLUFF		
		MAT.	INST.	TOTAL	MAT.	INST.	TOTAL	MAT.	INST.	TOTAL	MAT.	INST.	TOTAL	MAT.	INST.	TOTAL	MAT.	INST.	TOTAL
02	SITE CONSTRUCTION	73.7	99.2	93.3	77.9	99.8	94.7	76.7	83.9	82.2	101.7	99.0	99.6	76.7	83.9	82.2	79.9	83.9	83.0
03100	CONCRETE FORMS & ACCESSORIES	103.4	68.2	73.0	104.4	79.2	82.6	98.7	48.2	55.1	83.1	55.6	59.4	93.0	52.9	58.4	75.0	52.8	55.9
03200	CONCRETE REINFORCEMENT	103.6	82.3	91.5	100.9	80.9	89.5	95.5	72.8	82.6	91.4	54.5	70.4	95.8	71.9	82.2	95.7	71.9	82.2
03300	CAST-IN-PLACE CONCRETE	96.8	73.7	87.2	89.7	84.7	87.6	84.7	52.0	71.2	89.2	63.7	78.7	85.7	52.0	71.8	85.9	52.0	71.9
03	CONCRETE	106.9	72.5	89.7	95.5	81.0	88.3	84.1	55.0	69.5	85.0	60.0	72.5	84.2	56.9	70.6	85.6	56.8	71.2
04	MASONRY	103.6	71.6	83.9	97.5	65.7	77.9	97.0	58.4	73.3	91.3	54.9	68.9	95.2	58.4	72.6	115.0	58.4	80.2
05	METALS	98.1	72.0	88.7	98.9	73.4	89.7	96.6	72.8	88.0	90.9	82.3	87.8	96.2	72.6	87.7	95.4	72.5	87.1
06	WOOD & PLASTICS	104.2	67.3	85.1	102.8	80.9	91.5	100.4	47.3	73.0	83.3	57.0	69.7	96.9	53.6	74.5	75.7	53.6	64.3
07	THERMAL & MOISTURE PROTECTION	109.7	72.3	91.9	107.1	71.3	90.1	98.3	54.5	77.5	108.0	59.3	84.9	96.9	55.1	77.1	96.8	55.1	77.0
08	DOORS & WINDOWS	103.0	69.3	94.9	97.6	77.7	92.8	98.1	50.0	86.5	99.5	54.9	88.8	98.1	54.2	87.5	93.5	54.2	84.0
09200	PLASTER & GYPSUM BOARD	91.5	66.3	76.1	94.5	80.2	85.8	97.6	46.6	66.4	102.1	56.1	74.0	97.6	53.0	70.4	91.2	53.0	67.9
095,098	CEILINGS & ACOUSTICAL TREATMENT	99.2	66.3	77.5	104.7	80.2	88.5	101.1	46.6	65.1	101.7	56.1	71.6	101.1	53.0	69.4	97.0	53.0	68.0
09600	FLOORING	97.3	61.3	88.6	100.9	61.6	91.4	118.7	74.4	107.9	79.3	53.2	73.0	120.0	74.4	108.9	107.2	74.4	99.3
097,099	WALL FINISHES, PAINTS & COATINGS	99.1	58.5	75.5	107.4	58.5	78.9	95.6	67.7	79.4	84.3	60.9	70.7	95.6	51.8	70.1	95.6	51.8	70.1
09	FINISHES	94.9	65.6	80.0	98.1	73.8	85.7	102.0	54.9	78.0	92.6	55.8	73.8	102.4	56.7	79.1	96.8	56.7	76.3
10 - 14	TOTAL DIV. 10000 - 14000	100.0	82.6	96.3	100.0	85.1	96.8	100.0	70.5	93.8	100.0	65.4	92.7	100.0	71.4	94.0	100.0	71.4	94.0
15	MECHANICAL	100.3	71.7	87.1	100.0	76.2	89.0	100.1	50.6	77.2	100.3	50.7	77.3	100.1	57.3	80.3	100.1	53.5	78.5
16	ELECTRICAL	100.0	55.3	69.3	101.6	64.1	75.8	95.6	70.0	78.0	100.7	57.1	70.8	96.7	72.7	80.2	94.5	72.7	79.5
01 - 16	WEIGHTED AVERAGE	100.2	71.2	86.1	98.4	75.8	87.4	96.3	61.6	79.4	95.6	62.5	79.5	96.2	64.0	80.6	95.9	63.3	80.1

City Cost Indexes

DIVISION		ARKANSAS TEXARKANA			CALIFORNIA ANAHEIM			CALIFORNIA BAKERSFIELD			CALIFORNIA FRESNO			CALIFORNIA LOS ANGELES			CALIFORNIA OAKLAND		
		MAT.	INST.	TOTAL	MAT.	INST.	TOTAL	MAT.	INST.	TOTAL	MAT.	INST.	TOTAL	MAT.	INST.	TOTAL	MAT.	INST.	TOTAL
02	SITE CONSTRUCTION	96.0	84.6	87.3	94.1	110.2	106.5	98.7	107.1	105.2	100.2	107.2	105.5	89.5	109.0	104.5	128.3	104.0	109.7
03100	CONCRETE FORMS & ACCESSORIES	84.1	46.1	51.3	105.2	119.2	117.2	96.3	118.9	115.8	101.8	121.7	118.9	108.8	118.7	117.3	107.7	136.7	132.7
03200	CONCRETE REINFORCEMENT	95.2	52.8	71.0	106.8	117.3	112.8	106.4	117.1	112.5	106.9	117.1	112.7	108.3	117.2	113.4	98.8	118.0	109.8
03300	CAST-IN-PLACE CONCRETE	85.8	49.7	70.9	102.3	116.3	108.1	97.7	115.1	104.9	106.9	115.7	110.5	89.4	113.5	99.3	127.9	118.4	124.0
03	CONCRETE	81.1	49.6	65.4	112.1	116.9	114.5	109.2	116.2	112.7	114.2	117.7	115.9	108.2	115.6	111.9	125.9	125.6	125.7
04	MASONRY	97.1	38.5	61.1	87.8	116.5	105.4	107.2	113.3	111.0	109.7	112.4	111.4	97.8	116.5	109.3	145.5	125.5	133.2
05	METALS	87.7	64.2	79.2	110.5	101.4	107.2	105.0	100.5	103.4	106.9	101.0	104.8	111.2	99.3	106.9	101.7	106.4	103.4
06	WOOD & PLASTICS	85.7	49.4	66.9	99.5	117.7	108.9	89.1	117.9	104.0	101.4	120.6	111.4	99.6	117.3	108.7	106.4	139.1	123.3
07	THERMAL & MOISTURE PROTECTION	97.7	48.5	74.4	121.8	116.7	119.4	105.9	110.1	107.9	102.3	111.2	106.5	114.2	114.6	114.4	111.7	130.0	120.4
08	DOORS & WINDOWS	98.5	47.0	86.1	105.8	115.1	108.1	104.6	113.4	106.7	104.6	115.3	107.2	99.1	114.8	102.9	108.6	129.7	113.7
09200	PLASTER & GYPSUM BOARD	94.2	48.7	66.4	94.8	118.4	109.3	94.8	118.4	109.3	94.0	121.3	110.7	88.9	118.4	106.9	89.1	140.0	120.2
095,098	CEILINGS & ACOUSTICAL TREATMENT	103.8	48.7	67.5	118.1	118.4	118.3	118.1	118.4	118.3	118.1	121.3	120.2	115.2	118.4	117.3	116.9	140.0	132.2
09600	FLOORING	110.1	50.8	95.7	123.1	110.5	120.0	118.6	73.3	107.6	119.3	143.2	125.1	115.7	110.5	114.4	108.8	120.7	111.7
097,099	WALL FINISHES, PAINTS & COATINGS	95.6	42.7	64.8	111.3	116.0	114.0	114.4	106.0	109.5	113.9	101.7	106.8	108.6	114.8	112.2	118.3	133.2	126.9
09	FINISHES	99.8	46.7	72.7	112.9	117.1	115.1	114.1	109.6	111.8	114.2	123.9	119.2	108.5	116.6	112.7	112.6	134.4	123.7
10 - 14	TOTAL DIV. 10000 - 14000	100.0	46.6	88.7	100.0	115.8	103.3	100.0	130.8	106.5	100.0	133.3	107.0	100.0	114.5	103.1	100.0	136.6	107.7
15	MECHANICAL	100.1	44.8	74.5	100.3	114.1	106.7	100.3	112.7	106.0	100.3	115.1	107.1	100.2	114.0	106.6	100.1	136.4	116.9
16	ELECTRICAL	97.7	47.6	63.2	95.0	109.9	105.2	95.4	94.2	94.6	94.1	99.6	97.9	109.8	113.6	112.4	113.6	130.5	125.2
01 - 16	WEIGHTED AVERAGE	94.9	51.5	73.8	104.7	113.0	108.7	104.1	108.5	106.3	105.2	112.0	108.5	104.3	112.9	108.5	110.7	126.3	118.3

DIVISION		CALIFORNIA OXNARD			CALIFORNIA REDDING			CALIFORNIA RIVERSIDE			CALIFORNIA SACRAMENTO			CALIFORNIA SAN DIEGO			CALIFORNIA SAN FRANCISCO		
		MAT.	INST.	TOTAL	MAT.	INST.	TOTAL	MAT.	INST.	TOTAL	MAT.	INST.	TOTAL	MAT.	INST.	TOTAL	MAT.	INST.	TOTAL
02	SITE CONSTRUCTION	99.8	104.8	103.6	105.0	106.0	105.8	92.0	107.9	104.1	95.9	110.5	107.1	100.6	102.1	101.8	130.3	110.5	115.2
03100	CONCRETE FORMS & ACCESSORIES	103.1	119.3	117.1	102.7	120.8	118.3	106.3	119.1	117.3	106.2	121.2	119.2	106.6	113.8	112.8	108.0	137.5	133.4
03200	CONCRETE REINFORCEMENT	106.4	117.1	112.5	103.1	117.2	111.1	105.4	117.1	112.1	99.9	117.2	109.8	110.5	116.9	114.1	112.6	118.4	115.9
03300	CAST-IN-PLACE CONCRETE	103.7	115.4	108.5	117.4	114.5	116.2	101.3	116.3	107.5	111.0	114.8	112.6	105.2	105.0	105.1	127.8	120.1	124.6
03	CONCRETE	112.6	116.5	114.6	123.0	116.8	119.9	111.5	116.8	114.2	117.6	117.1	117.4	116.1	110.5	113.3	127.6	126.6	127.1
04	MASONRY	112.4	112.4	112.4	113.8	112.1	112.7	85.3	113.7	102.8	117.2	112.1	114.1	105.4	110.9	108.7	145.8	130.6	136.4
05	METALS	104.4	100.8	103.1	108.9	100.3	105.8	110.5	101.1	107.1	97.7	100.3	98.6	110.6	100.2	106.8	106.8	107.9	107.2
06	WOOD & PLASTICS	98.1	117.9	108.3	99.2	120.6	110.3	99.5	117.7	108.9	101.1	120.7	111.2	100.1	111.1	105.8	106.4	139.3	123.4
07	THERMAL & MOISTURE PROTECTION	111.3	114.1	112.6	112.0	112.8	112.4	120.8	115.4	118.3	121.3	113.4	117.5	114.7	107.8	111.4	111.7	132.0	121.3
08	DOORS & WINDOWS	103.4	115.1	106.3	106.3	112.5	107.8	105.8	115.1	108.1	120.6	116.5	119.6	104.6	110.9	106.1	111.1	129.3	115.5
09200	PLASTER & GYPSUM BOARD	94.8	118.4	109.3	95.3	121.3	111.2	94.6	118.4	109.2	87.8	121.3	108.2	88.1	111.4	102.4	90.9	140.0	120.9
095,098	CEILINGS & ACOUSTICAL TREATMENT	118.1	118.4	118.3	127.6	121.3	123.4	116.7	118.4	117.8	125.0	121.3	122.5	117.9	111.4	113.6	123.4	140.0	134.4
09600	FLOORING	118.6	110.5	116.6	117.8	111.4	116.3	123.0	110.5	120.0	111.2	111.4	111.3	119.1	109.8	116.8	108.8	120.7	111.7
097,099	WALL FINISHES, PAINTS & COATINGS	113.9	109.3	111.2	113.9	114.3	114.1	111.3	116.0	114.0	116.7	114.3	115.3	111.9	116.0	114.3	118.3	134.1	127.5
09	FINISHES	113.9	116.4	115.2	115.9	119.4	117.7	112.4	117.1	114.8	113.1	119.5	116.4	111.5	113.2	112.4	114.1	134.6	124.6
10 - 14	TOTAL DIV. 10000 - 14000	100.0	116.1	103.4	100.0	132.7	106.9	100.0	115.8	103.3	100.0	132.8	106.9	100.0	115.2	103.2	100.0	137.3	107.9
15	MECHANICAL	100.3	114.1	106.7	100.3	113.1	106.2	100.2	114.1	106.6	100.0	118.2	108.4	100.1	114.1	106.6	100.1	159.3	127.5
16	ELECTRICAL	99.2	113.1	108.8	104.4	99.5	101.0	95.1	105.5	102.2	104.0	100.4	101.5	92.8	93.9	93.6	109.3	140.3	130.6
01 - 16	WEIGHTED AVERAGE	105.2	112.4	108.7	108.3	110.6	109.4	104.3	111.7	107.9	107.5	112.4	109.9	105.6	107.0	106.3	111.9	133.9	122.6

DIVISION		CALIFORNIA SAN JOSE			CALIFORNIA SANTA BARBARA			CALIFORNIA STOCKTON			CALIFORNIA VALLEJO			COLORADO COLORADO SPRINGS			COLORADO DENVER		
		MAT.	INST.	TOTAL	MAT.	INST.	TOTAL	MAT.	INST.	TOTAL	MAT.	INST.	TOTAL	MAT.	INST.	TOTAL	MAT.	INST.	TOTAL
02	SITE CONSTRUCTION	136.3	100.4	108.8	100.2	107.2	105.5	98.1	106.4	104.5	92.5	110.7	106.4	110.1	98.3	101.1	108.4	107.8	107.9
03100	CONCRETE FORMS & ACCESSORIES	105.5	136.9	132.6	103.8	119.1	117.0	97.5	121.6	118.3	108.1	136.2	132.3	93.4	76.2	78.6	103.8	81.3	84.4
03200	CONCRETE REINFORCEMENT	103.4	118.3	111.9	106.4	116.9	112.4	106.9	117.3	112.8	98.5	118.2	109.7	98.9	80.8	88.6	98.4	81.5	88.7
03300	CAST-IN-PLACE CONCRETE	123.2	118.8	121.4	103.3	115.2	108.2	103.2	115.6	108.3	114.8	117.1	115.7	110.8	85.6	100.4	103.7	85.9	96.3
03	CONCRETE	122.0	126.0	124.0	112.4	116.3	114.4	112.1	117.7	114.9	119.4	124.7	122.0	103.8	80.7	92.2	108.3	83.1	95.7
04	MASONRY	149.4	129.1	136.9	107.6	109.8	109.0	112.1	115.6	114.2	83.3	129.4	111.6	108.3	69.4	84.4	107.6	83.3	92.7
05	METALS	111.1	108.4	110.2	104.3	100.4	102.9	106.5	101.4	104.7	101.0	102.5	101.5	101.6	85.9	96.0	105.2	86.8	98.5
06	WOOD & PLASTICS	106.0	139.0	123.1	98.1	117.9	108.3	93.4	120.6	107.5	99.6	138.9	119.9	91.4	76.5	83.7	101.5	82.2	91.5
07	THERMAL & MOISTURE PROTECTION	107.9	130.8	118.7	107.0	112.0	109.4	111.4	113.4	112.4	125.6	129.9	127.7	107.5	81.4	95.1	106.8	84.8	96.4
08	DOORS & WINDOWS	95.4	129.7	103.7	104.6	115.1	107.2	103.9	116.5	106.9	122.0	129.0	123.7	99.2	82.1	95.1	98.6	85.2	95.3
09200	PLASTER & GYPSUM BOARD	95.3	140.0	122.6	94.8	118.4	109.3	95.7	121.3	111.3	88.2	140.0	119.9	89.2	75.7	81.0	97.4	81.9	87.9
095,098	CEILINGS & ACOUSTICAL TREATMENT	109.9	140.0	129.8	118.1	118.4	118.3	118.1	121.3	120.2	125.0	140.0	134.9	97.5	75.7	83.1	96.2	81.9	86.7
09600	FLOORING	115.5	120.7	116.8	118.6	99.1	113.8	118.6	111.4	116.8	115.5	120.7	116.8	111.8	94.1	107.5	112.8	86.6	106.4
097,099	WALL FINISHES, PAINTS & COATINGS	115.6	131.4	124.8	113.9	109.3	111.2	113.9	98.9	105.2	114.6	133.2	125.4	111.9	62.4	83.1	111.9	76.8	91.5
09	FINISHES	113.1	134.2	123.9	114.1	114.5	114.3	114.1	118.1	116.1	111.7	134.1	123.1	100.5	76.9	88.5	101.6	81.5	91.4
10 - 14	TOTAL DIV. 10000 - 14000	100.0	136.4	107.7	100.0	116.1	103.4	100.0	133.4	107.0	100.0	135.6	107.5	100.0	85.0	96.8	100.0	85.8	97.0
15	MECHANICAL	100.3	146.3	121.5	100.3	114.1	106.7	100.3	114.0	106.6	100.0	130.9	114.3	100.2	80.9	91.3	100.1	84.7	93.0
16	ELECTRICAL	109.9	137.7	129.0	91.9	106.0	101.6	103.4	109.1	107.4	99.8	117.9	112.2	95.2	85.7	88.7	97.3	90.4	92.6
01 - 16	WEIGHTED AVERAGE	110.2	129.7	119.7	104.4	110.8	107.5	105.7	113.0	109.3	106.0	123.6	114.6	101.3	82.4	92.1	102.5	87.4	95.2

City Cost Indexes

	DIVISION	COLORADO												CONNECTICUT					
		FORT COLLINS			GRAND JUNCTION			GREELEY			PUEBLO			BRIDGEPORT			BRISTOL		
		MAT.	INST.	TOTAL	MAT.	INST.	TOTAL	MAT.	INST.	TOTAL	MAT.	INST.	TOTAL	MAT.	INST.	TOTAL	MAT.	INST.	TOTAL
02	SITE CONSTRUCTION	121.3	101.5	106.2	131.4	100.1	107.5	106.3	100.3	101.7	122.7	94.9	101.4	107.0	105.4	105.8	106.1	105.4	105.6
03100	CONCRETE FORMS & ACCESSORIES	103.2	75.1	79.0	113.0	51.6	60.0	100.2	55.9	62.0	107.0	76.5	80.7	99.3	102.7	102.3	99.3	108.0	106.8
03200	CONCRETE REINFORCEMENT	103.1	80.4	90.2	108.9	80.2	92.5	102.7	58.1	77.2	101.5	80.7	89.7	111.9	119.7	116.3	111.9	119.6	116.3
03300	CAST-IN-PLACE CONCRETE	120.4	82.1	104.6	119.4	51.2	91.3	100.6	68.7	87.4	107.1	86.3	98.5	105.2	114.5	109.0	98.5	114.4	105.1
03	CONCRETE	123.0	78.9	101.0	120.0	57.8	88.9	105.7	61.3	83.5	108.6	81.1	94.9	110.0	109.9	109.9	106.7	112.1	109.4
04	MASONRY	121.3	69.1	89.3	148.4	56.5	91.9	114.9	47.7	73.6	107.9	67.3	82.9	107.4	115.1	112.2	96.8	115.1	108.1
05	METALS	101.0	83.6	94.7	101.5	81.5	94.3	101.0	66.8	88.6	103.0	86.9	97.2	99.0	112.9	104.0	99.0	112.7	103.9
06	WOOD & PLASTICS	101.2	76.2	88.3	108.6	53.0	79.8	98.1	56.8	76.7	103.0	76.8	89.4	100.9	98.6	99.7	100.9	106.1	103.6
07	THERMAL & MOISTURE PROTECTION	107.4	76.3	92.6	105.5	62.4	85.1	106.5	66.1	87.4	104.9	80.1	93.1	99.4	112.4	105.5	99.5	107.5	103.3
08	DOORS & WINDOWS	96.6	74.4	91.2	101.4	58.2	91.0	96.6	49.6	85.2	95.8	82.2	92.6	110.0	111.7	110.4	110.0	111.0	110.3
09200	PLASTER & GYPSUM BOARD	96.4	75.7	83.8	102.3	51.6	71.3	95.6	55.6	71.1	89.5	75.7	81.1	97.8	97.4	97.5	97.8	105.0	102.2
095,098	CEILINGS & ACOUSTICAL TREATMENT	90.7	75.7	80.8	93.8	51.6	66.0	90.7	55.6	67.6	104.7	75.7	85.6	102.6	97.4	99.1	102.6	105.0	104.2
09600	FLOORING	112.7	83.0	105.5	120.7	83.0	111.5	111.2	83.0	104.4	117.0	86.6	109.6	99.9	109.3	102.2	99.9	109.3	102.2
097,099	WALL FINISHES, PAINTS & COATINGS	111.9	61.9	82.8	122.6	36.3	72.5	111.9	38.5	69.2	122.6	57.8	84.9	95.6	111.5	104.8	95.6	110.7	104.4
09	FINISHES	101.1	75.2	87.9	106.8	55.7	80.7	99.8	58.7	78.8	105.3	75.9	90.3	101.2	103.9	102.6	101.2	108.1	104.8
10 - 14	TOTAL DIV. 10000 - 14000	100.0	71.6	94.0	100.0	63.9	92.4	100.0	65.1	92.6	100.0	85.9	97.0	100.0	112.9	102.7	100.0	113.9	102.9
15	MECHANICAL	100.1	83.3	92.3	100.1	40.2	72.4	100.1	77.2	89.5	100.1	79.2	90.4	99.8	104.6	102.0	99.8	104.5	102.0
16	ELECTRICAL	90.1	70.4	76.6	94.0	75.0	80.9	90.1	70.4	76.6	95.3	81.9	86.1	104.0	103.9	104.0	104.0	105.0	104.7
01 - 16	WEIGHTED AVERAGE	104.1	79.2	91.9	107.1	63.6	85.9	100.9	68.4	85.1	102.6	80.9	92.0	103.3	107.6	105.4	102.3	108.5	105.3

	DIVISION	CONNECTICUT																	
		HARTFORD			NEW BRITAIN			NEW HAVEN			NORWALK			STAMFORD			WATERBURY		
		MAT.	INST.	TOTAL	MAT.	INST.	TOTAL	MAT.	INST.	TOTAL	MAT.	INST.	TOTAL	MAT.	INST.	TOTAL	MAT.	INST.	TOTAL
02	SITE CONSTRUCTION	106.3	105.4	105.6	106.3	105.4	105.6	106.1	106.3	106.3	106.8	105.4	105.8	107.5	105.5	105.9	106.5	105.4	105.7
03100	CONCRETE FORMS & ACCESSORIES	98.4	108.0	106.6	99.7	108.0	106.9	99.1	108.2	106.9	99.3	102.9	102.4	99.3	103.1	102.6	99.3	108.1	106.9
03200	CONCRETE REINFORCEMENT	111.9	119.6	116.3	111.9	119.6	116.3	111.9	119.7	116.3	111.9	119.8	116.4	111.9	119.8	116.4	111.9	119.7	116.3
03300	CAST-IN-PLACE CONCRETE	98.2	114.4	104.9	100.1	114.4	106.0	101.8	118.7	108.8	103.5	119.7	110.2	105.2	119.8	111.2	105.2	114.4	109.0
03	CONCRETE	106.5	112.1	109.3	107.6	112.1	109.8	109.3	113.7	111.5	109.2	111.8	110.5	110.0	111.9	111.0	110.0	112.2	111.1
04	MASONRY	96.8	115.1	108.1	96.9	115.1	108.1	97.0	115.1	108.2	97.2	117.4	109.6	97.3	117.4	109.6	97.3	115.1	108.3
05	METALS	99.6	112.7	104.3	95.8	112.7	101.9	96.0	112.9	102.1	99.0	113.2	104.1	99.0	113.4	104.2	99.0	112.8	104.0
06	WOOD & PLASTICS	100.9	106.1	103.6	100.9	106.1	103.6	100.9	106.1	103.6	100.9	98.6	99.7	100.9	98.6	99.7	100.9	106.1	103.6
07	THERMAL & MOISTURE PROTECTION	98.3	107.5	102.7	99.5	109.5	104.3	99.6	110.1	104.6	99.5	113.7	106.2	99.5	113.7	106.2	99.5	109.5	104.3
08	DOORS & WINDOWS	110.0	111.0	110.3	110.0	111.0	110.3	110.0	115.7	111.4	110.0	111.7	110.4	110.0	111.7	110.4	110.0	115.7	111.4
09200	PLASTER & GYPSUM BOARD	97.8	105.0	102.2	97.8	105.0	102.2	97.8	105.0	102.2	97.8	97.4	97.5	97.8	97.4	97.5	97.8	105.0	102.2
095,098	CEILINGS & ACOUSTICAL TREATMENT	102.6	105.0	104.2	102.6	105.0	104.2	102.6	105.0	104.2	102.6	97.4	99.1	102.6	97.4	99.1	102.6	105.0	104.2
09600	FLOORING	99.9	109.3	102.2	99.9	109.3	102.2	99.9	109.3	102.2	99.9	109.3	102.2	99.9	109.3	102.2	99.9	109.3	102.2
097,099	WALL FINISHES, PAINTS & COATINGS	95.6	110.7	104.4	95.6	110.7	104.4	95.6	104.7	100.8	95.6	111.5	104.8	95.6	111.5	104.8	95.6	109.6	103.7
09	FINISHES	101.2	108.1	104.7	101.2	108.1	104.8	101.3	107.5	104.4	101.2	103.9	102.6	101.3	103.9	102.6	101.1	108.0	104.6
10 - 14	TOTAL DIV. 10000 - 14000	100.0	113.9	102.9	100.0	113.9	102.9	100.0	113.9	102.9	100.0	112.9	102.7	100.0	113.1	102.8	100.0	113.9	102.9
15	MECHANICAL	99.8	104.5	102.0	99.8	104.5	102.0	99.8	104.6	102.0	99.8	104.6	102.0	99.8	104.6	102.0	99.8	104.6	102.0
16	ELECTRICAL	103.2	105.3	104.7	104.0	105.0	104.7	104.0	105.0	104.7	104.0	101.4	102.2	104.0	133.9	124.5	103.2	103.9	103.7
01 - 16	WEIGHTED AVERAGE	102.2	108.5	105.3	101.9	108.5	105.1	102.1	109.0	105.5	102.6	107.7	105.1	102.7	113.2	107.8	102.6	108.5	105.5

	DIVISION	D.C.			DELAWARE			FLORIDA											
		WASHINGTON			WILMINGTON			DAYTONA BEACH			FORT LAUDERDALE			JACKSONVILLE			MELBOURNE		
		MAT.	INST.	TOTAL	MAT.	INST.	TOTAL	MAT.	INST.	TOTAL	MAT.	INST.	TOTAL	MAT.	INST.	TOTAL	MAT.	INST.	TOTAL
02	SITE CONSTRUCTION	103.0	90.4	93.4	87.0	113.7	107.5	120.4	87.1	94.9	105.6	73.6	81.1	120.5	87.5	95.2	128.5	87.3	96.9
03100	CONCRETE FORMS & ACCESSORIES	97.0	79.3	81.7	100.4	102.0	101.8	96.6	70.6	74.2	93.9	65.1	69.0	96.3	59.2	64.3	90.3	72.2	74.7
03200	CONCRETE REINFORCEMENT	99.2	93.3	95.8	98.5	96.2	97.2	95.8	89.0	91.9	95.8	80.6	87.1	95.8	62.4	76.8	96.8	89.1	92.4
03300	CAST-IN-PLACE CONCRETE	112.6	85.5	101.4	76.8	91.7	82.9	91.0	74.0	84.0	95.5	69.7	84.8	91.9	58.6	78.2	105.7	75.9	93.4
03	CONCRETE	105.1	85.6	95.4	97.5	98.5	98.0	88.9	76.4	82.7	90.9	71.0	81.0	89.3	61.3	75.3	98.3	77.8	88.1
04	MASONRY	90.2	82.0	85.1	100.7	88.0	92.9	86.7	70.2	76.6	86.9	63.4	72.5	86.1	56.7	68.0	83.0	73.6	77.2
05	METALS	96.9	110.2	101.7	97.8	116.3	104.5	99.1	95.5	97.8	98.5	93.3	96.6	98.7	85.1	93.8	107.1	95.8	103.0
06	WOOD & PLASTICS	94.8	79.8	87.0	104.3	104.1	104.2	95.1	71.9	83.1	90.1	66.8	78.1	95.1	59.8	76.8	88.6	71.9	80.0
07	THERMAL & MOISTURE PROTECTION	94.1	83.6	89.1	99.5	103.7	101.5	96.5	73.6	85.7	96.5	68.6	83.3	96.8	63.9	81.2	96.8	76.0	86.9
08	DOORS & WINDOWS	101.9	85.9	98.0	97.4	104.4	99.1	100.7	69.5	93.2	98.3	65.1	90.3	100.7	56.3	90.0	99.9	72.9	93.4
09200	PLASTER & GYPSUM BOARD	102.3	79.0	88.1	105.9	104.1	104.8	107.2	71.5	85.4	106.7	66.3	82.0	107.2	59.1	77.8	104.2	71.5	84.2
095,098	CEILINGS & ACOUSTICAL TREATMENT	103.8	79.0	87.5	101.1	104.1	103.1	95.1	71.5	79.5	95.1	66.3	76.1	95.1	59.1	71.3	91.0	71.5	78.2
09600	FLOORING	96.3	88.3	94.4	80.3	93.1	83.4	115.9	74.2	105.8	115.9	70.8	105.0	115.9	58.1	101.9	112.6	74.2	103.3
097,099	WALL FINISHES, PAINTS & COATINGS	105.7	92.4	97.9	87.1	99.7	94.4	109.2	87.5	96.6	105.6	60.6	79.4	109.2	58.4	79.6	109.2	102.4	105.2
09	FINISHES	97.0	81.9	89.3	97.6	100.1	98.9	106.9	73.1	89.7	104.9	65.2	84.7	106.9	58.6	82.3	105.1	75.7	90.1
10 - 14	TOTAL DIV. 10000 - 14000	100.0	94.1	98.8	100.0	109.6	102.0	100.0	82.2	96.2	100.0	82.3	96.3	100.0	73.8	94.5	100.0	83.6	96.5
15	MECHANICAL	100.1	89.2	95.0	100.2	106.9	103.3	99.9	70.5	86.3	99.9	68.5	85.4	99.9	59.3	81.1	99.9	72.5	87.2
16	ELECTRICAL	97.4	96.4	96.7	99.8	96.2	97.3	96.7	65.6	75.3	96.7	83.6	87.7	96.2	70.3	78.4	96.4	72.8	80.2
01 - 16	WEIGHTED AVERAGE	99.2	90.1	94.8	98.5	102.8	100.6	98.7	75.0	87.2	97.9	73.5	86.0	98.7	66.6	83.1	100.8	77.7	89.5

City Cost Indexes

DIVISION		FLORIDA																	
		MIAMI			ORLANDO			PANAMA CITY			PENSACOLA			ST. PETERSBURG			TALLAHASSEE		
		MAT.	INST.	TOTAL	MAT.	INST.	TOTAL	MAT.	INST.	TOTAL	MAT.	INST.	TOTAL	MAT.	INST.	TOTAL	MAT.	INST.	TOTAL
02	SITE CONSTRUCTION	105.0	73.5	80.9	121.1	86.8	94.8	135.6	84.6	96.6	133.2	86.8	97.7	121.4	86.6	94.7	121.9	86.2	94.6
03100	CONCRETE FORMS & ACCESSORIES	93.7	65.1	69.1	96.4	61.9	66.6	95.2	32.6	41.2	84.7	59.8	63.2	93.7	55.6	60.8	96.3	45.7	52.6
03200	CONCRETE REINFORCEMENT	95.8	80.6	87.2	95.8	84.5	89.4	100.0	61.2	77.9	102.5	61.5	79.1	99.2	70.4	82.8	95.8	61.8	76.4
03300	CAST-IN-PLACE CONCRETE	93.0	70.1	83.5	98.9	73.8	88.6	96.6	39.7	73.1	96.6	62.5	82.5	102.9	63.5	86.7	95.2	55.1	78.7
03	CONCRETE	89.7	71.2	80.4	92.8	71.7	82.3	97.5	42.3	69.9	96.3	62.7	79.5	95.1	62.8	78.9	91.0	54.0	72.5
04	MASONRY	84.4	63.9	71.8	87.3	70.2	76.8	90.9	33.0	55.3	88.1	59.5	70.5	132.4	58.9	87.2	87.6	46.2	62.2
05	METALS	99.0	93.1	96.9	108.4	93.7	103.1	97.4	71.3	88.0	97.3	84.7	92.8	101.1	87.3	96.1	99.0	83.5	93.4
06	WOOD & PLASTICS	90.1	66.8	78.1	95.1	60.6	77.3	93.8	32.8	62.2	82.8	60.6	71.3	92.1	55.6	73.3	95.1	44.1	68.7
07	THERMAL & MOISTURE PROTECTION	99.5	70.1	85.6	96.9	72.5	85.3	97.1	36.1	68.2	96.8	60.5	79.6	96.5	57.0	77.7	97.0	53.4	76.3
08	DOORS & WINDOWS	98.3	65.7	90.4	100.7	61.5	91.2	98.4	30.7	82.0	98.3	57.7	88.5	99.4	53.4	88.3	100.7	47.0	87.7
09200	PLASTER & GYPSUM BOARD	106.7	66.3	82.0	110.0	59.9	79.4	105.7	31.2	60.2	101.8	59.9	76.2	105.3	54.8	74.4	107.2	42.8	67.8
095,098	CEILINGS & ACOUSTICAL TREATMENT	95.1	66.3	76.1	95.1	59.9	71.9	89.7	31.2	51.1	89.7	59.9	70.0	89.7	54.8	66.6	95.1	42.8	60.6
09600	FLOORING	123.5	71.7	110.9	115.9	74.2	105.8	115.3	22.3	92.8	109.5	61.8	97.9	114.3	61.6	101.5	115.9	45.2	98.8
097,099	WALL FINISHES, PAINTS & COATINGS	105.6	60.6	79.4	109.2	67.0	84.7	109.2	29.8	63.1	109.2	67.8	85.1	109.2	56.5	78.5	109.2	48.1	73.6
09	FINISHES	107.3	65.4	86.0	107.3	64.2	85.3	106.3	29.9	67.5	103.9	61.0	82.0	105.1	56.6	80.3	107.0	45.0	75.4
10 - 14	TOTAL DIV. 10000 - 14000	100.0	82.2	96.2	100.0	80.6	95.9	100.0	55.7	90.7	100.0	62.3	92.0	100.0	65.2	92.7	100.0	72.9	94.3
15	MECHANICAL	99.9	72.2	87.1	99.9	65.2	83.9	99.9	30.3	67.7	99.9	60.3	81.6	99.9	60.1	81.5	99.9	48.0	75.9
16	ELECTRICAL	97.2	78.5	84.4	97.3	53.1	66.9	94.9	39.7	57.0	100.4	64.3	75.6	97.1	57.4	69.8	97.3	49.2	64.2
01 - 16	WEIGHTED AVERAGE	98.1	73.4	86.1	100.8	69.5	85.6	99.8	43.6	72.5	99.4	66.1	83.2	102.1	64.3	83.7	99.2	56.2	78.2

DIVISION		FLORIDA			GEORGIA														
		TAMPA			ALBANY			ATLANTA			AUGUSTA			COLUMBUS			MACON		
		MAT.	INST.	TOTAL	MAT.	INST.	TOTAL	MAT.	INST.	TOTAL	MAT.	INST.	TOTAL	MAT.	INST.	TOTAL	MAT.	INST.	TOTAL
02	SITE CONSTRUCTION	121.7	86.6	94.8	105.9	76.4	83.3	110.3	95.2	98.8	106.2	92.8	96.0	105.8	76.5	83.4	106.6	94.4	97.3
03100	CONCRETE FORMS & ACCESSORIES	97.9	55.8	61.6	95.9	49.9	56.2	97.7	79.8	82.2	94.4	60.2	64.9	97.5	63.8	68.4	94.9	64.5	68.6
03200	CONCRETE REINFORCEMENT	95.8	70.4	81.4	95.8	92.2	93.7	102.9	93.8	97.7	108.7	85.2	95.3	95.8	92.4	93.9	98.2	92.4	94.9
03300	CAST-IN-PLACE CONCRETE	100.7	63.6	85.4	97.2	47.4	76.7	99.0	75.2	89.2	93.6	55.8	78.0	96.8	48.0	76.7	95.5	51.7	77.5
03	CONCRETE	93.8	62.9	78.4	91.9	58.7	75.3	94.6	80.6	87.6	90.9	63.7	77.3	91.8	65.0	78.4	91.3	66.6	79.0
04	MASONRY	87.7	58.9	70.0	89.0	37.6	57.4	90.9	72.1	79.3	91.2	47.4	64.3	88.9	41.4	59.7	103.7	45.2	67.7
05	METALS	102.0	87.5	96.8	96.4	92.5	95.0	92.3	81.2	88.3	91.1	74.8	85.2	96.8	93.1	95.5	91.7	93.5	92.4
06	WOOD & PLASTICS	96.7	55.6	75.5	94.4	50.1	71.5	98.6	82.4	90.3	95.4	62.8	78.5	96.2	69.0	82.1	100.7	67.7	83.6
07	THERMAL & MOISTURE PROTECTION	96.8	58.0	78.4	96.6	56.2	77.5	92.4	77.0	85.1	91.8	57.2	75.4	96.3	59.1	78.7	95.2	63.3	80.1
08	DOORS & WINDOWS	100.7	53.7	89.3	98.3	55.5	88.0	98.3	78.9	93.6	94.5	61.2	86.5	98.3	66.0	90.5	96.7	66.3	89.4
09200	PLASTER & GYPSUM BOARD	107.2	54.8	75.1	107.2	49.0	71.6	110.1	82.3	93.1	109.2	62.0	80.3	107.2	68.5	83.6	114.6	67.2	85.7
095,098	CEILINGS & ACOUSTICAL TREATMENT	95.1	54.8	68.5	95.1	49.0	64.7	100.9	82.3	88.6	102.4	62.0	75.7	95.1	68.5	77.6	87.5	67.2	74.1
09600	FLOORING	115.9	61.6	102.7	115.9	37.7	96.9	87.2	72.3	83.6	86.0	48.0	76.8	115.9	43.9	98.5	90.5	44.3	79.3
097,099	WALL FINISHES, PAINTS & COATINGS	109.2	56.5	78.5	105.6	48.1	72.2	94.1	78.2	84.9	94.1	45.7	66.0	105.6	46.1	71.0	107.2	56.4	77.7
09	FINISHES	106.9	56.6	81.3	105.0	46.3	75.0	93.5	78.3	85.7	93.0	56.5	74.4	104.9	58.4	81.1	91.6	59.7	75.3
10 - 14	TOTAL DIV. 10000 - 14000	100.0	75.9	94.9	100.0	73.1	94.3	100.0	82.0	96.2	100.0	69.8	93.6	100.0	75.4	94.8	100.0	76.9	95.1
15	MECHANICAL	99.9	60.3	81.6	99.9	53.4	78.4	100.1	80.3	91.0	100.1	50.6	77.2	99.9	43.7	73.9	99.9	49.2	76.4
16	ELECTRICAL	96.2	57.4	69.5	90.7	66.0	73.8	94.0	84.0	87.1	96.1	54.1	67.2	92.0	46.7	60.8	90.1	66.2	73.7
01 - 16	WEIGHTED AVERAGE	99.9	64.7	82.8	97.5	60.3	79.4	96.4	81.4	89.1	95.2	60.9	78.5	97.6	58.7	78.7	95.9	65.7	81.2

DIVISION		GEORGIA			HAWAII			IDAHO										
		SAVANNAH			VALDOSTA			HONOLULU			BOISE			LEWISTON			POCATELLO		
		MAT.	INST.	TOTAL	MAT.	INST.	TOTAL	MAT.	INST.	TOTAL	MAT.	INST.	TOTAL	MAT.	INST.	TOTAL	MAT.	INST.	TOTAL
02	SITE CONSTRUCTION	106.4	77.6	84.3	116.6	76.8	86.1	127.9	108.2	112.8	82.7	101.5	97.1	88.0	95.2	93.5	84.3	101.5	97.5
03100	CONCRETE FORMS & ACCESSORIES	95.7	60.1	65.0	81.4	51.8	55.9	106.9	150.1	144.2	101.5	87.3	89.3	112.9	76.6	81.6	101.5	87.0	89.0
03200	CONCRETE REINFORCEMENT	101.5	85.5	92.4	101.7	51.0	72.8	106.4	126.7	118.0	100.3	81.6	89.6	113.3	99.0	105.2	100.6	81.1	89.5
03300	CAST-IN-PLACE CONCRETE	93.8	56.4	78.4	95.1	57.1	79.5	190.4	128.6	165.0	101.2	95.3	98.8	109.7	93.0	102.8	100.4	95.1	98.2
03	CONCRETE	90.9	65.0	78.0	95.9	54.4	75.6	155.7	136.7	146.2	104.2	88.8	96.5	116.1	86.6	101.4	103.9	88.6	96.2
04	MASONRY	92.3	59.0	71.8	95.3	53.3	69.5	132.9	133.4	133.2	138.3	83.1	104.4	138.7	91.3	109.5	133.1	74.9	97.4
05	METALS	97.1	90.9	94.9	96.6	79.3	90.4	116.4	110.3	114.2	113.2	81.3	101.7	96.7	89.9	94.3	113.5	80.4	101.6
06	WOOD & PLASTICS	94.5	59.1	76.2	79.3	48.8	63.5	102.2	154.3	129.2	96.3	86.6	91.3	102.0	71.0	86.0	96.3	86.6	91.3
07	THERMAL & MOISTURE PROTECTION	96.7	59.0	78.9	96.4	62.2	80.2	113.7	132.2	122.5	98.1	84.2	91.5	168.2	84.1	128.3	98.3	77.7	88.6
08	DOORS & WINDOWS	98.3	58.5	88.7	93.8	45.2	82.1	109.0	143.5	117.4	95.5	81.2	92.0	115.8	75.3	106.0	95.5	75.6	90.7
09200	PLASTER & GYPSUM BOARD	107.2	58.4	77.3	101.0	47.7	68.4	111.9	156.0	138.9	88.2	86.0	86.9	152.1	70.1	102.0	88.2	86.0	86.9
095,098	CEILINGS & ACOUSTICAL TREATMENT	95.1	58.4	70.9	89.7	47.7	62.0	118.1	156.0	143.1	104.7	86.0	92.3	151.2	70.1	97.7	104.7	86.0	92.3
09600	FLOORING	115.9	56.6	101.5	107.7	45.2	92.5	119.4	131.9	122.5	103.8	65.3	94.5	143.7	95.9	132.1	104.1	65.3	94.7
097,099	WALL FINISHES, PAINTS & COATINGS	105.6	57.2	77.4	105.6	41.7	68.5	113.9	150.0	134.9	108.7	60.3	80.6	136.5	75.5	101.0	108.7	63.2	82.2
09	FINISHES	105.1	59.0	81.6	101.0	48.9	74.4	118.6	148.1	133.6	98.6	80.6	89.4	167.2	79.1	122.3	98.7	80.9	89.6
10 - 14	TOTAL DIV. 10000 - 14000	100.0	71.5	94.0	100.0	70.1	93.7	100.0	129.1	106.2	100.0	84.1	96.7	100.0	92.9	98.5	100.0	84.1	96.7
15	MECHANICAL	99.9	54.2	78.8	99.9	47.5	75.7	100.3	123.6	111.1	100.1	84.8	93.0	101.5	90.7	96.5	100.1	84.7	93.0
16	ELECTRICAL	93.4	62.7	72.3	89.0	36.0	52.5	114.3	129.9	125.1	81.4	80.3	80.6	83.6	89.6	87.7	84.5	79.9	81.3
01 - 16	WEIGHTED AVERAGE	97.9	64.5	81.7	97.4	54.0	76.3	116.6	129.0	122.7	102.1	85.1	93.8	113.7	88.1	101.2	102.1	83.7	93.2

COST INDEXES

583

City Cost Indexes

DIVISION		IDAHO TWIN FALLS			ILLINOIS CHICAGO			ILLINOIS DECATUR			ILLINOIS EAST ST. LOUIS			ILLINOIS JOLIET			ILLINOIS PEORIA		
		MAT.	INST.	TOTAL	MAT.	INST.	TOTAL	MAT.	INST.	TOTAL	MAT.	INST.	TOTAL	MAT.	INST.	TOTAL	MAT.	INST.	TOTAL
02	SITE CONSTRUCTION	91.6	99.6	97.7	86.0	91.0	89.8	82.2	95.3	92.2	100.1	95.8	96.8	86.0	90.1	89.1	90.2	94.4	93.4
03100	CONCRETE FORMS & ACCESSORIES	104.9	41.2	49.9	105.9	130.5	127.1	100.3	104.5	103.9	92.6	109.5	107.2	106.9	130.7	127.4	98.4	108.2	106.9
03200	CONCRETE REINFORCEMENT	109.3	57.0	79.5	94.9	158.7	131.3	96.5	110.4	104.4	102.1	118.2	111.2	94.9	144.0	122.9	96.5	114.5	106.7
03300	CAST-IN-PLACE CONCRETE	103.0	51.0	81.6	103.3	130.2	114.4	96.2	93.2	95.0	90.1	112.3	99.2	103.2	114.9	108.0	94.1	105.9	98.9
03	CONCRETE	113.2	48.7	80.9	100.6	134.6	117.6	93.8	101.9	97.8	85.2	112.9	99.1	100.6	126.5	113.5	92.5	108.7	100.6
04	MASONRY	136.7	46.0	80.9	93.9	131.5	117.0	69.0	101.3	88.9	73.1	110.4	96.0	95.5	126.4	114.5	113.0	109.7	111.0
05	METALS	113.4	68.4	97.2	96.4	123.7	106.3	95.5	108.1	100.0	93.0	123.9	104.1	94.5	115.2	102.0	95.5	112.0	101.4
06	WOOD & PLASTICS	99.7	41.2	69.5	103.3	128.9	116.5	102.6	103.5	103.1	96.8	108.9	103.1	105.3	131.3	118.7	102.6	106.0	104.4
07	THERMAL & MOISTURE PROTECTION	99.4	53.6	77.7	99.3	128.7	113.3	97.0	97.5	97.2	92.0	106.0	98.6	99.1	125.6	111.7	96.8	105.2	100.8
08	DOORS & WINDOWS	98.8	42.0	85.1	104.1	136.4	111.9	99.0	102.4	99.8	88.2	116.1	95.0	103.9	134.2	111.2	99.0	106.4	100.8
09200	PLASTER & GYPSUM BOARD	87.8	39.1	58.1	99.4	129.4	117.7	106.2	103.3	104.4	103.5	108.8	106.7	97.7	132.0	118.7	106.2	105.8	106.0
095,098	CEILINGS & ACOUSTICAL TREATMENT	97.9	39.1	59.1	99.2	129.4	119.1	99.2	103.3	101.9	93.8	108.8	103.7	99.2	132.0	120.8	99.2	105.8	103.6
09600	FLOORING	105.4	65.3	95.7	85.1	126.3	95.1	101.3	96.6	100.2	111.0	61.5	99.0	84.7	122.0	93.7	101.3	102.0	101.5
097,099	WALL FINISHES, PAINTS & COATINGS	108.7	37.2	67.1	80.7	134.4	111.9	91.0	102.0	97.4	99.2	101.7	100.6	78.4	123.0	104.3	91.0	96.2	94.0
09	FINISHES	98.4	44.9	71.1	89.4	129.9	110.1	97.9	103.3	100.6	97.8	99.4	98.6	88.9	128.7	109.2	97.8	105.6	101.8
10-14	TOTAL DIV. 10000 - 14000	100.0	57.7	91.1	100.0	123.7	105.0	100.0	101.0	100.2	100.0	105.5	101.2	100.0	122.7	104.8	100.0	102.8	100.6
15	MECHANICAL	100.1	43.0	73.7	100.0	124.5	111.3	100.2	99.6	99.9	100.1	99.8	99.9	100.2	119.7	109.2	100.2	105.1	102.4
16	ELECTRICAL	87.0	37.5	53.0	101.1	130.7	121.4	100.6	83.6	88.9	96.9	102.1	100.5	100.7	91.9	94.7	99.5	93.1	95.1
01-16	WEIGHTED AVERAGE	104.2	51.9	78.8	98.2	125.4	111.4	95.9	98.3	97.1	93.5	105.7	99.4	98.0	115.2	106.4	98.4	103.6	100.9

DIVISION		ILLINOIS ROCKFORD			ILLINOIS SPRINGFIELD			INDIANA ANDERSON			INDIANA BLOOMINGTON			INDIANA EVANSVILLE			INDIANA FORT WAYNE		
		MAT.	INST.	TOTAL	MAT.	INST.	TOTAL	MAT.	INST.	TOTAL	MAT.	INST.	TOTAL	MAT.	INST.	TOTAL	MAT.	INST.	TOTAL
02	SITE CONSTRUCTION	90.4	94.7	93.7	88.7	95.3	93.8	88.0	94.7	93.1	78.5	95.3	91.4	84.0	130.5	119.6	88.8	94.9	93.5
03100	CONCRETE FORMS & ACCESSORIES	103.2	112.5	111.3	102.6	104.6	104.3	91.5	82.2	83.5	103.1	83.4	86.1	91.0	84.1	85.0	91.1	77.1	79.0
03200	CONCRETE REINFORCEMENT	96.5	141.5	122.2	96.5	110.4	104.4	96.8	86.9	91.1	88.8	86.7	87.6	97.3	85.8	90.7	96.8	83.2	89.0
03300	CAST-IN-PLACE CONCRETE	96.2	104.5	99.6	89.4	100.7	94.1	101.7	83.3	94.1	102.7	81.8	94.1	98.1	93.5	96.2	108.0	81.4	97.0
03	CONCRETE	94.0	115.2	104.6	90.6	104.6	97.6	96.7	84.0	90.3	104.2	83.0	93.6	107.0	87.8	97.4	99.7	80.4	90.1
04	MASONRY	83.9	114.5	102.7	67.7	104.8	90.5	92.4	83.6	87.0	92.2	81.9	85.9	88.3	85.8	86.8	89.7	83.2	85.7
05	METALS	95.5	123.7	105.7	95.5	108.9	100.3	102.1	92.7	98.7	97.4	77.4	90.2	90.7	89.6	90.3	102.1	91.0	98.1
06	WOOD & PLASTICS	102.6	110.1	106.4	102.4	103.5	103.0	101.9	82.0	91.6	116.9	82.7	99.2	92.7	82.5	87.4	101.5	75.2	87.9
07	THERMAL & MOISTURE PROTECTION	96.7	112.8	104.3	96.6	103.5	99.9	100.0	82.3	91.6	92.1	86.7	89.5	95.3	88.5	92.1	99.3	86.4	93.2
08	DOORS & WINDOWS	99.0	116.5	103.2	99.0	102.4	99.8	101.9	84.8	97.8	105.5	83.4	100.2	98.0	83.0	94.4	101.9	77.9	96.1
09200	PLASTER & GYPSUM BOARD	106.2	110.0	108.5	106.2	103.3	104.4	106.5	82.0	91.5	105.0	82.9	91.5	101.1	80.8	88.7	102.2	74.9	85.5
095,098	CEILINGS & ACOUSTICAL TREATMENT	99.2	110.0	106.3	99.2	103.3	101.9	93.8	82.0	86.0	83.7	82.9	83.2	95.1	80.8	85.7	93.8	74.9	81.3
09600	FLOORING	101.3	103.0	101.7	101.5	96.8	100.4	87.0	77.4	84.7	102.2	93.7	100.2	95.7	86.6	93.5	87.0	82.6	85.9
097,099	WALL FINISHES, PAINTS & COATINGS	91.0	111.8	103.1	91.0	101.8	97.3	90.9	77.5	83.1	90.2	90.3	90.3	96.5	89.0	92.2	90.9	77.5	83.1
09	FINISHES	97.8	110.2	104.1	97.9	103.0	100.5	91.7	81.0	86.3	96.9	86.1	91.4	96.2	84.9	90.4	90.9	77.7	84.1
10-14	TOTAL DIV. 10000 - 14000	100.0	112.0	102.5	100.0	100.6	100.1	100.0	89.2	97.7	100.0	89.3	97.7	100.0	92.3	98.4	100.0	86.5	97.2
15	MECHANICAL	100.2	106.8	103.2	100.2	99.4	99.8	99.8	85.1	93.0	99.8	86.8	93.8	100.0	87.7	94.3	99.8	84.0	92.5
16	ELECTRICAL	99.5	113.1	108.8	100.5	95.1	96.8	81.3	92.8	89.2	101.0	88.4	92.3	95.7	89.6	91.5	81.8	86.7	85.2
01-16	WEIGHTED AVERAGE	97.0	111.2	103.9	95.6	100.6	98.1	97.1	87.2	92.3	99.3	85.9	92.8	97.1	91.8	94.6	97.3	84.7	91.1

DIVISION		INDIANA GARY			INDIANA INDIANAPOLIS			INDIANA MUNCIE			INDIANA SOUTH BEND			INDIANA TERRE HAUTE			IOWA CEDAR RAPIDS		
		MAT.	INST.	TOTAL	MAT.	INST.	TOTAL	MAT.	INST.	TOTAL	MAT.	INST.	TOTAL	MAT.	INST.	TOTAL	MAT.	INST.	TOTAL
02	SITE CONSTRUCTION	88.6	96.8	94.9	88.0	98.7	96.2	78.3	95.5	91.5	88.1	95.1	93.4	85.1	130.7	120.0	84.8	94.3	92.1
03100	CONCRETE FORMS & ACCESSORIES	92.0	108.3	106.1	92.0	87.4	88.0	88.5	82.0	82.9	96.8	78.4	80.9	92.6	84.9	86.0	100.9	78.3	81.4
03200	CONCRETE REINFORCEMENT	96.8	102.8	100.2	96.1	86.6	90.7	98.2	86.6	91.7	96.8	79.7	87.0	97.3	84.2	89.8	100.1	82.2	89.9
03300	CAST-IN-PLACE CONCRETE	106.2	107.5	106.7	97.8	87.5	93.6	107.9	82.7	97.5	98.9	86.6	93.8	94.9	88.3	92.2	105.3	83.4	96.3
03	CONCRETE	98.9	107.0	102.9	99.1	86.8	92.9	102.2	83.8	93.0	91.2	83.2	87.2	106.8	86.1	96.4	97.7	81.2	89.5
04	MASONRY	93.8	101.9	98.8	99.6	85.1	90.7	94.3	83.6	87.7	87.1	83.3	84.7	95.2	83.8	88.2	107.3	79.9	90.4
05	METALS	102.1	104.1	102.8	104.1	80.3	95.5	99.0	92.8	96.8	102.1	105.7	103.4	91.4	89.2	90.6	89.6	86.4	88.4
06	WOOD & PLASTICS	100.9	109.8	105.5	99.2	87.5	93.1	103.6	81.8	92.3	102.1	76.8	89.0	95.2	84.5	89.6	105.2	76.6	90.4
07	THERMAL & MOISTURE PROTECTION	99.2	104.0	101.4	98.7	88.1	93.7	93.6	82.6	88.4	98.9	86.9	93.2	95.4	86.6	91.3	98.2	81.0	90.0
08	DOORS & WINDOWS	101.9	105.7	102.8	108.1	87.7	103.1	100.1	84.7	96.4	94.9	78.7	91.0	98.6	85.4	95.4	103.0	79.7	97.4
09200	PLASTER & GYPSUM BOARD	103.0	110.7	107.7	103.0	87.4	93.4	100.0	82.0	89.0	106.1	76.6	88.0	101.1	82.9	89.9	100.4	76.4	85.7
095,098	CEILINGS & ACOUSTICAL TREATMENT	93.8	110.7	104.9	92.4	87.4	89.1	85.1	82.0	83.0	93.8	76.6	82.4	95.1	82.9	87.0	112.6	76.4	88.7
09600	FLOORING	87.0	108.4	92.2	86.8	97.7	89.4	95.1	77.4	90.8	87.0	78.7	85.0	95.7	96.4	95.9	126.4	67.1	112.0
097,099	WALL FINISHES, PAINTS & COATINGS	90.9	91.0	91.0	90.9	90.4	90.6	90.2	77.5	82.8	90.9	83.3	86.5	96.5	84.9	89.8	111.4	77.0	91.4
09	FINISHES	91.2	106.7	99.1	91.2	89.7	90.5	93.8	81.0	87.3	91.7	78.5	84.9	96.2	87.1	91.5	112.6	75.1	93.5
10-14	TOTAL DIV. 10000 - 14000	100.0	108.4	101.8	100.0	91.1	98.1	100.0	88.6	97.6	100.0	85.5	96.9	100.0	93.8	98.7	100.0	86.6	97.2
15	MECHANICAL	99.8	100.9	100.3	99.8	91.0	95.7	99.7	85.0	92.9	99.8	81.3	91.3	100.0	85.9	93.5	100.4	86.2	93.8
16	ELECTRICAL	88.3	105.6	100.2	99.0	92.7	94.7	89.4	87.0	87.8	94.2	87.6	89.7	92.9	88.4	89.8	96.2	83.5	87.4
01-16	WEIGHTED AVERAGE	97.9	103.7	100.7	100.0	89.6	94.9	97.5	86.3	92.1	96.2	86.2	91.3	97.6	91.2	94.5	99.6	83.5	91.8

City Cost Indexes

IOWA

DIVISION		COUNCIL BLUFFS			DAVENPORT			DES MOINES			DUBUQUE			SIOUX CITY			WATERLOO		
		MAT.	INST.	TOTAL	MAT.	INST.	TOTAL	MAT.	INST.	TOTAL	MAT.	INST.	TOTAL	MAT.	INST.	TOTAL	MAT.	INST.	TOTAL
02	SITE CONSTRUCTION	88.1	91.9	91.0	83.6	99.4	95.7	77.1	101.3	95.6	83.0	92.0	89.9	92.5	96.6	95.7	84.1	95.9	93.1
03100	CONCRETE FORMS & ACCESSORIES	76.5	64.7	66.3	100.5	93.6	94.5	102.6	80.7	83.7	78.5	69.5	70.8	100.9	70.2	74.4	101.5	56.2	62.4
03200	CONCRETE REINFORCEMENT	102.2	80.2	89.7	100.1	96.1	97.8	100.1	78.5	87.8	98.7	82.0	89.2	100.1	68.5	82.1	100.1	81.9	89.7
03300	CAST-IN-PLACE CONCRETE	109.6	72.8	94.4	101.5	94.2	98.5	105.7	85.0	97.2	103.1	68.6	88.9	104.4	62.3	87.1	105.3	53.6	84.0
03	CONCRETE	99.4	71.2	85.3	95.8	94.3	95.0	98.0	82.2	90.1	94.0	72.3	83.1	97.3	67.8	82.5	97.7	61.1	79.4
04	MASONRY	107.6	75.0	87.6	106.9	84.7	93.3	101.4	81.8	89.3	108.3	73.5	86.9	101.5	68.1	81.0	101.6	62.6	77.6
05	METALS	94.9	84.9	91.3	89.6	94.9	91.5	89.5	86.8	88.5	88.1	85.5	87.2	89.6	80.1	86.2	89.6	85.0	87.9
06	WOOD & PLASTICS	79.5	62.9	70.9	105.2	92.4	98.6	106.8	79.0	92.4	81.8	67.4	74.3	105.2	70.0	87.0	105.9	56.9	80.6
07	THERMAL & MOISTURE PROTECTION	97.3	68.9	83.8	97.5	88.2	93.1	98.3	78.1	88.7	97.6	71.2	85.1	97.5	64.0	81.6	97.1	58.5	78.8
08	DOORS & WINDOWS	102.0	66.6	93.5	103.0	92.0	100.3	103.0	78.8	97.2	102.0	74.8	95.5	103.0	67.8	94.5	98.3	65.8	90.5
09200	PLASTER & GYPSUM BOARD	92.7	62.2	74.1	100.4	92.4	95.5	97.3	78.6	85.8	93.1	66.9	77.1	100.4	69.3	81.4	100.4	55.8	73.1
095,098	CEILINGS & ACOUSTICAL TREATMENT	108.5	62.2	78.0	112.6	92.4	99.3	111.3	78.6	89.7	108.5	66.9	81.1	112.6	69.3	84.0	112.6	55.8	75.1
09600	FLOORING	101.0	54.8	89.8	112.1	86.4	105.8	111.8	50.0	96.9	115.6	46.7	98.9	112.6	70.7	102.4	113.7	71.0	103.4
097,099	WALL FINISHES, PAINTS & COATINGS	102.5	58.1	76.7	106.9	101.9	104.0	106.9	83.7	93.4	110.1	61.1	81.6	108.2	71.1	86.6	108.2	42.2	69.8
09	FINISHES	101.8	61.1	81.0	107.6	93.0	100.2	106.1	74.5	90.0	107.1	63.3	84.7	108.2	70.4	89.2	108.3	57.0	82.1
10 - 14	TOTAL DIV. 10000 - 14000	100.0	81.5	96.1	100.0	91.8	98.3	100.0	88.3	97.5	100.0	84.0	96.6	100.0	84.2	96.7	100.0	78.4	95.4
15	MECHANICAL	100.4	79.7	90.8	100.4	100.0	100.2	100.4	89.5	95.3	100.4	87.0	94.2	100.4	85.7	93.6	100.4	54.8	79.3
16	ELECTRICAL	104.5	84.9	91.0	93.6	88.4	90.0	96.9	90.1	92.2	102.2	81.9	88.3	96.2	77.9	83.6	96.2	64.3	74.2
01 - 16	WEIGHTED AVERAGE	99.8	77.5	88.9	98.6	93.6	96.1	98.4	86.2	92.5	98.4	79.2	89.1	99.0	77.6	88.6	98.2	66.2	82.7

KANSAS / KENTUCKY

DIVISION		DODGE CITY			KANSAS CITY			SALINA			TOPEKA			WICHITA			BOWLING GREEN		
		MAT.	INST.	TOTAL	MAT.	INST.	TOTAL	MAT.	INST.	TOTAL	MAT.	INST.	TOTAL	MAT.	INST.	TOTAL	MAT.	INST.	TOTAL
02	SITE CONSTRUCTION	105.8	93.2	96.1	86.1	90.4	89.4	95.6	93.5	93.9	88.4	90.8	90.2	89.7	93.7	92.8	70.8	100.9	93.9
03100	CONCRETE FORMS & ACCESSORIES	95.8	50.7	56.9	102.4	90.3	92.0	90.8	53.4	58.5	101.3	51.6	58.5	98.0	63.6	68.3	82.2	85.3	84.9
03200	CONCRETE REINFORCEMENT	104.4	69.7	84.7	99.3	92.9	95.6	103.8	86.9	94.2	96.5	92.6	94.3	96.5	87.8	91.5	88.1	73.8	79.9
03300	CAST-IN-PLACE CONCRETE	111.7	65.8	92.8	87.6	93.1	89.9	96.9	56.0	80.0	91.3	60.8	78.7	85.8	64.3	76.9	89.6	85.9	88.0
03	CONCRETE	110.1	61.1	85.6	92.0	92.4	92.2	97.5	62.2	79.8	91.4	64.1	77.8	88.4	69.7	79.1	95.3	83.4	89.4
04	MASONRY	103.1	50.3	70.7	103.1	88.7	94.3	117.7	48.4	75.1	96.7	60.0	74.5	91.3	65.9	75.6	95.2	86.2	89.7
05	METALS	92.8	86.3	90.4	97.6	100.5	98.6	92.6	93.4	92.9	95.5	97.4	96.2	95.5	95.2	95.4	94.8	80.8	89.8
06	WOOD & PLASTICS	94.2	50.8	71.7	102.4	91.4	96.7	89.4	54.6	71.4	98.8	48.3	72.7	96.3	62.8	79.0	88.5	82.8	85.6
07	THERMAL & MOISTURE PROTECTION	97.9	56.7	78.4	95.6	91.3	93.6	97.3	61.7	80.4	97.2	72.8	85.6	96.8	71.4	84.8	86.2	84.6	85.4
08	DOORS & WINDOWS	98.9	54.7	88.2	97.7	88.5	95.5	98.8	59.0	89.2	99.0	64.0	90.5	99.0	68.3	91.6	97.7	77.7	92.9
09200	PLASTER & GYPSUM BOARD	104.1	48.8	70.3	100.8	90.8	94.6	102.8	52.8	72.3	106.2	46.3	69.6	106.2	61.3	78.7	97.3	82.5	88.3
095,098	CEILINGS & ACOUSTICAL TREATMENT	92.4	48.8	63.7	99.2	90.8	93.7	92.4	52.8	66.3	99.2	46.3	64.3	99.2	61.3	74.2	91.0	82.5	85.4
09600	FLOORING	100.6	59.3	90.6	90.7	63.3	84.0	98.6	43.3	85.2	102.2	55.1	90.8	101.3	79.4	96.0	91.7	73.1	87.2
097,099	WALL FINISHES, PAINTS & COATINGS	91.0	58.6	72.2	99.2	98.5	98.8	91.0	46.2	64.9	91.0	68.4	77.9	91.0	68.7	78.0	96.5	78.8	86.2
09	FINISHES	97.0	52.6	74.4	96.2	86.1	91.0	95.6	50.0	72.3	98.1	52.2	74.7	97.9	66.6	82.0	93.7	82.2	87.8
10 - 14	TOTAL DIV. 10000 - 14000	100.0	60.1	91.6	100.0	90.7	98.0	100.0	72.7	94.2	100.0	79.5	95.7	100.0	76.8	95.1	100.0	74.5	94.6
15	MECHANICAL	100.2	56.8	80.1	100.0	97.1	98.7	100.2	42.9	73.7	100.2	72.0	87.2	100.2	73.8	88.0	100.2	87.5	94.3
16	ELECTRICAL	95.2	60.0	71.0	104.5	78.8	86.8	94.7	74.2	80.6	101.4	72.2	81.4	99.3	74.2	82.1	93.2	90.4	91.3
01 - 16	WEIGHTED AVERAGE	99.6	63.0	81.8	98.0	90.2	94.2	98.3	63.7	81.5	97.4	71.4	84.8	96.6	75.6	86.4	95.6	86.6	91.2

KENTUCKY / LOUISIANA

DIVISION		LEXINGTON			LOUISVILLE			OWENSBORO			ALEXANDRIA			BATON ROUGE			LAKE CHARLES		
		MAT.	INST.	TOTAL	MAT.	INST.	TOTAL	MAT.	INST.	TOTAL	MAT.	INST.	TOTAL	MAT.	INST.	TOTAL	MAT.	INST.	TOTAL
02	SITE CONSTRUCTION	78.3	102.9	97.1	67.9	100.4	92.8	83.2	130.4	119.3	104.2	85.4	89.8	114.7	87.0	93.4	116.4	86.7	93.7
03100	CONCRETE FORMS & ACCESSORIES	93.0	73.1	75.8	91.6	82.2	83.5	86.7	75.4	77.0	78.3	50.9	54.7	98.7	56.2	62.0	98.8	57.1	62.8
03200	CONCRETE REINFORCEMENT	97.3	98.1	97.7	97.3	99.1	98.3	88.2	99.4	94.6	97.0	68.2	80.6	98.1	59.2	75.9	98.1	59.0	75.8
03300	CAST-IN-PLACE CONCRETE	96.7	76.0	88.2	93.6	80.6	88.2	92.4	87.8	90.5	97.9	48.9	77.7	88.6	58.8	76.3	93.8	60.5	80.1
03	CONCRETE	97.3	79.1	88.2	95.7	84.9	90.3	104.9	84.5	94.7	90.0	54.6	72.3	93.3	58.5	75.9	95.8	59.5	77.7
04	MASONRY	92.3	56.6	70.3	92.7	81.8	86.0	92.3	61.3	73.3	114.7	50.8	75.4	102.3	60.7	76.7	101.1	61.0	76.5
05	METALS	96.3	90.7	94.3	96.3	91.4	94.5	86.2	92.1	88.3	86.6	76.6	83.0	101.0	72.2	90.6	100.6	72.3	90.4
06	WOOD & PLASTICS	95.2	75.3	84.9	98.2	82.8	90.3	88.4	74.9	81.4	80.0	51.4	65.2	105.8	56.6	80.4	103.7	57.5	79.8
07	THERMAL & MOISTURE PROTECTION	95.8	68.9	83.0	86.4	83.0	84.8	95.3	76.1	86.2	98.3	55.5	78.0	99.3	62.3	81.7	101.9	62.3	83.1
08	DOORS & WINDOWS	98.6	82.9	94.8	98.6	87.3	95.9	95.9	83.3	92.9	100.2	56.9	89.7	104.6	56.7	93.1	104.6	59.0	93.6
09200	PLASTER & GYPSUM BOARD	101.1	73.4	84.1	103.9	82.5	90.8	96.8	73.0	82.3	92.2	50.8	66.9	105.1	56.1	75.1	105.1	56.9	75.7
095,098	CEILINGS & ACOUSTICAL TREATMENT	95.1	73.4	80.8	95.1	82.5	86.8	81.5	73.0	75.9	99.7	50.8	67.4	100.7	56.1	71.3	100.7	56.9	71.8
09600	FLOORING	96.6	50.8	85.5	95.1	77.8	90.9	94.0	73.1	88.9	107.4	58.2	95.5	110.7	70.2	100.9	110.9	61.3	98.9
097,099	WALL FINISHES, PAINTS & COATINGS	96.5	69.5	80.8	96.5	83.8	89.1	96.5	100.7	98.9	95.6	49.0	68.5	97.1	56.2	73.3	97.1	51.8	70.8
09	FINISHES	96.5	68.1	82.1	96.2	81.2	88.6	92.6	77.2	84.7	98.3	51.6	74.5	104.0	58.7	80.9	104.0	57.2	80.1
10 - 14	TOTAL DIV. 10000 - 14000	100.0	92.0	98.3	100.0	92.4	98.4	100.0	94.0	98.7	100.0	84.9	96.8	100.0	73.5	94.4	100.0	73.7	94.5
15	MECHANICAL	100.0	57.3	80.3	100.0	83.5	92.4	100.2	63.0	83.0	100.1	45.3	74.7	100.2	51.6	77.7	100.2	63.6	83.2
16	ELECTRICAL	94.0	60.5	70.9	94.0	90.4	91.5	92.4	89.0	90.0	95.5	55.1	67.7	96.4	60.1	71.4	96.4	63.1	73.5
01 - 16	WEIGHTED AVERAGE	96.9	71.9	84.7	96.1	87.2	91.8	95.5	83.1	89.5	96.9	58.1	78.0	100.7	62.1	81.9	101.0	65.0	83.5

City Cost Indexes

DIVISION		LOUISIANA									MAINE								
		MONROE			NEW ORLEANS			SHREVEPORT			AUGUSTA			BANGOR			LEWISTON		
		MAT.	INST.	TOTAL	MAT.	INST.	TOTAL	MAT.	INST.	TOTAL	MAT.	INST.	TOTAL	MAT.	INST.	TOTAL	MAT.	INST.	TOTAL
02	SITE CONSTRUCTION	104.2	85.2	89.7	118.5	89.1	96.0	102.8	85.1	89.2	87.2	100.9	97.7	87.1	103.2	99.5	85.3	103.2	99.0
03100	CONCRETE FORMS & ACCESSORIES	77.6	51.7	55.3	97.7	64.2	68.8	100.1	54.8	61.0	97.8	39.0	47.0	91.4	90.6	90.7	97.9	90.3	91.4
03200	CONCRETE REINFORCEMENT	96.0	68.3	80.2	98.1	69.6	81.8	95.5	68.2	80.0	91.1	40.9	62.5	91.1	117.9	106.4	111.9	117.9	115.3
03300	CAST-IN-PLACE CONCRETE	97.9	53.8	79.7	92.9	62.5	80.4	96.5	57.9	80.6	81.6	48.9	68.1	81.6	74.4	78.6	90.8	74.3	84.0
03	CONCRETE	89.8	56.6	73.2	95.3	65.3	80.3	90.0	59.3	74.7	100.5	44.5	72.5	99.5	89.3	94.4	102.8	89.2	96.0
04	MASONRY	109.6	52.6	74.6	103.2	61.4	77.5	102.6	47.4	68.7	93.6	43.6	62.7	111.1	63.7	82.0	94.7	63.7	75.6
05	METALS	86.6	76.1	82.8	103.3	78.1	94.2	87.4	75.6	83.2	95.1	75.1	87.9	94.9	86.9	92.0	98.1	86.6	94.0
06	WOOD & PLASTICS	79.1	51.9	65.0	99.5	65.2	81.8	102.6	56.5	78.7	99.8	37.3	67.5	93.1	93.2	93.1	99.8	93.2	96.4
07	THERMAL & MOISTURE PROTECTION	98.2	57.9	79.1	102.7	65.9	85.3	97.4	57.1	78.3	99.8	42.3	72.5	99.6	64.2	82.8	99.5	64.2	82.8
08	DOORS & WINDOWS	100.2	62.0	91.0	103.4	68.7	95.0	98.1	59.7	88.8	106.8	45.3	92.0	106.8	82.1	100.8	110.0	82.1	103.3
09200	PLASTER & GYPSUM BOARD	91.7	51.3	67.0	104.2	64.9	80.2	97.6	56.0	72.2	96.7	34.1	58.4	93.9	91.7	92.6	98.2	91.7	94.3
095,098	CEILINGS & ACOUSTICAL TREATMENT	99.7	51.3	67.8	99.2	64.9	76.6	101.1	56.0	71.4	94.4	34.1	54.6	93.1	91.7	92.2	102.6	91.7	95.4
09600	FLOORING	107.0	50.6	93.3	111.4	60.1	98.9	118.5	53.1	102.6	99.9	50.9	88.0	97.5	50.9	86.2	99.9	50.9	88.0
097,099	WALL FINISHES, PAINTS & COATINGS	95.6	50.6	69.4	99.0	65.9	79.8	95.6	49.0	68.5	95.6	28.2	56.4	95.6	39.0	62.7	95.6	39.0	62.7
09	FINISHES	98.1	50.7	73.9	104.2	63.4	83.4	102.9	53.5	77.7	98.4	39.1	68.2	97.0	78.1	87.3	99.8	78.1	88.7
10 - 14	TOTAL DIV. 10000 - 14000	100.0	70.9	93.9	100.0	75.7	94.9	100.0	71.2	93.9	100.0	64.5	92.5	100.0	87.7	97.4	100.0	87.5	97.4
15	MECHANICAL	100.1	54.6	79.0	100.2	65.4	84.1	100.1	58.2	80.7	99.8	30.8	67.9	99.8	85.0	93.0	99.8	85.0	93.0
16	ELECTRICAL	98.6	59.0	71.4	96.7	67.7	76.8	96.6	69.6	78.1	104.0	86.1	91.7	100.6	86.2	90.7	104.0	53.1	69.0
01 - 16	WEIGHTED AVERAGE	96.8	60.7	79.2	101.4	69.1	85.7	96.9	63.3	80.5	99.5	56.9	78.8	99.8	84.4	92.3	100.8	78.7	90.0

DIVISION		MAINE			MARYLAND						MASSACHUSETTS								
		PORTLAND			BALTIMORE			HAGERSTOWN			BOSTON			BROCKTON			FALL RIVER		
		MAT.	INST.	TOTAL	MAT.	INST.	TOTAL	MAT.	INST.	TOTAL	MAT.	INST.	TOTAL	MAT.	INST.	TOTAL	MAT.	INST.	TOTAL
02	SITE CONSTRUCTION	84.7	103.2	98.9	94.5	92.3	92.8	84.7	88.1	87.3	92.3	110.2	106.0	89.9	106.0	102.3	88.9	106.2	102.1
03100	CONCRETE FORMS & ACCESSORIES	97.1	90.3	91.3	101.1	76.2	79.6	88.4	77.9	79.3	103.9	131.3	127.6	103.5	114.5	113.0	103.5	110.9	109.9
03200	CONCRETE REINFORCEMENT	111.9	117.9	115.3	97.2	93.1	94.9	85.9	79.2	82.1	110.7	141.3	128.1	111.9	141.0	128.5	111.9	128.0	121.1
03300	CAST-IN-PLACE CONCRETE	84.5	74.3	80.3	90.0	80.7	91.5	82.4	66.4	76.9	105.2	142.0	120.4	100.2	135.7	114.8	96.8	138.9	114.2
03	CONCRETE	99.7	89.2	94.4	100.4	82.2	91.3	85.7	75.4	80.6	113.3	136.0	124.6	110.4	126.2	118.3	108.8	123.3	116.0
04	MASONRY	92.6	63.7	74.8	88.1	74.6	79.8	92.3	77.7	83.3	113.9	142.6	131.5	109.5	135.5	125.5	109.2	138.3	127.1
05	METALS	98.2	86.6	94.0	95.9	102.6	98.3	95.0	95.4	95.1	100.9	123.7	109.1	98.0	120.4	106.1	98.0	115.8	104.4
06	WOOD & PLASTICS	99.8	93.2	96.4	101.2	77.3	88.8	88.5	77.1	82.6	104.0	130.6	117.8	103.0	110.8	107.1	103.0	106.2	104.7
07	THERMAL & MOISTURE PROTECTION	99.4	64.2	82.7	96.9	80.7	89.2	96.1	77.0	87.1	100.1	138.1	118.1	99.9	130.9	114.6	99.8	125.1	111.8
08	DOORS & WINDOWS	110.0	82.1	103.3	92.3	86.7	91.0	91.6	78.0	88.3	103.7	131.2	110.3	103.6	120.5	107.7	103.6	112.9	105.9
09200	PLASTER & GYPSUM BOARD	98.2	91.7	94.3	100.9	77.2	86.4	97.5	77.2	85.1	98.2	130.4	117.9	98.2	109.9	105.4	98.2	104.8	102.2
095,098	CEILINGS & ACOUSTICAL TREATMENT	102.6	91.7	95.4	89.7	77.2	81.4	91.2	77.2	81.9	99.5	130.4	119.9	102.6	109.9	107.4	102.6	104.8	104.0
09600	FLOORING	99.9	50.9	88.0	91.6	81.7	89.2	86.7	78.3	84.7	99.7	152.1	112.4	100.2	152.1	112.8	99.9	152.1	112.6
097,099	WALL FINISHES, PAINTS & COATINGS	95.6	39.0	62.7	98.2	80.2	87.7	98.2	47.8	68.9	96.3	139.3	121.2	95.6	125.7	113.1	95.6	125.7	113.1
09	FINISHES	99.9	78.1	88.7	91.1	77.1	84.0	88.7	74.6	81.5	99.0	136.3	118.0	99.7	122.2	111.2	99.6	119.6	109.8
10 - 14	TOTAL DIV. 10000 - 14000	100.0	87.5	97.4	100.0	85.5	96.9	100.0	88.1	97.5	100.0	126.4	105.6	100.0	122.3	104.7	100.0	122.8	104.8
15	MECHANICAL	99.8	85.0	93.0	99.9	86.2	93.6	99.9	85.3	93.2	99.8	122.0	110.1	99.8	101.1	100.8	99.8	102.2	100.9
16	ELECTRICAL	104.2	53.1	69.1	101.9	94.5	96.8	98.0	85.6	89.5	99.1	125.1	116.9	98.7	101.8	100.8	98.1	101.8	100.6
01 - 16	WEIGHTED AVERAGE	100.3	78.7	89.8	96.8	86.7	91.9	94.0	82.7	88.5	102.6	128.0	115.0	101.5	115.0	108.1	101.2	113.6	107.3

DIVISION		MASSACHUSETTS																	
		HYANNIS			LAWRENCE			LOWELL			NEW BEDFORD			PITTSFIELD			SPRINGFIELD		
		MAT.	INST.	TOTAL	MAT.	INST.	TOTAL	MAT.	INST.	TOTAL	MAT.	INST.	TOTAL	MAT.	INST.	TOTAL	MAT.	INST.	TOTAL
02	SITE CONSTRUCTION	86.6	106.0	101.5	90.7	106.0	102.5	89.8	106.0	102.2	88.6	106.2	102.0	90.9	104.9	101.6	90.1	105.0	101.5
03100	CONCRETE FORMS & ACCESSORIES	93.2	110.4	108.0	103.4	114.5	113.0	99.7	114.5	112.5	103.5	110.9	109.9	99.7	95.1	95.7	99.8	103.0	102.5
03200	CONCRETE REINFORCEMENT	89.7	124.9	109.8	111.0	126.3	119.7	111.9	125.8	119.8	111.9	128.0	121.1	92.9	106.2	100.5	111.9	119.4	116.2
03300	CAST-IN-PLACE CONCRETE	91.3	138.0	110.6	100.4	138.2	116.0	91.8	138.2	110.9	94.0	138.9	112.5	100.1	105.2	102.2	94.0	110.9	101.0
03	CONCRETE	100.4	122.2	111.3	110.4	124.4	117.4	109.9	124.1	112.5	107.4	123.3	115.3	102.6	100.4	101.5	102.0	108.3	105.1
04	MASONRY	108.5	138.4	126.9	108.5	138.4	126.9	94.7	129.5	116.1	109.0	138.3	127.0	95.4	106.4	102.1	94.9	111.4	105.0
05	METALS	94.7	113.9	101.6	95.7	114.8	102.6	95.6	110.9	101.2	98.0	115.8	104.4	95.5	98.3	96.5	98.0	104.1	100.2
06	WOOD & PLASTICS	92.7	105.8	99.5	103.0	110.8	107.1	101.9	110.8	106.5	103.0	106.2	104.7	101.9	92.5	97.0	101.9	101.3	101.6
07	THERMAL & MOISTURE PROTECTION	99.3	125.0	111.5	99.8	132.0	115.1	99.6	128.3	113.2	99.8	125.1	111.8	99.6	104.9	102.1	99.6	108.0	103.6
08	DOORS & WINDOWS	99.7	111.9	102.6	103.6	116.6	106.7	110.0	116.6	111.6	103.6	112.9	105.9	110.0	97.5	107.0	110.0	105.7	109.0
09200	PLASTER & GYPSUM BOARD	93.0	104.8	100.2	98.2	109.9	105.4	98.2	109.9	105.4	98.2	104.8	102.2	98.2	91.0	93.8	98.2	100.1	99.4
095,098	CEILINGS & ACOUSTICAL TREATMENT	93.1	104.8	100.8	102.6	109.9	107.4	102.6	109.9	107.4	102.6	104.8	104.0	102.6	91.0	94.9	102.6	100.1	101.0
09600	FLOORING	96.3	152.1	109.8	99.9	152.1	112.6	99.9	152.1	112.6	99.9	152.1	112.6	100.2	112.2	103.1	99.7	117.3	104.0
097,099	WALL FINISHES, PAINTS & COATINGS	95.6	125.7	113.1	95.6	125.7	113.1	95.6	125.7	113.1	95.6	125.7	113.1	95.6	94.1	94.7	96.8	94.1	95.3
09	FINISHES	95.7	119.3	107.7	99.6	122.2	111.1	99.6	122.2	111.1	99.6	119.6	109.8	99.7	97.8	98.7	99.6	104.5	102.1
10 - 14	TOTAL DIV. 10000 - 14000	100.0	121.7	104.6	100.0	122.4	104.7	100.0	122.4	104.7	100.0	122.8	104.8	100.0	107.7	101.6	100.0	109.7	102.0
15	MECHANICAL	99.8	102.1	100.9	99.8	105.2	102.3	99.8	117.6	108.0	99.8	102.2	100.9	99.8	88.3	94.5	99.8	98.4	99.2
16	ELECTRICAL	93.9	101.8	99.3	103.2	105.9	105.0	104.0	105.8	105.3	99.1	101.8	100.9	104.0	84.0	90.2	104.0	91.6	95.4
01 - 16	WEIGHTED AVERAGE	98.2	113.2	105.5	101.4	115.7	108.4	100.3	116.7	108.2	101.1	113.6	107.2	100.5	95.9	98.3	100.8	102.6	101.6

City Cost Indexes

DIVISION		MASSACHUSETTS WORCESTER			MICHIGAN ANN ARBOR			MICHIGAN DEARBORN			MICHIGAN DETROIT			MICHIGAN FLINT			MICHIGAN GRAND RAPIDS		
		MAT.	INST.	TOTAL	MAT.	INST.	TOTAL	MAT.	INST.	TOTAL	MAT.	INST.	TOTAL	MAT.	INST.	TOTAL	MAT.	INST.	TOTAL
02	SITE CONSTRUCTION	89.8	105.9	102.2	81.2	93.9	90.9	80.9	94.0	91.0	98.2	95.8	96.4	69.6	93.2	87.7	83.8	88.0	87.0
03100	CONCRETE FORMS & ACCESSORIES	100.3	113.7	111.8	97.1	116.9	114.2	97.0	120.6	117.4	98.8	120.6	117.6	98.3	98.9	98.8	95.9	79.0	81.3
03200	CONCRETE REINFORCEMENT	111.9	137.0	126.2	93.8	119.2	108.3	93.8	119.6	108.5	93.2	119.5	108.2	93.8	118.7	108.0	97.3	91.5	94.0
03300	CAST-IN-PLACE CONCRETE	91.8	132.5	108.6	92.5	122.3	104.8	90.4	124.2	104.3	99.2	124.2	109.5	93.1	105.4	98.2	96.9	89.5	93.8
03	CONCRETE	100.9	123.8	112.4	89.9	119.7	104.8	88.8	122.0	105.4	93.3	120.4	106.8	90.3	105.8	98.0	97.6	84.5	91.0
04	MASONRY	94.7	132.3	117.8	99.2	115.8	109.4	99.0	123.1	113.8	97.9	123.1	113.4	99.2	103.7	102.0	96.7	58.9	73.5
05	METALS	98.0	114.7	104.0	100.2	126.4	109.6	100.3	127.2	110.0	100.8	103.1	101.6	100.3	123.7	108.7	97.4	84.9	92.9
06	WOOD & PLASTICS	102.5	110.8	106.8	93.5	117.1	105.7	93.5	120.3	107.4	95.4	120.3	108.3	93.5	97.8	95.7	96.6	79.6	87.8
07	THERMAL & MOISTURE PROTECTION	99.6	119.3	108.9	94.0	116.4	104.6	93.1	124.8	108.2	91.5	124.8	107.3	92.3	99.5	95.7	91.7	67.1	80.1
08	DOORS & WINDOWS	110.0	119.6	112.3	97.8	113.9	101.7	97.8	115.6	102.1	97.4	115.7	101.8	97.8	101.1	98.6	93.3	73.8	88.6
09200	PLASTER & GYPSUM BOARD	98.2	109.9	105.4	107.9	115.8	112.7	107.9	119.2	114.8	107.9	119.2	114.8	107.9	95.9	100.6	101.1	72.9	83.9
095,098	CEILINGS & ACOUSTICAL TREATMENT	102.6	109.9	107.4	101.1	115.8	110.8	101.1	119.2	113.0	102.6	119.2	113.5	101.1	95.9	97.7	95.1	72.9	80.5
09600	FLOORING	99.9	143.8	110.6	90.4	121.8	98.0	89.9	123.5	98.1	90.1	123.5	98.2	90.1	88.3	89.7	95.3	52.1	84.8
097,099	WALL FINISHES, PAINTS & COATINGS	95.6	116.9	108.0	93.2	117.6	107.4	93.2	119.6	108.5	95.1	119.6	109.3	93.2	94.2	93.8	96.5	51.8	70.5
09	FINISHES	99.6	119.5	109.7	97.3	117.7	107.7	97.2	121.2	109.4	98.3	121.2	110.0	96.7	96.0	96.3	96.1	70.6	83.1
10 - 14	TOTAL DIV. 10000 - 14000	100.0	113.2	102.8	100.0	112.4	102.6	100.0	113.7	102.9	100.0	113.7	102.9	100.0	106.4	101.3	100.0	109.8	102.1
15	MECHANICAL	99.8	100.7	100.2	100.0	104.9	102.3	100.0	115.6	107.2	100.0	115.9	107.4	100.0	103.1	101.4	100.0	63.7	83.2
16	ELECTRICAL	104.0	97.2	99.3	95.2	104.1	101.3	95.2	115.4	109.1	96.5	115.4	109.5	95.2	105.3	102.1	93.5	72.6	79.1
01 - 16	WEIGHTED AVERAGE	100.6	111.9	106.1	97.0	111.1	103.9	96.9	117.0	106.7	98.1	114.8	106.2	96.7	103.8	100.1	96.7	74.4	85.9

DIVISION		MICHIGAN KALAMAZOO			MICHIGAN LANSING			MICHIGAN MUSKEGON			MICHIGAN SAGINAW			MINNESOTA DULUTH			MINNESOTA MINNEAPOLIS		
		MAT.	INST.	TOTAL	MAT.	INST.	TOTAL	MAT.	INST.	TOTAL	MAT.	INST.	TOTAL	MAT.	INST.	TOTAL	MAT.	INST.	TOTAL
02	SITE CONSTRUCTION	84.4	88.4	87.4	88.6	93.0	92.0	82.5	88.3	87.0	71.9	92.9	88.0	82.3	103.3	98.4	82.0	109.3	102.9
03100	CONCRETE FORMS & ACCESSORIES	95.0	92.2	92.5	100.7	96.9	97.4	95.9	89.4	90.3	97.1	95.6	95.8	97.5	121.4	118.1	98.2	139.9	134.2
03200	CONCRETE REINFORCEMENT	97.3	87.9	91.9	93.8	118.5	107.9	97.9	92.8	95.0	93.8	118.2	107.7	95.6	108.1	102.7	95.8	119.4	109.2
03300	CAST-IN-PLACE CONCRETE	98.6	110.7	103.6	92.5	102.5	96.6	96.3	108.2	101.2	91.3	102.7	96.0	109.5	115.1	111.8	107.9	127.3	115.9
03	CONCRETE	100.8	97.0	98.9	90.2	103.8	97.0	95.6	95.8	95.7	89.3	103.3	96.3	96.7	117.2	106.9	97.5	131.7	114.6
04	MASONRY	99.5	84.7	90.4	92.7	98.4	96.2	96.5	83.0	88.2	100.6	92.3	95.5	108.0	127.3	119.9	108.1	137.0	125.8
05	METALS	97.5	86.0	93.4	99.5	123.1	108.0	95.6	87.9	92.8	100.3	122.4	108.3	96.8	125.8	107.2	98.8	133.7	111.4
06	WOOD & PLASTICS	95.2	92.0	93.5	96.9	96.8	96.9	92.9	88.7	90.7	93.5	95.8	94.7	104.2	119.6	112.2	104.7	139.8	122.9
07	THERMAL & MOISTURE PROTECTION	91.4	88.5	90.1	93.6	96.8	95.1	90.6	81.5	86.3	92.8	93.8	93.3	99.0	126.7	112.2	98.9	135.1	116.1
08	DOORS & WINDOWS	93.3	80.4	90.2	97.8	100.6	98.4	92.6	87.9	91.4	97.8	100.0	98.3	101.0	127.7	107.5	102.2	149.4	113.6
09200	PLASTER & GYPSUM BOARD	101.1	85.7	91.7	110.7	95.0	101.1	92.3	82.3	86.2	107.9	93.9	99.3	100.6	121.0	113.1	100.6	141.8	125.7
095,098	CEILINGS & ACOUSTICAL TREATMENT	95.1	85.7	88.9	101.1	95.0	97.0	100.6	82.3	88.5	101.1	93.9	96.3	90.9	121.0	110.8	90.9	141.8	124.5
09600	FLOORING	95.3	60.4	86.8	90.4	77.2	87.2	94.4	75.2	89.8	90.4	70.2	85.5	112.1	129.2	116.2	109.2	118.1	111.4
097,099	WALL FINISHES, PAINTS & COATINGS	96.5	89.5	92.5	93.2	89.5	91.1	95.7	75.1	83.8	93.2	90.0	91.3	99.2	111.2	106.2	106.0	122.4	115.5
09	FINISHES	96.0	85.5	90.7	98.2	91.9	95.0	94.9	84.7	89.7	97.1	89.7	93.3	102.3	122.1	112.4	101.8	135.1	118.8
10 - 14	TOTAL DIV. 10000 - 14000	100.0	114.1	103.0	100.0	105.3	101.1	100.0	113.2	102.8	100.0	104.8	101.0	100.0	110.1	102.1	100.0	116.1	103.4
15	MECHANICAL	100.0	91.8	96.2	100.0	94.0	97.2	99.7	94.9	97.5	100.0	95.5	97.9	99.8	109.1	104.1	99.7	119.5	108.9
16	ELECTRICAL	93.1	85.0	87.5	93.9	100.7	98.6	93.4	79.8	84.0	97.6	88.1	91.0	102.3	104.4	103.8	103.4	101.4	102.0
01 - 16	WEIGHTED AVERAGE	97.2	89.0	93.2	96.8	99.8	98.3	95.7	88.5	92.2	96.9	96.8	96.9	99.6	115.0	107.1	100.1	123.6	111.6

DIVISION		MINNESOTA ROCHESTER			MINNESOTA SAINT PAUL			MINNESOTA ST. CLOUD			MISSISSIPPI BILOXI			MISSISSIPPI GREENVILLE			MISSISSIPPI JACKSON		
		MAT.	INST.	TOTAL	MAT.	INST.	TOTAL	MAT.	INST.	TOTAL	MAT.	INST.	TOTAL	MAT.	INST.	TOTAL	MAT.	INST.	TOTAL
02	SITE CONSTRUCTION	81.3	102.5	97.6	84.2	103.7	99.1	77.3	105.3	98.7	107.8	87.1	91.9	112.2	86.8	92.8	104.2	86.8	90.9
03100	CONCRETE FORMS & ACCESSORIES	98.1	109.9	108.3	88.3	132.1	126.1	79.1	108.8	104.7	99.0	52.6	58.9	77.2	43.6	48.2	93.5	49.8	55.8
03200	CONCRETE REINFORCEMENT	95.6	118.2	108.5	92.2	119.3	107.7	103.4	118.7	112.2	95.8	68.7	80.4	103.8	52.9	74.8	95.8	57.4	73.9
03300	CAST-IN-PLACE CONCRETE	105.8	105.0	105.5	108.8	123.7	115.0	95.7	117.9	104.8	105.7	54.0	84.4	106.2	47.3	81.9	103.7	50.2	81.7
03	CONCRETE	94.9	110.6	102.8	98.9	127.1	113.0	88.1	114.8	101.4	96.3	57.9	77.1	98.7	48.7	73.7	94.9	53.3	74.1
04	MASONRY	107.8	105.4	106.3	119.5	137.0	130.2	110.4	116.0	113.9	89.5	47.0	63.4	132.4	44.6	78.4	92.4	44.6	63.0
05	METALS	96.7	130.9	109.0	95.4	133.4	109.1	96.5	133.0	109.7	98.1	87.3	94.2	96.4	80.7	90.8	98.0	82.5	92.4
06	WOOD & PLASTICS	104.7	108.1	106.5	94.5	129.5	112.6	85.2	103.4	94.6	99.0	53.1	75.3	75.6	43.1	58.8	92.8	51.2	71.3
07	THERMAL & MOISTURE PROTECTION	98.8	107.9	103.1	98.7	133.9	115.4	96.8	124.2	109.8	96.5	54.5	76.6	96.5	49.5	74.2	96.2	51.1	74.8
08	DOORS & WINDOWS	101.0	132.3	108.6	98.1	143.8	109.1	94.7	129.7	103.2	98.3	56.2	88.1	97.7	45.8	85.2	98.7	51.3	87.3
09200	PLASTER & GYPSUM BOARD	100.6	109.2	105.9	97.0	131.2	117.9	87.7	104.4	97.9	108.0	52.2	73.9	101.0	41.8	64.8	108.0	50.2	72.7
095,098	CEILINGS & ACOUSTICAL TREATMENT	90.9	109.2	103.0	88.2	131.2	116.6	69.0	104.4	92.3	95.1	52.2	66.8	89.7	41.8	58.1	95.1	50.2	65.5
09600	FLOORING	111.8	92.7	107.2	104.7	118.1	107.9	100.2	129.2	107.3	115.9	49.4	99.8	106.3	42.2	90.8	115.9	48.8	99.7
097,099	WALL FINISHES, PAINTS & COATINGS	101.8	107.0	104.8	106.0	116.8	112.3	106.9	122.4	115.9	105.6	43.7	69.6	105.6	43.1	69.2	105.6	43.1	69.2
09	FINISHES	102.4	106.9	104.7	99.6	128.4	114.3	93.5	113.1	103.5	105.0	50.7	77.3	100.2	43.0	71.0	105.0	48.9	76.4
10 - 14	TOTAL DIV. 10000 - 14000	100.0	105.9	101.2	100.0	114.3	103.0	100.0	106.6	101.4	100.0	68.1	93.3	100.0	65.9	92.8	100.0	67.0	93.0
15	MECHANICAL	99.7	98.8	99.3	99.7	119.4	108.9	99.7	110.7	104.8	99.9	63.5	83.1	99.9	43.8	73.9	99.9	43.4	73.8
16	ELECTRICAL	102.2	94.5	96.9	101.3	104.2	103.3	98.9	104.2	102.6	96.7	56.1	68.8	95.6	46.7	62.0	97.2	46.7	62.5
01 - 16	WEIGHTED AVERAGE	99.3	106.4	102.7	99.5	121.7	110.3	96.2	113.4	104.6	98.9	62.4	81.1	100.5	53.3	77.6	98.7	55.1	77.5

City Cost Indexes

DIVISION		MISSISSIPPI MERIDIAN			MISSOURI CAPE GIRARDEAU			MISSOURI COLUMBIA			MISSOURI JOPLIN			MISSOURI KANSAS CITY			MISSOURI SPRINGFIELD		
		MAT.	INST.	TOTAL	MAT.	INST.	TOTAL	MAT.	INST.	TOTAL	MAT.	INST.	TOTAL	MAT.	INST.	TOTAL	MAT.	INST.	TOTAL
02	SITE CONSTRUCTION	103.1	86.8	90.6	93.9	93.3	93.4	98.1	94.5	95.4	99.2	95.9	96.7	92.0	94.4	93.9	90.6	93.6	92.9
03100	CONCRETE FORMS & ACCESSORIES	75.5	42.3	46.9	83.8	84.5	84.4	85.5	76.0	77.3	105.5	69.5	74.4	105.0	99.7	100.5	103.6	58.6	64.8
03200	CONCRETE REINFORCEMENT	102.5	56.9	76.5	106.0	91.6	97.8	100.4	104.4	102.7	104.0	76.6	88.3	98.8	105.4	102.6	96.5	104.4	101.0
03300	CAST-IN-PLACE CONCRETE	100.2	50.2	79.6	85.1	92.7	88.2	86.1	75.1	81.6	102.9	77.6	92.4	95.6	99.4	97.2	97.4	71.8	86.8
03	CONCRETE	91.4	49.9	70.7	85.9	90.5	88.2	81.7	83.0	82.3	97.8	75.0	86.4	94.0	101.2	97.6	94.6	73.3	83.9
04	MASONRY	89.2	36.4	56.7	114.3	83.0	95.1	130.9	75.5	96.9	94.7	61.8	74.5	100.1	99.7	99.9	86.5	73.8	78.7
05	METALS	96.4	81.7	91.1	93.7	116.1	101.8	93.0	120.1	102.8	95.8	94.8	95.4	99.2	110.5	103.2	96.3	105.0	99.4
06	WOOD & PLASTICS	73.9	41.3	57.0	85.3	84.2	84.7	89.4	72.7	80.8	102.7	69.9	85.7	102.8	99.4	101.1	102.6	54.9	78.0
07	THERMAL & MOISTURE PROTECTION	95.9	48.0	73.2	95.1	86.8	91.2	92.2	79.6	86.2	98.5	68.1	84.0	97.3	99.8	98.5	96.4	73.2	85.4
08	DOORS & WINDOWS	97.7	45.1	85.0	96.6	85.0	93.8	93.2	92.9	93.1	94.4	77.6	90.4	98.8	100.8	99.3	99.0	70.2	92.0
09200	PLASTER & GYPSUM BOARD	100.1	40.0	63.4	98.0	83.3	89.0	100.9	71.5	82.9	108.6	68.6	84.2	104.0	99.1	101.0	106.2	53.1	73.8
095,098	CEILINGS & ACOUSTICAL TREATMENT	89.7	40.0	56.9	92.6	83.3	86.4	93.8	71.5	79.1	98.5	68.6	78.8	104.0	99.1	100.7	99.2	53.1	68.8
09600	FLOORING	105.2	34.3	88.0	99.0	90.1	96.8	107.9	75.0	99.9	118.9	60.6	104.8	96.8	98.1	97.2	110.4	60.6	98.3
097,099	WALL FINISHES, PAINTS & COATINGS	105.6	42.5	68.9	100.8	90.2	94.6	99.2	71.5	83.1	92.6	53.3	69.7	97.4	104.8	101.7	93.5	76.3	83.5
09	FINISHES	99.1	40.2	69.1	94.0	85.5	89.6	96.3	73.1	84.5	105.8	65.7	85.3	100.6	99.7	100.1	101.1	59.3	79.8
10 - 14	TOTAL DIV. 10000 - 14000	100.0	65.8	92.8	100.0	81.5	96.1	100.0	95.8	99.1	100.0	86.3	97.1	100.0	95.9	99.1	100.0	84.1	96.6
15	MECHANICAL	99.9	43.4	73.8	100.1	96.9	98.6	100.1	97.8	99.0	100.3	63.2	83.1	100.1	99.5	99.8	100.2	72.7	87.5
16	ELECTRICAL	95.0	61.5	72.0	95.3	113.3	107.7	93.8	85.9	88.4	89.2	69.2	75.4	103.4	104.4	104.1	100.2	75.0	82.9
01 - 16	WEIGHTED AVERAGE	96.7	54.8	76.3	96.3	96.2	96.2	96.4	89.4	93.0	98.0	73.4	86.0	99.0	101.1	100.0	97.7	76.7	87.5

DIVISION		MISSOURI ST. JOSEPH			MISSOURI ST. LOUIS			MONTANA BILLINGS			MONTANA BUTTE			MONTANA GREAT FALLS			MONTANA HELENA		
		MAT.	INST.	TOTAL	MAT.	INST.	TOTAL	MAT.	INST.	TOTAL	MAT.	INST.	TOTAL	MAT.	INST.	TOTAL	MAT.	INST.	TOTAL
02	SITE CONSTRUCTION	93.4	91.5	91.9	93.3	96.6	95.8	82.7	98.2	94.6	88.6	96.9	95.0	92.1	97.7	96.4	93.8	97.7	96.8
03100	CONCRETE FORMS & ACCESSORIES	104.9	75.2	79.3	101.4	109.2	108.1	94.4	85.7	86.9	80.4	88.5	87.4	100.8	86.3	88.3	100.9	86.8	88.7
03200	CONCRETE REINFORCEMENT	97.6	99.0	98.4	97.0	108.8	103.7	100.1	104.9	102.9	108.6	104.9	106.5	100.1	104.9	102.9	103.7	99.9	101.5
03300	CAST-IN-PLACE CONCRETE	95.6	94.3	95.0	85.1	112.3	96.3	116.3	84.1	103.0	119.7	85.3	105.5	126.5	71.2	103.7	128.9	83.7	110.3
03	CONCRETE	93.8	87.5	90.6	86.0	111.3	98.6	102.6	89.1	95.9	102.3	90.7	96.5	108.1	84.8	96.5	109.7	88.4	99.1
04	MASONRY	99.4	88.2	92.5	96.4	110.6	105.1	122.6	96.2	106.4	121.0	96.9	106.2	124.7	99.5	109.2	121.9	96.5	106.3
05	METALS	97.8	107.1	101.2	96.1	125.3	106.7	98.4	99.4	98.8	97.7	99.1	98.2	98.8	99.2	98.9	98.0	96.7	97.5
06	WOOD & PLASTICS	103.4	71.7	87.0	103.4	107.8	105.7	96.6	86.0	91.1	83.8	91.1	87.6	105.0	86.0	95.2	105.2	86.0	95.2
07	THERMAL & MOISTURE PROTECTION	98.0	88.9	93.7	95.1	105.5	100.1	98.1	87.8	93.2	97.3	86.2	92.0	98.3	86.4	92.7	98.3	87.7	93.3
08	DOORS & WINDOWS	99.5	84.2	95.8	93.6	112.7	98.2	102.7	85.8	98.7	99.3	88.5	96.7	103.0	85.8	98.9	102.4	84.7	98.2
09200	PLASTER & GYPSUM BOARD	109.4	70.5	85.6	104.8	107.7	106.5	100.4	86.1	91.6	93.8	91.4	92.3	100.4	86.1	91.6	99.9	86.1	91.4
095,098	CEILINGS & ACOUSTICAL TREATMENT	102.6	70.5	81.4	96.7	107.7	103.9	112.6	86.1	95.1	109.9	91.4	97.7	112.6	86.1	95.1	109.9	86.1	94.2
09600	FLOORING	99.5	97.1	98.9	106.7	101.6	105.4	110.5	63.1	99.0	102.1	50.2	89.5	110.5	68.1	100.2	110.5	66.0	99.7
097,099	WALL FINISHES, PAINTS & COATINGS	93.2	88.4	90.4	100.8	109.6	105.9	102.5	78.4	88.5	102.5	60.5	78.1	102.5	68.1	82.5	102.5	75.6	86.8
09	FINISHES	101.8	79.5	90.4	98.3	107.5	103.0	106.7	80.4	93.3	102.4	78.3	90.1	106.8	80.6	93.5	106.3	81.4	93.6
10 - 14	TOTAL DIV. 10000 - 14000	100.0	88.6	97.6	100.0	106.3	101.3	100.0	84.7	96.8	100.0	84.8	96.8	100.0	85.6	97.0	100.0	76.5	95.0
15	MECHANICAL	100.3	88.2	94.7	100.1	108.1	103.8	100.4	90.4	95.7	100.4	87.8	94.5	100.4	91.6	96.3	100.4	90.5	95.8
16	ELECTRICAL	102.8	81.5	88.1	99.0	116.5	111.0	95.0	84.6	87.9	106.5	80.7	88.8	94.9	82.3	86.2	94.9	82.3	86.2
01 - 16	WEIGHTED AVERAGE	99.0	87.7	93.5	96.2	110.6	103.2	101.5	89.8	95.8	101.3	88.6	95.1	102.8	89.3	96.2	102.6	88.9	96.0

DIVISION		MONTANA MISSOULA			NEBRASKA GRAND ISLAND			NEBRASKA LINCOLN			NEBRASKA NORTH PLATTE			NEBRASKA OMAHA			NEVADA CARSON CITY		
		MAT.	INST.	TOTAL	MAT.	INST.	TOTAL	MAT.	INST.	TOTAL	MAT.	INST.	TOTAL	MAT.	INST.	TOTAL	MAT.	INST.	TOTAL
02	SITE CONSTRUCTION	72.8	96.9	91.3	93.1	90.7	91.3	84.2	90.7	89.2	93.9	90.4	91.2	76.0	89.4	86.2	64.2	102.3	93.4
03100	CONCRETE FORMS & ACCESSORIES	84.8	83.2	83.4	99.6	57.3	63.1	105.1	50.2	57.8	99.5	55.4	61.4	100.1	74.5	78.0	101.1	96.9	97.5
03200	CONCRETE REINFORCEMENT	110.5	110.3	110.4	105.0	78.9	90.2	96.5	78.7	86.4	106.4	78.7	90.6	101.0	80.3	89.2	108.4	117.3	113.5
03300	CAST-IN-PLACE CONCRETE	88.2	81.7	85.5	112.4	64.4	92.6	101.5	67.0	87.3	112.4	60.1	90.9	105.4	75.8	93.2	117.8	94.1	108.0
03	CONCRETE	81.1	88.1	84.6	104.2	65.2	84.7	96.7	63.0	79.8	104.4	63.0	83.7	98.0	76.2	87.1	113.4	99.5	106.5
04	MASONRY	147.4	91.2	112.9	104.7	54.5	73.8	94.6	51.8	68.3	89.9	47.7	64.0	100.4	78.9	87.2	136.3	84.2	104.3
05	METALS	97.8	101.0	99.0	92.9	88.0	91.1	95.8	87.6	92.9	93.1	91.8	92.6	98.4	79.4	91.6	103.9	101.6	103.1
06	WOOD & PLASTICS	88.6	85.7	87.1	98.9	53.6	75.5	104.2	44.3	73.2	98.7	53.6	75.4	99.6	75.6	87.2	95.4	96.7	96.1
07	THERMAL & MOISTURE PROTECTION	96.4	92.8	94.7	96.7	63.0	80.7	97.1	55.5	77.4	96.6	54.5	76.6	92.5	75.2	84.3	104.8	91.6	98.6
08	DOORS & WINDOWS	99.4	87.9	96.6	92.5	56.7	83.9	98.1	53.8	87.4	91.8	62.1	84.6	101.8	70.8	94.3	95.1	105.4	97.6
09200	PLASTER & GYPSUM BOARD	95.5	85.8	89.6	103.2	51.7	71.7	106.2	42.1	67.0	103.5	51.7	71.8	111.9	75.3	89.5	86.9	96.4	92.7
095,098	CEILINGS & ACOUSTICAL TREATMENT	109.9	85.8	94.0	92.4	51.7	65.6	99.2	42.1	61.6	93.8	51.7	66.0	138.8	75.3	96.9	97.9	96.4	96.9
09600	FLOORING	104.0	78.5	97.8	99.5	42.1	85.6	101.3	43.8	87.4	99.5	43.8	86.0	130.2	52.7	111.4	105.2	83.4	99.9
097,099	WALL FINISHES, PAINTS & COATINGS	102.5	68.1	82.5	91.0	49.9	67.1	91.0	50.3	67.3	91.0	60.3	73.1	146.4	60.3	96.3	108.7	93.7	100.0
09	FINISHES	102.5	80.5	91.3	96.1	52.4	73.8	98.2	47.4	72.3	96.4	53.0	74.2	123.9	68.6	95.7	97.0	93.3	95.2
10 - 14	TOTAL DIV. 10000 - 14000	100.0	72.4	94.2	100.0	82.3	96.3	100.0	81.1	96.0	100.0	66.1	92.8	100.0	83.3	96.5	100.0	120.9	104.4
15	MECHANICAL	100.4	82.9	92.3	100.2	84.4	92.9	100.2	84.4	92.9	100.2	82.1	91.8	99.8	83.2	92.2	100.2	96.9	98.7
16	ELECTRICAL	103.1	85.4	90.9	91.0	52.5	64.5	100.5	52.5	67.5	94.3	52.5	65.6	88.3	86.2	86.9	92.7	91.9	92.1
01 - 16	WEIGHTED AVERAGE	99.5	87.8	93.8	97.5	68.7	83.5	97.8	67.0	82.8	96.9	67.3	82.5	100.5	80.1	90.6	102.0	96.5	99.3

City Cost Indexes

DIVISION		NEVADA LAS VEGAS			NEVADA RENO			NEW HAMPSHIRE MANCHESTER			NEW HAMPSHIRE NASHUA			NEW HAMPSHIRE PORTSMOUTH			NEW JERSEY CAMDEN		
		MAT.	INST.	TOTAL	MAT.	INST.	TOTAL	MAT.	INST.	TOTAL	MAT.	INST.	TOTAL	MAT.	INST.	TOTAL	MAT.	INST.	TOTAL
02	SITE CONSTRUCTION	64.3	103.9	94.7	64.6	102.3	93.5	90.0	102.2	99.3	92.1	102.2	99.8	86.2	101.7	98.0	95.5	106.3	103.8
03100	CONCRETE FORMS & ACCESSORIES	97.2	110.0	108.3	101.6	97.0	97.6	99.4	72.6	76.2	99.8	72.6	76.3	85.7	69.5	71.8	99.8	123.5	120.2
03200	CONCRETE REINFORCEMENT	100.1	126.3	115.0	100.1	125.8	114.8	111.9	99.3	104.7	111.9	99.3	104.7	89.5	99.2	95.0	111.9	113.5	112.8
03300	CAST-IN-PLACE CONCRETE	112.1	112.0	112.1	120.0	94.1	109.3	95.1	89.5	92.8	92.8	89.5	91.4	88.0	85.6	87.0	80.7	122.0	97.7
03	CONCRETE	109.3	113.2	111.2	113.5	101.1	107.3	105.1	83.5	94.3	103.9	83.5	93.7	95.1	80.8	88.0	98.0	119.9	108.9
04	MASONRY	130.3	98.5	110.8	136.4	84.2	104.3	96.5	94.9	95.6	96.6	94.9	95.6	91.0	88.4	89.4	88.8	123.0	109.8
05	METALS	104.6	107.5	105.6	104.4	104.8	104.6	98.0	89.3	94.8	98.0	89.3	94.8	94.7	88.0	92.2	97.8	102.0	99.3
06	WOOD & PLASTICS	90.4	108.5	99.8	95.9	96.7	96.3	102.0	67.1	83.9	101.9	67.1	83.9	87.3	67.1	76.8	101.9	123.4	113.0
07	THERMAL & MOISTURE PROTECTION	104.4	102.2	103.3	104.8	91.6	98.6	99.4	96.5	98.0	99.8	96.5	98.2	99.3	97.0	98.2	99.2	118.6	108.4
08	DOORS & WINDOWS	95.5	114.0	99.9	95.5	105.4	97.9	110.0	74.3	101.4	110.0	74.3	101.4	111.0	67.4	100.5	110.0	116.6	111.6
09200	PLASTER & GYPSUM BOARD	87.4	108.6	100.3	88.2	96.4	93.2	98.2	64.8	77.8	98.2	64.8	77.8	91.3	64.8	75.1	98.2	122.9	113.3
095,098	CEILINGS & ACOUSTICAL TREATMENT	104.7	108.6	107.2	104.7	96.4	99.2	102.6	64.8	77.6	102.6	64.8	77.6	93.1	64.8	74.4	102.6	122.9	116.0
09600	FLOORING	105.2	99.3	103.8	105.2	83.4	99.9	100.2	101.9	100.6	99.9	101.9	100.4	94.7	101.9	96.5	99.9	123.2	105.6
097,099	WALL FINISHES, PAINTS & COATINGS	108.7	113.2	111.3	108.7	93.7	100.0	95.6	96.4	96.1	95.6	96.4	96.1	95.6	47.3	67.5	95.6	144.1	123.8
09	FINISHES	98.2	108.0	103.2	98.4	93.5	95.9	99.8	79.5	89.4	100.0	79.5	89.5	95.2	72.2	83.5	100.3	126.0	113.4
10 - 14	TOTAL DIV. 10000 - 14000	100.0	112.6	102.7	100.0	120.9	104.4	100.0	93.6	98.7	100.0	93.6	98.7	100.0	91.0	98.1	100.0	118.1	103.8
15	MECHANICAL	100.2	111.9	105.6	100.2	97.0	98.7	99.8	86.1	93.5	99.8	86.1	93.5	99.8	82.5	91.8	99.8	118.3	108.4
16	ELECTRICAL	94.8	110.2	105.3	92.7	91.9	92.1	104.2	82.3	89.1	104.0	82.3	89.1	100.7	82.3	88.0	104.0	120.3	115.2
01 - 16	WEIGHTED AVERAGE	101.5	108.6	105.0	102.3	97.0	99.7	101.3	86.8	94.2	101.2	86.8	94.2	98.3	83.8	91.3	100.1	117.5	108.6

DIVISION		NEW JERSEY ELIZABETH			NEW JERSEY JERSEY CITY			NEW JERSEY NEWARK			NEW JERSEY PATERSON			NEW JERSEY TRENTON			NEW MEXICO ALBUQUERQUE		
		MAT.	INST.	TOTAL	MAT.	INST.	TOTAL	MAT.	INST.	TOTAL	MAT.	INST.	TOTAL	MAT.	INST.	TOTAL	MAT.	INST.	TOTAL
02	SITE CONSTRUCTION	110.3	107.0	107.8	95.5	106.8	104.1	114.1	106.7	108.4	108.5	106.7	107.1	96.6	106.9	104.5	83.9	110.1	103.9
03100	CONCRETE FORMS & ACCESSORIES	111.1	125.6	123.6	99.8	126.5	122.9	98.1	126.6	122.7	98.6	126.6	122.8	98.9	124.0	120.6	99.5	73.3	76.9
03200	CONCRETE REINFORCEMENT	86.2	131.2	111.9	111.9	131.4	123.0	111.9	131.4	123.0	111.9	131.4	123.0	111.9	125.1	119.4	100.8	72.3	84.6
03300	CAST-IN-PLACE CONCRETE	88.9	128.7	105.3	80.7	130.8	101.4	90.9	130.8	107.3	102.8	130.7	114.3	90.8	124.8	104.8	110.2	79.5	97.6
03	CONCRETE	101.4	126.7	114.0	98.0	127.6	112.8	102.9	127.9	115.4	108.8	127.9	118.3	102.9	123.3	113.1	108.6	76.4	92.5
04	MASONRY	110.2	125.3	119.5	88.8	125.3	111.3	98.2	125.3	114.9	93.4	125.3	113.0	89.7	122.8	110.1	111.4	74.2	88.5
05	METALS	94.4	111.9	100.7	97.9	109.7	102.2	97.9	112.9	103.3	97.9	112.9	103.3	98.7	107.0	101.7	105.6	92.2	100.8
06	WOOD & PLASTICS	117.6	126.5	122.2	101.9	126.2	114.5	103.8	126.5	115.5	103.8	126.5	115.5	102.0	123.1	112.9	96.3	74.8	85.2
07	THERMAL & MOISTURE PROTECTION	100.1	121.9	110.4	99.2	122.9	110.4	99.5	121.9	110.1	99.9	122.1	110.4	98.1	121.7	109.3	103.6	79.5	92.2
08	DOORS & WINDOWS	111.4	126.6	115.0	110.0	126.8	114.1	116.5	126.6	118.9	116.5	126.6	118.9	110.0	123.6	113.3	95.4	76.1	90.8
09200	PLASTER & GYPSUM BOARD	101.2	126.1	116.4	98.2	125.8	115.1	98.2	126.1	115.3	98.2	126.1	115.3	98.2	122.7	113.2	91.1	73.4	80.3
095,098	CEILINGS & ACOUSTICAL TREATMENT	93.1	126.1	114.9	102.6	125.8	117.9	102.6	126.1	118.1	102.6	126.1	118.1	102.6	122.7	115.8	104.7	73.4	84.1
09600	FLOORING	104.5	66.4	95.3	99.9	133.2	108.0	100.2	133.2	108.2	99.9	133.2	108.0	100.2	123.3	105.8	105.2	72.6	97.3
097,099	WALL FINISHES, PAINTS & COATINGS	95.5	144.2	123.8	95.6	144.2	123.8	95.5	144.2	123.8	95.5	144.2	123.8	95.6	144.2	123.8	108.7	67.2	84.6
09	FINISHES	101.2	116.0	108.7	100.3	129.7	115.3	100.8	129.9	115.7	100.4	129.9	115.5	100.4	125.8	113.3	99.2	72.5	85.6
10 - 14	TOTAL DIV. 10000 - 14000	100.0	99.5	99.9	100.0	100.4	100.1	100.0	100.4	100.1	100.0	100.4	100.1	100.0	117.8	103.8	100.0	82.3	96.3
15	MECHANICAL	99.8	116.5	107.6	99.8	127.8	112.8	99.8	129.1	113.4	99.8	127.9	112.8	99.8	125.4	111.7	100.1	77.5	89.6
16	ELECTRICAL	98.8	115.7	110.4	105.7	133.7	124.9	105.4	121.9	116.8	105.7	118.9	114.8	104.2	127.6	120.3	87.2	77.8	80.8
01 - 16	WEIGHTED AVERAGE	101.7	117.4	109.3	100.3	124.2	111.9	102.8	122.7	112.5	103.1	122.0	112.3	100.9	121.4	110.9	100.6	81.2	91.2

DIVISION		NEW MEXICO FARMINGTON			NEW MEXICO LAS CRUCES			NEW MEXICO ROSWELL			NEW MEXICO SANTA FE			NEW YORK ALBANY			NEW YORK BINGHAMTON		
		MAT.	INST.	TOTAL	MAT.	INST.	TOTAL	MAT.	INST.	TOTAL	MAT.	INST.	TOTAL	MAT.	INST.	TOTAL	MAT.	INST.	TOTAL
02	SITE CONSTRUCTION	89.3	110.1	105.2	96.0	85.5	88.0	94.2	110.1	106.4	83.1	110.1	103.8	72.8	107.7	99.5	96.6	90.4	91.8
03100	CONCRETE FORMS & ACCESSORIES	99.5	73.4	76.9	96.9	71.3	74.8	99.5	73.3	76.8	99.5	73.3	76.9	98.4	88.4	89.8	104.3	79.5	82.9
03200	CONCRETE REINFORCEMENT	110.1	72.3	88.5	104.0	66.2	82.5	109.3	67.0	85.2	108.1	72.3	87.7	100.8	87.8	93.4	99.8	87.5	92.8
03300	CAST-IN-PLACE CONCRETE	110.9	79.6	97.9	96.4	68.7	85.0	102.1	79.5	92.8	104.2	79.5	94.0	83.7	104.0	92.1	106.0	92.6	100.5
03	CONCRETE	112.2	76.4	94.3	88.5	70.1	79.3	112.7	75.3	94.0	106.5	76.4	91.5	100.1	95.0	97.6	101.8	88.0	94.9
04	MASONRY	127.2	74.2	94.6	111.8	63.8	82.3	127.7	74.2	94.8	116.3	74.2	90.4	90.6	100.0	96.3	106.5	84.8	93.1
05	METALS	104.8	92.2	100.3	98.9	79.7	92.0	104.8	89.8	99.4	104.8	92.2	100.2	100.1	109.0	103.3	95.2	121.2	104.6
06	WOOD & PLASTICS	96.3	74.8	85.2	89.3	73.2	81.0	96.3	74.8	85.2	96.3	74.8	85.2	100.6	85.8	92.9	110.4	77.0	93.1
07	THERMAL & MOISTURE PROTECTION	104.6	79.5	92.7	89.9	70.5	80.7	104.9	79.5	92.8	103.9	79.5	92.3	91.1	95.1	93.0	100.8	86.8	94.2
08	DOORS & WINDOWS	98.4	76.1	93.0	87.6	73.6	84.2	94.4	74.5	89.6	94.5	76.1	90.1	98.1	82.8	94.4	93.1	77.0	89.2
09200	PLASTER & GYPSUM BOARD	86.4	73.4	78.5	88.5	73.4	79.3	86.4	73.4	78.5	86.4	73.4	78.5	102.4	85.0	91.8	108.1	75.7	88.3
095,098	CEILINGS & ACOUSTICAL TREATMENT	95.1	73.4	80.8	99.9	73.4	82.5	95.1	73.4	80.8	95.1	73.4	80.8	97.9	85.0	89.4	97.9	75.7	83.2
09600	FLOORING	105.2	72.6	97.3	138.1	62.6	119.8	105.2	72.6	97.3	105.2	72.6	97.3	84.8	93.7	87.0	94.6	92.2	94.0
097,099	WALL FINISHES, PAINTS & COATINGS	108.7	74.8	89.0	102.1	67.2	81.8	108.7	67.2	84.6	108.7	67.2	84.6	81.8	82.2	82.0	84.1	79.2	81.3
09	FINISHES	97.1	73.3	85.0	114.7	69.2	91.5	97.5	72.5	84.8	96.9	72.5	84.4	95.6	88.2	91.8	96.1	81.2	88.5
10 - 14	TOTAL DIV. 10000 - 14000	100.0	82.3	96.3	100.0	77.6	95.3	100.0	82.3	96.3	100.0	82.3	96.3	100.0	95.0	98.9	100.0	93.4	98.6
15	MECHANICAL	100.1	77.5	89.6	100.4	76.9	89.5	100.1	77.4	89.6	100.1	77.5	89.6	100.3	92.5	96.7	100.5	87.2	94.4
16	ELECTRICAL	84.5	77.8	79.9	85.7	65.7	71.9	85.7	77.8	80.3	87.2	77.8	80.8	101.9	96.1	97.9	102.8	80.1	87.2
01 - 16	WEIGHTED AVERAGE	102.0	81.3	91.9	97.7	72.7	85.6	101.8	80.7	91.6	100.2	81.2	90.9	97.9	96.3	97.1	99.0	88.2	93.7

City Cost Indexes

	DIVISION	NEW YORK																	
		BUFFALO			HICKSVILLE			NEW YORK			RIVERHEAD			ROCHESTER			SCHENECTADY		
		MAT.	INST.	TOTAL	MAT.	INST.	TOTAL	MAT.	INST.	TOTAL	MAT.	INST.	TOTAL	MAT.	INST.	TOTAL	MAT.	INST.	TOTAL
02	SITE CONSTRUCTION	101.6	93.3	95.2	116.3	131.1	127.7	139.5	128.6	131.1	116.6	131.1	127.7	77.3	107.6	100.5	73.0	107.8	99.6
03100	CONCRETE FORMS & ACCESSORIES	101.7	115.0	113.2	89.0	156.2	147.0	109.7	171.5	163.1	93.6	156.3	147.7	101.4	98.8	99.2	103.8	89.2	91.2
03200	CONCRETE REINFORCEMENT	101.6	103.9	102.9	101.1	177.9	144.9	105.1	181.2	148.5	103.0	177.9	145.7	102.5	89.8	95.2	99.5	87.8	92.8
03300	CAST-IN-PLACE CONCRETE	115.5	120.9	117.8	99.1	156.7	122.8	129.6	163.4	143.5	97.3	156.7	121.8	109.9	105.1	107.9	97.3	105.2	100.6
03	CONCRETE	106.2	114.1	110.1	105.8	158.9	132.3	127.6	168.6	148.1	104.7	159.0	131.8	112.0	100.4	106.2	107.0	95.8	101.4
04	MASONRY	107.2	119.0	114.4	108.8	159.6	140.0	110.9	167.5	145.7	115.7	159.6	142.7	103.1	101.2	101.9	91.2	102.1	97.9
05	METALS	100.0	95.7	98.4	109.8	137.9	119.9	105.9	141.0	120.8	109.8	138.0	120.0	103.3	110.5	105.9	100.1	109.0	103.3
06	WOOD & PLASTICS	103.4	114.9	109.4	90.2	157.9	125.2	112.4	172.6	143.6	94.9	157.9	127.5	99.9	100.0	99.9	107.0	85.8	96.0
07	THERMAL & MOISTURE PROTECTION	100.8	108.6	104.5	104.7	150.4	126.4	107.0	160.6	132.4	104.7	150.4	126.4	97.6	99.0	98.3	91.2	96.0	93.5
08	DOORS & WINDOWS	93.9	106.6	96.9	90.6	159.5	107.2	97.8	168.4	114.8	90.6	159.5	107.2	99.5	93.5	98.0	98.1	82.8	94.4
09200	PLASTER & GYPSUM BOARD	96.7	114.9	107.8	97.5	159.7	135.5	105.8	174.7	147.9	98.8	159.7	136.0	98.0	99.8	99.1	102.4	85.0	91.8
095,098	CEILINGS & ACOUSTICAL TREATMENT	92.2	114.9	107.2	81.5	159.7	133.1	106.5	174.7	151.5	81.5	159.7	133.1	96.5	99.8	98.7	97.9	85.0	89.4
09600	FLOORING	89.9	119.5	97.1	94.3	106.8	97.3	99.0	166.3	115.3	95.6	106.8	98.3	79.7	101.7	85.1	84.8	93.7	87.0
097,099	WALL FINISHES, PAINTS & COATINGS	84.6	118.1	104.1	106.7	145.9	129.5	97.5	151.8	129.1	106.7	150.7	132.3	81.8	105.1	95.4	81.8	82.2	82.0
09	FINISHES	92.7	116.9	105.0	105.6	145.2	125.8	108.8	168.8	139.4	106.1	145.7	126.3	93.9	100.4	97.2	95.3	88.7	91.9
10 - 14	TOTAL DIV. 10000 - 14000	100.0	106.4	101.3	100.0	138.0	108.0	100.0	144.8	109.5	100.0	138.1	108.0	100.0	96.1	99.2	100.0	95.9	99.1
15	MECHANICAL	99.8	97.1	98.6	99.7	145.2	120.8	100.4	163.8	129.7	99.7	145.2	120.8	99.8	92.8	96.6	100.3	93.7	97.2
16	ELECTRICAL	99.5	99.5	99.5	102.4	157.8	140.5	111.1	177.9	157.0	103.6	170.2	149.4	104.5	95.4	98.2	102.3	96.1	98.0
01 - 16	WEIGHTED AVERAGE	99.7	104.8	102.2	102.8	149.1	125.3	108.5	161.9	134.5	103.3	151.3	126.6	101.0	99.4	100.2	98.9	97.0	98.0

	DIVISION	NEW YORK															NORTH CAROLINA		
		SYRACUSE			UTICA			WATERTOWN			WHITE PLAINS			YONKERS			ASHEVILLE		
		MAT.	INST.	TOTAL	MAT.	INST.	TOTAL	MAT.	INST.	TOTAL	MAT.	INST.	TOTAL	MAT.	INST.	TOTAL	MAT.	INST.	TOTAL
02	SITE CONSTRUCTION	95.2	107.6	104.7	71.0	106.8	98.4	79.5	108.7	101.8	128.5	126.1	126.7	138.9	125.9	129.0	101.3	72.9	79.5
03100	CONCRETE FORMS & ACCESSORIES	103.1	86.1	88.5	104.3	76.2	80.1	83.8	84.0	84.0	110.5	125.7	123.6	109.5	135.7	132.1	93.0	45.4	52.0
03200	CONCRETE REINFORCEMENT	100.8	86.0	92.3	100.8	83.0	90.6	101.4	67.2	81.9	98.4	176.6	143.0	102.3	176.6	144.7	92.7	43.5	64.6
03300	CAST-IN-PLACE CONCRETE	97.3	95.6	96.6	89.9	95.0	92.0	104.7	97.1	101.6	114.1	128.3	120.0	127.8	129.3	128.4	95.7	53.1	78.1
03	CONCRETE	105.1	90.7	97.9	103.5	85.6	94.6	117.1	87.2	102.2	114.6	135.3	125.0	126.6	140.1	133.3	96.9	49.9	73.4
04	MASONRY	96.6	95.8	96.1	89.9	91.4	90.8	91.0	101.1	97.2	105.2	127.8	119.0	110.6	129.8	122.4	77.5	38.9	53.8
05	METALS	99.9	107.2	102.6	98.2	105.9	101.0	98.3	106.7	101.3	99.6	134.1	112.0	106.4	134.2	116.4	93.9	80.4	89.1
06	WOOD & PLASTICS	107.0	84.0	95.1	107.0	72.2	89.0	85.8	79.7	82.6	114.6	123.2	119.0	113.2	135.5	124.7	94.3	46.2	69.4
07	THERMAL & MOISTURE PROTECTION	99.0	94.2	96.8	91.2	91.5	91.3	91.5	95.4	93.3	107.9	134.2	120.4	108.3	138.1	122.4	96.1	46.7	72.7
08	DOORS & WINDOWS	95.9	79.1	91.9	98.1	73.9	92.3	98.1	79.1	93.5	94.4	140.8	105.6	97.9	147.6	109.9	93.1	44.1	81.3
09200	PLASTER & GYPSUM BOARD	102.4	83.1	90.6	102.4	71.0	83.2	96.4	78.7	85.6	102.4	123.7	115.4	105.5	136.4	124.4	100.3	44.3	66.1
095,098	CEILINGS & ACOUSTICAL TREATMENT	97.9	83.1	88.1	97.9	71.0	80.1	97.9	78.7	85.2	86.1	123.7	110.9	105.2	136.4	125.8	88.2	44.3	59.2
09600	FLOORING	86.4	85.6	86.2	84.8	85.7	85.1	77.4	80.9	78.2	95.1	165.3	112.1	94.7	165.3	111.8	92.4	48.1	81.6
097,099	WALL FINISHES, PAINTS & COATINGS	87.1	86.2	86.6	81.8	86.2	84.4	81.8	85.8	84.1	95.0	150.7	127.4	95.0	150.7	127.4	106.5	45.7	71.2
09	FINISHES	96.7	85.5	91.0	95.3	78.0	86.5	92.6	82.8	87.6	102.4	135.2	119.1	106.9	142.9	125.2	92.6	45.8	68.7
10 - 14	TOTAL DIV. 10000 - 14000	100.0	95.1	99.0	100.0	91.5	98.2	100.0	95.8	99.1	100.0	129.7	106.3	100.0	133.3	107.0	100.0	66.7	93.0
15	MECHANICAL	100.3	89.1	95.1	100.3	88.6	94.9	100.3	86.1	93.7	100.5	119.5	109.3	100.5	121.5	110.2	100.1	50.1	77.0
16	ELECTRICAL	102.3	85.3	90.6	102.2	84.7	90.2	102.2	87.1	91.8	95.5	130.9	119.9	108.6	141.0	130.9	99.6	47.7	63.9
01 - 16	WEIGHTED AVERAGE	99.7	92.2	96.1	98.1	89.3	93.8	99.5	91.7	95.7	102.6	129.7	115.8	107.6	134.1	120.5	95.6	53.1	75.0

	DIVISION	NORTH CAROLINA																	
		CHARLOTTE			DURHAM			FAYETTEVILLE			GREENSBORO			RALEIGH			WILMINGTON		
		MAT.	INST.	TOTAL	MAT.	INST.	TOTAL	MAT.	INST.	TOTAL	MAT.	INST.	TOTAL	MAT.	INST.	TOTAL	MAT.	INST.	TOTAL
02	SITE CONSTRUCTION	101.6	72.9	79.6	101.3	82.9	87.2	99.4	82.9	86.8	101.2	82.9	87.2	102.4	82.9	87.5	102.3	72.9	79.8
03100	CONCRETE FORMS & ACCESSORIES	101.5	45.5	53.2	95.2	45.7	52.5	90.3	45.7	51.9	95.2	45.7	52.5	98.3	45.7	52.9	94.8	45.7	52.5
03200	CONCRETE REINFORCEMENT	93.1	43.5	64.8	93.1	57.7	72.9	92.2	57.7	72.5	93.1	57.6	72.9	93.1	57.6	72.9	93.4	57.7	73.0
03300	CAST-IN-PLACE CONCRETE	98.1	53.1	79.6	96.0	53.2	78.3	92.6	53.2	76.4	95.1	53.2	77.8	101.5	53.2	81.6	95.3	53.2	77.9
03	CONCRETE	98.1	50.0	74.1	96.6	52.8	74.7	94.2	52.8	73.5	96.2	52.7	74.5	99.6	52.7	76.2	96.8	52.8	74.8
04	MASONRY	79.2	38.9	54.5	77.7	38.9	53.9	81.5	38.9	55.4	77.5	38.9	53.8	84.5	38.9	56.5	67.8	38.9	50.1
05	METALS	95.2	80.4	89.9	95.3	86.1	92.0	94.2	86.1	91.3	96.0	86.1	92.4	95.3	86.1	92.0	94.3	86.1	91.4
06	WOOD & PLASTICS	104.4	46.2	74.3	96.8	46.2	70.6	91.1	46.2	67.9	96.8	46.2	70.6	100.5	46.2	72.4	96.2	46.2	70.3
07	THERMAL & MOISTURE PROTECTION	96.1	47.8	73.2	96.8	47.4	73.3	96.4	47.4	73.1	96.8	47.4	73.3	96.6	46.8	72.9	96.1	47.4	73.0
08	DOORS & WINDOWS	97.1	44.1	84.3	97.1	48.3	85.3	93.2	48.3	82.4	97.1	48.3	85.3	93.9	48.3	82.9	93.3	48.3	82.4
09200	PLASTER & GYPSUM BOARD	105.8	44.3	68.2	105.8	44.3	68.2	100.5	44.3	66.2	105.8	44.3	68.2	108.6	44.3	69.3	101.0	44.3	66.3
095,098	CEILINGS & ACOUSTICAL TREATMENT	92.2	44.3	60.6	92.2	44.3	60.6	89.5	44.3	59.7	92.2	44.3	60.6	92.2	44.3	60.6	89.5	44.3	59.7
09600	FLOORING	95.8	48.1	84.2	96.0	48.1	84.4	92.5	48.1	81.7	96.0	48.1	84.4	96.0	48.1	84.4	93.3	48.1	82.3
097,099	WALL FINISHES, PAINTS & COATINGS	106.5	45.7	71.2	106.5	45.7	71.2	106.5	45.7	71.2	106.5	45.7	71.2	106.5	45.7	71.2	106.5	45.7	71.2
09	FINISHES	95.2	45.8	70.0	95.4	45.8	70.1	92.7	45.8	68.8	95.4	45.8	70.1	95.8	45.8	70.3	93.2	45.8	69.0
10 - 14	TOTAL DIV. 10000 - 14000	100.0	66.7	93.0	100.0	73.5	94.4	100.0	73.5	94.4	100.0	66.7	93.0	100.0	73.5	94.4	100.0	73.5	94.4
15	MECHANICAL	100.1	50.2	77.0	100.1	50.4	77.1	100.1	50.4	77.1	100.1	50.4	77.1	100.1	50.4	77.1	100.1	50.4	77.1
16	ELECTRICAL	99.1	42.8	60.4	97.5	45.6	61.8	92.7	45.6	60.3	98.6	45.6	62.1	98.2	45.6	62.0	99.3	45.6	62.4
01 - 16	WEIGHTED AVERAGE	96.9	52.4	75.2	96.5	55.1	76.4	95.0	55.1	75.6	96.6	54.9	76.3	97.0	55.1	76.6	95.2	54.1	75.2

City Cost Indexes

		NORTH CAROLINA			NORTH DAKOTA											OHIO		
	DIVISION	WINSTON-SALEM			BISMARCK			FARGO			GRAND FORKS			MINOT			AKRON		
		MAT.	INST.	TOTAL	MAT.	INST.	TOTAL	MAT.	INST.	TOTAL	MAT.	INST.	TOTAL	MAT.	INST.	TOTAL	MAT.	INST.	TOTAL
02	SITE CONSTRUCTION	101.6	82.9	87.3	85.1	96.4	93.8	85.1	96.4	93.8	93.0	95.8	95.2	90.7	96.4	95.1	102.4	109.4	107.8
03100	CONCRETE FORMS & ACCESSORIES	96.7	45.5	52.5	93.2	58.2	63.0	93.8	58.9	63.7	89.6	51.9	57.1	84.1	58.0	61.6	99.9	102.3	102.0
03200	CONCRETE REINFORCEMENT	93.1	43.5	64.8	108.6	67.6	85.2	100.1	67.6	81.6	107.0	61.2	80.9	110.7	67.6	86.1	99.1	99.7	99.4
03300	CAST-IN-PLACE CONCRETE	98.2	53.1	79.6	104.9	63.3	87.8	107.4	66.0	90.3	104.9	61.7	87.1	104.9	61.6	87.1	101.2	109.8	104.7
03	CONCRETE	97.8	50.0	73.9	98.0	62.6	80.3	98.2	63.8	81.0	102.8	58.0	80.4	101.3	61.9	81.6	100.3	103.2	101.8
04	MASONRY	77.8	38.9	53.9	107.3	60.5	78.6	107.4	60.1	78.3	107.3	52.9	73.9	107.7	60.6	78.8	93.4	103.6	99.7
05	METALS	95.2	80.5	89.9	95.6	74.5	88.0	95.5	74.5	87.9	95.6	71.0	86.7	95.9	74.3	88.1	91.8	83.5	88.8
06	WOOD & PLASTICS	96.8	46.2	70.6	85.3	54.5	69.4	85.3	55.4	69.8	81.9	50.1	65.5	76.4	54.5	65.1	95.2	101.5	98.4
07	THERMAL & MOISTURE PROTECTION	96.8	47.4	73.3	98.8	59.3	80.1	99.3	60.5	80.9	99.6	55.9	78.9	99.2	59.1	80.2	106.4	102.9	104.7
08	DOORS & WINDOWS	97.1	44.5	84.4	103.4	51.5	90.9	103.4	52.0	91.0	103.4	47.6	90.0	103.6	51.5	91.0	107.2	101.1	105.7
09200	PLASTER & GYPSUM BOARD	105.8	44.3	68.2	117.4	53.6	78.4	117.4	54.5	78.9	116.5	49.1	75.3	114.8	53.6	77.4	105.0	100.8	102.4
095,098	CEILINGS & ACOUSTICAL TREATMENT	92.2	44.3	60.6	143.1	53.6	84.1	143.1	54.5	84.6	143.1	49.1	81.1	143.1	53.6	84.1	99.5	100.8	100.4
09600	FLOORING	96.0	48.1	84.4	111.0	55.3	97.5	110.8	54.5	97.1	109.3	54.5	96.0	106.6	55.5	94.2	107.9	102.7	106.6
097,099	WALL FINISHES, PAINTS & COATINGS	106.5	45.7	71.2	102.5	45.9	69.6	102.5	42.2	67.5	102.5	35.1	63.3	102.5	38.4	65.3	103.9	122.1	114.5
09	FINISHES	95.4	45.8	70.1	115.2	55.8	84.9	115.0	55.7	84.8	115.0	50.1	81.9	113.7	55.0	83.8	104.4	104.5	104.5
10 - 14	TOTAL DIV. 10000 - 14000	100.0	66.7	93.0	100.0	61.2	91.8	100.0	61.2	91.8	100.0	57.8	91.1	100.0	61.2	91.8	100.0	102.4	100.5
15	MECHANICAL	100.1	50.2	77.0	100.6	69.5	86.2	100.6	76.1	89.3	100.6	55.5	79.7	100.6	59.4	81.5	100.1	105.3	102.5
16	ELECTRICAL	98.6	45.6	62.1	96.9	70.8	79.0	96.8	50.9	65.3	101.7	69.4	79.5	106.8	70.8	82.0	98.0	93.8	95.1
01 - 16	WEIGHTED AVERAGE	96.7	53.9	75.9	100.8	68.1	84.9	100.8	66.1	84.0	102.0	62.4	82.7	102.0	65.9	84.4	99.8	100.9	100.4

		OHIO																
	DIVISION	CANTON			CINCINNATI			CLEVELAND			COLUMBUS			DAYTON			LORAIN		
		MAT.	INST.	TOTAL	MAT.	INST.	TOTAL	MAT.	INST.	TOTAL	MAT.	INST.	TOTAL	MAT.	INST.	TOTAL	MAT.	INST.	TOTAL
02	SITE CONSTRUCTION	102.6	108.4	107.0	78.6	111.0	103.4	102.5	109.0	107.5	84.4	100.4	96.6	77.4	111.0	103.1	101.8	108.7	107.1
03100	CONCRETE FORMS & ACCESSORIES	99.9	90.7	91.9	96.8	90.5	91.3	100.0	109.3	108.0	94.4	90.6	91.1	96.7	87.2	88.5	99.9	91.5	92.7
03200	CONCRETE REINFORCEMENT	99.1	84.3	90.7	93.4	88.9	90.8	99.5	100.0	99.8	103.2	88.2	94.7	93.4	85.1	88.6	99.1	99.7	99.5
03300	CAST-IN-PLACE CONCRETE	102.2	106.8	104.1	90.0	92.5	91.1	99.3	118.0	107.0	90.0	98.3	93.4	82.0	92.7	86.4	96.3	98.6	97.3
03	CONCRETE	100.8	94.2	97.5	93.2	90.9	92.0	99.4	109.3	104.4	95.0	92.3	93.6	89.3	88.2	88.7	97.9	94.7	96.3
04	MASONRY	94.3	93.9	94.0	87.6	98.4	94.2	98.0	114.2	108.0	92.7	97.6	95.7	87.0	93.3	90.9	90.2	103.4	98.3
05	METALS	91.8	76.7	86.4	94.8	89.0	92.7	93.2	86.9	90.9	96.3	82.4	91.3	94.1	78.8	88.6	92.3	84.1	89.4
06	WOOD & PLASTICS	95.2	89.8	92.4	97.7	88.2	92.8	94.4	106.7	100.7	95.5	89.2	92.2	99.9	85.8	92.6	95.2	87.9	91.4
07	THERMAL & MOISTURE PROTECTION	106.9	98.5	102.9	89.9	98.0	93.7	105.1	115.9	110.2	101.7	97.9	99.9	95.3	94.0	94.7	106.7	105.4	106.1
08	DOORS & WINDOWS	101.4	83.1	97.0	97.2	85.8	94.5	95.6	104.7	97.8	99.8	87.5	96.8	97.5	84.1	94.2	101.4	94.5	99.7
09200	PLASTER & GYPSUM BOARD	105.0	88.7	95.0	102.5	88.2	93.7	104.3	106.1	105.4	102.9	88.6	94.1	102.5	85.7	92.2	105.0	86.7	93.8
095,098	CEILINGS & ACOUSTICAL TREATMENT	99.5	88.7	92.4	90.9	88.2	89.1	98.2	106.1	103.4	96.5	88.6	91.3	92.4	85.7	88.0	99.5	86.7	91.1
09600	FLOORING	108.1	88.9	103.4	92.9	103.7	95.5	107.6	112.6	108.9	97.1	95.8	96.8	95.4	91.1	94.4	108.1	110.2	108.6
097,099	WALL FINISHES, PAINTS & COATINGS	103.9	95.0	98.7	101.5	97.5	99.2	103.9	122.5	114.7	97.7	100.7	99.4	101.5	95.0	97.7	103.9	122.5	114.7
09	FINISHES	104.5	90.2	97.2	93.7	93.4	93.6	104.1	111.2	107.7	97.9	92.4	95.0	94.7	88.4	91.5	104.5	97.7	101.0
10 - 14	TOTAL DIV. 10000 - 14000	100.0	92.4	98.4	100.0	91.0	98.1	100.0	107.3	101.5	100.0	94.0	98.7	100.0	89.5	97.8	100.0	101.6	100.3
15	MECHANICAL	100.1	90.4	95.6	99.7	94.6	97.4	100.1	109.5	104.5	100.1	96.8	98.6	100.9	90.4	96.0	100.1	95.9	98.1
16	ELECTRICAL	96.5	97.5	97.2	95.8	87.9	90.4	97.1	105.8	103.1	102.1	89.4	93.4	94.5	90.1	91.5	96.5	96.4	96.4
01 - 16	WEIGHTED AVERAGE	99.2	92.9	96.1	95.3	93.9	94.6	98.7	107.3	102.9	97.9	93.1	95.5	95.2	90.9	93.1	98.6	97.2	97.9

		OHIO									OKLAHOMA								
	DIVISION	SPRINGFIELD			TOLEDO			YOUNGSTOWN			ENID			LAWTON			MUSKOGEE		
		MAT.	INST.	TOTAL	MAT.	INST.	TOTAL	MAT.	INST.	TOTAL	MAT.	INST.	TOTAL	MAT.	INST.	TOTAL	MAT.	INST.	TOTAL
02	SITE CONSTRUCTION	78.0	110.0	102.5	83.7	101.4	97.2	102.3	109.4	107.8	110.2	89.2	94.1	105.5	90.8	94.3	95.9	85.2	87.7
03100	CONCRETE FORMS & ACCESSORIES	96.7	88.7	89.8	94.4	103.9	102.6	99.9	92.8	93.7	94.3	46.5	53.1	98.6	64.1	68.8	100.4	42.2	50.2
03200	CONCRETE REINFORCEMENT	93.4	85.0	88.6	103.2	88.4	94.8	99.1	93.3	95.8	95.3	84.9	89.4	95.5	84.9	89.5	95.3	45.5	66.9
03300	CAST-IN-PLACE CONCRETE	86.0	92.2	88.5	90.0	108.0	97.4	100.2	108.6	103.7	96.5	57.9	80.6	93.4	57.9	78.8	86.9	44.4	69.3
03	CONCRETE	91.2	88.7	89.9	94.9	101.8	98.3	99.9	97.5	98.7	91.8	57.8	74.8	88.4	65.5	76.9	83.8	44.8	64.3
04	MASONRY	87.3	92.5	90.5	103.0	101.6	102.1	93.5	99.8	97.4	102.0	60.1	76.2	97.6	60.1	74.6	110.6	51.0	73.9
05	METALS	94.1	78.6	88.5	96.1	86.3	92.6	91.8	81.3	88.0	93.4	68.4	84.4	95.4	68.5	85.7	93.3	64.4	82.9
06	WOOD & PLASTICS	99.9	88.6	94.0	95.5	105.1	100.4	95.2	90.8	92.9	96.5	43.9	69.3	100.4	67.6	83.4	102.6	43.4	72.0
07	THERMAL & MOISTURE PROTECTION	95.3	94.0	94.7	103.6	106.4	105.0	107.0	100.3	103.8	98.6	64.6	82.5	98.4	67.0	83.5	98.2	51.8	76.2
08	DOORS & WINDOWS	97.5	80.2	93.3	99.7	97.6	99.2	101.4	91.7	99.0	96.6	57.2	87.1	98.1	70.0	91.3	96.5	42.7	83.5
09200	PLASTER & GYPSUM BOARD	102.5	88.6	94.0	102.9	105.0	104.1	105.0	89.7	95.6	95.2	43.3	63.5	97.8	67.6	79.3	96.9	42.6	63.7
095,098	CEILINGS & ACOUSTICAL TREATMENT	92.4	88.6	89.9	96.5	105.0	102.1	99.5	89.7	93.0	92.9	43.3	60.2	101.1	67.6	79.0	92.9	42.6	59.7
09600	FLOORING	95.4	91.1	94.4	96.3	74.1	90.9	108.1	95.9	105.1	115.6	62.3	102.7	118.7	62.3	105.0	119.9	50.2	103.0
097,099	WALL FINISHES, PAINTS & COATINGS	101.5	95.0	97.7	97.7	109.4	104.5	103.9	103.0	103.4	95.6	66.5	78.7	95.6	66.5	78.7	95.6	42.3	64.6
09	FINISHES	94.7	89.8	92.2	97.6	98.9	98.3	104.6	93.9	99.1	100.9	49.9	74.9	103.3	63.7	83.1	101.5	43.9	72.1
10 - 14	TOTAL DIV. 10000 - 14000	100.0	89.5	97.8	100.0	98.9	99.8	100.0	99.7	99.9	100.0	66.7	93.0	100.0	69.8	93.6	100.0	39.2	87.2
15	MECHANICAL	100.9	89.9	95.8	100.1	104.2	102.0	100.1	96.3	98.4	100.1	69.1	85.8	100.1	69.1	85.8	100.1	35.6	70.3
16	ELECTRICAL	94.5	89.4	91.0	102.8	98.1	99.6	96.5	92.0	93.4	95.0	70.8	78.4	97.7	65.1	75.3	94.4	41.9	58.3
01 - 16	WEIGHTED AVERAGE	95.4	90.6	93.1	98.5	99.7	99.1	99.0	95.7	97.4	97.6	65.7	82.1	97.8	68.5	83.6	96.8	48.8	73.4

City Cost Indexes

DIVISION		OKLAHOMA						OREGON											
		OKLAHOMA CITY			TULSA			EUGENE			MEDFORD			PORTLAND			SALEM		
		MAT.	INST.	TOTAL	MAT.	INST.	TOTAL	MAT.	INST.	TOTAL	MAT.	INST.	TOTAL	MAT.	INST.	TOTAL	MAT.	INST.	TOTAL
02	SITE CONSTRUCTION	105.8	91.4	94.8	102.6	86.2	90.0	99.4	106.4	104.8	108.5	106.4	106.9	100.5	106.4	105.0	99.6	106.4	104.8
03100	CONCRETE FORMS & ACCESSORIES	100.1	55.0	61.2	99.9	55.8	61.8	103.2	110.8	109.8	97.5	110.7	108.9	104.8	111.2	110.3	104.7	111.1	110.2
03200	CONCRETE REINFORCEMENT	95.5	84.9	89.5	95.5	84.9	89.5	109.3	105.5	107.1	100.9	105.5	103.5	104.5	105.8	105.2	110.4	105.8	107.8
03300	CAST-IN-PLACE CONCRETE	93.4	57.0	78.4	94.6	56.5	78.9	101.2	109.8	104.7	104.5	109.7	106.6	108.6	109.9	109.2	104.9	109.9	107.0
03	CONCRETE	88.5	61.2	74.8	89.1	62.5	75.8	111.7	109.1	110.4	119.0	109.0	114.0	114.9	109.4	112.1	113.8	109.3	111.6
04	MASONRY	99.5	62.7	76.8	96.8	62.8	75.9	112.3	106.4	108.7	109.1	106.4	107.4	113.6	111.9	112.6	114.3	111.9	112.8
05	METALS	95.6	68.4	85.8	96.0	84.6	91.9	95.1	103.7	98.2	94.7	103.6	97.9	93.8	104.4	97.7	95.7	104.4	98.8
06	WOOD & PLASTICS	102.6	54.1	77.5	101.2	55.7	77.7	98.2	109.8	104.2	92.2	109.8	101.3	100.1	109.8	105.1	100.1	109.8	105.1
07	THERMAL & MOISTURE PROTECTION	97.6	66.7	82.9	98.1	64.9	82.4	111.1	100.7	106.1	111.9	99.7	106.1	110.8	108.3	109.6	110.8	108.3	109.6
08	DOORS & WINDOWS	98.1	62.7	89.5	98.1	63.0	89.6	101.8	112.6	104.4	104.5	112.6	106.5	99.2	112.6	102.4	101.1	112.6	103.9
09200	PLASTER & GYPSUM BOARD	97.6	53.7	70.8	97.6	55.3	71.7	94.8	110.0	104.1	92.4	110.0	103.2	94.0	110.0	103.8	96.0	110.0	104.6
095,098	CEILINGS & ACOUSTICAL TREATMENT	101.1	53.7	69.8	101.1	55.3	70.9	118.1	110.0	112.8	116.7	110.0	112.3	118.1	110.0	112.8	129.0	110.0	116.5
09600	FLOORING	118.7	62.3	105.0	118.5	64.8	105.4	117.1	104.5	114.1	114.7	104.5	112.2	117.1	102.0	113.5	117.1	104.5	114.1
097,099	WALL FINISHES, PAINTS & COATINGS	95.6	66.5	78.7	95.6	62.3	76.3	119.6	82.9	98.3	119.6	70.1	90.9	119.6	82.9	98.3	119.6	82.9	98.3
09	FINISHES	103.4	56.4	79.4	102.8	57.5	79.7	113.9	106.6	110.2	113.1	105.2	109.0	113.8	106.2	109.9	116.0	106.6	111.2
10 - 14	TOTAL DIV. 10000 - 14000	100.0	69.1	93.5	100.0	71.9	94.1	100.0	110.2	102.2	100.0	110.1	102.1	100.0	110.2	102.2	100.0	110.2	102.2
15	MECHANICAL	100.1	70.5	86.4	100.1	62.8	82.8	100.3	112.5	105.9	100.3	112.5	105.9	100.3	112.5	105.9	100.3	112.5	105.9
16	ELECTRICAL	96.5	70.8	78.8	97.7	54.6	68.1	104.2	102.9	103.3	109.8	93.6	98.6	104.0	112.4	109.8	103.8	102.9	103.2
01 - 16	WEIGHTED AVERAGE	97.9	68.2	83.4	97.8	65.3	82.0	103.8	107.3	105.5	105.3	105.5	105.4	103.8	109.7	106.7	104.4	108.2	106.2

DIVISION		PENNSYLVANIA																	
		ALLENTOWN			ALTOONA			ERIE			HARRISBURG			PHILADELPHIA			PITTSBURGH		
		MAT.	INST.	TOTAL	MAT.	INST.	TOTAL	MAT.	INST.	TOTAL	MAT.	INST.	TOTAL	MAT.	INST.	TOTAL	MAT.	INST.	TOTAL
02	SITE CONSTRUCTION	94.1	107.7	104.6	98.5	107.7	105.6	94.9	108.4	105.3	82.8	106.3	100.8	101.9	95.5	97.0	101.3	110.3	108.2
03100	CONCRETE FORMS & ACCESSORIES	102.9	107.8	107.1	81.8	91.8	90.4	101.3	101.9	101.8	94.3	87.0	88.0	104.0	131.2	127.5	101.2	103.5	103.2
03200	CONCRETE REINFORCEMENT	100.8	108.5	105.2	97.8	98.3	98.1	99.8	98.2	98.9	100.8	93.2	96.4	100.2	123.3	113.3	102.6	107.3	105.2
03300	CAST-IN-PLACE CONCRETE	89.1	99.5	93.4	99.3	92.2	96.4	97.7	95.2	96.7	87.4	92.5	89.5	94.6	124.4	106.9	97.5	101.4	99.1
03	CONCRETE	99.8	106.2	103.0	94.2	94.7	94.4	93.2	100.2	96.7	98.2	91.7	95.0	106.8	126.9	116.8	95.1	104.8	99.9
04	MASONRY	95.0	95.7	95.4	96.8	82.2	87.8	88.1	98.4	94.4	92.4	88.5	90.0	89.9	125.5	111.8	89.2	102.3	97.2
05	METALS	98.1	125.0	107.8	91.8	118.9	101.6	91.9	120.8	102.3	99.9	117.6	106.3	100.1	122.1	108.0	98.6	126.1	108.5
06	WOOD & PLASTICS	107.0	110.5	108.8	81.7	93.1	87.6	102.5	102.1	102.3	99.0	85.8	92.2	107.7	131.3	119.9	103.0	103.7	103.3
07	THERMAL & MOISTURE PROTECTION	99.0	112.1	105.3	97.9	93.2	95.7	98.0	100.2	99.1	102.3	105.1	103.6	100.1	131.1	114.8	95.0	103.5	99.0
08	DOORS & WINDOWS	95.9	108.0	98.8	90.3	101.4	93.0	90.4	99.3	92.6	95.9	95.5	95.9	97.4	134.0	106.2	92.4	111.5	97.0
09200	PLASTER & GYPSUM BOARD	102.4	110.5	107.4	88.8	92.6	91.1	96.8	101.9	99.9	102.4	85.0	91.8	98.8	132.1	119.2	94.3	103.4	99.8
095,098	CEILINGS & ACOUSTICAL TREATMENT	97.9	110.5	106.2	92.4	92.6	92.5	97.9	101.9	100.5	97.9	85.0	89.4	99.9	132.1	121.1	96.8	103.4	101.1
09600	FLOORING	86.4	87.1	86.5	80.7	69.1	77.9	87.5	90.4	88.2	86.6	83.2	85.8	82.3	136.6	95.5	96.5	103.9	98.3
097,099	WALL FINISHES, PAINTS & COATINGS	87.1	102.2	95.9	82.7	104.0	95.1	90.9	88.0	89.2	87.1	88.9	88.1	89.0	138.8	117.9	92.0	116.0	105.9
09	FINISHES	96.7	103.7	100.3	92.1	88.4	90.2	96.9	98.3	97.6	95.7	85.8	90.6	96.9	133.2	115.4	98.6	104.6	101.7
10 - 14	TOTAL DIV. 10000 - 14000	100.0	103.9	100.8	100.0	100.9	100.2	100.0	105.5	101.2	100.0	93.6	98.6	100.0	124.9	105.3	100.0	105.2	101.1
15	MECHANICAL	100.3	102.7	101.4	99.7	88.3	94.4	99.7	98.6	99.2	100.3	93.4	97.1	99.9	124.7	111.4	99.8	100.4	100.1
16	ELECTRICAL	102.2	89.1	93.2	87.6	102.2	97.6	90.0	88.4	88.9	101.3	81.1	87.4	97.5	126.7	117.5	95.0	102.2	100.0
01 - 16	WEIGHTED AVERAGE	98.7	103.4	101.0	94.6	96.6	95.6	94.8	100.3	97.5	98.1	93.6	95.9	99.6	123.9	111.4	97.0	106.0	101.4

DIVISION		PENNSYLVANIA									PUERTO RICO			RHODE ISLAND			SOUTH CAROLINA		
		READING			SCRANTON			YORK			SAN JUAN			PROVIDENCE			CHARLESTON		
		MAT.	INST.	TOTAL	MAT.	INST.	TOTAL	MAT.	INST.	TOTAL	MAT.	INST.	TOTAL	MAT.	INST.	TOTAL	MAT.	INST.	TOTAL
02	SITE CONSTRUCTION	99.4	113.5	110.2	94.6	107.7	104.7	83.5	106.4	101.0	125.2	94.5	101.7	84.5	105.6	100.6	93.6	82.0	84.7
03100	CONCRETE FORMS & ACCESSORIES	103.5	93.1	94.6	103.1	93.9	95.1	81.6	87.5	86.7	98.9	21.4	32.1	102.5	111.2	110.0	95.2	44.6	51.6
03200	CONCRETE REINFORCEMENT	98.5	108.2	104.0	100.8	106.5	104.0	99.8	93.1	96.0	201.2	14.5	94.7	111.9	119.3	116.1	93.1	65.8	77.6
03300	CAST-IN-PLACE CONCRETE	69.2	98.8	81.4	93.1	95.3	94.0	87.7	93.6	90.1	109.8	36.2	79.4	86.1	118.9	99.6	81.5	55.6	70.8
03	CONCRETE	94.0	99.5	96.7	101.8	98.2	100.0	100.0	92.3	96.2	114.4	26.1	70.2	103.4	114.9	109.2	89.5	54.4	72.0
04	MASONRY	91.9	91.1	91.4	95.3	100.9	98.8	93.7	88.9	90.8	221.3	19.4	97.2	104.3	119.2	113.4	84.7	42.6	58.8
05	METALS	97.1	124.8	107.1	99.9	124.3	108.7	97.3	117.5	104.6	129.6	34.4	95.3	98.0	110.2	102.4	95.2	85.7	91.8
06	WOOD & PLASTICS	107.7	91.0	99.1	107.0	91.9	99.2	89.0	85.5	87.2	94.2	21.1	56.4	102.9	109.1	106.1	96.8	44.1	69.6
07	THERMAL & MOISTURE PROTECTION	100.1	108.8	104.2	98.9	108.0	103.2	97.4	106.2	101.6	168.9	27.1	101.7	98.5	112.4	105.1	96.3	50.5	74.6
08	DOORS & WINDOWS	97.4	97.4	97.4	95.9	103.3	97.7	92.8	95.7	93.5	160.0	18.4	125.8	103.6	111.1	105.4	97.1	46.9	85.0
09200	PLASTER & GYPSUM BOARD	105.9	90.5	96.5	102.4	91.4	95.7	97.9	84.7	89.8	225.4	18.2	98.8	98.2	108.2	104.3	105.8	42.2	66.9
095,098	CEILINGS & ACOUSTICAL TREATMENT	101.1	90.5	94.1	97.9	91.4	93.6	89.7	84.7	86.4	364.2	18.2	135.9	102.6	108.2	106.3	92.2	42.2	59.2
09600	FLOORING	80.5	85.9	81.8	86.4	93.8	88.2	82.4	83.2	82.6	201.6	19.9	157.6	99.9	126.2	106.3	96.0	47.6	84.3
097,099	WALL FINISHES, PAINTS & COATINGS	87.1	95.1	91.8	87.1	98.2	93.5	87.1	88.9	88.1	209.3	18.4	98.3	95.6	121.2	110.5	106.5	46.0	71.3
09	FINISHES	97.5	91.2	94.3	96.7	93.7	95.2	92.3	86.2	89.2	242.3	20.9	129.4	99.5	115.1	107.5	95.6	44.8	69.7
10 - 14	TOTAL DIV. 10000 - 14000	100.0	101.0	100.2	100.0	101.5	100.3	100.0	94.3	98.8	100.0	24.9	84.1	100.0	112.7	102.7	100.0	66.1	92.8
15	MECHANICAL	100.2	102.6	101.3	100.3	93.2	97.0	100.3	94.4	97.6	106.9	18.8	66.1	99.8	102.0	100.9	100.1	52.6	78.1
16	ELECTRICAL	100.1	81.0	87.0	102.3	86.6	91.5	91.9	81.1	84.5	144.8	16.1	56.3	98.6	97.0	97.5	98.5	39.6	58.0
01 - 16	WEIGHTED AVERAGE	97.9	99.0	98.4	99.2	98.8	99.0	96.3	94.0	95.2	143.2	29.0	87.6	100.1	108.1	104.0	95.9	54.6	75.8

City Cost Indexes

		SOUTH CAROLINA											SOUTH DAKOTA					
	DIVISION	COLUMBIA			FLORENCE			GREENVILLE			SPARTANBURG			ABERDEEN			PIERRE		
		MAT.	INST.	TOTAL	MAT.	INST.	TOTAL	MAT.	INST.	TOTAL	MAT.	INST.	TOTAL	MAT.	INST.	TOTAL	MAT.	INST.	TOTAL
02	SITE CONSTRUCTION	93.3	82.0	84.6	104.4	82.0	87.2	99.4	81.6	85.8	99.2	81.6	85.7	81.5	95.2	92.0	79.9	95.2	91.6
03100	CONCRETE FORMS & ACCESSORIES	100.6	47.2	54.5	78.9	47.2	51.6	95.1	47.0	53.6	99.3	47.0	54.2	93.5	44.4	51.2	91.7	46.7	52.9
03200	CONCRETE REINFORCEMENT	93.1	65.5	77.4	92.8	65.8	77.4	92.7	46.4	66.3	92.7	46.4	66.3	107.2	57.5	78.9	106.7	64.5	82.6
03300	CAST-IN-PLACE CONCRETE	79.9	51.5	68.2	72.1	55.5	65.3	72.1	55.4	65.2	72.1	55.4	65.2	100.0	55.9	81.9	97.2	51.8	78.4
03	CONCRETE	89.0	54.1	71.6	91.3	55.5	73.4	90.0	51.8	70.9	90.3	51.8	71.1	96.1	52.2	74.2	93.8	53.1	73.5
04	MASONRY	83.7	37.3	55.2	71.2	42.6	53.6	69.3	42.6	52.9	71.2	42.6	53.6	108.5	57.9	77.4	105.8	48.8	70.8
05	METALS	95.2	85.0	91.5	94.2	85.5	91.0	94.0	78.7	88.5	94.0	78.7	88.5	107.4	72.7	94.9	107.3	75.0	95.7
06	WOOD & PLASTICS	103.3	48.0	74.7	79.3	48.0	63.1	96.3	48.0	71.3	101.0	48.0	73.6	96.4	42.1	68.3	94.5	45.5	69.1
07	THERMAL & MOISTURE PROTECTION	96.3	48.4	73.5	96.5	50.9	74.9	96.6	50.9	74.9	96.7	50.9	74.9	97.1	52.4	75.9	97.2	50.3	75.0
08	DOORS & WINDOWS	97.1	49.0	85.5	93.2	49.0	82.5	93.2	44.4	81.4	93.2	44.4	81.4	98.5	45.8	85.8	102.0	49.8	89.4
09200	PLASTER & GYPSUM BOARD	105.8	46.2	69.4	96.7	46.2	65.8	101.2	46.2	67.6	102.4	46.2	68.1	101.7	40.8	64.5	100.6	44.3	66.2
095,098	CEILINGS & ACOUSTICAL TREATMENT	92.2	46.2	61.9	89.5	46.2	60.9	88.2	46.2	60.5	88.2	46.2	60.5	110.8	40.8	64.6	109.4	44.3	66.5
09600	FLOORING	95.8	47.6	84.1	87.0	47.6	77.5	93.6	48.5	82.7	95.3	48.5	83.9	111.4	68.4	100.9	110.5	48.7	95.5
097,099	WALL FINISHES, PAINTS & COATINGS	106.5	46.0	71.3	106.5	46.0	71.3	106.5	46.0	71.3	106.5	46.0	71.3	102.5	40.9	66.7	102.5	40.9	66.7
09	FINISHES	95.5	47.1	70.8	91.7	47.1	68.9	93.9	47.2	70.1	94.7	47.2	70.5	106.7	48.2	76.8	105.9	46.2	75.4
10 - 14	TOTAL DIV. 10000 - 14000	100.0	66.6	93.0	100.0	66.5	92.9	100.0	66.6	92.9	100.0	66.6	92.9	100.0	67.9	93.2	100.0	68.4	93.3
15	MECHANICAL	100.1	44.3	74.3	100.1	44.3	74.3	100.1	44.3	74.3	100.1	44.3	74.3	100.2	41.6	73.1	100.2	41.6	73.1
16	ELECTRICAL	99.6	41.7	59.8	95.4	23.8	46.2	98.9	37.9	57.0	98.9	37.9	57.0	102.5	48.1	65.1	96.1	56.4	68.8
01 - 16	WEIGHTED AVERAGE	95.9	53.2	75.1	94.2	51.0	73.1	94.4	52.0	73.8	94.7	52.0	73.9	101.2	55.7	79.1	100.6	56.5	79.1

		SOUTH DAKOTA						TENNESSEE											
	DIVISION	RAPID CITY			SIOUX FALLS			CHATTANOOGA			JACKSON			JOHNSON CITY			KNOXVILLE		
		MAT.	INST.	TOTAL	MAT.	INST.	TOTAL	MAT.	INST.	TOTAL	MAT.	INST.	TOTAL	MAT.	INST.	TOTAL	MAT.	INST.	TOTAL
02	SITE CONSTRUCTION	80.1	94.9	91.5	81.1	97.2	93.4	104.9	97.6	99.3	105.0	95.6	97.8	114.5	86.3	92.9	92.0	86.3	87.6
03100	CONCRETE FORMS & ACCESSORIES	104.8	42.6	51.1	91.7	44.9	51.4	95.4	52.5	58.4	86.2	38.0	44.6	79.2	54.7	58.1	94.3	54.8	60.2
03200	CONCRETE REINFORCEMENT	100.1	64.6	79.9	100.1	65.4	80.3	94.0	49.4	68.5	99.7	48.8	70.7	98.0	49.6	70.4	94.0	49.6	68.7
03300	CAST-IN-PLACE CONCRETE	96.5	48.7	76.8	100.5	56.1	82.2	103.0	54.6	83.0	101.7	38.8	75.8	82.8	61.0	73.8	96.7	55.9	79.9
03	CONCRETE	93.6	50.2	71.9	94.7	54.0	74.3	93.9	54.6	74.3	94.0	42.6	68.3	99.2	57.8	78.5	91.0	56.1	73.5
04	MASONRY	105.9	46.0	69.1	103.0	64.1	79.1	96.9	51.8	69.2	110.8	31.1	61.8	107.5	51.6	73.1	74.9	51.6	60.6
05	METALS	109.4	75.3	97.1	109.9	76.8	98.0	97.1	83.1	92.0	95.1	79.7	89.6	94.7	82.8	90.4	97.3	82.9	92.1
06	WOOD & PLASTICS	105.0	42.6	72.7	94.5	42.1	67.4	95.8	53.5	73.9	83.8	40.2	61.3	71.4	56.8	63.8	86.7	56.8	71.2
07	THERMAL & MOISTURE PROTECTION	97.7	49.2	74.7	97.2	56.0	77.6	103.0	54.5	80.0	100.7	40.3	72.0	97.3	55.3	77.4	94.9	54.8	75.8
08	DOORS & WINDOWS	103.0	48.2	89.8	103.0	47.9	89.7	100.7	52.8	89.2	101.7	44.1	87.8	97.0	56.2	87.1	93.7	56.2	84.6
09200	PLASTER & GYPSUM BOARD	101.6	41.3	64.7	101.6	40.8	64.5	95.1	52.7	69.2	98.3	38.9	62.0	104.7	56.0	74.9	110.4	56.0	77.2
095,098	CEILINGS & ACOUSTICAL TREATMENT	114.9	41.3	66.3	114.9	40.8	66.0	93.8	52.7	66.6	95.8	38.9	58.2	85.3	56.0	66.0	90.7	56.0	67.8
09600	FLOORING	110.5	57.3	97.6	110.5	79.3	102.9	99.4	59.1	89.7	90.7	28.6	75.7	92.7	60.8	84.9	98.2	60.8	89.1
097,099	WALL FINISHES, PAINTS & COATINGS	102.5	40.9	66.7	102.5	40.9	66.7	104.9	49.9	72.9	94.4	32.6	58.5	102.8	63.9	80.2	102.8	63.9	80.2
09	FINISHES	107.1	44.7	75.3	107.1	50.4	78.2	97.5	53.5	75.0	95.9	36.3	65.5	98.1	57.0	77.1	92.8	57.0	74.5
10 - 14	TOTAL DIV. 10000 - 14000	100.0	65.7	92.8	100.0	68.0	93.2	100.0	65.3	92.7	100.0	58.9	91.3	100.0	68.8	93.4	100.0	68.9	93.4
15	MECHANICAL	100.2	38.6	71.7	100.2	41.7	73.1	100.2	50.7	77.3	100.1	44.8	74.5	99.8	56.1	79.6	99.8	59.8	81.3
16	ELECTRICAL	96.9	56.4	69.1	95.8	72.1	79.5	103.5	63.9	76.3	99.3	36.6	56.2	90.9	39.1	55.3	100.4	62.8	74.5
01 - 16	WEIGHTED AVERAGE	101.3	54.9	78.7	101.2	61.6	81.9	98.9	62.1	81.0	98.8	49.2	74.7	98.3	58.9	79.1	95.0	63.3	79.6

		TENNESSEE						TEXAS											
	DIVISION	MEMPHIS			NASHVILLE			ABILENE			AMARILLO			AUSTIN			BEAUMONT		
		MAT.	INST.	TOTAL	MAT.	INST.	TOTAL	MAT.	INST.	TOTAL	MAT.	INST.	TOTAL	MAT.	INST.	TOTAL	MAT.	INST.	TOTAL
02	SITE CONSTRUCTION	98.7	92.1	93.7	95.6	100.3	99.2	103.0	85.1	89.3	103.1	86.1	90.1	88.1	85.9	86.4	96.7	87.4	89.6
03100	CONCRETE FORMS & ACCESSORIES	94.7	62.7	67.1	95.6	68.3	72.0	96.6	48.8	55.3	100.2	58.0	63.8	98.6	62.8	67.7	105.9	64.3	70.0
03200	CONCRETE REINFORCEMENT	92.3	64.2	76.3	95.0	63.1	76.8	95.5	59.9	75.2	95.5	59.5	75.0	94.0	59.9	74.6	94.8	54.0	71.6
03300	CAST-IN-PLACE CONCRETE	95.9	66.5	83.8	91.4	65.5	80.7	98.2	47.6	77.3	101.8	56.0	82.9	91.9	57.9	77.9	91.6	64.1	80.2
03	CONCRETE	89.9	66.2	78.1	89.4	68.0	78.7	90.6	51.5	71.0	92.6	58.4	75.5	82.6	61.1	71.8	88.7	63.0	75.9
04	MASONRY	84.2	62.5	70.8	85.5	64.9	72.9	100.2	53.0	71.2	104.0	47.8	69.4	97.2	53.9	70.6	103.3	65.3	80.0
05	METALS	96.5	93.9	95.6	99.7	91.8	96.9	96.6	70.3	87.1	96.6	70.0	87.0	95.7	69.6	86.3	95.9	70.7	86.8
06	WOOD & PLASTICS	93.4	64.5	78.5	90.9	70.5	80.4	97.2	49.5	72.5	100.4	60.6	79.8	98.2	66.2	81.6	107.6	66.1	86.1
07	THERMAL & MOISTURE PROTECTION	99.1	64.8	82.8	98.4	66.2	83.1	98.3	55.7	78.1	100.4	52.1	77.5	95.2	59.3	78.2	100.3	65.9	84.0
08	DOORS & WINDOWS	98.8	67.2	91.2	96.0	68.9	89.5	93.6	52.2	83.6	93.6	55.4	84.4	95.9	63.7	88.1	97.4	59.9	88.3
09200	PLASTER & GYPSUM BOARD	103.2	63.6	79.0	102.4	70.1	82.7	97.6	48.8	67.8	97.6	60.3	74.8	98.7	66.0	78.7	99.9	65.9	79.1
095,098	CEILINGS & ACOUSTICAL TREATMENT	89.2	63.6	72.3	87.7	70.1	76.1	101.1	48.8	66.6	101.1	60.3	74.2	90.6	66.0	74.3	108.4	65.9	80.3
09600	FLOORING	92.0	54.0	82.8	103.4	73.4	96.1	118.7	64.7	105.6	118.5	52.4	102.4	98.3	60.3	89.0	117.3	76.4	107.4
097,099	WALL FINISHES, PAINTS & COATINGS	93.5	64.8	76.8	109.1	60.2	80.7	94.4	64.6	77.1	94.4	46.1	66.3	95.0	51.8	69.9	91.0	62.2	74.3
09	FINISHES	93.2	60.9	76.7	101.7	68.6	84.8	102.8	53.3	77.6	102.8	55.6	78.8	93.9	61.3	77.3	98.7	66.5	82.3
10 - 14	TOTAL DIV. 10000 - 14000	100.0	71.7	94.0	100.0	73.1	94.3	100.0	68.3	93.3	100.0	70.7	93.8	100.0	71.3	93.9	100.0	74.9	94.7
15	MECHANICAL	100.0	66.7	84.6	100.0	66.2	84.4	100.1	43.1	73.7	100.1	59.3	81.2	99.7	67.7	84.9	100.0	65.7	84.1
16	ELECTRICAL	99.1	93.8	95.4	101.6	70.3	80.1	97.7	46.2	62.3	98.7	65.6	75.9	97.1	72.5	80.2	96.6	78.3	84.0
01 - 16	WEIGHTED AVERAGE	96.3	75.2	86.0	97.4	73.3	85.7	97.7	54.9	76.9	98.4	62.3	80.8	95.1	67.2	81.5	97.5	70.1	84.2

City Cost Indexes

DIVISION		TEXAS																	
		CORPUS CHRISTI			DALLAS			EL PASO			FORT WORTH			HOUSTON			LAREDO		
		MAT.	INST.	TOTAL	MAT.	INST.	TOTAL	MAT.	INST.	TOTAL	MAT.	INST.	TOTAL	MAT.	INST.	TOTAL	MAT.	INST.	TOTAL
02	SITE CONSTRUCTION	123.0	81.1	90.9	131.9	82.5	94.1	103.5	85.1	89.4	102.8	85.4	89.5	124.5	85.1	94.4	88.2	85.7	86.2
03100	CONCRETE FORMS & ACCESSORIES	102.2	47.0	54.6	95.4	62.6	67.1	97.3	57.5	63.0	99.3	62.2	67.3	92.8	71.6	74.5	90.2	47.1	53.0
03200	CONCRETE REINFORCEMENT	92.2	57.0	72.1	94.4	61.7	75.7	95.5	59.9	75.2	95.5	61.4	76.1	97.5	62.6	77.6	94.0	57.1	73.0
03300	CAST-IN-PLACE CONCRETE	108.9	53.6	86.1	102.1	62.4	85.7	100.5	43.1	76.8	96.5	57.6	80.5	94.4	72.4	85.3	83.9	60.4	74.2
03	CONCRETE	93.0	53.2	73.1	89.7	64.0	76.8	91.8	53.6	72.7	89.9	61.1	75.5	88.0	71.9	80.0	82.2	54.4	68.3
04	MASONRY	87.9	52.8	66.4	95.9	64.1	76.3	99.0	52.3	70.3	95.1	60.3	73.7	98.3	68.8	80.2	96.9	55.3	71.3
05	METALS	95.2	84.1	91.2	96.5	86.6	92.9	96.5	68.7	86.4	96.4	71.4	87.4	100.2	92.3	97.4	96.5	65.2	85.2
06	WOOD & PLASTICS	109.5	47.1	77.2	99.9	63.9	81.2	98.5	62.4	79.8	105.3	63.9	83.9	91.8	72.7	81.9	87.4	46.9	66.5
07	THERMAL & MOISTURE PROTECTION	97.9	54.1	77.1	93.8	64.6	79.9	97.9	58.9	79.4	98.8	60.0	80.4	92.7	69.6	81.7	94.1	57.0	76.5
08	DOORS & WINDOWS	102.9	47.2	89.5	106.6	60.2	95.4	93.6	55.7	84.5	87.1	60.2	80.6	105.7	68.6	96.8	96.5	45.9	84.3
09200	PLASTER & GYPSUM BOARD	99.7	46.1	66.9	100.8	63.7	78.1	97.6	62.1	75.9	97.6	63.7	76.9	98.2	72.5	82.5	96.5	46.1	65.7
095,098	CEILINGS & ACOUSTICAL TREATMENT	96.0	46.1	63.1	102.1	63.7	76.8	101.1	62.1	75.4	101.1	63.7	76.4	108.0	72.5	84.6	90.6	46.1	61.2
09600	FLOORING	110.4	57.0	97.4	110.5	73.0	101.4	118.7	66.7	106.1	153.9	58.4	130.7	104.7	65.5	95.2	95.1	57.0	85.9
097,099	WALL FINISHES, PAINTS & COATINGS	109.1	46.2	72.6	101.8	60.4	77.7	94.4	44.3	65.3	95.6	65.8	78.3	99.6	70.3	82.6	95.0	53.4	70.8
09	FINISHES	100.3	48.2	73.8	105.7	64.4	84.6	102.8	58.5	80.2	114.5	62.0	87.8	103.9	70.4	86.8	92.7	48.9	70.4
10 - 14	TOTAL DIV. 10000 - 14000	100.0	70.6	93.8	100.0	72.7	94.2	100.0	71.3	93.9	100.0	72.7	94.2	100.0	78.3	95.4	100.0	69.5	93.6
15	MECHANICAL	99.7	56.9	79.9	99.9	65.8	84.1	100.1	42.4	73.4	100.1	65.2	84.0	99.9	74.3	88.1	99.7	51.1	77.2
16	ELECTRICAL	94.0	58.8	69.8	98.9	68.6	78.0	96.5	61.7	72.6	96.5	66.3	75.7	98.5	71.6	80.0	97.1	69.0	77.8
01 - 16	WEIGHTED AVERAGE	98.3	60.0	79.7	100.0	69.1	85.0	97.7	58.6	78.7	97.8	66.4	82.6	99.8	75.0	87.7	95.0	59.8	77.8

DIVISION		TEXAS																UTAH		
		LUBBOCK			ODESSA			SAN ANTONIO			WACO			WICHITA FALLS			LOGAN		
		MAT.	INST.	TOTAL	MAT.	INST.	TOTAL	MAT.	INST.	TOTAL	MAT.	INST.	TOTAL	MAT.	INST.	TOTAL	MAT.	INST.	TOTAL
02	SITE CONSTRUCTION	133.7	83.3	95.1	103.0	85.6	89.7	87.7	90.1	89.6	101.7	85.4	89.2	102.5	85.1	89.2	96.0	99.8	98.9
03100	CONCRETE FORMS & ACCESSORIES	98.5	50.6	57.2	96.6	48.5	55.1	90.3	64.6	68.1	98.7	48.3	55.2	98.7	50.8	57.4	104.3	68.9	73.8
03200	CONCRETE REINFORCEMENT	94.8	59.5	74.7	97.7	59.3	75.8	94.0	58.2	73.6	95.5	59.7	75.1	95.5	51.6	70.5	100.6	75.7	86.4
03300	CAST-IN-PLACE CONCRETE	98.4	56.2	81.0	98.2	51.5	78.9	82.3	65.8	75.5	88.6	59.6	76.6	94.7	54.4	78.1	93.1	78.1	86.9
03	CONCRETE	89.4	56.3	72.8	90.8	52.6	71.7	81.4	64.4	72.9	86.0	55.4	70.7	89.0	53.4	71.2	112.3	73.7	93.0
04	MASONRY	99.7	44.1	65.5	100.2	44.2	65.7	96.7	62.3	75.6	97.3	56.3	72.1	97.7	54.2	71.0	122.1	72.5	91.6
05	METALS	100.7	86.3	95.5	96.8	69.3	86.9	97.1	71.0	87.7	96.6	70.0	87.0	96.5	70.7	87.2	105.3	78.3	95.6
06	WOOD & PLASTICS	100.5	52.0	75.4	97.2	49.3	72.4	87.4	65.7	76.2	104.1	45.8	74.0	104.1	51.8	77.1	90.1	67.4	78.4
07	THERMAL & MOISTURE PROTECTION	90.8	54.9	73.8	98.3	50.7	75.7	94.1	69.1	82.2	99.0	55.5	78.4	99.0	56.2	78.7	104.5	76.7	91.3
08	DOORS & WINDOWS	105.7	51.3	92.6	93.6	49.3	82.9	96.5	62.6	88.4	87.1	46.0	77.2	87.1	53.3	79.0	90.2	66.5	84.5
09200	PLASTER & GYPSUM BOARD	98.1	51.2	69.5	97.6	48.6	67.7	96.5	65.4	77.5	97.6	45.0	65.5	97.6	51.2	69.3	88.2	66.2	74.8
095,098	CEILINGS & ACOUSTICAL TREATMENT	103.8	51.2	69.1	101.1	48.6	66.5	90.6	65.4	73.9	101.1	45.0	64.1	101.1	51.2	68.2	104.7	66.2	79.3
09600	FLOORING	109.7	49.6	95.1	118.7	48.9	101.8	95.1	61.3	86.9	154.1	45.4	127.8	154.6	66.7	133.3	105.2	72.3	97.2
097,099	WALL FINISHES, PAINTS & COATINGS	106.4	41.4	68.6	94.4	41.4	63.6	95.0	53.4	70.8	95.6	42.7	64.8	97.5	67.7	80.1	108.7	57.7	79.1
09	FINISHES	104.4	49.0	76.1	102.8	47.4	74.5	92.7	62.6	77.4	114.5	46.2	79.7	114.8	55.4	84.5	100.0	67.8	83.6
10 - 14	TOTAL DIV. 10000 - 14000	100.0	69.1	93.5	100.0	70.3	93.7	100.0	74.9	94.7	100.0	70.3	93.7	100.0	68.7	93.4	100.0	78.5	95.5
15	MECHANICAL	99.7	54.2	78.7	100.1	46.4	75.3	99.7	73.3	87.5	100.1	62.9	82.9	100.1	52.0	77.9	100.1	74.4	88.2
16	ELECTRICAL	96.2	56.3	68.8	97.7	52.4	66.6	97.1	69.0	77.8	97.7	66.6	76.4	102.6	65.2	76.9	92.7	82.2	85.5
01 - 16	WEIGHTED AVERAGE	100.3	59.3	80.4	97.8	55.0	77.0	94.9	70.0	82.8	97.6	61.9	80.2	98.3	60.6	80.0	101.8	77.3	89.9

DIVISION		UTAH						VERMONT						VIRGINIA		
		OGDEN			PROVO			SALT LAKE CITY			BURLINGTON			RUTLAND			ALEXANDRIA		
		MAT.	INST.	TOTAL	MAT.	INST.	TOTAL	MAT.	INST.	TOTAL	MAT.	INST.	TOTAL	MAT.	INST.	TOTAL	MAT.	INST.	TOTAL
02	SITE CONSTRUCTION	83.3	99.8	95.9	92.3	97.8	96.5	82.9	99.7	95.8	83.3	101.5	97.3	83.3	101.5	97.3	112.6	83.8	90.5
03100	CONCRETE FORMS & ACCESSORIES	104.3	68.9	73.8	106.1	69.0	74.1	102.9	69.0	73.6	88.5	58.3	62.4	100.2	58.3	64.1	91.6	73.7	76.1
03200	CONCRETE REINFORCEMENT	100.6	75.7	86.4	108.8	75.7	89.9	98.5	75.7	85.5	111.9	66.6	86.1	111.9	66.6	86.1	82.7	91.3	87.6
03300	CAST-IN-PLACE CONCRETE	94.5	78.1	87.8	93.2	78.1	87.0	102.2	78.1	92.3	97.2	64.5	83.7	92.6	64.5	81.0	99.1	80.9	91.6
03	CONCRETE	101.2	73.7	87.4	111.7	73.7	92.7	104.6	73.7	89.2	105.3	62.5	83.9	103.9	62.5	83.2	96.6	80.6	88.6
04	MASONRY	115.3	72.5	89.0	127.6	72.5	93.7	133.9	72.5	96.2	96.3	70.1	80.2	84.0	70.1	75.4	83.9	70.7	75.7
05	METALS	105.7	78.3	95.8	102.7	78.3	93.9	104.5	78.3	95.1	98.6	72.6	89.2	98.0	72.6	88.8	95.2	97.4	96.0
06	WOOD & PLASTICS	90.1	67.4	78.4	92.0	67.4	79.3	90.2	67.4	78.4	88.6	56.9	72.2	102.5	56.9	78.9	95.7	74.4	84.7
07	THERMAL & MOISTURE PROTECTION	102.9	76.7	90.5	107.0	76.7	92.6	106.0	76.7	92.1	99.2	64.0	82.5	99.0	64.0	82.4	96.5	79.8	88.6
08	DOORS & WINDOWS	90.2	66.5	84.5	94.5	66.5	87.8	90.2	66.5	84.5	110.0	52.4	96.1	110.0	52.4	96.1	97.1	74.9	91.8
09200	PLASTER & GYPSUM BOARD	88.2	66.2	74.8	87.1	66.2	74.3	88.2	66.2	74.8	98.2	54.2	71.3	98.2	54.2	71.3	105.8	73.4	86.0
095,098	CEILINGS & ACOUSTICAL TREATMENT	104.7	66.2	79.3	96.5	66.2	76.5	104.7	66.2	79.3	102.6	54.2	70.7	102.6	54.2	70.7	92.2	73.3	79.8
09600	FLOORING	105.2	72.3	97.2	106.0	72.3	97.8	104.8	72.3	96.9	99.9	81.9	95.6	99.9	81.9	95.6	96.0	80.0	92.1
097,099	WALL FINISHES, PAINTS & COATINGS	108.7	57.7	79.1	108.7	67.0	84.5	108.7	67.0	84.5	95.6	46.9	67.3	95.6	46.9	67.3	117.5	77.8	94.4
09	FINISHES	98.9	67.8	83.0	99.0	68.9	83.6	99.2	68.9	83.7	99.1	61.2	79.7	99.0	61.2	79.7	96.1	74.9	85.3
10 - 14	TOTAL DIV. 10000 - 14000	100.0	78.5	95.5	100.0	78.5	95.5	100.0	78.5	95.5	100.0	96.7	99.3	100.0	96.7	99.3	100.0	86.0	97.0
15	MECHANICAL	100.1	74.4	88.2	100.1	74.4	88.2	100.3	74.4	88.3	99.8	60.1	81.4	99.8	60.1	81.4	100.1	84.6	92.9
16	ELECTRICAL	92.7	82.2	85.5	93.2	82.2	85.6	93.6	82.2	85.7	105.2	70.6	81.4	104.0	70.6	81.0	98.7	93.2	94.9
01 - 16	WEIGHTED AVERAGE	99.6	77.3	88.7	102.1	77.2	90.0	101.1	77.4	89.6	101.0	69.2	85.5	100.1	69.2	85.1	97.3	83.5	90.6

City Cost Indexes

DIVISION		VIRGINIA																	
		ARLINGTON			NEWPORT NEWS			NORFOLK			PORTSMOUTH			RICHMOND			ROANOKE		
		MAT.	INST.	TOTAL	MAT.	INST.	TOTAL	MAT.	INST.	TOTAL	MAT.	INST.	TOTAL	MAT.	INST.	TOTAL	MAT.	INST.	TOTAL
02	SITE CONSTRUCTION	123.0	83.4	92.7	105.2	85.0	89.7	104.4	86.1	90.4	103.6	84.6	89.1	105.8	85.7	90.4	101.9	81.7	86.4
03100	CONCRETE FORMS & ACCESSORIES	88.5	72.5	74.7	95.2	53.7	59.4	100.2	53.8	60.2	81.6	53.7	57.5	96.1	63.0	67.5	95.0	41.8	49.1
03200	CONCRETE REINFORCEMENT	93.3	77.9	84.5	93.1	76.4	83.6	93.1	76.4	83.6	92.8	78.7	84.8	93.1	76.6	83.7	93.1	77.8	84.4
03300	CAST-IN-PLACE CONCRETE	96.4	74.9	87.5	97.0	59.9	81.7	99.8	59.9	83.4	96.0	61.8	81.9	103.1	62.1	86.2	100.5	53.3	81.0
03	CONCRETE	104.2	75.5	89.8	97.1	61.8	79.5	98.9	61.9	80.4	95.7	62.9	79.3	100.2	66.7	83.5	98.8	54.5	76.7
04	MASONRY	95.0	68.2	78.6	86.0	57.9	68.7	93.7	57.9	71.7	90.7	56.2	69.5	82.5	55.8	66.1	85.6	39.8	57.5
05	METALS	94.2	92.4	93.5	95.3	91.6	93.9	94.5	91.7	93.5	94.5	91.5	93.4	95.3	92.2	94.1	95.1	87.1	92.2
06	WOOD & PLASTICS	91.3	74.4	82.6	96.8	54.1	74.7	102.9	54.1	77.7	82.1	54.1	67.6	98.1	65.8	81.4	96.8	41.5	68.2
07	THERMAL & MOISTURE PROTECTION	97.2	76.0	87.2	96.3	53.6	76.0	96.1	53.6	76.0	96.2	53.2	75.8	95.8	56.6	77.2	96.2	47.5	73.1
08	DOORS & WINDOWS	95.1	71.9	89.5	97.1	55.0	87.0	97.1	58.2	87.7	97.2	58.7	87.9	97.1	60.3	88.2	97.1	47.4	85.1
09200	PLASTER & GYPSUM BOARD	101.8	73.4	84.5	105.8	51.6	72.6	105.8	51.6	72.6	98.0	51.6	69.6	105.8	63.6	80.0	105.8	39.5	65.3
095,098	CEILINGS & ACOUSTICAL TREATMENT	89.5	73.4	78.9	92.2	51.6	65.4	92.2	51.6	65.4	92.2	51.6	65.4	92.2	63.6	73.3	92.2	39.5	57.4
09600	FLOORING	94.5	80.0	91.0	96.0	46.6	84.0	95.8	46.6	83.9	88.1	46.6	78.1	95.8	69.3	89.4	96.0	41.8	82.9
097,099	WALL FINISHES, PAINTS & COATINGS	117.5	93.3	103.4	106.5	49.7	73.5	106.5	64.3	82.0	106.5	43.6	69.9	106.5	65.0	82.4	106.5	38.8	67.2
09	FINISHES	95.4	76.2	85.6	95.4	51.4	72.9	95.3	53.1	73.8	91.7	50.7	70.8	95.2	64.6	79.6	95.2	40.9	67.5
10 - 14	TOTAL DIV. 10000 - 14000	100.0	85.0	96.8	100.0	76.7	95.1	100.0	76.8	95.1	100.0	76.8	95.1	100.0	84.2	96.7	100.0	70.7	93.8
15	MECHANICAL	100.1	83.2	92.3	100.1	68.1	85.3	100.1	65.2	83.9	100.1	65.2	83.9	100.1	67.7	85.1	100.1	47.6	75.8
16	ELECTRICAL	95.1	90.9	92.2	98.6	62.7	73.9	98.6	64.7	75.3	95.7	64.7	74.4	99.6	69.3	78.8	98.6	41.3	59.2
01 - 16	WEIGHTED AVERAGE	98.4	81.4	90.1	97.2	66.4	82.2	97.8	66.6	82.6	96.4	66.1	81.6	97.5	70.1	84.2	97.3	53.4	75.9

DIVISION		WASHINGTON																	
		EVERETT			RICHLAND			SEATTLE			SPOKANE			TACOMA			VANCOUVER		
		MAT.	INST.	TOTAL	MAT.	INST.	TOTAL	MAT.	INST.	TOTAL	MAT.	INST.	TOTAL	MAT.	INST.	TOTAL	MAT.	INST.	TOTAL
02	SITE CONSTRUCTION	93.3	118.0	112.2	97.2	90.7	92.2	97.3	115.7	111.4	98.0	90.7	92.4	96.1	118.0	112.9	108.8	103.4	104.7
03100	CONCRETE FORMS & ACCESSORIES	109.7	100.7	102.0	113.5	85.6	89.4	92.6	101.2	100.1	122.0	85.7	90.7	92.6	101.0	99.9	97.2	99.9	99.5
03200	CONCRETE REINFORCEMENT	107.1	95.6	100.6	101.1	92.7	96.4	101.1	95.7	98.0	101.8	92.9	96.7	101.1	95.6	98.0	104.0	95.5	99.2
03300	CAST-IN-PLACE CONCRETE	93.1	105.8	98.3	112.1	89.6	102.8	97.7	108.2	102.0	116.1	89.7	105.2	95.7	108.0	100.8	106.9	104.4	105.9
03	CONCRETE	100.3	100.9	100.6	109.2	88.2	98.7	102.3	102.0	102.2	111.7	88.2	99.9	101.3	101.8	101.6	112.9	100.3	106.6
04	MASONRY	129.6	104.9	114.5	113.8	87.2	97.4	123.1	105.6	112.3	115.5	90.2	99.9	123.0	105.6	112.3	124.7	105.7	113.0
05	METALS	99.3	88.8	95.5	93.1	85.4	90.4	100.8	90.9	97.2	92.9	85.7	90.3	100.3	88.8	96.2	98.8	91.5	96.2
06	WOOD & PLASTICS	111.3	99.0	105.0	103.2	84.6	93.6	93.9	99.0	96.5	113.5	84.6	98.5	92.6	99.0	95.9	90.7	98.6	94.7
07	THERMAL & MOISTURE PROTECTION	107.4	100.5	104.2	171.3	84.3	130.0	107.2	101.5	104.5	169.0	84.9	129.1	107.0	101.3	104.3	110.5	97.1	104.1
08	DOORS & WINDOWS	101.5	97.2	100.5	115.7	81.1	107.3	101.7	97.7	100.7	115.4	80.2	106.9	102.1	97.7	101.0	98.5	96.2	98.0
09200	PLASTER & GYPSUM BOARD	96.8	98.6	97.9	123.8	84.1	99.5	91.6	98.6	95.9	126.8	84.1	100.7	94.2	98.6	96.9	95.4	98.6	97.4
095,098	CEILINGS & ACOUSTICAL TREATMENT	110.9	98.6	102.8	122.2	84.1	97.0	112.0	98.6	103.2	122.2	84.1	97.0	113.7	98.6	103.8	122.0	98.6	106.6
09600	FLOORING	123.6	107.9	119.8	112.5	53.0	98.1	115.3	107.9	113.5	116.0	79.7	107.2	116.0	107.9	114.0	121.4	97.6	115.6
097,099	WALL FINISHES, PAINTS & COATINGS	115.0	91.5	101.3	117.6	76.4	93.6	115.0	91.5	101.3	117.6	80.9	96.3	115.0	91.5	101.3	123.1	75.3	95.3
09	FINISHES	115.9	100.9	108.3	129.2	77.8	103.0	112.8	101.0	106.8	130.8	83.8	106.8	113.6	101.0	107.2	114.0	96.6	105.1
10 - 14	TOTAL DIV. 10000 - 14000	100.0	104.2	100.9	100.0	92.4	98.4	100.0	104.5	100.9	100.0	92.4	98.4	100.0	104.4	100.9	100.0	99.2	99.8
15	MECHANICAL	100.1	105.2	102.5	100.7	99.9	100.4	100.1	111.8	105.5	100.7	89.8	95.7	100.2	102.3	101.2	100.4	101.4	100.8
16	ELECTRICAL	106.3	93.6	97.6	104.2	96.3	98.8	106.0	105.6	105.7	94.7	88.0	90.1	106.0	99.0	101.2	119.1	108.3	111.6
01 - 16	WEIGHTED AVERAGE	104.1	101.4	102.8	108.4	90.1	99.5	103.8	104.9	104.4	108.3	87.8	98.3	103.7	102.0	102.9	106.1	101.1	103.7

DIVISION		WASHINGTON			WEST VIRGINIA												WISCONSIN		
		YAKIMA			CHARLESTON			HUNTINGTON			PARKERSBURG			WHEELING			EAU CLAIRE		
		MAT.	INST.	TOTAL	MAT.	INST.	TOTAL	MAT.	INST.	TOTAL	MAT.	INST.	TOTAL	MAT.	INST.	TOTAL	MAT.	INST.	TOTAL
02	SITE CONSTRUCTION	99.3	117.3	113.1	100.6	86.0	89.4	102.1	86.4	90.1	106.6	86.0	90.9	107.3	85.1	90.3	82.2	102.3	97.6
03100	CONCRETE FORMS & ACCESSORIES	95.1	96.9	96.6	103.5	95.5	96.6	95.9	97.8	97.5	84.4	94.2	92.8	86.6	90.6	90.1	95.9	95.5	95.6
03200	CONCRETE REINFORCEMENT	103.6	94.8	98.6	93.1	85.7	88.8	93.1	92.1	92.5	91.8	97.6	95.1	91.1	93.1	92.2	103.7	99.3	101.2
03300	CAST-IN-PLACE CONCRETE	102.2	90.9	97.6	94.9	108.7	100.6	102.3	109.6	105.3	95.6	97.9	96.6	95.6	101.9	98.2	97.8	92.3	95.5
03	CONCRETE	106.8	93.8	100.3	96.6	99.0	97.8	99.8	101.5	100.6	99.8	96.8	98.3	99.8	95.9	97.8	95.0	95.1	95.1
04	MASONRY	114.9	87.1	97.8	84.4	98.1	92.8	85.8	102.1	95.8	74.1	93.7	86.1	96.1	92.2	93.7	93.6	97.6	96.1
05	METALS	99.2	85.4	94.2	95.3	101.8	97.6	95.3	104.5	98.6	94.1	105.7	98.3	94.2	104.2	97.8	94.0	96.8	95.0
06	WOOD & PLASTICS	95.3	99.0	97.2	106.2	94.9	100.4	96.8	96.8	96.8	84.6	92.6	88.8	86.7	89.5	88.2	106.9	94.5	100.5
07	THERMAL & MOISTURE PROTECTION	107.3	85.5	96.9	96.3	93.0	94.7	96.5	94.7	95.6	96.6	90.7	93.8	96.8	89.9	93.5	96.7	89.7	93.4
08	DOORS & WINDOWS	101.7	85.9	97.9	98.3	85.8	95.3	97.1	88.2	95.0	97.7	86.3	95.0	98.6	90.3	96.6	104.3	91.3	101.2
09200	PLASTER & GYPSUM BOARD	94.3	98.6	96.9	105.8	94.6	99.0	105.8	96.5	100.1	99.7	92.2	95.1	100.1	89.0	93.3	112.5	94.8	101.7
095,098	CEILINGS & ACOUSTICAL TREATMENT	109.6	98.6	102.4	92.2	94.6	93.8	92.2	96.5	95.1	89.5	92.2	91.3	89.5	89.0	89.2	108.5	94.8	99.5
09600	FLOORING	117.3	79.7	108.2	95.8	104.6	97.9	95.8	114.2	100.3	91.1	97.9	92.7	92.0	103.1	94.7	92.6	72.3	87.7
097,099	WALL FINISHES, PAINTS & COATINGS	115.0	80.9	95.2	106.5	100.2	102.8	106.5	95.1	99.9	106.5	104.8	105.5	106.5	93.3	98.8	91.7	80.7	85.3
09	FINISHES	113.5	92.2	102.7	95.4	98.1	96.7	95.2	100.9	98.1	92.7	96.0	94.4	93.0	93.2	93.1	101.1	89.4	95.1
10 - 14	TOTAL DIV. 10000 - 14000	100.0	101.5	100.3	100.0	99.4	99.9	100.0	100.2	100.1	100.0	99.5	99.9	100.0	95.6	99.1	100.0	96.1	99.2
15	MECHANICAL	100.2	100.4	100.3	100.1	89.1	95.0	100.1	93.0	96.8	100.1	95.8	98.1	100.1	93.3	97.0	100.2	90.8	95.9
16	ELECTRICAL	110.8	96.3	100.8	98.6	77.9	84.3	98.6	90.8	93.2	99.4	97.2	97.8	94.6	91.0	92.1	105.0	88.2	93.5
01 - 16	WEIGHTED AVERAGE	104.2	95.9	100.2	97.2	91.6	94.4	97.4	96.0	96.7	96.4	95.4	96.0	97.6	93.1	95.4	98.6	93.3	96.0

City Cost Indexes

DIVISION		WISCONSIN																	
		GREEN BAY			KENOSHA			LA CROSSE			MADISON			MILWAUKEE			RACINE		
		MAT.	INST.	TOTAL	MAT.	INST.	TOTAL	MAT.	INST.	TOTAL	MAT.	INST.	TOTAL	MAT.	INST.	TOTAL	MAT.	INST.	TOTAL
02	SITE CONSTRUCTION	84.4	98.1	94.9	89.2	101.7	98.8	76.6	102.3	96.2	84.4	105.0	100.2	85.7	93.5	91.7	84.8	105.5	100.7
03100	CONCRETE FORMS & ACCESSORIES	115.7	95.2	98.1	109.9	103.8	104.7	78.6	95.6	93.3	97.1	100.6	100.1	100.0	108.3	107.2	96.8	103.8	102.9
03200	CONCRETE REINFORCEMENT	100.1	91.2	95.0	111.7	96.8	103.2	102.2	89.8	95.1	104.6	90.1	96.4	104.6	97.0	100.3	104.6	96.8	100.1
03300	CAST-IN-PLACE CONCRETE	101.2	97.9	99.8	109.9	100.4	106.0	88.0	91.9	89.6	99.1	98.4	98.8	99.1	104.2	101.2	99.1	100.0	99.4
03	CONCRETE	96.7	95.5	96.1	104.0	101.2	102.6	86.1	93.3	89.7	96.9	97.8	97.4	97.3	103.8	100.6	96.9	101.1	99.0
04	MASONRY	114.5	98.7	104.8	100.4	108.8	105.6	92.8	99.9	97.1	104.5	103.3	103.8	104.4	112.9	109.6	104.5	108.8	107.1
05	METALS	96.6	93.8	95.6	96.7	98.3	97.3	94.0	93.2	93.7	97.0	94.9	96.2	98.6	88.6	95.0	97.0	98.3	97.5
06	WOOD & PLASTICS	122.6	94.5	108.1	117.3	102.6	109.7	88.3	94.5	91.5	108.1	100.2	104.0	111.4	107.3	109.3	108.4	102.6	105.4
07	THERMAL & MOISTURE PROTECTION	99.4	90.1	95.0	96.0	101.9	98.8	95.9	89.5	92.9	95.2	98.0	96.6	94.6	105.1	99.5	95.8	101.4	98.4
08	DOORS & WINDOWS	102.8	87.9	99.2	99.7	101.6	100.1	104.3	85.3	99.7	104.6	97.7	102.9	104.5	104.2	104.4	104.6	101.6	103.9
09200	PLASTER & GYPSUM BOARD	96.4	94.8	95.4	97.4	103.2	100.9	106.3	94.8	99.3	107.3	100.7	103.3	107.3	107.9	107.7	107.3	103.2	104.8
095,098	CEILINGS & ACOUSTICAL TREATMENT	103.5	94.8	97.8	85.8	103.2	97.3	103.1	94.8	97.6	94.8	100.7	98.7	94.8	107.9	103.4	94.8	103.2	100.3
09600	FLOORING	111.5	102.2	109.2	108.5	92.0	104.5	85.3	102.2	89.4	91.9	97.1	93.1	94.6	109.4	98.2	91.9	92.0	91.9
097,099	WALL FINISHES, PAINTS & COATINGS	102.5	72.8	85.3	103.4	96.3	99.3	91.7	80.7	85.3	92.6	92.2	92.4	95.2	104.9	100.8	92.6	97.0	95.2
09	FINISHES	105.0	94.4	99.6	100.9	101.4	101.1	96.5	95.6	96.0	97.8	99.4	98.6	98.9	108.5	103.8	97.9	101.4	99.7
10 - 14	TOTAL DIV. 10000 - 14000	100.0	95.0	98.9	100.0	101.1	100.2	100.0	96.1	99.2	100.0	96.9	99.4	100.0	102.9	100.6	100.0	101.1	100.2
15	MECHANICAL	100.6	90.7	96.0	100.4	98.4	99.5	100.2	91.6	96.2	100.1	94.8	97.6	100.1	99.0	99.6	100.1	98.4	99.3
16	ELECTRICAL	96.1	89.3	91.5	93.3	99.4	97.5	105.4	88.2	93.6	92.1	95.4	94.4	92.9	101.5	98.8	91.0	99.2	96.7
01 - 16	WEIGHTED AVERAGE	100.4	93.3	97.0	99.4	101.0	100.2	96.6	93.5	95.1	98.7	98.0	98.3	99.2	101.6	100.3	98.6	101.3	99.9

DIVISION		WYOMING									CANADA								
		CASPER			CHEYENNE			ROCK SPRINGS			CALGARY, ALBERTA			EDMONTON, ALBERTA			HAMILTON, ONTARIO		
		MAT.	INST.	TOTAL	MAT.	INST.	TOTAL	MAT.	INST.	TOTAL	MAT.	INST.	TOTAL	MAT.	INST.	TOTAL	MAT.	INST.	TOTAL
02	SITE CONSTRUCTION	84.1	99.2	95.7	84.0	99.2	95.7	87.3	98.8	96.1	107.1	97.4	99.7	108.6	97.4	100.0	110.7	97.6	100.7
03100	CONCRETE FORMS & ACCESSORIES	102.0	50.2	57.3	102.0	50.2	57.3	108.0	35.5	45.4	119.2	80.4	85.7	116.5	80.4	85.3	118.6	104.4	106.3
03200	CONCRETE REINFORCEMENT	107.2	54.2	77.0	100.6	54.4	74.2	109.3	54.0	77.8	161.1	64.3	105.9	161.1	64.3	105.9	174.4	97.3	130.4
03300	CAST-IN-PLACE CONCRETE	104.9	65.0	88.4	104.9	65.0	88.4	105.3	48.9	82.1	162.8	82.8	129.8	163.6	82.8	130.3	157.2	106.8	136.4
03	CONCRETE	106.9	56.9	81.9	106.1	56.9	81.5	107.9	44.8	76.4	146.0	78.8	112.4	146.3	78.8	112.5	144.9	103.6	124.3
04	MASONRY	109.0	45.9	70.2	108.0	48.5	71.4	175.8	37.5	90.8	146.7	81.4	106.5	146.7	81.4	106.6	148.1	111.6	125.8
05	METALS	105.1	66.7	91.2	106.3	66.9	92.1	105.1	65.4	90.8	100.8	83.0	94.4	100.8	83.0	94.4	100.9	95.9	99.1
06	WOOD & PLASTICS	96.8	48.7	71.9	95.8	48.7	71.4	103.7	34.5	67.9	116.2	79.7	97.3	111.9	79.7	95.3	117.0	103.8	110.1
07	THERMAL & MOISTURE PROTECTION	104.5	61.9	84.3	104.5	62.6	84.6	105.2	47.1	77.7	104.3	80.8	93.1	104.2	80.7	93.1	104.3	101.6	103.0
08	DOORS & WINDOWS	94.3	50.0	83.6	95.0	50.0	84.2	101.3	42.4	87.1	92.3	75.6	88.3	92.3	75.6	88.3	92.3	99.2	94.0
09200	PLASTER & GYPSUM BOARD	86.7	46.9	62.3	88.2	46.9	62.9	90.1	32.2	54.7	147.0	79.1	105.5	144.4	79.1	104.5	209.2	104.0	144.9
095,098	CEILINGS & ACOUSTICAL TREATMENT	96.5	46.9	63.7	104.7	46.9	66.5	96.5	32.2	54.1	102.4	79.1	87.0	102.4	79.1	87.0	102.4	104.0	103.4
09600	FLOORING	105.2	49.6	91.7	105.2	72.9	97.4	107.5	48.6	93.2	132.6	83.1	120.6	132.6	83.1	120.6	132.6	108.8	126.8
097,099	WALL FINISHES, PAINTS & COATINGS	108.7	65.7	83.7	108.5	65.7	83.7	108.7	39.9	68.7	109.2	78.1	91.1	109.3	78.1	91.2	109.2	110.3	109.8
09	FINISHES	97.4	50.9	73.7	99.0	55.6	76.9	98.6	37.5	67.4	120.4	81.0	100.3	120.1	81.0	100.2	129.5	106.3	117.6
10 - 14	TOTAL DIV. 10000 - 14000	100.0	77.6	95.3	100.0	77.6	95.3	100.0	59.8	91.5	100.0	89.0	97.7	100.0	84.6	96.8	100.0	110.0	102.1
15	MECHANICAL	100.1	69.3	85.9	100.1	52.5	78.1	100.1	37.1	70.9	102.5	73.3	89.0	102.5	73.3	89.0	102.5	102.2	102.4
16	ELECTRICAL	92.6	63.1	72.4	92.6	62.2	71.7	91.6	53.6	65.5	113.7	80.3	90.8	113.7	80.3	90.8	127.0	102.0	109.8
01 - 16	WEIGHTED AVERAGE	100.3	63.9	82.6	100.6	61.4	81.5	105.3	50.7	78.7	111.8	80.9	96.8	111.8	80.8	96.7	113.8	102.8	108.4

DIVISION		CANADA																	
		LONDON, ONTARIO			MONTREAL, QUEBEC			OTTAWA, ONTARIO			QUEBEC, QUEBEC			TORONTO, ONTARIO			VANCOUVER, B C		
		MAT.	INST.	TOTAL	MAT.	INST.	TOTAL	MAT.	INST.	TOTAL	MAT.	INST.	TOTAL	MAT.	INST.	TOTAL	MAT.	INST.	TOTAL
02	SITE CONSTRUCTION	110.7	97.4	100.5	91.1	97.8	96.2	111.0	97.4	100.6	92.6	97.8	96.6	111.7	98.1	101.2	109.3	98.9	101.4
03100	CONCRETE FORMS & ACCESSORIES	118.5	99.0	101.7	128.8	96.1	100.6	116.8	100.1	102.4	128.8	96.3	100.7	120.3	112.3	113.4	111.0	100.7	102.1
03200	CONCRETE REINFORCEMENT	122.9	95.4	107.2	159.2	97.5	124.0	174.4	95.3	129.3	147.2	97.5	119.1	170.7	101.9	131.5	161.1	97.8	125.0
03300	CAST-IN-PLACE CONCRETE	157.2	104.0	135.3	140.1	102.7	124.7	159.6	104.0	136.6	151.5	102.8	131.4	164.5	114.9	144.0	159.6	104.6	136.9
03	CONCRETE	138.5	100.1	119.2	135.3	98.5	116.9	146.0	100.4	123.2	139.5	98.6	119.1	148.2	110.8	129.5	150.2	101.3	125.7
04	MASONRY	147.6	108.3	123.4	143.9	97.6	115.5	148.6	107.8	123.5	144.0	97.6	115.5	149.0	121.4	132.0	146.4	104.6	120.7
05	METALS	99.2	95.4	97.8	101.1	93.8	98.4	99.9	95.4	98.9	101.1	94.0	98.5	100.9	99.3	100.3	101.1	95.2	99.0
06	WOOD & PLASTICS	117.0	97.7	107.0	131.0	96.3	113.1	116.0	99.7	107.5	131.0	96.3	113.1	118.2	111.2	114.6	103.8	98.6	101.1
07	THERMAL & MOISTURE PROTECTION	104.3	99.4	102.0	104.5	99.3	102.1	104.2	98.5	101.5	104.5	99.3	102.1	104.6	109.9	107.1	104.1	100.9	102.5
08	DOORS & WINDOWS	93.2	94.8	93.6	92.3	85.2	90.6	92.3	95.8	93.1	92.3	90.4	91.9	91.4	106.1	95.0	92.3	96.1	93.2
09200	PLASTER & GYPSUM BOARD	209.2	97.7	141.1	143.8	96.3	114.8	264.5	99.7	163.8	214.8	96.3	142.4	143.0	111.6	123.8	154.3	98.6	120.3
095,098	CEILINGS & ACOUSTICAL TREATMENT	102.4	97.7	99.3	102.4	96.3	98.4	102.4	99.7	100.6	102.4	96.3	98.4	102.4	111.6	108.5	102.4	98.6	99.9
09600	FLOORING	132.6	108.8	126.8	132.6	109.6	127.0	132.6	107.2	126.4	132.6	109.6	127.0	132.6	115.3	128.4	132.6	107.1	126.4
097,099	WALL FINISHES, PAINTS & COATINGS	109.2	110.3	109.8	109.2	102.6	105.4	109.2	102.7	105.5	109.2	102.6	105.4	112.3	120.3	116.9	109.2	116.0	113.2
09	FINISHES	129.5	102.3	115.6	119.0	99.8	109.2	137.9	102.1	119.6	129.8	99.8	114.5	119.7	114.2	116.9	122.1	103.1	112.4
10 - 14	TOTAL DIV. 10000 - 14000	100.0	108.4	101.8	100.0	94.4	98.8	100.0	106.1	101.3	100.0	94.4	98.8	100.0	112.8	102.7	100.0	105.3	101.1
15	MECHANICAL	102.5	98.1	100.5	102.5	89.0	96.3	102.5	98.4	100.6	102.5	89.0	96.3	102.5	108.6	105.3	102.5	100.2	101.5
16	ELECTRICAL	127.0	98.2	107.2	123.6	88.9	99.7	123.6	99.2	106.8	123.6	88.9	99.7	127.0	108.1	114.0	131.0	100.3	109.9
01 - 16	WEIGHTED AVERAGE	112.8	99.7	106.4	110.6	94.0	102.5	114.6	99.9	107.4	112.3	94.2	103.5	113.1	108.9	111.1	113.7	100.5	107.3

City Cost Indexes

DIVISION		CANADA WINNIPEG, MANITOBA																
		MAT.	INST.	TOTAL	MAT.	INST.	TOTAL	MAT.	INST.	TOTAL	MAT.	INST.	TOTAL	MAT.	INST.	TOTAL	MAT.	INST.	TOTAL
02	SITE CONSTRUCTION	108.4	96.3	99.2															
03100	CONCRETE FORMS & ACCESSORIES	119.3	79.2	84.7															
03200	CONCRETE REINFORCEMENT	161.1	64.8	106.2															
03300	CAST-IN-PLACE CONCRETE	146.7	79.3	118.9															
03	CONCRETE	131.1	77.1	104.1															
04	MASONRY	145.5	75.7	102.6															
05	METALS	100.8	82.0	94.0															
06	WOOD & PLASTICS	116.0	80.4	97.6															
07	THERMAL & MOISTURE PROTECTION	104.3	79.7	92.6															
08	DOORS & WINDOWS	92.3	71.7	87.3															
09200	PLASTER & GYPSUM BOARD	135.2	79.9	101.4															
095,098	CEILINGS & ACOUSTICAL TREATMENT	102.4	79.9	87.6															
09600	FLOORING	132.6	76.1	118.9															
097,099	WALL FINISHES, PAINTS & COATINGS	109.3	65.9	84.1															
09	FINISHES	118.6	77.6	97.7															
10 - 14	TOTAL DIV. 10000 - 14000	100.0	74.6	94.6															
15	MECHANICAL	102.5	81.8	92.9															
16	ELECTRICAL	128.8	85.1	98.8															
01 - 16	WEIGHTED AVERAGE	110.8	81.5	96.5															

Location Factors

Costs shown in *Means cost data publications* are based on National Averages for materials and installation. To adjust these costs to a specific location, simply multiply the base cost by the factor and divide by 100 for that city. The data is arranged alphabetically by state and postal zip code numbers. For a city not listed, use the factor for a nearby city with similar economic characteristics.

STATE/ZIP	CITY	MAT.	INST.	TOTAL
ALABAMA				
350-352	Birmingham	96.6	77.2	87.2
354	Tuscaloosa	96.3	61.1	79.1
355	Jasper	97.4	57.5	77.9
356	Decatur	96.3	65.6	81.3
357-358	Huntsville	96.2	68.6	82.8
359	Gadsden	96.9	65.0	81.4
360-361	Montgomery	97.2	63.9	81.0
362	Anniston	95.9	52.5	74.8
363	Dothan	96.9	57.2	77.6
364	Evergreen	96.1	58.7	77.9
365-366	Mobile	97.1	68.0	83.0
367	Selma	96.4	58.7	78.1
368	Phenix City	97.2	62.9	80.5
369	Butler	96.6	57.2	77.4
ALASKA				
995-996	Anchorage	134.5	114.1	124.5
997	Fairbanks	130.1	117.0	123.7
998	Juneau	132.9	114.2	123.8
999	Ketchikan	144.2	114.2	129.6
ARIZONA				
850,853	Phoenix	99.6	78.4	89.3
852	Mesa/Tempe	99.6	68.7	84.6
855	Globe	100.5	69.4	85.4
856-857	Tucson	98.4	75.8	87.4
859	Show Low	100.7	71.3	86.4
860	Flagstaff	102.8	74.8	89.2
863	Prescott	100.2	71.2	86.1
864	Kingman	98.7	72.7	86.1
865	Chambers	98.8	70.8	85.2
ARKANSAS				
716	Pine Bluff	95.9	63.3	80.1
717	Camden	94.3	45.9	70.7
718	Texarkana	94.9	51.5	73.8
719	Hot Springs	93.4	45.3	70.0
720-722	Little Rock	96.2	64.0	80.6
723	West Memphis	95.6	62.5	79.5
724	Jonesboro	95.6	62.5	79.5
725	Batesville	94.3	56.9	76.1
726	Harrison	95.8	56.9	76.9
727	Fayetteville	92.8	39.5	66.9
728	Russellville	94.4	54.9	75.2
729	Fort Smith	96.3	61.6	79.4
CALIFORNIA				
900-902	Los Angeles	104.3	112.9	108.5
903-905	Inglewood	100.6	111.1	105.7
906-908	Long Beach	102.4	111.1	106.6
910-912	Pasadena	101.3	111.1	106.1
913-916	Van Nuys	105.3	110.9	108.1
917-918	Alhambra	104.0	111.2	107.5
919-921	San Diego	105.6	107.0	106.3
922	Palm Springs	102.3	110.3	106.2
923-924	San Bernardino	99.8	110.6	105.0
925	Riverside	104.3	111.7	107.9
926-927	Santa Ana	102.1	111.0	106.4
928	Anaheim	104.7	113.0	108.7
930	Oxnard	105.2	112.4	108.7
931	Santa Barbara	104.4	110.8	107.5
932-933	Bakersfield	104.1	108.5	106.3
934	San Luis Obispo	106.2	110.7	108.4
935	Mojave	102.7	108.2	105.4
936-938	Fresno	105.2	112.0	108.5
939	Salinas	107.1	116.2	111.6
940-941	San Francisco	111.9	133.9	122.6
942,956-958	Sacramento	107.5	112.4	109.9
943	Palo Alto	105.5	129.2	117.0
944	San Mateo	108.7	126.1	117.2
945	Vallejo	106.0	123.6	114.6
946	Oakland	110.7	126.3	118.3
947	Berkeley	110.2	125.7	117.7
948	Richmond	110.0	124.3	116.9
949	San Rafael	111.9	127.2	119.3
950	Santa Cruz	110.9	116.3	113.6

STATE/ZIP	CITY	MAT.	INST.	TOTAL
CALIFORNIA (CONT'D)				
951	San Jose	110.2	129.7	119.7
952	Stockton	105.7	113.0	109.3
953	Modesto	105.8	113.8	109.7
954	Santa Rosa	107.2	127.1	116.9
955	Eureka	108.5	111.7	110.0
959	Marysville	107.3	110.8	109.0
960	Redding	108.3	110.6	109.4
961	Susanville	108.4	110.7	109.5
COLORADO				
800-802	Denver	102.5	87.4	95.2
803	Boulder	100.7	68.4	85.0
804	Golden	103.3	82.0	92.9
805	Fort Collins	104.1	79.2	91.9
806	Greeley	100.9	68.4	85.1
807	Fort Morgan	101.5	81.0	91.5
808-809	Colorado Springs	101.3	82.4	92.1
810	Pueblo	102.6	80.9	92.0
811	Alamosa	105.1	70.0	88.0
812	Salida	105.1	70.1	88.0
813	Durango	105.7	65.9	86.3
814	Montrose	103.9	63.2	84.1
815	Grand Junction	107.1	63.6	85.9
816	Glenwood Springs	104.9	76.1	90.9
CONNECTICUT				
060	New Britain	101.9	108.5	105.1
061	Hartford	102.2	108.5	105.3
062	Willimantic	102.6	107.1	104.8
063	New London	99.2	108.7	103.9
064	Meriden	101.9	106.9	104.3
065	New Haven	102.1	109.0	105.5
066	Bridgeport	103.3	107.6	105.4
067	Waterbury	102.6	108.5	105.5
068	Norwalk	102.6	107.7	105.1
069	Stamford	102.7	113.2	107.8
D.C.				
200-205	Washington	99.2	90.1	94.8
DELAWARE				
197	Newark	99.2	102.7	100.9
198	Wilmington	98.5	102.8	100.6
199	Dover	99.1	102.7	100.9
FLORIDA				
320,322	Jacksonville	98.7	66.6	83.1
321	Daytona Beach	98.7	75.0	87.2
323	Tallahassee	99.2	56.2	78.2
324	Panama City	99.8	43.6	72.5
325	Pensacola	99.4	66.1	83.2
326,344	Gainesville	100.2	62.0	81.6
327-328,347	Orlando	100.8	69.5	85.6
329	Melbourne	100.8	77.7	89.5
330-332,340	Miami	98.1	73.4	86.1
333	Fort Lauderdale	97.9	73.5	86.0
334,349	West Palm Beach	96.8	70.2	83.9
335-336,346	Tampa	99.9	64.7	82.8
337	St. Petersburg	102.1	64.3	83.7
338	Lakeland	98.8	64.4	82.1
339,341	Fort Myers	98.2	61.4	80.3
342	Sarasota	100.0	61.8	81.4
GEORGIA				
300-303,399	Atlanta	96.4	81.4	89.1
304	Statesboro	96.7	35.7	67.0
305	Gainesville	95.5	53.5	75.1
306	Athens	94.6	69.0	82.2
307	Dalton	96.6	36.8	67.5
308-309	Augusta	95.2	60.9	78.5
310-312	Macon	95.9	65.7	81.2
313-314	Savannah	97.9	64.5	81.7
315	Waycross	97.9	49.7	74.4
316	Valdosta	97.4	54.0	76.3
317	Albany	97.5	60.3	79.4
318-319	Columbus	97.6	58.7	78.7

Location Factors

STATE/ZIP	CITY	MAT.	INST.	TOTAL
HAWAII				
967	Hilo	115.9	129.0	122.3
968	Honolulu	116.6	129.0	122.7
STATES & POSS.				
969	Guam	211.9	60.7	138.4
IDAHO				
832	Pocatello	102.1	83.7	93.2
833	Twin Falls	104.2	51.9	78.8
834	Idaho Falls	101.3	62.9	82.6
835	Lewiston	113.7	88.1	101.2
836-837	Boise	102.1	85.1	93.8
838	Coeur d'Alene	113.4	63.2	89.0
ILLINOIS				
600-603	North Suburban	98.0	120.9	109.1
604	Joliet	98.0	115.2	106.4
605	South Suburban	98.0	119.6	108.5
606	Chicago	98.2	125.4	111.4
609	Kankakee	94.4	104.2	99.2
610-611	Rockford	97.0	111.2	103.9
612	Rock Island	94.9	100.2	97.5
613	La Salle	96.4	99.7	98.0
614	Galesburg	96.0	101.4	98.6
615-616	Peoria	98.4	103.6	100.9
617	Bloomington	95.2	103.4	99.2
618-619	Champaign	99.1	101.9	100.4
620-622	East St. Louis	93.5	105.7	99.4
623	Quincy	95.3	95.4	95.4
624	Effingham	94.6	99.3	96.9
625	Decatur	95.9	98.3	97.1
626-627	Springfield	95.6	100.6	98.1
628	Centralia	92.8	104.4	98.4
629	Carbondale	92.4	100.7	96.4
INDIANA				
460	Anderson	97.1	87.2	92.3
461-462	Indianapolis	100.0	89.6	94.9
463-464	Gary	97.9	103.7	100.7
465-466	South Bend	96.2	86.2	91.3
467-468	Fort Wayne	97.3	84.7	91.1
469	Kokomo	95.5	86.0	90.9
470	Lawrenceburg	93.4	84.3	89.0
471	New Albany	95.2	80.1	87.9
472	Columbus	97.4	86.2	91.9
473	Muncie	97.5	86.3	92.1
474	Bloomington	99.3	85.9	92.8
475	Washington	95.6	89.9	92.8
476-477	Evansville	97.1	91.8	94.6
478	Terre Haute	97.6	91.2	94.5
479	Lafayette	97.2	84.6	91.1
IOWA				
500-503,509	Des Moines	98.4	86.2	92.5
504	Mason City	96.9	65.2	81.4
505	Fort Dodge	97.2	59.5	78.8
506-507	Waterloo	98.2	66.2	82.7
508	Creston	97.7	71.6	85.0
510-511	Sioux City	99.0	77.6	88.6
512	Sibley	97.9	57.1	78.0
513	Spencer	100.2	56.1	78.7
514	Carroll	96.7	63.0	80.3
515	Council Bluffs	99.8	77.5	88.9
516	Shenandoah	96.7	57.3	77.5
520	Dubuque	98.4	79.2	89.1
521	Decorah	97.8	60.9	79.9
522-524	Cedar Rapids	99.6	83.5	91.8
525	Ottumwa	97.9	75.3	86.9
526	Burlington	96.7	76.2	86.7
527-528	Davenport	98.6	93.6	96.1
KANSAS				
660-662	Kansas City	98.0	90.2	94.2
664-666	Topeka	97.4	71.4	84.8
667	Fort Scott	97.1	70.2	84.0
668	Emporia	96.8	64.4	81.0
669	Belleville	98.8	63.0	81.4
670-672	Wichita	96.6	75.6	86.4
673	Independence	98.6	59.1	79.4
674	Salina	98.3	63.7	81.5
675	Hutchinson	93.7	58.0	76.3
676	Hays	98.1	63.0	81.1
677	Colby	98.9	63.0	81.4

STATE/ZIP	CITY	MAT.	INST.	TOTAL
KANSAS (CONT'D)				
678	Dodge City	99.6	63.0	81.8
679	Liberal	97.9	52.4	75.8
KENTUCKY				
400-402	Louisville	96.1	87.2	91.8
403-405	Lexington	96.9	71.9	84.7
406	Frankfort	96.1	76.4	86.5
407-409	Corbin	94.7	50.7	73.3
410	Covington	95.6	93.9	94.8
411-412	Ashland	93.8	101.9	97.7
413-414	Campton	95.7	50.5	73.7
415-416	Pikeville	96.7	70.9	84.1
417-418	Hazard	94.9	50.7	73.4
420	Paducah	93.6	91.1	92.4
421-422	Bowling Green	95.6	86.6	91.2
423	Owensboro	95.5	83.1	89.5
424	Henderson	93.1	94.1	93.6
425-426	Somerset	92.6	50.5	72.1
427	Elizabethtown	92.2	87.0	89.7
LOUISIANA				
700-701	New Orleans	101.4	69.1	85.7
703	Thibodaux	101.9	68.3	85.6
704	Hammond	98.7	67.0	83.3
705	Lafayette	100.8	61.6	81.7
706	Lake Charles	101.0	65.0	83.5
707-708	Baton Rouge	100.7	62.1	81.9
710-711	Shreveport	96.9	63.3	80.5
712	Monroe	96.8	60.7	79.2
713-714	Alexandria	96.9	58.1	78.0
MAINE				
039	Kittery	97.4	59.7	79.1
040-041	Portland	100.3	78.7	89.8
042	Lewiston	100.8	78.7	90.0
043	Augusta	99.5	56.9	78.8
044	Bangor	99.8	84.4	92.3
045	Bath	98.8	54.7	77.4
046	Machias	98.3	66.8	83.0
047	Houlton	98.5	65.3	82.3
048	Rockland	97.3	67.6	82.9
049	Waterville	98.9	56.9	78.5
MARYLAND				
206	Waldorf	97.3	78.3	88.1
207-208	College Park	97.2	82.0	89.8
209	Silver Spring	96.6	82.1	89.5
210-212	Baltimore	96.8	86.7	91.9
214	Annapolis	96.5	83.8	90.3
215	Cumberland	93.7	82.8	88.4
216	Easton	95.4	51.2	73.9
217	Hagerstown	94.0	82.7	88.5
218	Salisbury	95.9	58.5	77.7
219	Elkton	92.8	74.5	83.9
MASSACHUSETTS				
010-011	Springfield	100.8	102.6	101.6
012	Pittsfield	100.5	95.9	98.3
013	Greenfield	98.7	100.1	99.4
014	Fitchburg	97.1	111.5	104.1
015-016	Worcester	100.6	111.9	106.1
017	Framingham	96.7	117.2	106.7
018	Lowell	100.3	116.7	108.2
019	Lawrence	101.4	115.7	108.4
020-022, 024	Boston	102.6	128.0	115.0
023	Brockton	101.5	115.0	108.1
025	Buzzards Bay	95.9	113.2	104.3
026	Hyannis	98.2	113.2	105.5
027	New Bedford	101.1	113.6	107.2
MICHIGAN				
480,483	Royal Oak	94.7	108.5	101.4
481	Ann Arbor	97.0	111.1	10?
482	Detroit	98.1	114.8	1??
484-485	Flint	96.7	103.8	
486	Saginaw	96.9	96.8	
487	Bay City	96.5	96.7	
488-489	Lansing	96.8	99.8	
490	Battle Creek	96.8	91.8	
491	Kalamazoo	97.2	8?	
492	Jackson	95.5		
493,495	Grand Rapids	96.7		
494	Muskegon	95.7		

Location Factors

STATE/ZIP	CITY	MAT.	INST.	TOTAL
MICHIGAN (CONT'D)				
496	Traverse City	94.9	75.2	85.3
497	Gaylord	96.4	81.3	89.1
498-499	Iron Mountain	98.4	91.7	95.2
MINNESOTA				
550-551	Saint Paul	99.5	121.7	110.3
553-555	Minneapolis	100.1	123.6	111.6
556-558	Duluth	99.6	115.0	107.1
559	Rochester	99.3	106.4	102.7
560	Mankato	97.3	105.4	101.3
561	Windom	96.1	85.4	90.9
562	Willmar	95.3	93.1	94.3
563	St. Cloud	96.2	113.4	104.6
564	Brainerd	97.1	104.4	100.7
565	Detroit Lakes	99.7	93.4	96.6
566	Bemidji	98.9	101.5	100.2
567	Thief River Falls	98.0	94.1	96.1
MISSISSIPPI				
386	Clarksdale	96.9	35.0	66.8
387	Greenville	100.5	53.3	77.6
388	Tupelo	98.4	44.5	72.2
389	Greenwood	98.4	37.5	68.8
390-392	Jackson	98.7	55.1	77.5
393	Meridian	96.7	54.8	76.3
394	Laurel	98.3	38.8	69.4
395	Biloxi	98.9	62.4	81.1
396	McComb	96.7	36.1	67.2
397	Columbus	98.3	44.0	71.9
MISSOURI				
630-631	St. Louis	96.2	110.6	103.2
633	Bowling Green	96.2	93.7	95.0
634	Hannibal	94.9	92.3	93.6
635	Kirksville	97.3		90.6
636	Flat River	97.3		97.7
637	Cape Girardeau	96.3	96.2	96.2
638	Sikeston	94.9	90.0	92.5
639	Poplar Bluff	94.4	88.8	91.7
640-641	Kansas City	99.0	101.1	100.0
644-645	St. Joseph	99.0	87.7	93.5
646	Chillicothe	96.3	75.0	85.9
647	Harrisonville	95.8	96.8	96.3
648	Joplin	98.0	73.4	86.0
650-651	Jefferson City	95.0	88.9	92.0
652	Columbia	96.4	89.4	93.0
653	Sedalia	95.9	87.6	91.9
654-655	Rolla	94.7	85.2	90.0
656-658	Springfield	97.7	76.7	87.5
MONTANA				
590-591	Billings	101.5	89.8	95.8
592	Wolf Point	102.1	88.8	95.7
593	Miles City	100.0	89.7	95.0
594	Great Falls	102.8	89.3	96.2
595	Havre	100.8	89.2	95.1
596	Helena	102.6	88.9	96.0
597	Butte	101.3	88.6	95.1
598	Missoula	99.5	87.8	93.8
599	Kalispell	98.8	86.8	93.0
NEBRASKA				
680-681	Omaha	100.5	80.1	90.6
683-685	Lincoln	97.8	67.0	82.8
	Columbus	96.6	48.5	73.2
	Norfolk	99.1	66.9	83.4
	Grand Island	97.5	68.7	83.5
		97.2	58.5	78.4
		96.9	53.5	75.8
		96.9	67.3	82.5
			46.0	74.2
			41.8	71.7
		101.5	108.6	105.0
		102.3	88.1	95.4
		102.3	97.0	99.7
		102.0	96.5	99.3
		100.9	87.7	94.5
		101.2	86.8	94.2
		101.3	86.8	94.2

STATE/ZIP	CITY	MAT.	INST.	TOTAL
NEW HAMPSHIRE (CONT'D)				
032-033	Concord	99.0	86.8	93.1
034	Keene	98.3	60.2	79.7
035	Littleton	98.3	66.6	82.9
036	Charleston	97.7	56.7	77.8
037	Claremont	96.8	56.7	77.3
038	Portsmouth	98.3	83.8	91.3
NEW JERSEY				
070-071	Newark	102.8	122.7	112.5
072	Elizabeth	101.7	117.4	109.3
073	Jersey City	100.3	124.2	111.9
074-075	Paterson	103.1	122.0	112.3
076	Hackensack	100.1	122.7	111.1
077	Long Branch	99.7	119.0	109.1
078	Dover	100.5	119.7	109.8
079	Summit	100.6	116.0	108.1
080,083	Vineland	98.0	118.2	107.8
081	Camden	100.1	117.5	108.6
082,084	Atlantic City	99.0	118.4	108.5
085-086	Trenton	100.9	121.4	110.9
087	Point Pleasant	100.4	119.6	109.7
088-089	New Brunswick	100.8	119.7	110.0
NEW MEXICO				
870-872	Albuquerque	100.6	81.2	91.2
873	Gallup	102.0	81.2	91.9
874	Farmington	102.0	81.3	91.9
875	Santa Fe	100.2	81.2	90.9
877	Las Vegas	100.1	81.2	90.9
878	Socorro	99.6	81.2	90.6
879	Truth/Consequences	99.6	75.4	87.8
880	Las Cruces	97.7	72.7	85.6
881	Clovis	100.3	80.7	90.8
882	Roswell	101.8	80.7	91.6
883	Carrizozo	102.7	81.2	92.2
884	Tucumcari	101.3	80.7	91.3
NEW YORK				
100-102	New York	108.5	161.9	134.5
103	Staten Island	104.8	153.6	128.5
104	Bronx	102.0	156.0	128.3
105	Mount Vernon	102.9	131.5	116.8
106	White Plains	102.6	129.7	115.8
107	Yonkers	107.6	134.1	120.5
108	New Rochelle	103.6	131.4	117.1
109	Suffern	103.5	126.7	114.8
110	Queens	103.0	156.9	129.2
111	Long Island City	104.9	156.9	130.2
112	Brooklyn	105.2	156.9	130.3
113	Flushing	105.8	156.9	130.6
114	Jamaica	103.7	157.9	130.1
115,117,118	Hicksville	102.8	149.1	125.3
116	Far Rockaway	105.9	156.9	130.7
119	Riverhead	103.3	151.3	126.6
120-122	Albany	97.9	96.3	97.1
123	Schenectady	98.9	97.0	98.0
124	Kingston	103.4	115.7	109.3
125-126	Poughkeepsie	102.5	122.2	112.1
127	Monticello	101.7	113.2	107.3
128	Glens Falls	93.3	93.0	93.1
129	Plattsburgh	99.1	86.9	93.2
130-132	Syracuse	99.7	92.2	96.1
133-135	Utica	98.1	89.3	93.8
136	Watertown	99.5	91.7	95.7
137-139	Binghamton	99.0	88.2	93.7
140-142	Buffalo	99.7	104.8	102.2
143	Niagara Falls	98.2	107.0	102.5
144-146	Rochester	101.0	99.4	100.2
147	Jamestown	97.3	89.7	93.6
148-149	Elmira	96.6	90.3	93.6
NORTH CAROLINA				
270,272-274	Greensboro	96.6	54.9	76.3
271	Winston-Salem	96.7	53.9	75.9
275-276	Raleigh	97.0	55.1	76.6
277	Durham	96.5	55.1	76.4
278	Rocky Mount	95.2	40.8	68.7
279	Elizabeth City	95.8	43.4	70.3
280	Gastonia	96.7	52.0	75.0
281-282	Charlotte	96.9	52.4	75.2
283	Fayetteville	95.0	55.1	75.6
284	Wilmington	95.2	54.1	75.2
285	Kinston	93.7	40.4	67.8

Location Factors

STATE/ZIP	CITY	MAT.	INST.	TOTAL
NORTH CAROLINA (CONT'D)				
286	Hickory	93.9	39.4	67.4
287-288	Asheville	95.6	53.1	75.0
289	Murphy	95.4	38.4	67.6
NORTH DAKOTA				
580-581	Fargo	100.8	66.1	84.0
582	Grand Forks	102.0	62.4	82.7
583	Devils Lake	102.1	60.3	81.7
584	Jamestown	101.9	61.3	82.1
585	Bismarck	100.8	68.1	84.9
586	Dickinson	102.7	60.3	82.1
587	Minot	102.0	65.9	84.4
588	Williston	101.2	60.3	81.3
OHIO				
430-432	Columbus	97.9	93.1	95.5
433	Marion	95.3	89.0	92.2
434-436	Toledo	98.5	99.7	99.1
437-438	Zanesville	95.4	87.2	91.4
439	Steubenville	97.2	98.4	97.8
440	Lorain	98.6	97.2	97.9
441	Cleveland	98.7	107.3	102.9
442-443	Akron	99.8	100.9	100.4
444-445	Youngstown	99.0	95.7	97.4
446-447	Canton	99.2	92.9	96.1
448-449	Mansfield	96.7	92.7	94.7
450	Hamilton	95.3	92.3	93.8
451-452	Cincinnati	95.3	93.9	94.6
453-454	Dayton	95.2	90.9	93.1
455	Springfield	95.4	90.6	93.1
456	Chillicothe	94.9	97.2	96.0
457	Athens	98.4	75.3	87.1
458	Lima	99.0	89.7	94.5
OKLAHOMA				
730-731	Oklahoma City	97.9	68.2	83.4
734	Ardmore	96.1	67.6	82.2
735	Lawton	97.8	68.5	83.6
736	Clinton	97.6	65.7	82.1
737	Enid	97.6	65.7	82.1
738	Woodward	96.2	65.8	81.4
739	Guymon	97.5	38.0	68.6
740-741	Tulsa	97.8	65.3	82.0
743	Miami	94.9	71.1	83.3
744	Muskogee	96.8	48.8	73.4
745	Mcalester	94.4	61.1	78.2
746	Ponca City	95.0	66.7	81.2
747	Durant	95.0	66.7	81.2
748	Shawnee	96.7	64.9	81.2
749	Poteau	94.0	68.0	81.4
OREGON				
970-972	Portland	103.8	109.7	106.7
973	Salem	104.4	108.2	106.2
974	Eugene	103.8	107.3	105.5
975	Medford	105.3	105.5	105.4
976	Klamath Falls	106.3	104.9	105.6
977	Bend	105.1	107.3	106.2
978	Pendleton	98.5	106.2	102.2
979	Vale	96.0	98.2	97.1
PENNSYLVANIA				
150-152	Pittsburgh	97.0	106.0	101.4
153	Washington	94.6	105.5	99.9
154	Uniontown	94.7	103.9	99.2
155	Bedford	95.8	97.3	96.5
156	Greensburg	96.0	104.3	100.0
157	Indiana	94.6	102.2	98.3
158	Dubois	96.2	97.9	97.0
159	Johnstown	95.8	99.4	97.6
160	Butler	92.8	104.0	98.2
161	New Castle	92.7	104.0	98.2
162	Kittanning	93.3	105.7	99.4
163	Oil City	92.7	99.0	95.8
164-165	Erie	94.8	100.3	97.5
166	Altoona	94.6	96.6	95.6
167	Bradford	96.4	99.0	97.7
168	State College	96.1	98.0	97.0
169	Wellsboro	97.2	92.9	95.1
170-171	Harrisburg	98.1	93.6	95.9
172	Chambersburg	96.7	92.9	94.9
173-174	York	96.3	94.0	95.2
175-176	Lancaster	95.3	91.7	93.6

STATE/ZIP	CITY	MAT.	INST.	TOTAL
PENNSYLVANIA (CONT'D)				
177	Williamsport	93.6	88.0	90.9
178	Sunbury	96.0	93.5	94.8
179	Pottsville	95.1	94.5	94.8
180	Lehigh Valley	96.9	105.4	101.0
181	Allentown	98.7	103.4	101.0
182	Hazleton	96.0	96.0	96.0
183	Stroudsburg	96.0	107.4	101.6
184-185	Scranton	99.2	98.8	99.0
186-187	Wilkes-Barre	95.7	96.4	96.0
188	Montrose	95.2	98.1	96.6
189	Doylestown	95.4	117.6	106.2
190-191	Philadelphia	99.6	123.9	111.4
193	Westchester	96.8	113.9	105.1
194	Norristown	95.5	119.5	107.2
195-196	Reading	97.9	99.0	98.4
PUERTO RICO				
009	San Juan	143.2	29.0	87.6
RHODE ISLAND				
028	Newport	99.8	108.1	103.8
029	Providence	100.1	108.1	104.0
SOUTH CAROLINA				
290-292	Columbia	95.9	53.2	75.1
293	Spartanburg	94.7	52.0	73.9
294	Charleston	95.9	54.6	75.8
295	Florence	94.2	51.0	73.1
296	Greenville	94.4	52.0	73.8
297	Rock Hill	94.5	38.0	67.0
298	Aiken	95.6	38.6	67.8
299	Beaufort	96.2	42.5	70.1
SOUTH DAKOTA				
570-571	Sioux Falls	101.2	61.6	81.9
572	Watertown	100.1	55.7	78.5
573	Mitchell	98.9	55.5	77.8
574	Aberdeen	101.2	55.7	79.1
575	Pierre	100.6	56.5	79.1
576	Mobridge	99.7	55.7	78.3
577	Rapid City	101.3	54.9	78.7
TENNESSEE				
370-372	Nashville	97.4	73.3	85.7
373-374	Chattanooga	98.9	62.1	81.0
375,380-381	Memphis	96.3	75.2	86.0
376	Johnson City	98.3	58.9	79.1
377-379	Knoxville	95.0	63.3	79.6
382	Mckenzie	97.3	38.9	68.9
383	Jackson	98.8	49.2	74.7
384	Columbia	95.7	54.9	75.9
385	Cookeville	97.0	37.6	68.1
TEXAS				
750	Mckinney	99.5	63.6	82.1
751	Waxahackie	99.5	63.9	82.2
752-753	Dallas	100.0	69.1	85.0
754	Greenville	99.7	45.0	73.1
755	Texarkana	98.8	54.7	77.4
756	Longview	99.3	47.5	74.1
757	Tyler	99.9	58.5	79.8
758	Palestine	95.8	48.8	72.9
759	Lufkin	96.9	55.0	76.5
760-761	Fort Worth	97.8	66.4	82.6
762	Denton	98.1	59.2	79.2
763	Wichita Falls	98.3	60.6	80.0
764	Eastland	97.4	47.7	73.2
765	Temple	95.9	56.8	76.9
766-767	Waco	97.6	61.9	80.2
768	Brownwood	98.3	44.3	72.1
769	San Angelo	97.9	51.1	75.2
770-772	Houston	99.8	75.0	87.
773	Huntsville	98.7	47.8	7
774	Wharton	100.3	52.4	
775	Galveston	98.1	74.4	
776-777	Beaumont	97.5	70.1	
778	Bryan	95.0	68.7	
779	Victoria	100.4	56.9	
780	Laredo	95.0	59.8	
781-782	San Antonio	94.9	70	
783-784	Corpus Christi	98.3		
785	Mc Allen	98.8		
786-787	Austin	95.1		

Location Factors

STATE/ZIP	CITY	MAT.	INST.	TOTAL
TEXAS (CONT'D)				
788	Del Rio	98.2	37.0	68.4
789	Giddings	95.2	47.7	72.1
790-791	Amarillo	98.4	62.3	80.8
792	Childress	97.9	57.9	78.4
793-794	Lubbock	100.3	59.3	80.4
795-796	Abilene	97.7	54.9	76.9
797	Midland	101.1	56.6	79.4
798-799,885	El Paso	97.7	58.6	78.7
UTAH				
840-841	Salt Lake City	101.1	77.4	89.6
842,844	Ogden	99.6	77.3	88.7
843	Logan	101.8	77.3	89.9
845	Price	102.4	57.6	80.6
846-847	Provo	102.1	77.2	90.0
VERMONT				
050	White River Jct.	99.8	42.9	72.1
051	Bellows Falls	98.2	46.6	73.1
052	Bennington	98.5	42.6	71.3
053	Brattleboro	99.0	46.9	73.6
054	Burlington	101.0	69.2	85.5
056	Montpelier	98.2	69.1	84.0
057	Rutland	100.1	69.2	85.1
058	St. Johnsbury	100.0	49.5	75.4
059	Guildhall	98.4	48.9	74.3
VIRGINIA				
220-221	Fairfax	97.7	81.6	89.9
222	Arlington	98.4	81.4	90.1
223	Alexandria	97.3	83.5	90.6
224-225	Fredericksburg	96.2	73.1	84.9
226	Winchester	97.0	60.9	79.5
227	Culpeper	96.7	62.0	79.8
228	Harrisonburg	97.1	54.1	76.2
229	Charlottesville	97.3	65.2	81.7
230-232	Richmond	97.5	70.1	84.2
233-235	Norfolk	97.8	66.6	82.6
236	Newport News	97.2	66.4	82.2
237	Portsmouth	96.4	66.1	81.6
238	Petersburg	97.1	70.1	84.0
239	Farmville	96.5	48.3	73.1
240-241	Roanoke	97.3	53.4	75.9
242	Bristol	96.2	52.5	74.9
243	Pulaski	95.9	47.7	72.4
244	Staunton	96.7	52.4	75.1
245	Lynchburg	96.7	55.8	76.8
246	Grundy	96.2	47.7	72.6
WASHINGTON				
980-981,987	Seattle	103.8	104.9	104.4
982	Everett	104.1	101.4	102.8
983-984	Tacoma	103.7	102.0	102.9
985	Olympia	103.6	102.0	102.8
986	Vancouver	106.1	101.1	103.7
988	Wenatchee	104.9	91.2	98.3
989	Yakima	104.2	95.9	100.2
990-992	Spokane	108.3	87.8	98.3
993	Richland	108.4	90.1	99.5
994	Clarkston	107.6	89.6	98.8
WEST VIRGINIA				
247-248	Bluefield	95.0	85.1	90.2
249	Lewisburg	96.5	86.2	91.5
	Charleston	97.2	91.6	94.4
	Martinsburg	96.5	56.1	76.8
	Huntington	97.4	96.0	96.7
		94.7	89.2	92.0
		97.6	93.1	95.4
		96.4	95.4	96.0
			91.7	94.1
			91.7	94.3
			91.7	94.4
			92.2	94.2
			87.2	91.8
			90.1	93.2
		99.2	101.6	100.3
		99.4	101.0	100.2
		98.6	101.3	99.9
		98.5	97.5	98.0
		98.7	98.0	98.3

STATE/ZIP	CITY	MAT.	INST.	TOTAL
WISCONSIN (CONT'D)				
538	Lancaster	96.9	82.5	89.9
539	Portage	95.2	97.8	96.5
540	New Richmond	96.9	93.6	95.3
541-543	Green Bay	100.4	93.3	97.0
544	Wausau	96.1	93.1	94.6
545	Rhinelander	99.5	89.3	94.5
546	La Crosse	96.6	93.5	95.1
547	Eau Claire	98.6	93.3	96.0
548	Superior	96.8	98.9	97.8
549	Oshkosh	96.7	90.7	93.8
WYOMING				
820	Cheyenne	100.6	61.4	81.5
821	Yellowstone Nat'l Park	100.0	55.1	78.2
822	Wheatland	101.7	54.3	78.7
823	Rawlins	103.3	50.3	77.5
824	Worland	100.8	50.7	76.4
825	Riverton	102.0	53.9	78.6
826	Casper	100.3	63.9	82.6
827	Newcastle	100.5	50.3	76.1
828	Sheridan	101.6	59.2	81.0
829-831	Rock Springs	105.3	50.7	78.7
CANADIAN FACTORS (reflect Canadian currency)				
ALBERTA				
	Calgary	111.8	80.9	96.8
	Edmonton	111.8	80.8	96.7
	Fort McMurray	110.4	80.8	96.0
	Lethbridge	110.5	80.9	96.1
	Lloydminster	110.4	80.8	96.0
	Medicine Hat	110.5	80.9	96.1
	Red Deer	110.2	80.9	96.0
BRITISH COLUMBIA				
	Kamloops	110.6	97.3	104.2
	Prince George	114.1	97.3	106.0
	Vancouver	113.7	100.5	107.3
	Victoria	114.1	97.3	106.0
MANITOBA				
	Brandon	111.9	81.5	97.1
	Portage la Prairie	111.1	81.5	96.7
	Winnipeg	110.8	81.5	96.5
NEW BRUNSWICK				
	Bathurst	109.9	72.0	91.5
	Dalhousie	109.9	72.0	91.5
	Fredericton	110.6	77.2	94.4
	Moncton	109.5	72.0	91.3
	Newcastle	109.9	72.0	91.5
	Saint John	110.9	77.2	94.5
NEWFOUNDLAND				
	Corner Brook	115.9	71.5	94.3
	St. John's	115.9	71.5	94.3
NORTHWEST TERRITORIES				
	Yellowknife	108.1	73.3	91.1
NOVA SCOTIA				
	Dartmouth	111.0	79.5	95.7
	Halifax	111.0	81.0	96.4
	New Glasgow	111.0	79.5	95.7
	Sydney	108.2	79.5	94.3
	Yarmouth	111.0	79.5	95.7
ONTARIO				
	Barrie	115.7	98.1	107.2
	Brantford	113.5	103.9	108.8
	Cornwall	114.1	99.2	106.9
	Hamilton	113.8	102.8	108.4
	Kingston	114.1	98.9	106.7
	Kitchener	108.4	97.4	103.0
	London	112.8	99.7	106.4
	North Bay	113.5	97.0	105.5
	Oshawa	113.6	100.3	107.1
	Ottawa	114.6	99.9	107.4
	Owen Sound	115.8	96.6	106.5
	Peterborough	113.5	99.0	106.4
	Sarnia	112.7	104.6	108.8
	St. Catharines	107.3	97.3	102.4
	Sudbury	107.2	97.0	102.3

Location Factors

STATE/ZIP	CITY	MAT.	INST.	TOTAL
ONTARIO (CONT'D)				
	Thunder Bay	109.0	97.2	103.3
	Toronto	113.1	108.9	111.1
	Windsor	107.6	100.2	104.0
PRINCE EDWARD ISLAND				
	Charlottetown	113.1	67.0	90.7
	Summerside	113.3	67.0	90.8
QUEBEC				
	Cap-de-la-Madeleine	111.3	94.2	103.0
	Charlesbourg	111.3	94.2	103.0
	Chicoutimi	110.0	93.9	102.2
	Gatineau	108.3	94.0	101.4
	Laval	110.5	94.0	102.5
	Montreal	110.6	94.0	102.5
	Quebec	112.3	94.2	103.5
	Sherbrooke	110.5	94.0	102.5
	Trois Rivieres	111.3	94.2	103.0
SASKATCHEWAN				
	Moose Jaw	108.2	75.1	92.1
	Prince Albert	107.6	74.9	91.7
	Regina	108.9	75.1	92.5
	Saskatoon	107.8	74.9	91.8
YUKON				
	Whitehorse	108.1	73.3	91.2

Abbreviations

A	Area Square Feet; Ampere	Cab.	Cabinet	d.f.u.	Drainage Fixture Units
ABS	Acrylonitrile Butadiene Stryrene; Asbestos Bonded Steel	Cair.	Air Tool Laborer	D.H.	Double Hung
A.C.	Alternating Current; Air-Conditioning; Asbestos Cement; Plywood Grade A & C	Calc	Calculated	DHW	Domestic Hot Water
		Cap.	Capacity	Diag.	Diagonal
		Carp.	Carpenter	Diam.	Diameter
		C.B.	Circuit Breaker	Distrib.	Distribution
		C.C.A.	Chromate Copper Arsenate	Dk.	Deck
A.C.I.	American Concrete Institute	C.C.F.	Hundred Cubic Feet	D.L.	Dead Load; Diesel
AD	Plywood, Grade A & D	cd	Candela	DLH	Deep Long Span Bar Joist
Addit.	Additional	cd/sf	Candela per Square Foot	Do.	Ditto
Adj.	Adjustable	CD	Grade of Plywood Face & Back	Dp.	Depth
af	Audio-frequency	CDX	Plywood, Grade C & D, exterior glue	D.P.S.T.	Double Pole, Single Throw
A.G.A.	American Gas Association			Dr.	Driver
Agg.	Aggregate	Cefi.	Cement Finisher	Drink.	Drinking
A.H.	Ampere Hours	Cem.	Cement	D.S.	Double Strength
A hr.	Ampere-hour	CF	Hundred Feet	D.S.A.	Double Strength A Grade
A.H.U.	Air Handling Unit	C.F.	Cubic Feet	D.S.B.	Double Strength B Grade
A.I.A.	American Institute of Architects	CFM	Cubic Feet per Minute	Dty.	Duty
AIC	Ampere Interrupting Capacity	c.g.	Center of Gravity	DWV	Drain Waste Vent
Allow.	Allowance	CHW	Chilled Water; Commercial Hot Water	DX	Deluxe White, Direct Expansion
alt.	Altitude			dyn	Dyne
Alum.	Aluminum	C.I.	Cast Iron	e	Eccentricity
a.m.	Ante Meridiem	C.I.P.	Cast in Place	E	Equipment Only; East
Amp.	Ampere	Circ.	Circuit	Ea.	Each
Anod.	Anodized	C.L.	Carload Lot	E.B.	Encased Burial
Approx.	Approximate	Clab.	Common Laborer	Econ.	Economy
Apt.	Apartment	C.L.F.	Hundred Linear Feet	EDP	Electronic Data Processing
Asb.	Asbestos	CLF	Current Limiting Fuse	EIFS	Exterior Insulation Finish System
A.S.B.C.	American Standard Building Code	CLP	Cross Linked Polyethylene	E.D.R.	Equiv. Direct Radiation
Asbe.	Asbestos Worker	cm	Centimeter	Eq.	Equation
A.S.H.R.A.E.	American Society of Heating, Refrig. & AC Engineers	CMP	Corr. Metal Pipe	Elec.	Electrician; Electrical
		C.M.U.	Concrete Masonry Unit	Elev.	Elevator; Elevating
A.S.M.E.	American Society of Mechanical Engineers	CN	Change Notice	EMT	Electrical Metallic Conduit; Thin Wall Conduit
		Col.	Column		
A.S.T.M.	American Society for Testing and Materials	CO_2	Carbon Dioxide	Eng.	Engine, Engineered
		Comb.	Combination	EPDM	Ethylene Propylene Diene Monomer
Attchmt.	Attachment	Compr.	Compressor		
Avg.	Average	Conc.	Concrete	EPS	Expanded Polystyrene
A.W.G.	American Wire Gauge	Cont.	Continuous; Continued	Eqhv.	Equip. Oper., Heavy
AWWA	American Water Works Assoc.	Corr.	Corrugated	Eqlt.	Equip. Oper., Light
Bbl.	Barrel	Cos	Cosine	Eqmd.	Equip. Oper., Medium
B. & B.	Grade B and Better; Balled & Burlapped	Cot	Cotangent	Eqmm.	Equip. Oper., Master Mechanic
		Cov.	Cover	Eqol.	Equip. Oper., Oilers
B. & S.	Bell and Spigot	C/P	Cedar on Paneling	Equip.	Equipment
B. & W.	Black and White	CPA	Control Point Adjustment	ERW	Electric Resistance Welded
b.c.c.	Body-centered Cubic	Cplg.	Coupling	E.S.	Energy Saver
B.C.Y.	Bank Cubic Yards	C.P.M.	Critical Path Method	Est.	Estimated
BE	Bevel End	CPVC	Chlorinated Polyvinyl Chloride	esu	Electrostatic Units
B.F.	Board Feet	C.Pr.	Hundred Pair	E.W.	Each Way
Bg. cem.	Bag of Cement	CRC	Cold Rolled Channel	EWT	Entering Water Temperature
BHP	Boiler Horsepower; Brake Horsepower	Creos.	Creosote	Excav.	Excavation
		Crpt.	Carpet & Linoleum Layer	Exp.	Expansion, Exposure
B.I.	Black Iron	CRT	Cathode-ray Tube	Ext.	Exterior
Bit.; Bitum.	Bituminous	CS	Carbon Steel, Constant Shear Bar Joist	Extru.	Extrusion
Bk.	Backed			f.	Fiber stress
Bkrs.	Breakers	Csc	Cosecant	F	Fahrenheit; Female; Fill
Bldg.	Building	C.S.F.	Hundred Square Feet	Fab.	Fabricated
Blk.	Block	CSI	Construction Specifications Institute	FBGS	Fiberglass
Bm.	Beam			F.C.	Footcandles
Boil.	Boilermaker	C.T.	Current Transformer	f.c.c.	Face-centered Cubic
B.P.M.	Blows per Minute	CTS	Copper Tube Size	f'c.	Compressive Stress in Concrete; Extreme Compressive Stress
BR	Bedroom	Cu	Copper, Cubic		
Brg.	Bearing	Cu. Ft.	Cubic Foot	F.E.	Front End
Brhe.	Bricklayer Helper	cw	Continuous Wave	FEP	Fluorinated Ethylene Propylene (Teflon)
Bric.	Bricklayer	C.W.	Cool White; Cold Water		
Brk.	Brick	Cwt.	100 Pounds	F.G.	Flat Grain
Brng.	Bearing	C.W.X.	Cool White Deluxe	F.H.A.	Federal Housing Administration
Brs.	Brass	C.Y.	Cubic Yard (27 cubic feet)	Fig.	Figure
Brz.	Bronze	C.Y./Hr.	Cubic Yard per Hour	Fin.	Finished
Bsn.	Basin	Cyl.	Cylinder	Fixt.	Fixture
Btr.	Better	d	Penny (nail size)	Fl. Oz.	Fluid Ounces
BTU	British Thermal Unit	D	Deep; Depth; Discharge	Flr.	Floor
BTUH	BTU per Hour	Dis.;Disch.	Discharge	F.M.	Frequency Modulation; Factory Mutual
B.U.R.	Built-up Roofing	Db.	Decibel		
BX	Interlocked Armored Cable	Dbl.	Double	Fmg.	Framing
c	Conductivity, Copper Sweat	DC	Direct Current	Fndtn.	Foundation
C	Hundred; Centigrade	DDC	Direct Digital Control	Fori.	Foreman, Inside
C/C	Center to Center, Cedar on Cedar	Demob.	Demobilization	Foro.	Foreman, Outside

Abbreviations

Fount.	Fountain	J.I.C.	Joint Industrial Council	M.C.P.	Motor Circuit Protector
FPM	Feet per Minute	K	Thousand; Thousand Pounds; Heavy Wall Copper Tubing, Kelvin	MD	Medium Duty
FPT	Female Pipe Thread			M.D.O.	Medium Density Overlaid
Fr.	Frame	K.A.H.	Thousand Amp. Hours	Med.	Medium
F.R.	Fire Rating	KCMIL	Thousand Circular Mils	MF	Thousand Feet
FRK	Foil Reinforced Kraft	KD	Knock Down	M.F.B.M.	Thousand Feet Board Measure
FRP	Fiberglass Reinforced Plastic	K.D.A.T.	Kiln Dried After Treatment	Mfg.	Manufacturing
FS	Forged Steel	kg	Kilogram	Mfrs.	Manufacturers
FSC	Cast Body; Cast Switch Box	kG	Kilogauss	mg	Milligram
Ft.	Foot; Feet	kgf	Kilogram Force	MGD	Million Gallons per Day
Ftng.	Fitting	kHz	Kilohertz	MGPH	Thousand Gallons per Hour
Ftg.	Footing	Kip	1000 Pounds	MH, M.H.	Manhole; Metal Halide; Man-Hour
Ft. Lb.	Foot Pound	KJ	Kiljoule	MHz	Megahertz
Furn.	Furniture	K.L.	Effective Length Factor	Mi.	Mile
FVNR	Full Voltage Non-Reversing	K.L.F.	Kips per Linear Foot	MI	Malleable Iron; Mineral Insulated
FXM	Female by Male	Km	Kilometer	mm	Millimeter
Fy.	Minimum Yield Stress of Steel	K.S.F.	Kips per Square Foot	Mill.	Millwright
g	Gram	K.S.I.	Kips per Square Inch	Min., min.	Minimum, minute
G	Gauss	kV	Kilovolt	Misc.	Miscellaneous
Ga.	Gauge	kVA	Kilovolt Ampere	ml	Milliliter, Mainline
Gal.	Gallon	K.V.A.R.	Kilovar (Reactance)	M.L.F.	Thousand Linear Feet
Gal./Min.	Gallon per Minute	KW	Kilowatt	Mo.	Month
Galv.	Galvanized	KWh	Kilowatt-hour	Mobil.	Mobilization
Gen.	General	L	Labor Only; Length; Long; Medium Wall Copper Tubing	Mog.	Mogul Base
G.F.I.	Ground Fault Interrupter			MPH	Miles per Hour
Glaz.	Glazier	Lab.	Labor	MPT	Male Pipe Thread
GPD	Gallons per Day	lat	Latitude	MRT	Mile Round Trip
GPH	Gallons per Hour	Lath.	Lather	ms	Millisecond
GPM	Gallons per Minute	Lav.	Lavatory	M.S.F.	Thousand Square Feet
GR	Grade	lb.; #	Pound	Mstz.	Mosaic & Terrazzo Worker
Gran.	Granular	L.B.	Load Bearing; L Conduit Body	M.S.Y.	Thousand Square Yards
Grnd.	Ground	L. & E.	Labor & Equipment	Mtd.	Mounted
H	High; High Strength Bar Joist; Henry	lb./hr.	Pounds per Hour	Mthe.	Mosaic & Terrazzo Helper
		lb./L.F.	Pounds per Linear Foot	Mtng.	Mounting
H.C.	High Capacity	lbf/sq.in.	Pound-force per Square Inch	Mult.	Multi; Multiply
H.D.	Heavy Duty; High Density	L.C.L.	Less than Carload Lot	M.V.A.	Million Volt Amperes
H.D.O.	High Density Overlaid	Ld.	Load	M.V.A.R.	Million Volt Amperes Reactance
Hdr.	Header	LE	Lead Equivalent	MV	Megavolt
Hdwe.	Hardware	LED	Light Emitting Diode	MW	Megawatt
Help.	Helpers Average	L.F.	Linear Foot	MXM	Male by Male
HEPA	High Efficiency Particulate Air Filter	Lg.	Long; Length; Large	MYD	Thousand Yards
		L & H	Light and Heat	N	Natural; North
Hg	Mercury	LH	Long Span Bar Joist	nA	Nanoampere
HIC	High Interrupting Capacity	L.H.	Labor Hours	NA	Not Available; Not Applicable
HM	Hollow Metal	L.L.	Live Load	N.B.C.	National Building Code
H.O.	High Output	L.L.D.	Lamp Lumen Depreciation	NC	Normally Closed
Horiz.	Horizontal	L-O-L	Lateralolet	N.E.M.A.	National Electrical Manufacturers Assoc.
H.P.	Horsepower; High Pressure	lm	Lumen		
H.P.F.	High Power Factor	lm/sf	Lumen per Square Foot	NEHB	Bolted Circuit Breaker to 600V.
Hr.	Hour	lm/W	Lumen per Watt	N.L.B.	Non-Load-Bearing
Hrs./Day	Hours per Day	L.O.A.	Length Over All	NM	Non-Metallic Cable
HSC	High Short Circuit	log	Logarithm	nm	Nanometer
Ht.	Height	L.P.	Liquefied Petroleum; Low Pressure	No.	Number
Htg.	Heating	L.P.F.	Low Power Factor	NO	Normally Open
Htrs.	Heaters	LR	Long Radius	N.O.C.	Not Otherwise Classified
HVAC	Heating, Ventilation & Air-Conditioning	L.S.	Lump Sum	Nose.	Nosing
		Lt.	Light	N.P.T.	National Pipe Thread
Hvy.	Heavy	Lt. Ga.	Light Gauge	NQOD	Combination Plug-on/Bolt on Circuit Breaker to 240V.
HW	Hot Water	L.T.L.	Less than Truckload Lot		
Hyd.;Hydr.	Hydraulic	Lt. Wt.	Lightweight	N.R.C.	Noise Reduction Coefficient
Hz.	Hertz (cycles)	L.V.	Low Voltage	N.R.S.	Non Rising Stem
I.	Moment of Inertia	M	Thousand; Material; Male; Light Wall Copper Tubing	ns	Nanosecond
I.C.	Interrupting Capacity			nW	Nanowatt
ID	Inside Diameter	M^2CA	Meters Squared Contact Area	OB	Opposing Blade
I.D.	Inside Dimension; Identification	m/hr; M.H.	Man-hour	OC	On Center
I.F.	Inside Frosted	mA	Milliampere	OD	Outside Diameter
I.M.C.	Intermediate Metal Conduit	Mach.	Machine	O.D.	Outside Dimension
In.	Inch	Mag. Str.	Magnetic Starter	ODS	Overhead Distribution System
Incan.	Incandescent	Maint.	Maintenance	O.G.	Ogee
Incl.	Included; Including	Marb.	Marble Setter	O.H.	Overhead
Int.	Interior	Mat; Mat'l.	Material	O & P	Overhead and Profit
Inst.	Installation	Max.	Maximum	Oper.	Operator
Insul.	Insulation/Insulated	MBF	Thousand Board Feet	Opng.	Opening
I.P.	Iron Pipe	MBH	Thousand BTU's per hr.	Orna.	Ornamental
I.P.S.	Iron Pipe Size	MC	Metal Clad Cable	OSB	Oriented Strand Board
I.P.T.	Iron Pipe Threaded	M.C.F.	Thousand Cubic Feet	O. S. & Y.	Outside Screw and Yoke
I.W.	Indirect Waste	M.C.F.M.	Thousand Cubic Feet per Minute	Ovhd.	Overhead
J	Joule	M.C.M.	Thousand Circular Mils	OWG	Oil, Water or Gas

Abbreviations

Oz.	Ounce	S.	Suction; Single Entrance; South	THHN	Nylon Jacketed Wire
P.	Pole; Applied Load; Projection	SCFM	Standard Cubic Feet per Minute	THW.	Insulated Strand Wire
p.	Page	Scaf.	Scaffold	THWN;	Nylon Jacketed Wire
Pape.	Paperhanger	Sch.; Sched.	Schedule	T.L.	Truckload
P.A.P.R.	Powered Air Purifying Respirator	S.C.R.	Modular Brick	T.M.	Track Mounted
PAR	Weatherproof Reflector	S.D.	Sound Deadening	Tot.	Total
Pc., Pcs.	Piece, Pieces	S.D.R.	Standard Dimension Ratio	T-O-L	Threadolet
P.C.	Portland Cement; Power Connector	S.E.	Surfaced Edge	T.S.	Trigger Start
P.C.F.	Pounds per Cubic Foot	Sel.	Select	Tr.	Trade
P.C.M.	Phase Contract Microscopy	S.E.R.;	Service Entrance Cable	Transf.	Transformer
P.E.	Professional Engineer; Porcelain Enamel; Polyethylene; Plain End	S.E.U.	Service Entrance Cable	Trhv.	Truck Driver, Heavy
		S.F.	Square Foot	Trlr	Trailer
		S.F.C.A.	Square Foot Contact Area	Trlt.	Truck Driver, Light
Perf.	Perforated	S.F.G.	Square Foot of Ground	TV	Television
Ph.	Phase	S.F. Hor.	Square Foot Horizontal	T.W.	Thermoplastic Water Resistant Wire
P.I.	Pressure Injected	S.F.R.	Square Feet of Radiation		
Pile.	Pile Driver	S.F. Shlf.	Square Foot of Shelf	UCI	Uniform Construction Index
Pkg.	Package	S4S	Surface 4 Sides	UF	Underground Feeder
Pl.	Plate	Shee.	Sheet Metal Worker	UGND	Underground Feeder
Plah.	Plasterer Helper	Sin.	Sine	U.H.F.	Ultra High Frequency
Plas.	Plasterer	Skwk.	Skilled Worker	U.L.	Underwriters Laboratory
Pluh.	Plumbers Helper	SL	Saran Lined	Unfin.	Unfinished
Plum.	Plumber	S.L.	Slimline	URD	Underground Residential Distribution
Ply.	Plywood	Sldr.	Solder		
p.m.	Post Meridiem	SLH	Super Long Span Bar Joist	US	United States
Pntd.	Painted	S.N.	Solid Neutral	USP	United States Primed
Pord.	Painter, Ordinary	S-O-L	Socketolet	UTP	Unshielded Twisted Pair
pp	Pages	sp	Standpipe	V	Volt
PP; PPL	Polypropylene	S.P.	Static Pressure; Single Pole; Self-Propelled	V.A.	Volt Amperes
P.P.M.	Parts per Million			V.C.T.	Vinyl Composition Tile
Pr.	Pair	Spri.	Sprinkler Installer	VAV	Variable Air Volume
P.E.S.B.	Pre-engineered Steel Building	spwg	Static Pressure Water Gauge	VC	Veneer Core
Prefab.	Prefabricated	Sq.	Square; 100 Square Feet	Vent.	Ventilation
Prefin.	Prefinished	S.P.D.T.	Single Pole, Double Throw	Vert.	Vertical
Prop.	Propelled	SPF	Spruce Pine Fir	V.F.	Vinyl Faced
PSF; psf	Pounds per Square Foot	S.P.S.T.	Single Pole, Single Throw	V.G.	Vertical Grain
PSI; psi	Pounds per Square Inch	SPT	Standard Pipe Thread	V.H.F.	Very High Frequency
PSIG	Pounds per Square Inch Gauge	Sq. Hd.	Square Head	VHO	Very High Output
PSP	Plastic Sewer Pipe	Sq. In.	Square Inch	Vib.	Vibrating
Pspr.	Painter, Spray	S.S.	Single Strength; Stainless Steel	V.L.F.	Vertical Linear Foot
Psst.	Painter, Structural Steel	S.S.B.	Single Strength B Grade	Vol.	Volume
P.T.	Potential Transformer	sst	Stainless Steel	VRP	Vinyl Reinforced Polyester
P. & T.	Pressure & Temperature	Sswk.	Structural Steel Worker	W	Wire; Watt; Wide; West
Ptd.	Painted	Sswl.	Structural Steel Welder	w/	With
Ptns.	Partitions	St.; Stl.	Steel	W.C.	Water Column; Water Closet
Pu	Ultimate Load	S.T.C.	Sound Transmission Coefficient	W.F.	Wide Flange
PVC	Polyvinyl Chloride	Std.	Standard	W.G.	Water Gauge
Pvmt.	Pavement	STK	Select Tight Knot	Wldg.	Welding
Pwr.	Power	STP	Standard Temperature & Pressure	W. Mile	Wire Mile
Q	Quantity Heat Flow	Stpi.	Steamfitter, Pipefitter	W-O-L	Weldolet
Quan.; Qty.	Quantity	Str.	Strength; Starter; Straight	W.R.	Water Resistant
Q.C.	Quick Coupling	Strd.	Stranded	Wrck.	Wrecker
r	Radius of Gyration	Struct.	Structural	W.S.P.	Water, Steam, Petroleum
R	Resistance	Sty.	Story	WT., Wt.	Weight
R.C.P.	Reinforced Concrete Pipe	Subj.	Subject	WWF	Welded Wire Fabric
Rect.	Rectangle	Subs.	Subcontractors	XFER	Transfer
Reg.	Regular	Surf.	Surface	XFMR	Transformer
Reinf.	Reinforced	Sw.	Switch	XHD	Extra Heavy Duty
Req'd.	Required	Swbd.	Switchboard	XHHW; XLPE	Cross-Linked Polyethylene Wire Insulation
Res.	Resistant	S.Y.	Square Yard		
Resi.	Residential	Syn.	Synthetic	XLP	Cross-linked Polyethylene
Rgh.	Rough	S.Y.P.	Southern Yellow Pine	Y	Wye
RGS	Rigid Galvanized Steel	Sys.	System	yd	Yard
R.H.W.	Rubber, Heat & Water Resistant; Residential Hot Water	t.	Thickness	yr	Year
		T	Temperature; Ton	Δ	Delta
rms	Root Mean Square	Tan	Tangent	%	Percent
Rnd.	Round	T.C.	Terra Cotta	~	Approximately
Rodm.	Rodman	T & C	Threaded and Coupled	∅	Phase
Rofc.	Roofer, Composition	T.D.	Temperature Difference	@	At
Rofp.	Roofer, Precast	T.E.M.	Transmission Electron Microscopy	#	Pound; Number
Rohe.	Roofer Helpers (Composition)	TFE	Tetrafluoroethylene (Teflon)	<	Less Than
Rots.	Roofer, Tile & Slate	T. & G.	Tongue & Groove; Tar & Gravel	>	Greater Than
R.O.W.	Right of Way				
RPM	Revolutions per Minute	Th.; Thk.	Thick		
R.R.	Direct Burial Feeder Conduit	Thn.	Thin		
R.S.	Rapid Start	Thrded	Threaded		
Rsr	Riser	Tilf.	Tile Layer, Floor		
RT	Round Trip	Tilh.	Tile Layer, Helper		

Index

A

Abandon catch basin 22
Abatement asbestos 102
ABS DWV pipe 207
Absorber vibration 129
Absorption chiller 472
Absorption water chillers 311
A/C coils furnace 296
   fan coil system 486, 487
   packaged DX 484, 485
   packaged DXVAV 485
   rooftop DX 484
   rooftop DXVAV 484
   self contained 481, 482
   self contained VAV 481, 482
Access door duct 357, 358
   floor 78
   road and parking area 11
Accessories duct 357
   fuel oil 273
Accessory formwork 50
   masonry 54
Accumulator refrigerant 271
Acoustical ceiling 94
   sealant 68
Adapter fill 273
   polybutylene 217, 218
Adjustable frequency drives 396
Adsorptive desiccant 279
   filter 279
Aerator 44
Aerial lift 14
Aftercooler air-cooled 277
   compressor 277
   separator 277
   water-cooled 277
Ahu built up 475
   built up VAV 476
   central station 477
   central station VAV 478
   rooftop 479
   rooftop VAV 480
Air & dirt eliminator 245
   balancing 384
   cell pipe-covering 132
   cleaner electronic 380
   cock 248
   compressor 14
   compressor control system . . . 109
   compressor package 276
   condition and ventilation . 542, 543
   conditioner central station . . 321
   conditioner cooling & heating . 324
   conditioner direct expansion . 337
   conditioner fan coil 336
   conditioner gas heat 324
   conditioner packaged terminal 324
   conditioner portable 326
   conditioner receptacle 392
   conditioner removal 120
   conditioner rooftop 324,325
   conditioner self-contained . . . 326
   conditioner thru-wall 324
   conditioner ventilating 347
   conditioner window 326
   conditioner wiring 394
   conditioning 540, 544, 546
   conditioning and ventilation . 545, 546
   conditioning computer 323
   conditioning fan 364
   conditioning pump 272
   conditioning ventilating . . 108, 314, 315, 317, 318, 321, 324, 326, 327, 352, 353, 355, 357, 359, 362, 365, 368-370, 372-374, 376-378, 381, 382
   control 244
   cooled belt drive condenser . 315
   cooled condensing unit 316
   cooled direct drive condenser . 315
   curtain 72
   curtain door 72
   dryer adsorption 278
   dryer compressor 277
   dryer deliquescent 277
   dryer refrigeration 278
   dryer regenerative 278
   eliminator 244
   extractor 357
   filter 279, 379
   filter industrial 379
   filter regulator lubricator 279
   filter roll type 380
   filtration 102
   fitting boiler 245
   fitting compression tank . . . 245
   handler heater 297
   handler modular 296
   handling fan 364
   handling unit 320, 321
   hose 14
   line couplers 279
   make-up unit 323
   register 374
   return grille 372
   sampling 102
   scoop 245
   supply grille 371
   supply pneumatic control . . . 109
   supply register 374
   tool 14
   tube system 118
   unit induction 360
   vent automatic 245
Air-conditioner air cooled . . 444, 447, 454-458
   chilled water 444-446, 458
   direct expansion 448-457
   fan coil 444-447, 456, 457
   multizone 326, 450, 451
   penthouse 326
   self-contained 452-455
   single zone 448, 449
   split system 456, 457
   water cooled . . . 446, 452, 453, 458
Air-conditioning air cooled 444
   computer 458
   drug store 451
Airfoil fan centrifugal 365
Airless sprayer 102
Alarm burglar 116
   detection 116
   residential 394
   sprinkler 116
   standpipe 116
   vent 274
All fuel chimney 305-308
   service jacket insulation 137
Aluminum corrugated pipe 505
   diffuser 369
   diffuser perforated 369
   downspout 66
   ductwork 347
   flashing 66, 269
   flex duct 348
   grille 371
   handrail 78
   louver 78, 373
   nail 62
   operating louver 374
   pipe 40, 41
   register 374
   rivet 58
   salvage 121
Anchor chemical 56
   epoxy 56
   expansion 56
   hollow wall 56
   jute fiber 57
   lead screw 57
   machinery 58
   pipe conduit casing 38
   plastic screw 57
   selfdrilling 56
   wedge 57
Anchors drilling 56
Angle framing steel 58
   valve 236
Anti-freeze inhibited 270
Anti-siphon device 267
   water 267
Apartment bldg. heat fin-tube . . 420
   S.F. & C.F. 400
Appliance 86
   plumbing 285
   residential 82, 394
Approach ramp 78
Aquastats 109
Architectural woodwork 63
Area clean-up 103
Arena sport S.F. & C.F. 405
Asbestos abatement 102
   demolition 102
   disposal 104
   removal 101, 102
   removal process 529
As-built drawings 18
Ash conveyor 84
Asphalt flashing 66
Asphaltic binder 43
   concrete 20, 43
   pavement 43
   wearing course 43
Aspirator dental 91
Atomizer water 103
Attic ventilation fan 368
Auditorium S.F. & C.F. 400
Auger 13
Automatic timed thermostat . . . 111
   washing machine 87
Automotive equipment 83
   exhaust system 459
   lift 83
   sales S.F. & C.F. 400
Autopsy equipment 90
Axial flow fan 364

B

Backer rod 68
Backfill 25, 26
   trench 27
Backflow preventer 266
   preventer solar 106
Backhoe excavation 26
   trenching 500
Bag disposable 103
   glove 101
Balance valve circuit setter 248
Balancing air 384
   hydronic 265
   water 385
Baler 84
Ball check valve 237
   joint weld end 266
   valve 228, 236, 241

   valve grooved joint 188
Band vent-chimney 310
Bank air-conditioning . 444, 446, 448, 450, 452, 454, 456
   S.F. & C.F. 400
Barometric damper 300
Barricade 12
   tape 12
Barrier separation 102
Barriers and enclosures 12
Baseboard diffuser 371
   expansion coupling 250
   fin-tube heating 420
   heat 338
   heat electric 340
   register 375
Basket type strainer 261
Bath steam 96
   whirlpool 91
Bathroom exhaust fan 367
   heater & fan 367
Bathtub removal 120
Bead board insulation 95
Beam attachment welded 121
   clamp 122
   soldier 25
Bedding pipe 26
Bellow expansion joint pipe . . . 250
Bellows expansion joints 251
Belt driven fan 366
Bibb hose 248
Binder asphaltic 43
Bituminous coating 47
Blanket insulation 131
Blasting 26
Block manhole 42
   manhole/catch basin 510
Blowdown boiler 286
Blower 365
   pneumatic tube 118
Boiler 286
   air fitting 245
   blowdown 286
   coal fired 294
   coal-fired hydronic 418
   control system 108
   cut off low water 383
   demolition 120
   drain 248
   electric 287, 288, 415
Boiler electric HW 466
Boiler electric steam 287,467
Boiler feed pump 280
   fire tube 293
   fossil fuel 291
   fossil fuel hydronic 418
   gas fired 288-290
   gas pulse 290
   gas-fired hydronic 418
   gas/oil combination 290, 291
   general 120
   hot water 287, 289-294
   hot-water electric 415
   hydronic fossil fuel 416-419
   insulation 131
   oil fired 291, 292
   oil-fired hydronic 417, 418
   scotch marine 293
   solid fuel 294
   steam 288-293
   steam electric 414
   stoker fired 294
   wall hung 290
   water gauge 383
   water tube 294
Bolt & gasket set 172

Index

toggle ... 57
Bond performance ... 7, 519
Boom lift ... 14
Booster coil ... 334
   fan ... 364
Boot spiral duct register ... 357
Booth painting ... 83
Boring horizontal ... 28
   service ... 28
Bowling alley air-conditioning ... 444, 446, 448, 450, 452, 454, 456
Box buffalo ... 32
   culvert precast ... 505
   storage ... 9
   vent ... 54
Boxes VAV ... 360
Brace cross ... 59
Bracket pipe ... 121
Braided bronze hose ... 252
   steel hose ... 253
Brass pipe ... 141
   pipe fitting ... 141
   salvage ... 121
   screw ... 63
   valve ... 227, 228
Break glass station ... 116
Breaker vacuum ... 267
Breather cap fuel ... 274
Breeching draft inducer ... 311
   insulation ... 131
Brick catch basin ... 42
   chimney simulated ... 79
   manhole/catch basin ... 510
   vent ... 373
   vent box ... 54
Bridge sidewalk ... 10
Bronze body strainer ... 263
   butterfly valve ... 228
   valve ... 228
Broom finish concrete ... 51
Bucket crane ... 13
Buffalo box ... 32
Builder's risk insurance ... 514
Building air conditioning ... 444
   air-conditioning ... 444, 446-458
   permit ... 7
   portable ... 9
   sitework ... 510
   temporary ... 9
Built-in range ... 86
Bulldozer ... 25
Burglar alarm ... 116
Burner gas conversion ... 286
   gas/oil ... 286
   gun type ... 286
   industrial ... 286
   jet ... 286
   oil ... 287
   residential ... 286
Butterfly valve ... 20, 30, 232
   valve bronze ... 228
   valve grooved joint ... 189
Butyl caulking ... 69

C

Cabinet convector heater ... 343
   laboratory ... 90
   unit heater ... 339, 343, 344
Cabinet unit heater ... 465
Cable heating ... 340
   jack ... 16
Calcium silicate insulation ... 131
Cap vent ... 274
Carbon monoxide detector ... 109

Carousel compactor ... 84
Carpet computer room ... 78
Carrier pipe ... 122, 126
Case refrigerated ... 82
Cased evaporator coils ... 296
Casing anchor pipe conduit ... 38
Cast in place concrete ... 50
   iron fitting ... 170
   iron manhole cover ... 42
   iron pipe fitting ... 160
   iron radiator ... 338
   iron trap ... 268
Catch basin ... 42
   basin and manhole ... 510
   basin brick ... 42
   basin frame and cover ... 43
   basin masonry ... 42
   basin precast ... 42
   basin removal ... 22
Catwalk ... 11
Caulking ... 68, 69
   polyurethane ... 69
   sealant ... 68
Ceiling diffuser ... 369, 370
   drilling ... 56
   fan ... 365
   heater ... 87
   integrated ... 94
   polystyrene ... 95
   radiant ... 94
   register ... 374
   suspended ... 94
Cellular glass insulation ... 133
Cement plaster ... 95
Central station air handling unit ... 321
Central station heat pump ... 483
Central vacuum ... 82
Centrifugal airfoil fan ... 365
Centrifugal chiller ... 471
Centrifugal fan ... 364, 366
Centrifugal fan system ... 491
Centrifugal in-line motor ... 281
   liquid chiller ... 312
   pump ... 14, 280
Chain hoist ... 16
   link fence ... 12
   saw ... 15
   trencher ... 13
Chair dental ... 91
   hydraulic ... 91
Chamber decontamination ... 102
Channel framing steel ... 59
   slotted framing ... 59
Charge powder ... 58
Charges disposal ... 104
Chart recorders ... 112
Chase formwork ... 50
Check lift valve ... 229
   swing valve ... 228, 234
   valve ... 20, 30, 238, 240-242
   valve ball ... 237
   valve grooved joint ... 189
   valve silent ... 231, 234, 240, 243
   valve wafer ... 231, 235, 243
Chemical anchor ... 56
   feeder shot ... 296
   media filter ... 379
   toilet ... 15
Chiller reciprocating ... 470
Chiller water ... 311, 312
Chimney ... 54
   accessories ... 79
   all fuel ... 305-308
   cap ventilator ... 378
   concrete ... 54
   draft inducer ... 311

foundation ... 51
   high temperature vent ... 309
   metal ... 54, 79, 309, 310
   positive pressure ... 305-308
   prefabricated metal ... 305-308
   rain cap ... 378
   simulated brick ... 79
   vent ... 301
   vent fitting ... 308
   vent oval ... 304
Chlorination system ... 97
Church S.F. & C.F. ... 400
Chute linen ... 118
   package ... 118
   refuse ... 118
   rubbish ... 23
Chutes ... 118
C.I.P. concrete ... 50
Circuit sensor ... 246, 247
   setter ... 248
Circular saw ... 15
Circulating pump ... 270
City hall S.F. & C.F. ... 405
Clamp beam ... 122
   pipe ... 45, 46, 121, 129
   riser ... 122
Clamshell bucket ... 13
Classroom heating & cooling ... 337
Clay pipe ... 35, 41
Clean room ... 94
   tank ... 21
Cleaning internal pipe ... 45
   up ... 18
Clean-up area ... 103
Clevis pipe hanger ... 124
Climbing crane ... 15
   jack ... 16
Clock timer ... 394
Clothes dryer commercial ... 82
   dryer residential ... 87
Club country S.F. & C.F. ... 400
Coal burning furnace ... 298
   fired boiler ... 294
   stoker ... 287
Coalescent filter ... 279
Coating bituminous ... 47
   epoxy ... 47
   polyethylene ... 47
Cock air ... 248
   gauge ... 249
   pet ... 248
   steam ... 248
Cocks gas ... 227
Coil booster ... 334
   condenser ... 332, 333
   cooling ... 332
   heating ... 322
   water ... 332, 333
Cold storage door ... 72
Collar spin-in duct ... 358
   storm ... 308
Collection systems dust ... 88
Collector solar energy system ... 104, 105
College S.F. & C.F. ... 401
Combination device ... 391
Comfort station ... 96
Commercial air-conditioner ... 444
   air-conditioning ... 445, 446, 448-456, 458
   fin tube ... 338
   fin-tube heating ... 421
   heater electric ... 341
Commercial kitchen exhaust ... 463
Commercial kitchen vent ... 85
   refrigeration ... 95

unit-heater heating ... 422
Commissioning ... 18
Common nail ... 62
Community center S.F. & C.F. ... 401
Compact fill ... 26
Compaction soil ... 25
Compactor ... 84
   earth ... 13
   waste ... 84
Companion flange ... 172
Compensation workers' ... 7
Component control ... 107-109
   pneumatic ... 113, 114
Components furnace ... 296
Compression joint fitting ... 149
   tank ... 252
Compressor accessory ... 277
   aftercooler ... 277
   air ... 14
   air package ... 276
   filter ... 279
   reciprocating ... 276
   reciprocating hermetic ... 276
   refrigerant ... 276
   refrigeration ... 276
   tank mounted ... 109
Computer air conditioning ... 323
   floor ... 78
   room carpet ... 78
Concrete asphaltic ... 20, 43
   broom finish ... 51
   cast in place ... 50
   catch basin ... 42
   chimney ... 54
   C.I.P. ... 50
   core ... 22
   demo ... 23
   drill ... 22, 388
   drilling ... 56
   finish ... 51
   float finish ... 51
   footing ... 51
   formwork ... 50
   foundation ... 51
   hole cutting ... 389
   hole drilling ... 388
   insert hanger ... 123
   manhole ... 42
   monolithic finish ... 51
   patching ... 51
   pipe ... 33, 34, 40
   placing ... 51
   ready mix ... 50
   reinforcement ... 50
   removal ... 22
   structural ... 50
   tank ... 98
Condensate pump ... 271
   steam meter ... 273
Condenser ... 315
   air cooled belt drive ... 315
   air cooled direct drive ... 315
   coil ... 332, 333
   evaporative ... 316, 317
   pad ... 319
Condensing unit air cooled ... 316
   unit water cooled ... 317
Conditioner portable air ... 326
   roof top air ... 325
Conditioning ventilating air ... 317, 318, 321, 324, 326, 352, 353, 355, 357, 370, 372-374, 376, 377, 381, 382
Conduit & fitting flexible ... 395
   fitting pipe ... 37
   pipe ... 36
   prefabricated pipe ... 36, 37, 39, 40

Index

Connection motor 395
Connector flexible 252
   water copper tubing 252
Console heat pump 483
Construction aids 9
   cost index 7
   management fee 6
   photography 8
   sump hole 24
   temporary 9
Contained pipe 223
Contaminated soil 21
Contingencies 6
Contractor equipment 13
   overhead 18
   pump 15
Control & motor starter 397
   air 244
   component 107-109, 113
   component electronic 115
   component relay 110
   component sensor 110
   component transmitter 111
   draft 300
   electronic 115
   flow check 254
   hand-off-automatic 396
   pneumatic 110, 113, 114
   radiator supply 254
   station 396
   station stop-start 396
   system 107, 108
   system air compressor 109
   system boiler 108
   system cooling 108
   system electronic 108
   system split system 108
   system VAV box 109
   system ventilator 108
   valve heating 112, 254
   water feeder 383
   water level 383
Control-component 112
Controller pneumatic 114
Controls D.C. 108
   DDC 107
Convector cabinet heater 343
   cover 62
   heating unit 338
Conveyor ash 84
   door 73
Cooking equipment 85
   range 86
Cooler 95
   beverage 85
   door 72
   evaporator 317
Cooling & heating A/C 324
   coil 332
   computer 458
   control system 108
   electric furnace 298
   tower 314
   tower louver 373
   towers fiberglass 315
   towers stainless 315
Copper ACR tubing 144
   drum trap 268
   DWV tubing 143
   fitting 145
   flashing 66, 269
   grooved joint 153
   pipe 142, 143
   rivet 58
   salvage 121
   tubing oxygen class 143

Core drill 22
Cork insulation 95
Corridor air-conditioning . 444, 446,
      448 , 450, 452, 454, 456
Corrosion resis. backflow preven 266
   resistant duct fitting 353
   resistant fan 366
   resistant tank 251
Corrosive fume resistant duct .349-352
Corrugated metal pipe 40, 41
   pipe 40
Cost mark-up 8
Country club S.F. & C.F. 400
Couplers air line 279
Coupling expansion 250
   mechanical 182
   plastic 216
Course wearing 43
Courthouse S.F. & C.F. 401
Cover convector 62
   manhole 43
   stair tread 9
Covering calcium silicate pipe . 134,
                      135
   protection saddle 127
CPVC pipe 207
   valve 236, 237
Crane bucket 13
   climbing 15
   crawler 15
   hydraulic 16
   material handling 15
   tower 15
Crawler crane 15
Crematory 84
Crew 6
Cross brace 59
Crosses 28
Crossflow fuel heater 275
Cubic foot building cost 400
Culvert box precast 505
   end 34
   reinforced 33
Curb prefabricated 319
Curtain air 72
   damper fire 358
Cut in sleeve 29
   in valve 29
   off low water boiler 383
   pipe groove 189
Cutoff feeder combination 383
Cutting steel 23
   torch 15

D

Dairy case 82
Damper 359, 361, 362
   barometric 300
   fire round 359
   foundation vent 54
   louver type fire 359
   multi-blade 361
   relief 363
   round 363
   vent 301
Darkroom equipment 87
Dasher hockey 97
DDC controls 107, 108
Deck roof 63
Decontamination chamber 102
   enclosure 102
   equipment 103
Decorator device 391
   switch 391

Deep well pump 33
Dehumidifier 86
   selfcontained 329
Delicatessen case 82
Deliquescent desiccant 278
Demo concrete 23
Demolition 20, 120
   asbestos 102
   boiler 120
   ductwork 120
   enclosure 103
   furnace 120
   HVAC 120
   pavement 22
   plumbing 120
   selective 23
   site 22
Dental equipment 91
Department store air-cond. .. 447, 448
   store S.F. & C.F. 401
Derrick 16
   guyed 16
   stiffleg 16
Desiccant dryer 278, 279
Detection alarm 116
   leak 116
   probe leak 116
   system 116
   tape 20
Detector carbon monoxide 109
   check valve 30
   meter water supply 265
   smoke 116
   temperature rise 116
Device anti-siphon 267
   combination 391
   decorator 391
   GFI 392
   receptacle 392
   residential 390
Dewater 17, 23, 24
   pumping 24
Dewatering 23
Diaphragm pump 15
Dielectric union 249, 250
Diffuser aluminum 369
   baseboard 371
   ceiling 369, 370
   floor 371
   linear 369
   opposed blade damper 369
   perforated aluminum 369
   pump suction 264, 265
   rectangular 370
   sidewall 371
   steel 371
   supply/return combination . 370
   T-bar mount 370
Dimensionaire ceiling 94
Dimmer switch 390
Direct expansion A/C 337
Discharge hose 14
Dishwasher 85, 86
Dispenser product 83
Disposable bag 103
Disposal asbestos 104
   charges 104
   garbage 87
   waste 84
Distiller water 90
Ditching 23
Diverter valve 232, 239, 242
Door air curtain 72
   bell residential 393
   cold storage 72
   conveyor 73

cooler 72
   duct access 358
   fire 116
   grille 372
   refrigerator 72
   special 72
   steel louver 373
Dormitory S.F. & C.F. 401
Double wall pipe 35, 219
Downspout 66
   aluminum 66
   painting 76
   steel 66
Dozer 25
Draft control 300
   damper vent 301
   inducer breeching 311
   inducer chimney 311
Drain boiler 248
   pipe 24
Drainage pipe 33, 34, 41, 208
   site 42
   trap 268
Drainers liquid 255, 256
Drawings as-built 18
Drill & tap main 31
   concrete 22, 388
   core 22
   main 31
   rock 26
Drilling anchors 56
   ceiling 56
   concrete 56
   horizontal 28
   plaster 56
Driveway removal 22
Drug store air-conditioning . 445, 447,
      449 , 453, 455, 457
Drum trap copper 268
Dryer adsorption air 278
   clothes 87
   commercial clothes 82
   compressor air 277
   deliquescent air 277
   desiccant 278, 279
   receptacle 392
   refrigeration air 278
   regenerative air 278
   residential 82
   system 278
   vent 87
Drywall nail 62
Dual wall hdpe pipe 223
Duck tarpaulin 11
Duct access door 357
   accessories 357
   collar spin-in 358
   connection fabric flexible 358
   corrosive fume resistant . 349-352
   fiberglass 349-352
   fitting plastic 353
   fitting spiral 354-356
   flexible insulated 348
   flexible noninsulated 347
   furnace 299
   grille 78
   heater electric 330, 331
   humidifier 330
   HVAC 347
   insulation 131
   liner 132
   mechanical 347
   mixing box 360
   plastic 352
   preformed spiral 353, 355
   rigid fiberglass 349-352

Index

rigid plastic 352
silencer 363
smoke exhaust 349-352
sound trap 363
Ducted evaporative cooler 322
Ductile iron fitting 31
   iron pipe 31
   iron pipe cement lined 30
Ductless split system 328
Ductwork .. 347, 348, 353, 354, 356, 544
   aluminum 347
   demolition 120
   fabric coated flexible 347
   fabricated 347
   flexible 348
   galvanized 347
   laboratory 90
   rectangular 347
   removal 120
   rigid 347
Dump charge 23
   truck 13, 27
Dumpster 23
Duplex style strainer 262, 263
Dust collection systems 88
   partition 23
DWV pipe ABS 207
   PVC pipe 207
   tubing copper 143
DX coil evaporator 317-319

E

Earth compactor 13
   vibrator 25, 26
Earthwork 25
   equipment 13, 25
Education religious S.F. & C.F. ... 404
Eggcrate framed grille 372
   grille lay-in 373
   T-bar grille 373
Ejector pump 284
Elbow pipe 185
Elderly housing S.F. & C.F. 402
Electric baseboard heater 340
   boiler 287, 288, 414
   cabinet heater 343
   duct heater 330, 331
   fuel oil heater 274
   furnace 297
   generator 14
   heater 87
   heating 340
   hot air heater 342
   induction unit 360
   infra-red heater 340
   log 79
   pool heater 295
   unit heater 341-343
Electrical fee 6
   installation drilling 388
   knockout 389
   laboratory 90
Electronic air cleaner 380
   control 115
   control component 115
   control system 108
Electrostatic precipitator 85
Elementary school S.F. & C.F. ... 404
Elevated water tank 97
Elevator construction 16
   fee 6
Eliminator air 244
   air & dirt 245

Employer liability 7
Encapsulation 104
   pipe 104
Enclosure decontamination 102
   demolition 103
End culvert 34
Energy circulator air solar 104
   recovery wheel 346
   system air purger solar 106
   system air vent solar 106
   system balancing valve solar . 106
   system control valve solar ... 106
   system controller solar 105
   system expansion tank solar . 106
   system gauge pressure solar . 106
   system heat exchanger solar . 105
   system storage tank solar ... 105
   system vacuum relief solar .. 106
Engine exhaust elimination 381
Engineering fee 6
   fees 522
Enthalpy control 109
   recovery wheel 346
Entrainment eliminator 268
Epoxy anchor 56
   coating 47
   grout 51
   resin fitting 209
Equipment 13, 82
   darkroom 87
   dental 91
   earthwork 13, 25
   feedwater 296
   formwork 50
   foundation 50
   general 8
   hospital 90
   insurance 7
   laboratory 90
   laundry 82
   medical 90
   rental 13, 15, 525
   service station 83
   swimming pool 97
   waste handling 83
Escutcheon 259
Estimate electrical heating 340
Estimating 514
Ethylene glycol 270
Evaporative condenser 316, 317
   cooler 322
   cooler ducted 322
Evaporator coils cased 296
   cooler 317
   DX coil 317-319
   freezer 318
Excavation 25, 27
   backhoe 26
   hand 25, 26
   hauling 27
   machine 25
   structural 26
   trench 26, 27, 500-502
Excavator rental 13
Exchanger heat 320
Exhaust elimination engine 381
Exhaust fume hood 460
Exhaust granular fan 366
   hood 87, 90
   hood kitchen 86
   registers 376
   system 381
   vent 377
   vent damper 358
Exhaust-elimination garage 381
Exhauster industrial 366

roof 368
Expansion anchor 56
   coupling 250
   coupling baseboard 250
   grooved joint 188
   joint 250
   joints bellows 251
   shield 56
   tank 251
   tank steel 251
Exterior sprinkler 44
Extra work 8
Extractor air 357
   fitting 273
   industrial 82
Eye rod linked 125
   rod welded 125

F

Fabricated ductwork 347
Factor 6
Factory air-conditioning 445, 447, 449, 451, 455, 457
   S.F. & C.F. 401
Fan 364, 366
   air conditioning 364
   air handling 364
   attic ventilation 368
   axial flow 364
   bathroom exhaust 367
   belt driven 366
   booster 364
   ceiling 365
   centrifugal 364, 366
   coil air conditioning 336
   corrosion resistant 366
   exhaust granular 366
   house ventilation 369
   induced draft 311
   in-line 364
   kitchen exhaust 367
   low sound 364
   paddle 394
   propeller exhaust 366
   residential 367, 394
   roof 368
   utility set 366, 368
   vane-axial 365
   ventilation 394
   ventilator 364
   wall exhaust 369
   wiring 394
Fastener timber 62
   wood 63
Fastenings metal 56
Faucet & fitting 282
Fee engineering 6
Feeder control water 383
   cut-off combination 383
Feedwater equipment 296
Fence chain link 12
   plywood 12
   temporary 12
   wire 12
Fiberglass tank 99
Fiberglass cooling tower 473
Fiberglass cooling towers 315
   duct 349-352
   insulation 131
   panel 9
   pipe covering 136, 137
   reinforced ceiling 95
Fiberglass storage tank 494, 495
Fiberglass tank 100

Field office 9
   office expense 9
Fill 25, 26
   adapter 273
   box fuel 273
   cap fuel 273
   gravel 26
Film equipment 87
Filter air 279, 379
   chemical media 379
   compressor 279
   dryer refrigerant 271
   grille 372
   mechanical media 380
   odor 279
   oil 273
   regulator lubricator air 279
   swimming pool 97
Filtration air 102
   equipment 97
Fin tube commercial 338
   tube radiation 338
   tube radition 338
Finish concrete 51
   nail 62
   refrigeration 95
Fire alarm 116
   call pullbox 116
   damper curtain type 358
   damper louver type 359
   damper round 359
   door 116
   horn 116
   protection kitchen 85
   signal bell 116
   station S.F. & C.F. 401
   tube boiler 293
Fired burner gas 286
   pool heater oil 295
Fireplace free standing 79
   prefabricated 78
Fireproofing 67
Firestopping 67
Fitting adapter pipe 147
   cast iron 170
   cast iron pipe 162, 163
   compression joint 149
   compression tank air 245
   copper 145
   copper pipe 147
   epoxy resin 209
   extractor 273
   flanged 171, 172
   flare joint 150
   forged steel pipe 177, 179, 180
   gas vent 302, 303
   grooved joint ... 153, 185, 187, 188
   grooved joint pipe 185
   grooved unit 153, 186
   insert pipe 218
   malleable iron 164
   malleable iron pipe 167, 169, 180, 182
   mechanical joint pipe 180, 182
   monoflow 256
   pipe .. 160, 161, 165, 171, 172, 176
   pipe conduit 37
   plastic 209
   plastic pipe 213, 217
   polybutylene pipe 217, 218
   PVC 211
   PVC pipe 219
   refrigeration 150
   stainless steel 195
   stainless steel pipe .. 195-197, 199, 201, 202, 204

Index

steel 160, 170
steel carbon pipe 175
steel pipe 177, 179-181
tee 31
vent chimney . . . 302, 303, 308, 309
water pipe 31
weld 176
weld joint pipe 177
welded steel pipe 181
wye 31
Fittings HDPE 221
Fixed blade louver 374
roof tank 98
Fixture plumbing 33, 270, 280
removal 121
residential 393
Flange companion 172
gasket 172
grooved joint 187
stainless steel 200
steel 187
steel weld-on 179
threaded 172, 173
Flanged fitting 171, 172
pipe 155, 158
valve 240
Flare joint fitting 150
Flashing 66
aluminum 66, 269
asphalt 66
copper 66, 269
galvanized 269
masonry 66
neoprene 270
stainless 66
vent 269
vent chimney 303, 309
Flatbed truck 13
Flex duct aluminum 348
Flexible conduit & fitting 395
connector 252
duct connection fabric 358
ductwork 348
ductwork fabric coated 347
insulated duct 348
metal hose 252
Float finish concrete 51
valve 254
Floater equipment 7
Floating roof tank 99
Floodlight 14
Floor access 78
cleaning 18
diffuser 371
heating radiant 344-346
nail 62
patching concrete 51
pedestal 78
refrigeration 95
register 375
Flow check control 254
fitting hydronic 256
measurement 246-248
meter 247, 248, 265
meter circuit sensor 246
Flue chimney metal 301
prefab metal 301
Fluid heat transfer 105
Foam pipe covering 139
urethane 69
Food service equipment 85
Foot valve 237, 273
valve oil 273
Footing concrete 51
Forged valves 241
Formwork accessory 50

chase 50
concrete 50
equipment 50
sleeve 50
structural 50
Fossil fuel boiler 289, 291-294
fuel burner 286
fuel hydronic heating 416-418
Foundation chimney 51
concrete 51
equipment 50
mat 51
vent 54
Fountain 44
lighting 44
Framing lightweight 58
metal 58
steel angle 58
steel channel 59
Fraternal club S.F. & C.F. 400
Fraternity house S.F. & C.F. 402
Freezer 82, 85, 95
door 72
evaporator 318
Freezestat 110
Fuel breather cap 274
burner fossil 286
fill box 273
fill cap 273
heater crossflow 275
heater safety type 275
heater steam 275
heater suction bell 275
oil gauge 274
oil heater 274, 275
oil heater electric 274
oil pump 273, 281
oil pump & motor set 273
oil specialties 273
oil valve 274
spill container 273
storage tank 98
system components 273
tank 99, 100
Fume elimination welding 382
hood 90
Funeral home S.F. & C.F. 402
Furnace A/C coils 296
coal burning 298
components 296
demolition 120
duct 299
electric 297
gas 299
gas fired 297, 298
heating & cooling 298
hot air 297
oil fired 298, 299
type humidifier 330
wall 300
wood coal/oil combination . . . 298
Fused pipe 31
Fusible oil valve 274

G

Gage stick wood 273
Galvanized ductwork 347
flashing 269
Garage exhaust system 459
exhaust-elimination 381
S.F. & C.F. 402
Garbage disposal 87
Gas boiler HW 468
boiler steam 469

Gas cocks 227
connector 252
conversion burner 286
fired boiler 288-290
fired burner 286
fired furnace 297, 298
fired infra-red heater 340
fired pool heater 295
fired space heater 299, 300
furnace 299
heat air conditioner 324
incinerator 84
log 79
meter residential 275
pipe 36
pressure regulators 257
pulse combustion 290
regulator 257
safety valve 287
station piping 35
station tank 99, 100
stop valve 227
vent 301
vent fitting 302, 303
Gasket & bolt set 172
flange 172
neoprene 69
Gas/oil boiler HW 469
boiler steam 469
Gas/oil burner 286
combination boiler 291
steam boiler 291
Gasoline pump 83
tank 99
Gas-service polyethlene pipe 507
Gate box 29
valve 20, 29, 233, 239, 241, 243
valve grooved-joint 189
valve soldered 229
Gauge cock 249
fuel oil 274
oil tank 274
pressure 110
pressure/vacuum 110
vacuum 110
water 383
General equipment 8
Generator construction 14
steam 90
GFI receptacle 392
Gland seal pipe conduit 36
Glass insulation cellular 133
lined water heater 87
Glassware washer 90
Globe valve . . . 21, 230, 233, 240, 242, 243
Glove bag 101
Glycol ethylene 270
Gradall 13
Grading 28
Grass sprinkler 44
Gravel fill 26
pack well 32
Gravity syphon ventilator 377
Grille air return 372
aluminum 371
convector cover 62
door 372
duct 78
eggcrate framed 372
filter 372
lay-in T-bar 373
painting 76
plastic 373
transfer 373
Grinder pump system 284

Groove cut labor steel 189
roll labor steel 190
Grooved joint check valve 189
joint copper 153
joint expansion 188
joint fitting 153, 185, 187, 188
joint flange 187
joint pipe 183
joint plug valve 189
joint strainer 188
joint suction diffuser 188
reducer 186
unit fitting 153, 186
Ground box hydrant 282
post hydrant 283
Grout epoxy 51
topping 51
underlayment 51
Guardrail temporary 12
Guide pipe 123
Gutter swimming pool 97
Guyed derrick 16
Gymnasium S.F. & C.F. 402
Gypsum lath nail 62

H

Hand excavation 25-27
Handling rubbish 23
unit air 321
waste 83
Hand-off-automatic control 396
Handrail aluminum 78
Hanger & support pipe 121-123, 125, 128, 129
assembly pipe 126
clevis 124
pipe 121
plastic pipe 129
rod coupling 125
rod pipe 125
rod socket 128
roll 126
split ring 124
trapeze 126
U-bolt pipe 129
U-hook 129
vibration absorbing 129
Hanging lintel 59
Hauling excavation 27
truck 28
Hazardous waste handling 84
HDPE fittings 221
pipe 31
Header pipe 17
Heat baseboard 338
coil steam 335, 336
electric baseboard 340
exchanger 320
exchanger plate 320
exchanger residential 346
loss 538
pump 328
Heat pump central station 483
pump console 483
Heat pump residential 395
pump rooftop 329
radiant 94, 340
recovery 84, 85
recovery air to air 346
recovery ventilation 346
recovery wheel 346
steam 335
temporary 8
trace system 341

Index

transfer fluid 105
transfer package 285
Heater & fan bathroom 367
  air handler 297
  cabinet convector 343
  ceiling mount unit 343
  contractor 14
  electric 87
  electric commercial 341
  electric hot air 343
  electric residential 341
  electric wall 341
  floor mounted space 299
  fuel oil 274, 275
  infra-red 340
  infra-red quartz 342
  raceway 341
  sauna 95
  swimming pool 295
  swimming pool solar 439
  terminal 339, 340
  unit 299, 300, 337, 340
  valance 340
  vertical discharge 342
  wall mounted 343
  warm air 299
  water 87
Heating .. 98, 105, 131, 135-137, 139,
    274, 286-292, 295, 299, 300, 302, 303,
    308 -310, 320, 338, 340, 537
  & cooling classroom 337
  cable 340
  coal hydronic 416, 418
  coil 322
  coil tank 275
  commercial bldg 422
  control transformer 342
  control valve 254
  electric 340
  estimate electrical 340
  fin-tube 418
  fin-tube residential 420
  fixture 418
  gas-fired hydronic 416
  hot air 297, 299, 337, 338, 340
  hot water solar 429
  hydronic 287-289, 291, 338
  hydronic electric 415
  hydronic fossil fuel .. 416, 418, 419
  industrial 287
  insulation 131
  oil hydronic 416, 418
  oil-fired hydronic 417
  panel radiant 342
  solar-energy 439, 441, 443
  subsoil 97
  system hot-water 416
  system oil fired 420, 422
  systems 536
  terminal unit-heater .. 416, 417
  unit convector 338
Heavy duty shoring 9, 10
High impact pipe 205
  school S.F. & C.F. 404, 405
  temperature vent chimney 310
Historical Cost Index 7
Hockey dasher 97
Hoist and tower 16
  automotive 83
  contractor 16
  lift equipment 15
  personnel 16
Hole cutting electrical 389
  drilling electrical 388
Hollow wall anchor 56
Hood exhaust 90

fire protection 85
fume 90
range 87
Hook type pipe outlet 182
Hopper refuse 118
Horizontal boring 28
  drilling 28
Horizontal heat pump 483
Horn fire 116
Hose adapter vapor 273
  air 14
  bibb 248
  bibb sillcock 282
  braided bronze 252
  braided stainless steel ... 253
  braided steel 253
  discharge 14
  metal flexible 252
  suction 14
  water 14
Hospital equipment 90
  S.F. & C.F. 402
Hot air furnace 297
  air heater electric 342
  air heating . 297, 299, 337, 338, 340
  water boiler 287, 289-294
  water coil 334
  water heating 287, 338
  water-steam exchange ... 285, 320
  water-water exchange 320
Hot-water heating forced 420
  solar-energy 425, 427, 429, 431,
      433, 435, 437, 441, 443
House fan whole 367
  ventilation fan 369
Housing elderly S.F. & C.F. . 402
  public S.F. & C.F. 402
HTRW 101
Humidifier 87
  duct 330
  furnace type 330
  room 329
Humidifiers 329
HVAC demolition 120
  duct 347
  piping specialties .. 251, 253, 257,
      262 -264, 268, 383
  removal 120
Hydrant ground box 282
  ground post 283
  removal 22
  wall 282
  water 282
Hydraulic chair 91
  crane 16
  jack 16
  oil pump 281
  pressure test 385
Hydrocumulator 283
Hydronic balancing 265
  expansion coupling 250
  flow fitting 256
  heating 287-289, 291, 338

I

Ice rink 97
  skating equipment 97
  skating rink S.F. & C.F. .. 403
Icemaker 85, 87
Impact wrench 14
Incinerator gas 84
  municipal 84
  waste 84
Increaser vent chimney 310

Index construction cost 7
Indicator thermoflo 265
Induced draft fan 311
Induction air unit 360
  unit electric 360
  unit vertical 360
Industrial air filter 379
  burner 286
  equipment 88
  exhauster 366
  heating 287
Inert gas 21
Infra-red heater 340
  heater gas fired 340
  quartz heater 342
In-line fan 364
Insert fitting PVC 218
  hanger concrete 123
Insulated protectors ADA 131
Insulation 131, 132, 136, 137, 139
  all service jacket 137
  blanket 131
  boiler 131
  breeching 131
  calcium silicate 131
  duct 131
  fiberglass 131
  pipe 132, 139
  polyethylene 132, 138
  polyolefin 139
  refrigeration 95
  removal 101
  subsoil 97
  vapor proof 133
  vaporproof 139
  water heater 131
Insurance 7, 514
  builder risk 7
  equipment 7
  public liability 7
Intake-exhaust louver 374
  vent 378
Integrated ceiling 94
Interval timer 391
Intrusion system 116
Invert manhole 43
Inverted bucket steam trap .. 259
Investigation subsurface 22
Iron body valve 232, 234
Irrigation system 44

J

J hook clamp w/nail 129
Jack cable 16
  hydraulic 16
Jacking 28
Jail S.F. & C.F. 403
Jet burner 286
  water system 33
Jetting pump 17
Job condition 6
Joint expansion 250
  sealer 68
  weld end ball 266
Jute fiber anchor 57

K

Kettle 85
Kitchen equipment 85
  equipment fee 6
  exhaust fan 367
  exhaust hood 86

heat recovery 346
smoke control 86
vent commercial 85
ventilation 85
Knockout electrical 389

L

Labels pipe & valve 130
Labor index 7
  pipe groove roll 190
  pipe threading st steel ... 194
  pipe threading steel 159
  pipe weld stainless steel . 194
  pipe weld steel 159
Laboratory cabinet 90
  equipment 90
  research S.F. & C.F. 404
  sink 90
Ladder 14
  rolling 11
Lag screw 58
  screw shield 57
Lagging 25
Landfill fees 23
Lateral-o-let 179
Latex caulking 69
Laundry equipment 82
Lavatory removal 121
Lead flashing 66
  salvage 121
  screw anchor 57
Leak detection 116
  detection probe 116
  detection tank 116
Liability employer 7
Library air-conditioning 444, 446, 448,
      452, 454
  S.F. & C.F. 403
Lift aerial 14
  automotive 83
Light dental 91
  stand 102
  temporary 8
  tower 14
  underwater 97
Lighting ceiling 94
  fountain 44
  outdoor 14
  outlet 393
  residential 393
Lightning suppressor 390
Lightweight framing 58
Linear diffuser 369
Linen chute 118
  collector 118
Liner duct 132
Link seals pipe 259
Linked eye rod 125
Lintel hanging 59
Liquid chiller centrifugal .. 312
  drainers 255, 256
Loadcenter residential 390
Loader vacuum 103
Locking receptacle 393
Log electric 79
  gas 79
Louver 78, 373
  aluminum 78, 373
  aluminum operating 374
  coating 374
  cooling tower 373
  door steel 373
  fixed blade 374
  intake-exhaust 374

Index

midget 78
  mullion type 374
  redwood 63
  silencer 364
  ventilation 63
Lube equipment 83
Lubricated plug valve 238
Lubricating oil 316
Luminaire ceiling 94

M

Machine excavation 25
  screw 58
  welding 15
Machinery anchor 58
Magnetic motor starter 397
Maintenance 18
Make-up air unit 322
  air unit rooftop 323
Malleable iron fitting 164
  iron pipe fitting ... 164, 165
Management fee construction 6
Manhole 42
  concrete 42
  cover 43
  invert 43
  precast 42
  raise 43
  removal 22
  step 43
Marker pipe 130
Mark-up cost 8
Masonry accessory 54
  catch basin 42
  flashing 66
  manhole 42
  nail 62
  saw 15
  ventilator 54
Masons scaffolding 11
Mat foundation 51
Material handling system ... 118
  hoist 16
  index 7
Measurement flow 246-248
Meat case 82
Mechanical 530
  coupling 182
  duct 347
  equipment demolition 120
  fee 6
  media filter 380
Medical center air-conditioner 457
  center air-conditioning .. 445, 447,
                        449, 451, 453, 455
  clean copper tubing 143
  clinic S.F. & C.F. 403
  equipment 90
  office S.F. & C.F. 403
Melting snow 340
Metal chimney 54, 79, 309, 310
  ductwork 347
  fastenings 56
  flexible hose 252
  flue chimney 301
  framing 58
  pipe 30, 41
Meter flow 247, 248
  steam condensate 273
  venturi flow 265
  water supply 265
  water supply detector 265
  water supply domestic 265
Meters 275

Microtunneling 28
Miscellaneous pumps 281
Mixing box duct 360
  boxes VAV 360, 361
  valve 256
Modification to cost 6
Modular air handler 296
Modulating damper motorized 362
Monoflow fitting 256
Monolithic finish concrete . 51
Monoxide detector carbon ... 109
Mortuary refrigeration 90
Motel S.F. & C.F. 403
Motor actuated valve 112
  centrifugal in-line 281
  connection 395
  controlled valve 112
  generator 14
  starter 397
  starter & control 397
  starter enclosed & heated 397
  starter magnetic 397
  starter w/circuit protector 397
  starter w/fused switch ... 397
Motorized modulating damper 362
  zone valve 112
Mount vibration absorbing .. 130
Mounting board plywood 63
Movie S.F. & C.F. 405
Mullion type louver 374
Multi-blade damper 361
Multipurpose valve 235
Multizone air cond. rooftop 325
  air-conditioner 326
Municipal incinerator 84
Mushroom ventilator stationary 378
Mylar tarpaulin 11

N

Nail 62
Needle valve 237
Neoprene flashing 270
  gasket 69
Nondestructive pipe testing 385
Nonfusible oil valve 274
Nursing home S.F. & C.F. ... 403
Nut threaded 125

O

Observation well 32
Occupancy determinations ... 521
Odor filter 279
Off highway truck 14
Office air-conditioner 457
  air-conditioning . 445, 447, 449, 451,
                        453, 455
  field 9
  floor 78
  medical S.F. & C.F. 403
  S.F. & C.F. 403
  trailer 9
Oil burner 287
  filter 273
  fired boiler 291, 292
  fired furnace 298, 299
  fired heating 292
  fired infra-red heater ... 340
  fired pool heater 295
  fired space heater 300
  heater temporary 14
  lubricating 316
  pump fuel 281

pump hydraulic 281
  refrigeration 316
  shut-off valve 274
  specialties fuel 273
  storage tank 98
  valve fuel 274
Oils 316
Omitted work 8
Opening roof frame 59
Operating cost equipment ... 13
OSHA testing 103
Outdoor lighting 14
Outlet heater tank 275
  lighting 393
Oval chimney vent 304
Oven 85, 86
Overhaul 23, 28
Overhead & profit 8
  and profit 518
  contractor 18
Overtime 7, 8, 520
Oxygen lance cutting 23

P

P trap 268
  trap refrigeration 272
Package chute 118
Packaged ventilation 85
Packaging waste 104
Pad condenser 319
  prefabricated 319
  vibration absorbing 130
Paddle blade air circulator 365
  fan 394
Painting booth 83
  downspout 76
  grille 76
  pipe 76
  swimming pool 97
Paints & coatings 76
Pan shower 66
Panel fiberglass 9
  fire 116
  radiant heat 342
Parking garage S.F. & C.F. . 402
  lots paving 44
Partition dust 23
  refrigeration 95
Patching asphalt 20
  concrete 51
  concrete floor 51
  concrete wall 51
Pavement asphaltic 43
  demolition 22
  replacement 44
Paving parking lots 44
Pedestal floor 78
Penthouse air-conditioner .. 326
  roof louver 374
Perforated aluminum pipe ... 40
  pipe 40, 41
  PVC pipe 41
Performance bond 7, 519
Permit building 7
Personal respirator 102
Personnel hoist 16
Pet cock 248
Pex tubing 344
Photography 8
  construction 8
  time lapse 8
Pickup truck 15
Pigtail steam syphon 249
Pile driving 525

H 25
  steel sheet 24
  wood sheet 24
Piling sheet 24, 528
Pillow tank 99
Pin powder 58
Pipe 42
  & fittings 36, 38, 39, 121, 122, 124,
    125, 127, 129, 142, 143, 145, 147, 154
    -156, 158-161, 163, 165, 167, 170, 173,
    175, 179-181, 183-185, 190, 191, 193
    -197, 199, 201-208, 213, 216, 218, 228,
    230, 233, 234, 236-240, 242, 250, 266,
                         267, 269, 282
  & valve labels 130
  aluminum 40, 41
  and fittings 531
  bedding 26, 503
  bedding trench 27
  bellow expansion joint ... 250
  bituminous coated metal .. 505
  bituminous fiber 505
  bituminous fiber perforated 508
  brass 141
  carrier 126
  clamp 45, 46, 121
  clamp plastic 129
  clay 35, 41
  cleaning internal 45
  concrete 33, 34, 40
  concrete nonreinforced ... 505
  concrete porous wall 508
  concrete reinforced 505
  conduit 36
  conduit gland seal 36
  conduit prefabricated .. 36, 37, 39, 40
  conduit system 36, 37
  contained 223
  copper 142, 143
  copper in trench 509
  corrugated 40
  corrugated metal 40, 41
  corrugated metal coated .. 505
  corrugated metal plain ... 505
  corrugated oval arch coated 506
  corrugated oval arch plain 506
  covering 132, 136, 137
  covering calcium silicate 134, 135
  covering fiberglass .. 136, 137
  covering foam 139
  CPVC 207
  double wall 35, 219
  drain 24
  drainage 33, 34, 41
  dual wall HDPE 223
  ductile iron 31
  ductile iron in trench ... 509
  DWV ABS 207
  DWV PVC 207
  elevated installation 120
  encapsulation 104
  fitting . 160, 161, 165, 171, 172, 176
  fitting adapter 147
  fitting brass 141
  fitting cast iron ... 160, 162, 163
  fitting copper 145, 147
  fitting DWV 213
  fitting flanged cast iron 170
  fitting forged steel . 177, 179, 180
  fitting grooved joint 185
  fitting insert 218
  fitting malleable iron 165, 167, 169,
                         180, 182
  fitting mechanical joint 180, 182
  fitting plastic ... 211, 213, 216, 217

Index

Entry	Page
fitting polybutylene	217, 218
fitting PVC	219
fitting stainless steel	195-197, 199, 201, 202, 204
fitting steel	173, 177, 179-181
fitting steel carbon	173, 175
fitting weld joint	177
fitting weld steel	173
fitting welded steel	181
flanged	155, 158
fused	31
gas	36
groove roll labor	190
grooved joint	183
hanger	121
hanger & support	121-123, 125, 128, 129
hanger assembly	126
hanger clevis	124
hanger plastic	129
hanger rod	125
hanger strap	128
hanger U-bolt	129
HDPE	31
header	17
high impact	205
in trench	505-509
inspection internal	45
insulation	132, 139
insulation removal	101
iron ductile mech joint	509
iron ductile tyton joint	509
link seals	259
marker	130
metal	41
metal perforated coated	508
painting	76
perforated	41
perforated aluminum	40
plastic	205-208
plastic, FRP	204
polyethylene	36
polyethylene gas-service	507
polyethylene in trench	509
polyvinyl chloride	506
PVC	34, 207
PVC in trench	509
PVC pressure	206
quick coupling	17
reinforced concrete	33
relay	24
removal	121
repair	45, 46
ring hanger	123
rodding	45
roll single	126
sewage	33, 34, 41
shock absorber	268
specialties	257, 266
stainless steel	190-192, 195
steel	40, 41, 153, 155-158, 160
steel gas-service	507
steel tarred & wrapped	507
subdrainage	40
subdrainage plastic	41
support framing	59
support saddle	127
testing	385
testing nondestructive	385
threading st steel labor	194
threading steel labor	159
vitrified clay	34, 41
vitrified clay perforated	508
vitrified clay premium joint	506
weld joint	154, 157, 190, 191, 193, 194
weld stainless steel labor	194
weld steel labor	159
wrapping	47
X-ray	385
Piping drainage & sewage	505
excavation	500
gas station	35
gas-service	507
specialties	383
specialties HVAC	251, 253, 257, 262-264, 268, 383
subdrainage	40
to 10' high	532
valve	20
water distribution	509
Pit excavation	26
sump	23
Placing concrete	51
Plank scaffolding	10
Plant power S.F. & C.F.	404
screening	13
Plaster cement	95
drilling	56
Plastic convector cover	62
coupling	216
duct	352
duct fitting	353
fitting	209
Plastic, FRP pipe	204
Plastic grille	373
pipe	205-208
pipe clamp	129
pipe fitting	209, 211, 213, 216
pipe hanger	129
pipe polyethylene	219
pipe weld	222
roof ventilator	366
screw anchor	57
trap	269
valve	236
Plate heat exchanger	320
Plate heat exchanger	474
Platform trailer	15
Plating zinc	62
Plenum silencer	364
Plug valve	238
valve grooved joint	189
Plumbing	266
appliance	285
demolition	120
fixture	33, 270, 280
fixtures removal	120
laboratory	90
Plywood fence	12
mounting board	63
sheathing, roof and walls	63
sidewalk	12
Pneumatic air operated valve	112
component	113, 114
control	110, 113, 114
control system	108
controller	114
ejector pumps	84
stud driver	58
thermostat	111
tube	114, 115
tube fitting	114, 115
tube system	118
Police station S.F. & C.F.	404
Polybutylene adapter	217, 218
Polyethylene coating	47
insulation	132, 138
pipe	36, 208
plastic pipe	219
tarpaulin	11
Polyolefin insulation	139
Polypropylene trap	269
valve	236
Polystyrene ceiling	95
insulation	95
Polysulfide caulking	69
Polyurethane caulking	69
Polyvinyl tarpaulin	11
Pool heater electric	295
heater gas fired	295
swimming	96
swimming S.F. & C.F.	405
Porous wall concrete pipe	508
Portable air compressor	14
building	9
heater	14
Positive pressure chimney	305-308
Post office S.F. & C.F.	404
Powder actuated tool	58
charge	58
pin	58
Power plant S.F. & C.F.	404
Power roof vent	492
Power temporary	8
Precast catch basin	42
manhole	42
manhole/catch basin	510
Precipitator electrostatic	85
Pre-engineered structure	96
Prefabricated comfort station	96
fireplace	78
metal chimney	305-308
Pressure booster pump	283
gauge	110
pipe	206
pipe PVC	206
regulator	257, 258
regulator oil	257
regulator steam	258
regulators gas	257
relief valve	230
test	385
test hydraulic	385
valve relief	230
Pressure/temperature safety plug	259
Pressure/vacuum gauge	110
Prison S.F. & C.F.	403
Prisons work in	6
Process air handling fan	364
Produce case	82
Product dispenser	83
piping	35
Project overhead	8
sign	12
size modifier	547
Propeller exhaust fan	366
Propylene glycol	270
Protection winter	9
worker	103
Protectors ADA insulated	131
Psychrometric table	541
P&T relief valve	230
Public housing S.F. & C.F.	402
Pulse boiler gas	290
Pump	33, 280, 283
air conditioning	272
centrifugal	280
circulating	270
condensate	271
condensate return	271
contractor	23
diaphragm	15
Pump double suction	462
end suction	462
Pump fuel oil	273
gasoline	83
heat	328
jetting	17
operator	24
pressure booster	283
rental	15
rotary	281
sewage ejector	284
shallow well	33
submersible	15, 285
sump	87, 285
trash	15
turbine	281
vacuum	272
water	14, 17, 32
wellpoint	17
Pumping	23
dewater	24
Pumps miscellaneous	281
pneumatic ejector	84
vacuum	280
Putlog scaffolding	11
PVC DWV pipe	207
fitting	211
insert fitting	218
pipe	31, 34, 205, 207
pipe perforated	41
SDR35 pipe	506
trap	269
valve	236, 237

Q

Entry	Page
Quartz heater	342

R

Entry	Page
Raceway	388-390
for heater cable	341
Radiant ceiling	94
floor heating	344-346
heat	94, 340
heating panel	342
Radiation fin tube	338
Radiator air vent	246
cast iron	338
supply control	254
supply valve	254
thermostat control sys.	109
Rain cap chimney	378
Raise manhole	43
manhole frame	43
Raised floor	78
Ramp approach	78
temporary	11
Range cooking	86
hood	87
receptacle	393
Reach-in cooler evaporator	317
Ready mix concrete	50, 523
Receptacle air conditioner	392
device	392
dryer	392
GFI	392
locking	393
range	393
telephone	393
television	393
weatherproof	392
Reciprocating chiller	470
Reciprocating compressor	276
hermetic compressor	276
water chiller	313
Recorders chart	112
Recovery heat	85
wheel energy	346

Index

Rectangular diffuser 370
   ductwork 347
Reducer grooved 186
Redwood louver 63
   tank 99
Refrigerant accumulator 271
   compressor 276
   filter dryer 271
   removal 104
Refrigerated case 82
Refrigeration 95
   commercial 82, 85
   compressor 276
   equipment 97
   fitting 150
   floor 95
   insulation 95
   mortuary 90
   oil 316
   P trap 272
   panel fiberglass 95
   partition 95
   sightglass 272
   specialties 271
   tubing 144
   valves 272
   walk-in 95
Refrigerator door 72
Refuse chute 118
   hopper 118
Register air supply 374
   baseboard 375
   boot spiral duct 357
   return 376
   steel 375
   wall 376
Registers exhaust 376
Regulator gas 257
   oil pressure 257
   pressure 257, 258
   steam pressure 258
Reinforced concrete pipe ... 33, 34
   culvert 33
Reinforcement concrete 50
Reinforcing steel 50
Relay control component 110
   pipe 24
Relief damper 363
   vent ventilator 378
   water vacuum 267
Religious education S.F. & C.F. .. 404
Remote oil gauge 274
Removal air conditioner 120
   asbestos 101, 102
   catch basin 22
   concrete 22
   driveway 22
   hydrant 22
   insulation 101
   lavatory 121
   pipe 121
   pipe insulation 101
   plumbing fixtures 120
   refrigerant 104
   sidewalk 23
   tank 21
   urinal 121
   water closet 121
   water fountain 121
Rental equipment 13, 15
   equipment rate 13
   generator 14
Repair pipe 45, 46
Replacement pavement 44
Residential alarm 394
   appliance 82, 394

burner 286
   device 390
   door bell 393
   dryer 82
   fan 367, 394
   fin-tube heating 420
   fixture 393
   gas meter 275
   heat exchanger 346
   heat pump 395
   heater electric 341
   lighting 393
   loadcenter 390
   roof jack 367
   service 390
   smoke detector 394
   switch 390
   transition 367
   ventilation 368
   wall cap 367
   washer 82
   water heater 394
   wiring 390, 394
Resistant duct fitting corrosion .. 353
Respirator 103
   personal 102
Restaurant air-conditioner 457
   air-conditioning . 445, 447, 449, 451,
                453, 455
   S.F. & C.F. 404
Retail store S.F. & C.F. 404
Return register 376
Ridge vent 78
Rigid fiberglass duct 349-352
   plastic duct 352
Ring hanger pipe 123
Rink ice 97
Riser clamp 122
Rivet 58
   aluminum 58
   copper 58
   steel 58
   tool 58
Road temporary 11
Rock drill 26
Rod backer 68
   coupling hanger 125
   pipe hanger 125
   socket 128
   threaded 125
   tie 59
Roll hanger 126
   type air filter 380
Roller compaction 25
   sheepsfoot 25
Rolling ladder 11
   tower 11
Roof exhauster 368
   flashing vent chimney .. 309
   frame opening 59
   jack residential 367
   sheathing 63
   specialties, prefab ... 66
   ventilator 377
   ventilator plastic 366
   rooftop air cond. 324, 448-451
   heat pump 329, 483
Rooftop make-up air unit 323
   multizone 450, 451
   multizone air cond. ... 325
   single zone 448, 449
   space heater 299
Room clean 94
   humidifier 329
Rotary pump 281
Rotating chimney cap 378

Round damper 363
   damper fire 359
   diffuser 370
   fire damper 359
   volume damper 363
Rubber pipe insulation 139
Rubbish chute 23
   handling 23

S

Saddle covering protection 127
   pipe support 127
Safety equipment laboratory ... 90
   plug pressure/temperature ... 259
   type fuel heater 275
   valve gas 287
Salamander 15
Sales tax 7, 519
Sampling air 102
Sand backfill 503
Sauna 95
Saw 15
   chain 15
   circular 15
   masonry 15
Scaffold specialties 10
Scaffolding 524
   masons 11
   plank 10
   putlog 11
   tubular 9
School air-conditioner 457
   air-conditioning . 445, 447, 449, 451,
                453, 455
   elementary S.F. & C.F. ... 404
   S.F. & C.F. 404
Scoop air 245
Scotch marine boiler 293
Screening plant 13
Screw brass 63
   lag 58
   machine 58
   sheet metal 62
   steel 63
   wood 63
Sealant caulking 68
   tape 69
Sealer joint 68
Sediment strainer Y valve 238
Selective demolition 23
Self propelled crane 16
Self-closing relief valve 230
Self-contained air conditioner .. 326
Selfcontained dehumidifier 329
Selfdrilling anchor 56
Semi-steel valve 238
Sensor circuit 246, 247
   control component 110
Separation barrier 102
Separator 268
   aftercooler 277
Service boring 28
   residential 390
   station equipment 83
Setter circuit 248
Sewage ejector pump 284
   pipe 33, 34, 41
   piping 505
   sitework 510
Sheathing 63
   roof 63
Sheathing, roof & walls plywood .. 63
Sheathing wall 63
Sheepsfoot roller 25

Sheet metal screw 62
   metal transitions 349
   piling 24, 528
Sheeting 24
   open 25
   tie back 25
   wale 24
   wood 24, 527
Shell/tube heat exchanger 474
Shelving refrigeration 95
Shield expansion 56
   lag screw 57
Shift work 7
Shipping door 72
Shock absorber 268
   absorber pipe 268
Shoring 24
   heavy duty 9, 10
   slab 10
   temporary 25
Shot chemical feeder 296
Shower pan 66
Shredder compactor 84
Shut-off valve oil 274
Sidewalk bridge 10
   removal 23
   temporary 12
Sidewall diffuser 371
Sightglass refrigeration 272
Sign 12
   project 12
Signal bell fire 116
Silencer duct 363
   louver 364
   plenum 364
Silent check valve . 231, 234, 240, 243
Silicone caulking 69
Sillcock hose bibb 282
Simplex style strainer 261
Single pipe roll 126
   zone rooftop unit 324
Sink darkroom 88
   laboratory 90
   removal 121
Site demolition 22
   drainage 42
   improvement 44
   preparation 22
   utility 503
Sitework manhole 510
Skating rink S.F. & C.F. 403
Slab on grade 51
   shoring 10
Sleeve 29, 259
   cut in 29
   formwork 50
   wall 259
Slotted framing channel 59
   pipe 34
Small tools 11
Smoke control kitchen 86
   detector 116
   exhaust duct 349-352
   vent chimney 301
Smokestack 54
Snow melting 340
Snubber valve 249
Social club S.F. & C.F. 400
Socket hanger rod 128
Socket-o-let 179
Softener water 87
Soil compaction 25
   compactor 13
   decontamination 21
   tamping 25

615

Index

Solar backflow preventer 106
   drain back hot water 428
   draindown hot-water 431
   energy 104-106
   energy circulator air 104
   energy system . 425, 427, 433, 435, 437
   energy system air purger 106
   energy system air vent 106
   energy system balancing valve 106
   energy system collector 105
   energy system control valve ... 106
   energy system controller 105
   energy system expansion tank . 106
   energy system gauge pressure . 106
   energy system heat exchanger 105
   energy system storage tank ... 105
   energy system thermometer .. 106
   energy system vacuum relief .. 106
   heating 529
   system solenoid valve 106
Solar-energy pool heater 439
   space-heater 437
   water-heater 441, 443
Solder 142
Soldier beam 25
Solid fuel boiler 294
   fuel furnace 298
Sorority house S.F. & C.F. 402
Sound trap duct 363
Space heater floor mounted 299
   heater rental 14
   heater rooftop 299
   heater suspended 300
Special door 72
   purpose room 94
   systems 341-343
Specialties & accessories, roof 66
   piping 383
   piping HVAC ... 250-253, 257, 262 -264, 268
   refrigeration 271
   scaffold 10
Spill container fuel 273
Spinner ventilator 377
Spiral duct fitting 354-356
   duct preformed 353, 355
   duct register boot 357
Split ring hanger 124
   system control system 108
   system ductless 328
Splitter damper assembly 362
Sport arena S.F. & C.F. 405
Spotter 28
Spray substrate 103
Sprayer airless 102
Spread footing 51
Sprinkler alarm 116
   grass 44
   system 44
Square foot building cost 400
Stainless cooling towers 315, 473
   duct 347
   flashing 66
   steel fitting 195
   steel flange 200
   steel pipe 190-195
   steel spiral duct 357
   steel valve 242, 243
Stair temporary protection 9
Stand heat pump 319
Standpipe alarm 116
   steel 98
Starter board & switch 397
   motor 397
Station control 396

Stationary ventilator 377
   ventilator mushroom 378
Steam bath 96
   bath residential 96
   boiler 288-293
   boiler electric 287
   boiler system 414
   cock 248
   condensate meter 273
   fuel heater 275
   heat 335
   heat coil 335, 336
   heating commercial 421
   humidifier 330
   jacketed kettle 85
   pressure reducing valve 257
   pressure valve 254
   radiator valve 255
   regulator pressure 258
   syphon pigtail 249
   trap 260, 261
Steel cooling tower 473
Steel cutting 23
   diffuser 371
   downspout 66
   expansion tank 251
   fitting 160
   flange 179, 187
   oval arch corrugated pipe 506
   pipe 40, 41, 153, 155-158, 160
   pipe downspout 67
   pipe fitting 160, 173
   pipe stainless 193, 194
   register 375
   reinforcing 50
   rivet 58
   salvage 121
   screw 63
   sheet pile 24
   spiral duct 353, 355
   spiral duct stainless 357
   standpipe 98
Steel storage tank 496, 497
Steel structural 58
   tank 98, 251
   valve 238, 240, 242
   well screen 32
Step manhole 43
Sterilizer dental 91
   medical 90
Stiffleg derrick 16
Stoker coal 287
   fired boiler 294
Stop-start control station 396
Storage box 9
   building S.F. & C.F. 405
   door cold 72
   tank 97-100
Store air-conditioning . 444, 451, 452, 454, 456
   department S.F. & C.F. 401
   retail S.F. & C.F. 404
Storm collar 308
Stove 79
   wood burning 79
Strainer basket type 261, 262
   bronze body 263
   duplex style 262, 263
   grooved joint 188
   simplex style 261
   Y type 263
   Y type iron body 264
Strap pipe hanger 128
Strapless outlet 182
Strip footing 51
Structural concrete 50

   excavation 26
   formwork 50
   steel 58
Structure pre-engineered 96
Stud driver pneumatic 58
Subcontractor O & P 8
Subcontractors 530
Subdrainage pipe 40
   piping 40, 508
   plastic pipe 41
   system 40
Submersible pump .. 15, 32, 281, 285
Subsoil heating 97
Subsurface investigation 22
Suction bell fuel heater 275
   diffuser grooved joint 188
   diffuser pump 264, 265
   hose 14
Sump hole construction 24
   pit 23
   pump 87, 285
Supermarket air-conditioning ... 445, 447, 449, 453, 455, 457
   air-condtioning 451
   S.F. & C.F. 405
Supply/return comb. diffuser 370
Support framing pipe 59
   pipe 121
Suppressor lightning 390
Surfactant 103
Suspended ceiling 94
   space heater 300
Suspended space heater 464
Suspension mounted heater ... 299
Swimming pool 96, 97
   pool equipment 97
   pool heater 295
   pool painting 97
   pool S.F. & C.F. 405
Swimming-pool heater 439
Swing check valve 228
Switch decorator 391
   dimmer 390
   residential 390
Switchboard electric 396
Syphon ventilator rotary 377
System boiler fossil fuel 417
   boiler gas fired 416
   boiler hot-water 415
   boiler oil fired 419
   boiler steam 414
   catch basin/manhole 510
   components fuel 273
   control 107, 108
   dryer 278
   exhaust 381
   fin-tube heating 414
   fin-tube radiation 418, 419
   grinder pump 284
   heat trace 341
   heating 437
   heating hydronic 414-421
   hydronic fossil fuel 416, 418
   irrigation 44
   manhole/catch basin 510
   pipe conduit 36, 37, 39
   sitework catch basin 510
   sitework trenching 500
   solar-energy ... 425, 427, 429, 431, 433, 435, 437, 439, 441
   solenoid valve solar 106
   sprinkler 44
   subdrainage 40
   tube 118
   VAV box control 109

T

Tag valve 131
Tamper 14, 26
Tamping soil 25
Tank clean 21
   compression 252
   corrosion resistant 251
   darkroom 87
   disposal 21
   expansion 251
   fibergalss 99
   fixed roof 98
   gasoline 99
   gauge oil 274
   heating coil 275
   leak detection 116
   outlet heater 275
   pillow 99
   removal 23
   steel 98, 251
   storage 97-100
   water 98-100
   water storage solar 106
   water-heater storage 285
Tape barricade 12
   detection 20
   sealant 69
   underground 20
Tapping main 31
Tarpaulin 9, 11
   duck 11
   mylar 11
   polyethylene 11
   polyvinyl 11
Tavern air-conditioning 444, 446, 448, 450, 452, 454, 456
Tax 7
   sales 7
   social security 7
   unemployment 7
Taxes 519
T-bar grille eggcrate 373
   grille lay-in 373
   mount diffuser 370
Tee fitting 31
   pipe 185
Telephone exchange S.F. & C.F. . 405
   receptacle 393
Television receptacle 393
Temperature relief valve 230
   rise detector 116
Tempering valve 231
   valve water 256
Temporary building 9
   construction 8, 9, 12
   facility 12
   fence 12
   guardrail 12
   heat 8
   light 8
   oil heater 14
   shoring 25
   toilet 15
   utility 8
Terminal A/C packaged 324
   heater 339, 340
Terne coated flashing 66
Test pressure 385
   well 32
Testing OSHA 103
   pipe 385
Theater S.F. & C.F. 405
Thermoflo indicator 265
Thermometer 111
   solar energy system 106

Index

well 182
Thermometers 111
Thermostat 111
   automatic timed 111
   contactor combination 342
   control sys. radiator 109
   integral 341
   pneumatic 111
   wire 395
Threaded flange 172, 173
   nut 125
   rod 125
Threading labor st steel 194
   labor steel 159
Thread-o-let 178
Thru-the-wall A/C 490
   heat pump 490
Thru-wall air conditioner 324
Tie back sheeting 25
   rod 59
Tile ceiling 94
Timber fastener 62
Time lapse photography 8
Timer clock 394
   interval 391
   switch ventilator 369
Toggle bolt 57
Toilet chemical 15
   temporary 15
   trailer 15
Tool air 14
   powder actuated 58
   rivet 58
Tools small 11
Topping grout 51
Torch cutting 15, 23
Tower cooling 314
   crane 15, 525
   hoist 16
   light 14
   rolling 11
Town hall S.F. & C.F. 405
Tractor 25, 26
   truck 15
Trailer office 9
   platform 15
   toilet 15
   truck 13, 15
Training 18
Transfer grille 373
Transformer heating control 342
Transition residential 367
Transitions sheet metal 349
Transmission of heat 539
Transmitter control component . . 111
Trap cast iron 268
   drainage 268
   inverted bucket steam 259
   plastic 269
   polypropylene 269
   PVC 269
   steam 260, 261
Trapeze hanger 126
Trash pump 15
Trench backfill 27, 503, 504
   excavation 26, 27
   utility 26
Trencher 13
   chain 13
Trenching 23, 501
Truck dump 13
   flatbed 13
   hauling 28
   mounted crane 16
   off highway 14
   pickup 15

rental 13
tractor 15
trailer 13, 15
vacuum 15
Trucking 28
Tube fitting pneumatic 114, 115
   pneumatic 114, 115
   system 118
   system air 118
   system pneumatic 118
Tubing copper 142
   copper type K 509
   pex 344
Tubular scaffolding 9
Tunneling 28
Turbine pump 281
   water meter 266
Turnbuckle 125
Turning vanes 364

U

U-bolt pipe hanger 129
U-hook hanger 129
Underdrain 42
Undereave vent 78
Underground piping 503
   storage tank removal 526
   tape 20
Underlayment grout 51
Underwater light 97
Unemployment tax 7
Union dielectric 249, 250
Unit gross area requirements . . . 520
   heater 299, 300, 337, 340
   heater cabinet 339, 343, 344
   heater electric 341-343
   water cooled condensing 317
Urethane foam 69
   insulation 95
Urinal removal 121
Utensil washer medical 90
Utility excavation 500
Utility fan system 491
Utility pump 280
   set fan 366, 368
   structure 510
   temporary 8
   trench 26

V

Vacuum breaker 267
   central 82
   cleaning 82
Vacuum dust collection 461
Vacuum gauge 110
   loader 103
   pumps 272, 280
   relief water 267
   truck 15
Valance heater 340
Valve . . . 227-230, 234, 236-238, 240,
                         242, 243, 270
   angle 236
   balancing & shut off 255
   ball 228, 236, 241
   brass 227
   bronze 228
   bronze butterfly 228
   butterfly 20, 232
   check 20, 30, 238, 240-242
   check lift 229
   check swing 234

circuit setter balance 248
control solar energy system . . . 106
CPVC 236
cut in 29
diverter 232, 239, 242
flanged 240
float 254
foot 237, 273
foot oil 273
fusible oil 274
gas stop 227
gate 20, 29, 233, 239, 241, 243
globe . . 21, 230, 233, 240, 242, 243
grooved joint butterfly 189
grooved-joint gate 189
heating control 112, 254
hot water radiator 254
iron body 232, 234
lubricated plug 238
materials 533-535
mixing 256
motor actuated 112
motor controlled 112
motorized zone 112
multipurpose 235
needle 237
nonfusible oil 274
oil shut-off 274
piping 20
plastic 236
plug 238
pneumatic air operated 112
polypropylene 236
PVC 236
radiator supply 254
relief pressure 230
selection considerations 533
semi-steel 238
snubber 249
soldered gate 229
stainless steel 242, 243
steam pressure 254
steam radiator 255
steel 238, 240, 242
swing check 228
tag 131
tempering 231
water pressure 230
water tempering 256
Y sediment strainer 238
Y-check 238
zone 266
Valves forged 241
   refrigeration 272
Vane-axial fan 365
Vanes turning 364
Vapor barrier 95
   hose adapter 273
   proof insulation 133
Vaporproof insulation 139
Variable air volume 326
   frequency drives 396
   volume damper 362
VAV boxes 360
   mixing boxes 360, 361
   terminal 488, 489
   terminals 488
Vent air automatic 246
   alarm 274
   automatic air 245
   box 54
   brick 373
   cap 274
   chimney 301, 303, 309
   chimney all fuel 308
   chimney fitting . 302, 303, 308, 309

chimney flashing 303, 309
chimney high temperature . . . 309, 310
chimney increaser 310
damper 301
damper exhaust 358
draft damper 301
dryer 87
exhaust 377
flashing 269
foundation 54
intake-exhaust 378
metal chimney 301
radiator air 246
relief 378
ridge 78
ridge strip 78
silencer 364
Vent-chimney band 310
Ventilating air conditioning . 108, 314,
   315, 327, 347, 359, 362, 365, 368, 369,
                                       378
Ventilation 542
   fan 394
   garage 381
   heat recovery 346
   kitchen 85
   louver 63
   packaged 85
   residential 368
Ventilator 377
   chimney cap 378
   control system 108
   fan 364
   gravity syphon 377
   masonry 54
   mushroom stationary 378
   relief vent 378
   roof 377
   rotary syphon 377
   spinner 377
   timer switch 369
Venturi flow meter 265
Vertical induction unit 360
Vibration absorber 129
   absorbing pad 130
Vibrator earth 25, 26
   plate 25
Vibratory equipment 13
Vitrified clay pipe 34, 41
Vocational school S.F. & C.F. . . . 405
Volume control damper 362
   damper round 363

W

Wafer check valve 231, 235, 243
Wale sheeting 24
Wales 24
Walk-in cooler evaporator 317
   refrigeration 95
Wall cap residential 367
   exhaust fan 369
   furnace 300
   heater 87
   heater electric 341
   hung boiler 290
   hydrant 282
   patching concrete 51
   register 376
   sheathing 63
   sleeve 259
Warehouse S.F. & C.F. 405, 406
Warm air heater 299

Index

Washer commercial 82
    darkroom 87
    residential 82, 87
Washers 124
Washing machine automatic 87
Waste compactor 84
    disposal 84
    handling 83
    handling equipment 83
    incinerator 84
    packaging 104
Water anti-siphon 267
    atomizer 103
    balancing 385
    chiller 311, 312
    chillers absorption 311
    closet removal 121
    coil 332, 333
    coil hot 334
    copper tubing connector ... 252
    distiller 90
    distribution system 30
    fountain removal 121
    gauge 383
    gauge boiler 383
    heater 87, 394
    heater insulation 131
    heater removal 121
    heater wrap kit 131
    heating hot 287, 338
    hose 14
    hydrant 282
    level control 383
    meter turbine 266
    piping polyethylene 509
    pressure relief valve 230
    pressure valve 230
    pump 14, 17, 32, 87, 270
    pumping 23
    softener 87
    storage solar tank 106
    supply detector meter 265
    supply domestic meter 265
    supply meter 265
    tank 98-100
    tank elevated 97
    tempering valve 231, 256
    tube boiler 294
    vacuum relief 267
    well 32
Water-heater solar-energy .. 425, 427,
               429, 431, 433, 435, 437
    storage tank 285
Wearing course 43
Weatherproof receptacle 392
Wedge anchor 57
Weld fitting 176
    joint pipe . 154, 157, 190, 191, 193,
                          194
    plastic pipe 222
    X-ray 385
Welded beam attachment 121
    eye rod 125
Welding fume elimination 382
    labor stainless-steel 194
    labor steel 159
    machine 15
Weld-o-let 178
Weld-on flange steel 179
Well 24, 32
    gravel pack 32
    pump deep 33
    pump shallow 33
    screen steel 32
    thermometer 182
    water 32

Wellpoint 24
    equipment 16
    header pipe 17
    pump 17
Wellpoints 526
Wheelbarrow 15
Whirlpool bath 91
Whole house fan 367
Window air conditioner 326
Winter protection 9
Wire fence 12
    thermostat 395
Wiring air conditioner 394
    fan 394
    methods 388
    residential 390, 394
Wood burning furnace 298
    coal/oil combination furnace .. 298
    fastener 63
    gage stick 273
    screw 63
    sheet piling 527
    sheeting 24, 527
    tank 99
Woodwork architectural 63
Work extra 8
    in prisons 6
Worker protection 103
Workers' compensation 7
Workers' compensation 515-517
Wrapping pipe 47
Wrench impact 14
Wye fitting 31

X

X-ray pipe 385
    weld 385

Y

Y sediment strainer valve 238
    type iron body strainer 264
Y-check valve 238
YMCA club S.F. & C.F. 401

Z

Zinc plating 62
Zone valve 266

Notes

Notes

Notes

Notes

Notes

Notes

Notes

Notes

Notes

Notes

Notes

Notes

CMD Group...

R.S. Means Company, Inc., a CMD Group company, which is a division of Cahners Business Information, the leading provider of construction cost data in North America, supplies comprehensive construction cost guides, related technical publications and educational services.

CMD Group, a leading worldwide provider of total construction information solutions, is comprised of three synergistic product groups crafted to be the complete resource for reliable, timely and actionable project, product and cost data. In North America, CMD Group encompasses:
- Architects' First Source
- Construction Market Data (CMD)
- Manufacturer's Survey Associates (MSA)
- R.S. Means
- CMD Canada
- BIMSA/Mexico
- Worldwide, CMD Group includes Byggfakta Scandinavia (Denmark, Estonia, Finland, Norway and Sweden) and Cordell Building Information Services (Australia).

First Source for Products, available in print, on the Means CostWorks CD, and on the Internet, is a comprehensive product information source. In alliance with The Construction Specifications Institute (CSI) and Thomas Register, Architects' First Source also produces CSI's SPEC-DATA® and MANU-SPEC® as well as CADBlocks.℠ Together, these products offer commercial building product information for the building team at each stage of the construction process.

CMD Exchange, a single electronic community where all members of the construction industry can communicate, collaborate, and conduct business, better, faster, easier.

Construction Market Data provides complete, accurate and timely project information through all stages of construction. Construction Market Data supplies industry data through productive leads, project reports, contact lists, market penetration analysis and sales evaluation reports. Any of these products can pinpoint a county, look at a state, or cover the country. Data is delivered via paper, e-mail or the Internet.

CMD Group Canada serves the Canadian construction market with reliable and comprehensive information services that cover all facets of construction. Core services include: Buildcore, product selection and specification tools available in print and on the Internet; CMD Building Reports, a national construction project lead service; CanaData, statistical and forecasting information; Daily Commercial News, a construction newspaper reporting on news and projects in Ontario; and Journal of Commerce, reporting news in British Columbia and Alberta.

Manufacturers' Survey Associates is a quantity survey and specification service whose experienced estimating staff examines project documents throughout the bidding process and distributes edited information to its clients. Material estimates, edited plans and specifications and distribution of addenda form the heart of the Manufacturers' Survey Associates product line, which is available in print and on CD-ROM.

Clark Reports is the premier provider of industrial construction project data, noted for providing earlier, more complete project information in the planning and pre-planning stages for all segments of industrial and institutional construction.

Byggfakta Scandinavia AB, founded in 1936, is the parent company for the leaders of customized construction market data for Denmark, Estonia, Finland, Norway and Sweden. Each company fully covers the local construction market and provides information across several platforms including subscription, ad-hoc basis, electronically and on paper.

Cordell Building Information Services, with its complete range of project and cost and estimating services, is Australia's specialist in the construction information industry. Cordell provides in-depth and historical information on all aspects of construction projects and estimation, including several customized reports, construction and sales leads, and detailed cost information among others.

For more information, please visit our web site at www.cmdg.com.

CMD Group Corporate Offices
30 Technology Parkway South #100
Norcross, GA 30092-2912
(800) 793-0304
(770) 417-4002 (fax)
info@cmdg.com
www.cmdg.com

Means Project Cost Report

By filling out and returning the Project Description, you can receive a discount of $20.00 off any one of the Means products advertised in the following pages. The cost information required includes all items marked (✔) except those where no costs occurred. The sum of all major items should equal the Total Project Cost.

$20.00 Discount per product for each report you submit.

DISCOUNT PRODUCTS AVAILABLE—FOR U.S. CUSTOMERS ONLY—STRICTLY CONFIDENTIAL

Project Description (No remodeling projects, please.)

✔ Type Building _____ Owner _____
✔ Location _____ Architect _____
   Capacity _____ General Contractor _____
✔ Frame _____ ✔ Bid Date _____
✔ Exterior _____ Typical Bay Size _____
✔ Basement: full ☐ partial ☐ none ☐ crawl ☐ ✔ Labor Force: _____ % Union _____ % Non-Union
✔ Height in Stories _____ ✔ Project Description (Circle one number in each line)
✔ Total Floor Area _____   1. Economy 2. Average 3. Custom 4. Luxury
   Ground Floor Area _____   1. Square   2. Rectangular 3. Irregular 4. Very Irregular
✔ Volume in C.F. _____
   % Air Conditioned _____ Tons _____
   Comments _____

	✔ Total Project Cost		$
A	✔ General Conditions		$
B	✔ Site Work		$
BS	Site Clearing & Improvement		
BE	Excavation	(C.Y.)	
BF	Caissons & Piling	(L.F.)	
BU	Site Utilities		
BP	Roads & Walks Exterior Paving	(S.Y.)	
C	✔ Concrete		$
C	Cast in Place	(C.Y.)	
CP	Precast	(S.F.)	
D	✔ Masonry		$
DB	Brick	(M)	
DC	Block	(M)	
DT	Tile	(S.F.)	
DS	Stone	(S.F.)	
E	✔ Metals		$
ES	Structural Steel	(Tons)	
EM	Misc. & Ornamental Metals		
F	✔ Wood & Plastics		$
FR	Rough Carpentry	(MBF)	
FF	Finish Carpentry		
FM	Architectural Millwork		
G	✔ Thermal & Moisture Protection		$
GW	Waterproofing-Dampproofing	(S.F.)	
GN	Insulation	(S.F.)	
GR	Roofing & Flashing	(S.F.)	
GM	Metal Siding/Curtain Wall	(S.F.)	
H	✔ Doors and Windows		$
HD	Doors	(Ea.)	
HW	Windows	(S.F.)	
HH	Finish Hardware		
HG	Glass & Glazing	(S.F.)	
HS	Storefronts	(S.F.)	

J	✔ Finishes		$
JL	Lath & Plaster	(S.Y.)	
JD	Drywall	(S.F.)	
JM	Tile & Marble	(S.F.)	
JT	Terrazzo	(S.F.)	
JA	Acoustical Treatment	(S.F.)	
JC	Carpet	(S.Y.)	
JF	Hard Surface Flooring	(S.F.)	
JP	Painting & Wall Covering	(S.F.)	
K	✔ Specialties		$
KB	Bathroom Partitions & Access.	(S.F.)	
KF	Other Partitions	(S.F.)	
KL	Lockers	(Ea.)	
L	✔ Equipment		$
LK	Kitchen		
LS	School		
LO	Other		
M	✔ Furnishings		$
MW	Window Treatment		
MS	Seating	(Ea.)	
N	✔ Special Construction		$
NA	Acoustical	(S.F.)	
NB	Prefab. Bldgs.	(S.F.)	
NO	Other		
P	✔ Conveying Systems		$
PE	Elevators	(Ea.)	
PS	Escalators	(Ea.)	
PM	Material Handling		
Q	✔ Mechanical		$
QP	Plumbing	(No. of fixtures)	
QS	Fire Protection (Sprinklers)		
QF	Fire Protection (Hose Standpipes)		
QB	Heating, Ventilating & A.C.		
QH	Heating & Ventilating	(BTU Output)	
QA	Air Conditioning	(Tons)	
R	✔ Electrical		$
RL	Lighting	(S.F.)	
RP	Power Service		
RD	Power Distribution		
RA	Alarms		
RG	Special Systems		
S	✔ Mech./Elec. Combined		$

Product Name _____
Product Number _____
Your Name _____
Title _____
Company _____
   ☐ Company
   ☐ Home   Street Address _____
City, State, Zip _____
☐ Please send _____ forms.

Please specify the Means product you wish to receive. Complete the address information as requested and return this form with your check (product cost less $20.00) to address below.

R.S. Means Company, Inc.,
Square Foot Costs Department
P.O. Box 800
Kingston, MA 02364-9988

R.S. Means Company, Inc...
a tradition of excellence in Construction Cost Information and Services since 1942.

For more information visit Means Web Site at www.rsmeans.com

Table of Contents
Annual Cost Guides, Page 2
Reference Books, Page 6
Seminars, Page 11
Consulting Services, Page 13
Electronic Data, Page 14
New Titles, Page 15
Order Form, Page 16

Book Selection Guide

The following table provides definitive information on the content of each cost data publication. The number of lines of data provided in each unit price or assemblies division, as well as the number of reference tables and crews is listed for each book. The presence of other elements such as an historical cost index, city cost indexes, square foot models or cross-referenced index is also indicated. You can use the table to help select the Means' book that has the quantity and type of information you most need in your work.

Unit Cost Divisions	Building Construction Costs	Mechanical	Electrical	Repair & Remodel.	Square Foot	Site Work Landsc.	Assemblies	Interior	Concrete Masonry	Open Shop	Heavy Construc.	Residential	Light Commercial	Facil. Construc.	Plumbing	Western Construction Costs
1	1098	586	667	862		1089		491	987	1096	1132	416	621	1612	659	1091
2	3141	1467	509	1881		9149		852	1655	3090	5621	1064	984	5015	1826	3114
3	1516	91	71	768		1321		198	1937	1510	1494	268	233	1378	43	1514
4	871	27	0	688		751		636	1174	853	651	313	396	1124	0	849
5	1857	177	174	955		803		985	667	1824	1056	779	809	1849	95	1841
6	1388	97	84	1314		493		1336	339	1379	612	1532	1473	1411	59	1772
7	1365	180	91	1329		490		546	451	1365	353	805	1036	1357	188	1364
8	1899	60	0	1904		337		1821	726	1880	9	1133	1184	2032	0	1900
9	1650	50	0	1473		188		1746	327	1598	113	1251	1390	1845	50	1644
10	991	58	31	565		217		822	196	995	0	250	467	996	249	991
11	1158	435	213	622		151		917	34	1052	95	116	237	1221	371	1051
12	352	0	0	48		228		1536	30	344	0	72	68	1538	0	343
13	1384	1221	498	587		443		1075	106	1358	302	237	562	2094	1083	1339
14	364	43	0	261		36		330	0	364	40	9	18	362	17	362
15	2510	14742	763	2243		1809		1477	80	2515	2028	1087	1553	12312	10902	2545
16	1517	541	10191	1200		879		1292	62	1535	916	706	1290	9872	481	1459
17	421	0	0	421		0		421	421	460	0	421	421	421	0	421
Totals	23482	19775	13292	17121		18384		16508	9192	23218	14422	10459	12742	46439	16023	23600

Assembly Divisions	Building Construction Costs	Mechanical	Electrical	Repair & Remodel.	Square Foot	Site Work Landsc.	Assemblies	Interior	Concrete Masonry	Open Shop	Heavy Construc.	Residential	Light Commercial	Facil. Construc.	Plumbing	Western Construction Costs
1		0	0	202	131	738	775	0	689		752	841	131	89	0	
2		0	0	40	33	35	48	0	49		0	584	32	0	0	
3		0	0	443	1114	0	3064	228	1245		0	1546	805	174	0	
4		0	0	714	1382	0	3201	255	1273		0	2050	1200	26	0	
5		0	0	288	227	0	459	0	0		0	1081	223	25	0	
6		0	0	1006	843	0	1295	1723	152		0	1077	738	330	0	
7		0	0	42	89	0	179	159	0		0	709	33	71	0	
8		2257	149	998	1689	0	2718	926	0		0	1686	1189	1114	2084	
9		0	1383	355	367	0	1320	306	0		0	238	358	319	0	
10		0	0	0	0	0	0	0	0		0	0	0	0	0	
11		0	0	365	465	0	724	153	0		0	0	466	201	0	
12		501	160	539	84	2392	699	0	698		541	0	84	119	663	
Totals		2758	1692	4992	6424	3165	14482	3750	4106		1293	9812	5259	2468	2747	

Reference Section	Building Construction Costs	Mechanical	Electrical	Repair & Remodel.	Square Foot	Site Work Landsc.	Assemblies	Interior	Concrete Masonry	Open Shop	Heavy Construc.	Residential	Light Commercial	Facil. Construc.	Plumbing	Western Construction Costs
Tables	150	46	84	66	4	84	223	59	83	146	55	49	70	86	49	148
Models					102							32	43			
Crews	408	408	408	389		408		408	408	391	408	391	391	389	408	408
City Cost Indexes	yes	yes	yes	yes	yes	yes	yes	yes	yes	yes	yes	yes	yes	yes	yes	yes
Historical Cost Indexes	yes	yes	yes	yes	yes	yes	yes	yes	yes	yes	yes	no	yes	yes	yes	yes
Index	yes	yes	yes	yes	no	yes	yes	yes	yes	yes	yes	yes	yes	yes	yes	yes

Annual Cost Guides

For more information visit Means Web Site at www.rsmeans.com

Means Building Construction Cost Data 2001

Available in Both Softbound and Looseleaf Editions

The "Bible" of the industry comes in the standard softcover edition or the looseleaf edition.

Many customers enjoy the convenience and flexibility of the looseleaf binder, which increases the usefulness of *Means Building Construction Cost Data 2001* by making it easy to add and remove pages. You can insert your own cost information pages, so everything is in one place. Copying pages for faxing is easier also. Whichever edition you prefer, softbound or the convenient looseleaf edition, you'll get the *DesignBuildIntelligence* newsletter at no extra cost.

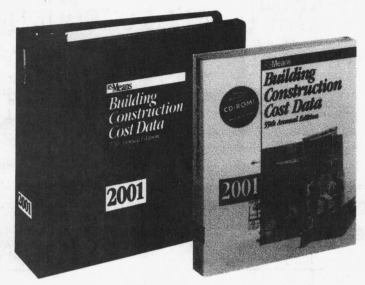

$89.95 per copy, Softbound
Catalog No. 60011

$116.95 per copy, Looseleaf
Catalog No. 61011

Means Building Construction Cost Data 2001

Offers you unchallenged unit price reliability in an easy-to-use arrangement. Whether used for complete, finished estimates or for periodic checks, it supplies more cost facts better and faster than any comparable source. Over 23,000 unit prices for 2001. The City Cost Indexes cover over 930 areas, for indexing to any project location in North America. Order and get the *DesignBuildIntelligence* newsletter sent to you FREE. You'll have year-long access to the Means Estimating **Hotline** FREE with your subscription. Expert assistance when using Means data is just a phone call away.

$89.95 per copy
Over 650 pages, illustrated, available Oct. 2000
Catalog No. 60011

Means Building Construction Cost Data 2001

Metric Version

The Federal Government has stated that all federal construction projects must now use metric documentation. The *Metric Version* of *Means Building Construction Cost Data 2001* is presented in metric measurements covering all construction areas. Don't miss out on these billion dollar opportunities. Make the switch to metric today.

$89.95 per copy
Over 650 pages, illustrated, available Nov. 2000
Catalog No. 63011

For more information
visit Means Web Site
at www.rsmeans.com

Annual Cost Guides

Means Mechanical Cost Data 2001

- HVAC • Controls

Total unit and systems price guidance for mechanical construction . . . materials, parts, fittings, and complete labor cost information. Includes prices for piping, heating, air conditioning, ventilation, and all related construction.

Plus new 2001 unit costs for:
- Over 2500 installed HVAC/controls assemblies
- "On Site" Location Factors for over 930 cities and towns in the U.S. and Canada
- Crews, labor and equipment

$89.95 per copy
Over 600 pages, illustrated, available Oct. 2000
Catalog No. 60021

Means Plumbing Cost Data 2001

Comprehensive unit prices and assemblies for plumbing, irrigation systems, commercial and residential fire protection, point-of-use water heaters, and the latest approved materials. This publication and its companion, *Means Mechanical Cost Data*, provide full-range cost estimating coverage for all the mechanical trades.

$89.95 per copy
Over 500 pages, illustrated, available Nov. 2000
Catalog No. 60211

Means Electrical Cost Data 2001

Pricing information for every part of electrical cost planning: More than 15,000 unit and systems costs with design tables; clear specifications and drawings; engineering guides and illustrated estimating procedures; complete labor-hour and materials costs for better scheduling and procurement; the latest electrical products and construction methods.
- A Variety of Special Electrical Systems including Cathodic Protection
- Costs for maintenance, demolition, HVAC/ mechanical, specialties, equipment, and more

$89.95 per copy
Over 450 pages, illustrated, available Oct. 2000
Catalog No. 60031

Means Electrical Change Order Cost Data 2001

You are provided with electrical unit prices exclusively for pricing change orders—based on the recent, direct experience of contractors and suppliers. Analyze and check your own change order estimates against the experience others have had doing the same work. It also covers productivity analysis and change order cost justifications. With useful information for calculating the effects of change orders and dealing with their administration.

$89.95 per copy
Over 450 pages, available Oct. 2000
Catalog No. 60231

Means Facilities Maintenance & Repair Cost Data 2001

Published in a looseleaf format, *Means Facilities Maintenance & Repair Cost Data* gives you a complete system to manage and plan your facility repair and maintenance costs and budget efficiently. Guidelines for auditing a facility and developing an annual maintenance plan. Budgeting is included, along with reference tables on cost and management and information on frequency and productivity of maintenance operations.

The only nationally recognized source of maintenance and repair costs. Developed in cooperation with the Army Corps of Engineers.

$199.95 per copy
Over 600 pages, illustrated, available Dec. 2000
Catalog No. 60301

Means Square Foot Costs 2001

It's Accurate and Easy To Use!

- **Updated 2001 price information,** based on nationwide figures from suppliers, estimators, labor experts and contractors.
- "How-to-Use" Sections, with **clear examples** of commercial, residential, industrial, and institutional structures.
- Realistic graphics, offering true-to-life illustrations of building projects.
- Extensive information on using square foot cost data, including **sample estimates** and **alternate pricing methods.**

$99.95 per copy
Over 450 pages, illustrated, available Oct. 2000
Catalog No. 60051

Annual Cost Guides

For more information visit Means Web Site at www.rsmeans.com

Means Repair & Remodeling Cost Data 2001
Commercial/Residential

You can use this valuable tool to estimate commercial and residential renovation and remodeling.

Includes: New costs for hundreds of unique methods, materials and conditions that only come up in repair and remodeling. PLUS:
- Unit costs for over 16,000 construction components
- Installed costs for over 90 assemblies
- Costs for 300+ construction crews
- Over 930 "On Site" localization factors for the U.S. and Canada.

$79.95 per copy
Over 600 pages, illustrated, available Nov. 2000
Catalog No. 60041

Means Facilities Construction Cost Data 2001

For the maintenance and construction of commercial, industrial, municipal, and institutional properties. Costs are shown for new and remodeling construction and are broken down into materials, labor, equipment, overhead, and profit. Special emphasis is given to sections on mechanical, electrical, furnishings, site work, building maintenance, finish work, and demolition. More than 45,000 unit costs plus assemblies and reference sections are included.

$219.95 per copy
Over 1150 pages, illustrated, available Nov. 2000
Catalog No. 60201

Means Residential Cost Data 2001

Contains square foot costs for 30 basic home models with the look of today—plus hundreds of custom additions and modifications you can quote right off the page. With costs for the 100 residential systems you're most likely to use in the year ahead. Complete with blank estimating forms, sample estimates and step-by-step instructions.

Means Light Commercial Cost Data 2001

Specifically addresses the light commercial market, which is an increasingly specialized niche in the industry. Aids you, the owner/designer/contractor, in preparing all types of estimates, from budgets to detailed bids. Includes new advances in methods and materials. Assemblies section allows you to evaluate alternatives in the early stages of design/planning.

Over 13,000 unit costs for 2001 ensure you have the prices you need... when you need them.

$79.95 per copy
Over 550 pages, illustrated, available Dec. 2000
Catalog No. 60171

$79.95 per copy
Over 600 pages, illustrated, available Dec. 2000
Catalog No. 60181

Means Assemblies Cost Data 2001

Means Assemblies Cost Data 2001 takes the guesswork out of preliminary or conceptual estimates. Now you don't have to try to calculate the assembled cost by working up individual components costs. We've done all the work for you.

Presents detailed illustrations, descriptions, specifications and costs for every conceivable building assembly—240 types in all—arranged in the easy-to-use UniFormat system. Each illustrated "assembled" cost includes a complete grouping of materials and associated installation costs including the installing contractor's overhead and profit.

Means Site Work & Landscape Cost Data 2001

Means Site Work & Landscape Cost Data 2001 is organized to assist you in all your estimating needs. Hundreds of fact-filled pages help you make accurate cost estimates efficiently.

Updated for 2001!
- Demolition features—including ceilings, doors, electrical, flooring, HVAC, millwork, plumbing, roofing, walls and windows
- State-of-the-art segmental retaining walls
- Flywheel trenching costs and details
- Updated Wells section
- Thousands of landscape materials, flowers, shrubs and trees

$149.95 per copy
Over 550 pages, illustrated, available Oct. 2000
Catalog No. 60061

$89.95 per copy
Over 600 pages, illustrated, available Oct. 2000
Catalog No. 60281

For more information
visit Means Web Site
at www.rsmeans.com

Annual Cost Guides

Means Open Shop Building Construction Cost Data 2001

The latest costs for accurate budgeting and estimating of new commercial and residential construction... renovation work... change orders... cost engineering. *Means Open Shop BCCD* will assist you to...
- Develop benchmark prices for change orders
- Plug gaps in preliminary estimates, budgets
- Estimate complex projects
- Substantiate invoices on contracts
- Price ADA-related renovations

$89.95 per copy
Over 650 pages, illustrated, available Nov. 2000
Catalog No. 60151

Means Heavy Construction Cost Data 2001

A comprehensive guide to heavy construction costs. Includes costs for highly specialized projects such as tunnels, dams, highways, airports, and waterways. Information on different labor rates, equipment, and material costs is included. Has unit price costs, systems costs, and numerous reference tables for costs and design. Valuable not only to contractors and civil engineers, but also to government agencies and city/town engineers.

$89.95 per copy
Over 450 pages, illustrated, available Nov. 2000
Catalog No. 60161

Means Building Construction Cost Data 2001
Western Edition

This regional edition provides more precise cost information for western North America. Labor rates are based on union rates from 13 western states and western Canada. Included are western practices and materials not found in our national edition: tilt-up concrete walls, glu-lam structural systems, specialized timber construction, seismic restraints, landscape and irrigation systems.

$89.95 per copy
Over 600 pages, illustrated, available Dec. 2000
Catalog No. 60221

Means Heavy Construction Cost Data 2001
Metric Version

Make sure you have the Means industry standard metric costs for the federal, state, municipal and private marketplace. With thousands of up-to-date metric unit prices in tables by CSI standard divisions. Supplies you with assemblies costs using the metric standard for reliable cost projections in the design stage of your project. Helps you determine sizes, material amounts, and has tips for handling metric estimates.

$89.95 per copy
Over 450 pages, illustrated, available Nov. 2000
Catalog No. 63161

Means Construction Cost Indexes 2001

Who knows what 2001 holds? What materials and labor costs will change unexpectedly? By how much?
- Breakdowns for 305 major cities.
- National averages for 30 key cities.
- Expanded five major city indexes.
- Historical construction cost indexes.

$198.00 per year/$49.50 individual quarters
Catalog No. 60141

Means Interior Cost Data 2001

Provides you with prices and guidance needed to make accurate interior work estimates. Contains costs on materials, equipment, hardware, custom installations, furnishings, labor costs . . . every cost factor for new and remodel commercial and industrial interior construction, including updated information on office furnishings, plus more than 50 reference tables. For contractors, facility managers, owners.

$89.95 per copy
Over 550 pages, illustrated, available Oct. 2000
Catalog No. 60091

Means Concrete & Masonry Cost Data 2001

Provides you with cost facts for virtually all concrete/masonry estimating needs, from complicated formwork to various sizes and face finishes of brick and block, all in great detail. The comprehensive unit cost section contains more than 8,500 selected entries. Also contains an assemblies cost section, and a detailed reference section which supplements the cost data.

$83.95 per copy
Over 450 pages, illustrated, available Nov. 2000
Catalog No. 60111

Means Labor Rates for the Construction Industry 2001

Complete information for estimating labor costs, making comparisons and negotiating wage rates by trade for over 300 cities (United States and Canada). With 46 construction trades listed by local union number in each city, and historical wage rates included for comparison. No similar book is available through the trade.

Each city chart lists the county and is alphabetically arranged with handy visual flip tabs for quick reference.

$199.95 per copy
Over 300 pages, available Dec. 2000
Catalog No. 60121

Reference Books

For more information visit Means Web Site at www.rsmeans.com

Project Scheduling & Management for Construction
New 2nd Edition
By David R. Pierce, Jr.

A comprehensive, yet easy-to-follow guide to construction project scheduling and control—from vital project management principles through the latest scheduling, tracking, and controlling techniques. The author is a leading authority on scheduling with years of field and teaching experience at leading academic institutions. Spend a few hours with this book and come away with a solid understanding of this essential management topic.

$64.95 per copy
Over 250 pages, illustrated, Hardcover
Catalog No. 67247A

Residential & Light Commercial Construction Standards
A Unique Collection of Industry Standards That Define Quality in Construction

For Contractors & Subcontractors, Owners, Developers, Architects & Engineers, Attorneys & Insurance Personnel

Compiled from the nation's major building codes, professional associations, and product manufacturer institutes, this unique resource enables you to:
- Set a standard for subcontractors and employees
- Protect yourself against defect claims
- Resolve disputes
- Answer client questions with authority

$59.95 per copy
Over 500 pages, illustrated, Softcover
Catalog No. 67322

The Building Professional's Guide to Contract Documents
New 3rd Edition
By Waller S. Poage, AIA, CSI, SAVE

This comprehensive treatment of Contract Documents is an important reference for owners, design professionals, contractors, and students.
- Structure your Documents for Maximum Efficiency
- Understand the Roles and Responsibilities of Construction Professionals
- Improve Methods of Project Delivery

$64.95 per copy, 400 pages
Diagrams and construction forms, Hardcover
Catalog No. 67261A

Cyberplaces II: The Internet Guide for AECs & Facility Managers 2nd Edition
By Paul Doherty

With an added section on the New Economy and e-commerce—such as AEC arbitrage, aggregate purchasing, auctions, e-supply chains, and more. Plus the best Knowledge Management and other Internet tools, including PDAs and the newest Web software for scheduling and project management.

The **Web site** keeps you current on the best Internet tools and applications for building design, construction and management—with workshops to put it all into practice.

$59.95 per copy
Over 700 pages, illustrated, Softcover
Catalog No. 67317A

Value Engineering: Practical Applications
... For Design, Construction, Maintenance & Operations
By Alphonse Dell'Isola, PE, leading authority on VE in construction

A tool for immediate application—for engineers, architects, facility managers, owners and contractors. Includes: Making the Case for VE—The Management Briefing, Integrating VE into Planning and Budgeting, Conducting Life Cycle Costing, Integrating VE into the Design Process, Using VE Methodology in Design Review and Consultant Selection, Case Studies, VE Workbook, and a Life Cycle Costing program on disk.

$79.95 per copy
Over 450 pages, illustrated, Hardcover
Catalog No. 67319

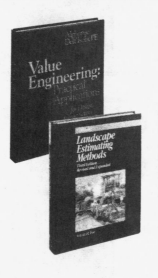

Means Landscape Estimating Methods
3rd Edition
By Sylvia H. Fee

This revised edition offers expert guidance for preparing accurate estimates for new landscape construction and grounds maintenance. Includes a complete project estimate featuring the latest equipment and methods, and **two totally new chapters**—Life Cycle Costing and Landscape Maintenance Estimating.

$62.95 per copy
Over 300 pages, illustrated, Hardcover
Catalog No. 67295A

For more information visit Means Web Site at www.rsmeans.com

Reference Books

Preventive Maintenance Guidelines for School Facilities
NEW! By John C. Maciha

This new publication gives you a complete program for school maintenance that ensures: sustained security, safety, property integrity, facility acceptance, user satisfaction, and reasonable ongoing expenditures.

Includes schedules for weekly, monthly, half-yearly and annual maintenance. 3-ring binder with hard copy forms, electronic forms on disk, and laminated wall chart.

$149.95 per copy
Over 150 pages, Hardcover
Catalog No. 67326

Means Environmental Remediation Estimating Methods By Richard R. Rast

Guidelines for estimating 50 standard remediation technologies. Use it to: prepare preliminary budgets, develop detailed estimates, compare costs and solutions, estimate liability, review quotes, negotiate settlements. A valuable support tool for *Means Envitonmental Remediation Unit Price* and *Assemblies books.*

$99.95 per copy
Over 600 pages, illustrated, Hardcover
Catalog No. 64777

Means Facilities Maintenance Standards By Roger W. Liska, PE, AIC

Unique features of this one-of-a-kind working guide for facilities maintenance

A working encyclopedia that points the way to solutions to every kind of maintenance and repair dilemma. With a labor-hours section to provide productivity figures for over 180 maintenance tasks. Included are ready-to-use forms, checklists, worksheets and comparisons, as well as analysis of materials systems and remedies for deterioration and wear.

$159.95 per copy, 600 pages, 205 tables, checklists and diagrams, Hardcover
Catalog No. 67246

HVAC: Design Criteria, Options, Selection
Expanded Second Edition

By William H. Rowe III, AIA, PE

Includes Indoor Air Quality, CFC Removal, Energy Efficient Systems and Special Systems by Building Type. Helps you solve a wide range of HVAC system design and selection problems effectively and economically. Gives you clear explanations of the latest ASHRAE standards.

$84.95 per copy
Over 600 pages, illustrated, Hardcover
Catalog No. 67306

The Facilities Manager's Reference
By Harvey H. Kaiser, PhD

The tasks and tools the facility manager needs to accomplish the organization's objectives, and develop individual and staff skills. Includes Facilities and Property Management, Administrative Control, Planning and Operations, Support Services, and a complete building audit with forms and instructions, widely used by facilities managers nationwide.

$86.95 per copy, over 250 pages,
with prototype forms and graphics, Hardcover
Catalog No. 67264

Facilities Maintenance Management
By Gregory H. Magee, PE

Now you can get successful management methods and techniques for all aspects of facilities maintenance. This comprehensive reference explains and demonstrates successful management techniques for all aspects of maintenance, repair and improvements for buildings, machinery, equipment and grounds. Plus, guidance for outsourcing and managing internal staffs.

$86.95 per copy
Over 280 pages with illustrations, Hardcover
Catalog No. 67249

Cost Planning & Estimating for Facilities Maintenance

In this unique book, a team of facilities management authorities shares their expertise at:
- Evaluating and budgeting maintenance operations
- Maintaining & repairing key building components
- Applying *Means Facilities Maintenance & Repair Cost Data* to your estimating

With the special maintenance requirements of the 10 major building types.

$82.95 per copy
Over 475 pages, Hardcover
Catalog No. 67314

The ADA in Practice
By Deborah S. Kearney, PhD

Helps you meet and budget for the requirements of the Americans with Disabilities Act. Shows how to do the job right by understanding what the law requires, allows, and enforces. Includes a "buyers guide" for 70 ADA-compliant products.

*Winner of IFMA's "Distinguished Author of the Year" award.

See page 9 for another ADA resource.

$72.95 per copy
Over 600 pages, illustrated, Softcover
Catalog No. 67147A

Reference Books

For more information visit Means Web Site at www.rsmeans.com

Basics for Builders: Plan Reading & Material Takeoff
By Wayne J. DelPico

For Residential and Light Commercial Construction

A valuable tool for understanding plans and specs, and accurately calculating material quantities. Step-by-step instructions and takeoff procedures based on a full set of working drawings.

$35.95 per copy
Over 420 pages, Softcover
Catalog No. 67307

Planning & Managing Interior Projects New 2nd Edition, By Carol E. Farren, CFM

Addresses changes in technology and business, guiding you through commercial design and construction from initial client meetings to post-project administration. Includes: evaluating space requirements, alternative work models, telecommunications and data management, and environmental issues.

$69.95 per copy
Over 400 pages, illustrated, Hardcover
Catalog No. 67245A

Superintending for Contractors:
How to Bring Jobs in On-time, On-Budget
By Paul J. Cook

This book examines the complex role of the superintendent/field project manager, and provides guidelines for the efficient organization of this job. Includes administration of contracts, change orders, purchase orders, and more.

$35.95 per copy
Over 220 pages, illustrated, Softcover
Catalog No. 67233

Means Forms for Contractors
For general and specialty contractors

Means editors have created and collected the most needed forms as requested by contractors of various-sized firms and specialties. This book covers all project phases. Includes sample correspondence and personnel administration.
Full-size forms on durable paper for photocopying or reprinting.

$79.95 per copy
Over 400 pages, three-ring binder, more than 80 forms
Catalog No. 67288

Total Productive Facilities Management
By Richard W. Sievert, Jr.

The best of today's cost and project management, quality, and value engineering principles. Includes Benchmarking, Value Engineering, and Scheduling.

$79.95 per copy
Over 270 pages, illustrated, Hardcover
Catalog No. 67321

Means Heavy Construction Handbook

Informed guidance for planning, estimating and performing today's heavy construction projects. Provides expert advice on every aspect of heavy construction work including hazardous waste remediation and estimating. To assist planning, estimating, performing, or overseeing work.

$74.95 per copy
Over 430 pages, illustrated, Hardcover
Catalog No. 67148

Estimating for Contractors:
How to Make Estimates that Win Jobs
By Paul J. Cook

Estimating for Contractors is a reference that will be used over and over, whether to check a specific estimating procedure, or to take a complete course in estimating.

$35.95 per copy
Over 225 pages, illustrated, Softcover
Catalog No. 67160

HVAC Systems Evaluation
By Harold R. Colen, PE

You get direct comparisons of how each type of system works, with the relative costs of installation, operation, maintenance and applications by type of building. With requirements for hooking up electrical power to HVAC components. Contains experienced advice for repairing operational problems in existing HVAC systems, ductwork, fans, cooling coils, and much more!

$84.95 per copy
Over 500 pages, illustrated, Hardcover
Catalog No. 67281

Quantity Takeoff for Contractors:
How to Get Accurate Material Counts
By Paul J. Cook

Contractors who are new to material takeoffs or want to be sure they are using the best techniques will find helpful information in this book, organized by CSI MasterFormat division.

Now $17.98 per copy, limited quantity
Over 250 pages, illustrated, Softcover
Catalog No. 67262

Roofing: Design Criteria, Options, Selection
By R.D. Herbert, III

This book is required reading for those who specify, install or have to maintain roofing systems. It covers all types of roofing technology and systems. You'll get the facts needed to intelligently evaluate and select both traditional and new roofing systems.

Now $31.48 per copy
Over 225 pages with illustrations, Hardcover
Catalog No. 67253

For more information visit Means Web Site at www.rsmeans.com

Reference Books

Means Estimating Handbook

This comprehensive reference covers a full spectrum of technical data for estimating, with information on sizing, productivity, equipment requirements, codes, design standards and engineering factors.
Means Estimating Handbook will help you: evaluate architectural plans and specifications, prepare accurate quantity takeoffs, prepare estimates from conceptual to detailed, and evaluate change orders.

$99.95 per copy
Over 900 pages, Hardcover
Catalog No. 67276

Means Repair and Remodeling Estimating Third Edition
By Edward B. Wetherill & R.S. Means

Focuses on the unique problems of estimating renovations of existing structures. It helps you determine the true costs of remodeling through careful evaluation of architectural details and a site visit.
New section on disaster restoration costs.

$69.95 per copy
Over 450 pages, illustrated, Hardcover
Catalog No. 67265A

Successful Estimating Methods:
From Concept to Bid
By John D. Bledsoe, PhD, PE

A highly practical, all-in-one guide to the tips and practices of today's successful estimator. Presents techniques for all types of estimates, *and* advanced topics such as life cycle cost analysis, value engineering, and automated estimating.
Estimate spreadsheets available at Means Web site.

$64.95 per copy
Over 300 pages, illustrated, Hardcover
Catalog No. 67287

Means Productivity Standards for Construction
Expanded Edition (Formerly Man-Hour Standards)

Here is the working encyclopedia of labor productivity information for construction professionals, with labor requirements for thousands of construction functions in CSI MasterFormat.
Completely updated, with over 3,000 new work items.

$159.95 per copy
Over 800 pages, Hardcover
Catalog No. 67236A

Means Electrical Estimating Methods Second Edition

Expanded version includes sample estimates and cost information in keeping with the latest version of the CSI MasterFormat. Contains new coverage of Fiber Optic and Uninterruptible Power Supply electrical systems, broken down by components and explained in detail. A practical companion to *Means Electrical Cost Data.*

$64.95 per copy
Over 325 pages, Hardcover
Catalog No. 67230A

Means Mechanical Estimating Methods Second Edition

This guide assists you in making a review of plans, specs and bid packages with suggestions for takeoff procedures, listings, substitutions and pre-bid scheduling. Includes suggestions for budgeting labor and equipment usage. Compares materials and construction methods to allow you to select the best option.

$64.95 per copy
Over 350 pages, illustrated, Hardcover
Catalog No. 67294

Means Scheduling Manual
Third Edition
By F. William Horsley

Fast, convenient expertise for keeping your scheduling skills right in step with today's cost-conscious times. Covers bar charts, PERT, precedence and CPM scheduling methods. Now updated to include computer applications.

$64.95 per copy
Over 200 pages, spiral-bound, Softcover
Catalog No. 67291

Maintenance Management Audit

This annual audit program is essential for every organization in need of a proper assessment of its maintenance operation. Downloadable electronic forms from Means Web Site allow managers to identify and correct problems and enhance productivity.

Now $32.48 per copy, limited quantity
125 pages, spiral bound, illustrated, Hardcover
Catalog No. 67299

Means Square Foot Estimating Methods Second Edition
By Billy J. Cox and F. William Horsley

Proven techniques for conceptual and design-stage cost planning. Steps you through the square foot cost process, demonstrating faster, better ways to relate the design to the budget. Now updated to the latest version of UniFormat.

$69.95 per copy
Over 300 pages, illustrated, Hardcover
Catalog No. 67145A

Means ADA Compliance Pricing Guide

Gives you detailed cost estimates for budgeting modification projects, including estimates for each of 260 alternates. You get the 75 most commonly needed modifications for ADA compliance, each an assembly estimate with detailed cost breakdown.

$72.95 per copy
Over 350 pages, illustrated, Softcover
Catalog No. 67310

Reference Books

For more information visit Means Web Site at www.rsmeans.com

Means Unit Price Estimating Methods
2nd Edition
$59.95 per copy
Catalog No. 67303

Legal Reference for Design & Construction
By Charles R. Heuer, Esq., AIA
Now $54.98 per copy, limited quantity
Catalog No. 67266

Means Plumbing Estimating Methods
New 2nd Edition
By Joseph J. Galeno & Sheldon T. Greene
$59.95 per copy
Catalog No. 67283A

Structural Steel Estimating
By S. Paul Bunea, PhD
Now $39.98 per copy, limited quantity
Catalog No. 67241

Business Management for Contractors
How to Make Profits in Today's Market
By Paul J. Cook
Now $17.98 per copy, limited quantity
Catalog No. 67250

Understanding Legal Aspects of Design/Build
By Timothy R. Twomey, Esq., AIA
$79.95 per copy
Catalog No. 67259

Risk Management for Building Professionals
By Thomas E. Papageorge, RA
Now $29.98 per copy, limited quantity
Catalog No. 67254

Contractor's Business Handbook
By Michael S. Milliner
Now $21.48 per copy, limited quantity
Catalog No. 67255

Basics for Builders: How to Survive and Prosper in Construction
By Thomas N. Frisby
$34.95 per copy
Catalog No. 67273

Successful Interior Projects Through Effective Contract Documents
By Joel Downey & Patricia K. Gilbert
Now $34.98 per copy, limited quantity
Catalog No. 67313

Understanding Building Automation Systems
By Reinhold A. Carlson, PE & Robert Di Giandomenico
$39.98 per copy, limited quantity
Catalog No. 67284

Illustrated Construction Dictionary, Condensed
$59.95 per copy
Catalog No. 67282

Managing Construction Purchasing
By John G. McConville, CCC, CPE
Now $31.48 per copy, limited quantity
Catalog No. 67302

Hazardous Material & Hazardous Waste
By Francis J. Hopcroft, PE, David L. Vitale, M. Ed., & Donald L. Anglehart, Esq.
Now $44.98 per copy, limited quantity
Catalog No. 67258

Fundamentals of the Construction Process
By Kweku K. Bentil, AIC
Now $34.98 per copy, limited quantity
Catalog No. 67260

Construction Delays
By Theodore J. Trauner, Jr., PE, PP
$59.95 per copy
Catalog No. 67278

Basics for Builders: Framing & Rough Carpentry
By Scot Simpson
$24.95 per copy
Catalog No. 67298

Interior Home Improvement Costs
New 7th Edition
$19.95 per copy
Catalog No. 67308B

Exterior Home Improvement Costs
New 7th Edition
$19.95 per copy
Catalog No. 67309B

Concrete Repair and Maintenance Illustrated
By Peter H. Emmons
$69.95 per copy
Catalog No. 67146

How to Estimate with Metric Units
Now $24.98 per copy, limited quantity
Catalog No. 67304

For more information
visit Means Web Site
at www.rsmeans.com

Seminars

Developing Facility Assessment Programs

This two-day program concentrates on the management process required for planning, conducting, and documenting the physical condition and functional adequacy of buildings and other facilities. Regular facilities condition inspections are one of the facility department's most important duties. However, gathering reliable data hinges on the design of the overall program used to identify and gauge deferred maintenance requirements. Knowing where to look . . . and reporting results effectively are the keys. This seminar is designed to give the facility executive essential steps for conducting facilities inspection programs.

Inspection Program Requirements • Where is the deficiency? • What is the nature of the problem? • How can it be remedied? • How much will it cost in labor, equipment and materials? • When should it be accomplished? • Who is best suited to do the work?

Note: Because of its management focus, this course will not address trade practices and procedures.

Repair and Remodeling Estimating

Repair and remodeling work is becoming increasingly competitive as more professionals enter the market. Recycling existing buildings can pose difficult estimating problems. Labor costs, energy use concerns, building codes, and the limitations of working with an existing structure place enormous importance on the development of accurate estimates. Using the exclusive techniques associated with Means' widely acclaimed **Repair & Remodeling Cost Data**, this seminar sorts out and discusses solutions to the problems of building alteration estimating. Attendees will receive two intensive days of eye-opening methods for handling virtually every kind of repair and remodeling situation . . . from demolition and removal to final restoration.

Mechanical and Electrical Estimating

This seminar is tailored to fit the needs of those seeking to develop or improve their skills and to have a better understanding of how mechanical and electrical estimates are prepared during the conceptual, planning, budgeting and bidding stages. Learn how to avoid costly omissions and overlaps between these two interrelated specialties by preparing complete and thorough cost estimates for both trades. Featured are order of magnitude, assemblies, and unit price estimating. In combination with the use of **Means Mechanical Cost Data**, **Means Plumbing Cost Data** and **Means Electrical Cost Data**, this seminar will ensure more accurate and complete Mechanical/Electrical estimates for both unit price and preliminary estimating procedures.

Unit Price Estimating

This seminar shows how today's advanced estimating techniques and cost information sources can be used to develop more reliable unit price estimates for projects of any size. It demonstrates how to organize data, use plans efficiently, and avoid embarrassing errors by using better methods of checking.

You'll get down-to-earth help and easy-to-apply guidance for:
- making maximum use of construction cost information sources
- organizing estimating procedures in order to save time and reduce mistakes
- sorting out and identifying unusual job requirements to improve estimating accuracy.

Square Foot Cost Estimating

Learn how to make better preliminary estimates with a limited amount of budget and design information. You will benefit from examples of a wide range of systems estimates with specifications limited to building use requirements, budget, building codes, and type of building. And yet, with minimal information, you will obtain a remarkable degree of accuracy.

Workshop sessions will provide you with model square foot estimating problems and other skill-building exercises. The exclusive Means building assemblies square foot cost approach shows how to make very reliable estimates using preliminary budget and design information.

Scheduling and Project Management

This seminar helps you successfully establish project priorities, develop realistic schedules, and apply today's advanced management techniques to your construction projects. Hands-on exercises familiarize participants with network approaches such as the Critical Path Method. Special emphasis is placed on cost control, including use of computer-based systems. Through this seminar you'll perfect your scheduling and management skills, ensuring completion of your projects *on time* and *within budget*. Includes hands-on application of **Means Building Construction Cost Data.**

Facilities Maintenance and Repair Estimating

With our Facilities Maintenance and Repair Estimating seminar, you'll learn how to plan, budget, and estimate the cost of ongoing and preventive maintenance and repair for all your buildings and grounds. Based on R.S. Means' groundbreaking cost estimating book, this two-day seminar will show you how to decide to either contract out or retain maintenance and repair work in-house. In addition, you'll learn how to prepare budgets and schedules that help cut down on unplanned and costly emergency repair projects. Facilities Maintenance and Repair Estimating crystallizes what facilities professionals have learned over the years, but never had time to organize or document. This program covers a variety of maintenance and repair projects, from underground storage tank removal, roof repair and maintenance, exterior wall renovations, and energy source conversions, to service upgrades and estimating energy-saving alternatives.

Managing Facilities Construction and Maintenance

In you're involved in new facility construction, renovation or maintenance projects and are concerned about getting quality work done on time and on or below budget, in Means' seminar **Managing Facilities Construction and Maintenance** you'll learn management techniques needed to effectively plan, organize, control and get the most out of your limited facilities resources.

Learn how to develop budgets, reduce expenditures and check productivity. With the knowledge gained in this course, you'll be better prepared to successfully sell accurate project budgets, timing and manpower needs to senior management . . . plus understand how to evaluate the impact of today's facility decisions on tomorrow's budget.

Call 1-800-448-8182 for more information

Seminars

2001 Means Seminar Schedule

For more information visit Means Web Site at www.rsmeans.com

Location	Dates
Las Vegas, NV	March 12 - 15
Washington, DC	April 23 - 26
Denver, CO	May 21-24
San Francisco, CA	June 11 - 14
Cape Cod, MA	September TBD
Washington, DC	October 1 - 4
San Diego, CA	October TBD
Atlantic City, NJ	November TBD
Orlando, FL	November 5-8

Note: Call for exact dates and details

Registration Information

Register Early... Save up to $150! Register 30 days before the start date of a seminar and save $150 off your total fee. *Note: This discount can be applied only once per order.*

How to Register Register by phone today! Means toll-free number for making reservations is: **1-800-448-8182**.

Individual Seminar Registration Fee $875 To register by mail, complete the registration form and return with your full fee to: Seminar Division, R.S. Means Company, Inc., 63 Smiths Lane, Kingston, MA 02364.

Federal Government Pricing All Federal Government employees save 25% off regular seminar price. Other promotional discounts cannot be combined with Federal Government discount.

Team Discount Program Two to four seminar registrations: $760 per person—Five or more seminar registrations: $710 per person—Ten or more seminar registrations: Call for pricing.

Consecutive Seminar Offer One individual signing up for two separate courses at the same location during the designated time period pays only $1,400. You get the second course for only $525 (**a 40% discount**). Payment must be received at least ten days prior to seminar dates to confirm attendance.

Refunds Cancellations will be accepted up to ten days prior to the seminar start. There are no refunds for cancellations postmarked later than ten working days prior to the first day of the seminar. A $150 processing fee will be charged for all cancellations. Written notice or telegram is required for all cancellations. Substitutions can be made at any time before the session starts. **No-shows are subject to the full seminar fee.**

AACE Approved Courses The R.S. Means Construction Estimating and Management Seminars described and offered to you here have each been approved for 14 hours (1.4 recertification credits) of credit by the AACE International Certification Board toward meeting the continuing education requirements for re-certification as a Certified Cost Engineer/Certified Cost Consultant.

AIA Continuing Education R.S. Means is registered with the AIA Continuing Education System (AIA/CES) and is committed to developing quality learning activities in accordance with the CES criteria. R.S. Means seminars meet the AIA/CES criteria for Quality Level 2. AIA members will receive (28) learning units (LUs) for each two day R.S. Means Course.

Daily Course Schedule The first day of each seminar session begins at 8:30 A.M. and ends at 4:30 P.M. The second day is 8:00 A.M.–4:00 P.M. Participants are urged to bring a hand-held calculator since many actual problems will be worked out in each session.

Continental Breakfast Your registration includes the cost of a continental breakfast, a morning coffee break, and an afternoon break. These informal segments will allow you to discuss topics of mutual interest with other members of the seminar. (You are free to make your own lunch and dinner arrangements.)

Hotel/Transportation Arrangements R.S. Means has arranged to hold a block of rooms at each hotel hosting a seminar. To take advantage of special group rates when making your reservation be sure to mention that you are attending the Means Seminar. You are of course free to stay at the lodging place of your choice. (**Hotel reservations and transportation arrangements should be made directly by seminar attendees.**)

Important Class sizes are limited, so please register as soon as possible.

Registration Form

Call 1-800-448-8182 x5115/to register or FAX 1-800-632-6732. Visit our web site www.rsmeans.com

Please register the following people for the Means Construction Seminars as shown here. Full payment or deposit is enclosed, and we understand that we must make our own hotel reservations if overnight stays are necessary.

☐ Full payment of $ _____ enclosed.

☐ Bill me

Name of Registrant(s)
(To appear on certificate of completion)

_____

_____

P.O. #: _____
GOVERNMENT AGENCIES MUST SUPPLY PURCHASE ORDER NUMBER

Firm Name _____

Address _____

City/State/Zip _____

Telephone No. _____ Fax No. _____

E-Mail Address _____

Charge our registration(s) to: ☐ MasterCard ☐ VISA ☐ American Express ☐ Discover

Account No. _____ Exp. Date _____

Cardholder's Signature _____

| Seminar Name | City | Dates |

Please mail check to: R.S. MEANS COMPANY, INC., 63 Smiths Lane, P.O. Box 800, Kingston, MA 02364 USA

Consulting Services Group

Solutions For Your Construction and Facilities Cost Management Problems

We are leaders in the cost engineering field, supporting the unique construction and facilities management costing challenges of clients from the Federal Government, Fortune 1000 Corporations, Building Product Manufacturers, and some of the World's Largest Design and Construction Firms.

Developers of the Dept. of Defense Tri-Services Estimating Database

Recipient of SEARS 1997 Chairman's Award for Innovation

Research...

- **Custom Database Development**—Means expertise in construction cost engineering and database management can be put to work creating customized cost databases and applications.
- **Data Licensing & Integration**—To enhance applications dealing with construction, any segment of Means vast database can be licensed for use, and harnessed in a format compatible with a previously developed proprietary system.
- **Cost Modeling**—Pre-built custom cost models provide organizations that expend countless hours estimating repetitive work with a systematic time-saving estimating solution.
- **Database Auditing & Maintenance**—For clients with in-house data, Means can help organize it, and by linking it with Means database, fill in any gaps that exist and provide necessary updates to maintain current and relevant proprietary cost data.

Estimating...

- **Estimating Service**—Means expertise is available to perform construction cost estimates, as well as to develop baseline schedules and establish management plans for projects of all sizes and types. Conceptual, budget and detailed estimates are available.
- **Benchmarking**—Means can run baseline estimates on existing project estimates. Gauging estimating accuracy and identifying inefficiencies improves the success ratio, precision and productivity of estimates.
- **Project Feasibility Studies**—The Consulting Services Group can assist in the review and clarification of the most sound and practical construction approach in terms of time, cost and use.
- **Litigation Support**—Means is available to provide opinions of value and to supply expert interpretations or testimony in the resolution of construction cost claims, litigation and mediation.

Training...

- **Core Curriculum**—Means educational programs, delivered on-site, are designed to sharpen professional skills and to maximize effective use of cost estimating and management tools. On-site training cuts down on travel expenses and time away from the office.
- **Custom Curriculum**—Means can custom-tailor courses to meet the specific needs and requirements of clients. The goal is to simultaneously boost skills and broaden cost estimating and management knowledge while focusing on applications that bring immediate benefits to unique operations, challenges, or markets.
- **Staff Assessments and Development Programs**—In addition to custom curricula, Means can work with a client's Human Resources Department or with individual operating units to create programs consistent with long-term employee development objectives.

For more information and a copy of our capabilities brochure, please call 1-800-448-8182 and ask for the Consulting Services Group, or reach us at www.rsmeans.com

MeansData™

CONSTRUCTION COSTS FOR SOFTWARE APPLICATIONS
Your construction estimating software is only as good as your cost data.

Software Integration

A proven construction cost database is a mandatory part of any estimating package. We have linked MeansData™ directly into the industry's leading software applications. The following list of software providers can offer you MeansData™ as an added feature for their estimating systems. Visit them on-line at *www.rsmeans.com/demo/* for more information and free demos. Or call their numbers listed below.

3D International
713-871-7000 budwit@3di.com

ACT
Applied Computer Technologies
Facility Management Software
919-859-1335 patw@srs.net

AEPCO, Inc.
301-670-4642 blueworks@aepco.com

ArenaSoft ESTIMATING
888-370-8806 info@arenasoft.com

Ares Corporation
650-401-7100
sales@arescorporation.com

ASSETWORKS, Inc.
Facility Management Software
800-659-9001 info@assetworks.com

Benchmark, Inc.
800-393-9193 sales@benchmark-inc.com

BSD
Building Systems Design, Inc.
888-273-7638 bsd@bsdsoftlink.com

cManagement
800-945-7093 sales@cmanagement.com

CDCI
Construction Data Controls, Inc.
800-285-3929 sales@cdci.com

CMS
Computerized Micro Solutions
800-255-7407 cms@proest.com

CONAC GROUP
800-663-2338 sales@concac.com

ESTIMATING SYSTEMS, Inc.
800-967-8572 pulsar@capecod.net

G2 Estimator
A Div. of Valli Info. Syst., Inc.
800-657-6312 info@g2estimator.com

GEAC COMMERCIAL SYSTEMS, Inc.
800-554-9865 info@geac.com

GRANTLUN CORPORATION
480-897-7750 support@grantlun.com

HCI SYSTEMS, Inc.
800-750-4424 info@hcisystems.com

IQ BENECO
801-565-1122 mbrown@beneco.com

Lugs International
888-682-5573 info@lugs.com

MC²
Management Computer Controls
800-225-5622 vkeys@mc2-ice.com

PRISM COMPUTER CORPORATION
Facility Management Software
800-774-7622 nalbadri@prismcc.com

PYXIS TECHNOLOGIES
888-841-0004 ldenigris@ebid98.com

QUEST SOLUTIONS, Inc.
800-452-2342 info@questsolutions.com

RICHARDSON ENGINEERING SERVICES, Inc.
480-497-2062 info@resi.net

SANDERS SOFTWARE, Inc.
800-280-9760 hsander@vallnet.com

STN, Inc.
Workline Maintenance Systems
800-321-1969 ask@stn.com

TIMBERLINE SOFTWARE CORP.
800-628-6583
product.info@timberline.com

TMA SYSTEMS, Inc.
Facility Management Software
800-862-1130 sales@tmasys.com

US COST, Inc.
800-372-4003
usc-software@compuserve.com

VERTIGRAPH, Inc.
800-989-4243
info-request@vertigraph.com

WENDLWARE
714-895-7222 sales@corecon.com

WINESTIMATOR, Inc.
800-950-2374 sszabo@winest.com

DemoSource™ One-stop shopping for the latest cost estimating software for just $19.95. This evaluation tool includes product literature and demo diskettes for ten or more estimating systems, all of which link to MeansData™. Call 1-800-334-3509 to order.

**FOR MORE INFORMATION ON ELECTRONIC PRODUCTS CALL
1-800-448-8182 OR FAX 1-800-632-6732.**

MeansData™ is a registered trademark of R.S. Means Co., Inc., *A CMD Group Company*.

For more information
visit Means Web Site
at www.rsmeans.com

New Titles

Means Illustrated Construction Dictionary, 3rd Edition
New Updated Edition with interactive CD-ROM

Long regarded as the Industry's finest, the *Means Illustrated Construction Dictionary* is now even better. With the addition of over 1,000 new terms and hundreds of new illustrations, it is the clear choice for the most comprehensive and current information.

The companion CD-ROM that comes with this new edition adds many extra features: larger graphics, expanded definitions, and links to both CSI MasterFormat numbers and product information.

- 19,000 construction words, terms, phrases, symbols, weights, measures, and equivalents
- 1,000 new entries
- 1,200 helpful illustrations
- Contemporary coverage of professional & trade terms
- Easy-to-use format, with thumbtabs

$99.95 per copy
Over 790 pages, illustrated, Hardcover
Catalog No. 67292A

Historic Preservation: Project Planning & Estimating

- *Managing Historic Restoration, Rehabilitation and Preservation Building Projects and*
- *Determining and Controlling Their Costs*

By Swanke Hayden Connell Architects

- Applicable standards and financing
- Site survey and documentation methods
- Protecting finishes and features
- Hazardous materials
- Identifying and qualifying specialty contractors
- Machanical/Electrical systems upgrades
- Restoring old building materials
- How to assemble a reliable, detailed estimate and budget for future maintenance

$99.95 per copy
Over 450 pages, Hardcover
Catalog No. 67323

Facilities Operations & Engineering Reference
An all-in-one technical reference for planning & managing facility projects & solving day-to-day operations problems

Ten major sections address:

- **Management Skills**
- **Economics** (budgeting/cost control, financial analysis, VE, etc.)
- **Civil Engineering & Construction Practices**
- **Maintenance** (detailed staffing guidance and job descriptions, CMMS, planning/scheduling, training, work orders, preventive/predictive maintenance)
- **HVAC & Energy Efficiencies** (optimizing energy use: heating, cooling, lighting, water)
- **Mechanical and Electrical Engineering**
- **Instrumentation & Controls**
- **Environment, Health & Safety**

Produced jointly by R.S. Means and
the Association for Facilities Engineering.

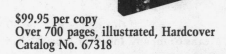

$99.95 per copy
Over 700 pages, illustrated, Hardcover
Catalog No. 67318

2001 Order Form

ORDER TOLL FREE 1-800-334-3509
OR FAX 1-800-632-6732.

Qty.	Book No.	COST ESTIMATING BOOKS	Unit Price	Total
	60061	Assemblies Cost Data 2001	$149.95	
	60011	Building Construction Cost Data 2001	89.95	
	61011	Building Const. Cost Data-Looseleaf Ed. 2001	116.95	
	63011	Building Const. Cost Data-Metric Version 2001	89.95	
	60221	Building Const. Cost Data-Western Ed. 2001	89.95	
	60111	Concrete & Masonry Cost Data 2001	83.95	
	50141	Construction Cost Indexes 2001	198.00	
	60141A	Construction Cost Index-January 2001	49.50	
	60141B	Construction Cost Index-April 2001	49.50	
	60141C	Construction Cost Index-July 2001	49.50	
	60141D	Construction Cost Index-October 2001	49.50	
	60311	Contr. Pricing Guide: Framing/Carpentry 2001	36.95	
	60331	Contr. Pricing Guide: Resid. Detailed 2001	36.95	
	60321	Contr. Pricing Guide: Resid. Sq. Ft. 2001	39.95	
	64021	ECHOS Assemblies Cost Book 2001	149.95	
	64011	ECHOS Unit Cost Book 2001	99.95	
	54001	ECHOS (Combo set of both books)	214.95	
	60231	Electrical Change Order Cost Data 2001	89.95	
	60031	Electrical Cost Data 2001	89.95	
	60201	Facilities Construction Cost Data 2001	219.95	
	60301	Facilities Maintenance & Repair Cost Data 2001	199.95	
	60161	Heavy Construction Cost Data 2001	89.95	
	63161	Heavy Const. Cost Data-Metric Version 2001	89.95	
	60091	Interior Cost Data 2001	89.95	
	60121	Labor Rates for the Const. Industry 2001	199.95	
	60181	Light Commercial Cost Data 2001	79.95	
	60021	Mechanical Cost Data 2001	89.95	
	60151	Open Shop Building Const. Cost Data 2001	89.95	
	60211	Plumbing Cost Data 2001	89.95	
	60041	Repair and Remodeling Cost Data 2001	79.95	
	60171	Residential Cost Data 2001	79.95	
	60281	Site Work & Landscape Cost Data 2001	89.95	
	60051	Square Foot Costs 2001	99.95	
		REFERENCE BOOKS		
	67147A	ADA in Practice	72.95	
	67310	ADA Pricing Guide	72.95	
	67298	Basics for Builders: Framing & Rough Carpentry	24.95	
	67273	Basics for Builders: How to Survive and Prosper	34.95	
	67307	Basics for Builders: Plan Reading & Takeoff	35.95	
	67261	Bldg. Prof. Guide to Contract Documents-3rd Ed.	64.95	
	67312	Building Spec Homes Profitably	29.95	
	67250	Business Management for Contractors	17.98	
	67146	Concrete Repair & Maintenance Illustrated	69.95	
	67278	Construction Delays	59.95	
	67255	Contractor's Business Handbook	21.48	
	67314	Cost Planning & Est. for Facil. Maint.	82.95	
	67317A	Cyberplaces: The Internet Guide-2nd Ed.	59.95	
	67230A	Electrical Estimating Methods-2nd Ed.	64.95	
	64777	Environmental Remediation Est. Methods	99.95	
	67160	Estimating for Contractors	35.95	
	67276	Estimating Handbook	99.95	
	67249	Facilities Maintenance Management	86.95	

Qty.	Book No.	REFERENCE BOOKS (Cont.)	Unit Price	Total
	67246	Facilities Maintenance Standards	$159.95	
	67264	Facilities Manager's Reference	86.95	
	67318	Facilities Operations & Engineering Reference	99.95	
	67301	Facilities Planning & Relocation	109.95	
	67231	Forms for Building Const. Professional	94.95	
	67288	Forms for Contractors	79.95	
	67260	Fundamentals of the Construction Process	34.98	
	67258	Hazardous Material & Hazardous Waste	44.98	
	67148	Heavy Construction Handbook	74.95	
	67323	Historic Preservation: Proj. Planning & Est.	99.95	
	67308C	Home Improvement Costs-Interior Projects	19.95	
	67309C	Home Improvement Costs-Exterior Projects	19.95	
	67304	How to Estimate with Metric Units	24.98	
	67306	HVAC: Design Criteria, Options, Select.-2nd Ed.	84.95	
	67281	HVAC Systems Evaluation	84.95	
	67282	Illustrated Construction Dictionary, Condensed	59.95	
	67292A	Illustrated Construction Dictionary, w/CD-ROM	99.95	
	67295A	Landscape Estimating-3rd Ed.	62.95	
	67266	Legal Reference for Design & Construction	54.98	
	67299	Maintenance Management Audit	32.48	
	67302	Managing Construction Purchasing	31.48	
	67294	Mechanical Estimating-2nd Ed.	64.95	
	67245A	Planning and Managing Interior Projects-2nd Ed.	69.95	
	67283A	Plumbing Estimating Methods-2nd Ed.	59.95	
	67326	Preventive Maint. Guidelines for School Facil.	149.95	
	67236A	Productivity Standards for Constr.-3rd Ed.	159.95	
	67247A	Project Scheduling & Management for Constr.	64.95	
	67262	Quantity Takeoff for Contractors	17.98	
	67265A	Repair & Remodeling Estimating-3rd Ed.	69.95	
	67322	Residential & Light Commercial Const. Stds.	59.95	
	67254	Risk Management for Building Professionals	29.98	
	67291	Scheduling Manual-3rd Ed.	64.95	
	67145A	Square Foot Estimating Methods-2nd Ed.	69.95	
	67287	Successful Estimating Methods	64.95	
	67313	Successful Interior Projects	34.98	
	67233	Superintending for Contractors	35.95	
	67321	Total Productive Facilities Management	79.95	
	67284	Understanding Building Automation Systems	39.98	
	67259	Understanding Legal Aspects of Design/Build	79.95	
	67303	Unit Price Estimating Methods-2nd Ed.	59.95	
	67319	Value Engineering: Practical Applications	79.95	

MA residents add 5% state sales tax
Shipping & Handling**
Total (U.S. Funds)*

Prices are subject to change and are for U.S. delivery only. *Canadian customers may call for current prices. **Shipping & handling charges: Add 7% of total order for check and credit card payments. Add 9% of total order for invoiced orders.

Send Order To: ADDV-1001

Name (Please Print) _____

Company _____

☐ Company
☐ Home Address _____

_____

City/State/Zip _____

Phone # _____ P.O. # _____

Mail To: R.S. Means Company, Inc., P.O. Box 800, Kingston, MA 02364-0800   (Must accompany all orders being billed)